The Evolution and Classification of the Cretaceous Ammonites and their Jurassic Progenitors

This book is a unique text that illustrates and describes all the nominal genera of Cretaceous ammonites and their Jurassic progenitors on the basis of their type species.

There is no judgement as to validity, and the authors leaves it to the reader to subjectively recognize or synonymize those taxa as they see fit, based on the premises and philosophies to which they ascribe. The author argues that the taxonomy offered by the *Treatise*, the fundamental reference work, is no longer appropriate and genera need to be redefined according to testable hypotheses. The purpose of this book, therefore, is redescribe the type species of every nominal genus and to use these to construct a new phylogenetic classification not hampered by preconceived opinion. It emphasizes the importance of "trifling characters" for classification, and that taxonomic rank is not optional but is predetermined by evolutionary position. A new higher classification is introduced for the group using the higher taxa proposed for the phylogenetic classification of the Bivalvia. Hypothesized relationships are depicted by the use of cladograms, recognizing these portray only affiliations among the sample and not necessarily close taxonomic bond.

This book will be of great interest to palaeontologists, geologists, stratigraphers, biogeographers and interested students of geology, and paleology. It will also make a good addition to universities, museums, and public libraries.

The Evolution and Classification of the Cretaceous Ammonites and their Jurassic Progenitors

Michael Robert Cooper

CRC Press
Taylor & Francis Group
Boca Raton London New York

CRC Press is an imprint of the
Taylor & Francis Group, an **informa** business

Designed cover image: Michael Robert Cooper

First published 2025
by CRC Press/Balkema
4 Park Square, Milton Park, Abingdon, Oxon, OX14 4RN

and by CRC Press/Balkema
2385 NW Executive Center Drive, Suite 320, Boca Raton FL 33431

CRC Press/Balkema is an imprint of the Taylor & Francis Group, an informa business

A catalogue record for this book is available from the British Library

ISBN: 978-1-032-76279-1 (hbk)
ISBN: 978-1-032-76282-1 (pbk)
ISBN: 978-1-003-47784-6 (ebk)

DOI: 10.1201/9781003477846

Typeset in CIDFont+F1
by KnowledgeWorks Global Ltd.

Publisher's note: This book has been prepared from camera-ready copy provided by the authors.

"*Satisfaction of one's curiosity is one of the greatest sources of happiness in life*" (Linus Pauling).

CONTENTS

PREFACE

"Let there be no emnity among seekers after truth" (Aristotle).

This book is the culmination of over 50 years of work on Cretaceous ammonites, punctuated by temporary excursions into other fields of palaeontology. My passion for the field developed under the enthusiasm and mentorship of Norman M. Savage (Eugene, Oregon) and my formative years in ammonitology were guided by the commitment, dedication and expertise of W.J. Kennedy (Oxford) and, to a lesser extent, the late C.W. Wright and J.M. Hancock, to each of whom I owe unrepayable gratitude. Early on, however, I realized there were differences with my mentors, particularly with regard to the role of phylogeny in classification and the discrimination of genera. Eventually this led to a series of publications, completely ignored but not refuted, in which I attempted to show that the prevailing classification, based on Linnaean methodology involving unification on the basis of primitive characters (plesiomorphies), is utilitarian and unnatural. I advocated a Darwinian approach in which grouping is based on acquired characteristics (apomorphies). This publication represents an extension and culmination of this work.

Personal experience showed that most taxonomists, especially in their formative years, cram species into whichever taxa are on offer. Thus the taxa available for selection is crucial and, in this respect, the *Treatise* has always played a leading role, as it was designed to. However, heavy synonymy for the sake of taxonomic simplicity creates "*categories of convenience*", many of which I disagree with, since "*taxonomic lumping*" masks phylogeny and hinders the achievement of a natural classification;

"Simplicity itself is not necessarily virtuous, particularly if it obscures true relationships between entities in the natural world" (Daniel Lieberman).

Importantly, as more and more new taxa are discovered, and evolutionary relationships increasingly clarified, the need to resurrect subjective synonyms becomes ever more necessary, especially if taxonomy is to replicate the evolutionary process. Hence, a higher taxonomy based on phylogenetic principles will be as dynamic as the field of research itself, a disturbing thought to some.

Since subjective synonyms remain available to those who find use for them, here I have attempted to illustrate and redescribe (not diagnose) all nominal genera on the basis of their type species, without prejudice as to validity, and to incorporate them into hypothetical arrangements. It is left for readers:

to subjectively recognize or synonymize those taxa they see fit, based on the premises and philosophies to which they ascribe.

Although I have not handled as many Cretaceous ammonites as some of my more illustrious colleagues, and my experience of Jurassic and earliest Cretaceous Boreal elements is bleak, in consulting the primary literature and drawing the ammonites for this book I have achieved a familiarity that, I believe, is acceptable for this work, for I fully ascribe to the opinion that

".... false views, if supported by some evidence, do little harm, for everyone takes a salutary pleasure in proving their falseness; and when this is done, one path towards error is closed and the road to truth is often at the same time opened" (C. Darwin, 1871. *The descent of man*).

Obviously, though, this work is a compilation of the enormous efforts of others, whose names fill the reference list - it rests on the shoulders of giants.

The cladograms presented here are offered as a starting point to provoke testing and correction. As this work has been compiled over 30 years without financial, institutional, drafting, artistic, secretarial, indexing or proofreading assistance, I take full responsibility for any errors that may occur.

Since the evolution of some groups transcended the Jurassic-Cretaceous boundary, in order to present a phylogenetic classification it became necessary to include Jurassic progenitors. This situation is particularly acute among phylloceratids, perisphinctid, axtaxioceratids and haploceratids, and their treatment is thus both irregular and perhaps incomplete, but is offered as an insight into ancestral relationships.

M. R. Cooper
New Germany, South Africa
November, 2023

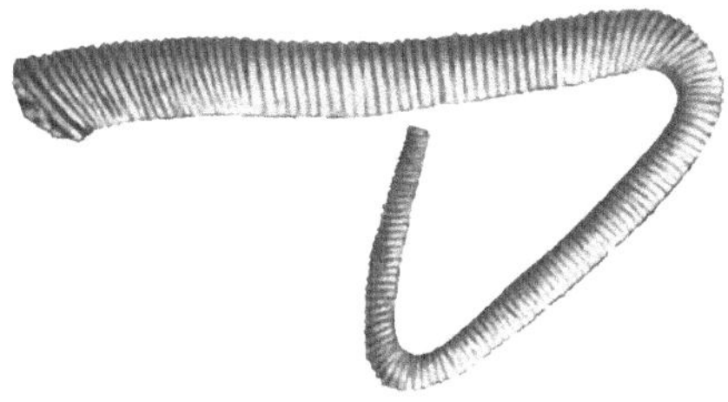

INTRODUCTION

In offering a new classification of the Cretaceous ammonites and their Jurassic progenitors, it is important that colleagues understand the premises and philosophy on which this is based.

My formative years were during a period of taxonomic rationalization when, for reasons not totally clear to me, the thrust was towards reducing the number of taxa to an absolute minimum by heavy synonymy, by "*taxonomic lumping*". As my early papers showed, it is easy to get caught up in this taxonomic hype. Particularly noteworthy is how ready I was to synonymize taxa about which I knew nothing. But I was not alone. A perusal of the literature showed that many subjective synonyms were unfamiliar taxa from distant parts of the globe and that valid taxa had a decidedly eurocentric bias. Interesting in this respect also is how, since most Cretaceous ammonitologists were concentrated in Western Europe, there was a tendency to seek ancestor-descendent relationships among the known faunas of Western Europe, as if evolution occurred only in that part of the globe.

During the 70s and 80s the prevailing methodology among ammonitologists, myself included, was to regard each important phyletic line as a "genus" and, where necessary, to horizontally subdivide this line into "subgenera". While, philosophically, there was nothing wrong with this approach, many of the subgenera were closer to each other than to the nominate subgenus. It was only later in my career that I was able to identify the flaw in this process. Almost invariably, synonymy and unification were based upon assigning greatest taxonomic weight to primitive characters (plesiomorphies). This methodology, *Linnaean taxonomy*, leads to horizontal groupings in which evolutionary dichotomies are obfuscated or ignored, and heavy synonymy a natural consequence.

This taxonomic rationalization was promulgated by the latest *Treatise*, albeit recognizedly

> *"... compiled over three and a half decades within an earlier convention"* (Wright, 1996: xviii).

Nonetheless, many of the taxa admitted by the *Treatise* were horizontal groupings unified by primitive characters, comprising stem members of different phyletic lines.

Linnaean vs Darwinian genera

Fundamental to any classification is the concept of a genus. However genera are subjective taxa, many established in the era prior to Darwinian and Hennigian methodology, before primitive and derived characters were discriminated, when classifications were utilitarian rather than phylogenetic. To achieve a natural classification, therefore, all genera *and their subjective synonyms* must be re-evaluated using testable hypotheses.

The greatest influence on ammonite systematics was the publication of the first *Treatise* (Arkell *et al.*, 1957). Quite obviously this compelling synthesis was required to be "*simple*", with the number of higher taxa reduced to manageable proportions by subjective synonymy, particularly of taxonomic inconveniences. To a large extent this procedure was followed by its successor (Wright *et al.*, 1996). Conveniently, but unfortunately, for most of the last 250 years biological classification has been largely

> *"... a scheme for arranging together those living objects which are most alike, and for separating those which are most unlike"* (Darwin, 1859: 372).

This philatelic approach uses shared primitive characters to unite and create higher taxa, thereby assigning plesiomorphies greatest taxonomic weight in assessing relationships. Unfortunately, Linnaean taxonomy is capable of recognizing only major (basal) dichotomies in the evolutionary tree, i.e. high-ranking taxa (subfamilies, families, orders etc.). It cannot discriminate low-ranking taxa, e.g. tribes, genera and subgenera which, therefore, require *a priori* reasoning for their recognition. Since they are created in a mysterious fashion, Linnaean taxa have to be dictated by leaders in the field. Rightly or wrongly their views are protected by the *Matthew effect* (Merton, 1968) whereby the work of high profile researchers receives greater attention than equivalent work by lesser-known colleagues. However,

> *"The world progresses, year by year, century by century, as the members of the younger generation find out what was wrong among the things that their elders said. So, you must always be skeptical - always think for yourself"* (Linus Pauling).

Traditionally Linnaeanists have regarded the differences between genera as a magnitude greater than those separating species. As a result the literature is replete with claims that

> *"... the differences are not considered to warrant generic separation"*, *".. the differences are regarded as specific only"*, or *"... generic separation seems unnecessary"*.

The simple reality is that Linnaean genera are subjective and untestable; they have to be accepted on faith. However, to be scientifically valid, they should satisfy the *Falsification Principle* of Karl Popper (1959); they must be testable and thus able to be proven false.

Darwin (1859: 378) appreciated that, in order to be natural,

"*....the arrangement of the groups in each class ... must be strictly genealogical*".

This guidance, a profound truism, has been overlooked (not read) by countless taxonomists who have continued to group like with like. Stated simply a natural classification is one that *accurately* replicates the evolutionary history of the group. The goal of Darwinian (phylogenetic) taxonomy therefore is to identify all the branches of the evolutionary tree and to replicate these in an appropriate classification. It is a search for evolutionary lineages, groups with

"*... propinquity of descent*" (Darwin, 1859: 372).

Darwinian taxonomy is the only methodology capable of detecting the finer (more recent) twigs of the evolutionary tree, and of objectively determining genera and other low-ranking taxa. Importantly, it produces classifications that are testable. Classifications not strictly rooted in genealogy are nothing more than "*convenient fiction*" (Dawkins, 2004). Unfortunately Darwinian methodology poses severe problems for traditionalists schooled in Linnaean taxonomy.

Linnaean genera are dissimilar, broadly based, and encompass many of the minor branches of the evolutionary tree. Since they obfuscate details of the evolutionary process, Linnaean genera correspond to the subtribes, tribes and even subfamilies of Darwinian taxonomy.

Darwinian genera, on the other hand, can be very similar, narrowly defined and encompass relatively few species. Because they define discrete twigs (*clades*) on the evolutionary tree, Darwinian taxonomy results in a multiplicity of genera and is characterized by "*taxonomic splitting*". As Darwinian genera record steps in the evolutionary process, they are not separated by quantum leaps in phenotype but simply by the acquisition of a new character (apomorphy) that is phylogenetically meaningful. It may seem insignificant, but one should not ignore the

"*... importance, for classification, of trifling characters*" (Darwin, 1859: 375).

In Darwinian taxonomy there is no intrinsic difference between genera and species, merely their position within the phylogenetic scheme. Genera with large numbers of species are "*categories of convenience*", since any cursory study will show that in such groups some species are always closer than others, and the finer branches of the evolutionary tree have been ignored.

Over a period of 50 years I have sought the maximum number of species descended from a single ancestor; this number is only 6 (in the case of *Eogaudryceras*). This suggests that in most cases genera with more than 6 species are "*categories of convenience*" that encompass more than one phyletic line (Darwinian genus). Obviously ancestor-descendent relationships are not always easily resolved and, for pragmatic reasons, a genus may comprise more than 6 species. Those, however, that contain dozens and even hundreds of species (as in the insect world) ignore phylogeny and are unnatural conveniences. Certainly it is incumbent upon systematists to incessantly beg the question of every taxon, "*Which was its ancestor?*"

The implication of Darwinian taxonomy to the classification of Cretaceous ammonites is that far more taxa are required to reflect the evolution of the group than admitted by prevailing classifications. Obviously though the resulting

"*...plethora of new higher taxa*" (Wright, 1996: xviii)

will be deplored by those who favour "*categories of convenience*" with preconceived limits, an approach preferred by those who view ammonites primarily as stratigraphical tools. As a result workers who have attempted to reflect phylogeny in their classification, e.g. Hyatt, Spath, van Hoepen, Casey, have been castigated for their taxonomic excesses. However, Darwin (1859: 434) offered a cautionary note about

"*... the blindness of preconceived opinion*"

since,

"*... in phylogenetic taxonomy, rank is not discretionary, it is pre-determined by evolutionary position*" (Cooper, 2015).

Evolution is a complex process and hence natural (phylogenetic) classifications must be complex also. Taxonomic splitting is mandatory if genealogy is to be replicated accurately, and it should be appreciated that the more refined the taxonomy, the greater the value of the organisms for biostratigraphical resolution.

Primitive and derived characters

The characters of all organisms are *biological*

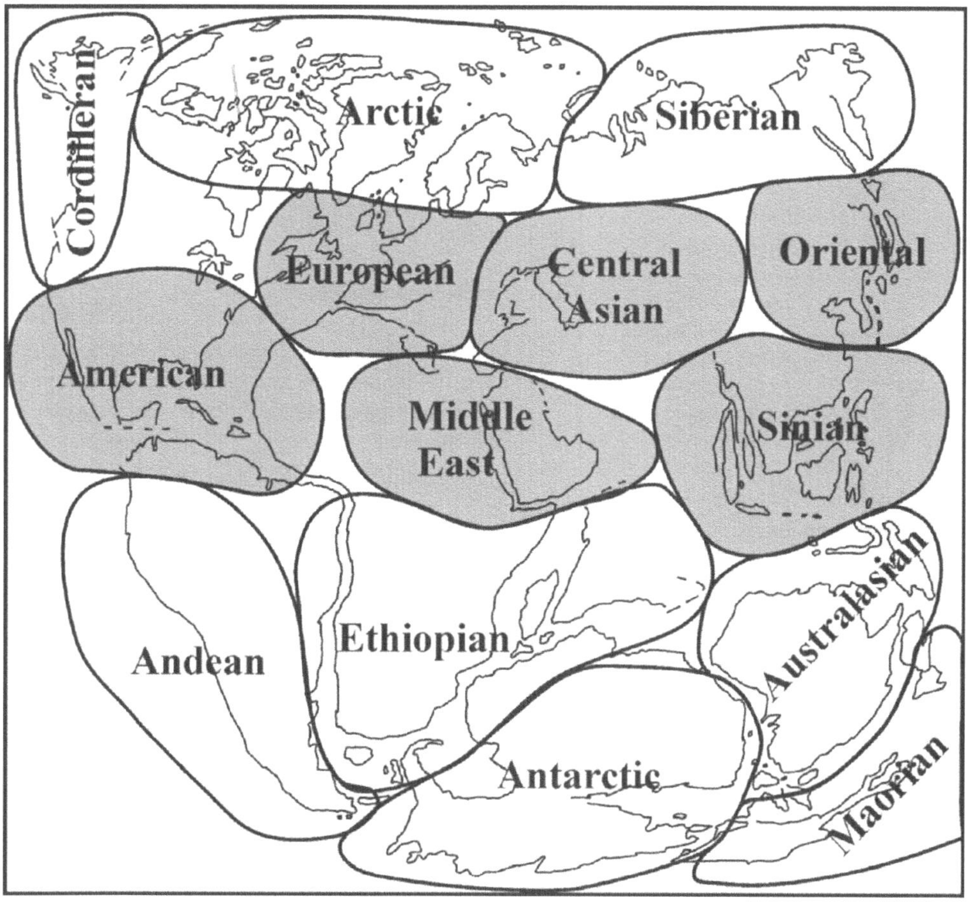

Fig. 1. Late Jurassic-Early Cretaceous palaeobiogeographical provinces. The Tethyan Realm is shaded grey, with the Boreal Realm to the north and the Gondwanic (= Austral Realm) to the south (after Cooper, 2015).

isolating mechanisms designed to facilitate recognition by their own kind and to inhibit interbreeding with closest relatives. Thus evolution is the creation of differences, and consistent disparities can be established only where inbreeding is averted and gene flow inhibited.

In determining relationships it is important to appreciate that the characters of all organisms fall into two fundamental groups:

- *primitive characters* (plesiomorphies) inherited unchanged from the ancestor, and
- *derived characters* (apomorphies) that distinguish it from its ancestor.

In any organism primitive characters are overwhelmingly predominant and, since they are inherited, are invaluable in tracking ancestry. Mostly plesiomorphies are recognized by the *Commonality Principle*:

the most widespread characters
are likely to be the more primitive.

Derived characters on the other hand constitute a minor component of the features of an organism and, since they are idiosyncratic, do not record distant relationships. Very often they are analogous characters that have arisen in unrelated organisms due to adaptation to similar habits and habitats. Such *convergence* (homoeomorphy) is so common and widespread that

"... We must not, therefore, in classifying,
trust to resemblance" (C. Darwin, 1859: 373).

Character weighting

Species are a mosaic of primitive and derived characters, and the taxonomic weighting of these two character sets greatly influences classification. Since primitive characters dominate, grouping like with like assigns plesiomorphies greatest taxonomic weight, thereby obfuscating the finer branches of the evolutionary tree. As, however, evolution is

"... descent with modification"
(C. Darwin, 1859: 112),

derived characters (*apomorphies*) should carry greatest weight in taxonomic placement.

Phenotypic discontinuities

Darwin (1859: 388) appreciated that

"... if every form which has ever lived on this earth were suddenly to reappear ... it would be quite impossible to give definitions by which each group could be distinguished from other groups, as all would blend together by steps as fine as those between the finest existing varieties".

As such all evolutionary relationships are naturally paraphyletic. Fortunately for taxonomists this evolutionary continuum is compartmentalized by phenotypic discontinuities resulting from parts of the evolutionary/stratigraphical record not having been preserved, or still to be discovered. Such discontinuites are a measure of the number of taxa missing, the quantity to be inferred by the magnitude of the discontinuity. Cladistically the gaps between adjacent branches of a cladogram are all phenotypic discontinuities of different magnitude, and each may cover anything from a species to a class.

Minor phenotypic discontinuities involve species and genera, and ancestral relationships remain discernible and obvious. Such discontinuities separate genera and, where they involve only a number of species, are the basis for recognition of monotypic genera. As a guidance,

"The more aberrant any form is, the greater must be the number of connecting forms which .... have been exterminated and utterly lost" (Darwin, 1859: 386).

Moderate phenotypic discontinuities are indicated where ancestral relationships are not obvious but may be reasonably inferred; thus the omissions involve subtribes and tribes.

Major phenotypic discontinuities involve subfamilies and families (and higher taxa) and ancestral relationships are obscure. They create *cryptogenic taxa* whose evolutionary divergence is recorded by quantum shifts in phenotype. They, together with taxonomic lumping (which obfuscates the finer branches of the evolutionary tree), are largely responsible for the hypotheses of *quantum evolution* (Simpson, 1944) and *punctuated equilibrium* (Eldredge & Gould, 1972).

For a natural classification to be achieved, taxonomic rank cannot be preconceived since the evolutionary significance of characters can never be anticipated. In particular the

"... importance, for classification of trifling characters" (Darwin, 1859: 375)

is paramount. Attention to detail provides the only means of disentangling the "*inextricable web*" of evolution, exposing numerous convergences and frequent recapitulation.

Cladistic analysis

Cladograms displaying inferred relationships have become a preferred basis for depicting shared ancestry. However cladograms have a serious and completely overlooked deficiency when used in the construct of classifications. They exclude and disregard extinct and unknown taxa. Such omissions result in cladograms placing taxa adjacent that, actually, may be well removed phylogenetically. Hence, due to phenotypic discontinuities, cladograms portray much closer evolutionary relationships that actually exist and, at best, offer only a loose guide to affiliation among the sample available, and not necessarily close evolutionary proximity or taxonomic bond. As a first attempt, the hypothetical relationships offered here are designed to provoke close scrutiny and correction, for the writer is firmly of the opinion that

".... false views, if supported by some evidence, do little harm, for everyone takes a salutary pleasure in proving their falseness" (C. Darwin, 1871. *The descent of man*).

Palaeobiogeography

All organisms are geographically restricted and the full distribution of extinct species may never be known. Following Servais *et al.* (2013) *endemic taxa* are restricted to a particular geographical region, *pandemic taxa* to several geographical regions and *cosmopolitan taxa* have a worldwide distribution. Taxonomic lumping has led to the belief that most ammonite genera are cosmopolitan but this is not so. As with most organisms ammonites are provincialized, and the majority are pandemic with a significant number endemic; very few, if any, are truly cosmopolitan.

Pangaeal fragmentation commenced in the Jurassic and by the mid-Cretaceous biogeography had become highly complex. There were a multitude of drifting continents, islands of varying shapes and sizes, a vast increase in shoreline lengths, an enormous expansion in shallow epeiric seas, diversion of ocean currents, new migration routes and barriers to migration. The evolution of Cretaceous ammonites, therefore, occurred in a highly dynamic biogeographical scenario, and evolutionary rates in many group were tachytelic. In particular, rapid plate motion during the Late Jurassic marked the opening of the Atlantic and Indian Oceans, the spread of which facilitated widespread dispersal and further radiation. On the other hand, end-Jurassic regression created isolated basins and high degrees of provincialism and endemism.

Palaeobiogeography is greatly influenced by taxonomy. During phases of taxonomic lumping, groups are attributed a very wide often cosmopolitan distribution, whereas at times of taxonomic splitting patterns of provincialism and endemism emerge.

Based on infaunal bivalves (trigoniids), a good indicator of local bottom conditions, the writer identified a number of Jurassic-Early Cretaceous faunal provinces prior to the break-up of Pangaea (**Fig. 1**). Their application to ammonite palaeobiogeography is open to question but is offered here for testing.

Rifting of the central Atlantic commenced in the Late Triassic with graben formation and volcanism, and the Atlantic and Pacific oceans were connected through the Hispanic Corridor in the Hettangian (Sha, 2019). Rifting of the Indian Ocean commenced in the Middle Jurassic, with marine Toarcian in Kenya (Cox, 1965), Bathonian in Tanzania (Aitken, 1959) and Oxfordian in Madagascar (Collignon, 1959). Rifting in the South Atlantic began in the late Jurassic, with volcanics and thick Aptian salt deposits overlain by Late Aptian marine strata (Bengtson *et al.*, 2018). Seafloor spreading commenced north of the Walvis Ridge at the end of the Aptian (~112 Ma) (Torsvik *et al.*, 2009), and opening of the North Atlantic was an Early Tertiary (~56 Ma) event (Upton, 2021). This spreading created additional biogeographical provinces/boundaries whose relationships weakened as continents drifted further apart in the Cretaceous.

TAXONOMIC CATEGORIES

The only taxonomic units formally recognized below the genus level are subgenus, species and subspecies (ICZN, 2000). However the evolutionary record is more complex than this and, as a result, phylogenetic taxonomy requires far more taxonomic categories (*cf.* Carter *et al.*, 2011).

Demes

Demes are the only objective biological taxon, unfortunately not recognized by the ICZN. They are

discrete naturally occurring populations, the members of which recognize one another and which prefer to breed among themselves.

Demes are genetically distinct populations whose members remain capable of interbreeding with members of other demes. Their boundaries are vague, marked by zones of hybridization, and the progeny of hybrids remain fertile. Demes largely conform to the "*varieties*" of Darwin & Wallace (1858). Palaeontologically they are either dismissed as slightly different, or are treated as ecotypes, varieties or subspecies depending on the preferences of the taxonomist concerned.

Subspecies

Subspecies are the smallest formally recognized taxonomic unit. They are "*incipient species*" (Darwin, 1859: 450) and, as such, represent an *advanced* stage in the speciation process. However, since they are subjective, their status vacillates for, during phases of taxonomic "*lumping*", many species are downgraded to subspecies and during phases of "*splitting*" many subspecies are upgraded to species so as to satisfy the notions of the taxonomist concerned.

Perspicaciously Mayr (1942) observed that

"Every subspecies that was ever carefully analyzed was found to be composed of a number of genetically distinct populations [= demes]. *It is, in many cases, entirely dependent upon the judgement of the individual taxonomist how many of these populations are to be included in one subspecies".*

A major failing of the prevailing subspecies concept is that ICZN (2000) rules require allocation to the nominate species irrespective of ancestry, thus advocating the creation of "*evolutionary fans*". Phylogenetically implicit to such fans is that all subspecies are descend from the nominate subspecies, when even a cursory study shows this is not true. Phylogenetically

a subspecies is a group of demes that define a discrete twiglet on the evolutionary tree.

Heterochronous subspecies are often used to connect species in the fossil record. In such cases the species to which they are assigned is based on majority (primitive) characters. As a result most heterochronous subspecies are referred to the ancestral (parent) species even though differentiating (derived) characters favour assignment to the descendent species. Phylogenetically many subspecies are assigned to the wrong species. Discrimination of primitive and derived characters is required for the correct specific allocation of subspecies.

Species

The "*species concept*" has filled countless tomes but, in reality, it is as true today as when first penned,

> "... *how entirely vague and arbitrary is the distinction between species and varieties"* (Darwin, 1859).

The *biological species concept* controlled classification for the last half of the 20th Century and was the basis for species recognition. According to this notion a species is:

> "... *an interbreeding population of individuals which, under natural conditions, does not interbreed with its closest relatives"* (Mayr, 1963).

This definition emphasises a shared gene pool, gene isolation and reproductive nonviability. Unfortunately the cornerstone of this hypothesis, breeding exclusivity, was demonstrably erroneous at the day of inception. The clause "*under natural conditions*" was designed to circumvent known interbreeding in captivity but this is irrelevant; organisms are either genetically viable or not. Morever countless observations before and since have shown that "*good species*" can and do interbreed (*cf.* Darwin, 1859). In fact wherever mate selection is inhibited, either in the wilds or in captivity, hybridization is almost habitual.

Phylogenetically

> *a species is a group of subspecies that share a common ancestor and define a discrete twig on the evolutionary tree.*

Although many workers (Spath, Wright, Casey, Owen) have suspected and even suggested hybridization among ammonites, it is nearly impossible to prove but, as with most organisms, is highly likely.

Mostly species discrimination among ammonites is based on type and strength of ornament, density of ribbing, degree of involution and shell inflation. As shown by collections from the Western Interior of North America, where there is tight stratigraphical control, sutural details can be important also. Unfortunately indiscriminate mixing of samples from different stratigraphical levels obfuscates such subtle evolutionay changes, diminishing the value of trifling characters for species recognition. In this respect it is worth emphasizing the exceptional importance of the Western Interior of North America to the evolutionary study of Cretaceous ammonites. The atypical environment seems to have favoured the rapid evolution of certain groups, particularly among the Turrilitidina. Moreover the semi-isolated basin, together with rapid synorogenic (molasse) sedimentation and frequent bentonite markers, makes ancestor-descendent relationships unusually obvious; it is an "evolutionary laboratory". The same is true for some of the restricted Boreal basins of the Early Cretaceous (Sverdrup, northern Siberia, Russian platform).

Subgenus

The subgenus is a subjective taxon with the same status as genus. Its recognition is arbitrary and discretionary. In the past many workers, the writer included, used subgenera as a means of indicating relationships. While this approach served to restrict the number of genera and broadly preserve the *Treatise* classification of 1957, it did not reduce the number of genus-level taxa. It merely increased the number of subgenera and resulted in a cumbersome written nomenclature of trinomens and quadrinomens. Once inferred relationships are displayed on a cladogram there seems little need for subgenera, moreso since generally they have been unified on the basis of primitive characters. They are not used here.

Genus

Traditionally species are grouped on the basis of similarity into genera, but

> "*.... generic limits are not defined by any rules .... for practical reasons, the lumping of several distinct species-groups into a single genus is currently in vogue* (Fry & Keith, 2004: 456).

Such is the fickle grounds on which Linnaean genera are constructed. It constitutes "*the tyrrany of taxonomy*" (Cooper, 2015). Phylogenetically however

> *a genus is a group of species that share a common recent ancestor and define a discrete twig on the evolutionary tree.*

It is a twig (*clade*) comprising all the species down to the last significant evolutionary dichotomy, the stem of which is defined by extinct species among surviving genera and undiscovered species among fossils.

Among ammonites, generally but not consistently, acquisition or loss of tubercles has been accorded generic rank; this principle is adhered to here since it represents an evolutionary novelty that is important to understanding phylogeny. Modern collections from the Western Interior of North America have shown that *progenetic dwarfs* are common offshoots of normal sized stock and usually these are separated at the genus level.

Intraspecific variation

Morphological features are plastic. In some groups shell form is fairly consistent, e.g. Lytoceratidina, but where dimorphism occurs may be extreme, e.g. Ancyloceratidina. Since the pioneering studies of Haas (1946) and Reeside & Cobban (1960), modern investigations have served to emphasize the polymorphism of very many ammonite populations, but this should not be assumed to be invariant.

Intraspecific variation may be *clinal* (continuous) or *discontinuous* (disjunct), in which latter case it becomes possible to distinguish *morphotypes* (deme, varieties). Prior to the 1970's, according to the philosophy of the time, these morphotypes were frequently accorded specific or subspecific distinction. Normally intraspecific variation in ammonites ranges from compressed and weakly sculptured to inflated and strongly ornamented end members.

Far from polymorphism being "*useless and directionless*" (Arnould-Saget, 1956), polytypy is the birthground of new species. Morphotypes represent the first tentative step in the speciation process. Indeed, it is more than probable that the gamodemes from which new species arise are not random genomic samples but, rather, are dominated by morphotypes with a discrete gene pool. Consequently very many new species are not created by randon gene mutations but rather by neutral genes becoming advantageous under new conditions. Certainly morphotype recognition and discrimination is fundamental to *population analysis* and species definition, irrespective of nomenclatural status, and hence has an important role to play in taxonomy.

Many claims of bewildering intraspecific variation are based on allochthonous faunas from condensed deposits, reworked remanié, scattered surface scree and unlocalized material from substantial thicknesses of strata and different ammonite zones. Often they are the basis for subjective synonymy. Theoretically however intraspecific variation can be established only among strictly contemporaneous populations, i.e. from the same bedding plane. Tight stratigraphical control and adequate sample size (practical experiments suggest at least 50 specimens) are prerequisites for the satisfactory determination of intraspecific variation and hence, also, species discrimination.

The Late Campanian of Western Europe is subdivided into three zones (*marroti*, *polyplocum*, *hyatti*) whereas under perfect conditions, in the Western Interior of North America, 16 ammonite zones are discriminated. It could be argued, therefore, that the faunas of the *polyplocum* Zone of Western Europe represent faunal mixing of at least 12 zones of the Western Interior. This has implications for the alleged intraspecific variation of long-ranging species such as *Hoploscaphites constrictus* (J. Sowerby).

In many cases in Western Europe, the richest best preserved faunas, i.e. those most suited to population analysis, comprise phosphatized remanié associated with stratigraphical discontinuites, e.g. the Cambridge Greensand. Attempts to document intraspecific variation, species definition and synonymy on the basis of these winnowed faunas, that derive from several horizons, have led to flawed conclusions within some groups (scaphitids) but not others (puzosiids), emphasizing different rates of evolution within different groups under the same conditions.

Maturation

One of the problems facing ammonitologists is determining whether or not an ammonite is a small species, or a juvenile of a much larger species, particularly where there is substantial ontogenetic change. Sexual maturity in normally coiled ammonites is indicated most obviously by scaphitoid uncoiling of the adult body chamber, i.e. egression of the umbilical seam. Usually this is accompanied by crowding and simplification of sutures and, in many cases, there is also a change in ornament, e.g. weakening or loss of ribbing. Often apertural adornments, e.g. lappets, rostra, are a feature.

Dimorphism

Modern studies have shown that very many, if not most, ammonite species are dimorphic and that probably the differences are sexual. However, many dimorphs remain as distinct species or even different genera.

Although, now, dimorphism is widely recognized and acknowledged, the challenge remains of convincingly matching dimorphs. This is not easy.

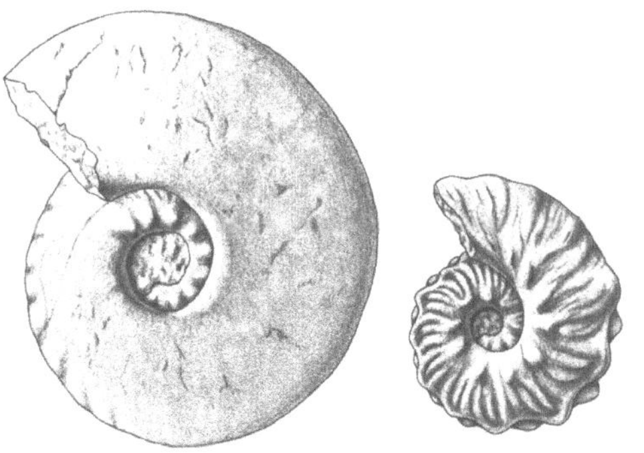

Fig. 2. Sexual dimorphism in *Callihoplites.*

No sooner had *Labeceras-Myloceras* and *Costidiscus-Macroscaphites* been declared dimorphs than papers appeared disputing these claims. It is particularly difficult in heteromorphs where both coiling and ornament may be substantially different, and has led to separate phylogenies for macroconchs and microconchs.

Many ammonite species had been collected for countless years without dimorphism becoming apparent. Either they are not dimorphic or adults were ecological segregated, some populations comprising macroconchs and others microconchs. Thus in Western Europe, phosphatized Albian greensand faunas are dominated by microconchs and juveniles, whereas macroconchs are preserved mainly in clay and malmstone facies. The former are typical of shallow sediment-starved high-energy environments, whereas that latter record offshore low-energy environments with relatively high rates of sediment input.

Due to the vagaries of selective pressure, there is no reason to believe that evolutionary change in dimorphic lineages occured at the same rate. Among living ploceid birds males evolved far more rapidly than females. Thus tachytelic dimorphs are likely to be subdivided into more taxa than bradytelic dimorphs. In the Puzosiidae macroconch genera outnumber microconch genera. In such cases taxonomic discrimination should be based upon the more rapidly evolving dimorph, not all lumped into the slowly evolving dimorph.

The most widespread and oft claimed dimorphism involves size in which microconchs are smaller than macroconchs, with little or no morphological difference. As noted by Scherzinger *et al.* (2015), microconchs are paedomorphic by progenesis, maturing at an earlier developmental stage and smaller size than corresponding macroconchs. All too often sexual dimorphism has been claimed by the discovery of unusually small or unusually large individuals. However this can be environmently induced and size dimorphism can be confirmed only by population analysis of an adequate sample, and proven bimodality. Some studies have shown substantial size overlap in dimorphs (Kisevev & Rogov, 2018).

Phenotypic dimorphism involves a change in adult morphology. Very often macroconchs lose ornament in later growth and often become smooth, e.g. *Callihoplites*, *Surites*. Microconchs normally retain strong ornament to maturity and acquire apertural adornments, e.g. lappets. In such cases there can be little doubt that *sexual dimorphism* is involved.

The determination of dimorphs usually requires matching of microconchs with the inner whorls (early growth stages) of corresponding macroconch within contemporaneous populations. However within heterochronous populations similarity of inner whorls normally reflects ancestry.

Sexual dimorphism may be exhibited also by macroconchs becoming more inflated, e.g. *Rogersites*, or more compressed, e.g. *Neogastroplites*. In many oxyconic microconchs with a narrow tabulate to acute venter, e.g. *Proplacenticeras*, the corresponding more inflated macroconchs have a broadly rounded to subtabulate venter. In many pachydiscids and kossmaticeratids microconchs develop ventrolateral tubercles and have been separated at the genus level.

Changes in mode of coiling have been invoked as evidence for dimorphism in ancyloceratoids. Small aspinocones, subaspinocones and ancylocones having been interpreted as microconchs and larger co-existent criocones as macroconchs. Most scaphite macroconchs have an umbilical bulge at the base of the adult body chamber. In some baculites microconchs have a down-curved aperture, whereas macroconchs have a trumpet-shaped aperture.

Similarity of inner whorls led Cooper (1981) to suggest that *Olcostephanus* may have been a protandric hermaphrodite, changing from an earlier male morphology to a later female morphology, a process well known in gastropods.

Suture vs morphology

One of the greatest areas of dispute in ammonite taxonomy has been the emphasis placed on different aspects of the shell for classification. The

"*British school*" long emphasized morphology and regarded sutural differences as subordinate, whereas Russian and German workers place great store on the suture and its ontogenetic development.

The problems posed by sutures are numerous. Like all characters, there must be intraspecific variation, complicated by individual variation produced by the position of tubercles and ribs. Moreover there is significant ontogenetic change so that different growth stages have different sutures, complexity increasing with age. An added problem is the lack of documentation of intraspecific variation itself for most species. Moreover there is no valid reason to believe that because one species displays sutural plasticity, all species do.

Since all aspects of an organism are subject to evolution, *all characters* are available and *should be used* for systematic purposes. In this respect it is important to appreciate that the characters of an organism all do not evolve at the same rate. At any one point in time, different characters are undergoing stasis, bradytely, horotely or tachytely.

In groups that perfected their shell design early, e.g. phylloceratids, engonoceratids and placenticeratids, morphological change is minimal and sutural evolution predominant. On the other hand, at times of adaptive radiation, changes in shell morphology predominate and sutural change is masked by phenotypic radiation. Hence the criteria for classification change with time and it is an error to perpetually emphasize one character at the expense of another, even within one lineage. All characters are important to the classification of an organism, including non-apparent ones such as habitat, ecology and behaviour.

One of the reasons for dismissing sutural details as unimportant is the claim of bewildering variability. However, W.A. Cobban's meticulous work based on bed-by-bed collecting of the rapidly evolving scaphitid and baculitid lineages of the Western Interior of North America, demonstrated that sutural characteristics can be species specific. Certainly sutural characteristics are invaluable for the detection of convergence, e.g. in separating *Cunningtoniceras* and *Euomphaloceras*. The writer believes that the basic design (*bauplan*) of the suture, and its ontogenetic development, is of prime taxonomic importance. The fundamentality of the simplified goniatite suture to their classification emphasizes the importance of basic design. Ceratites and ammonites simply add sublobes and frills to the goniatite bauplan.

Synonymy

It is standard practise for ammonitologists to criticize the taxonomic excesses of their predecessors. Workers in the late 20th Century frequently could be heard bemoaning the "*excesses*" of Buckman, Hyatt, Spath, van Hoepen, Casey and others. However these workers all were victims of the conventions of their time, just as many present workers have fallen foul to "*lumping*". In fact as one watches the pendulum swing it is hard not to gain the impression that the ammonites themselves have little to do with taxonomy. Perhaps I do my colleagues a grave injustice but, as with so many aspects of life, it seems that too many

see what they want to see, and disregard the rest.

The rationalization of taxa in the first *Treatise* (Wright, 1957), many of which were said to be "cosmopolitan", was invaluable for facilitating interregional correlation and biostratigraphy. However as the faunas from different parts of the world became better known it became increasingly apparant that the faunas from different regions were not the same, they were provincialized. The latest *Treatise* (Wright, 1996) attempted to maintain the higher taxonomy established by its predecessor, and to regulate the number of genera on offer by heavy synonymy. But subjective synonymy is a personal perspective that, all too often, is a means of dispensing with taxonomic inconveniences and preserving the *status quo*. However, despite prevailing views, subjective synonyms remain available for those who find use for them and, as a result, all nominal genera are redescribed here.

An important point in synonymy involves original definition. Many subjective synonyms have been promulgated because their initial separation was deemed "*unnecessary*", "*unwarranted*" or "*insufficient for generic separation*". Such subjective appelations imply predetermined taxonomic limits. However, once a specimen is designated holotype/lectotype, *all its characters* become available for rediagnosis, not merely those used in the initial discriminations. Thus characters originally overlooked may, under review, become critically important.

A significant difference between Darwinian and Linnaean taxonomy relates to synonymy. In Linnaean taxonomy, subjective synonyms are a judgement based on preconceived limits placed on a taxon by the worker involved. Once determined, such synonyms are deemed buried and forgotten, their synonymy merely to be replicated by workers unconcerned with the problem, thereby adding unjustified credibility.

In Darwinian taxonomy subjective synonyms require constant re-evaluation as new taxa are added, and the original concepts of higher taxa broaden. Indeed, as more species are described, it becomes increasingly certain that more subjective synonyms will require resurrection and it is an error to believe that, once synonymized, a taxon can be dismissed forever.

Sea-level change, anagenesis and cladogenesis

The evolution of shallow marine organisms, and their dispersal, is closely linked to fluctuations in sea level. Eustatic transgressions and the creation of shallow epeiric seas are the stimulus for habitat expansion, range extension, reduced competition, tachytelic rates of evolution, adaptive radiation and cladogenesis (Cooper, 1977). Regressive phases on the other hand are associated with habitat reduction, niche destruction, increased competition, within-basin evolution and periods of anagenesis. Thus *cladogenesis* (adaptive radiation) and *anagenesis* (species replacement) alternate in time and are a feature of most long ranging ammonite lineages.

Whereas many ammonite genera have pandemic distributions, others display high degrees of provincialism. All ammonite groups evolved within particular sedimentary basins and hence their early history is likely to comprise endemic lineages created by *within-basin evolution*, followed by *out-of-basin migration* accompanying highstands of sea level. Obviously high degrees of endemism are a feature of evolution within restricted basins.

There are workers who believe that a strict phylogenetic classification must comprise only monophyletic taxa. However paraphyly is the fundamental aspect of biological evolution since, theoretically, there is an unbroken chain from the first virus to the most recently formed species. The fossil record shows that paraphyly is most detectable during periods of anagenesis and is obfuscated by periods of cladogenesis. It should be appreciated also that, due to recapitulation, an apomorphic character may appear within a lineage more than once due to the switching on and off of certain genes.

Taxonomic rank

Following the lead provided by Carter *et al.* (2011) for the Bivalvia, the following taxonomic ranks are available for the phylogenetic reclassification of ammonites:

Subclass — *-ia*
Infraclass — *-ia*
Cohort — *-omorphi*
Subcohort — *-ioni*
Infrasubcohort — *-idia*
Megaorder — *-ata*
Superorder — *-iformii*
Order — *-ida*
Suborder — *-idina*
Hyporder — *-oidei*
Minorder — *-oitei*
Superfamily — *-oidea*
Epifamily — *-oidae*
Series — *-itae*
Family — *-idae*
Subfamily — *-inae*
Tribe — *-ini*
Subtribe — *-ina*

Methodology

Here the type species of nearly all nominate genera are figured and redecribed (not diagnosed), without judgement as to validity. In this way the descriptions of the type species only, and not referred species, are keyed in to the cladograms offered. It is left to the discretion of individual workers to select the taxa they favour. Although the illustrations are based on particular specimens, generally the types, in some cases they have been restored and reconstructed to varying degrees in classic d'Orbigny-esque fashion, particularly where original illustrations are poor and material fragmentary. Sadly in an academic world of "*publish or perish*" some ammonitologists, perhaps in their rush to publish yet another paper, would prefer to provide unwhitened or coloured illustrations followed by lengthy diagnoses and descriptions, seemingly unaware that one good picture

"... paints a thousand words"
(Frederick Barnard, *Printer's ink*, 1921).

While every effort has been made to ensure accurate reconstruction, *all illustrations* herein should be regarded as hypothetical depictions of non-existent specimens; in cases of doubt the onus is on the researcher to consult the primary literature. Moreover, in order to achieve conformity, and to facilitate comparison, illustrations have been reorientated and, in very many cases, reversed. Since the ammonite shell is bilaterally symmetrical this is not viewed as a problem. Unfortunately the quality of my illustrations is, to a large extent, controlled by the original figures that, in some cases, are very poor.

Like all organisms the adult size of ammonites is a fundamental attribute of the species. However normal size varies with environment and can be very different from maximum size, and is greatly complicated by size dimorphism. Moreover holotypes may represent any growth stage and hence, in very many perhaps most cases, the normal and absolute adult sizes are unknown. As such the size of figured types is often meaningless. Here specimens are figured to approximately the same size, without indication of enlargment or reduction. The apparent adult size (diameter) suggested in the descriptions is as follows:

Gigantic >750 mm
Enormous - 500-749 mm
Very large - 350-499 mm
Large - 200-349 mm
Moderately large - 150-199 mm
Medium sized - 100-149 mm
Small - 50-99 mm
Very small - 20-49 mm
Micromorph <20 mm

I continue to use the sutural terminology of Kullmann & Wiedmann (1970). The sutures illustrated are external sutures taken from the literature, with the first lateral lobe (L) shaded. For obvious reasons it is not possible to illustrate all sutures at the same stage of ontogenetic development and, hence, the simplicity/complexity of the sutures may reflect the growth stage of the individual and is not necessarily good grounds for discrimination. Irrespective of ontogenetic stage, all sutures are illustrated at more or less the same size so as to facilitate comparison of the overall structure (*bauplan*) which is believed to be significant. Internal sutures are not illustrated, although in some cases these are known to have taxonomic value.

As regards authorship, here I follow the first *Treatise* (Arkell *et al.*, 1957) in listing surnames as the authors would have. I find the practise of dropping prefixes to names, e.g. Orbigny instead of d'Orbigny, Grossouvre instead of de Grossouvre and Hoepen instead of van Hoepen, disrespectful to those workers who would not have done so themselves.

With regard to content, pre-Cretaceous genera are variously treated. I considered an introduction to perisphinctoids necessary because they are the progenitors of later groups that are important in the Cretaceous. So too with phylloceratids, haploceratids and ataxioceratids. In those groups that cross the Cretaceous-Jurassic boundary treatment is inconsistent and not every member may be dealt with. The ataxioceratids were particularly problematic due to the blurred boundary with their progenitors (perisphinctids) and their descendents (dorsoplanitids and berriasellids). In some cases I was dragged uncomfortably further back into the Jurassic by tracking evolutionary lineages that are relevant to the origin of Cretaceous groups.

The purpose of this book is to illustrate and redescribe all validly introduced genera without judgement. The taxa on offer is not inhibited by personal preference. Even though many may be junior synonyms, all subjective synonyms are available to those who find use for them. It will become apparent that many species deviate from the description of the type species, and that the concept of many genera has broadened considerably over time with the discovery of new species, in many cases to unacceptable levels. Thus it is left to individual workers to recognize and synonymize those taxa that best serve their motives. I hope this work will be seen as an initial attempt to achieve a natural classification.

Finally, although the latest *Treatise* is authored by "C.W. Wright with J.H. Calloman (*sic*) and M.K. Howarth", the contributions of the latter two are very small. Callomon coauthored only the Craspeditidae, and Howarth dealt only with the Tetragonitoidea while providing a list of Cretaceous phylloceratids and lytoceratids. As a result most of the views concerning Cretaceous ammonites and their relationships are those of C.W. Wright alone and so, in order to recognize his immense contribution, in this publication this work is cited as Wright (1996). The same is true for the first *Treatise* (Arkell *et al.*, 1957) in which the Cretaceous ammonites were authored almost entirely by C.W. Wright (Arkell co-authored only a very few groups).

ACKNOWLEDGEMENTS

I am indebted to all those colleagues and librarians, too numerous to mention individually, that over the last 50 years, have provided me with reprints and PDF files of their work. Especially, though, I should like to thank Dr N. Clark (Glasgow), Dr G. Delanoy (Nice), Prof. W.J. Kennedy (Oxford), Dr H.C. Klinger (Cape Town), Dr N. Landman (New York), Dr I. López-Palomino (Hidalgo), Prof. V.V. Mitta (Moscow) and Dr J. Vermeulen (Moriez) who all went out of their way to provide me with literature I was unable to access for completion of this work.

My views have been greatly enlightened by 30 years of provocative discussion and collaboration with my good friend Dr Hugh G. Owen at the Natural History Museum (London), not that he necessarily agreed with all my views, and am grateful to the curators of that institute who, over many years, facilitated my study of their collections. Prof. T.P. Poulton (Calgary) and Dr H. Parent (Rosario) offered valuable comment on the manuscript, and Ms Nancy Rebecca of Taylor and Francis who provided editorial assistance.

Finally I am deeply indebted to my family, in particular Brenda, for their patience, encouragement, love, unwavering support and understanding for my neglect and the countless hours I spent in front of the computer and at the art desk (diningroom table).

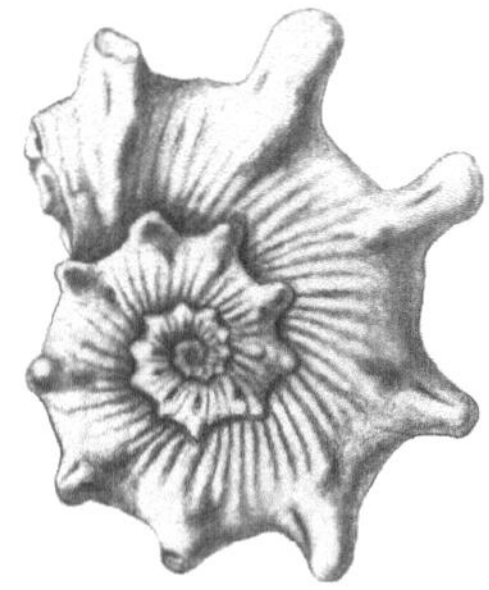

Phylum MOLLUSCA Linnaeus, 1758

Discussion. Monophyletic derivatives of Lophotrochozoa (Hyanalynch *et al.*, 1995); comprises the clades Aplacophora and Eumollusca. *Cambr. - Rec.*

Clade EUMOLLUSCA Ruppert, Fox & Barnes, 2004

Discussion. Comprises the subphyla Polyplacophora and Conchifera. *Cambr. - Rec.*

Subphylum CONCHIFERA Gegenbaur, 1878

Discussion. Monophyletic derivatives of Polyplacophora; comprises the superclasses Monoplacophora, Cyrtosoma and Diasoma. *Cambr. - Rec.*

Superclass CYRTOSOMA Runnegar & Pojeta, 1974

(= Rhacopoda Ruppert *et al.*, 2004)
(= Visceroconcha Mizzaro-Wimmer & Salvini-Plawin, 2001)
(= Viscelloconcha Tsuchiya, 2008)

Discussion. Monophyletic derivatives of Monoplacophora (Hengsbach, 1981); comprises the classes Gastropoda and Cephalopoda. *Cambr. - Rec.*

Class CEPHALOPODA Cuvier, 1797

(= Tetrabranchiata Owen, 1832 *pars*)

Discussion. Monophyletic derivatives of Gastropoda; typically with hyponome, circum-oral tentacles, chitinous beaks and image-forming eyes; comprises the subclasses Palcephalopoda and Neocephalopoda. *Cambr. - Rec.*

Subclass NEOCEPHALOPODA Lehmann & Hillmer, 1980

(= Dibranchiata Owen, 1832 *pars*)
(= Angusteradulata Lehmann, 1967)

Discussion. Monophyletic derivatives of Palcephalopoda, close to Endoceratoidea; characteristically with small subspherical protoconch (1.5-3.0mm) separated from rest of shell by constriction and without cicatrix; 9 tentacles; radula with 7 rows of teeth. Comprises the megaorders Orthocerata, Ammonita and Octopoda. *L. Cambr. - Rec.*

Megaorder AMMONITA Agassiz, 1840

(*nom. transl.* herein *ex* order Ammonitida)

Discussion. Monophyletic derivatives of Orthocerata; comprises the superorders Goniatitiformii, Ceratitiformii and Ammonitiformii; typically with chambered external shell with folded internal septa (differentiated suture line) and narrow ventrally positioned siphuncle. *U. Gedinn. - Danian.*

Superorder AMMONITIFORMII Zittel, 1884

(*nom. transl.* herein *ex* order Ammonitida)

Discussion. Monophyletic descendents of Ceratitiformii (Hyatt, 1889); characteristically with ammonitic suture and, primitively, a lituid internal lobe. Comprises the orders Phylloceratida, Lytoceratida, Ammonitida, Stephanoceratida (Stephanoceratidina + Cardioceratidina), Perisphinctida, Acanthoceratida and Ancyloceratida. *U. Scyth. - Danian.*

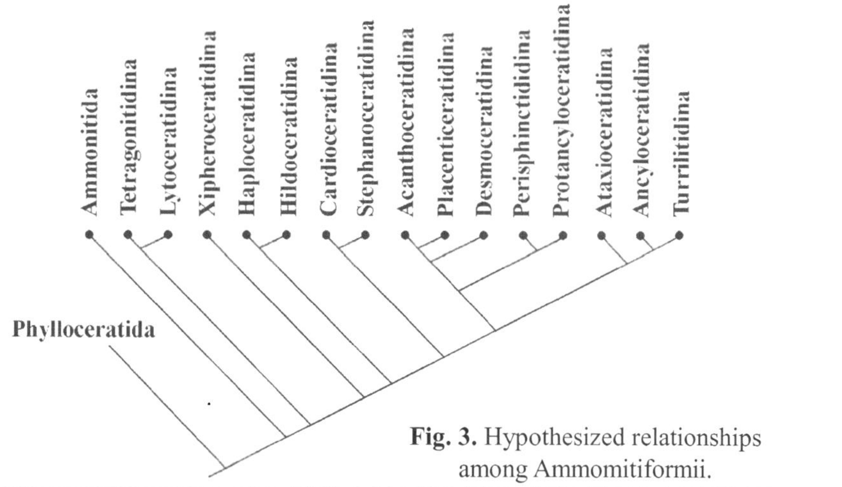

Fig. 3. Hypothesized relationships among Ammomitiformii.

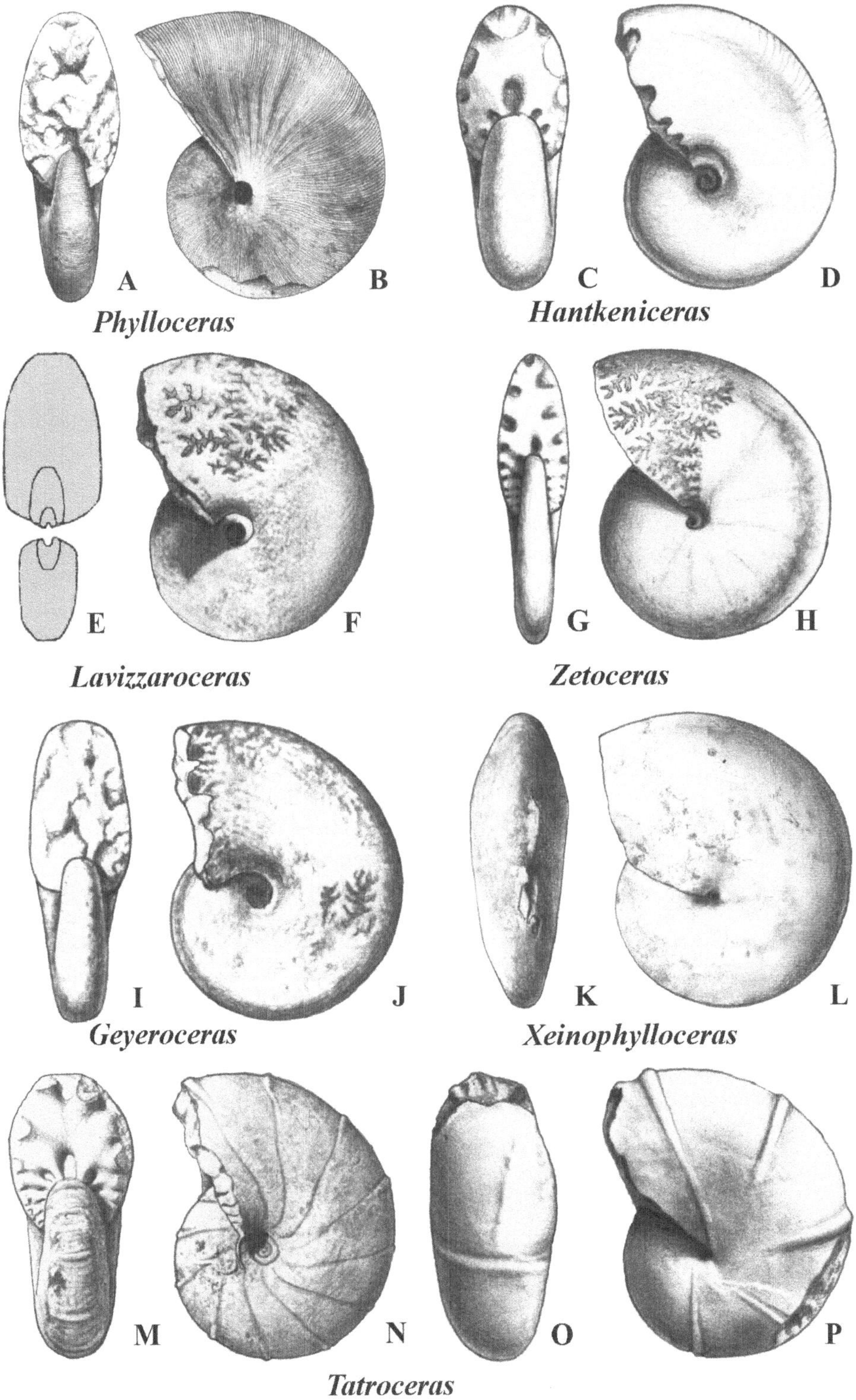

Fig. 4. Phylloceratidae.

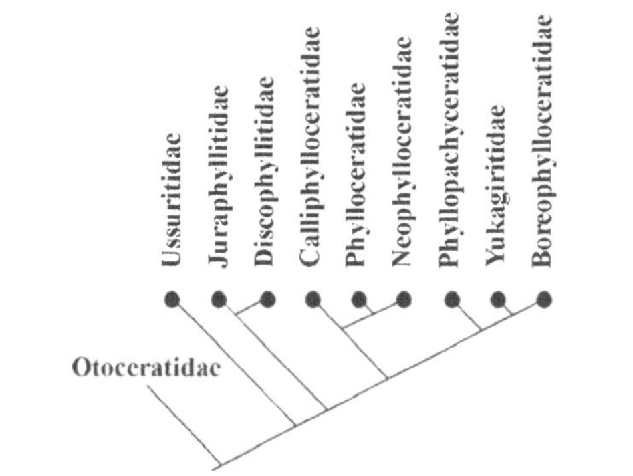

Fig. 5. Hypothesized relationships among Phylloceratida.

Order PHYLLOCERATIDA Hyatt, 1900

Discussion. Monophyletic derivatives of Ceratitida (Hyatt, 1889); comprises only Phylloceratidina, the single ammonite group to survive the Triassic-Jurassic extinction event. *U. Scyth. - U. Maastr.*

Suborder PHYLLOCERATIDINA Hyatt, 1900

(*nom. transl.* Schindewolf 1923 *ex* Phylloceratida, *nom. correct.* herein)

Discussion. Monophyletic derivatives of Ceratitidina (Arkell, 1950), close to Otoceratidae; ammoniticones with thin test, a retrochoanitic septal neck to the siphuncle in early to middle growth, later becoming amphichoantitic (Tanabe *et al.*, 1993), and a simple unfrilled lituid dorsal lobe to the suture. Comprises the superfamilies Ussuritoidea, Phylloceratoidea, Phyllopachyceratoidea and Boreophylloceratoidea. *U. Scyth. - U. Maastr.*

Superfamily PHYLLOCERATOIDEA Zittel, 1884

(*nom. transl.* Hyatt 1900 *ex* Phylloceratida, *nom. correct.* Arkell, 1952)

Discussion. Monophyletic derivatives of Ussuritoidea (Rollier, 1922); typically very involute, compressed, finely lirate to smooth, with narrowly rounded to subtabulate and occasionally keeled venter, with or without constrictions and quinquelobate diphyllic to polyphyllic suture commonly with phylloid folioles and a descending series of numerous auxiliary elements. Due to conservative shell form, evolutionary change involves mainly sutural modification and alteration in details of ornament that, for the most part, are subtle. Comprises the families Juraphyllitidae (not treated here), Phylloceratidae, Calliphylloceratidae and Neophylloceratidae. A persistent and conservative stock, commonest in the deeper water low-energy environments of the Tethyan Realm. *U. Scyth. - U. Maastr.*

Family PHYLLOCERATIDAE Zittel, 1884

Discussion. Monophyletic derivatives of Discophyllitidae (Rollier, 1922); typically unconstricted, with circum-umbilical depression and diphyllic or tetraphyllic E/L to the suture, but tetraphyly may have evolved more than once (Joly, 1977). Comprises only Phylloceratinae. *Hettang. - Valang.*

Subfamily Phylloceratinae Zittel, 1884

Discussion. A monophyletic taxon with an origin in Discophyllitinae, close to *Rhacophyllites* (Rollier, 1922); typically large, compressed, with broad weakly convex to flattened flanks, narrowly rounded to subtabulate venter, and strong sharp dense fine flexuous lirae. *Hettang. - Valang.*

Phylloceras SUESS, 1865 [*Ammonites heterophyllus* J. Sowerby, 1820; OD]. Large; very involute, compressed, with small funnel-shaped circum-umbilical depression surrounding deep pit-like umbilicus, convex flanks, narrowly rounded venter and strongly compressed oval whorl section; uniform fine dense flexuous prorsiradiate lirae arise at edge of circum-umbilical depression and cross venter with slight forward arch; weak radial pleats with bundled lirae may develop around umbilicus in later growth; suture with shallow E, narrow stemmed diphyllic or triphyllic E/L, broad subsymmetrically bifid L, triphyllic L/U_2 that protrudes well beyond E/L, and pro-

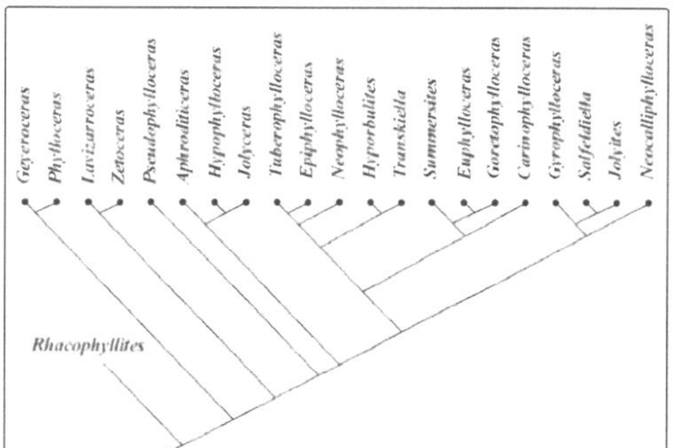

Fig. 6. Hypothesized relationships among Phylloceratidae and Neophylloceratidae.

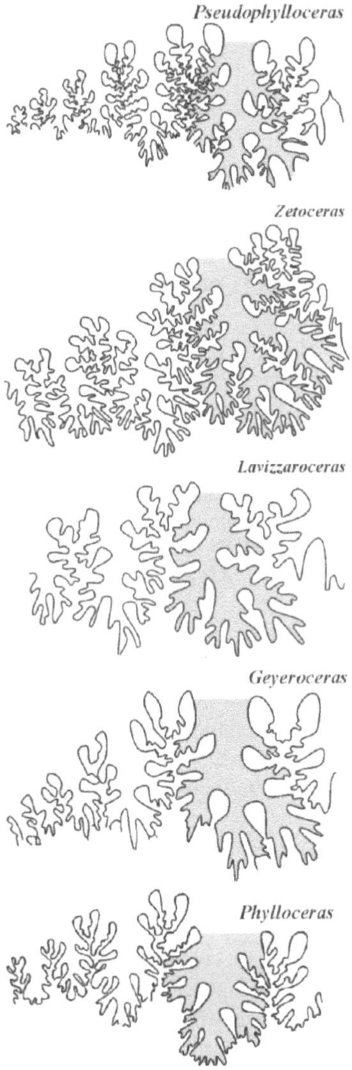

Fig. 7. Phylloceratoid sutures.

nounced phylloid terminations to folioles. *Hettang. - Kimmer.* **Fig. 4A-B** - *P. heterophyllum* (J. Sowerby).

= *Rhacoceras* Hyatt, 1867 (obj.).

= *Plylloceras* Kilian, 1892 (*laps. cal.*)

= *Heterophylloceras* Kovacs, 1939 (obj.).

Geyeroceras HYATT, 1900 [*Ammonites cylindricus* J. de C. Sowerby, 1831; OD]. Moderately large; like *Phylloceras* but with steeper umbilical wall, broad subtabulate venter, subtrapezoidal whorl section and unornamented shell; suture with shallow E and oblique diphyllic E/L; internal suture with 4 monophyllic saddles. *L. Hettang.* **Fig. 4I-J** - *G. cylindricum* (J. de C. Sowerby), juv.

Zetoceras KOVÁCS, 1939 [*Ammonites zetes* d'Orbigny, 1850; OD]. Moderately large, with very compressed whorl section, flattened outer flanks, narrowly rounded venter, subtrigonal whorl section, indistinct liration and suture with tetraphyllic saddles; the original protograph infers

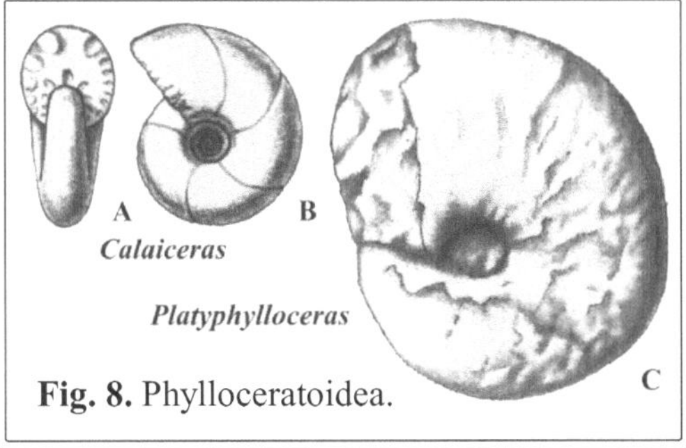

Fig. 8. Phylloceratoidea.

numerous straight feeble prorsiradiate constrictions. *Sinem. - Valang.* **Fig. 4G-H** - *Z. zetes* (d'Orbigny).

Platyphylloceras REPIN, 1998 [*P. kedonicum*; OD]. Microconch to very small; differs from *Phylloceras* in its flattened subparallel flanks, slightly wider umbilicus with steep walls and nepionic characteristics. *U. Bajoc.* **Fig. 8C** - *P. kedonicum* REPIN.

= *Platypachyceras* REPIN, 1998 (*nom.van.*).

Lavizarroceras KOVÁCS, 1939 [*Ammonites lavizzari* von Hauer, 1854; OD]. Small to medium sized; like *Zetoceras* but more inflated, with subtabulate venter and less divided suture with L/U_2 protruding slightly beyond E/L. *Pleins.* **Fig. 4E-F** - *L. lavizzari* (von Hauer).

Family CALLIPHYLLOCERATIDAE Spath, 1927

(*nom. transl.* herein *ex* Calliphylloceratinae)

Discussion. The sister taxon of Phylloceratidae (Joly & Furters, 2007); typically involute, inflated, with convex flanks, rounded to subtabulate venter, and unornamented shell save for growth striae and periodic constrictions indicated across the venter by labial ridges that persist to maturity. A predominantly Jurassic taxon comprising Calliphylloceratinae, Ptychophylloceratinae and Holcophylloceratinae. *Hettang. - Alb.*

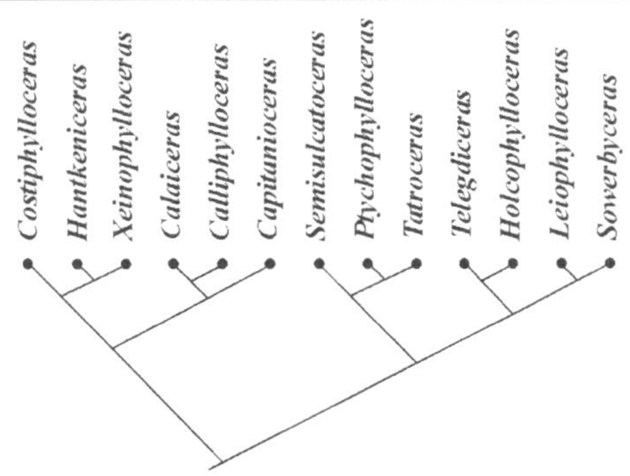

Fig. 9. Hypothesized relationships among Calliphylloceratidae.

Fig. 10. *Costiphylloceras.*

Subfamily Calliphylloceratinae Spath, 1927

Discussion. The sister taxon of Phylloceratinae, but with immediate ancestor uncertain; typically without lirae or riblets; suture with shallow E, oblique diphyllic E/L and trifid L. A waning group in the Early Cretaceous. *Hettang. - Tithon.*

Calliphylloceras SPATH, 1927 [*Phylloceras disputabile* Zittel, 1869; OD]. Medium sized; involute, compressed, with broad gently convex flanks converging to narrowly rounded venter and oval whorl section; shell smooth with periodic constrictions indicated by labial ridges across venter; internal mould with 4-9 strong sigmoidal constrictions; suture with shallow E, oblique triiphyllic E/L, large L much deeper than E, triphyllic L/U_2 and 5 auxiliary elements. *Hettang. - Tithon.* **Fig. 12E-F** - *C. disputabile* (Zittel).

Calaiceras KOVÁCS, 1939 [*Ammonites calais* Meneghini, 1881; OD]. Based on a juvenile; relatively evolute, inflated, with steep umbilicak wall, broad low weakly convex venter and subcircular whorl section; internal mould with 6 concave constrictions that project strongly forward near venter; suture complex. *Sinem. - Pleinsb.* **Fig. 8A-B** - *C. calais* (Meneghini).

Hantkeniceras KOVÁCS, 1939 [*Phylloceras hantkeni* Schloenbach, 1867; OD]. Large; like *Calaiceras* but with flatter flanks and compressed oval whorl section; shell smooth, with growth striae that curve forward near venter; suture relatively simple, with L/U_2 as large as E/L. *U. Pliens.* **Fig. 4C-D** - *H. hankeni* (Schloenbach).

Adult of *Calaiceras* (*fide* Blau & Meister, 1991).

Capitanioceras KOVÁCZ, 1939 [*Phylloceras capitanei* Catullo, 1847; OD]. Like *Calliphylloceras* but with more numerous straight prorsiradiate constrictions without inflexion at midflank. *Sinem.* **Fig. 13** - *C. capitanei* (Catullo).

Costiphylloceras JOLY, 2000 [*C. argelliezi*; OD]. Medium sized; like *Calliphylloceras* but with radial folds superimposed with falciform lirae; periodic falciform constriction difficult to detect, absent at large size. *Toarc. - M. Bajoc.* **Fig. 10** - *C. argelliezi* Joly.

Xeinophylloceras BUCKMANN, 1921 [*X. xeinus*; OD]. Large; like *Calliphylloceras* but more involute, with narrower venter, maximum width close to midflank and subrhomboidal whorl section; constrictions obsolete in later growth. *Toarc.* **Fig. 4K-L** - *X. xeinum* Buckmann.

Subfamily Ptychophylloceratinae Collignon, 1937

Discussion. Monophyletic derivative of Calliphylloceratinae, close to *Calliphylloceras* (Rulleau, 1997); mostly small to medium sized, rather inflated, with broadly rounded venter, periodic periumbilical constrictions and circum-umbilical depression; suture with deep E, asymmetrically trifid L and phylloid folioles. *Toarc. - Apt.*

Ptychophylloceras SPATH, 1927 [*Phylloceras feddeni* Waagen, 1875; OD]. Medium sized; like *Calliphyll-*

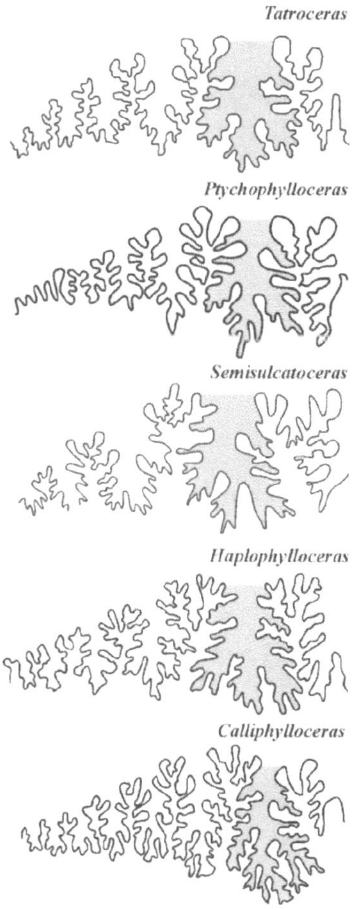

Fig. 11. Calliphylloceratid sutures.

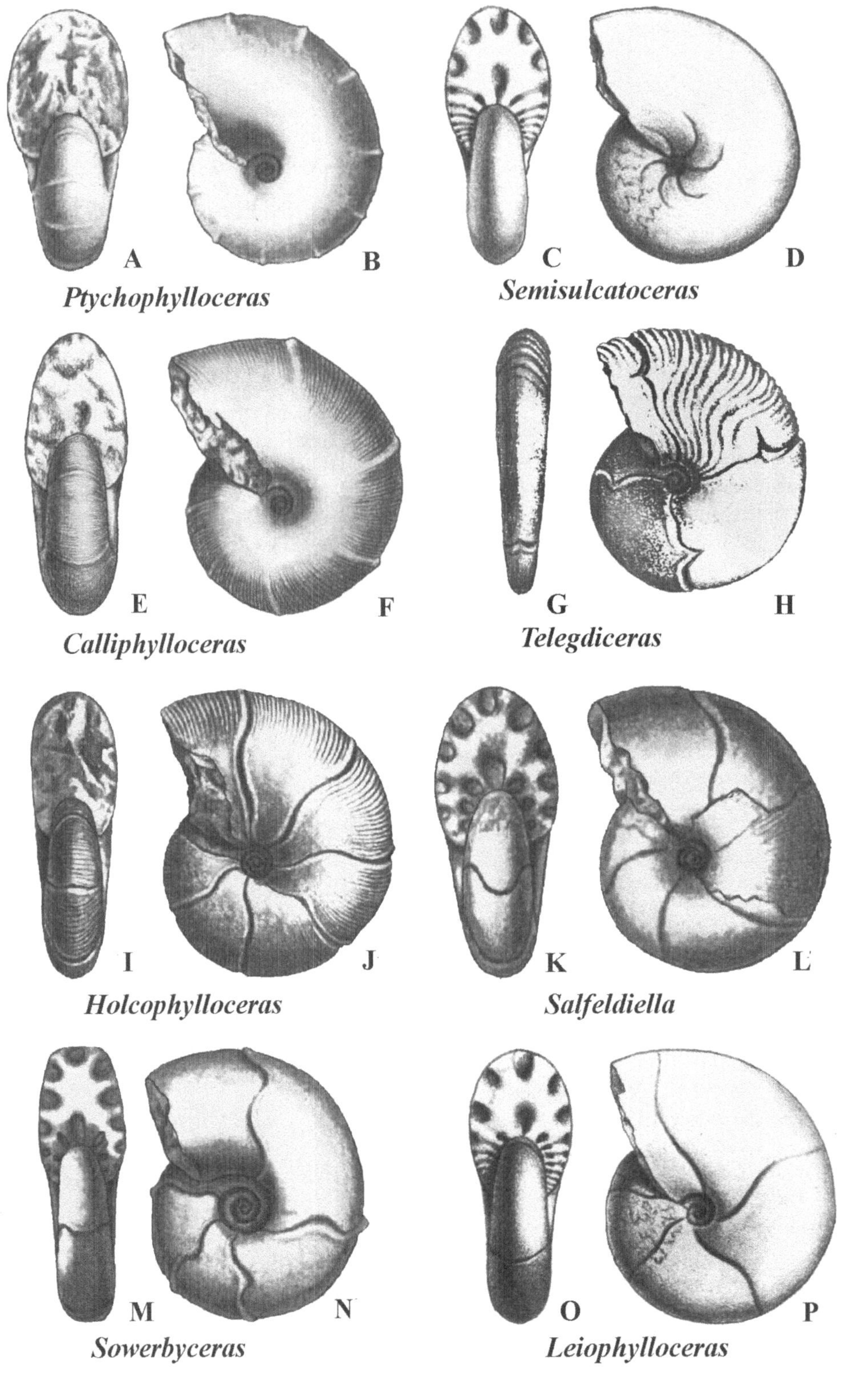

Fig. 12. Calliphylloceratinae.

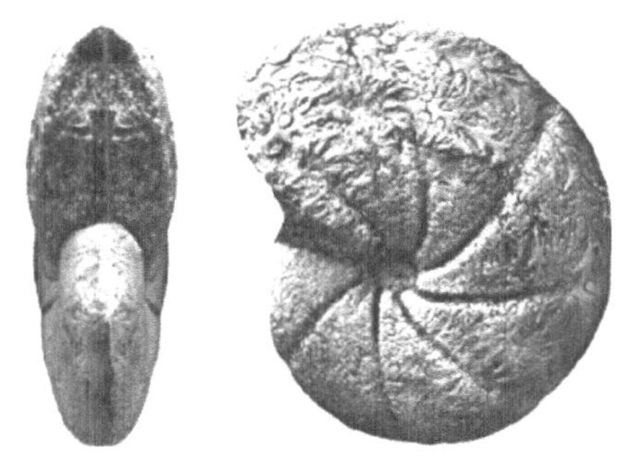

Fig. 13. *Capitanioceras.*

oceras but more inflated, smooth, with broadly rounded venter, gentle umbilical slope and funnel-shaped circum-umbilical depression; 3-8 labial ridges on internal mould; peristome with lateral lappets and shallow dorsal sinus; suture with diphyllic saddles, E as deep as L, and 8 auxiliary elements. *Bajoc. - Haut.* **Fig. 12A-B** - *P. feddeni* (Waagen).

= *Neumayriceras* Sorrentino, 1942 (*non* Rollier 1909).

Tatroceras KOVÁCS, 1939 [*Ammonites tatricus* von Pusch, 1837; OD]. Medium sized; like *Ptychophylloceras* but with occluded umbilicus, large circum-umbilical depression, maximum width on outer flanks and broadly rounded venter with fewer labial ridges. *Toarc.-Oxford.* **Fig. 4M-N** - *T. euphyllum* (Neumayr); **Fig. 4O-P** - *T. tatricum* (Pusch).

= *Tatrophylloceras* Beznosov, 1957 (obj.).

Semisulcatoceras JOLY, 2000 [*Ammonites semisulcatus* d'Orbigny, 1841; OD]. Very small; like *Ptychophylloceras* but with compressed subelliptical whorl section and constrictions reduced to peri umbilical crescents. *Berrias.-Apt.* **Fig. 12C-D** - *S. semisulcatum* (d'Orbigny).

Subfamily Holcophylloceratinae Druschicz, 1956

Discussion. Probably monophyletic derivatives of Calliphylloceratinae (Rulleau, 1998), but an origin in Phylloceratinae has been suggested also (Besnozov & Mikhailova, 1984); somewhat open umbilicus, with convex flanks, rounded to subtabulate venter, very finely lirate to "smooth" shell, prominent falcoid to sigmoidal constrictions to internal mould, sometimes accompanied by flares, and a suture with shallow E and diphyllic saddles. A waning group in the Lower Cretaceous. *Toarc. - U. Alb.*

Telegdiceras KOVÁCS, 1939 [*Phylloceras ultramontanum* Zittel, 1869; OD]. Small, compressed, involute, with flat flanks, narrowly rounded venter and compressed elliptical whorl section; 4 prominent biconcave prorsiradiate constrictions affect shell as well as internal mould and are linguoid over venter; in later growth with sinusoidal ribs strongest across venter; suture with broad E, narrow oblique bifid E/L, broad trifid L much deeper than E, narrow bifid L/U_2 and 6 auxiuliary elements. *Toarc.-Bajoc.* **Fig. 12G-H** - *T. ultramontanum* (Zittel).

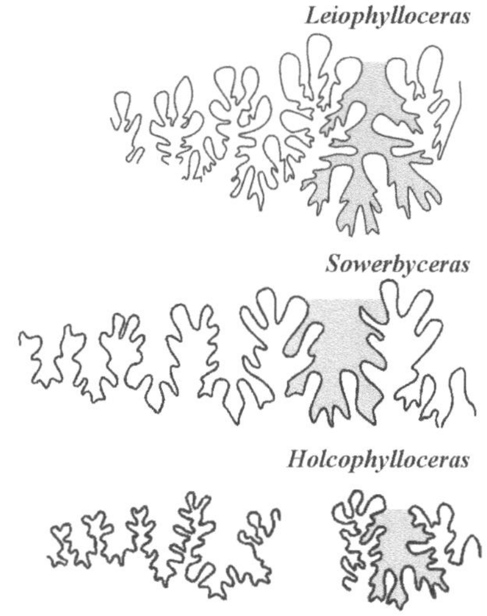

Fig. 14. Holcophylloceratine sutures.

Holcophylloceras SPATH, 1927 [*Phylloceras mediterranneum* Neumayr, 1871, = *Ammonites zignodianus* d'Orbigny, 1848; OD]. Like *Telegdiceras* but moderately large, more inflated, with 6-8 prominent biconcave prorsiradiate constrictions which form chevrons over venter; lirae parallel constrictions but are effaced from inner flank, strengthening on adult body chamber; microconch aperture with ventral and lateral lappets; suture with bipartite heads to saddles of internal suture. *Bajoc. - Alb.* **Fig. 12I-J** - *H. zignodianum* (d'Orbigny).

Sowerbyceras PARONA & BONARELLI, 1895 [*Ammonites tortisulcatus* d'Orbigny, 1841; OD]. Very large; like *Holcophylloceras* but more evolute, compressed, with steep umbilical wall and stepped umbilicus, flattened flanks and venter, subrectangular whorl section, and deep sigmoidal constrictions (marked by raised ribs) that are linguoid across venter; suture with E as large as L, E/L diphyllic and L/U_2 triphyllic. *Bathon. - Valang.* **Fig. 12M-N** - *S. tortisulcatum* (d'Orbigny), juv.

= *Martelliceras* Sorrentino, 1942 (*non* Schindewolf, 1925).

= *Sowerbiceras* Khimshiashvili, 1957 (*nom. van.*).

Leiophylloceras ARKADIEV, 2002 [*Ammonites calypso* d'Orbigny, 1842; OD]. Medium sized; like *Holcophylloceras* but with convex flanks, oval whorl section, fewer stronger prorsiradiate constrictions per whorl and no intervening riblets; suture with splayed trifid L much deeper than E and phylloid folioles to saddle elements. *L. Berrias.* **Fig. 12O-P** - *L. calypso* (d'Orbigny).

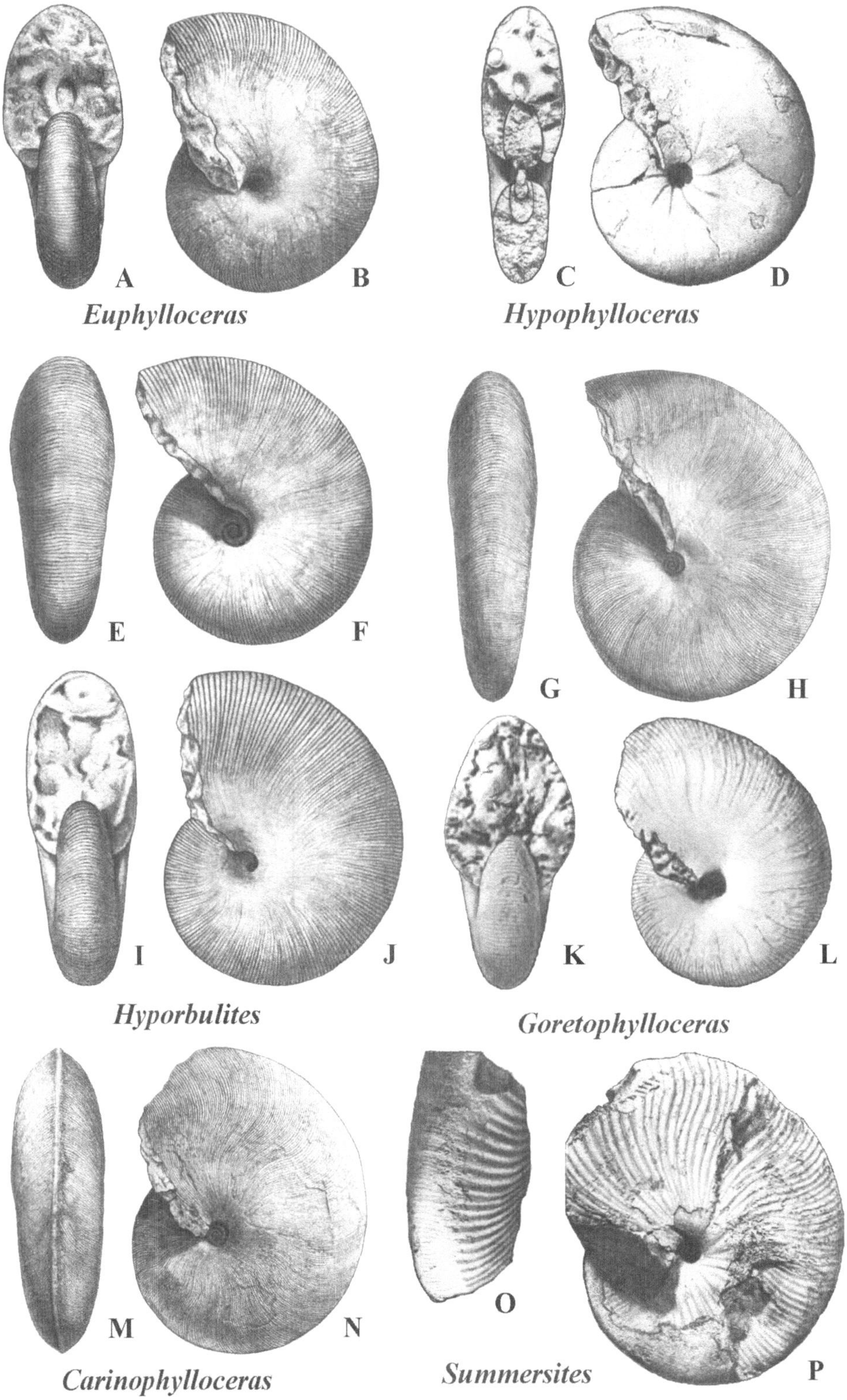

Fig. 15. Phylloceratidae.

Fig. 16. Phylloceratid sutures.

Family Hypophylloceratidae Spath, 1927

(*nom. transl.* herein *ex* Hypophylloceratinae)

Discussion. A monophyletic group with an origin in Phylloceratidae (Spath, 1927); typically with prorsiradiate constrictions to internal mould at least in early to middle growth and more complex suture. Although Beznosov & Mikhailova (1984) looked to Holcophylloceratinae for the origin of *Pseudophylloceras*, constrictions have arisen more than once in phylloceratoids. Comprises the subfamilies Hypophylloceratinae, Neophylloceratinae and Neocalliphylloceratinae. *Bajoc. - Coni.*

Subfamily Hypophylloceratinae Spath, 1927

= Pseudophylloceratinae Besnosov, 1958

Discussion. A monophyletic group with an origin in Phylloceratinae (Spath, 1927), close to *Zetoceras* (Murphy & Rhodda, 2007); typically with constrictions to internal mould and relatively complex suture with di- or tetraphyllic E/L. A rising group in the Cretaceous. *U. Bajoc. - Coni.*

Pseudophylloceras Besnosov, 1957 [*Ammonites kudernatschi* von Hauer, 1854; OD]. Medium sized; moderately inflated, with convex flanks, oval whorl section, reduced circum-umbilical depression, narrowly rounded venter and oval whorl section; bundles of coarse and fine lirae (2-5), striate on lower flank; upper flank with flexuous slightly prorsiradiate lirae that curve forward near periphery and cross venter transversely; internal mould with prominent straight prorsiradiate constrictions; suture with short E, oblique tetraphyllic E/L, narrow splayed trifid L and tetraphyllic L/U_2 and phylloid folioles. *U. Bajoc. - Kimmer.* **Fig. 17** - *P. kudernatschi* (von Hauer).

Hypophylloceras Salfeld, 1924 [*Phylloceras onoense* Stanton, 1895; OD]. Moderately large; like *Pseudophylloceras* but more involute (U<8%) and flat sided, with elliptical whorl section, smaller circum-umbilical depression, geniculate to biconcave constrictions in early to middle growth (D<80 mm), finer denser lirae that are not bundled, and more incised suture with diphyllic E/L and L/U_2, latter becoming polyphyllic. *Tithon. - Coni.* **Fig. 15C-D** - *H. knoxvillense* (Stanton); **Fig. 15G-H** - *H. onoense* (Stanton).

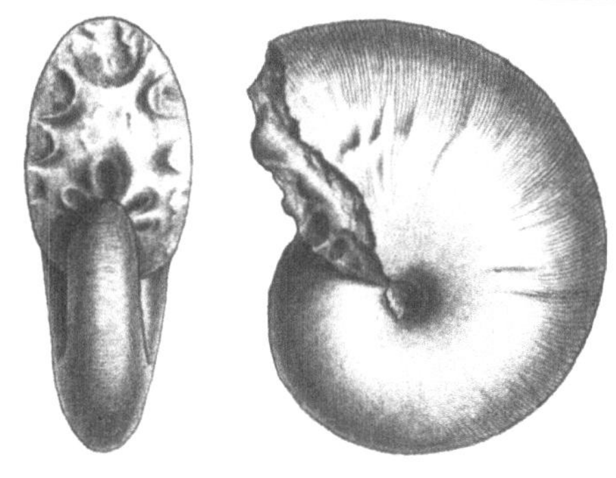

Fig. 17. *Pseudophylloceras*.

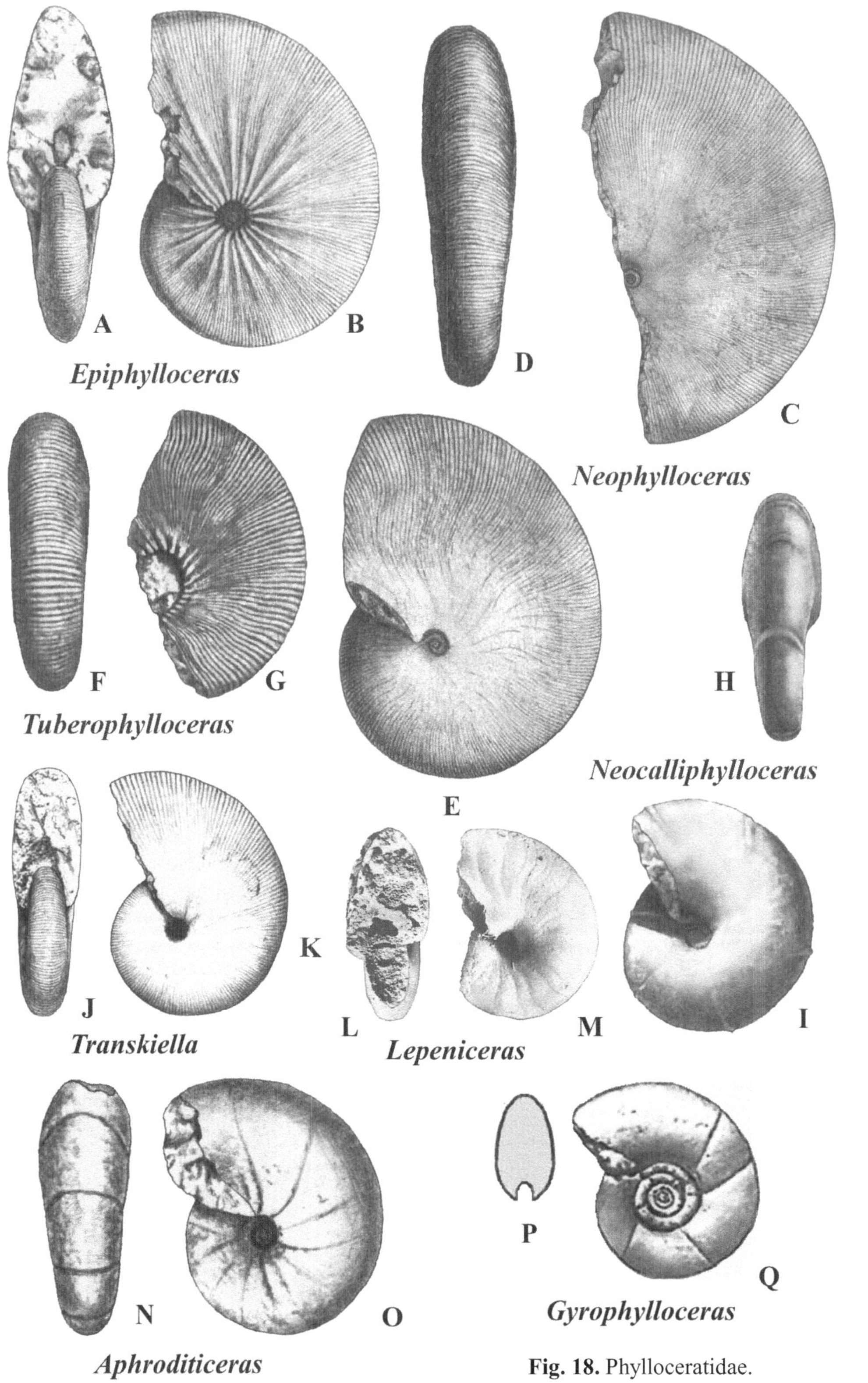

Fig. 18. Phylloceratidae.

Euphylloceras DRUSHCHITS, 1953 [*Ammonites ponticuli* Rousseau, 1842; OD]. Large; like *Hypophylloceras* but inflated, with convex flanks, broader venter, oval whorl section, coarser lirae, early loss of constrictions and suture with U_2 much smaller than L, 8 auxiliary lobes, indented I with lateral branches and 3-lobed proseptum. *Haut. - M. Alb.* **Fig. 15A-B** - *E. ponticuli* (Rousseau); **Fig. 15E-F** - *E. velledae* (d'Orbigny).

Jolyceras VERMEULEN, 2005 [*Phylloceras paquieri* Sayn, 1896; OD]. Like *Hypophylloceras* but with prominent geniculate constrictions in early growth, later with 5-6 straight constrictions, weak across rounded venter; superficially similar to *Salfeldiella*. *L. Apt. - L. Alb.* **Fig. 22G-H** - *J. paquieri* (Sayn).

Summersites MURPHY & RHODDA, 2006 [*S. summersi*; OD.] Medium sized; like *Euphylloceras* but with vertical umbilical wall, angulated edge and coarse broad adorally biconcave riblets on flanks that split into lirae over venter. *U. Apt.* **Fig. 15O-P** - *S. summersi* Murphy & Rhodda.

Carinophylloceras KLINGER, WIEDMANN & KENNEDY, 1975 [*C. collignoni*; OD]. Moderately large; like *Euphylloceras* but with fastigiate venter to inner whorls, later with distinct siphonal keel and lirae forming chevrons over venter; suture with tetraphyllic L/U_2 in later growth. *M. Alb.* **Fig. 15M-N** - *C. collignoni* Klinger, Wiedmann & Kennedy.

Goretophylloceras COLLIGNON, 1949 [*Plylloceras goreti* Kilian, 1892, = *Ammonites fortunei* Honnorat-Bastide, 1892; OD]. Small; like *Euphylloceras* but more inflated, with convergent flanks, narrowly rounded venter and oval to subtrigonal whorl section; constrictions vestigial to obsolete; fine lirae grouped in fasciculate bundles, but internal mould smooth; suture with shallow E and diphyllic saddles with phylloid terminal folioles to E/L and L/U_2. *U. Valang. - U. Alb.* **Fig. 15K-L** - *G. fortunei* (Honnorat-Bastide).

Aphroditiceras MAHMOUD, 1955 [*Phylloceras aphrodite* Fallot & Termier, 1923; OD]. Small; like *Hypophylloceras* but with up to 10 straight prorsiradiate constrictions per whorl, fine flexuous striae and irregular low folds (internal mould smooth); suture with bulbous folioles to saddles. *U. Alb.* **Fig. 18N-O** - *A. aphrodite* (Fallot & Termier).

Lepeniceras RABRENOVIC, 1989 [*L. lepensis*; OD]. Like *Goretophylloceras* but with wider umbilicus and numerous sigmoidal striae that in later growth bifurcate and trifurcate from the umbilical edge. *Uppermost Barrem.* **Fig. 18L-M** - *L. lepensis* Rabrenovic.

Subfamily Neocalliphylloceratinae Joly, 2000

Discussion. Monophyletic derivatives of Hypophylloceratinae, close to *Goretophylloceras* (Joly, 2002); typically small, compressed, flat sided, finely

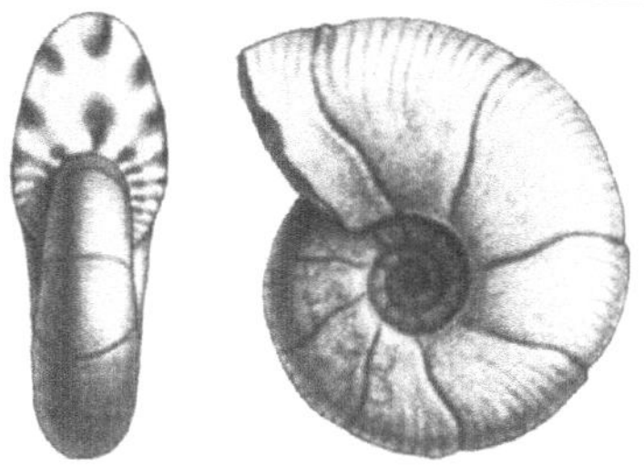

Fig. 19. *Jolyites.*

lirate, with prorsiradiate constrictions marked by labial ridge over venter persisting to middle growth, and relatively simple suture. *L. Apt. - U. Alb.*

Gyrophylloceras HOWARTH, 2018 [*Phylloceras lateumbilicatum* Pervinquière, 1907; OD]. Very small; evolute (U=37%), with narrowly rounded venter and slightly compressed oval whorl sectionin early growth, later with subcircular whorl section; internal mould smooth, with 5 straight prorsiradiate constrictions, slightly sinuous later; shell with very fine liration in referred species; suture with trifid L much deeper than E and retracted suspensive lobe. *L. Apt. - Lower L. Alb.* **Fig. 18P-Q** - *G. lateumbilicatum* (Pervinquière).

= *Gyrophyllites* Wiedmann, 1963 (*non* Glocker, 1841, ichnotaxon).

Salfeldiella SPATH, 1927 [*Ammonites guettardi* Raspail, 1831; OD]. Small; moderately inflated, involute, with convex flanks, narrowly rounded venter and oval to subtrigonal whorl section; 5-8 prorsiradiate falciform constrictions (both on test and internal mould) project strongly forward on outer flank and linguoid over venter; suture relatively simple, with shallow E, diphyllic or triphyllic E/L and L/U_2, and trifid L deeper than E. *Barrem. - U. Apt.* **Fig. 12K-L** - *S. guettardi* (Raspail).

***Jolyites* gen. nov.** [*Ammonites belus* d'Orbigny, 1841; OD herein]. Very small; like *Salfeldiella* but more evolute and compressed, with constrictions that recurve posteriorly on outer flank and are transverse, not linguoid, over venter; suture with diphyllic saddles with bulbous folioles and trifid L deeper than E. *M.-U. Apt.* **Fig. 19** - *J. belus* (d'Orbigny).

Neocalliphylloceras BESAIRIE, 1936 [*Calliphylloceras (N.) collignoni*; OD]. Small; compressed, with flat subparallel flanks, elliptical whorl section and shell surface without lirae; constrictions marked externally by labial ridges across venter. *M. - U. Alb.* **Fig. 18H-I** - *N. collignoni* (Besairie).

Subfamily Neophylloceratinae Joly, 1993

Discussion. Monophyletic derivatives of Hypophylloceratinae, close to *Euphylloceras* (Collignon,

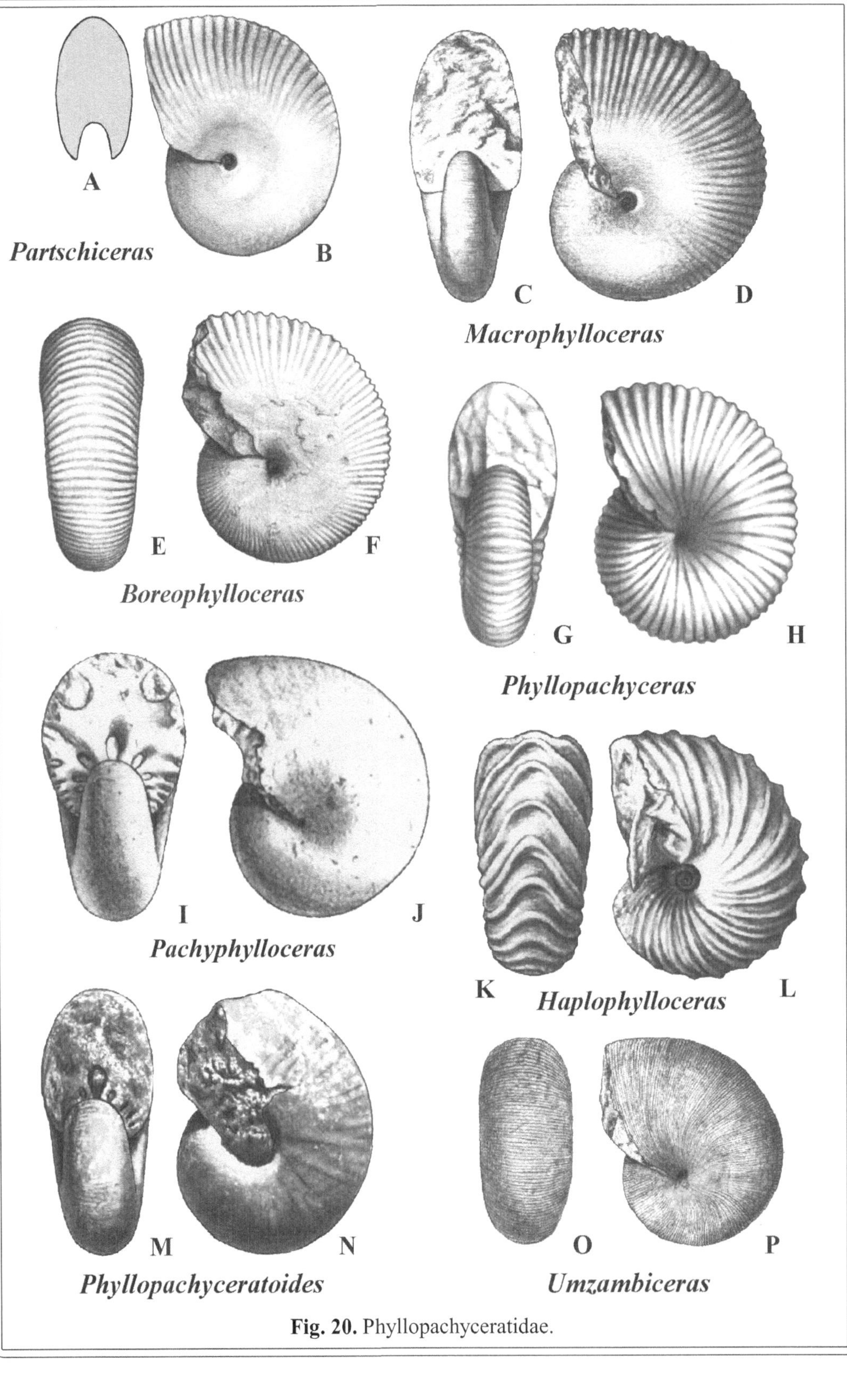

Fig. 20. Phyllopachyceratidae.

1956); strongly compressed, unconstricted, with flattened flanks and relatively complex suture with tetraphyllic saddles. Comprises the tribes Hyporbulitini and Neophylloceratini. *Haut. - U. Maastr.*

Tribe Hyporbulitini Joly, 1993
(*nom. transl.* herein *ex* Hyporbulitinae)

Diagnosis. Typically with moderately incised suture with phylloid folioles; constrictions lacking. *Haut. - L. Camp.*

Hyporbulites BREISTROFFER, 1947 [*Phylloceras seresitense* Pervinquière, 1907; OD]. Moderately large; deep umbilicus with steep umbilical wall, narrowly rounded venter, subelliptical whorl section, prorsiradiate growth striae on inner flank becoming rectiradiate to slightly rursiradiate on outer flank; lirae become flat topped in later growth; internal mould smooth; suture with 7 or more auxiliary elements. *Haut. - Coni.* **Fig. 15I-J** - *H. seresitense* (Pervinquière).

***Transkiella* gen. nov.** [*Phylloceras woodsi* van Hoepen, 1921, p.3; OD herein]. Small to medium sized; like *Hyporbulites* but more compressed, with flat parallel flanks, narrowly rounded venter, compressed elliptical whorl section and prorsiradiate lirae on outer flank. *Sant. - L. Camp.* **Fig. 18J-K** - *T. woodsi* (van Hoepen).

Tribe Neophylloceratini Joly, 1993
(*nom. transl.* herein *ex* Neophylloceratinae)

Diagnosis. Suture extremely incised, with multiphyllic saddles (terminations with >4 folioles). Monophyletic derivatives of Hyporbulitinae, close to *Hyporbulites* (Joly, 1993); *Turon. - U. Maastr.*

Neophylloceras SHIMIZU, 1934 [*Ammonites (Scaphites?) ramosus* Meek, 1857; OD]. Large; strongly compressed, very involute (U<10%), high whorled, with rounded umbilical shoulder, broad flat subparallel flanks, narrowly rounded venter, elliptical whorl section and thin test; very fine dense biconcave prorsiradiate lirae, weak or effaced from innere flank, crossing venter transversely; suture complex, finely incised, tetraphyllic, with deeply bipartite saddles, and minute phylloid terminal folioles; very faint shallow constrictions to internal mould in early growth. *Turon. - Maastr.* **Fig. 18C-E** - *N. ramosum* (Meek).

Epiphylloceras COLLIGNON, 1956 [*Phylloceras surya* Forbes, 1846; OD]. Moderately large; like *Neophylloceras* but with flat convergent flanks, narrowly arched venter, subtrigonal whorl section, well developed slightly prorsiradiate primary ribs on lower flank that give rise to fasciculate bundles of rectiradiate riblets coarser and less flexuous than in *Neophylloceras*, not projecting forward near venter; suture with very shallow E and auxiliary elements running in straight line to umbilical seam. *M. Camp. - U. Maastr.* **Fig. 18A-B** - *E. surya* (Forbes). A homoeomorph of *Costiphylloceras*.
= *Paraphylloceras* Shimizu, 1935 (*non* Salfeld, 1919) (obj.).

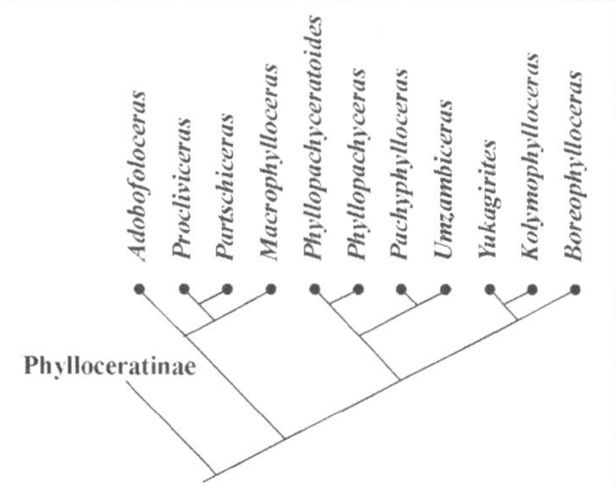

Fig. 21. Hypothesized relationships among Phyllopachyceratoidea and Boreophylloceratoidea.

***Tuberophylloceras* gen. nov.** [*Epiphylloceras cottreaui* Collignon, 1956, p.25; OD herein]. Small; like *Epiphylloceras* but more evolute, with subparallel flanks, narrow evenly rounded venter, compressed elliptical whorl section and fine flexuous riblets arising in fasciculate bundles from pinched umbilical bullae. *U. Maastr.* **Fig. 18F-G** - *T. cottreaui* (Collignon).

Superfamily PHYLLOPACHYCERATOIDEA COLLIGNON, 1937
(*nom. transl.* herein *ex* Phyllopachyceratinae)

Discussion. A seemingly monophyletic group descended from Phylloceratoidea. Mostly small, inflated, very involute, with ribs across the venter in middle to later growth; constrictions present on early whorls of primitive representatives; suture with long narrow tetraphyllic main saddles and trifid L. Comprises only the family Phyllopachyceratidae. *Hettang. - U. Maastr.*

Family PHYLLOPACHYCERATIDAE COLLIGNON, 1937
(*nom. transl.* Beznosov, 1958 *ex* Phyllopachyceratinae)

Discussion. Monophyletic derivatives of Calliphylloceratidae; the group appeared simultaneously with the first Phylloceratinae in the Hettangian, and the two evolved in parallel until the end of the Cretaceous. Generally small, inflated, ribbed and lacking constrictions (restricted to inner whorls of primitive members only). Commonest in the Tethyan Realm, but never abundant. *Hettang. - U. Maastr.*

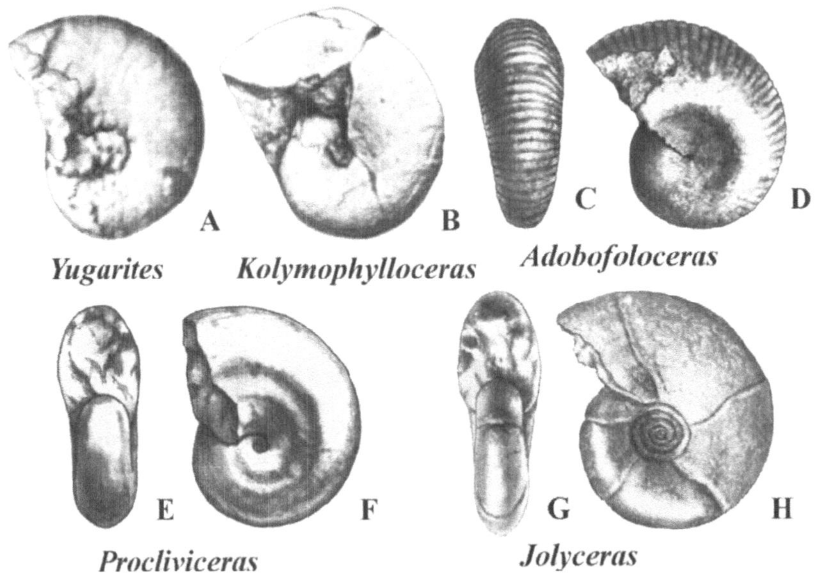

Fig. 22. Phylloceratidina.

Subfamily Phyllopachyceratinae Collignon, 1937

Discussion. Monophyletic derivatives of uncertain ancestry, perhaps Ptychophylloceratinae close to *Ptychophylloceras*. Matsumoto (1959) and Murphy & Rhodda (2006) have questioned the relationship between *Partschiceras* and *Phyllopachyceras*; generally small, very involute and inflated, and commonly ribbed across venter in middle to later growth. Comprises the tribes Partschiceratini and Phyllopachyceratini. *Hettang. - U. Maastr.*

Tribe Partschiceratini nov.

Type genus: *Partschiceras* Fucini, 1923.

Diagnosis. Small, involute, compressed, high whorled, with strong ribs across venter in middle to later growth; early whorls constricted; suture with phylloid folioles, shallow E, slender diphyllic saddles, oblique E/L and retracted suspensive lobe. *Hettang. - Alb.*

Partschiceras FUCINI, 1923 [*Ammonites partschi* Stur, 1851 (*non* Klipstein 1843), = *Partschiceras monestieri* Breistroffer, 1949, = *Ammonites striatocostatus* Meneghini, 1853; SD Spath, 1927]. Small; very involute (U=8-14%), compressed, with indistinct umbilical wall, broad rather flat flanks, rounded venter, oval whorl section and inner whorls with conspicuous constrictions; weak distinct ribs arise at midflank and cross venter transversely; suture with very short E, diphyllic main saddles, large trifid L much deeper than E and at least 5 retracted auxiliary elements. *Hettang. - Apt.* **Fig. 20A-B** - *P. striatocostatum* (Meneghini).

= *Partschiphylloceras* Roman, 1938 (obj.).

Procliviceras FUCINI, 1923 [*Phylloceras proclive* Rosenberg, 1909; OD]. Very small; like *Partschiceras* but with oval whorl section and rursiradiate ribbing on outer flank. *Pleins.* **Fig. 22E-F** - *P. proclive* (Rosenberg), juv.

= *Proclivioceras* Roman, 1938 (*nom. van.*).

Macrophylloceras SPATH, 1927 [*Phylloceras ptychostoma* Bennett, 1868; OD]. Medium sized; like *Partschiceras* but with subvertical umbilical wall, well developed umbilical shoulder, subparallel flanks, no circumumbilical depression and broad venter; inner whorls with very finely lirae visible across venter, but developing coarse strong ventral ribs on adult body chamber; suture with moderately broad saddles. *Tithon. – U. Alb.* **Fig. 20C-D** - *M. ptychostomum* (Bennett).

Tribe Phyllopachyceratini Collignon, 1937

(*nom. transl.* herein *ex* Phyllopachyceratinae)

Type genus: *Phyllopachyceras* Spath, 1925.

Diagnosis. Typically small, inflated, with prominent circum-umbilical depression, subcircular whorl section and strong rigid ribs in middle to later growth; suture with deep penetrating E, straight E/L, tetraphyllic saddles, and suspensive lobe of 8-10 elements; constrictions lacking. Monophyletic derivatives of Partschiceratini, close to *Partschiceras*; *Toarc. - U. Maastr.*

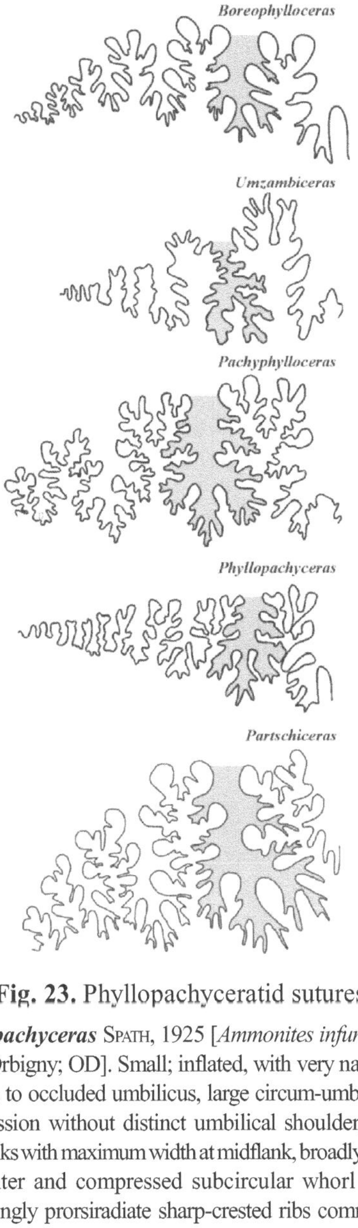

Fig. 23. Phyllopachyceratid sutures.

Phyllopachyceras SPATH, 1925 [*Ammonites infundibulum* d'Orbigny; OD]. Small; inflated, with very narrow pit-like to occluded umbilicus, large circum-umbilical depression without distinct umbilical shoulder, convex flanks with maximum width at midflank, broadly rounded venter and compressed subcircular whorl section; strongly prorsiradiate sharp-crested ribs commence in early growth (15 mm diameter), becoming strong at edge of circum-umbilical depression where they recurve slightly to alternate long and short on outer flank; ribs strongest over venter, forming slight forward convexity, and superimposed with fine lirae; suture with deep penetrating E, long narrow tetraphyllic E/L and L/U_2, trifid L shorter than E, and at least 6 auxiliary elements. *Bajoc. - Apt.* **Fig. 20G-H** - *P. infundibulum* (d'Orbigny).

Adabofoloceras JOLY, 1977 [*Phyllopachyceras adobofolense* Collignon; OD]. Very small; like *Phyllopachyceras* but with flattened or slightly convex sides, weak ribs and suture with E much shallower than L. *Bathon.- Oxford.* **Fig. 22C-D** - *A. adabofolense* (Collignon).

= *Adobofoloceras* Howarth (*in* Wright, 1996) (*nom. van.*)

Haplophylloceras SPATH, 1925 [*Phylloceras strigile* Blanford, 1903; OD].Small; early whorls smooth, with gentle umbilical slope and rounded venter; later venter flattens and strong simple prorsiradiate ribs develop which sharpen on crossing venter with linguoid tongue; suture with saddle endings tending to become multifid. *Tithon. - Berrias.* **Fig. 20K-L** - *H. strigile* (Blanford).

Phyllopachyceratoides **gen. nov.** [*Phyllopachyceras betiokyense* Collignon, 1963, p.7; OD herein]. Small; like *Phyllopachyceras* but more evolute, with subvertical umbilical wall, flattened flanks, compressed oval whorl section and faint blunt rigid ribs that tend to efface over the venter. *Alb.* **Fig. 20M-N** - *P. betiokyensis* (Collignon).

Pachyphylloceras **gen. nov.** [*Ammonites forbesianus* d'Orbigny, 1850, p.211; OD herein]. Small; like *Phyllopachyceras* but strongly inflated, more involute, with smaller circum-umbilical depression, broader rounder venter, equidimensional semicircular whorls, weak ribbing restricted to outer flank and venter of adult body chamber only, and suture more incised with up to 10 auxiliary elements and retracted suspensive lobe. *U. Valang. - U. Maastr.* **Fig. 20I-J** - *P. forbesianum* (d'Orbigny).

Umzambiceras **gen. nov.** [*Phylloceras umzambiense* van Hoepen, 1920, p.142; OD herein]. Very small; like *Neophyllopachyceras* but flat sided, with fasciculate bundles of strong lirae that increase by intercalation and bifurcation; lirae prorsiradiate on inner flank, recurving to become rectiradiate on outer flank; ribs lacking; suture distinctive, with very shallow E, much reduced E/L, broad bifid L, trifid L/U_2 and at least 7 auxiliary elements. *Sant.* **Fig. 20O-P** - *U. umzambiense* (van Hoepen).

Superfamily BOREOPHYLLOCERATOIDEA Alexeev & Repin, 1995

Discussion. This superfamily differs from Phylloceratoidea in that the siphuncle becomes ventral by the beginning of the 1st whorl and, at the same time, the septal neck is prochoanitic. In addition it shows accelerated sutural development with the formation of an 8-lobed suture. However Berriasian *Boreophylloceras* barely differs from some Phyllopachyceratinae in gross morphology, and is distinguished by features of the nepionic stages that are not well documented in many other phylloceratids. Questionable linking of Early Jurassic Yukagiritidae further complicates the taxon that seems to be endemic to the Siberian Province of the Boreal Realm. *Hettang. - Berrias.*

Family YUKAGIRITIDAE Repin, 1998

Discussion. Assignment to Boreophylloceratoidea is based upon nepionic characters, i.e. siphuncle becoming ventral in first whorl and septal necks prochoanitic. Referred species are very small (D<23 mm) and may be based on stunted dwarfs (an associated *Phylloceras* is only 14.3 mm in diameter), making taxonomic discrimination difficult. Typically involute with rounded venter and moderately convex flanks; ornament varies from fine striae to distinct thin ribs. Sutural complexity increases by division of the internal lateral lobe (I) and simultaneous separation of new umbilical lobes. *Hettang. - U. Pliensb.*

Kolymophylloceras REPIN, 1998 [*K. turomchense*; OD]. Small, very involute, with flattened oval whorl section; flanks moderately convex, venter rounded; protoconch rounded, large; siphuncle ventral from end of first whorl; 3rd and all succeeding septal necks prochoanitic and short. *Hettang.* **Fig. 22B** - *P. turomchense* REPIN.

Yukagirites REPIN, 1998 [*Y. molodoensis*; OD]. Small, involute, with weakly convex flanks and narrow rounded venter; flanks with numerous striated riblets; at end of first whorl siphuncle becomes ventral; septal necks prochoanitic, from 9-10th septa with long pillars. *U. Sinem. - U. Pliensb.* **Fig. 22A** - *Y. molodoensis* REPIN.

Family BOREOPHYLLOCERATIDAE Alexeev & Repin, 1995

Discussion: A monotypic taxon, typically with a very large protoconch and caecum, sharply defined primary constriction, siphuncle ventrally positioned from the beginning of the first whorl and the eight-lobed suture formed suddenly, immediately following the five-lobed suture, without intermediate changes. *Berrias.*

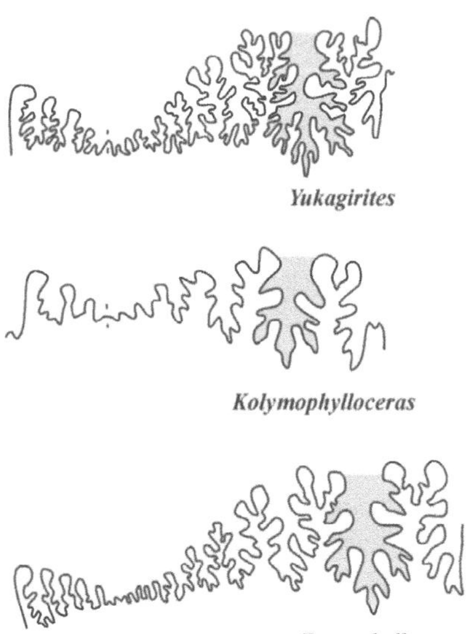

Fig. 24. Boreophylloceratoid sutures.

Boreophylloceras ALEKSEEV & REPIN, 1998 [*Phyllopachyceras praeinfundibulum* Voronez, 1962; OD]. Small; moderately inflated and very involute, with weakly convex to almost flat flanks and broadly rounded venter; in early growth with fine flexuous striae only; from 20-30 mm diameter, fine ribs appear on outer flank and cross venter; in later growth ribs prominent, but inner flanks still smooth; sutural complexity increases by division of inner lateral lobe (I) and separation of umbilical lobes. *Berrias.* **Fig. 20E-F** - *B. densicostatum* (Igolnikov).

Order LYTOCERATIDA Hyatt, 1889

(*nom. transl.* Teichert, 1988 *ex* Lytoceratina)
= Lithoceratida Mikhailova, 1979 (*nom. van.*).

Discussion. Monophyletic derivatives of Psiloceratida (Wiedmann, 1970); comprises Lytoceratidina and Tetragonitidina. *Hettang. - U. Maastr.*

Suborder LYTOCERATIDINA Hyatt, 1889

(*nom. correct.* Arkell, 1950 *pro* Lytoceratinae, *nom. correct.* herein *pro* Lytoceratina)

Discussion. Monophyletic derivatives of Psiloceratidina (Wiedmann, 1970) with prochoanitic septal necks throughout most of ontogeny (Tanabe *et al.*, 1993) and non-lituid dorsal lobe (Kullmann & Wiedmann, 1970); comprises only Lytoceratoidea. *Hettang. - Maastr.*

Superfamily LYTOCERATOIDEA Neumayr, 1875

(*nom. transl.* Buckmann 1894 *ex* Lytoceratidae, *nom. correct.* Dommergues 2002 *pro* Lytocerataceae).

Discussion. Monophyletic derivatives of Arietitoidea (Hoffmann & Keupp, 2006); comprises Derolytoceratidae, Nannolytoceratidae and Lytoceratidae, of which only the last extends into the Cretaceous and is dealt with here. Small to very large serpenticones, with or without ribs and crinkled flares. A conservative stock, mostly on the decline by the beginning of the Cretaceous; commonest in the Tethyan and Pacific Realms, scarse in the Boreal Realm. *Hettang. - Cenom.*

Family LYTOCERATIDAE Neumayr, 1875

= Thysanoidea Hyatt, 1867
= Lithoceratidae Mikhailova, 1979 (*nom. van.*).

Discussion. Monophyletic derivatives of Derolytoceratidae (Hoffmann & Keupp, 2006); typically moderately large to very large serpenticones with rounded whorls, periodic flares that are often crinkled and associated with constrictions, fine to coarse rectiradiate ribbing in some, and suture with 5 lobes, few auxiliary elements and large dorsal lobe.

A diverse Jurassic group, in the Cretaceous represented by the subfamilies Lytoceratinae, Ammonoceratitinae and Hemilytoceratinae. Hoffman's (2010, 2015) concepts of the family are non-phyletic. *Sinem. - Cenom.*

Subfamily Lytoceratinae Neumayr, 1875

(*nom. transl.* Spath, 1927 *ex* Lytoceratidae)

Discussion. Monophyletic derivatives of Derolytoceratidae, close to *Derolytoceras* (Hoffmann & Keupp, 2006) (**Fig. 26F**); typically serpenticone with rapidly expanding whorls, periodic flares and delicate fimbriate ornament, rarely with bold ribs in Cretaceous representatives. As interpreted here, *Lytoceras* is wholly Jurassic and does not extend into the Cretaceous. A waning group in the Lower Cretaceous. *Hettang. – Cenom.*

Lytoceras SUESS, 1865 [*Ammonites fimbriatus* J. SOWERBY, 1817; OD]. Moderately large serpenticone with rounded venter, subcircular whorl section and simple, annular, more or less fimbriate ribbing; suture with narrow short E, oblique narrow stemmed bifid E/L, splayed asymmetrically bifid L and narrow stemmed bifid L/U$_2$. *L. Pliens. - L. Toarc.* **Fig. 26A-B** - *L. fimbriatum* (J. Sowerby).

Thysanoceras HYATT, 1867 [*Ammonites cornucopia* Young & Bird, 1822; OD]. Very large; whorls rounded, slightly depressed, with rounded venter and flattened flanks in maturity; 18-20 prominent, annular, undulating crinkled flares; shell covered with transverse and longitudinal striae, that cross each other and produce a successive series of small quadratic cellular depressions in the shell texture. *Toarc. - Oxford.* **Fig. 26E** - *T. cornucopium* (Young & Bird).

Thysanolytoceras BUCKMAN, 1905 [*Ammonites eudesianus* d'Orbigny, 1846; OD]. Very large; evolute, with rapidly expanding rounded whorls and oval to subcircular whorl section; numerous crenulated flares extend halfway up flank, separated by fine thread-like ribs and fimbriate growth striae; spiral striae may produce recticulate pattern to shell surface; fimbriate ornament declining in later growth; constrictions lacking; highly incised suture with shallow E with elongate median prong, oblique bifid E/L and widely splayed asymmetrically trifid L much deeper than E. *Aalen. - Bathon.* **Fig. 26G-H** - *T. eudesianum*

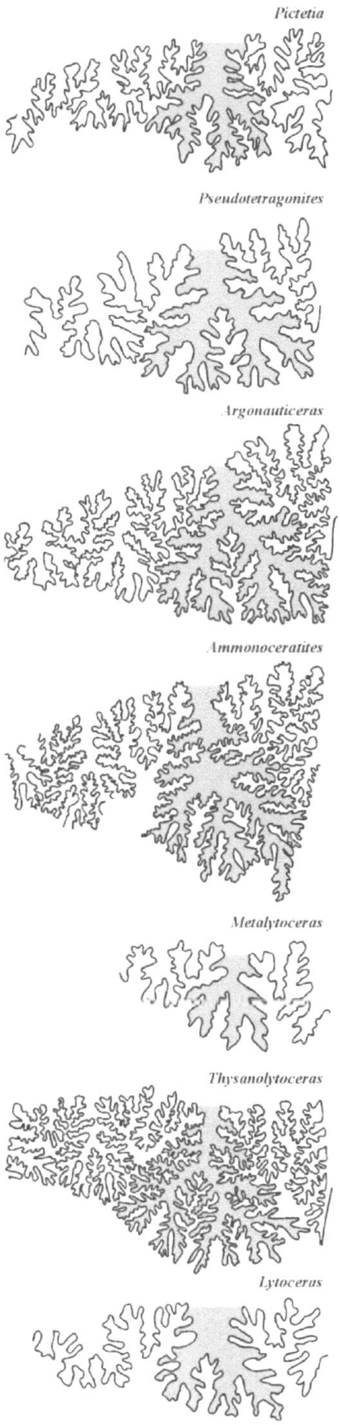

Fig. 25. Lytoceratine sutures.

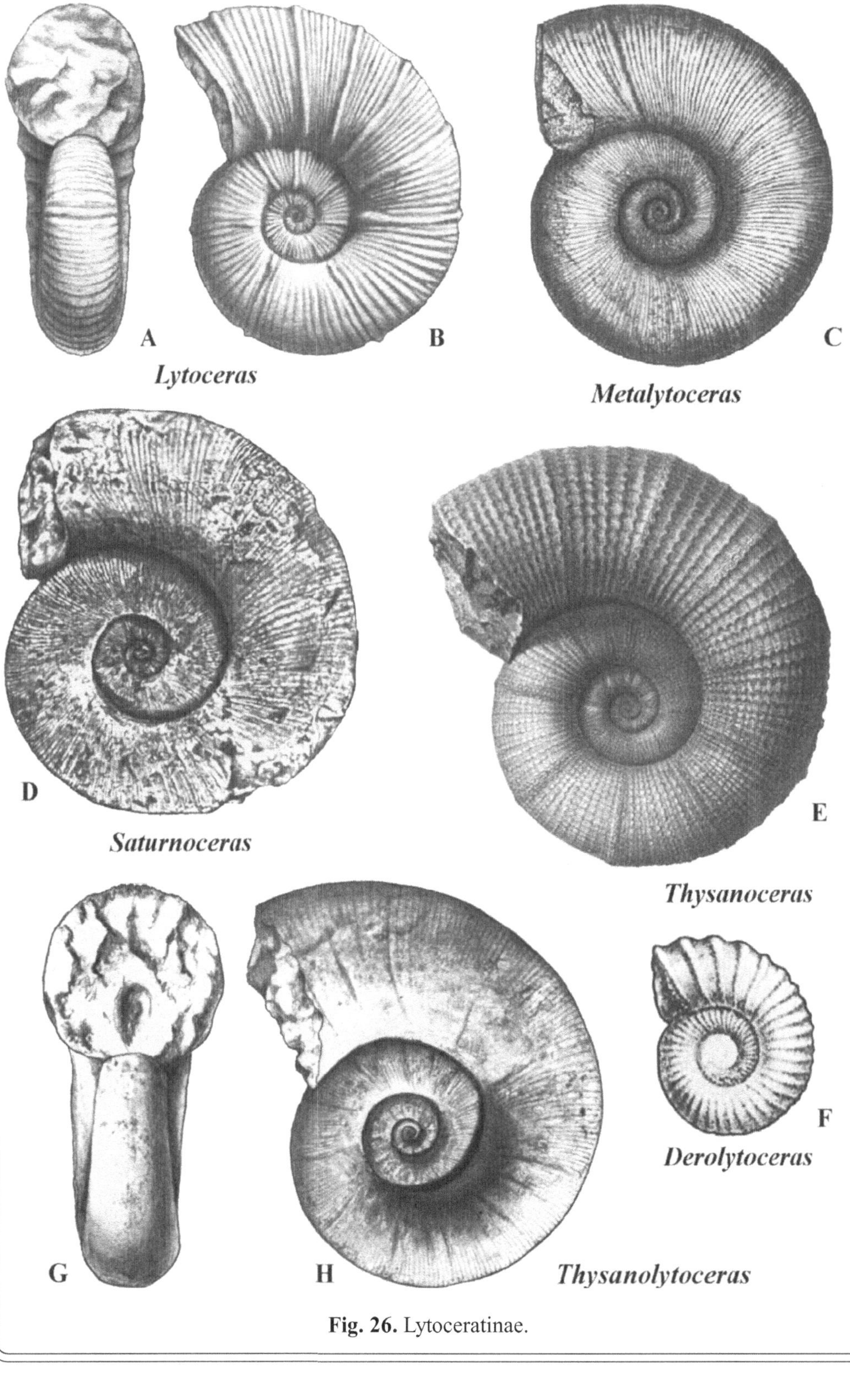

Fig. 26. Lytoceratinae.

Fig. 27. Hypothesized relationships among later Lytoceratidae.

(d'Orbigny); **Fig. 28H** - *T. cinctum* Besnosov. Microconch *Trachylytoceras* (*fide* Rulleau, 1998).

Subfamily Ammonoceratitinae Brunnschweiler, 1966

(*nom. transl.* herein *ex* Ammonoceratitidae)

Discussion. Monophyletic derivatives of Lytoceratinae, close to *Thysanolytoceras*; very large with rapidly expanding whorls and dense fine crinkled ribs and striae; flares restricted to early growth; suture with shallow E and strongly oblique E/L. *M. Valang. - Cenom.*

Saturnoceras ANDERSON, 1938 [*Lytoceras (S.) saturnale*; OD]. Very large; like *Thysanolytoceras* but with depressed whorl section, regular fine ribs and lack of flares. *M. - U. Valang.* **Fig. 26D** - *S. saturnale* (Anderson).

Hoffman (2015) regarded the genus a *nomen nudum* introduced without definition. However Anderson (1938: 146) noted that it differed from "... most of the following [lytoceratids] in its more robust form, in its more restrained sculpture, and in the cross-section of its whorls".

Metalytoceras SPATH, 1927 [*Lytoceras triboleti* Uhlig, 1902; OD]. Small (?juv.); like *Saturnoceras* but compressed, with uniformly expanding whorls and fine prorsiradiate ribs that split into riblets over venter; suture with E almost as long as L, and E/L not oblique. *Valang.* **Fig. 26C** - *M. triboleti* (Uhlig).

Ammonoceratites BOWDITCH, 1822 [*Ammonites lamarcki*; OD]. Very large serpenticone (U=38-42%) with almost circular whorl section; prominent flares and constrictions in earliest growth only; later dense fine crenulated prorsiradiate ribs and striae, convex forward over venter; suture with massive cruciform septal lobe. *U. Apt. - U. Alb.* **Fig. 28C-E** - *A. mahadeva* (Stoliczka).

= *Ammonoceratites* Rafinesque, 1815 (*nom. nud.*).

Ammonoceras LAMARCK, 1822 [*A. glossoidea*; OD]. Based on fragments of a gigantic Indian ammonite, perhaps *A. mahadeva* (Stoliczka); a *nomen dubium*. *Alb.*

Argonauticeras ANDERSON, 1938 [*Lytoceras argonautarum*; OD] Large; like *Ammonoceratites* but with narrower umbilicus, somewhat flattened convergent flanks, subtabulate venter, subtrapezoidal whorl section and very fine, generally non-crenulate lirae in maturity, transverse across venter. *U. Apt.* **Fig. 28F-G** - *A. argonautarum* (Anderson).

Pseudotetragonites DRUSHCHITS, 1956 [*P. kudrjavzevi*; OD]. Small (?juv.); like *Saturnoceras* but with crenulated lirae to early whorls. *U. Apt.* **Fig. 28A-B** - *P. kudrjavzevi* Druschicz.

Subfamily Hemilytoceratinae Spath, 1927

Discussion. Monophyletic derivatives of Lytoceratinae, close to *Thysanolytoceras* (Roman, 1938), radiating in the Early Cretaceous; characteristically medium sized serpenticones with slowly expanding rounded whorls, numerous non-crinkled flares and, often, strong regular riblets; suture relatively simple with E almost as deep as L and E/L slightly oblique to straight. *L. Tithon. - U. Apt.*

Hemilytoceras SPATH, 1927 [*Ammonites immanis* Oppel, 1865; OD]. Small; smooth inner whorls like *Pterolytoceras*; later with depressed whorls and closely spaced, concave, funnel-shaped lamellar flares. *Bajoc. - U. Tithon.* **Fig. 31G-H** - *H. immane* (Oppel); **Fig. 30I-J** - *H. fraasi* (Dacquè).

The defining characters are based upon early growth stages and do not persist to maturity.

Pterolytoceras SPATH, 1927 [*Ammonites exoticus* Oppel, 1863; OD]. Medium sized to moderately large; very evolute (U=44-46%), with slowly expanding circular whorls barely in contact; distant rectiradiate to slightly-prorsiradiate crinkled flares separated by very fine, dense, crinkled lirae so as to appear "smooth"; flares distant on septate whorls, becoming crowded on adult body chamber; spiral threads may create reticulate ornament; shell thin and internal mould smooth. *L. Tithon. - Berrias.* **Fig. 31A-B** - *P. exoticum* (Oppel).

Eulytoceras SPATH, 1927 [*Ammonites inaequalicostatus* d'Orbigny, 1840; OD]. Medium sized; like *Pterolytoceras* but with regular distant primary ribs separated by 1-4 weaker secondaries and striae; constrictions lacking. *L. Haut.- L. Apt.* **Fig. 31C** - *E. inaequalicostatum* (d'Orbigny).

Biasaloceras DRUSCHICZ, 1953 [*B. sauclum* Druschicz; OD]. Medium sized; inner whorls like *Eulytoceras* with ribs separated by groups of striae; outer whorls with strong fine closely spaced ribs crossing venter transversely; sparse constrictions on internal mould. *Haut. - Barrem.* **Fig. 31D-E** - *B. sauclum* (Druschicz).

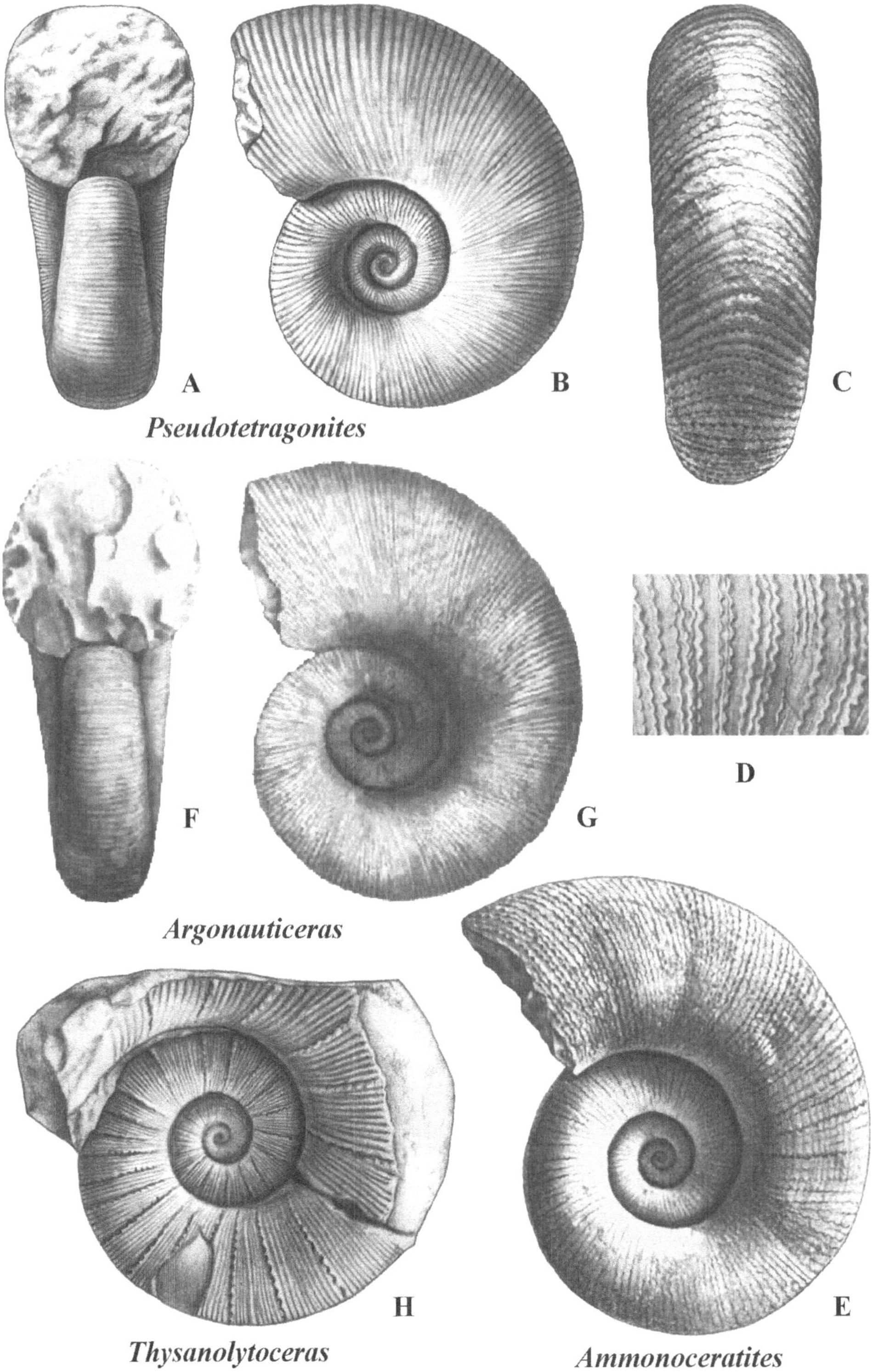

Fig. 28. Lytoceratinae.

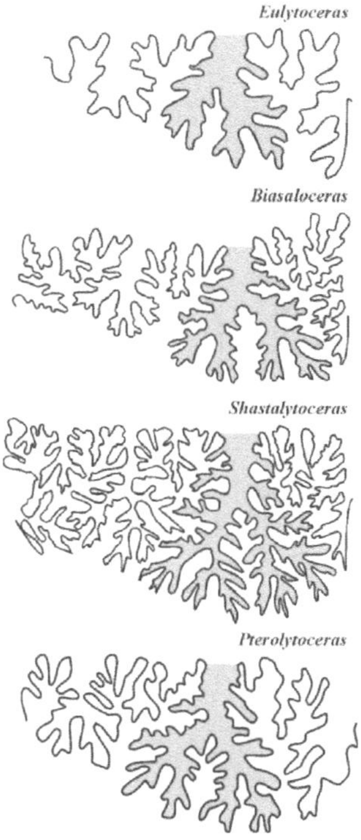

Fig. 29. Hemilytoceratine sutures.

***Shastalytoceras* gen. nov.** [*Ammonites batesi* Trask, 1855, p.39; OD herein]. Moderately large; like *Biasaloceras* but with uniform strong fine flexuous prorsiradiate riblets throughout, without strengthened primaries, flares or constrictions. *Haut. - Barrem.* **Fig. 31F** - *S. batesi* (Trask).

Suborder TETRAGONITIDINA Schindewolf, 1968

(*nom. correct.* herein *pro* Tetragonitina)

Discussion. Monophyletic derivatives of Lytoceratidina (Schindewolf, 1968), retaining a septal lobe but with sexlobate primary suture; the most derived of all ammonites. Comprises only Tetragonitoidea. *U. Tithon. - Maastr.*

Superfamily TETRAGONITOIDEA Hyatt, 1900

(*nom. transl.* Wiedmann, 1962 *ex* Tetragonitidae)

Discussion. Monophyletic derivatives of Lytoceratoidea (Wright., 1957); comprises Protetragonitidae, Gaudryceratidae and Tetragonitidae. *U. Tithon. - U. Maastr.*

Family PROTETRAGONITIDAE Spath, 1927

Discussion. Monophyletic derivatives of Lytoceratidae, close to *Nannolytoceras* (Roman, 1938); characteristically small unornamented serpenticones with subcircular to oval whorls, frequent constrictions, and simple quadrilobate suture with deep E, narrow stemmed saddles, widely splayed bifid L and 1-2 auxiliary elements. Inclusion of Protetragonitidae in Lytoceratinae emphasizes primitive characters. *U. Tithon. - U. Alb.*

Protetragonites HYATT, 1900 [*Ammonites quadrisulcatus* d'Orbigny, 1841; OD]. Very small; serpenticone with subcircular whorl section and flattened flanks; every quadrant deep slightly prorsiradiate weakly curved constrictions fronted by strong collar; ribs lacking; suture simple, with narrow deep penetrating E split by prominent median saddle, narrow bifid oblique E/L and large splayed bifid L. *U. Tithon. - U. Alb.* **Fig. 30A-B** - *P. quadrisulcatus* (d'Orbigny).

Leptotetragonites SPATH, 1927 [*Ammonites honnoratianus* d'Orbigny, 1841; OD]. Small; like *Protetragonites* but with flattened flanks, compressed elliptical whorl section and more (9-10) constrictions per whorl that are concave forward on flank and linguoid over venter. *Valang.* **Fig. 30E-F** - *L. honnoratianus* (d'Orbigny).

Hemitetragonites SPATH, 1927 [*Ammonites crebrisulcatus* Uhlig, 1883; OD]. Small; like *Protetragonites* but more inflated, slightly more involute, with somewhat flattened

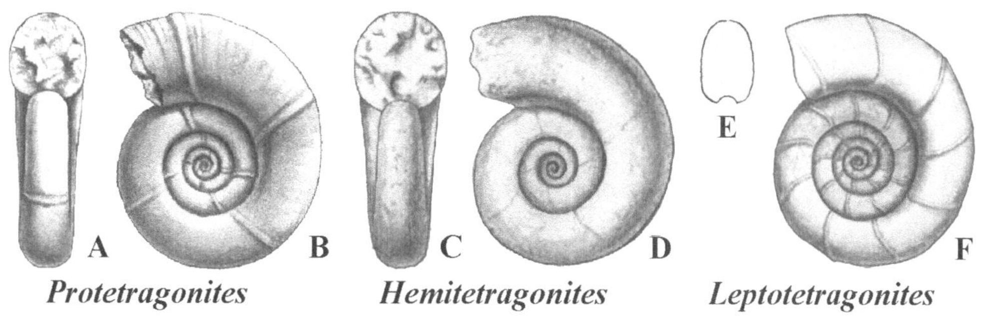

Fig. 30. Protetragonitidae.

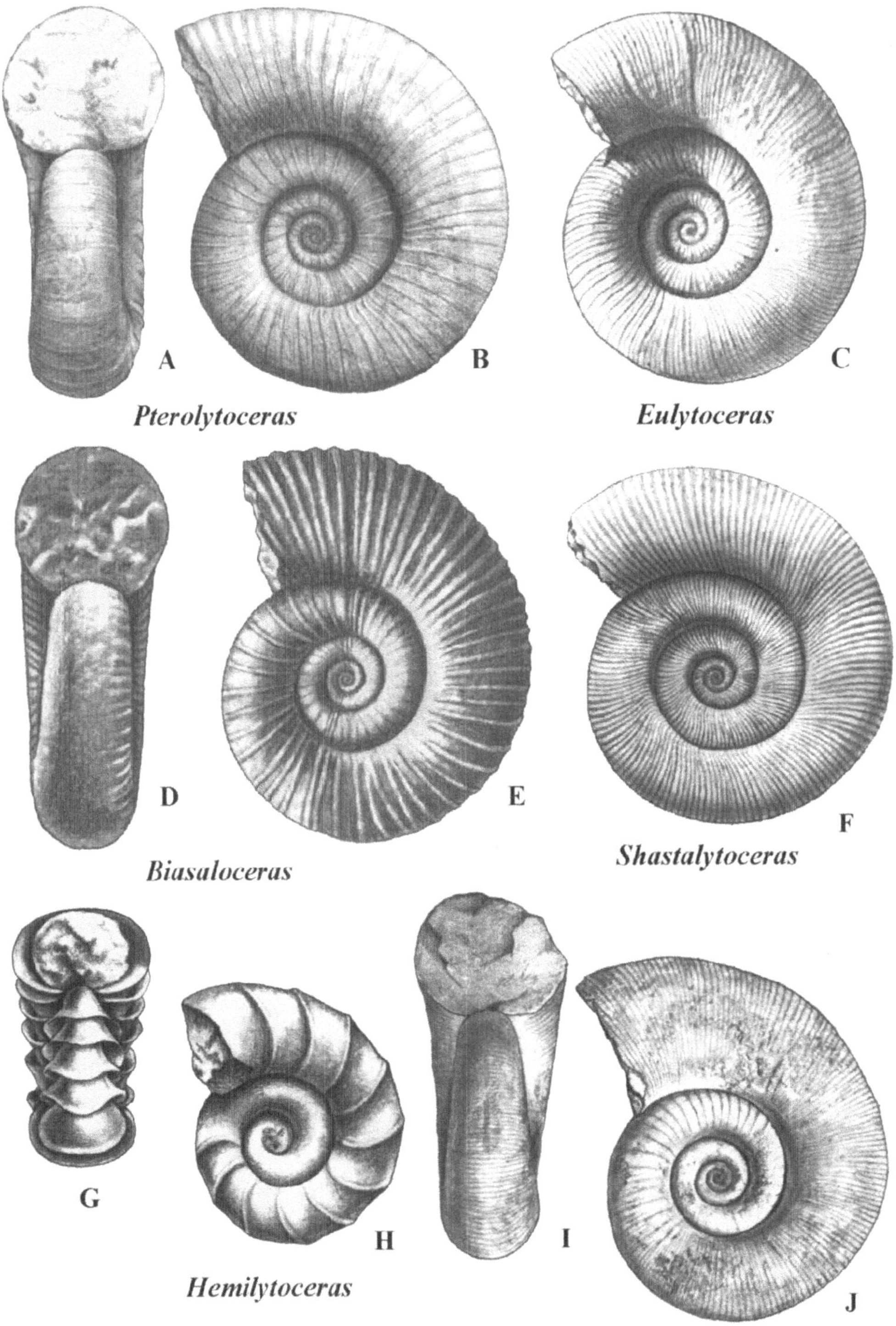

Fig. 31. Hemilytoceratinae.

flanks and venter, subquadratic whorl section and weaker more numerous constrictions; suture like *Leptotetragonites* but more incised, with L/U_2 smaller than E/L and E as deep as L. *Haut. - M. Alb.* **Fig. 30C-D** - *H. crebrisulcatus* (d'Orbigny).

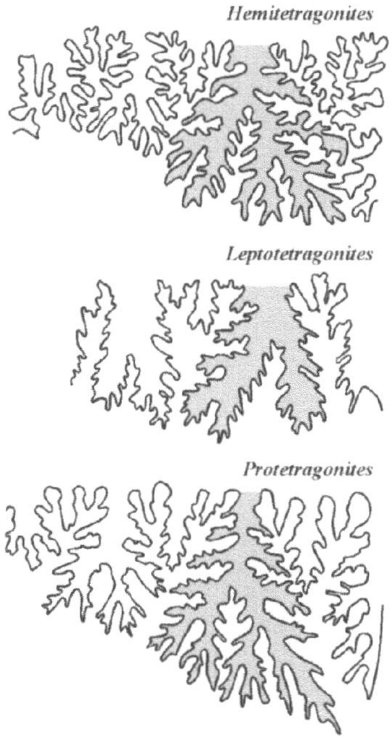

Fig. 32. Protetragonitid sutures.

Family GAUDRYCERATIDAE Spath, 1927

Discussion: Monophyletic derivatives of Protetragonitidae (Wiedmann, 1962) whose acme was in the Late Cretaceous. The most primitive representative, *Eogaudryceras*, is the rootstock of 5 main lines of gaudryceratid evolution, all of which evolved in parallel and, here, are accorded subfamily rank; these are Gaudryceratinae, Anagaudryceratinae, Hypogaudryceratinae, Kossmatellinae and Gabbioceratinae. *Barrem. - U. Maastr.*

Subfamily Gaudryceratinae Spath, 1927

(*nom. transl.* Wright, 1957 *ex* Gaudryceratidae)

Discussion. Monophyletic descendents of Protetragonitidae, close to *Hemitetragonites* (Wiedmann, 1962); typically evolute, compressed, with rounded venter and whorls increasing rapidly in height, wiry flexuous branching riblets, periodic collar ribs (constrictions) and a strongly retracted suspensive lobe to the suture. Commonest in the deeper water environments of the Tethyan Realm. The group represents an intriguing evolutionary radiation that has been masked by taxonomic lumping. Comprises the tribes Gaudryceratini and Vertebritini. *Barrem. - U. Maastr.*

Tribe Gaudryceratini Spath, 1927

(*nom. transl.* herein *ex* Gaudryceratinae)
Type genus: *Gaudryceras* de Grossouvre, 1894.

Diagnosis. A paraphyletic taxon typically with subtabulate (primitive) to rounded venter across which lirae do not split. *Barrem.- M. Camp.*

Eogaudryceras SPATH, 1927 [*Ammonites numidus* Coquand, 1880; OD]. Small; moderately evolute (U=25-36%), compressed, with trapezoidal early whorls; frequent periodic flexuous collar ribs in early to middle growth; in later growth with rounded venter and oval whorl section; dense fine flexuous single or branching riblets cross venter transversely and may become flat topped and band like in maturity; suture with deep narrow E with long tapering median process, narrow stemmed bifid E/L, widely splayed symmetrically bifid L as deep as E and retracted suspensive lobe with at least 3 auxiliary elements. *Barrem. - U. Alb.* **Fig. 33A-B** - *E. numidum* (Coquand); **Fig. 33C-E** - *E. italicum* Wiedmann & Dieni.

Gaudryceras DE GROSSOUVRE, 1894 [*Ammonites mitis* von Hauer, 1866; OD]. Moderately large; like *Eogaudryceras* but more evolute (U=42-52%, decreasing to 32% in maturity), with flattened flanks, broadly rounded venter and subelliptical whorl section; riblets branch at various levels on flank, particularly in later growth; inner whorls with frequent prominent broad flexuous collar ribs that are linguoid over venter but later split into riblets and are barely distinguishable from intervening riblets at least across venter; fine riblets retained to maturity; suture like *Eogaudryceras* but more indented. *Cenom. - M. Camp.* **Fig. 33R-S** - *G. mite* (von Hauer).
= *Gaudriceras* Lehmann, 1981 (*laps. cal.*).

Mesogaudryceras SPATH, 1927 [*Ammonites leptonema* Sharpe, 1855; OD]. Small; like *Gaudryceras* but more involute (U=30-32%) and high whorled, with convergent flanks, narrowly arched venter, subtrigonal whorl section, and well spaced non-bifurcating flexuous lirae which are linguoid across venter. *Cenom.* **Fig. 33T-U** - *M. leptonemum* (Sharpe).

Pseudogaudryceras SHIMIZU, 1934 [*Gaudryceras tenuiliratum* var. *infrequens* Yabe, 1903, = *G. denseplicatum* Jimbo, 1894; OD]. Very large; like *Gaudryceras* but more inflated and with subquadratic early whorls; in middle to later growth whorls increase rapidly in height, umbilicus narrows and irregular closely spaced ribs appear covered in lirae. *Turon. - M. Camp.* **Fig. 33P-Q** - *P. infrequens* (Yabe); **Fig. 36I-J** - *P. denseplicatum* (Jimbo).

Hemigaudryceras SHIMIZU, 1934 [*Lytoceras (Gaudry-ceras) denmanense* Whiteaves, 1901; OD]. Medium sized; like *Gaudryceras* but with finer denser lirae that bifurcate and trifurcate close to umbilical shoulder and are replaced on adult body chamber by strong regular ribs that recurve

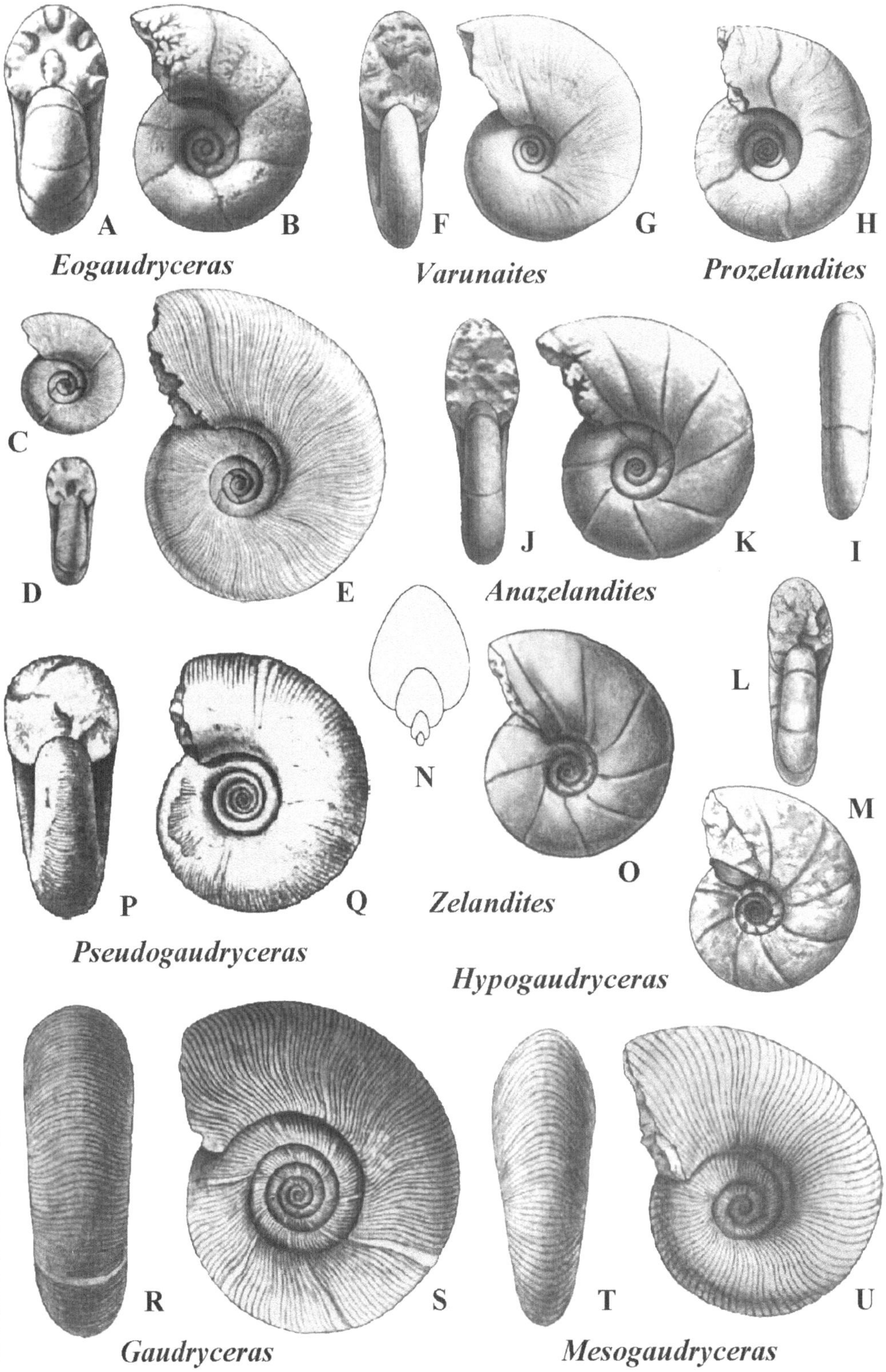

Fig. 33. Gaudryceratinae.

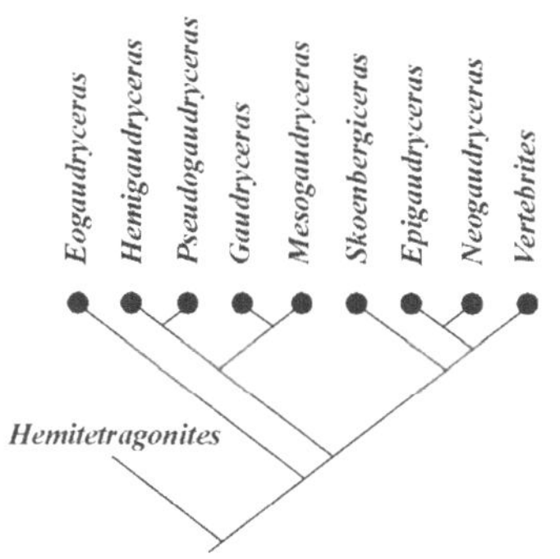

Fig. 34. Hypothesized relationships among Gaudryceratinae.

on outer flank. *U. Sant. - Camp.* **Fig. 33H** - *H. denmanense* (Whiteaves).

Tribe Vertebritini Wiedmann, 1962

(*nom. transl.* herein *ex* Vertebritinae)

Type genus: *Vertebrites* Marshall, 1926.

Diagnosis. Typically with lirae splitting into fine threads over venter. Monophyletic derivatives of Gaudryceratini, close to *Gaudryceras* (Marshall, 1925). *M. Cenom. - U. Maastr.*

***Skoenbergiceras* gen. nov.** [*G. stefaninii* Venzo, 1936, p.21; OD herein]. Medium sized; inner whorls serpenticone (U=47-51%), with sloping umbilical wall, abruptly rounded ventrolateral shoulders, flattened to weakly convex venter and very depressed subcoronate whorl section; fine dense flexuous prorsiradiate lirae split into very fine bundles at ventrolateral shoulder creating impression of "smooth" venter; in middle to later growth whorls more compressed and involute (U=34-37%) with rounded venter and lirae coarsening and becoming *Gaudryceras* like, without splitting over venter. *M. - U. Cenom.* **Fig. 36E-G** - *S. stefaninii* (Venzo).

Neogaudryceras SHIMIZU, 1934 [*Gaudryceras tenuiliratum* Yabe, 1903; OD]. Large to very large; like *Skoenbergiceras* but with rounded venter throughout and lirae not so fine across venter; adult body chamber with distant flexuous ribs separated by coarse lirae that do not split across venter; suture with dendritic saddles and extremely retracted suspensive lobe. *Coni. - L. Camp.* **Fig. 36C-D** - *N. tenuiliratum* (Yabe).

Epigaudryceras SHIMIZU, 1934 [*Lytoceras striatum* Jimbo, 1894; OD]. Small; like *Neogaudryceras* but with even finer and denser hair-like lirae (absent on internal mould); narrow collar ribs increase in abundance with growth (up to 10 on adult body chamber) without intervening riblets; convergent towards *Anagaudryceras*. *U. Sant. - L. Camp.* **Fig. 36A-B** - *E. striatum* (Jimbo).

Vertebrites MARSHALL, 1926 [*V. murdochi* Marshall; OD]. Small; like inner whorls of *Skoenbergiceras* but more evolute (U=56-62%), with flattened venter retained to maturity; suture with 6 internal saddles. *Maastr.* **Fig. 36O-P** - *V. murdochi* Marshall.

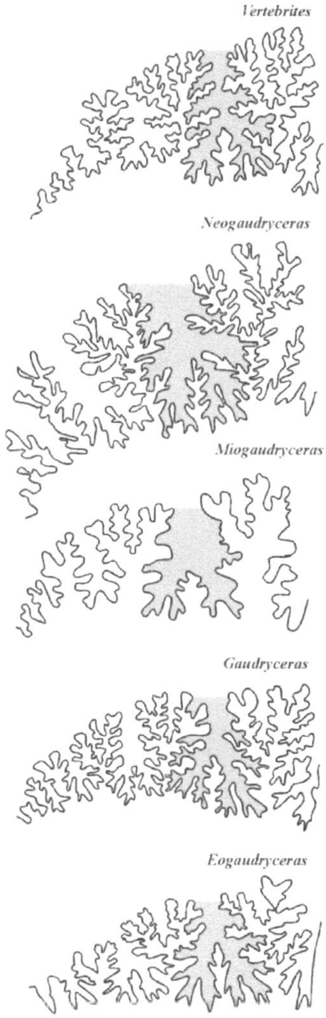

Fig. 35. Gaudryceratine sutures.

Subfamily Anagaudryceratinae nov.

Diagnosis: Monophyletic derivatives of Gaudryceratinae (Collignon, 1956), close to *Eogaudryceras* (Wright, 1997); characteristically with extremely fine dense lirae and numerous constrictions commonly separated by broad flat-topped "ribs". Comprises the tribes Anagaudryceratini and Parajaubertellini. *L. Alb. - U. Maastr.*

Tribe Anagaudryceratini nov.

Type genus: *Anagaudryceras* Shimizu, 1934.

Diagnosis: An ammoniticone in early to middle growth, with rounded umbilical edges. *U. Apt. - U. Maastr.*

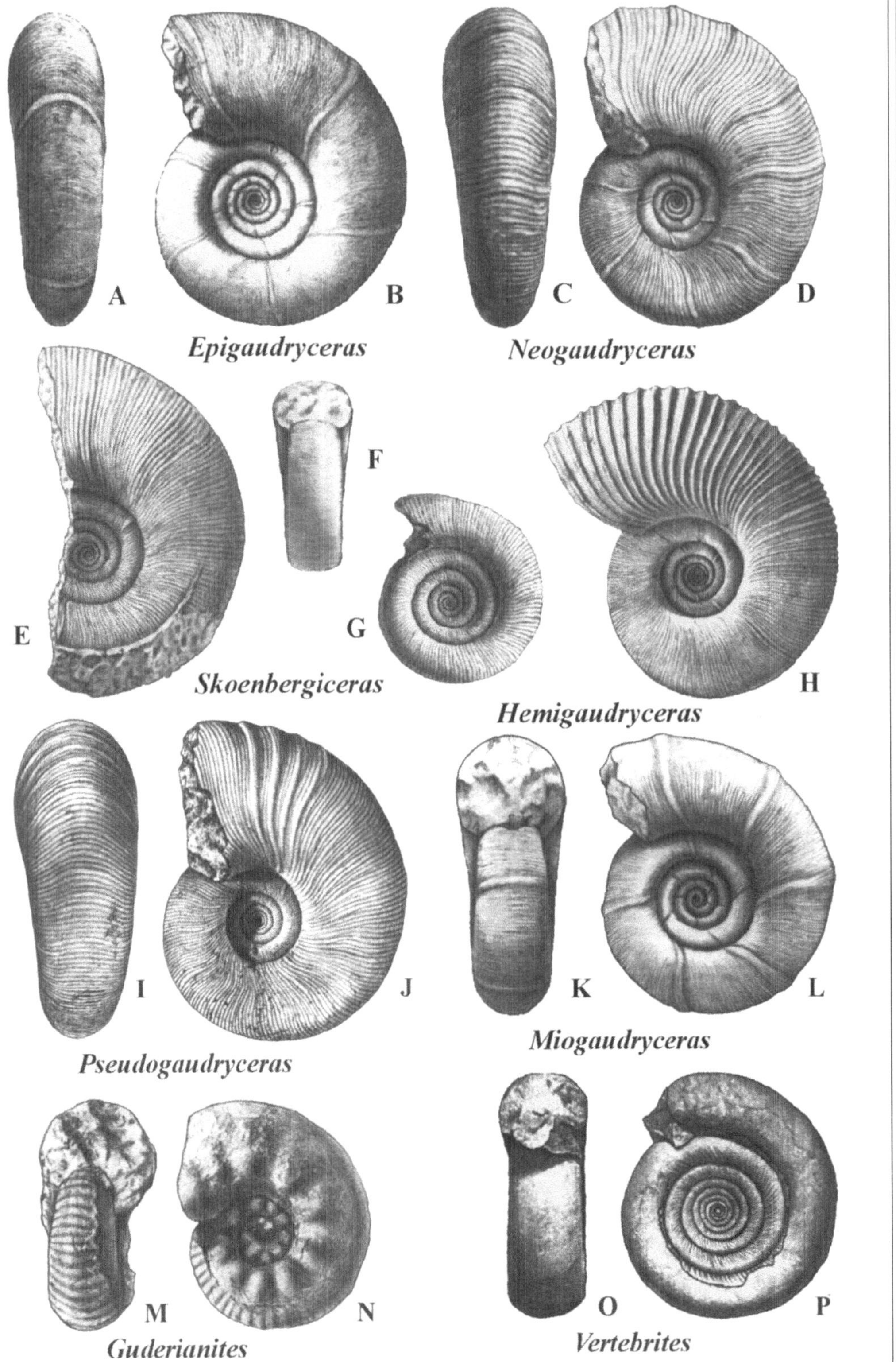

Fig. 36. Gaudryceratidae.

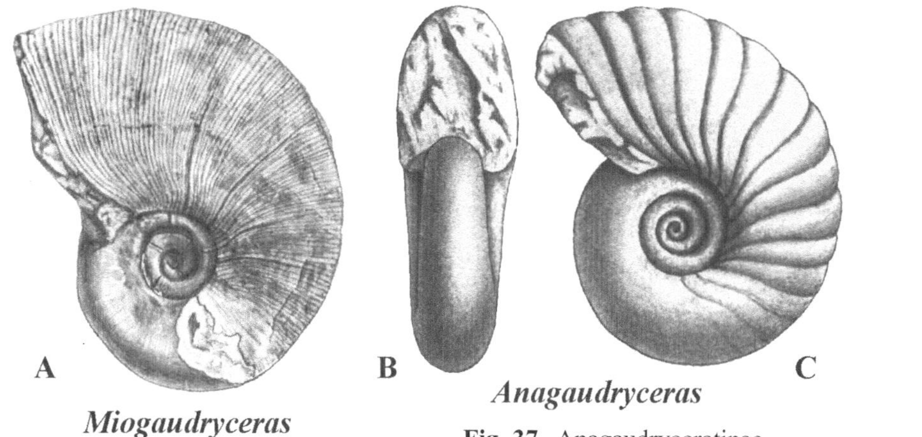

Fig. 37. Anagaudryceratinae.

***Drushchitzia* gen. nov.** [*Ammonites duvalianus* d'Orbigny, 1842, p.158; OD herein]. Small; evolute, compressed, with vertical umbilical wall, subparallel flanks, broadly rounded venter and subelliptical whorl section; numerous (14-22 per whorl) deep sinuous prorsiradiate constrictions convex over venter; shell surface finely lirate; suture with deep narrow E with penetrative median saddle, bifid narrow stemmed E/L and L/U_2, widely splayed bifid L as deep as E, and non-retracted suspensive lobe. *U. Apt.* **Fig. 40J** - *D. duvaliana* (d'Orbigny); **Fig. 40K-L** - *D. wintunia* (Anderson).

Anagaudryceras SHIMIZU, 1934 [*Ammonites sacya* Forbes, 1846; OD]. Medium sized; in early growth evolute (U=40-44%), with evenly rounded subcircular to depressed oval whorl section and periodic straight slightly prorsiradiate collar ribs (constrictions) that cross venter transversely; in middle to later growth whorls become compressed, umbilicus narrows (U=35%) and numerous flexuous constrictions (20-35 per whorl) develop preceded by broad flat-topped lirate collar; suture with retracted suspensive lobe. *L. Alb. - Cenom.* **Fig. 37B-C, 40H-I** - *A. sacya* (Forbes).

= *Anaguadryceras* Henderson, 1970 (*laps. cal.*).

Murphyella MATSUMOTO, 1972 [*M. enigma*; OD]. Small; like *Anagaudryceras* but with depressed subrounded whorls throughout, broad flat-topped collar ribs from early growth and straighter less flexuous constrictions. *L. Cenom.* **Fig. 40F-G** - *A. enigma* (Matsumoto).

Miogaudryceras MATSUMOTO, 1995 [*M. yokoi*; OD]. Small; inner whorls like *Anagaudryceras* but with reniform to subcircular whorl section increasing rapidly in height, 4 collar ribs per whorl, and dense very fine straight lirae; in maturity involute and high whorled, like *Mesogaudryceras*, with compressed suboval to subrectangular whorl section, coarser sparser lirae and weak narrow subcostae that are straight across flanks. *L. Cenom.* **Fig. 36K-L** - *M. pulchrum* (Crick); **Fig. 37A** - *M. yokoi* Matsumoto.

***Shimizuites* gen. nov.** [*Gaudryceras howarthi* Matsumoto, 1995, p.150, OD herein]. Small to medium sized, like inner whorls of *Anagaudryceras* but more involute (U=25-37%), with lirae that are prorsiradiate on inner flank and recurve on outer flank to cross venter transversely; only 4-5 narrow lirate collar ribs per whorl. *L. Cenom. - Coni.* **Fig. 43** - *S. howarthi* (Matsumoto).

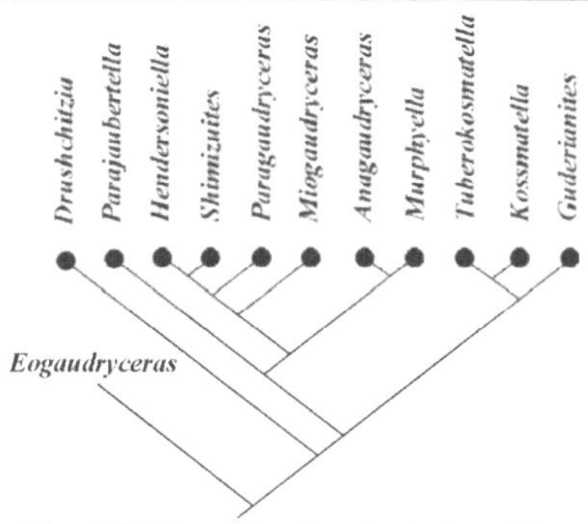

Fig. 38. Hypothesized relationships among some Anagaudryceratidae.

Fig. 39. *Paragaudryceras.*

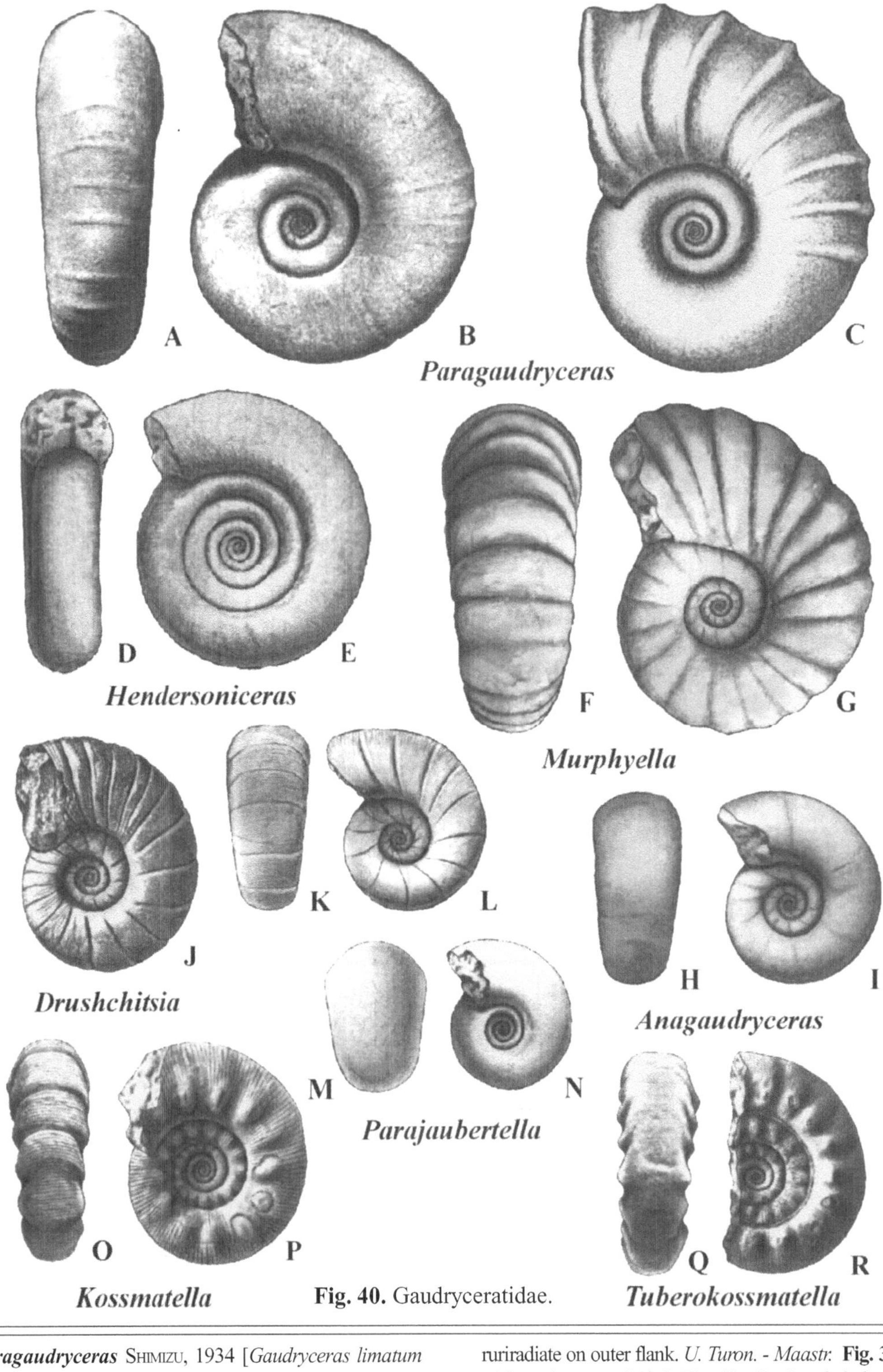

Fig. 40. Gaudryceratidae.

Paragaudryceras SHIMIZU, 1934 [*Gaudryceras limatum* Yabe, 1903; OD]. Medium sized; like *Shimizuites* but lacking constrictions; in later growth with strong ribs, prorsiradiate on lower flank and rectiradiate to slightly rursiradiate on outer flank. *U. Turon. - Maastr.* **Fig. 39, 40C** - *P. limatum* (Yabe); **Fig. 40A-B** - *P. mikobokense* (Collignon).

Fig. 41. Anagaudryceratine sutures.

***Hendersoniceras* gen. nov.** [*Anagaudryceras tennenti* Henderson, 1970, p.19; OD herein]. Small; like *Shimizuites* but serpenticone (U=47-56%), with lirae splitting into fine threads over venter as in *Vertebrites*. *Camp.* **Fig. 40D-E** - *H. tennenti* (Henderson).

Tribe Parajaubertellini nov.

Type genus: *Parajaubertella* Matsumoto, 1942.

Diagnosis. In early to middle growth a depressed cadicone. An offshoot of the main anagaudryceratine line, with an origin close to *Anagaudryceras*. Similarity to *Gabbioceras* is convergent. *U. Alb. - L. Cenom.*

Parajaubertella MATSUMOTO, 1942 [*P. kawakitana*; OD]. Large; like *Anagaudryceras* but strongly inflated, whorls increasing rapidly in height, with broad vertical umbilical wall to inner whorls, deep pit-like umbilicus, subangular umbilical edge, broadly rounded venter and depressed reniform whorl section; mature whorls more compressed, with fold ribs and adult body chamber shows scaphitoid uncoiling; suture with deep penetrating E, narrow stemmed bifid E/L and L/U_2, narrow bifid L not as deep as E and retracted suspensive lobe with 6 external lobes. *L. Cenom.* **Fig. 40M-N, 49D-E** - *P. kawakitana* Matsumoto.

Subfamily Kossmatellinae Breistroffer, 1953

Discussion. Monophyletic derivatives of Gaudryceratinae, close to *Eogaudryceras* (Collignon, 1956); typically small with lirate test and deep constrictions separated by swollen whorl segments; suture like *Gaudryceras* but with 5-6 auxiliary elements to retracted suspensive lobe. *U. Apt. - L. Cenom.*

Kossmatella JACOB, 1907 [*Ammonites agassizianus* Pictet, 1847; SD Roman, 1938). Very small; evolute (U=34-40%) with rounded and depressed inner whorls, later becoming compressed oval; numerous (13-17) rectiradiate constrictions separated by swollen whorl segments superimposed with straight rectiradiate lirae that cross venter uninterrupted; suture highly incised, with narrow stemmed saddles and splayed L as deep as E. *U. Apt. - L. Cenom.* **Fig. 40O-P** - *K. agassiziana* (Pictet).

***Tuberokossmatella* gen. nov.** [*Kossmatella schindewolfi* Wiedmann & Dieni, 1965, p.41; OD herein]. Very small; like *Kossmatella* but very evolute (U=48-50%), with subtabulate venter, trapezoidal whorl section, prominent umbilicolateral bullae and smooth venter; suture with L shorter than E. *M. - U. Alb.* **Fig. 40S-T** - *T. schindewolfi* (Wiedmann & Dieni).

Guderianites WIEDMANN, 1962 [*Kossmatella costata* Douvillé, 1916; OD]. Small; like *Kossmatella* but inflated, with convex flanks, subcircular whorl section and

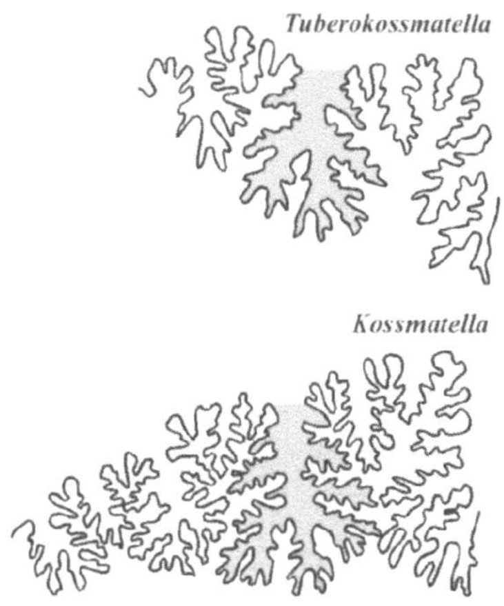

Fig. 42. Kossmatelline sutures.

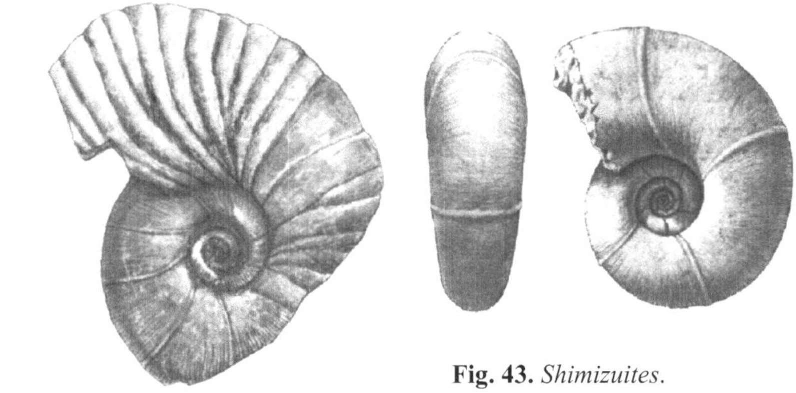

Fig. 43. *Shimizuites*.

coarse swollen umbilicolateral bullae weakly connected to fine rectiradiate ribs that cross venter transversely. *?L. Alb.* **Fig. 36M-N** - *G. costatus* (Douvillé).

Subfamily Hypogaudryceratinae Shimizu, 1934

Discussion. Monophyletic derivatives of Gaudryceratinae, close to *Eogaudryceras* (Wiedmann, 1962); typically small, involute, compressed, with narrowly rounded venter, very fine flexuous lirae and rectiradiate to prorsiradiate flexuous constrictions. Forms a well defined lineage parallel to Gaudryceratinae. *L. Alb. - U. Maastr.*

Prozelandites **gen. nov.** [*Desmoceras dozei* Fallot, 1885, p.235; OD herein]. Very small; compressed, moderately evolute (U=32-37%), with steep umbilical wall, broad weakly convex flanks, narrowly arched venter and compressed oval to subtrigonal adult whorl section; 4-5 almost rectiradiate sigmoidal constrictions per whorl that cross venter; very fine flexuous lirae; suture with narrow stemmed bifid saddles, widely splayed bifid L and at least 9 auxiliary elements to retracted suspensive lobe. *L. Alb. – Cenom.* **Fig. 33H-I** - *P. dozei* (Fallot).

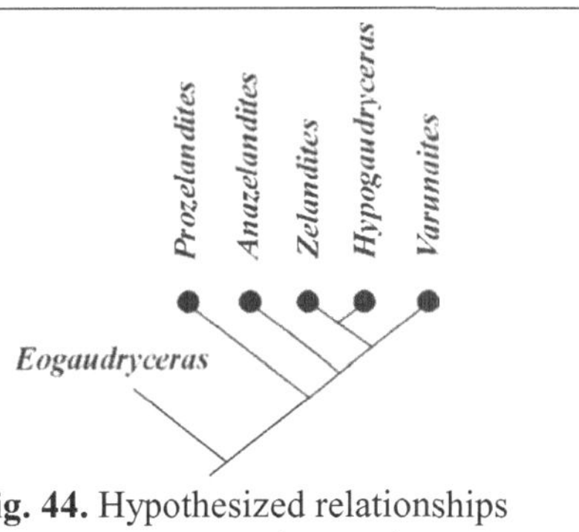

Fig. 44. Hypothesized relationships among Hypogaudryceratinae.

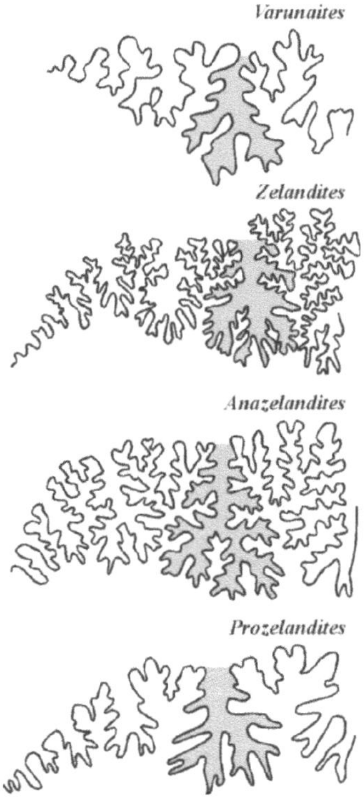

Fig. 45. Hypogaudryceratine sutures.

Anazelandites MATSUMOTO, 1938 [*Lytoceras (Gaudryceras) flicki* Pervinquière, 1907; OD]. Very small; like *Prozelandites* but more compressed and high whorled, with subelliptical whorl section, more numerous (10) straighter prorsiradiate constrictions, and more complex suture with fewer (6) auxiliary elements. *Upper U. Alb.* **Fig. 33J-K** - *A. flicki* (Pervinquière).

Hypogaudryceras SHIMIZU, 1934 [*Desmoceras kawanoi* Jimbo, 1894; M]. Very small; like *Anazelandites* but with narrower umbilicus, maximum width at midflank, and less inclined constrictions that project strongly forward near venter. *Sant. – M. Camp.* **Fig. 33L-M** - *H. kawanoi* (Jimbo).

Zelandites MARSHALL, 1926 [*Z. kaiparaensis*; OD]. Very small; like *Hypogaudryceras* but more inflated, with rhomboidal whorl section, more constrictions (9-12) that weaken over siphonal line, and more complex suture. *Camp.* **Fig. 33N-O** - *Z. kaiparaensis* Marshall.

= *Zealandites* Shimizu, 1935 (*nom. van.*).

The holotype of *Z. kaiparaensis* is lost. Henderson (1977) designated a neotype that is very similar to *Hypogaudryceras* but does not conform to the holotype.

Varunaites SHIMIZU, 1926 [*Ammonites varuna* Forbes, 1846; OD]. Very small; like *Hypogaudryceras* but more involute (U=21%), with distinct umbilical shoulder, flattened flanks and fewer constrictions (4 per whorl) that recurve at midflank and are effaced from venter; suture with L much deeper than E and suspensive lobe in an ascendent straight line. *U. Maastr.* **Fig. 33F-G** - *V. varuna* (Forbes).

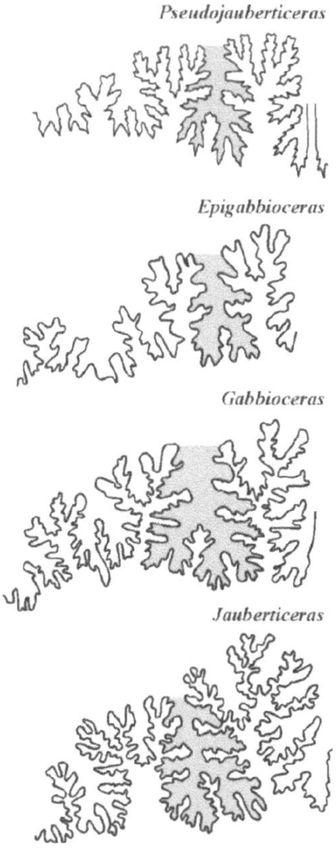

Fig. 47. Gabbioceratine sutures.

Subfamily Gabbioceratinae Breistroffer, 1953

Discussion. Monophyletic descendents of Gaudryceratinae, close to *Eogaudryceras* (Collignon, 1956); despite being derived from *Eogaudryceras* independent of *Eotetragonites*, and appearing in the mid-Aptian prior to the appearance of *Eotetragonites* (the ancestor of *Tetragonites*), modern workers have consistently assigned Gabbioceratinae to Tetragonitidae, but the similarities are due to shared ancestry. Typically small, evolute, highly inflated with strongly depressed whorls, broad flat sloping umbilical wall, funnel-shaped umbilicus, more or less conspicuous ventrolateral angulations, broadly rounded to subtabulate venter, collared constrictions and lirate test. *M. Apt. - L. Cenom.*

Gabbioceras HYATT, 1900 [*Ammonites batesi* Gabb, 1869, *non* Trask 1855, = *Gabbioceras angulatum* Anderson, 1902; OD]. Very small; in early growth with depressed reniform whorl section, becoming round and oval in later growth; inclined umbilical wall passing into broadly rounded venter without flanks; penultimate whorl with prominent mid-lateral angulation exposed in umbilicus; fine dense prorsiradiate lirae on umbilical wall; shallow prorsiradiate collared constrictions, particularly on adult body chamber, gently curved across venter; external suture with only 2 saddles, and median saddle of E prolonged into narrow tapering process; internal suture with 3 lobes. *M. Apt. - L. Alb.* **Fig. 46A-B** - *G. angulatum* Anderson; **Fig. 46C-D** - *G. lamberti* (Breistroffer).

Jauberticeras JACOB, 1907 [*Ammonites jaubertianus* d'Orbigny, 1850; OD]. Very small; like *Gabbioceras*, but with almost flat venter so that sharp lateral angulations become ventrolateral and not exposed in umbilicus that

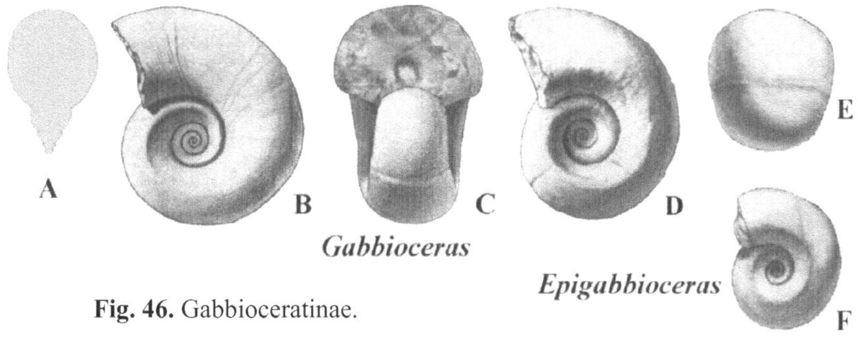

Fig. 46. Gabbioceratinae.

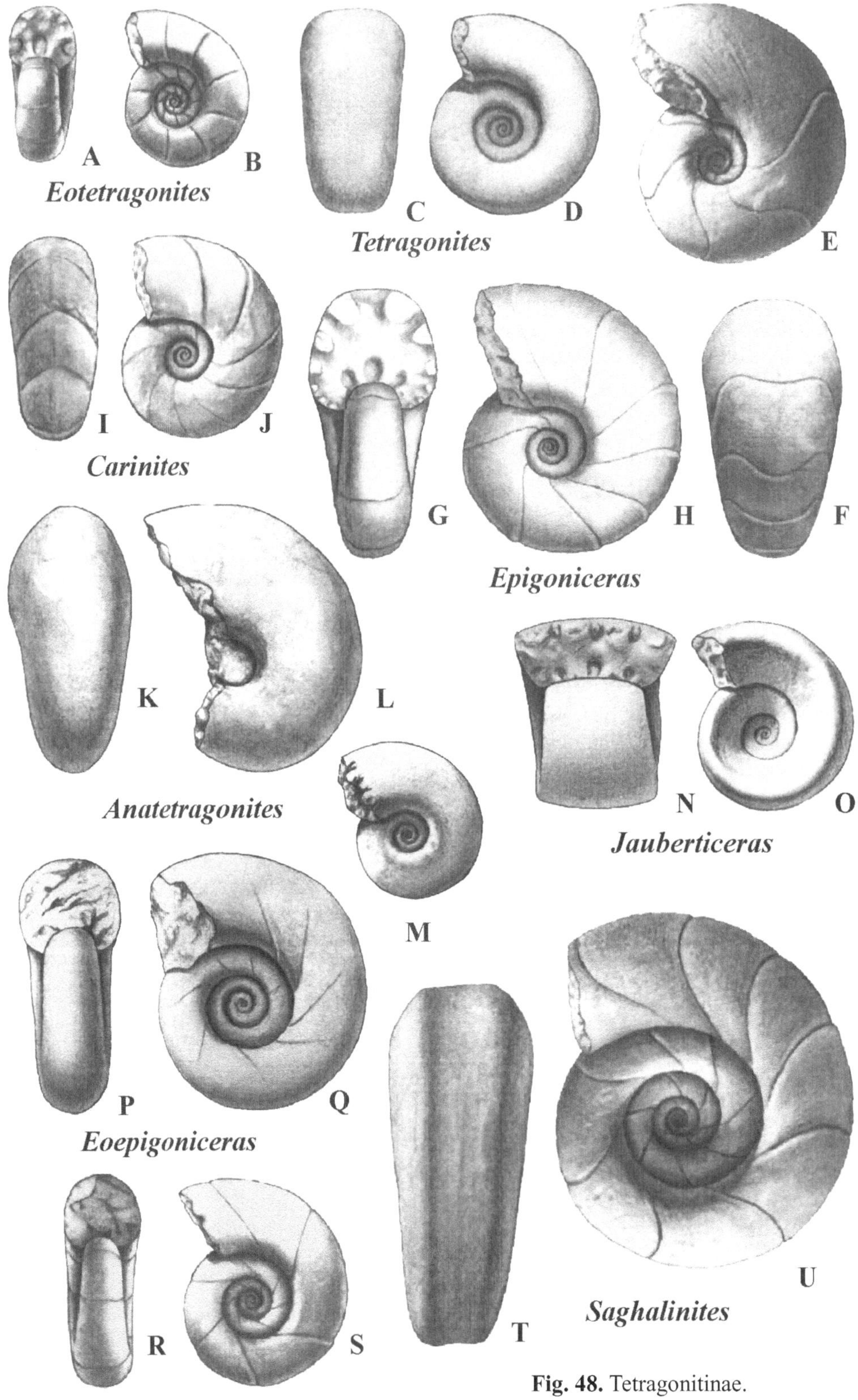

Fig. 48. Tetragonitinae.

becomes a smooth sided conical funnel; whorl section strongly depressed and trapezoidal throughout; internal suture with 5 lobes. *U. Apt. - U. Alb.* **Fig. 48N-O** - *J. jaubertianum* (d'Orbigny).

= *Jaubertella* Jacob, 1908 (obj.).

***Epigabbioceras* gen. nov.** [*Ammonites michelinianum* d'Orbigny, 1850, p.124; OD herein]. Very small; like *Gabbioceras* but more inflated, cadicone; 3 saddles to both external and internal suture, narrow subrectangular L and U, and median saddle of E not prolonged into tapering process. *Upper L. Albian - Lower L. Cenom.* **Fig. 46E-F** - *E. michelinianum* (d'Orbigny).

Pseudojauberticeras SHARIKADZE, 2015 [*P. egoiani*; OD]. Very small cadicone with moderately wide umbilicus, strongly convex flanks, very broad gently convex venter and depressed reniform whorl section; in early to middle growth unornamented; last half whorl with fine uniform ribs, single or bifurcating as wide as the interspaces, prorsiradiate on lower flank, recurving on outer flank and crossing venter transversely; one bifid saddle to external and internal sutures, narrow subrectangular L and U, and median saddle of E prolonged into tapering process. *L. Alb.* **Fig. 49A-C** - *P. egoiani* Sharikadze.

Family TETRAGONITIDAE Hyatt, 1900

Discussion. Monophyletic derivatives of Gaudryceratidae (Breistroffer, 1947); characteristically with sexlobate primary suture. Comprises only the subfamily Tetragonitinae. *U. Apt. - U. Maastr.*

Subfamily Tetragonitinae Hyatt, 1900

(*nom. transl.* Wright 1957 *ex* Tetragonitidae)

Discussion. Monophyletic derivatives of Gaudryceratinae, close to *Eogaudryceras* (Breistroffer, 1947); primitively unornamented with, in early to middle growth, a subtabulate venter, depressed quadratic to trapezoidal whorl section, prominent prorsiradiate constrictions which are transverse or form a sinus over the venter, and a suture with prominent septal lobe and non-retracted suspensive lobe. Early representatives share many primitive characters with early Gaudryceratinae but later the two groups are distinct. Comprises the tribes Tetragonitini and Pseudophyllitini. A conservative group, commonest in the Tethyan Realm, particularly in deeper water environments. *U. Apt. - U. Maastr.*

Tribe Tetragonitini Hyatt, 1900

(*nom. transl.* herein *ex* Tetragonitinae)

Type genus: *Tetragonites* Kossmat, 1895.

Diagnosis. Characteristically small to medium sized, with whorls showing moderate expansion, flattened flanks and venter in early growth, quadratic to trapezoidal whorl section and mostly regular prorsiradiate constrictions that are concave across the venter. *U. Apt. - U. Maastr.*

Eotetragonites BREISTROFFER, 1947 [*E. raspaili* Breistroffer; OD]. Very small; evolute, compressed, with flattened subparallel flanks, subtabulate venter, slowly expanding subrectangular whorls and more or less distinct ventrolateral shoulders; 8-10 strong, rather rigid, prorsiradiate constrictions recurve on outer flank and cross venter with slight forward convexity; ornament of growth striae only; involution may increase with growth, and adult body chamber retains broad flattened venter and subrectangular whorl section; suture with large asymmetrically bifid E/L, large deeply incised, symmetrically bifid L and 3 auxiliary elements. *U. Apt. - M. Cenom.* **Fig. 48A-B** - *E. raspaili* Breistroffer.

Tetragonites KOSSMAT, 1895 [*Ammonites timotheanus* Pictet, 1847; OD]. Very small; like *Eotetragonites* but with very depressed subrectangular whorl section and constrictions lost early; suture rather simple, with narrow stemmed asymmetrically bifid to trifid external saddles, narrow stemmed bifid L shorter than or as deep as E, and slightly retracted suspensive lobe. *Upper U. Apt. - U. Cenom.* **Fig. 48C-D** - *T. timotheanus* (Pictet).

The type species of *Tetragonites* is not well known, and concepts of the genus rest largely with better known *T. subtimotheanus* Wiedmann (**Fig. 48E-F**). The latter

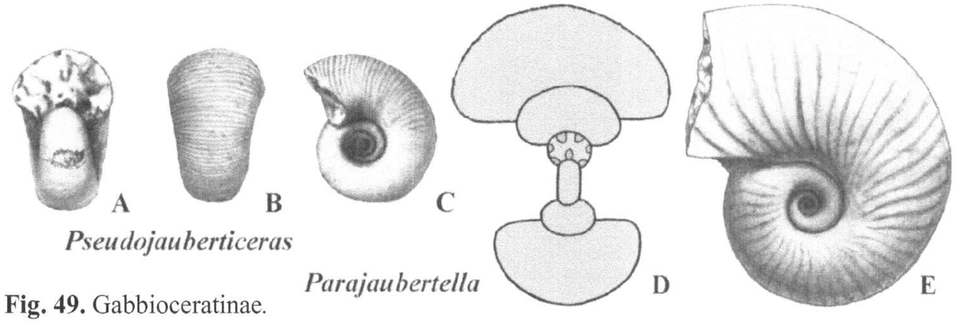

Fig. 49. Gabbioceratinae.

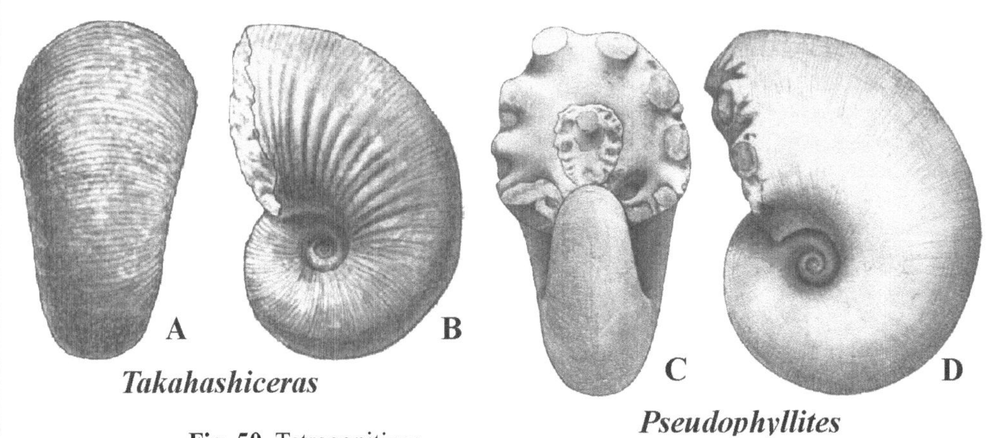

Fig. 50. Tetragonitinae.

differs from *T. timotheanus* in being larger, more involute (U=20-30%) with convergent flanks, trapezoidal whorl section, rounded venter in later growth; prominent collared constrictions persist into middle growth and are strongly prorsiradiate and flexuous across the flanks and concave over the venter.

Carinites WIEDMANN, 1973 [*Tetragonites spathi* Fabre, 1940; OD]. Very small; like *Tetragonites* but more involute, with faint low blunt siphonal ridge and persistent constrictions that form chevrons over venter. *Cenom. - Camp.* **Fig. 47G-H** - *C. spathi* (Fabre).

Epigoniceras SPATH, 1925 [*Ammonites epigonus* Kossmat, 1895; OD]. Small; like *Tetragonites* but more involute and high whorled, with straighter constrictions that may be poorly developed on shell, rounder venter in later growth that may have blunt siphonal ridge, and more complex suture with strongly retracted suspensive lobe. *L. Turon. - U. Camp.* **Fig. 47I-J** - *E. epigonum* (Kossmat).

= *Neoepigoniceras* Shimizu, 1935 [*T. schmidti* Shimizu; OD] (*nom. nud.*).

Eoepigoniceras SHIMIZU, 1935 [*Lytoceras (Tetragonites) kingianus* Kossmat, 1895; OD]. Very small; like *Epigoniceras* but more evolute (U=45%), with whorls

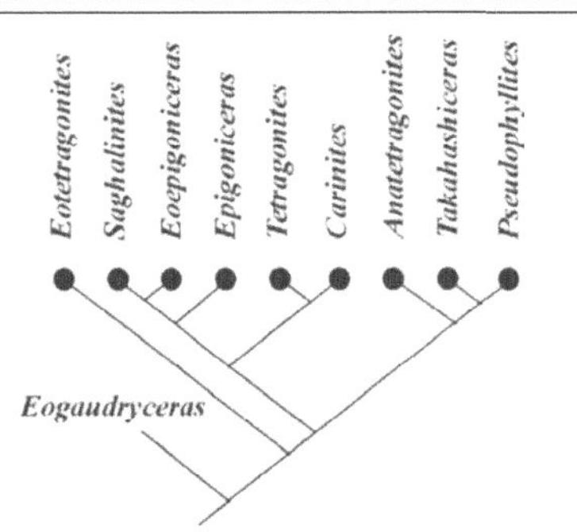

Fig. 51. Hypothesized relationships among Tetragonitidae.

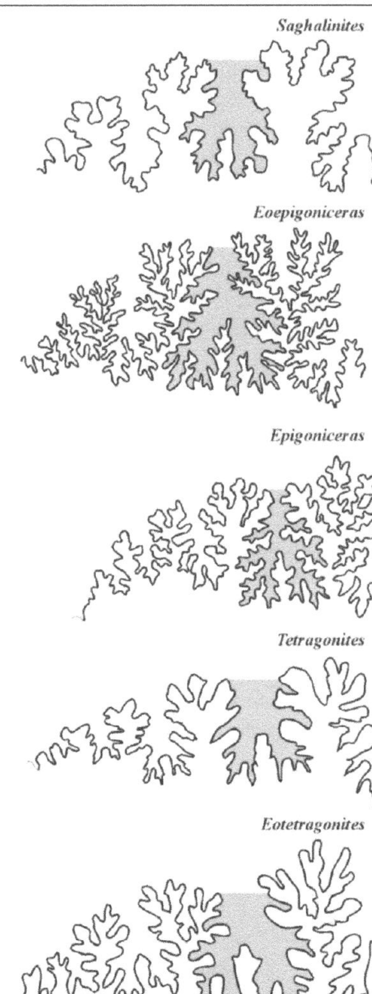

Fig. 52. Tetragonitid sutures.

retaining uniform expansion, convex flanks that do not converge towards broadly rounded venter, subrounded to subquadratic whorl section that is equidimensional in later growth, and constrictions that are transverse over venter. *Sant. - L. Camp.* **Fig. 48P-Q** - *E. kingianum* (Kossmat); **Fig. 48R-S** - *E. nuperum* (van Hoepen).

Saghalinites WRIGHT & MATSUMOTO, 1954 [*Ammonites cala* Forbes, 1846; OD]. Small; inner whorls like *Eoepigoniceras* but with constrictions that recurve strongly on outer flank; later with almost flat convergent flanks, subcarinate ventrolateral shoulders and flat to shallowly concave venter with low blunt siphonal ridge. *U. Camp. - Maastr.* **Fig. 48M-O** - *S. cala* (Forbes).

Tribe Pseudophyllitini nov.

Type genus: *Pseudophyllites* Kossmat, 1895.

Diagnosis. Typically lacking constrictions and with whorls expanding rapidly in height; in later growth involute, with narrowly rounded venter, subtrigonal whorl section, fold-like ribs and lirae, and strigation; suture with strongly retracted suspensive lobe with 5 or more auxiliary lobes. Monophyletic derivatives of Tetragonitini, close to *Tetragonites* (Wright, 1957) with which it evolved in parallel. *M. Alb. - U. Maastr.*

Anatetragonites **gen. nov.** [*Ammonites jurinianus* Pictet, 1848, p.297; OD herein]. Medium sized; relatively involute (U=16-28%), inflated; in early growth with depressed subtrapezoidal whorl section, later becoming high whorled and compressed with narrowly rounded venter and oval whorl section; constrictions lacking throughout; suture relatively complex with asymmetrical L and regularly descending suspensive lobe of 5-6 elements. *M. Alb. - L. Cenom.* **Fig. 48K-M** - *A. jurinianus* (Pictet).

Takahashiceras HOFFMAN & HOWARTH, 2015 [*Takahashia eureka* Matsumoto, 1984; OD]. Medium sized; inflated, relatively involute, with rapidly expanding high whorls and broadly rounded venter; moderately strong convex radial ribs superimposed with crinkled lirae, crossing venter with slight concavity; suture like *Pseudophyllites*. *L. Cenom.* **Fig. 50A-B** - *T. eureka* (Matsumoto).

= *Takahashia* Matsumoto, 1984 (*non* Cockerell, 1896, Homoptera)

= *Takahashiella* Cooper, 2012 (*non* Borchsenius, 1964, Hemiptera)

Pseudophyllites KOSSMAT, 1895 [*Ammonites indra* Forbes, 1846; OD]. Large; innermost whorls like *Anatetragonites*; whorls expanding rapidly, becoming involute and high whorled with compressed oval and then subtrigonal whorl section; venter well rounded; in middle growth delicate folds appear on flank and venter, coarsening with growth; later whorls with distinct ornament of low flat ribs and lirae; at very large size strigation produces reticulate ornament; constrictions lacking at all stages; suture intricately divided from early stage, with retracted suspensive lobe of 6 or more auxiliary elements. *U. Coni. - U. Maastr.* **Fig. 50C-D** - *P. indra* (Forbes).

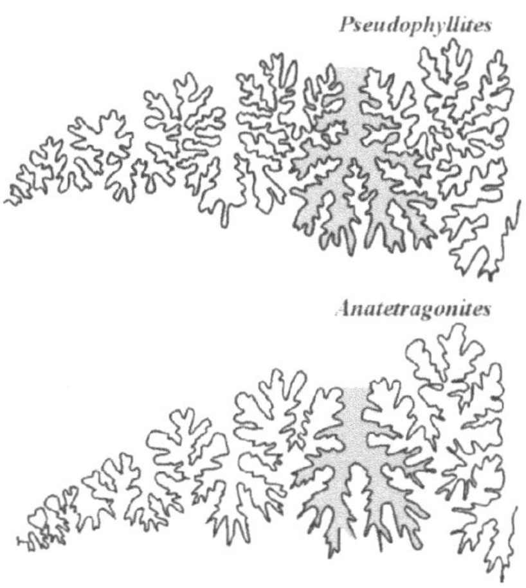

Fig. 53. Tetragonitid sutures.

Order AMMONITIDA Zittel, 1884

Discussion. Monophyletic derivatives of Phylloceratida (Hyatt, 1889), close to Ussuritoidea; a paraphyletic taxon comprising the suborders Psiloceratidina, Xipheroceratidina, Haploceratidina and Hildoceratidina. *Hettang. - L. Sant.*

Suborder HAPLOCERATIDINA Besnosov & Mikhailova, 1984

(*nom. correct.* herein *pro* Haploceratina)

Discussion. Monophyletic derivatives of Xipheroceratidina (Besnosov & Mikhailova, 1983; Ventura & Bilotta, 2008) with an undivided U_1 and proliferation of umbilical lobes (Wiedmann & Kullman, 1981); comprises the superfamilies Oppelioidea and Haploceratoidea. *L. Bajoc. - L. Sant.*

Superfamily OPPELIOIDEA Douvillé, 1890

(= Oppelacea Buckman, 1919; *nom. transl. et correct.* Arkell, 1951 *ex* Oppelidae; *nom. correct.* herein *pro* Oppeliaceae).

Discussion. Monophyletic derivatives of Hammatoceratoidea (Arkell, 1957); comprises Oppeliidae, Taramelliceratidae and Aconeceratidae. A long ranging and conservative group of weakly ornamented compressed oxycones and platycones generally provincialized and conspicuously dimorphic. *L. Bajoc. - L. Sant.*

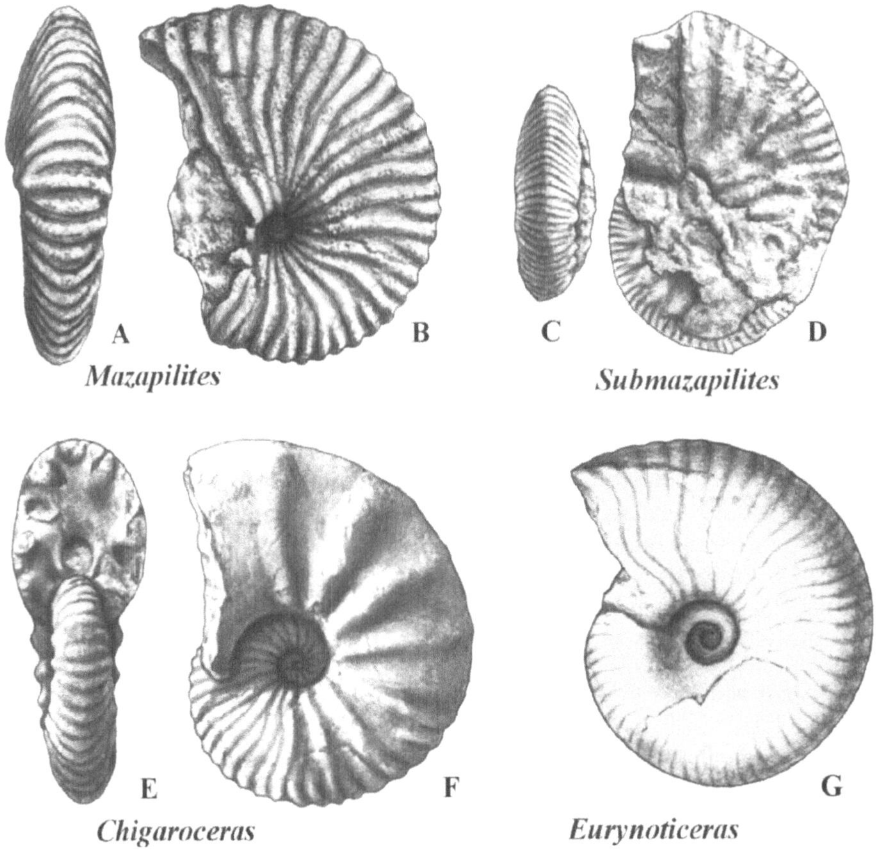

Fig. 54. Mazapilitinae.

Family TARAMELLICERATIDAE Spath, 1928

Discussion. Monophyletic derivatives of Oppeliidae (Roman, 1938); comprises the subfamilies Taramelliceratinae and Mazapelitinae, of which only the latter is dealt with here. *U. Callov.* - *U. Tithon.*

Subfamily Mazapilitinae Spath, 1928

Discussion. Monophyletic derivatives of Taramelliceratinae, close to *Taramelliceras* (Burckhardt, 1919); typically involute, compressed, with narrowly rounded venter in early to middle growth, geniculate ribs that branch on the outer flank, and relatively simple suture. *L. Kimmer.* - *U. Tithon*

Eurynoticeras CANAVARI, 1897 [*E. paparellii*; SD Roman, 1938] [**m**]. Small; compressed, involute, high whorled, with broadly rounded venter and subelliptical whorl section; weak low blunt geniculate ribs strengthen and bifurcate or trifurcate on outer flank; adult aperture with lateral lappets; suture relatively simple, with narrow E, narrow bifid E/L, and large broad subsymmetrical L deeper than E. *L.* - *M. Kimmer*. **Fig. 54G** - *E. paparellii* Canavari.

Mazapilites BURCKHARDT, 1919 [*M. symonensis*; OD]. Small; like *Eurynoticeras* but more involute, with vertical umbilical wall, convex inner flanks, flat convergent outer flanks, oval whorl section, stronger ribbing across flanks and more incised suture; in early growth with siphonal tubercles and weak lateral keels. *L. Tithon*. **Fig. 54A-B** - *M. symonensis* Burckhardt.

= *Mazapylites* Roman, 1938 (*nom. van.*).

Submazapilites CANTÚ-CHAPA, 1963 [*S. mexicanus*; OD]. Small; like *Mazapilites* but with denser ribbing bifurcating from lateral tubercles, frequent intercalatories and feeble siphonal nodes. *U. Tithon., ?Berrias.* **Fig. 54C-D** - *S. mexicanus* Cantú-Chapa.

Chigaroceras HOWARTH, 1992 [*C. banikense*; OD]. Medium sized; inner whorls like *Mazapilites* but slightly more evolute and inflated, with undercut umbilical wall, feeble umbilical bullae and ribs branching at distinct midlateral tubercles; in later growth with rounder flank, broader

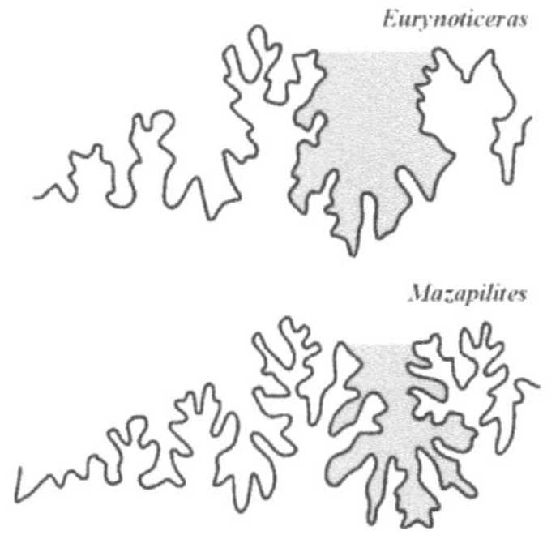

Fig. 55. Mazapilitine sutures.

venter and ornament reduced to broad distant fold-like ribs that do not cross venter; suture complex and highly divided *U. Tithon.* **Fig. 54E-F** - *C. banikense* Howarth.

Family STREBLITIDAE Spath, 1923

Discussion. Monophyletic derivatives of Taramelliceratidae (Donovan *et al.*, 1981); typically weakly ornamented small to medium sized oxycones and platycones; suture with a descending series of auxiliary elements. Comprises the subfamilies Ochetoceratinae and Streblitinae, only some of which are dealt with here to emphasize their relevance to aconeceratid ancestry. *L. Kimmer. – Sant.*

Subfamily Ochetoceratinae Spath, 1928

Discussion. Monophyletic derivatives of Taramelliceratinae, close to *Neocampylites* (Rogov, 2001), only some of which are treated here. Mostly oxycones with falcoid ribs, with or without lateral groove. *U. Callov. - U. Valang.*

Suboxydiscites POULTON, ZEISS & JELETZKY, 1988 [*Ochetoceras (S.) manningense*; OD]. Small; strongly compressed oxycone with very narrow umbilicus, sharp umbilical edge, and finely serrated, non-undulating siphonal keel; inner flanks with feeble distant prorsiradiate ribs, outer flank with stronger crescents; microconchs may have mid-lateral spiral groove. *Kimmer.* **Fig. 57U** - *S. manningensis* (Poulton, Zeiss & Jeletzky).

Oxylenticeras SPATH, 1950 [*O. lepidum*; OD]. Small; compressed oxycone with very narrow (juvenile) to occluded (adult) umbilicus, convex flanks, sharp venter and lanceolate whorl section; sigmoidal growth striae that curve sharply backwards on approaching venter; in middle to later growth with distant low convex undulations on outer flank, parallel to striae; suture highly complex, with deep E and very broad E/L. *Upper L. - U. Tithon.* **Fig. 57C-D** - *O. lepidum* Spath.

Subfamily Streblitinae Spath, 1925

(*nom. transl.* Spath, 1928 *ex* Streblitidae)

Discussion. Monophyletic derivatives of Ochetoceratinae, close to *Strebliticeras* (Douvillé, 1913); suture typically very complex with shallow E and L/U_2 not protruding beyond E/L. *L. Kimmer. - U. Valang.*

Streblites HYATT, 1900 [*Ammonites tenuilobatus* Oppel, 1862; OD]. Medium sized; compressed involute oxycone with rounded venter and sublanceolate whorl section; inner flank with distant curved strongly prorsi-radiate ribs weakly connected to arcuate ribs on outer flank with weak ventrolateral bullae; falcoid main ribs with feeble lateral tubercles, separated by fine secondaries across venter. *L. Kimmer.* **Fig. 57G-H** - *S. tenuilobatus* (Oppel).

Pseudoppelia A. LEANZA, 1946 [*P. oxynota*; OD]. Small; like *Streblites* but more compressed, with narrower venter, plications round the umbilicus, unornamented flanks and a simpler suture. *L. Kimmer.* **Fig. 57D-E** - *P. oxynota* A. Leanza.

Apparently an abraded *Streblites*.

Semiformiceras SPATH, 1925 [*Ammonites fallauxi* Oppel, 1870; OD] [**m**]. Very small to small; evolute, compressed, with flattened subparallel flanks, evenly rounded venter and subelliptical to subrectangular whorl section; sparse umbilical and lateral tubercles connected by feeble ribs; in early to middle growth venter with minute median serrations; adult body chamber with oblique ventrolateral clavi and prominent siphonal groove. *L. - M. Tithon.* **Fig. 57A-B** - *S. fallauxi* (Oppel).

Parastreblites DONZE & ÉNAY, 1961 [*Oppelia tenuilobata* var. *circumnodosa* Fontannes, 1879; OD][**M**]. Medium sized; like *Streblites* but more involute, with stronger main ribs, lack of fine secondaries over venter, and simpler less incised suture with deep E and L not splayed. *L. Tithon.* **Fig. 57A-B** - *P. circumnodosus* (Fontannes).

Substreblites SPATH, 1925 [*Ammonites zonarius* Oppel, 1865; OD] [**M**]. Small; strongly compressed, very

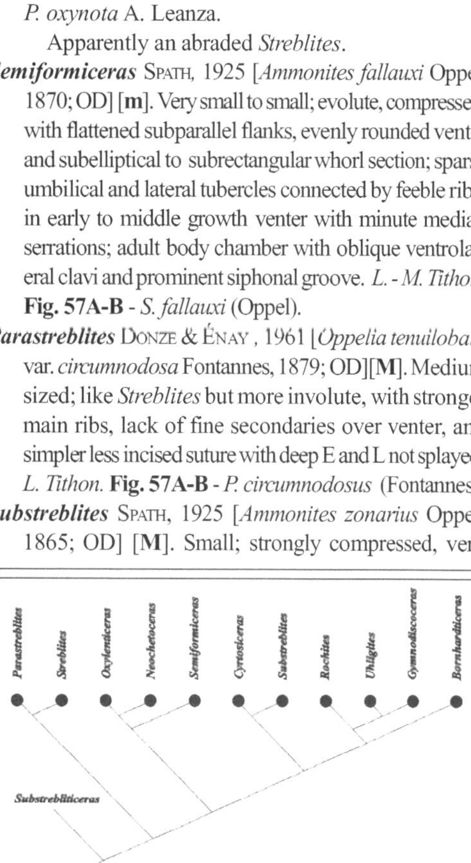

Fig. 56. Hypothesized relationships among Streblitinae.

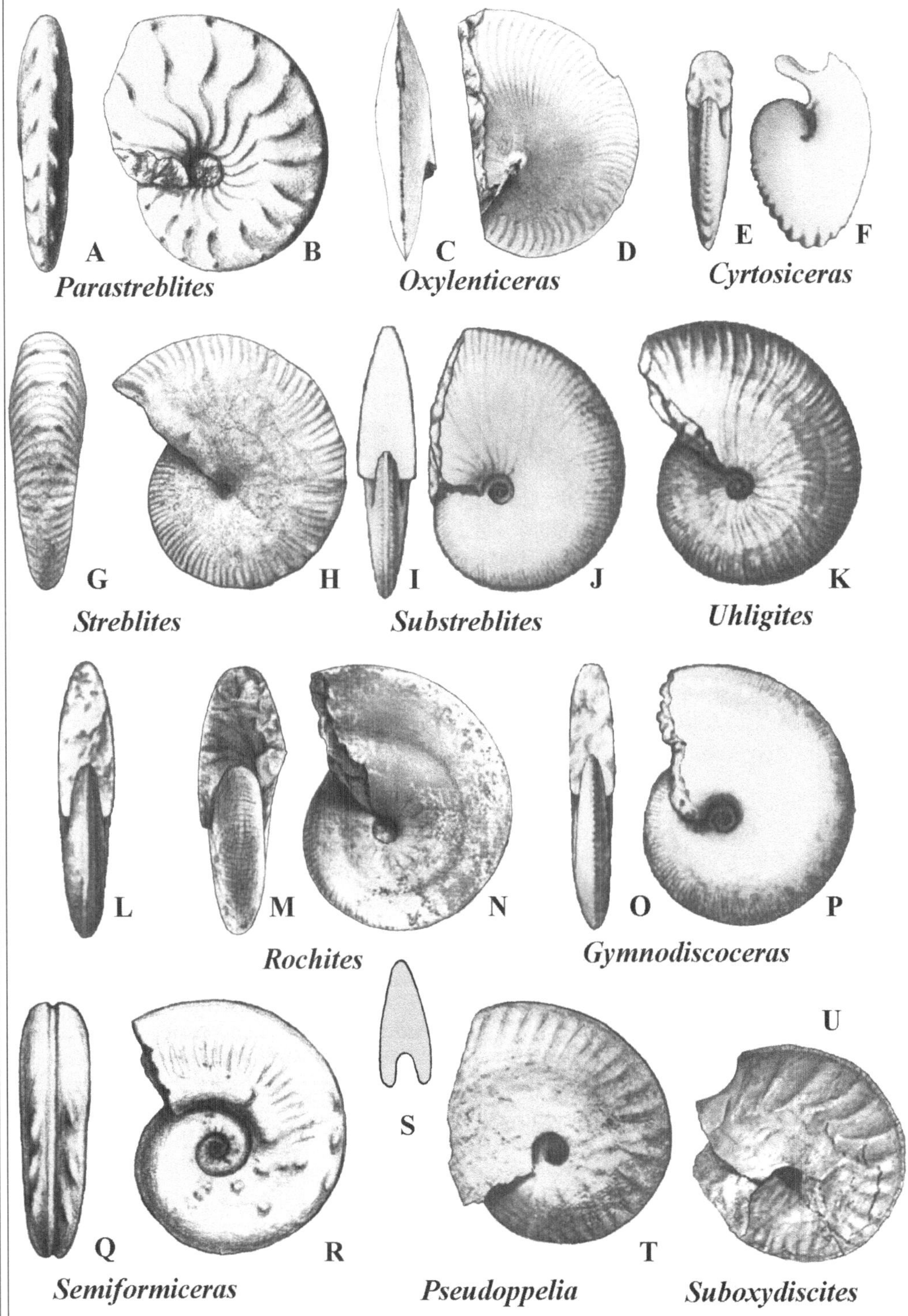

Fig. 57. Oppelioidea.

involute, with steep umbilical wall, subangular shoulder, broad gently convex convergent flanks, rounded venter with low smooth tabulate siphonal keel and sublanceolate whorl section; very weak prorsiradiate ribs on inner flank connected to fine rectiradiate ribs on outer flank; very complex suture with shallow E, oblique narrow stemmed bifid E/L, widely splayed bifid L much deeper than E, and descending series of auxilliaries. *Tithon. - U. Valang.* **Fig. 57I-J** - *S. zonarius* (Oppel).

Cyrtosiceras HYATT, 1900 [*Ammonites macrotelus* Oppel, 1865; OD] [**m**]. Very small; like *Substreblites* but with occluded umbilicus, crenulate venter, adult body chamber showing geniculate uncoiling, long lateral lappets and simple suture. *U. Tithon. - Berrias.* **Fig. 57E-F** - *C. macrotelum* (Oppel).

= *Meunieria* Gregorio, 1922 (*non* Kiefer, 1904).

Uhligites KILIAN, 1913 [*Streblites kraffti* Uhlig, 1903; SD Roman, 1938]. Moderately large; like *Substreblites* but with narrowly rounded venter, convex flanks, oval whorl section, finely serrated siphonal keel, numerous regular falciform ribs continuous across flank, fewer secondary ribs over venter, and suture with exceptionally large L. *L. Tithon. - U. Valang.* **Fig. 57K** - *U. kraffti* (Uhlig).

Gymnodiscoceras SPATH, 1925 [*Oppelia acucincta* Blanford, 1903; OD]. Small; like *Uhligites* but with weaker ornament and simpler suture. *L. - M. Tithon.* **Fig. 57O-P** - *G. acucinctum* (Blanford).

Rochites COLLIGNON, 1960 [*R. rochi*; OD]. Small; like *Uhligites* but more inflated, with weaker ribbing and shallow spiral groove at midflank; suture very complex, with massive elements. *L. Tithon.* **Fig. 57L-N** - *R. rochi* Collignon.

Bornhardticeras BÖHM & RIEDEL, 1933 [*Placenticeras discoidale* Müller, 1900; M] [**M**]. Large, discoidal, very involute, with narrowly rounded venter and sublanceolate whorl section; in early growth with falcate ribs and growth striae with two rows of lateral tubercles, later smooth; suture with long very narrow tapering trifid lobes, deep narrow accessory lobe to E/L and L much deeper than E. *U. Haut.* **Fig. 59B-D** - *B. discoidale* (Müller).

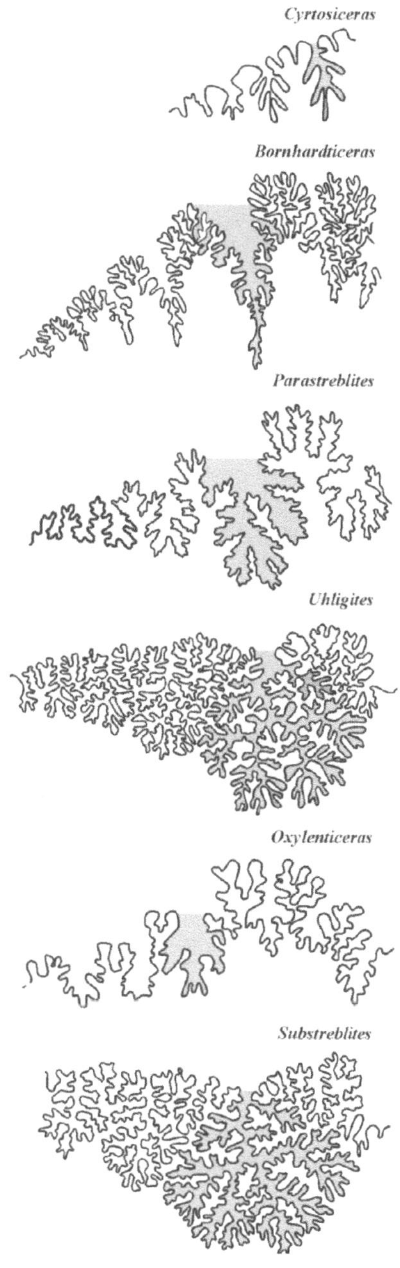

Fig. 58. Oppeliid sutures.

Family ACONECERATIDAE Spath, 1923

Discussion. Monophyletic derivatives of Streblitidae (Wright, 1957, 1996); typically weakly ornamented small oxycones and platycones; suture with a descending series of auxiliary elements. Comprises Aconeceratinae and Binneyitinae. *U. Haut. – Sant.*

Subfamily Aconeceratinae Spath, 1923

(*nom. transl.* Wright, 1957 *ex* Aconeceratidae)

Discussion. A paraphyletic taxon of keeled oxycone derivatives of Streblitinae (Wright, 1957), close to *Substreblites*; mostly with ribs forming crescents on the outer flank, a keeled fastigiate venter and suture with, in most cases, L/U_2 larger than and protruding beyond E/L. Dimorphism marked; microconchs with lateral lappets, long ventral rostrum and a pronounced midlateral spiral groove. Whereas juveniles abound in offshore pyritic-clay facies, adults are not uncommon in very shallow water environments, suggesting ecological partitioning. *U. Haut. - U. Alb.*

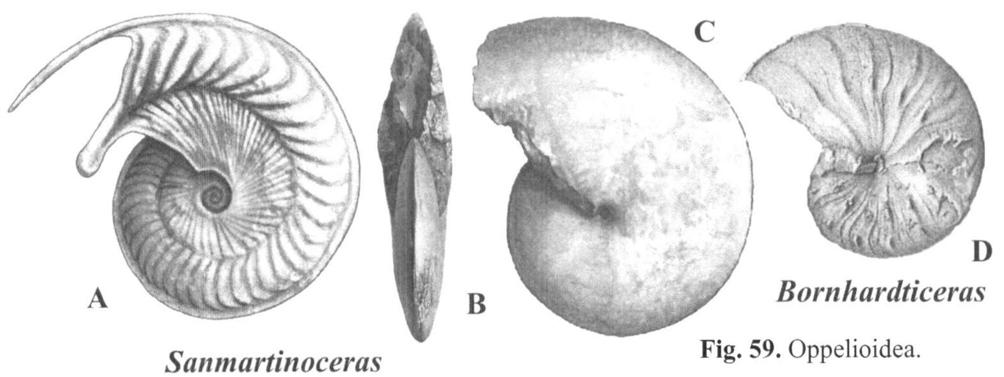

Fig. 59. Oppelioidea.

Protaconeceras CASEY, 1954 [*Oppelia patagoniensis* Favre, 1908; OD]. Medium sized compressed oxycone; inner whorls with narrow pit-like umbilicus, steep umbilical wall, evenly rounded shoulder, broad subparallel flanks, compressed elliptical whorl section, and siphonal keel coarsely crenulated in juvenile stages; very weak flat falcate ribs project strongly forward near venter; suture with very narrow shallow E, narrow somewhat oblique, narrow stemmed, asymmetrically trifid E/L, very broad trifid L much deeper than E, and broad asymmetrically bifid L/U_2 much larger than and extending beyond E/L. *U. Haut.* **Fig. 64C-D** - *P. patagoniense* (Favre).

Aconeceras HYATT, 1903 [*Ammonites nisus* d'Orbigny; OD] [**m**]. Very small; like *Protaconeceras* but high whorled, with convergent flanks, narrow acute venter, finely denticulate siphonal keel and more complex suture with deep E and narrow L. *U. Barrem. - L. Alb.* **Fig. 64A-B** - *A. nisus* (d'Orbigny).

= *Adolphia* Stolley, 1907 (obj.).

= *Adolphites* Hennig, 1932 (*nom. van.*).

Doridiscus CASEY, 1961 [*D. rotulus*; OD]. Very small; like *Protaconeceras* but more evolute, with flatter flanks, very low minutely beaded keel and simpler suture with much-reduced E/L. *L. Apt.* **Fig. 64G-H** - *D. rotulus* Casey.

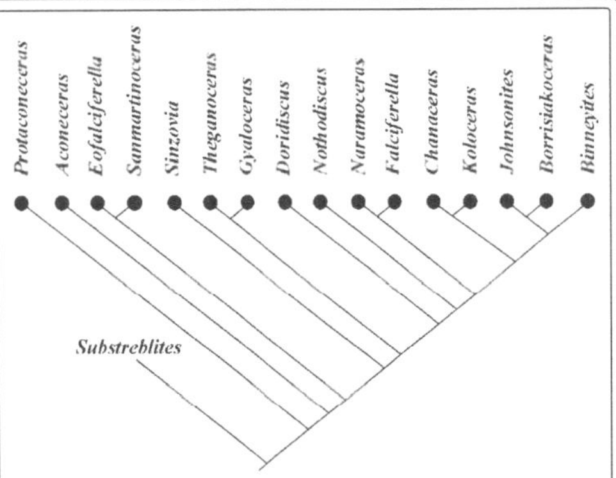

Fig. 60. Hypothesized relationships among Aconeceratidae.

Sanmartinoceras BONARELLI, 1921 [*S. patagonicum*; M] [**m**]. Small (m) to medium sized (M); like *Aconeceras* but larger, somewhat more inflated, with very weak prorsiradiate ribs on inner flanks, distinct spiral groove at midflank, prominent low broad crescents on outer flank, and suture with wider L; juveniles unornamented. *U. Apt. - L. Alb.* **Fig. 59A** - *S. groenlandicum* Rozenkrantz; **Fig. 64E-F** - *S. patagonicum* Bonarelli.

Many of the species assigned here have lateral lappets and are microconchs.

Eofalciferella BRUNNSCHWEILER, 1959 [*E. condoni*; OD]. Very small; like *Sanmartinoceras* but with high, floored, entire keel; maximum width of whorls near ventrolateral shoulders; lower half of flanks flat or concave, with conspicuous midlateral spiral groove; outer flank with broad, very low, biconcave falcoid ribs and, on body chamber, with strigae; suture simple, like *Falciferella*. *Apt.* **Fig. 61** - *E. condoni* Brunnschweiler.

Sinzovia SAZONOVA, 1958 [*S. sazonovae* Wright, = *Ammonites trautscholdi* Sinzow *non* Oppel, 1862; OD]. Small; like *Sanmartinoceras* but with subparallel flanks, very low siphonal keel and lack of midlateral spiral groove. *L. Apt. - L. Alb.* **Fig. 64I-J** - *S. sazonovae* Wright.

Theganoceras WHITEHOUSE, 1926 [*Oppelia scalata* von Koenen; OD]. Very small; like *Sinzovia* but with broader venter, indistinct ventrolateral shoulders and stronger denser ribbing that is acutely falcate and arises from

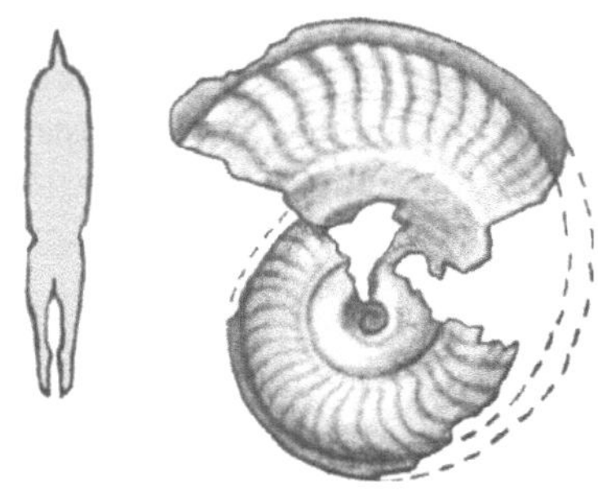

Fig. 61. *Eofalciferella*.

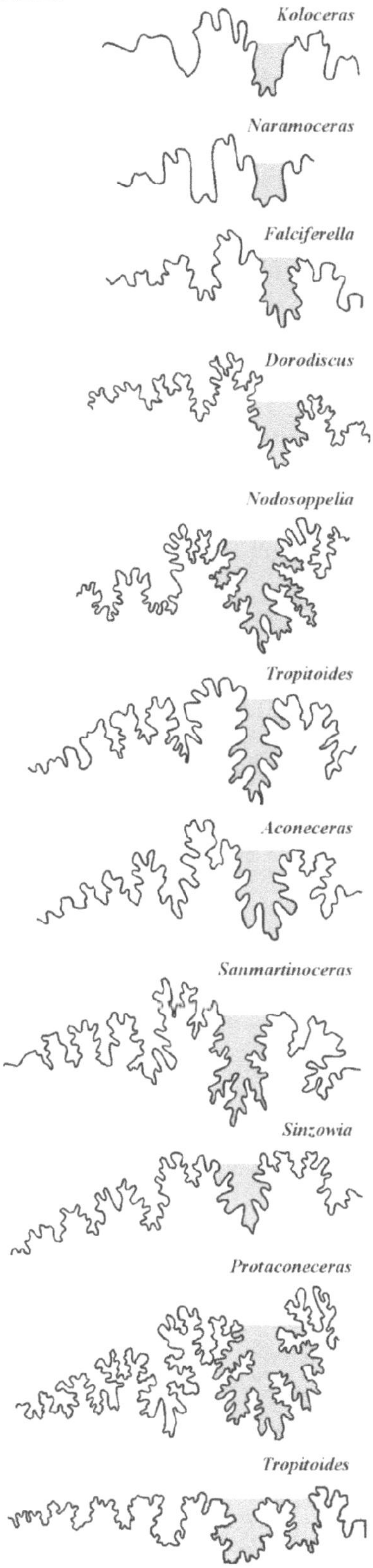

Fig. 62. Aconeceratine sutures.

umbilicus. *L. Apt.* **Fig. 64K-L** - *T. scalatum* (von Koenen).

***Nodosoppelia* gen. nov.** [*Sanmartinoceras (Theganoceras) nodosum* Kennedy & Klinger, 1979, p.107; OD herein]. Medium sized; like *Theganoceras* but with prominent umbilical bullae; inner whorls with broad weakly falcate ribs with incipient ventrolateral clavi; ribs decline in later growth when comma-shaped bullae give rise to bundles of falcate striae. *U. Barrem.* **Fig. 64O-Q** - *N. nodosa* (Kennedy & Klinger).

Gyaloceras WHITEHOUSE, 1927 [*G. smithi*; OD] [**M**]. Medium sized; like *Aconeceras* but more inflated, with wider umbilicus and rounder venter to adult body chamber. *U. Apt., ?U. Alb.* **Fig. 64M-N** - *G. smithi* Whitehouse.

Nothodiscus CASEY, 1962 [*N. planus*; OD]. Very small; very involute platycone with flat flanks, narrow tabulate venter and sharply rounded ventrolateral edges; feeble keel becomes obsolete in later growth; falcoid growth striae only. *U. Apt.* **Fig. 68H-I** - *N. planus* Casey.

Falciferella CASEY, 1954 [*F. milbournei*; OD]. Very small; like *Nothodiscus* but keel not so persistent and with spiral groove to midflank; suture like *Aconeceras* but simpler and with fewer auxiliary elements. *M. - U. Alb.* **Fig. 68A-B** - *F. milbournei* Casey.

Naramoceras McNAMARA, 1985 [*Falciferella breadeni* Brunnschweiler, 1959; OD]. Micromorph; like *Falciferella* but more evolute, with different whorl profile and simpler suture. *M. - Low U. Alb.* **Fig. 68G** - *N. breadeni* (Brunnschweiler).

Koloceras RICCARDI, AGUIRRE & MEDINA, 1987 [*K. talenkanum*; OD]. Micromorph; like *Naramoceras* but with tabulate to rounded venter and intermittent feeble siphonal keel; suture with trifid lobes. *U. Alb.* **Fig. 68C-D** - *K. talenkanum* Riccardi, Aguirre & Medina.

Chanaceras CHAO, 1978 [*C. oxynotum*; OD]. Very small unornamented platycone with rounded venter and persistent siphonal keel; suture with asymmetrically bifid E/L and trifid L not as deep as E. *Upper M. - Lower U. Alb.* **Fig. 63** - *C. oxynotum* Chao.

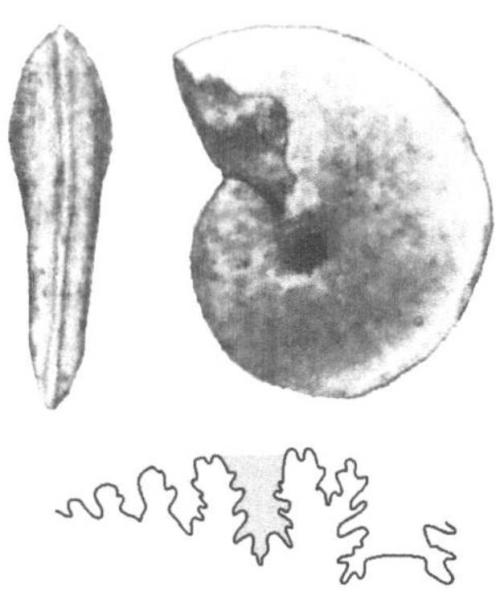

Fig. 63. *Chanaceras*.

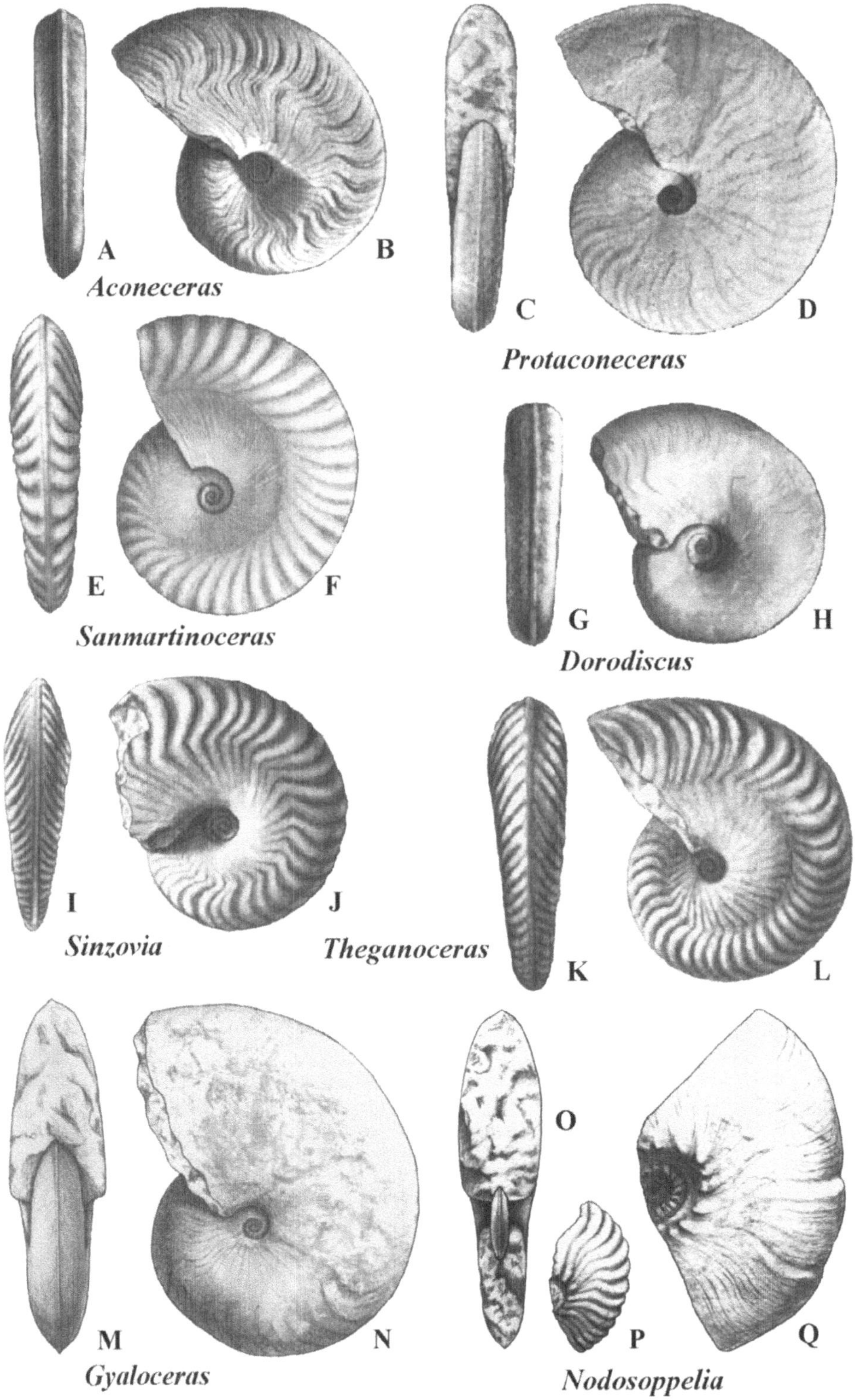

Fig. 64. Aconeceratidae.

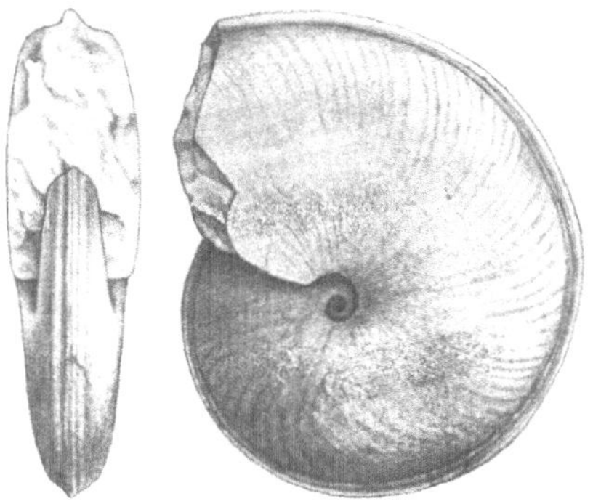

Fig. 65. *Tropitoides*.

Subfamily Tropitoidinae nov.

Diagnosis. A cryptogenic monotypic taxon, superficially resembling Aconeceratinae but separated by a long time gap, with a suture closer to Placenticeratidae. Diagnosis as for genus. *Turon*.

Tropitoides SPATH, 1925 [*Ammonites obesus* Stoliczka, 1865; OD]. Medium sixed; compressed oxycone with tiny umbilicus and gently convex flanks converging to narrow fastigiate venter with bluntly rounded siphonal keel; low sinuous ribs, weak on inner flank and not broadening significantly ventrally, terminate in small ventrolateral clavi that parallel siphonal line; adult whorls smooth; suture with shallow E, bifid E/L, very broad asymmetrically bifid L much deeper than E and 5 auxilliary elements. *Turon*. **Fig. 65** - *T. obesus* (Stoliczka).

Subfamily Binneyitinae Reeside, 1927

(*nom. transl.* Donovan *et al.*, 1981 *ex* Binneyitidae)

Discussion. Monophyletic derivatives of Aconeceratinae (Casey, 1961), close to *Koloceras*; characteristically weakly ornamented micromorph platycones with subtabulate noncarinate venter and greatly simplified suture. *M. Cenom. - L. Sant.*

Borissjakoceras ARKHANGELSKY, 1916 [*B. mirabilis*; OD]. Micromorph to very small; moderately evolute platycone with flat flanks and evenly rounded to flattened venter; flanks unornamented or with thin closely spaced falcoid ribs that may develop ventrolateral tubercles; body chamber shows scaphitoid uncoiling; suture with narrow bifid lobes and much broader bifid or trifid saddles. *M. Cenom. - L. Turon*. **Fig. 68E-F** - *B. mirabile* (Arkhangelsky).

= *Borrisiakoceras* Wright, 1957 (*nom. van.*).

Johnsonites COBBAN, 1961 [*J. sulcatus*; OD]. Very small; like *Borrisjakoceras* but very involute with flat to

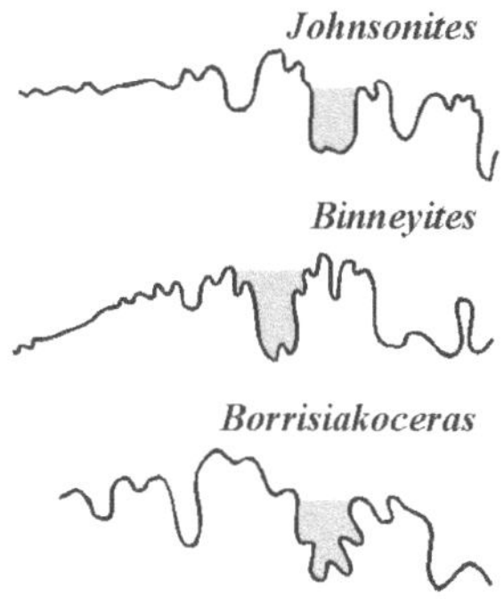

Fig. 66. Binneyitine sutures.

shallowly sulcate venter and weak falcoid ribs; suture with very wide deeply divided E/L. *U. Cenom*. **Fig. 68L-M** - *J. sulcatus* Cobban.

Binneyites REESIDE, 1927 [*B. parkensis*; OD]. Very small; like *Johnsonites* but larger, with tabulate venter, distant short prorsiradiate ribs across venter of adult body chamber and fine biconcave riblets near aperture; suture with very narrow entire finger-like E/L and very wide shallowly bifid L. *L. Turon. - L. Sant.* **Fig. 68J-K** - *B. parkensis* Reeside.

Superfamily HAPLOCERATOIDEA Zittel, 1884

(*nom. transl.* Arkell, 1957 *ex* Haploceratidae)

Discussion. A monophyletic group probably with an origin in Oppelioidea (Douvillé, 1913), although Hildoceratoidea remain an option (Wright, 1996); comprises Strigoceratidae, Lissoceratidae and Haploceratidae, of which only the last family extends into the Cretaceous and is dealt with here. *L. Bajoc. - U. Haut.*

Family HAPLOCERATIDAE Zittel, 1884

Discussion. A monophyletic group with an origin in Taramelliceratidae (Donovan *et al.*, 1981); comprises only the subfamily Haploceratinae. Sexual dimorphism conspicuous. Typically small to medium sized, compressed, weakly ornamented, with flattish flanks and rounded to subtabulate venter. A declining group in the Lower Cretaceous; comprises the subfamilies Haploceratinae and Glochiceratinae. *M. Bajoc. – U. Haut.*

Subfamily Haploceratinae Zittel, 1884

(*nom. transl.* Douvillé 1884 *ex* Haploceratidae)

Discussion. Questionably monophyletic derivatives of Taramelliceratinae, close to *Metahaploceras*

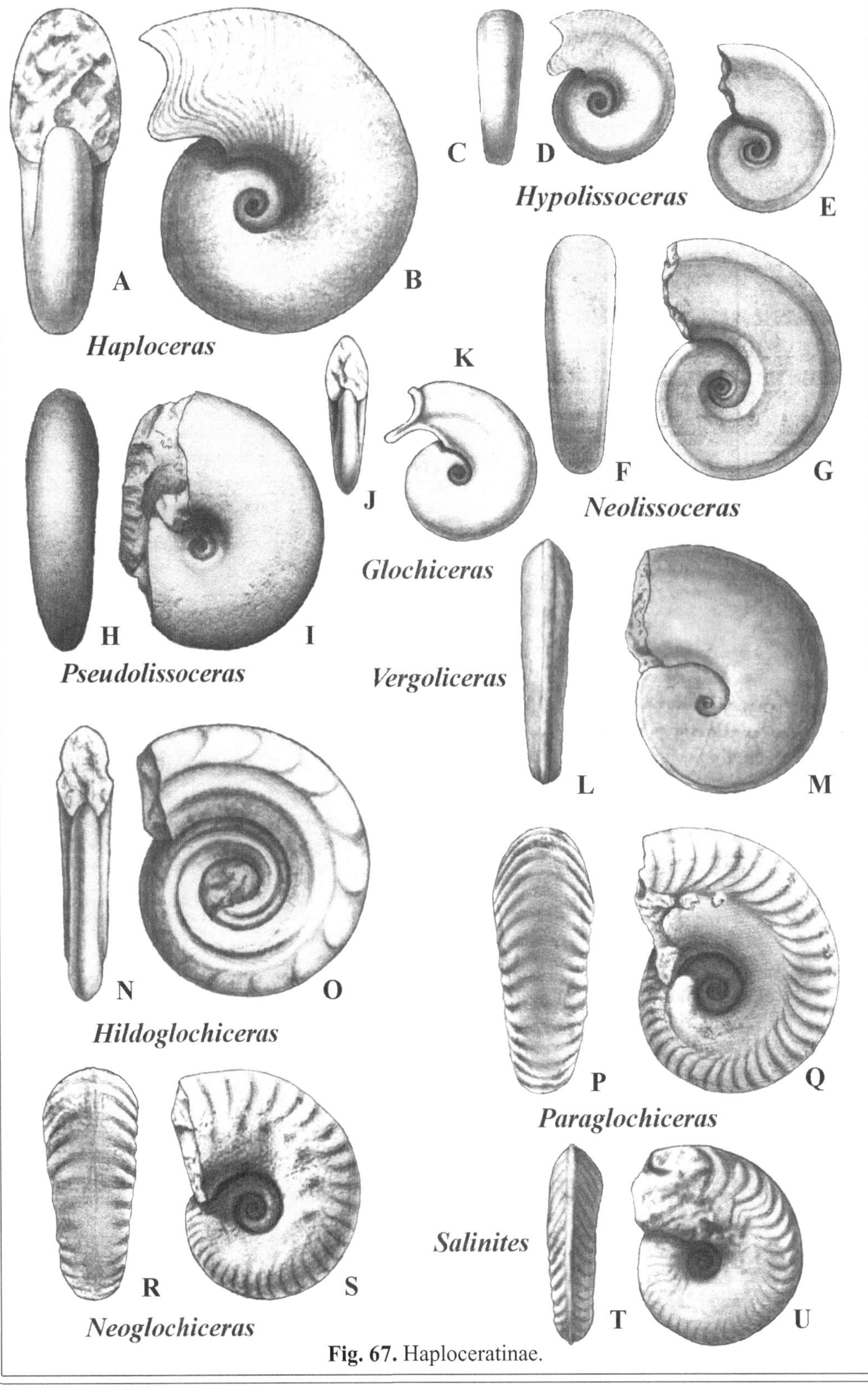

Fig. 67. Haploceratinae.

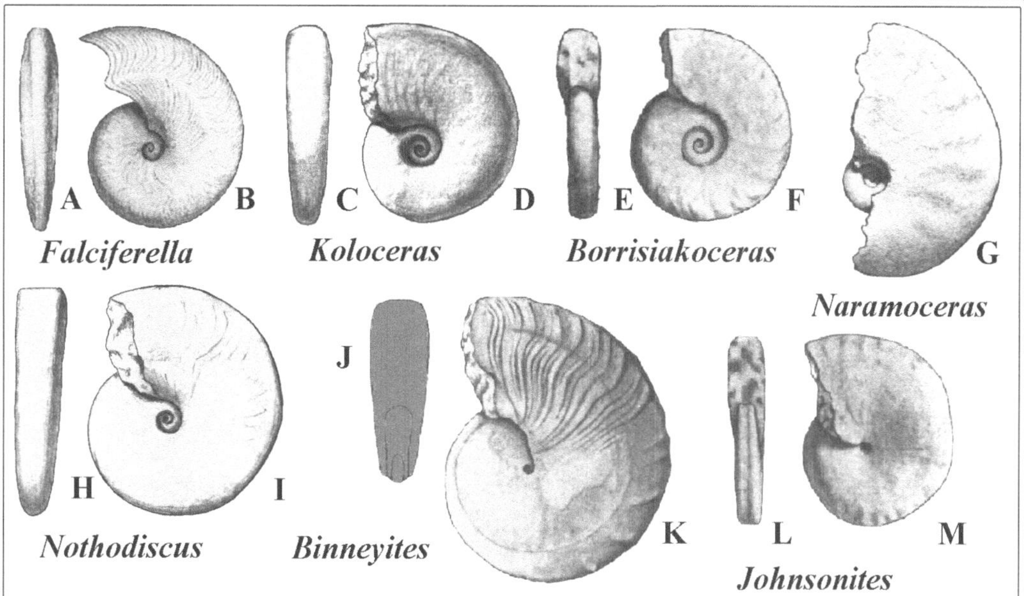

Fig. 68. Binneyitidae.

(Hölder, 1955). *Neolissoceras* is of cryptic origin (Donovan *et al.*, 1981); in morphology and suture it is similar to *Lissoceras* (Lissoceratidae) but a long time gap (*M. Oxford. - L. Tithon.*) separates the two. Strongly dimorphic, microconchs mostly rather involute platycones with subtabulate venter, feeble ventral folds to body chamber and aperture with ventral rostrum and rudimentary lateral lappets. Suture with bifid L/U_2 larger than E/L and 4-5 accessory lobes. *Upper U. Kimmer. - U. Haut.*

Haploceras ZITTEL, 1870 [*Ammonites elimatus* Oppel, 1865; SD Spath, 1923] [**M**]. Small; moderately evolute (U=26-34%), with shallow umbilicus, flanks converging to narrowly rounded venter and compressed oval whorl section; final aperture with broad short lateral lappets. *U. Kimmer. - Berrias.* **Fig. 67A-B** - *H. elimatum* (Oppel).

Volanites ÉNAY & CECCA, 1986 [*Ammonites verruciferus* Zittel, 1869; OD] [**M**]. Like *Haploceras* but more inflated, with convex flanks, broadly rounded venter and ventral constriction and flare at the sinuous adult peristome. *L. - M. Tithon.* **Fig. 70A** - *V. verruciferus* (Zittel).

Hypolissoceras BREISTROFFER, 1947 [*Ammonites carachtheis* Zeuschner, 1846; OD] [**m**]. Very small; like *Haploceras* but with flat sided whorls, lateral lappets and weak ribs across subtabulate venter of adult body chamber. *Tithon. - Berrias.* **Fig. 67C-D** - *H. carachtheis* (Zeuschner).

Generally believed to be based on microconchs of either or both of *Haploceras* and *Volanites*.

= *Neoglochiceras* Patrulius & Avram, 1976 (*non* Collignon, 1960).

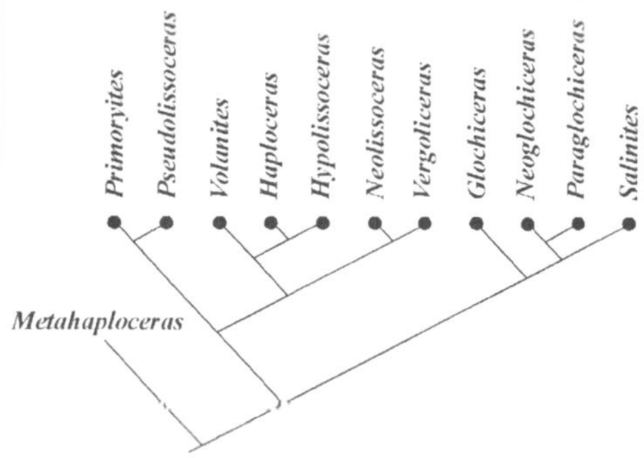

Fig. 69. Hypothesized relationships among Haploceratinae.

Pseudolissoceras SPATH, 1925 [*Neumayria zitteli* Burckhardt, 1903; OD]. Very small; like *Haploceras* but more compressed and high whorled, with narrower venter, more inclined umbilical wall and simpler suture with broader L and L/U_2 not protruding beyond E/L. *M. - U. Tithon.* **Fig. 67H-I** - *P. zitteli* (Burckhardt).

Primorytes KHUDOLAY, 1960 [*P. primoryense*; OD]. Very involute, compressed, with high oval whorl section, narrowly arched venter; flanks smooth, unornamented; siphuncle off-centre (?pathological); suture more complex than *Pseudolissoceras*, with bulbous folioles, narrow-stemmed asymmetrically bifid E/L, wide L and L/U_2 protruding beyond E/L. *L. Tithon.* **Fig. 70B-D** - *P. primoryensis* Khudolay.

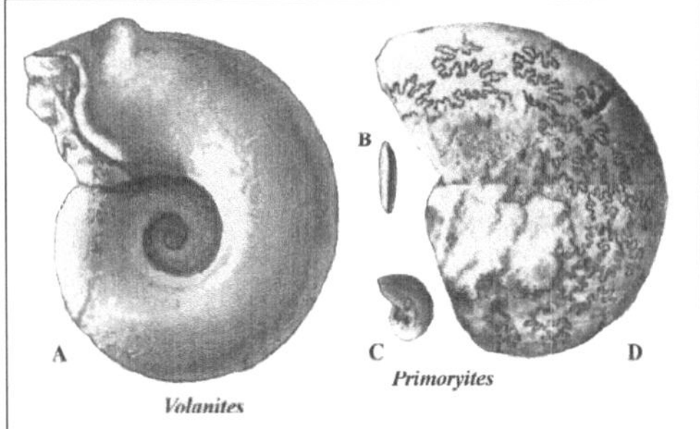

Fig. 70. Haploceratinae.

Neolissoceras SPATH, 1923 [*Ammonites grasianus* d'Orbigny, 1841; OD]. Very small; like *Pseudolissoceras* but with subangular umbilical shoulder, broad flat parallel flanks,well defined ventrolateral shoulders, subtabulate venter and subrectangular whorl section; microconch with lateral lappets and long ventral rostrum; suture like *Haploceras* but with broader L. *U. Tithon. – U. Haut.* **Fig. 67E-G** - *N. grasianum* (d'Orbigny).

Vergoliceras ATROPOS & REBOULET, 1996 [*Haploceras salinarium* Uhlig, 1887; OD]. Very small; like *Neolissoceras* but more involute and compressed, with rounded umbilical shoulder, gently convex flanks, indistinct ventrolateral shoulders, narrowly rounded venter and blunt low rounded siphonal keel from early growth that is prolonged into a long oblique rostrum at adult aperture; adult body chamber shows marked scaphitoid uncoiling and 7mcroconchs have long lateral lappets. *L. - U. Valang.* **Fig. 67L-M** - *C. salinarium* (Uhlig).

= *Carinites* Atropos & Reboulet, 1995 (*non* Wiedmann, 1973).

Subfamily Glochiceratinae Hyatt, 1900

(*nom. transl.* herein *ex* Glochiceratidae)

Discussion. Monophyletic derivatives of Haploceratinae, close to *Haploceras*; typically with more or less prominent median lateral groove; derived forms with coarse rursiradiate arcuate ribs on outer flank and even lateral tubercles or siphonal keel. *Oxford. - Berrias.*

Glochiceras HYATT, 1900 [*Ammonites nimbatus* Oppel, 1863; OD] [**m**]. Micromorph; involute, compressed, with convex flanks, subacute venter and oval whorl section; unornamented save for median lateral groove and large lateral lappets to adult aperture. *Oxford. - Kimmer.* **Fig. 67J-K** - *G. nimbatum* (Oppel).

Paraglochiceras COLLIGNON, 1959 [*P. giraudoni*; OD] [**M**]. Small; like *Glochiceras* but more inflated, with wider umbilicus, broadly rounded venter, oval whorl section, weak lateral groove and numerous rursiradiate arcuate

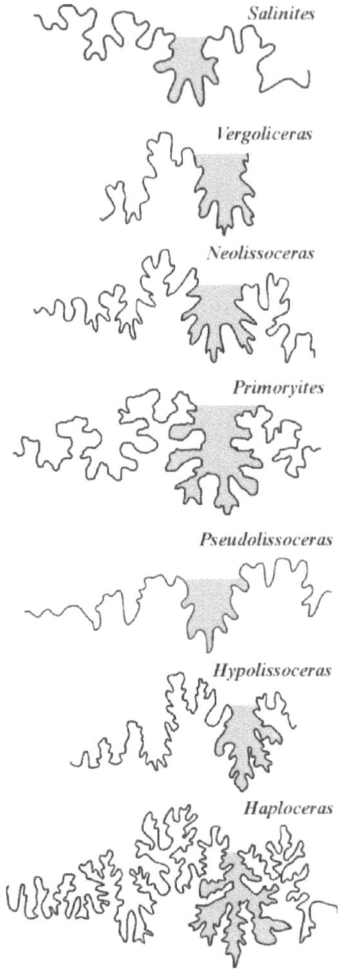

Fig. 71. Haploceratid sutures.

ribs on outer flank. *M. Kimmer. - L. Tithon.* **Fig. 67P-Q** - *P. giraudoni* Collignon.

Hildoglochiceras SPATH, 1924 [*Hecticoceras latistrigatum* Uhlig, 1903; OD] [**m**]. Small; like *Paraglochiceras* but very evolute, with flat flanks, narrow subacute venter, deep spiral groove at midflank and more distant rursiradiate ribs on outer flank; aperture with lateral lappets *U. Kimmer. - M. Tithon.* **Fig. 67N-O** - *H. latistrigatum* (Uhlig).

Neoglochiceras COLLIGNON, 1960 [*Paraglochiceras (N.). tuberosum* Collignon, 1960; SD herein] [**M**]. Small; like *Paraglochiceras* but with arcuate ribs arising in 2-3's from blunt midlateral tubercles. *L. Tithon.* **Fig. 67R-S** - *N. tuberosum* (Collignon).

Salinites CANTÚ-CHAPA, 1968 [*Hildoglochiceras grossicostatum* Imlay, 1942; OD]. Very small; like *Paraglochiceras* but more involute and compressed, with weak prorsiradiate ribs on lower flank, prominent siphonal keel and ascendent suture with bifid L and broad L/U_2. *?M. Tithon., U. Tithon. - Berrias.* **Fig. 67T-U** - *S. grossicostatus* (Imlay).

Order PERISPHINCTIDA Houša, 1965

Discussion. Monophyletic descendents of Stephanoceratidina (Roman, 1938); comprises the suborders Perisphinctidina and Protancyloceratidina. *U. Bajoc. - L. Apt.*

Suborder PERISPHINCTIDINA Houša, 1965

(*nom. transl.* Besnosov & Mikhailova, 1983 *ex* Perisphinctida, *nom. correct.* herein)

Discussion. Monophyletic descendents of Stephanoceratidina (Roman, 1938); comprises the superfamilies Perisphinctoidea, Aspidoceratoidea and Idoceratoidea.Unfortunately classification of this group has been hindered by the belief that

"... *too many genera defeats one of the principle objectives of binomial nomenclature: namely, to indicate relationships between species by grouping like species under the same name*" (Arkell, 1950: 354).

This Linnaean approach resulted in taxonomic lumping, overlooking the "*trifling characters*" so important to unravelling phylogeny. It led Arkell (1950: 356) to suggest that

"... *A logical classification of the Perisphinctaceae is perhaps an impossibility*".

The prevailing arrangements within the suborder (Klein, 2005) remain largely utilitarian. *U. Bajoc. - L. Berrias.*

Superfamily PERISPHINCTOIDEA Steinmann, 1890

Discussion. Monophyletic derivatives of Stephanoceratoidea (Roman, 1938). Sexual dimorphism is conspicuous. The complex evolutionary history of the group has still to be worked out, in particular the boundary with ataxioceratids is blurred. Comprises the families Parkinsoniidae, Perisphinctidae, Zigzagiceratidae, Morphoceratidae, Tulitidae, Rieneckiidae, Pachyceratidae and Aulacostephanidae, of which only Perisphinctidae and Zigzagiceratidae touched upon here. Faunal provicialism is conspicuous, particularly in the Late Jurassic. *L. Bajoc. - L. Berrias.*

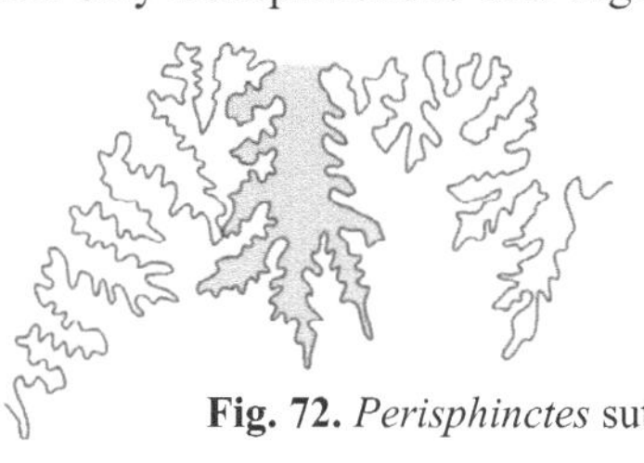

Fig. 72. *Perisphinctes* sutures.

Family ZIGZAGICERATIDAE Buckmann, 1920

Discussion. Monophyletic derivatives of Leptosphinctinae (Westermann, 1993); comprises the subfamilies Zigzagiceratinae, Grossouvriinae, Proplanulitinae, Kinkeliceratinae and Sivajiceratinae. *L. Bathon. - U. Callov.*

Subfamily Grossouvriinae Spath, 1931

Discussion: Derivatives of Zigzagoceratinae (Arkell, 1957), close to *Wagnericeras* (Mangold *et al.*, 2012). Small (**m**) to very large (**M**), moderately evolute to serpenticone. Early whorls densely ribbed with frequent prominent constrictions; in middle growth single and biplicate ribs more distant, thick, raised and often subbullate; suture with slightly retracted suspensive lobe; dimorphic. *U. Bathon. - L. Oxford.*

Alligaticeras BUCKMAN, 1923 (*Ammonites alligatus* Leckenby, 1859; OD]. Small to medium sized, very evolute, with flattened flanks, quadratic whorls, regular biplicate ribs, strong constrictions form chevrons over venter, and heavy parabolic nodes. *U. Callov.* **Fig. 74** - *A. alligatus* (Leckenby).

Subgrossouvria SPATH, 1924 (*Perisphinctes aberrans* Waagen, 1875; OD] [**M**]. Small (**m**) to moderately large (**M**); serpenticone with low rounded venter and slightly compressed oval intercostal whorl section; strong prorsiradiate primary ribs, dense and as wide as interspaces in early growth, becoming distant later, flaring to form crescentic umbilicolateral bullae that split into weak indistinct secondaries over the venter in maturity; microconchs like inner whorls of macroconch, with lappets. *U. Callov.* **Fig. 73H-I** - *S. aberrans* (Waagen).

The specimen figured by Arkell *et al.* (1957) as *S. aberrans* is actually the holotype of *Indosphinctes spirorbis* (Waagen) (Dutta & Bardhan, 2015).

Indosphinctes SPATH, 1930 [*Ammonites calvus* J. de C. Sowerby, 1840: OD] [**M**]. Medium sized (m) to very large (M); compressed, moderately evolute to evolute, wih densely ribbed early whorls; in middle to later growth flat sided, with strong primary ribs bifurcating or trifurcating at midflank or higher, with intercalatories, weakening over venter; in later growth primaries reduced to umbilicolateral bullae; constrictions indistinct, but parabolic ribs frequent at beginning of body chamber;

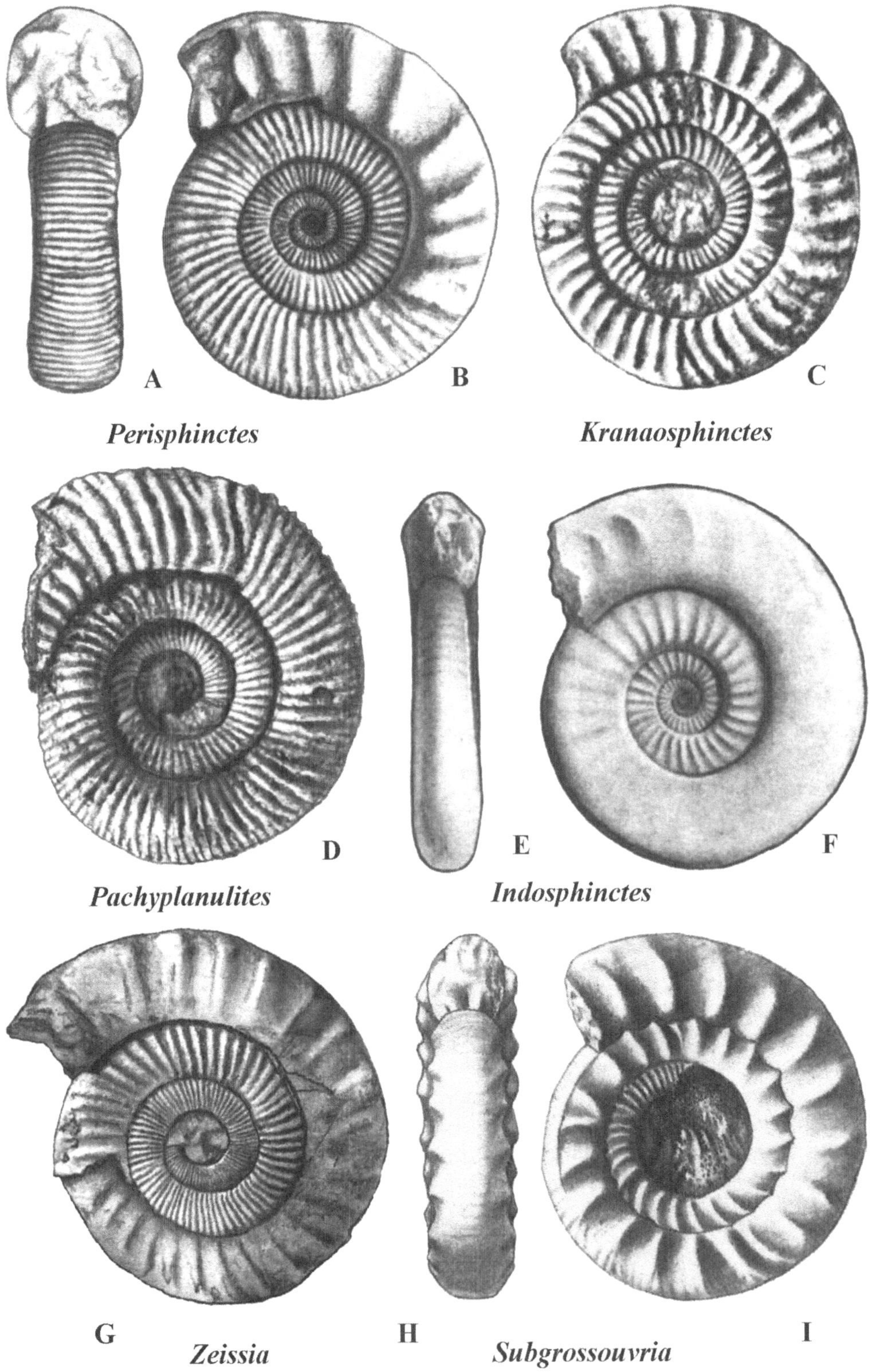

Fig. 73. Perisphinctidae, Vinalesphinctidae.

suture complex; microconch ribbed throughout, ribs tending to be rursiradiate, with lappets. *L. Callov.* **Fig. 73E-F** - *I. indicus* (Siemiradski).

The microconch is *Elatmites* Shevyrev, 1960 [*Perisphinctes submutatus* Nikitin, 1881; OD].

Family PERISPHINCTIDAE Steinmann, 1890

(*nom. transl.* Hyatt, 1900 *ex* Perisphinctinae)

Discussion. Questionably monophyletic derivatives of Zigzagiceratidae but, given the large number of genera placed here (Énay& Howarth, 2019), the family is not monophyletic. Typically evolute planulates with sharp biplicate ribs and complex suture with retracted suspensive lobe. Comprises the subfamilies Perisphinctinae and Epipeltoceratinae, only the former dealt with here. *U. Callov. - Kimmer.*

Subfamily Perisphinctinae Steinmann, 1890

Discussion. Derivatives of Zigzagiceratinae (Melendez *et al.*, 2009) or Grossouvrinae (Énay& Howarth, 2019); typically serpenticone planulates with biplicate ribs, with or without constrictions; suture with strongly retracted suspensive lobe. Predominantly Tethyan. *U. Callov. - Kimmer.*

A number of genera are variously assigned to Perisphinctinae or Ataxioceratinae, and some have even been treated as subgenera of *Perisphinctes* (Arkell, 1957). A problem is that ataxioceratine microconchs preserve the biplicate ribbing of ancestral perisphinctids, whereas macroconchs have bundled ataxioceratid ribbing. In addition evolution was more rapid in microconchs, e.g. *Perisphinctes* is claimed to have 3 different microconchs (*Otosphinctes*, *Dichotomosphinctes*, *Dichotomoceras*) (Meléndez *et al.*, 1984). Only a few genera are dealt with here so as to provide an introduction to a group that includes the progenitors of some of the most important Cretaceous ammonites.

Perisphinctes WAAGEN, 1869 [*Ammonites variocostatus* Buckland, 1836; ICZ 1954 (Opinion 303) [**M**]. Very large; compressed, evolute, with flattened flanks, subtabulate venter and quadratic whorl section; in early to middle growth with numerous sharp prorsiradiate primary ribs which branch at ventrolateral shoulder and loop over venter; adult body chamber with robust distant primary ribs; constrictions lacking; suture with subrectangular bifid E/L, narrow trifid L as deep as E and strongly retracted suspensive lobe. *U. Oxford.* **Fig. 73A-B** - *P. variocostatus* (Buckland).

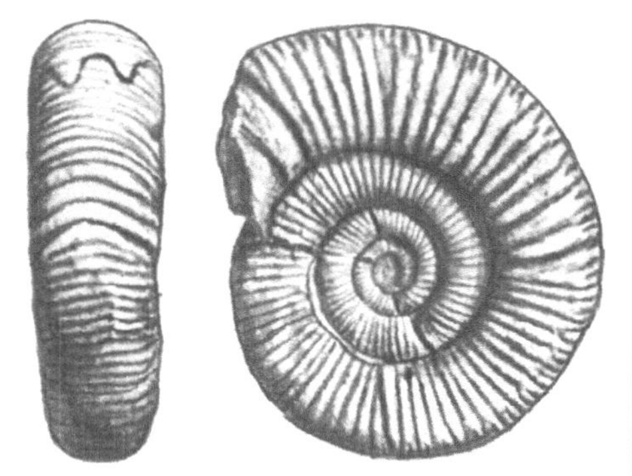

Fig. 74. *Alligaticeras.*

Kranaosphinctes BUCKMAN, 1921 [*K. kranaus*; OD] [**M**]. Very large to gigantic, with rounded to depressed whorls; coarse simple curved primary ribs split into finer secondaries over venter but these become obsolete in later growth; periodic deep constrictions and simple peristome. *Oxford.* **Fig. 73C** - *K. kranaus* Buckman.

The microconch is *Otosphinctes* Buckman (1926).

Pachyplanulites SPATH, 1930 [*Perisphinctes subevolutus* Waagen, 1869; OD] [**M**]. Moderately large, evolute with depressed subcircular whorl section in early growth, venter moderatey arched later; prorsiradiate biplicate flank costae with occasional single ribs and rare triplicates; ribs cross venter transversely on inner whorls, later with forward convexity; rib density increases with growth, with three prorsiradiate constrictions per whorl. *U. Oxford.* **Fig. 73D** - *P. subevolutus* (Waagen).

Superfamily ASPIDOCERATOIDEA Zittel, 1895

(*nom. transl.* Parent *et al.*, 2020 *ex* Aspidoceratidae)

= Aspidoceratiformes Buck, 1958 (*nom. nud.*).

Discussion. Derivatives of Perisphinctoidea (Waagen, 1875); comprises the families Aspidoceratidae, Physodoceratidae and Peltoceratidae (Parent *et al.*, 2020), of which only the first two are dealt with here. Mostly inflated with rounded to quadratic whorls, broadly rounded to subtabulate venter and umbilical tubercles, with or without lateral spines; constrictions lacking; short to very short body

Fig. 75. *Subgrossouvria* suture.

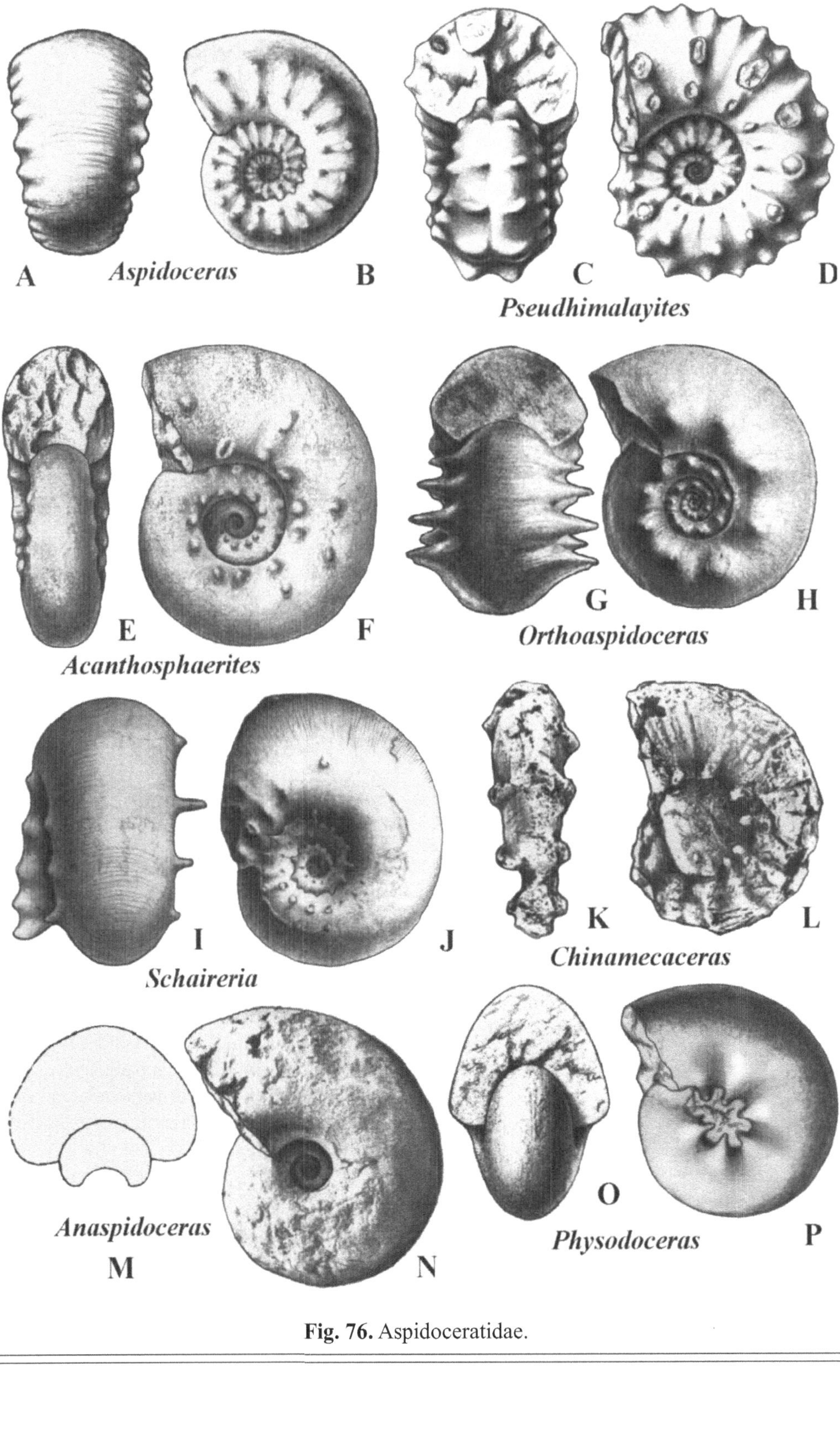

Fig. 76. Aspidoceratidae.

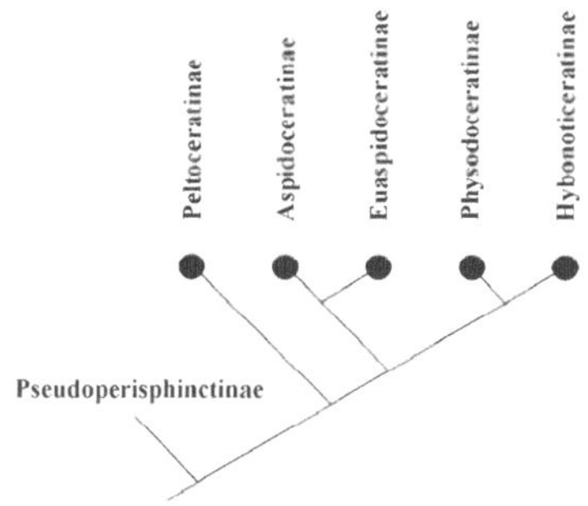

Fig. 77. Hypothesized relationships among Aspidoceratidae.

chamber; jaws calcified laevaptychi; suture with narrow E, rectangular E/L and trifid L as deep as E. *Upper M. Callov. - L. Berrias.*

Family ASPIDOCERATIDAE Zittel, 1895

Discussion. Paraphyletic derivatives of Zigzagiceratidae (Waagen, 1875); characterized by generally involute globular shells with 1-2 rows of spinose tubercles and weak ornament, at least on outer whorls. Predominantly Jurassic, comprising the subfamilies Epipeltoceratinae, Euaspidoceratinae and Aspidoceratinae; only the last is dealt with here. *Upper M. Callov. - L. Berrias.*

Subfamily Aspidoceratinae Zittel, 1895

(*nom. transl.* Salfeld, 1919 *ex* Aspidoceratidae)

Discussion. Monophyletic derivatives of Grossouvriinae (Énay & Howarth, 2019), close to *Alligaticeras*; typically inflated with broadly rounded smooth venter, depressed reniform to coronate whorl section, generally weak ribbing with umbilical and/or lateral spines; in early growth umbilical tubercles sometimes penetrate umbilicus. Commonest in the Tethyan Realm, but in strong decline by the beginning of the Cretaceous. *U. Oxford. - L. Berrias.*

Acanthosphaerites ROLLIER, 1922 [*Ammonites acanthicus* Oppel, 1863; OD] [**M**]. Large; like *Aspidoceras* but more evolute, with compressed oval whorl section, narrower venter, reduced tuberculation and suture with broader E/L and L. *U. Oxford. - L. Tithon.* **Fig. 76E-F** - *A. acanthicus* (Oppel).

The type species is part of a phyletic line (*cf.* Checa, 1985), including *A. sesquinodosus* (Fontannes), *A. uninodosus* (Toula) and *A. linaresi* (Checa), discrete from *Aspidoceras*.

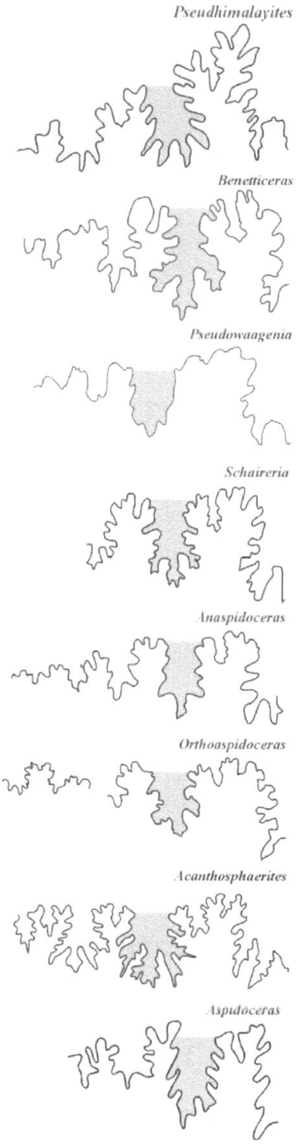

Fig. 78. Aspidoceratid sutures.

Aspidoceras ZITTEL, 1868 [*Ammonites rogoznicensis* Zejszner, 1846; M] [**M**]. Moderately large; fairly evolute, globular, with subvertical umbilical wall, broadly-rounded venter and strongly depressed reniform whorl section; two rows of lateral tubercles, outer near midflank and quickly becoming obsolete in some species; adult body chamber with reduced tuberculation and low ribs or folds over venter; suture with deep narrow bifid E, asymmetrically bifid E/L, narrow trifid L almost as deep as E, and bifid L/U_2. *L. Kimmer. - L. Berrias.* **Fig. 76A-B** - *A. rogoznicense* (Zejszner).

Chinamecaceras CANTÚ-CHAPA, 2006 [*C. maldonadoi*; OD]. Small; like *Aspidoceras* but with very compressed (juv.) to subquadratic (ad.) whorl section, sharp rectiradiate ribs with long umbilical bullae that project into

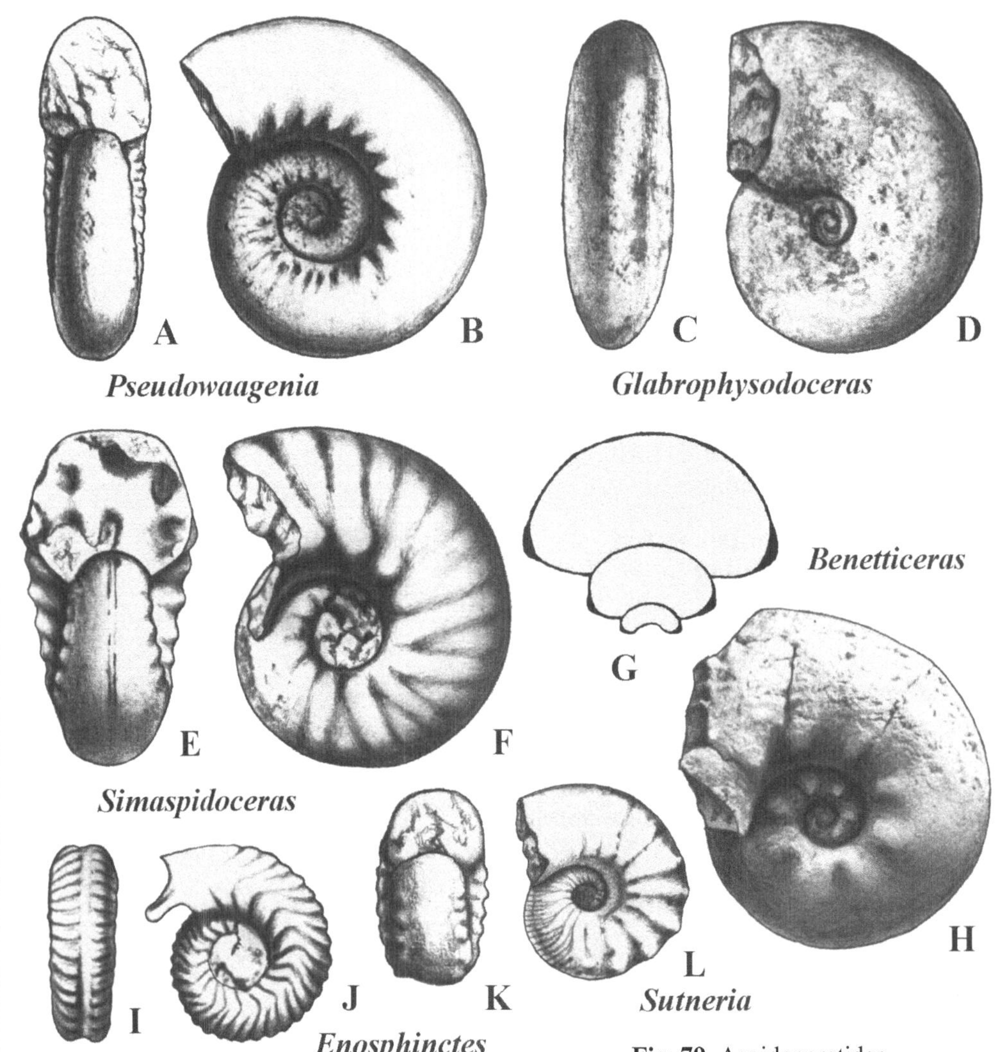

Fig. 79. Aspidoceratidae.

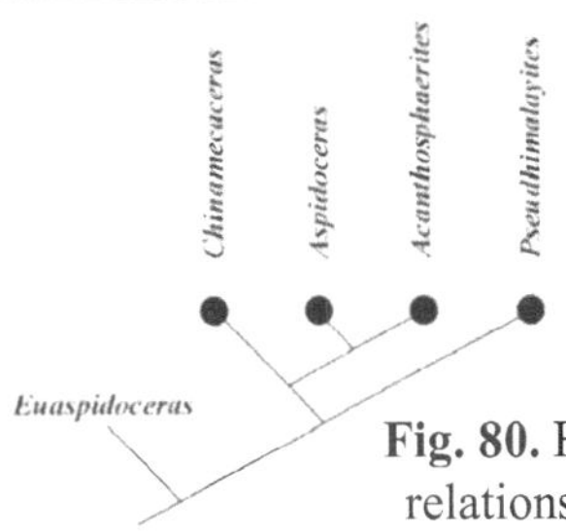

Fig. 80. Hypothesized relationships among Aspidoceratinae.

umbilicus; in early growth ribs trituberculate (umbilical, lateral, ventrolateral); in maturity ribs weaken and only umbilical tubercles persist; suture with shallow simple elements. *U. Tithon.* **Fig. 76K-L** - *C. maldonadoi* Cantú-Chapa.

Pseudhimalayites SPATH, 1925 [*Aspidoceras steinmanni* Haupt, 1907, = *Cosmoceras subpretiosum* Uhlig, 1878; OD]. Medium sized; like *Aspidoceras* but more inflated, with depressed subpentagonal whorl section, swollen septispinose outer lateral tubercles, and weak ribs on outer flank swelling to form ventral bullae separated by shallow siphonal groove. *L.- M. Tithon.* **Fig. 76C-D** - *P. subpretiosus* (Uhlig).

Family PHYSODOCERATIDAE Zittel, 1895

Discussion. Monophyletic derivatives of Aspidoceratidae (Roman, 1938), comprising the subfamilies Hybonoticeratinae and Physodoceratinae; only the latter is dealt with here. *U. Callov. - L. Berrias.*

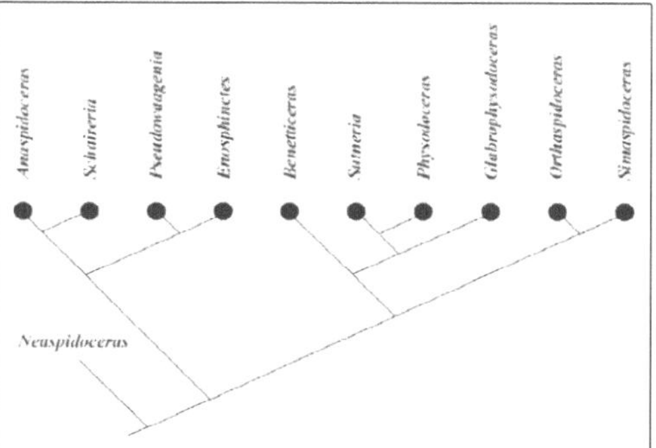

Fig. 81. Hypothesized relationships among Physodoceratinae.

Subfamily Physodoceratinae Zittel, 1925

(*nom. transl.* Checa, 1985 *ex* Physodoceratidae)
= Sutneriinae Zeiss, 1984

Discussion. Monophyletic derivatives of Euaspidoceratinae (Roman, 1938), close to *Neaspidoceras* gr. *radisense* (Checa, 1985), and thus phyletically distinct from Aspidoceratinae; typically with one row of tubercles/spines that vary in position from umbilical to mid-lateral; microconchs micromorph, with lateral lappets. Commonest in the Tethyan Realm, but in strong decline by the beginning of the Cretaceous. *U. Oxford. - L. Berrias.*

Pseudowaagenia SPATH, 1931 [*Ammonites haynaldi* Herbich, 1868; OD]. Small; evolute, compressed, with gently convex flanks converging to rounded venter and oval whorl section; vertical umbilical wall with subangular shoulder ornamented by numerous small spinose umbilical tubercles that are directed into umbilicus; in early to middle growth with distant small high-lateral tubercles and feeble ribs that become obsolete in later growth. *U. Oxford. - M. Kimm.* **Fig. 79A-B** - *P. haynaldi* (Herbich).

Enosphinctes SCHINDEWOLF, 1925 [*Sutneria subeumela* Schneid, 1914; OD] [**m**]. Micromorph; evolute, compressed, with gently convex flanks, low rounded venter with prominent siphonal groove, and subquadratic whorl section; strong prorsiradiate primary ribs reflexed at midflank where they give rise to paired rursiradiate ribs, the point of furcation sometimes marked by a lateral tubercle; aperture with lateral lappets. *Lower U. Kimmer.* **Fig. 79I-J** - *E. subeumelus* (Schneid).

Perhaps a microconch of *Pseudowaagenia*.

Physodoceras HYATT, 1900 [*Ammonites circumspinosus* Quenstedt, 1858; OD] [**M**]. Medium sized; moderately involute, globular, with rounded to depressed oval whorl section; smooth save for umbilical spines that protrude into umbilicus. *U. Oxford. - L. Kimmer.* **Fig. 76O-P** - *P. circumspinosum* (Quenstedt).

Glabrophysodoceras SCOTT, 1943 [*G. abyssineanum*; OD]. Like *Physodoceras* but compressed, high whorled, with broad flat flanks and subelliptical whorl section; inner whorls smooth and *Haploceras*-like; in later growth blunt feeble umbilical bullae give rise to distant weak ribs. *?U. Oxford., L. Kimmer.* **Fig. 79C-D** - *G. abyssineanum* Scott.

Benetticeras CHECA, 1985 [*B. benetti*; OD] [**M**]. Small; like *Physodoceras* but with subvertical convex umbilical wall, more depressed whorl section and larger umbilical bullae that do not penetrate the umbilicus. *U. Oxford. - L. Kimmer.* **Fig. 79G-H** - *B. benetti* Checa.

Sutneria ZITTEL, 1884 [*Nautilus platynotus* Reinecke, 1818; SD Munier-Chalmas, 1892] [**m**]. Micromorph; inflated, involute, with depressed reniform whorl section; in early to middle growth numerous weak prorsiradiate primaries give rise to bundles of 3-4 fine rectiradiate secondaries that cross venter uninterrupted; in later growth ribs weaken over venter; adult body chamber with strong distant rectiradiate primary ribs that terminate in ventrolateral tubercles; aperture with lateral lappets. *L. Kimmer.* **Fig. 79K-L** - *S. platynota* (Reinecke).

Perhaps a microconch of *Physodoceras*.

Anaspidoceras VIGH, 1984 [*Ammonites neoburgensis* Oppel, 1863; OD] [**M**]. Medium sized; like *Schaireria* but non-tuberculate and involute; adult body chamber shows scaphitoid uncoiling; suture with L almost as deep as E. *L. Tithon.* **Fig. 76M-N** - *A. neoburgense* (Oppel).

Simaspidoceras SPATH, 1925 [*Aspidoceras argobbae* Dacqué, 1905; OD]. Moderately large; fairy involute, with subvertical umbilical wall, flattened flanks converging to subtabulate venter and subpentagonal whorl section; blunt umbilical bullae give rise to 1-2 feeble prorsiradiate ribs that form weak ventrolateral swellings; venter smooth; suture complex. *L. Kimmer.* **Fig. 79E-F** - *S. argobbae* (Dacqué).

Orthaspidoceras SPATH, 1925 [*Ammonites orthocera* d'Orbigny, 1850; OD]. Large; inflated, globose, with moderately wide umbilicus, depressed coronate whorl section and swollen laterally-directed umbilicolateral spines. *M. Kimmer. - Tithon.* **Fig. 76G-H** - *O. orthocerum* (d'Orbigny).

Schaireria CHECA, 1985 [*Aspidoceras avellanum* Zittel, 1870; OD] [**M**]. Medium sized; like *Pseudowaagenia* but with two rows of tubercles (lateral and umbilicolateral); suture as in *Aspidoceras*. *U. Kimmer. - L. Berrias.* **Fig. 76I-J** - *S. depressa* (Steuer).

= *Shaireria* Howarth, 1998 (*nom. van*).

Superfamily IDOCERATOIDEA Spath, 1924

(*nom. transl.* herein *ex* Idoceratidae)

Discussion. Comprises the paraphyletic families Vinalesphinctidae (Oxfordian) and Idoceratidae (Kimmeridgian) with an origin in Perisphinctoidea (Geyssant, 1985; Vašíček & Skupien, 2016). Very

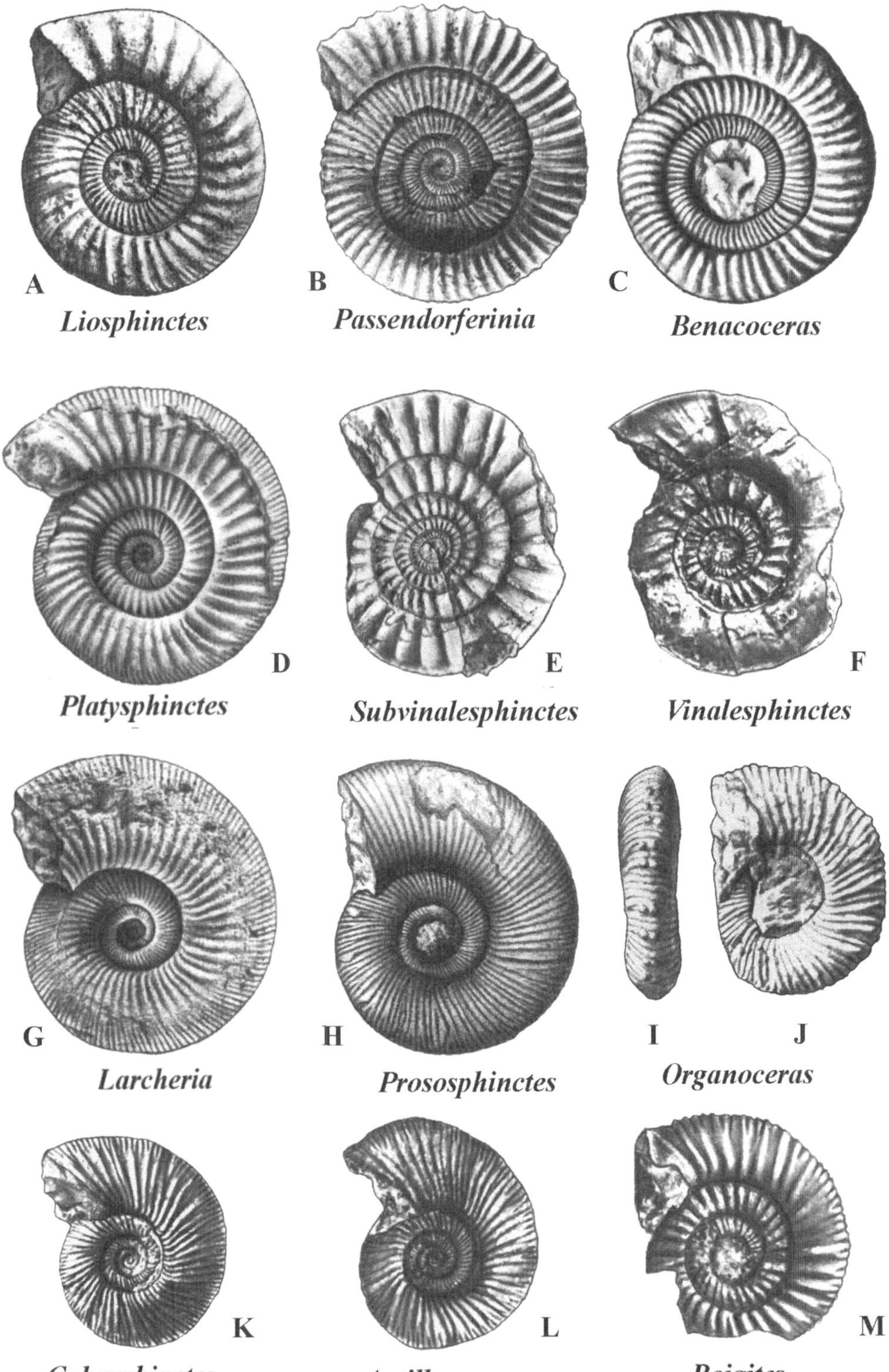

Fig. 82. Idoceratoidea.

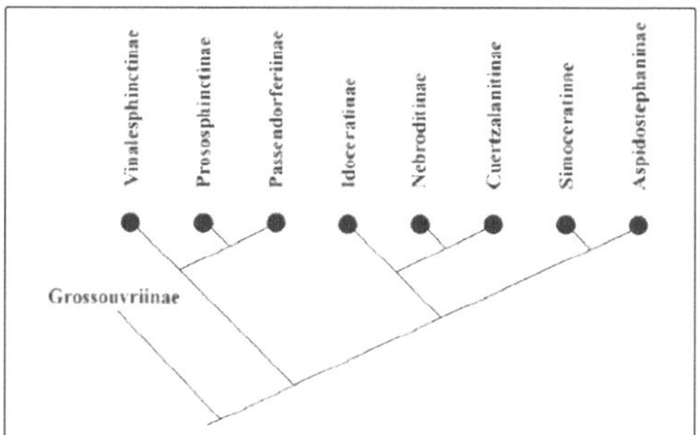

Fig. 83. Hypothesized relationships among Idoceratoidea and Simoceratidae.

evolute to serpentcone, often with flattened venter; biplicate and single ribs predominant, branching high on flank; constrictions and parabolic ribs frequent; dimorphic, microconchs lappeted. An important group, the progenitor of Ataxioceratoidea. *L. Oxford. - U. Tithon.*

Family VINALESPHINCTIDAE Meléndez & Myczynski, 1987

(*nom. transl.* herein *ex* Vinalesphinctinae)

Discussion. A Tethyan group derived from Perisphinctidae (Arkell, 1956); typically with biplicate and single ribs, rare intercalatories and parabolae (Meléndez *et al.*, 2009). Comprises the subfamilies Passendorferiinae, Prososphinctinae and Vinalesphinctinae. *L. - U. Oxford.*

Subfamily Passendorferiinae Meléndez, 1989

Discussion. A monophyletic group descended from Grossouvriinae, close to *Alligaticeras* (Meléndez *et al.*, 2006), largely confined to the southern margin of Tethys. Typically serpenticones with subcircular to subquadratic whorl sections, rectiradiate simple and biplicate ribs branching on the upper flank. Lapetted microconchs (D=80-120 mm) isocostate, densely ribbed, typically with parabolic nodes on final part of adult phragmocone. Macroconchs large to very large (D>240 mm), slightly variocostate, with thick uniform columnar ribs in maturity with incipient ventrolateral tubercles. All these features create a strong resemblance to Kimmeridgian *Nebrodites*, with which they have long been grouped and to which they may be linked phyletically by Late Oxfordian *Geyssantia* (Meléndez *et al.*, 2006). There are two main phyletic lines associated with the genera *Passendorferia* and *Sequeirosia*. *L. - U. Oxford.*

Passendorferia BROCHWICZ-LEWINSKI, 1973 [*P. teresiformis*; OD] [**M**]. Small (**m**) to very large (**M**), serpenticone; initially with subcircular whorls, in maturity subquadratic with a subtabulate venter; initially densicostate with rectiradiate to slightly prorsiradiate biplicate ribs, occasionally single, becoming more distant in middle to later growth (D>55 mm), narrower than interspaces; rib furcation at subtuberculate ventrolateral shoulder, uninterrupted over venter with slight forward arch; early whorls heavily constricted. *M. - U. Oxford.* **Fig. 82B** - *P. teresiformis* (Brochwicz-Lewinski.

Enayites Brochwicz-Lewinski & Rózak, 1976 [*Passendorferia (E.) gygi*; OD] [**m**]. Very small to small, very evolute, with subcircular whorl section and alternating simple and biplicate rectiradiate ribs crossing venter transversely; oblique constrictions frequent, and with peristomal collar and lateral lappets. *L. - U. Oxford.* **Fig. 87D** - *E. gygi* (Brochwicz-Lewinski & Rózak).

A microconch of some *Passendorferia*.

Sequeirosia MELÉNDEZ, 1984 (*Passendorferia brochwiczi* Sequeiros, 1977; OD] [**M**]. Large, evolute, with flattish subparallel flanks and subquadratic whorl section; angular umbilical edge; dense fine rectiradiate to slightly prorsiradiate ribs, barely modified on adult body chamber; numerous constrictions. *M. Oxford.* **Fig. 87A-B** - *S. brochwiczi* (Sequeiros).

As interpreted by Meléndez *et al.* (2009), different species of *Sequeirosia* evolved from different species of *Passendorferia* and hence the genus is polyphyletic.

Gemmellarites MELÉNDEZ, 1989 (*Ammonites trichopocus* Gemmellaro, 1875; OD] [**m**]. Very small, like *Enayites* but with subquadratic whorl section, finer denser ribs branching on upper third of flank, and two constrictions per whorl. *M. Oxford.* **Fig. 87C** - *G trichopocus* (Gemmellaro).

A microconch of some *Sequeirosia*.

Neumannia GLOWNIAK, 1989 (*Perisphinctes cyrilli* Neumann, 1907; OD] [**m**]. Like *Sequeirosia* but with depressed subrectangular whorl section, ribs effaced from

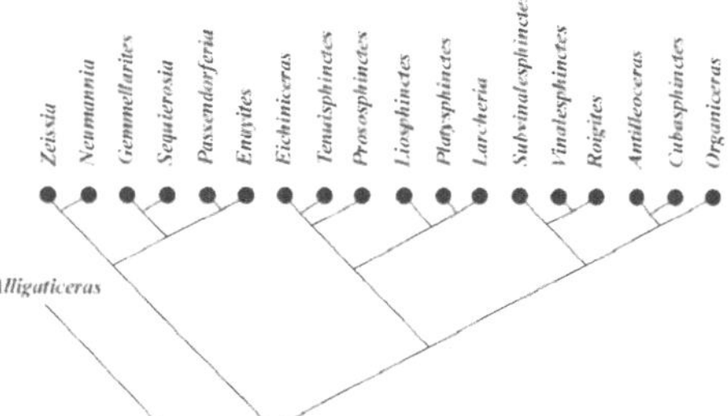

Fig. 84. Hypothesized relationships among Vinalesphinctidae.

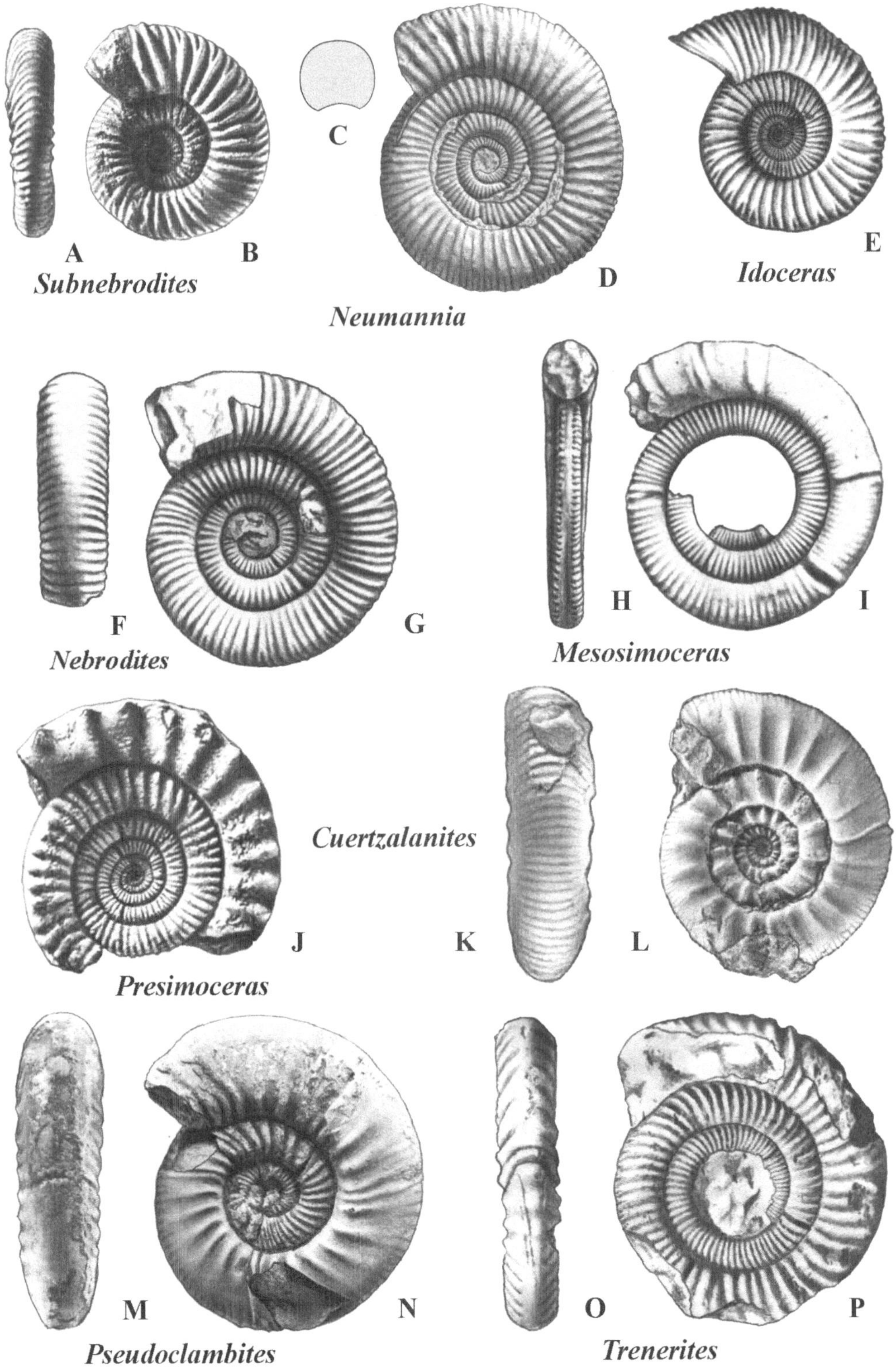

Fig. 85. Idoceratidae, Simoceratidae.

venter and fewer simple ribs. *M. Oxford.* **Fig. 86** - *N. cyrilli* (Neumann).

Zeissia ÉNAY, 2009 [*Z. simoceroides*; OD]. Medium sized; phragmocone with rounded venter and subrectangular whorl section; strong, dense, prorsiradiate ribs, biplicate and single, weak over venter; body chamber with strong distant rectiradiate ribs thickening ventrally but effaced from venter. *M. Oxford.* **Fig. 83G** - *Z. simoceroides* Énay.

Subfamily Prososphinctinae Glowniak, 2012

Discussion. Derivatives of Passendorferiinae (Glowniak, 2012). Typically discoidal, evolute, compressed with subelliptical whorl section; ribs mostly biplicate and triplicate with intercalatories across the venter, but occasionally biplicate; macroconchs with smooth or weakly ribbed adult body chamber. *L. - U. Oxford.*

Liosphinctes BUCKMANN, 1925 [*L. apolipon*; OD]. Like *Passendorferia* but with narrower umbilicus, finer denser secondary ribbing that weakens on adult body chamber where ribs frequently triplicate; periodic constrictions. *L. - U. Oxford.* **Fig. 82A** - *L. apolipon* Buckmann.

Prososphinctes SCHINDEWOLF, 1925 [*Perisphinctes mazuricus* Bukowski, 1887; OD][**M**]. Large, moderately involute high whorled, with dense fine biplicate and single ribs, often alternating, and periodic parabolic ribs; occasional constrictions. *L. Oxford.* **Fig. 82H** - *P. mazuricus* (Bukowski).

Larcheria TINANT, 1961 [*L. larcheri*; OD] [**M**]. Compressed, involute and high whorled with convergent flanks and oval whorl section; dense fine slightly prorsiradiate primary ribs weakly connected to 2-3 fine secondaries that flex forward and cross venter with forward convexity; branching occurs above midflank. *U. Oxford.* **Fig. 82G** - *L. schilli* (Oppel).

Platysphinctes TINANT, 1961 [*Perisphinctes (P.) perplanatus*; OD]. Small (**m**) to large (**M**); like *Liosphinctes* but becoming more involute with growth; whorl section trapezoidal to oval with flat convergent flanks; primary ribs split into 2's and 3's at distinct ventrolateral shoulder, separated by intercalatories; ornamental changes associated with frequent constrictions; flanks becoming smooth near adult aperture save for weak blunt primaries at umbilical edge. *M. Oxford.* **Fig. 82D** - *P. tinanti* Gygi.

Tenuisphinctes GYGI, 1998 [*T. (T.) kruegeri*; OD] [**M**]. Medium sized; like *Platysphinctes* but innermost whorls densely ribbed, becoming progressively more wider spaced and finally evanescing on adult body chamber; suture more complicated than *Platysphinctes*. *L. Oxford.* **Fig. 87F** - *T. kruegeri* Gygi.

Eichiniceras GYGI, 1998 [*Tenuisphinctes (E.) rolandi*; OD] [**m**]. Small; the microconch of *Tenuisphinctes* (*fide* Gygi, 1998). *L. Oxford.* **Fig. 87E** - *E. rollandi* (Gygi).

Subfamily Vinalesphinctinae Meléndez & Myczynski, 1987

Discussion. Derivatives of Prosospinctinae (Glowniak, 2012). Macroconchs with weak to obsolete ribbing in later growth; microconchs with parabolic nodes. Largely endemic to the Carribean Subprovince. *Oxford.*

Vinalesphinctes SPATH, 1931 [*V. rogoi*; OD] [**M**]. Large with simple peristome; evolute to very evolute, with variable whorl section from oval to quadratic and trapezoidal; early whorls densicostate, with prorsiradiate single and biplicate ribs, later sometimes with triplicates and intercalatories, crossing venter transversely; ribbing obsolete in middle to later growth; numerous deep collared constrictions. *Oxford.* **Fig. 82F** - *V. niger* Spath.

Subvinalesphinctes WIERZBOWSKI, 1976 [*Perisphinctes (Ampthillia?) corrali* Chudolay & Furrazola, 1968; OD]. [**M**]. Like *Vinalesphinctes* but ribs remaining strong on flanks to maturity. *Oxford.* **Fig. 82E** - *S. corrali* (Chudolay & Furrazola).

Cubasphinctes CHUDOLAY & FURRAZOLA, 1968 [*Perisphinctes (C.) jaworskii*; OD]. Ribbing irregular with several simple primaries and constrictions, associated with polyschizotome, polyfurcated and flared ribs; microconchs with lappets. *Oxford.* **Fig. 82K** - *C. jaworskii* (Chudolay & Furrazola.

Antilloceras WIERZBOWSKI, 1976 [*Perisphinctes antillarum* Jaworski, 1940; OD] [**m**]. Small to medium sized; peristome with lappets; perhaps the microconch of *Cubasphinctes*. *Oxford.* **Fig. 82L** - *A. antillarum* (Jaworski).

Roigites WIERZBOWSKI, 1976 [*Prososphinctes subconsociatus* Spath, 1931; OD] [**m**]. Small, inner whorls like *Vinalesphinctes*, aperture lappetted; comprises microconchs of *Vinalesphinctes* and, perhaps, *Subvinalesphinctes*. *Oxford.* **Fig. 82M** - *R. subconsociatus* (Spath).

Organoceras MYCZYNSKI, 1976 [*O. znoskoi*; OD] [**m**]. Medium sized; evolute, with flat flanks, irregular prorsiradiate biplicate ribbing and parabolic nodes to adult

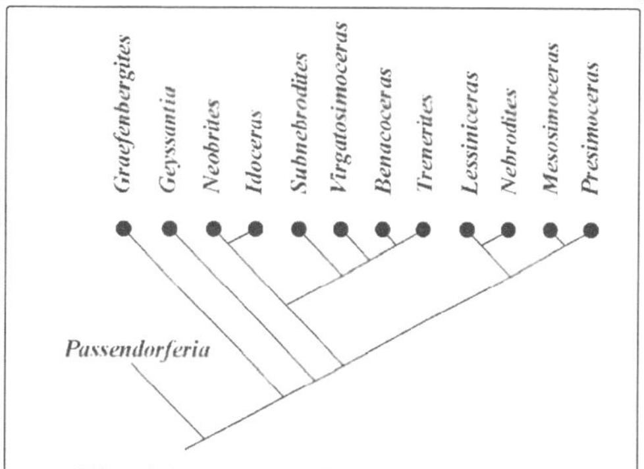

Fig. 86. Hypothesized relationships among Idoceratidae.

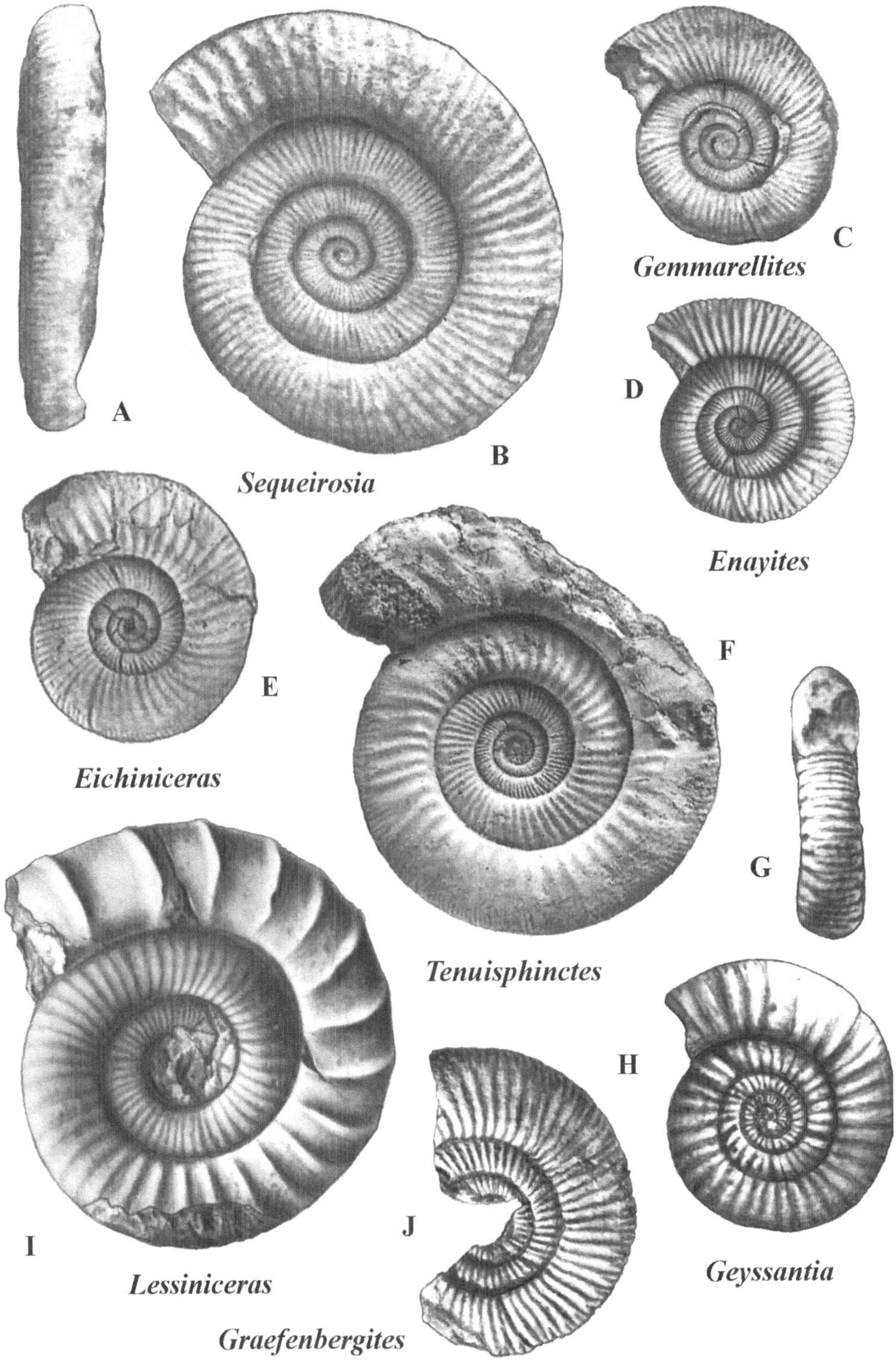

Fig. 87. Vinalesphinctinae, Nebroditinae.

body chamber. *Oxford*. **Fig. 82I-J** - *O. znoskoi* Myczynski.

Family IDOCERATIDAE Spath, 1924

(*nom. transl.* herein *ex* Idoceratinae)

Discussion. A predominantly Kimmeridgian Tethyan group, derivatives of Vinalesphinctidae (Brochwicz-Lewinski, 1973); typically with biplicate and single ribs, rare intercalatories, and periodic parabolic ribs (Meléndez *et al.*, 2009). Comprises the subfamilies Nebroditinae, Idoceratinae and Cuertzalanitinae. *Uppermost Oxford. - L. Kimmer.*

Subfamily Nebroditinae Cantú-Chapa, 2009

(*nom. transl.* herein *ex* Nebroditidae)

Discussion. A monophyletic taxon with an origin in Vinalesphinctinae, close to *Vinalesphinctes*; typically serpenticones with subquadratic whorls, flattened venter and weakly curved simple and infrequent bifurcating rectiradiate primary ribs, subtuberculate at ventrolateral edges and subtransverse over venter; irregular constrictions; suture with narrow bifid E/L, very narrow asymmetrically bifid L deeper than E, narrow L/U_2 and retracted suspensive lobe. Predominantly Mediterranean. *Uppermost Oxford. - L. Kimmer.*

Graefenbergites SCHAIRER & SCHLAMPP, 2003 [*Perisphinctes idoceratoides* Dorn, 1930; OD] [**m**]. Medium sized (**m**) to moderately large (**M**); compressed serpenticones with flat flanks, narrowly arched venter, strong regular slightly prorsiradiate biplicate ribs that form chevrons over venter, and periodic constrictions; macroconchs with triplicate and polygyrate ribs also. *U. Oxford.* **Fig. 87J** - *G. idoceratoides* (Dorn).

Nebrodites BURCKHARDT, 1910 [*Simoceras agrigentinus* Gemmellaro, 1872; SD Spath, 1925] [**m**]. Small; compressed serpenticone with flattened flanks, subtabulate venter and subquadratic whorl section; numerous rectiradiate ribs, mostly simple but also biplicate, branching very high on flank; ribs weak or effaced from siphonal line; periodic constrictions. *U. Oxford. - L. Kimmer.* **Fig. 85F-G** - *N. agrigentinus* (Gemmellaro).

Geyssantia MELÉNDEZ, 1989 [*G. geyssanti*; OD]. Medium sized serpenticone with rounded subcircular whorl section and crassicostate ornament with coarse and large ribs on flank and venter, becoming smooth at an early stage; ribbing effaced from venter; constrictions prominent. *Uppermost Oxford. - L. Kimmer.* **Fig. 87G-H** - *G. geyssanti* Meléndez.

Mesosimoceras SPATH, 1925 [*Simoceras cavouri* Gemmellaro, 1872; OD]. Medium sized; like *Nebrodites* but more evolute and compressed, with secondary ribs terminating in small ventrolateral nodes on either side of smooth siphonal band; constrictions bordered by flared collars; ribs fading on adult body chamber. *Uppermost Oxford. - L. Kimmer.* **Fig. 85H-I** - *M. cavouri* (Gemmellaro).

Lessiniceras PARVIA, BENETTI & MINETTI, 1987 [*Perisphinctes ptychodes* Neumayr, 1873; OD] [**M**]. Large, very evolute, with fine prorsiradiate single ribs, occasionally bifid, on phragmocone; coarse distant ribbing on adult body chamber. *L. Kimmer.* **Fig. 87I** - *L. ptychodes* (Neumayr).

Presimoceras SARTI, 1990 [*Ammonites herbichi* von Hauer, 1866; OD] [**M**]. Moderately large; ribs simple, coarse, distant in middle to later growth, crossing venter uninterrupted. *L. Kimmer.* **Fig. 85J** - *P. herbichi* (von Hauer).

Subfamily Idoceratinae Spath, 1924

(*nom. transl.* Khimshiashvili *et al.*, 1958 *ex* Idoceratidae)

Discussion. A seemingly monophyletic group with an origin in Nebroditinae, close to *Nebrodites* (Brochwicz-Lewinski, 1973). Mostly fairly coarsely ribbed, with frequent simple ribs, prominent constrictions and secondary ribs that project forward and are weak or effaced across the venter; suture with U_2 much smaller than L. An Early Kimmeridgian radiation in the Submediterranean Subprovince of Tethys. *Uppermost Oxford. - L. Kimmer.*

Subnebrodites SPATH, 1925 [*S. planulata nom. nov. pro Ammonites planula* Quensted, 1888; OD] [**m**]. Small, with convergent flanks, fairly narrow subtabulate venter and subtrapezoidal to oval and elliptical whorl section; prorsiradiate biplicate ribs, frequently separated by intercalatory, branching above midflank and projecting foward near venter; ribbing becomes coarse and distant in later growth; 3-6 oblique constrictions per whorl; suture with square bifid E/L, narrow L not as deep as E, and very

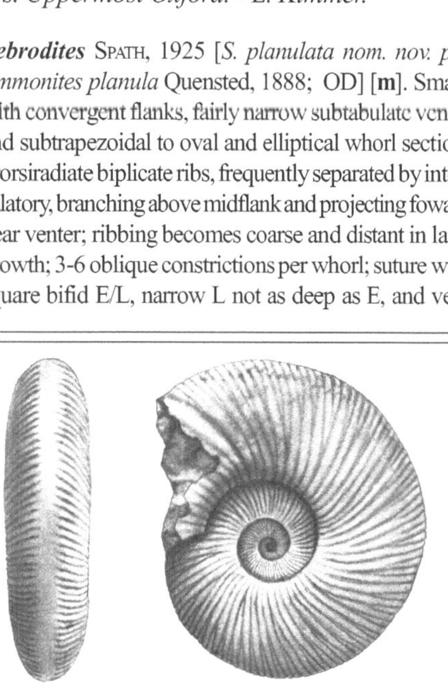

Fig. 88. *Idoceras*.

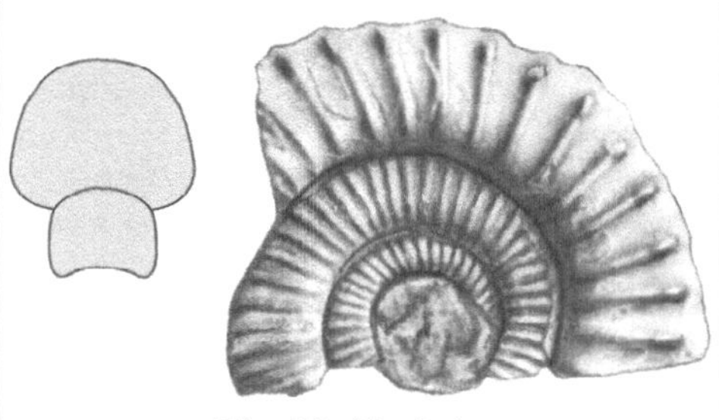

Fig. 89. *Neobrites*.

small U_2. *Uppermost Oxford. - Lower L. Kimmer.* **Fig. 85A-B** - *S. planula* Spath.

Idoceras BURCKHARDT, 1906 [*Ammonites balderus* Oppel, 1863; SD Spath, 1925]. Small to large; very evolute, compressed, with coarse dense biplicate and single ribs projecting forward near venter and weak or effaced from siphonal line; ribs may form chevrons across venter in later growth. Predominantly Central American. *Upper U. Oxford. - L. Kimmer.* **Fig. 85E** - *I. laxevolutum* (Hehl); **Fig. 88** - *I. densicostatum* Imlay.

= *Paraidoceras* Cantú-Chapa, 1992 (obj.).

Neobrites ALBRITTON, 1937 [*N. nodosocostatus* Burckhardt, 1912; OD] [**M**]. Like *Subnebrodites* but more inflated, with subquadratic whorl section and fewer ribs that become coarse and distant in later growth with well developed ventrolateral bullae; ribs interrupted across siphonal line in early growth, but crossing venter in maturity. *L. Kimmer.* **Fig. 89** - *N. nodosocostatus* (Burckhardt).

Trenerites SARTI, 1993 [*Simoceras evolutum* Gemmellaro, 1876; OD] [**M**]. Medium sized (**m**) to large (**M**) serpenticone with flat flanks and compressed subrectangular whorl section; ribs mostly simple, occasionally biplicate in early growth, crossing venter or interrupted by smooth siphonal band; deep constrictions throughout. *L. Kimmer.* **Fig. 85O-P** - *T. evolutus* (Gemmellaro).

Benacoceras SPATH, 1925 [*Simoceras heteroplocum* Gemmellaro, 1877; OD] [**M**]. Large serpenticone with strong narrow mainly biplicate ribs effaced across venter in later growth; secondary ribs very short and projected. *L. Kimmer.* **Fig. 82C** - *B. heteroplocum* (Gemmellaro).

Subfamily Cuetzalanitinae Cantú-Chapa, 2009

(*nom. correct.* herein *ex* Cuetzalaninae)

Discussion. Descendents of Nebroditinae (Cantú-Chapa, 2009). Evolute, compressed, with subrounded whorls and moderately broad venter; simple distant rectiradiate biplicate and triplicate primary ribs separated by intercalatories, with or without constrictions. *U. Tithon.*

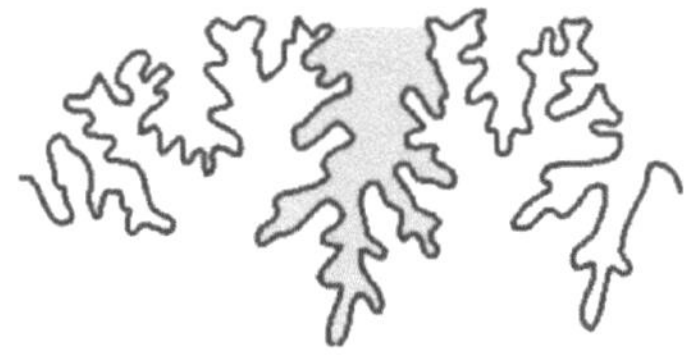

Fig. 90. *Subnebrodites* suture.

Cuetzalanites CANTÚ-CHAPA, 2009 [*C. tonatuiloides*; OD]. Small, evolute, compressed, with subrounded whorls, moderately broad venter, simple distant rectiradiate biplicate primary ribs splitting into 2-3 secondaries across venter, separated by 2-3 intercalatoriesy. *Lower U. Tithon.* **Fig. 85K-L** - *C. tonatuiloides* Cantú-Chapa.

Pseudoclambites SPATH, 1925 [*P. aenigmaticus*; OD]. Medium sized; evolute, compressed, with subtrapezoidal whorl section, gently convex flanks converging to broad shallowly sulcate venter with rounded ventrolateral shoulders; 4-5 constrictions per whorl parallel to primary ribs; in early growth flanks with sharp prorsiradiate ribs, with 2-3 secondaries per primary, arising from umbilical clavi; ribs weaken in middle to later growth. *U. Tithon.* **Fig. 85M-N** - *P. araensis* Howarth.

The genus was created for poorly preserved material from Bihendula (Somalia), the type species based on a body chamber, its early growth stages unknown; here it is interpreted on the basis of *P. araensis* Howarth.

Suborder PROTANCYLOCERATIDINA Vermeulen, 2005

(*nom. correct.* herein *pro* Protancyloceratina)

Discussion. A seemingly monophyletic group with an origin in Perisphinctidina (Checca, 1997; Schweigert & Zeiss, 1998); superficially similar Jurassic Spiroceratoidea are believed to originate in Hildoceratina (Donovan *et al.,* 1981). Primary suture quinquelobate, later quadrilobate. Predominantly Tethyan. Comprises the superfamilies Protancyloceratoidea, Bochianitoidea and Ptychoceratoidea. *L. Tithon. – L. Apt.*

Superfamily PROTANCYLOCERATOIDEA Breistroffer, 1947

(*nom. transl.* Dimitrova, 1970 *ex* Protancyloceratidae)

Discussion. A monophyletic group with an origin in Idoceratoidea (Checca, 1997; Schweigert & Zeiss, 1998); small to medium sized helicones, ellipticones and ancylocones, generally with strong simple ribs

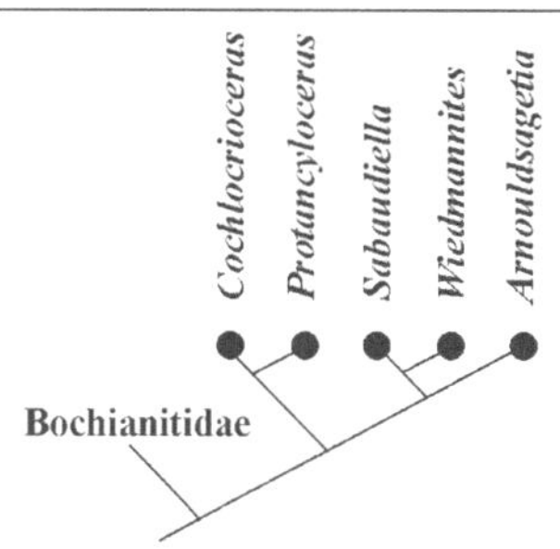

Fig. 91. Hypothesized relationships among Protancyloceratinae.

weak or interrupted across the venter; some constricted. Comprises the families Protancyloceratidae, Leptoceratidae and Leptoceratoididae. *L. Tithon. – L. Apt.*

Family PROTANCYLOCERATIDAE Breistroffer, 1947

Discussion. A seemingly monophyletic group with an origin in Vinalesphinctidae (Schweigert & Zeiss, 1998); shell variable, from loose helicone to criocone and ancylocone, with oblique ribs tending to form chevrons over the venter and generally interrupted by a siphonal groove, with or without ventrolateral tubercles; simple suture with bifid saddles, trifid lobes and reduced U. Comprises the subfamilies Protancyloceratinae and Wiedmannitinae. *L. Tithon. - Barrem.*

Subfamily Protancyloceratinae Breistroffer, 1947

Discussion. Seemingly monophyletic derivatives of uncertain origin, either Passendorferiinae, close to *Passendorferia*, or Idoceratinae close to *Berckhemeria* (a microconch of *Presimoceras*) (Schweigert & Zeiss, 1996). Small to medium sized very loose helicones and aspinocones with simple annular ribs coarsening on the adult body chamber. *L. Tithon. - U. Berrias.*

Protancyloceras SPATH, 1924 [*Ancyloceras guembeli* Oppel, 1868; OD]. Medium sized very loose helicone with subcircular whorl section in middle growth; coarse distant prorsiradiate ribs on outer flank form blunt oblique ventrolateral clavi separated by deep ventral groove. *L. Tithon.* **Fig. 92A-B** - *P. guembeli* (Oppel).

As shown by Wierzbowski (1990), the genus *Protancyloceras* covers a number of different phyletic lines that require discrimination.

Due to poor knowledge of the type species, *P. kurdistanensis* Spath (**Fig. 92N-P**) has come to typify the genus (Wright, 1956, 1996). It is a very loose helicone with compressed oval whorl section later becoming circular with a shallowly impressed dorsum; dense prorsiradiate ribs terminate along a faint smooth siphonal band, connected across the sulcate dorsum by fine striae that become concave adorally; adult body chamber with coarse annular ribs, narrower than interspaces, becoming increasingly prorsiradiate anteriorly; ventral lappet to peristome. A second specimen lacks the ventral groove.

Cochlocrioceras SPATH, 1950 [*C. turriculatum*; OD]. Very small; loose helicone with coarse distant ribs interrupted across venter in early to middle growth, but forming chevrons later. *Upper L. - Lower U. Tithon.* **Fig. 92I-J** - *C. turriculatum* Spath.

***Arnouldsagetia* gen. nov.** [*Protancyloceras punicum* Arnould-Saget, 1951, p.114; OD herein]. Very small; seemingly a loose criocone with compressed oval whorl section and smooth dorsum; sharp rectiradiate ribs curve strongly forward on outer flank and form acute chevrons over venter interrupted by a narrow siphonal groove; suture relatively simple, with asymmetrical deeply bifid saddles and narrow trifid L. *M. - U. Berrias.* **Fig. 92C-D** - *A. punica* (Arnould-Saget).

Subfamily Wiedmannitinae nov.

Diagnosis. Very small criocones with more or less well developed fibulate ribbing and ventrolateral spines; monophyletic derivatives of Protancyloceratinae. *Berrias. - L. Barrem.*

Wiedmannites CANTÚ-CHAPA, 2009 [*W. fibuloidis*; OD]. Based on fragment suggesting a very small criocone or ellipticone; flat flanks and fine flexuous ribs looped to small sharp ventrolateral tubercles; suture with broad asymmetrically bifid E/L and narrow trifid L deeper than E.. *Berrias. - U. Valang.* **Fig. 92G** - *W. fibuloidis* Cantú-Chapa.

The type species is barely separable from older *W. bicostatus* (Arnould-Saget).

Sabaudiella VASÍCEK & HOEDEMAEKER, 2003 [*Ancyloceras sabaudiana* Pictet & de Loriol, 1858; OD]. Small gently expanding ancylocone with loose criocone spire, long almost straight proversum with fine rectiradiate ribs narrower than interspaces, all with ventrolateral spines; ribbing more distant on U-shaped flexus and retroversum, with periodic fibulation; suture relatively complex, with bifid saddles and asymmetrically trifid L. *Uppermost Haut. - L. Barrem.* **Fig. 94F** - *S. argosiensis* Vašíček & Hoedemaeker.

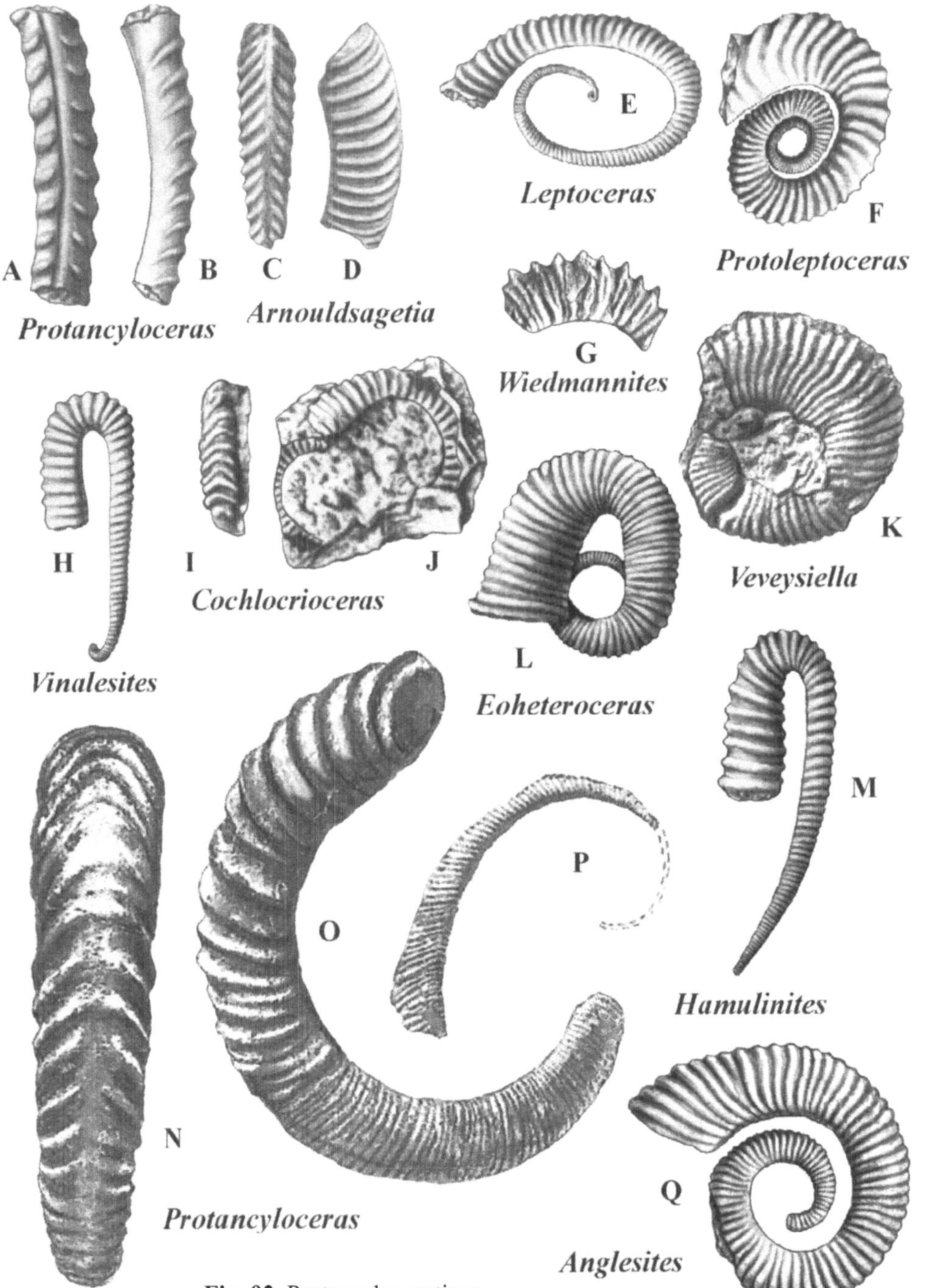

Fig. 92. Protancyloceratinae.

Family LEPTOCERATIDAE Manolov, 1962

(*nom. transl.* herein ex Leptoceratinae)

Discussion. Monophyletic derivatives of Protancyloceratidae (Wright, 1996); typically micromorphs with very simple sutures and annular ribbing uninterrupted and transverse across the venter. Comprises the subfamilies Leptoceratinae, Leptoceratoidinae and Karsteniceratinae. *L. Tithon. - L. Apt.*

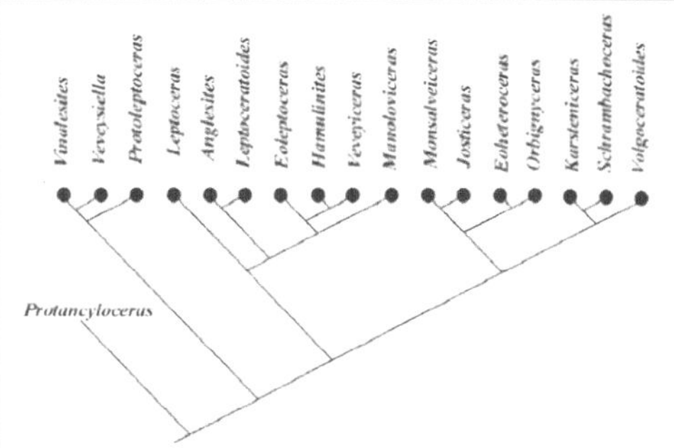

Fig. 93. Hypothesized relationships among Leptoceratidae.

Subfamily Leptoceratinae Manolov, 1962

Discussion. Monophyletic micromorph derivatives of Protancyloceratinae, close to *Protancyloceras gracile* (Oppel) (Wierzbowski, 1990); typically small to very small strongly ribbed criocones, ellipticones and aspinocones of only a few whorls with non-tuberculate annular ribs transverse and uninterrupted across rounded venter; simple suture with broad square relatively indented L. *L. Tithon. - L. Valang.*

Vinalesites THIEULOY, 1966 [*Hamulina rosariensis* Imlay, 1942; OD]. Micromorph ancylocone with parallel shafts and U-shaped flexus; proversum with simple prorsiradiate ribs that coarsen on flexus and retroversum. *L. - U. Tithon., ?U. Berrias.* **Fig. 92H** - *V. rosariensis* (Imlay).
= *Pseudoanahamulina* Judoley & Furrazola-Bermudez, 1968 (obj.).

Leptoceras UHLIG, 1883 [*Ancyloceras brunneri* Ooster, 1860; SD Roman, 1938]. Loose slowly expanding ellipticone of just over one whorl, with simple uniform non-tuberculate rectiradiate annular ribs narrower than interspaces; body chamber expanding rapidly, with constricted aperture; suture simple, with symmetrically bifid E/L, trifid L as deep as E and tiny V-shaped U much shallower than narrow I. *U. Berrias. - L. Valang.* **Fig. 92E** - *L. brunneri* (Ooster).

Protoleptoceras NIKOLOV, 1966 [*P. jelevi*, = *Ancyloceras studeri* Ooster, 1860; OD]. Like *Leptoceras* but larger; spire a tightly coiled criocone of two and a half whorls with strong, slightly rursiradiate annular ribs, broader than interspaces; elliptical body chamber barely separated from spire, increasing rapidly in height, with coarse flexed ruriradiate ribs and compressed elliptical whorl section; suture with shallower E and narrow trifid L. *U. Berrias. - L. Valang.* **Fig. 92F** - *P. studeri* (Ooster).
= *Proleptoceras* Vašíček & Hoedemaker, 2003 (*nom. van.*).

Subfamily Leptoceratoidinae Thieuloy, 1966

Discussion. Seemingly monophyletic derivatives of Leptoceratinae (Avram, 1999), but separated by a stratigraphical gap and perhaps homoeomorphs (Vašíček & Wiedmann, 1994; Wright, 1996). Typically the suture is simple, with a relatively narrow E/L, barely trifid to entire L, and large deep U. Predominantly Tethyan where they are commonest in offshore argillaceous facies, but there are too many genera to represent a single phyletic line. *L. Barrem.- L. Apt.*

Leptoceratoides THIEULOY, 1966 [*Crioceras (Leptoceras) pumilum* Uhlig, 1883; OD]. Micromorph; initial loose criocone followed by planispiral whorl; flexuous single rectiradiate to rursiradiate ribs; periodic strong constrictions backed, in later growth, by flared collar. *Barrem.* **Fig. 94E** - *L. pumilis* (Uhlig).

Veveysiceras VAŠÍČEK & WIEDMANN, 1993 [*Ancyloceras escheri* Ooster, 1860; OD]. Like *Leptoceratoides* but with initial loose planispire followed by an ellipticone with rather tight bends; ribbing dense and fine without strengthened primaries; weak constrictions may occur on retroversum; suture poorly known, but with barely indented saddles, weakly trifid L and much reduced U. *Lower L. Barrem.* **Fig. 94A** - *V. escheri* (Ooster).

Hamulinites PAQUIER, 1900 [*Hamulina munieri* Nicklès, 1894; OD]. Like *Veveysiceras* but a hamulinicone with initial planispire followed by more or less straight proversum; uniform strong coarse ribs; suture like *Karsteniceras* but more indented, with broader L/U. *L. Barrem.* **Fig. 92M** - *H. munieri* (Nicklès).
= *Wrightites* Manolov, 1962 (obj.).

Josticeras VAŠÍČEK & KLAJMON, 1998 [*J. wiedmanni*; OD]. A loose ellipticone like *Eoleptoceras* but larger, with body chamber uncoiling perpendicular to penultimate shaft and expanding in height, with alternating fine and strong ribs; suture like *Veleziceras*. *L. Barrem.* **Fig. 94C** - *J. wiedmanni* Vašíček & Klajmon.

Eoheteroceras VAŠÍČEK & WIEDMANN, 1993 [*E. silesiacum*; OD]. Initial more or less tight helix enclosed by curved proversum and hook; ribbing relatively coarse, simple, annular; a *Heteroceras* homoeomorph. *L. Barrem.* **Fig. 92L** - *E. silesiacum* Vašíček & Wiedmann.

Veveysiella BUSNARDO, CHAROLLAIS, WIEDMANN & CLAVET, 2003 [*V. vibiscumi*; OD]. Small evolute planispire with contiguous whorls increasing markedly in height; whorls with flattened flanks and compressed, subrectangular whorl section; simple fine rectiradiate ribs, broader than interspaces, flexing forward on outer flank in later growth and crossing venter transversely. *Barrem.* **Fig. 92K** - *V. vibiscumi* Busnardo, Charollais, Wiedmann & Clavel.

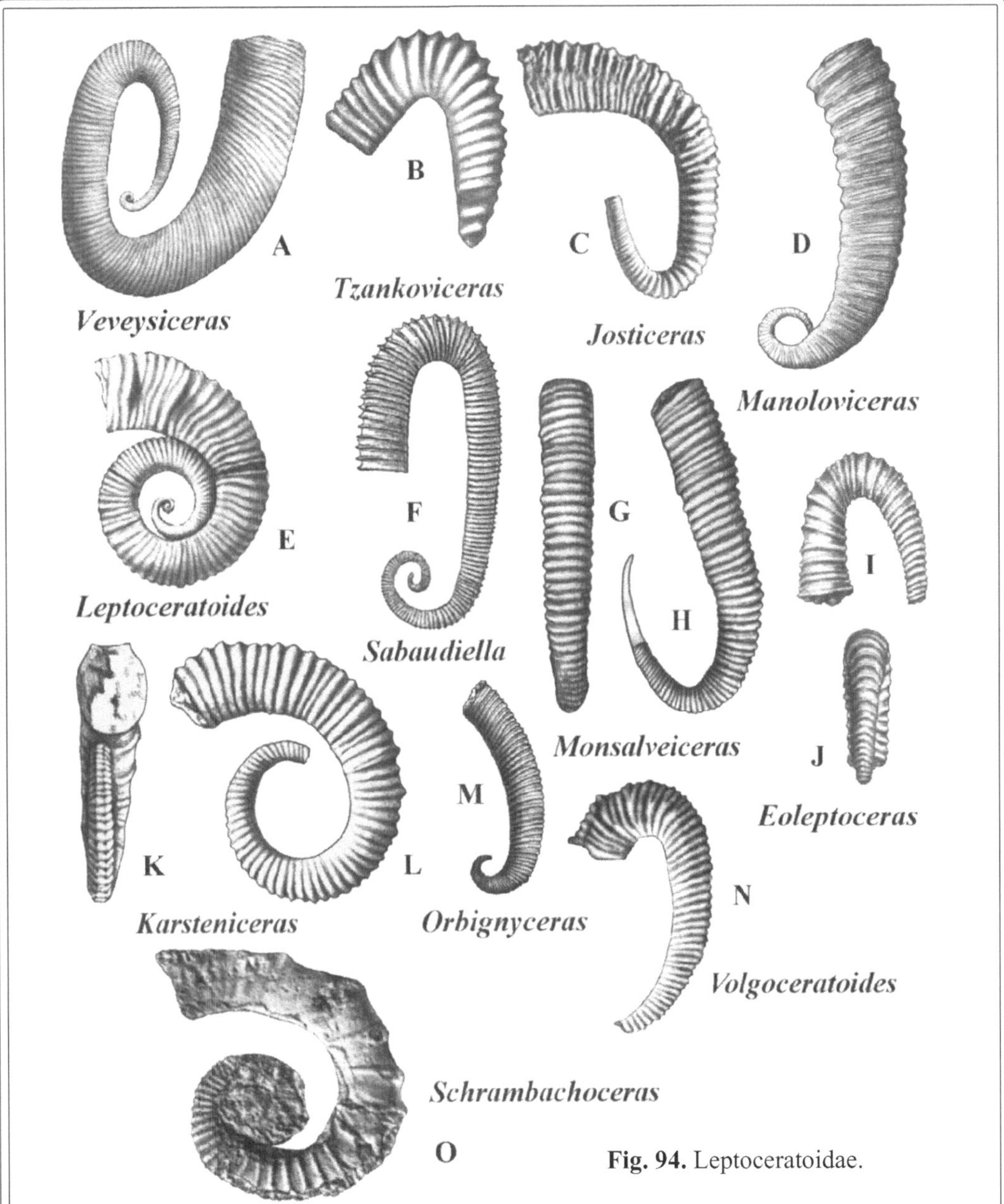

Fig. 94. Leptoceratoidae.

Eoleptoceras MANOLOV, 1962 [*Crioceras (Leptoceras) parvulum* Uhlig, 1883; OD]. Like *Hamulinites* but with more open hook and less regular ribbing to retroversum. *Barrem.* **Fig. 94I-J** - *E. parvulum* (Uhlig).

Tzankoviceras MANOLOV, 1962 [*Crioceras (Leptoceras) assimile* Uhlig, 1883; OD]. Like *Hamulinites* but larger and more robust, with sharper more distant ribs that are projected on phragmocone, and a more complex suture with narrow lobes. *?U. Barrem.* **Fig. 94B** - *T. assimile* (Uhlig).

Manoloviceras VAŠÍČEK & WIEDMANN, 1993 [*Hemibaculites saharievae* Manolov, 1962; OD]. Initial helix loose and open, followed by rapidly expanding gently curved final shaft; simple prorsiradiate ribs, at first dense and fine, coarsening with age; ribs and interspaces covered with fine growth lines; suture like *Karsteniceras* but more indented, with longer saddles, weakly digitate L and trifid U. *L. Barrem.* **Fig. 94D** - *M. saharievae* (Manolov).

Orbignyceras ROYO Y GÓMEZ, 1945 [*O. veleziense*; OD]. Known only from straight to gently curved rapidly expanding shafts with elliptical whorl section, simple non-tuberculate ribs and weak constrictions; like *Hamulinites* but, seemingly, without final hook and with

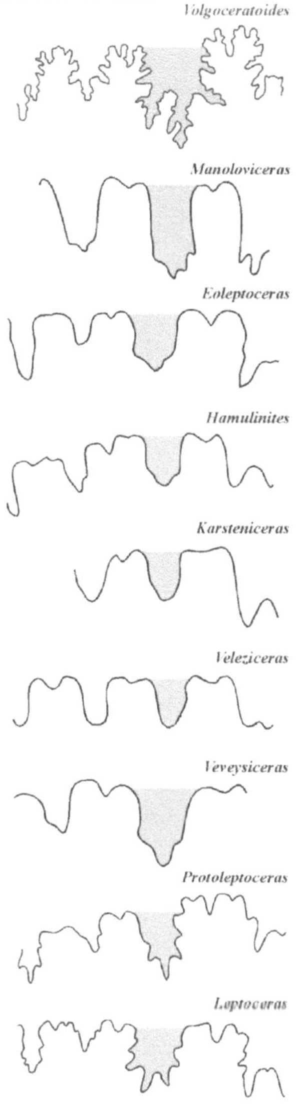

Fig. 95. Leptoceratid sutures.

simpler suture with small median saddle to E and undivided lobes. *Barrem.* **Fig. 94M** - *V. veleziense* (Royo y Gómez).

= *Veleziceras* Wright, 1957 (obj.).

Monsalveiceras Kakabadze & Hoedemaeker, 1997 [*M. monsalvense*; OD]. Like *Orbignyceras* but shell hamiticone, with closer spaced constrictions separated by only 2 ribs. *Apt.* **Fig. 94G-H** - *M. monsalvense* Kakabadze & Hoedemaeker.

Subfamily Karsteniceratinae Immel, 1987

Discussion. Questionably monophyletic derivatives of Leptoceratoidinae; criocones and hamulinicones with ventrolateral tubercles. *L. Valang. - L. Apt.*

Schrambachoceras Boorová, Skupien, Vašíček & Lobitzer, 2015 [*Protancyloceras obscurocostatum* Vašícek & Hoedemaker, 2003; OD]. Very small criocone; phragmocone with dense simple convex forward rursiradiate ribs with occasional ventrolateral tubercles; on body chamber ribs becoming very distant, some uniting in prominent ventrolateral tubercles; finally with weak dense fine ribs similar to phragmocone; suture similar to *Arnoldsagetia. L. Valang.* **Fig. 94O** - *S. obscurocostatum* (Vašíček & Hoedemaker).

Anglesites Delanoy & Busnardo, 2007 [*Crioceras puzosianus* d'Orbigny, 1842; OD]. Small criocone, with broad flattish flanks, subtabulate venter and compressed subtrapezoidal whorl section; fine simple straight to flexuous ribs pass uninterrupted over venter with more or less marked ventrolateral bullae; suture simple, similar to *Veveysiceras. Upper U. Barrem.* **Fig. 92Q** - *A. puzosianus* (d'Orbigny).

Karsteniceras Royo y Gómez, 1945 [*Ancyloceras beyrichi* Karsten, 1858; OD]. Very small; loose criocone with low trochospiral initial coil followed by subelliptical whorls; whorl section oval to subcircular, with subtabulate to shallowly grooved venter; initially smooth, followed by simple prorsiradiate ribs with ventrolateral bullae; in middle and later growth ribs differentiate into stronger primaries separated by 1-4 secondaries with occasional looping and periodic constrictions; suture very simple, barely incised, with broad bifid saddles and wide barely indented L. *Barrem. - L. Apt.* **Fig. 94K-L** - *K. beyrichi* (Karsten).

Volgoceratoides Mikhailova & Baraboshkin, 2002 [*V. schilovkensis*; OD]. Micromorph; hamulicone with initial loose criocone followed by arcuate proversum and flexus; proversum with simple ribs with lateral and ventral tubercles; on flexus ribs biplicate and intercalated, with loss of ventral tubercles; retroversum with comb of tubercles; suture simple, with shallow E, relatively narrow asymmetrically bifid E/L, broad trifid L much deeper than E, reduced trifid U and trifid I. *L. Apt.* **Fig. 94N** - *V. schilovkensis* Mikhailova & Baraboshkin.

Superfamily BOCHIANITOIDEA Spath, 1922

(*nom. transl.* Klein *et al.*, 2007 *ex* Bochianitidae)

Discussion. A monophyletic group with an origin in Protancyloceratoidea (Wright, 1996; Arkadiev *et al.*, 2011). A straight baculicone, smooth or with prorsiradiate ribs, sometimes tuberculate, or a ribbed hamulinicone or ptychocone. Comprises the families Bochianitidae and Anahamulinidae (Vermeulen *et al.*, 2020). *U. Tithon. - L. Apt.*

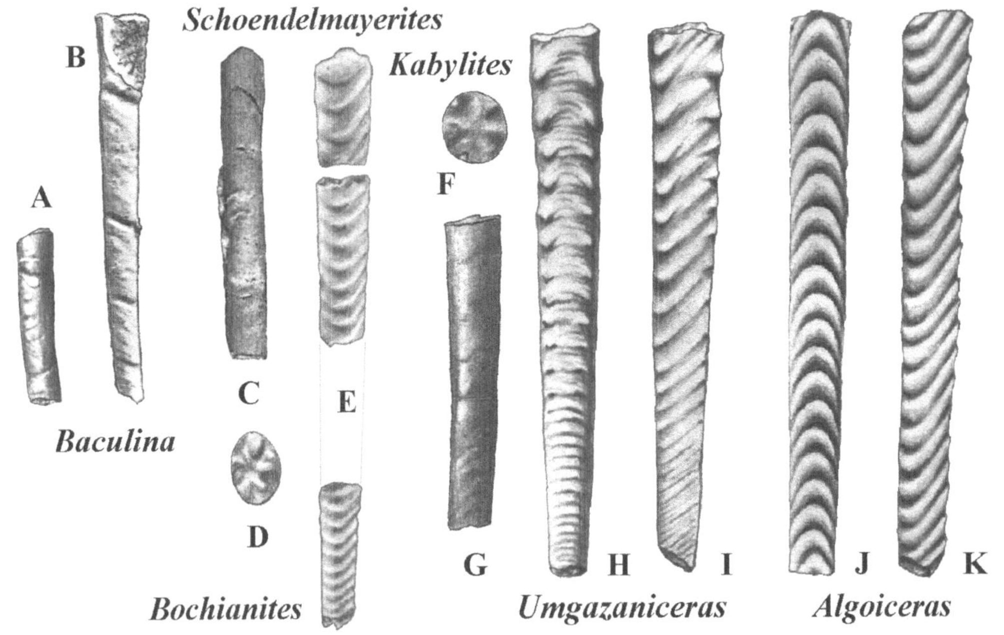

Fig. 96. Bochianitidae.

Family BOCHIANITIDAE Spath, 1922

(*nom. transl.* Breistroffer, 1947 *ex* Bochianitinae)

Discussion. The sister taxon of Protancyloceratidae, comprising straight shafted homoeomorphs of Late Cretaceous Baculitidae; smooth or with prorsiradiate annular ribs and simple suture. Comprises the subfamilies Bochianitinae and Baculininae. *U. Tithon. - L. Apt.*

Subfamily Bochianitinae Spath, 1922

(*nom. transl.* herein *ex* Bochianitidae)

Discussion. Monophyletic derivatives Protancyloceratinae (Thieuloy, 1977), close to *P. gracile* (Wierbowski) (Arkadiev *et al.*, 2011). Small to medium sized baculicones, smooth or with strong prorsiradiate ribs, sometimes tuberculate. Suture relatively simple, generally with E as deep as L, bifid saddles and trifid L. *U. Tithon. - L. Apt.*

Bochianites LORY, 1898 [*Baculites neocomiensis* d'Orbigny, 1842; OD]. Small baculicone with slowly tapering straight shaft and subcircular to oval (?deformed) whorl section; earliest stages with growth striae only, later with variably developed arcuate ribs, strongest on lower flank, prorsiradiate on outer flank and convex across venter, sometimes forming chevrons; incised suture with narrow trifid L and very reduced U. *U. Tithon. - Barrem.* **Fig. 96E** - *B. neocomiensis* (d'Orbigny).

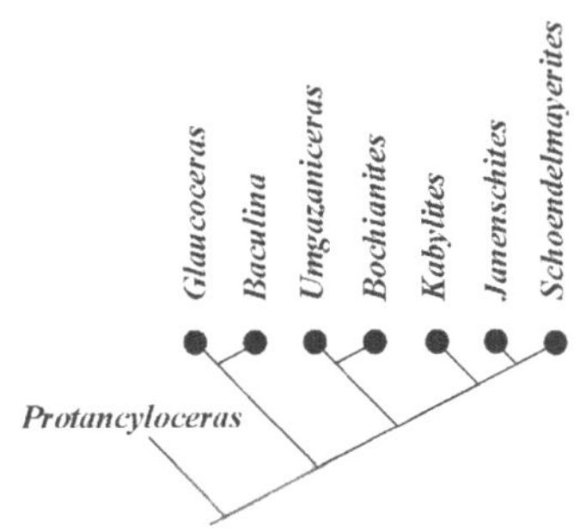

Fig. 97. Hypothesized relationships among Bochianitoidea.

***Arkadieviceras* gen. nov.** [*Bochianites? ambiguus* Arkadiev, Rogov & Perminov, 2011, p.392; OD herein]. Moderately long baculicone; straight to slightly curved, very slowly expanding, with bold slightly prorsiradiate ribs, interrupted ventrally by barely visible smooth band; periodic well developed constrictions. *Berrias.* **Fig. 99A** - *A. ambiguum* (Arkadiev, Rogov & Perminov).

***Algoiceras* gen. nov.** [*Hamites africanus* Tate, 1867, p.150; OD herein). Small; elliptical (juv.) to subtrigonal (ad.) whorl section with, in all but earliest stages, flattened dorsum; ribs coarse, prorsiradiate, annular in early growth and transverse across dorsum, sometimes with incipient dorsolateral bullae; in later growth with tabulate venter across which ribs broaden developing asymmetrical

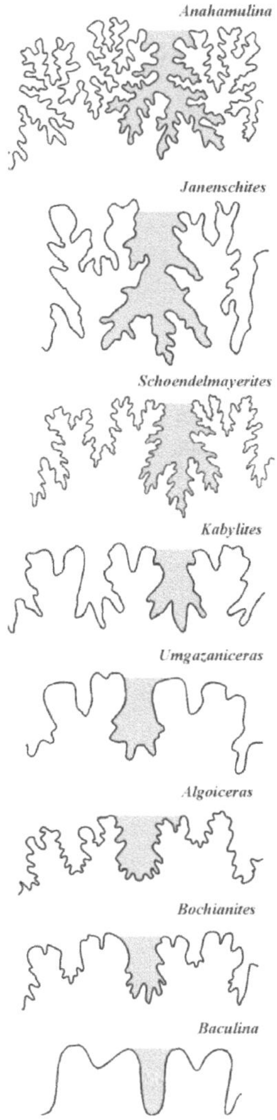

Fig. 98. Bochianitid sutures.

profiles (steepest anteriorly); suture with very broad, deeply bifid L/U. *U. Valang.* **Fig. 96J-K** - *A. africanum* (Tate).

Umgazaniceras KLINGER & KENNEDY, 1979 [*U. thieuloyi*; OD]. Moderately long baculicone, with fine prorsiradiate riblets and striae in early growth, and periodic shallow constrictions; later with tabulate dorsum, strong arcuate dorsolateral bullae weakly connected to oblique ventrolateral clavi joined in middle to later growth across venter by looped ribs; suture with L asymmetrically trifid L. *U. Valang.* **Fig. 96H-I** - *U. thieuloyi* Klinger & Kennedy.
= *Umganaziceras* Vermeulen, 2006 (*nom. van.*).

Upper Valanginian *Bochianites glaber* Kitchin is very small, slowly expanding, intially with a circular whorl section becoming elliptical, with very fine prorsiradiate growth striae across the flank that are linguoid over the venter and convex forward across the dorsum (Cooper, 1981), like the early growth stage of *U. thieuloyi*.

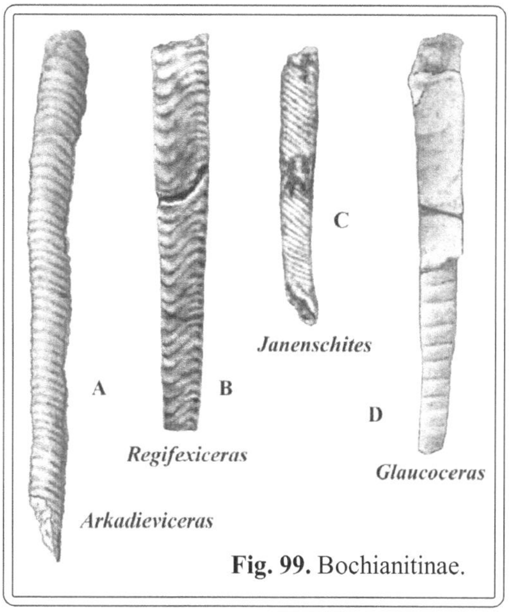

Fig. 99. Bochianitinae.

***Regifexiceras* gen. nov.** [*Bochianites nodosocostatus* Mandov, 1971, p.104; OD herein]. Like *Bochianites* but with compressed elliptical whorl section, differentiated ribbing with principle ribs at 40° to axis, separated by an intercalatory in middle growth, all ribs with ventrolateral clavi. *Lower U. Valang.* **Fig. 99B** - *R. nodosocostatum* (Mandov).

Janenschites DURAND DELGA, 1954 [*Bochianites janenschi* Zwierzycki, 1914; OD]. Slowly expanding baculicone with strong fine prorsiradiate ribbing, as wide as the interspaces; suture like *Schoendelmayerites* but simpler, with long central lobule to E, L as deep as E, and narrow stemmed L/U. *Barrem.* **Fig. 99C** - *J. janenschi* (Zwierzycki).

Kabylites DURAND DELGA, 1954 [*Bochianites superstes* Pervinquière, 1910; OD]. Like *Bochianites* but with faint, distant, prorsiradiate folds on flanks; suture with subequal narrow stemmed bifid saddles and trifid lobes, with large well developed U. *Barrem. - L. Apt.* **Fig. 96F-G** - *K. superstes* (Pervinquière).

Subfamily Baculininae Gill, 1871

(*nom. transl.* herein *ex* Baculinidae)

Discussion. Monophyletic derivatives of Bochianitinae; small unornamented baculicones with periodic deep constrictions and relatively simple sutures. *U. Valang. - U. Haut.*

Baculina D'ORBIGNY, 1850 [*B. rouyana*; OD]. Very small; shaft straight to gently curved, slowly expanding, with subcircular whorl section; regular broad deep prorsiradiate constrictions separated by indistinct folds, weak prorsiradiate ribs and striae; very simple suture with entire bifid saddles and unincised lobes. *U. Valang.* **Fig. 96A-**

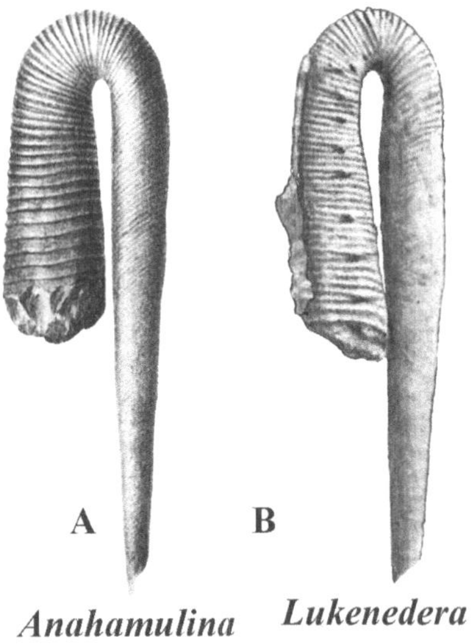

Fig. 100. Anahamulinidae.

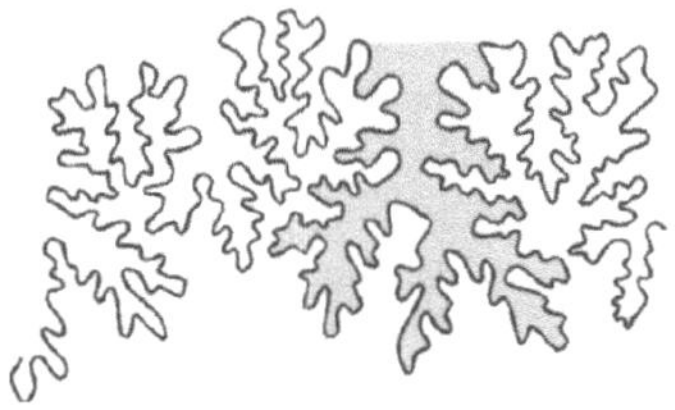
Fig. 101. *Anahamulina* suture.

B - *B. rouyana* d'Orbigny.
= *Baculina* d'Orbigny, 1849 (*nom. nud.*).

***Schoendelmayerites* gen. nov.** [*Bochianites oosteri* Sarasin & Schöndelmayer, 1902, p.179; OD herein]. Shaft smooth, slowly expanding; suture incised with narrow E, long narrow bifid E/L, trifid L deeper than E, and very broad asymmetrically bifid L/U. *L. Haut.* **Fig. 96C-D** - *S. oosteri* (Sarasin & Schöndelmayer).

Glaucoceras MATAMALES-ANDREU, JUAREZ-RUIZ, PICTET & MORENO-BEDMAR [*Baculites renevieri* Ooster, 1860; OD]. Small baculicone with subcircular whorl section; strong slightly prorsiradiate constrictions, narrower than interspaces, weakening across dorsum. *U. Haut.* **Fig. 99D** - *G. renevieri* (Ooster).

Superfamily PTYCHOCERATOIDEA Gill, 1871

(*nom. transl.* Klein *et al.*, 2007 *ex* Ptychoceratidae)

Discussion. Monophyletic derivatives of Bochianitoidea (Breistroffer, 1952; Vermuelen, 2006); hamulinicones and ptychocones with complex suture. Comprises the families Anahamulinidae and Ptychoceratidae. *Uppermost Haut. - L. Barrem.*

Family ANAHAMULINIDAE Breistroffer, 1952

Discussion. Monophyletic derivatives of Bochianitinae (Breistroffer, 1952; Vermuelen, 2006); small to medium sized with separated proversum and retroversum, the latter relatively long and expanding rapidly in height; flexus and retroversum strongly ribbed, with or without tubercles. *Uppermost Haut. - L. Barrem.*

Anahamulina HYATT, 1900 [*Hamulina subcylindrica* d'Orbigny, 1850; OD]. Medium sized; proversum and retroversum subparallel and close together, the latter expanding rapidly; proversum with fine prorsiradiate ribs that strengthen on flexus and become distant and slightly prorsiradiate on retroversum; suture with L only slightly deeper than E so that E/L is not oblique, and U smaller than I. *Uppermost Haut. - basal Barrem.* **Fig. 100A** - *A. subcylindrica* (d'Orbigny).

***Lukenedera* gen. nov.** [*Hamites (Hamulina) lorioli* Uhlig, 1883, p.212; OD herein]. Proversum like *Anahamulina*; retroversum touching proversum at aperture; retroversum with fine rursiradiate ribbing with periodic prominent dorsolateral tubercles giving rise to biplicate and triplicate riblets; flexus with prorsiradiate varix. *L. Barrem.* **Fig. 100B** - *L. lorioli* (Uhlig).

Family PTYCHOCERATIDAE Gill, 1871

Discussion. Seemingly monophyletic derivatives of Anahamulinidae (Breistroffer, 1952; Vermuelen *et al.*, 2020); small to medium sized, with three parallel shafts that may or may not be in contact, generally slowly expanding and with strong ribbing restricted to flexus and retroversum; suture with bifid or trifid L. *U. Haut. - Upper U. Apt.*

Euptychoceras BREISTROFFER, 1952 [*Ptychoceras meyrati* Ooster, 1860; OD]. Small; three subparallel tapering shafts, first and second barely or not touching, retroversum just overlapping proversum; first two shafts smooth or with weak ribs that may become scale-like on retroversum; suture with trifid L. *U. Haut. - Barrem.* **Fig. 102G** - *E. meyrati* (Ooster).

Assignment of 3-shafted *Euptychoceras* to single shafted Bochianitidae (Klein *et al.*, 2007) is not followed here.

Ptychoceras D'ORBIGNY, 1842 [*P. emericianus*; SD Diener, 1925]. Small; strongly prorsiradiate growth striae recurve dorsolaterally to cross venter transversely; flexus unornamented or with feeble ribs; fairly long retroversum with distant asymmetrical rursiradiate to rectiradiate ribs with angular to rounded crests, effacing from lower flank and dorsum; suture with deep penetrating E, narrow

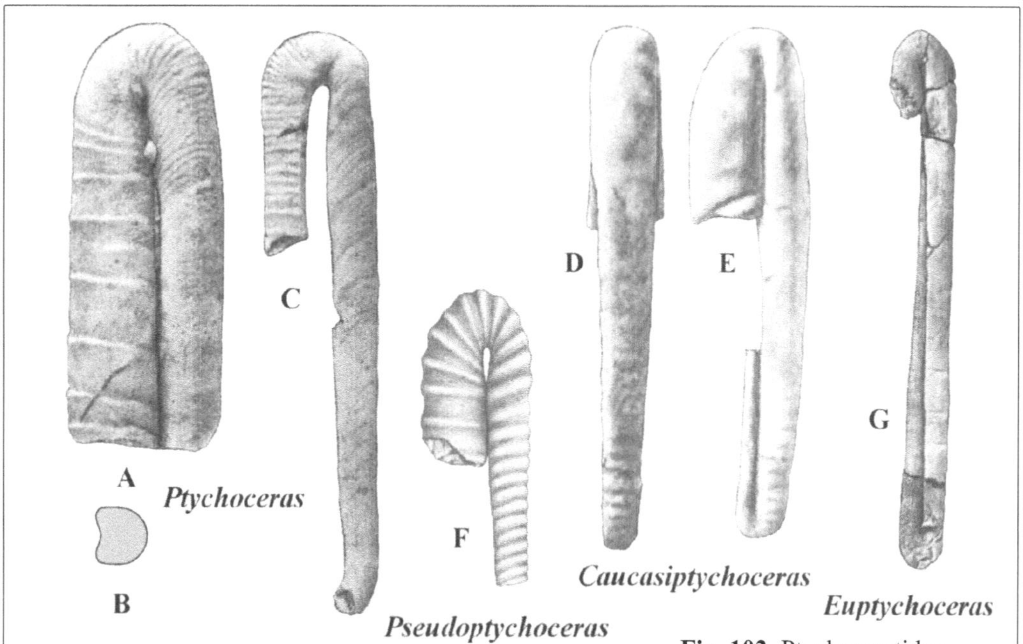

Fig. 102. Ptychoceratidae.

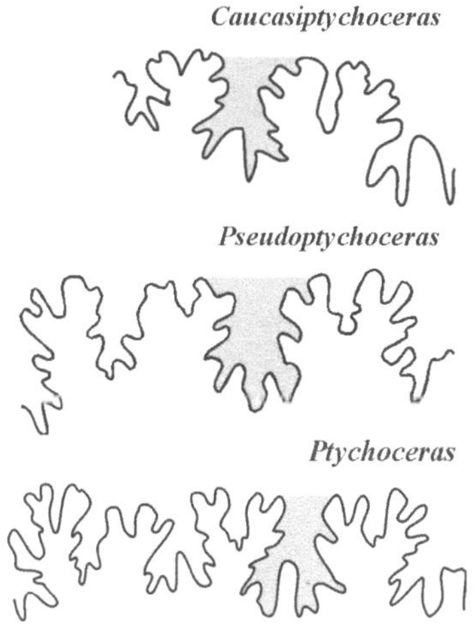

Fig. 103. Ptychoceratid sutures.

stemmed bifid saddles, relatively wide bifid L, reduced U and narrow deep trifid I. *Upper Barrem. - Lower U. Apt.* **Fig. 102A-B** - *P. emericianum* (d'Orbigny).

Ptychoceras puzosianum (d'Orbigny) (**Fig. 102C**) has the final two shafts adpressed, the retroversum with a sulcate dorsum and depressed reniform whorl section, weak rectiradiate ribs to the flexus that has a weak varix at its start, and distant rursiradiate ribs to the retroversum.

***Caucasiptychoceras* gen. nov.** [*Ptychoceras renngarteni* Egoian, 1968, p.182; OD herein). Moderately large (L_{max}=120 mm) with 3 parallel shafts in tight adpressed contact, and tiny keyhole aperture at flexus; proversum slowly expanding, with faint, blunt ribs, coarsening on moderately long, rapidly expanding retroversum with oblique apertural constriction; suture with narrow bifid L/U. *Uppermost Apt.* **Fig. 102D-E** - *C. renngarteni* (Egoian).

Pseudoptychoceras ETAYO SERNA, 1979 [*P. gilberti*; OD]. Like *Caucasiptychoceras* but much smaller, with prominent, blunt, rounded, slightly prorsiradiate annular ribs, broader than deep interspaces, slightly rursiradiate and distant on retroversum; suture with broad rectangular saddles, trifid L and very broad asymmetrically bifid U. *Upper U. Apt.* **Fig. 102F** - *P. gilberti* Etayo Serna.

Superfamily TOXOCERATOIDEA Gill, 1871

(*nom. transl.* herein *ex* Toxoceratidae)

Discussion. Seemingly monophyletic derivatives of Protancyloceratoidea (Vermeulen, 2006). Mostly large bituberculate criocones and ancylocones with trifid L; comprises the families Aegocrioceratidae and Toxoceratidae. *U. Valang.- U. Apt.*

Family AEGOCRIOCERATIDAE Vermeulen, 2006

Discussion. A seemingly monophyletic group thought to be derivatives of Protancyloceratidae (Vermeulen, 2006; Klein *et al.*, 2007), but large size, criocone coiling and ungrooved venter are derived

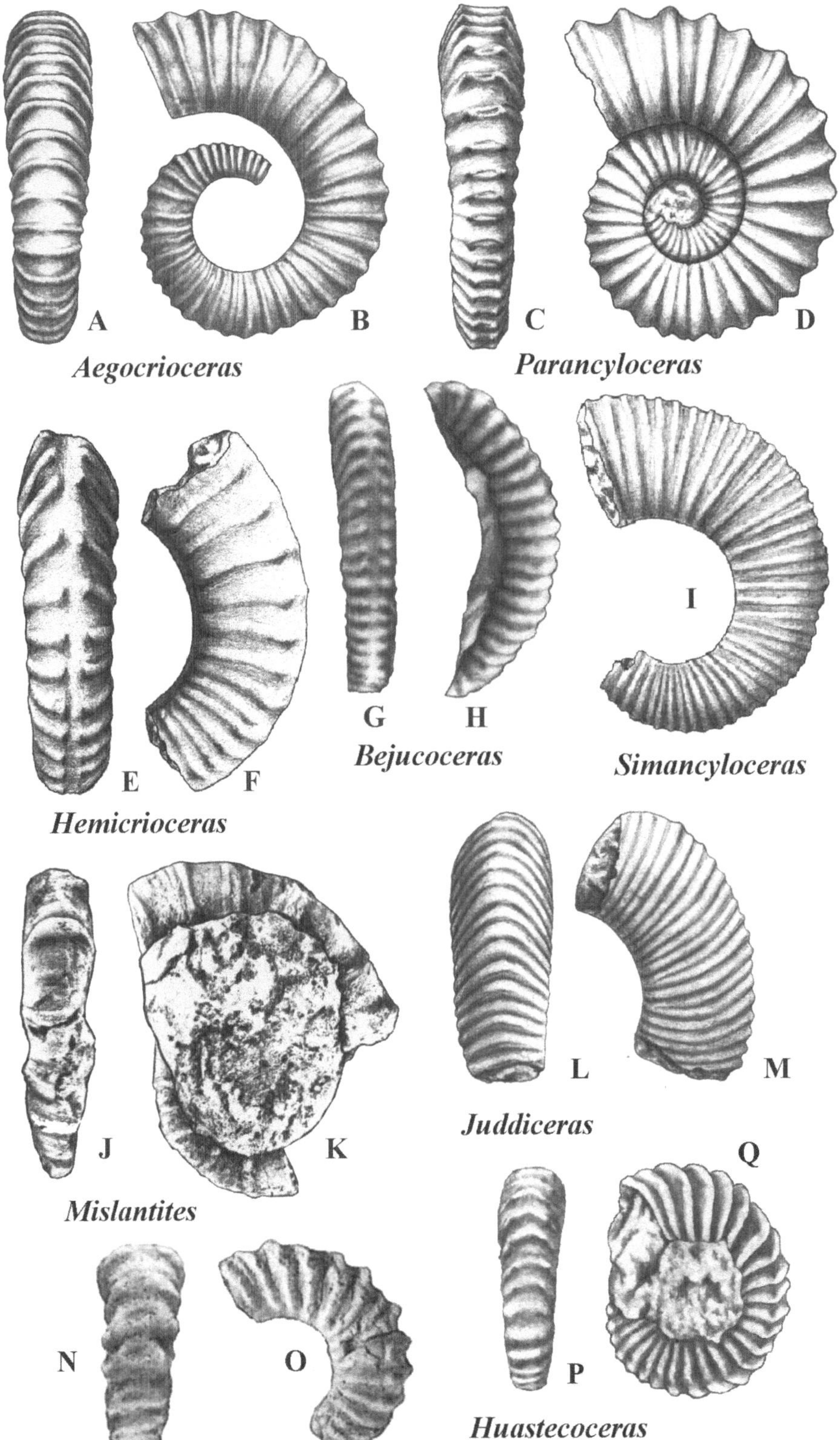

Fig. 104. Aegocrioceratidae.

characters shared with Toxoceratidae. An origin in *Crioceratites* (Wienkopf *et al.*, 2021) seems unlikely. Mostly medium sized to enormous, loose to tight criocones and ancylocones with strong, simple, well spaced ribs that generally coarsen in maturity, with or without umbilical and single or double ventrolateral tubercles; suture with narrow trifid L; predominantly Boreal. Comprises the subfamilies Aegocrioceratinae, Aspinoceratinae and Annuloceratinae. *U. Valang. - U. Apt.*

Subfamily Aegocrioceratinae Vermeulen, 2006.

Discussion. Small to large criocones and serpenticones with strong simple mostly bituberculate ribs, descended from Protancyloceratinae. *U. Valang.- U. Apt.*

Juddiceras SPATH, 1924 [*Crioceras curvicosta* von Koenen, 1902; OD]. Large; known only from gently curved fragments suggesting a loose criocone; whorl section compressed oval with flattened dorsum; strong simple non-tuberculate ribs, occasionally bifurcating close to dorsolateral shoulder, project forward on outer flank and linguoid over rounded venter. *U. Valang.* **Fig. 104L-M** - *J. curvicostum* (von Koenen).

***Vonkoenenites* gen. nov.** [*Crioceras curvicostata* von Koenen, 1902: 327, pl. 50, figs. 2-3; OD herein]. Large loose criocone with compressed suboval whorl section; simple concave forward ribs of variable strength, narrower than interspaces, stronger ribs with small ventrolateral tubercles; in maturity ribs uniform, distant. *U. Valang.* **Fig. 106I-J** - *V. vonkoeneni* sp. nov.

Aegocrioceras SPATH, 1924 [*Hamites capricornu* Roemer, 1841; OD]. Medium sized; loose to tight criocone with compressed elliptical whorl section and somewhat flattened venter in later growth; regular strong single slightly prorsiradiate ribs, narrower than interspaces, weakening across siphonal line with or without ventrolateral bullae in early growth; suture with broad based E/L, very narrow trifid L much deeper than E, and moderately developed U. *Upper L. - Lower U. Haut.* **Fig. 104A-B** - *A. bicarinatum* (Young & Bird).

Bejucoceras CANTÚ-CHAPA, 1976 [*B. simplecostatum*; OD]. Small; loosely coiled criocone expanding slowly in height, with strong simple bituberculate ribs interrupted across venter. *Lower U. Haut.* **Fig. 104G-H** - *B. simplecostatum* Cantú-Chapa.

Misantlites CANTÚ-CHAPA, 2009 [*M. reyesi*; OD]. Small; like *Aegocrioceras* but with stronger more distant ribs that are flared and swollen over venter, weak across flanks and with blunt umbilical bullae; feeble intercalatories separate primary ribs across venter. *Haut.* **Fig. 104J-K** - *M. reyesi* Cantú-Chapa.

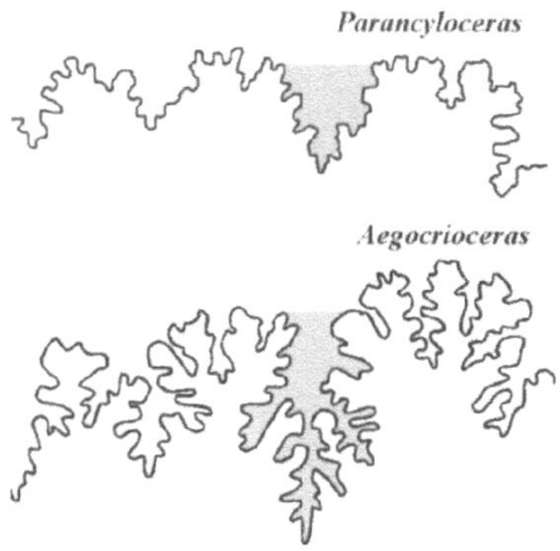

Fig. 105. Aegocrioceratid sutures.

Parapedioceras COLLIGNON, 1962 [*P. colcanapi*; OD]. Medium sized; known only by body chamber flexus; whorl section compressed rectangular, simple distant slightly convex rursiradiate ribs with feeble inner and outer ventrolateral tubercles. *L. Barrem.* **Fig. 106A** - *P. colcanapi* Collignon.

Hemicrioceras SPATH, 1924 [*Crioceras rude* von Koenen, 1902; OD]. Medium sized, fragments suggesting a loose criocone; like *Parancyloceras* but larger, with compressed elliptical whorl section; subdued umbilical bullae to long ribs, weak inner and outer ventrolateral tubercles and ribs interrupted across narrow venter; suture highly incised, with narrow trifid L at midflank. *U. Barrem.* **Fig. 104E-F** - *H. rude* (von Koenen).

Simancyloceras KEMPER, 1973 [*S. stolleyi*; OD]. Medium sized; like *Hemicrioceras* but a smaller ancylocone with finer, straight, slightly prorsiradiate non-tuberculate ribs that in later growth are separated by 1-2 long intercalatories. *U. Barrem.* **Fig. 104I** - *S. stolleyi* Kemper.

Some referred species have feeble umbilical and ventrolateral bulges on flexus.

Parancyloceras SPATH, 1924 [*Crioceras bidentatum* von Koenen, 1902; OD]. Small serpenticone, with flattened venter, rectiradiate ribs and small ventrolateral tubercles joined across venter by looped ribs; some with weak lower ventrolateral tubercles. *U. Barrem.* **Fig. 104C-D** - *P. bidentatum* (von Koenen).

Huastecoceras CANTÚ CHAPA, 1976 [*Crioceras trispinosoides* Burckhardt, 1925; OD]. Small, like a miniature *Parancyloceras*; evolute, planispirally coiled, with flattened flanks and venter, and subquadratic whorl section; strong well spaced primary ribs with incipient umbilical bullae and well developed lateral and doubled ventrolateral tubercles; ribs looped across venter or not, with occasional non-tuberculate intercalatories, where they form chevrons. *Uppermost L. Apt. - Lowermost U. Apt.* **Fig. 104N-Q** - *H. trispinosoides* (Burckhardt).

Subfamily Aspinoceratinae nov.

Diagnosis. Medium sized ancylocones, with fine ribs arising in bundles commonly from umbilical

Fig. 106. Aegocrioceratidae.

bullae, with or without constrictions. Seemingly monophyletic derivatives of Aegocrioceratinae with which they evolved in parallel; precise ancestor unknown but closest to *Juddiceras*. *U. Valang. - L. Barrem.*

Aspinoceras ANDERSON, 1938 [*A. hamlini*; OD]. Small, moderately inflated with flattened dorsum; initial criocone with circular whorl section followed by curved proversum and flexus with compressed elliptical whorl section; ribs commonly bifurcating from dorsolateral shoulder, but also single or bifurcating higher up, non-tuberculate, weak across dorsum, coarsening on body chamber; two deep narrow constrictions. *U. Valang.* **Fig. 106B** - *A. hamblini* Anderson.

Hoplocrioceras SPATH, 1924 [*Hamites phillipsi* Phillips, 1829; OD]. Medium sized ancylocone with flattened flanks, compressed subelliptical whorl section and curved proversum; ribs arise in pairs from subdued umbilico-lateral bullae, separated by an intercalatory, and transverse over venter. *Upper U. Haut. - L. Barrem.* **Fig. 106F** - *H. phillipsi* (Phillips).

Subaspinoceras THOMEL, DELANOY & AUTRAN, 1987 [*Ancyloceras mulsanti* Astier, 1851; OD]. Medium sized; like

Hoplocrioceras but with large initial criocone followed by short gently curved J-shaped body chamber on which ribbing remains fine; initial part of adult body chamber with ventrolateral clavi. *Upper U. Haut.* **Fig. 106C** - *S. mulsanti* (Astier).

Fissicostaticeras KAKABADZE & HOEDEMAEKER, 2010 [*Hamites (Ammonites) fissicostatus* Roemer, 1841; OD]. Moderately large; like *Hoplocrioceras* but whorls barely separated, increasing more rapidly in height, with flattened convergent flanks, tabulate venter and trapezoidal whorl section; comma-shaped umbilicolateral bullae strengthen with growth and give rise to fasciculate bundles of ribs separated by 2-3 intercalatories; in later growth (D>120 mm) trituberculate main ribs appear, separated by 2-3 non-tuberculate secondaries; in maturity only distant trituberculate main ribs remain. *L. Barrem.* **Fig. 106D-E** - *F. fissicostatum* (Roemer).

Subfamily Annuloceratinae nov.

Diagnosis. A monotypic taxon, perhaps with an origin in Aegocrioceratinae (Vermeulen, 2006) but separated by a significant phenotypic discontinuity. Diagnosis as for genus. *L. Barrem.*

Annuloceras MURPHY, RODDA & EMBREE, 1995 [*A. summersi*; OD]. Initial loosely coiled helix followed by a perpendicularly orientated slowly expanding criocone with annular sharp-crested rectiradiate to slightly rursiradiate ribs becoming coarser and more distant on adult body chamber; suture with narrow trifid L. *Barrem.* **Fig. 106K** - *A. summersi* Murphy, Rodda & Embree.

According to Vermeulen (2006) the helicoid early whorls are convergent towards Heteroceratidae and suggested an origin close to *Paraancyloceras*.

Family TOXOCERATIDAE Gill, 1871

Discussion. A seemingly monophyletic taxon with an origin in Aegocrioceratidae. Typically large to very large slowly expanding ancylocones with simple bituberculate or quadrituberculate ribs; ribs may attenuate across venter.

Controversy has long surrounded this taxon since the types of *Toxoceras* were presumed lost. Both Autran *et al.* (1986) and Busnardo *et al.* (2007) concluded its ornament and suture were "*indistinguishable*" from *Crioceratites*, and rejected the genus as composite and artificial. Since then the types have been rediscovered and Ebbo & Delanoy (1999) resurrected the genus, regarding it a senior synonym of *Moutoniceras*. Comprises the subfamilies Toxoceratinae and Shasticrioceratinae. *Upper L. Barrem. - Lowermost Apt.*

Subfamily Toxoceratinae Gill, 1871

(*nom. transl.* herein *ex* Toxoceratidae)

Discussion. Small to gigantic, loose slowly expanding ancylocone with strong mostly simple bituberculate ribs to spire with body chamber uncoiling. *Uppermost L. Barrem. - U. Barrem.*

Toxoceras D'ORBIGNY, 1842 [*T. requienianum*; SD Diener, 1925]. Large to very large, fragments indicating an open criocone; in middle growth gently curved whorls slowly expanding, with flat flanks and compressed subrectangular to oval whorl section; uniform, bluntly rounded, almost rectiradiate primary ribs, as wide as or broader than interspaces, with prominent rounded ventrolateral tubercles separated by non-tuberculate secondary; suture with very large, widely splayed, trifid L deeper than E, and U much smaller than trifid I. *Uppermost L. Barrem. - U. Barrem.* **Fig. 108D-H** - *T. requienianum* d'Orbigny.

Moutoniceras SARKAR, 1954 [*Toxoceras moutonianum* d'Orbigny, 1850; OD]. Gigantic (up to 2.2 m); a loose criocone followed by a long curved proversum with compressed elliptical whorl section, open flexus and short retroversum; spire with rectiradiate to slightly prorsiradiate and sinuous ribs with ventrolateral bullae; proversum with simple uniform non-tuberculate ribs; body chamber with coarse distant simple slightly prorsiradiate to slightly rursiradiate ribs with dorsolateral and high lateral tubercles that evanesce on the retroversum; constrictions present in early growth; suture with small U and shallow I. *Uppermost L. Barrem.* **Fig. 108A-C** - *M. moutonianum* (d'Orbigny).

Subfamily Shasticrioceratinae Bert, Bersac & Canut, 2019

(*nom. transl.* herein *ex* Shasticrioceratidae)

Discussion. Derivatives of Toxoceratinae close to *Moutoniceras*; large criocones with whorls almost in contact, broad gently convex flanks, tabulate venter, compressed trapezoidal whorl section, and mostly single flexuous ribs with ventrolateral clavi, sometimes also inner ventrolateral clavi. *M. Barrem. - Lowermost Apt.*

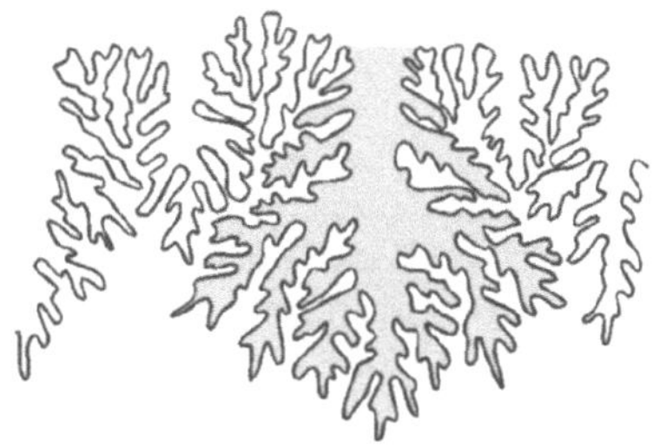

Fig. 107. *Toxoceras* suture.

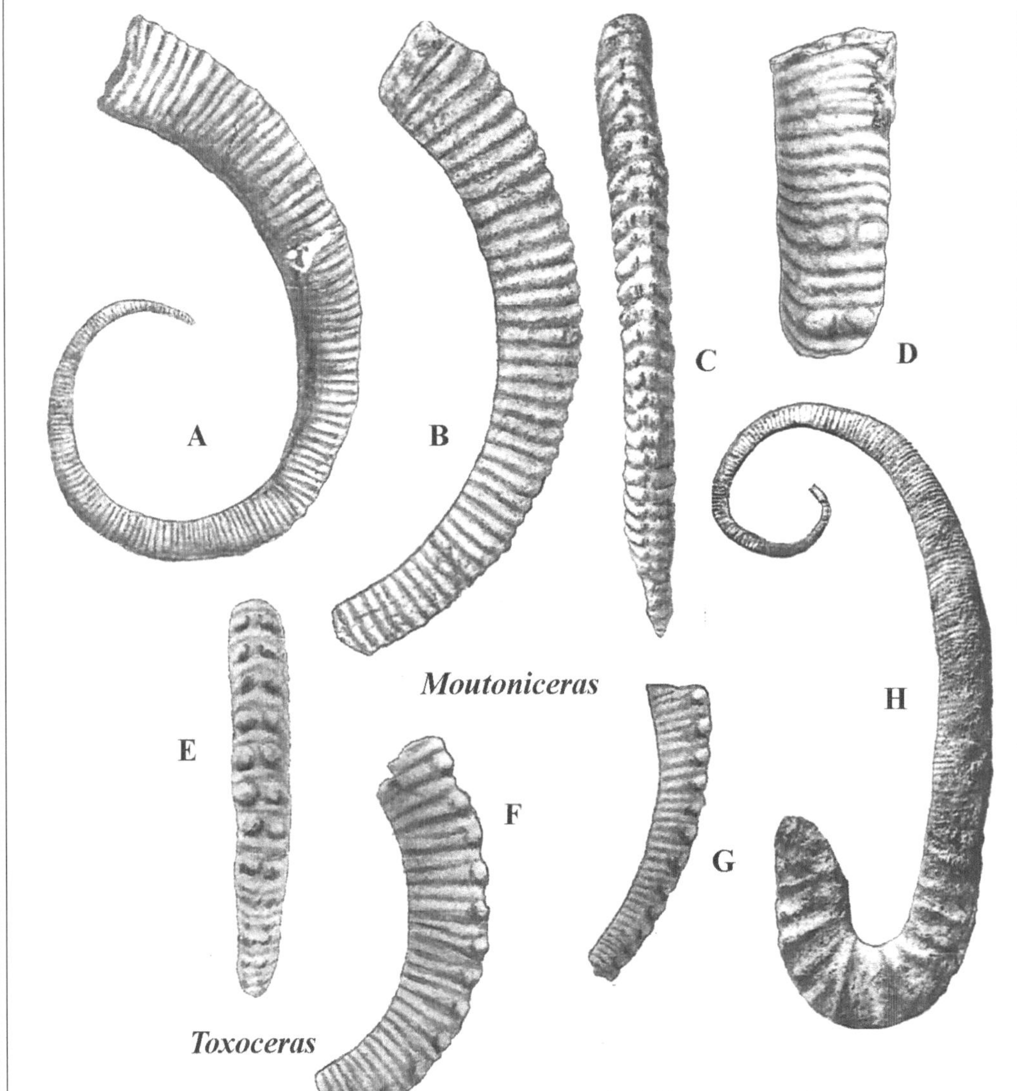

Fig. 108. Toxoceratidae.

Bert *et al.* (2019) introduced the family for criocones with doubled ventrolateral tubercles, but these are not a feature of the type species.

Shasticrioceras ANDERSON, 1938 [*S. poniente*; SD Wright, 1957]. Large; initially tight criocone with whorls almost in contact, followed by uncoiled straight proversum; almost flat flanks with maximum width just above umbilical shoulder, becoming more convex with growth; costally with tabulate venter and compressed trapezoidal whorl section at all stages; in early growth with periodic constrictions, strong fine biconcave to sigmoidal simple ribs, occasionally intercalated or bifurcating, coarsening with age and weakly crossing the venter transversely; all ribs flattening and broadening ventrally, with ventrolateral clavi that strengthen with growth; ribs coarse and distant on adult body chamber; suture florid, with broad necked asymmetrically trifid L. *M. Barrem. - Lower L. Apt.* **Fig. 106G-H** - *S. poniente* (Anderson).

= *Backraceiceras* Mourgues, 2007 (*nom. nud.*).

Order ACANTHOCERATIDA Hyatt, 1900

Discussion. A monophyletic group with an origin in Perisphinctida (Roman, 1938); comprises the suborders Desmoceratidina, Placenticeratidina and Acanthoceratidina. *L. Kimmer. - U. Maastr.*

Suborder DESMOCERATIDINA Ruzhencev, 1962a

(*nom. correct.* herein *pro* Desmoceratida)
= Desmoceratida Houša, 1965
= Olcostephanina Kvantaliani, Lominadze, Topchishvili & Sharikadze, 1999

Discussion. Monophyletic derivatives of Perisphinctidina. Comprises the superfamilies Olcostephanoidea, Silesitoidea, Desmoceratoidea and Pachydiscoidea. *L. Kimmer. - U. Maastr.*

Superfamily OLCOSTEPHANOIDEA Haug, 1910

(*nom. transl.* Pavlov, 1892 *ex* Olcostephanidae)

Discussion. A monophyletic group descended from Idoceratoidea (Brochwicz-Lewinski, 1973); comprises the families Simoceratidae, Spiticeratidae and Olcostephanidae. Generally replaced by Polyptychitidae in the Boreal Realm. *L. Kimmer. - U. Haut.*

Family SIMOCERATIDAE Spath, 1924

Discussion. A Late Jurassic group typical of the Mediterrannean Subprovince, with an origin in Passendorferiidae (Brochwicz-Lewinski, 1973). Mostly very evolute to serpenticone, compressed, with perisphinctoid early whorls; later with coarse simple ribbing that may be interrupted over the venter. Comprises the subfamilies Simoceratinae and Aspidostephaninae. *L. Kimmer. - U. Tithon.*

Subfamily Simoceratinae Spath, 1924

= Chignahuapitinae Cantu-Chapa 2009

Discussion. A seemingly monophyletic group with an origin in Nebroditinae, close to *Nebrodites* (Brochwicz-Lewinski, 1973; Olóriz, 1978). Mostly very evolute to serpenticone, compressed, with perisphinctoid early whorls; later with coarse simple distant ribs. *L. Kimmer. - U. Tithon.*

Ceratosphinctes ZIEGLER, 1959 [*Ammonites septenarius* Quenstedt, 1858; OD]. Very small to small; very evolute, compressed, with well spaced simple low crescentic ribs terminating in prominent ventrolateral tubercles; suture with narrow E, broad asymmetrically bifid E/L and narrow trifid L much shorter than E. *L. - M. Kimmer.* **Fig. 111I** - *C. amatitlaensis* Olóriz & Villasenor.

Lytogyroceras SPATH, 1925 [*Ammonites fasciatus* Quenstedt, 1848; OD]. Compressed serpenticone with rounded umbilical shoulder, broad gently convex flanks and narrowly rounded venter; flanks unornamented but for 2 deep collared constrictions per whorl; adult peristome collared. *Kimmer. - Tithon.* **Fig. 114A-B** - *L. fasciatum* (Quenstedt).

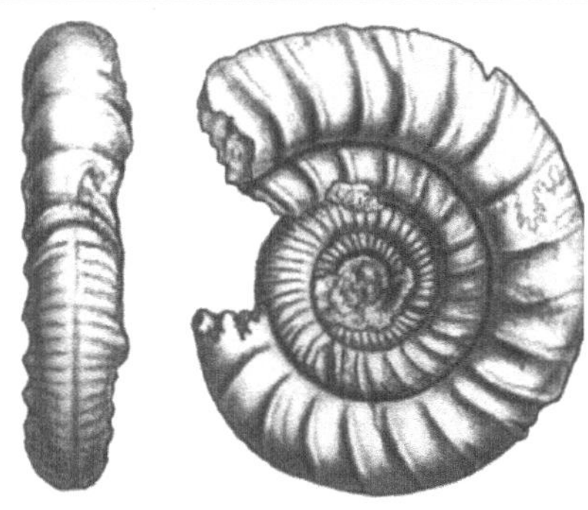

Fig. 109. *Virgatosimoceras.*

Virgatosimoceras SPATH, 1925 [*Simoceras rothpletzi* Schneid, 1915; OD]. Serpenticone, with prominent, distant, biplicate and triplicate ribs; constrictions bordered adorally by conspicuous collar; venter with siphonal groove, becoming smooth and ungrooved on outer whorl; dimorphic, microconchs isocostate, macroconchs variocostate. *L. Tithon.* **Fig. 109** - *V. rothpletzi* (Schneid).

Simoceras ZITTEL, 1870 [*Ammonites biruncinatus* Quenstedt, 1847; SD Fischer, 1882]. Medium sized, evolute, with compressed flat to concave flanks, subrectangular whorl section; flank costae biplicate in early growth; in middle to later growth weak umbilical bullae give rise to 1-2 weak low broad ribs that terminate in strong ventrolateral clavi; ribs weaken in later growth. *Tithon.* **Fig. 111A** - *S. biruncinatum* (Quenstedt).

Volanoceras GEYSSANT, 1985 [*Ammonites volanense* Oppel, 1863; SD Spath, 1925][**M**]. Large to very large; like *Simoceras* but serpenticone with slightly compressed subquadratic whorls; inner flanks feebly ornamented in early growth, with simple and biplicate ribs; in middle growth small umbilical bullae appear connected to prominent ventrolateral tubercles (pointed or clavate) by broad weak low ribs; tubercles alternate across sunken venter; dimorphic, microconchs very small. *L. Tithon.* **Fig. 111B-C** - *V. volanense* (Oppel).

Pseudovolanoceras CECCA, 2002 [*Ammonites aesinensis* Meneghini, 1885; OD]. Medium sized (**M**) to small (**m**); in early growth ribs simple, concave forwards, with ventrolateral clavi that are pleated in later growth separated by "longitudinal furrows"; umbilical bullae

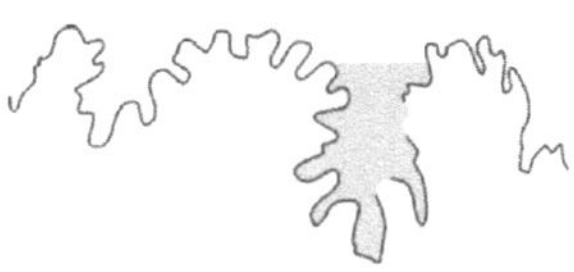

Fig. 110. *Pseudovolanoceras* suture.

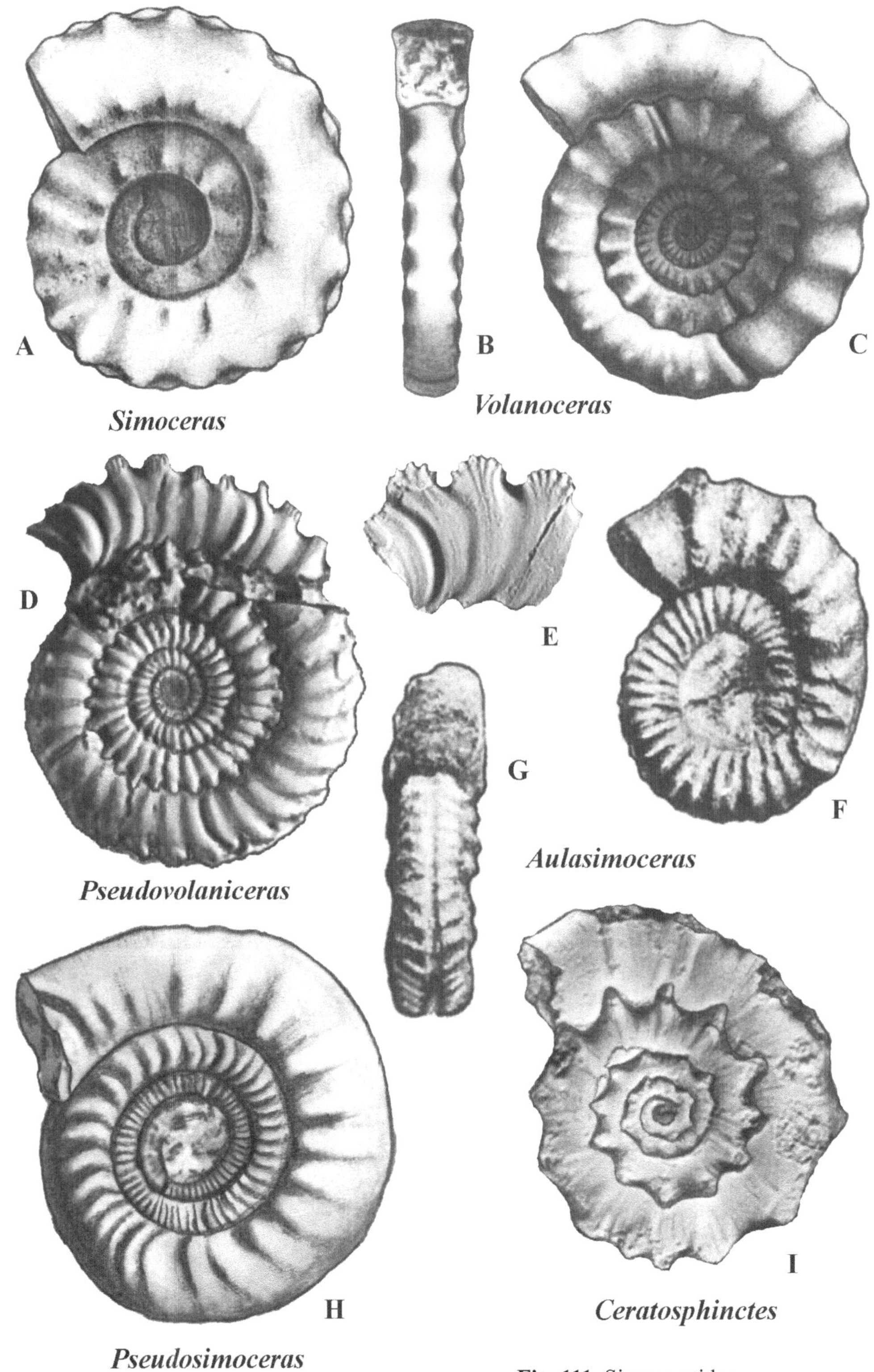

Fig. 111. Simoceratidae.

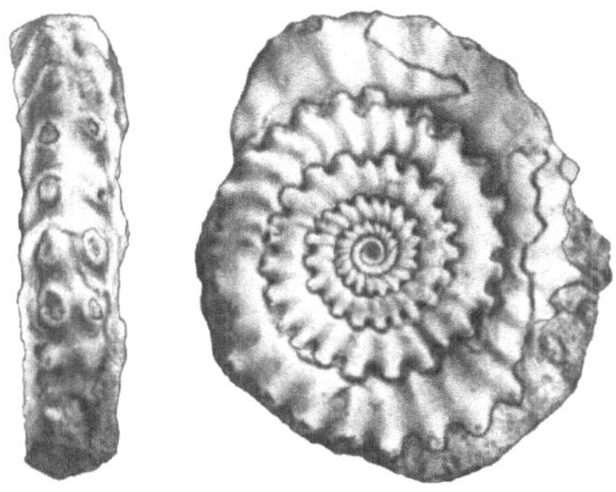

Fig. 112. *Chignahuapites.*

appear in middle growth when ribs weaken at midflank; constrictions variable, better developed in later growth, parallel to ribbing; suture with narrow E, subrectangular asymmetrically bifid E/L, narrow trifid L deeper than E, and broad multifid L/U_2. *L. Tithon.* **Fig. 111D-E** - *P. aesinense* (Meneghini).

Chignahuapites CANTÚ-CHAPA, 2009 [*C. chignahuapensis*; OD] (**m**). Very small; like *Pseudovolaniceras* but with fewer ribs, earlier appearance of clavi, development of umbilical bullae in later growth, and weaker more distant ribs in maturity; 2-3 constrictions per whorl; microconchs lappeted. *L. Tithon.* **Fig. 112** - *C. chignahuapensis* Cantú-Chapa.

Cordubiceras OLÓRIZ & TAVERA, 1979 [*Simoceras (C.) geminatum*; OD] [**M**]. Medium sized; like *Volaniceras* but with looped flank costae to adult body chamber and frequent constrictions. *U. Tithon.* **Fig. 113E** - *C. geminatum* (Olóriz & Tavera).

According to Schweigert *et al.* (2002) looped flank costae occur also in *Volaniceras*.

Baeticoceras GEYSSANT, ÉNAY & BUSNARDO, 1979 [*B. baeticum* Geyssant, 1979; OD] [**m**]. Medium sized; like *Cordubiceras* but smaller, with periodic constrictions and auriculate peristome. *U. Tithon.* **Fig. 113F-G** - *B. baeticum* Geyssant.

Based on a microconch of *Cordubiceras geminatum* (Geyssant) (*fide* Tavera, 1985).

Aulasimoceras Spath, 1931 [*Waagenia auberti* Pervinquière, 1907; OD]. Small, very evolute, flat sided, with subtabulate venter and compressed rectangular whorl section; phragmocone with strong, rigid, slightly rursiradiate simple ribs, occasionally bifurcating from umbilical shoulder interrupted across venter by deep siphonal groove; adult body chamber with coarse, simple distant ribs with ventrolateral bullae. *Tithon.* **Fig. 111F-G** - *A. auberti* (Pervinquière).

Subfamily Aspidostephaninae nov.

Diagnosis. Inflated, with broad smooth venter and subrectangular to lanceolate whorl section; persistent umbilical bullae give rise to ribs that may have ventrolateral tubercles but evanesce in later growth; up to 4 contrictions per whorl. Derivatives of Simoceratinae (Cantu-Chapa, 2009); *U. Tithon. - L. Valang.*

Aspidostephanus SPATH, 1925 [*Holcostephanus depressus* Steuer, 1897; OD]. Medium sized; very evolute, with depressed subrectangular whorl section in early growth becoming oval in maturity as venter arches; inner whorls with robust pillar-like ribs bearing umbilical and ventrolateral tubercles; in maturity ribs and ventrolateral tubercles obsolete, leaving only umbilical bullae; suture with broad based rectangular E/L with superficial subdivisions. *?U. Tithon., Berrias.* **Fig. 122O-P** - *A. depressus* (Steuer).

Praenegreliceras COLLIGNON, 1960 [*Spiticeras (P.) hourcqui* Collignon, 1960; SD herein]. Small; very compressed serpenticone with subtabulate venter and compressed rectangular whorl section; ornament restricted to periumbilical bullae and weak constrictions; suture like *Negreliceras. U. Tithon.* **Fig. 117E-F** - *P. hourqui* (Collignon).

Cantianiceras FARAONI, FLORES, MARINI, PALLINI & PEZONNI 1997 [*C. cantianense*; OD]. Small; like *Aspidostephanus* but with strongly compressed sublanceolate whorl section, two narrow shallow constrictions per whorl, and last part of adult body chamber with ribs branching from umbilical bullae, the adapical rib with a high-lateral tubercle from which 3 secondary ribs project forward. *L. Valang.* **Fig. 117C-D** - *C. cantianense* Faraoni, Flores, Marini, Pallini & Pezonni.

Howarthiceras ZHAKAROV & ROGOV, 2020 [*Groebericeras? baileyi* Imlay & Jones, 1970; OD]. Large; much like *Paskentites* but with flat flanks converging to narrowly arched venter and compressed elliptical whorl section; innermost whorls (D<30 mm) with fairly strong mostly simple rectiradiate ribs that flex forward on upper flank and arch forward, weakening across siphonal line; weak to distinct umbilical bullae are weakly connected to ribs, singly or in pairs, some of which branch at midflank and form weak chevrons across venter; 3-4 constrictions per whorl, strongest in early growth; in maturity ribs strong, simple, without umbilical bullae. *U. Tithon.* **Fig. 117I-J** - *H. baileyi* (Imlay & Jones).

Paskentites IMLAY & JONES, 1970 [*P. paskentaensis*; OD]. Very large; coronate inner whorls with tiny ventrolateral tubercles from which secondary ribs project strongly forward over venter forming weak chevrons, with weak constrictions; earliest whorls may have siphonal groove; outer whorls characterized by loss of tuberculation, weakening of primary ribs, development of biplicate ribs and strong projection of secondary ribs to form chevrons over venter. *L. Valang.* **Fig. 117G-H** - *P. paskentaensis* Imlay & Jones.

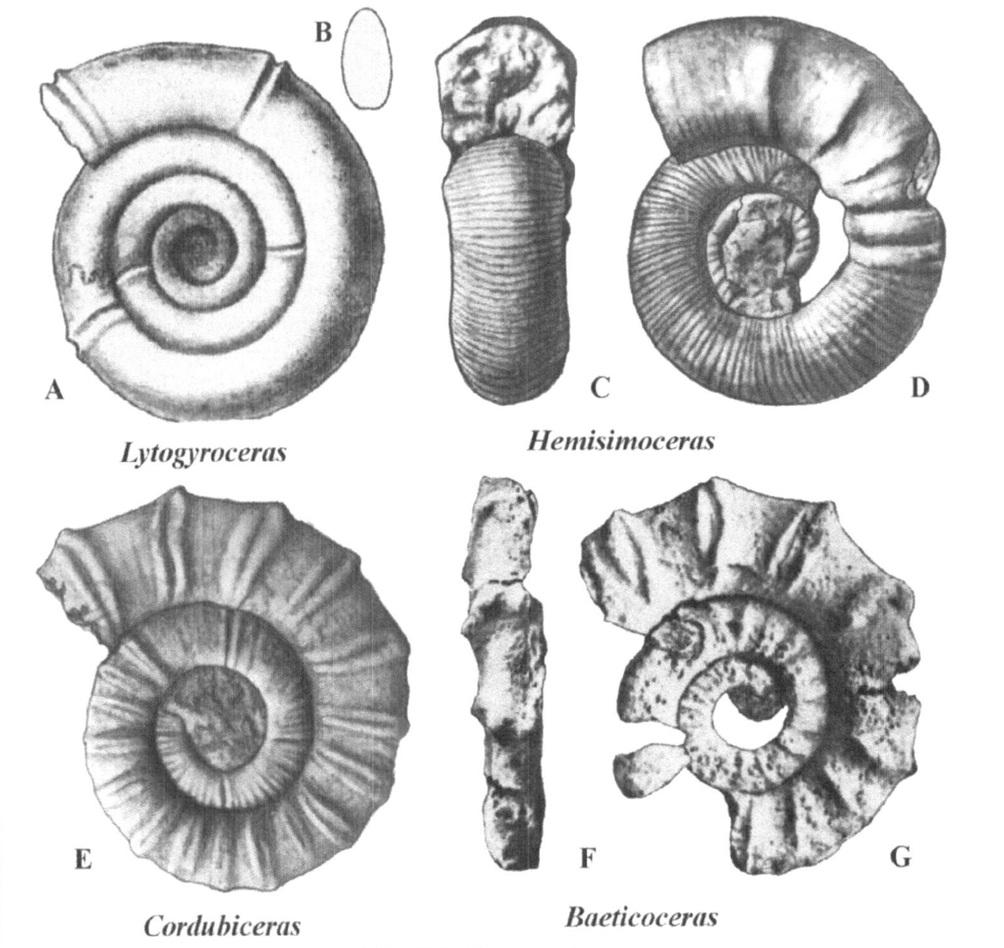

Fig. 113. Simoceratidae.

Family SPITICERATIDAE Spath, 1924

(*nom. trans.* Cantú-Chapa, 2012 *ex* Spiticeratinae)

Discussion. A monophyletic group with an origin in Simoceratidae (Olóriz & Tavera, 1979; Énay& Cariou, 1999). Mostly evolute compressed with bundled ribbing arising from umbilical bullae, with periodic contsrictions oblique to ribbing; comprises the subfamilies Groebericeratinae and Spiticeratinae. *U. Tithon. - L. Haut.*

Subfamily Groebericeratinae Cantú-Chapa, 2012

Discussion. A monophyletic group with an origin in Simoceratinae (Énay & Cariou, 1999; Cecca, 1999), although Howarth (1998) looked to *Pseudoclambites* (Cuetzalanitinae) for its progenitor. Small to medium sized, moderately evolute to evolute, compressed, with narrow subrounded to subacute venter; primary ribs alternating with some secondary ribs, both weaken at midflank, even on outer flank where they project forwards; periodic concave constrictions cut across ribbing behind; small tubercles in different positions on the flanks, disappearing in middle to later growth; suture with broad E, wide shallowly bifid subquadratic E/L with prominent auxiliary saddle, shallowly bifid quadratic L/U_2 and narrow asymmetrically trifid L shorter than E. *U. Tithon. - Berrias.*

Simospiticeras OLÓRIZ & TAVERA, 1979 [*S. lojense*; OD]. Medium sized; compressed, evolute, with subquadratic to subtrigonal whorl section and, in middle growth, almost keeled venter; ornament poorly known due to poor preservation; prominent distant umbilical tubercles seem to weaken in later growth, 2-3 deep curved constrictions

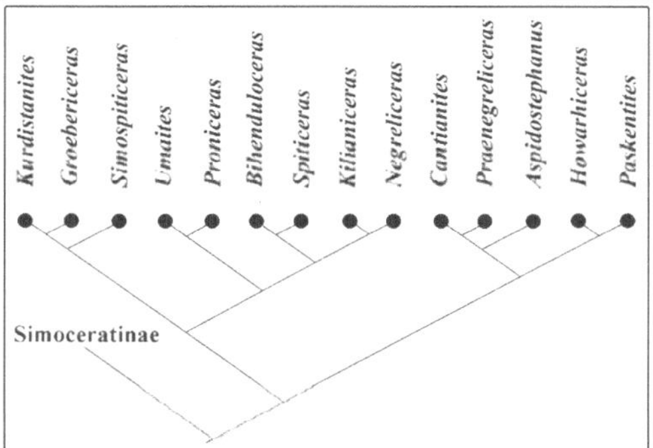

Fig. 114. Hypothesized relationships among Spiticeratidae.

per whorl; fine ribs form chevrons over venter. *Lower U. Tithon.* **Fig. 117A-B** - *S. lojense* Olóriz & Tavera.

Groebericeras A. LEANZA, 1945 [*G. bifrons*; OD]. Small; inner whorls like *Proniceras* but with tuberculation subdued or absent; later compressed and involute, with subacute venter, subtrigonal whorl section and smooth adult body chamber. *L.- U. Berrias.* **Fig. 116A, 122K-L** - *G. bifrons* Leanza.

Kurdistanites CANTÚ-CHAPA, 2012 [*Ammonites rocardi* Pomel, 1899; OD] Small to medium sized, compressed, moderately evolute (U=26%), with flat flanks, rounded venter and compressed elliptical whorl section; umbilical bullae not connected to simple prorsiradiate ribs on outer flank, convex forward over venter, becoming obsolete in later growth; earliest whorls with umbilical bullae; 2-4 slightly flexuous prorsiradiate constrictions per whorl; suture with rectangular saddles and narrow asymmetrically trifid L. *Berrias.* **Fig. 116B, 117K-L** - *K. rocardi* (Pomel).

Subfamily Spiticeratinae Spath, 1924

(*nom. transl.* Arkell & Wright, 1957 *ex* Spiticeratidae)

Discussion. Derivatives of Groebericeratinae, close to *Simospiticeras* (Olóriz & Tavera, 1979). Typically evolute, compressed, with regular (often every 180°) prominent straight oblique collared constrictions that cut ribbing behind, well developed umbilical bullae giving rise to fasciculate bundles of ribs; lateral tubercles where present lost early; suture with narrow asymmetrically bifid E/L, narrow trifid L and retracted auxiliary lobe. Dimorphic; microconchs with lateral lappets. *U. Tithon. - Berrias.*

Proniceras BURCKHARDT, 1919 [*Ammonites pronus* Oppel, 1865; SD Roman, 1938] [**m**]. Small, evolute with compressed oval whorl section; prominent umbilical bullae from which arise 2-3 curved slightly prorsiradiate ribs separated by an intercalatory; ribs form obtuse chevrons over venter and weaken across siphonal line; ornament may decline on adult body chamber; suture with L as deep as E. *Upper U. Tithon.* **Fig. 122A-B** - *P. pronum* (Oppel).

= *Prospiticeras* DJANÉLIDZE, 1922 [*nom. nud.*].

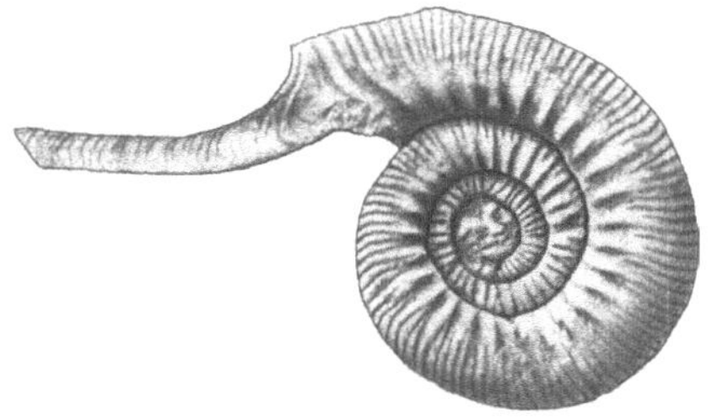

Fig. 115. *Spiticeras* microconch.

Umaites SPATH, 1931 [*U. rajnathi*; OD] [**M**]. Medium sized; very evolute, compressed, with wide shallow umbilicus, steeply inclined to overhanging umbilical wall, flat to gently convex flanks and compressed elliptical whorl section; inner whorls like *Proniceras* with at least 2 deep adorally concave constriction per whorl (constrictions lost >100 mm diameter); prominent umbilicolateral bullae strengthen with growth and give rise to bundles (up to 6) fine sharp secondaries with irregular intercalatories; secondary ribs flex forward to cross venter with linguoid projection that increases with growth; suture moderately complex with short squat E/L and deeply incised narrow trifid L. *U. Tithon.* **Fig. 122M-N** - *U. rajnathi* Spath.

Said to be the macroconch of *Proniceras* (Shome & Bardhan, 2009), but the sutures are different.

Spiticeras UHLIG, 1903 [*Ammonites spitiensis* Blanford, 1863; SD Roman, 1938]. Small to medium sized; like *Proniceras* but larger; whorl section subtrigonal in maturity, with narrowly rounded venter; inflated innermost whorls with umbilical and lateral tubercles, latter lost early; umbilical bullae weakly connected to bundles of 3-4 ribs that may bifurcate on flanks; ribs convex forward over venter. *Berrias.* **Fig. 115** - *S. mirum* (Retowski); **Fig. 122E-F** - *S. spitiense* (Blanford).

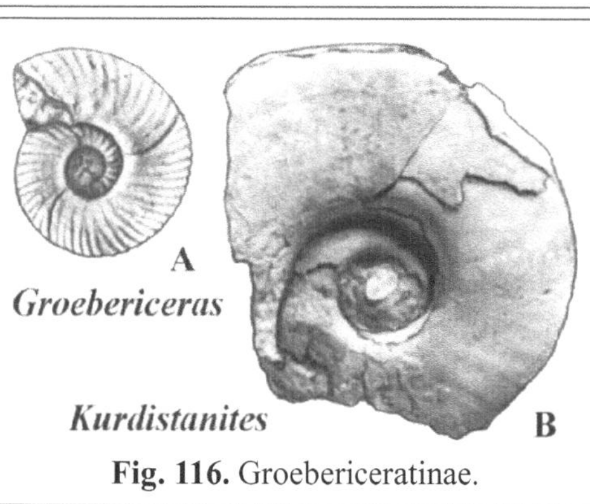

Fig. 116. Groebericeratinae.

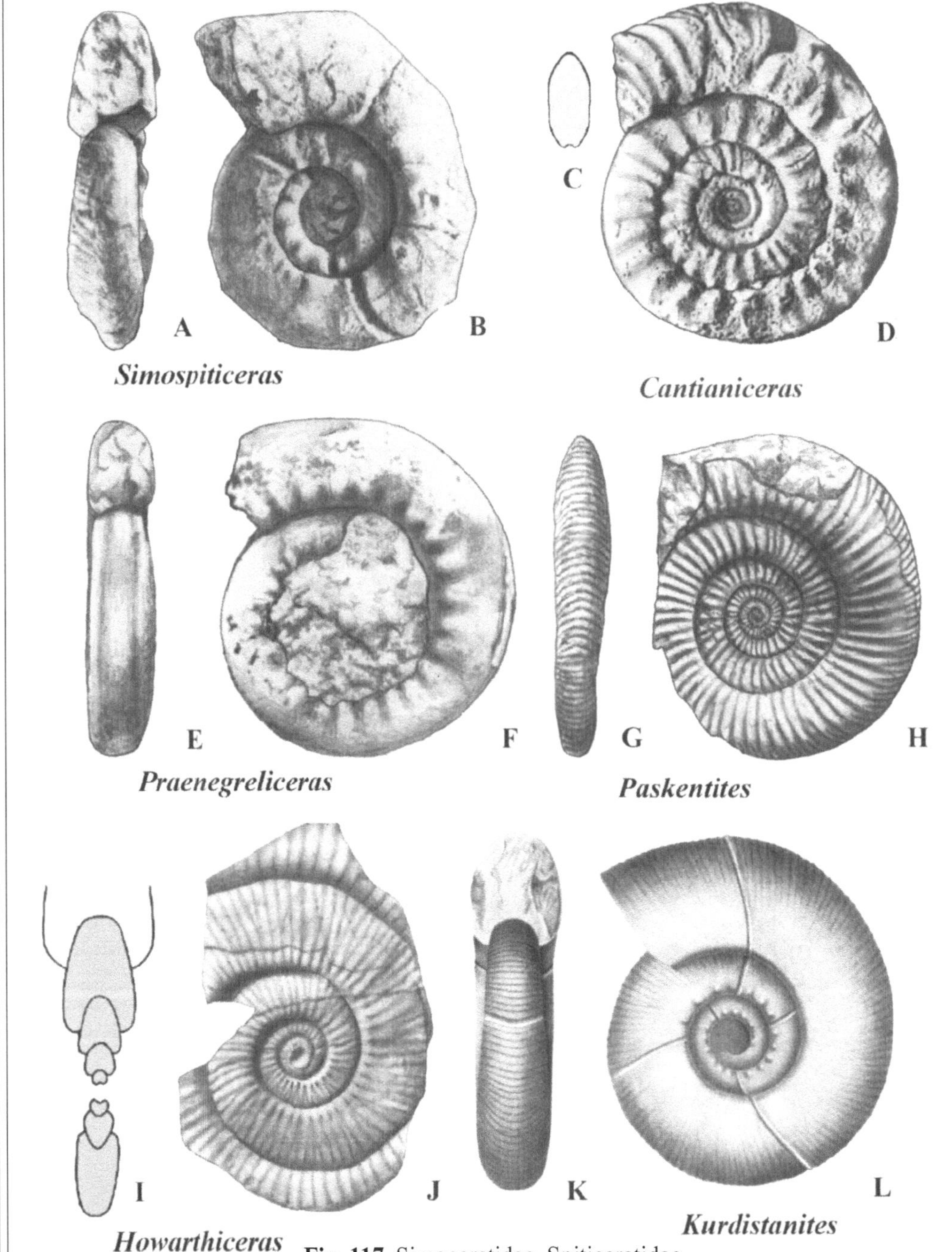

Fig. 117. Simoceratidae, Spiticeratidae.

Bihenduloceras SPATH, 1925 [*B. gregoryi*; M]. Medium sized; like *Spiticeras* but more evolute, with flat convergent flanks, tabulate venter, distinct ventrolateral shoulders, trapezoidal whorl section, larger rounder umbilical tubercles and ribs transverse or slightly concave over venter with some weakening across siphonal line. *U. Tithon.* **Fig. 122C-D** - *B. gregoryi* Spath.

Djanelidzeites CANTÚ-CHAPA, 2012 [*Ammonites mojsvari* Uhlig, 1903; OD] [**M**]. Moderately large; like *Spiticeras* but more densely and finely ribbed and suture with broad shallowly bifid E/L and L/U_2, and narrow trifid lobes. *Berrias.* **Fig. 118** - *D. mojsvari* (Uhlig).

Kilianiceras DJANÉLIDZE, 1922 [*Stephanoceras damesi* Steuer, 1897; SD Roman, 1938] [**M**]. Medium sized;

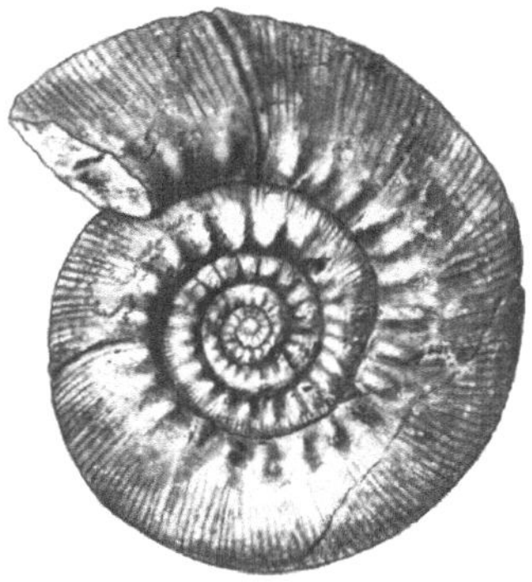

Fig. 118. *Djanelidzeites*.

like *Spiticeras* but larger, more evolute, with broadly rounded venter, depressed reniform whorl section, numerous umbilicolateral bullae and mid-lateral tubercles that persist into later growth. *Berrias*. **Fig. 122I-J** - *K. damesi* (Steuer).

Negreliceras DJANÉLIDZE, 1922 [*Ammonites negreli* Matheron, 1880; SD Roman, 1938] [**m**]. Small; like *Spiticeras* but compressed, with flat subparallel flanks and elliptical whorl section; lateral tubercles absent at all stages; dense fine ribs effaced from inner flank; ribbing declines with growth and adult body chamber may be smooth. *Berrias*. **Fig. 122G-H** - *N. negreli* (Matheron).

Family OLCOSTEPHANIDAE Haug, 1910

= Holcostephanidae Haug, 1910 (incorrect original spelling, ICZN Opinion 575).

Discussion. A monophyletic group with an origin in Spiticeratidae (Roman, 1938); comprises the subfamilies Olcostephaninae, Saynoceratinae, Taraisitinae, Capeloitinae and Provalanginitinae. Generally replaced by Polyptychitidae in the Boreal Realm. *U. Tithon?, L. Valang. - L. Haut.*

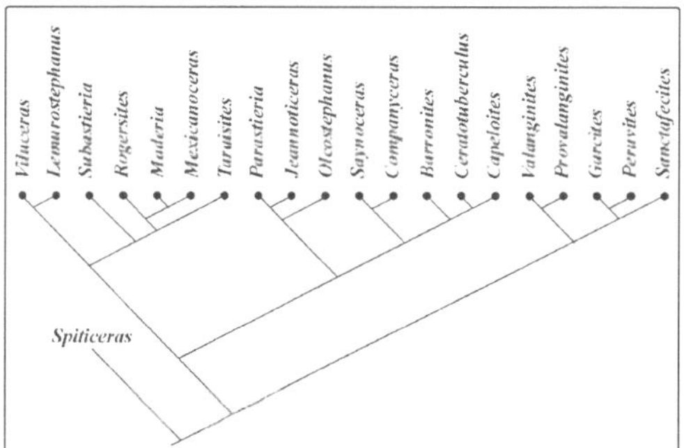

Fig. 119. Hypothesized relationships among Olcostephanidae.

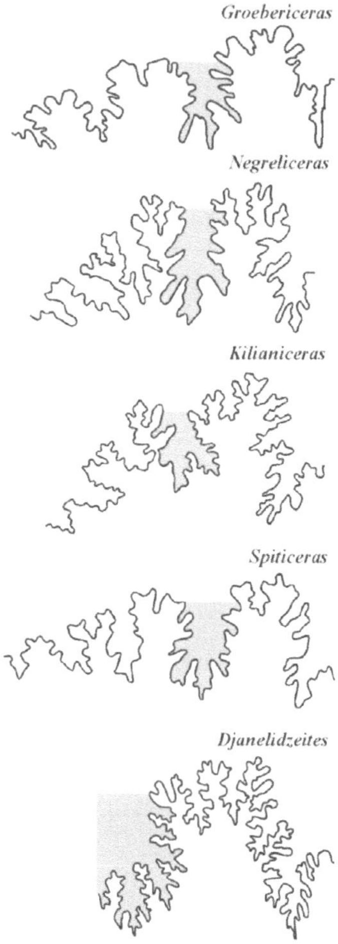

Fig. 120. Spiticeratid sutures.

Subfamily Olcostephaninae Haug, 1910

Discussion. Monophyletic derivatives of Spiticeratinae, close to *Spiticeras* (Roman, 1938). Generally medium sized to large, with ribs arising in fasciculate bundles from prominent umbilical tubercles with primary ribs on the umbilical wall;

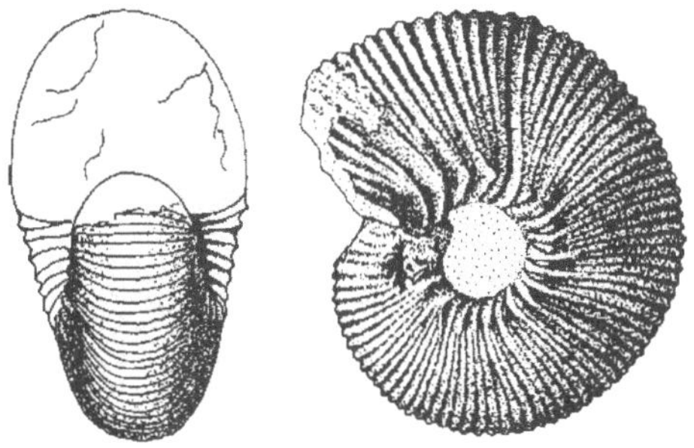

Fig. 121. *Olcostephanus ventricosus*.

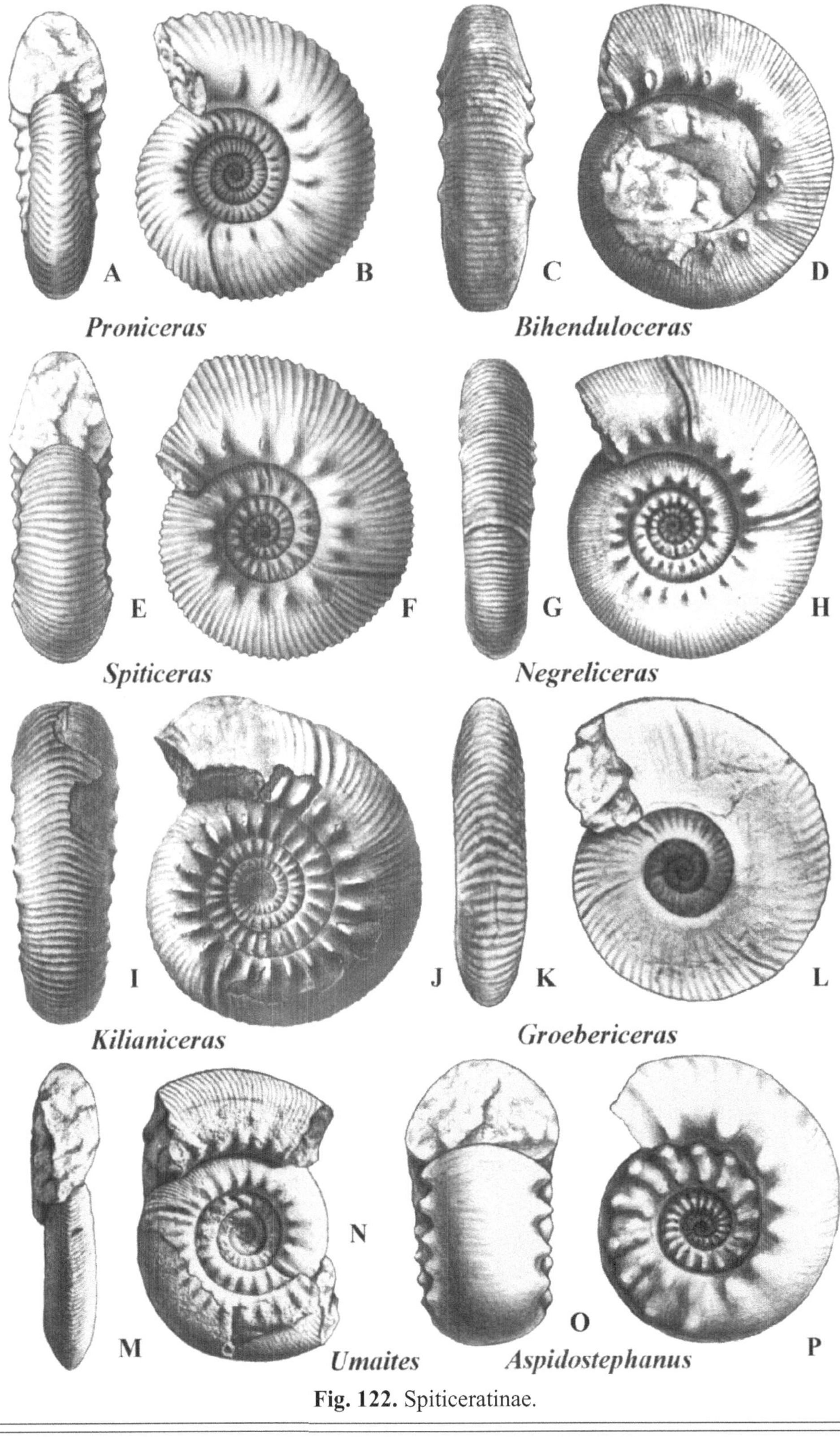

Fig. 122. Spiticeratinae.

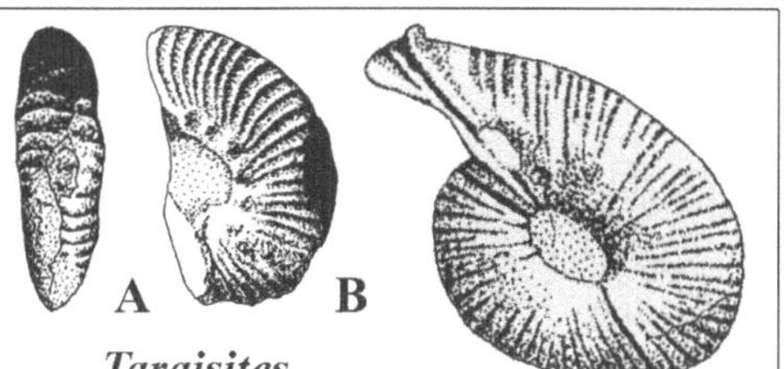

Fig. 123. Olcostephaninae.

prorsiradiate ribs cross the venter transversely and, in many, are cut obliquely by collared varices. Suture with long slender saddles, trifid L and non-retracted suspensive lobe. Sexual dimorphism conspicuous, with lappetted microconchs. *L. Valang. - L. Haut.*

Satoites CANTÚ-CHAPA, 1966 (*nom. nud.*). The genus had no type species. Very small, globular, involute (U=26%); fine rectiradiate ribs arise in pairs from minute umbilical bullae and bifurcate above midflank; 1-2 deep prorsiradiate constrictions per whorl, backed by prominent collar forming chevron over venter; ventral lappet to aperture. *?Berrias*. **Fig. 123C**.

Olcostephanus NEUMAYR, 1875 [*Ammonites astierianus* d'Orbigny, 1840; OD] [**M**]. Small; compressed, moderately evolute, with flattish flanks, rounded venter and compressed subelliptical whorl section; prominent primary ribs terminate in well developed umbilical bullae that give rise to 2-4 rectiradiate to prorsiradiate secondaries which cross venter transversely and uninterrupted, with collared constrictions (varices); suture with deep narrow E, long thin saddles with thin folioles and lobules, and narrow trifid L. *L. Valang. - M. Haut.* **Fig. 121** - *O. ventricosus* (von Koenen); **Fig. 126A-B** - *O. astierianus* (d'Orbigny);

= *Holcostephanus* Sayn, 1889 (ICZN rejected name)
= *Astieria* Pavlow, 1892 (ob.)

Rogersites SPATH, 1924 [*R. modderensis*; OD] [**M**]. Small (**m**) to large (**M**); like *Olcostephanus* but macroconchs highly inflated cadicone with steep to overhanging umbilical wall, deep narrow crater-like umbilicus and strongly depressed reniform whorl section; ribs relatively coarse; generally only two collared constrictions per whorl that are lost on outer whorls of macroconchs; lappetted microconchs with depressed subcoronate whorl section and relatively coarse ribs. *Lower U. Valang*. **Fig. 126I-J** - *R. modderensis* Spath (**M**). **Fig. 128C-D** - *R. baini* (Sharpe) (**m**).

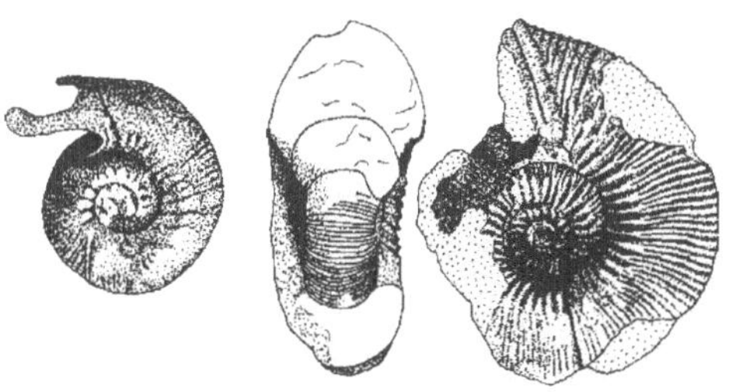

Fig. 124. *Jeannoticeras frequens* (**m** & **M**)

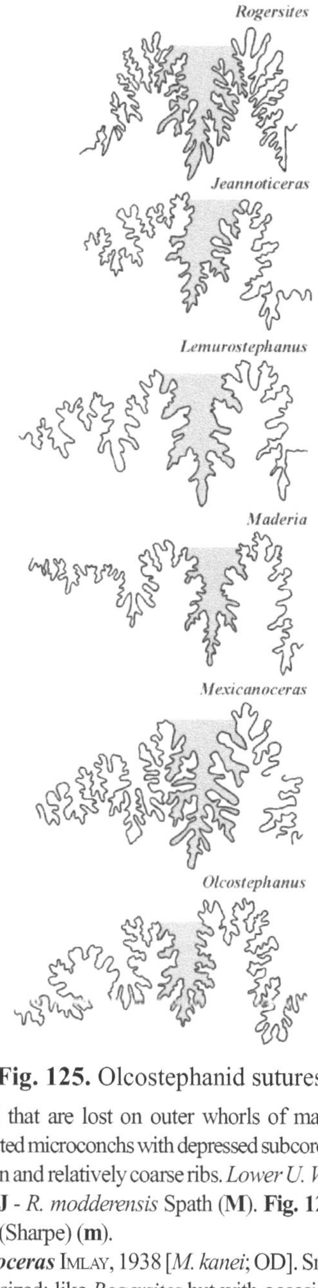

Fig. 125. Olcostephanid sutures.

Mexicanoceras IMLAY, 1938 [*M. kanei*; OD]. Small to medium sized; like *Rogersites* but with occasional rib bifurcation, no constrictions and venter with smooth siphonal band or shallow groove. *L. Haut*. **Fig. 126E-F** - *M. kanei* Imlay.

Lemurostephanus THIEULOY, 1977 [*Holcostephanus madagascariensis* Lemoine, 1906; OD] [**M**]. Very small (**m**) to small (**M**); like *Rogersites* but very evolute (U=40-45%), with inclined umbilical wall, broad gently convex venter and depressed subcoronate whorl section. *L. Valang. - L. Haut.* **Fig. 126K** - *L. madagascariensis* (Lemoine) (**M**); **Fig. 128G-H** - *L. mittrei* (d'Orbigny) (**m**).

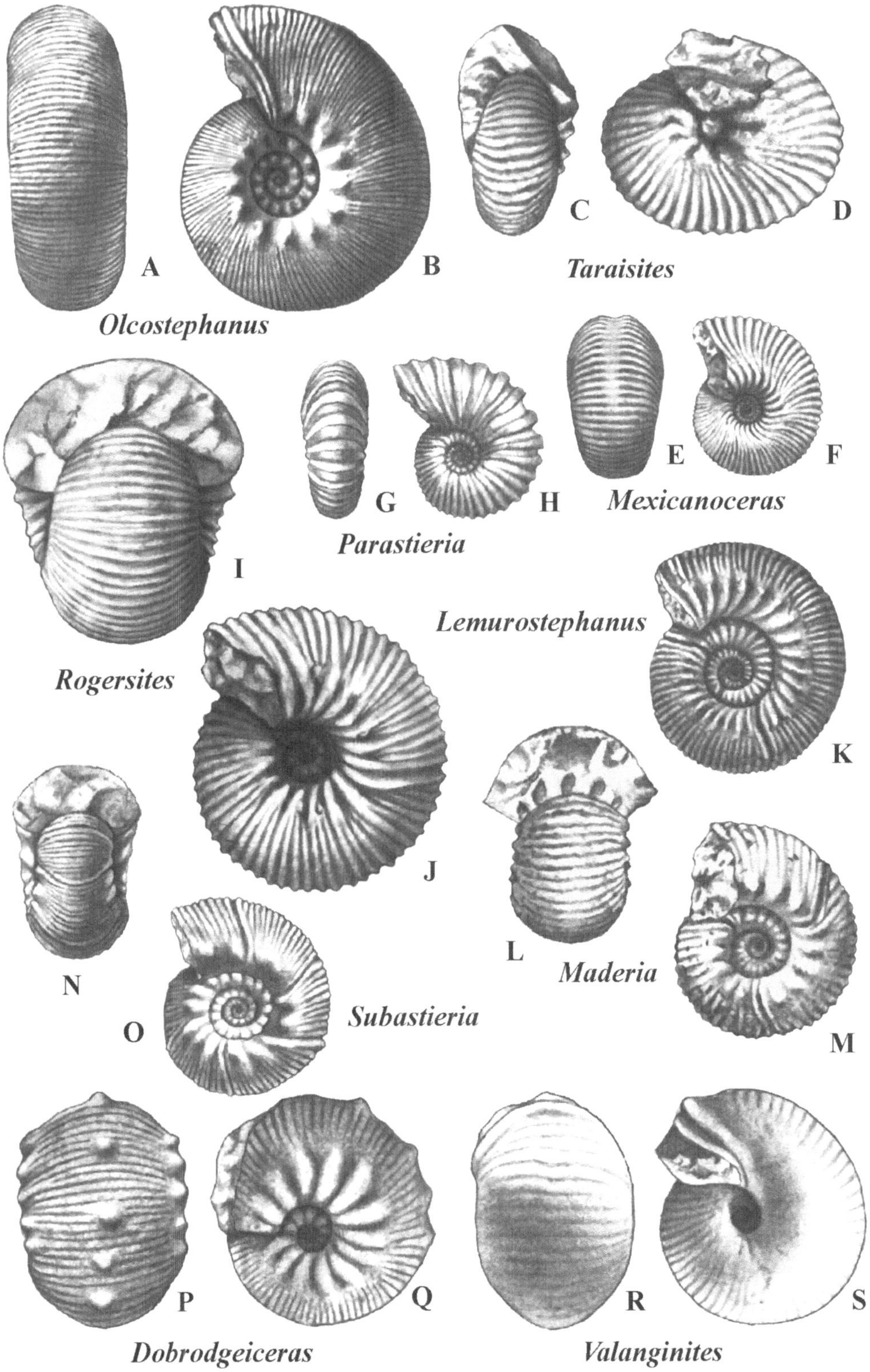

Fig. 126. Olcostephanidae.

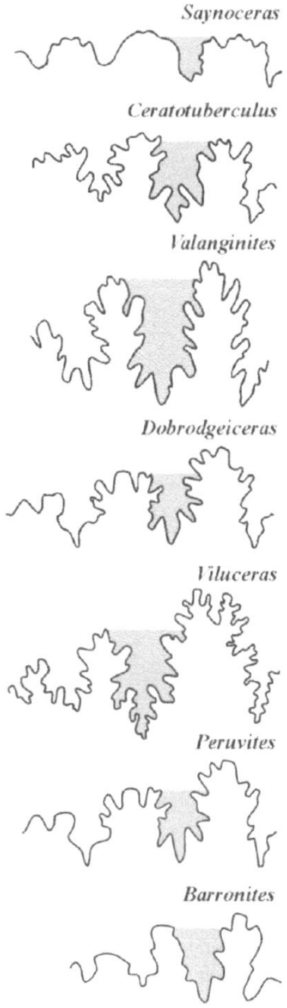

Fig. 127. Olcostephanid sutures.

Viluceras Aguirre-Urreta & Rawson, 1999 [*Simbirskites permolestus* A. Leanza, 1958; OD]. Medium sized; like *Lemurostephanus* but serpenticone, with steeper umbilical wall, and well developed flares and constrictions in maturity. *Valang.* **Fig. 128E-F** - *V. permolestum* (Leanza).

Maderia Imlay, 1938 [*M. coronata*; OD]. Adult size unknown, based on pyritic nuclei; like *Subastieria* but with coarser ribbing, fewer constrictions, less depressed whorls and thinning of ribs over siphonal line of outer whorls; suture with non-retracted suspensive lobe. *L. Haut.* **Fig. 126L-M** - *M. coronata* Imlay.

Subastieria Spath, 1923 [*Olcostephanus (Astieria) sulcosus* Pavlow, 1892; OD]. Very small; like microconchs of *Rogersites* but with broad strongly inclined umbilical wall and coronate whorl section, so umbilical tubercles shift to midlateral position, and 6 strongly prorsiradiate constrictions per whorl. *L. Haut.* **Fig. 126N-O** - *S. sulcosa* (Pavlow).

Jeannoticeras Thieuloy, 1965 [*Ammonites jeannoti* d'Orbigny, 1841; OD]. Small (**m**) to large (**M**); like *Olcostephanus* but more evolute and compressed, with numerous fine umbilical bullae giving rise to bundles of 2-3 thin almost rectiradiate ribs, with occasional intercalatories; with or without constrictions. *Lower L. Haut.* **Fig. 124** - *J. frequens* (Zwierzycki); **Fig. 128A-B** - *J. jeannoti* (d'Orbigny).

Subfamily Taraisitinae Cantú Chapa, 1966

Discussion. A monotypic taxon with an origin in Olcostephaninae. Diagnosis as for genus. *U. Valang.*

Taraisites Cantú Chapa, 1966 [*T. bosei*; = *Astieria* aff. *baini* Böse, 1923: OD]. Small, like microconch of *Rogersites* but compressed with sharp prorsiradiate ribs arising in triplicates from prominent umbilical bullae and recurving on outer flank to cross venter transversely. *U. Valang.* **Fig. 123A-B** - *T. bosei* Cantú Chapa.

The choice of the type species was unfortunate since some referred species, e.g. *T. bangei* (Böse) (**Fig. 126C-D**), are very involute and inflated, quite unlike either *Rogersites* or *Olcostephanus*. Others have been included in the synonymy of *O. symonensis* (Böse), paratypes of which have only 2 ribs per umbilical bulla (González-Arriola *et al.*, 2014).

Subfamily Saynoceratinae Cantú Chapa, 2012

Discussion. A monophyletic group with an origin in early Olcostephaninae, close to *Olcostephanus* (Pervinquière, 1907); typically very small, inflated, with fibulate ribbing and large ventral tubercles in later growth, with or without lateral tubercles. *Upper L. Valang. - L. Haut.*

Companyceras **gen. nov.** [*Saynoceras contestanum* Company, 1985, p.120; OD herein]. Micromorph; inflated, very involute, with depressed polygonal whorl section and tabulate venter; bundled fine rectiradiate to slightly rursiradiate ribs cross venter uninterrupted but loop to, and between, periodic midlateral and ventrolateral tubercles that alternate over venter. *Upper L. Valang.* **Fig. 128I-J** - *C. contestanum* (Company).

Saynoceras Munier-Chalmas, 1894 [*Ammonites verrucosus* d'Orbigny, 1841; OD]. Very small; like *Companyceras* but more evolute, with coarser ribs and stronger tubercles; suture with broad shallow feebly incised E/L. *Low U. Valang.* **Fig. 128K-L** - *S. verrucosum* (d'Orbigny).

Baronnites Bulot, Company & Thieuloy, 1990 [*Saynoceras hirsutum* Fallot & Termier, 1923; OD]. Micromorph; inflated, moderately involute with subtrapezoidal whorl section; after initial smooth stage, prominent umbilical tubercles give rise to 3 rectiradiate ribs that develop ventrolateral bullae; ribs weak across venter in later growth when tubercles become spinose. *Upper L. Valang.* **Fig. 128O** - *B. hirsutum* (Fallot & Termier).

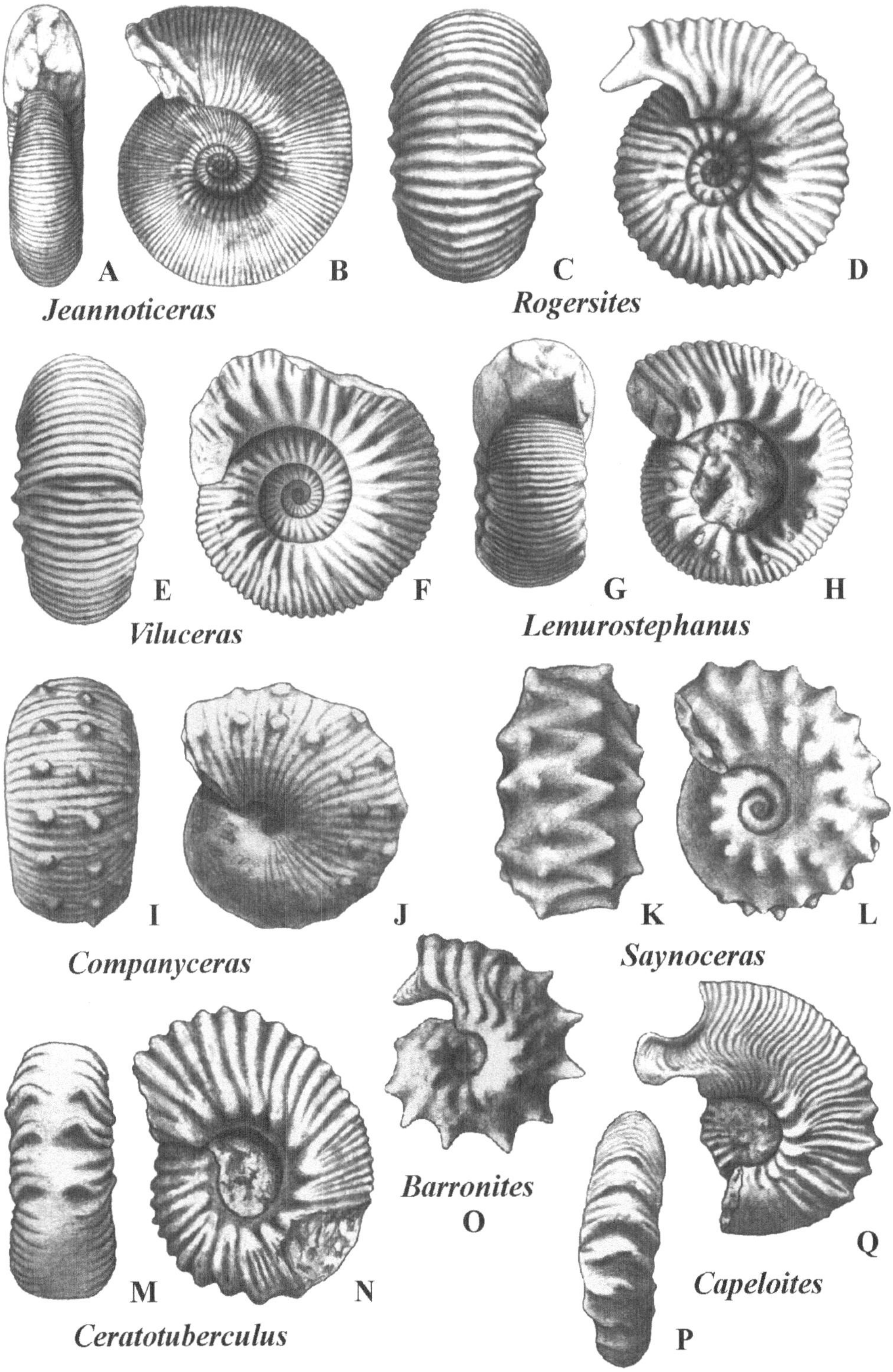

Fig. 128. Olcostephanidae.

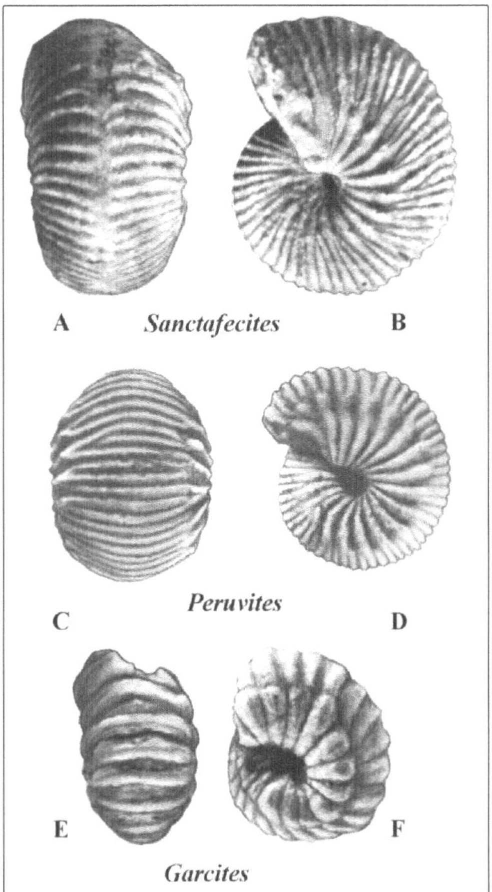

Fig. 129. Provalanginitinae.

Ceratotuberculus IMLAY, 1938 [*C. casitensis*; OD]. Very small; inner whorls like *Olcostephanus* but, after 3rd whorl, with high thick ventrolateral bullae and more or less grooved venter; flank ribs loop to or zigzag between tubercles that are opposite over venter. *L. Haut.* **Fig. 128M-N** - *C. casitensis* Imlay.

Subfamily Capeloitinae Cantú-Chapa, 2009

Discussion. A monophyletic taxon of uncertain ancestry, perhaps Saynoceratini close to *Barronites*; typically small lappetted microconchs with compressed, moderately evolute whorls, rounded venter and markedly discrepent ornament in later growth. *Upper U. Valang. - L. Haut.*

Capeloites LISSÓN, 1937 [*C. larozai*, = *Ammonites perelegans* Matheron, 1878; OD]. Very small; inner whorls like *Ceratotuberculus* but outer whorls superficially resembling *Jeannoticeras*, with weakening of umbilical bullae, loss of siphonal tubercles and fine flexuous ribs passing uninterrupted over venter. *Upper U. Valang. - basal L. Haut.* **Fig. 128P-Q** - *C. perelegans* (Matheron).

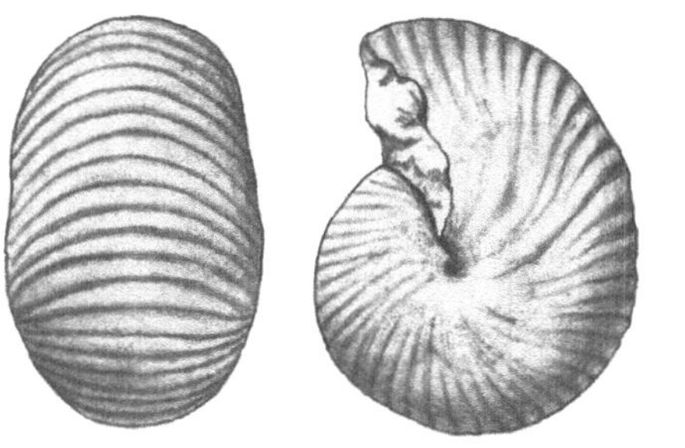

Fig. 130. *Provalanginites.*

Parastieria SPATH, 1923 [*Acantoceras? peltoceratoides* Pavlow, 1892; OD][**m**]. Very small; inner whorls like *Jeannoticeras* but with rather distant straight to slightly sinuous flat-topped ribs on first half of body chamber. *L. Haut.* **Fig. 126G-H** - *P. peltoceratoides* (Pavlow).

Subfamily Provalanginitinae Fatmi, 1972

= Garcitinae Cantú Chapa, 2012

Discussion. A monophyletic group with an origin in Olcostephaninae, but immediate antecedent unknown. Small to very small globose sphaerocones with very narrow umbilicus, broadly rounded to subtabulate venter and depressed subcoronate whorl section; primary ribs terminate in prominent lateral to ventrolateral nodes, from which bundles of 2-4 strong prorsiradiate ribs arise, sometimes bifurcating; ribs transverse across venter, continuous or interruped by smooth siphonal band; siphonal nodes may develop in later growth. *L. - U. Valang.*

Provalanginites FATMI, 1972 [*P. rhodesi*; OD]. Small; very involute sphaerocone with depressed semicircular whorl section; ribs moderately dense, strongly prorsiradiate, simple, bifurcating or trifurcating at midflank, crossing venter transversely; adult body chamber shows scaphitoid uncoiling, with contracted aperture. *?U. Tithon., L. Valang.* **Fig. 130** - *P. rhodesi* Fatmi.

Valanginites KILIAN, 1910 [*Ammonites nucleus* Roemer, 1841; SD Roman, 1938]. Very small; like *Provalanginites* but more inflated, with weaker rectiradiate ribs that often are absent from inner flank, and flared guttered collar to subtrigonal adult aperture. *L. - U. Valang.* **Fig. 126R-S** - *V. nucleus* (Roemer).

= *Rotundites* Stolley, 1937 (*nom. nud.*).

Dobrodgeiceras NIKOLOV, 1963 [*D. ventrotuberculatum*, = *Holcostephanus wilfridi* Karakasch, 1902; OD]. Very small, with low broadly rounded venter; prominent umbilicolateral bullae give rise to bundles of ribs looped

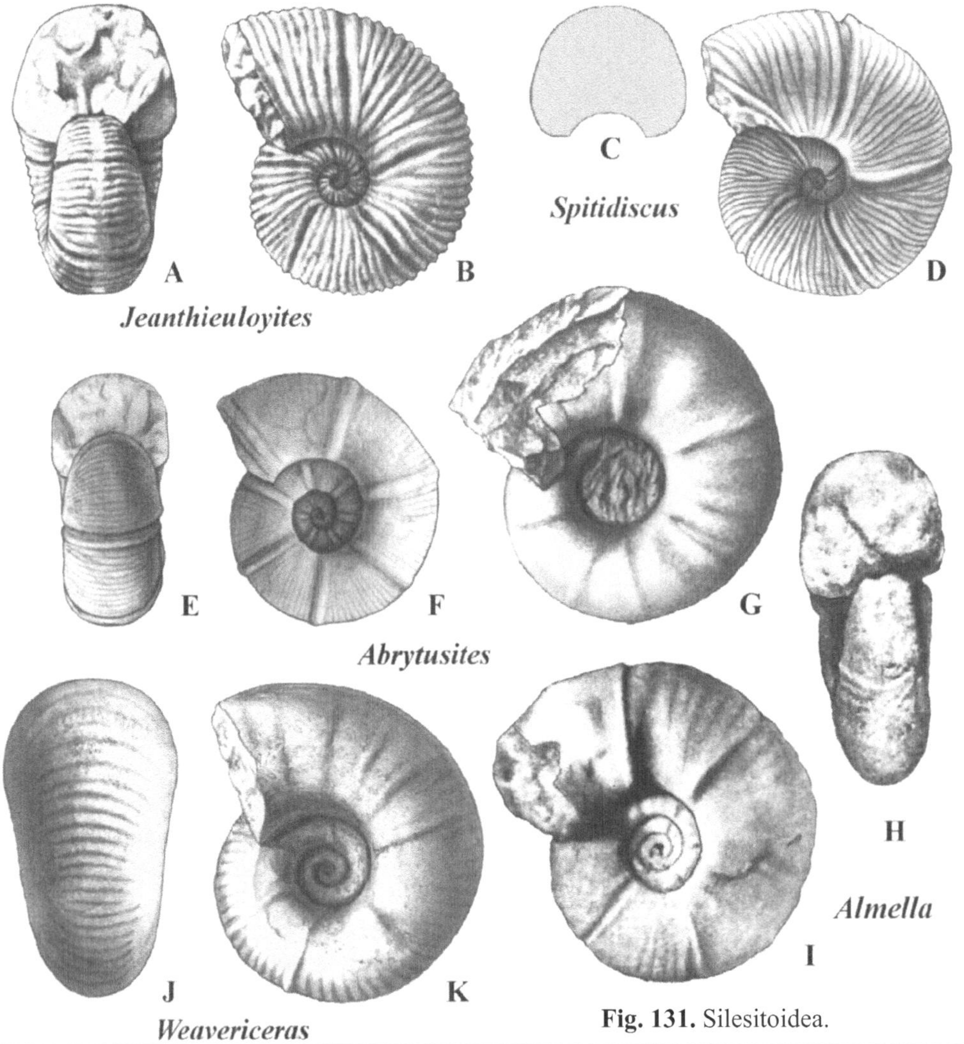

Fig. 131. Silesitoidea.

across venter, and with periodic siphonal tubercles in later growth; suture with long narrow E, narrow E/L and wider trifid L as deep as E. *L. - Lower U. Valang.* **Fig. 126P-Q** - *D. wilfridi* (Roemer).

= *Dobrodgeites* Nikolov, 1962 (*non* Kittl, 1908).

Peruvites CANTÚ-CHAPA, 2012 [*Sphaeroceras peruvianum* Lissón, 1938; OD]. Like *Dobrodgeiceras* but without siphonal nodes and with very high subvertical umbilical wall; suture with narrow E, long subrectangular multifid E/L, narrow trifid L, and broad L/U_2 shorter than E. *U. Valang.* **Fig. 129C-D** - *P. peruvianum* (Lissón).

Garcites CANTÚ-CHAPA, 2001 [*G. potrerensis*; OD]. Like *Peruvites* but with sloping umbilical wall, shorter, coarser primary ribs arising singly or in pairs from mid-lateral tubercles and crossing venter transversely without looping; venter may be subtabulate. *L. Valang. - L. Haut.* **Fig. 129E-F** - *G. potrerensis* Cantú-Chapa.

The ornament has been coarsened by pyrite encrustation.

Sanctafecites Etayo-Serna, 1985 [*Ammonites sanctafecinus* d'Orbigny, 1842; OD]. Like *Peruvites* but larger; in later growth primary ribs terminate in lateral tubercles and give rise to secondaries in triplicates that are interrupted across the siphonal line by a shallow groove. *Lower U. Valang.* **Fig. 129A-B** - *S. sanctafecinus* (d'Orbigny).

Superfamily SILESITOIDEA Hyatt, 1900

(*nom. transl.* Vermeulen, 2007 *ex* Silesitidae)

Discussion. Monophyletic derivatives of Olcostephanoidea (Kilian, 1910; Roman, 1938); comprises the families Holcodiscidae, Abrytusitidae and Silesitidae. *U. Valang. - M. Alb.*

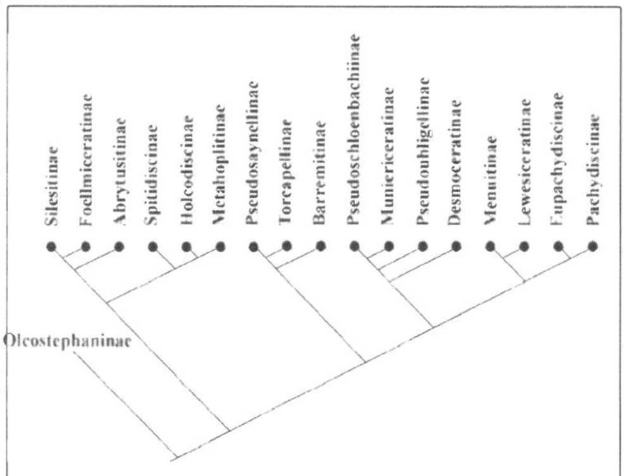

Fig. 132. Hypothesized relationships among Desmoceratidina.

Family HOLCODISCIDAE Spath, 1923

Discussion. Monophyletic derivatives of Olcostephanidae (Thieuloy, 1972). Mostly relatively small to micromorph, with sharp fine biplicate ribs, constrictions in many and relatively simple suture with trifid L. Comprises the subfamilies Spitidiscinae, Holcodiscinae and Metahoplitinae. *U. Valang. - Barrem.*

Subfamily Spitidiscinae Vermeulen & Thieuloy, 1999

Discussion. Monophyletic derivatives of Olcostephaninae, close to *Mexicanoceras*. Typically small, rather inflated and involute, with biplicate ribs that tend to weaken over the siphonal line, and with frequent straight constrictions. *U. Valang. - U. Haut.*

Jeanthieuloyites COOPER, 1981 [*Rogersites quinquestriatus* Besairie, 1936; OD]. Small; strongly inflated, involute, with deep umbilicus, convergent flanks, subtabulate venter and depressed subtrapezoidal whorl section; slightly prorsiradiate biplicate ribs (rarely simple) branch above midflank and cross subtabulate venter transversely with weakening over siphonal line; 5-6 deep oblique collared constrictions per whorl. *U. Valang. - L. Haut.* **Fig. 131A-B** - *J. quinquestriatus* (Besairie).

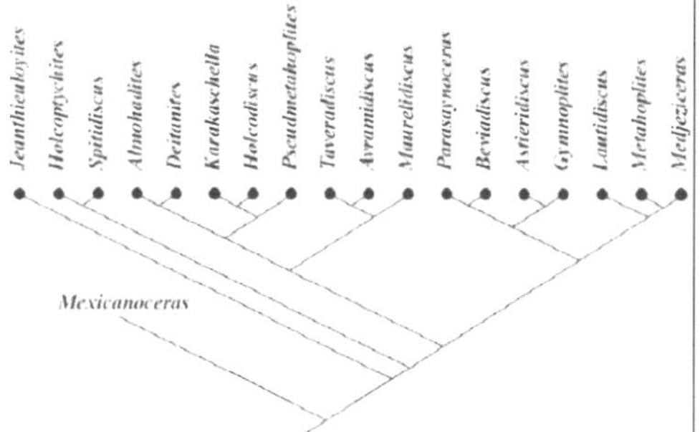

Fig. 133. Hypothesized relationships among Holcodiscidae.

Spitidiscus KILIAN, 1910 [*Ammonites rotula* J. de C. Sowerby, 1827; SD Roman, 1938]. Small; like *Jeanthieuloyites* but more finely ribbed, with rounded venter and ribs bifurcating sometimes more than once at various levels, and crossing venter uninterrupted with forward convexity; suture with long rectangular asymmetrically bifid saddles and moderately wide trifid L almost as deep as E. *U. Valang. - U. Haut.* **Fig. 131C-D** - *S. rotula* (J. de C. Sowerby).

Holcoptychites GERTH, 1921 [*Polyptychites neuquensis* Douvillé, 1910; SD Wright, 1957]. Medium sized, dimorphic; evolute to moderately involute; in early growth with flattened flanks and subquadratic whorl section, later with rounded venter and compressed oval whorl section; in early to middle growth with 7 straight slightly prorsiradiate constrictions per whorl; biplicate or triplicate ribs, occasionally single, arise from incipient (juv.) to strong (ad.) umbilical bullae; ribs strong across siphonal line; outer whorls of macroconch with enlarged umbilical tubercles and blunt low ribs. *M. Haut.* **Fig. 134A-B** - *H. neuquensis* (Douvillé) [**m**].

Subfamily Holcodiscinae Spath, 1923

(*nom. transl.* Vermeulen & Thieuloy, 1999 *ex* Holcodiscidae)

Discussion. Monophyletic derivatives of Spitidiscinae, close to *Spitidiscus* (Vermeulen, 1996); micromorph (often based on nuclei) to moderately large, finely ribbed, with lateral and sometimes ventrolateral tubercles. Comprises the tribes Holcodiscini and Deitanitini. *L. Barrem.*

Tribe Holcodiscini Spath, 1923

(*nom. transl.* herein *ex* Holcodiscinae)

Type genus: *Holcodiscus* Uhlig, 1882.

Discussion. Moderately involute, with fine biplicate ribs crossing venter uninterrupted. *L. Barrem.*

Maurelidiscus VERMEULEN, 2007 [*Ammonites intermedius* d'Orbigny, 1841; OD]. Small to moderately large; fairly involute but showing excentric uncoiling; broad flat subparallel flanks, well rounded venter and compressed elliptical whorl section; fine slightly flexuous, prorsiradiate riblets bifurcate at various levels on flank, with an occasional intercalatory, curving forward to cross venter with convex arch; 6-7 regularly spaced constrictions per whorl parallel ribbing, backed by swollen collar that arises at midflank; suture with massive asymmetrically bifid L and trifid U. *L. Barrem.* **Fig. 134M-N**

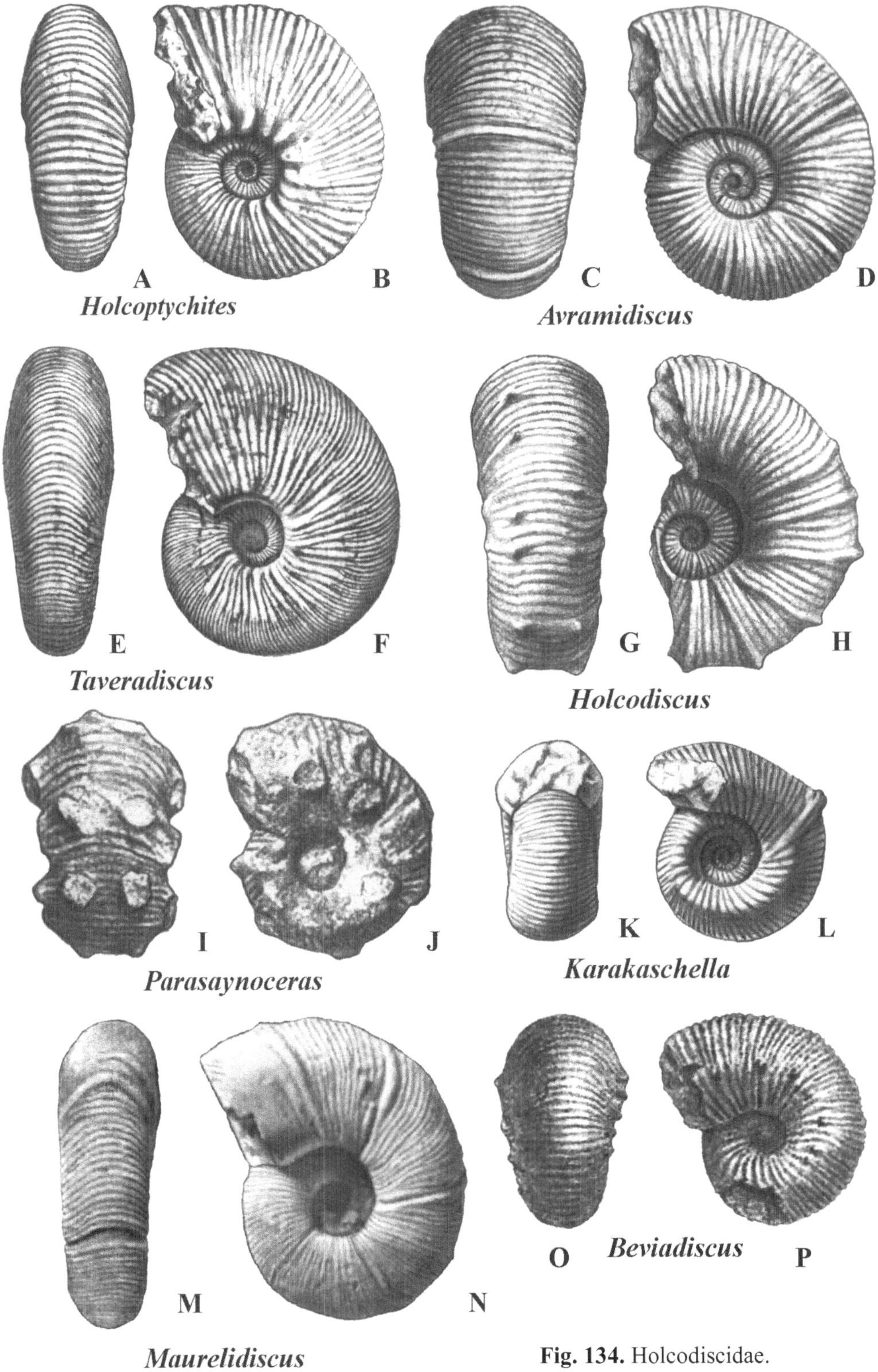

Fig. 134. Holcodiscidae.

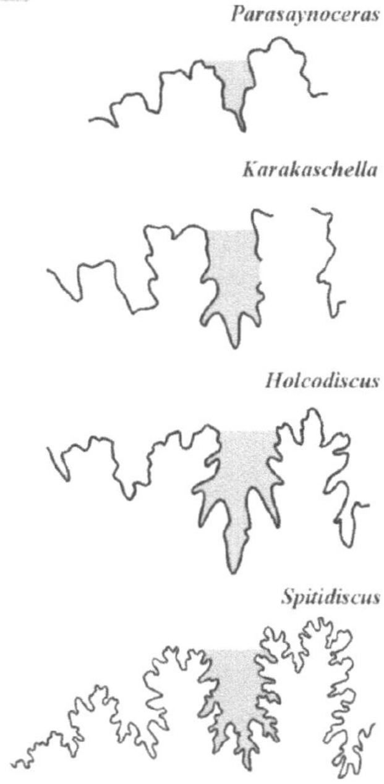

Fig. 135. Holcodiscid sutures.

- *M. intermedius* (d'Orbigny).

A homoeomorph of *Kennicottia*.

Avramidiscus VERMEULEN, 1996 [*Ammonites gastaldianus* d'Orbigny, 1850; OD]. Small; moderately involute, inflated, with broadly rounded venter and depressed whorl section; straight relatively thin biplicate ribs with minute lateral tubercle at point of furcation, crossing venter transversely; 5-8 straight collared constrictions. *L. Barrem.* **Fig. 134C-D** - *A. gastaldianus* (d'Orbigny).

Holcodiscus UHLIG, 1882 [*Ammonites caillaudianus* d'Orbigny, 1850; OD]. Very small; relatively evolute, with somewhat flattened flanks and venter, and rounded subquadratic whorl section; about 8 parabolic ribs per half whorl splitting across venter, with lateral and ventral tubercles, separated by fine single or biplicate non-tuberculate secondaries; suture with prominent median saddle to E, long narrow trifid E/L, asymmetrically trifid L much deeper than E and bifid L/U_2. *L. Barrem.* **Fig. 134G-H** - *H. caillaudianus* (d'Orbigny).

Pseudometahoplites VERMEULEN, 2007 [*Ammonites geronimae* Hermite, 1879; OD]. Micromorph; moderately evolute with rapidly inflating whorls, convex flanks, broad gently rounded venter and depressed subhexagonal whorl section; strong well spaced prorsiradiate primary ribs terminate in lateral spines that give rise to 2-3 fine secondaries that frequently loop to nodes or clavi that alternate irregularly and do not correspond, crossing venter uninterrupted; rare constrictions in early growth. *L. Barrem.* **Fig. 136I-J** - *P. geronimae* (Hermite).

Karakaschella **gen. nov.** [*Astieria cadoceroides* Karakasch, 1907, p.71; OD herein]. Micromorph; cadicone with sloping umbilical wall, subangular umbilical edge positioned at midflank and depressed subcoronate to reniform whorl section; numerous strong prorsiradiate primary ribs form weak bullae at umbilical edge and give rise to 2-3 secondaries; adult body chamber with closely spaced parabolic ribs with paired ventral bullae; suture like *Holcodiscus* but simpler; a homoeomorph of *Subastieria* (Olcostephanidae). *L. Barrem.* **Fig. 134K-L** - *K. cadoceroides* (Karakasch).

= *Neoastieria* Obata & Matsukawa, 1984 (*non* Egoian, 1969).

Parasaynoceras BREISTROFFER, 1947 [*Ammonites horridus* d'Orbigny, 1850; OD]. Very small; like inner whorls of *Karakaschella* but with very large spinose lateral and ventral tubercles covering several riblets; suture simple, with narrow L. *Barrem.* **Fig. 134I-J** - *P. horridum* (d'Orbigny).

Taveraidiscus VERMEULEN, 1996 [*Ammonites hugii* Ooster, 1860; OD]. Very small; like *Jeanthieuloyites* but compressed, high whorled with broad flanks, evenly rounded venter and elliptical whorl section; numerous fine, prorsiradiate primary ribs terminate at midflank in feeble lateral tubercles that give rise to 2-3 finer secondaries that flex forward and cross venter with slight forward convexity. *Lower L. Barrem.* **Fig. 134E-F** - *T. alcoyensis* (Nicklès).

Tribe Deitanitini nov.

Type genus: *Deitanites* Company, Fozy, Sandoval & Tavera, 2006.

Diagnosis. Typically very evolute with coarse ribbing interrupted across the venter, with or without ventrolateral tubercles. Monophyletic derivatives of Holcodiscini, close to *Holcodiscus* (Company *et al.*, 2006). *L. Barrem.*

Almohadites WIEDMANN, 1966 [*A. subcamelinus*, = *Ammonites camelinus* d'Orbigny, 1850; OD]. Very small to small; evolute, compressed, with equidimensional whorls, rounded intercostally, subquadratic costally; strong simple robust ribs obliquely cut by prominent collared constrictions, posterior collar with strong ventral bullae; smooth grooved venter in early growth; suture very simple, with broad saddles. *L. Barrem.* **Fig. 143U-V** - *A. camelinus* (d'Orbigny).

Deitanites COMPANY, FOZY, SANDOVAL & TAVERA, 2006 [*D. labatlanensis*; OD]. Micromorph; like *Almohadites* but very evolute, with strong rectiradiate biplicate ribs, rarely single or triplicate, with ventrolateral tubercles. *L. Barrem.* **Fig. 143S-T** - *D. labatlanensis* Company, Fozy, Sandoval & Tavera.

Subfamily Metahoplitinae Tzankov & Breskovski, 1982

(*nom. transl.* herein *ex* Metahoplitidae)

Discussion. Monophyletic derivatives of Holcodiscinae; typically compressed involute high whorled

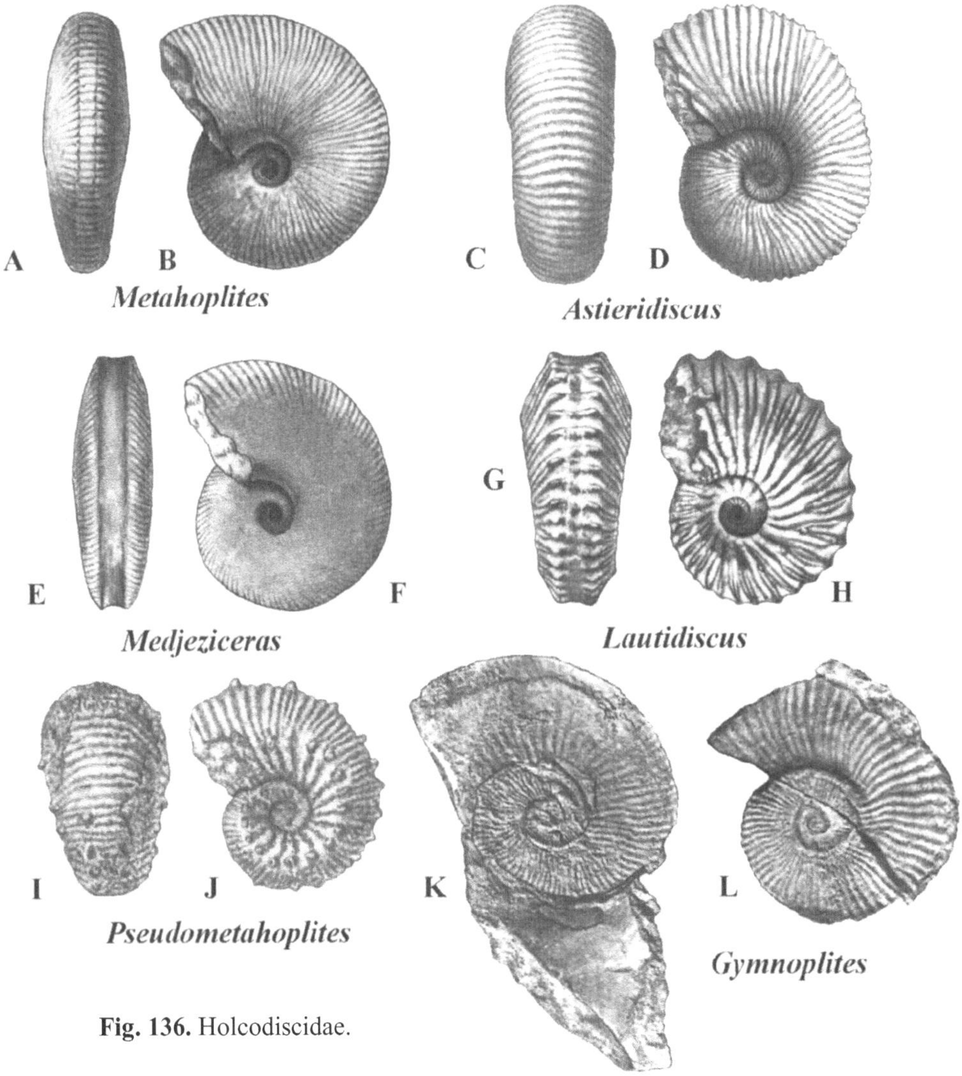

Fig. 136. Holcodiscidae.

micromorphs, with flattened flanks, narrow tabulate venter, fine ribbing and tiny ventrolateral tubercles; suture like *Holcodiscus* but simpler. Comprises the tribes Metahoplitini and Astieridiscini. *M. Haut. - Barrem.*

Tribe Astieridiscini Tzankov & Breskovski, 1982

(*nom. transl.* herein *ex* Astieridiscidae)
Type genus: *Astieridiscus* Kilian, 1910.

Discussion. Monophyletic derivatives of Holcodiscinae, close to *Avramidiscus*; typically without constrictions or tubercles. *M. Haut. - Barrem.*

Astieridiscus KILIAN, 1910 [*Holcodiscus morleti* Kilian, 1889; SD Roman, 1938]. Small; like inner whorls of *Avramidiscus*; later with rounded whorls that increase rapidly in height, and fine ribbing that is mostly single on adult body chamber. *L. Barrem.* **Fig. 136C-D** - *A. morleti* (Kilian).

Gymnoplites SPATH, 1933 [*G. simplex*; OD]. Based on crushed specimens; like *Astieridiscus* but larger, with ribs more flexed and smooth adult body chamber. *L. Barrem.* **Fig. 136K-L** - *G. simplex* Spath.

Beviadiscus VERMEULEN, 2007 [*Holcodiscus astieriformis* Sayn, 1890; OD; OD]. Very small; moderately involute, globose, with a steep umbilical wall, broadly rounded venter and subrounded to depressed subelliptical whorl section; thin, simple, slightly prorsiradiate to rectiradiate primary ribs terminate in small lateral spines from which fine secondary ribs bifurcate or trifurcate passing transversely over venter. *L. Barrem.* **Fig. 134O-P** - *B. astieriformis* (Sayn).

Tribe Metahoplitini Tzankov & Breskovski, 1982

(*nom. transl.* herein *ex* Metahoplitidae)
Type genus: *Metahoplites* Spath, 1924.

Discussion. Very small to small, strongly compressed, involute, with tabulate to sulcate venter with sharp edges or ventrolateral bullae, with or without lateral tubercles and constrictions. *L. Barrem.*

Lautidiscus **gen. nov.** [*Ammonites diversecostatus* Coquand, 1880, p.19; OD herein]. Small; moderately involute, compressed, high whorled, with convex flanks, subtabulate to slightly sulcate venter and trapezoidal whorl section; prorsiradiate primary ribs terminate in mid-lateral bullae from which 2-3 secondaries loop to ventrolateral bullae; in early to middle growth ribs interrupted across venter, later zigzagging. *L. Barrem.* **Fig. 136G-H** - *L. diversecostatus* (Coquand).

Metahoplites SPATH, 1924 [*Ammonites henoni* Coquand, 1880; M]. Very small; like *Lautidiscus* but with tabulate venter, very fine flexuous ribs that branch from umbilical shoulder, non-fibulate ribbing and tiny ventrolateral tubercles; in later growth ribs cross venter transversely. *L. Barrem.* **Fig. 136A-B** - *M. henoni* (Coquand).

Medjeziceras BUSNARDO & DAVID, 1957 [*Spitidiscus (M.) collignoni*; OD]. Very small; like *Metahoplites* but with concave venter and smooth sharp ventrolateral edges in later growth; constrictions occur throughout. *L. Barrem.* **Fig. 136E-F** - *M. collignoni* (Busnardo & David).

Family ABRYTUSITIDAE Breskovski, 1977

(*nom. correct.* Wright & Kennedy, 1984 *pro* Abritusitinae; *nom. transl.* Vermeulen, 2007)

Discussion. Paraphyletic derivatives of Holcodiscidae (Thieuloy, 1972); typically medium sized to moderately large, fairly evolute, inflated with flat flanks, elliptical to subquadratic whorl section, regular collared constrictions and strong non-tuberculate ribs that may be lost in later growth or entirely. Comprises only the subfamilyAbrytusitinae, descended from Holcodiscidae. *U. Valang. - L. Alb.*

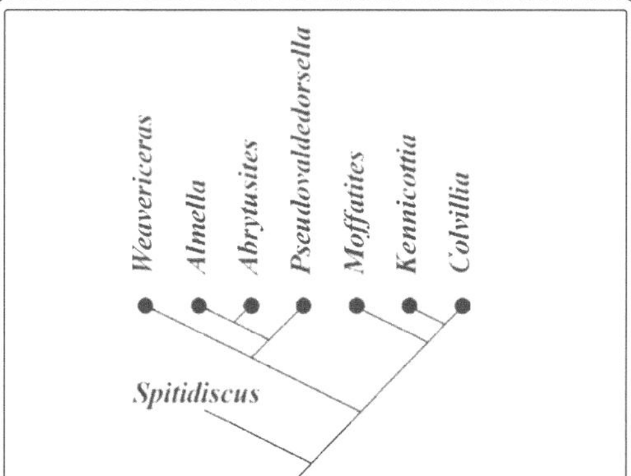

Fig. 137. Hypothesized relationships among Abrytusitinae.

Subfamily Abrytusitinae Breskovski, 1977

Discussion. Monophyletic derivatives of Spitidiscinae (Thieuloy, 1972). Mostly large, with prominent, straight, rectiradiate to slightly prorsiradiate primary ribs separated by fine secondaries that may evanesce in later growth; suture with deep narrow E, long rectangular bifid E/L and narrow trifid L. Comprises the tribes Abrytusitini and Kennicottiini. *?U. Valang., L. Haut. - L. Barrem.*

Tribe Abrytusitini Breskovski, 1977

Type genus: *Abrytusites* Nikolov & Breskovski, 1969.

Discussion. Relatively large with blunt rounded ribs single or bifurcating low on the flank in maturity, and suture with L much deeper than E; endemic to the Mediterranean Subprovince. *?U. Valang., L. Haut. - Barrem.*

Abrytusites NIKOLOV & BRESKOVSKI, 1969 [*Pachydiscus neumayri* Haug, 1889; OD]. Medium sized to moderately large; evolute; inner whorls like *Spitidiscus*, with flattish flanks, depressed subquadratic whorl section and thick rectiradiate primary ribs that cross venter transversely, separated by 10-12 fine single secondary ribs that are indistinct on inner flank; in maturity secondary ribbing obsolete and primary ribs poorly defined and with weak umbilical bullae. *?U. Valang., L. Haut. - Barrem.* **Fig. 131E-F** - *A. julianyi* (Honnorat-Bastide); **Fig. 131G** - *A. neumayri* (Haug).

Weavericeras H. LEANZA & WIEDMANN, 1980 [*Latidorsella vacaensis* Weaver, 1931; OD]. Moderately large; like *Abrytusites* but very inflated, with very deep umbilicus, vertical umbilical wall, subangular umbilical edge, broadly rounded venter and depressed reniform to semicircular whorl section; indistinct primary ribs separated on outer flank and venter by coarse low weak secondaries; faint constrictions present; suture with broad asymmetrically bifid E/L and deep penetrating trifid L and U_2. *Upper L. Haut.* **Fig. 131J-K** - *W. vacaense* (Weaver).

Almella GORN, 1969 [*A. almensis*; OD]. Introduced simultaneously with *Abrytusites* to include the type species of the latter; moderately large and evolute; in early growth flanks converging to well rounded venter and whorl section compressed oval, later with subparallel flanks

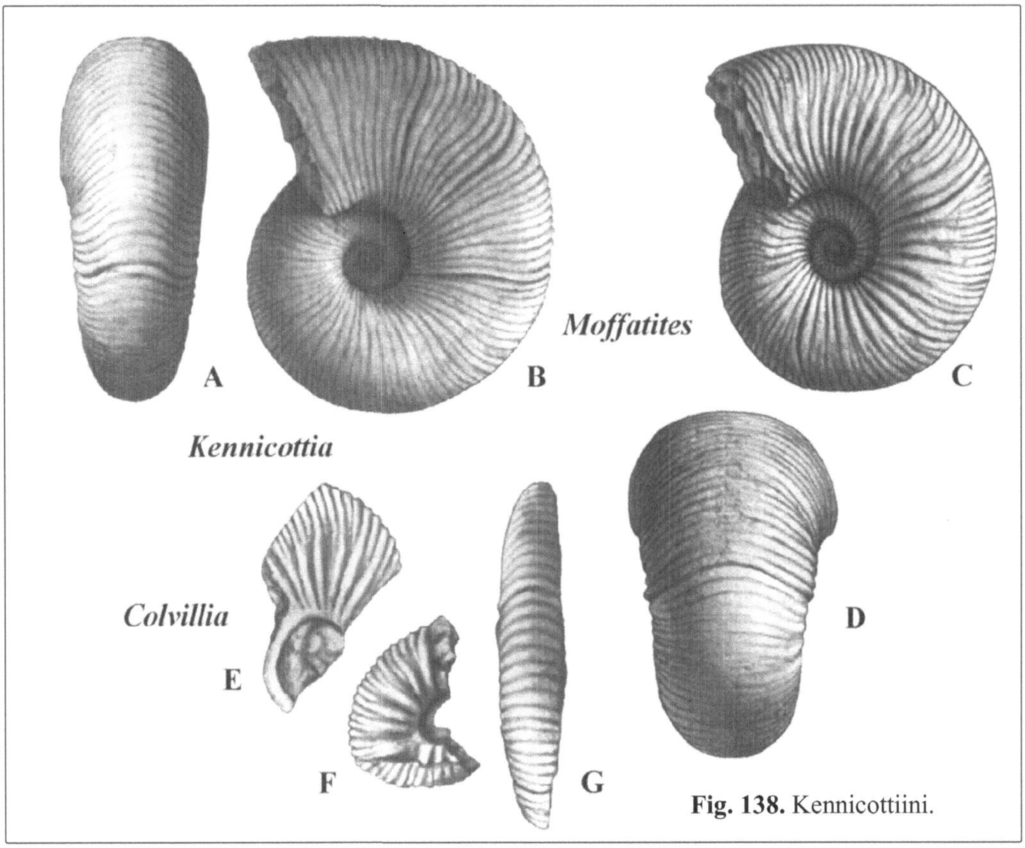

Fig. 138. Kennicottiini.

and equidimensional, quadratic whorl section; 6 strong, regular, rectiradiate primary ribs per whorl fronted by a collar, constrictions weakening and broadening with growth, separated by 6-8 feeble rectiradiate ribs; suture with very narrow E, high rectangular trifid E/L and shallow asymmetrically bifid L/U_2 and partially retracted suspensive lobe. *L. Barrem.* **Fig. 131H-I** - *A. almensis* Gorn.

Pseudovaldedorsella CHECCA, FARAONI & MARINI, 1998 [*Desmoceras crassidorsatum* Karakasch, 1907; OD]. Medium sized to very large, with subquadratic to subcircular whorl section; 11 prorsiradiate constrictions, anterior margins strongly raised on lower half of flank with umbilical notch; suture with narrow E, narrow bifid E/L as deep as E and trifid L. *Barrem.* **Fig. 139** - *P. crassidorsatum* (Karakasch).

Juveniles barely separable from *Davidiceras* (Barremitinae) but with long deep E to suture, like Desmoceratidae.

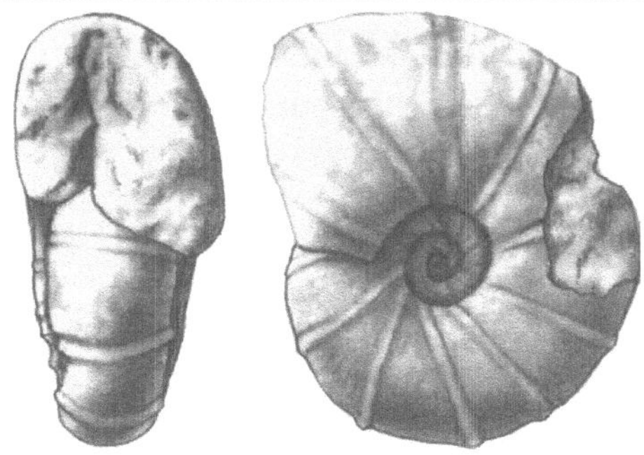

Fig. 139. *Pseudovaldedorsella.*

Tribe Kennicottiini nov.

Type genus: *Kennicottia* Imlay, 1959.

Diagnosis. Ornament of long sharp primary ribs that bifurcate or trifurcate on the upper flank, often with triangular profile, and suture with L much deeper than E; endemic to the Cordilleran Province. Monophyletic derivatives of Holcodiscini, close to *Maurelidiscus*. *L. Alb.*

Kennicottia IMLAY, 1959 [*K. bifurcata*; OD]. Small, with vertical umbilical wall, subquadratic whorl section in later growth, weak constriction with poorly developed collars, and strong primary ribs that regularly bifurcate or trifurcate on upper flank, projecting strongly forward near venter and weakening over siphonal line; suture with almost trifid L. *L. Alb.* **Fig. 138A-B** - *K. bifurcata* Imlay.

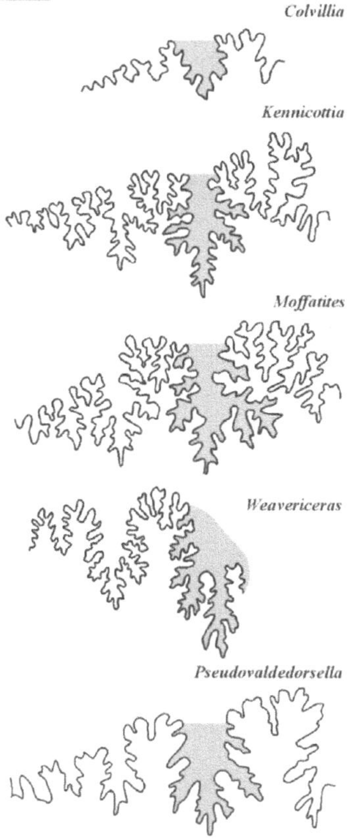

Fig. 140. Abrytusitine sutures.

Moffatites IMLAY, 1959 [*M. robustus*; OD]. Small; like *Kennicottia* but more inflated and evolute, with depressed coronate whorl section in later growth, strong robust primary ribs (triangular in profile) that bifurcate or trifurcate at ventrolateral shoulder with fine secondary riblets crossing venter with moderate forward convexity and tending to weaken over siphonal line. *L. Alb.* **Fig. 138C-D** - *M. robustus* Imlay.

Colvillia IMLAY, 1961 [*C. kenti*; OD]. Small; strongly compressed and rather involute, with flattened flanks, narrowly arched venter and compressed oval whorl section; strong slightly flexuous prorsiradiate primary ribs (triangular in profile) trifurcate or less commonly bifurcate (with associated intercalatory) on outer third of flank where they curve strongly forward and cross venter uninterrupted; 4 prorsiradiate constrictions per whorl, strongest over venter; suture with short broad E, broad trifid L and 5 auxiliary elements. *L. Alb.* **Fig. 138E-G** - *C. kenti* Imlay.

Colvillia was considered a desmoceratid (Imlay, 1961) and assigned to Cleoniceratidae (Wright, 1996), but the latter have a broad asymmetrically bifid L to the suture. The triangular profile to the ribs of *Colvillia* is an unique derived character (autapomorphy) shared with *Moffatites* from the same faunal province, to which it seems closely related.

Family SILESITIDAE Hyatt, 1900

Discussion. A monophyletic group with an origin in Abrytusitidae (Vermeulen, 2007). Characteristically serpenticone with oval to coronate whorl section, periodic constrictions and fine sharp biplicate ribs frequently with tubercle at point of rib furcation. Comprises the subfamilies Foellmiceratinae and Silesitinae. *Upper L. Barrem. - M. Alb.*

Subfamily Foellmiceratinae nov.

Diagnosis. Typically very small to small, evolute, with feeble distant ribs effaced from venter, occasionally absent, and periodic deep constrictions. Monophyletic derivatives of Abrytusitinae close to *Puezalpella* (Vermeulen *et al.*, 2018). *Upper L. Barrem. - L. Alb.*

Foellmiceras VERMEULEN, LAZARIN, LÉPINAY, LEROY & MASCARELLI, 2015 [*Ammonites vulpes* Coquand *in* Matheron, 1880; OD]. Medium sized, very evolute, compressed, with weakly convex flanks, rounded venter and compressed oval whorl section; early whorls smooth, later with thick, well separated, rectiradiate to slightly prorsiradiate ribs that attenuate or evanesce across venter particularly in early growth; deep wide constrictions, generally fronted by varix, linguoid or forming chevrons over venter; suture little incised, with trifid L and asymmetrically bifid U. *Upper L. - lower U. Barrem.* **Fig. 141B** - *F. vulpes* (Coquand).

Platysilesites VERMEULEN, LAZARIN, LÉPINAY, LEROY & MASCARELLI, 2016 [*P. duvali*; OD]. Like *Foellmiceras* but smaller, with narrower umbilicus, more compressed whorls, narrowly rounded venter and stronger more prominent constrictions. *Upper L. - lower U. Barrem.* **Fig. 141C** - *P. duvali* Vermeulen, Lazarin, Lépinay, Leroy & Mascarelli.

Haplobrancoceras AVRAM, 1990 [*H. subquadratum*; OD]. Very small, evolute, with steep umbilical wall, flat subparallel flanks, broad low slightly convex venter and subquadratic whorl section; flank unornamented or with weak sigmoidal striae and 5 straight prorsiradiate constrictions per whorl crossing venter transversely; suture simple, with narrow E, rectangular feebly bifid E/L and relatively broad trifid L. *Upper L. Barrem.* **Fig. 143O-P** - *H. subquadratum* Avram.

According to Vermeulen *et al.* (2018) based upon indeterminate juveniles, thus a *nomen dubium*, but *Foellmiceras anglesense* Vermeulen *et al.* (**Fig. 141E**) is smooth to its largest known diameter of 80 mm, with only 4 constrictions per whorl (see also *Tingriceras* (Brancoceratidae).

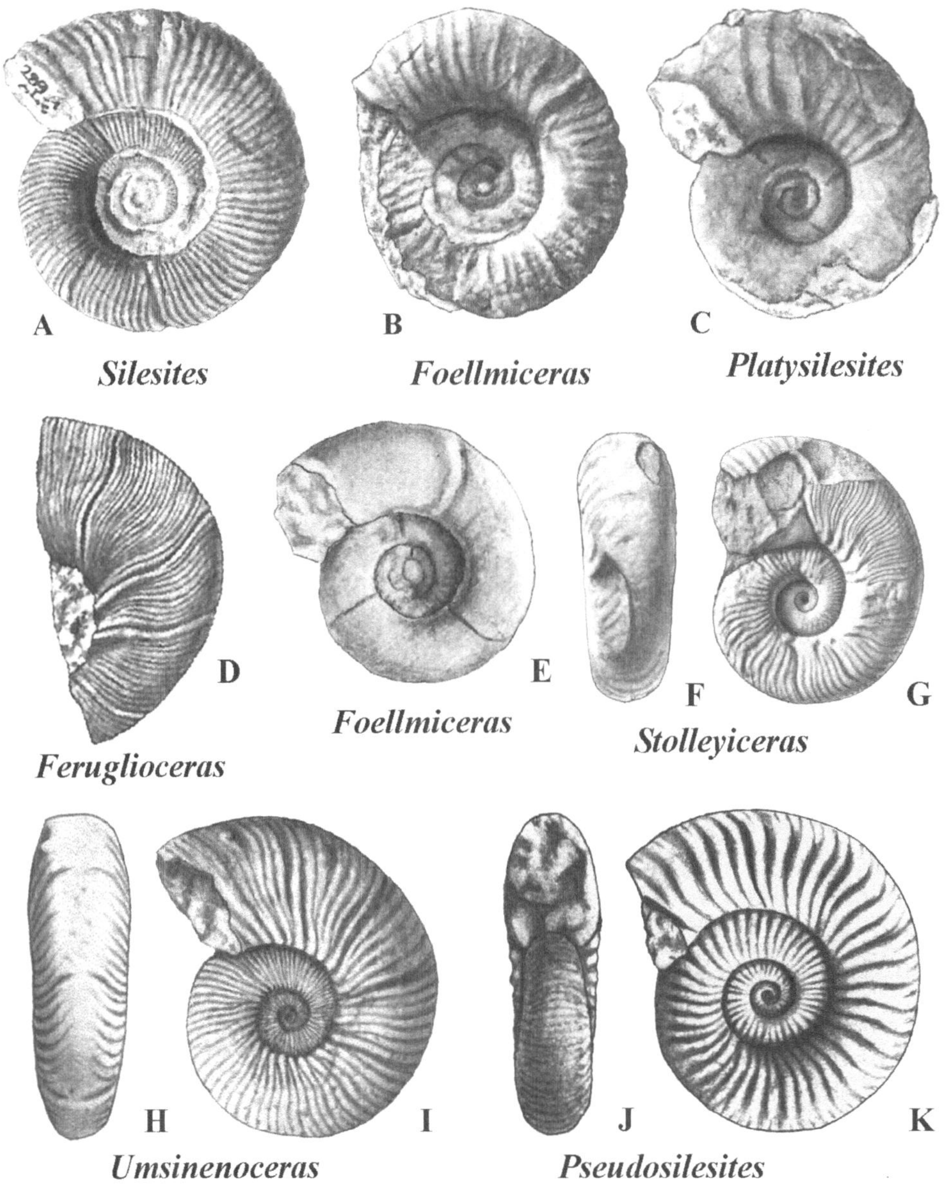

Fig. 141. Silesitidae.

Hameimaceras LATIL, 2011 [*H. zghalae*; OD]. Micromorph; inflated, moderately evolute, with convex flanks and low rounded venter; 6-9 deep sigmoidal constrictions form chevrons over venter, sometimes flared, separated by 2-3 simple ribs effaced from venter; suture simple. Superficially similar to *Puzosiella*. *L. Alb.* **Fig. 143M-N** - *H. zghalae* Latil.

Subfamily Silesitinae Hyatt, 1900

Discussion. Small to medium sized, evolute, with strong ribbing, with or without umbilical and/or ventrolateral tubercles, and frequent collared constrictions. Comprises the tribes Silesitini and Neosilesitini. *U. Barrem. - M. Alb.*

Tribe Silesitini Hyatt, 1903

(*nom. transl.* herein *ex* Silesitidae)

Type genus: *Silesites* Uhlig, 1883.

Diagnosis. Small to medium sized serpenticones with flattened flanks and fine biplicate ribbing continuous across venter. Monophyletic derivatives of Foellmicer-

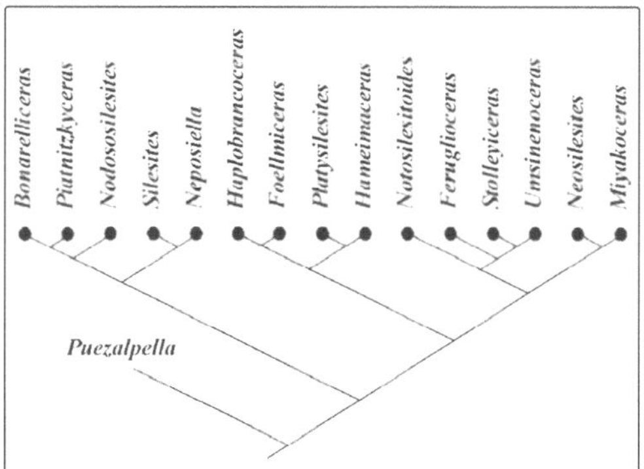

Fig. 142. Hypothesized relationships among Silesitidae.

atinae, close to *Foellmiceras* (Vermeulen, 2007). *Barrem.* - *L. Alb.*

Silesites UHLIG, 1883 [*Ammonites seranonis* d'Orbigny, 1841; OD]. Medium sized; like *Foellmiceras* but with fine wiry rectiradiate ribs that curve forwards near venter and are linguid or form chevrons over it; rare short intercalatories across venter increase in number in maturity; 3-5 deep fairly wide constrictions per whorl, their width increasing sharply in maturity, becoming pronounced and subbullate across siphonal line; suture with retracted auxiliary elements. *U. Barrem.* **Fig. 141A** - *S. seranonis* (d'Orbigny).

= *Beneckeia* Uhlig, 1882 (*non* Mojsisovics, 1882).

Nodososilesites VERMEULEN, LAZARIN, LÉPINAY, LEROY & MASCARELLI, 2016 [*N. sylvestreae*; OD]. Like *Silesites* but with oval whorl section, venter becoming narrowly rounded in later growth; strong simple rectiradiate primary ribs project forward near venter, some bifurcating at ventrolateral shoulder; in later growth alternate primaries have small pointed ventrolateral tubercle that give rise to paired secondaries; periodic deep constrictions linguoid over venter; suture relatively simple, with narrow bifid saddles and trifid lobes, L as deep as E, and a few non-retracted auxiliaries. *U. Barrem.* **Fig. 132A-B** - *N. sylvestreae* Vermeulen, Lazarin, Lépinay, Leroy & Mascarelli.

Piatnitzkyceras MEDINA & RICCARDI, 2005 [*P. bonarelli*; OD]. Small; like *Silesites* but with ribs forming weak umbilical bullae, occasionally bifurcating below mid-flank and with only sporadic small ventrolateral clavi; adult peristome constricted; suture with large trifid L deeper than E. *L. Alb.* **Fig. 143 I-J** - *P. bonarelli* Medima & Riccardi.

Bonarelliceras MEDINA & RICCARDI, 2005 [*Kossmaticeras meseticum* Bonarelli, 1921; OD]. Small; like *Piatnitzkyceras* but more involute, with rounder whorls, almost rectiradiate ribs that bifurcate at midflank (or joined by intercalatory) often from prominent mid-lateral tubercles, and less marked constrictions. *M. Alb.* **Fig. 143K-L** - *B. meseticum* (Bonarelli).

Feruglioceras A. LEANZA, 1967 [*F. piatnitzkyi*; OD]. Very small; moderately evolute, compressed, high whorled, with subvertical umbilical wall, broad gently convex flanks and subtabulate venter; numerous flexuous prorsiradiate constrictions separated by 8-16 sharp wiry riblets branching from sporadic umbilical bullae in early growth, and again on flank, sometimes with lateral tubercles. *L. Alb.* **Fig. 141D** - *F. piatnitzkyi* A. Leanza.

Stolleyiceras MEDINA & RICCARDI, 2005 [*Silesites desmoceratoides* Stolley, 1912; OD]. Very small; relatively involute becoming more evolute with growth, with steep umbilical wall and subrounded to compressed subrectangular whorl section, compression increasing with growth; ribbing changes from fine and almost rectiradiate on inner whorls, to coarse and flexuous in middle growth and finer again with constrictions in maturity; at end of adult body chamber ribs arise in bundles from umbilical bullae and weaken over venter. *L. Alb.* **Fig. 141F-G** - *S. desmoceratoides* (Stolley).

Tribe Neosilesitini nov.

Type genus: *Neosilesites* Breistroffer, 1951.

Diagnosis. Monophyletic derivatives of Silesitini, typically with ventrolateral tubercles. *U. Apt.* - *L. Alb.*

Umsinenoceras KENNEDY, WRIGHT & KLINGER, 1979 [*U. inguatuberculatum*; OD]. Small, evolute, with prorsiradiate ribs reduced to delicate striae over venter and constrictions interrupted over siphonal line on internal mould; in later growth with irregular ventrolateral nodes joined by looped ribs; on adult body chamber ribs strongly concave and venter rounded; suture with retracted suspensive lobe. *Lower M. Alb.* **Fig. 141H-I** - *U. linguatuberculatum* Kennedy, Wright & Klinger.

Umsinenoceras was assigned to Puzosiinae by Wright (1996) and Medina & Riccardi (2005) but is much like *Stolleyceras* in early to middle growth. Its derived characters (wiry ribbing, subtabulate venter across which ribbing weakens or is effaced, sporadic ventrolateral nodes and retracted suspensive lobe) are features of early Kossmaticeratidae.

Neoastieria EGOIAN, 1969 [*N. reliqua*; OD]. Small, with depressed coronate whorl section, broadly convex venter and prorsiradiate main ribs with sharp oblique ventrolateral spines; ribs cross venter transversely; a homoeomorph of *Karakaschella* (Holcodiscidae). *U. Apt.* **Fig. 143C-D** - *N. reliqua* Egoian.

Neosilesites BREISTROFFER, 1951 [*Silesites seranonis* var. *balearensis* Fallot, 1920; OD]. Very small; like *Neoastieria* but more compressed, with subquadratic whorl section, rectiradiate primaries and small ventrolateral nodes; ribbing becoming more distant in later growth;

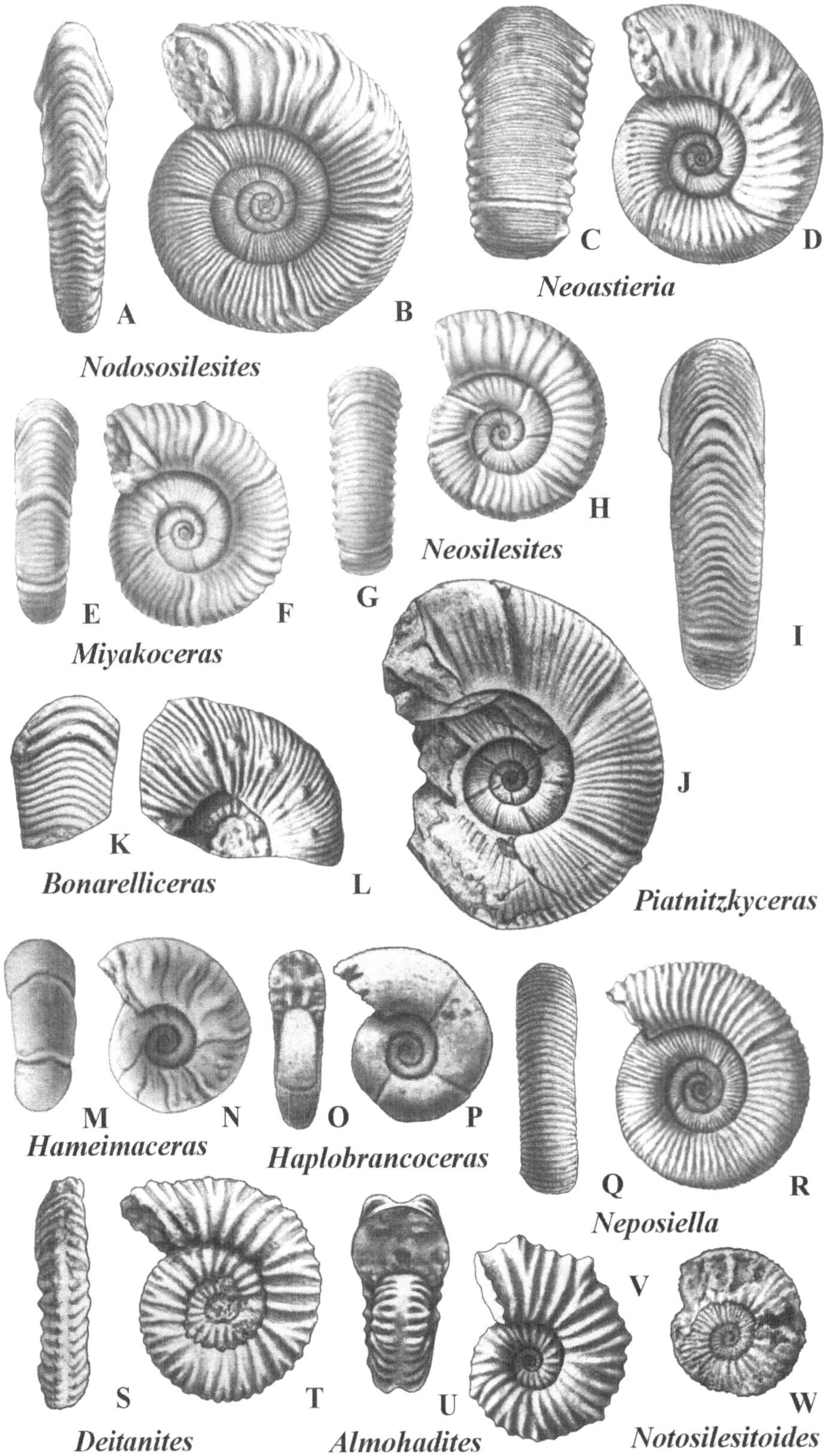

Fig. 143. Silesitidae.

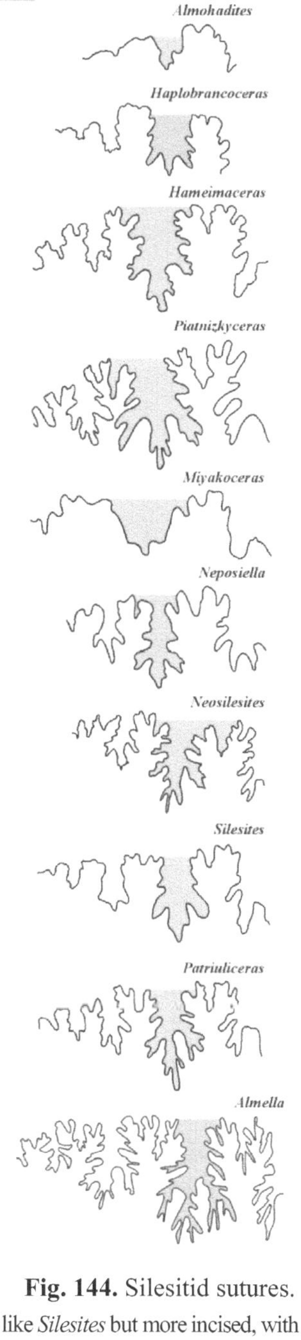

Fig. 144. Silesitid sutures.

suture like *Silesites* but more incised, with broader E/L. *U. Apt.* **Fig. 143G-H** - *N. balearensis* (Fallot).

Miyakoceras OBATA, 1967 [*M. tanohatense*; OD]. Small; like *Neosilesites* but smaller, with rounded venter, coarser more distant primary ribs that do not split across venter, and occasional ventrolateral tubercles; collared constrictions form weak chevrons over venter; very simple suture with broad E/L, L and L/U_2. *U. Apt.* **Fig. 143E-F** - *M. tanohatense* Obata.

Pseudosilesites EGOIAN, 1969 [*P. seranoniformis*; OD]. Very small, evolute, compressed; in later growth with ribs splitting into two fine secondaries, commonly separated by an intercalatory, and crossing venter transversely; adult body chamber with deep flexuous constrictions; suture deeply incised, with narrow stemmed saddles and trifid L. *U. Apt.* **Fig. 141J-K** - *P. seranoniformis* Egoian.

Notosilesitoides MATSUKAWA, SENDON, MATEER, SATO & OBATA, 2012 [*N. philippinensis*; OD]. Micromorph; evolute, with gently convex flanks, rounded venter and slightly depressed subquadratic whorl section; flanks with paired rectiradiate ribs, broader than interspaces, and periodic constrictions. *U. Apt.* **Fig. 143W** - *N. philippinensis* Matsukawa, Sendon, Mateer, Sato & Obata.

Neposiella MAHMOUD, 1953 [*Silesites nepos* Douvillé, 1917; OD]. Small; like *Neosilesites* but with fewer constrictions, denser finer sharp ribbing that does not coarsen and becomes distant in later growth, and with feeble ventrolateral nodes. *L. Alb.* **Fig. 143Q-R** - *N. nepos* (Douvillé).

= *Neposiella* Breistroffer, 1952 (*nom. nud.*).

Superfamily DESMOCERATOIDEA Zittel, 1895

Discussion. Monophyletic derivatives of Silesitoidea (Thieuloy, 1972); comprises the families Barremitidae, Desmoceratidae and Muniericeratidae. *L. Haut. - M. Camp.*

Family BARREMITIDAE BRESKOVSKI, 1977

(*nom. transl.* Vermeulen, 1996 *ex* Barremitinae)

Discussion. Monophyletic derivatives of Holcodiscidae (Thieuloy, 1972); mostly involute, compressed, weakly ornamented and constricted; suture relatively simple, with bifid saddles, large asymmetrically bifid L and a straight declining series of auxilliary elements. Genera are poorly discriminated, but the interpretation of this family offered by Klein & Vašíček (2011) is not followed here. Comprises the subfamilies Barremitinae, Torcapellinae and Pseudosaynellinae. *L. Haut. - U. Apt.*

Subfamily Barremitinae Breskovski, 1977

Discussion. Monophyletic derivatives of Spitidiscinae, close to *Spitidiscus* (Thieuloy, 1972); generally small to medium sized, compressed, high whorled and relatively involute with frequent generally prorsiradiate constrictions and weak to obsolete ribbing; suture with narrow bifid saddles and

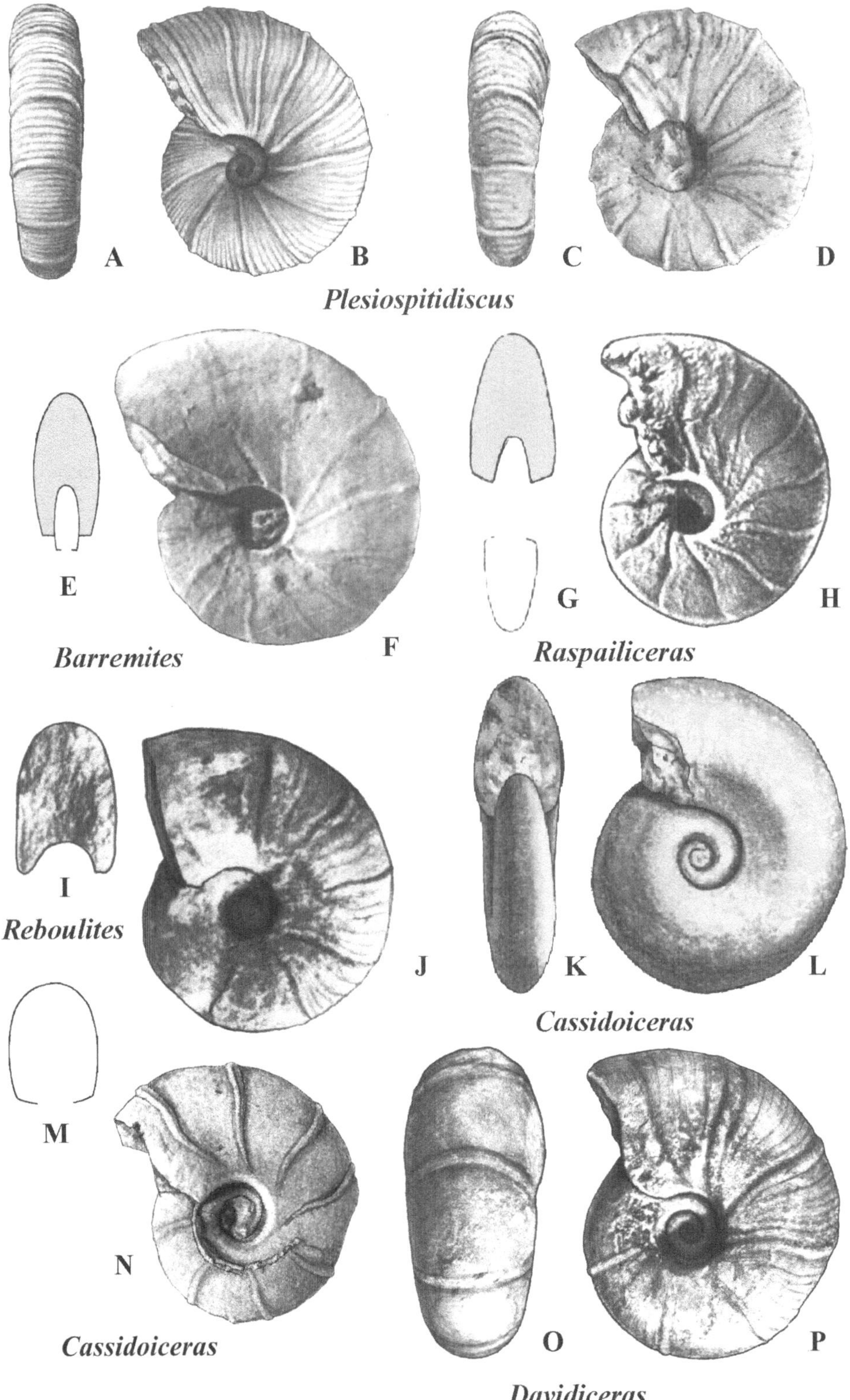

Fig. 145. Barremitidae.

asymmetrically bifid to trifid L deeper than E.

Insufficient knowledge of ontogenetic change, together with taxonomic lumping, have served to confuse relationships within this complex Barremian radiation. *L. Haut. - L. Apt.*

Tribe Barremitini Breskovski, 1977
Type genus: *Barremites* Kilian, 1913.

Discussion. Strongly compressed, involute and high whorled with subtabulate to narrowly rounded venter and subrectangular to sublanceolate whorl section; early whorls unconstricted in most, later with up to 12 almost straight to slightly sinuous prorsiradiate constrictions per whorl, with or without intervening riblets. *L. Haut. - L. Apt.*

Plesiospitidiscus BREISTROFFER, 1947 [*Ammonites ligatus* d'Orbigny, 1841; OD]. Medium sized, with convex flanks, maximum width at midflank, broad venter, and straight rigid prorsiradiate constrictions that parallel ribbing and form deep notches at umbilical edge; constrictions evanesce in later growth; suspensive lobe of suture not retracted (but retracted in *P. fasciger*). *L. - U. Haut.* **Fig. 145A-B** - *P. fasciger* (Thieuloy); **Fig. 145C-D** - *P. ligatus* (d'Orbigny).

Nabdalsiceras BRESKOVSKI, 1977 [*Ammonites nabdalsa* Coquand, 1880; OD]. Small to medium sized; moderately involute, flat sided with narrowly rounded to subtabulate venter and subrectangular whorl section; 6-7 prominent, deep, wide, straight prorsiradiate constrictions per whorl, notching the umbilical edge; ribs lacking; suture like *Plesiospitidiscus* with deep E. *Barrem.* **Fig. 146I-J** - *N. nabdalsum* (Coquand).

Montanesiceras BRESKOVSKI, 1977 [*Barremites (Cassidoiceras) chaputi tschuprenensis* Dimitrova, 1967; OD] [**M**]. Medium sized, involute, adult body chamber showing scaphitoid uncoiling, with circum-umbilical depression in later growth; in early growth with biconcave prorsiradiate constrictions separated over venter by feeble ribs; adult body chamber with numerous flexuous prorsiradiate constrictions; suture highly incised, with trifid L. *U. Haut. - L. Barrem.* **Fig. 146A-B** - *M. tschuprenense* (Dimitrova); **Fig. 146C-E** - *M. breskovskii* Vašíček, Rabrenovic, Radulovic, Radulovic & Mojsic.

Trimontioniceras BRESKOVSKI, 1977 [*Barremites charrierianus* Drušcic, 1960; OD]. Small; moderately involute, with weakly convex umbilical wall, flat flanks, broadly rounded venter and subtrapezoidal whorl section; flexuous constrictions strongest across venter, crossing with forward convexity; in later growth constrictions backed by short bold rib with faint riblets in interspaces. *Barrem.* **Fig. 149A-B** - *T. charrierianum* (Drušcic).

Patruliusiceras AVRAM, 1990 [*P. crenelatum*; OD]. Small, based on juveniles (D<30 mm); shell moderately evolute, smooth, with 6-7 straight shallow constrictions each accompanied by indistinct rib; ribs and constrictions project forward over venter. *L. Barrem.* **Fig. 149C-D** - *P. crenelatum* Avram.

Raspailiceras WRIGHT, 1956 [*Ammonites cassida* Raspail, 1831; OD]. Small to medium sized, involute, compressed, with sloping umbilical wall without circum-umbilical depression; broad gently convex convergent flanks, narrowly rounded venter and strongly compressed oval whorl section; earliest whorls smooth, later with 5-9 bold, prorsiradiate sigmoidal constrictions per whorl commencing at umbilical edge; suture with long narrow E, narrow stemmed symmetrically bifid saddles and symmetrically trifid L deeper than E. *Barrem.* **Fig. 145G-H** - *R. cassidum* (Raspail).

Paralectotypes more compressed with a narrowly rounded to subtabulate venter and steeper umbilical wall.

= *Raspailites* Wright, 1955 (*non* Jeannet, 1951).

Barremites KILIAN, 1913 [*Ammonites difficilis* d'Orbigny, 1841; M]. Small to medium sized, involute but umbilical seam egressing in maturity; strongly compressed, with vertical umbilical wall, angular umbilical edge, flat to weakly convex flanks, distinct circum-umbilical depression, narrowly arched venter and compressed sublanceolate whorl section; earliest whorls unornamented, later with 10-12 prorsiradiate biconcave constrictions per whorl; suture with narrow shallow E, oblique narrow stemmed bifid E/L, asymmetrically bifid L much deeper than E, and regularly descending suspensive lobe of 5 elements. *U. Barrem.* **Fig. 145E-F** - *B. difficilis* (d'Orbigny).

The lectotype was said to be a juvenile of *strettostoma* "type" (Gauthier, 2006).

= *Falloticeras* Breskovski, 1977 (*non* Parona & Bonarelli, 1897).

Pseudobarremites VERMEULEN, LAZARIN, LÉPINAY, LEROY & MASCARELLI, 2014 [*P. raybaudae*; OD]. Like *Barremites* but with primary ribs better developed on inner flank, deeper constrictions and fine falcate ribs of various length that persist to maturity. *Uppermost Haut. - basal Barrem.* **Fig. 146O-P** - *P. raybaudae* Vermeulen, Lazarin, Lépinay, Leroy & Mascarelli.

Barremitites AVRAM, 1997 [*Haploceras strettostoma* Uhlig, 1883; OD]. Very small, involute, strongly compressed, with vertical umbilical wall, angular umbilical edge, gently convex subparallel inner flanks, shallowly concave outer flank, narrow subtabulate venter and strongly compressed subtrigonal whorl section; spiral depression with feeble crescents; after 20 mm diameter 5 strongly prorsiradiate, falcate constrictions per half whorl; suture with shallow E, narrow stemmed bifid E/L and very broad asymmetrically bifid L. *Uppermost Barrem. - L. Apt.* **Fig. 146G-H** - *B. strettostoma* (Uhlig).

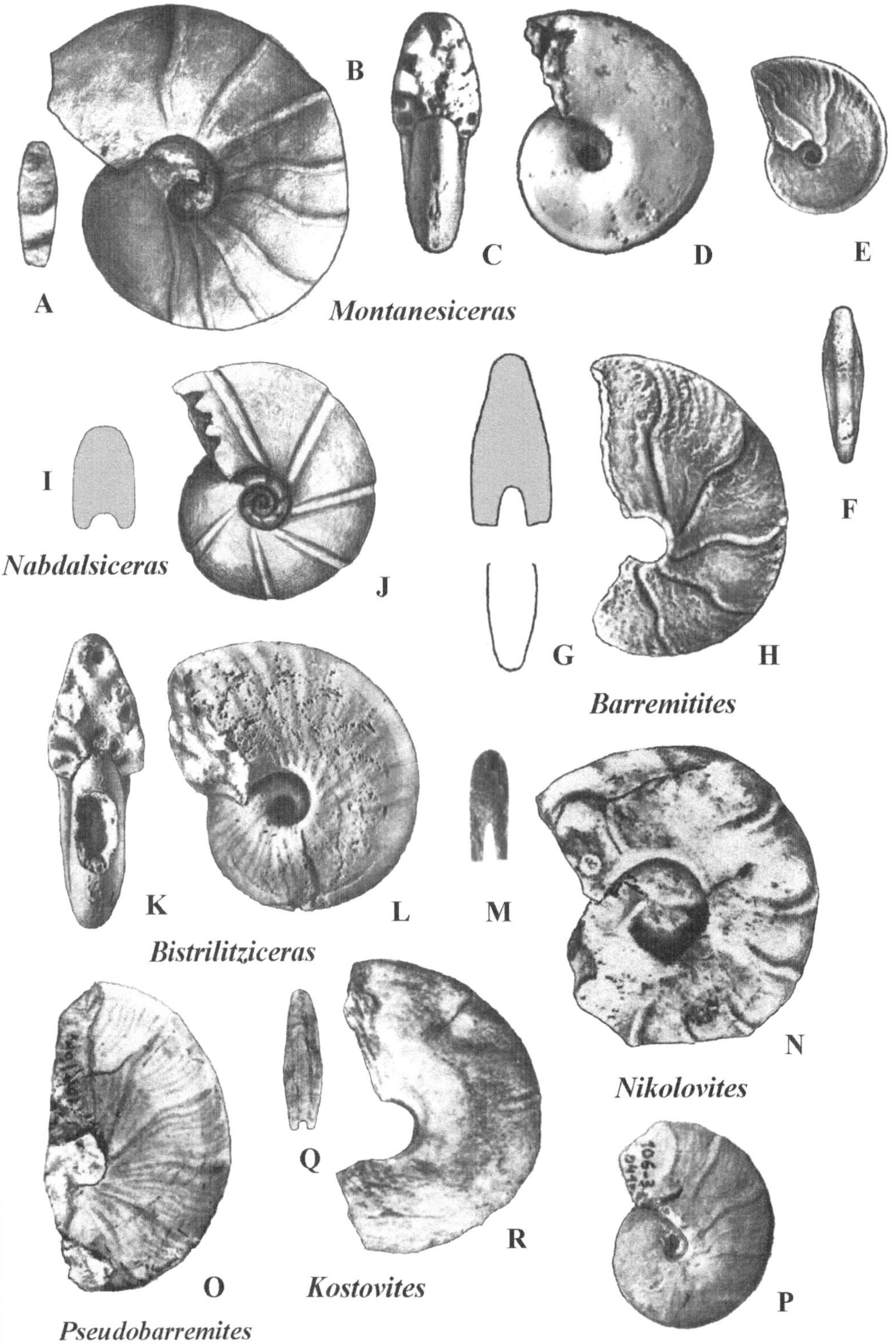

Fig. 146. Barremitinae.

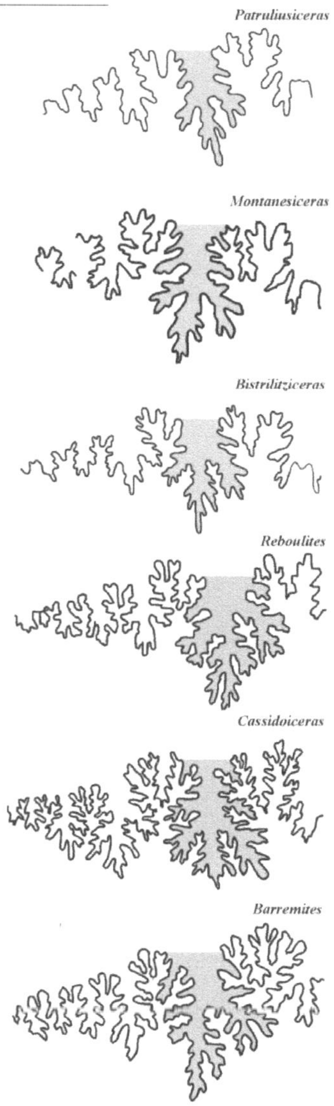

Fig. 147. Barremitid sutures.

Tribe Cassidoiceratini Breskovski, 1977
Type genus: *Cassidoiceras* Dimitrova, 1967.

Discussion. Like Barremitini but more evolute and inflated, with convex flanks and rounded venter; numerous prorsiradiate, more or less sigmoidal collar ribs. *U. Barrem. - L. Apt.*

Cassidoiceras DIMITROVA, 1967 [*Haploceras cassidoides* Uhlig, 1883; OD]. Medium sized; moderately evolute, with relatively flat flanks, maximum width at midflank, with shallow circum-umbilical depression, rounded venter and subelliptical whorl section; umbilical wall perpendicular with angular umbilical edge; earliest whorls without constrictions, but later with 8-12 relatively shallow, weak collar ribs per whorl (on internal mould) forming crescents on outer flank; suture with long narrow E much deeper than trifid L and symmetrically bifid L/U_2. *U. Barrem.* **Fig. 145K-L** - *C. chaputi* (Dimitrova); **145M-N** - *C. cassidoides* (Uhlig).

= *Cassidoides* Dimitrova, 1967 (*nom. van.*).

Nikolovites BRESKOVSKI, 1977 [*Barremites (Cassidoiceras) fegirensis* Dimitrova, 1967; OD]. Moderately involute but showing scaphitoid uncoiling, with subangular umbilical edge, flat parallel flanks, well rounded venter and very compressed elliptical whorl section; outer whorl with about 10 crescentic collar ribs on outer flank. *L. Barrem.* **Fig. 146M-N** - *N. fegirensis* (Dimitrova).

Kostovites BRESKOVSKI, 1977 [*Barremites difficilis dimboviciorensis* Breskovski, 1966; OD] [**M**]. Small, involute, discoidal, with narrowly rounded venter, sharp umbilical edge, and compressed oval whorl section; 8-12 sinuous collar ribs per whorl, strongest over venter where they are convex fowards, flanked on both sides by constrictions. *Barrem.* **Fig. 146Q-R** - *K. dimboviciorensis* (Breskovski).

Davidiceras VERMEULEN, 2003 [*Ammonites potieri* Matheron, 1880; OD]. Small, like *Reboulites* but more inflated, with slightly convex subparallel flanks, broad well rounded venter and compressed subquadratic whorl section; 8 slightly prorsiradiate collared constrictions per whorl crossing venter with slight forward arch, separated by feeble sinuous ribs. *L. Barrem.* **Fig. 145O-P** - *D. potieri* (Matheron).

Reboulites DIMITROVA, 1967 [*Puzosia issarpayensis* Kilian & Reboul, 1915; OD]. Small, with broad flattened parallel flanks, low rounded venter and compressed elliptical whorl section; in middle to later growth with 6 slightly prorsiradiate sinuous collared constrictions per half whorl, separated by feeble sinuous ribs. *Upper L. - U. Barrem.* **Fig. 145I-J** - *R. issarpayensis* (Kilian & Reboul).

= *Rebulites* Breskovski, 1977 (*nom. van.*).

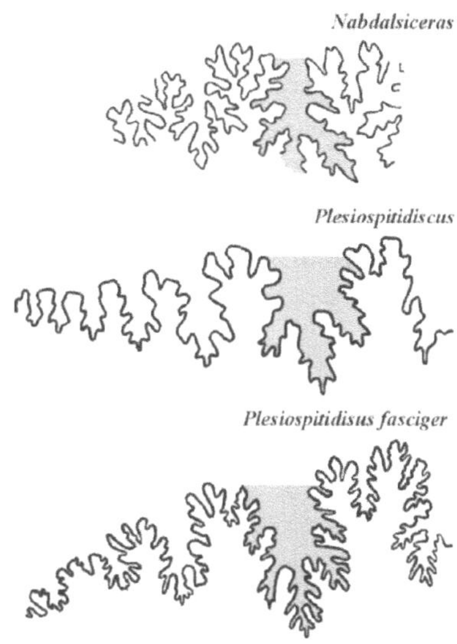

Fig. 148. Barremitid sutures.

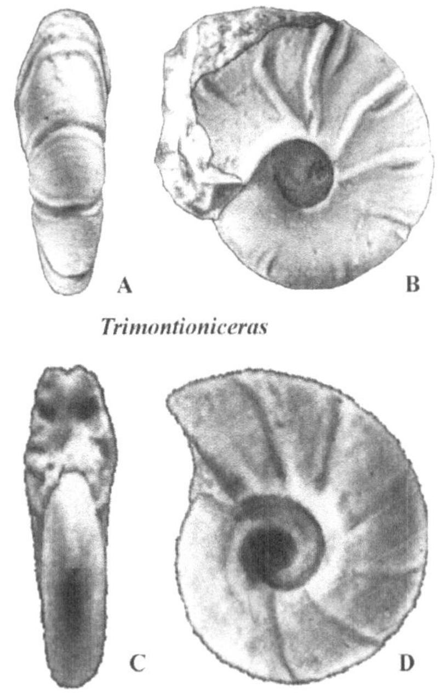

Fig. 149. Barremitidae.

Bistrilitzeras BRESKOVSKI, 1980 [*Barremites (B.) balkanicus* Manolov, 1975; OD]. Small, moderately involute platycone with vertical umbilical wall and sharp umbilical edge; convex inner flank, concave outer flank, narrowly rounded venter and subtrigonal whorl section; prorsiradiate, slightly sigmoidal growth striae, bundled in early growth, parallelled by 10-12 feeble constrictions, the last 4-5 strongest; suture with bifid saddles and trifid lobes. *U. Barrem.* **Fig. 146K-L** - *B. balkanicum* (Manolov).

Miodesmoceras WRIGHT, 1955 [*Haploceras lechinum* Uhlig, 1883; OD]. Medium sized; strongly compressed, evolute, with wide umbilicus, rounded umbilical shoulder, flat flanks and strongly arched venter; ornament of sigmoidal growth striae, some stronger than others; constrictions lacking. *L. Barrem.* **Fig. 150** - *M. lechinum* (Uhlig).

Tribe Mogharaeceratini nov.

Type genus: *Mogharaeceras* Breistroffer, 1940.

Diagnosis. A monotypic taxon with an origin close to *Pseudobarremites*, but set well apart by a distinctive suture. Diagnosis as for genus. *U. Barrem.*

Mogharaeceras BREISTROFFER, 1940 [*Knemiceras priscum* Douvillé, 1917; OD]. Small, involute, compressed, high whorled, with very narrow sulcate bicarinate venter; flanks with weak falcoid riblets and growth striae, and crescents on outer flank; suture distinctive, with long narrow saddles, deeply bifid E/L, very broad bifid L much deeper than E, and ascending auxiliary lobe. *U. Barrem.* **Fig. 151** - *M. priscum* (Douvillé).

Fig. 150. *Miodesmoceras*.

Subfamily Torcapellinae Breskovski, 1977

(*nom. correct.* Thieuloy & Bulot, 1992 *ex* Torcapelinae)

Discussion. Monophyletic derivatives of Barremitinae (Wright, 1957), the inner whorls of *T. serbiensis* (**Fig. 153A-B**) close to *Bistrilitzeras*. Typically oxycone with narrowly rounded to acute or grooved venter, low flexuous ribs to the inner whorls, smooth outer whorls and wide deep penetrating L to the suture. *U. Haut. - Barrem.*

Thieuloyceras BRESKOVSKI, 1980 [*T. densecostatus*; OD]. Small, involute, compressed, with vertical umbilical wall, gently convex flanks, narrow venter and sublanceolate whorl section; strong prorsiradiate primary rib in circumumbilical depression split into 2-3 fine sigmoidal ribs a few of which bifurcate higher on flank, forming chevrons across venter. *U. Haut.* **Fig. 153E-F** - *T. densecostatum* Breskovski.

Subtorcapella VERMEULEN, 2002 [*S. defayae*; OD].Small; compressed, involute with steep umbilical wall, broad

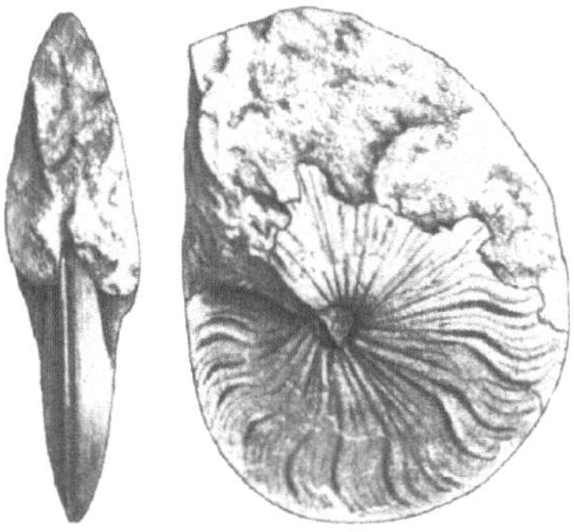

Fig. 151. *Mogharaeceras*.

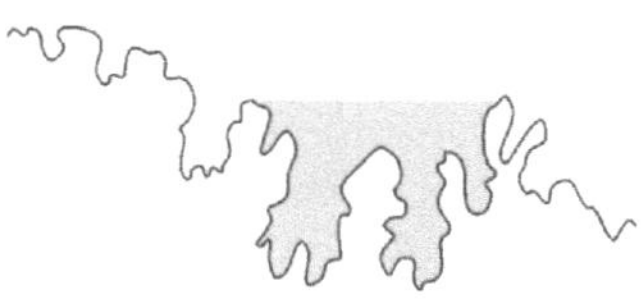

Fig. 152. *Mogharaeceras* suture.

flat subparallel flanks, evenly rounded venter and subelliptical whorl section; early whorls unornamented; in later growth middle and outer flanks with weak simple distant ribs that project forward and are linguoid over venter. *L. Barrem.* **Fig. 153C-D** - *S. defayae* Vermeulen.

Torcapella BUSNARDO, 1970 [*Ammonites fabrei* Torcapel, 1884; OD]. Large to very large; like *Subtorcapella* but with maximum width at midflank and strong fine dense falcoid ribs, single or branching from umbilicus, persisting to maturity. *L. - U. Barrem.* **Fig. 153A-B** - *T. serbiensis* Vašícek, Rabrenovic, Radulovic, Radulovic & Mojsic; **Fig. 153G-H** - *T. fabrei* (Torcapel).

Subfamily Pseudosaynellinae Casey, 1961

Discussion. A monotypic taxon with an origin in Barremitinae, close to *Barremitites* (Dimitrova, 1967) but superficially at least more like Torcapellinae; oxycones with acute venter, lanceolate whorl section, coarse low falcate ribs in early to middle growth, weakening later, and suture with a curved suspensive lobe of at least 6 elements. *L. - U. Apt.*

Pseudosaynella SPATH, 1923 [*Ammonites bicurvatus* Michelin, 1838; OD]. Based on early growth stage, but size dimorphic and macroconchs large; moderately inflated oxycone with acute venter and lanceolate whorl section; inner whorls with strong coarse flexuous ribs strongest on outer flank, some of which bifurcate, where they form crescents; suture with shallow E, bifid E/L, asymmetrically trifid L much deeper than E, and descendiing series of 6 auxilliary saddles; a homoeomorph of *Saynella*. *L. - U. Apt.* **Fig. 155** - *P. bicurvata* (Michelin).

Family DESMOCERATIDAE Zittel, 1895

Discussion. A monophyletic group with an origin in Barremitidae (Wright, 1957)); typically rather involute and inflated, with periodic constrictions, weak to strong ribs and, in some, a ventral keel; suture with auxiliaries to suspensive lobe in a descending straight line. Comprises the subfamilies Desmoceratinae and Desmophyllitinae. *U. Apt. - U. Maastr.*

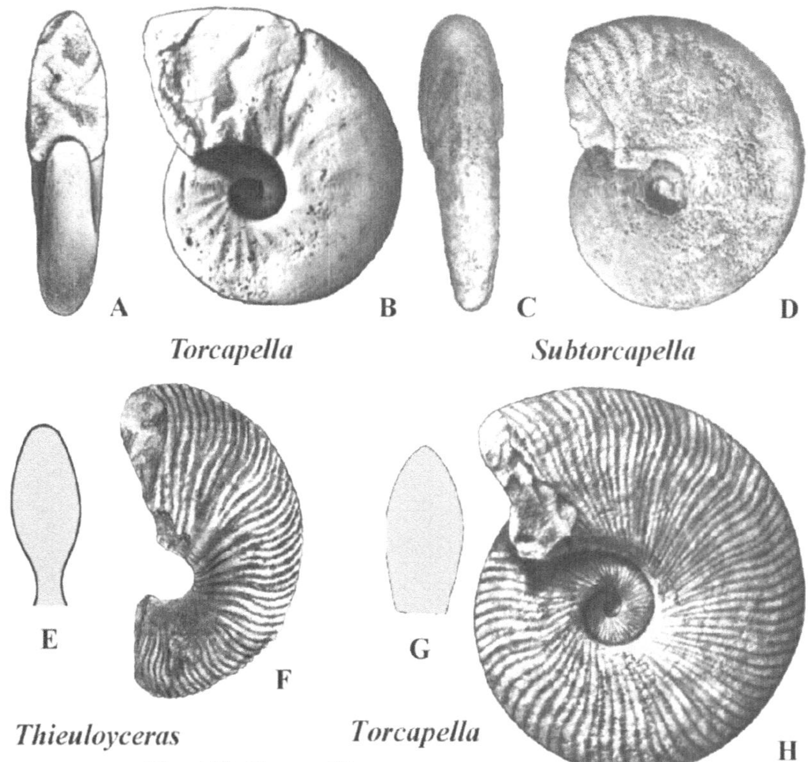

Fig. 153. Torcapellinae.

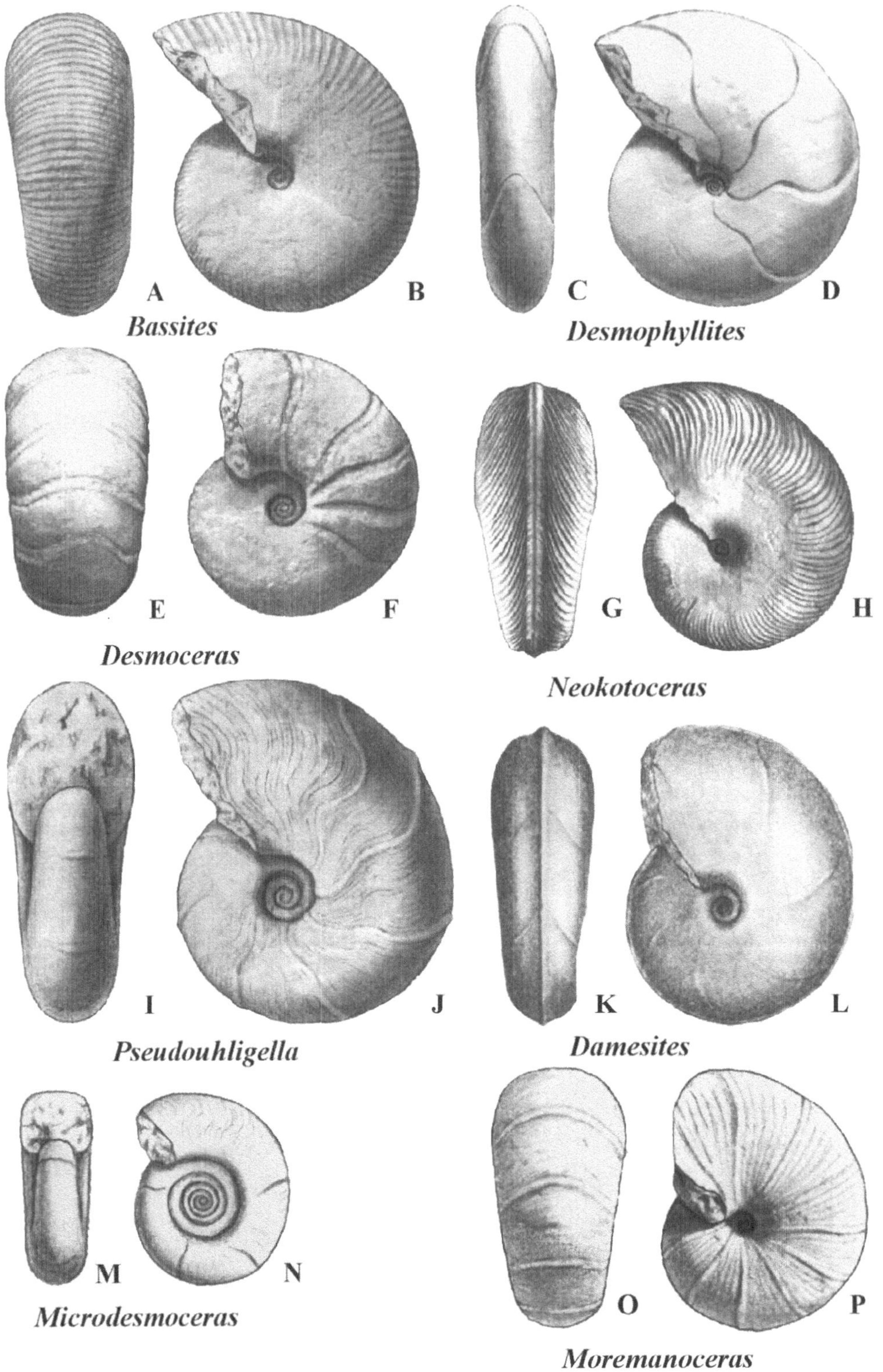

Fig. 154. Desmoceratidae.

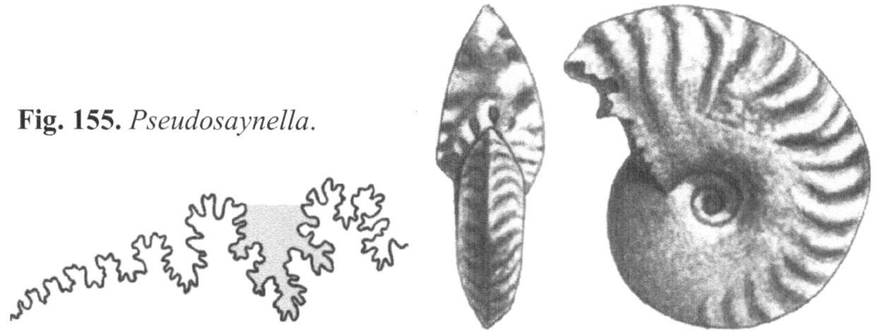

Fig. 155. *Pseudosaynella.*

Subfamily Desmoceratinae Zittel, 1895

Discussion. A monophyletic taxon with an origin in Barremitinae (Wright, 1957), close to *Davidiceras*; characteristically inflated and moderately involute with periodic flexuous constrictions, weak to obsolete ribs, and suture with narrow stemmed asymmetrically bifid saddles and auxilliaries in regularly descending series. *U. Apt. - Coni.*

Valdedorsella BREISTROFFER, 1947 [*Desmoceras akuschaense* Anthula, 1899; OD]. Small to medium sized; moderately inflated, with strong flexuous constrictions backed by primary ribs that cross venter with forward convexity, separated by single and bifurcating secondaries; carries strong ornament of inner whorls to middle and later growth; suture with broad narrow stemmed E/L, rather narrow asymmetrically bifid L deeper than E and a regularly descending series of auxiliary lobes. *U. Apt.* **Fig. 156A-B** - *V. getulina* (Coquand); **Fig. 156C-D** - *V. falcistriata* (Anthula).

Desmoceras ZITTEL, 1885 [*Ammonites latidorsatus* Michelin, 1838; SD Böhm, 1895]. Small; inflated, involute, with rounded to somewhat flattened flanks, broadly rounded to subtabulate venter, and depressed to slightly compressed reniform to subquadratic and elliptical whorl section; periodic flexuous constrictions bend forward near venter crossing with forward convexity; ribs lacking, only flexuous growth striae; suture with deep narrow E, broad E/L and narrow trifid L as deep as E. *U. Apt. - M. Cenom.* **Fig. 154E-F** - *D. latidorsatum* (Michelin).
= *Latidorsella* Jacob, 1907 (obj.).

Phyllodesmoceras SPATH, 1925 [*Ammonites valdedorsatus* Reynés, 1876; OD]. Very small; like *Desmoceras* but with circuar whorl section and highly complex suture with profound E and phylloid terminations to folioles. *Alb.* **Fig. 157E-F** - *P. valdedorsatum* (Reynés).

Lunatodorsella BREISTROFFER, 1947 [*Puzosia chirichensis* Pervinquière, 1907; OD]. Micromorph; like *Desmoceras* but smaller, strongly inflated cadicone with subvertical umbilical wall, very depressed subcoronate whorl section, 5 straight prorsiradiate constrictions per whorl and very simple suture with large E, weakly bifid saddles and trifid lobes; superficially similar to inner whorls of *Parajaubertella* (Gaudryceratidae). *Upper U. Alb.* **Fig. 157I-J** - *L. chirichensis* (Pervinquière).

Microdesmoceras MATSUMOTO & MURAMOTO, 1972 [*M. tetragonum*; OD]. Very small; evolute, outer whorls with flat flanks, broadly rounded venter, tetragonal whorl section; weak constrictions on internal mould, sigmoidal growth striae on test; adult body chamber with scaphitoid uncoiling; suture like *Desmoceras* but simpler, with very broad L; superficially similar to *Tetragonites*. *L. Cenom.* **Fig. 154M-N** - *M. tetragonum* Matsumoto & Muramoto.

Moremanoceras COBBAN, 1971 [*Tragodesmoceras scotti* Moreman, 1942; OD]. Small, like *Desmoceras* but more

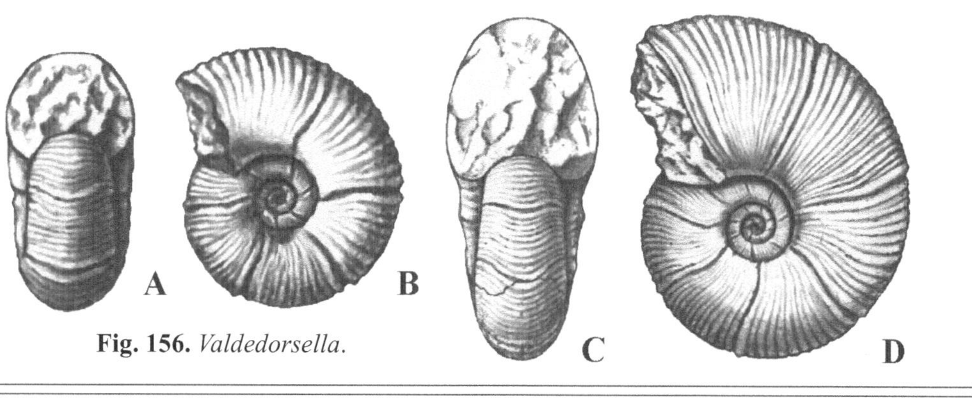

Fig. 156. *Valdedorsella.*

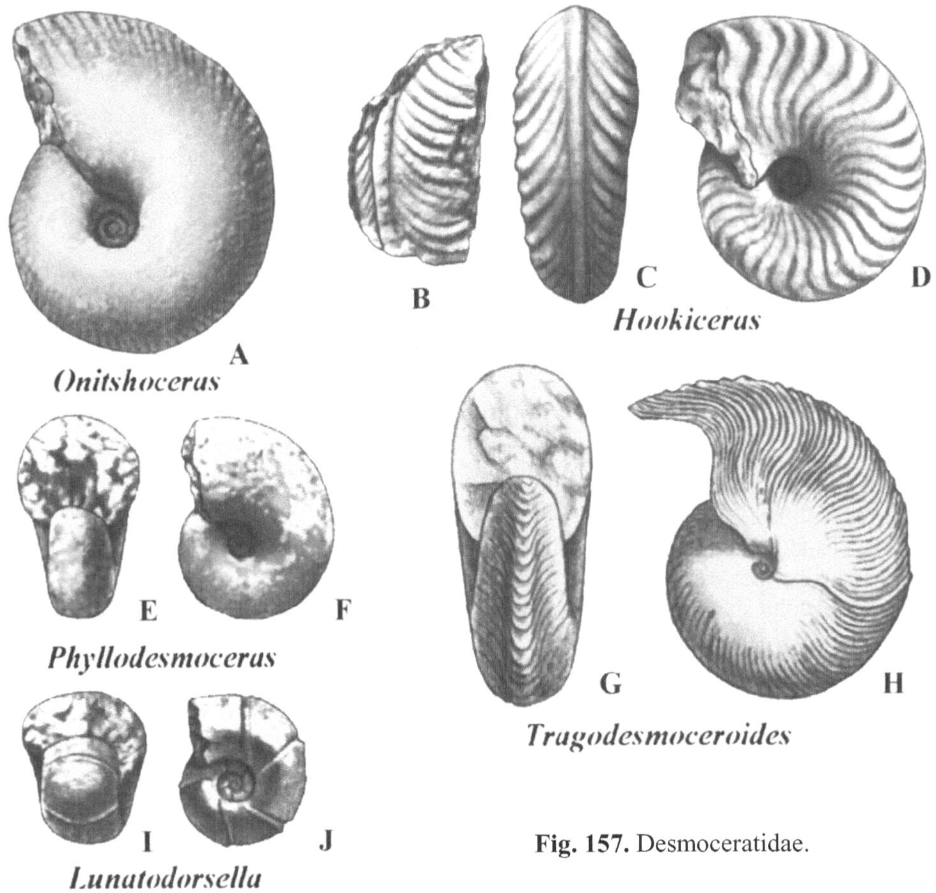

Fig. 157. Desmoceratidae.

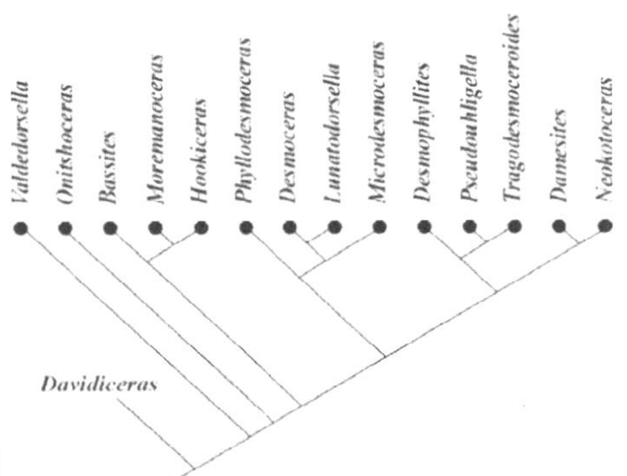

Fig. 158. Hypothesized relationships among Desmoceratidae

involute, with fewer straight prorsiradiate collared constrictions that flex forward near venter; ribs parallel constrictions and strongest on outer flanks; adult body chamber with numerous strong distant bluntly rounded rectiradiate ribs that strengthen across venter; suture simple, with narrow E/L, fairly wide L longer than E, fairly wide L/U_2, and 3-5 short auxiliary lobes. *Upper U. Cenom.* **Fig. 154O-P** - *M. scotti* (Moreman).

Bassites COBBAN, 1987 [*B. reesidei*; OD]. Medium sized; like *Moremanoceras* but with weak dense flat ribs and simple suture with bifid L; constrictions lacking. *L. Turon.* **Fig. 154A-B** - *B. reesidei* Cobban.

Onitshoceras REYMENT, 1954 [*O. matsumotoi*; OD]. Small; inner whorls like *Bassites*; later with subquadratic whorl section and thin irregular flexuous prorsiradiate ribs arising from umbilical shoulder and strengthening over venter; suture like *Desmoceras* but with very narrow deep penetrating L and rectangular L/U_2. *Coni.* **Fig. 157A** - *O. matsumotoi* Reyment.

Subfamily Desmophyllitinae nov.

Diagnosis. Typically involute and compressed, with high whorls, flat subparallel flanks, rounded venter, sigmoidal to biconcave constrictions that project strongly fowards on venter and, in some, a blunt siphonal keel. A monophyletic group with an origin in Desmoceratinae, close to *Desmoceras* (Collig-

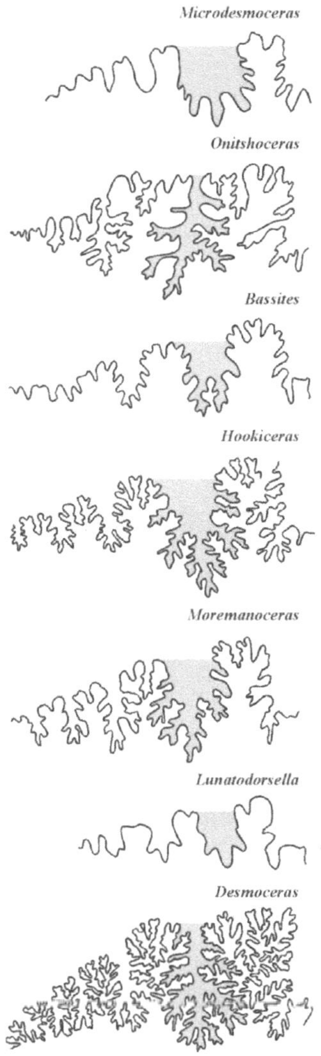

Fig. 159. Desmoceratine sutures.

non, 1961), with which they evolved in parallel. *M. Alb. - U. Maastr.*

Pseudouhligella MATSUMOTO, 1938 [*Desmoceras dawsoni* var. *japonica* Yabe, 1904; OD]. Small; like *Desmoceras* but compressed and high whorled, with flat subparallel flanks, compressed oval to elliptical whorl section, biconcave constrictions that project strongly forward near venter and cross it with low collar posteriorly; bundled branching growth striae follow course of constrictions, and there may be weak riblets across venter in later growth; suture like *Desmoceras* but with L not as deep as E. *M. Alb. - L. Turon.* **Fig. 154I-J** - *P. japonica* (Yabe).

Hookiceras **gen. nov.** [*Moremanoceras costatum* Cobban, Hook & Kennedy, 1989, p.19; OD herein]. Small; like *Pseudouhligella* but with blunt low weakly crenulated siphonal keel; typically with numerous falcate single ribs that are weak on inner flank and strongly projected over venter. *M. - Lower U. Cenom.* **Fig. 157B-D** - *H. costatum* (Cobban, Hook & Kennedy).

Tragodesmoceroides MATSUMOTO, 1942 [*T. subcostatus*; OD]. Like *Pseudouhligella* but smaller, with very narrow pit-like umbilicus, gently convex flanks, broadly rounded venter, and depressed (juv.) to compressed (ad.) whorl section; fine sigmoidal ribs appear in middle growth parallelled by infrequent constrictions; in later growth ribs effaced from inner flank but strong and sharp on outer flank and linguoid over venter; adult aperture with ventral rostrum and lateral lappets; suture with long narrow elements. *Turon. - L. Sant.* **Fig. 157G-H** - *T. subcostatus* Matsumoto.

Desmophyllites SPATH, 1929 [*Desmoceras larteti* Seunes, 1891; SD Spath, 1921]. Small to medium sized; like *Pseudouhligella* but with very narrow umbilicus; suture with broad deep E, broad E/L, broad trifid L as deep as E, narrow stemmed bifid L/U_2 and some 15 auxiliary elements. *Sant. - U. Maastr.* **Fig. 154C-D** - *D. diphylloides* (Forbes).

= *Schlueteria* de Grossouvre, 1894 (*non* Fritsch, 1887).

= *Schluetericeras* Collignon, 1938 (*non* Hyatt, 1903).

Damesites MATSUMOTO, 1942 [*Desmoceras damesi* Jimbo, 1894; ICZN Opinion 555]. Small; like *Pseudouhligella* but more involute, with low blunt siphonal keel that strengthens with growth and bordered by shallow sulci; flexuous growth striae, weak on inner flank, project strongly forward on venter parallelled by weak constrictions that may be lost in later growth; longitudinal striae present in well preserved material; suture like *Desmophyllites* with narrow bifid E/L and at least 5 auxilliary elements. *Cenom. - L. Camp.* **Fig. 154K-L** - *D. damesi* (Jimbo).

= *Damesites* Matsumoto, 1938 (*nom. nud.*).

= *Kotoceras* Yabe, 1927 (ICZN rejected name, No. 1264).

Neokotoceras ANDERSON, 1958 [*N. fresnoense*, = *Damesites hetonaiensis* Matsumoto, 1942; OD]. Moderately large; like *Damesites* but more compressed and involute (U=6-8%), with very narrow crater-like umbilicus, sharp fine siphonal keel without bounding sulci, "smooth" or with very fine riblets remarkably projected on venter; constrictions straight across flank, projecting strongly forward near venter, declining with growth and sometimes obsolete in maturity; highly incised suture with narrow L and retracted suspensive lobe. *Camp. - L. Maastr.* **Fig. 154G-H** - *N. hetonaiense* (Matsumoto).

Family MUNIERICERATIDAE Wright, 1952

Discussion. Monophyletic derivatives of Desmoceratidae (Wright, 1952); characteristically involute,

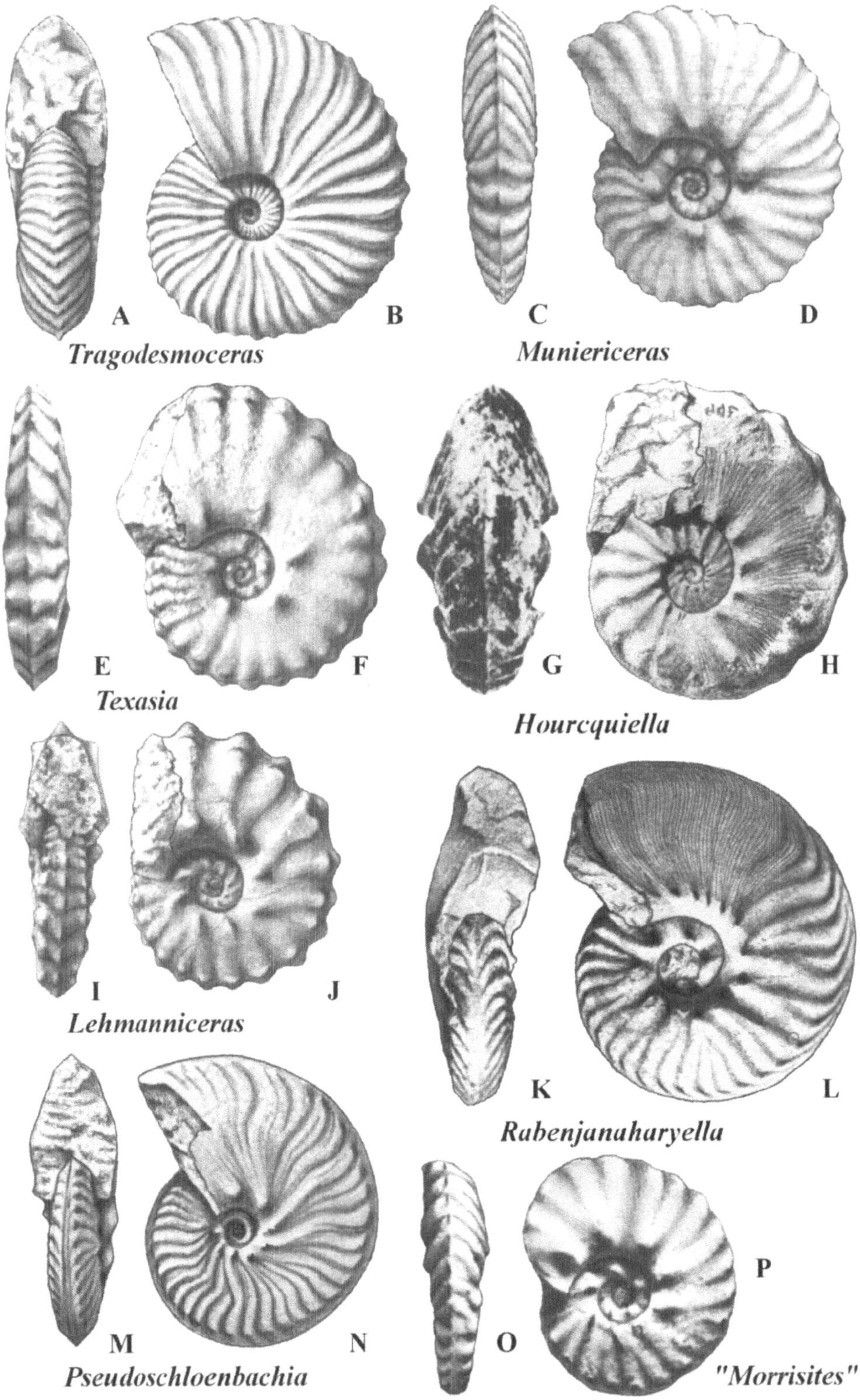

Fig. 160. Munierceratinae.

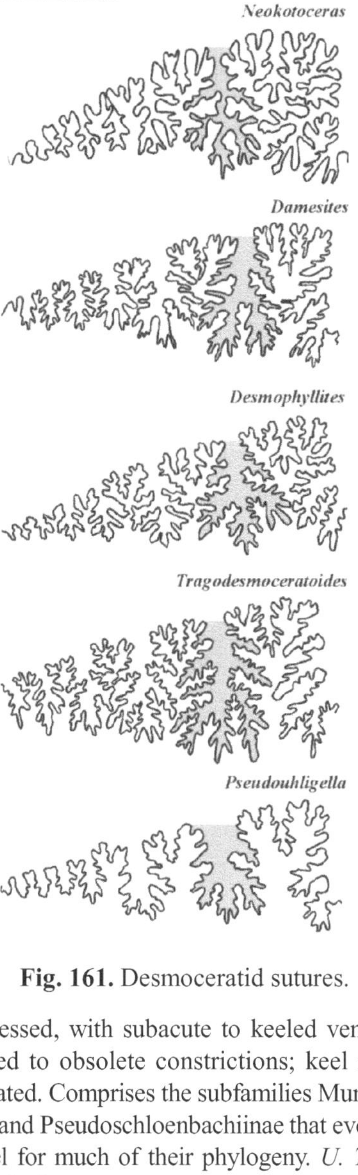

Fig. 161. Desmoceratid sutures.

compressed, with subacute to keeled venter and subdued to obsolete constrictions; keel may be crenulated. Comprises the subfamilies Muniericeratinae and Pseudoschloenbachiinae that evolved in parallel for much of their phylogeny. *U. Turon. - Upper L. Camp.*

Subfamily Muniericeratinae Wright, 1952

(*nom. transl.* herein *ex* Muniericeratidae)

Discussion. Monophyletic derivatives of Desmophyllitinae, close to *Hookiceras*; characteristically compressed with a subacute to keeled venter and subdued to obsolete constrictions. *U. Turon. - L. Camp.*

Tragodesmoceras SPATH, 1922 [*Desmoceras clypealoides* Leonhard, 1897; OD]. Medium sized; involute, compressed, high whorled, with weakly convex flanks, very narrowly arched to subacute venter and sublanceolate whorl section; umbilical wall subvertical on inner whorls, inclined in maturity; strong sinuous ribs differentiated into primaries and bifurcating secondaries, arise at umbilical edge and project forward on outer flank, forming chevrons over venter; prominent flexuous constrictions in maturity when venter rounds, with siphonal nodes; suture very incised, with broad short E, asymmetrically bifid E/L and broadly trifid L. *U. Turon. - Coni.* **Fig. 160A-B** - *T. carlislense* Cobban; **Fig. 164** - *T. clypealoides* (Leonhard).

Praemuniericeras COLLIGNON, 1966 [*P. proprium*; OD]. Small, involute, with steep umbilical wall, compressed and high whorled, with broad convergent flanks, fastigiate venter and subtrigonal whorl section; slightly prosocline umbilical bullae giving rise to bold flexuous prorsiradiate ribs with occasional intercalatories, and ventrolateral bullae projecting forwards onto venter; siphonal keel with clavi more or less corresponding to ribs; constrictions lacking. *Upper U. Coni.* **Fig. 165** - *P. proprium* Collignon.

In *P. lapparentiforme* (Collignon) the flank costae arise in triplicates in later growth.

"Morrisites" VAN HOEPEN, 1968 [*M. louwi*; OD]. Small, poorly known; like *Praemuniericeras* but with periodic deep constrictions, coarser more distant long and short ribs that are linguoid over fastigiate venter and with small siphonal clavi. *M. Coni.* **Fig. 160O-P** - *M. louwi* van Hoepen.

non *Morrisites* Buckman (1921, Ammonoidea, *nec* Gibson 1955, Arthropoda).

Tuberodiscoides COLLIGNON, 1966 [*T. termierorum*; OD]. Small, with flat convergent flanks, trapezoidal whorl section, fastigiate venter, coarse sparse ribs with few intercalatories, very prominent umbilical bullae separated by deep concave interspaces, and prominent ventrolateral clavi and siphonal nodes. *M. Sant.* **Fig. 166** - *T. carinatus* (Collignon); **Fig. 169A-B** - *T. termierorum* Collignon.

Muniericeras DE GROSSOUVRE, 1894 [*M. lapparenti*; OD]. Small; like *Praemuniericeras* but with rounded umbilical

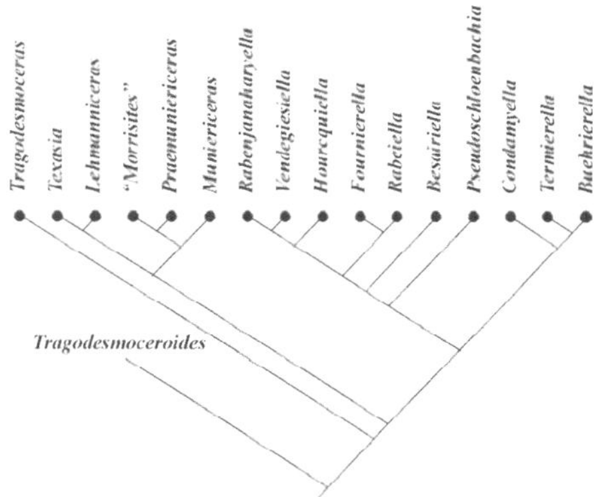

Fig. 162. Hypothesized relationships amng Muniericeratidae.

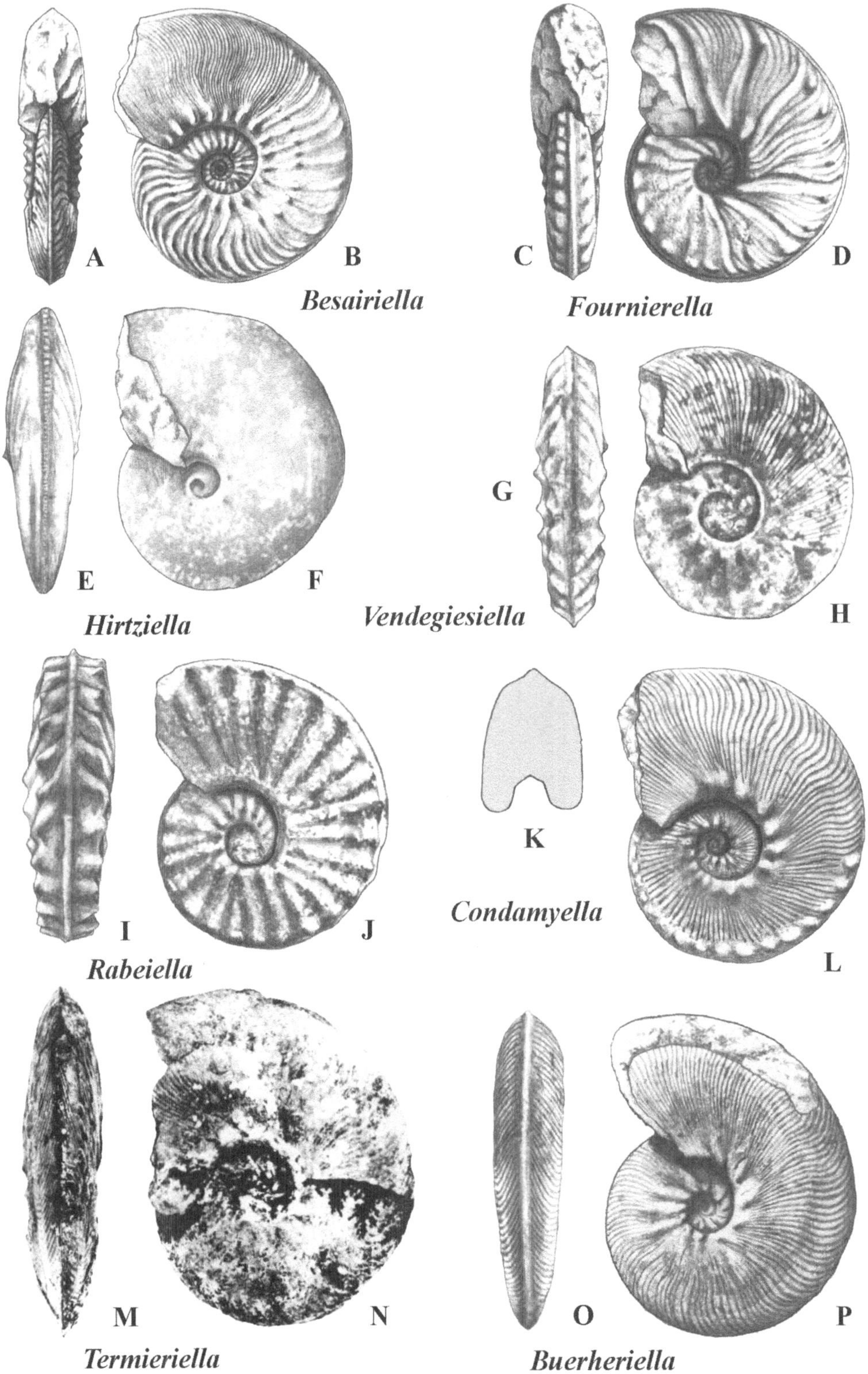

Fig. 163. Pseudoschloenbachiinae.

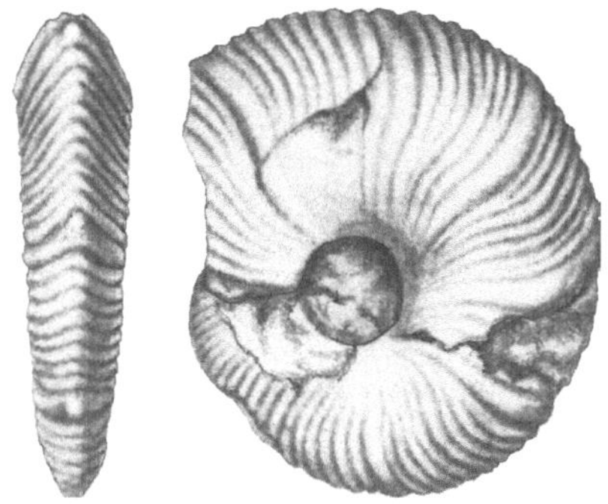

Fig. 164. *Tragodesmoceras.*

wall, convergent outer flanks, sharp fastigiate venter coarsely crenulated by ribbing, sparse (5-7) generally strong umbilical bullae that give rise to 2-3 flexuous ribs separated by 1-2 intercalatories. *L. - M. Sant.* **Fig. 160C-D** - *M. lapparenti* (de Grossouvre).

Lehmaniceras COLLIGNON, 1966 [*L. sornayi*; OD]. Small; like *Praemuniericeras* but more coarsely ornamented, with conspicuous ventrolateral and siphonal clavi that correspond in number. *L.- M. Sant.* **Fig. 160I-J** - *L. cricki* (Spath).

Texasia REESIDE, 1932 [*Ammonites dentatocarinatus* Roemer, 1852; SD Wright, 1957]. Medium sized; like *Lehmaniceras* but more compressed and high whorled, with strongly fastigiate venter, fewer (5) umbilical bullae, straight rigid ribs and frequent intercalatories; suture with deep narrow E, broad narrow stemmed trifid E/L, narrow trifid L deeper than E, and moderately broad narrow stemmed bifid L/U_2. *L. Camp.* **Fig. 160E-F** - *T. dentatocarinata* (Roemer).

Subfamily Pseudoschloenbachiinae van Hoepen, 1968

= Pseudoschloenbachiinae Collignon, 1969

Discussion. Monophyletic oxycone derivatives of Muniericeratinae (Wright, 1996), close to *Praemuniericeras*; typically strongly compressed and involute, with fastigiate venter, prominent siphonal keel that may be crenulated; strong umbilical bullae give rise to bundles of flat sinuous ribs, occasionally with lateral tubercles, that may be replaced by fasciculate bundles of falciform riblets on adult body chamber. *U. Coni. - Upper L. Camp.*

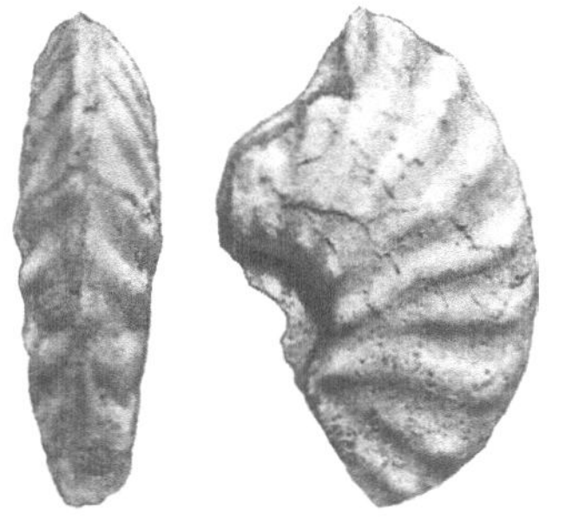

Fig. 165. *Praemunericeras.*

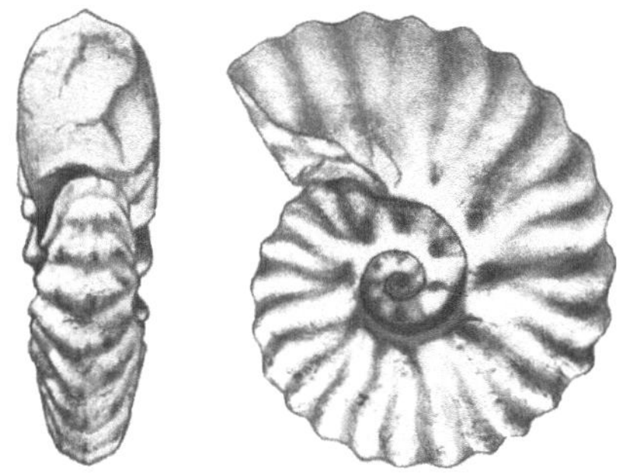

Fig. 166. *Tuberodiscoides.*

Pseudoschloenbachia SPATH, 1921 [*Ammonites umbulazi* Baily, 1855; OD]. Small to medium sized; strongly compressed oxycone with raised fastigiate venter and sharp siphonal keel; prominent pointed umbilical tubercles give rise to pairs of low prorsiradiate falcate ribs that bifurcate and broaden ventrally; in later growth long ribs tend to be single separated by up to 3 intercalatories of different length; suture with narrow stemmed bifid E/L and moderately wide trifid *L. U. Coni. - Lower L. Camp.* **Fig. 160M-N** - *P. umbulazi* (Baily).

Fournierella COLLIGNON, 1966 [*Schloenbachia fournieri* de Grossouvre, 1894; OD]. Small to medium sized; like *Pseudoschloebachia* but with broader venter, subelliptical whorl section, prosocline umbilical bullae, straighter ribs that weaken across flanks, fewer intercalated ribs, strong ventrolateral clavi, numerous broad shallow sinuous constrictions and sinuous bundled costellae to adult body chamber. *M. Sant. - L. Camp.* **Fig. 163C-D** - *F. fournieri* (de Grossouvre).

Vendegiesiella COLLIGNON, 1969 [*Pseudoschloenbachia (V.) spinosa*; OD]. Medium sized; like *Fournierella* but with strong distant ribs in early to middle growth and irregular mid-lateral tubercles; constrictions lacking. *U. Sant. - L. Camp.* **Fig. 163G-H** - *V. spinosa* (Collignon).

Besairiella COLLIGNON, 1969 [*Pseudoschloenbachia (B.) besairiei*; OD]. Moderately large; like *Vendegiesiella* but with prominent umbilical tubercles connected to low-lateral tubercles by strong primary ribs from which slightly prorsiradiate secondaries bifurcate or trifurcate and terminate in small oblique ventrolateral clavi that project forward on venter; rib bundles separated by 1-2 intercalatories; adult body chamber with umbilical bullae giving rise to fine sigmoidal costellae. *L. Camp.* **Fig. 163A-B** - *B. blanfordiana* (Stoliczka).

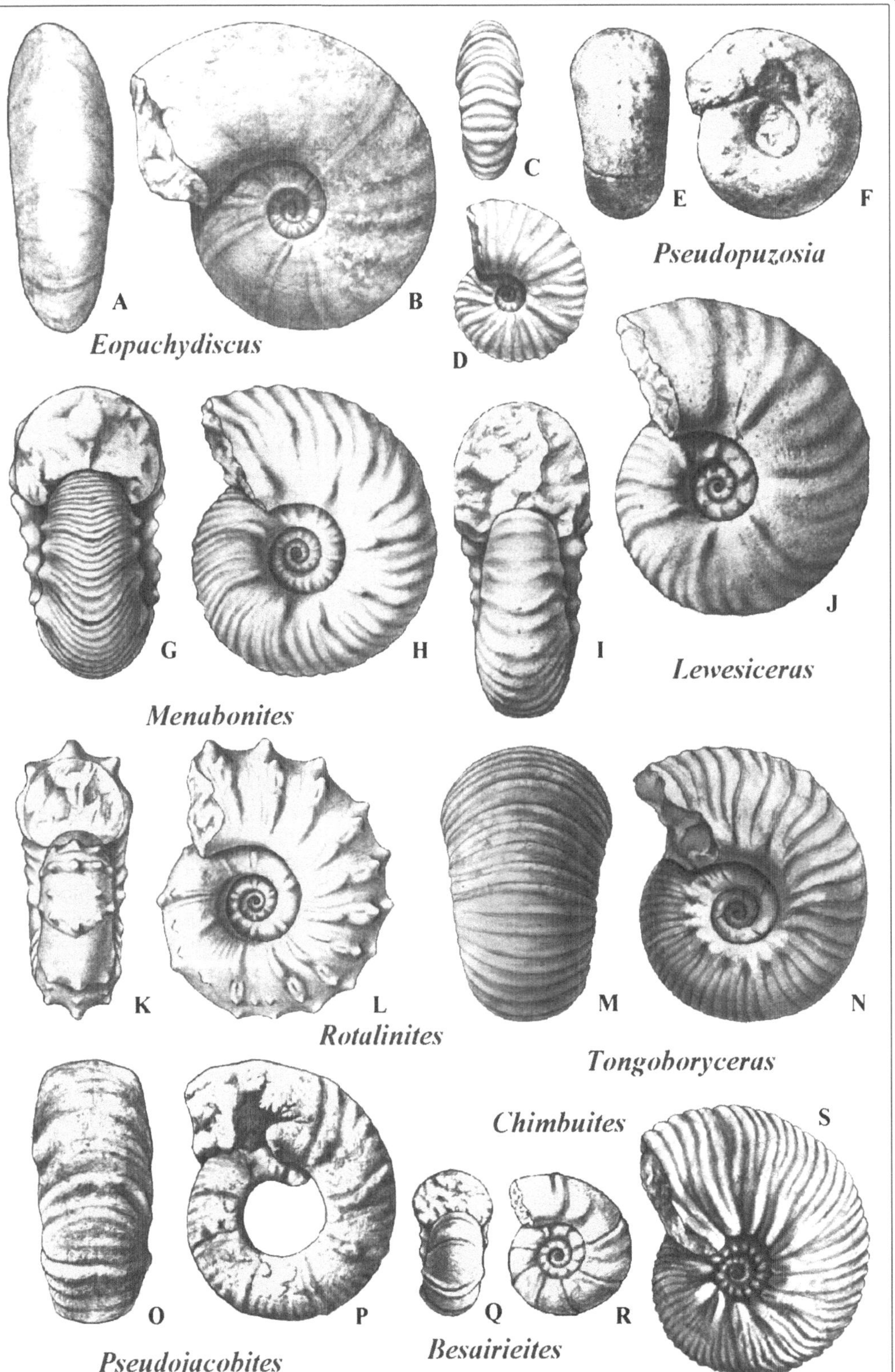

Fig. 167. Lewesiceratinae.

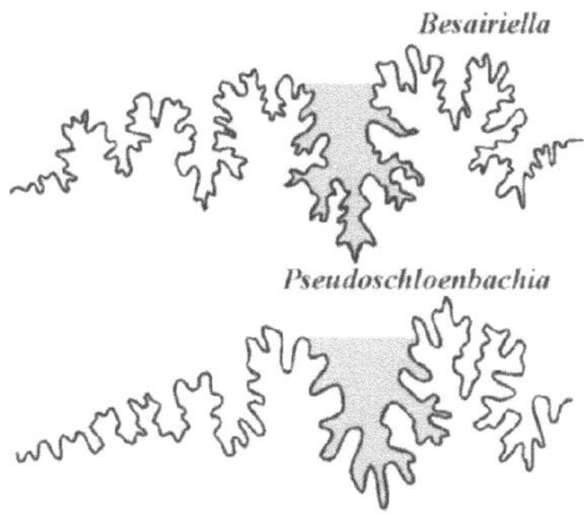

Fig. 168. Muniericeratid sutures.

Rabeiella COLLIGNON, 1969 [*Pseudoschloebachia (R.) orthogonia*; OD]. Medium sized, evolute, flat sided with subparallel flanks, broad initially fastigiate venter becoming flat later, and subrectangular to subtrapezoidal whorl section; strong sparse umbilical bullae give rise to pairs of coarse rigid slightly prorsiradiate ribs projecting forward on venter; on adult body chamber umbilical tubercles weaken, ventrolaterals evanesce, ribs become fine and sigmoidal; siphonal keel finely serrated; suture with long slender trifid L. *L. Camp.* **Fig. 163I-J** - *R. orthogonia* (Collignon).

= *Rabiella* Wright, 1996 (*nom.van.*).

Hourcquiella COLLIGNON, 1969 [*Pseudoschloenbachia (H.) bererensis* (Hourcq MS); OD]. Moderately large, with ogival whorl section; prominent initially prosocline umbilical bullae become rounded later; flanks with mostly single flat low broad prorsiradiate ribs with ventrolateral clavi, projecting strongly forward on venter and weakening with growth; in maturity coarse ribbing abruptly replaced by bundles of fine sigmoidal costellae; siphonal clavi prominent, not corresponding exactly with ribs; suture with massive L. *L. Camp.* **Fig. 160G-H** - *H. bererensis* (Collignon).

= *Hourcqiella* Wright, 1996 (*nom. van.*).

Rabenjanaharyella COLLIGNON, 1969 [*Pseudoschloenbachia (R.) rhomboidalis*; OD]. Moderately large; much like *Hourcquiella* but with strongly convergent flanks, fastigiate venter and subtrigonal whorl section; large to very large umbilical tubercles give rise to pairs of coarse rectiradiate ribs, separated by an intercalatory, that curve strongly forwards on outer flank; adult body chamber with fine sinuous costellae. *L. Camp.* **Fig. 160K-L** - *R. rhomboidalis* (Collignon).

Condamyella COLLIGNON, 1969 [*Pseudoschloenbachia (C.) condamyi*; OD]. Large; like *Pseudoschloenbachia* but more inflated, with broader venter, subparallel flanks, subquadratic whorl section; numerous prominent umbilical bullae give rise to broad folds superimposed with fasciculate bundles of fine falciform riblets that loop to strong oblique ventrolateral clavi that become obsolete on adult body chamber. *L. Camp.* **Fig. 163K-L** - *C. condamyi* (Collignon).

Buehreriella COLLIGNON, 1969 [*Pseudoschloenbachia (B.) buehreri*; OD]. Large; like *Condamyella* but more compressed, with higher whorls and narrowly rounded venter; ventrolateral tubercles lacking and flanks with strongly falciform riblets. *L. Camp.* **Fig. 163O-P** - *B. buehreri* (Collignon).

= *Buehrierella* Wright, 1996 (*nom. van.*).

Hirtziella COLLIGNON, 1969 [*Pseudoschloenbachia (H.) inornata*; OD]. Medium sized, compressed, discoidal, without ventrolateral shoulders and extremely small, pointed, prosocline umbilical bullae in early growth; unornamented save for occasional hint of flank costae but without relief; siphonal keel very finely denticulate, denticles becoming widely spaced on adult body chamber. *Upper L. Camp.* **Fig. 163E-F** - *H. inornata* (Collignon).

Termieriella COLLIGNON, 1969 [*Pseudoschloenbachia (T.) lenticularis*; OD]. Medium sized to moderately large, with large umbilical bullae that become triangular in middle growth and exaggerated on adult body chamber; flanks convex, without ventrolateral shoulders; adult body chamber with bundles of extremely fine falciform riblets; suture with finely divided bifid E/L with divergent branches and massive L. *Upper L. Camp.* **Fig. 163M-N** - *T. lenticularis* (Collignon).

Superfamily PACHYDISCOIDEA Spath, 1922

(*nom. transl.* herein *ex* Pachydiscidae)

Discussion. Monophyletic derivatives of Desmoceratoidea (Roman, 1938). Mostly involute to moderately evolute, compressed, with strong ribs arising from umbilical bullae, with siphonl tubercles in some; constrictions prominent in primitive representatives. Sexual dimorphism conspicuous, microconchs with ventrolateral tubercles. Comprises the families Menuitidae and Pachydiscidae. *U. Alb. - U. Maastr.*

Family MENUITIDAE nov.

Diagnosis. Evolute, moderately to strongly inflated, often constricted, with umbilical bullae and simple ribs. Sexual dimorphism conspicuous but not widely documented, micro-conchs with bituberculate venter in maturity. Monophyletic descendents of Desmoceratidae (Roman, 1938); comprises the subfamilies Lewesiceratinae and Menuitinae. *U. Alb. - U. Maastr.*

Subfamily Lewesiceratinae nov.

Diagnosis. Typically evolute, strongly inflated (at least in early growth), with prominent constrictions

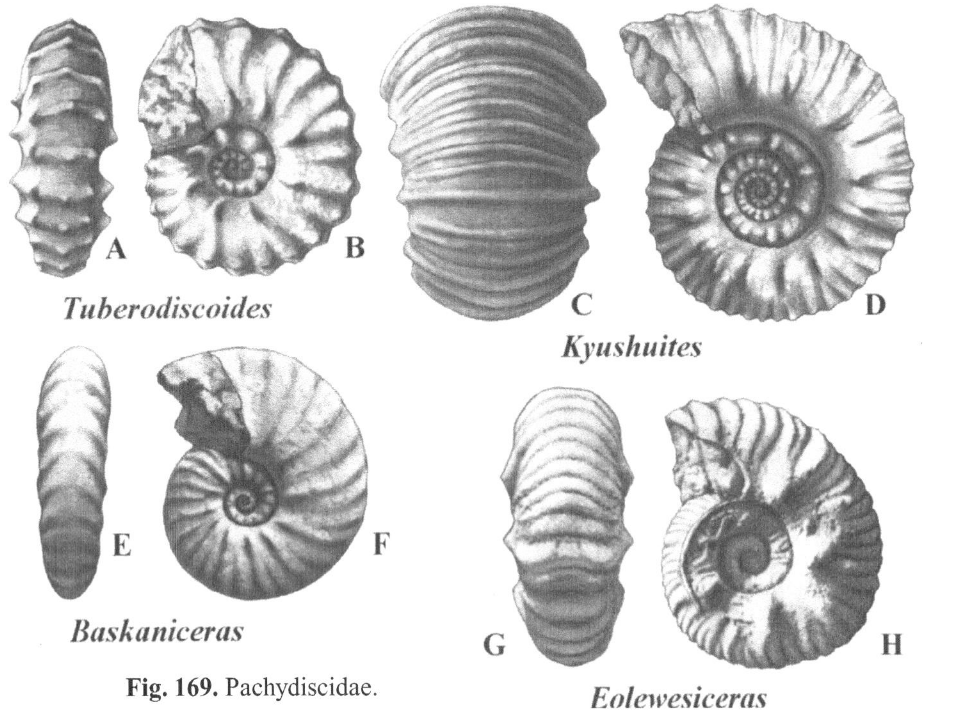

Fig. 169. Pachydiscidae.

backed by a collar and small umbilical tubercles giving rise to 2-3 ribs; ribs and constrictions linguoid over venter; complex suture with long narrow E/L and trifid L deeper than E. Monophyletic descendents of Pseudouhligellinae, close to *Pseudouhligella*. *U. Alb. - Coni.*

Eopachydiscus WRIGHT, 1955 [*Pachydiscus laevicaniculatus* Lasswitz, 1904, = *Ammonites marcianus* Shumard, 1854; OD]. Medium sized; compressed, rather involute, with steep convex umbilical wall, well rounded umbilical shoulder, more or less convex flanks, narrowly rounded venter and oval whorl section in all but earliest growth stages; 8-12 strong straight to slightly sinuous collar ribs (constrictions) separated by up to 3 coarse ribs; umbilical bullae sometimes present; ribs and constrictions become increasingly projected on shoulders with age; outer whorls smooth or with distant bar-like folds on inner flank; suture with broad bifid E/L and broad trifid L much deeper than E. *U. Alb.* **Fig. 167A-D** - *E. marcianus* (Shumard).

Baskaniceras WRIGHT & KENNEDY, 1984 [*B. deshayesitoides*; OD]. Very small; like juvenile *Eopachydiscus* but more compressed, with subelliptical whorl section, straighter ribs tending to alternate long and short and weakening over siphonal line; suture simple. *L. Cenom.* **Fig. 169E-F** - *B. deshayesitoides* Wright & Kennedy.

***Eolewesiceras* gen. nov.** [*Lewesiceras cenomanense* Wiedmann, 1962, p.667; OD herein]. Very small; moderately involute with flattened flanks, rounded venter and subpentagonal whorl section; prominent umbilical tubercles give rise to 1-2 strong sinuous ribs, the posterior one strongest, separated by 3-7 finer secondaries; ribs coarsen with growth and subtransverse over venter; suture with asymmetrically trifid L as deep as E. *L. - M. Cenom.* **Fig. 169G-H** - *E. cenomanense* (Wiedmann).

Lewesiceras SPATH, 1939 [*Ammonites peramplus* Mantell, 1822; OD]. Moderately large; inflated, evolute, inner whorls like *Eolewesiceras* but with anterior rib of bundles strongest and fronted by prominent constriction; 2-4 distant secondaries on outer flank; ribs project forward on outer flank, sometimes with hint of tubercle at point of flexure, and linguoid over venter; with growth ribs become more uniform, then weaken and evanesce over venter; adult body chamber smooth save for weak lateral bulges; suture like *Eopachydiscus* but with longer saddles, smaller narrower E/L and larger better developed U_2. *Turon.* **Fig. 167I-J** - *L. mantelli* Spath.

Pseudojacobites SPATH, 1922 [*Pachydiscus farmeri* Crick, 1910; OD]. Medium sized; like *Lewesiceras* but more evolute (U= 41%), inner and middle whorls with depressed reniform whorl section and hint of fine rectiradiate ribs; outer whorls with slightly flattened flanks and subquadratic adult whorl section, with 8 well marked rectiradiate constrictions (backed by collar) on internal mould, crossing venter almost transversely; in middle and later growth broad rounded ventrolateral tubercles appear irregularly, and collars to constrictions develop

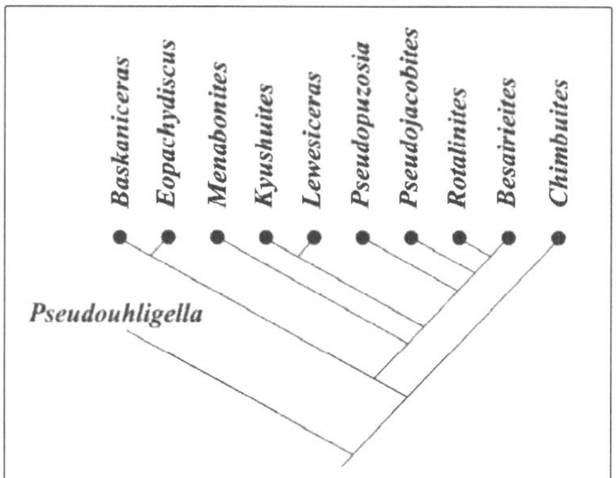

Fig. 170. Hypothesized relationships among Lewesiceratinae.

low rounded ventral tubercles on either side of siphonal line; siphonal tubercles present, but absent from holotype; suture like *Lewesiceras* but with much longer narrower elements, deeper E, and narrower asymmetrically bifid L. *U. Turon.* **Fig. 167O-P** - *P. farmeri* (Crick), juv.

= *Psuedojacobites* Matsumoto, 1990 (*laps. cal.*).

Pseudopuzosia SPATH, 1926 [*Desmoceras marlowense* Noble, 1911; OD]. Small, based on a poorly preserved internal mould of a juvenile; like *Pseudojacobites* but with narrower umbilicus and finer denser ribbing on inner whorls. *U. Turon.* **Fig. 167E-F** - *P. marlowensis* (Noble).

Rotalinites SHIMIZU, 1935 [*Ammonites rotalinus* Stoliczka, 1864; OD]. Like *Pseudojacobites* but more involute (U=36%), with rounder flanks in later growth, prominent umbilicolateral bullae connected by looped ribs to small ventrolateral tubercles and well developed siphonal tubercles; secondary ribs and constrictions lacking on outer whorls; suture like *Pseudojacobites* but with trifid L. *U. Turon.- Cont.* **Fig. 167K-L** - *R. rotalinus* (Stoliczka).

Menabonites HOUŠA, 1967 [*Pachydiscus anapadensis* Kossmat, 1895; OD]. Medium sized; like *Pseudojacobites* but more involute (U=31%), with rounder venter, ribs and constrictions strongly projected on outer flank and linguoid over venter, finer ribs to inner whorls, and no evidence of siphonal tubercles (although postulated); suture like *Rotalinites*. *Turon. - Coni.* **Fig. 167G-H** - *M. anapandensis* (Kossmat).

***Kyushuites* gen. nov.** [*Lewesiceras kawashitai* Matsumoto, 1979, p.32; OD herein]. Medium sized; like *Menabonites* but with coronate whorl section, strong regular umbilicolateral bullae that give rise to pairs of strong rectiradiate ribs, separated by intercalatories, that cross broad low venter transversely, and suture with longer narrower saddles. *Turon.* **Fig.169C-D** - *K. kawashitai* (Matsumoto).

Chimbuites CASEY & GLAESSNER, 1958 [*C. sinuosocostatus*; OD]. Small; an aberrant pachydiscid if, indeed, it is one; involute, high whorled, with flattened flanks and swollen crowded umbilical bullae giving rise to fasciculate bundles of coarse low rounded flexuous prorsiradiate

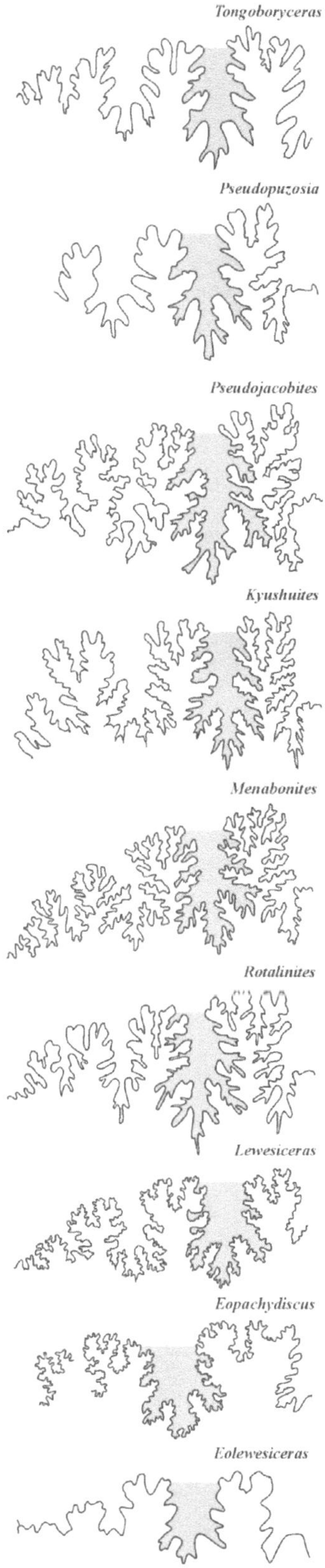

Fig. 171. Lewesiceratine sutures.

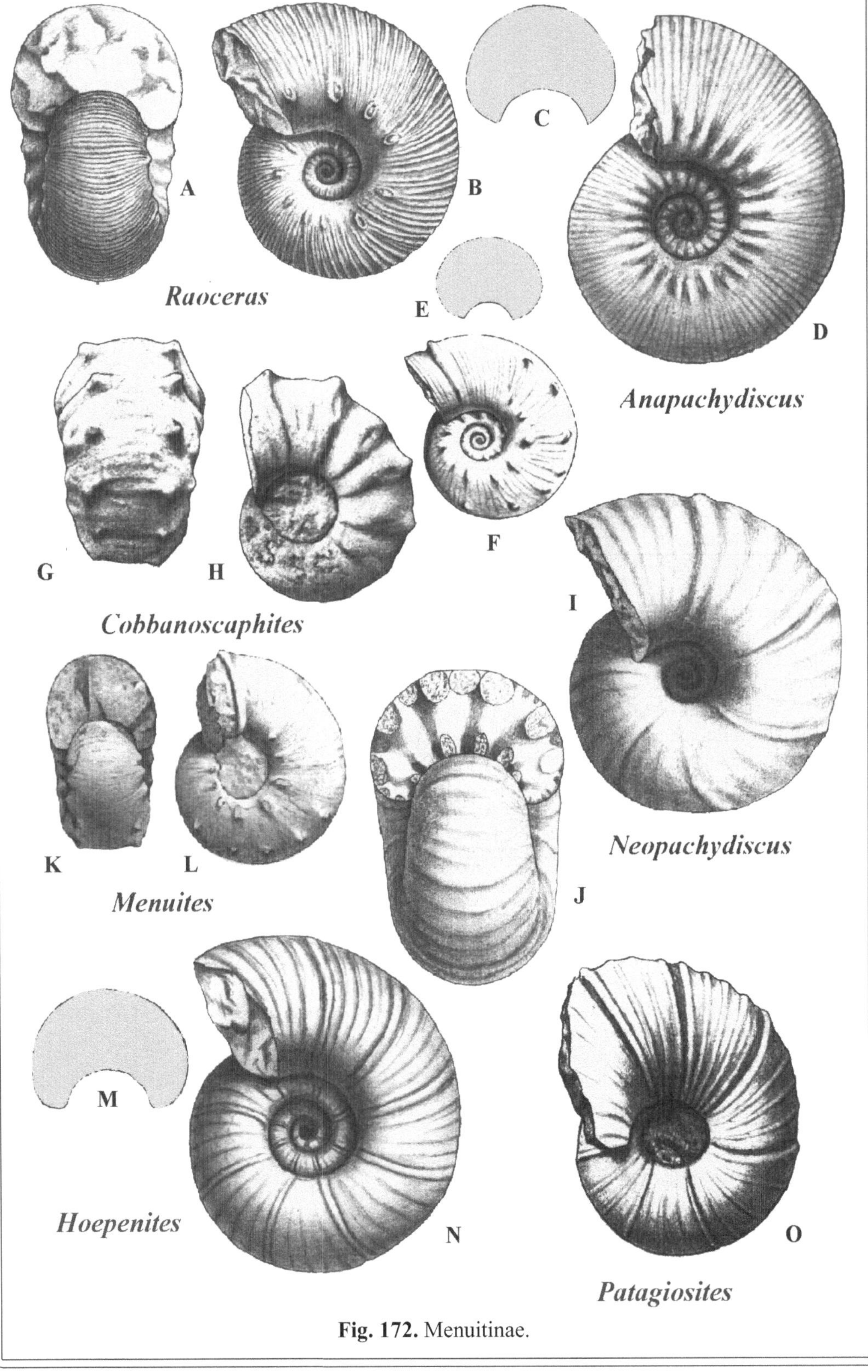

Fig. 172. Menuitinae.

ribs that project forward near venter and are separated by intercalatories. *U. Cenom.* **Fig. 167S** - *C. sinuoso-costatus* Casey & Glaessner.

Subfamily Menuitinae nov.

Diagnosis. Typically inflated, with depressed whorl section and relatively fine ribbing. It is now recognized that bituberculate *Menuites* is a microconch of some *Anapachydiscus*. Monophyletic descendents of Lewesiceratinae, close to *Lewesiceras*. *U. Turon. - U. Maastr.*

Tongoboryceras HOUŠA, 1967 [*Lewesiceras tongoboryense* Collignon, 1952; OD]. Small; evolute, inner whorls like *Lewesiceras*; in middle growth, strong rounded rectiradiate primary ribs arise from weak umbilical bullae and are convex forward across flank, projecting strongly on outer flank to cross venter with broad convexity; 1-2 intervening secondaries as strong as primaries arise at various levels on flank; suture like *Rotalinites*, with long relatively narrow elements. *U. Turon. - Coni.* **Fig. 167M-N** - *T. tongoboryense* (Collignon).

Raoceras AYYASAMI, 1993 [*Ammonites koluturensis* Stoliczka, 1865; OD]. Small; like *Tongoboryceras* but more inflated and involute (U=27%), with rounder flanks and deeper umbilicus; dense fine crowded somewhat irregular ribbing, with rectiradiate primaries arising in pairs from small distant umbilical bullae, separated by branching intercalatories of different length; ribs curve gently forward on outer flank to cross venter with weak convexity; fine ornament persists to maturity; suture with narrow deep E, long narrow asymmetrically bifid E/L, long narrow trifid L deeper than E, and long narrow asymmetrically bifid L/U_2. *M. Coni.* **Fig. 172A-B** - *R. koluturense* (Stoliczka).

Anapachydiscus YABE & SHIMIZU, 1926 [*Pachydiscus (Parapachydiscus) fascicostatus* Yabe, 1921; OD] [**M**]. Moderately large; like *Raoceras* but with more inclined umbilical wall and numerous septispinate umbilical bullae that give rise to bundles of 3-4 fine rectiradiate ribs separated by 2-4 intercalatories; ribs with slight forward curvature near venter; adult body chamber with strong distant flared ribs; microconch small, with finely ribbed inner whorls, distant umbilicolateral spines, prominent opposite ventrolateral tubercles in maturity, and collared adult aperture; suture like *Raoceras* but more incised. *Sant. - U. Camp.* **Fig. 172C-F** - *A. fascicostatus* (Yabe), **M** and **m**.

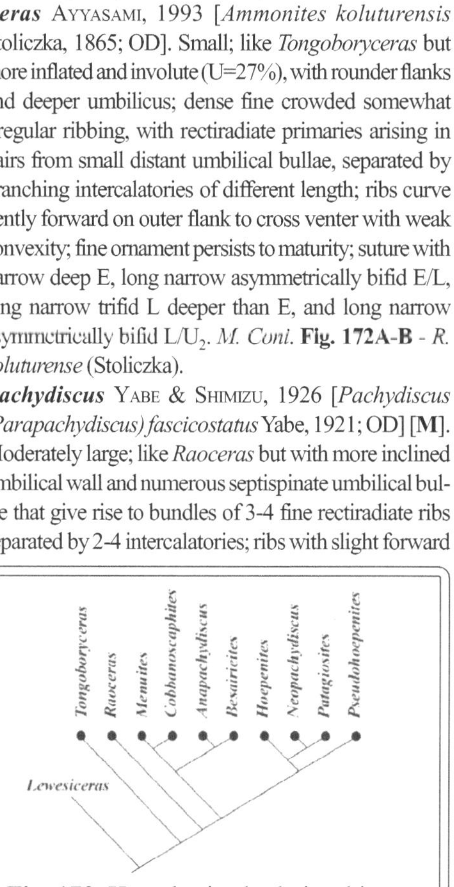

Fig. 173. Hypothesized relationships among Menuitinae.

Neopachydiscus YABE & SHIMIZU, 1926 [*Pachydiscus naumanni* Yokoyama, 1890; OD] [**M**]. Moderately large; like *Anapachydiscus* but with slightly narrower umbilicus (U=22-26%), broader flatter venter, depressed subquadratic whorl section, fewer umbilical bullae, coarser sparser ribbing, and strong constrictions persisting to large size. *Sant.- Camp.* **Fig. 172I-J** - *N. naumanni* (Yokoyama).

Menuites SPATH, 1922 [*Ammonites menu* Forbes, 1846; OD] [**m**]. Small; like inner whorls of *Anapachydiscus* but unribbed, with flatter venter and ventrolateral tubercles that alternate across venter; suture incised, with deep narrow E, broad narrow stemmed asymmetrically bifid E/L, very narrow deep penetrating L, and narrow stemmed asymmetrically bifid L/U_2. *U. Maastr.* **Fig. 172K-L** - *M. menu* (Forbes).

A microconch of some *Anapachydiscus*.

Cobbanoscaphites COLLIGNON, 1969 [*C. menabensis*; OD] [**m**]. Small; like *Menuites* but more inflated, with strong ribs looped between lateral and ventrolateral tubercles where they are opposite, and looped again across the venter; macroconch unknown. *M. Camp.* **Fig. 172G-H** - *C. menabensis* Collignon.

Besairieites COLLIGNON, 1931 [*Kossmaticeras pseudorotalinus* Collignon, 1931 [**m**], = *Pachydiscus boulei* Collignon (**M**); M]. Small; macroconch like *Neopachydiscus* but smaller (D<100 mm), with umbilicolateral bullae that persist to maturity; constrictions feeble to obsolete in later growth; microconchs very small, with umbilicolateral and ventrolateral bullae; suture with narrow E/L and trifid L as deep as E. *L. Santon.* **Fig. 167Q-R** - *B. boulei* (Collignon) juv.

= *Besaireites* Wright, 1996 (*nom. van.*).

Hoepenites COLLIGNON, 1952 [*Pachydiscus patagonicus* Paulcke, 1907; OD] [**M**]. Small; like *Neopachydiscus* but with finer more numerous ribs, 9-10 strong collared concave forward constrictions that cross venter with forward convexity and suture with narrower lobes; umbilical bullae lacking. *L. Camp.* **Fig. 172M-N** - *H. patagonicus* (Paulcke).

Patagiosites SPATH, 1953 [*Ammonites patagiosus* Schlüter, 1867; OD] [**M**]. Small to medium sized; poorly known, type deformed; like *Hoepenites* but more involute and with more rigid constrictions. *Camp.* **Fig. 172O** - *P. patagiosus* (Schlüter).

Pseudohoepenites **gen. nov.** [*Pachydiscus antecursor* van Hoepen, 1921, p.24; OD herein]. Small; like *Hoepenites* but evolute (U=33-36%) with slightly depressed rounded

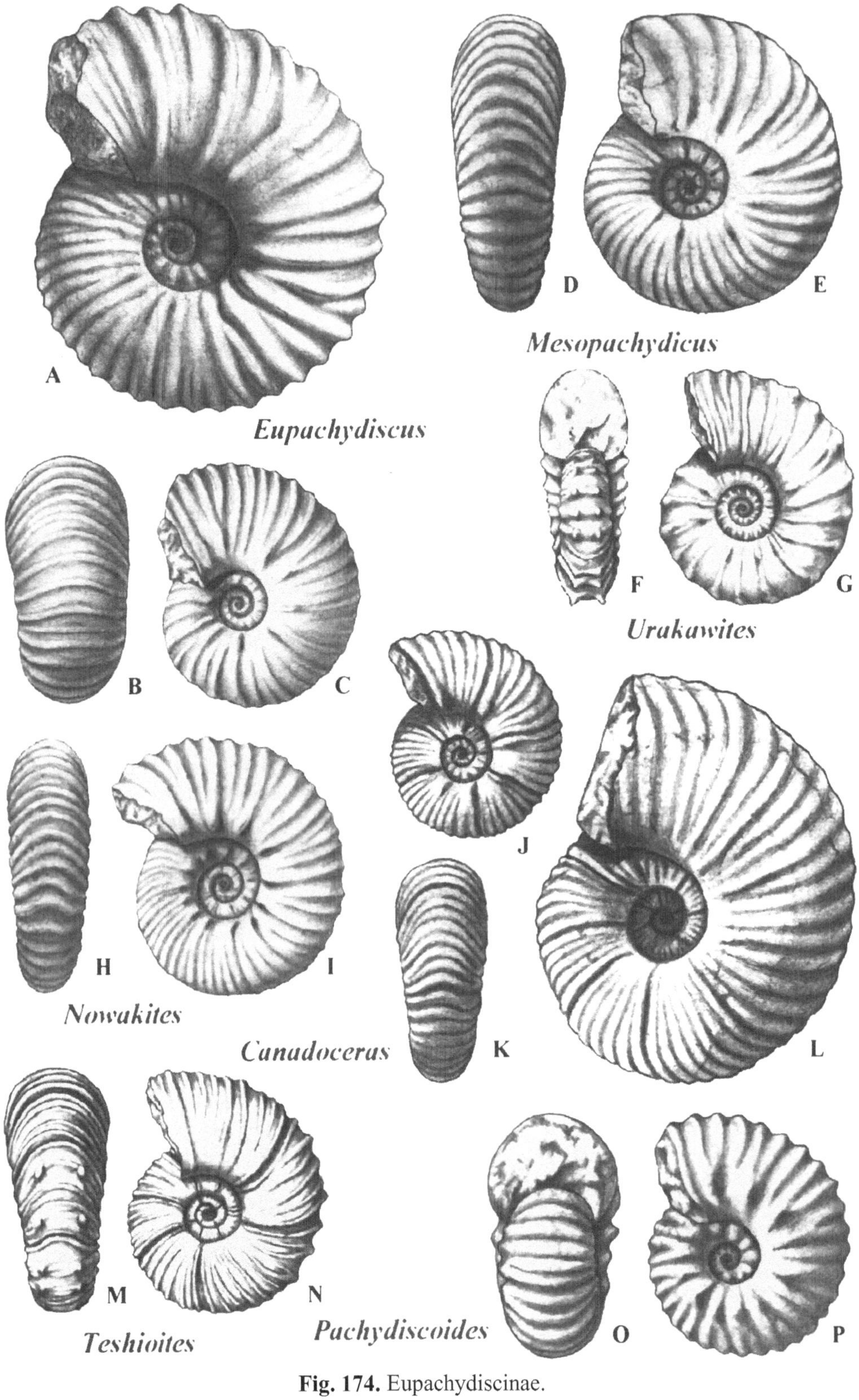

Fig. 174. Eupachydiscinae.

subquadratic whorl section; single sigmoidal ribs, separated by 1-3 intercalatories of various length, arise from incipient to feeble umbilical bullae and cross venter with weak convexity; ornament declines in later growth to delicate crowded ribs and striae; constrictions sigmoidal; suture like *Anapachydiscus* but simpler, with broader asymmetrically bifid L. *U. Sant.* **Fig. 175** - *P. antecursor* (van Hoepen).

Family Pachydiscidae Spath, 1922

Discussion. Monophyletic derivatives of Menuitidae; mostly involute, compressed and high whorled; primitively with strong ribs and constrictions, later without constrictions and ribs tending to weaken, especially across the flank. Comprises the subfamilies Eupachydiscinae and Pachydiscinae. *Coni. - U. Maastr.*

Subfamily Eupachydiscinae nov.

Diagnosis. Strongly ribbed monophyletic derivatives of Lewesiceratinae, close to *Tongoboryceras* (Matsumoto, 1990) evolving in parallel with Menuitinae; typically inflate and coarsely ribbed, but ribs may evanesce in maturity. *Coni. - U. Camp.*

Eupachydiscus Spath, 1922 [*Ammonites isculensis* Redtenbacher, 1873; OD] [**M**]. Moderately large; like *Tongoboryceras* but with coarser stronger ribs that are straight across flank and do not project strongly forward over venter; ribbing coarsens and strengthens in later growth. *Coni. - Camp.* **Fig. 174A-C** - *E. isculensis* (Redtenbacher).

Nowakites Spath, 1922 [*Pachydiscus carezi* de Grossouvre, 1894; OD] [**M**]. Small to medium sized; like *Eupachydiscus* but with narrowly rounded umbilical shoulder, convergent outer flanks, ovate whorl section, prominent collared constrictions in early to middle growth, densely

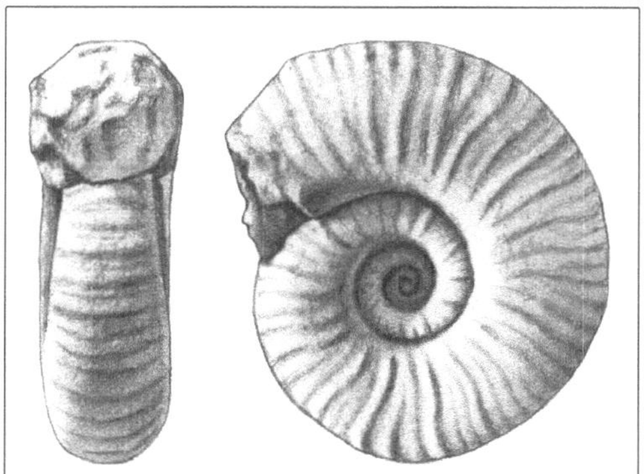

Fig. 175. *Pseudohoepenites.*

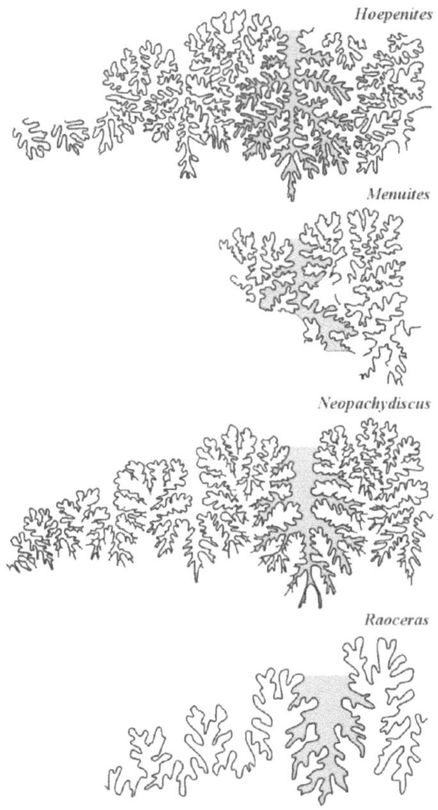

Fig. 176. Menuitine sutures.

ribbed inner whorls and, on outer whorls, finer narrower ribs that do not increase in strength on adult body chamber; microconchs with ventrolateral tubercles connected across venter by looped ribs. *Coni. - M. Sant.* **Fig. 174H-I** - *N. carezi* (de Grossouvre).

Pachydiscoides Spath, 1922 [*Sonneratia janeti* de Grossouvre, 1894; OD]. Small; moderately involute, inflated with depressed subcircular whorl section; prominent umbilical bullae give rise to strong rectiradiate ribs in pairs, with an occasional intercalatory, that cross venter transversely. *Coni.* **Fig. 174O-P** - *P. janeti* (de Grossouvre).

Mesopachydiscus Yabe & Shimizu, 1926 [*Pachydiscus haradai* Jimbo, 1894; OD] [**M**]. Medium sized; like *Eupachydiscus* but with more numerous umbilical bullae, coarser stronger ribbing in middle growth with 3-4 intercalatories between primary ribs, more compressed outer whorls with subovate whorl section, and ribs that curve forward on outer flank to cross venter with forward convexity. *L. - U. Camp.* **Fig. 174D-E** - *E. haradai* (Jimbo).

Urakawites Matsumoto, 1955 [*Pachydiscus rotalinoides* Yabe, 1915; OD][**m**]. Small; like *Mesopachydiscus* but with sparser ribbing and strong simple bituberculate ribs to adult body chamber. *Camp.* **Fig. 174F-G** - *U. rotalinoides* (Yabe).

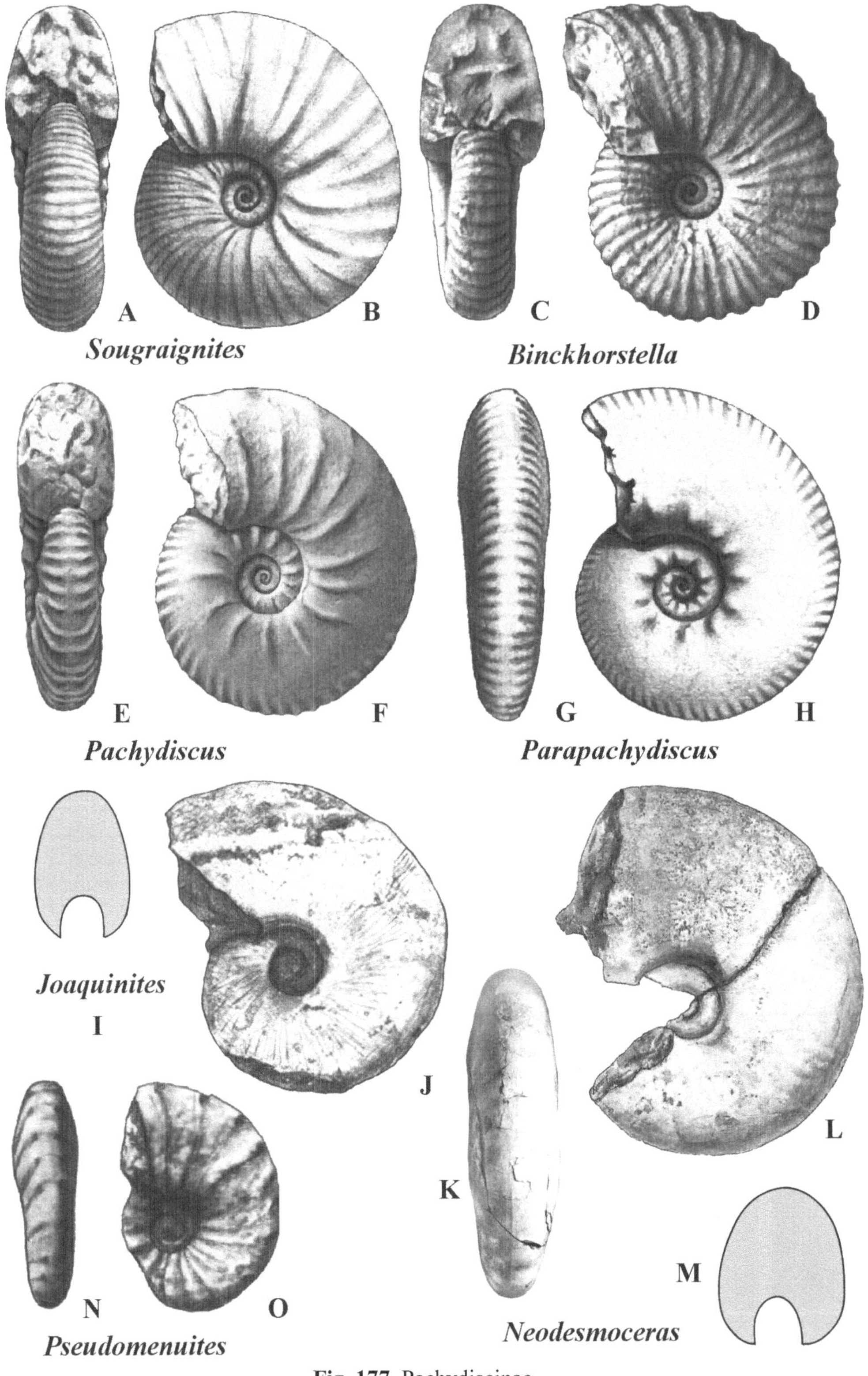

Fig. 177. Pachydiscinae.

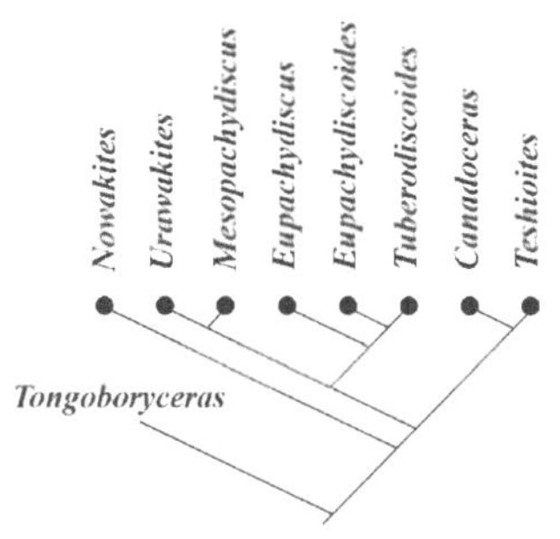

Fig. 178. Hypothesized relationships among Eupachydiscinae.

Canadoceras SPATH, 1922 [*Ammonites newberryanus* Meek, 1876; OD] [**M**]. Moderately large; like *Nowakites* but with up to 8 deep concave constrictions per whorl, each backed by a bullate collar, separated by 3-8 finer secondaries that occasionally bifurcate; shell becomes more compressed and involute (U=25-28%) with growth, with weakening of umbilical bullae to primary ribs and ribs more widely spaced; constrictions obsolete on outer whorls of macroconchs when ribbing becomes uniform and fine; suture with broad E, deep trifid L, narrow L/U_2 and 4-5 auxiliary saddles that descend in straight line. *Camp.* **Fig. 174J-L** - *C. newberryanum* (Meek).

=*Pseudopachydiscus* Yabe & Shimizu, 1926 (*nom. nud.*).

Teshioites MATSUMOTO, 1955 [*T. ryugasensis*; OD] [**m**]. Small; like inner whorls of *Canadoceras* but with more distant ribs, no umbilical bullae and secondary ribs looped to ventrolateral clavi on adult body chamber. *Camp.* **Fig.174M-N** - *T. ryugasensis* Matsumoto.

A microconch of *Canadoceras*.

Subfamily Pachydiscinae Spath, 1922

Discussion. Monophyletic descendents of Eupachydiscinae, close to *Nowakites*; mostly rather involute and compressed, with high whorls, prominent umbilicolateral bullae and generally weak ribs that tends to efface across the flanks and siphonal line. *U. Sant. - U. Maastr.*

***Sougraignites* gen. nov.** [*Pachydiscus jeani* de Grossouvre, 1894, p.187; OD herein] [**M**]. Medium sized; compressed and rather involute (U=18%), with somewhat flattened flanks converging to evenly rounded venter; inner whorls with distant feebly bullate primary ribs (10 per whorl) that give rise to 1-3 narrow fine secondaries separated by 2-5 long and short intercalatories; ribs flex forward and strengthen on outer flank, crossing venter with feeble convexity; beyond 70 mm diameter ornament coarsens, with primaries developing elongate umbilicolateral bullae separated by 1-2 long or short secondaries. *U. Sant.* **Fig. 177A-B** - *S. jeani* (de Grossouvre).

Pachydiscus ZITTEL, 1884 [*Ammonites neubergicus* von Hauer, 1858; SD de Grossouvre, 1894] [**M**]. Medium sized to moderately large; like *Sougraignites* but more compressed and high whorled, with a narrower venter; weak umbilical bullae of early growth become long strong umbilicolateral bullae (14-17 per whorl) on outer whorls that are weakly connected, if at all, to strong slightly prorsiradiate ribs on outer flank, separated by 2-3 short intercalated ribs of similar strength; ribs cross venter transversely but weaken markedly, or are effaced, from siphonal line, forming weak ventrolateral bullae that outnumber umbilicals 3-4:1; highly incised suture with deep narrow E, long narrow bifid E/L, narrow trifid L as deep as E and extremely retracted suspensive lobe. *U. Camp. - L. Maastr.* **Fig. 177E-F** - *P. neubergicus* (von Hauer).

Pseudomenuites MATSUMOTO, 1955 [*Pachydiscus ambiguus* de Grossouvre, 1894; OD][**m**]. Small; like *Pachydiscus* but with stronger ribs across flanks and ventrolateral clavi to the adult body chamber. *M. Camp.* **Fig. 177N-O** - *P. ambiguus* (de Grossouvre).

Assumed to be a microconch of *Pachydiscus*.

Parapachydiscus HYATT, 1900 [*Ammonites gollevillensis* d'Orbigny, 1850; OD][**M**]. Medium sized; like *Pachydiscus* but more compressed and involute (U=23-27%), with flattened convergent outer flanks and subtabulate venter; ribbing obsolete on flanks, leaving only

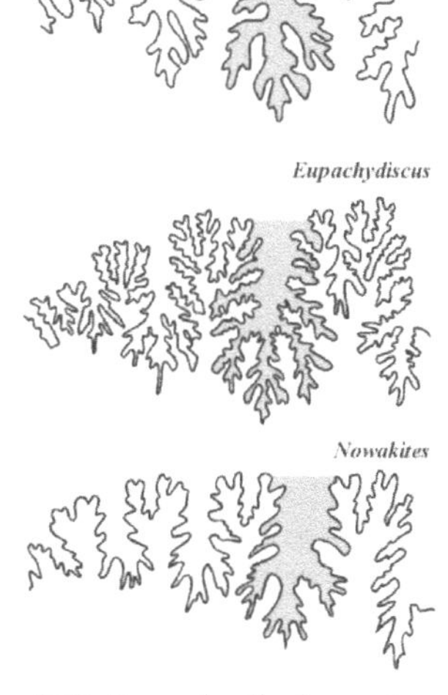

Fig. 179. Eupachydiscine sutures.

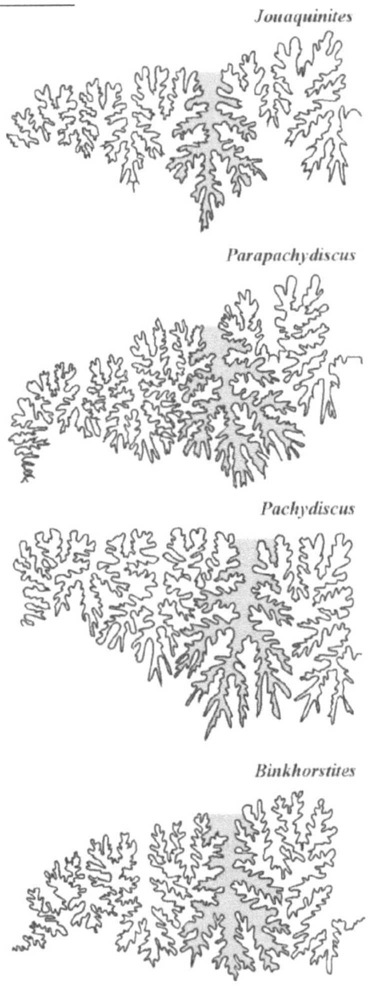

Fig. 181. Pachydiscine sutures.

short umbilical bullae (9-11 per whorl) weakly connected if at all to strong prorsiradiate ventrolateral ribs that persist to relatively large size and outnumber the umbilica bullae by 8:1; siphonal line smooth; suture like *Pachydiscus* but with shorter E so that E/L is oblique in later growth. *U. Maastr.* **Fig. 177G-H** - *P. gollevillensis* (d'Orbigny).

Lack of ventrolateral bullae distinguishes this genus from *Pachydiscus*.

***Binckhorstella* gen. nov.** [*Ammonites colligatus* Binckhorst, 1861, p.25; OD herein] [**M**]. Moderately large; like *Sougraignites* but more inflated, particularly inner whorls, and with sparser ribbing arising singly or in pairs separated by 1-3 intercalatories; umbilical tubercles lost in middle growth when inner flanks become smooth, restricting ribs to outer flank and venter; ribs sometimes weakening over siphonal line and may persist to large size, evanescing in maturity when outer whorls become smooth. *L. Camp. - L. Maastr.* **Fig. 177C-D** - *B. colligatus* (Binkhorst).

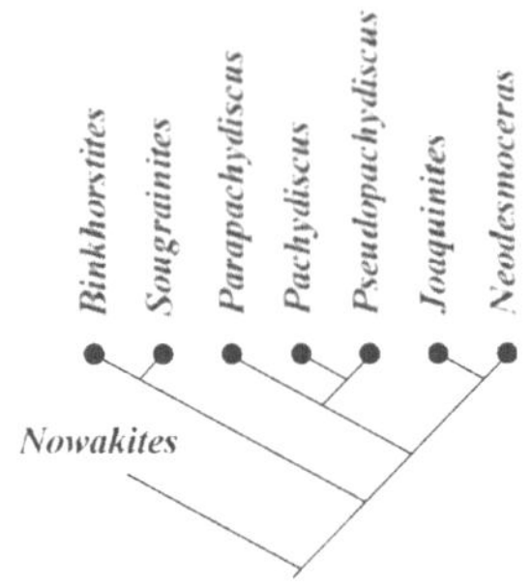

Fig. 180. Hypothesized relationships among Pachydiscinae.

Neodesmoceras MATSUMOTO, 1947 [*Pachydiscus (N.) japonicus*; OD] [**M**]. Large to very large; inner whorls like *Binckhorstella*; ribbing weakens in middle growth and is obsolete later; microconchs with ventrolateral tubercles; suture very complex, with long narrow L. *U. Maastr.* **Fig. 177L-M** - *N. japonicum* (Matsumoto).

= *Neodesmoceras* Matsumoto, 1938 (*nom. nud.*).

Joaquinites ANDERSON, 1958 [*J. fascicostatus*, = *Pachydiscus subcompressus* Matsumoto, 1954; OD]. Moderately large; like *Neodesmoceras*, fairly evolute (U=21%), with rounded umbilical edge, flattened slightly convergent flanks and subtabulate venter; flanks with bundles (>4) of fine sinuous costal threads arising from numerous (c. 25 per whorl) small umbilical bullae, some threads splitting at midflank but fading across venter; in later growth whorls almost smooth, with faint umbilical bulges and infrequent feeble flexiradiate long ribs. *L. Maastr.* **Fig. 177I-K** - *J. subcompressus* (Matsumoto).

Suborder PLACENTICERATIDINA Hyatt, 1900

(*nom. transl. et correct.* herein *pro* Placenticeratida) = Hoplitida Spath, 1922.

Discussion. Monophyletic derivatives of Desmoceratidina (Spath, 1922); comprises the superfamilies Puzosioidea and Placenticeratoidea. *Uppermost Haut. - U. Maastr.*

Superfamily PUZOSIOIDEA Spath, 1922

(*nom. transl.* Cooper & Greyling, 1998 *ex* Puzosiidae)

Discussion. Monophyletic derivatives of Desmoceratoidea (Spath, 1922); generally evolute, compressed with frequent constrictions and suture with a retracted suspensive lobe; comprises the families Puzosiidae, Cleoniceratidae and Kossmaticeratidae. *Uppermost Haut. - U. Maastr.*

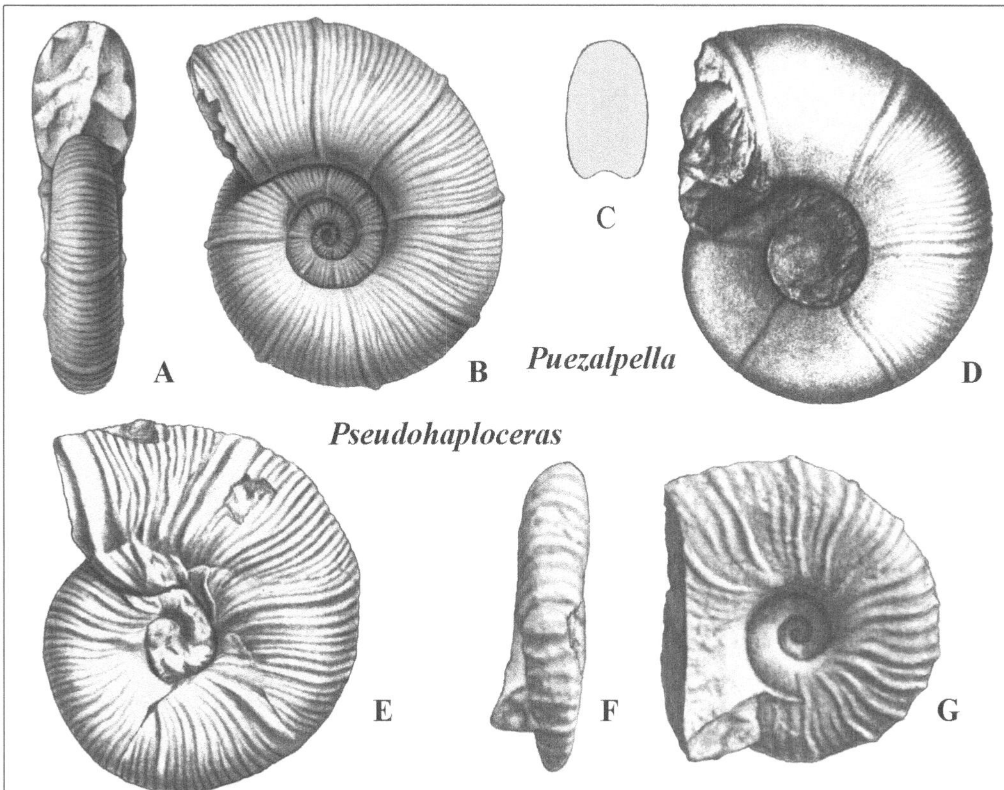

Fig. 182. Puzosiidae.

Family PUZOSIIDAE Spath, 1922

Discussion. Seemingly monophyletic derivatives of Desmoceratidae (Spath, 1922; Klein & Vašíček, 2011). Early members of this lineage are poorly discriminated from Desmoceratidae but, typically, the suspensive lobe of puzosiids is retracted and that of desmoceratids forms a regularly descending series. *Pseudohaploceras* is very similar to *Valdedorsella*, with a non-retracted suspensive lobe, but its wide umbilicus and compressed whorls are puzosiid. *Anapuzosia* has fine ribbing and constrictions transverse over the venter, as in *Pseudohaploceras*, whereas *Melchiorites* has unribbed early whorls and constrictions forming chevrons across the venter as in *Puzosia*. Being close to the basal dichotomy of two important phyletic lines it is not surprising that early puzosiids and desmoceratids are poorly discriminated and taxonomic placement relies heavily on character weighting.

Typically puzosiids are compressed and evolute, with frequent regular constrictions marked by strong ribs on the test and, except in early growth, paralleled by generally subordinate ribbing that is strongest over the rounded venter. Other than in early representatives the suture has a trifid L and strongly retracted suspensive lobe. Comprises the subfamilies Anapuzosiinae, Puzosiinae, Hauericeratinae and Silesitoidinae. *Uppermost Haut. - U. Maastr.*

Subfamily Anapuzosiinae nov.

Diagnosis. Medium sized to large, with fairly wide umbilicus, regular constrictions that cross the venter transversely and do not flex forward on the outer flank, separated by strong sinuous flank costae, often bifurcating, that arise from the umbilical shoulder. Primitively the suture has an asymmetrically bifid L and non-retracted suspensive lobe, but later with a trifid L and strongly retracted suspensive lobe like *Puzosia*. Derivatives of Desmoceratinae close to *Valdedorsella*. *Uppermost Haut. - L. Sant.*

Puezalpella DIMITROVA, 1967 [*Desmoceras uhligi* Haug, 1889; OD]. Small, evolute, compressed, with steep umbilical wall, gently convex flanks, rounded venter and elliptical whorl section; 5-6 prorsiradiate constrictions separated by ribs that are weak across inner flank and

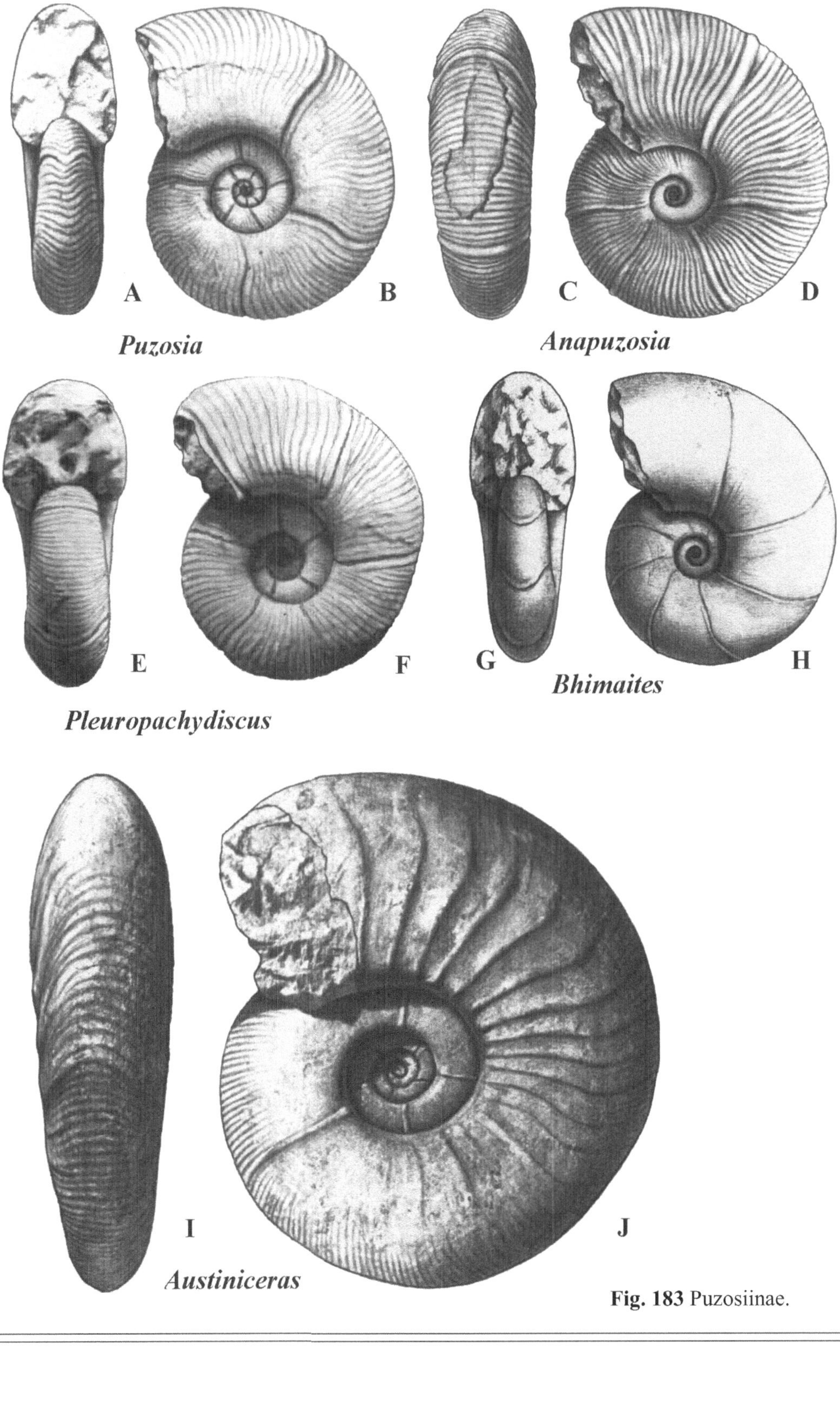

Fig. 183 Puzosiinae.

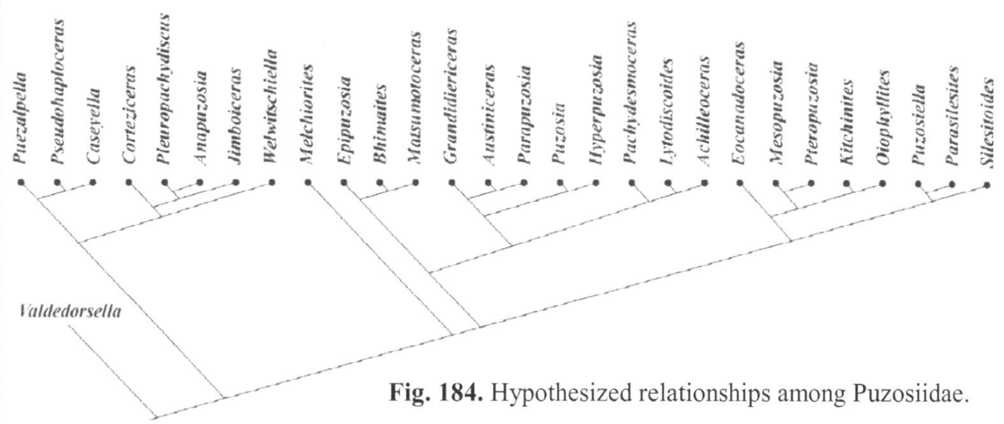

Fig. 184. Hypothesized relationships among Puzosiidae.

cross venter with a broad shallow convexity. *Uppermost Haut. - basal Barrem.* **Fig. 182C-D** - *P. uhligi* (Haug).

Pseudohaploceras HYATT, 1900 [*Ammonites liptoviensis* Zeuschner, 1856; OD]. Medium sized; moderately involute, compressed, with broad weakly convex flanks, rounded venter and oval to subelliptical whorl section; 7-8 regular flexuous rectiradiate constrictions per whorl, transverse across venter, separated by fine flexuous ribs that arise on umbilical shoulder and commonly bifurcate on flank; suture with deep asymmetrically bifid L. *Barrem - U. Apt.* **Fig. 182A-B** - *P. matheroni* (d'Orbigny); **Fig. 182E** - *P. liptoviense* (Zeuschner).
= *Pleurohaploceras* Riedel, 1938 (*nom. van.*).

Caseyella CANTÚ-CHAPA, 1976 [*Uhligella reesidei* Humphrey, 1949; OD]. Very small; involute, compressed, high whorled, initially with compressed subelliptical whorls, later squarish; numerous flexuous primary ribs backed by constrictions that weaken with growth, separated by 1-2 rarely 3 simple intercalatories arising at or slightly below midflank, strengthening on outer flank and weakening over venter; inner flanks smooth so ribs not exposed in umbilicus. *M. Apt.* **Fig. 182F-G** - *P. reesidei* (Humphrey).

Corteziceras ETAYO SERNA, 1979 [*C. cortezi*; OD]. Small; similar to *Caseyella* but more inflated and evolute, with fewer constrictions, stronger, coarse, simple, flexuous ribs arising at various levels on flank leaving umbilical shoulder smooth. *U. Apt.* **Fig. 187** - *C. cortezi* Etayo-Serna.

Pleuropachydiscus HYATT, 1900 [*Ammonites hoffmanni* Gabb, 1869; OD]. Medium sized; moderately involute, inflated, with broadly rounded venter and depressed semi-elliptical whorl section in middle growth; 6 flexuous constrictions per whorl cross venter transversely, separated by fine ribs; suture with E as deep as L, straight E/L and L/U_2 protruding beyond E/L. *L. Alb.* **Fig. 183E-F** - *P. hoffmanni* (Gabb).

Anapuzosia MATSUMOTO, 1954 [*Puzosia buenaventura* Anderson, 1938; OD] [**M**]. Medium sized to large; inner whorls much like *Pleuropachydiscus* but more involute, with compressed whorls, narrowly rounded venter and more frequent bifurcation of flank riblets; adult body chamber with distant strong rectiradiate ribs; suture with deep E and L/U_2 projecting beyond E/L. *L. Alb. - U. Cenom.* **Fig. 183C-D** - *A. multicostatum* Renz; **Fig. 186A-B** - *A. dibleyi* (Spath); **Fig. 186C** - *A. buenaventura* (Anderson) (**M**).

Jimboiceras MATSUMOTO, 1954 [*Desmoceras planulatiforme* Jimbo, 1894; OD] [**m**]. Like *Anapuzosia* but more inflated, with convex flanks, less flexuous ribs most of which branch or are intercalated at midflank; straight prorsiradiate constrictions marked by broad flared ribs that flex forward on outer flank. *Turon. - L. Sant.* **Fig. 189C-D** - *J. planulatiforme* (Jimbo).
= *Jimboiceras* Shimizu, 1935 (*nom. nud.*).
= *Jimboireras* Matsumoto, 1988 (*laps. cal.*).

Welwitschiella **gen. nov.** [*Puzosia garajuana* Maury, 1930, p.289; OD herein]. Medium sized; compressed, evolute, with rounded venter that becomes subtabulate in later growth; about 7 straight prorsiradiate constrictions per whorl split by wedge-shaped rib across venter; ribs coarse, simple, strong, transverse over venter and effaced from lower flank. *Lower M. Alb.* **Fig. 190** - *W. garajuana* (Maury).

Subfamily Puzosiinae Spath, 1922

= Melchioritinae Breskovski, 1977 (*nom. correct.* Wright & Kennedy, 1984 *pro* Melcioritinae).

Discussion. Monophyletic derivatives of Anapuzosiinae, close to *Pseudohaploceras*, easily distinguished by suture with strongly retracted suspensive lobe with 4 elements. Dimorphism marked; microconchs compressed, evolute, with broad flattish flanks, evenly arched venter, periodic collared constrictions, fine flexuous ribs strongest across outer flank and venter, and with lateral lappets. Macro-

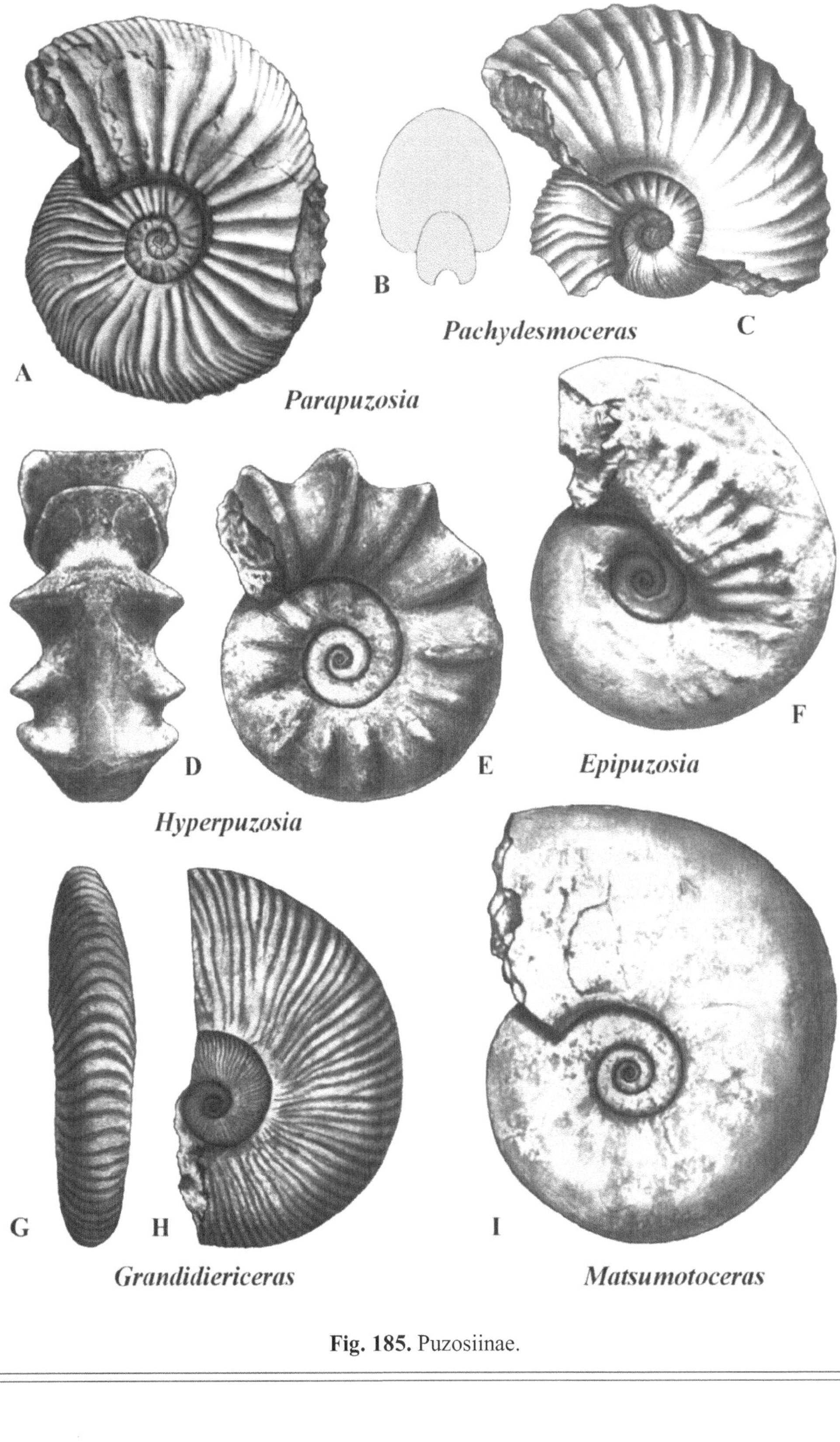

Fig. 185. Puzosiinae.

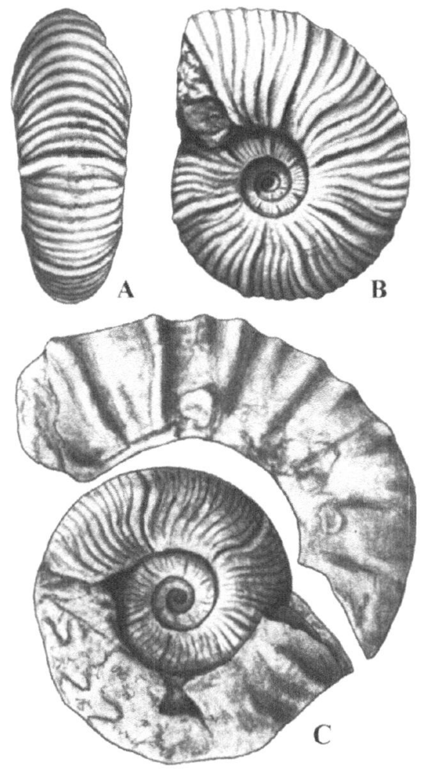

Fig. 186. *Anapuzosia.*

conchs enormous to gigantic, including the largest known ammonites, generally with coarse robust ribs, lateral flares, tubercles and/or spines to the adult body chamber. Suture with trifid L and retracted suspensive lobe. Macroconch diversity suggests more rapid evolution than microconchs and, hence, better for tracking evolution and taxonomic discrimination. Comprises the tribes Puzosiini and Kitchinitini. *U. Barrem. - Maastr.*

Tribe Puzosiini Spath, 1922

(*nom. transl.* herein *ex* Puzosiinae)

Type genus: *Puzosia* Bayle, 1875.

Diagnosis. Microconchs mainly with ribs strongest across outer flank and venter, primitively forming chevrons. *U. Barrem.- U. Maastr.*

Melchiorites SPATH, 1923 [*Ammonites melchioris* Tietze, 1872; M]. Small; like *Pseudohaploceras* but without ribs except feebly over venter in later growth; deep almost straight prorsiradiate constrictions flex forward on outer flank and are linguoid over venter; suture like *Pseudohaploceras* but simpler and with slightly retracted suspensive lobe of 4 elements; some may have feeble umbilical bullae. *U. Barrem. - L. Alb.* **Fig. 192** - *M. melchioris* (Tietze).

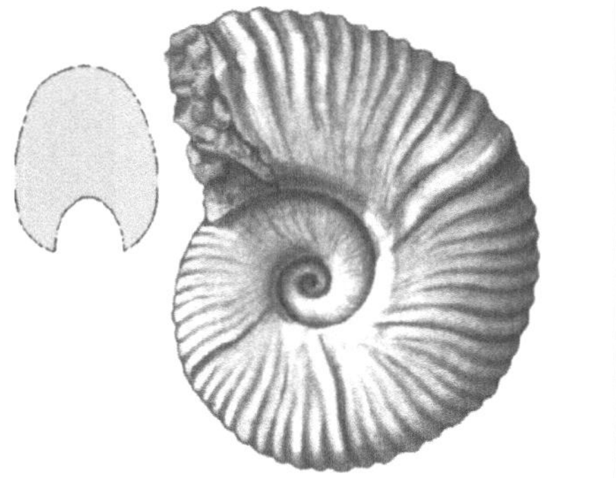

Fig. 187. *Corteziceras.*

Puzosia BAYLE, 1878 [*Ammonites planulatus* J. de C. Sowerby, 1827; SD Douvillé, 1879]. Small (**m**) to very large (**M**); compressed, moderately evolute, with vertical umbilical wall, evenly rounded umbilical shoulder, broad weakly convex flanks, fairly narrow evenly rounded venter and compressed oval whorl section; 4-8 sinuous constrictions per whorl, the number decreasing with growth, backed by ventral collar, projecting strongly fowards on upper flank and forming chevrons over venter; strong fine simple prorsiradiate ribs arise above umbilical shoulder leaving innermost flank essentially unornamented; body chamber of medium sized to large macroconchs almost smooth, save for distant low radial bulges to inner flank and 1-2 horns; suture much incised, with relatively shallow E, narrow stemmed bifid E/L and L/U_2, long narrow asymmetrically trifid L with deep penetrating accessory saddles. *L. Alb - Cenom.* **Fig. 183A-B** - *P. mayoriana* (d'Orbigny); **Fig. 193** - *P.* sp. macroconch; **Fig. 194** - *P. compressa* Kosmat.

Hyperpuzosia MATSUMOTO, 1988 [*H. tamon*; OD] [**M**]. Very large; inner whorls like *Puzosia*; compressed, evolute, with nearly smooth inner flank, fine riblets on outer flank

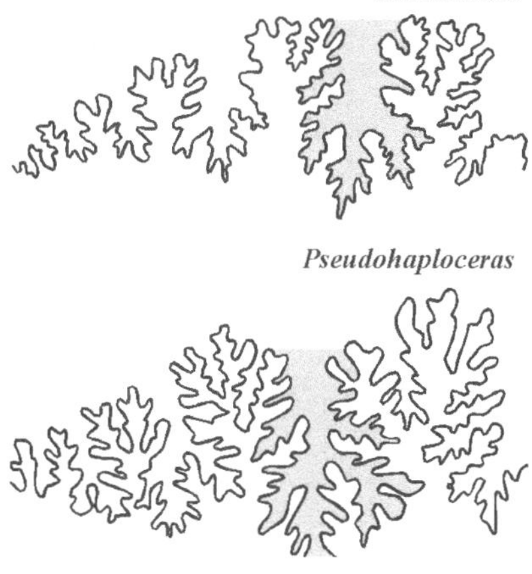

Fig. 188. Puzosiid sutures.

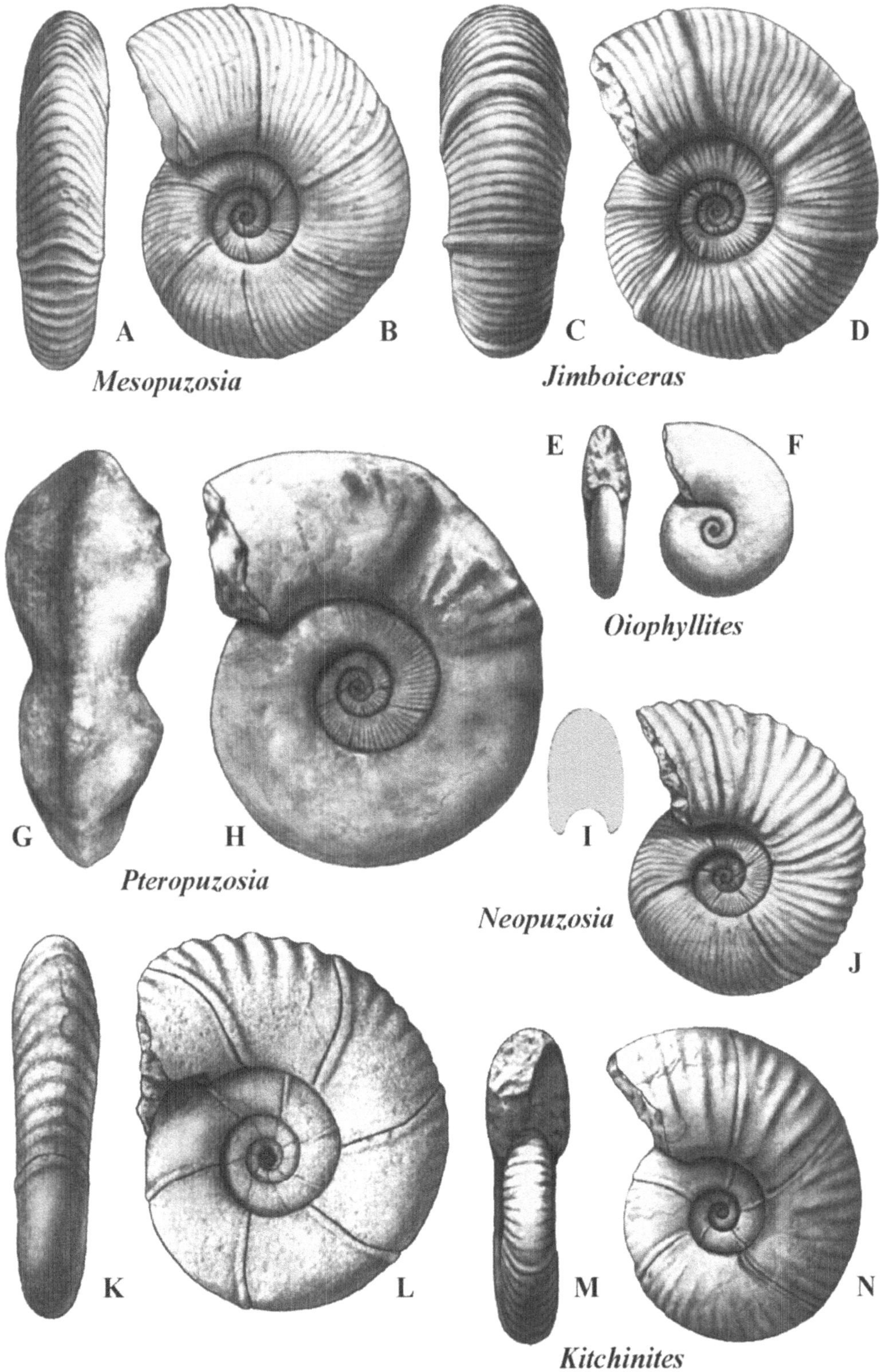

Fig. 189. Puzosiinae.

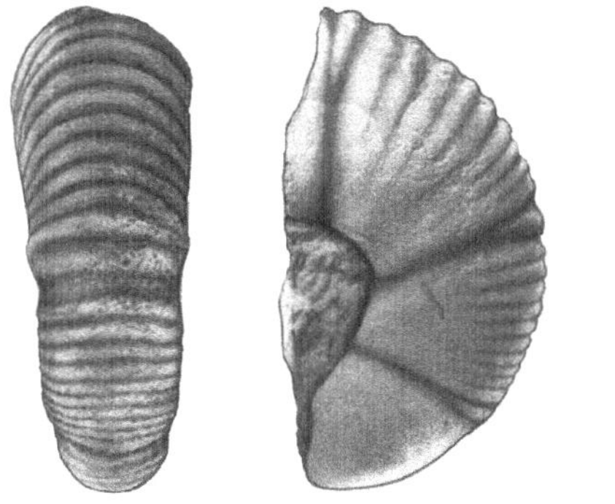

Fig. 190. *Welwitschiella*.

and venter, and periodic nearly rectiradiate collared constrictions; last part of phragmocone and adult body chamber with broad slightly fastigiate venter and strong distant flared ribs with ventrolateral nodes or horns. *Upper L.- M. Alb.* **Fig. 185D-E** - *H. tamon* Matsumoto.

Achilleoceras VAN HOEPEN, 1951 [*A. erasmusi*; OD] [**M**]. Gigantic; inner whorls with frequent long, rectiradiate, gently flexuous major ribs (constrictions) separated by numerous crowded minor ribs of unequal length; ribs flex forward near venter; adult body chamber with prominent cock's-comb crest and irregular subhorizontal lateral and ventrolateral spines. *U. Alb.* **Fig. 195C-D** - *A. erasmusi* van Hoepen.

Pachydesmoceras Spath, 1922 [*Ammonites denisonianus* Stoliczka, 1865; OD] [**M**]. Gigantic, becoming more evolute with growth; subvertical umbilical wall, gently convex flanks, rounded venter and broadly ovate whorl section becoming subcircular in maturity; in early to middle growth with long straight rectiradiate to slightly prorsiradiate main ribs that flex strongly forward near venter, separated by 2-5 intercalatories of various length; adult macroconch body chamber with single robust distant ribs without horns; suture with very shallow E. *U. Alb. - Coni.* **Fig. 185B-C** - *P. kossmati* Matsumoto (**m**); **Fig. 197** - *P. denisonianum* (Stoliczka) (**M**).

Lytodiscoides SPATH, 1922 [*Pachydiscus conduciensis* Choffat, 1903; OD] [**M**]. Enormous; inner whorls like *Pachydesmoceras*; adult body chamber with sparse irregular fine ribs and very long horizontally protruding umbilicolateral and ventrolateral spines. *U. Alb. - L. Cenom.* **Fig. 195A-B** - *L. conduciensis* (Choffat).

= *Lytodescoides* Breskovski, 1977 (*laps. cal.*).

Bhimaites MATSUMOTO, 1954 [*Ammonites bhima* Stoliczka, 1865; OD]. Small to medium sized; like *Puzosia*, but higher whorled and more involute (U=23-30%), with nearly flat flanks, narrowly arched venter and 7-8 almost straight constrictions that flex forward near venter; ribs feeble over venter. *U. Alb. - Turon.* **Fig. 183G-H** - *B. bhima* (Stoliczka).

Epipuzosia MATSUMOTO, 1988 [*E. maya* Matsumoto & Kawashita, 1988; OD]. Enormous; inner whorls like

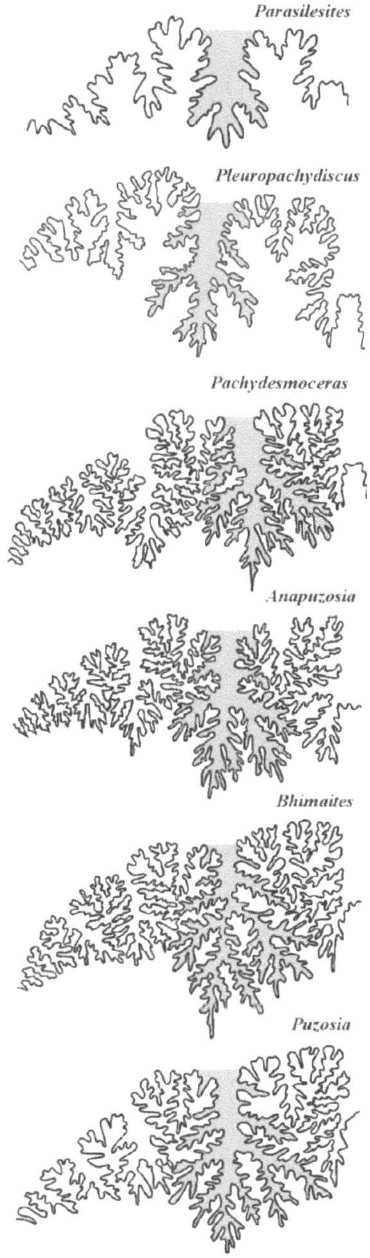

Fig. 191. Puzosiine sutures.

Bhimaites, almost smooth save for weak infrequent flares; adult body chamber shows scaphitoid uncoiling, with strongly undercut, overhanging umbilical edge and weak rectiradiate ribs terminating in rounded ventrolateral nodes. *L. - M. Turon.* **Fig. 185F** - *E. maya* Matsumoto & Kawashita.

Matsumotoceras VAN HOEPEN, 1968 [*M. donlisteri*; OD] [**M**]. Enormous; fairly narrow umbilicus (U=26%), vertical umbilical wall, subangular umbilical shoulder and oval whorl section; constrictions lacking on outer whorls; fine concave forward riblets on outer flank with

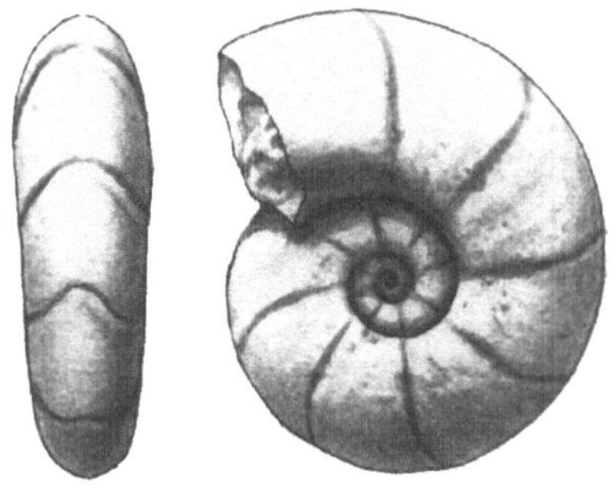

Fig. 192. *Melchiorites.*

broad low radial bulges to lower two-thirds of flanks on final whorl. *L. Coni.* **Fig. 185I** - *M. donlisteri* van Hoepen.

Klinger & Kennedy (2014) tentatively included *Matsumotoceras* in the synonymy of *Pachydesmoceras* but the macroconchs are markedly different, and its narrow umbilicus places it closer to *Epipuzosia*.

Austiniceras SPATH, 1922 [*Ammonites austeni* Sharpe, 1855; OD]. Very large, like *Puzosia* but more compressed and involute, with narrowly arched venter, fewer (4-5) almost straight narrow primary ribs (constrictions) that arise from weak umbilical bullae and flex forward near venter, and more numerous (up to 30) ribs intervening between constrictions on outer flank; adult body chamber with numerous strong primary ribs that parallel constrictions but weaken on outer flank and do not cross venter. *L. Cenom. - Camp.* **Fig. 183I-J** - *A. austeni* (Sharpe).

Parapuzosia NOWAK, 1913 [*Sonneratia daubreei* de Grossouvre, 1894; SD Spath, 1922]. Like *Austiniceras* but larger, with numerous strong, closely spaced, almost straight to slightly flexuous primary ribs from early growth; primary ribs branch on outer flank, separated by 2-3 secondary ribs; ribbing may evanesce on adult body chamber in some. *Coni. - L. Camp.* **Fig. 185A** - *P. daubreei* (de Grossouvre); **Fig. 196** - *P. bererensis* (Collignon).

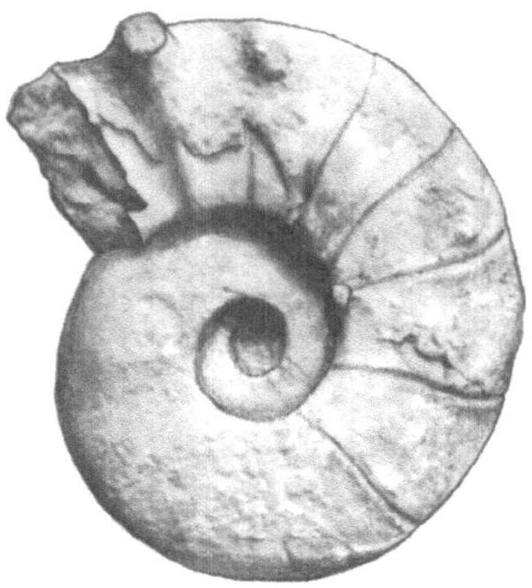

Fig. 193. *Puzosia* macroconch.

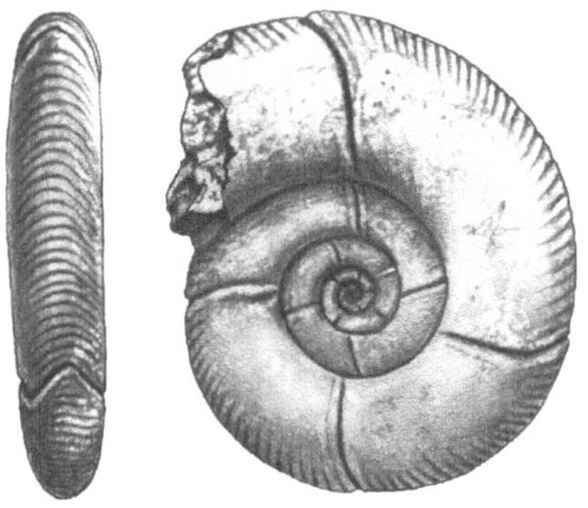

Fig. 194. *Puzosia compressa* Kossmat.

Grandidiericeras COLLIGNON, 1961 [*G. grandidierorum*; OD]. Moderately large, compressed; like *Austiniceras* but with dense round ribs arising singly or in pairs from umbilical shoulder, with occasional bifurcation higher up; adult body chamber with prominent distant umbilical bullae; macroconchs with smooth inflated adult body chamber. *Coni. - Camp.* **Fig. 185G-H** - *G. grandidierorum* Collignon.

Tribe Kitchinitini nov.

Type genus: *Kitchinites* Spath, 1922.

Diagnosis. Small to medium sized; primitively moderately inflated with strong ribbing arising from umbilical shoulder, straight prorsiradiate constrictions that flex forward on outer flank, and suture with E/L projecting beyond L/U_2; primitively much like *Puzosia* but becoming more compressed with time to homoeomorph Hauericeratinae. A seemingly monophyletic group with an origin in Puzosiini, close to *Puzosia*, although this was questioned by Maeda (1993). *L. Turon. - U. Maastr.*

Eocanadoceras ANDERSON, 1958 [*Puzosia (E.) hannai*; OD]. Large, evolute, compressed, with vertical umbilical wall, convex flanks sloping towards rounded venter and compressed oval whorl section; rectiradiate primary ribs arise from feeble bullae, sometimes splitting in 2-3 branches on outer flank where they curve forward to cross venter with slight forward convexity, separated by thinner secondary ribs that arise at mid-flank; 6 broad deep constrictions parallel ribbing. *Middle Turon.* **Fig. 199** - *E. hannai* (Anderson).

Mesopuzosia MATSUMOTO, 1954 [*M. pacifica*; OD]. Medium sized [**m**] to large [**M**]; like *Puzosia* but generally more compressed, with flatter flanks, oval to subelliptical whorl section and straighter primary ribs (constrictions) that flex forward near venter, separated by variable number of uniform finer bifurcating sigmoidal secondaries that

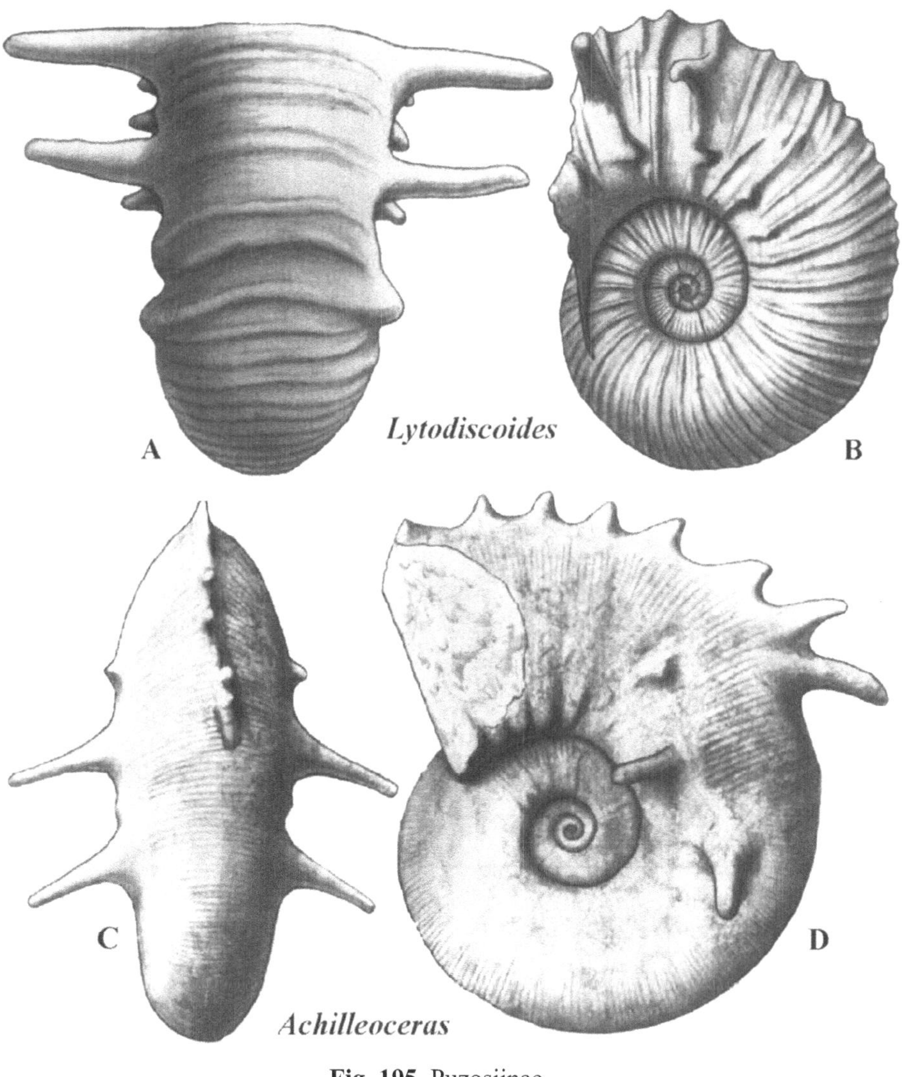

Fig. 195. Puzosiinae.

arise from umbilical shoulder or by bifurcation on flank; constrictions evanesce on outer whorls; adult body chamber of macroconch smooth, generally with low undulations on flank. *U. Turon. - Camp.* **Fig. 189A-B** - *M. pacifica* Matsumoto [**m**].

Pteropuzosia MATSUMOTO, 1988 [*P. kawashitai* Matsumoto; OD] [**M**]. Enormous; inner whorls like *Mesopuzosia* but adult body chamber smooth, with fastigiate venter and huge auriculate lateral flares that may form ventrolateral nodes with superimposed ribs. *L. - M. Turon.* **Fig. 189G-H** - *P. kawashitai* Matsumoto.

Kitchinites SPATH, 1922 [*Holcodiscus pondycherryanus* Kossmat, 1897; OD] [**m**]. Small, moderately evolute (U=31%), with compressed elliptical whorl section; ribbing largely confined to outer flank on phragmocone, so they are not visible in the umbilicus; outer whorl with

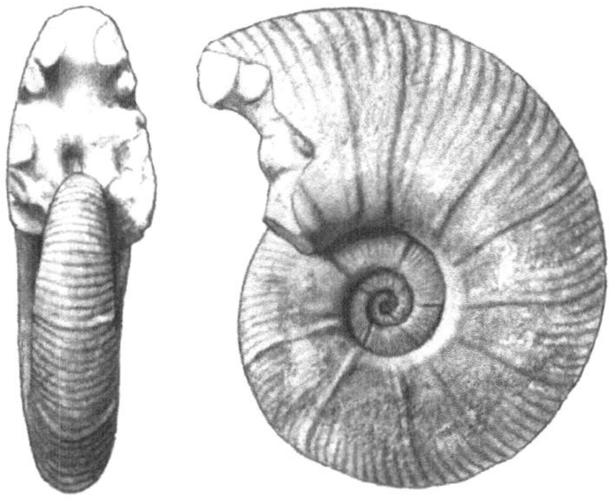

Fig. 196. *Parapuzosia.*

Fig. 197. *Pachydesmoceras.*

low blunt straight almost rectiradiate main ribs that arise close to umbilical shoulder and flex slightly forward on outer flank; weak ribs cross venter with broad convexity; long ribs separated by 1-2 intercalatories of equal strength on outer flank and, very occasionally, may bifurcate; 5 collared constrictions per whorl; macroconchs superficially resemble *Hauericeras*. *Camp. - L. Maastr.* **Fig. 189K-L** - *K. densicostatus* (Marshall) (**M**); **Fig. 189M-N** - *K. pondycherryanus* (Kossmat) (**m**).

Fig. 198. Puzosiine sutures.

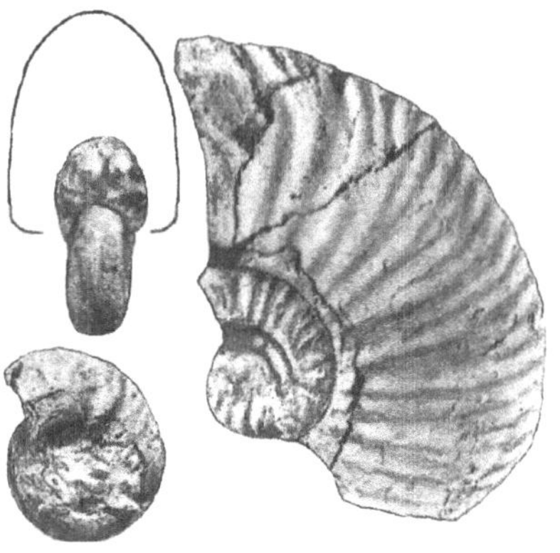

Fig. 199. *Eocanadoceras.*

Oiophyllites SPATH, 1953 [*O. decipiens*; OD]. Very small, based on an unornamented juvenile; moderately involute with rounded umbilical shoulder, flat convergent flanks, narrowly rounded venter and compressed ovate whorl section; suture with subphylloid folioles. *U. Camp.* **Fig. 189E-F** - *O. decipiens* Spath.

The only other species of *Oiophyllites*, *O. angolensis* Spath, is based on the innermost whorls of *Kitchinites* and this may be true also for *O. decipiens*, although it could be a juvenile *Hauericeras* (Wright, 1996).

Subfamily Hauericeratinae Matsumoto, 1938

Discussion. Monophyletic derivatives of Puzosiinae, close to *P. compressa* Kossmat (**Fig. 194**); primitively only moderately evolute but later very evolute and strongly compressed, with flattened flanks, keeled fastigiate venter, lanceolate whorl section and conspicuous concave and biconcave constrictions; fine prorsiradiate ribs across venter in primitive members only. Suture with prominent adventive lobe to E/L in primitive members, reducing in derived members. Sexual dimorphism is conspicuous, microconch with lateral lappets and macroconch with sigmoidal aperture. *M. Turon. - U. Maastr.*

Mossamedites COOPER, 2003 [*Puzosia serratocarinata* Kennedy & Cobban, 1988; OD]. Small, moderately evolute (U=25-33%), compressed, with flattened flanks, low fastigiate venter, blunt siphonal keel, and compressed sublanceolate whorl section; flanks with weak sigmoidal growth striae; outermost flanks with short, fine, uniform, prorsiradiate riblets that form chevrons over venter and serve to crenulate keel; periodic constrictions marked by short V-shaped labial collar on venter. *M. Turon.* **Fig. 200A-B** - *M. serratocarinatus* (Kennedy & Cobban).

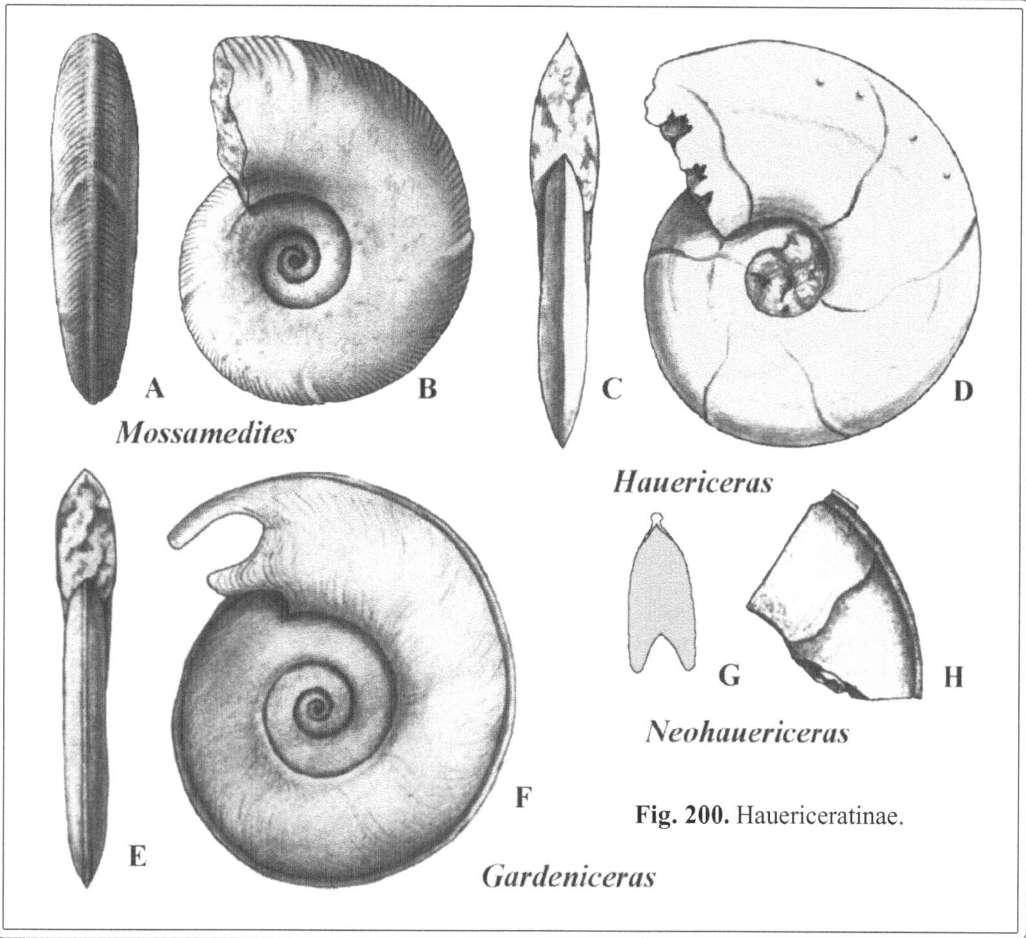

Fig. 200. Hauericeratinae.

Hauericeras DE GROSSOUVRE, 1894 [*Ammonites pseudogardeni* Schlüter, 1872; OD]. Moderately large, like *Mossamedites* but more involute (U=21-29%), with angular umbilical edge, shallow narrow biconcave constrictions that project strongly forward near venter; last part of phragmocone (?and body chamber) with distant weak ventrolateral nodes and oblique riblets below which is a shallow spiral groove; suture with narrow stemmed saddles, asymmetrically bifid L, with L/U_2 level with E/L and a gradually descending series of auxiliaries. *Sant.* **Fig. 200C-D** - *H. pseudogardeni* (Schlüter).

= *Schlueteria* Rollier, 1922 (*non* Fritsch, 1887).

= *Pseudogardenia* Tomlin, 1930 (obj.).

= *Hsuericeras* Matsumoto *et al.*, 1990 (*laps. cal.*)

Gardeniceras MATSUMOTO & OBATA, 1956 [*Ammonites gardeni* Baily, 1855; OD]. Medium sized; like *Hauericeras* but more evolute (U=37-50%) and compressed, with subvertical umbilical wall of inner whorls becoming inclined in later growth; venter with specialized elevated siphonal keel with excavated sides in middle to adult growth; ribs lacking, with ornament of flexuous growth striae only; strong concave forward constrictions to internal moulds in later growth; suture like *Hauericeras* but with L/U_2 protruding beyond E/L and strongly retracted suspensive lobe. *L. Sant. - L. Camp.* **Fig. 200E-F** - *G. angustum* (Yabe); **Fig. 201** - *G. gardeni* (Baily).

***Neogardeniceras* gen. nov.** [*Ammonites rembda* Forbes, 1846, p.111; OD herein]. Small; like *Hauericeras* with biconcave to sinuous constrictions, but very evolute and with undercut siphonal keel; suture with broader shallower E. *U. Maastr.* **Fig. 200G-H** - *N. rembda* (Forbes).

Subfamily Silesitoidinae Breistroffer, 1953

Discussion. A monophyletic taxon with an origin in Anapuzosiinae, close to *Pseudohaploceras*. *Puzosiella* is like a micromorph *Parasilesites* but *P. kilianiforme* (Fallot) (**Fig. 203C-D**) shows features of *Puzosia*. Very small to small, very evolute to serpenticone, compressed, with frequent constrictions; suture with deep narrow E, narrow stemmed bifid saddles, trifid L deeper than E and declining but not retracted suspensive lobe. *U. Apt. - L. Alb.*

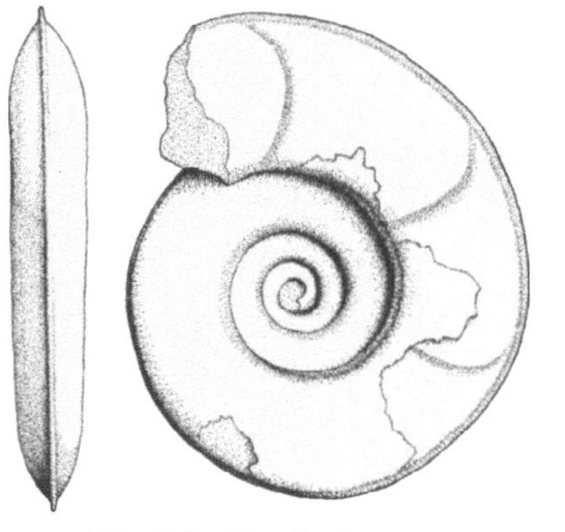

Fig. 201. *Gardeniceras.*

Silesitoides SPATH, 1925 [*Silesites escragnollensis* Jacob, 1908; OD]. Very small; in early growth evolute with flattened flanks, slightly depressed subrectangular whorl section and 4-5 prominent constrictions which form chevrons over venter; subsequently serpenticone, with 20-22 straight, blunt, widely spaced rectiradiate ribs that cross venter transversely; adult aperture contracted. *U. Apt.* **Fig. 203A-D** - *S. escragnollensis* (Jacob).

= *Jacobella* Passendorfer, 1930 (*non* Jeannet, 1908).

Parasilesites IMLAY, 1959 [*P. bullatus*; OD]. Very small, very evolute, compressed, with barely flexuous prorsiradiate constrictions backed by bullate collar; ribbing strengthens on adult body chamber but weak over siphonal line; suture simple with 3 auxiliary elements. *L. - M. Alb.* **Fig. 204A-B** - *P. bullatus* Imlay; **Fig. 204C-D** - *P. kilianiforme* (Fallot).

= *Pseudosilesitoides* Breistroffer, 1951 (*nom. nud.*).

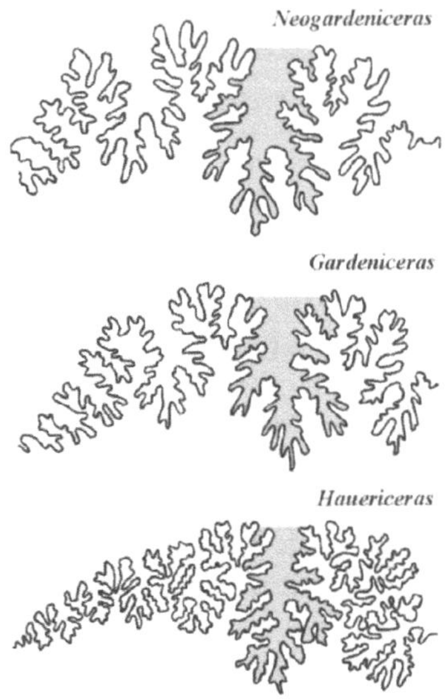

Fig. 202. Hauericeratine sutures.

Puzosiella EGOIAN, 1969 [*P. minuta*; OD]. Like *Parasilesites* but micromorph; 7 deep constrictions per whorl linguoid over venter, separated by fine riblets. *U. Apt.* **Fig. 204E-G** - *P. minuta* Egoian.

Szives (2007) included *P. minuta* in the synonymy of *Puzosia nolani* var. *kilianiformis* Fallot (1910) that, if correct, becomes the type species. However it is clear from the specific name that this genus was regarded a genuine micromorph.

Family BEUDANTICERATIDAE Breistroffer, 1953

(*nom. transl.* herein *ex* Beudanticeratinae)

Discussion. Monophyletic derivatives of Puzosiidae (Wright, 1957, 1996), comprising the subfamilies Beudanticeratinae and Rapidoplacenticeratinae. Mostly involute, compressed and high whorled, with strong (primitive) to feeble (derived) flexuous ornament, often with shallow constrictions. *L. - U. Alb.*

Subfamily Beudanticeratinae Breistroffer, 1953

= Beidanticeratinae Breskovski, 1977 (*laps. cal.*).
= Eodesmoceratinae Spath, 1923 (*nom. dub.*)

Discussion. A seemingly monophyletic group with an origin in Anapuzosiinae, close to *Pseudohaploceras*; typically involute, compressed and high whorled with prorsiradiate constrictions, weak ribs and a suture with shallow E and large asymmetrically bifid L; superficially similar to some Barremitinae. *M. Apt. - lower U. Alb.*

Zuercherella CASEY, 1954 [*Desmoceras zuercheri* Jacob & Tobler, 1906; OD]. Small, involute, compressed, with broad convex flanks, rounded venter and compressed oval whorl section; 5-10 falciform constrictions per whorl backed by strong labial collar over venter, with distinct fine ribs on outer flank and venter; suture with rather wide E, narrow E/L, relatively broad asymmetrically bifid L much deeper than E, narrow stemmed asymmetrically trifid L/U_2, and at least 6 auxiliary elements. *L. - M. Alb.*

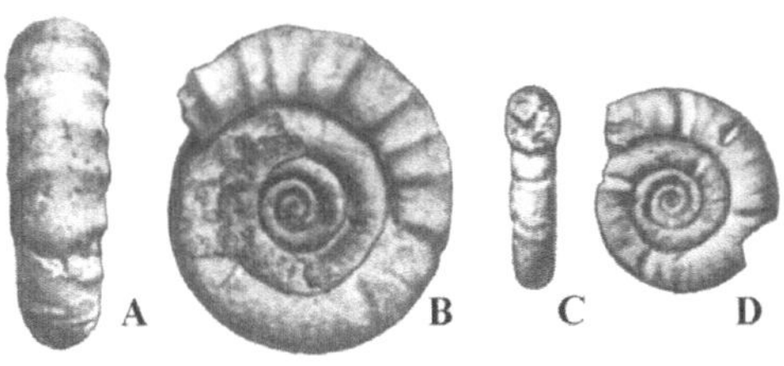

Fig. 203. *Silesitoides.*

Fig. 204. Puzosiidae.

Fig. 206C-D - *Z. zuercheri* (Jacob & Tobler).

= *Zucherella* Wright, 1957 (incorrect emend.).

Kislovodskoceras POLKOVOY, 2021 [*K. kislovodskense*; OD]. Small, involute (U=19-22%), with narrow flattened flanks, subtabulate venter, compressed subtrapezoidal whorl section, angular umbilical edge, 10-14 flexuous constrictions per whorl linguoid over venter, parallel to growth striae; suture with asymmetrically trifid L. *M. Apt.* **Fig. 206P-Q** - *K. kislovodskense* Polkovoy.

Uhligella JACOB, 1907 [*Desmoceras clansayense* Jacob, 1905; SD Kilian, 1907). Small (**m**) to medium sized (**M**); moderately involute, with steep umbilical wall, convex flanks, rounded venter and compressed oval whorl section; in early to middle growth prominent umbilical bullae give rise to coarse, flexuous, slightly prorsiradiate ribs, separated by 2-4 intercalatories on outer flank, crossing venter with slight forward arch; weak shallow constrictions; ribs weakening or effaced on adult body chamber of macroconch. *L. - M. Alb.* **Fig. 208** - *U. clansayensis* (Jacob).

Beudanticeras HITZEL, 1905 [*Ammonites beudanti* Brongniart, 1822; OD]. Small; like *Zuercherella* but more involute (U=12-16%), higher whorled, with flatter flanks, narrowerly rounded venter and compressed subtrigonal to sublanceolate whorl section; sparse prorsiradiate constrictions, mainly in later growth, marked by labial ridges across venter; ribbing very weak to obsolete; suture highly incised, with asymmetrically bifid L and L/U_2. *M. - Lower U. Alb.* **Fig. 206A-B** - *B. beudanti* (Hitzel).

Beudanticeras dupinianum (**Fig. 206E-F**) with its wider umbilicus and weak ribbing provides the link to *Zuercherella*, but with linguoid constrictions over venter.

= *Buedanticeras* Spath, 1923 (*laps. cal.*).

= *Beaudanticeras* Kennedy, 1993 (*nom. van*).

Pseudorbulites CASEY, 1961 [*Uhligella convergens* Jacob, 1908; OD]. Small; like *Beudanticeras* but more involute, whorls increasing rapidly in height, with biconcave constrictions, highly dissected suture and obsolete ribbing. *L. Alb.* **Fig. 206R-S** - *P. convergens* (Jacob).

= *Pseudorbulites* Breistroffer, 1953 (*nom. nud.*).

Grantziceras IMLAY, 1961 [*Beudanticeras (G) multiconstrictum*, = *Desmoceras affine* Whiteaves, 1893; OD]. Small, like *Beudanticeras* but with inclined umbilical wall, indistinct umbilical shoulder and funnel-shaped umbilicus; variable number (up to 18) of asymmetrical (steepest anteriorly) falciform constrictions to internal mould, at large size marked by labial collar over venter; shell surface with bundled growth striae; adult body chamber shows scaphitoid uncoiling; suture intricate, with asymmetrically bifid E/L, wide trifid L much deeper than E, and descending series of 5-6 auxiliary lobes. *L. Alb.* **Fig. 206M-N** - *G. affine* (Imlay).

= *B. (Grantziceras)* Imlay, 1960 (*nom. nud.*).

Boliteceras WHITEHOUSE, 1928 [*Ammonites daintreei* Etheridge, 1872; OD]. Medium sized, with steep umbilical wall, distinct subrounded umbilical shoulder, slightly convergent flanks, broadly arched venter, somewhat depressed whorl section and whorls not increasing rapidly in height; numerous low broad closely spaced sigmoidal ribs continuous across venter; internal mould with periodic constrictions; suture with fairly broad bifid saddles and trifid lobes. *Upper L. - Lower M. Alb.* **Fig. 206O** - *B. daintreei* (Etheridge).

Cophinoceras WHITEHOUSE, 1928 [*C. ogilviei*; OD]. Moderately large; like *Beudanticeras* but more inflated, with convergent flanks, arched venter and compressed oval whorl section; 8 prorsiradiate, nearly straight collar ribs (constrictions) per whorl, separated by 8-10 short ribs. *U. Alb.* **Fig. 206I-J** - *C. ogilvei* Whitehouse.

= *Beudantiella* Breistroffer, 1947 (unnecess. replacement name).

Eodesmoceras SPATH, 1923 [*Ammonites celestini* Pictet & Campiche, 1860; OD]. Small, based on a stratigraphically

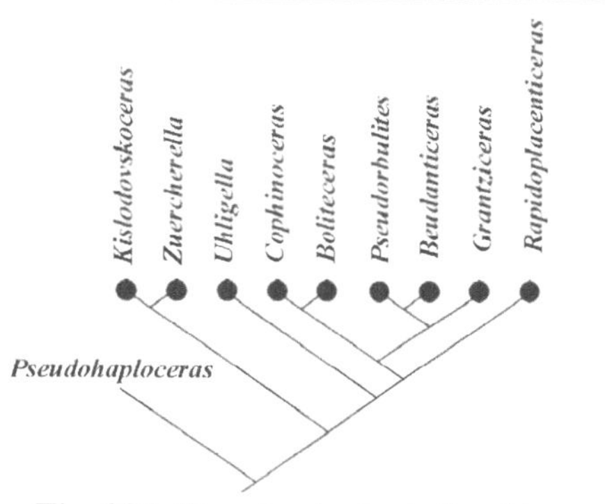

Fig. 205. Hypothesized relationships among Beudanticeratidae.

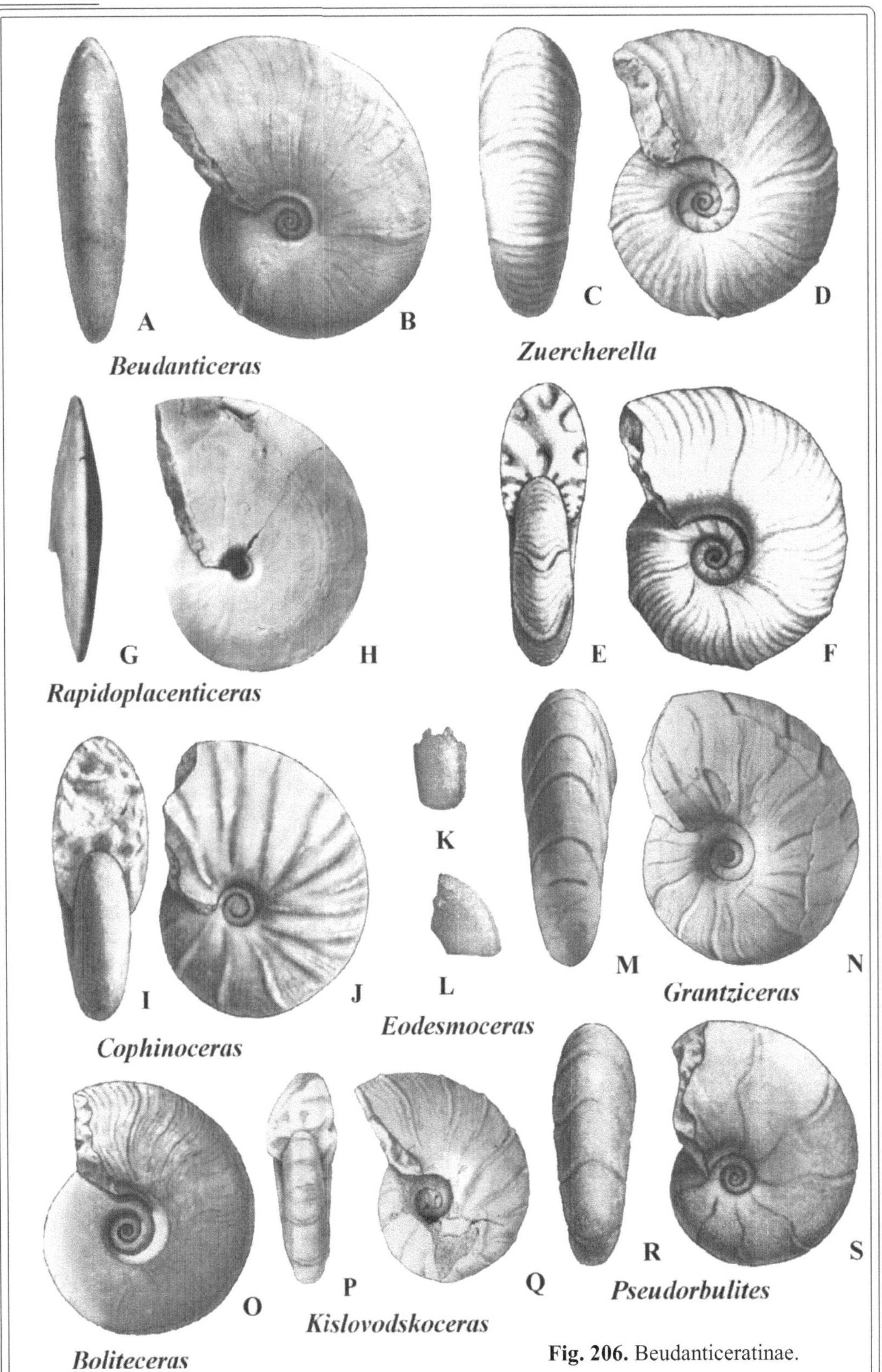

Fig. 206. Beudanticeratinae.

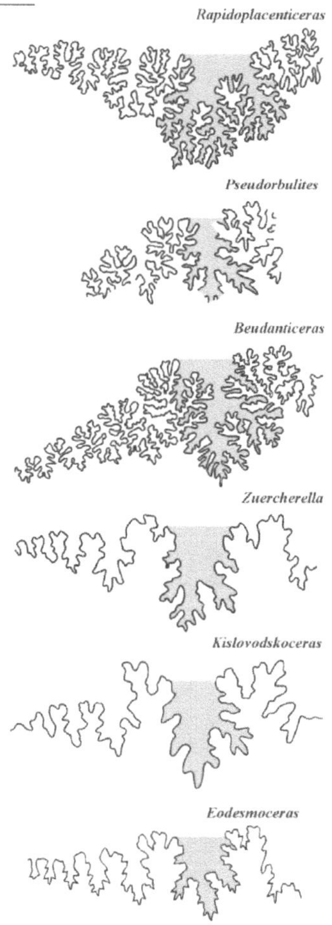

Fig. 207. Beudanticeratid sutures.

misplaced fragment of juvenile *"Puzosia"*, together with fragments of *Desmoceras* and *Oxytropidoceras*, erroneously attributed to the Valanginian (Busnardo & Thieuloy, 1989). The holotype clearly is not an *Oxytropidoceras*, and the suture does not have the retracted suspensive lobe of *Beudanticeras*, but is similar to *Zuercherella*. It is a *nomen dubium*. *?Alb.* **Fig. 206K-L** - *E. celestini* (Pictet & Campiche).

Subfamily Rapidoplacenticeratinae Alabushev, 1997

Discussion. A problematic taxon variously assigned to Placenticeratidae (McLearn, 1972; Alabushev 1988; Alabushev & Wiedmann, 1997; Jagt-Yazykova, 2011) or Beudanticeratinae (Wright, 1996; Riccardi & Medina, 2002; Kennedy & Klinger, 2012), with the type genus considered a subjective junior synonym of *Proplacenticeras* by Jagt-Yazykova (2011). However it is also similar to *Eocleoniceras* but more compressed and with an acute venter. Diagnosis as for genus. *U. Alb. - L. Cenom.*

Rapidoplacenticeras ALABUSHEV, 1988 [*Proplacenticeras sutherlandbrowni* McLearn, 1972; OD). Moderately large oxycone with high steep umbilical wall, subangular umbilical edge, very narrowly rounded venter and lack of ornament save for growth striae; suture with descending trifid E/L and ascendent suspensive lobe. *U. Alb. - L. Cenom.* **Fig. 206G-H** - *R. sutherlandbrowni* (McLearn).

Family CLEONICERATIDAE Whitehouse, 1926

Discussion. Paraphyletic derivatives of Puzosiidae (Owen, 1988). Mostly involute, compressed, with a subtabulate venter; umbilical bullae give rise to single or bifurcating ribs, with or without intercalatories; primitively with constrictions and trifid L, but later unconstricted and with asymmetrically bifid L. Comprises the subfamilies Vnigriceratinae, Cleoniceratinae, Lemuroceratinae and Gastroplitinae. *U. Barrem. - L. Cenom.*

Subfamily Vnigriceratinae Savel'ev, 1972

Discussion. Monophyletic derivatives of Anapuzosiinae, close to *Pseudohaploceras*. Well rounded subquadratic (**m**) to moderately high whorl section (**M**), with rounded to slightly flattened venter; comma-shaped umbilical bullae give rise to bold prorsiradiate ribs, singly or in pairs, that flex forward near the venter and thicken on the ventrolateral shoulders, sometimes weakening over the siphonal line; constrictions may be present; dimorphic, ribbing strongly reduced on body chamber of macroconchs. *U. Barrem. - Lower L. Alb.*

Callizoniceras SPATH, 1923 [*Desmoceras hoyeri* von Koenen, 1902; OD]. Very small; moderately involute, with broad gently convex flanks, rounded venter and compressed oval to subelliptical whorl section; periodic

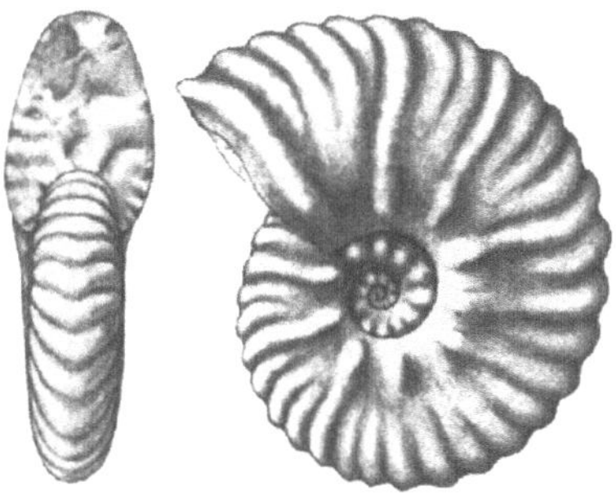

Fig. 208. *Uhligella.*

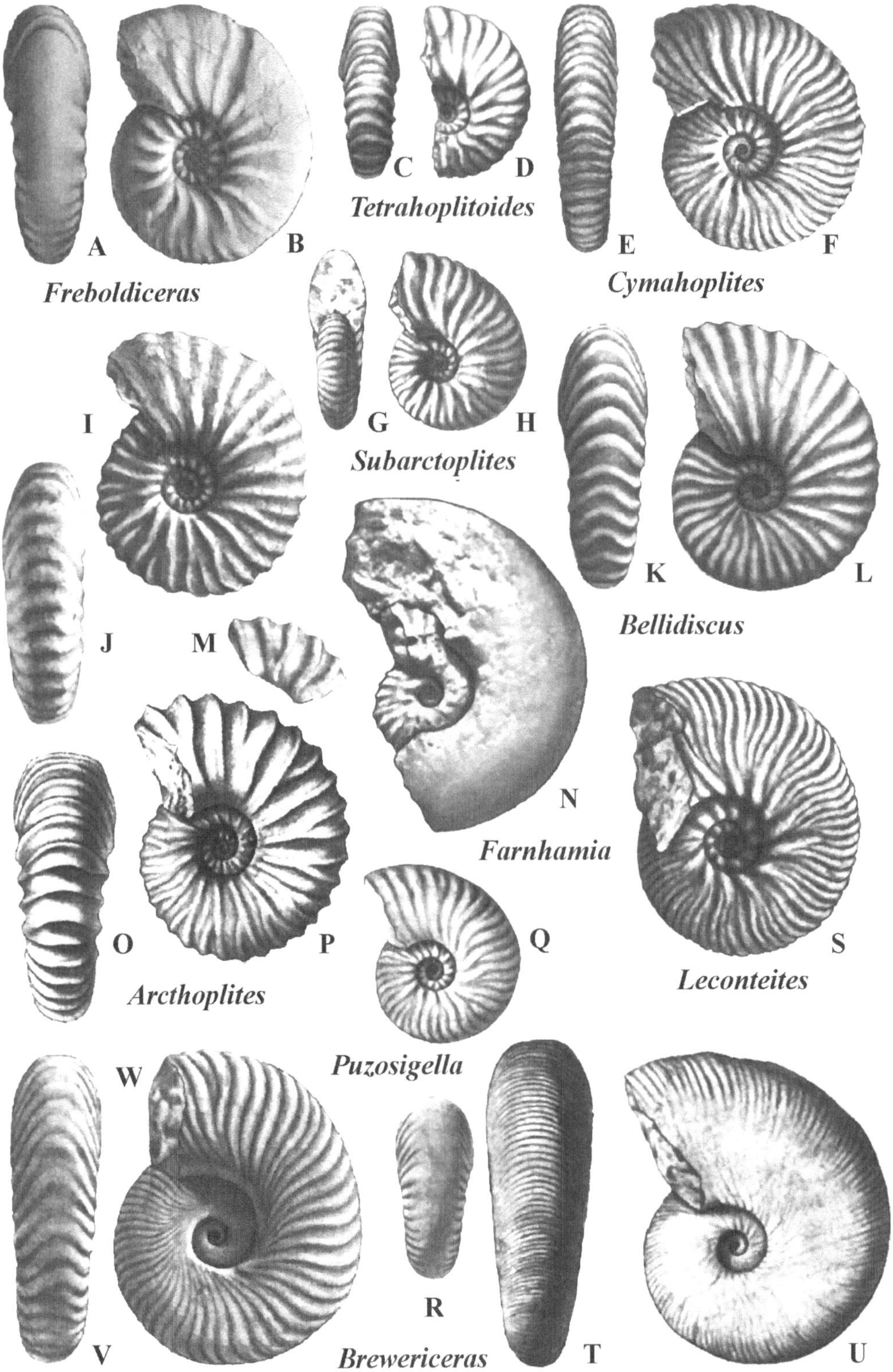

Fig. 209. Cleoniceratidae.

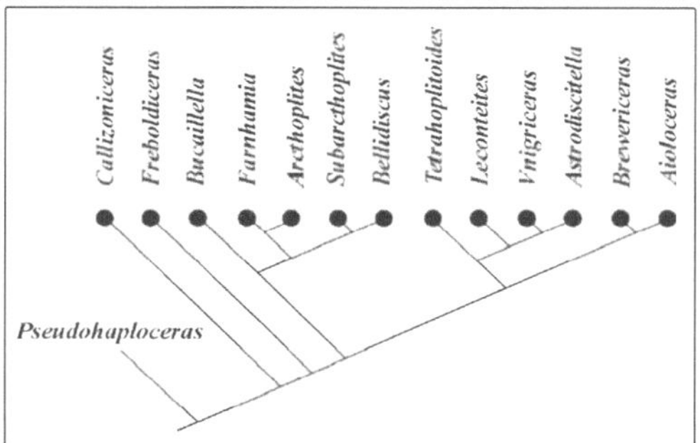

Fig. 210. Hypothesized relationships among Vnigriceratinae.

constrictions separated by rather coarse flexuous ribs that bifurcate on outer flank, or alternate long and short in later growth; suture with deep penetrating E, rather narrow asymmetrically bifid E/L, broad trifid L as deep as E, narrow asymmetrically bifid L/U_2, and few non-retracted auxiliaries. *U. Barrem. - U. Apt.* **Fig. 211** - *C. hoyeri* (von Koenen).

Freboldiceras IMLAY, 1959 [*F. singulare*; OD]. Small; like *Callizoniceras* but with regularly spaced, long, curved, umbilicolateral bullae giving rise to 1-2 weak ribs separated on outer flank by 1-2 intercalatories; ribs effaced from venter; adult body chamber with broad shallow constrictions; suture like *Callizoniceras*. *Lowermost Alb.* **Fig. 209A-B** - *F. singulare* Imlay.

= *Freboldites* Savel'ev, 1973 (*nom. van.*).

Subarcthoplites CASEY, 1954 [*Lemuroceras belli* McLearn, 1945; OD]. Small to medium sized; involute, with weakly convex flanks that converge towards narrowly arched venter and compressed oval whorl section; non-biconcave ribs bifurcate low on flank, separated by an intercalatory, weakening over venter in later growth; occasional constrictions; adult body chamber almost smooth, with broad low radial undulations. *Lower L. Alb.* **Fig. 209G-J** - *S. belli* (McLearn).

Bellidiscus SAVEL'EV, 1973 [*B. probus*; OD]. Very small (**m**) to moderately large (**M**); like *Subarcthoplites* but more involute and high whorled, with distinctly prorsi-radiate ribs tending to alternate long and short rather than forming Y's and crossing rounded venter with forward convexity. *Lower L. Alb.* **Fig. 209K-L** - *B. probus* Savel'ev.

Arcthoplites SPATH, 1925 [*Ammonites jachromensis* Nikitin, 1888; OD]. Small; like *Freboldiceras* but carrying strong ornament to maturity, with distant slightly prorsiradiate Y-shaped ribs bifurcating at midflank and remaining high and strong over venter; constrictions obsolete; suture like *Freboldiceras* but more incised and with squarer E/L. *Lower L. Alb.* **Fig. 209O-P** - *A. jachromensis* (Nikitin).

Farnhamia CASEY, 1960 [*F. farnhamensis*; OD] [**M**]. Very large; inner whorls like *Arcthoplites* with first rounded then subquadratic whorl section; strong blunt sparse ribs

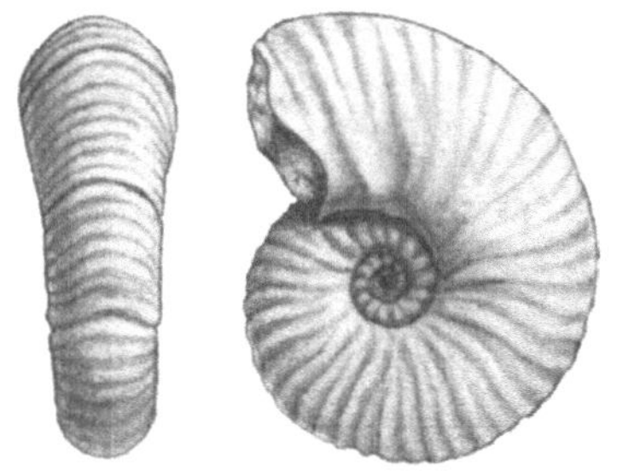

Fig. 211. *Callizoniceras.*

arise in 2's and 3's from umbilical bullae and flex forward near venter; in middle growth ribs may alternate long and short; adult body chamber smooth, with rounded venter and broad radial bulges to lower flank; suture with narrow E/L, broader trifid L, very narrow L/U_2 and 4 auxiliaries. *Lower L. Alb.* **Fig. 209M-N** - *F. farnhamensis* Casey.

Bucaillella DESTOMBES, JUIGNET & RIOULT, 1973 [*B. cayeuxi*; OD]. Small; like *Arcthoplites* but strongly inflated with depressed reniform whorl section; body chamber reduced in cross section and showing scaphitoid uncoiling; suture with broad shallowly bifid E/L, narrow trifid L and few auxiliary elements. *Lower L. Alb.* **Fig. 215I-J** - *B. cayeuxi* Destombes, Juignet & Rioult.

Leconteites CASEY, 1954 [*Desmoceras lecontei* Anderson, 1902; OD]. Small (**m**) to medium sized (**M**); moderately inflated with distinct umbilical bullae that give rise to 2-3 sinuous ribs, separated by sometimes branching intercalatories; ribs project forward on outer flank and in later growth tend to weaken on lower flank and across venter; constrictions present; suture with fairly broad E/L, narrow trifid L deeper than E, and narrow L/U_2. *Upper L. Alb.* **Fig. 209S** - *L. lecontei* (Anderson).

= *Leconteites* Breistroffer, 1951 (*nom. nud.*).

Tetrahoplitoides CASEY, 1954 [*Sonneratia stantoni* Anderson, 1902; OD]. Very small; like juvenile *Leconteites* but more coarsely ribbed, with ribs that thicken slightly on the ventrolateral shoulders and are strong and transverse across the venter; suture like *Leconteites*. *L. Alb.* **Fig. 209C-D** - *T. stantoni* (Anderson).

= *Coloboceras* Crickmay, 1927 (*non* Trouessart, 1889, Acari).

Vnigriceras SAVEL'EV, 1973 [*V. emendatum*; OD] [**m**]. Very small to small; like *Anadesmoceras* but with bundles of 2-3 low weak flexuous ribs arising from feeble umbilical bullae and separated by 1-2 intercalatories; ribbing persists to maturity but weakens across venter; adult body chamber with flat subparallel flanks, broad subtabulate venter, subrectangular whorl section and frequent flexuous constrictions. *Upper L. Alb.* **Fig. 212M-N** - *V. emendatum* (Savel'ev); **Fig. 214** - *V.* sp. (**m**).

Astrodiscitella **nom. nov.** [*Vnigriceras insegestum* Savel'ev; OD]. Very small to small; involute, compressed, high whorled with narrowly rounded venter in early to

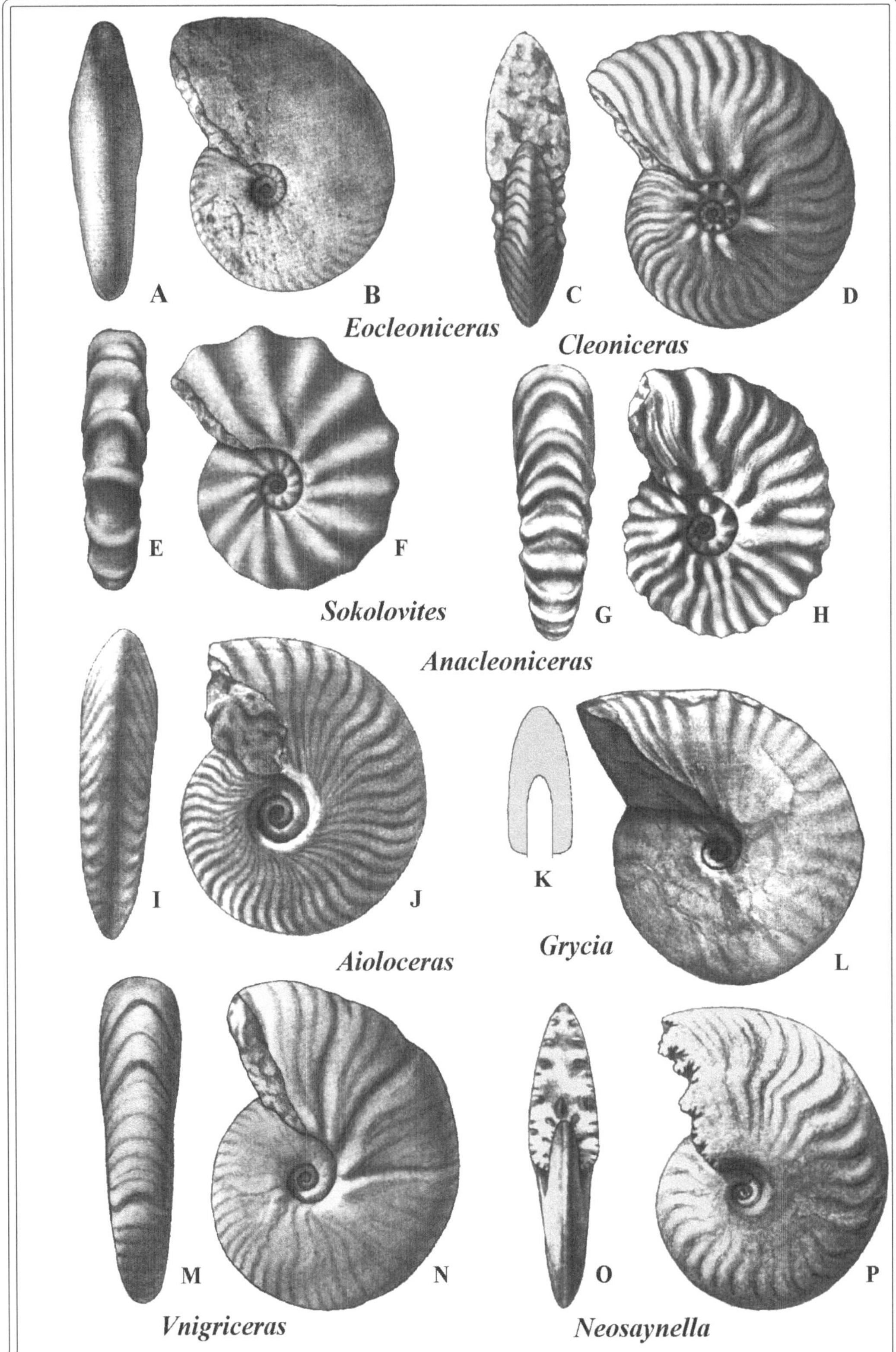

Fig. 212. Cleoniceratinae.

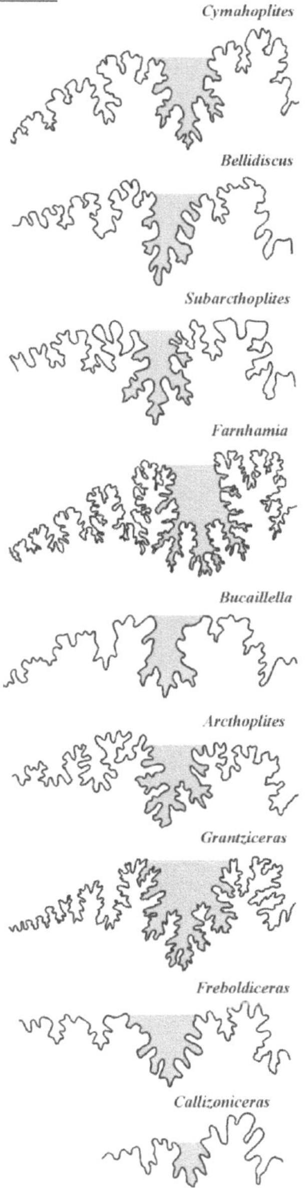

Fig. 213. Cleoniceratid sutures.

middle growth, becoming subtabulate in maturity; strong umbilical bullae give rise to prorsiradiate ribs in bundles of 2-3 with occasional intercalatories, linguoid over venter. *Lower L. Alb.* **Fig. 221A-B** - *A. insegestum* (Savel'ev).

=*Astrodiscus* Savel'ev, 1973 (*non* Schulze 1875, Foraminiferida).

Brewericeras CASEY, 1954 [*Ammonites breweri* Gabb, 1864; OD]. Medium sized; like *Leconteites* but with umbilical bullae lost early leaving only simple or bundled prorsiradiate falcate ribs with asymmetrical profiles (steepest posteriorly); ribs may bifurcate and joined by intercalatories across venter, breaking up into fine threads on body chamber of macroconchs; ribs vary considerably in strength, some individuals being almost smooth; distant labial ridges, noticeable on weakly ribbed forms, mark position of constrictions; suture like *Leconteites* but more deeply incised and interlocking, with shallower L. *Lower M. Alb.* **Fig. 209T-U (M), V-W (m)** - *B. breweri* (Gabb).

Aioloceras WHITEHOUSE, 1926 [*Cleoniceras argentinum* Bonarelli, 1921; OD]. Small to medium sized; like *Brewericeras* but more involute, with rounded umbilical edge and narrowly rounded venter throughout; umbilical bullae lacking; strong blunt flexuous ribs of inner whorls become sparser and thicker on adult body chamber where they are effaced from lower flank and venter; suture with very shallow E, oblique E/L and deep penetrating asymmetrically subtrifid L, and retracted suspensive lobe. *Upper L. Alb.* **Fig. 212I-J** - *A. argentinum* (Bonarelli).

=*Ailoceras* Casey, 1966 (*laps. cal.*).

Subfamily Lemuroceratinae Owen, 1988

Discussion. Monophyletic derivatives of Vnigriceratinae, close to *Vnigriceras* (Savel'ev, 1973); typically very small to moderately large, fairly evolute and compressed, with rather coarse prorsiradiate ribs branching on the outer flank. Sexual dimorphism conspicuous; macroconchs involute and high whorled, with a smooth adult body chamber, frequent constrictions and narrowly rounded venter; microconchs with short lappets. Predominantly Tethyan and Austral in distribution. *Upper L. - Lower M. Alb.*

Lemuroceras SPATH, 1942 [*Pseudohaploceras aburense*; OD]. Small (**m**) to medium sized (**M**); compressed, evolute, with wide shallow umbilicus, evenly rounded umbilical shoulder, flattened flanks, broadly rounded low venter and subquadratic whorl section; innermost whorls almost smooth, with broadly rounded venter; numerous rather distant prorsiradiate umbilicolateral bullae give rise to pairs of sharp, prorsiradiate, flexuous ribs separated by 1-2 long intercalatories; ribs rarely bifurcate and

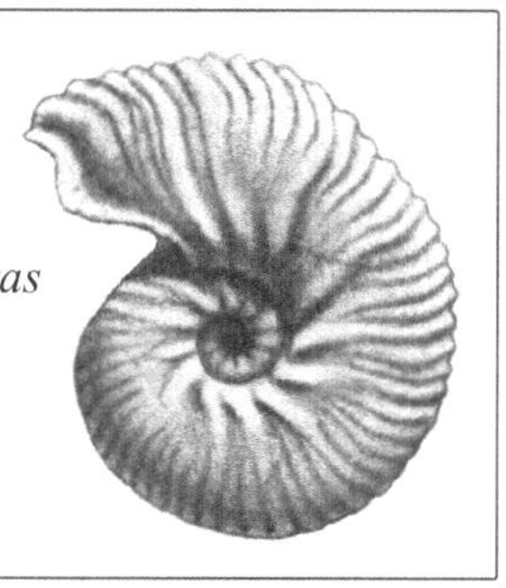

Fig. 214. *Vnigriceras* microconch.

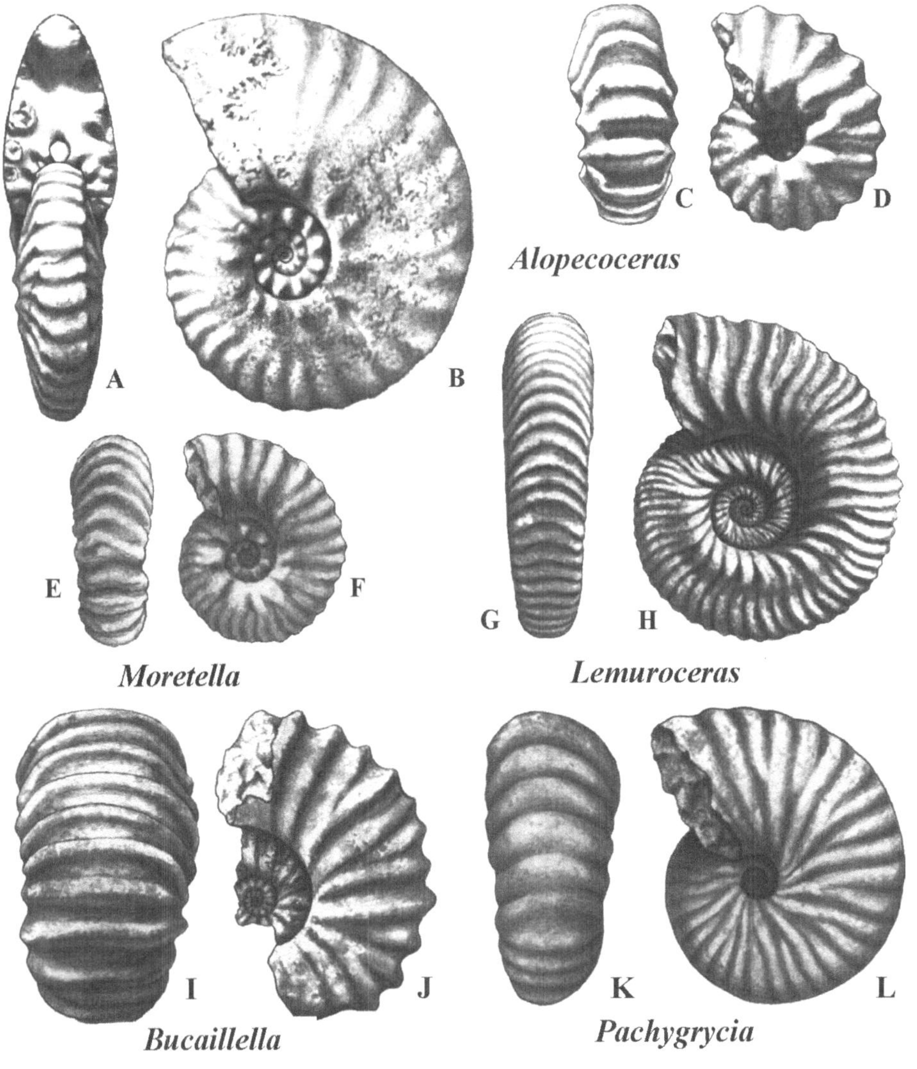

Fig. 215. Cleoniceratidae.

are slightly convex forward across venter with no sign of weakening; suture with rather wide E, square symmetrically bifid E/L, relatively narrow asymmetrically trifid L, asymmetrically bifid L/U_2 and three auxiliary lobes. *Upper L. Alb.* **Fig. 215G-H** - *L. aburense* (Spath).

Cymahoplites SPATH, 1922 [*Ammonites kerenskianus* Bogoslovsky, 1902; M]. Small (**m**) to moderately large (**M**); inner whorls like *Subarcthoplites* but with narrowly rounded to subtabulate venter, denser ribbing that may be effaced from venter, and hint of umbilical bullae; suture with broad subquadratic bifid E/L, very narrow deep asymmetrically bifid L, and moderately wide asymmetrically bifid L/U_2. *Upper L. Alb.* **Fig. 209E-F** - *C.* sp.; **Fig. 217A-B** - *C. kerenskianus* (Bogoslovsky).

= *Vjasemkiceras* Sazonova, 1961 (obj.).

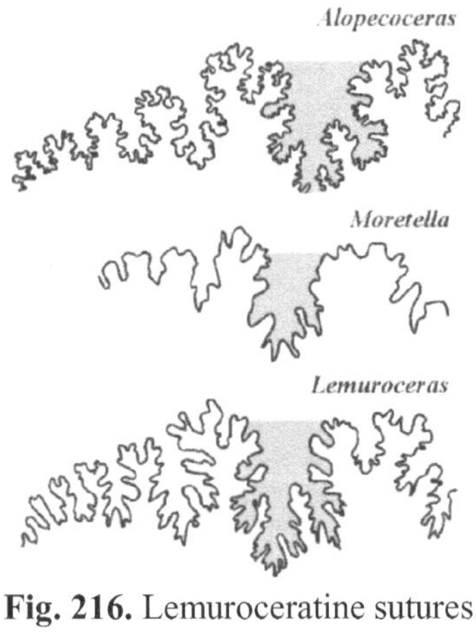

Fig. 216. Lemuroceratine sutures.

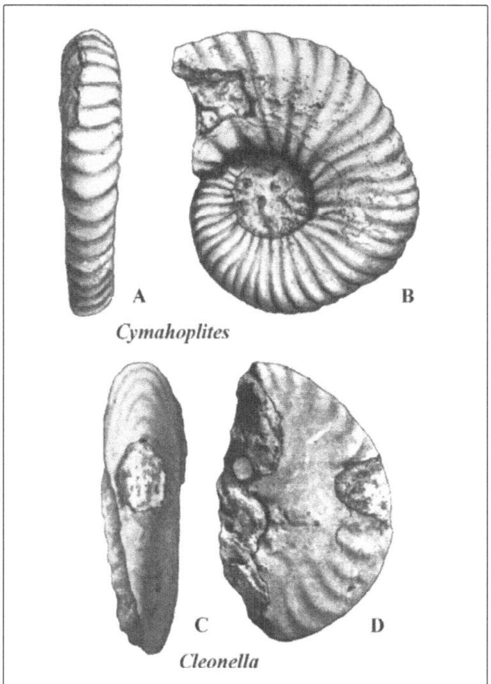

Fig. 217. Cleoniceratidae.

Anahoplitoides CASEY, 1961 [*Saynella splendens* var. *gigas* Sinzow, 1915; OD] [**M**]. Large; like *Cymahoplites* but more involute, with flattened flanks and venter, and compressed trapezoidal whorl section; prominent umbilical bullae and feeble ventrolateral bullae opposite across venter; macroconchs with coarse, bold, distant, prorsiradiate ribs in later growth effaced from venter. *L. Alb.* **Fig. 218** - *A. gigas* (Sinzow) (juv.).

Puzosigella CASEY, 1954 [*Pachydiscus sacramenticus* Anderson, 1902; OD]. Small; like *Leconteites* but more inflated, with rounder flanks and coarser ribbing. *Upper L. Alb.* **Fig. 209Q-R** - *L. sacramenticus* (Anderson).

Moretella COLLIGNON, 1963 [*Pseudosonneratia madagascariensis* Collignon, 1949; OD]. Small; inner whorls like *Lemuroceras* but with subquadratic whorl section, fewer umbilical bullae, straighter more rigid ribs that do not project strongly forward on outer flank and which thicken markedly over venter where they tend to form weak chevrons, more numerous intercalatories and very weak constrictions. *Upper L. Alb.* **Fig. 215E-F** - *M. madagascariensis* (Collignon).

Subfamily Cleoniceratinae Whitehouse, 1926

(*nom. transl.* Wright, 1952 *ex* Cleoniceratidae)

Discussion. Monophyletic derivatives of Vnigriceratinae, close to *Freboldiceras* (Owen, 1988). Compressed, involute, high whorled, with narrowly

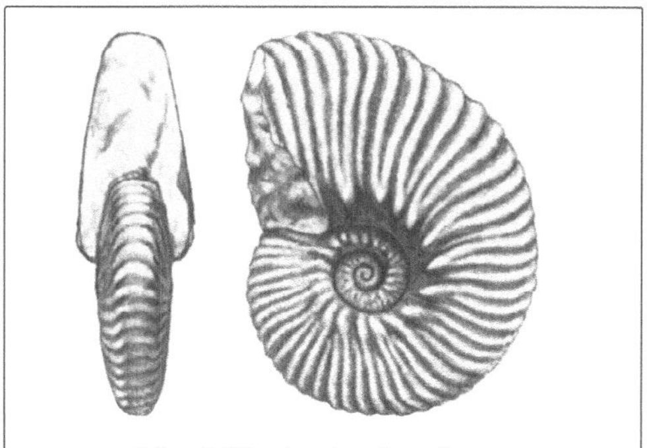

Fig. 218. *Anahoplitoides.*

rounded to tabulate venter and oval to lanceolate whorl section. Sexual dimorphism conspicuous but not well documented, macroconchs with smooth outer whorls and rounded venter. *L. - M. Alb.*

Anadesmoceras CASEY, 1954 [*A. strangulatum*; OD]. Small (**m**) to medium sized (**M**); moderately involute, discoidal, with distinct umbilical edge and flattened convergent flanks; in middle growth with narrowly arched to acute venter and sublanceolate whorl section; innermost whorls with hint of umbilical nodes and sigmoidal ribs, later only with bundled growth striae strongest on inner flank; adult body chamber with rounded venter, subelliptical whorl section and 2-3 sigmoidal constrictions that broaden ventrally; suture like *Cleoniceras*. *L. Alb.* **Fig. 221C** - *A. strangulatum* Casey.

Carloscasceresiceras ETAYO SERNA, 1979 [*C. caceresi*; OD]. Small; like *Anadesmoceras* but more evolute (U=20-30%), with subparallel flanks, evenly rounded venter and subelliptical whorl section; inner whorls with bundled sinuous to biconcave growth striae that project forward near venter; in maturity umbilical seam egresses (U=30-39%) and adult body chamber with very broad prominent sinuous constrictions backed by labial collar; suture with narrow stemmed bifid saddles, asymmetrically trifid L longer than E and feebly retracted auxiliaries. *L. - M. Alb.* **Fig. 221D-E** - *C. caceresi* Etayo Serna.

Differs from *Anadesmoceras* in the trifid L to its suture.

Eocleoniceras SAVEL'EV, 1992 [*Cleoniceras (E.) discors*; OD]. Small; involute, compressed with evenly rounded umbilical shoulder and convex flanks converging to narrowly arched venter; strong rounded flexuous prorsiradiate ribs arise in pairs from prominent umbilical bullae, separated by 1-2 intercalat-

Fig. 219. Partial suture of *Anahoplitoides*.

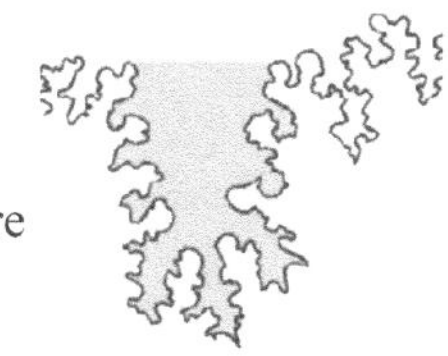

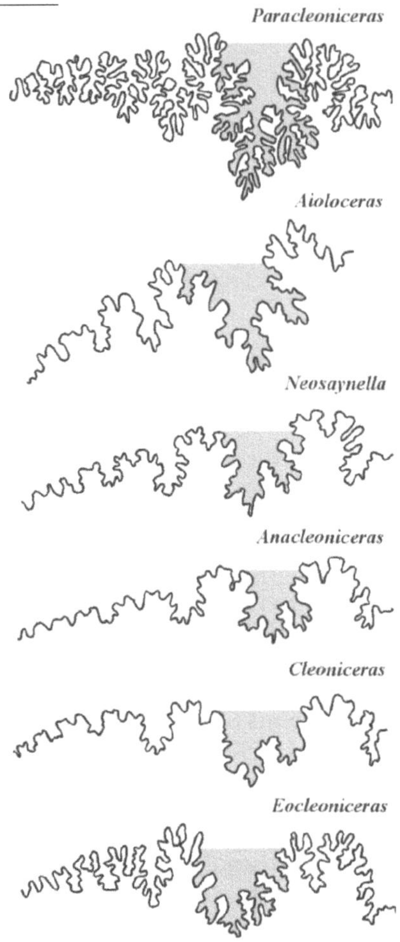

Fig. 220. Cleoniceratid sutures.

ories, sometimes bifurcating on outer flank; ribbing effaced from siphonal line and flanks in maturity; suture with broad bifid E/L, narrower trifid L deeper than E, and several auxiliary elements. *Lower L. Alb.* **Fig. 212A-B** - *E. discors* (Savel'ev).

Chikkimites **gen. nov.** [*Cleoniceras oberhauseri* Lukeneder, Suttner & Bertle, 2013; OD]. Small, with initially straight

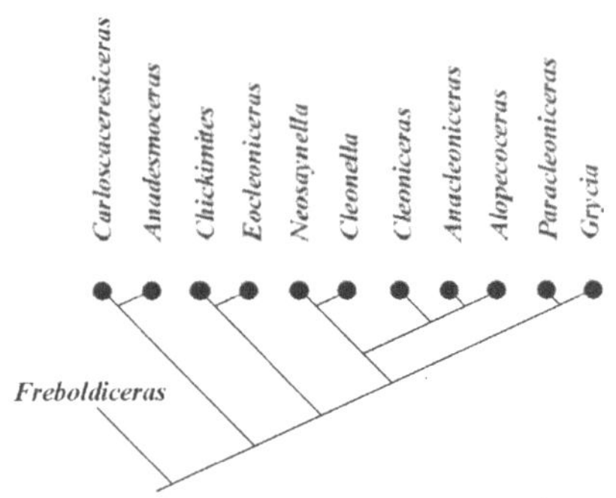

Fig. 222. Hypothesized relationships among Cleoniceratinae.

prorsiradiate ribs broadening ventrally, singly and in pairs; in middle growth strong umbilicolateral bullae give to thin falcate riblets in fasciculate bundles of 3 separated by a long intercalatory, interrupted over siphonal line; ornament coarsens in later growth and evanesces in maturity; suture with broad shallowly indented, symmetrically bifid E/L. *?U. Apt., L. Alb.* **Fig. 224** - *C. oberhauseri* (Lukeneder, Suttner & Bertle).

Cleoniceras PARONA & BONARELLI, 1897 [*Ammonites cleon* d'Orbigny, 1850; OD]. Like *Eocleoniceras* but more compressed and involute, with acute venter in later growth, sublanceolate whorl section, weaker umbilical bullae and feeble ribs on inner flank; adult body chamber smooth; suture like *Eocleoniceras* but with asymmetrically trifid E/L and broad asymmetrically bifid L. *L. - Lower M. Alb.* **Fig. 212C-D** - *C. cleon* (d'Orbigny).

Anacleoniceras MIRZOEV, 1969 [*A. caseyi*, = *Cleoniceras dimorphum* Casey, 1966; OD] [**m**]. Like *Cleoniceras* but with stronger ornament, prominent umbilical bullae, robust low falcoid ribs that persist onto adult body chamber where they pass strongly over flattened venter. *L. Alb.* **Fig. 212G-H** - *A. dimorphum* (Casey).

Alopecoceras KENNEDY & KLINGER, 1978 [*A. ankeritterae*; OD]. Small (**m**) to medium sized (**M**); in early growth inflated and coarsely ribbed with flat convergent flanks, subtabulate venter, trapezoidal whorl section, prominent

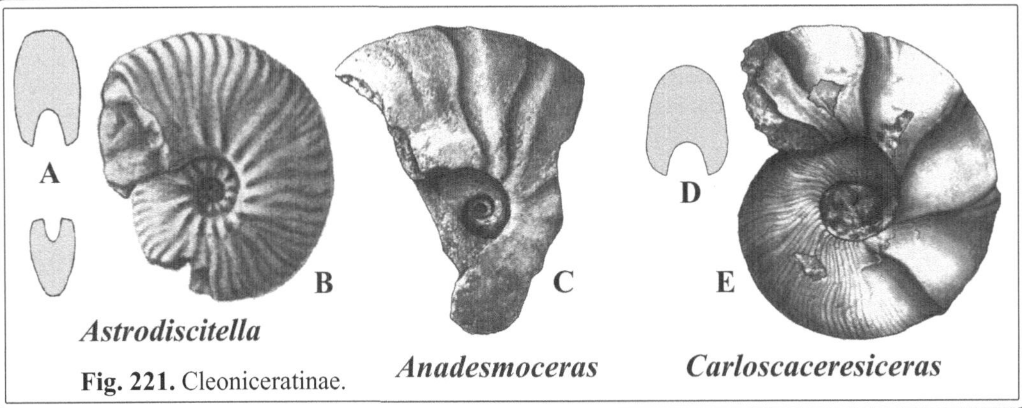

Fig. 221. Cleoniceratinae.

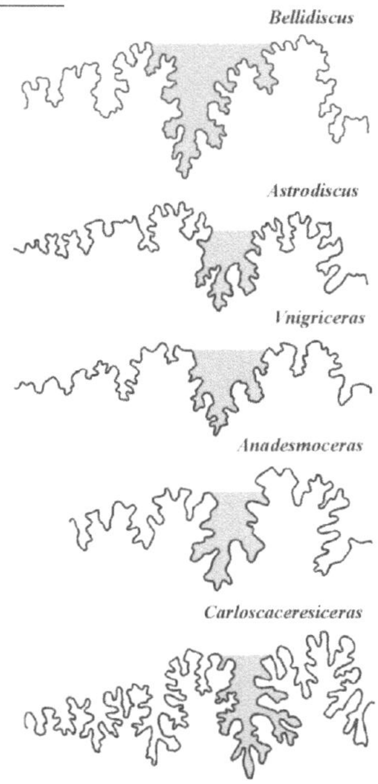

Fig. 223. Cleoniceratid sutures.

sparse rectiradiate umbilicolateral bullae give rise to coarse robust rectiradiate ribs with occasional intercalatories that form weak oblique ventrolateral bullae before crossing venter with slight forward curvature and weakening over siphonal line; outer whorls compressed, high whorled, rather involute, with narrowly rounded venter and sublanceolate whorl section, like macroconchs of *Lemuroceras* with only weak indistinct broad prorsiradiate falcate ribs and growth striae; suture with rather narrow asymmetrically bifid E/L, deep penetrating broad asymmetrically bifid L and rather narrow long asymmetrically bifid L/U_2. *Lower M. Alb.* **Fig. 215A-D** - *A. ankeritterae* Kennedy & Klinger.

Cleonella DESTOMBES, 1970 [*Cleoniceras dimorphum* Casey, 1966; OD]. Very small; like *Cleoniceras* but with rounded venter and ribs very weak on inner flank. *L. Alb.* **Fig. 217C-D** - *C. dimorpha* (Casey).

Neosaynella CASEY, 1954 [*Cleoniceras (N.) inornatum*; OD]. Like *Cleonella* but with tabulate venter to inner

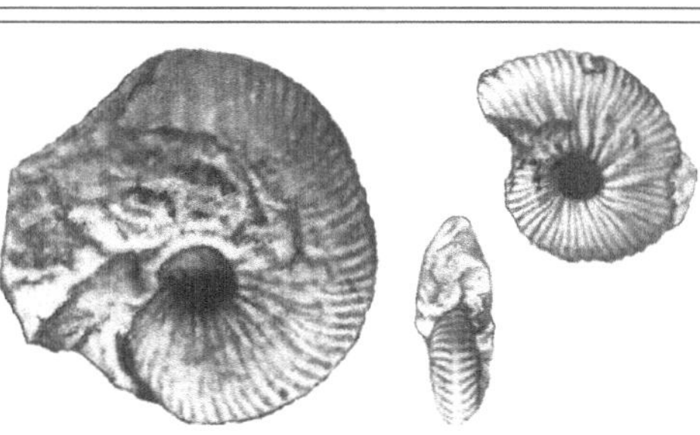

Fig. 224. *Chikkimites.*

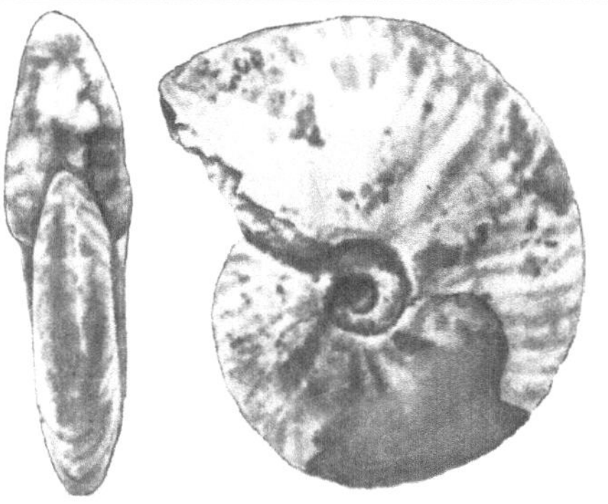

Fig. 225. *Paracleoniceras.*

whorls, sharp umbilical rim, loss of umbilical bullae, ribbing reduced to weak crescents on outer flank, and sharp venter in later growth. *L. Alb.* **Fig. 212O-P** - *N. inornata* (Casey).

Grycia IMLAY, 1961 [*Cleoniceras (G.) sablei*; OD]. Medium sized; like *Cleoniceras* but more involute, with subangular umbilical shoulder in later growth, flat outer flanks, indistinct coarse ribbing that tends to alternate long and short and strengthens with growth; scaphitoid uncoiling of adult body chamber. *L. Alb.* **Fig. 212K-L** - *G. sablei* (Imlay).

Paracleoniceras COLLIGNON, 1963 [*Aioloceras besairiei* Collignon, 1949; OD]. Small to medium sized; like *Grycia* but with vertical umbilical wall, coarser blunter ribs of variable strength, feeble constrictions on internal mould, and more complex suture with deeper E, straight E/L and bifid L, and barely retracted suspensive lobe. A homoeomorph of *Aioloceras*. *L. Alb.* **Fig. 225** - *P. besairiei* (Collignon).

A group of small strongly ornamented lowermost Albian cleoniceratids from Madagascar, centered around *"Pseudosonneratia" sakalava* Collignon (**Fig. 227**), are very similar to *Paracleoniceras* and, if not microconchs, belong to a new paedomorphic genus.

Subfamily Gastroplitinae Wright, 1952

Discussion. A seemingly monophyletic group with an origin in Cleoniceratinae, close to *Subarcthoplites* (Casey, 1964), or *Grycia* (Jeletzky, 1980), or *Arcthoplites* (Baraboshkin & Michailova, 2005). Sexual dimorphism marked, macroconchs up to 60 cm diameter, frequently with "smooth", compressed, sublanceolate outer whorls. Microconchs tend to be inflated, carrying coarse ornament to maturity, with scaphitoid uncoiling, apertural rostra, coarse asymmetrical ribs (steepest posteriorly), distinct ventrolateral shoulders and tabulate venter to adult body chamber. The suture is typically cleoniceratid, with

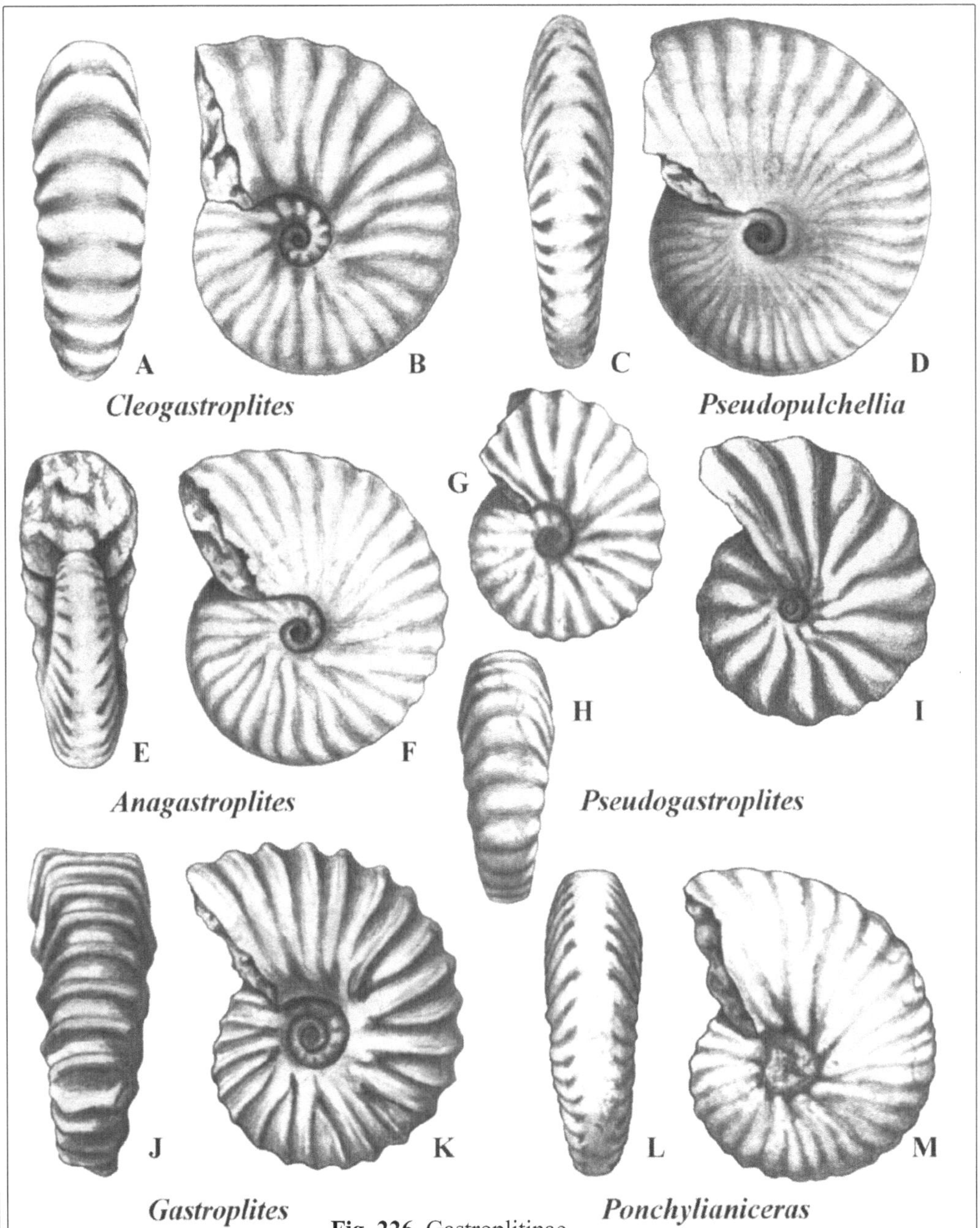

Fig. 226. Gastroplitinae.

a shallow E, subrectangular E/L and asymmetrically bifid L. Endemic to the Boreal Realm. *L. Alb. - L. Cenom.*

Cleogastroplites JELETZKY, 1980 [*C. aberrans*; OD]. Very small, involute, compressed; in early to middle growth like *Eocleoniceras* with somewhat flexuous prorsiradiate ribs bifurcating from weak umbilical bullae and interrupted across venter; in maturity ribs trifurcate from umbilical bullae and adult body chamber becomes subrectangular with tabulate venter; suture simple with shallow E, shallowly indented bifid E/L, broad subsymmetrically bifid L with cauliflower-like adventive saddle, narrow bifid L/U_2 with 4 auxiliary lobes. *L. Alb.* **Fig. 226A-B** - *C. aberrans* Jeletzky.

Pseudopulchellia IMLAY, 1961 [*P. pattoni*; OD] [**m**]. Medium sized; in early growth like *Cleogastroplites* with closely spaced bifurcating rib bundles but ribs coarsening in maturity, becoming sparse and single on adult

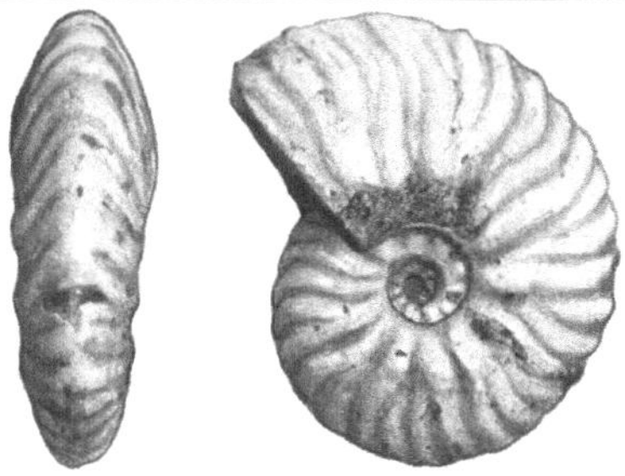

Fig. 227. *"Pseudosonneratia" sakalava* Collignon.

body chamber and effaced from lower flank; in later growth with steep sided, rounded to flat topped, fold-like ribs that broaden markedly towards venter, forming incipient ventrolateral tubercles and crossing venter as very broad weak undulations; suture more incised than *Cleogastroplites*. *L. Upper M. Alb.* **Fig. 226C-D** - *P. pattoni* Imlay.

Sokolovites CASEY, 1966 [*S. subdragunovi*; OD]. Medium sized; like *Cleogastroplites* but with narrowly rounded umbilical shoulder, broad flattish flanks, narrow subtabulate venter and subrectangular whorl section; prominent distant umbilical bullae give rise to pairs of low broad distant ribs that form weak ventrolateral bullae before crossing venter transversely; adult body chamber shows scaphitoid uncoiling; suture with subtrifid L. *L. Alb.* **Fig. 212E-F** - *S. subdragunovi* Casey.

Stelckiceras JELETZKY, 1980 [*Placenticeras (perezianum? var.) liardense* Whiteaves, 1889; OD] [**M**]. Very large; inner whorls like *Pseudopulchellia* but with early appearance of tabulate venter, ribs with triangular cross section, pronounced forward bend of ribs near venter and their weakening or effacement from venter; suture very incised, closely spaced and frequently overlapping, with extremely long narrow asymmetrically to symmetrically trifid L and trifid auxiliary lobes. *Upper M. Alb.* **Fig. 229B** - *S. liardense* (Whiteaves).

Perhaps a macroconch of *Pseudopulchellia* (Wright, 1996).

Anagastroplites JELETZKY, 1980 [*A. tozeri*; OD] [**m**]. Small; like *Cleogastroplites* but with subdued ornament, small umbilical bullae, thin ribs that weaken across subrounded venter and do not broaden ventrally, and suture with more asymmetrical L. *Upper M. Alb.* **Fig. 226E-F** - *A. tozeri* Jeletzky.

Pachygrycia JELETZKY & STELCK, 1981 [*P. canadensis*; OD]. Small; like *Anagastroplites* but more inflated, with convex flanks and broadly rounded venter over which ribbing remains strong. *L. Alb.* **Fig. 215K-L** - *P. canadensis* Jeletzky & Stelck.

Pseudogastroplites JELETZKY, 1980 [*P. arcticus*; OD] [**m**]. Small; strong, regular, round topped, bifurcating ribs broadening ventrally, becoming fold-like and transverse across venter; ribs as wide as or wider than interspaces; venter rounded in early to middle growth, becoming tabulate on adult body chamber; suture like *Cleogastroplites* but with deeper penetrating adventive saddle to L. *Upper M. - Lower U. Alb.* **Fig. 226G-I** - *P. arcticus* Jelezky.

Stotticeras JELETZKY, 1980 [*S. crowense*; OD] [**M**]. Large; in middle growth like *Pseudogastroplites* with ribs arising from umbilicolateral bullae, later with smooth outer whorls; suture with long, slender stemmed, extremely denticulate lobes and subsymmetrically bifid L. *M. Alb.* **Fig. 229A** - *S. crowense* (Jeletzky).

Irenicoceras WARREN & STELCK, 1958 [*I. bahani*; OD] [**M**]. Like *Stotticeras* but very compressed, involute, with more numerous, sharp crested, bifurcating ribs persisting to maturity and subdued ventrolateral bullae; suture with very shallow E and small oblique E/L. *?U. Alb.* **Fig. 229F** - *I. bahani* Warren & Stelck.

Gastroplites MCLEARN, 1930 [*Hoplites canadensis* Whiteaves, 1893; OD] [**m**]. Small (**m**) to moderately large (**M**); like *Pseudogastroplites* but with tabulate venter with distinct ventrolateral angles from middle growth; rather rigid, thin, sharp crested, asymmetrical ribs (steepest posteriorly) arise from umbilicolateral bullae, separated by wide interspaces; incipient ventrolateral tubercles and ribs convex forward across venter; macroconch with smooth outer whorls; suture simple, like *Pseudogastroplites*. *M. Alb.* **Fig. 226J-K** - *G. canadensis* (Whiteaves).

Paragastroplites IMLAY, 1961 [*Gastroplites spiekeri* McLearn, 1931; OD] [**m**]. Like *Gastroplites* but more involute and compressed, with convergent flanks and narrowly arched venter; distant single (with an intercalatory) or paired ribs arise from umbilical shoulder (without umbilical bullae) and flex forward on outer flank to terminate in small ventrolateral tubercles; ribbing very weak or effaced from venter in early to middle growth; suture very simple, almost ceratitic, with regular descend-

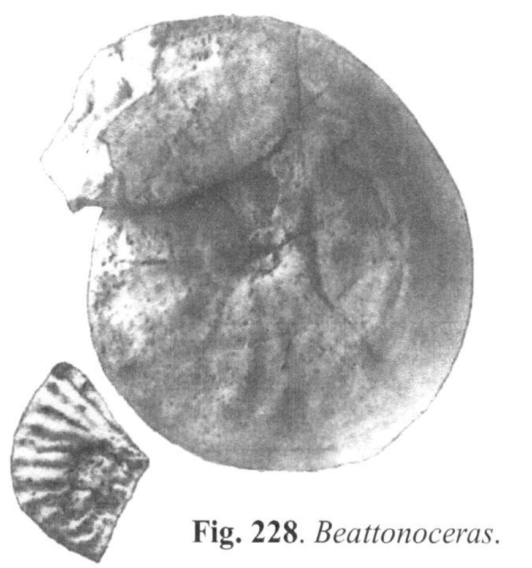

Fig. 228. *Beattonoceras*.

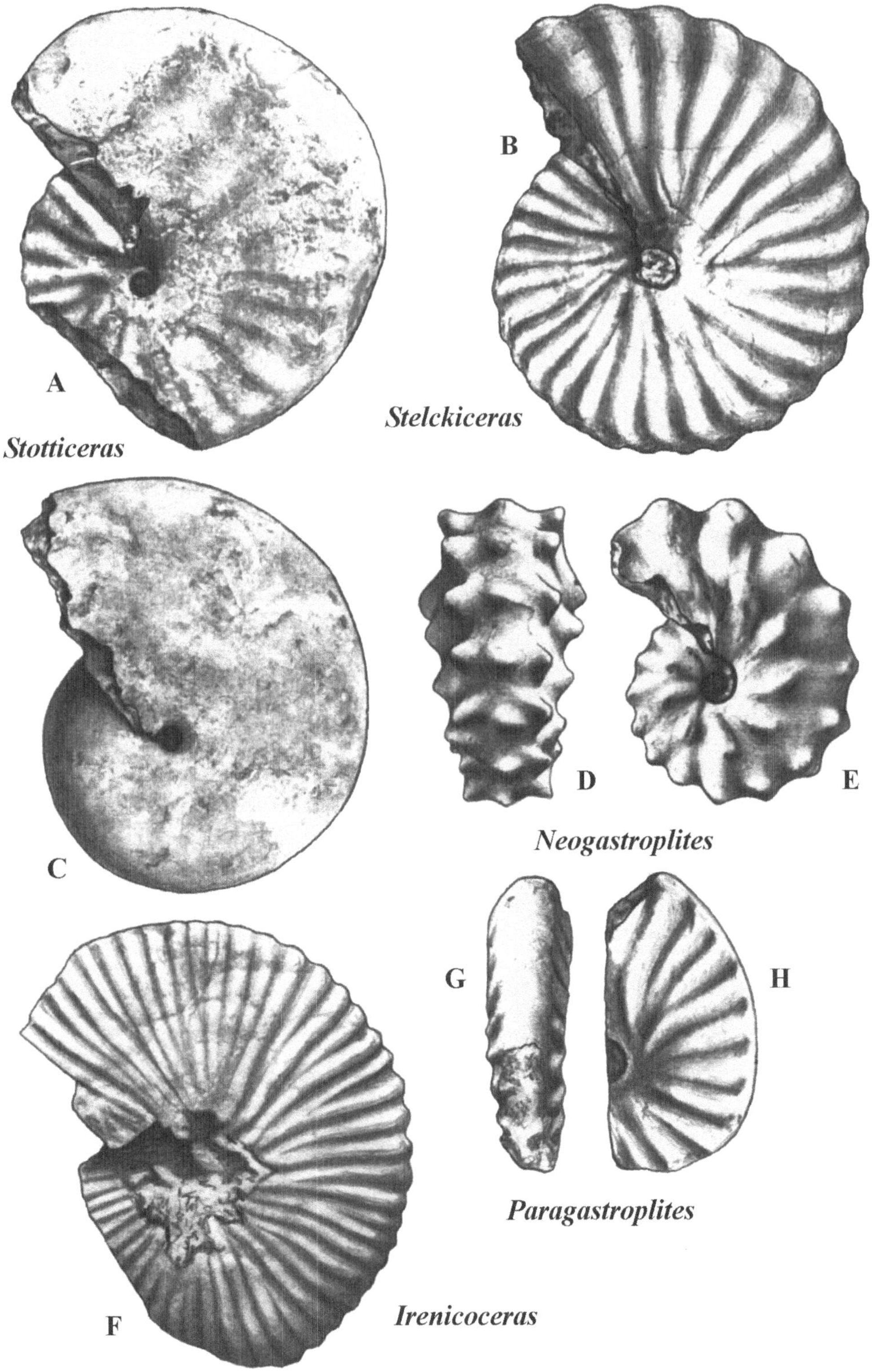

Fig. 229. Gastroplitinae.

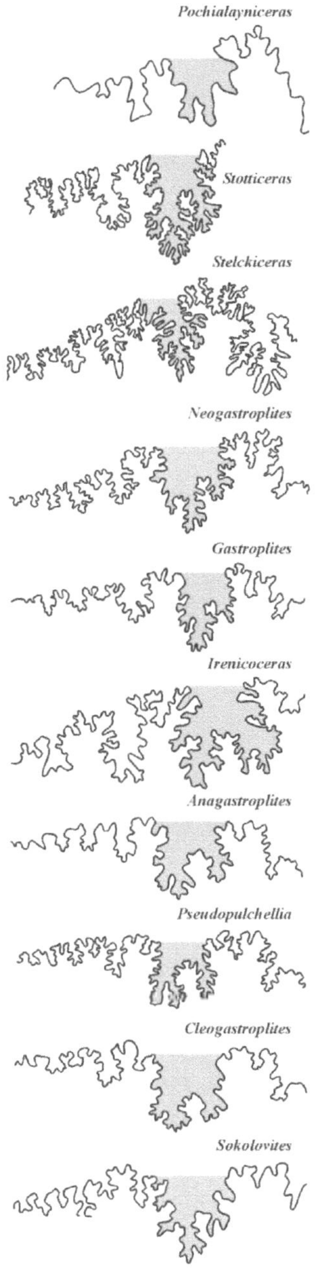

Fig. 230. Gastroplitine sutures.

ing series of 6 auxiliary lobes. *M. Alb.* **Fig. 229G-H** - *P. spiekeri* (McLearn).

Neogastroplites McLEARN, 1931 [*Buchiceras? cornutum* Whiteaves, 1885; OD] [**m**]. Small (**m**) to medium sized (**M**); like *Gastroplites* but more involute, with distinct umbilical edge, coarser ornament with ribs subordinate to tubercles, prominent low lateral (umbilical) and ventrolateral tubercles, and gently convex venter with distinct low siphonal bullae from early growth; macroconchs with smooth outer whorls; suture with moderately broad E, broad asymmetrically bifid E/L with inner margin tending to slope towards wide asymmetrically bifid L that is much deeper than E, and moderately wide asymmetrically bifid L/U_2. *U. Alb., ?L. Cenom.* **Fig. 229C (M), D-E (m)** - *N. cornutus* (Whiteaves).

Beattonoceras WARREN & STELCK, 1958 [*B. beattonense*; OD] [**M**]. Very large; very compressed, involute, with distinct ventrolateral nodes on penultimate whorl, absence of umbilical nodes, and suture like *Neogastroplites* with short broad stemmed lobes and asymmetrically trifid downard tapering L. *Upper L. Cenom.* **Fig. 228** - *B. beattonense* Warren & Stelck.

Pochialyniceras ABALUSHEV & ABALUSHEVA, 1988 [*Anahoplites yakounensis* McLearn, 1972; OD] [**m**]. Small (**m**) to medium sized (**M**); like inner whorls of *Cleogastroplites* but more involute, with subangular ventrolateral shoulders, narrow tabulate venter, oblique ventrolateral bullae, ribs weak or effaced from venter; suture with very asymmetrical L; macroconchs with "smooth" outer whorls; a homoeomorph of *Anahoplites. L. Alb.* **Fig. 226L-M** - *P. yakounense* (McLearn).

Family KOSSMATICERATIDAE Spath, 1922

Discussion. Monophyletic derivatives of Puzosiidae with bundled ribbing; comprises the subfamilies Marshallitinae, Kossmaticeratinae, Grossouvritinae and Brahmaitinae. Strongly dimorphic although dimorphs have not been satisfactorily discriminated; microconchs with lateral lappets. *U. Apt. - U. Maastr.*

Subfamily Marshallitinae Matsumoto, 1955

Discussion.Monophyletic derivatives of Puzosiinae, close to *Melchiorites* (Anderson, 1938); dense fine bundled ribbing arises from umbilical bullae and is truncated posteriorly by oblique constrictions. Comprises the tribes Marshallitini and Eogunnaritini. *U. Apt. - U. Turon.*

Tribe Marshallitini Matsumoto, 1955

(*nom. transl.* herein *ex* Marshallitinae)

Type genus: *Marshallites* Matsumoto, 1955.

Diagnosis. Generally compressed, more or less flat sided, with tendency for venter to flatten; some with ventrolateral nodes. *U. Apt. - L. Turon.*

Hulenites MATSUMOTO, 1955 [*Puzosia reesidei* Anderson, 1938; OD]. Very small; compressed, moderately evolute (U=26-31%), with low subvertical umbilical wall, broad gently convex to flattened flanks, evenly rounded venter that becomes somewhat flattened in maturity, and com-

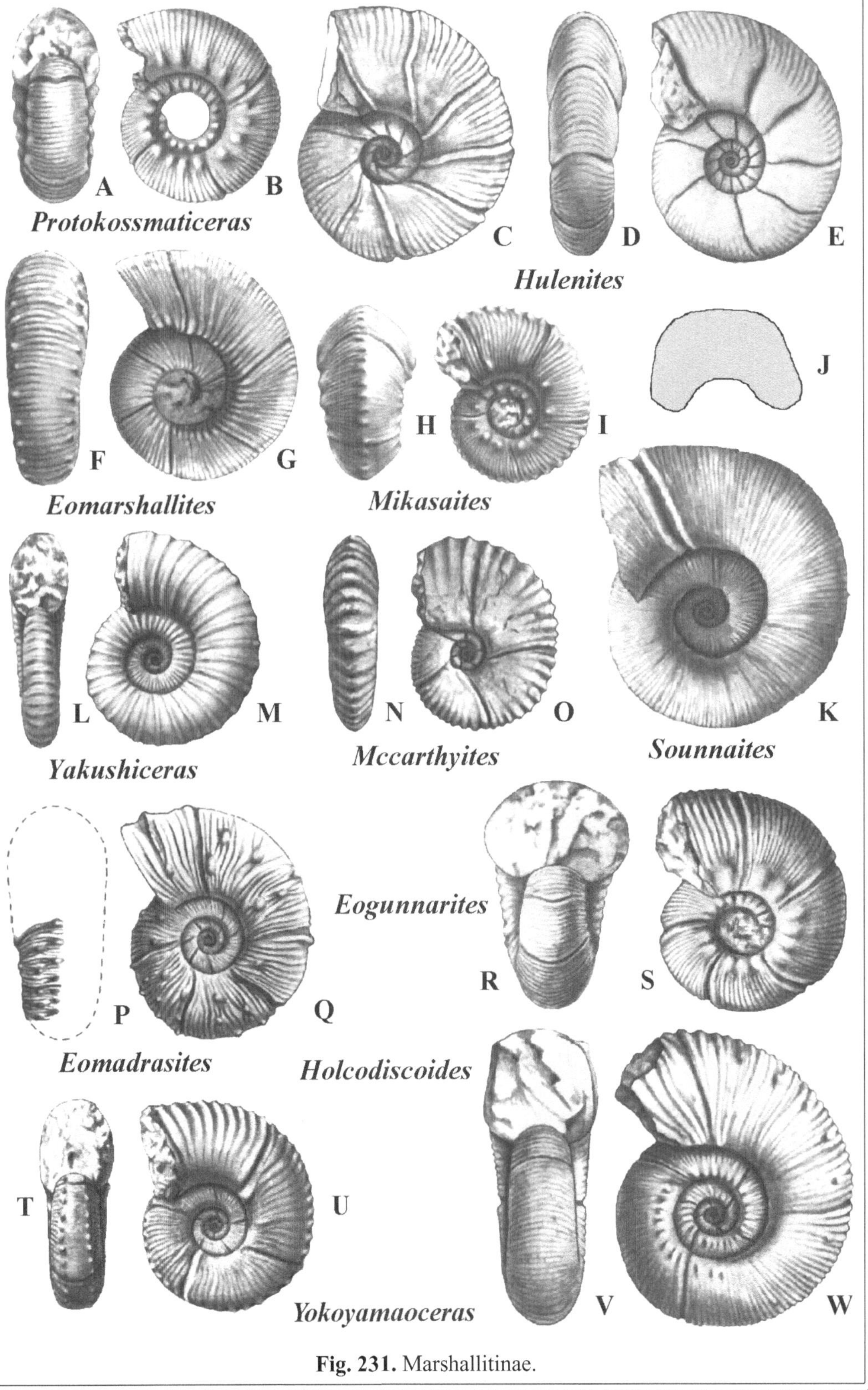

Fig. 231. Marshallitinae.

pressed subelliptical whorl section; fine sharp sigmoidal ribs commence on umbilical shoulder, frequently branch at various levels, and cross venter with forward projection and adoral sinus; more or less distinct spiral groove to midflank on internal moulds; 4-9 prorsiradiate flexuous constrictions flex forward near venter; suture with narrow bifid saddles, trifid L and retracted suspensive lobe. *L. Alb.* **Fig. 231C-E** - *H. reesidei* (Anderson).

Marshallites Matsumoto, 1955 [*M. compressus*; OD]. Small (**m**) to medium sized (**M**); like *Hulenites* but more evolute, with wiry ribs arising in bundles (2-3) from numerous incipient to weak umbilical bullae and sometimes branching on flank, more numerous (up to 14) irregularly spaced, weakly collared constrictions that cut obliquely across ribbing, and more complex suture with narrower stemmed saddles; outer whorls become flat sided, high whorled and with subtabulate venter; microconchs with lappets. *U. Apt. - U. Cenom.* **Fig. 233A-B** - *M. compressus* Matsumoto (**m**), **Fig. 233C-D** - *M. canadensis* (McLearn) (**M**).

Eomarshallites Medina & Rinaldi, 1986 [*E. espinosus*; OD]. Very small; like *Marshallites* but more evolute, with subquadratic whorls, straighter more rigid ribs and constrictions, and small ventrolateral nodes to inner whorls. *U. Alb.* **Fig. 231F-G** - *E. espinosus* Medina & Rinaldi.

Yeharites Matsumoto, 1991 [*Y. kobayashii* Matsumoto & Takahashi, 1991; OD]. Small; like *Marshallites* but more involute and with coarser straighter ribs, fewer constrictions and mostly single ribs in maturity. *U. Alb. - L. Cenom.* **Fig. 233E-F** - *Y. kobayashii* Matsumoto & Takahashi.

Holcodiscoides Spath, 1922 [*Ammonites cliveanus* Stoliczka, 1866; OD]. Small; like *Eomarshallites* but adult body chamber with subquadratic whorl section and sporadic ventrolateral tubercles. *Turon.* **Fig. 231V-W** - *H. clivianus* (Stoliczka).

Yakushiceras Matsumoto, 1991 [*Y. takahasii*; OD]. Very small; serpenticone, with compressed subovate whorl section and flattened venter; ribs predominantly single, dense on inner whorls, distant on outer, with feeble or no umbilical bullae; all ribs with subdued ventrolateral tubercles. *L. Cenom.* **Fig. 231L-M** - *Y. takahasii* Matsumoto.

Tribe Eogunnaritini nov.

Type genus: *Eogunnarites* Wright & Matsumoto, 1954.

Diagnosis. Generally with inflated whorls, depressed whorl section and rounded venter. *L. Cenom.*

Eogunnarites Wright & Matsumoto, 1954 [*Olcostephanus unicus* Yabe, 1904; OD]. Medium sized, moderately evolute; inner whorls with deep steep sided umbilicus, angular umbilical shoulder, convex flanks, broadly rounded venter and depressed reniform whorl section; fine sharp concave ribs arise in bundles of 3-4 from umbilical bullae separated by intercalatories, frequently bifurcating on flank; regular oblique constrictions truncate ribbing to the posterior. *L. Cenom.* **Fig. 231R-S** - *E. unicus* (Yabe).

= *Eogunnarites* Matsumoto, 1942 (*nom. nud.*).

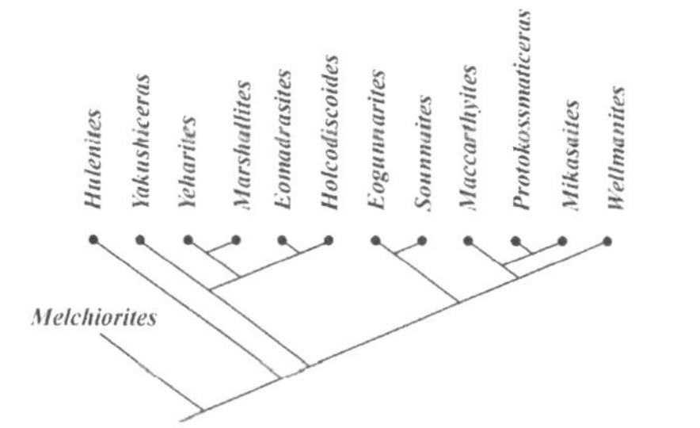

Fig. 232. Hypothesized relationships among Marshalliinae.

Sounnaites Matsumoto & Inoma, 1991 [*Eogunnarites alaskaensis* Matsumoto, 1959; OD] [**M**]. Moderately large; like *Eogunnarites* but more evolute in early growth, with flattened venter, subtrapezoidal whorl section and without umbilical bullae. *L. Cenom.* **Fig. 231J-K** - *S. alaskaensis* (Matsumoto).

Perhaps a macroconch of *Eogunnarites* (Wright, 1996).

Protokossmaticeras Collignon, 1964 [*P. madagascariense*; OD]. Very small; like *Eogunnarites* but very evolute, with narrowly rounded venter and trigonally ovate whorl section; a homoeomorph of *Spiticeras*. *L. Cenom.* **Fig. 231A-B** - *P. madagascariense* Collignon.

Mikasaites Matsumoto, 1956 [*M. orbicularis*; OD]. Very small; like *Protokossmaticeras* but more inflated, with ribs looping to siphonal bullae. *L. Cenom.* **Fig. 231H-I** - *M. orbicularis* Matsumoto.

Maccarthyites Matsumoto, 1959 [*M. gracilis*; OD]. Very small; compressed, rather involute, with steep umbilical walls, flattened flanks and fastigiate venter; periodic well-marked collared constrictions are prorsiradiate on inner flank becoming radial on outer part; small umbilical tubercles restricted to inner whorls; in early growth narrow distant rectiradiate ribs develop on outer flank and carry siphonal bullae; in maturity some ribs almost reach umbilical wall; suture similar to *Marshallites*. *L. Cenom.* **Fig. 231N-O** - *M. gracilis* Matsumoto.

Wellmanites Wright, 1957b [*W. zelandicus*; OD]. Small; like *Eogunnarites* but with somewhat flattened flanks, feeble ribs confined to outer flank and 1-3 high lateral tubercles between constrictions; in some species adult body chamber with irregular long lateral, ventrolateral and siphonal spines; suture with narrow trifid L. *L. Cenom.* **Fig. 233G-I** - *W. zelandicus* Wright.

Eomadrasites Matsumoto, 1955 [*E. nipponicus*; OD]. Small; like inner whorls of *Sounnaites* but not so inflated, with wiry ribs looping between irregular lateral, ventro-

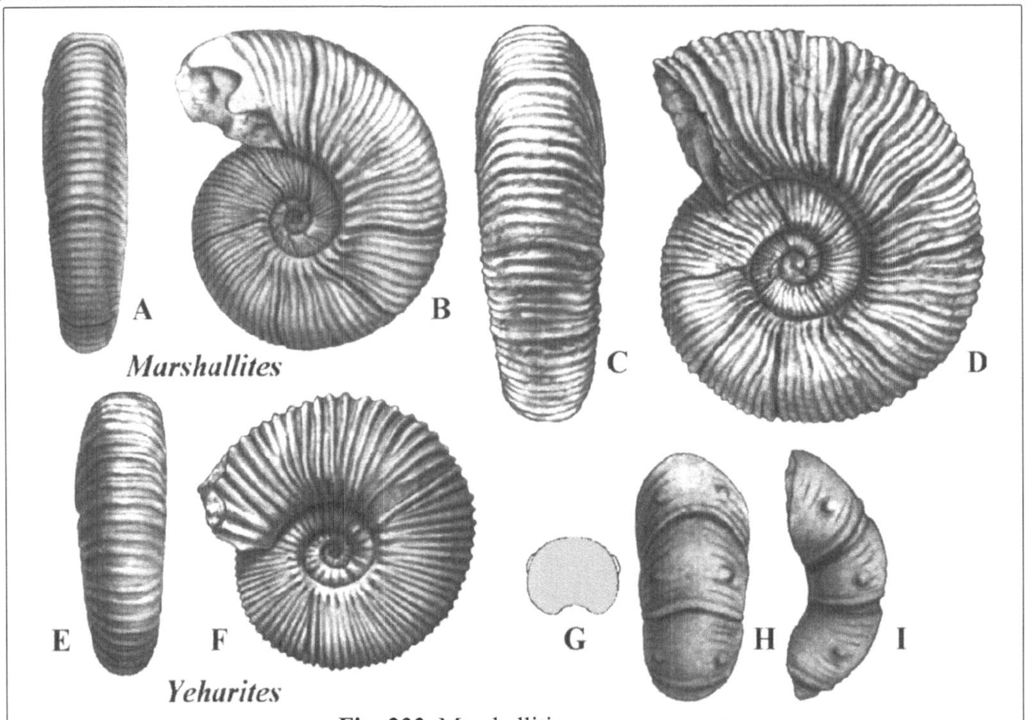

Fig. 233. Marshallitinae.

lateral and siphonal tubercles. *Cenom.* **Fig. 231P-Q** - *E. nipponicus* Matsumoto.

= *Eomadrasites* Matsumoto, 1954 (*nom. nud.*).

= *Jacobitoides* MATSUMOTO, 1954 (*nom. nud.*).

Subfamily Kossmaticeratinae Spath, 1922

Discussion. A paraphyletic taxon comprising derivatives of Marshallitinae, close to *Marshallites*; typically evolute, compressed, mostly with relatively coarse bundled ribbing and oblique collared constrictions that project forward across the venter. Predominantly Gondwanic in distribution where the group radiated. *U. Turon. - U. Maastr.*

Kossmaticeras DE GROSSOUVRE, 1901 [*Ammonites theobaldianus* Stoliczka, 1866; SD Diener, 1925]. Medium sized; compressed, evolute, with slightly compressed oval whorl section and rounded venter throughout ontogeny; in early to middle growth fine straight to slightly flexuous prorsiradiate ribs arise singly from umbilical shoulder, separated by an intercalatory, and bifurcate on outer flank with only single ribs exposed in umbilicus; umbilical bullae lacking; 4-5 prominent oblique collared constrictions per whorl that flex forward on outer flank; suture with narrow bifid saddles, trifid L deeper than E, and retracted suspensive lobe. *U. Turon. - Coni.* **Fig. 235A-B** - *K. theobaldianum* (Stoliczka).

= *Pseudoholcodiscus* Kilian & Reboul, 1909 (*nom. nud.*).

= *Madrasites* Kilian & Reboul, 1909 (obj.).

Yokoyamaoceras WRIGHT & MATSUMOTO, 1954 [*Holcodiscus kotoi* Jimbo, 1894; OD] [**m**]. Very small; like *Holcodiscoides* but more coarsely ribbed with ribs effaced from venter in middle growth; constrictions linguiform over venter; in maturity bituberculate ribs alternate long and short, projecting across venter; blunt siphonal tubercles may develop. *Coni. - Sant.* **Fig. 231T-U** - *Y. kotoi* (Jimbo).

= *Yokoyamaoceras* Shimizu, 1935 (*nom. nud.*).

Neopuzosia MATSUMOTO, 1954 [*K. japonicus* Spath, 1922; OD][**M**]. Small; superficially similar to *Mesopuzosia* but more compressed, with constrictions that cut across ribbing and are strongly projected on outer flank; microconch evolute, with gently convex flanks, narrowly rounded venter and compressed subelliptical whorl section; frequent well marked, gently sigmoidal, collared constrictions project forward on outer flank and are linguoid over venter; early whorls almost smooth; last part of phragmocone with fine, weak, sigmoidal ribs and growth striae, ribs becoming stronger on adult body chamber; suture with narrow stemmed saddles, rather wide E, deeply bifid E/L, deep penetrating asymmetrically trifid L deeper than E, and retracted suspensive lobe. *U. Coni. - L. Camp.* **Fig. 189I-J** - *N. japonicus* (Spath).

According to Maeda (1993) the macroconch of *Yokoyamaoceras*.

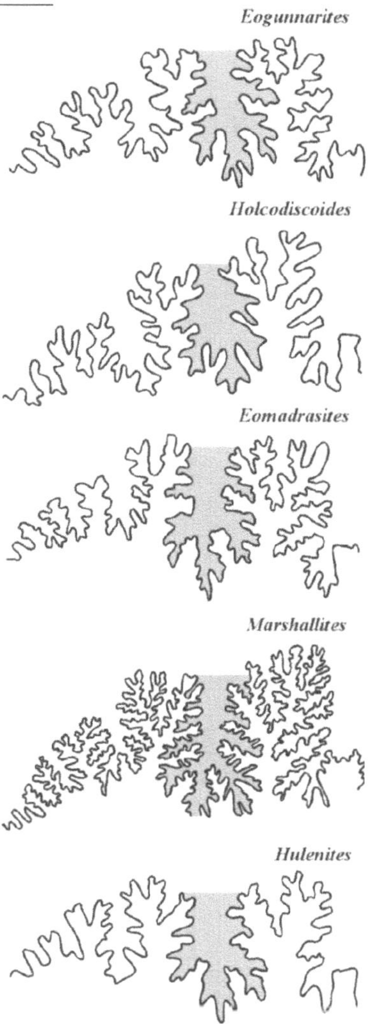

Fig. 234. Marshallitine sutures.

Natalites COLLIGNON, 1954 [*Madrasites natalensis* Spath, 1922, = *Holcodiscus africanus* van Hoepen, 1920; OD]. Small; like *Kossmaticeras* but more compressed, with flatter flanks, higher whorls, narrower venter, pronounced umbilical tubercles giving rise to bundles (2-3) of ribs, and coarser ornament in maturity; constrictions project forward on outer flank. *Sant. - L. Camp.* **Fig. 235C-D** - *N. africanus* (van Hoepen).

Karapadites COLLIGNON, 1954 [*Holcodiscus karapadensis* Kossmat, 1897; OD]. Like *Natalites* but with broad flat flanks, comma-shaped umbilical bullae and sparser, broader, lower, coarser ribs that are effaced from inner flank. *Sant. - Camp.* **Fig. 235E-F** - *K. karapadensis* (Kossmat).

Subfamily Grossouvritinae nov.

Diagnosis. Like Kossmaticeratinae but prorsiradiate constrictions straight, not flexing forward markedly and not linguoid over the venter; prorsiradiate ribs generally rigid, sometimes fine or tuberculate. Largely restricted to the cold circum-polar waters of the Austral Realm. Comprises the tribes Maoritini and Jacobitini. *U. Turon. - .L. Maastr.*

Tribe Maoritini nov.

Type genus: *Maorites* Marshall, 1926.

Diagnosis. Like *Kossmaticeras* but with finer ribbing generally arising from umbilical bullae, and straight non-flexed constrictions. *U. Camp. - Maastr.*

***Epikossmaticeras* gen. nov.** [*Kossmaticeras japonicum* Matsumoto, 1955, p.150; OD herein]. Small; like *Kossmaticeras* but with more numerous (6-7), straight, prorsiradiate constrictions, denser broad low flexuous ribs that are mostly single but occasionally bifurcate from umbilical shoulder, and lack of rib furcation on outer flank. *Sant.* **Fig. 237** - *E. japonicum* (Matsumoto).
= *Neopuzosia* Matsumoto, 1954 (*nom. nud.*).

Kaiparaites MATSUMOTO, 1955 [*Madrasites multicostatus* Marshall, 1926; OD]. Small; like *Natalites* but more evolute (U=32-40%), with more numerous umbilical bullae that are not comma shaped, broader evenly rounded venter, fewer less inclined constrictions that do not project forward on outer flank, ribbing that remains fine in maturity, and very strongly retracted suspensive lobe. *U. Camp. - Maastr.* **Fig. 235G-H** - *K. multicostatus* (Marshall).

Caledonites COLLIGNON, 1977 [*C. neocaledonicus*; OD]. Medium sized; like *Kaiparaites* but with more numerous constrictions that are lost early; umbilical bullae may become exaggerated in later growth when ribbing coarsens. *Camp.* **Fig. 235I** - *C. neocalidonicus* Collignon.

Gunnarites KILIAN & REBOUL, 1909 [*Olcostephanus antarcticus* Weller, 1903; SD Diener, 1925]. Medium sized; like *Kaiparaites* but larger, more inflated, with denser more rectiradiate ribs crenulated by minute tubercles, and broadly rounded venter across which ribs become flat topped. *U. Camp. - L. Maastr.* **Fig. 235J-K** - *G. zelandicus* (Marshall).

Grossouvrites KILIAN & REBOUL, 1909 [*Ammonites gemmatus* Hupé, 1854; OD]. Large; like densicostate *Kaiparaites* but more involute and high whorled, with constrictions lost early (<30 mm diamtere); ribs straight rigid wiry, narrower than interspaces; umbilical tubercles strengthen with growth and, on adult macroconchs, become rounded and exaggerated; suture very complex. *U. Camp.* **Fig. 238A-C** (A, adult macroconch; B-C, juv.) - *G. gemmatus* (Hupé).
= *Grossouvreites* Wright, 1957 (*nom. van.*).

Maorites MARSHALL, 1926 [*Kossmaticeras tenuicostatum* Marshall, 1917; OD]. Moderately large; compressed, involute, with broad flat flanks and rounded venter; early

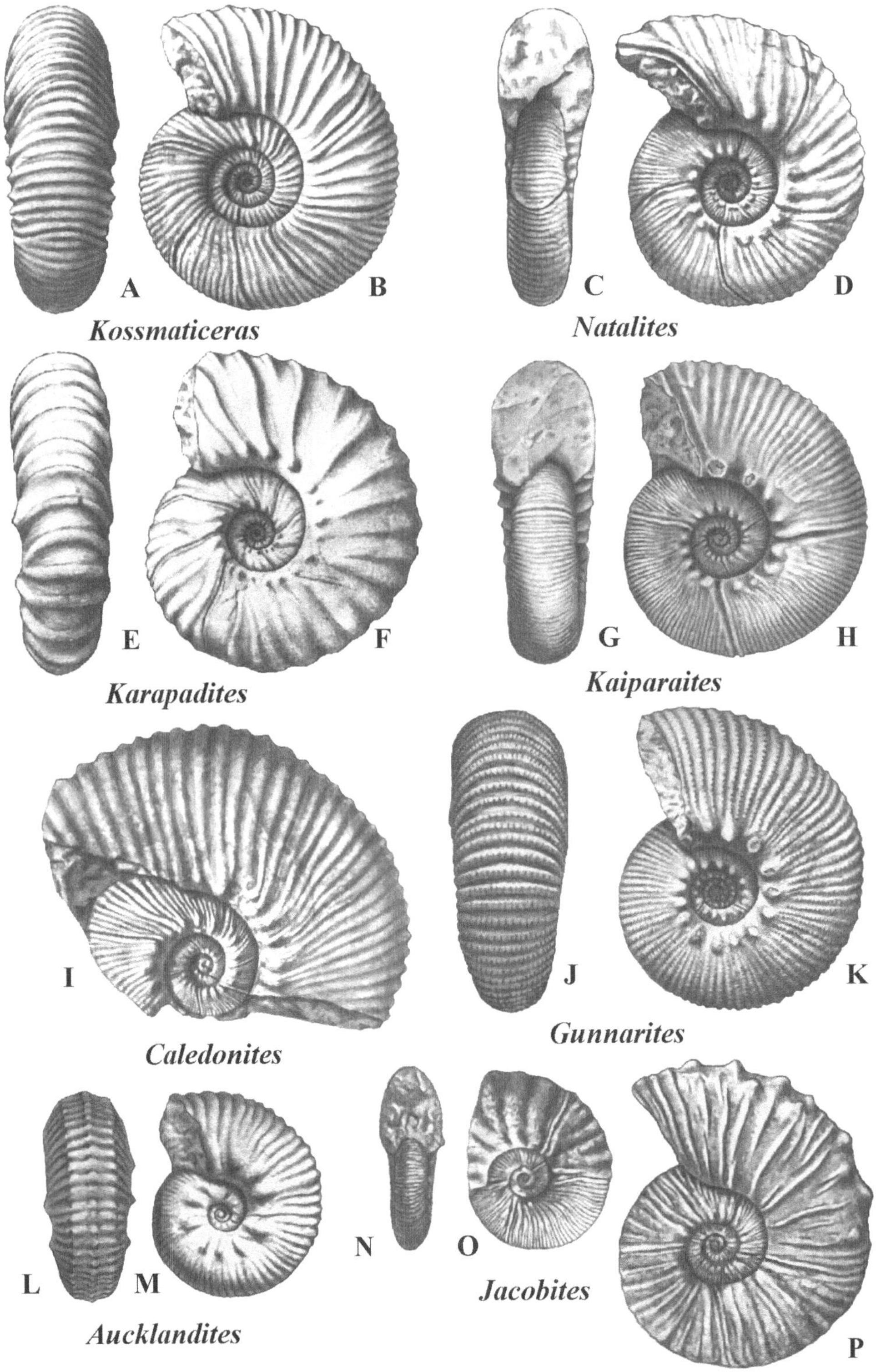

Fig. 235. Kossmaticeratinae.

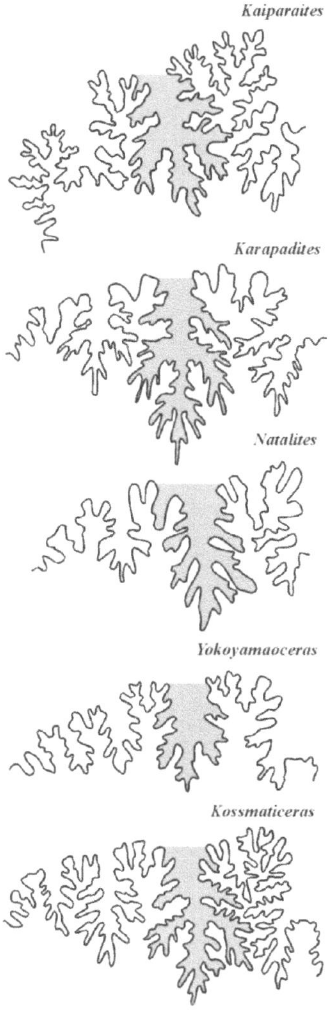

Fig. 236. Kossmaticeratid sutures.

whorls like *Kaiparaites*; in middle growth umbilical bullae give rise to bundles of dense, fine, sinuous, prorsiradiate riblets, bifurcating several times on flanks, separated by non-tuberculate secondaries; 5 strongly prorsiradiate, flexuous, collared constrictions tend to be slightly biconcave, cutting obliquely across ribbing to the posterior; in maturity umbilical bullae obsolete and adult body chamber develops weak folds; suture highly complex. *U. Camp. - L. Maastr.* **Fig. 238G-H** - *M. tenuicostatus* (Marshall).

Maorites densicostatus (Kilian & Reboul) (**Fig. 238E-F**) differs from the type in being strongly compressed, with narrowly arched venter; perhaps a microconch.

Tribe Jacobitini nov.

Type genus: *Jacobites* Kilian & Reboul, 1909.

Diagnosis. Characteristically with densicostate inner whorls, early loss of constrictions and sparsicostate

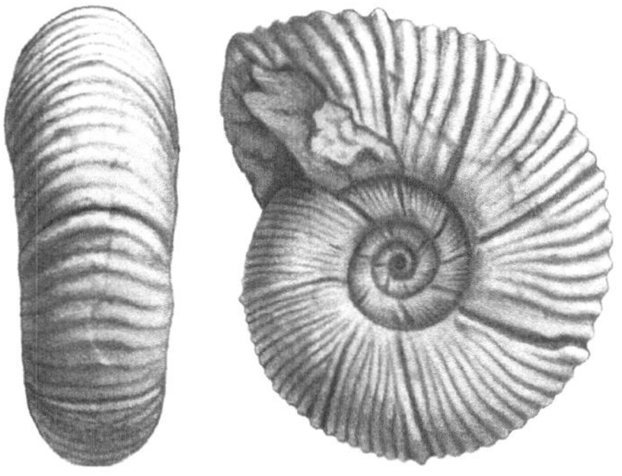

Fig. 237. *Epikossmaticeras*.

outer whorls that, in derived members, become multituberculate with a fastigiate venter. A short lived offshoot of Grossouvritini, close to *Kaiparaites*. *U. Camp. - Maastr.*

Jacobites KILIAN & REBOUL, 1909 [*J. anderssoni*; OD]. Large, inner whorls like *Caledonites*; in middle growth ornament changes dramatically, ribs become distant, irregular and discontinuous, venter becomes fastigiate, and irregularly developed spinose ventrolateral and siphonal tubercles appear. *Camp.* **Fig. 235N-P** - *J. anderssoni* (Kilian & Reboul).

Neomadrasites MARSHALL, 1926 [*N. nodulosus*; OD] [**m**]. Very small; inflated, rather involute with depressed whorls, ventrolateral angulations and broadly rounded venter; prominent umbilical bullae give rise to 3-4 ribs, separated by 2-3 intercalatories, that loop to rounded ventrolateral nodes; ribs cross venter with forward convexity, and all carry weak siphonal nodes as do collared constrictions. *Camp.* **Fig. 240** - *N. nodulosus* (Marshall).

Aucklandites MARSHALL, 1927 [*Acanthoceras ultimum*; OD] (?**m**). Small; like *Jacobites* but with equidimensional whorls and flattened flanks; outer whorls with strong distant ribs all of which carry ventrolateral and siphonal tubercles, as do collared constrictions. *Camp.* **Fig. 235L-M** - *A. ultimatus* (Marshall).

Tainuia MARSHALL, 1926 [*T. aucklandica*; OD]. Moderately large; like *Jacobites* but larger, more evolute and compressed, with flatter flanks, outer whorls with single straight prorsiradiate ribs with up to 4 rows of small lateral tubercles; all ribs with ventrolateral and siphonal tubercles. *Camp.* **Fig. 238I-J** - *T. aucklandica* Marshall.

Subfamily Brahmaitinae Collignon, 1977

Discussion. Monophyletic derivatives of Kossmaticeratini, close to *Natalites*; typically very evolute, with subcircular whorl section and ribs generally effaced from venter in early to middle growth; derived members with

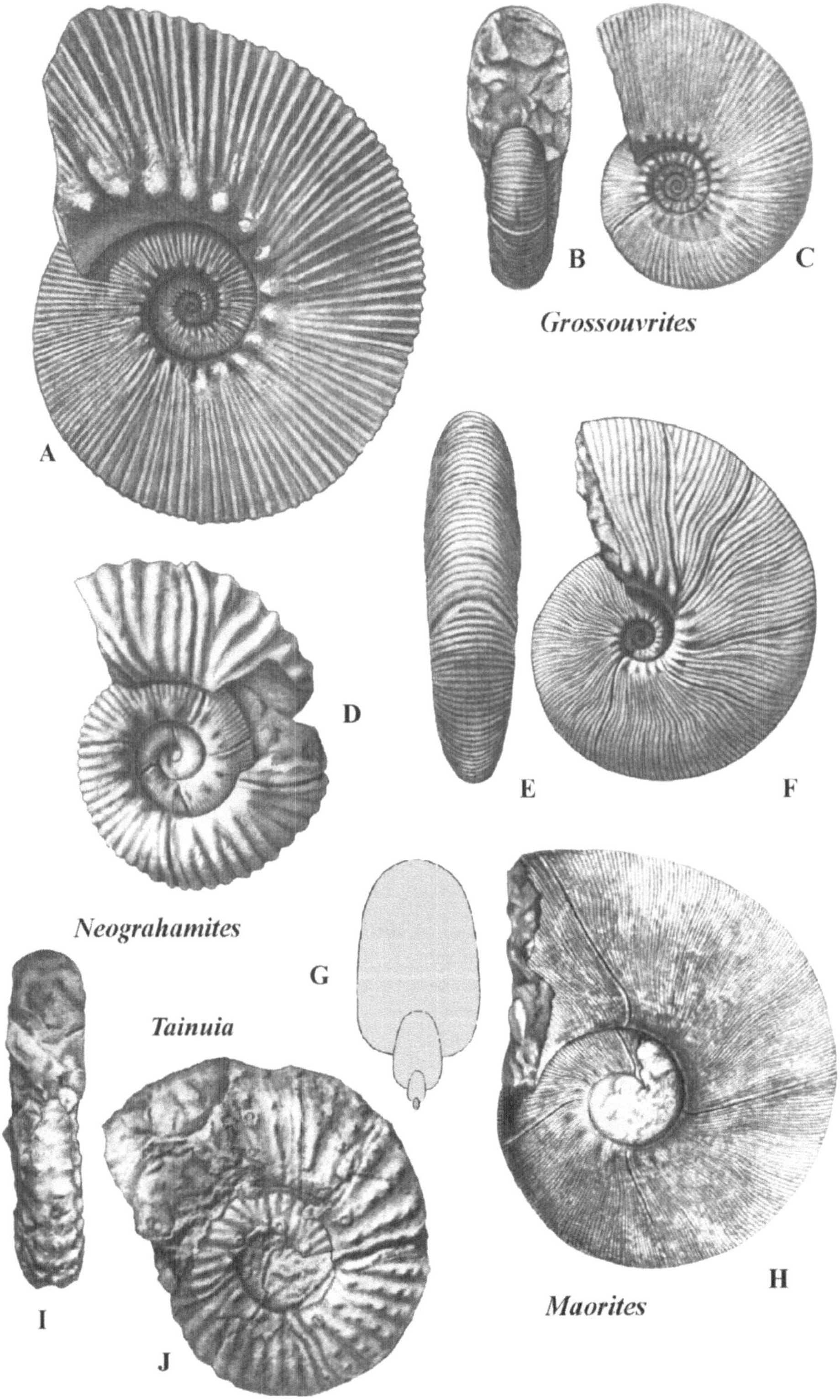

Fig. 238. Kossmaticeratinae.

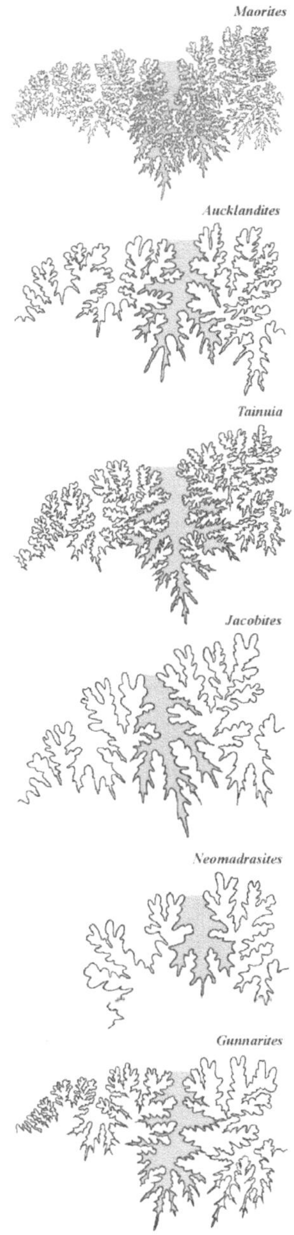

Fig. 239. Kossmaticeratine sutures.

siphonal tubercles to collared constrictions, and suture with long narrow saddles and lobes. Predominantly Austral in distribution but with incursions into the northern Tethyan. *U. Camp. - U. Maastr.*

Neograhamites SPATH, 1953 [*N. kiliani*; OD]. Small; compressed, evolute (U=43-45%), initially with circular whorl section becoming compressed oval in middle growth; inner whorls smooth but later (>20 mm diameter) with sparse (12/whorl) prominent umbilical bullae that give rise to bundles of 3-4 ribs separated by intercalatories and effaced from venter; deep collared prorsiradiate constrictions, about 4 per whorl, persist to maturity and project forward on venter; ribbing coarsens and simplifies in later growth when ribs cross venter uninterrupted; suture very complex, with slender saddles and deep trifid L. *U. Camp.* **Fig. 238D** - *N. taylori* Spath; **241I-J** - *N. kiliani* Spath.

***Eurokossmaticeras* gen. nov.** [*Ammonites brandti* Redtenbacher, 1873, p.106; OD herein]. Medium sized; like *Neograhamites* but with narrower umbilicus (U=30-42%), broader venter, ribs strong across venter and not coarsening in maturity; weak to obsolete constrictions; suture with long narrow saddles and lobes. *U. Camp. - L. Maastr.* **Fig. 241K-L** - *E. galicianum* (Favre).

Pseudokossmaticeras SPATH, 1922b [*Ammonites pacificus* Stoliczka, 1866; OD]. Small; like *Neograhamites* but more evolute (U=47-49%), with more numerous umbilical bullae and generally single ribs in middle to later growth; suture relatively simple; siphonal tubercles may be present on collars of constrictions in maturity. *L. Maastr.* **Fig. 241G-H** - *P. pacificum* (Stoliczka).

Brahmaites KOSSMAT, 1897 [*Ammonites brahma* Forbes, 1846; SD Yabe & Shimizu, 1924]. Medium sized; like *Pseudokossmaticeras* but more inflated, with rounder flanks, more depressed whorls, strong umbilical bullae, and ribs effaced from midflank; ribs rejuvenate in later growth, becoming strong, rectiradiate and distant, but still weak over siphonal line, with hint of ventrolateral bullae on adult body chamber, the last part of which is smooth and with apertural constriction; posterior collar of constrictions frequently with umbilical bullae and, at large size, a siphonal bulla; incised suture with narrow E, narrow stemmed asymmetrically bifid saddles, very narrow trifid L as deep as E, and retracted suspensive lobe with two saddles. *U. Maastr.* **Fig. 241C-F** - *B. brahma* (Forbes).

Anabrahmaites YABE & SHIMIZU, 1924 [*Ammonites vishnu* Forbes, 1846; OD]. Like *Brahmaites* but with equidimensional to slightly compressed whorls, smooth inner whorls, complete absence of umbilical bullae, weak distant annular ribs that strengthen and become distant in middle to later growth when they parallel constrictions, and siphonal tubercles that protrude forward to interrupt

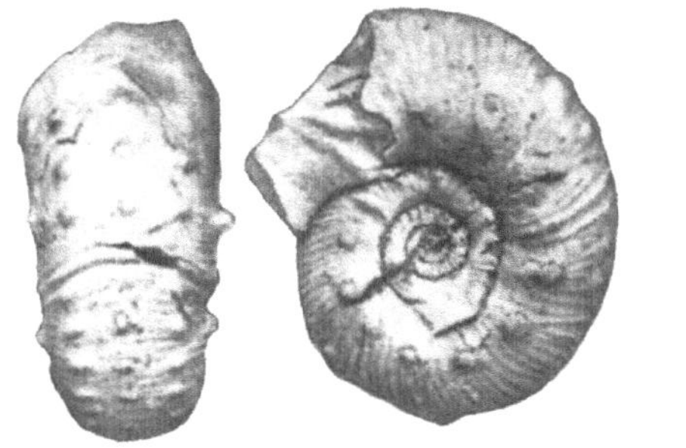

Fig. 240. *Neomadrasites*.

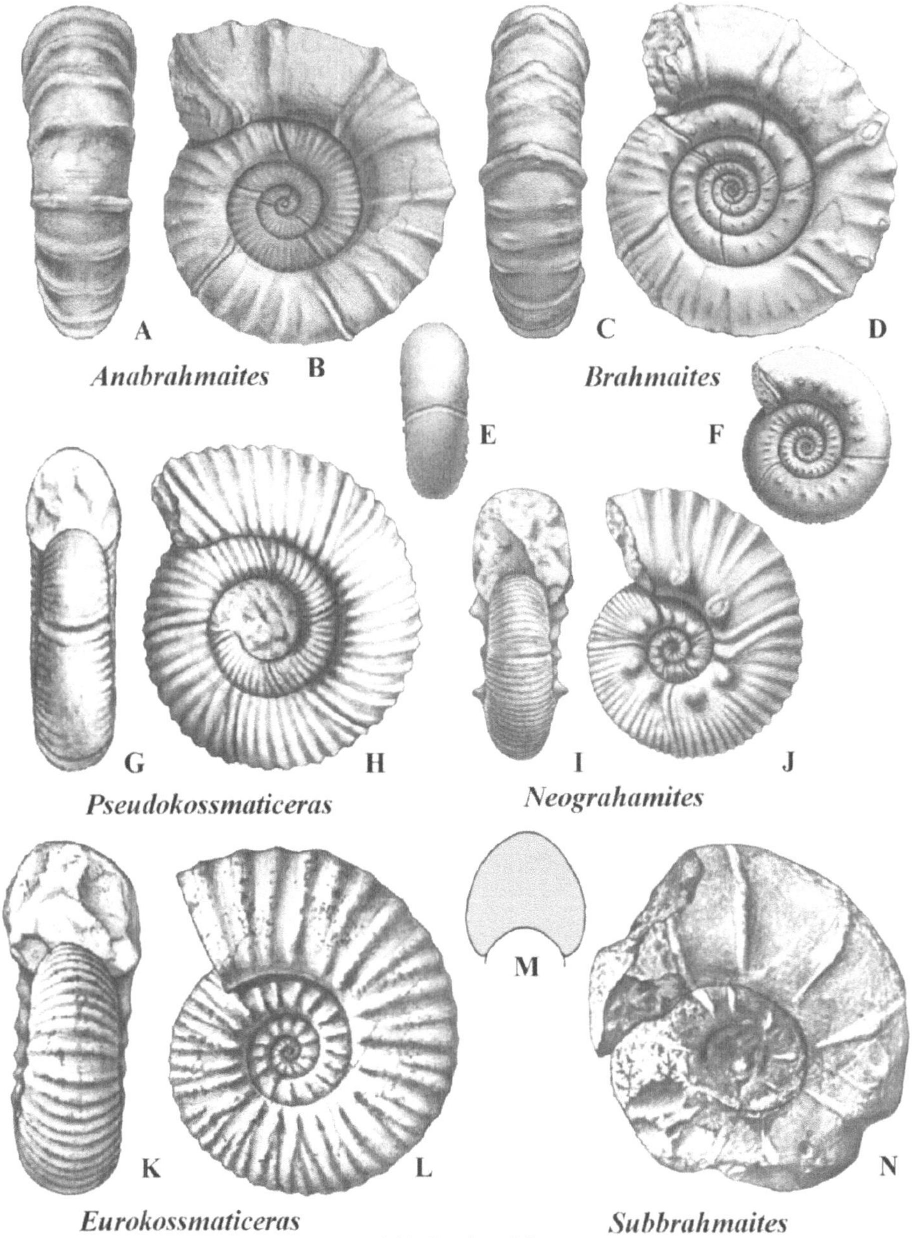

Fig. 241. Brahmaitinae.

the constriction; external suture with 3 saddles. *U. Maastr*. **Fig. 241A-B** - *A. vishnu* (Forbes).

Subbrahmaites YABE & SHIMIZU, 1924 [*S. sachalinensis*; OD]. Poorly known; like *Anabrahmaites* but with trigonally ovate whorl section, sparse distant asymmetrical (steepest anteriorly) flank ribs (10/whorl) with low mid-lateral bullae and 4 well developed saddles to external suture. *?Maastr*. **Fig. 241M-N** - *S. sachalinensis* Yabe & Shimizu.

Superfamily PLACENTICERATOIDEA Hyatt, 1900

(*nom. transl.* herein *ex* Placenticeratida)

Discussion. Monophyletic derivatives of Puzosioidea (Spath, 1925); comprises the families Sonneratiidae, Hoplitidae, Schloenbachiidae and Placenticeratidae. *L. Alb. - U. Maastr*.

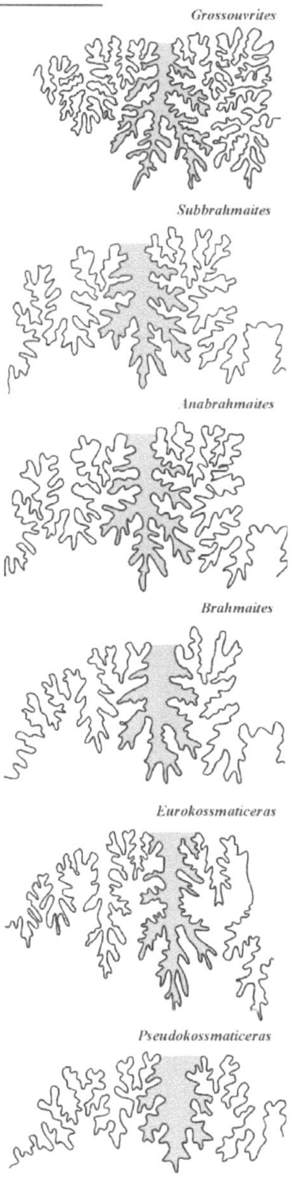

Fig. 242. Brahmaitine sutures.

Family SONNERATIIDAE Destombes, Juignet & Rioult, 1972

(*nom. transl.* Cooper & Owen, 2012 ex Sonneratiinae)

Discussion. A monophyletic taxon with an origin in Puzosiidae (Spath, 1925); typically moderately evolute with strong ribs arising in pairs from umbilical bullae, lack of constrictions and suture with asymmetrically trifid L and only 3-4 auxiliary lobes. Comprises the paraphyletic subfamilies Sonneratiinae and Otohoplitinae (Cooper & Owen, 2012). *L. Alb.*

Subfamily Sonneratiinae Destombes, Juignet & Rioult, 1972

Discussion. A monophyletic group with an origin in Beudanticeratinae, close to *Uhligella* (Spath, 1925); primitively with linguoid ribbing uninter rupted across rounded to subtabulate venter; derived members with ventrolateral tubercles that are opposite over the venter. *L. Alb.*

Eosonneratia SAVEL'EV, 1973 [*Sonneratia (E.) vnigri*; OD]. Small (**m**) to moderately large (**M**); relatively involute, moderately inflated, with gently convex flanks, narrowly rounded venter and oval whorl section; well developed umbilical bullae give rise to paired strong flexuous ribs with occasional intercalatories that form chevrons over venter; suture with shallow E, deeply bifid E/L and broad asymmetrically trifid L deeper than E. *Lower L. Alb.* **Fig. 243C-D** - *E. vnigri* (Savel'ev).

Globosonneratia SAVEL'EV, 1973 [*Sonneratia (G.) globulosa*, = *Sonneratia perinflata* Breistroffer, 1947; OD]. Small; like *Eosonneratia* but inflated cadicone with strongly depressed coronate whorl section, deep pit-like umbilicus, broadly rounded venter and rectiradiate to slightly rursiradiate paired ribs with or without intercalatories; suture relatively simple. *Lower L. Alb.* **Fig. 243A-B** - *G. perinflata* (Breistroffer).

Sonneratia BAYLE, 1878 [*Ammonites dutempleanus* d'Orbigny, 1850; M]. Small to medium sized; in early growth like *Eosonneratia*, later more evolute and inflated, with subhexagonal to subquadratic whorl section, broad subtabulate venter, straighter more distant ribs which are transverse across venter and long umbilicolateral bullae shifting to midflank in later growth; ribs may show incipient thinning across siphonal line; suture like *Eosonneratia* but with longer narrower elements. *L. Alb.* **Fig. 243G-H** - *S. dutempliana* (d'Orbigny).

Tetrahoplites CASEY, 1952 [*Sonneratia subquadrata* Sinzow, 1907; OD]. Small; like *Sonneratia* but with sparse more distant ribs distinctly weakened across siphonal line; adult body chamber subquadratic, with distant mostly single ribs crossing venter uninterrupted. *L. Alb.* **Fig. 243E-F** - *T. subquadratus* (Sinzow).

Hemisonneratia CASEY, 1952 [*Ammonites puzosianus* d'Orbigny, 1841; OD]. Small; barely separable from *Tetrahoplites* but with ribs forming distinct ventrolateral bullae in middle growth. *L. Alb.* **Fig. 243I-K** - *H. puzosiana* (d'Orbigny).

= *Hemisonneratia* Breistroffer, 1952 (*nom. nud.*).

Dragunoviceras COOPER & OWEN, 2012 [*Tetrahoplites dragunovi* Savel'ev, 1973; OD]. Small; like *Tetrahoplites* but with flat parallel flanks and compressed rectangular to subtrapezoidal whorl section; asymmetrical ribs (steepest posteriorly) interrupted across venter in early growth

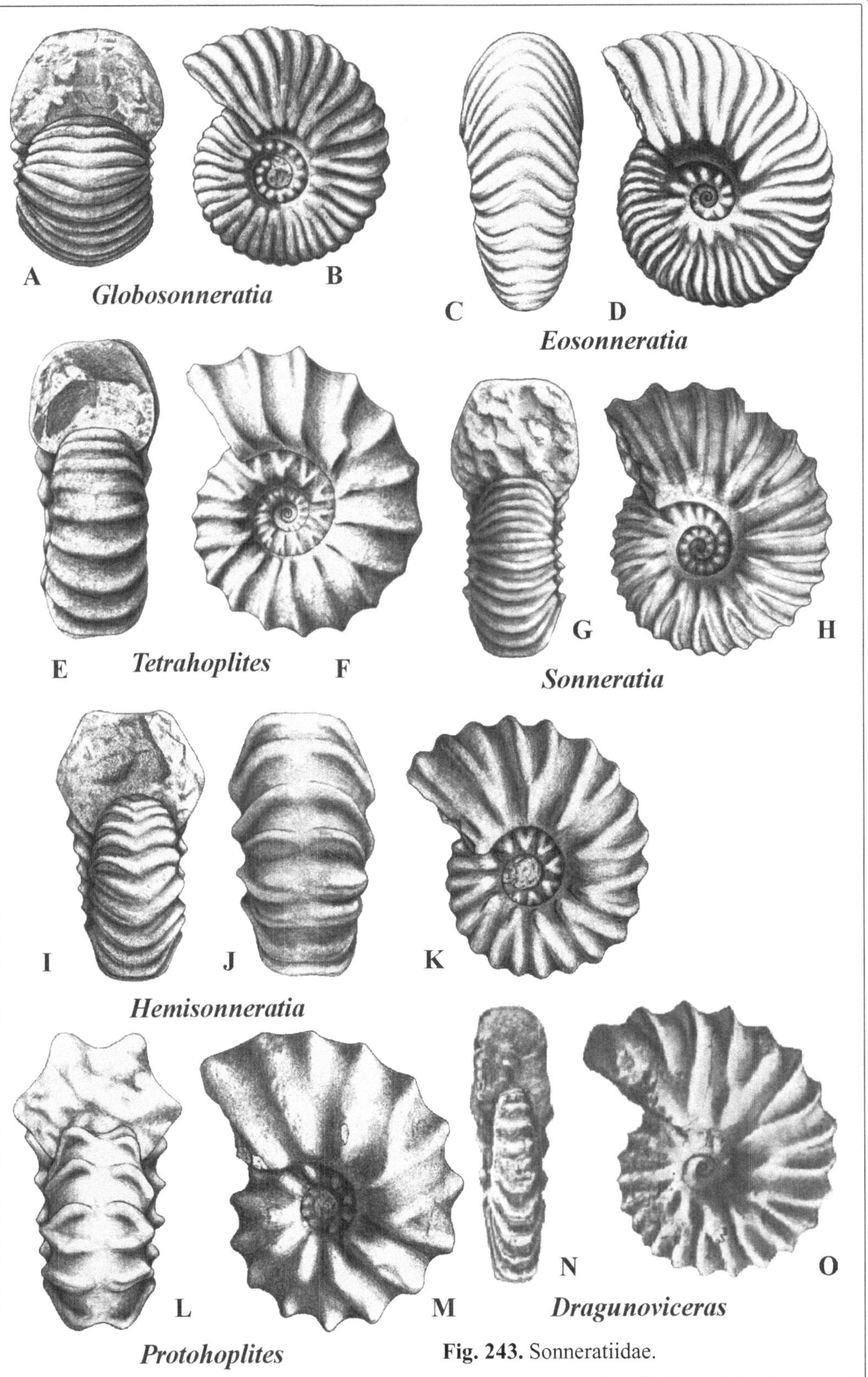

Fig. 243. Sonneratiidae.

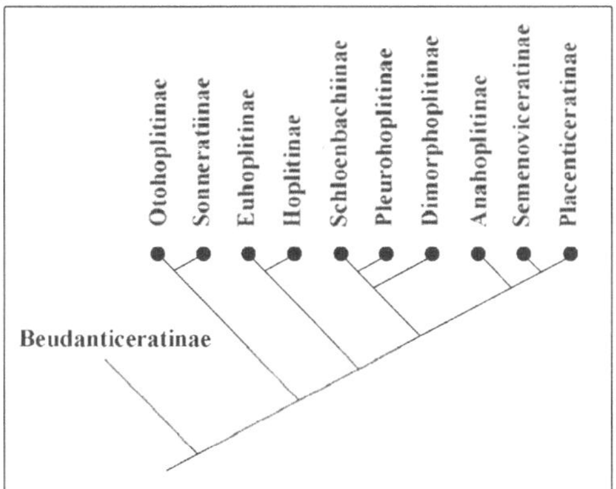

Fig. 244. Hypothesized relationships among Placenticeratoidea.

but, in middle to later growth, strong high and uninterrupted over venter; suture like *Sonneratia*. *L. Alb.* **Fig. 243N-O** - *D. dragunovi* (Savel'ev).

Protohoplites SPATH, 1923 [*Ammonites archiacanus* d'Orbigny, 1841; OD]. Small; like *Dragunoviceras* but more inflated, with bifid or single ribs and prominent to auriculate ventrolateral bullae. *L. Alb.* **Fig. 243L-M** - *P. archiacanus* (d'Orbigny).

Subfamily Otohoplitinae Savel'ev, 1992

Discussion. A paraphyletic group with an origin in Sonneratiinae, close to *Eosonneratia* (Owen, 1971). Typically with subtabulate venter at some stage of growth, ventrolateral tubercles that alternate across the venter and shallow ventral sulcus. Dimorphism conspicuous. *L. Alb.*

Caseyhoplites COOPER & OWEN, 2012 [*Otohoplites waltoni* Casey, 1965; OD]. Small; inflated, moderately evolute, with depressed subhexagonal whorl section, feebly concave venter, prominent umbilicolateral bullae, coarse ribs initially paired but later alternating long and short, and blunt conical to obliquely clavate ventrolateral tubercles that zigzag over venter; in middle growth forward projection of ventrolateral tubercles gradually reduces until, eventually, they are transverse and bullate; suture like *Sonneratia* but with shallowly bifid E/L and broader L; superficially similar to younger *Hemisonneratia*. *L. Alb.* **Fig 251A-B** - *C. waltoni* (Casey).

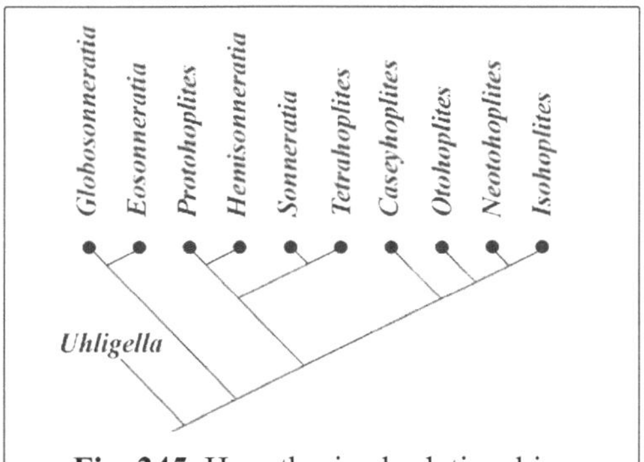

Fig. 245. Hypothesized relationships among Sonneratinae.

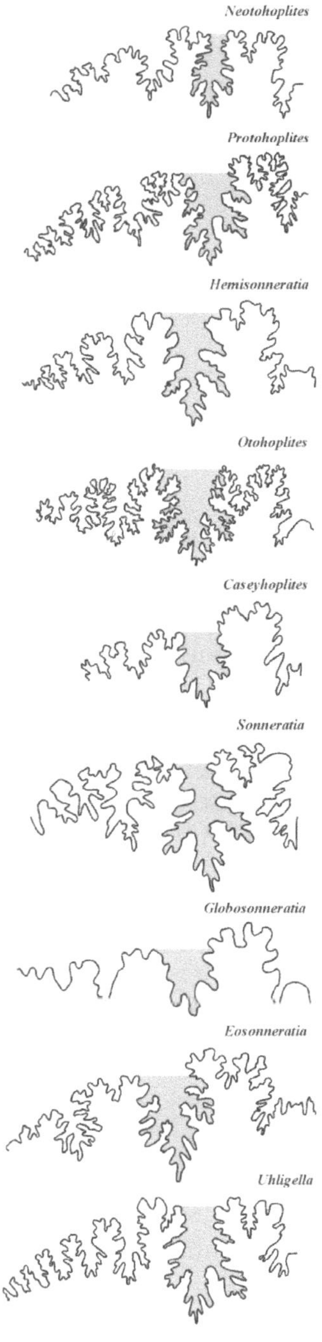

Fig. 246. Sonneratiine sutures.

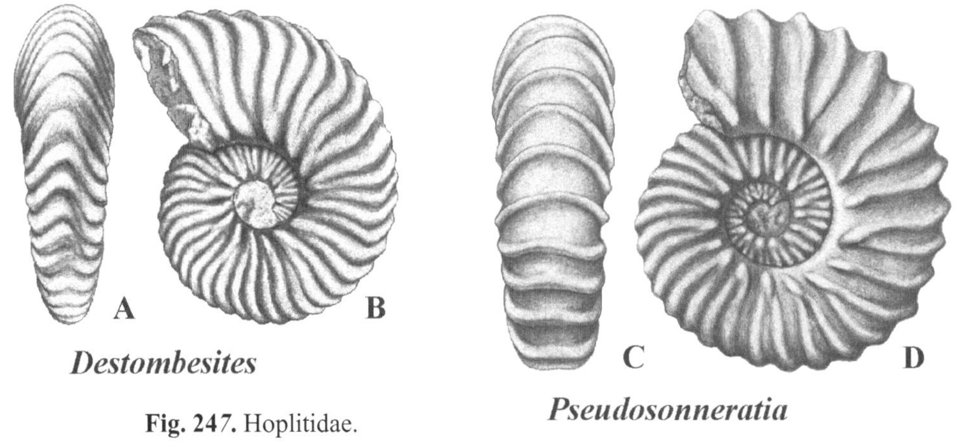

Fig. 247. Hoplitidae.

Otohoplites STEINMANN, 1925 [*Ammonites raulinianus* d'Orbigny, 1841; OD]. Small; like *Caseyhoplites* but more compressed, with broader flatter flanks, narrower shallowly sulcate venter, more numerous umbilical bullae, finer fibulate ribbing, prominently raised oblique ventrolateral clavi and suture with deeply bifid E/L. *L. Alb.* **Fig. 251C-D** - *O. raulinianus* (d'Orbigny).

Neotohoplites COOPER & OWEN, 2012 [*Otohoplites bulliensis* Destombes, 1970; OD]. Small; like *Otohoplites* but with ventrolateral clavi parallel to siphonal line and tilted outwards to overhang flanks; ribs with asymmetrical profile (steepest adapically); unornamented venter sulcate in early growth, later tabulate and finally fastigiate; a homoeomorph of *Dimorphoplites*. *Upper L. Alb.* **Fig. 251E-F** - *N. bulliensis* (Destombes, Juignet & Rioult).
= *Subotohoplites* BARABOSHKIN, 1991 (*nom. nud.*).

Family HOPLITIDAE Douvillé, 1890

Discussion. A monophyletic group with an origin in Sonneratiidae (Amedro, 1984); comprises the subfamilies Hoplitinae and Euhoplitinae that evolved in parallel with both Placenticeratidae and Schloenbachiidae. *Upper L. Alb. - L. Cenom.*

Subfamily Hoplitinae Douvillé, 1890

Discussion. A paraphyletic taxon with an origin in Otohoplitinae, close to *Neotohoplites* (Amedro, 1984). Typically compressed, flat sided, with strong flexuous frequently looped ribs, *en-echelon* tuberculate rib endings and a sulcate venter. Sexual dimorphism conspicuous but poorly documented, macroconchs becoming coarsely ribbed. *Upper L. - M. Alb.*

Destombesites COOPER & OWEN, 2011b [*Pseudosonneratia iserensis* Spath, 1925; OD]. Small; like *Eosonneratia* but compressed and flat sided, with sinuous ribs that flex strongly forward on outer flank and are linguoid over venter; in later growth becoming evolute; ribs with symmetrical rib profiles. *L. Alb.* **Fig. 247A-B** - *D. iserensis* (Spath).

Pseudosonneratia SPATH, 1925 [*P. typica*; OD]. Small (**m**) to moderately large (**M**); inner whorls like *Destombesite* but in later growth very evolute, with subquadratic whorl section, broader venter and strong distant ribs that cross venter transversely. *L. Alb.* **Fig. 247C-D** - *P. typica* Spath.

Isohoplites CASEY, 1954 [*Parahoplites steinmanni* Jacob, 1907; OD]. Small; like *Pseudosonneratia* but finely ribbed throughout, with trapezoidal whorl section and stiff to slightly flexuous ribs with asymmetrical profiles (steepest adapically); ribs with ventrolateral tubercles that alternate across venter in early growth but are opposite in maturity, and are weak across siphonal line; macroconchs with strong coarse ribs in maturity. *Uppermost L. Alb.* **Fig. 253A-B** - *I. steinmanni* (Jacob).

Amedroites COOPER & OWEN, 2011b [*Hoplites bennettianus* J. de C. Sowerby, 1826; OD]. Like *Isohoplites* but strongly inflated and hypertuberculate, with depressed trapezoidal whorl section, high sloping umbilical wall and sparse swollen umbilicolateral tubercles; coarse robust symmetrical ribs terminate in weak bullae that interfinger across venter to produce deep zigzagging ventral sulcus. *Uppermost L. - lowest M. Alb.* **Fig. 249** - *A. bennettianus* (J. de C. Sowerby).
= *Rudihoplites* Baraboshkin, 1991 (*nom. nud.*).

Hoplites NEUMAYR, 1875 [*Ammonites dentatus* J. Sowerby, 1821; ICZN Opinion 353]. Small (**m**) to medium sized (**M**); like *Isohoplites* but slightly more involute and high whorled, with distinct ventral sulcus, frequent intercalatory between paired ribs, asymmetrical rib profiles (steepest adorally), and prominent oblique rib terminations that alternate *en-echelon* across sulcate venter in maturity. *M. Alb.* **Fig. 253C-D** - *H. dentatus* (J. Sowerby).

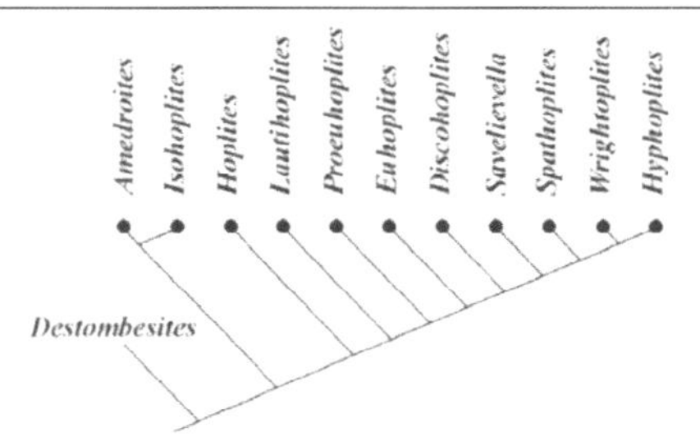

Fig. 248. Hypothesized relationships among Hoplitidae.

Lautihoplites BARABOSHKIN, 1996 [*Hoplites engersianus* Rouillier & Fachrenkohl., 1847; OD]. Small; innermost whorls like *Hoplites*; in middle growth with looped and zigzagging ribs on flanks and prominent oblique ventrolateral clavi; in later growth with bifurcating and intercalated ribs and elongate umbilicolateral bullae. *Lower M. Alb.* **Fig. 253E-F** - *L. similis* (Spath).

= *Odonthoplites* Breistroffer, 1947 (ICZN Rejected Name No. 296).

= *Lautihoplites* Baraboshkin, 1991 (*nom. nud.*).

Proeuhoplites COOPER & OWEN, 2011b [*Euhoplites loricatus* Spath, 1925; OD]. Like *Lautihoplites* but with looped ribbing from early growth and with ventrolateral clavi parallel to siphonal line. *M. Alb.* **Fig. 253G-H** - *P. loricatus* (Spath).

Subfamily Euhoplitinae Savel'ev, 1992

= Discohoplitinae Savel'ev, 1992.

Discussion. A monophyletic group with an origin in Hoplitinae, close to *Proeuhoplites* (Spath, 1926). Typically with fibulate ribbing, deep steep-sided siphonal channel to the venter, and suture with narrow E. *Upper M. - L. Cenom.*

Euhoplites SPATH, 1925 [*E. truncatus*; OD]. Small; like *Proeuhoplites* but with distinct siphonal channel from early growth and more complex suture with very broad L. *Lower U. Alb.* **Fig. 253I-J** - *E. truncatus* Spath.

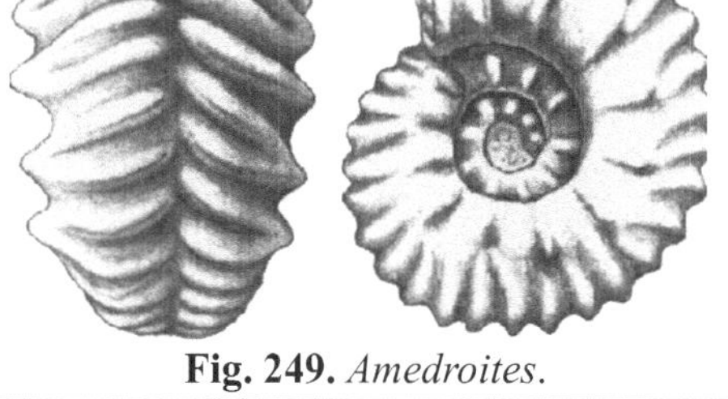

Fig. 249. *Amedroites.*

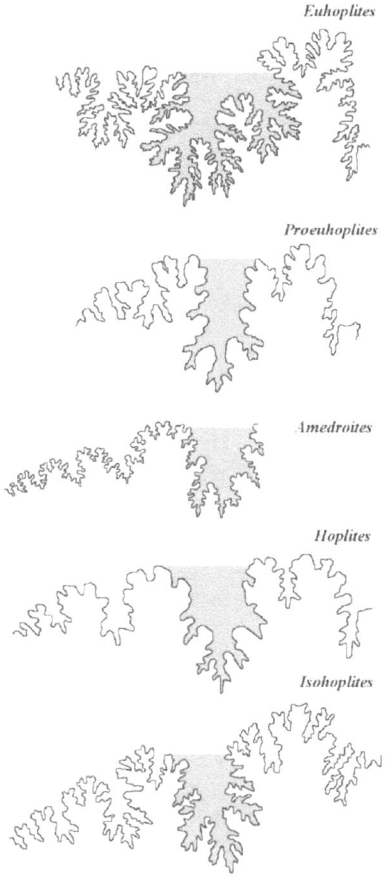

Fig. 250. Hoplitine sutures.

Discohoplites SPATH, 1925 [*Ammonites coelonotus* Seeley, 1865; OD]. Small; like *Euhoplites* but strongly compressed and high whorled, with narrowly rounded venter; dense fine ribs projecting strongly forward near venter, and simpler suture with shallow E and deep penetrating narrow bifid L; ventrolateral tubercles obsolete. *Upper U. Alb.* **Fig. 253K-L** - *D. coelonotus* (Seeley).

Savelievella COOPER & OWEN, 2011b [*Discohoplites varicosus* Spath, 1925; OD]. Very small; like *Discohoplites* but evolute, with subquadratic to rectangular whorls and coarse flexuous ribs; umbilical tubercles persist to maturity; suture simple, almost pseudoceratitic; superficially resembles *Hysteroceras* (Brancoceratidae). *Upper U. Alb.* **Fig. 253M-N** - *S. varicosus* (Spath).

Spathoplites COOPER & OWEN, 2011b [*Hyphoplites transitorius* Spath, 1925; OD]. Very small; like *Savelievella* but with wider umbilicus; smoothish spiral groove at mid-flank at join between "haft" and "blade" of sickle-shaped ribs; ribs with weak oblique ventrolateral clavi to all ribs, some stronger than others. *Upper U. Alb.* **Fig. 253O-P** - *S. transitorius* (Spath).

Wrightoplites COOPER & OWEN, 2011b [*Hyphoplites pylorus* Wright & Wright, 1949; OD]. Like *Spathoplites* but larger, with doubled ventrolateral tubercles and low keels

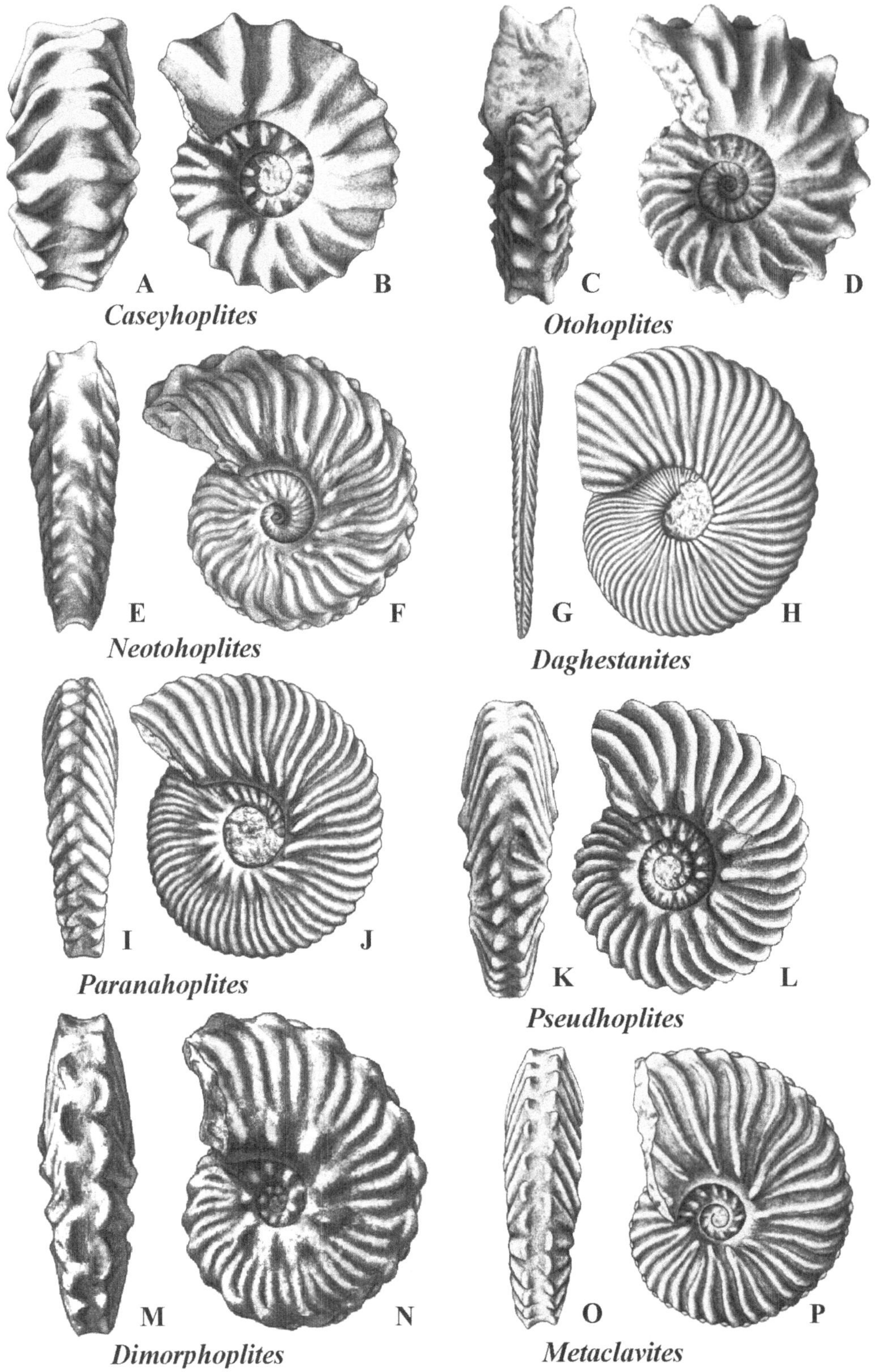

Fig. 251. Dimorphoplitinae.

bordering siphonal channel. *Upper U. Alb.* **Fig. 255M-N** - *W. pylorus* (Wright & Wright).

Hyphoplites SPATH, 1922 [*Ammonites falcatus* Mantell, 1822; OD]. Small; like *Wrightoplites* but more compressed, with strongly falcate ribs with narrow hafts and broad blades; ventrolateral tubercles lost on outer whorls. *L. Cenom.* **Fig. 255O-P** - *H. falcatus* (Mantell).

Family SCHLOENBACHIIDAE Parona & Bonarelli, 1897

(*nom. correct.* Wright & Wright, 1951 *ex* Schloenbachidae)

Discussion. A monophyletic group with an origin in Sonneratiidae; comprises the subfamilies Dimorphoplitinae, Pleurohoplitinae and Schloenbachiinae. *Lower M. Alb. - U. Cenom.*

Subfamily Dimorphoplitinae Cooper & Owen, 2011c

Discussion. A paraphyletic taxon with an origin in Otohoplitinae, close to *Isohoplites* (Cooper & Owen, 2011c); typically with prominent ventrolateral tubercles that alternate across the deeply sulcate venter and asymmetrical ribs that are steepest posteriorly. *Lower M. - Middle U. Alb.*

Daghestanites GLAZUNOVA, 1953 [*D. daghestanensis*; T]. Small to medium sized; involute, compressed, high whorled, with narrow sulcate venter; numerous elongate prorsiradiate umbilical bullae give rise to paired flexuous ribs terminating in ventrolateral nodes that alternate across venter; in later growth ribs may simplify, forming chevrons over venter (**m**), or body chamber may become weakly ornamented to "smooth" (**M**); superficially similar to *Anahoplites*. *Lower M. Alb.* **Fig. 251G-H** - *D. daghestanensis* Glazunova.

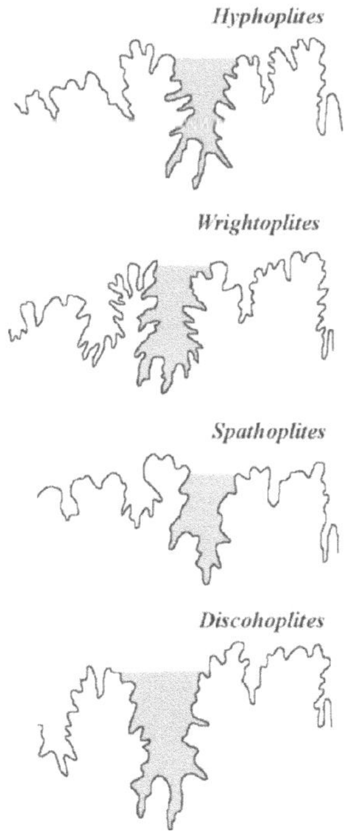

Fig. 252. Hoplitine sutures.

Pseudhoplites COOPER & OWEN, 2011C [*Dimorphoplites niobe* Spath, 1925; OD]. Small; like *Daghestanites* but more evolute, with coarser more distant ribs projecting forward on outer flank and terminating in low, blunt, oblique ventrolateral clavi that rise well above siphonal line; ribs distinctly asymmetrical, steepest adapically, with one connection of pairs weakening so ribs tend to become single with paired intercalatories and, eventually, alternating long and short; suture symmetrical with asymmetrically trifid L deeper than E. *Lower M. Alb.* **Fig. 251K-L** - *P. niobe* (Spath).

=*Praeanahoplites* Baraboshkin, 1991 (*nom. nud.*).

Dimorphoplites SPATH, 1925 [*Ammonites biplicatus* Mantell, 1822; OD]. Small; like *Pseudhoplites* but more involute, with fibulate ribbing and large *en-echelon* ventrolateral clavae parallel to siphonal line. *M. Alb.* **Fig. 251M-N** - *D. biplicatus* (Mantell).

Gazdaganites MIRZOEV, 1969 [*G. gazdaganensis*; OD]. Medium sized; like *Pseudhoplites* but larger, more involute, with elongate umbilicolateral bullae, coarser more flexuous ribs which do not project strongly forwards on outer flank and which form chevrons over subtabulate venter; suture with strongly bifid E/L and asymmetrically trifid L. *Upper M. - Lower U. Alb.* **Fig. 255A-B** - *G. gazdaganensis* Mirzoev.

Metaclavites CASEY, 1965 [*Hoplites denarius* var. *compressus* Parona & Bonarelli, 1897; OD]. Small; like *Gazdaganites* but more compressed, with finer less flexuous asymmetrical ribs that often arise in triplets from prorsiradiate umbilical bullae and terminate in distinct ventrolateral clavi parallel to siphonal line leaving smooth narrow tabulate venter; suture with deeply divided E/L. *Lower U. Alb.* **Fig. 251O-P** - *M. compressus* (Parona & Bonarelli).

Epihoplites SPATH, 1925 [*Ammonites denarius* J. de C. Sowerby, 1826; OD]. Small; like *Gazdaganites* but more evolute and inflated, with rounder flanks, sloping umbilical walls, smaller less elongate umbilical bullae and less flexuous ribs which frequently arise in triplets. *Lower U. Alb.* **Fig. 255C-D** - *E. denarius* (J. de C. Sowerby).

Subfamily Pleurohoplitinae Savel'ev, 1992

Discussion. A monophyletic group with an origin in Dimorphoplitinae, close to *Procallihoplites* (Cooper & Owen, 2011c); typically with raised to fasti-

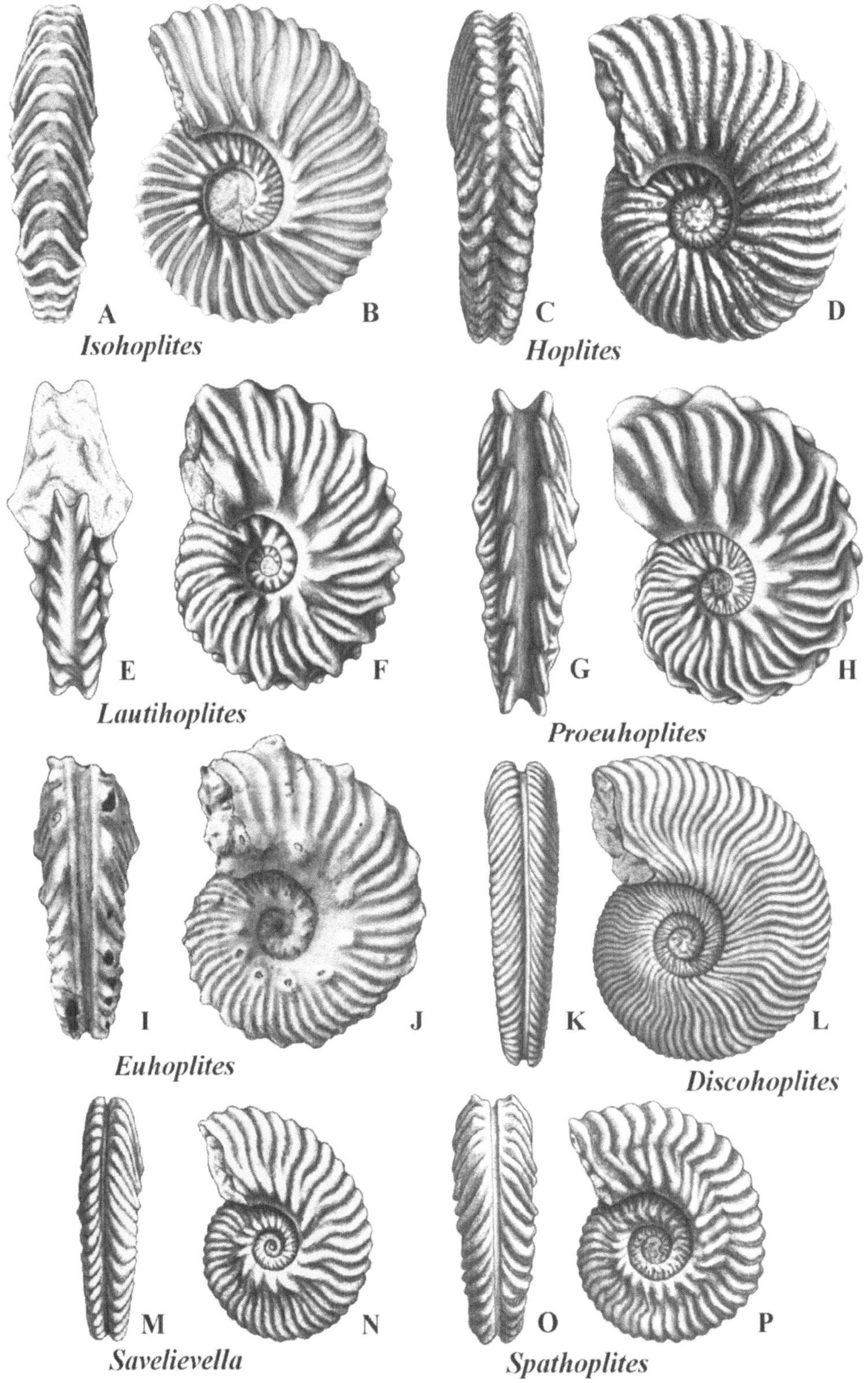

Fig. 253. Euhoplitinae.

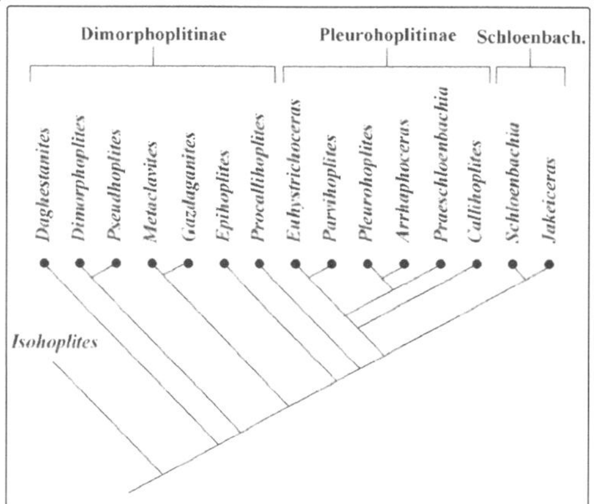

Fig. 254. Hypothesized relationships among Schloenbachiidae.

giate and keeled venter and generally prominent umbilical tubercles. *U. Alb.*

Procallihoplites COOPER & OWEN, 2011C [*Ammonites auritus* J. Sowerby, 1827; OD] [**m**]. Small (**m**) to medium sized (**M**); compressed, involute, high whorled, with strong sharp fibulate ribbing and plate-like ventrolateral clavi that parallel siphonal line and rise high above smooth deeply sulcate venter; macroconchs with smooth outer whorl and rounded venter; a homoeomorph of *Dimorphoplites*. *Middle U. Alb.* **Fig. 255I-J** - *P. auritus* (J. Sowerby) [**m**].

Callihoplites SPATH, 1925 [*Ammonites catillus* J. de C. Sowerby, 1827; OD] [**M**]. Small (**m**) to medium sized (**M**); like *Procallihoplites* but microconchs smaller, with fibulate rib endings forming low ventrolateral clavi bordering a weakly convex to fastigiate venter; adult macroconch like *Procallihoplites*. *Middle U. Alb.* **Fig. 255E-F** - *C. fuscus* Spath, (**m**); **Fig. 255G-H** - *C. catillus* (J. de C. Sowerby) (**M**); **Fig. 258N-O** - *C. vraconnensis* (Pictet & Campiche) (**m**).

Saltericeras ATABEKIAN, 1960 [*Ammonites salteri* Sharpe, 1856; OD] [**m**]. Types lost; long regarded a pathological *Schloenbachia* (Spath, 1925; Kennedy, 1971) but the small type is virtually indistinguishable from *Callihoplites vraconnensis* (Pictet & Campiche). It seems to be a reworked Late Albian remanié specimen. *U. Alb.* **Fig. 258L-M** - *S. salteri* (Sharpe).

Arrhaphoceras WHITEHOUSE, 1927 [*Ammonites woodwardi* Seeley, 1865; = *Ammonites studeri* Pictet & Campiche, 1860; OD]. Small to very small; evolute, inflated, with rounded to fastigiate venter, coronate whorl section, about 10 prominent swollen rounded umbilicolateral tubercles that give rise to 2-3 strong blunt straight rigid ribs that cross venter uninterrupted in early growth but later terminate in weak ventrolateral bullae leaving raised siphonal line smooth; microconchs with strong ventrolateral tubercles and flattened venter to adult body chamber. *Upper U. Alb.* **Fig. 256** - *A. studeri* (Pictet & Campiche); **Fig. 258A-B** - *A. woodwardi* (Seeley) (teratological).

Pleurohoplites SPATH, 1921 [*Ammonites renauxianus* d'Orbigny, 1840; OD] [**M**]. Small (**m**) to medium sized (**M**); inner whorls like *Arrhaphoceras*; later flat sided, compressed and high whorled, with smaller umbilical bullae and ribs that sweep forward to form chevrons on fastigiate venter with blunt siphonal ridge. *Upper U. Alb.* **Fig. 258C-D** - *P. renauxianus* (d'Orbigny).

Praeschloenbachia SCHOLZ, 1973 [*Schloenbachia (P.) briacensis*; OD]. Small; like *Arrhaphoceras* but with coarser more distant ribbing, strongly raised ventrolateral tubercles and distinct blunt siphonal keel; ribs tend to become single in later growth. *Uppermost U. Alb.* **Fig. 258E-F** - *P. briacensis* (Scholz).

Parvihoplites COOPER & OWEN, 2011C [*Euhystrichoceras constrictum* Spath, 1933; OD]. Very small; like *Praeschloenbachia* but with depressed polygonal adult whorls, occasional looped ribs, umbilical tubercles shifting to midlateral position in later growth, and well developed but irregular oblique ventrolateral clavi that, in later growth, occur on every second rib and, on adult body chamber, parallel siphonal line; in general rib terminations opposite across venter; suture very simple, with broad E/L and narrow asymmetrically bifid L much shallower than E. *L. Cenom.* **Fig. 258G-I** - *P. constrictus* (Spath).

Euhystrichoceras SPATH, 1923 [*Ammonites nicaisei* Coquand, 1862; OD]. Very small; inner whorls like *Parvihoplites*; outer whorls more compressed and evolute, with uniform ribs arising from small umbilical bullae that remain on shoulder and with feeble ventrolateral swellings; ribs simplify in later growth; suture with barely incised saddles and U-shaped L. *Upper U. Alb. - L. Cenom.* **Fig. 258J-K** - *E. nicaisei* (Coquand).

Subfamily Schloenbachiinae Parona & Bonarelli, 1897

(*nom. transl.* Cooper & Owen, 2011c *ex* Schloenbachiidae)

Discussion. Monophyletic derivatives of Pleurohoplitinae (Spath, 1942), close to *Callihoplites* (Owen, 1976); typically with fastigiate keeled venter and lateral tubercles. *L. - U. Cenom.*

Schloenbachia NEUMAYR, 1875 [*Ammonites varians* J. Sowerby, 1817; SD Douvillé, 1890]. Small (**m**) to medium sized (**M**); moderately evolute, compressed to inflated, high whorled, with strongly fastigiate venter and distinct thin high siphonal keel; ribs irregular, typically with tiny umbilical, prominent low lateral, bullate high lateral and prominent ventrolateral tubercles varying in

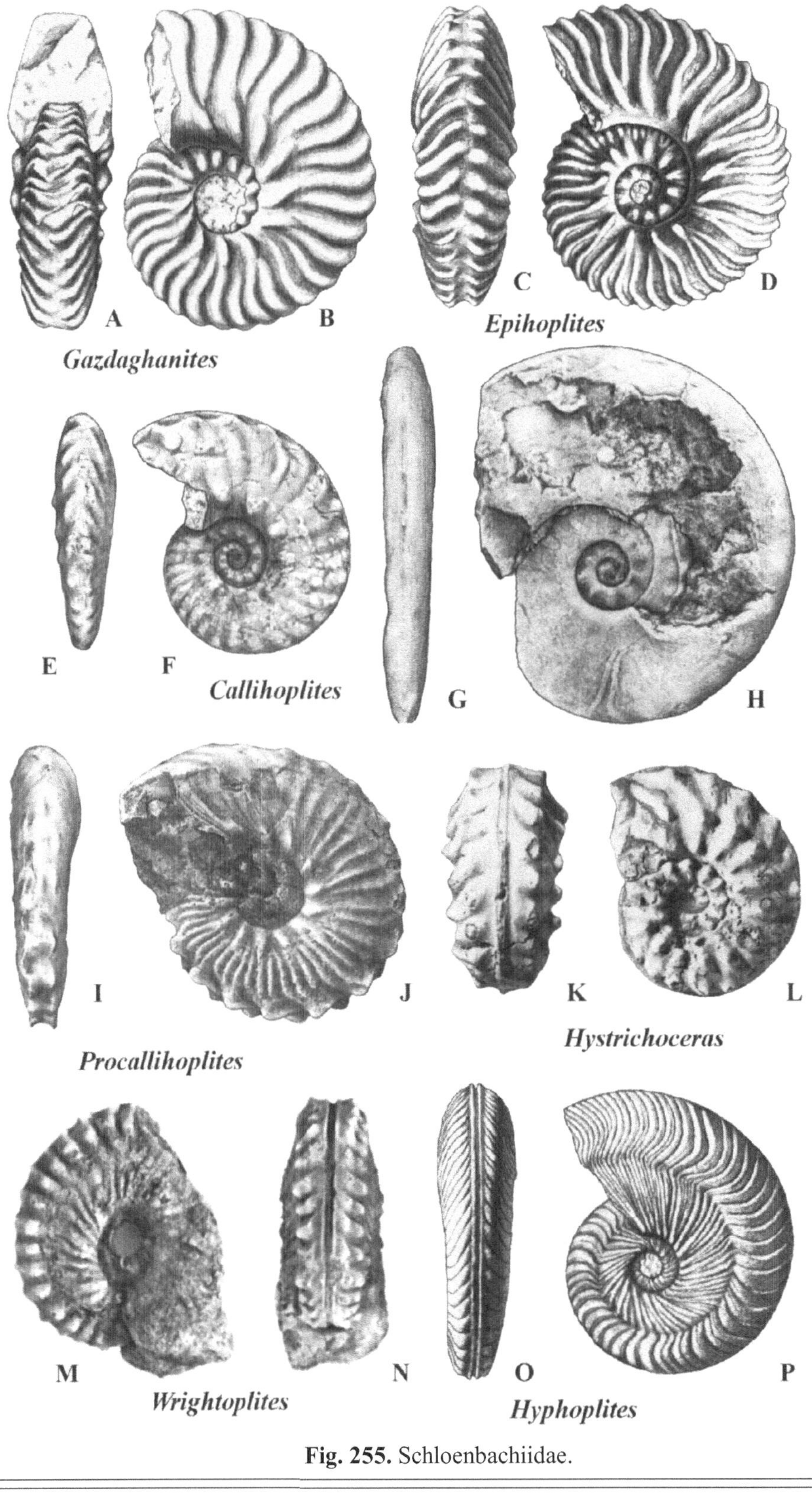

Fig. 255. Schloenbachiidae.

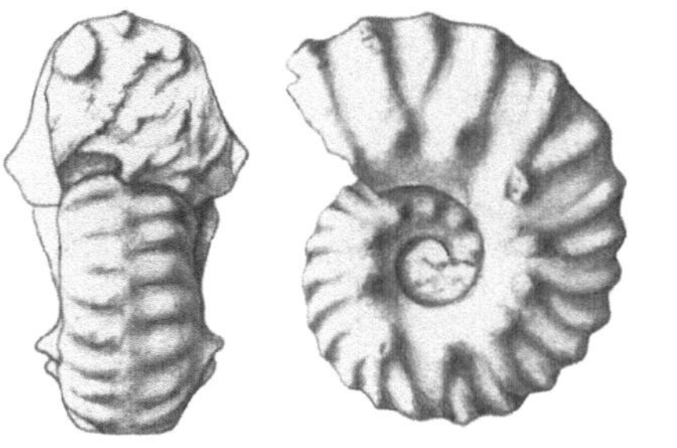

Fig. 256. *Arrhaphoceras.*

strength and number; suture with narrow lobes and trifid L deeper than E. *L.- U. Cenom.* **Fig. 258P-Q** - *S. varians* (J. Sowerby).

Hystrichoceras HYATT, 1900 [*Ammonites coupei* Brongniart, 1822; OD]. Small; like *Schloenbachia* but generally more inflated, with depressed coronate whorl section and prominent nodose mid-lateral tubercles; high lateral tubercles obsolete; suture perhaps with E deeper than L. *U. Cenom.* **Fig. 255K-L** - *H. coupei* (Brongniart).

Jakeiceras COOPER & OWEN, 2011C [*Schloenbachia glabra* Spath, 1924; OD]. Small; strongly compressed with weak falcoid ribs and growth striae arising from tiny pinched prorsiradiate umbilical bullae as in *Hyphoplites*; lateral tubercles absent; constrictions present, forming chevrons over venter. *L. Cenom.* **Fig. 258R-S** - *J. glabrum* (Spath).

Family PLACENTICERATIDAE Hyatt, 1900

Discussion. A monophyletic group with an origin in Sonneratiidae (Sarasin, 1897); comprises the subfamilies Anahoplitinae, Semenoviceratinae and Placenticeratinae, the first two restricted to the hoplitoid province, the latter pandemic. Typically involute, compressed, high whorled, with tabulate to sulcate venter; broad low flexuous to falcate ribs terminate in ventrolateral clavi; suture with very broad deeply divided L. *M. Alb. - U. Maastr.*

Subfamily Anahoplitinae Breistroffer, 1947

(*nom. transl.* Wright, 1996 *ex* Anahoplitidae)

Discussion. A monophyletic group with an origin in Otohoplitinae (Sarasin, 1897); characteristically involute, compressed and high whorled, with narrow tabulate to sulcate (primitive) and fastigiate (derived) venter. Restricted to the hoplitoid province. Sexual dimorphism conspicuous but poorly documented. Comprises the tribes Anahoplitini and Anaplacenticeratini. *M. Alb. - Cenom.*

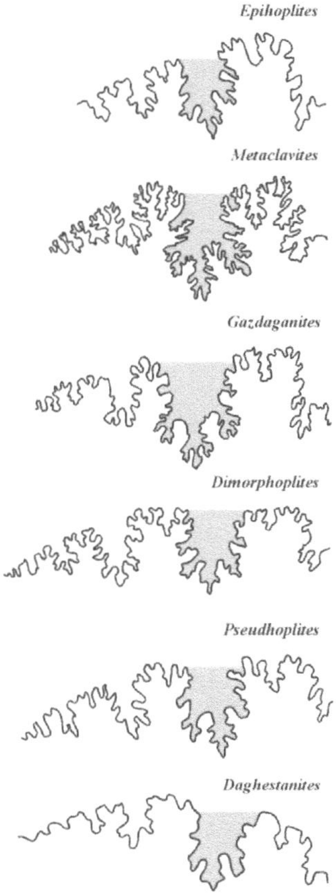

Fig. 257. Schloenbachiid sutures.

Tribe Anahoplitini Breistroffer, 1947

(*nom. transl.* Cooper & Owen, 2012 *ex* Anahoplitinae)

Diagnosis. Small; suture with moderately deep E and narrow L. *M. Alb. - Cenom.*

Anahoplites HYATT, 1900 [*Ammonites splendens* J. Sowerby, 1815; OD]. Small; strongly compressed, involute, with broad flat convergent flanks and narrow tabulate to shallowly sulcate venter; distinct umbilical bullae give rise to paired flexuous ribs separated by 1-2 intercalatories that terminate in small oblique ventrolateral clavi alternating across smooth venter; ribs with asymmetrical profiles (steepest adapically), sometimes weakening at midflank; suture asymmetrical, with displaced siphuncle and widely splayed bifid L much deeper than E. *M. - Lower U. Alb.* **Fig. 261A-B** - *A. planus* (J. Sowerby).

Lepthoplites SPATH, 1925 [*L. falcoides*; SD Spath 1928]. Small, like *Anahoplites* but with stronger sharper ribs that are not effaced from flanks, less clavate ventrolateral tubercles, fastigiate venter and simpler suture with broad subrectangular E/L and non-splayed asymmetrically trifid L. *Upper U. Alb.* **Fig. 261G-H** - *L. falcoides* Spath.

Sulcohoplites MICHAILOVA & SAVEL'EV, 1994 [*S. furcatus*; OD]. Medium sized; like *Anahoplites* but with distinctly

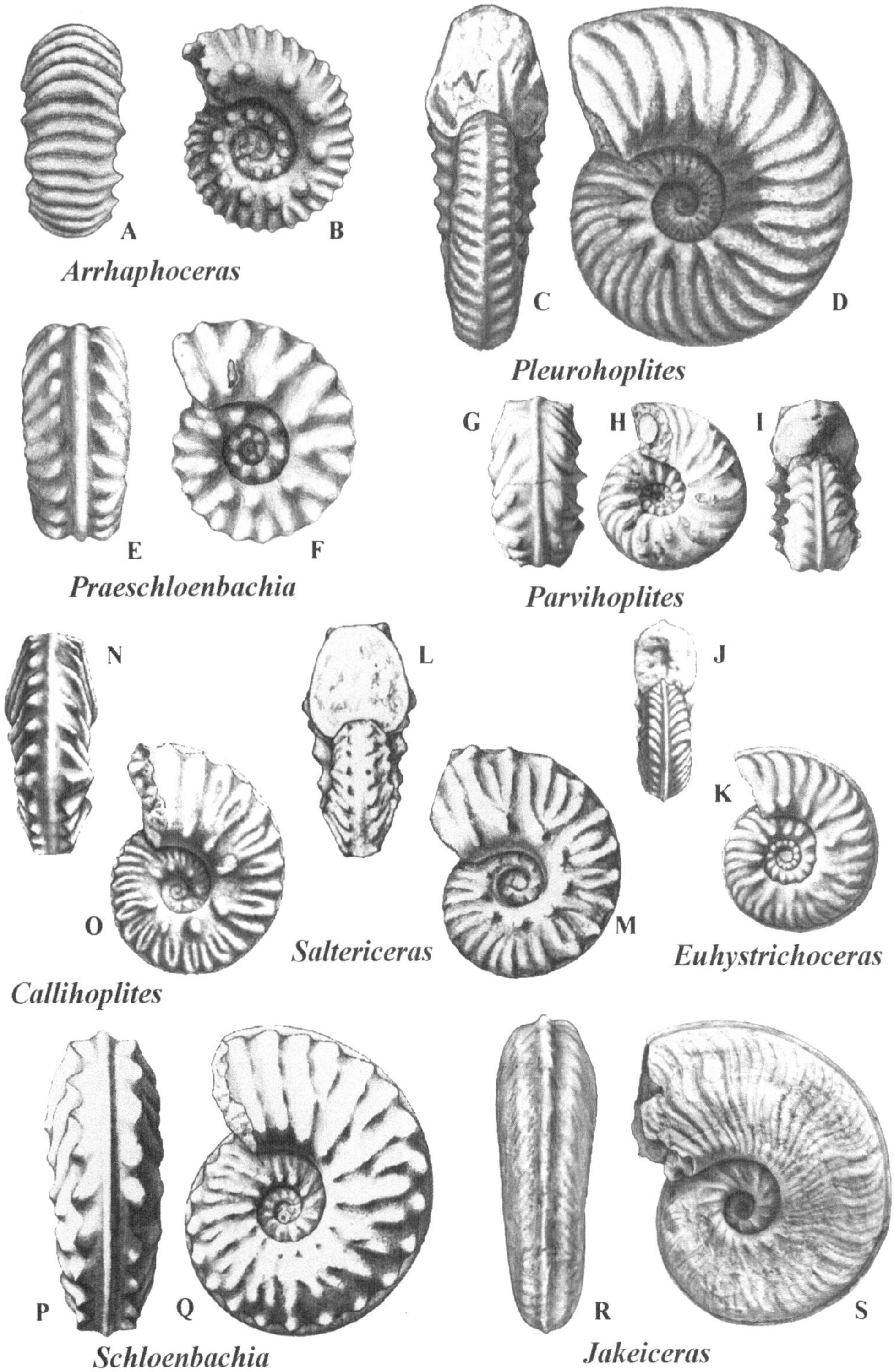

Fig. 258. Schloenbachiidae.

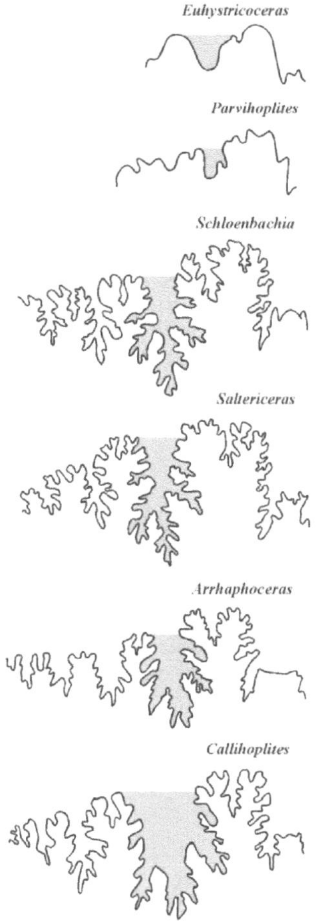

Fig. 259. Schloenbachiid sutures.

sulcate venter and numerous small umbilical bullae; suture symmetrical, with narrow E, broad deeply bifid E/L and large trifid L. *Uppermost M. - Lower U. Alb.* **Fig. 261C-D** - *S. furcatus* Michailova & Savel'ev.

Euhoplitoides COOPER & OWEN, 2010 [*Euhoplites inornatus* Spath, 1925; OD]. Small; inner whorls like *Anahoplites*; smooth outer whorls more inflated, not so high whorled with subquadratic whorl section, flexuous constrictions and conspicuously sulcate venter. *Lower U. Alb.* **Fig. 261E-F** - *E. inornatus* (Spath).

Tribe Anaplacenticeratini Il'in, 2000
(*nom. transl.* Cooper & Owen, 2012
ex Anaplacenticeratidae)

Diagnosis. Medium sized to moderately large, almost oxycone; suture with shallow E, deeply divided E/L and moderately (primitive) to very broad (derived) asymmetrically bifid L. *U. Alb. - Cenom.*

Hengestites CASEY, 1960 [*H. applanatus*; OD]. Moderately large; very compressed and high whorled, almost oxycone, with sharp umbilical edge; early whorls with coarse flexuous ribs and sulcate venter with nodate ventrolateral keels; suture with shallow E, deeply divided E/L with small outer element and very broad asymmetrically bifid L; a homoeomorph of *Placenticeras*. *Lower U. Alb.* **Fig. 261I-J** - *H. applanatus* Casey.

Anaplacenticeras IL'IN, 1959 [*A. turkmenense*; OD]. Medium sized; like *Hengestites* but with wider umbilicus and suture with subsymmetrical E/L. *Cenom.* **Fig. 263O-P** - *A. turkmenense* Il'in.

Subfamily Semenoviceratinae Mirzoev, 1967
(*nom. correct.* Cooper & Owen, 2011a
pro Semenovitinae)

Discussion. A paraphyletic taxon with an origin in Anahoplitinae, close to *Anahoplites* (Spath, 1926); transitional between Anahoplitinae and Placenticeratinae, retaining the coarse ornament of the former but suture with the broad asymmetrically bifid L of the latter. *Upper M. Alb. - M. Cenom.*

Neanahoplites COOPER & OWEN, 2011a [*Anahoplites daviesi* Spath, 1926; OD]. Small; like *Anahoplites* when young, but adult shell with slightly convex flanks and rather sparse broad asymmetrical (steepest adapically) ribs of low relief that project strongly forward near venter; ribs weakly connected to prosocline umbilical bullae, terminating in low oblique ventrolateral clavi that rise above tabulate venter and outnumber ribs; suture asymmetrical, with displaced siphuncle and broad shallowly bifid E/L. *Uppermost M. Alb.* **Fig. 263A-B** - *N. daviesi* (Spath).

Planihoplites SAVEL'EV, 1992 [*Hoplites uhligi* Semenov, 1899; OD]. Small; inner whorls like *Neanahoplites*; in middle to later growth with prominent umbilical bullae feebly connected to broad flat crescents on outer flank; ribs with ventrolateral clavi parallel to siphonal line; suture symmetrical, with very broad asymmetrically trifid L. *Lower U. Alb.* **Fig. 263C-D** - *P. uhligi* (Semenov).

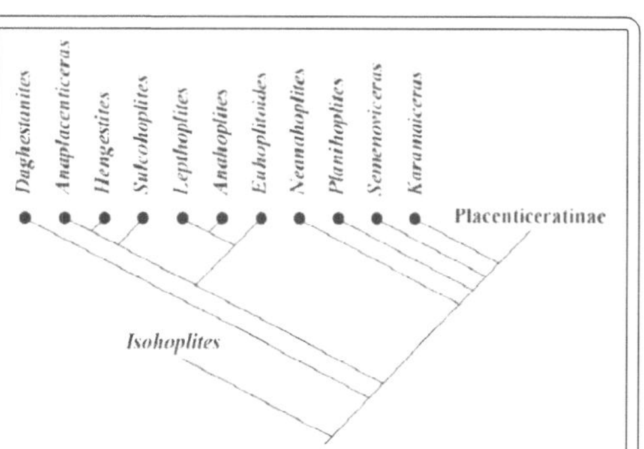

Fig. 260. Hypothesized relationships among primitive placenticeratids

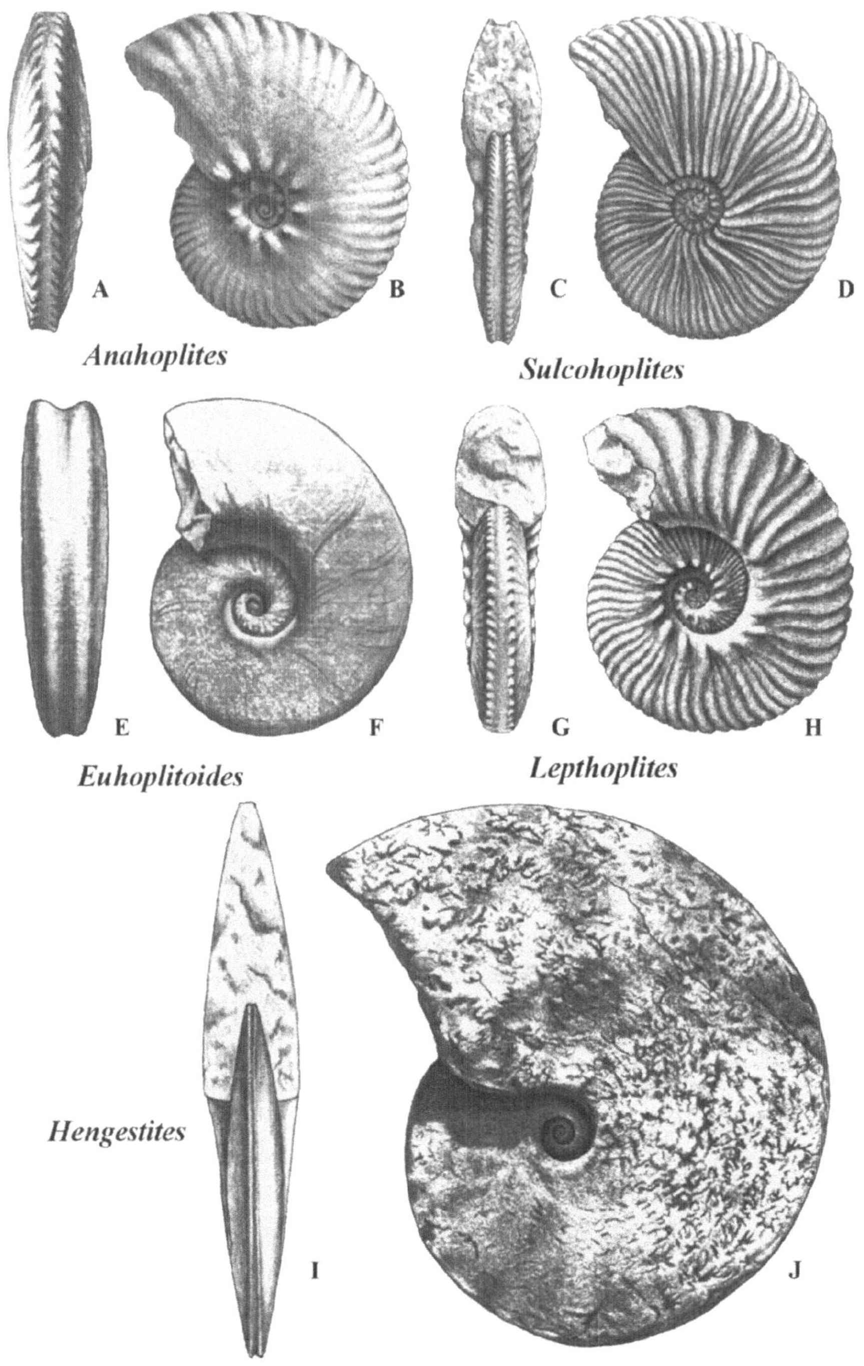

Fig. 261. Anahoplitinae.

Semenoviceras WRIGHT, 1996 [*Hoplites michalskii* Semenov, 1899; OD]. Small; like *Planihoplites* but more involute, with ribbing effaced from lower flank leaving rounded umbilical tubercles disconnected from broad low crescents on outer flank; suture with deeply bifid E/L. *Lower U. Alb*. **Fig. 263E-F** - *S. michalskii* (Semenov).

= *Semenovites* Glazunova, 1960 (*non* Tarbinskii, 1932).

Karamaiceras SOKOLOV, 1967 [*K. kolbajense*; OD]. Small; like *Karamaites* but more compressed, with inclined umbilical wall, funnel-shaped umbilicus, convergent flanks and narrow tabulate venter; ribs and tubercles obsolete;

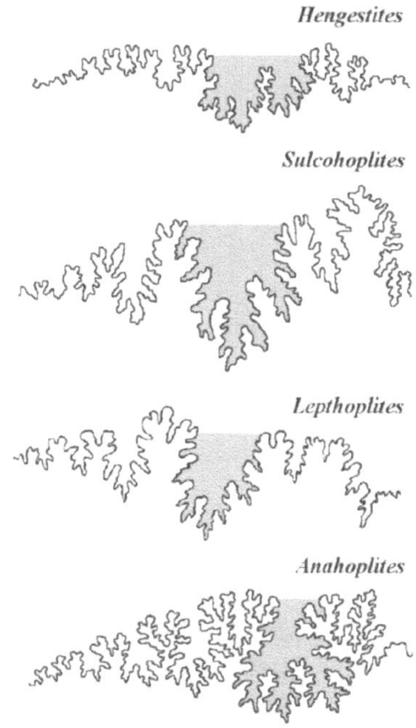

Fig. 262. Anahoplitine sutures.

suture with less deeply divided E/L and shorter median saddle to L. *Upper U. Alb.* **Fig. 263I-J** - *K. kolbajense* Sokolov.

Subfamily Placenticeratinae Hyatt, 1900

(*nom. transl.* Chiplonkar & Ghare, 1976 *ex* Placenticeratidae)

Discussion. A monophyletic group with an origin in Semenoviceratinae (Mirzoev, 1967), close to *Karamaiceras* (Il'in, 2000); mostly weakly ornamented oxycones with low broad crescents on outer flanks, ventrolateral clavi and suture with very broad bifid L and numerous auxiliary elements. Primitive representatives confined to the Upper Albian of the Boreal hoplitoid province where the group arose, their subsequent widespread dispersal occurring in the Cenomanian. Dimorphism conspicuous, macroconchs inflated with weaker ornament and rounded venter in maturity. Due to phenotypic conservativism sutural evolution is significant. Unfortunately the consistency of sutural differences at the population level, particularly when influenced by exasperating morphological plasticity, has still to be established. There is litle doubt that some of the nominal genera should be placed in synonymy, but also that lumping of all placenticeratids into a single genus is simplistic and unhelpful to understanding the evolution of the group. Comprises the tribes Baghiceratini, Placenticeratini, Metaplacenticeratini and Hoplitoplacenticeratini. *L. Cenom. - U. Maastr.*

Tribe Baghiceratini Chiplonkar & Ghare, 1976

(*nom. transl.* herein *ex* Baghiceratinae)
Type genus: *Baghiceras* Chiplonkar & Ghare, 1976.

Diagnosis. In early to middle growth with tabulate to sulcate venter and, often, lateral crescents to flanks. Suture with 4 auxiliary lobes. According to Il'in (2000) this lineage is phylogenetically distinct from that leading to *Placenticeras* but, likewise with an origin in *Karamaiceras*. It is distinguished on the basis of sutural characteristics. *L. Cenom. - U. Sant.*

Mediasiaceras IL'IN, 1975 [*M. beliakovae*; OD]. Small to medium sized; like *Karamaiceras* but more compressed and higher whorled with lanceolate whorl section, tiny umbilical tubercles, falcate striae and crescents on outer flank; ventrolateral clavi parallel siphonal line; suture little incised, with deep penetrating splayed bifid adventive lobe to E/L whose outer branch is smaller than the inner branch, and more protrusive digitate adventive saddle to L. *Cenom.* **Fig. 265A-B** - *M. beliakovae* Il'in.

= *Mediasiceras* Klinger & Kennedy, 1989 (*nom. van.*).

Proplacenticeras SPATH, 1926 [*Placenticeras fritschi* de Grossouvre, 1894; OD]. Medium sized; like *Mediasiaceras* but ventrolateral clavi obsolete and suture more incised, with cauliflower-like terminations to saddles, much broader deeply bifid E/L and deeply divided symmetrical L. *Upper U. Cenom. - Sant.* **Fig. 265F-G** - *P. fritschi* (de Grossouvre).

Parastantonoceras COLLIGNON, 1965 [*P. murphyi*; OD]. Medium sized; like *Proplacenticeras* but more inflated, with swollen inner ventrolateral nodes and prominent outer ventrolateral clavi that alternate across venter. *U. Turon. - M. Coni.* **Fig. 269A-B** - *P. murphyi* (Collignon).

Baghiceras CHIPLONKAR & GHARE, 1976 [*B. ambai*; OD] [**m**]. Medium sized; moderately evolute (U=27%), strongly inflated with broadly rounded venter and broadly elliptical to circular whorl section; 7-8 spinose umbilical tubercles give rise, mostly in pairs, to widely splayed ribs with inner and outer ventrolateral clavi alternating across venter; suture relatively simple. *Cenom. - Turon.* **Fig. 265I-J** - *B. ambai* Chiplonkar & Ghare.

Type included in the synonymy of *Placenticeras mintoi* (Vredenberg) (Jaitly & Ajani, 2013).

Malwiceras CHIPLONKAR & GHARE, 1976 [*M. variabile*; OD]. Medium sized (D=83 mm), rather inflated, moderately evolute (U=27-32%), with maximum width one-third up flank and compressed subquadratic to oval and subcircular whorl section; narrow tabulate venter in early growth becoming broadly rounded in maturity; 7-8 pointed umbilical tubercles, and ventrolateral clavi alternate

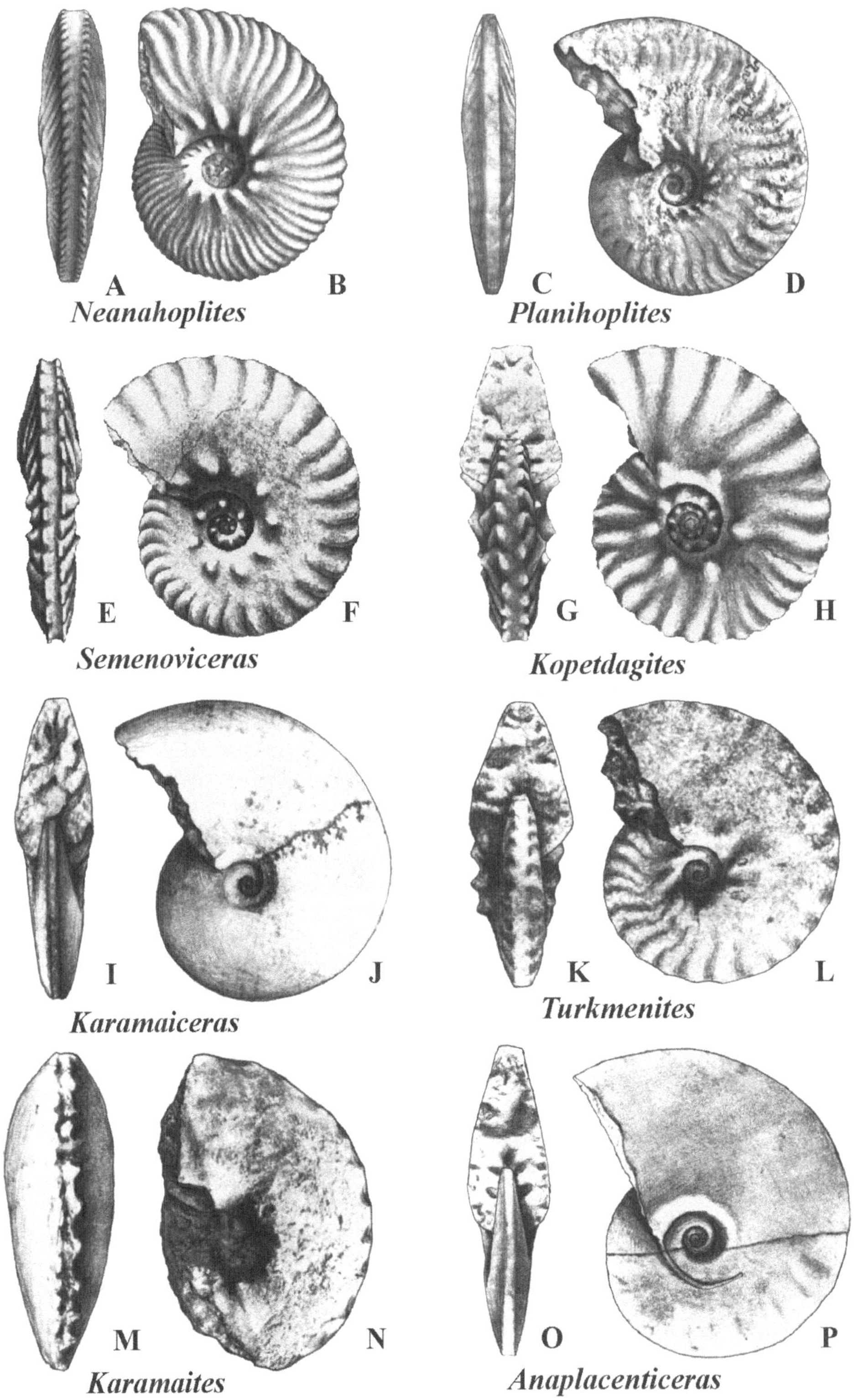

Fig. 263. Placenticeratinae.

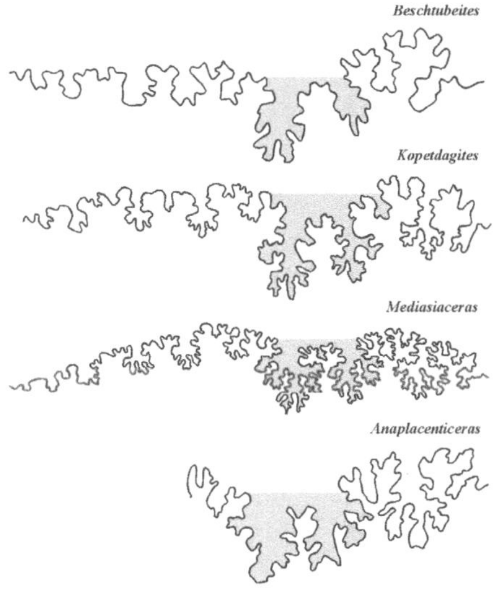

Fig. 264. Placenticeratid sutures.

across venter; suture with few elements and trifid L. *Cen.* - *Turon.* **Fig. 267** - *M. variabile* Chiplonkar & Ghare.

Placentoscaphites CHIPLONKAR & GHARE, 1976 [*P. dangerfieldi*; OD]. Medium sized; an elliptical shell with body chamber undergoing scaphitoid uncoiling and oval umbilicus; flat to weakly convex flanks converge to narrow tabulate venter; weak low ribs bifurcate from sparse umbilical bullae and terminate in small oblique outer ventrolateral clavi with smaller inner ventrolateral clavi; suture simple, with very narrow short E, rather narrow to broad inclined E/L, very broad bifid L and only two auxiliary elements. *Cenom.* - *Turon.* **Fig. 265H** - *P. dangerfieldi* Chiplonkar & Ghare.

Tribe Placenticeratini Hyatt, 1900

(*nom. transl.* herein *ex* Placenticeratinae)

Type genus: *Placenticeras* Meek, 1876.

Diagnosis. Suture with deeply bifid E/L, deeply bifid almost symmetrical L and 5-7 auxiliary lobes. *L. Cenom.* - *L. Camp.*

Turkmenites IL'IN, 1975 [*T. gouerdakensis* Luppov, 1963; OD]. Small; like *Karamaites* but suture with deeply bifid E/L and deeply bifid almost symmetrical L. *L. Cenom.* **Fig. 263K-L** - *T. gouerdakensis* Luppov.

Kopetdagites IL'IN, 1975 [*K. kopetdagensis*; OD]. Small to medium sized; inner whorls like *Turkmenites* but more involute and inflated, with fewer swollen rounded umbilical tubercles and broader ribs which terminate in slightly oblique ventrolateral clavi projecting onto venter; ribs tend to form weak lateral swelling at point of furcation, declining with growth; outer whorls almost smooth save for ventrolateral clavi; suture like *Karamaiceras* but more incised, with inner branch of E/L larger than outer branch. *M.* - *U. Cenom.* **Fig. 263G-H** - *K. kopetdagensis* Il'in.

Karamaites SOKOLOV, 1965 [*Placenticeras? grossouvrei* Semenov, 1899; OD]. Medium sized; like *Planihoplites* but more inflated with maximum width close to umbilical shoulder, convergent flanks, narrow venter and coarser broader ribs weakly connected to fewer (7-8) umbilical tubercles; suture with deeply bifid asymmetrical L and ascendent suspensive lobe. *U. Cenom.* **Fig. 263M-N** - *K. grossouvrei* (Semenov).

= *Karamaites* Sokolov, 1961 (*nom. nud.*).

Beschtubeites IL'IN, 1975 [*B. beschtubensis*; OD]. Small; like *Kopetdagites* but with more numerous ribs and lateral tubercles; ventrolateral clavi not oblique; suture like *Kopetdagites* but with outer branch of E/L longer than inner branch so that, together with deep penetrating accessory saddle to L, they form a regularly descending series; dividing lobe of L/U_2 still shallower than U_2. *L. Turon.* **Fig. 265C-D** - *B. beschtubensis* Il'in.

Placenticeras MEEK, 1876 [*Ammonites placenta* DeKay, 1828; OD]. Medium sized to moderately large; like *Proplacenticeras* but innermost whorls with closely spaced ventrolateral nodes alternating across venter and unornamented flanks; suture more incised, with longer thinner elements and dividing adventive lobe to L/U_2 as deep as or deeper than U_2. *Sant.* - *Camp.* **Fig. 269C-D** - *P. placenta* (De Kay).

= *Placentoceras* Meek, 1871 (*nom. oblit.*).

Diplacomoceras HYATT, 1900 [*Ammonites bidorsatus* Roemer, 1841; OD]. Medium sized; inner whorls like *Placenticeras*, but adult body chamber subquadratic with broad subtabulate venter and initially sparse conical inner ventrolateral tubercles that become auriculate and move up until level with venter; outer ventrolateral clavi form two entire keels on either side of siphonal line in later growth; suture like *Placenticeras* but with adventive saddle to L as long as inner branch of E/L and fewer auxiliary elements. *L. Camp.* **Fig. 269G-H** - *D. bidorsatum* (Roemer).

= *Diplacmoceras* Hyatt, 1903 (*nom. van.*).

Pseudoplacenticeras SPATH, 1926 [*Ammonites milleri* von Hauer, 1866; OD]. Like *Diplacomoceras* but less compressed in early growth, and without umbilical and inner ventrolateral tubercles or ventral groove; venter relatively broad with widely spaced ventrolateral clavi. *L. Camp.* **Fig. 270K** - *P. milleri* (von Hauer).

Stantonoceras JOHNSON, 1903 [*S. pseudocostatum*, = *Ammonites guadalupae* Roemer, 1849, = *Ammonites syrtalis* Morton, 1834; OD]. Small to moderately large; like *Placenticeras* but with evolute outer whorls with broad venter, quadratic whorl section, and very prominent spinose lateral and inner ventrolateral tubercles, the latter sometimes lost on adult body chamber; outer ventrolateral clavi (now ventral in position) form two crenulated keels on either side of narrow venter, notches greatly outnumbering inner ventrolateral tubercles. *U. Sant.* - *L. Camp.* **Fig. 269E-F** - *S. syrtale* (Morton).

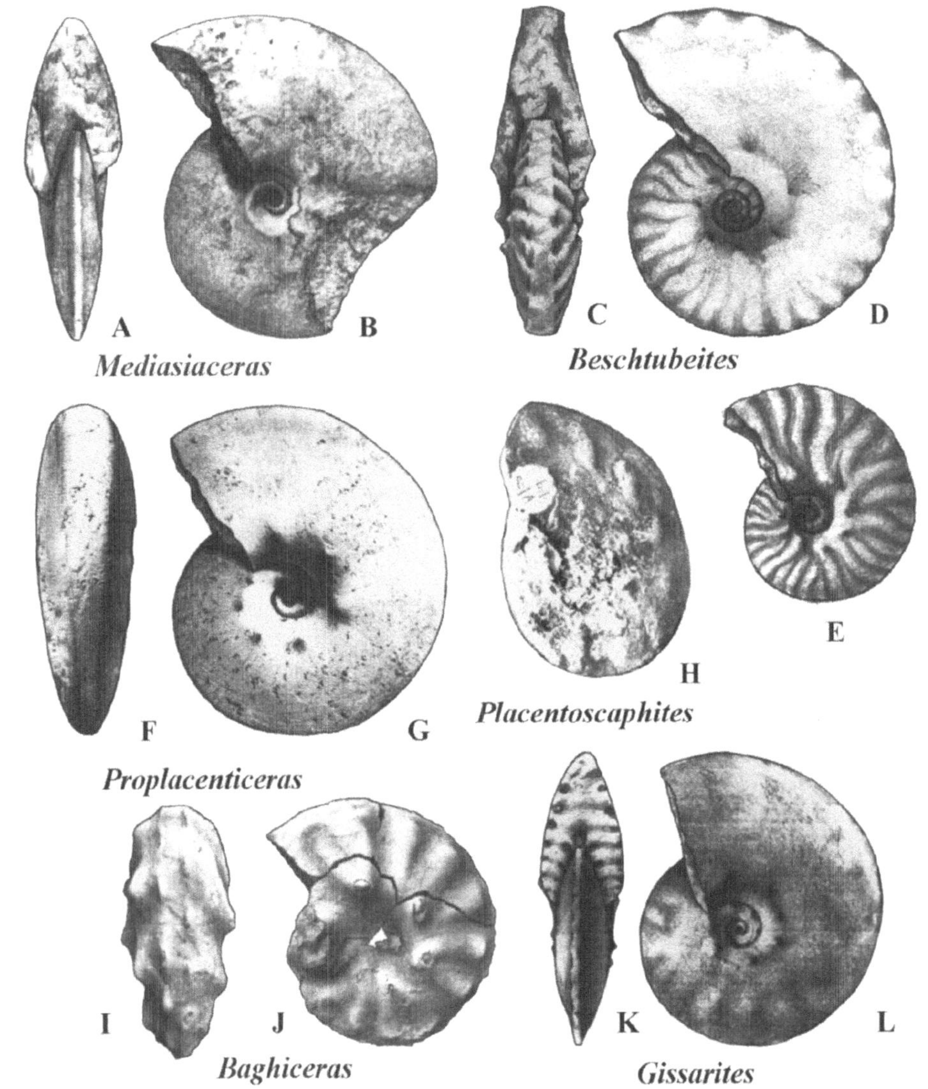

Fig. 265. Placenticeratinae.

Sancarlosia CHIPLONKAR & GHARE, 1976 [*Placenticeras sancarlosense* Hyatt, 1903; OD]. Moderately large; like *Stantonoceras* but with narrow venter, coarser ventrolateral clavi not forming keels and bullate umbilical tubercles migrating to high-lateral position in later growth; suture like *Proplacenticeras* but with more auxiliary elements. *Sant.* **Fig. 269I-J** - *S. sancarlosensis* (Hyatt).

Asiatostantonoceras IL'IN, 1975 [*Stantonoceras tagamense* Ilín, 1959; OD]. Small; in early growth like *Parastantonoceras* but with loss of high-lateral tubercles in later growth, when umbilicals shift to mid-lateral position; suture like *Proplacenticeras* but simpler. *U. Sant.* **Fig. 269K-L** - *A. tagamense* (Il'in).

Gissarites IL'IN, 1959 [*G. kysylchensis*; OD]. Medium sized; like *Placenticeras* but with tabulate venter remaining narrow throughout ontogeny; suture with short blunt branches to E and 6th lobe larger than 5th. *L. Camp.* **Fig. 265K-L** - *G. kysylchensis* Il'in.

Tribe Hoplitoplacenticeratini nov.

Type genus: *Hoplitoplacenticeras* Paulcke, 1907.

Diagnosis. Typically small with bold often flat topped ribs and doubled ventrolateral clavi that are opposite across the venter; suture with relatively narrow L, not deeply divided, and few auxiliary elements. Monophyletic derivatives of uncertain origin, perhaps Placenticeratini close to *Sancarlosia*. However given the very different sutures, Matsumoto (1985) suggested an origin in Pseudoschloen-

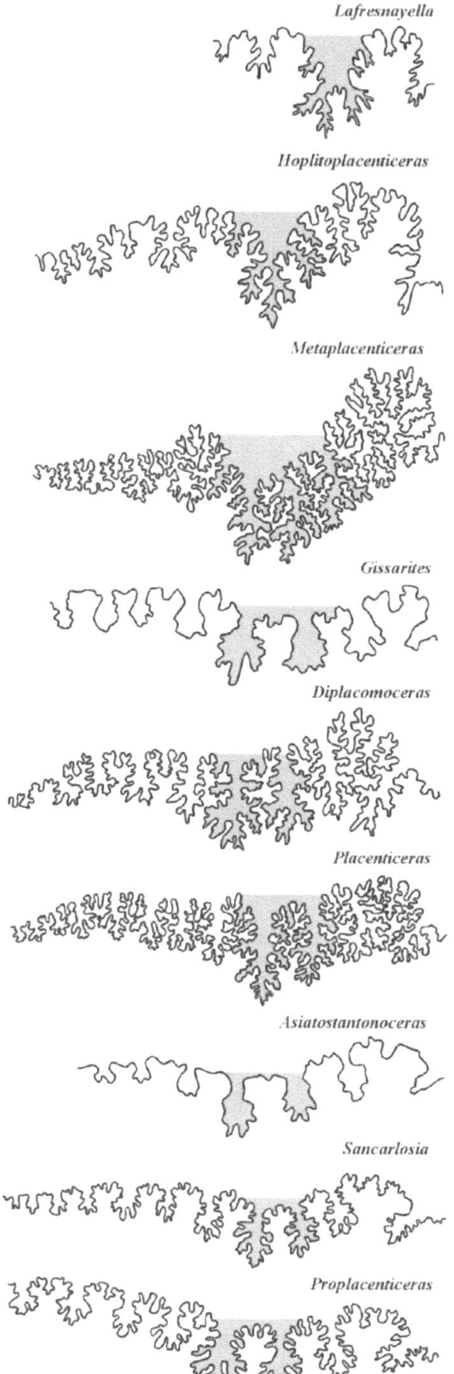

Fig. 266. Placenticeratid sutures.

bachiinae. *Lemfoerdiceras* shares the fine ribbing of *Metaplacenticeras* but the similarity may be convergent since the venters are completely different. *M. Camp. - U. Maastr.*

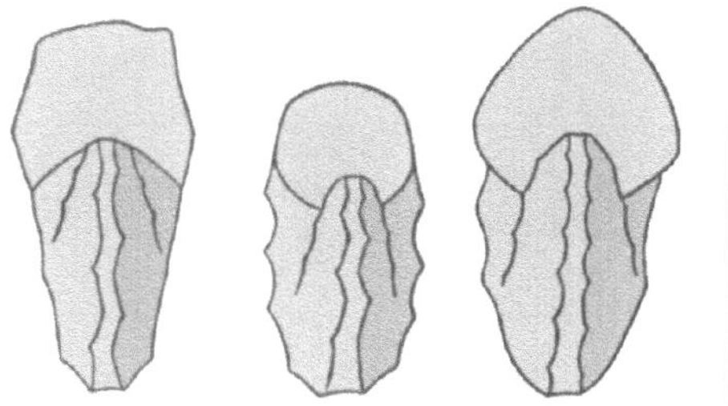

Fig. 267. *Malwiceras variabile* Chiplonkar & Ghare, variable whorl sections.

Hoplitoplacenticeras PAULCKE, 1907 [*Hoplites plasticus* Paulke; ICZN Opinion 554]. Small; rather evolute, compressed to inflated, with flat subparallel or convergent flanks, narrow tabulate venter and polygonal to trapezoidal whorl section; more or less prominent umbilical bullae give rise to strong prorsiradiate ribs, commonly bifurcating or alternating long and short, with inner and outer ventrolateral tubercles; ribs cross venter transversely so that ventrolaterals are opposite; suture with narrow E, broad deeply incised asymmetrically bifid E/L with sloping inner margin due to elongate accessory saddle, very narrow trifid L deeper than E, and multifid L/U_2. *M. - U. Camp.* **Fig. 270E-F** - *H. plasticum* (Paulcke).
= *Dechenoceras* Kayser, 1924 (*nom. nud.*).

Lafresnayella **gen. nov.** [*Ammonites lafresnayanus* d'Orbigny, 1841, p.326; OD herein]. Very small; like compressed *Hoplitoplacenticeras* but with more numerous umbilical bullae, flexuous flat topped prorsiradiate ribs that alternate long and short and broaden ventrally, paired ventrolateral clavi merging into a single clavus in later growth when umbilical bullae become obsolete; relatively simple suture with narrow L and without multiplicity of adventive and auxiliary elements. *U. Maastr.* **Fig. 270G-H** - *L. lafresnayana* (d'Orbigny).

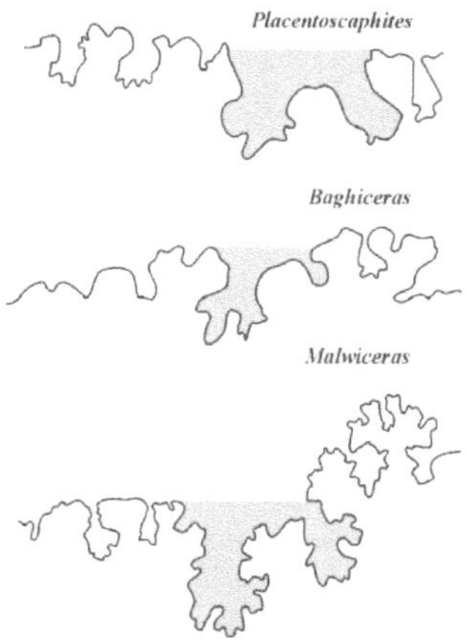

Fig. 268. Placenticeratid sutures.

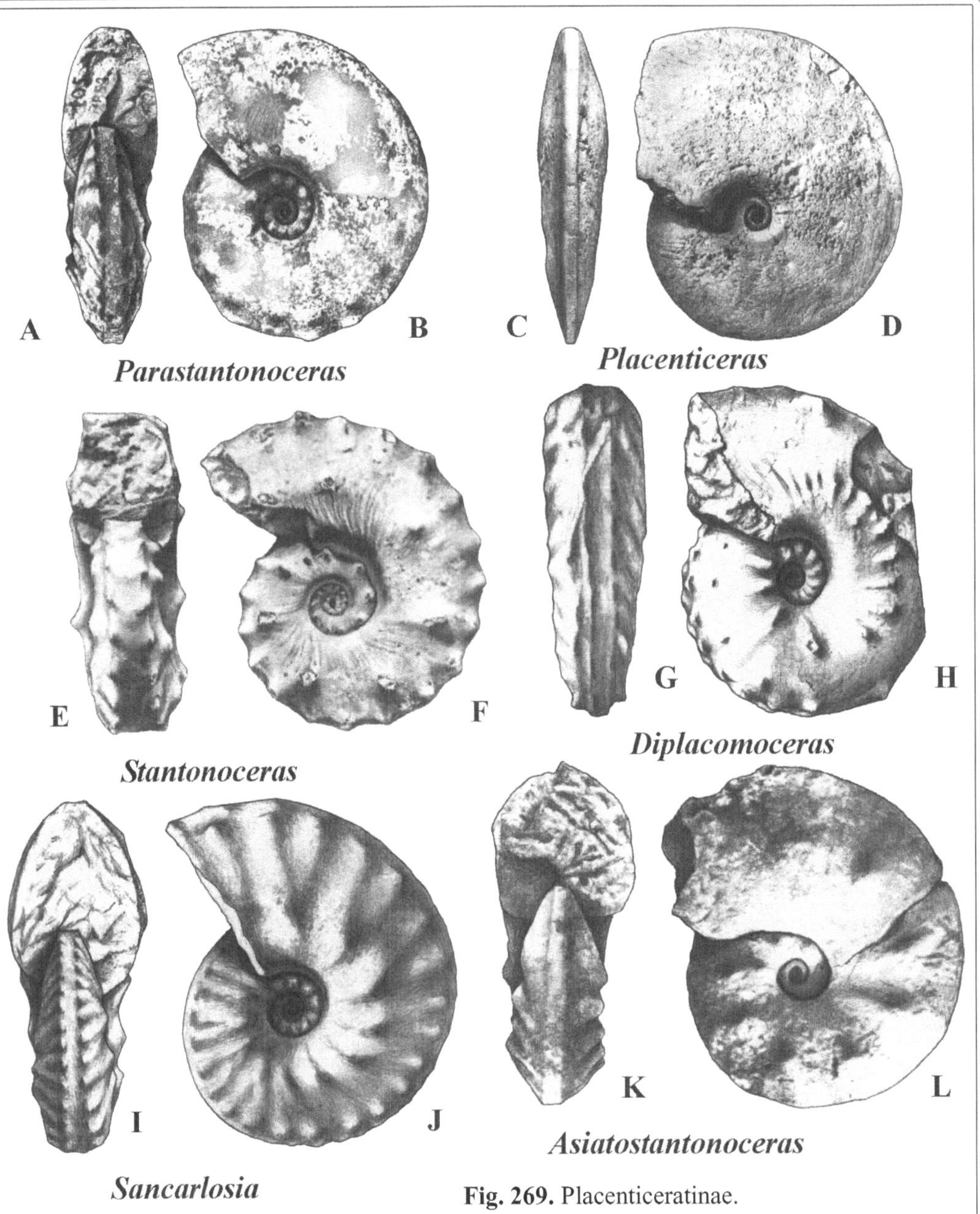

Fig. 269. Placenticeratinae.

Lemfoerdiceras KENNEDY, 1986 [*Ammonites lemfoerdensis* Schlüter, 1872; OD]. Small; like *Hoplitoplacenticeras* but higher whorled, with fine looped ribbing and midlateral and siphonal tubercles. *L. - U. Maastr.* **Fig. 270A-B** - *L. lemfoerdense* (Schlüter).

Tribe Metaplacenticeratini nov.

Type genus: *Metaplacenticeras* Spath, 1927.

Diagnosis. Venter tricarinate; a group of uncertain origin since the resemblance to *Placenticeras* may be convergent; fine falcoid ribbing is shared with *Lemfoerdiceras*; an ancestry in Pseudoschloenbachiinae has been suggested (Matsumoto, 1988), perhaps close to *Hirtziella*. *U. Camp.*

Metaplacenticeras SPATH, 1926 [*Placenticeras pacificum* Smith, 1900; OD]. Medium sized; inner whorls like *Placenticeras* but later with umbilical bullae from which bundles of 2-4 dense falcoid ribs arise and project strongly forward near narrow tricarinate venter; in maturity venter becomes tabulate to sulcate, bordered by finely nodate

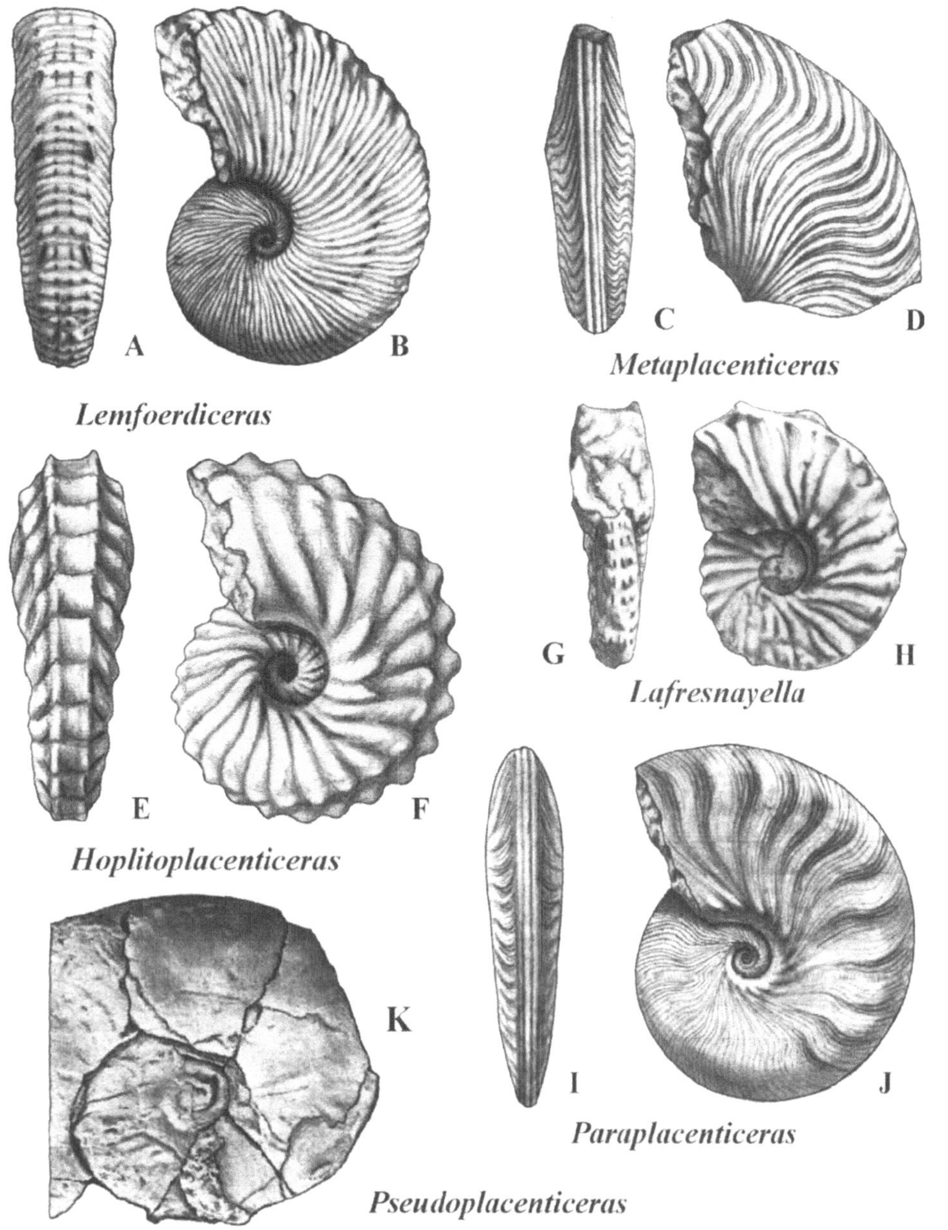

Fig. 270. Placenticeratidae.

keels; suture with very narrow stemmed E/L with rather short dividing lobe, very wide L with two deep penetrating accessory lobes, and U_2 deeper than dividing lobe to L/U_2; line of suture follows that of ribs. *U. Camp*. **Fig. 270C-D** - *M. pacificum* (Smith).

Paraplacenticeras MATSUMOTO, 1953 [*Placenticeras subtilistriatum* Jimbo, 1894; OD]. Small; like *Metaplacenticeras* but with more distant and less numerous ribs that do not show multiple branching. *U. Camp*. **Fig. 270I-J** - *P. subtilistriatum* (Jimbo).

Suborder ACANTHOCERATIDINA Hyatt, 1900

(*nom. transl.* Cooper, 1992 *ex* Acanthoceratida, *nom. correct.* herein)

Discussion. Monophyletic derivatives of Placenticeratidina (Brinkmann, 1937); comprises the hyporders Brancoceratoidei and Acanthoceratoidei. *L. Alb. - U. Maastr.*

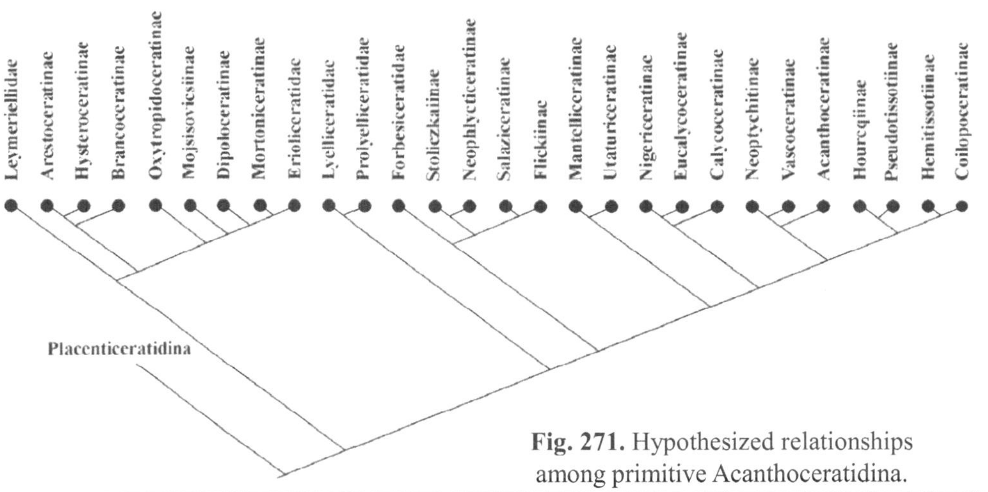

Fig. 271. Hypothesized relationships among primitive Acanthoceratidina.

Hyporder BRANCOCERATOIDEI nov.

Discussion. Monophyletic derivatives of Desmoceratoidei; comprises the superfamilies Brancoceratoidea, Lyelliceratoidea and Forbesiceratoidea. *Lowermost Alb. - U. Alb.*

Superfamily BRANCOCERATOIDEA Spath, 1900

(*nom. transl.* Cooper, 1992 *ex* Brancoceratidae)

Discussion. Monophyletic derivatives of Puzosioidea (Brinkmann, 1937); comprises the paraphyletic families Leymeriellidae, Brancoceratidae, Mojsisovicziidae, Mortoniceratidae and Erioliceratidae. *Lowermost Alb. - Uppermost Alb.*

Family LEYMERIELLIDAE Breistroffer, 1951

(*nom. transl.* Wright,1955 *ex* Leymeriellinae)

Discussion. A monophyletic group with an origin in Cleoniceratidae (Vnigriceratinae), close to *Callizoniceras* (Brinkmann, 1937); comprises the nominate family only, the sister taxon of Brancoceratidae. Very evolute, compressed, with simple or bifurcating flexuous ribs that are flat topped or grooved, either forming chevrons over the venter (primitive) or interrupted by a siphonal groove (derived). Typically Boreal in distribution. *L. Alb.*

Wollemaniceras BREISTROFFER, 1947 [*Desmoceras keilhacki* Wolleman, 1907; OD]. Very small; evolute, compressed, with rounded umbilical shoulder, flattened flanks, narrowly rounded venter and compressed oval whorl section increasing rapidly in height; numerous deep flexuous constrictions form chevrons across venter; flanks with simple strong rounded flexuous ribs, sometimes bifurcating at or above midflank, project forward on outer flank but effaced from venter; suture with rectangular E/L, narrow trifid L deeper than E, reduced U_2 and slightly retracted suspensive lobe. *Lowermost Alb.* **Fig. 275A-B, M** - *W. keilhacki* (Wolleman).

Large collections indicate substantial intraspecific variation within the type species (Kemper, 1982), that includes virtual absence of ribs and a uniform whorl expansion rate in early to middle growth.

Proleymeriella BREISTROFFER, 1947 [*Parahoplites schrammeni* Jacob, 1907; OD]. Small; like *Wollemaniceras* but with slightly convex flanks, coarser stronger ribs, some bifurcating, that form chevrons and may flatten across venter; suture with shallower E, markedly asymmetrical E/L and wide deep trifid L. *Lower L. Alb.* **Fig. 275C-D** - *P. schrammeni* (Jacob).

Leymeriella JACOB, 1907 [*Ammonites tardefurcatus* d'Orbigny, 1841; SD Spath, 1925]. Small; like *Proley-*

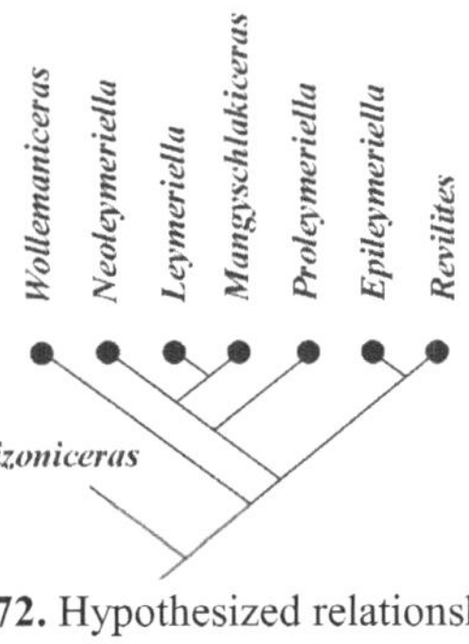

Fig. 272. Hypothesized relationships among Leymeriellidae.

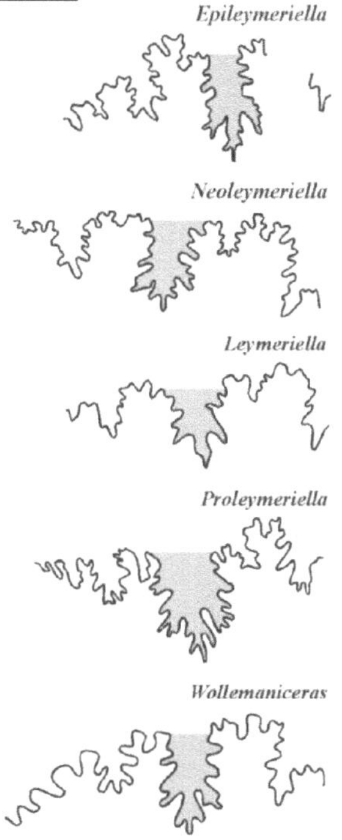

Fig. 273. Leymerielline sutures.

meriella but more compressed, with broad flat convergent flanks, narrow subtabulate venter, and compressed subrectangular whorl section; dense fine sharp simple almost straight to slightly sigmoidal prorsiradiate ribs broaden and flatten ventrally, and from just below midflank are shallowly grooved; all ribs with oblique ventrolateral clavi, opposite across the venter and separated by a smooth siphonal band; tubercles obsolete on adult body chamber when ribs and constrictions form chevrons over venter; suture with relatively broad almost square E/L and E almost as deep as trifid L. *Lower L. Alb.* **Fig. 275E-F** - *L. tardefurcata* (d'Orbigny).

***Mangyschlakiceras* gen. nov.** [*Leymeriella (L.) brevicostata* Savel'ev, 1973, p.226; OD herein]. Small, to 10 mm diameter like *Leymeriella* but thereafter with convergent flanks, compressed trapezoidal whorl section and simple, flexuous, sharp-crested ribs with oblique ventrolateral clavi standing high above venter. *Lower L. Alb.* **Fig. 275I-J** - *M. brevicostatum* (Savel'ev).

Neoleymeriella SAVEL'EV, 1973 [*Leymeriella consueta* Casey, 1957; OD]. Small; like *Leymeriella* but more coarsely ribbed and evolute, with umbilicolateral tubercles on inner whorls, blunt ventrolateral horns and subhexagonal costal whorl section; ribs extend onto umbilical wall. *Lower L. Alb.* **Fig. 275G-H** - *N. consueta* (Casey).

Epileymeriella BREISTROFFER, 1947 [*Parahoplites hitzeli* Jacob, 1907; OD]. Small; like *Proleymeriella* but more compressed and involute, with higher whorls and ribs branching from umbilical shoulder and then again higher on flank; suture like *Leymeriella* but with relatively long and narrow saddles and lobes, and L much deeper than E. *Lower L. Alb.* **Fig. 275N-O** - *E. hitzeli* (Jacob).

Revilites CASEY, 1978 [*Hoplites (Leymeriella) revili* Jacob, 1908; OD]. Small; like *Epileymeriella* but more evolute and densicostate, with fine riblets showing regular double bifurcation; suture like *Leymeriella*. *Lower L. Alb.* **Fig. 275K-L** - *R. revili* (Jacob).

Family BRANCOCERATIDAE Spath, 1900

= Hystatoceratidae Hyatt, 1900.

Discussion. A seemingly monophyletic group with an origin in Leymeriellidae (Casey, 1957). Generally very small, compressed, evolute, with incipient to strong siphonal keel; flexuous ribs arise from umbilical bullae, thickening on ventrolateral shoulders. Comprises the subfamilies Brancoceratinae, Hysteroceratinae and Arestoceratinae. *L. - U. Alb.*

Subfamily Brancoceratinae Spath, 1900

(*nom. transl.* Spath, 1934 *ex* Brancoceratidae)

Discussion. A paraphyletic taxon derived from Leymeriellidae (Casey, 1957), close to *Wollemaniceras*. Typically very small, evolute, compressed, with straight to flexuous blunt ribs that broaden ventrally, an incipient siphonal keel may be present at some stage, and suture with a trifid L deeper than E. *L. - M. Alb.*

Carinolytoceras WIEDMANN, 1962 [*C. carinatum*; OD]. Micromorph; compressed serpenticone with subcircular whorl section and tricarinate-bisulcate venter; flanks with 20 weak prorsiradiate constrictions per whorl; simple suture with deep narrow bifid E, broad weakly trifid E/L, shallow trifid L and broad barely indented L/U_2. *L. Alb.* **Fig. 276A-C** - *C. carinatum* Wiedmann.

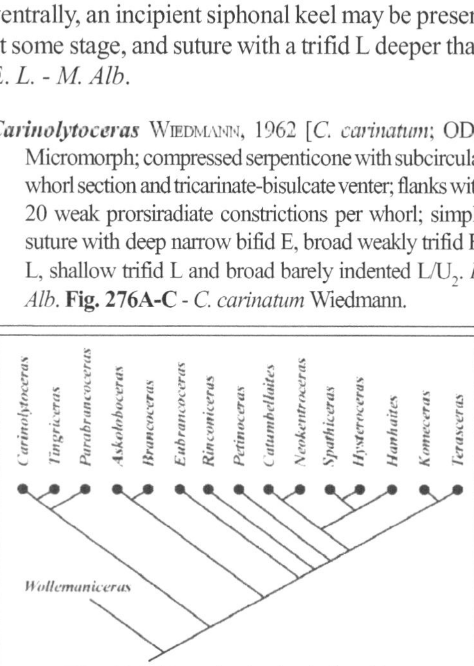

Fig. 274. Hypothesized relationships among Brancoceratidae.

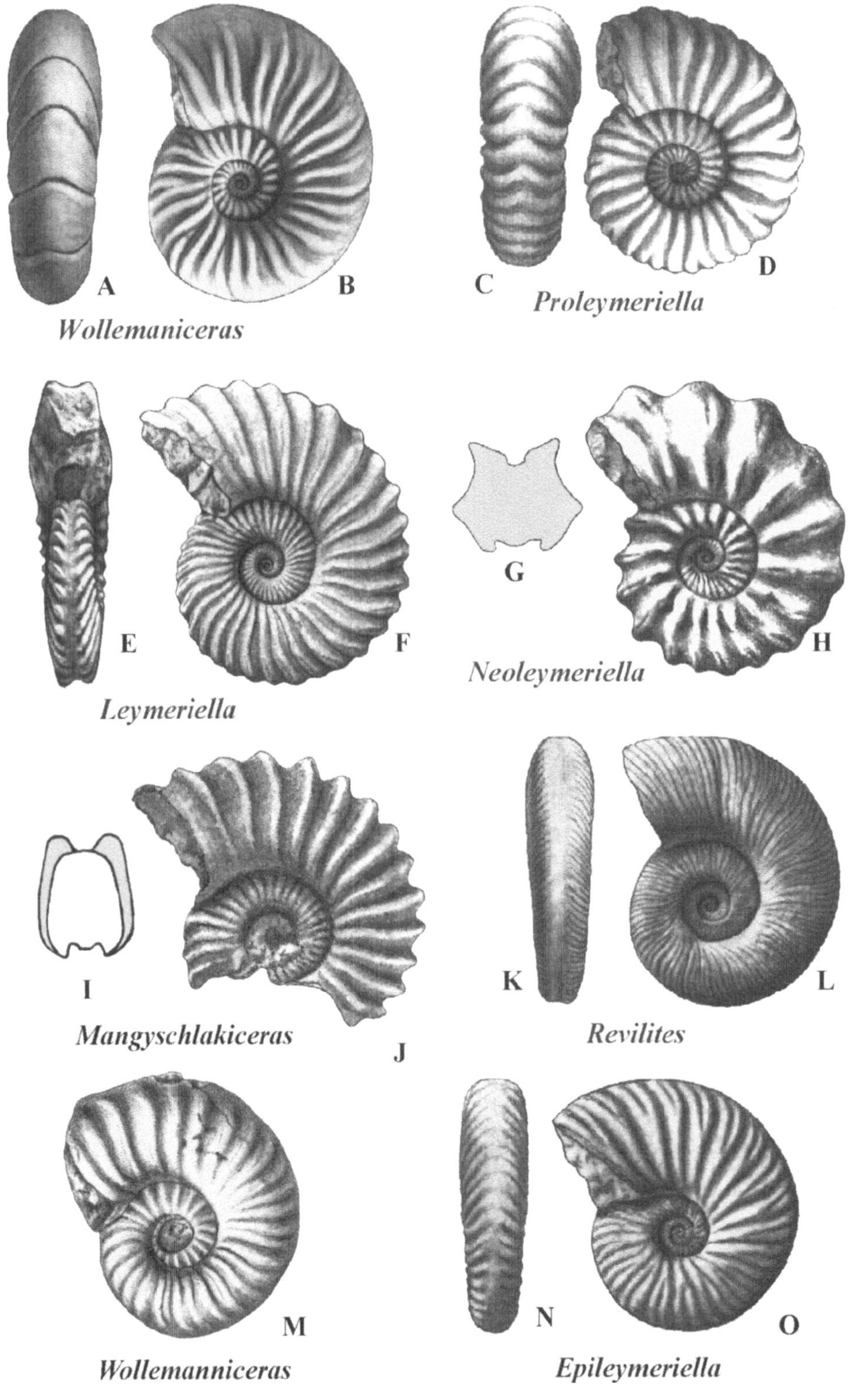

Fig. 275. Leymeriellidae.

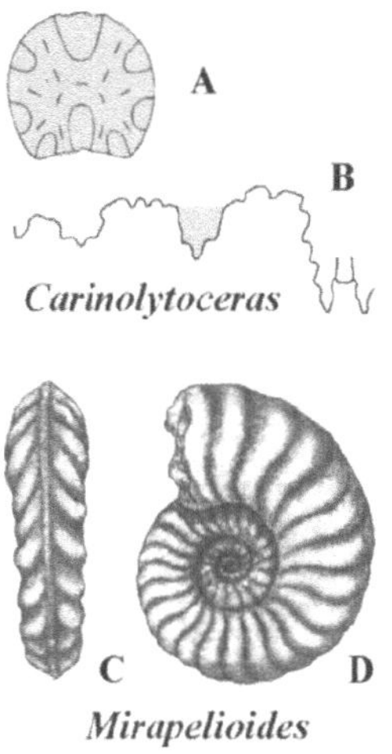

Fig. 276. Brancoceratinae.

Originally assigned to Lytoceratidae, but the lack of a septal lobe makes this untenable. Its resemblance to Barremian *Haplobrancoceras* is convergent, but its suture is very similar to Upper Albian *Tingriceras*.

Tingriceras CHAO, 1976 [*T. chanaense*; OD]. Very small; like *Parabrancoceras* but more involute and without constrictions (at least in later growth); type species with flattened venter, but broadly rounded in referred species (*T. tingriense*); suture simple, with deep E and small trifid L; a homoeomorph of *Tetragonites* but with different suture. *Lower U. Alb.* **Fig. 277AA-BB** - *T. tingriense* Chao; **Fig. 277CC-DD** - *T. chanaense* Chao.
= *Tringiceras* Chao, 1976 (*nom. van.*)

Parabrancoceras BREISTROFFER, 1951 [*Brancoceras besairiei* Collignon, 1949; OD]. Very small; like *Carinolytoceras* but with compressed oval whorl section and inner whorls with 5-6 rectiradiate constrictions separated by several broad low folds; outer whorls with distant broad ribs, at least until body chamber, separated by wide shallow constrictions on flanks. *L. Alb.* **Fig. 277E-H** - *P. besairiei* (Collignon).

Brancoceras STEINMANN, 1881 [*Ammonites senequieri* d'Orbigny, 1841; SD Hyatt, 1900]. Very small; like *Parabrancoceras* but with strong distant flexuous ribs in middle to later growth; constrictions lacking; ribs arise singly or in pairs from strong umbilical bullae, or alternate long and short, and thicken across venter that is crossed transversely; faint very delicate siphonal ridge present; in later growth with quadratic whorl section and sharp narrow rectiradiate single ribs; suture pseudoceratitic, with shallow E, broad rectangular shallowly bifid E/L, moderately wide incised L deeper than E, moderately wide bifid to entire L/U_2 and reduced U_2. *L. Alb.* **Fig. 277A-B** - *B. paronai* Collignon; **Fig. 278A-B** - *B. helcion* (Reynès).
= *Hystatoceras* Hyatt, 1900 (obj.).

Askoloboceras VAN HOEPEN, 1944 [*A. fastigatum*; OD]. Very small; like densicostate *Brancoceras* but ribs forming chevrons over venter in later growth when umbilical tubercles obsolete; relict apertures (old "mouth-edges") present. *Lower U. Alb.* **Fig. 277S-T** - *A. fastigatum* van Hoepen.

***Cinctibrancoceras* gen. nov.** [*Brancoceras magneti* Collignon, 1949, p.98; OD herein). Very small; in early growth compressed, high whorled, involute (U~20%) and smooth, with almost flat subparallel flanks, fastigiate venter and delicate siphonal keel; in middle to later growth umbilicus widens (U=33-39%), whorls more inflated with broadly convex flanks, venter rounds and deep constriction-like interspaces develop, separated by broad flat-topped wedge-shaped ribs that alternate with normal thin ribs; adult body chamber with narrow sharp prorsiradiate ribs separated by wide interspaces; suture pseudoceratitic, with narrow E, broad shallowly indented asymmetrically bifid E/L as deep as E with incised inner margin and entire outer margin, and entire remaining saddles. *L. Alb.* **Fig. 278C-D** - *C. magneti* (Collignon).

Eubrancoceras BREISTROFFER, 1951 [*Brancoceras aegoceratoides* Steinmann, 1881; OD]. Very small; like *Brancoceras* but with higher whorls, subtrapezoidal whorl section, single ribs only, no siphonal keel and more complex suture with narrower saddles and trifid L. *Lower M. Alb.* **Fig. 277C-D** - *E. aegoceratoides* (Steinmann).

***Mirapelioides* gen. nov.** [*Brancoceras? carinatum* Collignon, 1949, p.96; OD herein]. Very small; evolute, compressed with wide shallow umbilicus, broad convex flanks and fastigiate venter with strong raised siphonal keel; strong simple sigmoidal ribs with asymmetrical profiles project forward on outer flank and meet keel obliquely; suture with broad little incised saddles like *Brancoceras*. *L. Alb.* **Fig. 276C-D** - *M. carinatum* (Collignon).
= *Paroxytropidoceras* Breistroffer, 1951 (*nom. nud.*).

Rinconiceras ETAYO SERNA, 1979 [*R. rinconi*; OD]. Very small; like *Eubrancoceras* but with umbilical bullae in early growth and ribs that project forward on outer flank and do not thicken markedly across venter; in middle growth venter subfastigiate and ribs form chevrons over it; suture with L deeper than E and narrow U_2. *M. Alb.* **Fig. 277Q-R** - *R. rinconi* Etayo Serna.

Subfamily Hysteroceratinae nov.

Diagnosis. Typically very small to small, with a siphonal keel at least in early to middle growth, ribs that bifurcate from weak to strong umbilical bullae or alternate long and short, and a relatively complex suture with L as long as or shorter than E. A monophyletic group with an origin in Branco-ceratinae (Spath, 1924), close to *Eubrancoceras cricki* (Spath)

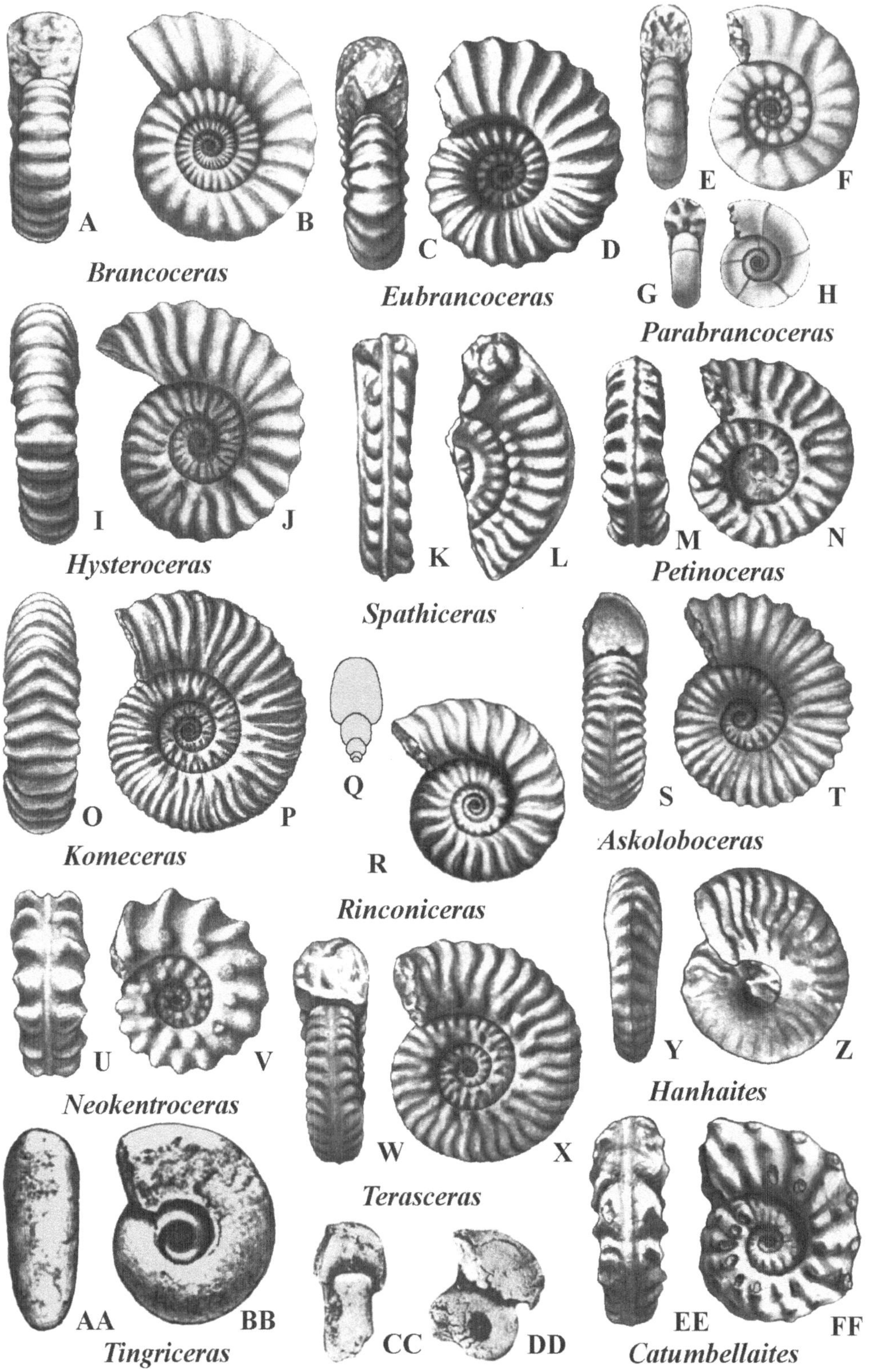

Fig. 277. Brancoceratidae.

that has traces of a siphonal keel on its innermost whorls. *U. Alb.*

Hysteroceras HYATT, 1900 [*Ammonites varicosus* J. de C. Sowerby, 1824; OD]. Very small; evolute (U=38-44%), compressed, with flattened flanks and venter; in early to middle growth with elliptical whorl section, fastigiate venter and prominent siphonal keel; after earliest smooth whorls umbilical bullae give rise to strong round sigmoidal ribs that alternate long and short, rarely bifurcating; on subquadratic adult body chamber ribs mostly long, broadening and thickening ventrally, crossing venter transversely; suture relatively simple, with broad shallowly bifid E/L and narrow trifid L almost as deep as E. *U. Alb.* **Fig. 277I-J** - *H. varicosum* (J. de C. Sowerby).

Spathiceras WHITEHOUSE, 1927 [*Hystrichoceras antipodeum* Etheridge, 1902; OD]. Small; like *Hysteroceras* but very compressed, evolute and flat sided, with dense uniform single rectiradiate ribs that form blunt ventrolateral bulges on either side of persistent siphonal keel. *U. Alb.* **Fig. 277K-L** - *S. antipodeum* (Etheridge).

The pyritized fauna from Point Charles, Darwin, occurs as loose beach-sorted ironstone pebbles derived by erosion of strata exposed on the sea floor (Henderson, 1990). The presence of *Myloceras cornucopia* Spath and *Mariella bergeri* (Brongniart) indicate a mixed assemblage from almost all levels in the Upper Albian, bar the basal *cristatum* Zone. There is no intrinsic way of knowing if the specimens of *Hysteroceras* gr. *subbinum* Spath figured by Cooper & Kennedy (1979) and Henderson (1990), that were used to synonymize *Spathiceras*, are in any way related to the lectotype that shows mainly straight simple ribs that are not effaced at midflank and ventrolateral bulges if not tubercles.

***Hanhaites* gen. nov.** [*Hysteroceras falcicostatum* Haas, 1942, p.35; OD herein]. Micromorph to very small; like *Hysteroceras* but more involute and strongly compressed; inner and middle whorls smooth; in later growth with closely spaced, flat topped, broad falcate ribs that form crescents on outer flank and are weakly if at all connected to the umbilical bullae; ribs asymmetrical, steepest posteriorly; siphonal keel thread like. *Lower U. Alb.* **Fig. 277Y-Z** - *H. falcicostatus* (Haas).

Komeceras VAN HOEPEN, 1944 (*K. acuticostatum*; OD). Very small; like densicostate *Hysteroceras* but with finer sharper ribs; siphonal keel obsolete in maturity when ribs form chevrons over venter. *Lower U. Alb.* **Fig. 277O-P** - *K. acuticostatum* van Hoepen.

Terasceras VAN HOEPEN, 1944 [*T. cariniferum*; OD]. Very small; like *Komeceras* but with stronger umbilical bullae, blunter more flexuous ribs and siphonal keel persisting onto all but last part of adult body chamber that develops subfastigiate venter across which ribs form chevrons. *Lower U. Alb.* **Fig. 277W-X** - *T. cariniferum* van Hoepen.
= *Terascoceras* Wright, 1996 (*nom. van.*).

Petinoceras VAN HOEPEN, 1944 [*P. recuperator*; OD]. Micromorph; outer whorls like *Hysteroceras* but with periodic *Dipoloceras*-like flares and distant ribs alternating long and short, not broadening ventrally. *Lower U. Alb.* **Fig. 277M-N** - *P. recuperator* van Hoepen.

Neokentroceras SPATH, 1921 [*N. curvicornu*; ICZN Opinion 1254]. Very small; like *Hysteroceras* but with exaggerated umbilical and ventrolateral tubercles, so that sparse ribbing is subordinate; ventrolaterals stand well above keel and may form backwardly directed horns; tuberculation weakens on adult body chamber when siphonal keel evanesces and ribs form broad folds over venter. *Lower U. Alb.* **Fig. 277U-V** - *N. curvicornu* Spath.

***Catumbellaites* gen. nov.** [*Neokentroceras trituberculatum* Howarth, 1965, p.352; OD herein]. Like *Neokentroceras* but with distinct high-lateral tubercle from middle growth, and weaker ventrolateral tubercles that become clavate on adult body chamber without forming horns. *U. Alb.* **Fig. 277EE-FF** - *C. trituberculatus* (Howarth).

Subfamily Arestoceratinae van Hoepen, 1942

(*nom. transl.* Cooper, 2018 *ex* Arestoceratidae)
= Cainoceratidae van Hoepen, 1942

Discussion. Monophyletic derivatives of Hysteroceratinae (Cooper, 2018), close to *Hysteroceras*; mostly small to medium sized,compressed, moder-

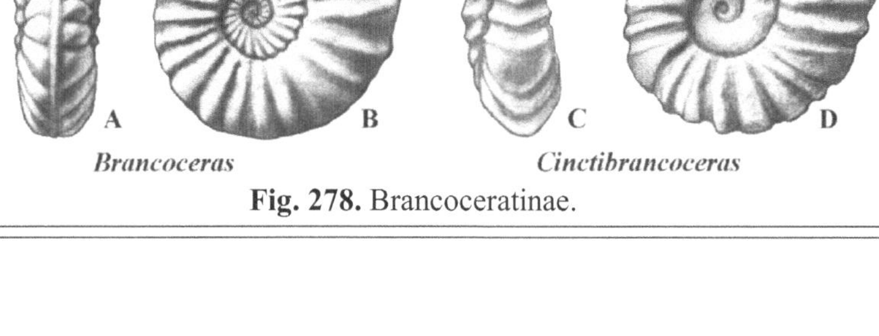

Fig. 278. Brancoceratinae.

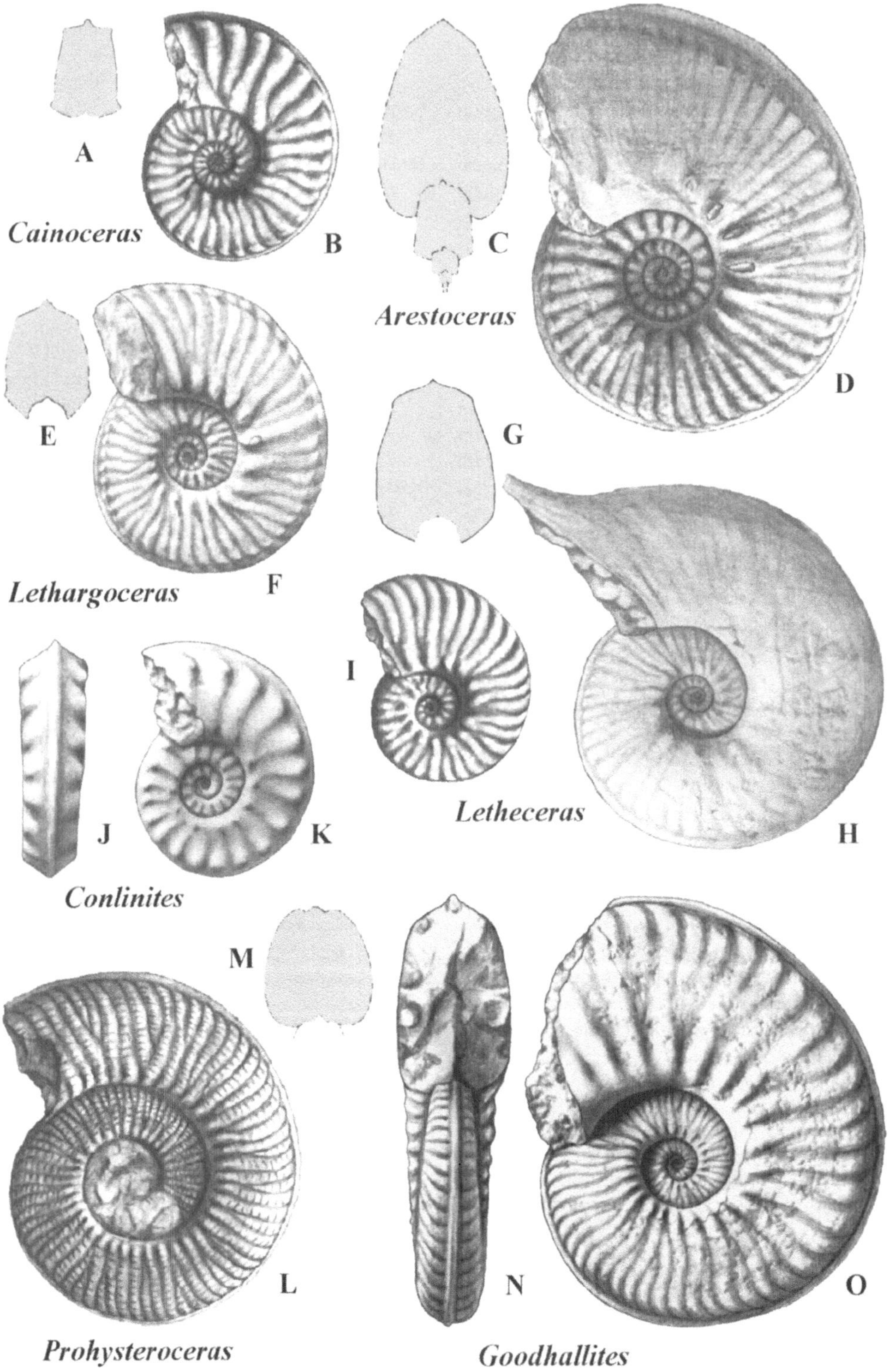

Fig. 279. Arestoceratinae.

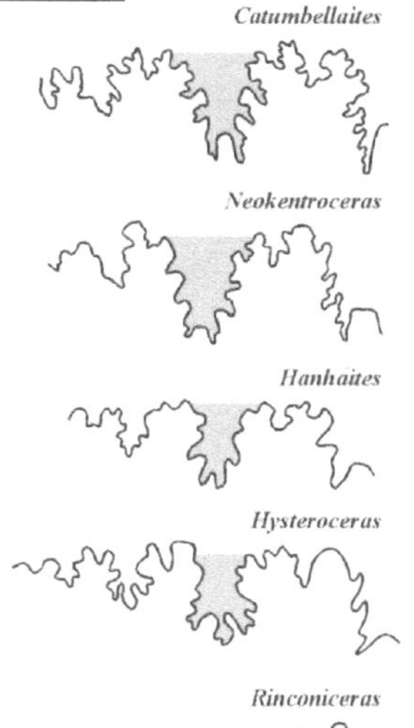

Fig. 280. Brancoceratid sutures.

ately involute, high whorled and bituberculate, with more or less fastigiate venter without bounding sulci to siphonal keel; spiral ornament variably developed; adult whorls show scaphitoid uncoiling. Dimorphism conspicuous but matching dimorphs a challenge; adult aperture with a straight oblique ventral rostrum or following spiral curvature of venter. *Middle U. Alb. - uppermost Alb.*

Goodhallites SPATH, 1932 [*Ammonites goodhalli* J. Sowerby, 1817; OD] [**M**]. Large; compressed, flat sided, high whorled, moderately evolute, with shallow umbilicus, broad subparallel flanks, fastigiate venter and high siphonal keel without bounding grooves; flexuous ribs arise singly or in pairs from small but distinct umbilical bullae and terminate in weak ventrolateral tubercles; suture with broad stemmed asymmetrically trifid E/L with well developed adventive lobe to inner margin, and narrow bifid L almost as deep as E. *Lower U. Alb.* **Fig. 279N-O** - *G. goodhalli* (J. Sowerby).

Prohysteroceras SPATH, 1921 [*P. wordiei* Spath, 1922; ICZN Opinion 1254]. Moderately large; like *Goodhallites* but very evolute, with convex flanks, rounded venter and low siphonal keel flanked by shallow grooves; umbilical and ventrolateral tubercles feeble; spiral ornament strong; suture with deep narrow E, quadratic bifid E/L, bifid L as deep as E, and narrow stemmed trifid L/U_2. *Middle U. Alb.* **Fig. 279L-M** - *P. wordiei* Spath.

Drakeoceras YOUNG, 1957 [*D. drakei*; OD]. Medium sized; inner whorls like *Goodhallites* but more inflated, becoming subquadratic in middle to later growth when sparse straight rectiradiate ribs tend to alternate long and short and well spaced outer lateral and ventral tubercles develop; ribs flex forward from outer laterals; a homoeomorph of *Angolaites* (Mortoniceratinae). *U. Alb.* **Fig. 304K-L** - *D. drakei* Young.

Conlinites KENNEDY, 2004b [*C. wrighti*; OD]. Micromorph; like inner whorls of *Drakeoceras* but with single ventrolateral tubercles. *Upper U. Alb.* **Fig. 279J-K** - *C. wrighti* Kennedy.

Arestoceras VAN HOEPEN, 1942 [*A. collinum*; OD][**M**]. Medium sized; like *Goodhallites* but more involute with only prominent umbilical bullae exposed in umbilicus; slightly convex flanks converge to narrow fastigiate venter with prominent siphonal keel; almost straight broad low rectiradiate ribs arise in pairs, or singly with associated intercalatory, from umbilical bullae and, up to 70 mm diameter, show faint mid-lateral bullae; all ribs with weak but distinct ventrolateral clavi; ornament weakening in later growth, when intercalatories become more frequent and venter markedly fastigiate. *Lower U. Alb.* **Fig. 279C-D** - *A. collinum* van Hoepen.

Aidoceras VAN HOEPEN, 1942 [*A. jubatum*; OD] [**M**]. Middle whorls like *Letheceras* but with more irregular ribbing; outer whorls show scaphitoid uncoiling and become very evolute; ribbing weak and indistinct on penultimate whorl, but strengthening on adult body chamber, as do ventrolateral clavi; umbilical tubercles weaken with

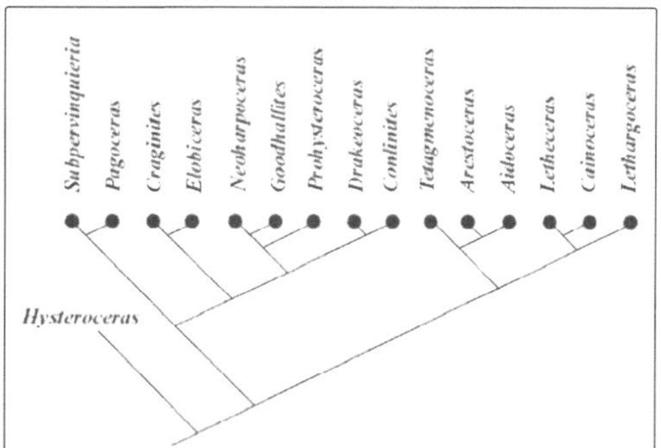

Fig. 281. Hypothesized relationships among Arestoceratinae.

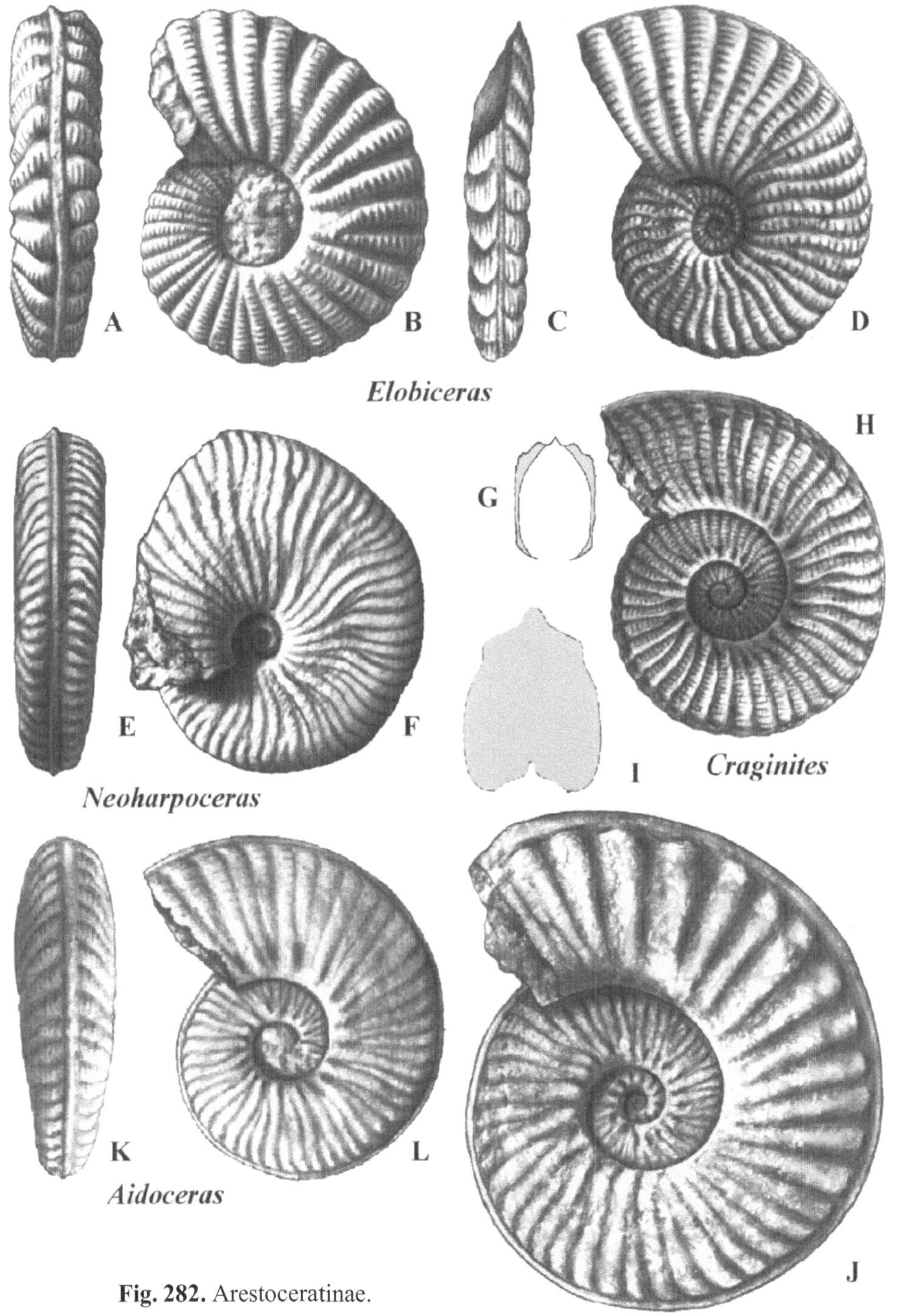

Fig. 282. Arestoceratinae.

growth, becoming obsolete in maturity; faint spiral ornament present. *Lower U. Alb.* **Fig. 282J-L** - *A. jubatum* van Hoepen.

Tetagmenoceras VAN HOEPEN, 1942 [*T. tumulosum* van Hoepen; SD Wright, 1996]. Moderately large; like *Aidoceras* but with distinct high-lateral tubercles from middle growth; siphonal keel lacks flanking grooves. *U. Alb.* **Fig. 299G-H** - *T. tumulosum* van Hoepen.

Letheceras VAN HOEPEN, 1942 [*L. complanatum*; OD] [**M**]. Large; like *Goodhallites* but with straighter coarser ribs in early growth and ornament declining later, becoming obsolete in maturity; suture with narrow stemmed bifid

Fig. 283. Arestoceratine sutures.

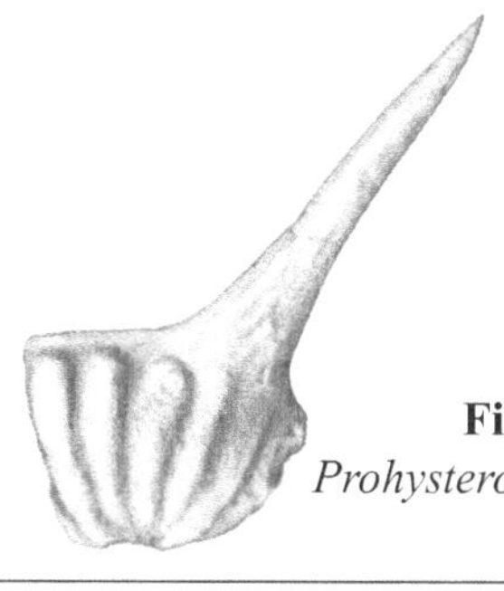

Fig. 284. *Prohysteroceras* rostrum.

saddles. *Lower U. Alb.* **Fig. 279G-I** - *L. complanatum* van Hoepen.

Cainoceras VAN HOEPEN, 1942 [*C. liberum*; OD] [**m**]. Small; like inner whorls of *Letheceras* but with broader more sinuous ribs that persist to maturity. *Lower U. Alb.* **Fig. 279A-B** - *C. liberum* van Hoepen.

Lethargoceras VAN HOEPEN, 1942 [*L. incommodum*; OD] [**m**]. Small; like *Cainoceras* but somewhat larger and more densicostate, with narrower umbilicus and weak mid-lateral tubercles in later growth. *Middle U. Alb.* **Fig. 279E-F** - *L. incommodum* van Hoepen.

Pagoceras VAN HOEPEN, 1951 [*P. amplificatum*; OD] [**M**]. Moderately large; like *Goodhallites* but smaller, more inflated, with rigid straight robust rectiradiate ribs arising in pairs from prominent umbilical bullae. *U. Alb.* **Fig. 299M-N** - *P. amplificatum* van Hoepen.

Poikiloceras VAN HOEPEN, 1951 [*P. firmum*; OD] [**m**]. Small; like *Pagoceras* but smaller; inner whorls with quadratic whorl section, almost tabulate venter and weak bounding sulci to siphonal keel; outer whorls compressed with fastigiate venter; coarse robust ribs generally arising in pairs from acute umbilical tubercles; all ribs with well developed midlateral bullae and prominent ventrolateral tubercles that, on internal mould, appear weakly doubled. *Lower U. Alb.* **Fig. 286** - *P. firmum* Van Hoepen.

Elobiceras SPATH, 1921 [*Schloenbachia elobiensis* Szajnocha, 1885; OD]. Moderately large; inner whorls like evolute *Neoharpoceras* but ribbing coarser and not trifurcating, with conspicuous spiral ornament; later ribs rather rigid, rectiradiate to slightly prorsiradiate, mostly single but with occasional intercalatories, broadening ventrally; in maturity with single widely spaced, markedly club-shaped ribs and fastigiate venter. *U. Alb.* **Fig. 282A-B** - *E. intermedium* Spath; **Fig. 282C-D** - *E. elobiense* (Szajnocha).

Craginites YOUNG, 1957 [*Schloenbachia leonensis* var. *serratescens* Cragin, 1893; OD]. Small; like inner whorls of *Elobiceras* but with rectangular whorl section, distinct umbilical bullae and doubled ventrolateral clavi. *Lower U. Alb.* **Fig. 282G-H** - *C. serratescens* (Cragin).

Neoharpoceras SPATH, 1921 [*Ammonites hugardianus* d'Orbigny; OD]. Medium sized; involute and high whorled, with compressed oval whorl section, obsolete tuberculation and frequent trifurcation of striated ribbing;

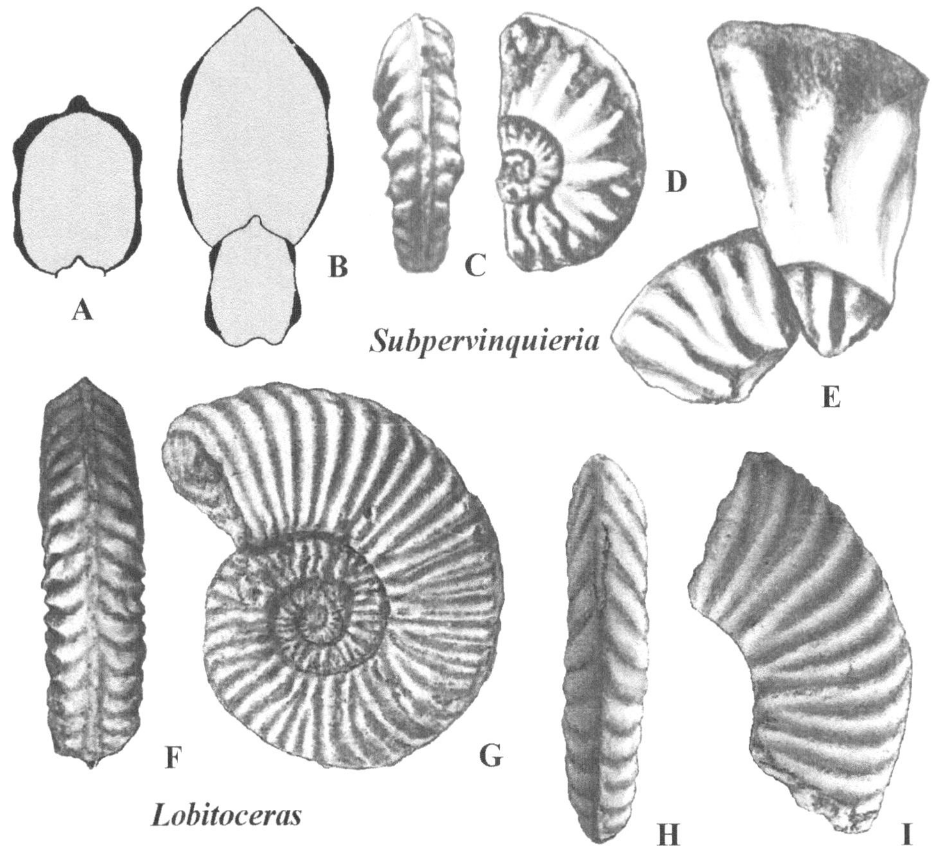

Fig. 285. Arestoceratinae.

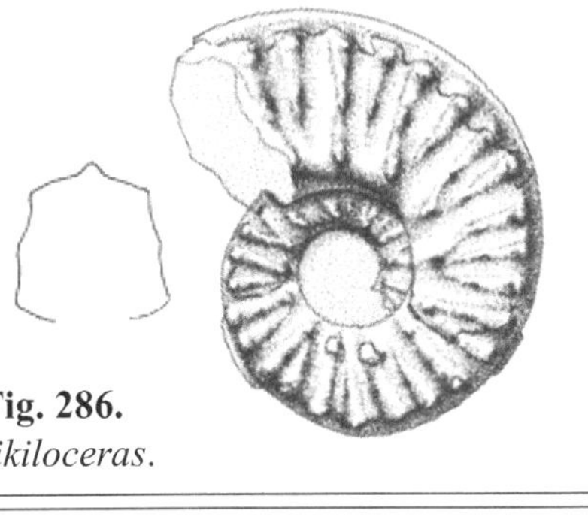

Fig. 286. *Poikiloceras.*

suture highly incised, with narrow L. *U. Alb.* **Fig. 282E-F** - *N. hugardianum* (d'Orbigny).

Lobitoceras COOPER, 2018 [*Elobiceras arietiforme* Spath, 1922; OD]. Moderately large, evolute, strongly compressed, with broad weakly convex to flattened flanks and sharp, strongly fastigiate, acute keeled venter; phragmocone with strong slightly prorsiradiate ribs arising singly or in pairs from small umbilical bullae, single ribs accompanied by an intercalatory, all with weak spiral ornament and prominent ventrolateral tubercles that project strongly forwards on venter; adult body chamber with strong, simple, almost straight ribs lacking umbilical bullae and spiral ornament, becoming increasingly distant towards aperture. *Middle U. Alb.* **Fig. 285F-I** - *L. arietiforme* (Spath).

Subpervinquieria MIRZOEV, 1969 [*Pervinquieria (S.) luppovi*; OD]. Based on a septate fragment, probably moderately large; evolute, compressed, with gently sloping umbilical wall, rounded umbilical shoulder, broad flat flanks, compressed rectangular whorl section and strong siphonal keel; inner whorls with robust flexuous rectiradiate ribs branching from umbilical bullae and projecting foward on venter from weak ventrolateral tubercles; in later growth distant ribs alternate long and short, long ribs with elongate umbilicolateral bullae and ventrolateral bullae showing faint division; suture with very broad squat E/L, narrower symmetrically bifid L as deep as E with short median saddle, and relatively small U_2. *U. Alb.* **Fig. 285A-E** - *S. gissarensis* (Mirzoev).

Although not mentioned in the description, Mirzoev's figured cross section shows a feeble lateral tubercle, while a referred species, *S. gissarensis* (Mirzoev), is bituberculate and shows weakening of ornament in later growth when it develops a strongly fastigiate venter. More than one species may be involved.

Family MOJSISOVICZIIDAE Hyatt, 1903

Discussion. A seemingly monophyletic group descended from Brancoceratidae. Comprises the subfamilies Oxytropidoceratinae and Mojsisovicziinae. *Lowermost L. - Middle U. Alb.*

Subfamily Oxytropidoceratinae nov.

Diagnosis. Medium sized to very large keeled oxycones seemingly with an origin in Leymeriellidae, close to *Wollemanniceras*. Strong simple to branching flexuous ribs, commonly with asymmetrical profiles (steepest anteriorly), and suture with narrow E, long bifid E/L with sloping inner margin with up to 5 accessory saddles, and narrow trifid L. There are indications of dimorphism but it may not be ubiquitous, and has not been proven. As interpreted here the subfamily comprises two tribes that, for the most part, evolved in parallel with brancoceratines, mojsisovicziines, dipoloceratines, mortoniceratines and arestoceratines throughout much of the Middle and Upper Albian. *L. - Middle U. Alb.*

Tribe Venezoliceratini nov.

Type genus: *Venezoliceras* Spath, 1920.

Diagnosis. Typically with strong sinuous ribs thickening ventrally and projecting forwards on the venter, rarely transverse; derived members with ventrolateral, then also lateral and finally umbilical tubercles as well. Derivatives of Leymeriellidae close to *Wollemanniceras*. *L. Albian - Middle U. Alb.*

Kollmanniceras **gen. nov.** [*Oxytropidoceras (Mirapelia) advena* KENNEDY (*in* KENNEDY *et al.* 2000, p.677; OD herein]. Medium sized; ribs simple, distant, straight to flexuous, prorsiradiate, narrower than interspaces, geniculated and projecting strongly forwards on outer flank, with low rounded to asymmetrical profiles; siphonal keel undercut at base. *L. - middle M. Alb.* **Fig. 288O** - *K. advenum* (Kennedy).

Mirapelia COOPER, 1982 [*Ammonites mirapeli* d'Orbigny, 1850; OD]. Medium sized to moderately large; ribs simple, coarse, flexuous, broadening ventrally and thickening on ventrolateral shoulders with asymmetrical profiles (steepest anteriorly); ribs may be transverse or projected forwards on venter. *Upper L. Alb. - lower U. Alb.* **Fig. 288M-N** - *M. mirapelianum* (d'Orbigny).

Mauryiceras **gen. nov.** [*Ammonites sergipensis* White, 1887, p.221; OD herein]. Medium sized, moderately inflated, evolute, with wide shallow umbilicus (U= 32-35%) in later growth, steep umbilical wall, compressed oval intercostal whorl section but subrectangular costally; ribs single, distant, narrower than interspaces, slightly flexuous and prorsiradiate, terminating in prominent oblique ventrolateral clavi that project forwards on tabulate venter with a high siphonal keel. *Lower M. Alb.* **Fig. 288I-J** - *M. sergipense* (White).

Venezoliceras SPATH, 1925 [*Oxytropidoceras venezolanum* Stieler, 1920; OD]. Medium sized to large, moderately evolute, with broad flat flanks converging to narrow subtabulate venter with high siphonal keel that is laterally undercut; for most of growth ribs simple, rectiradiate, narrower than interspaces, with low lateral and ventrolateral tubercles. *Upper M. Alb. - lower U. Alb.* **Fig. 288K-L** - *V. venezolanum* (Stieler).

= *Lophoceras* van Hoepen, 1931 (*non* Hyatt, 1893).

Tarfayites COLLIGNON, 1967 [*Oxytropidoceras (T.) bituberculatum*; OD]. Large; in early growth (D<65 mm) with fine prorsiradiates ribs generally alternating long and short, but also bifurcating from weak umbilical bullae, terminating in ventrolateral tubercles; lateral tubercles lacking; from D>65 mm, sporadic low lateral bullae appear on main ribs and some secondaries; in later growth ribs distant, single, narrower than interspaces, concave forward. *Lower U. Alb.* **Fig. 288D-E** - *T. bituberculatus* (Collignon).

= *Tarfayites* Collignon, 1963 (*nom. nud.*).

Laraiceras RENZ, 1968 [*Oxytropidoceras (L.) laraense*; OD]. Large; like *Venezoliceras* but primary ribs with strong umbilical and mediolateral tubercles, and oblique ventrolateral clavi; tuberculate ribs alternate across the venter. *Middle U. Alb.* **Fig. 288G-H** - *L. laraense* (Renz).

Tribe Oxytropidoceratini nov.

Type genus: *Oxytropidoceras* STIELER, 1920.

Diagnosis. Compressed oxycones with high siphonal keel, sinuous ribs bifurcating from incipient to strong umbilical bullae, and sometimes also on

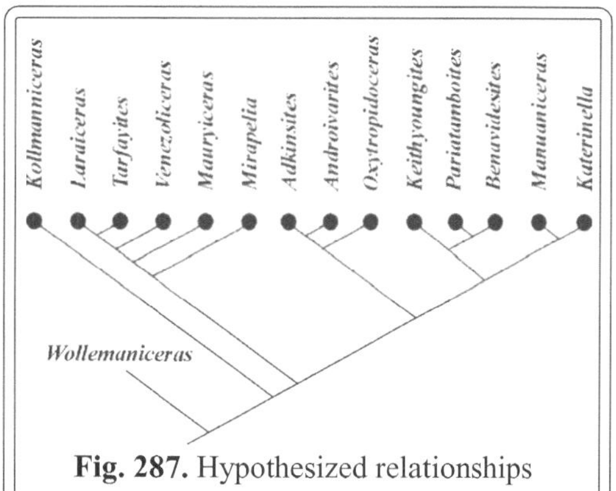

Fig. 287. Hypothesized relationships among Oxytropidoceratinae.

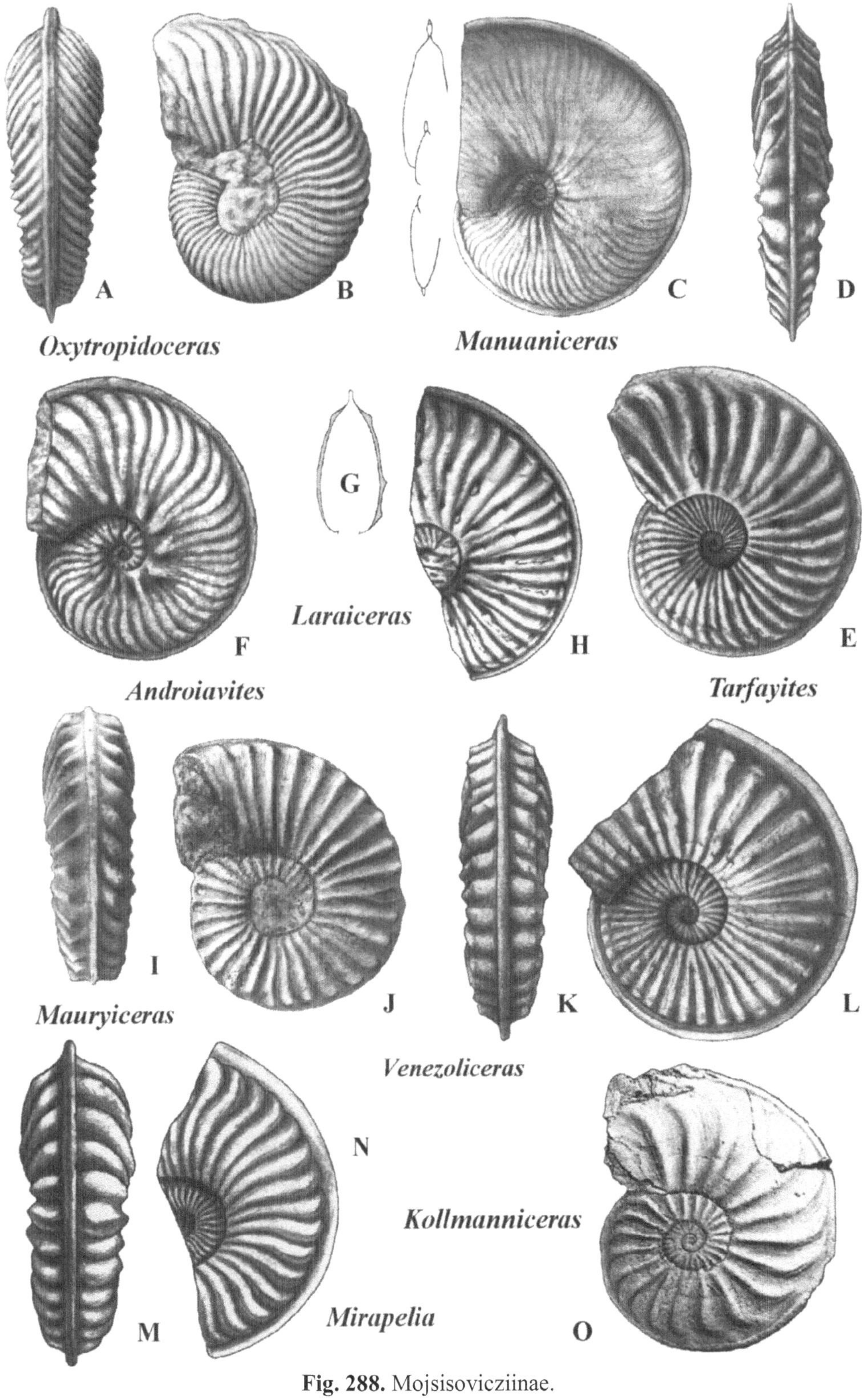

Fig. 288. Mojsisovicziinae.

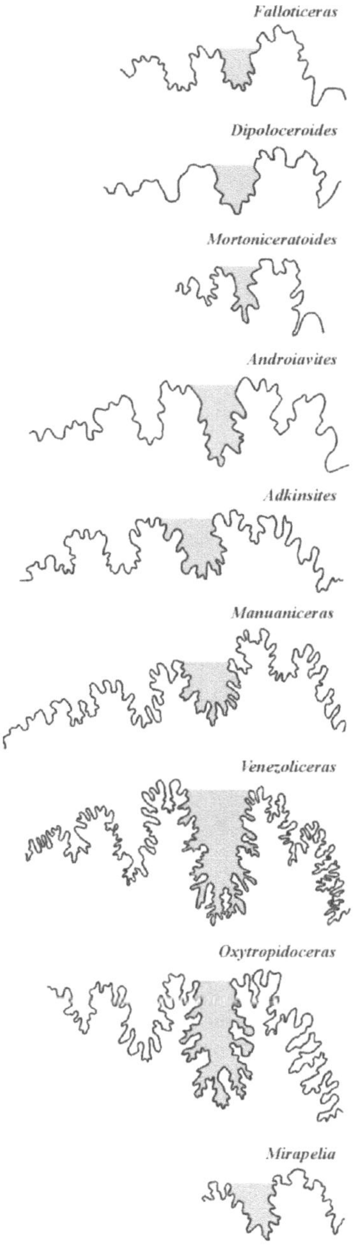

Fig. 289. Mojsisovicziid sutures.

the flank, with intercalatories; ribs with asymmetrical profiles. Monophyletic derivatives of Venezoliceratini, close to *Mirapelia*. *M. - Lower U. Alb.*

Oxytropidoceras STIELER, 1920 [*Ammonites roissyanus* d'Orbigny, 1841; OD]. Small to medium sized; phragmocone moderately evolute, with convex flanks, moderately wide subtabulate venter and high siphonal keel; fine, sharp, flexuous ribs arise from umbilical seam, with occasional intercalatories, bifurcating from slight thickenings on umbilical shoulder, broadening on ventrolateral shoulder and projecting strongly fowards on venter, coarsening markedly with growth; ribs with asymmetrical profiles (steepest adorally), simplifying on adult body chamber; suture with deep narrow L. *M. - Lower U. Alb.* **Fig. 288A-B** - *O. roissyanum* (d'Orbigny).

= *Pseudophacoceras* Spath, 1921 (obj.).

Androiavites COLLIGNON, 1936 [*Oxytropidoceras (A.) besairiei*; OD]. Medium sized; like *Oxytropidoceras* but with coarser ribs and frequent intercalatories, broader than interspaces; periodic strong umbilical bullae in early to middle growth, evanescing later; ribs not projected strongly forwards on venter and with low profiles that make asymmetry difficult to detect; ribs all long on adult body chamber. *Upper M.- Lower U. Alb.* **Fig. 288F** - *A. besairiei* (Collignon).

= *Androyavites* Collignon, 1963 (*laps. cal.*).

Adkinsites SPATH, 1931[*Ammonites belknapi* Marcou, 1858; OD]. Moderately large with well developed umbilical bullae, straight rectiradiate ribs in early growth with some intercalation and bifurcation, and weak ventrolateral clavi; in middle growth with almost straight single coarse ribs with steeper adoral slopes; suture like *Oxytropidoceras* but with broader squatter E/L and broader L. *M. - Lower U. Alb.* **Fig. 291F** - *A. belknapi* (Marcou).

An undescribed specimen from the Kiamichi Formation, Texas (**Fig. 290**), only referrable to *Adkinsites*, shows remarkable characters that may require a complete re-interpretation of the genus.

Tribe Pariatamboitini nov.

Type genus: *Pariatamboites* gen. nov.

Diagnosis. Oxytropidoceratines with flat-topped or rounded ribs, in early to middle growth broader than the interspaces, bifurcating or trifurcating on the flanks with occasional intercalatories; ribs may strengthen in maturity, with or without the development of low lateral tubercles, or may become

Fig. 290. *Adkinsites?* sp.

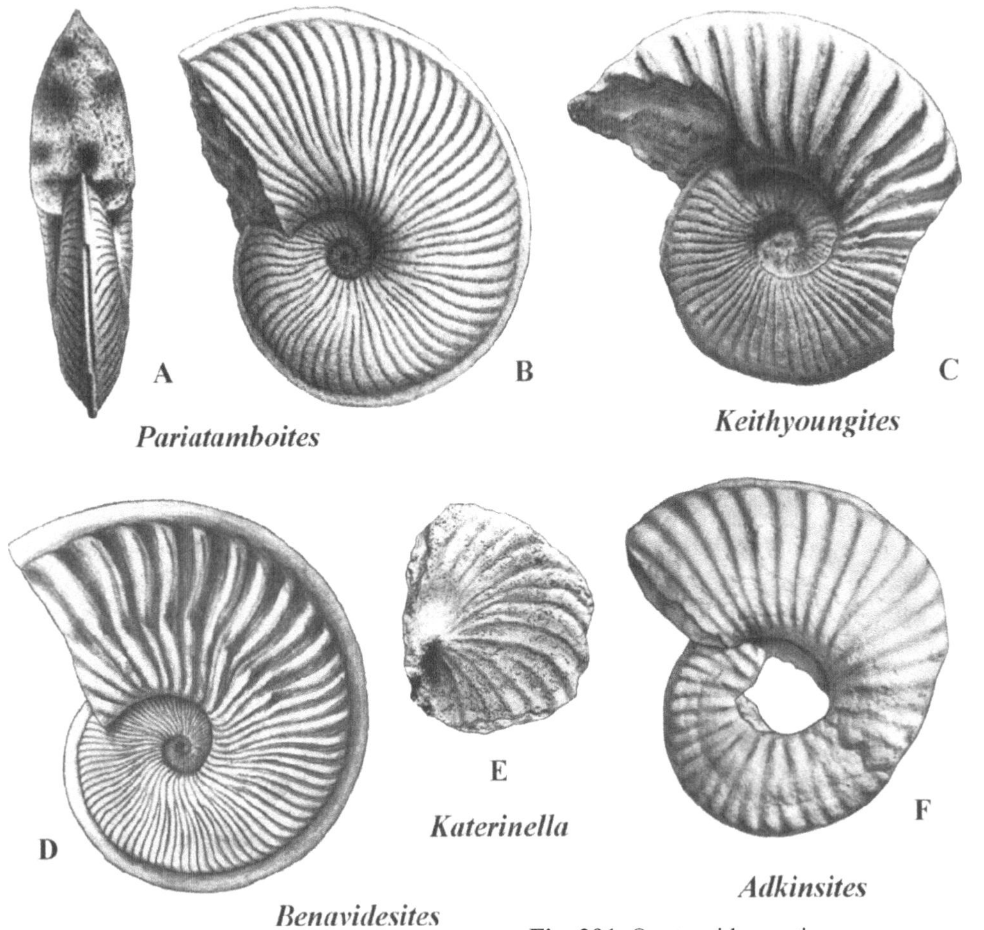

Fig. 291. Oxytropidoceratinae.

obsolete to leave a smooth adult body chamber. *Middle - Lower U. Alb.*

Pariatamboites **gen. nov.** [*Ammonites carbonarius* Gabb 1877, p.269; OD herein]. Large to very large, evolute, in middle to later growth with bifurcating ribs exposed in umbilicus; in early to middle growth with uniform, flat-topped, sinuous, slightly prorsiradiate ribs (even on the steinkern), separated by narrow interspaces, bifurcating on umbilical shoulder; in later growth most ribs single, persisting to maturity. *M. Alb.* **Fig. 291A-B** - *P. carbonarius* (Gabb).

Benavidesites COOPER, 1982 [*Venezoliceras harrisoni* Benavides-Cáceres, 1956; OD]. Moderately large; inner whorls like *Pariatamboites*; adult body chamber with strong simple prorsiradiate rounded ribs, flexed at midflank, with occasional long intercalatories, all with ventrolateral clavi; main ribs on last part of phragmocone and body chamber with prominent lateral bullae below midflank. *M. Alb.* **Fig. 291D** - *B. harrisoni* (Benavides-Cáceres).

Keithyoungites **gen. nov.** [*Manuaniceras powelli* YOUNG 1966, p.106; OD herein]. Moderately large; in early growth like *Pariatamboites*, but body chamber with strong, single, narrow, rounded, and relatively widely spaced ribs. *Lower U. Alb.* **Fig. 291C** - *K. powelli* (Young).

Manuaniceras SPATH, 1925 [*Pseudophacoceras manuanense* Spath, 1921; OD]. Moderately large to large; similar to *Pariatamboites* but phragmocone with weak but distinct umbilical bullae in early to middle growth, rounded ribs as wide as interspaces bifurcating and trifurcating on inner flank with frequent intercalatories, weakening or effaced from adult body chamber; siphonal keel laterally undercut; suture like *Adkinsites* with very broad E/L. *Low U. Alb.* **Fig. 288C** - *M. manuanense* (Spath).

Katerinella **gen. nov.** [*Oxytropidoceras (Manuaniceras) falcatum* Kennedy & Klinger, 2011, p.73; OD herein]. Medium sized, strongly compressed with tiny, nearly occluded umbilicus, very broad, flat, bifurcating, crescentic ribs with asymmetrical profiles (convex adapically and concave adorally). *Uppermost M. Alb.* or *Lower U. Alb.* **Fig. 291E** - *K. falcatae* (Kennedy & Klinger).

Subfamily Mojsisovicziinae Hyatt, 1903

(*nom. transl.* Wright, 1952 *ex* Mojsisovicziidae)

Discussion. Derivatives of Oxytropidoceratinae (Spath, 1922) close to *Mauryiceras*; typically small to medium sized, relatively inflated, evolute, with smooth keeled early whorls; in middle to later growth with broad subtabulate venter and simple coarse flexuous ribs with or without high lateral and ventrolateral tubercles. *M. Alb.*

Mojsisoviczia STEINMANN, 1881 [*Ammonites ventanillensis* Gabb, 1869; OD]. Medium sized; inner whorls smooth, moderately involute, with depressed subquadratic whorl section and low siphonal keel; in middle growth strong, single, rather distant, slightly sinuous, rursiradiate to rectiradiate ribs appear suddenly with prominent ventrolateral tubercles that project forward on venter; every second to third rib with inner ventrolateral bullae; adult body chamber shows scaphitoid uncoiling, conspicuous obliquely clavate ventrolateral tubercles and high siphonal keel. *Lower M. Alb.* **Fig. 293A-C** - *M. ventanillensis* (Gabb).

= *Mojsisovicsia* Gabb, 1877 (invalid subsequent spelling).

Dipoloceroides BREISTROFFER, 1947 [*Ammonites delaruei* d'Orbigny, 1841; OD]. Small; like *Mojsisoviczia* but smaller, with convex flanks, maximum width at midflank, more distant flared ribs some with slight asymmetrical profiles, especially in early growth; sporadic ribs with mid-lateral bullae, and occasional ribs stronger than others; bullate rib endings perpendicular to siphonal keel; some species with straight ventral rostrum that is oblique to coiling; suture simple, with subquadratic E/L and fairly broad trifid L. *M. Alb.* **Fig. 293H-I** - *D. delaruei* (d'Orbigny).

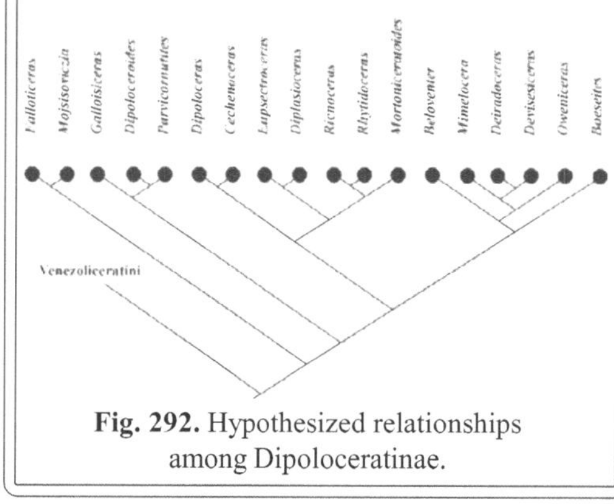

Fig. 292. Hypothesized relationships among Dipoloceratinae.

Falloticeras PARONA & BONARELLI, 1897 [*Ammonites proteus* d'Orbigny, 1842; OD]. Very small; like inner whorls of *Mojsisoviczia* but with generally depressed subquadratic whorl section and early whorls with simple broad prorsiradiate ribs projected onto venter to meet fine siphonal keel obliquely; ornament declines in later growth and last half of body chamber smooth; constrictions variably developed; suture very simple, like *Dipoloceroides* but with L not as deep as E. *Lower M. Alb.* **Fig 293D-E** - *F. proteus* (d'Orbigny).

Has been considered a microconch of *Mojsisoviczia* (Gebhard, 1983).

Galloisiceras **gen. nov.** [*Dipoloceras subdelaruei* Spath, 1931, p.355; OD herein]. Very small; like *Dipoloceroides* but more evolute, with simple sinuous ribs separated by deep interspaces that project strongly forwards on venter; lateral tubercles lacking and ventrolateral tubercles obsolete in later growth. *Upper M. Alb.* **Fig. 293J** - *G subdelaruei* (Spath).

Parvicornutites **gen. nov.** [*Ammonites cornutus* Pictet & Campiche, 1854, p.93; OD herein]. Very small, involute but with excentric uncoiling; depressed oval whorl section; lower to mid-flank of penultimate whorl smooth; outer whorl with simple, straight, slightly prorsiradiate ribs narrower than interspaces terminating in feeble oblique ventrolateral clavi; body chamber with periodic flared ribs with bulbous ventrolateral protuberances. *Upper M. Alb.* **Fig. 293F-G** - *P. cornutus* (Pictet & Campiche).

Family MORTONICERATIDAE Douvillé, 1912

(*nom. transl.* herein *ex* Mortoniceratinae)

Discussion. Monophyletic derivatives of Mojsisovicziidae (Spath, 1922). Small to very large, evolute, primitively bituberculate later trituberculate and quadrituberculate, with strong low siphonal keel and ribs mostly with spiral ornament. Comprises the subfamilies Dipoloceratinae and Mortoniceratinae. *Uppermost M. - uppermost U. Alb.*

Subfamily Dipoloceratinae Spath, 1921

(*nom. transl.* herein *ex* Dipoloceratidae)

= Cechenoceratidae van Hoepen, 1941

= Drepanoceratidae van Hoepen, 1941

Discussion. Monophyletic derivatives of Mojsisovicziinae, close to *Dipoloceroides* (Spath, 1922); typically with low siphonal keel, inflated subquadratic to rounded whorls, conspicuous spiral ornament, horizontal rostrum with down-curved tip (**Fig.**

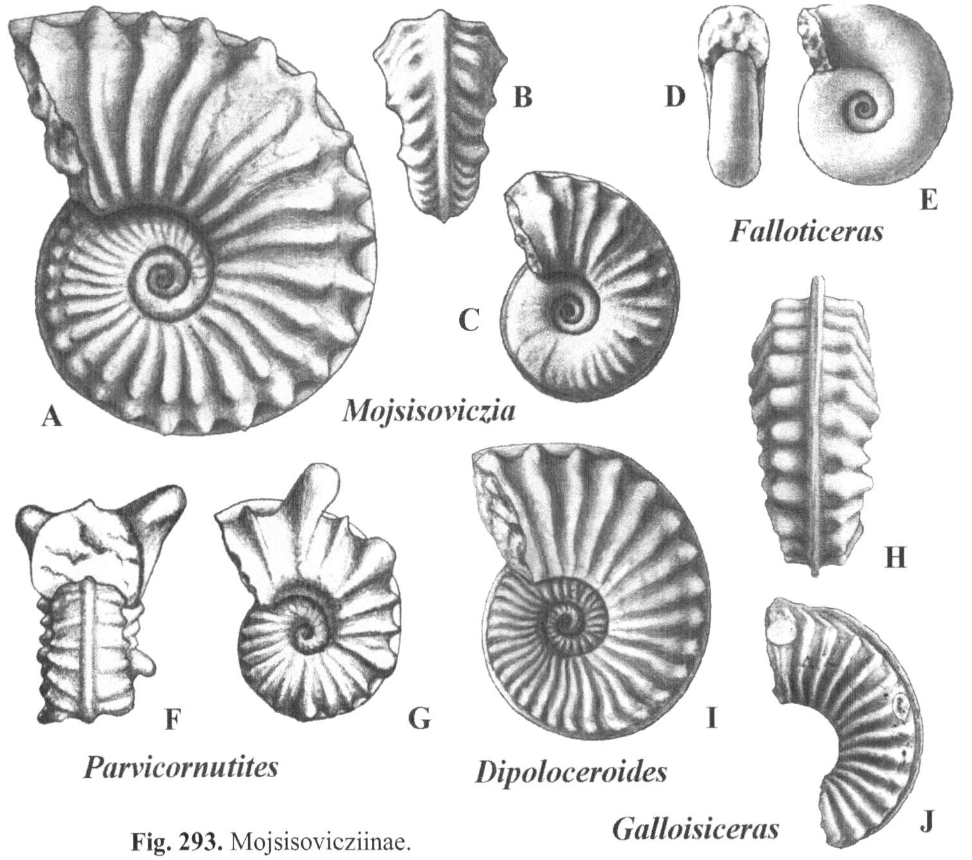

Fig. 293. Mojsisovicziinae.

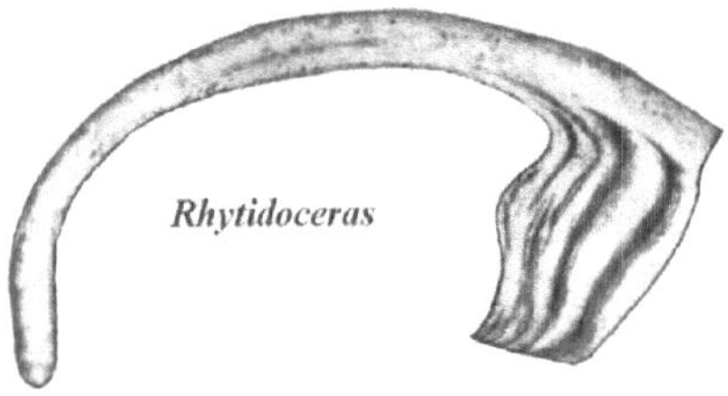

Fig. 294. Dipoloceratine rostrum.

294), and broad E/L to suture; some with flared ribs and alae. *Upper M. - Lower U. Alb.*

Dipoloceras HYATT, 1900 [*Ammonites cristatus* Brongniart, 1822; OD]. Small to medium sized; evolute, inflated; umbilical shoulder, flanks and venter rounded, with subcircular whorl section; innermost whorls smooth, following which distinct umbilical bullae give rise to 1-2 flexuous ribs; in middle growth flares appear, incorporating 2-3 flank ribs, and rib terminations rise above low siphonal keel to form weak oblique ventrolateral clavi that are steepest anteriorly; flared ribs quickly grow into alae that, on venter, give rise to paired ribs that curve gently forward to terminate at shallow grooves on either side of keel; alae separated by up to 3 ribs; posterior surface of alae may show weak spiral ornament. *Upper M. - Lower U. Alb.* **Fig. 297A-B** - *D. cristatum* (Brongniart).

Cechenoceras VAN HOEPEN, 1941 [*C. reversum*; OD]. Medium sized; like *Dipoloceras* but larger, without alate flares but with periodically strengthened ribs, flexuous rursiradiate ribs that may bifurcate on outer flank and strong spiral ornament; suture with narrower E/L and larger L. *Upper M. - Lower U. Alb.* **Fig. 297C-D** - *C. reversum* van Hoepen.

Rhytidoceras VAN HOEPEN, 1931 [*R. elegans*; OD]. Moderately large to large; like *Cechenoceras* but flat sided, with subquadratic whorl section and uniform ribbing without flared ribs or alae; suture with longer narrower saddles and lobes. *Upper M. - lower U. Alb.* **Fig. 297L-M** - *R. elegans* van Hoepen.

= ***Drepanoceras*** van Hoepen, 1931 (*non* Stein, 1878) (**Fig. 297N-O**) was distinguished by flat flanks to its early whorls, lack of ventrolateral tubercles and frequent rib bifurcation from the umbilical bullae.

Ricnoceras VAN HOEPEN, 1931 [*R. pandai*; OD]. Very small, based on a juvenile; like *Cechenoceras* but with flat sided subquadratic whorls with distinct umbilical bullae, and finer denser ribs that frequently bifurcate on outer flank; alate flares and spiral ornament lacking; adult stages unknown. *Upper M. Alb.* **Fig. 300** - *R. pandai* van Hoepen.

Mortoniceratoides COOPER, 1982 [*Mortoniceras rigidum* Spath, 1933; OD]. Small; evolute, with flattened flanks and subquadratic whorl section; straight rigid asymmetrical ribs (steepest anteriorly) that thicken ventrally, with distinct umbilical bullae; spiral ornament lacking; suture with narrow E/L and narrow trifid L shorter than E. *Lower U. Alb.* **Fig. 297H-I** - *M. rigidus* (Spath).

Diplasioceras VAN HOEPEN, 1946 [*D. fallax*; OD]. Small; inner whorls like *Mortoniceratoides* but with alate flares and spiral ornament; final whorl with mostly single flexuous rursiradiate ribs lacking spiral striation on internal mould. *Lower U. Alb.* **Fig. 297E-G** - *D. fallax* van Hoepen.

Eupsectroceras VAN HOEPEN, 1946 [*E. strigile*; OD]. Like *Diplasioceras* but with strong, distant, rigid, simple, slightly rursiradiate ribs in later growth, distinct ventrolateral bullae and strong spiral striations; alae lacking. *Lower U. Alb.* **Fig. 295** - *E. strigile* van Hoepen.

= *Euspectroceras* Wright, 1996 (*nom. van.*).

Mimeloceras VAN HOEPEN, 1944 [*Mimoceras binodosum* van Hoepen, 1941; SD Young, 1957]. Moderately large, like *Rhytidoceras* but more inflated; inner whorls with flat flanks, vertical umbilical wall and small umbilical bullae that remain on umbilical shoulder to large size; in later growth ribs straight, rigid and widely splayed, sometimes single. *Lower U. Alb.* **Fig. 297J-K, 298C-D** - *M. binodosum* (van Hoepen).

= *Mimoceras* van Hoepen, 1941 (*non* Hyatt, 1884).

Deiradoceras VAN HOEPEN, 1931 [*Inflaticeras prerostratum* Spath, 1921; OD]. Large; like *Mimeloceras* but more inflated, with depressed trapezoidal whorl section and enlarged umbilicolateral bullae. *Lower U. Alb.* **Fig. 298A-B** - *D. prerostratum* (Spath).

Oweniceras **gen. nov.** [*Mortoniceras (Pervinquieria) geometricum* Spath, 1932, p.395; OD herein]. Large, very evolute (U=40-57%) with umbilical seam egressing

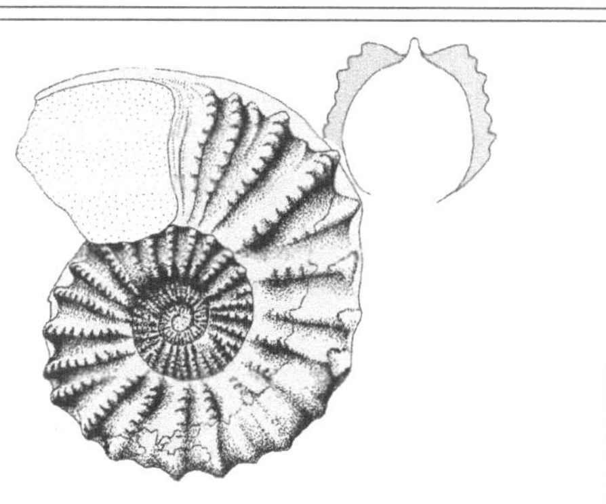

Fig. 295. *Eupsectroceras*.

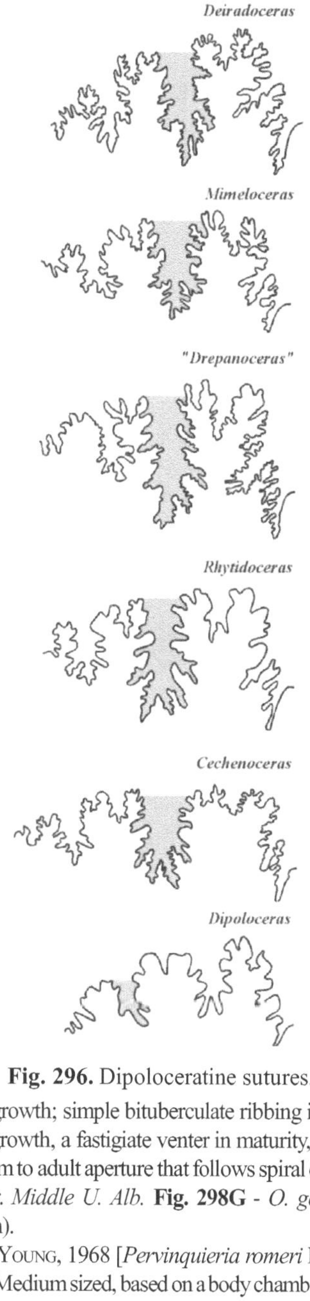

Fig. 296. Dipoloceratine sutures.

with growth; simple bituberculate ribbing in middle to later growth, a fastigiate venter in maturity, and a short rostrum to adult aperture that follows spiral curvature of venter. *Middle U. Alb.* **Fig. 298G** - *O. geometricum* (Spath).

Boeseites YOUNG, 1968 [*Pervinquieria romeri* Haas, 1942; OD]. Medium sized, based on a body chamber fragment; like *Deiradoceras* but in later growth with very sparse distant frequently single ribs with ventrolateral horns; suture like *Pervinquieria* but simpler, with broader L. *Lower U. Alb.* **Fig. 304G-H** - *B. romeri* (Haas).

At least some of the referred *Boeseites* species are *Deiradoceras*, but the type species seems different.

= *Boesites* Wright, 1996 (*nom. van.*) (*non* Miller & Furnish, 1940, Nautiloidea).

Jukesbrowneia **gen. nov.** [*Mortoniceras* (*Pervinquieria*) *potternense* Spath, 1932, p.399; OD herein]. Small, very evolute, with depressed subrectangular whorl section,

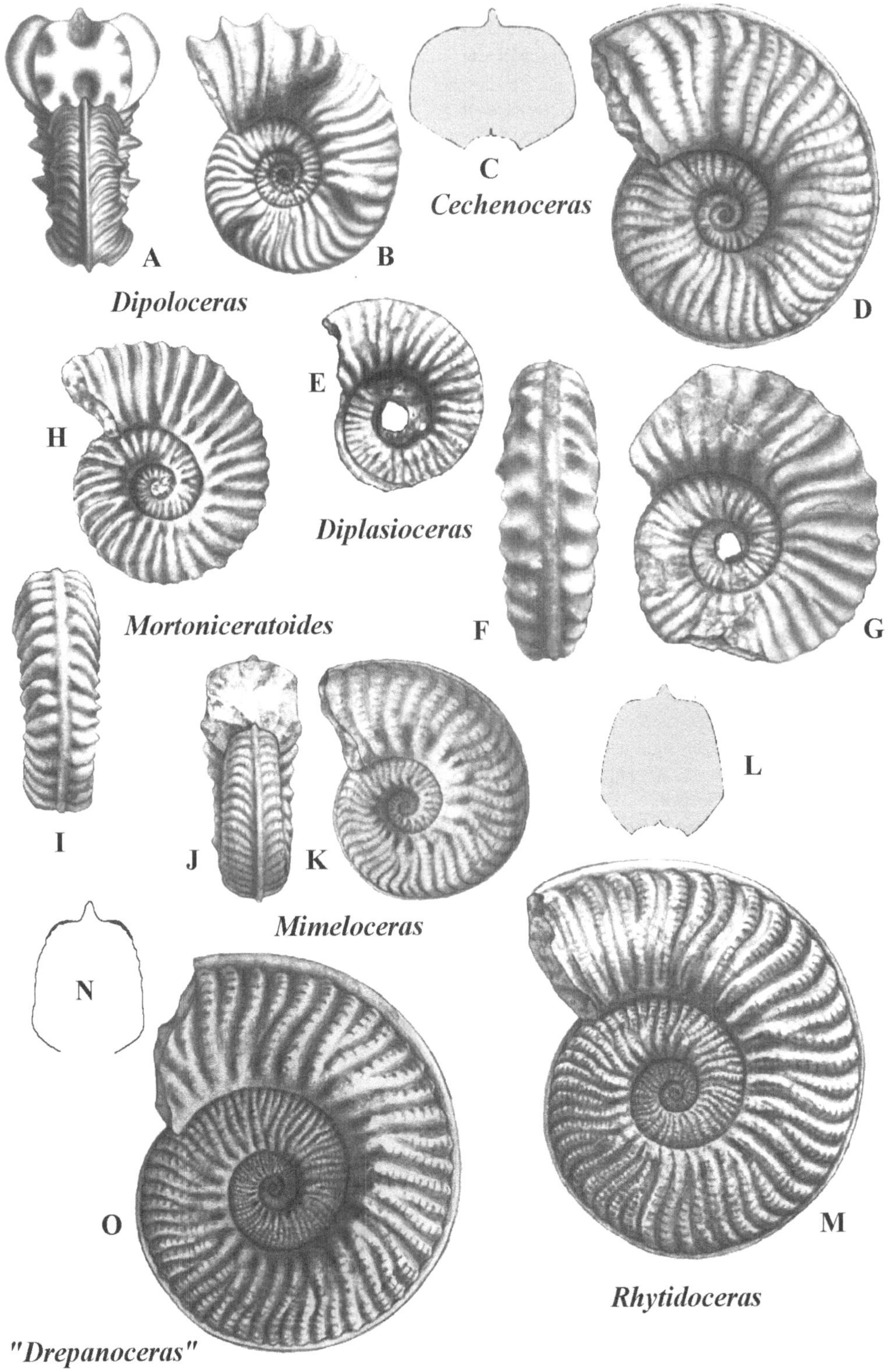

Fig. 297. Dipoloceratinae.

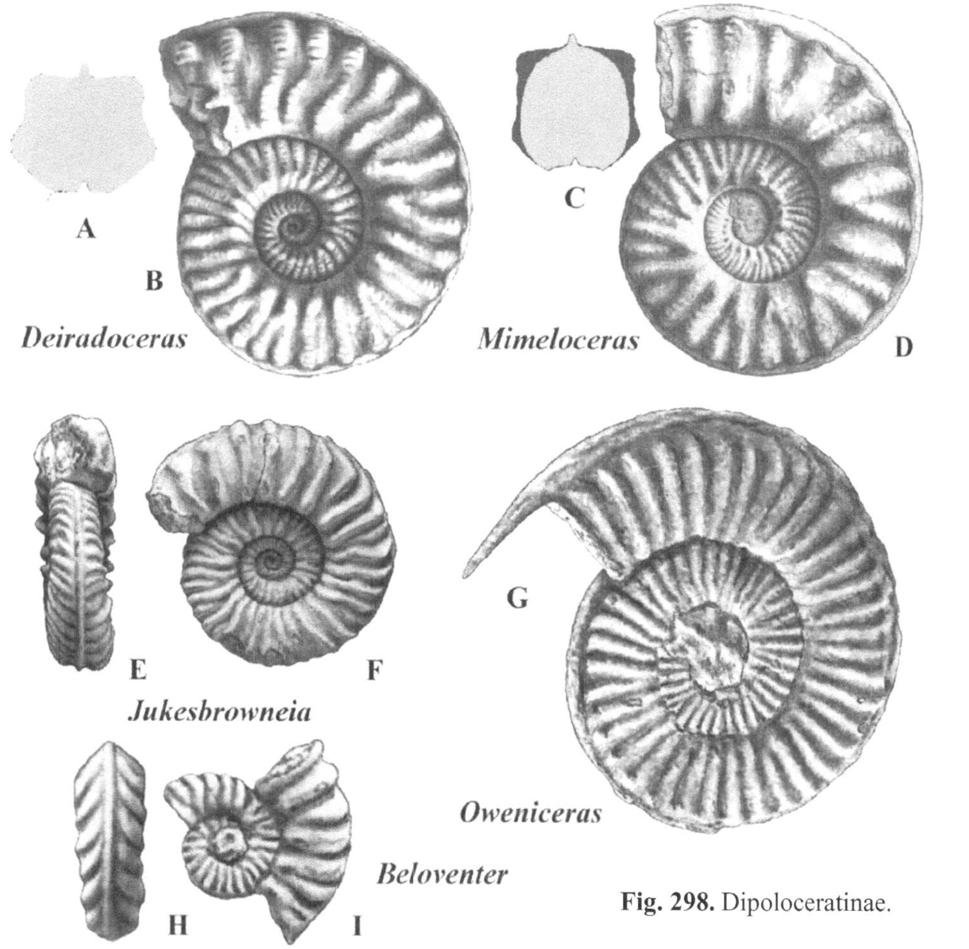

Fig. 298. Dipoloceratinae.

flat subparallel flanks, broad subtabulate keeled venter; on phragmocone feeble umbilical bullae give rise to flexuous rursiradiate ribs, generally in pairs with ventrolateral bullae, projecting forwards on venter; on adult body chamber bituberculate ribs single, distant, slightly rursiradiate to rectiradiate, narrower than interspaces. *Middle U. Alb.* **Fig. 298E-F** - *J. potternensis* (Spath).

Beloventer **gen. nov.** [*Mortoniceras (Deiradoceras) beloventer* Kennedy, Gale, Hancock, Crampton & Cobban, 1999, p.1111; OD herein]. Very small, evolute; phragmocone with inclined umbilical wall, depressed rectangular whorl section; small sharp umbilical bullae giving rise singly or in pairs, with occasional intercalatories, to strong, slightly flexuous, rursiradiate ribs, broader than interspaces, thickened on ventrolateral shoulders and projecting strongly forwards to form acute chevrons across venter; adult body chamber with quadratic whorl section, subparallel flanks, low fastigiate venter and coarse robust ribs with blunt umbilical bullae giving rise to straight, rectiradiate to feebly prorsiradiate ribs with ventrolateral thickening, sweeping forwards to form interrupted chevrons with strong sharp siphonal keel at apex. *Lower U. Alb.* **Fig. 298H-I** - *B. beloventer* (Kennedy, Gale, Hancock, Crampton & Cobban).

Subfamily Mortoniceratinae Douvillé, 1912

(*nom. transl.* Spath, 1934 *ex* Mortoniceratidae)
= Inflaticeratidae Spath, 1925
= Pervinquieriinae Spath, 1925

Discussion. A monophyletic group with an origin in Dipoloceratinae (Spath, 1932), close to *Mimeloceras*; typically evolute, with rectiradiate to prorsiradiate ribs bifurcating from umbilical bullae but becoming single in later growth, with or without midlateral bullae; ventrolateral bullae incipiently binodate (primitive), later strongly so (derived); spiral ornament generally conspicuous; adult aperture with straight oblique (primitive) or recurved

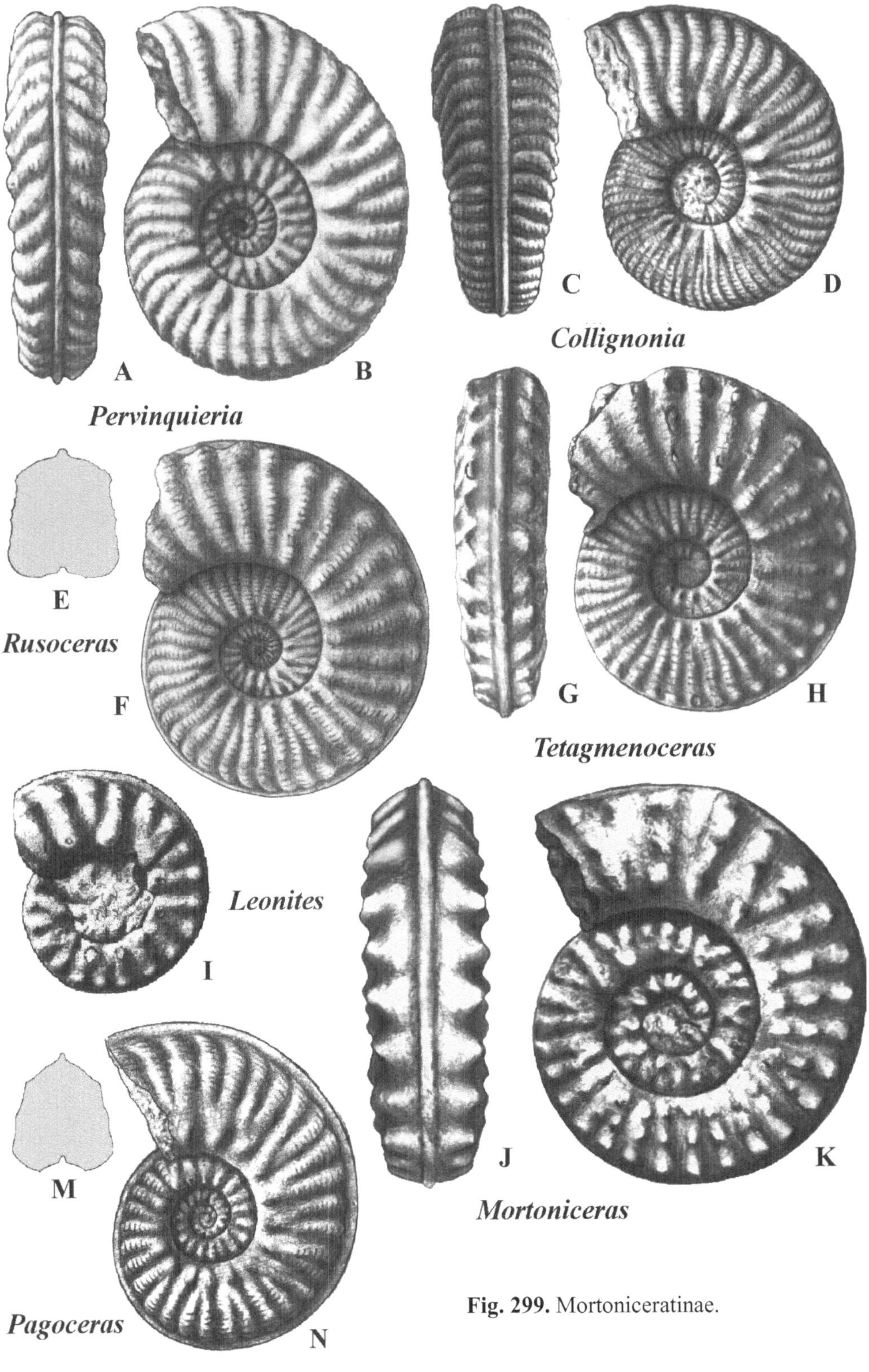

Fig. 299. Mortoniceratinae.

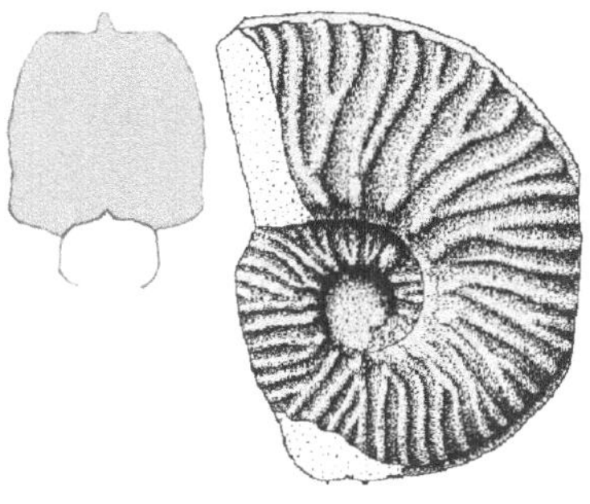

Fig. 300. *Ricnoceras.*

(derived) rostrum. Comprises the tribes Mortoniceratini and Durnovaritini. *U. Alb.*

Tribe Mortoniceratini Douvillé, 1912

(*nom. transl.* herein *ex* Mortoniceratinae)
Type genus: *Mortoniceras* Meek, 1876.

Diagnosis. Mortoniceratines with a straight oblique rostrum (**Fig. 302**); primitively with only umbilical tubercles prominent, later acquiring lateral and even doubled ventrolateral tubercles. *U. Alb.*

Collignonia van Hoepen, 1951[*Pervinquieria undulatocostata* van Hoepen, 1942; OD]. Medium sized; evolute, compressed, with shallow umbilicus, steep umbilical wall with rounded shoulder, flattened slightly convergent flanks and compressed subrectangular whorl section; slightly prorsiradiate ribs arise in pairs from numerous prominent umbilical bullae and thicken at ventrolateral shoulder without forming discrete tubercles; rib endings perpendicular to siphonal keel that is bounded by shallow grooves; beyond 50 mm diameter ribs develop weak mid-lateral bullae; suture with narrow deeply divided E/L and narrow L almost as deep as E. *U. Alb.* **Fig. 299C-D** - *C. vallifera* (van Hoepen).

Rusoceras van Hoepen, 1946 [*R. nothum*; OD]. Medium sized; like *Collignonia* but larger, more compressed and higher whorled, with convexoconcave flanks; ribs project forward on venter, with distinct ventrolateral tubercles in later growth and incipient midlateral swellings, but essentially bituberculate. *Lower U. Alb.* **Fig. 299E-F** - *R. nothum* van Hoepen.

Pervinquieria Böhm, 1910 [*Ammonites inflatus* J. Sowerby, 1817; OD]. Large to very large; like *Rusoceras* but more evolute with, in middle to later growth, distinct midlateral bullae and weakly doubled ventrolateral tubercles; straight oblique rostrum (**Fig. 302**); suture with little divided E/L and very narrow asymmetrically bifid L. *Middle U. Alb.* **Fig. 299A-B** - *P. inflata* (J. Sowerby).
= *Inflaticeras* Steiler, 1920 (obj.).

Ophryoceras van Hoepen, 1942 [*O. jugosum*; OD]. Large to very large; inner whorls like *Rusoceras* but with

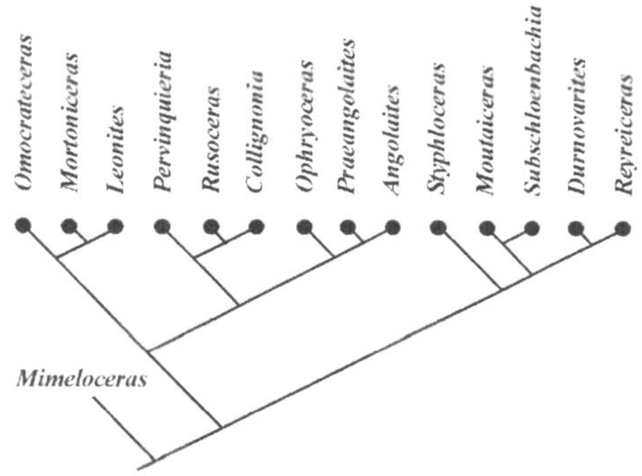

Fig. 301. Hypothesized relationships among Mortoniceratinae.

stronger ventrolateral tubercles and more widely splayed ribs; umbilical bullae obsolete in later growth when ribs become single, distant, and concave forwards; some species with distinct high-lateral tubercles; suture with broad based saddles. *Middle U. Alb.* **Fig. 303C-D** - *O. jugosum* van Hoepen.

Praeangolaites Collignon, 1979 [*P. galvaoi*; OD]. Medium sized; like *Ophryoceras* but very evolute (U=40-43%) and compressed, with feeble umbilical bullae in early growth, later obsolete; mainly single slightly prorsiradiate ribs, occasionally bifurcating from umbilical shoulder, terminate in prominent rounded ventrolateral tubercles that, in maturity, split weakly into approximated inner and outer ventrolaterals. *U. Alb.* **Fig. 303E-F** - *P. galvaoi* Collignon.

Angolaites Spath, 1932 [*Pervinquieria gregoryi* Spath, 1922; OD]. Medium sized; like *Praeangolaites* but with smaller distinctly paired ventrolateral tubercles and ribs more distant in later growth. *Upper U. Alb.* **Fig. 303I-J** - *A. gregoryi* (Spath).

Styphloceras van Hoepen, 1951 [*Pervinquieria nodosocostata* van Hoepen, 1942; OD]. Moderately large; like *Pervinquieria* but whorl section equidimensional or depressed at all stages, with prominent row of rounded midlateral tubercles and doubled ventrolaterals discrete in middle and later growth; ribs become single on last

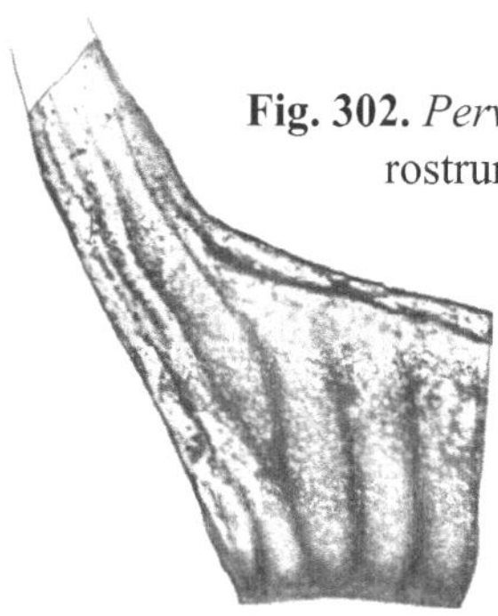

Fig. 302. *Pervinquieria* rostrum.

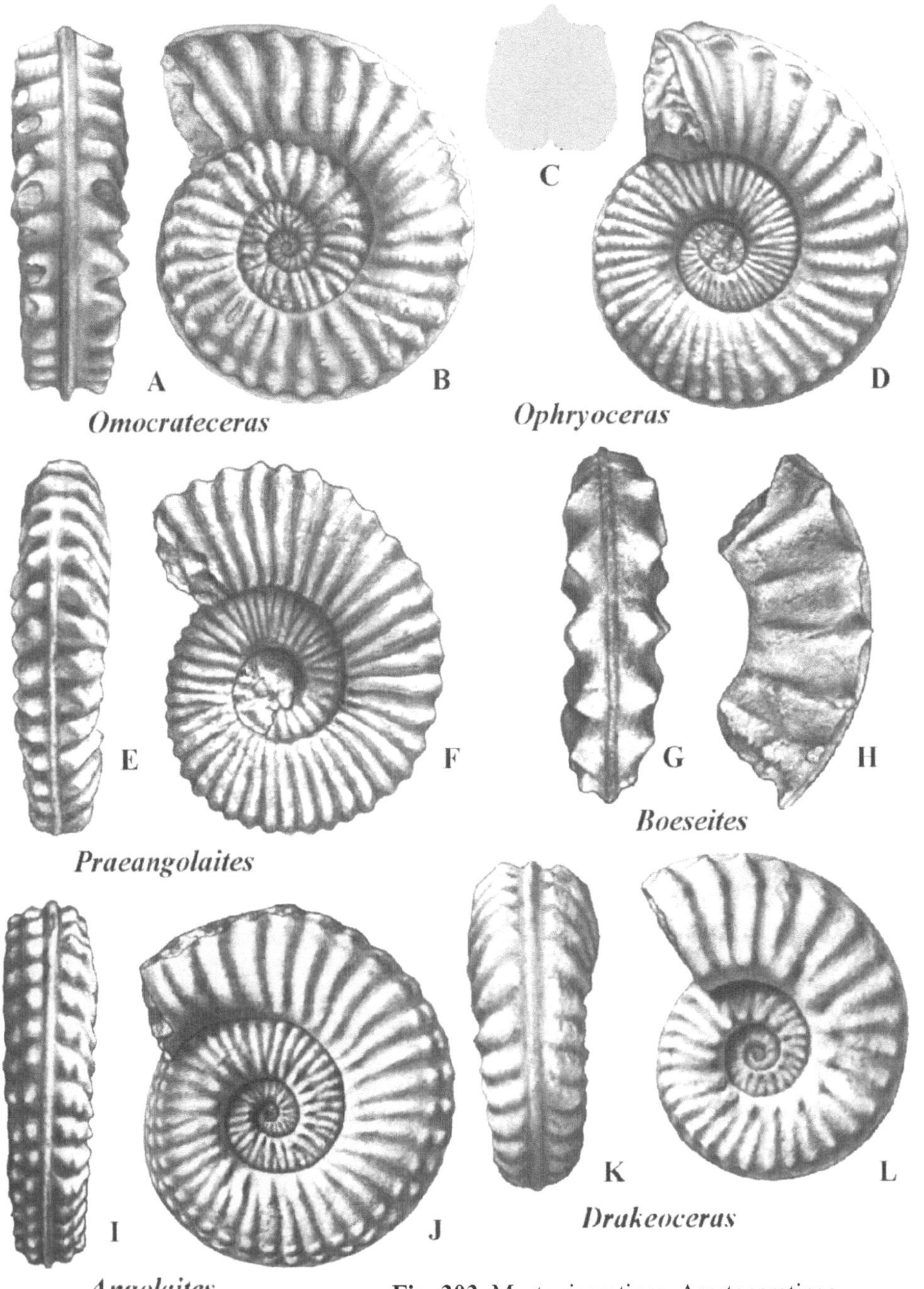

Fig. 303. Mortoniceratinae, Arestoceratinae.

half of adult body chamber. *Middle U. Alb*. **Fig. 307G-I** - *S. nodosocostatum* (van Hoepen).

Mortoniceras MEEK, 1876 [*Ammonites vespertinus* Morton, 1834; OD]. Very large to enormous; very evolute, almost serpenticone, compressed, with subquadratic whorl section; strong rectiradiate ribs arise singly or in divergent pairs from strong umbilical bullae; non-striated ribs with distinct mid-lateral tubercles from early growth, terminate in single ventrolateral tubercles that may be clavate; adult body chamber with strong single distant ribs, with weak umbilical and lateral tubercles, and slightly fastigiate venter. *U. Alb*. **Fig 299J-K** - *M. vespertinum* (Morton).

Leonites SPATH, 1932 [*Ammonites leonensis* Conrad, 1857; OD]. Based on a juvenile; evolute (U= 43%), with squarish whorls, prominent umbilical tubercles that give rise to 1-2 distant rectiradiate ribs with strong nodate

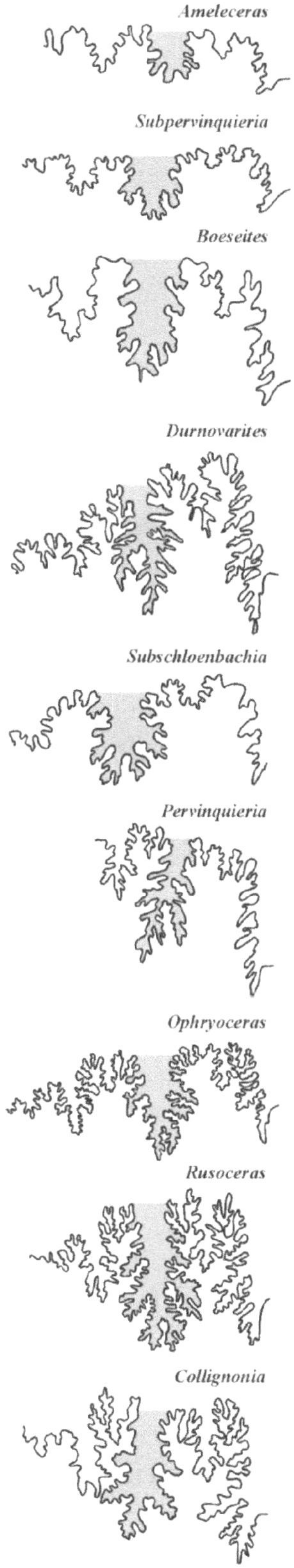

Fig. 304. Mortoniceratid sutures.

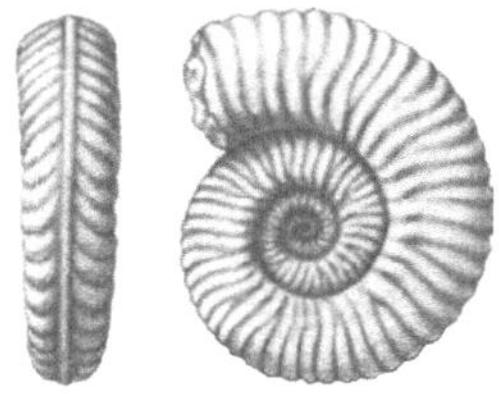

Fig. 305. *Cantabrigites.*

ventrolateral tubercles and weak lateral tubercles in later growth when ribs tend to alternate long and short. *U. Alb*. **Fig. 299I** - *L. leonensis* (Conrad).

A juvenile *Mortoniceras*.

Omocrateceras VAN HOEPEN, 1951 [*Pervinquieria (O.) planiventer*; OD]. Large; like *Mortoniceras* but with denser costation, especially on adult body chamber, and conspicuous spiral ornament. **Fig. 303A-B** - *M. planiventer* (van Hoepen).

= *Omocraterceras* van Hoepen, 1951 (*nom. van.*).

Cantabrigites SPATH, 1933 [*Mortoniceras (C.) cantabrigense* Spath, 1932; OD]. Very small, very evolute, compressed, with flat parallel flanks, rounded venter and persistent siphonal keel flanked by faint sulci; strong sharp flexuous ribs arising singly or in pairs from weak umbilical bullae, forming feeble ventrolateral swellings and projecting strongly forward on venter; suture with very deep E and shallow L. *Upper U. Alb*. **Fig. 305** - *C. cantabrigensis* (Spath). Superficially similar to both *Hysteroceras* and *Erioliceras* and assignment uncertain; perhaps a progenetic dwarf of *Pervinquieria*.

= *Cantabrigites* Spath, 1932 (*nom. nud.*).

= *Cantabridgites* Hayakawa, Kano & Katano, 1999 (*nom. van.*).

Tribe Durnovaritini nov.

Type genus: *Durnovarites* Spath, 1932.

Diagnosis. Typically with 4 rows of strong flank tubercles in early to middle growth and recoiled terminal rostrum. Monophyletic derivatives of Mortoniceratini (Spath, 1922), close to *Styphloceras*; *Upper U. Alb.*

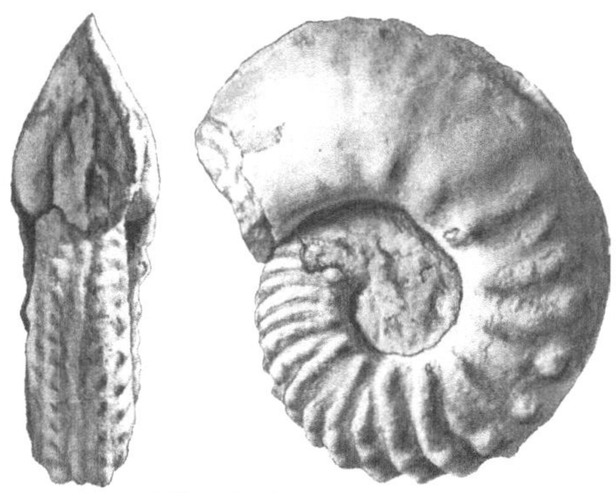

Fig. 306. *Moutaiceras.*

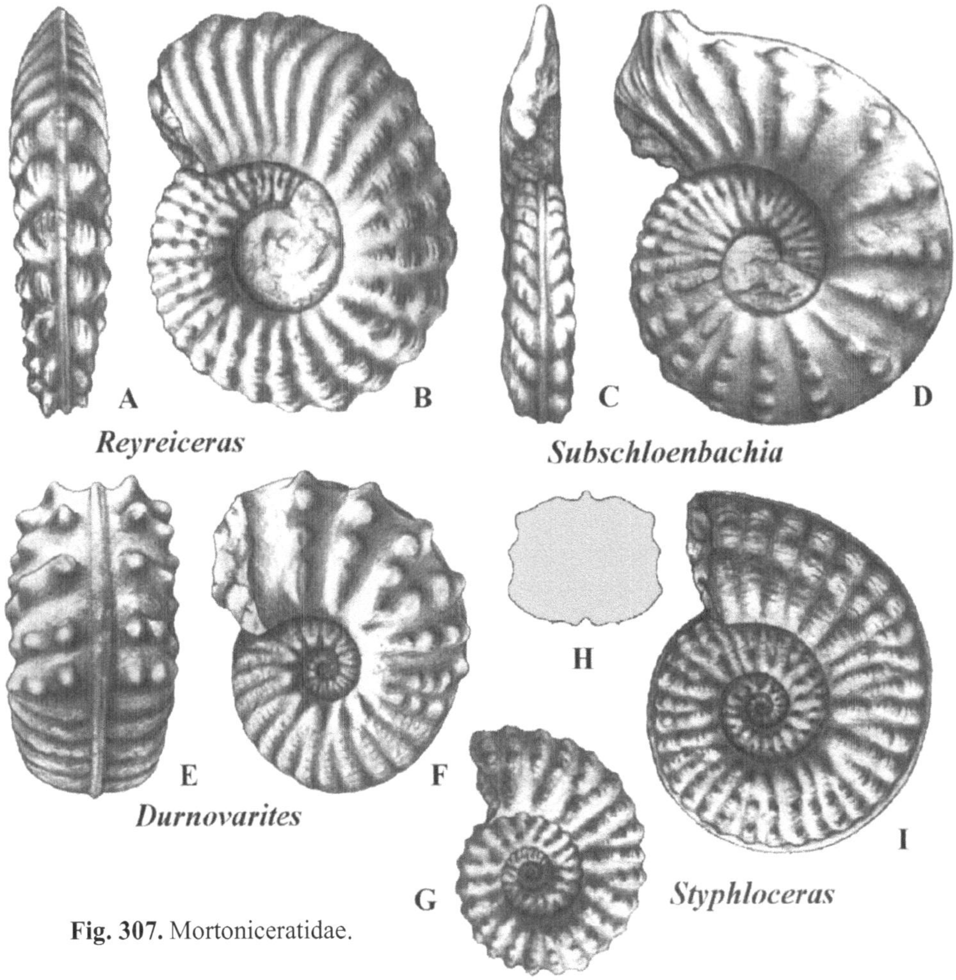

Fig. 307. Mortoniceratidae.

Subschloenbachia SPATH, 1921 [*Ammonites rostratus* J. Sowerby, 1817; OD]. Medium sized, evolute, compressed, with flat subparallel flanks and compressed rectangular whorl section; in early to middle growth with straight rectiradiate ribs arising in pairs from umbilical bullae, all with distinct lateral and inner and outer ventrolateral tubercles; in later growth ribs single and widely spaced, umbilical bullae obsolete; suture with broad subquadratic E/L and relatively broad asymmetrically bifid L. *Upper U. Alb*. **Fig. 307C-D** - *S. rostrata* (J. Sowerby).

Durnovarites SPATH, 1932 [*Subschloenbachia perinflata* Spath, 1922; M]. Small; like *Subschloenbachia* but smaller, more inflated, with depressed quadratic whorl section, narrower adult umbilicus and weaker tuberculation on phragmocone; suture with long narrow deeply bifid E/L and narrow L. *Upper U. Alb*. **Fig. 307E-F** - *D. perinflatus* (Spath).

Reyreiceras COLLIGNON, 1979 [*R. reyesi*; OD]. Medium sized; like *Subschloenbachia* but more evolute, with lanceolate non-tuberculate adult body chamber with simple recurved ribs that are strongly convex across flanks, and high acute siphonal keel; recurved adult rostrum; body chamber homoeomorphs *Oxytropidoceras* and has been identified as such. *Uppermost U. Alb*. **Fig 307A-B** - *R. reyrei* Collignon.

= *Reyericeras* Kennedy *et al.* 2005 (*nom. van.*).

Moutaiceras COOPER, 2018 [*Sharpeiceras goliath* Haas, 1942; OD]. Gigantic; inner whorls like *Subschloenbachia* but ornament weakening on lanceolate adult body chamber; adult aperture without recurved rostrum. *Uppermost Alb*. **Fig. 306** - *M. goliath* (Haas).

Family ERIOLICERATIDAE van Hoepen, 1955

Discussion. A seemingly monophyletic group of uncertain origin, either Hysteroceratinae or Mortoniceratinae; typically small, very evolute, flat sided,

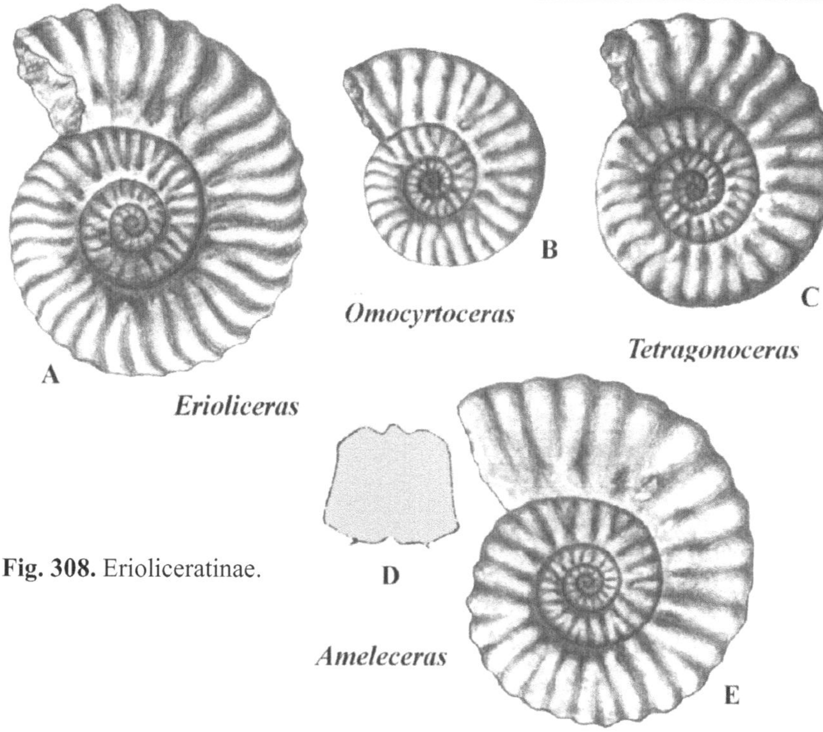

Fig. 308. Erioliceratinae.

with robust flexuous ribs arising singly or in pairs from prominent umbilical tubercles and sometimes weakening at midflank, distinct ventrolateral tubercles steepest posteriorly and siphonal keel obsolete on last part of adult body chamber. *U. Alb.*

Erioliceras VAN HOEPEN, 1955 [*E. tenuis*; OD]. Small; very evolute, compressed, with flat sided rectangular whorls; strong blunt sinuous ribs arise single or in pairs from umbilical tubercles and project forward on venter; low ventrolateral bullae markedly asymmetrical with steep sometimes overhanging adapical side. *Middle U. Alb.* **Fig. 308A** - *E. tenuis* van Hoepen.

Omocyrtoceras VAN HOEPEN 1955 [*O. errabundum*; OD]. Small; like *Erioliceras* but with quadratic to depressed whorl section, midlateral tubercles and strong siphonal keel standing above prominent ventrolateral tubercles. *Middle U. Alb.* **Fig. 308B** - *O. errabundum* van Hoepen.

Tetragonoceras VAN HOEPEN, 1955 [*T. robustum*; OD]. Small; like *Omocyrtoceras* but with lateral tubercles restricted to final whorl when ventrolateral tubercles become very prominent. *Middle U. Alb.* **Fig. 308C** - *T. robustum* van Hoepen.

Ameleceras VAN HOEPEN, 1942 [*A. abjectum*; OD]. Small; compressed, evolute (U=42%), with quadratic whorls, coarse low ribs arising singly or in pairs from prominent umbilical nodes, low blunt bipartite ventrolateral bullae, low siphonal keel flanked by shallow grooves, and keel becoming obsolete in maturity; spiral ornament lacking; suture simple, with broad bifid E/L and fairly wide trifid L not as deep as E. *U. Alb.* **Fig. 308D-E** - *A. abjectum* van Hoepen.

Superfamily LYELLICERATOIDEA Spath, 1921

(*nom. transl.* Cooper, 1992 *ex* Lyelliceratidae)

Discussion. A paraphyletic group with an origin in Brancoceratoidea (Spath, 1931); comprises the families Lyelliceratidae and Prolyelliceratidae. Mostly small, evolute, with umbilical bullae and trituberculate venter. *Upper M. Alb. - U. Cenom.*

Family LYELLICERATIDAE Spath, 1921

Discussion. A monophyletic group with an origin in Brancoceratidae, close to *Brancoceras* (Spath, 1931); typically evolute, medium sized, with strong generally simple ribs with ventrolateral and siphonal tubercles; some with umbilical and lateral tubercles also. *Upper L. - Lower M. Alb.*

Rauliniceras DOUVILLÉ, 1911 [*Hoplites gladiator* Bayle, 1878; OD]. Small; compressed, moderately evolute, with compressed subrectangular whorl section, distant simple rectiradiate ribs with elongate umbilicolateral bullae that terminate in weak lateral bullae and are feebly connected

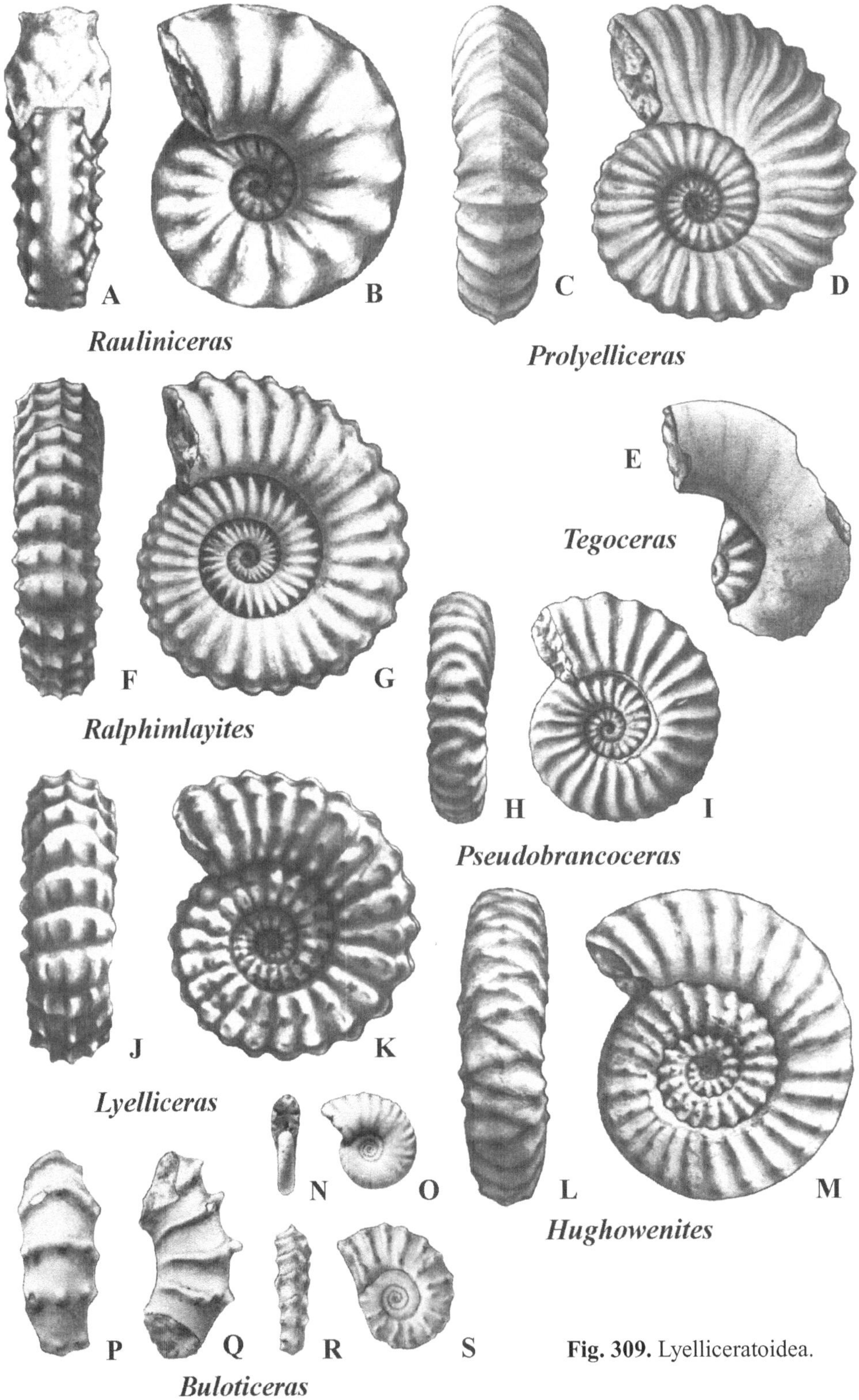

Fig. 309. Lyelliceratoidea.

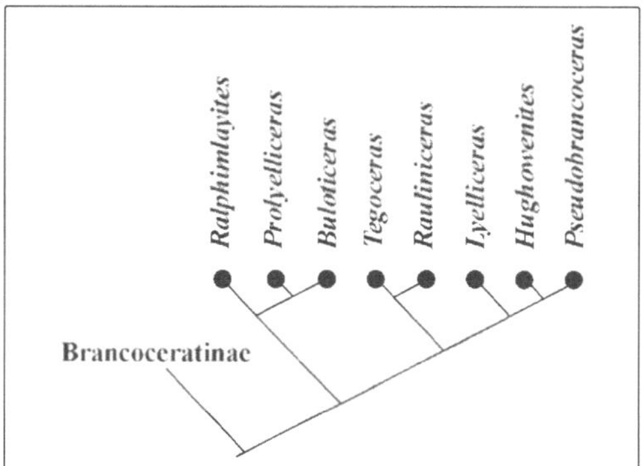

Fig. 310. Hypothesized relationships among Lyelliceratidae and Prolyelliceratidae.

to ventrolateral tubercles that alternate across venter; ventrolaterals may be nodate, when ribs zigzag across venter, or clavate when venter is smooth; strong ornament of inner whorls carried to maturity; suture with broad asymmetrically bifid E/L, deep narrow asymmetrically bifid L deeper than E, and shallowly bifid L/U_2. *Upper L. - Lower M. Alb.* **Fig. 309A-B** - *R. gladiator* (Bayle).

Tegoceras HYATT, 1903 [*Ammonites mosensis* d'Orbigny, 1841; OD]. Very small; inner whorls like *Raulíniceras*; later more evolute (U=45%), with strong rectiradiate umbilicolateral bullae exposed in umbilicus and impression of penultimate whorl showing subtabulate venter crossed by blunt zigzagging ribs; adult body chamber smooth, with oval whorl section, rounded venter and slight well spaced flank undulations that alternate on either side of venter; suture unlike *Raulíniceras*, with broad shallow E, very broad bifid E/L with finger-like folioles and narrow L. *Upper L. - Lower M. Alb.* **Fig. 309E** - *T. mosense* (d'Orbigny).

Lyelliceras SPATH, 1922 [*Ammonites lyelli* d'Orbigny, 1841; OD]. Small; evolute (U=45-50%), with polygonal whorl section, strong pointed umbilical tubercles, mid-lateral tubercles that persist to maturity, and inner ventrolateral and siphonal clavi. *Lower M. Alb.* **Fig. 309F-G** - *L. lyelli* (d'Orbigny).

= *Seunesiceras* Breistroffer & de Villoutreys, 1953 (obj.).

***Hughowenites* gen. nov.** [*Acanthoceras pseudolyelli* Parona & Bonarelli, 1897, p.99; OD]. Small; like *Lyelliceras* but very evolute, with primary ribs terminating at ventrolateral shoulder; secondary ribs zigzag across venter and siphonal tubercles greatly outnumber ventrolaterals. *Uppermost L. Alb.* **Fig. 309L-M** - *H. pseudolyelli* (Parona & Bonarelli).

Pseudobrancoceras KENNEDY, 2004a [*Ammonites versicostatus* Michelin, 1838; OD]. Small; like *Hughowenites* but smaller, with vestigial ventral tuberculation, no lateral tubercles and suture with broader L; a homoeomorph of *Brancoceras* (Brancoceratinae). *Lower M. Alb.* **Fig. 309H-I** - *P. versicostatum* (Michelin).

Family PROLYELLICERATIDAE Latil, Robert & Bulot, 2009

Discussion. Homoeomorphs of Lyelliceratidae, with which they evolved in parallel, seemingly with an independent origin in Brancoceratinae (Casey, 1957) close to *Parabrancoceras* (Latil, 2011). *L. Alb.*

Buloticeras LATIL, 2011 [*Prionotropis radenaci* Pervinquière, 1907; OD]. Very small to small, rather evolute, with polygonal whorl section; early whorls with more or less carinate venter and smooth flanks; in middle growth with simple, distant, flexuous rectiradiate ribs with umbilical bullae, rounded inner and outer ventrolateral tubercles and siphonal clavi, the latter becoming smaller and conical at end of body chamber; suture simple, with broad saddles and few elements. *L. Alb.* **Fig. 309N-S** - *B. radenaci* (Pervinquière).

Prolyelliceras SPATH, 1930 [*P. peruvianum*; OD]. Small; evolute (U=34%), compressed, with slightly convex

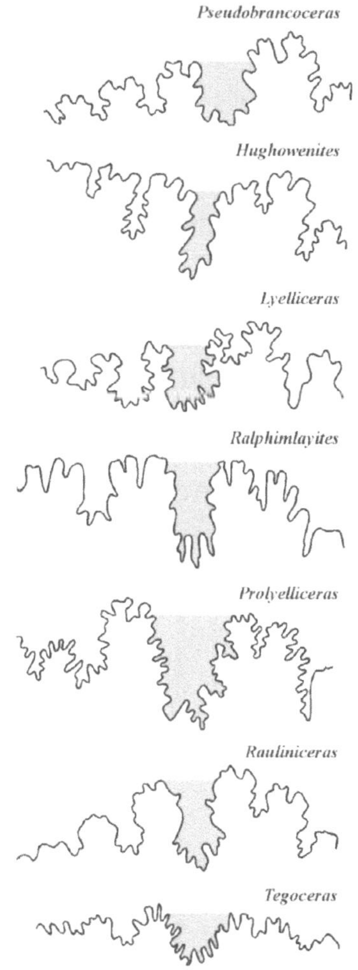

Fig. 311. Lyelliceratine sutures.

flanks, broadly rounded to fastigiate venter, and subquadratic to oval whorl section; strong simple straight rather distant prorsiradiate ribs broaden over venter and with ventrolateral and siphonal clavi; beyond 30 mm diameter tuberculation declines, becoming obsolete on adult body chamber when sinuous ribs form chevrons over venter; suture with L/U_2 projecting beyond E/L. *L. Alb.* **Fig. 309C-D** - *P. peruvianum* Spath.

Ralphimlayites ETAYO SERNA, 1979 [*Acanthoceras prorsocurvatum* Gerhardt, 1897; OD]. Small; like *Prolyelliceras* but more evolute (U=36-45%), with subquadratic adult whorl section, feeble umbilical bullae and rounded inner ventrolateral tubercles; flattening of ribs between inner and outer ventrolateral tubercles produces a spiral depression; inner ventrolateral and umbilical tubercles obsolete on adult body chamber. *L. Alb.* **Fig. 309F-G** - *R. prorsocurvatus* (Gerhardt).

Superfamily FORBESICERATOIDEA Wright, 1952

(*nom. transl.* Cooper, 1992 *ex* Forbesiceratidae)

Discussion. Monophyletic derivatives of Lyelliceratoidea (Wright, 1996). Mostly involute, compressed, with weak to strong ornament, primitively with siphonal tubercles and, sometimes, a siphonal keel. Comprises the families Forbesiceratidae, Flickiidae and Stoliczkaiidae. *U. Alb. - U. Cenom.*

Family FORBESICERATIDAE Wright, 1952

(*nom. transl.* Wright, 1955 *ex* Forbesiceratinae)

Discussion. A monophyletic group with an origin in Prolyelliceratidae (Wright, 1996). Typically strongly compressed, involute and high whorled, with subdued ventrolateral tubercles and nodate siphonal keel that may become entire or obsolete in middle to later growth. Comprises the subfamilies Neophlycticeratinae, Forbesiceratinae and Algericeratinae; widely distributed in the Tethyan Realm. *U. Alb. - U. Cenom.*

Subfamily Neophlycticeratinae Cooper, 1992

Discussion. Questionably monophyletic derivatives of Lyelliceratidae (Spath, 1922), since there are two phyletic lines that evolved in parallel, and the precise antecedents are unknown. Generally small to very small, with narrow umbilicus, compressed to inflated whorls, umbilical bullae giving rise to bundles of ribs with feeble ventrolateral bullae and prominent siphonal clavi; suture with more or less distinct adventive lobe to E/L. Comprises the tribes Neophlycticeratini and Cenisellini. *U. Alb. - L. Cenom.*

The phyletic line *Neophlycticeras - Faraudiella - Budaiceras* persisted throughout the Late Albian and most of the Early Cenomanian, and evolved in parallel with other members of the Neophlycticeratinae as well as the Stoliczkaiinae.

Tribe Neophlycticeratini Cooper, 1992

(*nom. transl.* herein *ex* Neophlycticeratinae)

Type genus: *Neophlycticeras* Spath, 1922.

Diagnosis. Typically with fastigiate venter and siphonal clavi persisting to maturity. *Upper Alb. - L. Cenom.*

Neophlycticeras SPATH, 1922 [*Ammonites brottianum* d'Orbigny, 1841; OD]. Small; involute, moderately inflated, with strongly convex flanks, evenly rounded

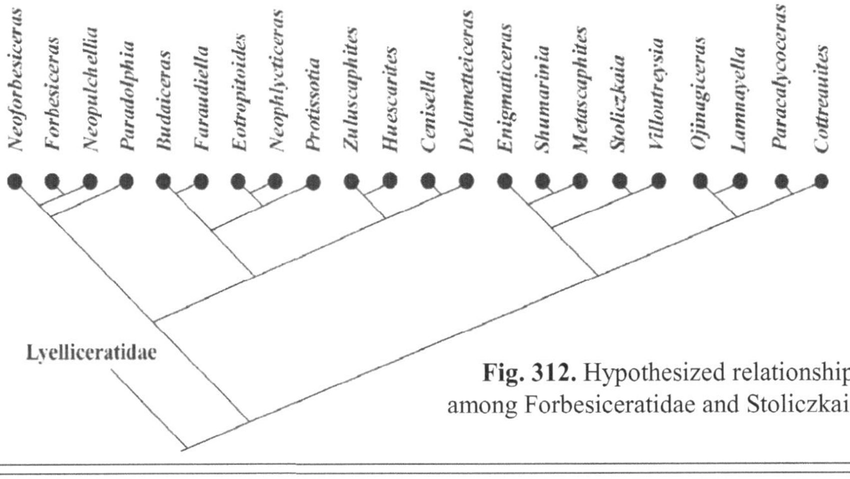

Fig. 312. Hypothesized relationships among Forbesiceratidae and Stoliczkaiidae.

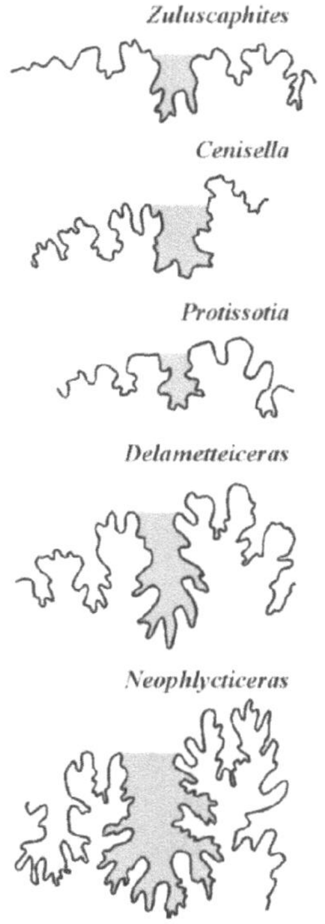

Fig. 313. Neophlycticeratine sutures.

venter and subovate whorl section; 2-3 strong low broad rather rigid rectiradiate to slightly rursiradiate ribs arise from blunt low umbilical bullae, with occasional intercalatories, and broaden ventrally, terminating in feeble ventrolateral tubercles that persist to end of adult body chamber, but are weak in maturity; prominent siphonal clavi flanked by smooth bands in early to middle growth; suture with deep narrow E with diverging branches, long rather narrow asymmetrically bifid E/L with narrow stem, deep rather narrow trifid L deeper than E, and narrow asymmetrically bifid narrow stemmed L/U_2. *Lower U. Alb.* **Fig. 314A-B** - *N. brottianum* (d'Orbigny).

Eotropitoides CASEY, 1965 [*Neophlycticeras jayeti* Breistroffer, 1953; OD] is based on a juvenile with feeble ribbing and siphonal clavi coalescing to form a weak low keel. *U. Alb.* **Fig. 314C-D** - *E. jayeti* (Breistroffer).

According to Kennedy & Delamette (1991) it is a *nomen nudum.*

= *Eotropidoites* Follmi, 1989 (*laps. cal.*).

Protissotia COLLIGNON, 1932 [*P. madagascariensis*, = *Ammonites itierianus* d'Orbigny, 1841; OD]. Micromorph; moderately involute, compressed, with flat flanks, fastigiate venter and scaphitoid uncoiling of adult body chamber; coarse round distant ribs arise singly or in pairs from weak umbilical bullae and carry prominent ventrolateral clavi matched in number by pronounced siphonal clavi; ribbing simplifies in maturity; adult aperture with short oblique rostrum; suture like *Neophlycticeras* but simpler, with rather broad shallow E, broad subquadratic E/L, narrow L, and moderately wide nonbifid L/U_2. *Lower U. Alb.* **Fig. 314I-J** - *P. itieriana* (d'Orbigny).

Faraudiella BREISTROFFER, 1947 [*Ammonites gardonica* Herbert & Munier-Chalmas, 1875, = *Ammonites blancheti* Pictet & Campiche, 1849; OD]. Small; like *Neophlycticeras* but more involute, compressed and high whorled, with more frequent intercalatories between long primary ribs; ribbing may weaken at midflank but crosses venter strongly forming obtuse chevrons; all ribs with weak ventrolateral and siphonal clavi that persist into middle growth; adult body chamber with low broad fold-like ribs, with or without feeble siphonal tubercles; suture with moderately wide E, narrow stemmed asymmetrically bifid E/L, narrow asymmetrically bifid L deeper than E, and narrow asymmetrically bifid L/U_2. *Upper U. Alb. - L. Cenom.* **Fig. 321A-B** - *F. blancheti* (Pictet & Campiche).

Budaiceras BÖSE, 1928 [*B. mexicanum*; OD]. Small; like *Faraudiella* but larger and more evolute, with ribbing effaced from venter, and tubercles to siphonal keel outnumbering ventrolaterals; ribs single in maturity and convex across flanks; ornament obsolete on last part of adult body chamber when venter rounds; suture simple, with broad E/L and very narrow L not as deep as E. *L. Cenom.* **Fig. 314R-S** - *B. mexicanum* Böse.

Tribe Cenisellini nov.

Type genus: *Cenisella* Delamette & Latil, 1989.

Diagnosis. Typically larger than neophlycticeratins with siphonal clavi evanescing in middle to later growth, leaving rounded venter over which ribs cross uninterrupted; ribbing fine, tending to evanesce in later growth, but in the type genus with strong umbilical bullae. *?Upper M. Alb., U. Alb. - L. Cenom.*

Delametteiceras **gen. nov.** [*Neophlycticeras rhodanense* Delamette, 1983, p.43; OD herein]. Very small; like *Neophlycticeras* but with narrower umbilicus, flat convergent flanks, compressed oval whorl section, fine ribbing that is weak on inner flank but strengthens on outer flank and venter; straight strongly prorsiradiate primary ribs branch both low and high on flank, separated by intercalatories, and feeble ventrolateral tubercles restricted to early growth; adult body chamber shows scaphitoid uncoiling and decline in ornament, becoming smooth; suture simpler, with short E, E/L with deep adventive lobe, and asymmetrically bifid L much deeper than E. *Lower U. Alb.* **Fig. 314P-Q** - *D. rhodanense* (Delamette).

Cenisella DELAMETTE & LATIL, 1989 [*Ammonites bonnetianus* Pictet, 1847; OD]. Small; like *Delametteiceras* but larger, more evolute, with subparallel flanks, distinct umbilical bullae that give rise to bundles of

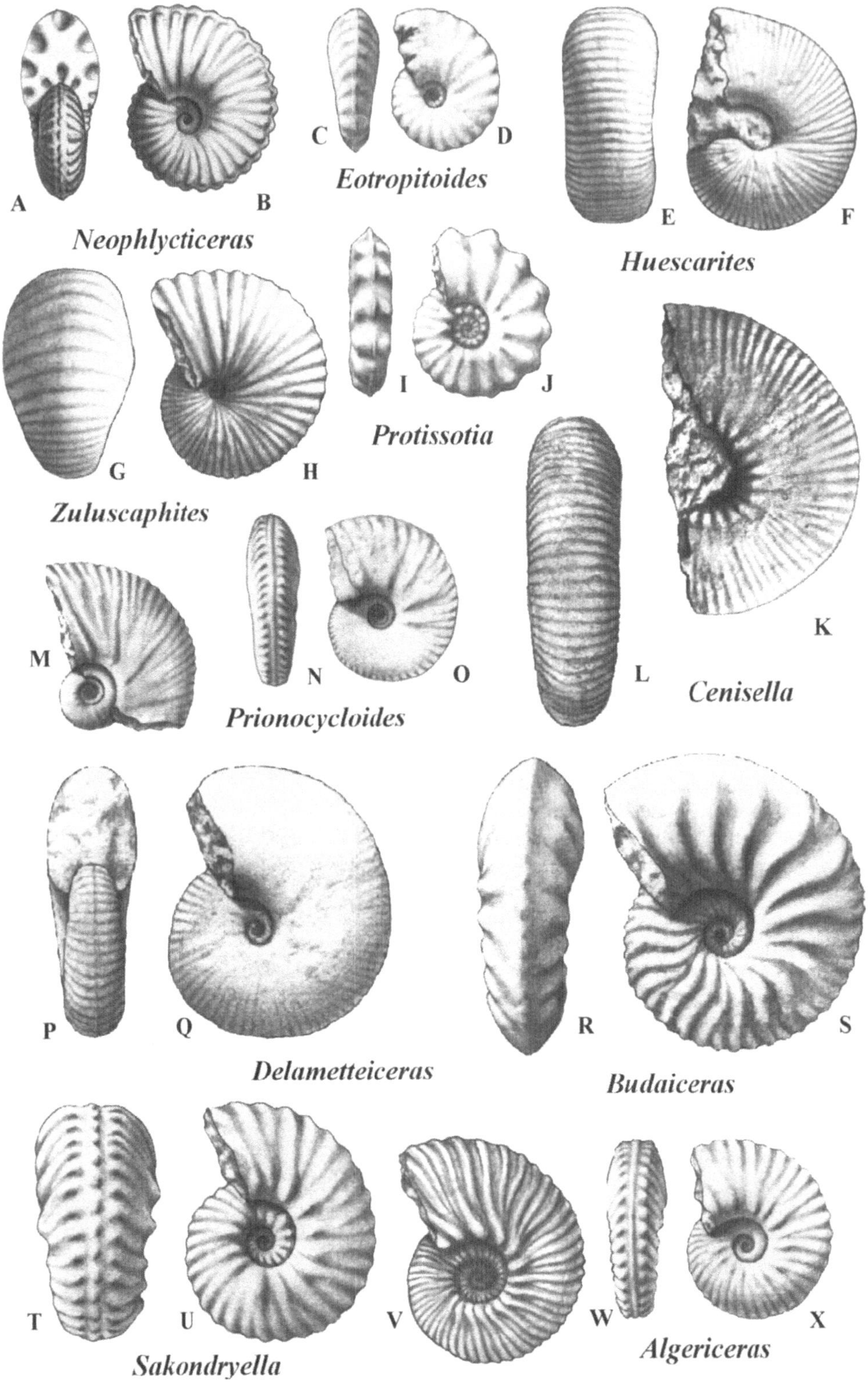

Fig. 314. Stoliczkaiidae, Flickiidae, Algericeratidae.

fine ribs that increase by bifurcation and trifurcation and persist to maturity; siphonal tubercles lost early. *Upper U. Alb.* **Fig. 314K-L** - *C. bonnetiana* (Pictet).

Zuluscaphites VAN HOEPEN, 1955 [*Z. orycteropusi*; OD]. Small; like *Faraudiella* but with rapidly inflating phragmocone, tiny deep umbilicus, broadly rounded venter and depressed reniform whorl section; inner whorls with almost straight slightly prorsiradiate narrow ribs that increase by bifurcation and intercalation and cross venter with broad convexity, with tiny siphonal tubercles; in middle growth ribs coarsen markedly, tending to alternate long and short on last part of phragmocone; suture simple, with broad bifid L, rather narrow shallow L and moderately wide bifid L/U_2. *Upper M. - Lower U. Alb.* **Fig. 314G-H** - *Z. orycteropusi* van Hoepen.

Van Hoepen (1955) emphasized the stratigraphic position of *Zuluscaphites* as upper Middle or low Upper Albian.

Huescarites LATIL, 1990 [*H. companyi*; OD]. Small; like *Zuluscaphites* but whorls not increasing so rapidly in width, more finely and weakly ribbed, and siphonal tubercles persisting longer; ornament may decline on adult body chamber. *Upper U. Alb.* **Fig. 314E-F** - *H. companyi* Latil.

Subfamily Forbesiceratinae Wright, 1952

Discussion. Monophyletic derivatives of Neophlycticeratinae, close to Cenisellini (Cooper, 1992). Primitively small, later very large, discoidal, very involute, with narrow tabulate venter sometimes with siphonal tubercles. *L. - U. Cenom.*

Paradolphia CASEY, 1965 [*P. prisca*; OD]. Small; involute, strongly compressed and high whorled with flat convergent flanks and narrow fastigiate venter with nodate siphonal keel; faint flexuous prorsiradiate ribs arise in bundles from feeble umbilical bullae and become strong and flat topped only near venter where they project forward; suture with bifid L and oblique incipient adventive lobe that divides E/L into two unequal parts, outer part being smaller; a homoeomorph of *Niceforoceras* (Subprionocyclinae). *Upper U. Alb. - L. Cenom.* **Fig. 316J-K** - *P. prisca* Casey.

Forbesiceras KOSSMAT, 1897 [*Ammonites largilliertianus* d'Orbigny, 1841; SD Diener, 1925]. Large; compressed oxycone with tiny umbilicus, broad gently convex flanks and narrow weakly fastigiate venter in early to middle growth, becoming acutely rounded later; numerous slightly flexuous ribs, rectiradiate to prorsiradiate on lower flank become stronger on outer flank where they curve forwards to meet small ventrolateral clavi before crossing venter transversely; in some venter may be smooth with

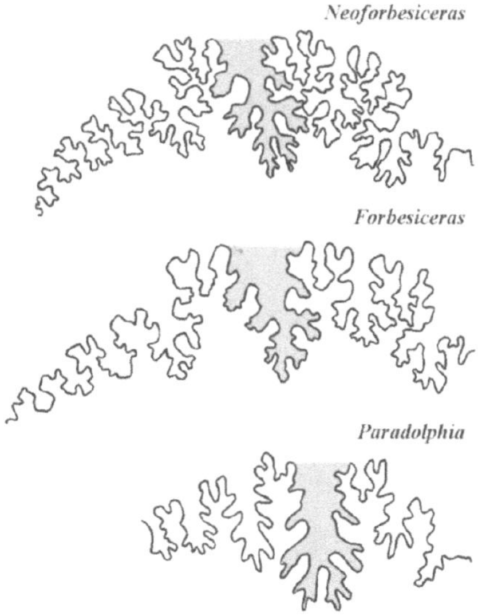

Fig. 315. Forbesiceratine sutures.

development of ventrolateral keels and low blunt siphonal ridge; suture with broad E, very broad deeply incised oblique E/L with deep penetrating adventive lobe, narrow bifid L almost as deep as E, long narrow asymmetrically bifid L/U_2, and descending series of bifid auxiliary saddles. *L. - U. Cenom.* **Fig. 316A-B** - *F. largilliertianum* (d'Orbigny).

= *Discoceras* Kossmat, 1897 (*non* Barrande, 1867).

= *Cenomanites* Haug, 1898 (obj.).

= *Ferbesiceras* Kennedy & Juignet, 1984 (*laps. cal.*).

Neopulchellia COLLIGNON, 1929 [*Pulchellia (N.) gignouxi*; SD Wright, 1996]. Based on compressed, very involute platycone with tabulate venter and single and intercalated prorsiradiate ribs strengthening ventrally and terminting in ventrolateral clavi. *L. Cenom.* **Fig. 316E-F** - *N. gignouxi* (Collignon).

Believed to represent the nucleus of a *Forbesiceras*.

Praeclariella **gen. nov.** [*Forbesiceras brundrettei* Young, 1958, p.289; OD herein]. Like *Forbesiceras* but smaller, with tabulate venter and coarse flexuous ribs bifurcating and trifurcating at various levels on flank. *M. Cenom.* **Fig. 316C-D** - *P. brundrettei* (Young).

Neoforbesiceras **gen. nov.** [*Ammonites obtectus* Sharpe, 1855, p.20; OD herein]. Very large, discoidal; like *Forbesiceras* but with discrepant flank ornament; inner flanks with radial ribs that terminate in small rounded midlateral tubercles; outer flank with numerous weak gently curved strongly rursiradiate ribs that terminate in ventrolateral clavi on distinct low keels without crossing venter; venter with weak siphonal tubercles in early to middle growth, and numerous transverse riblets later; ornament declines in maturity and outer whorls smooth, with lanceolate whorl section and acute venter; suture with phylloid folioles and strongly retracted suspensive lobe. *L?, M. Cenom.* **Fig. 316G-I** - *N. obtectum* (Sharpe).

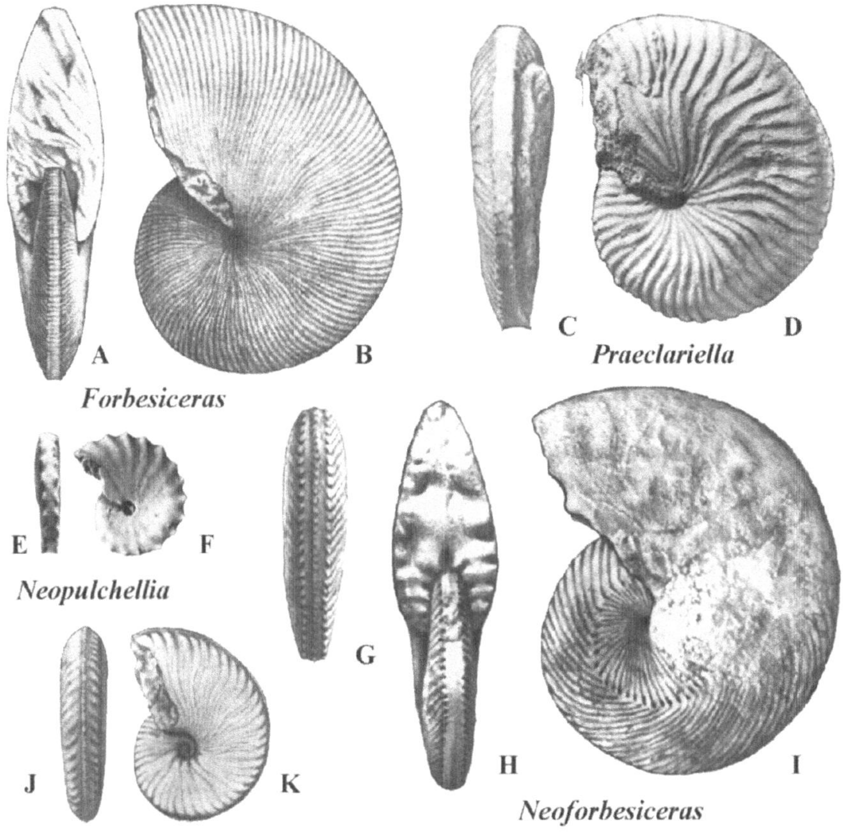

Fig. 316. Forbesiceratinae.

Subfamily Algericeratinae nov.

Diagnosis. Typically micromorph, strongly ribbed, with a crenulated to finely serrated siphonal keel, a character unknown among mortoniceratids, and simple suture. An enigmatic taxon, superficially similar to Mortoniceratinae (Kennedy & Wright, 1981), but perhaps with an origin in Forbesiceratinae, perhaps close to *Sakondryella*. *L. Cenom.*

Sakondryella COLLIGNON, 1964 [*Euhystrichoceras (S.) madagascariense*, = *E. remolinense* Böse, 1928; OD]. Very small, moderately evolute (U=24-33%), with convergent weakly convex to rounded flanks, broadly rounded to subtabulate venter and depressed trapezoidal to subquadratic whorl section; prominent sometimes massive umbilical bullae give rise to pairs of strong almost straight ribs separated by 1-2 intercalatories; all ribs terminate in small rounded oblique ventral tubercles separated by narrow smooth band from corresponding number of aperturally displaced rounded siphonal clavi on weak ridge; suture simple, with quadratic shallowly bifid subsymmetrical E/L, narrow L as deep as E, and rounded little incised L/U_2. *L. Cenom.* **Fig. 314T-U** - *S. remolinensis* (Böse).

Algericeras SPATH, 1925 [*Ammonites boghariensis* Coquand, 1880; OD]. Like *Sakondryella* but more evolute and flat sided, with compressed subrectangular whorl section, strong rigid prorsiradiate ribs bifurcating from numerous regular umbilical bullae, separated by single intercalatory, oblique ventrolateral tubercles and entire siphonal keel with microscopic serrations in maturity; suture like *Sakondryella* but more incised, with broader shallower L; a homoeomorph of *Cantabrigites* (Brancoceratidae). *L. Cenom.* **Fig. 312V** - *A. b. boghariense* (Coquand), **Fig. 314W-X** - *A. b. paucinodatum* Kennedy & Wright.

Prionocycloides SPATH, 1925 [*Ammonites proratus* Coquand, 1880; OD]. Like *Algericeras* but more involute, higher whorled, with more convex flanks, fewer poorly developed umbilical bullae, finer weaker bidichotomous ribs with more numerous intercalatories, and more incised suture with deeply bifid E/L. *L. Cenom.* **Fig. 314M-O** - *P. proratus* (Coquand).

Family FLICKIIDAE Adkins, 1928

(*nom. correct.* Wright, 1957 *pro* Flickidae)

Discussion. A monophyletic group with an origin

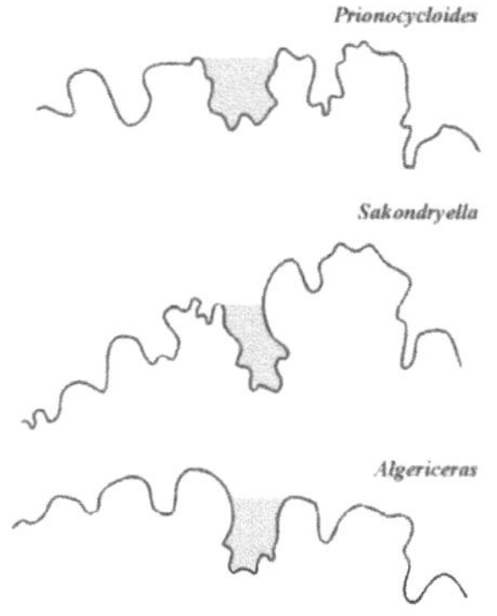

Fig. 317. Algericeratine sutures.

in Forbesiceratidae; micromorph to very small, typically weakly ornamented with a simplified suture. Comprises the subfamilies Salaziceratinae and Flickiinae. A widely distributed but uncommon component of Tethyan and Austral faunas. *Upper U. Alb. - U. Cenom.*

Subfamily Salaziceratinae Kennedy & Wright, 1984

Discussion. Monophyletic derivatives of Neophlycticeratinae, close to *Neophlycticeras* (Kennedy & Wright, 1984); typically with strong ribs and tubercles, and simple suture. Dimorphic, microconchs with lateral lappets. *Upper U. Alb. - L. Cenom.*

Salaziceras BREISTROFFER, 1936 [*Ammonites salazacensis* Herbert & Munier-Chalmas, 1875; OD]. Very small; inflated, moderately involute, subglobose, with deep conical umbilicus, inclined umbilical wall, convex flanks, broadly rounded venter, and depressed reniform whorl section; strong sparse umbilical bullae give rise to 1-2 strong rounded straight to feebly convex slightly rursiradiate ribs, with occasional intercalatories, that thicken markedly across venter; adult body chamber shows scaphitoid uncoiling; simple suture with broad E, broad bifid E/L, narrower little divided, rather deep L and moderately wide asymmetrically bifid L/U_2. *Upper U. Alb.* **Fig. 319E-F** - *S. salazacense* (Herbert & Munier-Chalmas).

= *Salazaciceras* Breistroffer, 1940 (*nom. van.*).

Noskytes SCHOLZ, 1979 [*Salaziceras (N.) bakonyense*; OD]. Very small; inner whorls like *Salaziceras* but with vestigial siphonal tubercles; adult body chamber with strong coarse ribbing and prominent ventrolateral horns joined across venter by looped ribs; suture rather incised, with deep E, trifid L much shorter than E and narrow-stemmed L/U_2. *Upper U. Alb.* **Fig. 319I** - *N. bakonyensis* (Scholz).

Interpreted as the microconch of *Salaziceras* (Wright, 1996).

= *Noskytes* Scholz, 1978 (*nom. nud.*).

Neosaynoceras BREISTROFFER, 1947 [*Saynoceras gazellae* Pervinquière, 1907; OD]. Micromorph; like *Salaziceras* but smaller, more inflated and globose; adult body chamber occludes umbilicus; inner whorls smooth, with periodic weak constrictions; later with rather weak distant prorsiradiate ribs with inner and outer ventrolateral bullae that become spinose on adult body chamber and are linked across venter by looped or zigzagging riblets or striae; outer ventrolateral tubercles tend to alternate across venter, but may be opposite; vestigial siphonal tubercles may be present. *L. Cenom.* **Fig. 319G-H** - *N. gazellae* (Pervinquière).

Subfamily Flickiinae Adkins, 1928

(*nom. transl.* Kennedy & Wright, 1984 *ex* Flickiidae)

Discussion. A monophyletic group with an origin in Salaziceratinae, close to *Salaziceras* (Wright & Kennedy, 1979); typically weakly ornamented micromorphs with goniatitic sutures. Comprises the tribes Flickiini and Litophragmatoceratini. *L. - U. Cenom.*

Tribe Flickiini Adkins, 1928

(*nom. transl.* herein *ex* Flickiinae)

Type genus: *Flickia* Pervinquière, 1907.

Discussion. Comprises ammoniticones and cadicones. *L. - U. Cenom.*

Ficheuria PERVINQUIÈRE, 1910 [*F. kiliani*; OD]. Very involute, globular, with deep conical umbilicus, angular umbilical edge, broadly rounded venter and depressed reniform whorl section; early whorls smooth or with sparse shallow constrictions; adult body chamber shows slight scaphitoid uncoiling, with low flat ribbon-like flexuous prorsiradiate ribs separated by deep interspaces; suture like *Salaziceras* but very simple, barely incised, with shallowly bifid E/L and L/U_2. *L. Cenom.* **Fig. 319C-D** - *F. kiliani* Pervinquière.

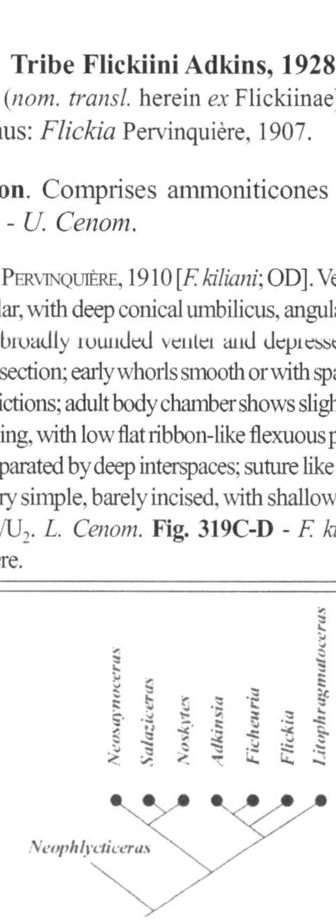

Fig. 318. Hypothesized relationships among Flickiidae.

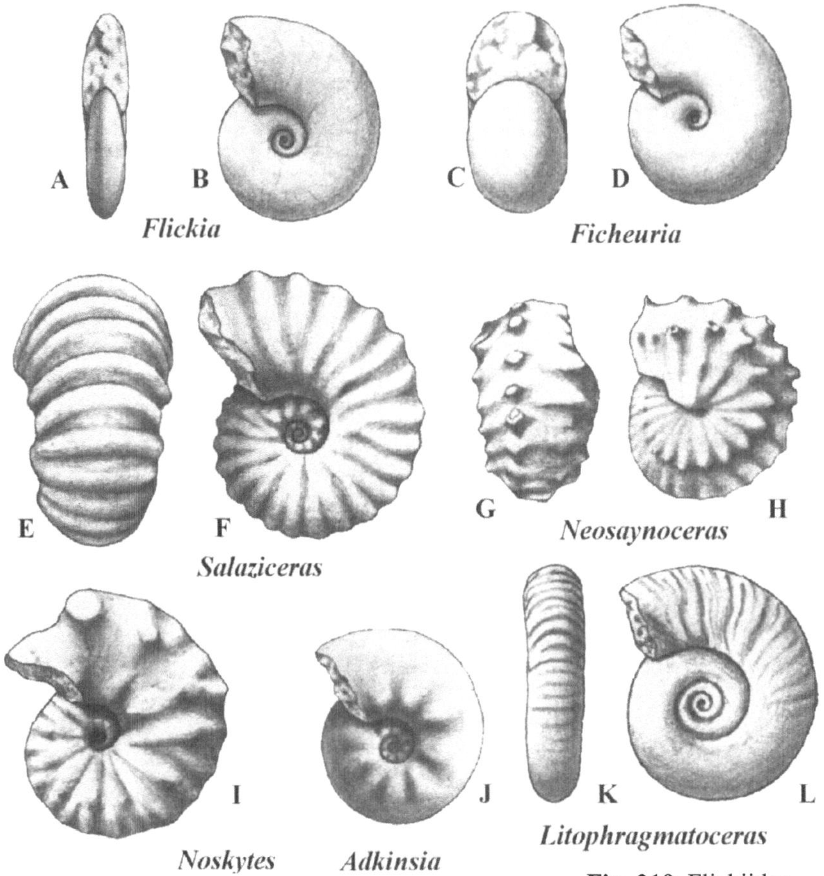

Fig. 319. Flickiidae.

Flickia PERVINQUIÈRE, 1907 [*F. simplex*; OD]. Inner whorls like *Ficheuria*, but outer whorls compressed and evolute, with low umbilical wall, rounded umbilical shoulder, flat subparallel flanks, narrowly rounded to fastigiate venter and simpler goniatitic suture with entire saddles and lobes, narrow E/L and L deeper than E. *L. Cenom.* **Fig. 319A-B** - *F. simplex* Pervinquière.

Adkinsia BÖSE, 1928 [*A. adkinsi*, = *Flickia? bosquensis* Adkins, 1920; OD]. Like *Flickia*, but more inflated, with deeper umbilicus; distinct umbilical bullae appear in middle growth and remain strong until adult body chamber when they weaken; single low broad convex prorsiradiate ribs appear on outer whorls but decline on outer flank and may, or may not, cross venter. *L. Cenom.* **Fig. 319J** - *A. bosquensis* Adkins.

Tribe Litophragmatoceratini nov.

Type genus: *Litophragmatoceras* Kennedy & Cobban, 1988.

Diagnosis. Weakly ornamented platycones with an origin in Flickiini, but separated by a significant phenotypic discontinuity. *Lower U. Cenom.*

Litophragmatoceras KENNEDY & COBBAN, 1988 [*L. incomptum*; OD]. Very small; like *Flickia* but an evolute platycone with crowded broad and narrow flexuous ribs separated by deep interspaces on adult body chamber; suture simple, like *Flickia* but with very broad E, and L as deep as E. *Lower U. Cenom.* **Fig. 319K-L** - *L. incomptum* Kennedy & Cobban.

Family STOLICZKAIIDAE Breistroffer, 1953

(*nom. transl.* herein *ex* Stoliczkaiinae)

Discussion. Monophyletic derivatives of Forbesiceratidae (Spath, 1922), typically very small to small with siphonal tubercles, if present, restricted to earliest growth; comprises only the subfamily Stoliczkaiinae. *Upper U. Alb. - L. Cenom.*

Subfamily Stoliczkaiinae Breistroffer, 1953

= Stoliczkaiellinae Cooper, 2012 (unnessecary replacement name).

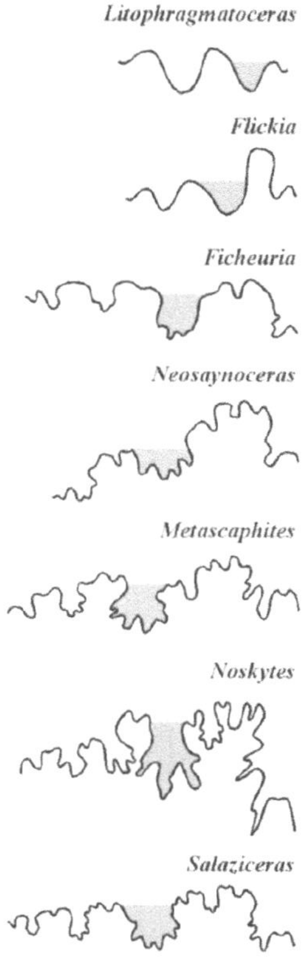

Fig. 320. Flickiid sutures.

Discussion. Monophyletic derivatives of Neophlycticeratinae (Spath, 1931; Casey, 1965) but the precise antecedent still uncertain, perhaps close to *Cenisella* (Cooper, 1992). Save for sutural differences, in earliest growth *Stoliczkaia* and *Forbesiceras* are very similar, particularly their narrow subtabulate venters and ventrolateral tubercles. Typically compressed and high whorled with narrow umbilicus, tabulate venter that rounds in later growth, prorsiradiate primary ribs separated by shorter secondaries with ventrolateral tubercles that evanesce in middle growth; if present siphonal tubercles are restricted to earliest growth. Suture acanthoceratid, with a subquadratic shallowly bifid E/L. *Upper U. Alb. - L. Cenom.*

Stoliczkaia NEUMAYR, 1875 [*Ammonites dispar* d'Orbigny, 1841; SD Diener, 1925]. Small; like *Faraudiella* but inner whorls more compressed and involute, with flat subparallel flanks, subtabulate to subfastigiate venter, falcate ribs with distinct ventrolateral clavi and feeble siphonal nodes; in early middle growth ventral tuberculation obsolete, venter rounds and ribbing becomes coarser and more rigid, tending to alternate long and short; ornament declines on adult body chamber and final portion is smooth. *Uppermost Alb.* **Fig. 321C-D** - *S. dispar* (d'Orbigny); **Fig. 323** - *S. tenuis* Renz.

non *Stoliczkaia* Boulenger, 1893 (Serpentes), an objective synonym of *Stoliczkia* Jerdon, 1870.

= *Stoliczkaiella* Cooper, 2012 (unnecessary replacement name).

Villoutreysia CASEY, 1965 [*Stoliczkaia (V.) villoutreysi*; OD]. Small; like *Stoliczkaia* but somewhat more inflated, with broad subtabulate venter and coarser ribbing that persists to maturity. *Uppermost Alb. - L. Cenom.* **Fig. 321S-T** - *V. villoutreysi* (Casey).

Shumarinaia MATSUMOTO & INOMA, 1975 [*Stoliczkaia (S.) hashimotoi*; OD]. Very small; like *Stoliczkaia* but smaller, relatively evolute, with sparse low broad slightly flexuous main ribs, either bifurcating at midflank or separated by intercalatory; ribs flex forward near venter and cross it without developing tubercles; suture simple; superficially resembles *Brancoceras* (Brancoceratidae). *Uppermost Alb. - L. Cenom.* **Fig. 321G-H** - *S. hashimotoi* (Matsumoto & Inoma).

Ojinagiceras COBBAN & KENNEDY, 1989 [*O. ojinagaense*; OD]. Micromorph; very involute, with compressed whorls and subparallel flanks; in early growth with fastigiate venter, strong ribs and ventrolateral tubercles; in middle growth tubercles decline and flexuous prorsiradiate ribs arise singly or in pairs from umbilical bullae, separated by intercalatories, and project forward on outer flank to cross venter with broad convexity; venter flattens in later growth; suture simple. *Upper L. Cenom.* **Fig. 321K-L** - *O. ojinagaense* Cobban & Kennedy.

= *Ojianagiceras* Cobban & Kennedy, 1989 (*laps. cal.*).

Metascaphites WIEDMANN, 1962 [*Scaphites? thomasi* Pervinquière, 1907; OD]. Micromorph, a progenetic dwarf perhaps of *Shumarinaia*; involute, compressed with flattened flanks, broadly rounded venter, compressed subquadratic whorl section, and maximum width close to ventrolateral shoulders; inner whorls unornamented; simple distant rectiradiate ribs, both long and short, appear on last part of final whorl that shows scaphitoid uncoiling, accompanied by ventrolateral bullae; ribs feeble across venter, with gentle forward arch. *Uppermost Alb., ?L. Cenom.* **Fig. 321H-I** - *M. thomasi* (Pervinquière).

Enigmaticeras KENNEDY, 2004b [*E. riceae*; OD]. Micromorph; compressed, involute, with flat subparallel flanks, narrow subtabulate venter and feeble ventrolateral folds to phragmocone; adult body chamber with delicate, prorsiradiate, biconvex, falcoid riblets and growth striae; suture like *Shumarinaia* but with deeper E and deeply bifid E/L; a homoeomorph of *Naramoceras* (Binneyitidae). *Uppermost Alb.* **Fig. 321O-P** - *E. riceae* Kennedy.

Lamnayella WRIGHT & KENNEDY, 1978 [*Stoliczkaia (L.) juigneti*; OD]. Small; in early growth like *Villoutreysia* with narrowly rounded fastigiate trituberculate venter; later with elongate umbilicolateral bullae passing imperceptibly into distant weakly flexuous prorsiradiate

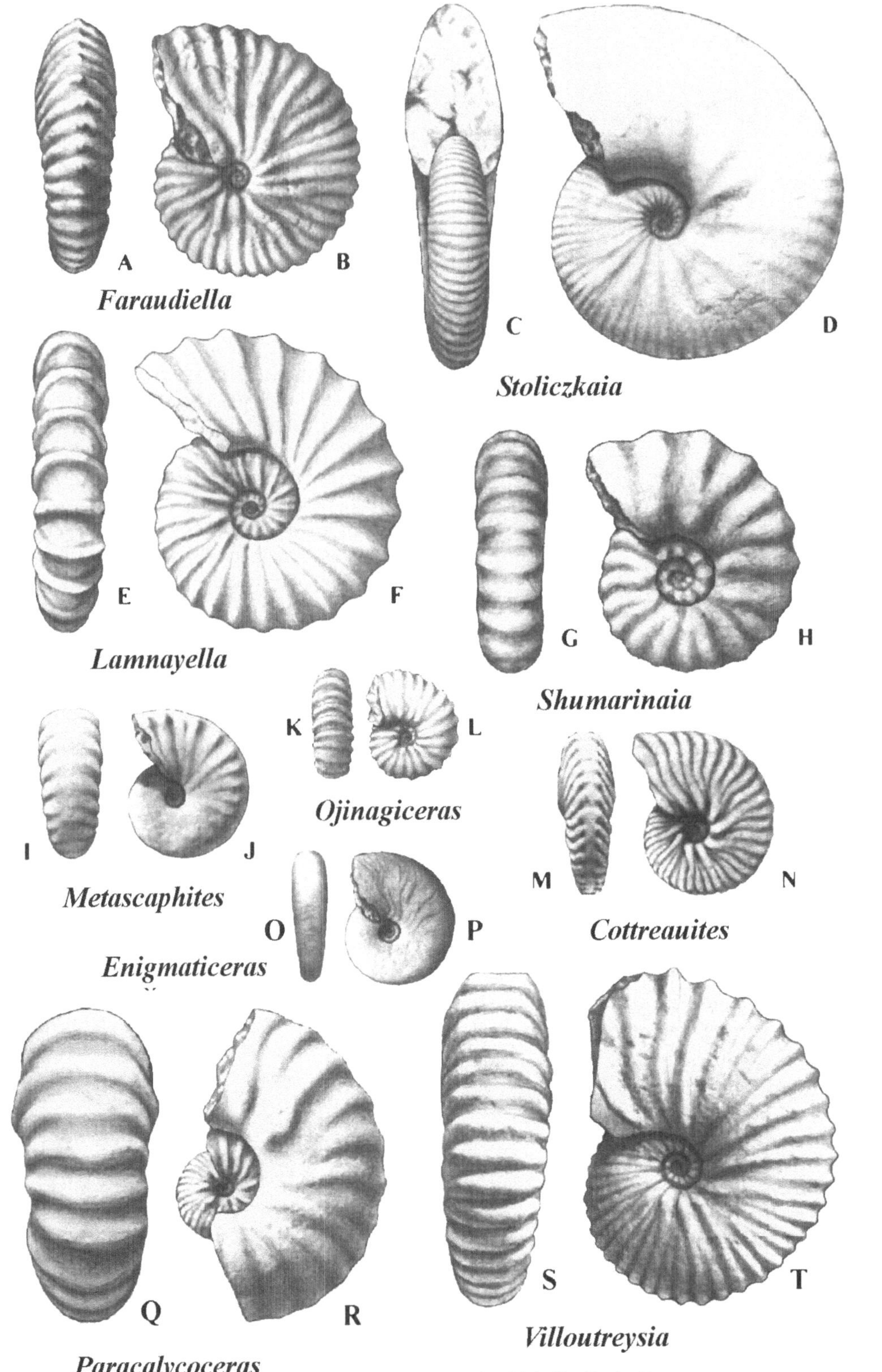

Fig. 321. Stoliczkaiinae.

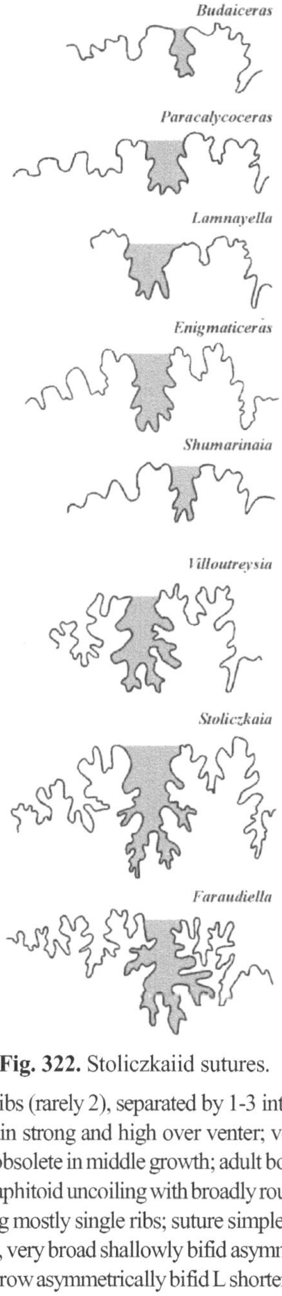

Fig. 322. Stoliczkaiid sutures.

primary ribs (rarely 2), separated by 1-3 intercalatories, that remain strong and high over venter; ventral tuberculation obsolete in middle growth; adult body chamber shows scaphitoid uncoiling with broadly rounded venter and strong mostly single ribs; suture simple, with rather narrow E, very broad shallowly bifid asymmetrical E/L, rather narrow asymmetrically bifid L shorter than E, and fairly narrow L/U_2. *Lower L. Cenom.* **Fig. 321E-F** - *L. juigneti* (Wright & Kennedy).

Paracalycoceras SPATH, 1925 [*Ammonites wiestii* Sharpe, 1857; OD]. Small; like inflated *Lamnayella* but inner whorls with 4-6 intercalatories between primary ribs, inner (weak) and outer ventrolateral tubercles, and distinct siphonal ridge that may be feebly tuberculate; inflated outer whorl with distant curved ribs, rursiradiate on outer flank; primary ribs with more or less prominent umbilical bullae, separated by 1-3 intercalatories. *L. Cenom.* **Fig. 321Q-R** - *P. wiestii* (Sharpe).

Fig. 323. *Stoliczkaia.*

Cottreauites COLLIGNON, 1929 [*Acanthoceras (Prionotropis) subvicinale* Boulé, Lemoine & Thevenin, 1907; OD]. Based on pyritic nuclei; inner whorls compressed, involute, high whorled, with broad flattened flanks converging to narrow fastigiate venter with weak siphonal clavi; prominent prorsiradiate primary ribs give rise to pairs strong flexuous ribs separated by single intercalatory; ribs flex strongly forward near venter; and carry weak inner and strong outer ventrolateral clavi. *L. Cenom.* **Fig. 321M-N** - *C. subvicinalis* (Boulé, Lemoine & Thevenin).

Said to be based on pyritic nuclei of *Forbesiceras*. = *Cottereauites* Roman, 1938 (*nom. van.*).

Hyporder ACANTHOCERATOIDEI de Grossouvre, 1894

(*nom. transl.* herein *ex* Acanthoceratoidea)

Discussion. Monophyletic derivatives of Brancoceratoidei (Spath, 1931), comprising the superfamilies Acanthoceratoidea, Mammitoidea and Sphenodiscoidea. *L. Cenom. - U. Maastr.*

Superfamily ACANTHOCERATOIDEA de Grossouvre, 1894

(*nom. transl. et correct.* Hyatt, 1900 *ex* Acanthoceratidés, *nom. transl.* Wright & Wright, 1951 *ex* Acanthoceratida)

Discussion. Monophyletic derivatives of Lyelliceratoidea (Spath, 1931) comprising the families Mantelliceratidae, Calycoceratidae, Acanthoceratidae and Vascoceratidae, primitively with bituberculate venter, then trituberculate, quadrituberculate and even quinquetuberculate, often with ventrolateral horns and keels, and finally sometimes nontuberculate. *L. Cenom. - L. Coni.*

Family MANTELLICERATIDAE Hyatt, 1903

Discussion. Monophyletic derivatives of Stoliczkaiidae (Spath, 1937), typically with quadrangular flat sided whorls, strong ribs and without siphonal tubercles; a paraphyletic taxon comprising the subfamilies Utaturiceratinae and Mantelliceratinae. *L. Cenom.*

Subfamily Utaturiceratinae Matsumoto, 1969

Discussion. Monophyletic derivatives of Stoliczkaiinae (Spath, 1937), close to *Stoliczkaia* although Wright & Kennedy (1984) looked to *Lamnayella*; typically large, compressed and high whorled, with a deeply incised florid suture, in some cases with phylloid folioles. Of widespread distribution in the Tethyan and Austral Realms. *L. Cenom.*

Utaturiceras WRIGHT, 1956 [*Ammonites vicinalis* Stoliczka, 1864; OD]. Moderately large; compressed, high whorled, involute, with broad convex flanks and narrow tabulate venter; delicate umbilical bullae give rise to crowded flexuous ribs, singly or in pairs, with long intercalatories; all ribs with feeble inner and stronger outer ventrolateral clavi, the former declining in later growth when incipient high-lateral bullae appear; suture florid with subphylloid folioles, like *Saxbyites* but with broader E/L, narrower L/U_2 that projects beyond E/L, and additional auxilliary elements. *L. Cenom.* **Fig. 330P-Q** - *U. vicinale* (Stoliczka).

Graysonites YOUNG, 1958 [*G. lozoi*, = *Mantelliceras wacoense* Böse, 1928; OD]. Very large; inner whorls like *Utaturiceras* but with compressed subrectangular to subquadratic whorl section, strong rather rigid ribs that alternate long and short and with bullate inner ventrolateral tubercles; in middle growth umbilical tubercle may split into two bullae on raised rib, and ventrolaterals coalesce to form horns; ribbing becomes coarser and more distant in later growth but without developing high-lateral bullae; suture like *Utaturiceras* with subphylloid terminations, rather shallow E, wide E/L, L deeper than E, and narrow L/U_2. *Lower L. Cenom.* **Fig. 330E-I** - *G. wacoensis* (Böse).

= *Graysonites* Adkins, 1933b (*nom. nud.*).

Tlahualiloceras KELLUM & MINTZ, 1962 [*T. tlahualiloense*;; OD]. Very large; like *Graysonites* but larger, more evolute, with umbilical bullae, prominent mid-lateral tubercles from early growth, single ribs in middle to later growth, and approximated but separate ventrolateral horns. *L. Cenom.* **Fig. 328** - *T. tlahualiloense* Kellum & Mintz; **Fig. 329** - *T. florencae* (Spath).

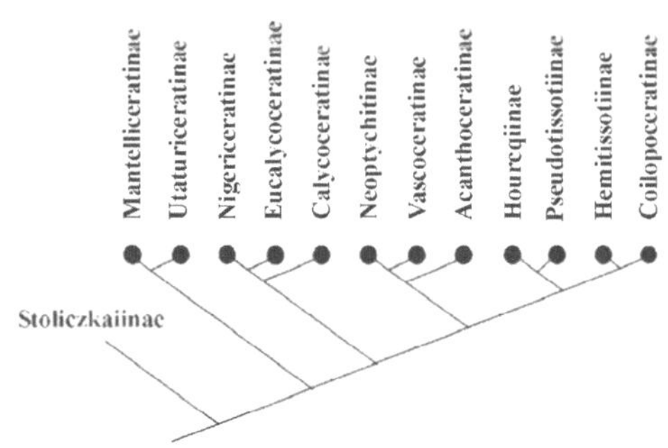

Fig. 324. Hypothesized relationships among Acanthoceratoidea.

Skoenbergites **gen. nov.** [*Acanthoceras (Mantelliceras) falloti* Collignon, 1931, p.81 (41); OD herein]. Small, very evolute, with quadratic whorl section; strong, simple rectiradiate ribs, narrower than or as wide as interspaces, with subspinose umbilical, lateral and inner and outer ventrolateral tubercles, the latter clavate and rising vertically well above the venter. *Lower M. Cenom.* **Fig. 332** - *S. falloti* (Collignon).

Sharpeiceras HYATT, 1903 [*Ammonites laticlavius* Sharpe, 1855; OD]. Very large to enormous; like *Tlahualiloceras* but with more closely spaced ribs, particularly on adult body chamber, broader ventral sulcus and, in later growth, with low siphonal ridge and feeble to distinct row of high-lateral (5th row) tubercles. *L. Cenom.* **Fig. 328** - *S. kikuae* Matsumoto & Kawashita; **Fig. 330R-S** - *S. laticlavium* (Sharpe);

Acanthoceras laticlavium var. *byzacenica* Pervinquière (**Fig. 325**), assigned by some to *Mhriliceras*, shows incipient low-lateral and high-lateral bullae, and a blunt siphonal ridge and seems to be transitional to *S. indicum* (Kossmat).

Mhriliceras KENNEDY & WRIGHT, 1985 [*Mammites lapparenti* Pervinquière, 1907; OD]. Small; moderately involute, with equidimensional to depressed pentagonal whorl section; robust straight, slightly prorsiradiate ribs alternate long and short; long ribs with prominent umbilical bullae and all ribs with inner and outer ventrolateral clavi; ribs cross venter transversely; suture with broad asymmetrically bifid E/L, narrow asymmetrically bifid L as deep

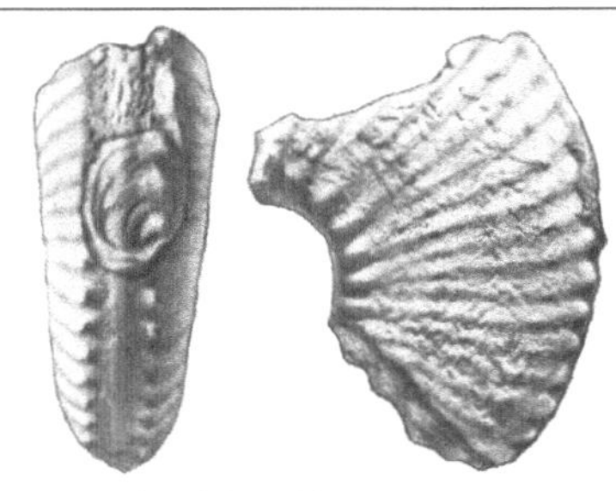

Fig. 325. *Sharpeiceras.*

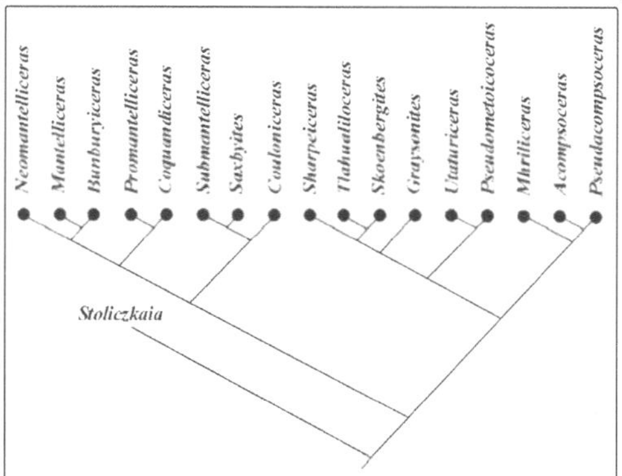

Fig. 326. Hypothesized relationships among Mantelliceratidae.

Fig. 328. *Thlahualiloceras.*

as E and shallow L/U_2. *L. Cenom.* **Fig. 333I-J** - *M. lapparenti* (Pervinquière).

***Pseudometoicoceras* gen. nov.** [*Metoicoceras fasciculatum* Collignon, 1964, p.151; OD herein]. Very small, strongly involute (U=12%), high whorled with flat subparallel flanks, smooth fastigiate venter and fine flexuous bituberculae ribs that form chevrons over venter. *L. Cenom.* **Fig. 335A-B** - *P. fasciculatum* (Collignon).

Kennedy & Wright (1985) interpreted the type as a microconch; a supposed macroconch fragment (**Fig. 333K-L**) is very similar to *Acompsoceras.*

Acompsoceras HYATT, 1903 [*Ammonites bochumensis* Schlüter, 1871, = *Ammonites renevieri* Sharpe, 1857; OD]. Large; compressed, rather involute and high whorled; inner whorls with broad gently convex flanks and narrow weakly fastigiate to tabulate venter, with compressed whorl section; rather distant rectiradiate ribs arise in pairs from prominent umbilical bullae and carry inner and outer ventrolateral clavi; ribs cross slightly raised venter transversely; with growth ribs weaken on flanks and lower ventrolaterals become obsolete, leaving strong umbilical bullae, subdued flank costae, distinct (outer) ventrolateral clavi and narrow fastigiate venter; outer whorls of macroconchs with rounded venter and umbilical bullae that give rise to short radial folds only; suture florid, with phylloid terminations to elements, very broad stemmed asymmetrically bifid E/L, narrow bifid L and narrow stemmed asymmetrically bifid L/U_2. *L. Cenom.* **Fig. 333M-N, P-Q** - *A. renevieri* (Sharpe).

Pseudacompsoceras SPATH, 1925 [*P. vectense* Spath, 1925, = *Ammonites inconstans* Schlüter, 1871; OD]. Large; like *Acompsoceras* but with distinct lateral tubercles; very early whorls with siphonal clavi positioned ahead of outer ventrolaterals; adult macroconch smooth. *Upper L. Cenom.* **Fig. 333O** - *P. inconstans* (Schlüter).

Subfamily Mantelliceratinae Hyatt, 1903

(*nom. transl.* Wright & Wright, 1951 *ex* Mantelliceratidae)

Discussion. Monophyletic derivatives of Stoliczkaiinae (Spath, 1937), close to *Utaturiceras*; typically strongly ribbed, variously with umbilical, lateral and ventrolateral tubercles carried to maturity. Comprises the tribes Mantelliceratini and Coquandiceratini. *L. Cenom.*

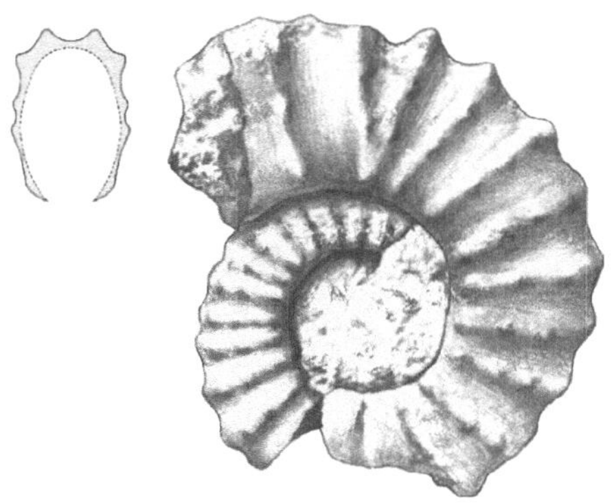

Fig. 327. *Sharpeiceras.*

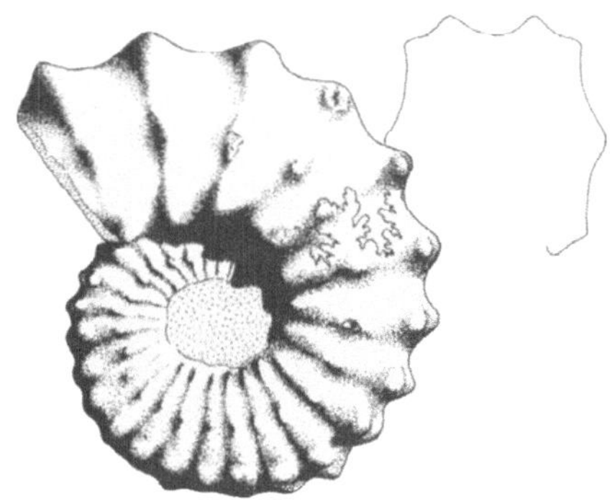

Fig. 329. *Tlahualiloceras.*

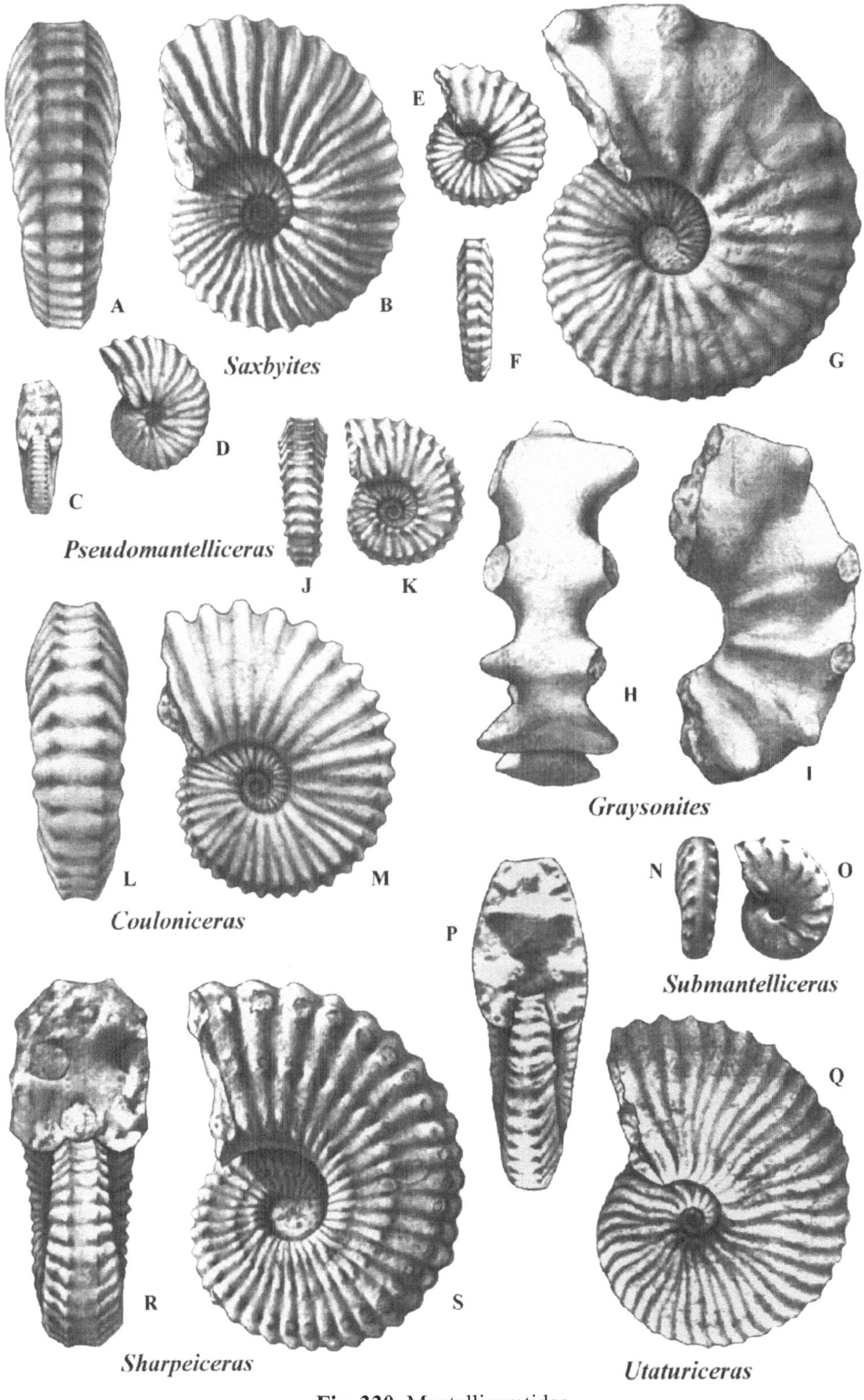

Fig. 330. Mantelliceratidae.

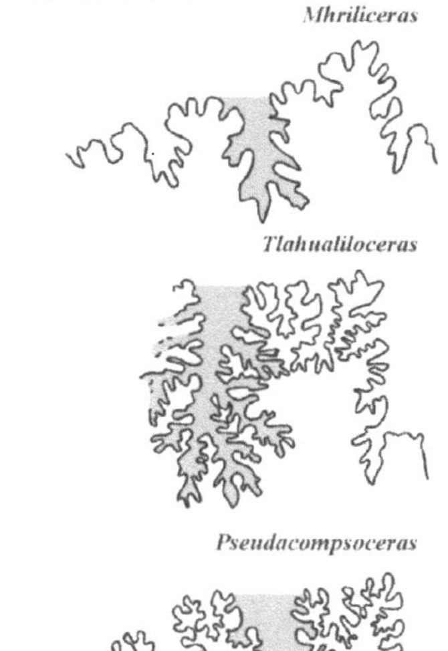

Fig. 331. Utaturiceratine sutures.

Tribe Mantelliceratini Hyatt, 1903

(*nom. transl.* herein *ex* Mantelliceratinae)

Type genus: *Mantelliceras* Hyatt, 1903.

Diagnosis. Typically small to medium sized, with umbilical and paired ventrolateral tubercles and, in derived members, lateral tubercles also. *L. Cenom.*

Submantelliceras SPATH, 1923 [*Ammonites aumalensis* Coquand, 1862; OD]. Micromorph; compressed, very involute, with flat subparallel flanks and broad subtabulate venter; earliest whorls smooth, followed by distant single prorsiradiate sigmoidal ribs with feeble umbilical bullae; ribs strengthen markedly on outer flank where separated by 1-2 intercalatories of equal strength; all ribs with weak high lateral bullae and prominent ventrolateral clavi; venter raised along siphonal line. *L. Cenom.* **Fig. 330N-O** - *S. aumalense* (Coquand).

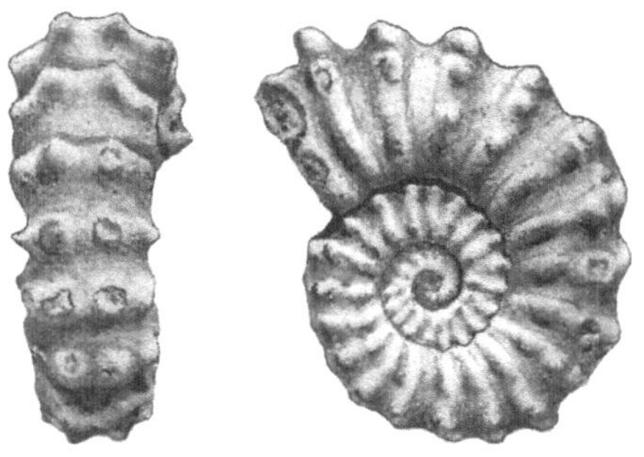

Fig. 332. *Skoenbergites.*

***Saxbyites* gen. nov.** [*Ammonites saxbii* Sharpe, 1857, p.45; OD herein]. Small to medium sized; compressed, involute, high whorled, with broad gently convex flanks and narrowly rounded to tabulate venter; strong rounded flexuous prorsiradiate ribs arise singly from umbilical bullae, separated by 1-2 intercalatories; all ribs with weak inner ventrolateral tubercles and prominent outer ventrolateral clavi joined across venter by low transverse ribs; suture with strongly retracted suspensive lobe. *L. Cenom.* **Fig. 330A-D** - *S. saxbii* (Sharpe).

Couloniceras BUSNARDO, 1966 [*Ammonites couloni* d'Orbigny, 1850; OD]. Medium sized to moderately large; like *Saxbyites* but more involute with, in maturity, straight rectiradiate ribs that alternate long and short; inner ventrolateral tubercles lost early (D<30 mm); in later growth umbilical bullae obsolete, outer ventrolateral clavi exaggerated and standing high above narrowly sulcate venter; on adult body chamber venter subtabulate and ventrolateral tubercles obsolete. *Upper L. Cenom.* **Fig. 330L-M** - *C. couloni* (d'Orbigny).

Promantelliceras THOMEL, 1972 [*Mantelliceras picteti* Hyatt, 1903; OD]. Medium sized; like *Saxbyites* but more evolute, compressed with narrow subtabulate venter, prominent lateral tubercles to main ribs and low outer ventrolateral clavi; tubercles obsolete on adult body chamber when ribs tend to alternate long and short. *L. Cenom.* **Fig. 333G-H** - *M. picteti* (Hyatt).

This name is available for the *picteti-dixoni* lineage.

Mantelliceras HYATT, 1903 [*Ammonites mantelli* J. Sowerby, 1814; ICZN Opinion 557]. Small to medium sized; like *Promantelliceras* but more inflated, with narrower umbilicus, depressed octagonal to polygonal whorl section, and regular persistent lateral tubercles. *L. Cenom.* **Fig. 333A-B** - *M. mantelli* (J. Sowerby).

Bunburyiceras THOMEL, 1972 [*Mantelliceras cantianum* Spath, 1926; OD]. Medium sized; like *Mantelliceras* but with broadly rounded venter and smooth weakly tuberculate ribs; inner ventrolateral tubercles lost early but lateral and umbilical tubercles persist to maturity; macroconchs commonly with rursiradiate ribs in later growth, homoeomorphing *Calycoceras*. *L. Cenom.* **Fig. 333C-D** - *B. cantianum* (Spath).

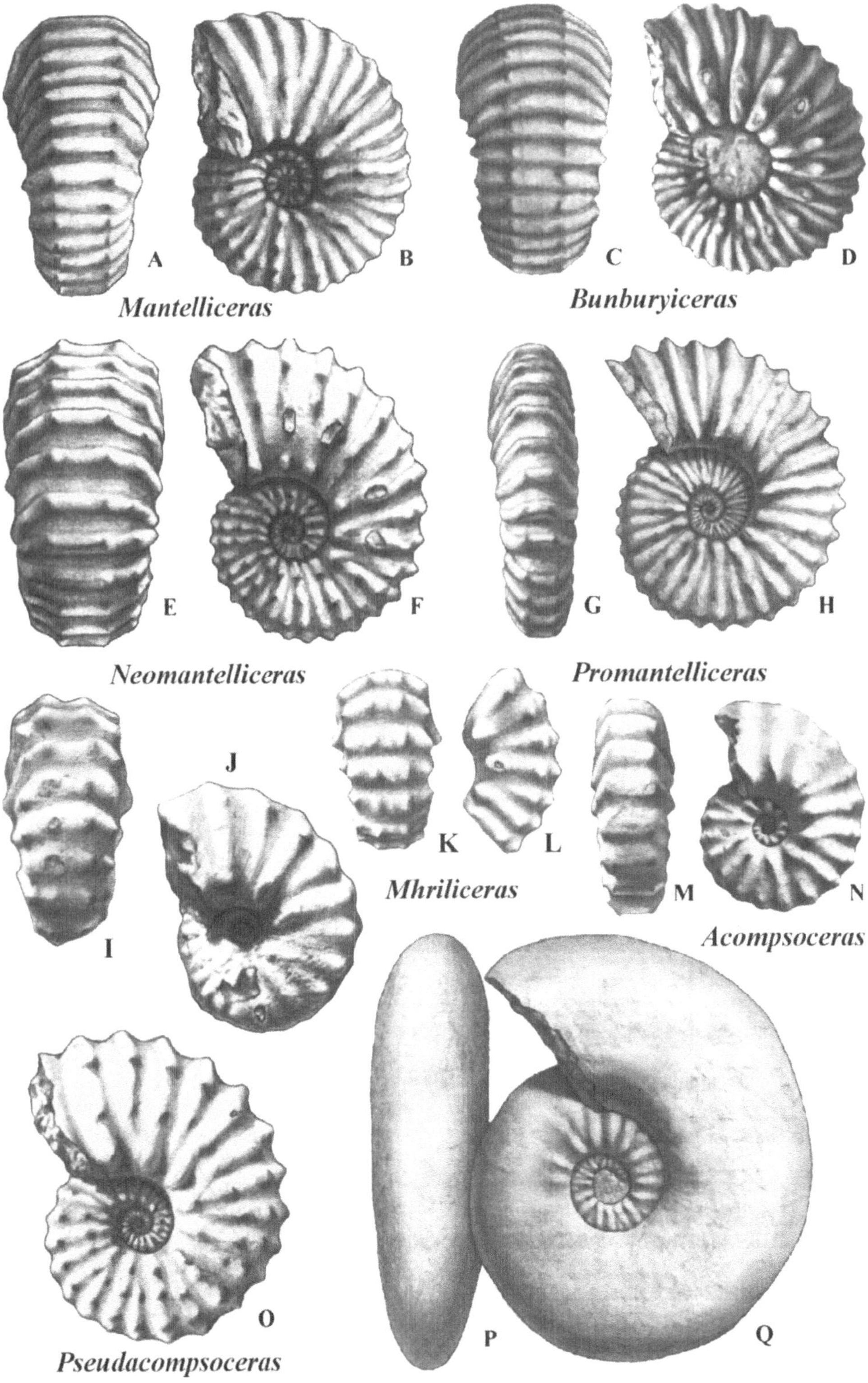

Fig. 333. Mantelliceratidae.

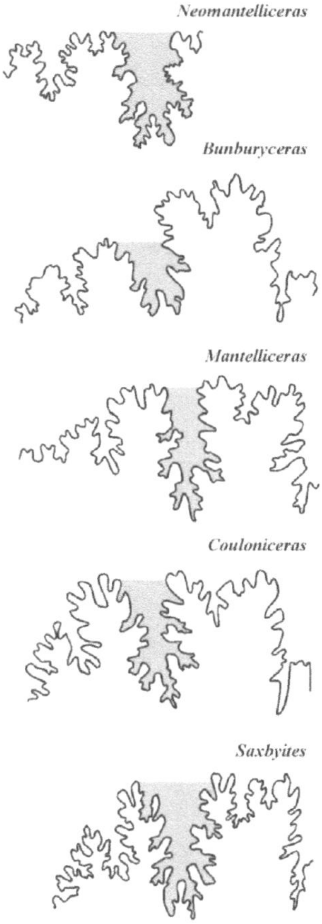

Fig. 334. Mantelliceratid sutures.

Neomantelliceras THOMEL, 1972 [*Ammonites mantelli tuberculata* Mantell, 1822; OD] [**m**]. Medium sized to moderately large; like *Mantelliceras* but more evolute, whorls increasing more rapidly in height, ribs becoming more distant, and with strong tuberculation retained to maturity. *L. Cenom.* **Fig. 333E-F** - *N. tuberculatum* (Mantell).

Tribe Coquandiceratini nov.

Type genus: *Coquandiceras* Kennedy, 2020.

Diagnosis. A monotypic taxon with an origin close to *Promantelliceras*, but separated by a significant phenotypic discontinuity. Diagnosis as for genus. *L. Cenom.*

Coquandiceras KENNEDY, 2020 [*Ammonites jubae* Coquand, 1880; OD]. Very small, very evolute; umbilical bullae give rise to primary ribs that may branch and, with additional intercalatories, are linked to outer ventrolateral clavi on either side of sunken venter that rounds in maturity; tubercles efface on adult body chamber when ribs cross transversely on the adapertural part; suture simple with minor incision to E/L. *L. Cenom.* **Fig. 335C-D** - *C. jubae* (Coquand).

Other species assigned here have both inner and outer ventrolateral tubercles.

Family CALYCOCERATIDAE nov.

Diagnosis. Monophyletic derivatives of Mantelliceratidae, characteristically with siphonal tubercles, at least in early to middle growth; comprises the subfamilies Calycoceratinae, Eucalycoceratinae and Nigericeratinae. *Upper L. Cenom. - L. Turon.*

Subfamily Calycoceratinae nov.

Diagnosis. Generally with depressed polygonal whorls in early growth becoming rounded later with early loss of outer ventrolateral and siphonal tubercles. Paraphyletic derivatives of Mantelliceratinae, close to *Promantelliceras*. *Upper L. - U. Cenom.*

Newboldiceras THOMEL, 1972 [*Acanthoceras newboldi* Kossmat, 1897; = *Acanthoceras rhotomagense* var. *asiaticum* Jimbo, 1894; OD]. Moderately inflated, evolute, with flat subparallel flanks, distinct ventrolateral facets, almost flat venter and polygonal to subquadratic and subcircular whorl section, whorls rapidly expanding; strongly and often densely ribbed, with strong rounded prorsiradiate primaries arising singly or in pairs from numerous crowded umbilical bullae; on early whorls all ribs with distinct inner and outer ventrolateral and siphonal tubercles; inner ventrolateral and siphonal tubercles weaken with growth, often becoming obsolete; suture like *Gentoniceras* but with larger deeper L and

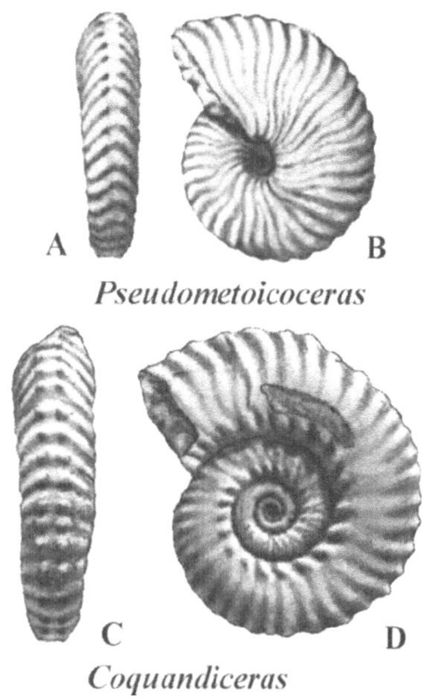

Fig. 335. Mantelliceratinae.

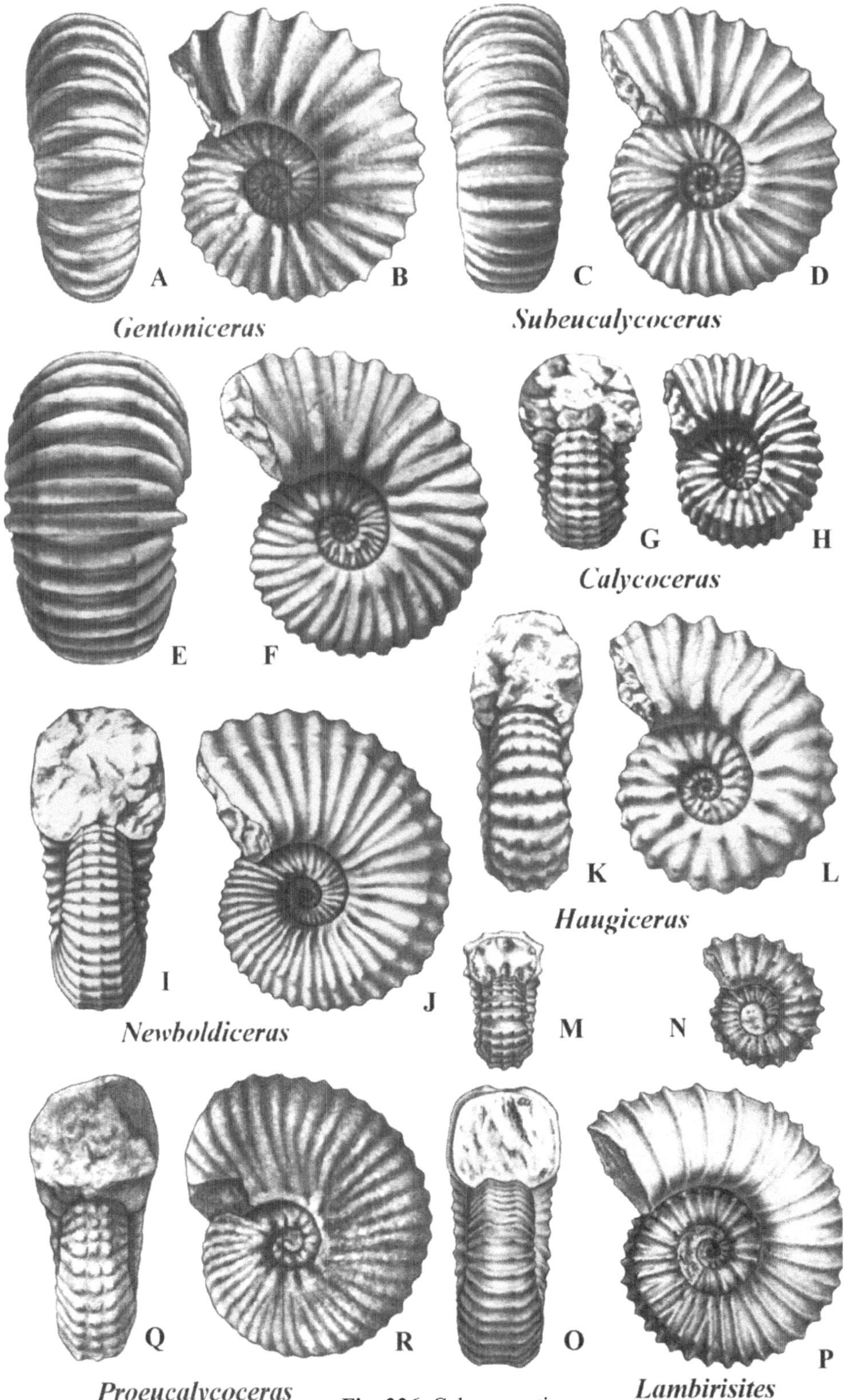

Fig. 336. Calycoceratinae.

U_2 above umbilical shoulder. *Uppermost L. - Lower U. Cenom.* **Fig. 336I-J** - *N. asiaticum* (Jimbo).

Gentoniceras THOMEL, 1972 [*Ammonites gentoni* Brongniart, 1822; OD]. Small to medium sized; inner whorls like *Newboldiceras*; outer ventrolateral and siphonal tubercles lost early, leaving strong ribs with umbilical and (inner) ventrolateral tubercles crossing venter uninterrupted in middle growth; in maturity lower ventrolaterals reduced to angulations, but umbilicals may or may not persist; suture with deep rectangular E, broad quadrate shallowly bifid E/L, narrow bifid L shorter than E, and small U_2 positioned on umbilical shoulder. *M. - Lower U. Cenom.* **Fig. 336A-B** - *G gentoni* (Brongniart).

Calycoceras HYATT, 1900 [*Ammonites navicularis* Mantell, 1822; ICZN Opinion 577]. Evolute, inflated; inner whorls like *Gentoniceras* but with broadly rounded to subtabulate venter and depressed reniform to subquadratic whorl section; strong rounded rursiradiate main ribs, separated by 1-2 intercalatories, arise singly or occasionally in pairs from well developed umbilical bullae (sometimes absent), with strong inner ventrolateral, small rounded to bullate outer ventrolateral and siphonal tubercles; in middle growth siphonal tubercles lost, ventrolaterals weaken and may also be lost with position marked by slight angulation in ribs; suture like *Gentoniceras* but with broader L. *Lower U. Cenom.* **Fig. 336E-H** - *C. naviculare* (Mantell).

= *Metacalycoceras* Spath, 1926 (obj.).

Subeucalycoceras THOMEL, 1972 [*Ammonites sarthacensis* Bayle, 1878; OD]. Like *Gentoniceras* but with flatter venter, subquadratic whorl section and wider adult umbilicus. **Fig. 336C-D** - *S. sarthacense* (Bayle).

= *Subeucalycoceras* Thomel, 1969 (*nom. nud.*).

Pseudacanthoceras THOMEL, 1969 [*Acanthoceras tapara* Wright, 1963; OD]. Like *Newboldiceras* but more compressed and evolute, with flexuous primary ribs separated by intercalatories, weak umbilical bullae and clavate ventrolateral and siphonal tubercles; in later growth siphonal clavi lost and ventrolateral clavi stand above shallowly sulcate venter; a homoeomorph of *Acanthoceras*. *M. Cenom.* **Fig. 340B-C** - *P. tapara* (Wright).

Proeucalycoceras THOMEL, 1972 [*Calycoceras (Eucalycoceras) besairiei* Collignon, 1937; OD]. Inner whorls like *Newboldiceras* but somewhat more compressed and involute, with square to rectangular whorl section; generally somewhat flat topped ribs arise from feeble to moderate umbilical tubercles, slightly twisted or not; inner ventrolateral tubercles weak or absent, outer ventrolaterals distinct and clavate; siphonal tubercles weak to moderate; just before body chamber, ornament changes suddenly to distant strong well rounded ribs with little or no angulation at position of tubercles and rounded quadratic whorl section. *Upper M. - Low U. Cenom.* **Fig. 336Q-R** - *P. besairiei* (Collignon); **Fig. 336** - *P. choffati* (Kossmat).

Mourreiceras THOMEL, 1972 [*Newboldiceras (M.) mourrei*; OD]. Large; like *Newboldiceras* but more evolute, with depressed whorl section; single rectiradiate ribs in early to middle growth, becoming flexuous and slightly rursiradiate later, narrower than the interspaces, with prominent umbilicolateral bullae and enlarged marginal (inner ventrolateral) bullae. *M. Cenom.* **Fig. 340A** - *M. mourrei* (Thomel).

In several characters a link to *Guerangericeras*.

Hourcqiceras COLLIGNON, 1939 [*Calycoceras (H.) hourcqi*; OD]. Moderately large, with very thick test; inner whorls like *Subeucalycoceras* with rigid prorsiradiate ribs arising singly and in pairs from umbilical bullae, but more robust and more evolute (U=40%); outer whorls with sparse, distant, rigid, simple ribs that flex forwards at inner ventrolateral tubercles, narrower than interspaces, with an occasional intercalatory and persistent constrictions;

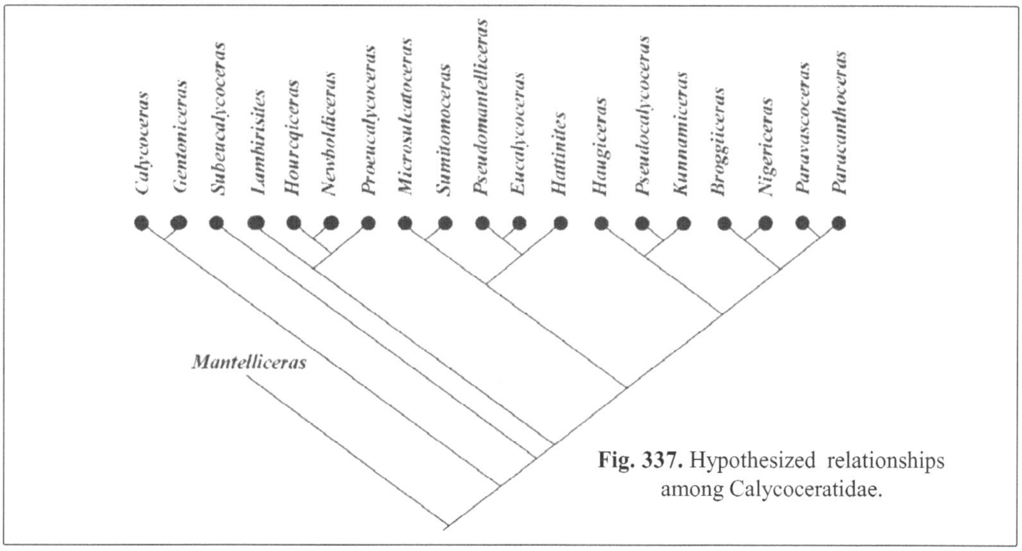

Fig. 337. Hypothesized relationships among Calycoceratidae.

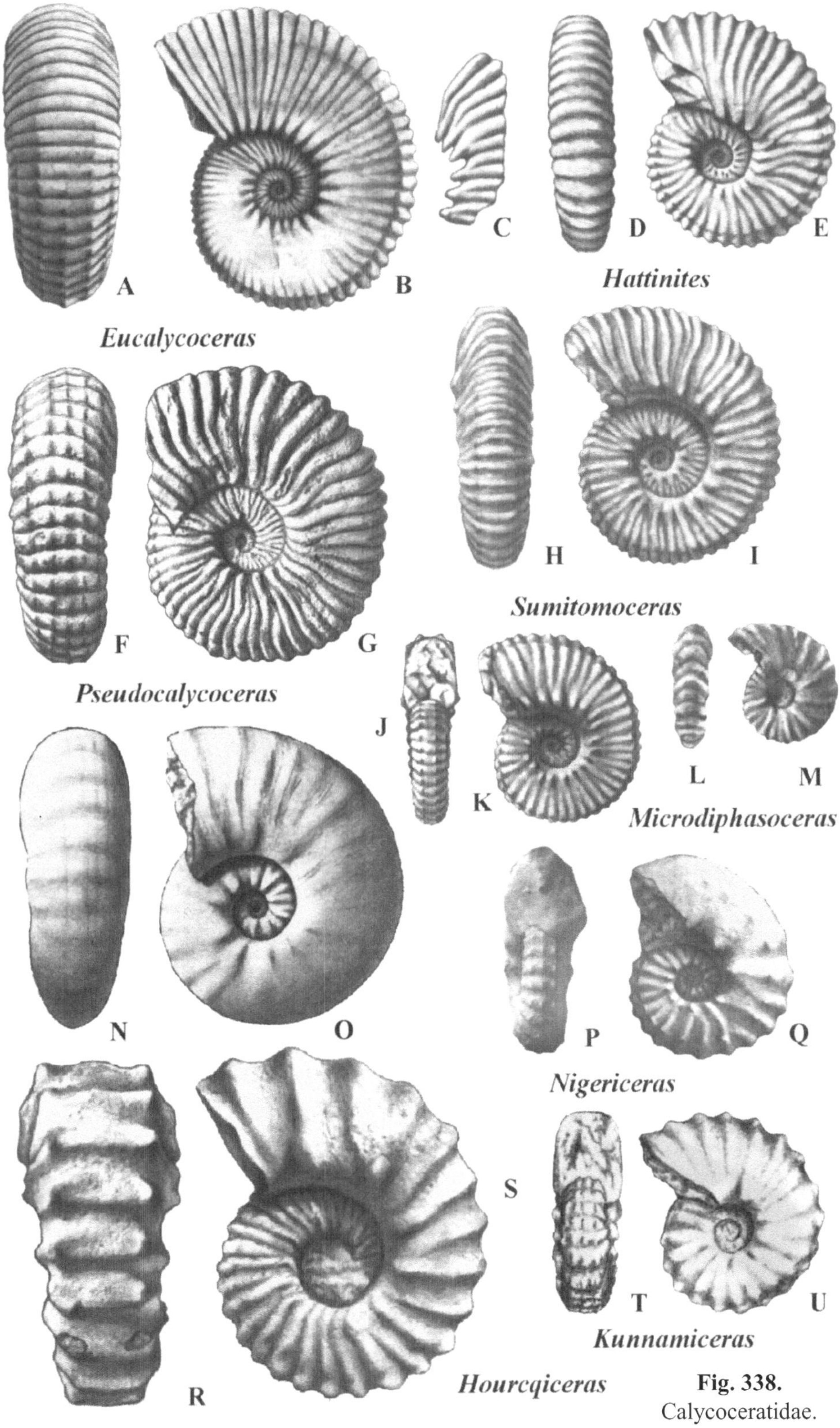

Fig. 338.
Calycoceratidae.

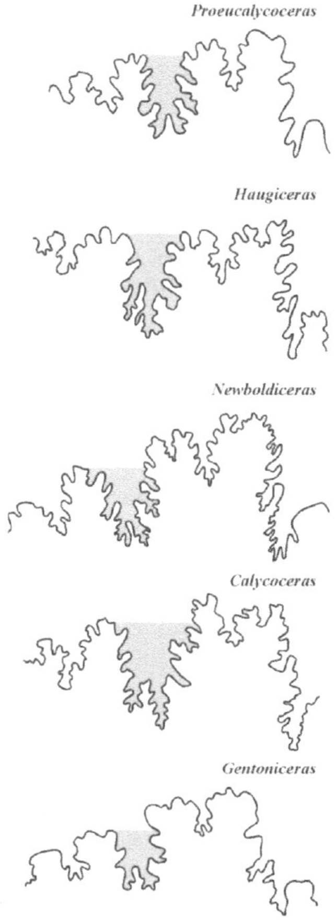

Fig. 339. Calycoceratine sutures.

tuberculation is feeble in middle to later growth when siphonal node tends to evanesce. *M. - Lower U. Cenom.* **Fig. 342** - *H. hourcqi* (Collignon).

Hourcqiceras canitaurinum (Haas) (**Fig. 337R-S**) has a subtabulate venter to the adult body chamber with short ventrolateral horns joined by feebly looped ribs.

***Lambirisites* gen. nov.** [*Ammonites coleroonensis* Stoliczka, 1866, p.71; OD herein]. Moderately large, very evolute; superficially similar to *Hourcqiceras* but inner whorls with single, well spaced rectiradiate ribs with umbilical, lateral, inner and outer ventrolateral and siphonal tubercles; in maturity umbilical, lateral and siphonal tubercles obsolete and body chamber with sulcate venter and single distant ribs some separated by a short intercalatory across venter; constrictions lacking. *M. Cenom.* **Fig. 336O-P** - *L. coleroonensis* (Stoliczka).

Subfamily Eucalycoceratinae nov.

Diagnosis. Generally small to medium sized, evolute and compressed, with persistent tuberculation; the flat flanks, ventrolateral facets and dense ribbing are features of juvenile *Newboldiceras* and *Proeucalycoceras*. A monophyletic taxon with an origin in Calycoceratinae, close to *Gentoniceras* (Wright & Kennedy, 1990). *Upper M. - U. Cenom.*

Pseudomantelliceras THOMEL, 1960 [*Acanthoceras (Mantelliceras) pervinquierei* Collignon, 1931; OD]. Based on a nucleus; evolute, compressed, with flat parallel flanks, distinct ventrolateral facets, subtabulate to slightly sunken venter and subrectangular whorl section; strong prorsiradiate ribs with pointed inner and outer ventrolateral tubercles; siphonal tubercles poorly developed in early stages but strengthening with growth, while inner ventrolaterals decline; suture with deeply bifid E/L and narrow L deeper than E; a homoeomorph of *Submantelliceras*. *U. Cenom.* **Fig. 330J-K** - *P. pervinquierei* (Collignon).

The type species may be a junior subjective synonym of *P. rowei* (Spath).

Eucalycoceras SPATH, 1923 [*Ammonites pentagonum* Jukes-Browne, 1896; OD]. Medium sized; inner whorls

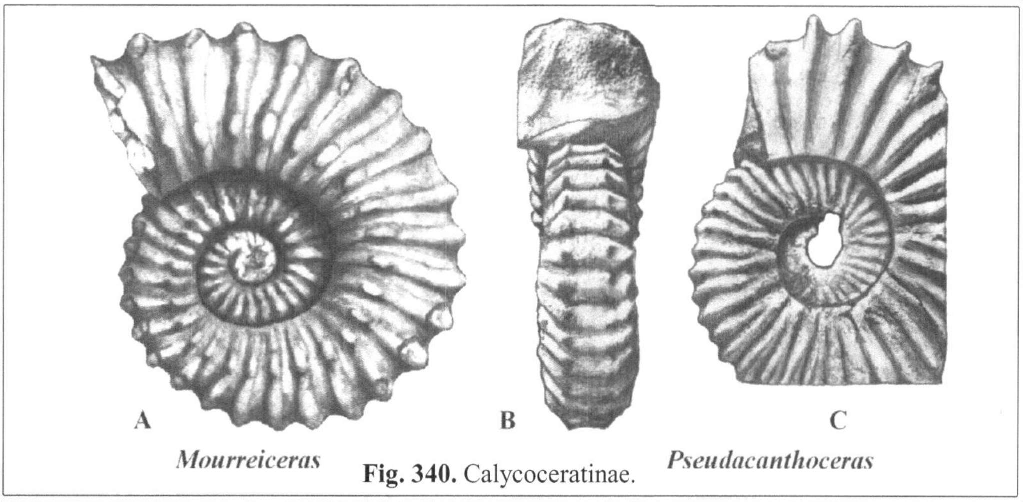

Fig. 340. Calycoceratinae.

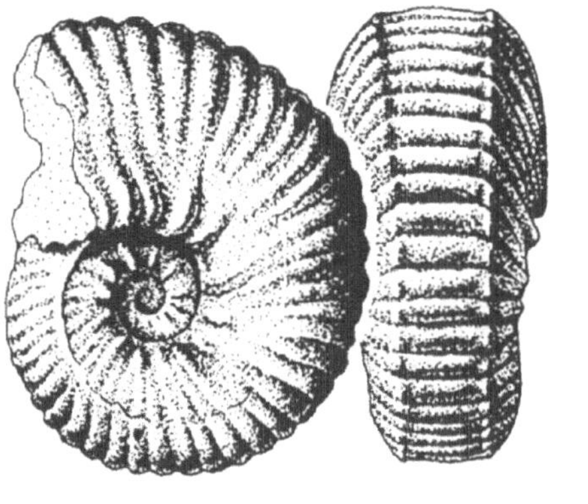

Fig. 341. *Proeucalycoceras.*

like *Pseudomantelliceras* but later more involute, with compressed high whorls, umbilical bullae protruding into umbilicus, ribs weak or effaced from midflank but strengthening over venter, bullate siphonal tubercles and crowded, flat topped, asymmetrical (steepest posteriorly) ribs to adult body chamber; suture relatively complex, with finger-like folioles and very narrow L/U_2. *Upper M. - U. Cenom.* **Fig. 338A-B** - *E. pentagonum* (Jukes-Browne).

Sumitomoceras MATSUMOTO, 1969 [*S. faustum*; OD]. Relatively small; like inner whorls of *Eucalycoceras* but with very early loss (D<15 mm) of siphonal tubercles; ventrolateral tubercles obsolete in later growth, leaving ribs to cross venter uninterrupted, and adult body chamber with constrictions; suture with rather narrow rectangular E/L, L almost as deep as E, and narrow bifid L/U_2. *Upper U. Cenom.* **Fig. 338H-I** - *S. faustum* Matsumoto.

Microsulcatoceras KENNEDY & COBBAN, 1990 [*M. puzosiiforme*; OD]. A progenetic dwarf, perhaps of *Summitomoceras*; strong ribs with very reduced tuberculation form chevrons over venter; primary ribs separated by 1-2 intercalatories. *Upper U. Cenom.* **Fig. 344H-I** - *M. puzosiiforme* Kennedy & Cobban.

Fig. 342. *Hourcqiceras.*

***Hattinites* gen. nov.** [*Eucalycoceras templetonense* Cobban, 1988, p.10; OD herein]. Like *Eucalycoceras* but more compressed, with slightly wider umbilicus (U=30-33%), narrow subtabulate venter to phragmocone, feebly ornamented flanks to inner whorls, prorsiradiate ribs that generally alternate long and short and recurve on outer flank, clavate siphonal tubercles, subspinose umbilical tubercles that protrude into the umbilicus, and ribs that broaden and flatten on adult body chamber but do not form belt-like ribs. *Upper M. Cenom.* **Fig. 338C-E** - *H. templetonensis* (Cobban).

Subfamily Nigericeratinae nov.

Diagnosis. Primitively medium sized with 5 rows of tubercles closely approximated on the venter; later with loss of ornament and smooth outer whorls. A monophyletic group with an origin in Calycoceratinae, close to *Proeucalycoceras*. *U. Cenom. - L. Turon.*

Pseudocalycoceras THOMEL, 1969 [*Ammonites harpax* Stoliczka, 1864; OD]. Inner whorls like *Hattinites*; outer whorls with coarser ribs that recurve on outer flank; ribs strong and elevated across venter with 5 rows of equidistant clavi, becoming finer and denser on adult body chamber; finally evanescing; suture with deep E, broad asymmetrically bifid E/L, narrow bifid L not nearly as deep as E, and narrow non-bifid L/U_2. *U. Cenom.* **Fig. 338F-G** - *P. harpax* (Stoliczka).

= *Neocalycoceras* Thomel, 1969 (*nom. nud.*).

Kunnamiceras GALE & KENNEDY, 2019 [*Ammonites tropicus* Stoliczka, 1864; OD]. Very small; like *Pseudocalycoceras* but tuberculation weakening in maturity leaving only ventrolateral angulations and coarse bold ribs crossing arched venter. *Lower U. Cenom.* **Fig. 338T-U** - *K. tropicum* (Stoliczka).

Haugiceras THOMEL, 1969 [*Acanthoceras haugi* Pervinquière, 1907; OD]. Small, evolute, compressed, with wide shallow umbilicus, subquadratic whorls and gently convex venter; coarse strong flexuous ribs arise singly (with associated intercalatory) or in pairs from prominent umbilical tubercles; all ribs with rounded inner and outer ventrolateral and siphonal tubercles; ribbing remains coarse and distant onto adult body chamber; suture with broad E/L and narrow trifid L not as deep as E. *U. Cenom.* **Fig. 336K-L** - *H. haugi* (Pervinquière).

Although questionably included in the synonymy of *Proeucalycoceras* (Wright & Kennedy, 1992), the similarities are confined mainly to primitive characters. As noted by Thomel (1969), there is a noteworthy resemblance to *Pseudocalycoceras angolaense* (Spath).

Nigericeras SCHNEEGANS, 1943 [*N. gignouxi*; OD]. Medium sized; inner whorls like *Pseudocalycoceras angolaense*

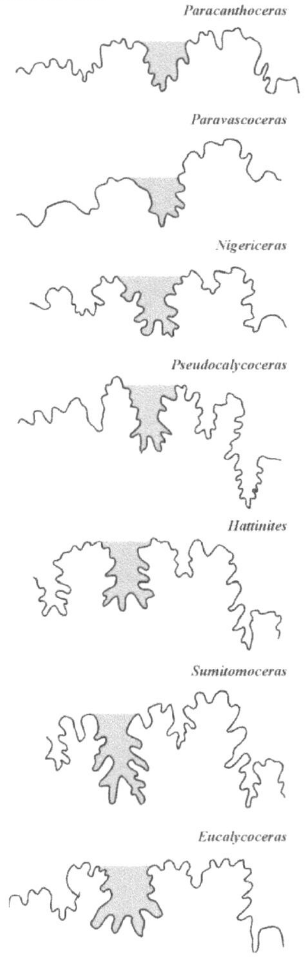

Fig. 343. Eucalycoceratine sutures.

(Spath) but ribs and tubercles weaken at various stages; where ribs persist, outer whorls with broad weak lateral bulges and faint spiral striation to venter; where ribs lost early, outer whorls smooth and indistinguishable from *Paravascoceras*; suture with narrow symmetrically bifid L as deep as E. *Upper U. Cenom.* **Fig. 338N-Q** - *N. gignouxi* Schneegans.

Paravascoceras FURON, 1935 [*Vascoceras cauvini* Chudeau, 1907; OD]. Like *Nigericeras* but more involute, without ribs or tubercles to inner whorls; adult body chamber moderately inflated, shows scaphitoid uncoiling, with numerous low folds across venter, broader than interspaces; suture simple, with parallel sided E, very broad shallowly bifid E/L, narrow asymmetrical L and broad undivided L/U_2. *Upper U. Cenom.* **Fig. 344C-E** - *P. cauvini* (Chudeau).

Ezilloella REYMENT, 1954 [*E. ezilloensis*; OD]. Poorly known; like *Vascoceras* but with fastigiate venter to inner whorls that later becomes broadly rounded with subquadratic whorl section; flanks with well spaced prominent bulges; inner whorls without ventral ornament but faint siphonal and clavate ventrolateral tubercles appear on adult whorls; suture with deep narrow E, broad weakly bifid E/L with inner side sloping towards moderately wide shallow trifid L as deep as E, and short narrow subrectangular L/U_2. *L. Turon.* **Fig. 351Q** - *E. ezilloensis* Reyment.

Paracanthoceras FURON, 1935 [*Vascoceras (P.) chevalieri*; OD]. Moderately large, involute with steep umbilical wall, convex convergent flanks, rounded venter and compressed oval whorl section; outer whorl with coarse, blunt, rounded ribs across venter, narrower than interspaces. *Upper U. Cenom.* **Fig. 344F-G** - *P. chevalieri* (Furon).

Subsequent morphometric analysis of *ex situ* desert-floor float (Schöbel, 1975) suggested that *Paracanthoceras* is part of the intraspecific variation of *P. cauvini* (Chudeau).

Broggiiceras BENAVIDES-CACÈRES, 1956 [*B. ohlssoni*; OD]. Like *Paravascoceras* but more inflated and more involute, with subvertical umbilical wall and few strong well spaced folds across venter to adult body chamber that shows scaphitoid uncoiling. *?L. Turon.* **Fig. 344A-B** - *P. ohlssoni* (Benavides-Cacères).

Family ACANTHOCERATIDAE de Grossouvre, 1894

(*nom. correct.* Hyatt, 1900 *pro* Acanthoceratidés)

Discussion. Monophyletic derivatives of Calycoceratidae, mostly with squarer whorls, more persistent tuberculation with the outer ventrolaterals remaining strong to maturity, and stronger more distant ribs. Comprises the subfamilies Acanthoceratinae and Protacanthoceratinae. *M. - U. Cenom.*

Subfamily Acanthoceratinae de Grossouvre, 1894

(*nom. transl.* Wright & Wright, 1951 *ex* Acanthoceratidae)

Discussion. A monophyletic group with an origin in Calycoceratinae, close to *Newboldiceras*; although some workers have looked to *Acompsoceras* for the origin of the group, the latter has a different florid suture. *M. Cenom.*

Acanthoceras NEUMAYR, 1875 [*Ammonites rhotomagensis* Brongniart, 1822; SD de Grossouvre, 1894]. Moderately large; very evolute, compressed; inflated inner whorls with flat flanks, broadly rounded venter, and depressed subquadratic to rectangular polygonal whorl section; coarse weak straight prorsiradiate ribs usually arise from umbilical bullae singly or in pairs, or may alternate long and short, and cross venter transversely; all ribs with conical inner ventrolateral nodes and outer ventrolateral and siphonal clavi; with growth whorl section tends to

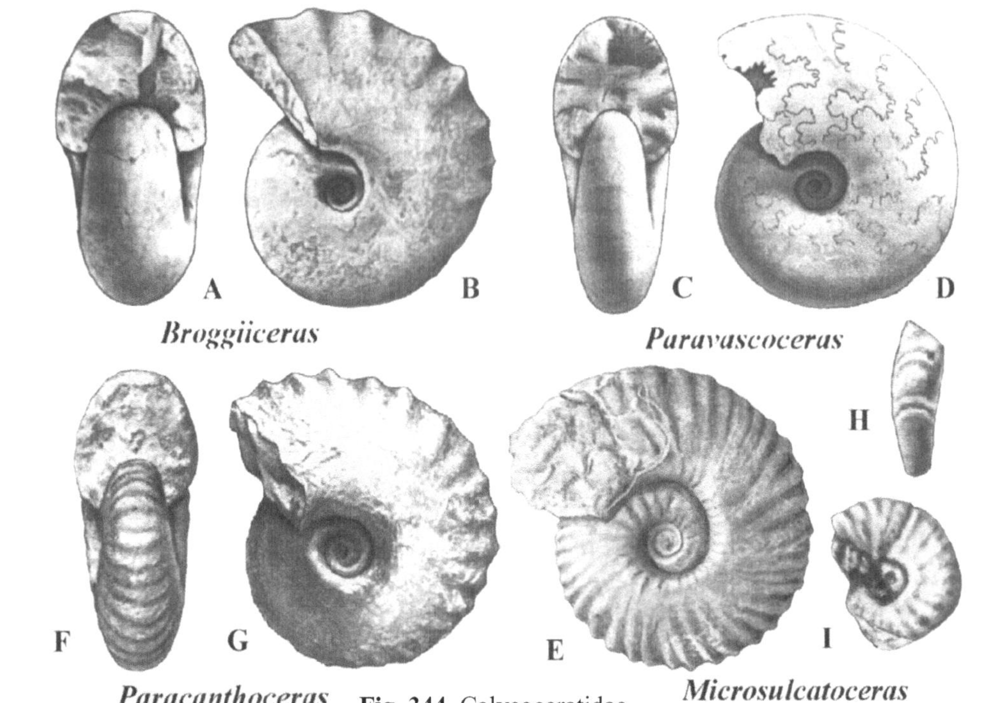

Fig. 344. Calycoceratidae.

become slightly compressed, rib furcation ceases and most ribs are long; in maturity siphonal clavi become obsolete and the narrow venter shallowly concave; suture with broad asymmetrically bifid E/L and narrow asymmetrically bifid L much shorter than E. *M. Cenom.* **Fig. 347A-B** - *A. rhotomagense* (Brongniart).

= *Metacanthoplites* Hyatt, 1900 (obj.).

Alternacanthoceras MARCINOWSKI, 1979 [*Protacanthoceras jukesbrownei* Spath, 1926; OD]. Like *Acanthoceras* but with more numerous secondary ribs to inner whorls, early loss of siphonal and inner ventrolateral tubercles, leading to distinctive depressed trapezoidal whorl section; mature whorls with broad almost "smooth" venter, blunt ventrolateral horns, coarse distant ribs that tend to alternate long and short, and strong umbilical bullae; suture more finely denticulate than *Acanthoceras* with L almost as deep as E. *Upper M. Cenom.* **Fig. 347I-J** - *A. jukesbrownei* (Spath).

***Phindaites* gen. nov.** [*Acanthoceras cornigerum* Crick, 1907, p.207; OD herein]. Medium sized, very evolute, with subquadratic whorl section; in early growth sharp umbilical bullae give rise to single gently flexuous prorsiradiate ribs that alternate with shorter intercalatories; all ribs with inner and outer ventrolateral angulations and feeble siphonal clavi; in later growth all ribs single, distant, umbilical bullae migrating to midflank, with development of large, outward and upward directed horns linked across venter by strong ribs. *M. Cenom.* **Fig. 348** - *P. cornigerus* (Crick).

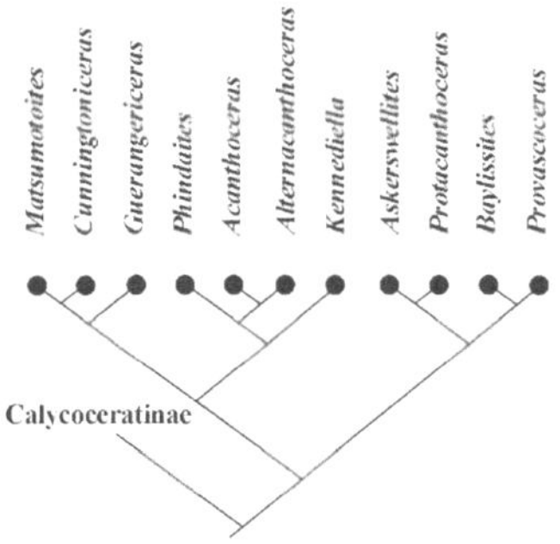

Fig. 345. Hypothesized relationships among Acanthoceratidae.

Guerangericeras THOMEL, 1972 [*Ammonites confusus* Guéranger, 1867; OD]. Like *Acanthoceras* but more evolute, with slightly rursiradiate simple ribs and robust subspinose lower ventrolateral tubercles impressed into umbilical wall of succeeding whorl. *Lower M. Cenom.* **Fig. 347M** - *G. confusum* (Guéranger).

Guerangericeras has been considered a synonym of both *Acanthoceras* and *Cunningtoniceras*; the name is available for members of the *Cunningtoniceras inerme* (Pervinquière) group that, itself, seems to be intermediate between *Acanthoceras* and *Cunningtoniceras*. The types are lost.

Cunningtoniceras COLLIGNON, 1937 [*Ammonites cunningtoni* Sharpe, 1855; OD]. Like *Acanthoceras* but generally

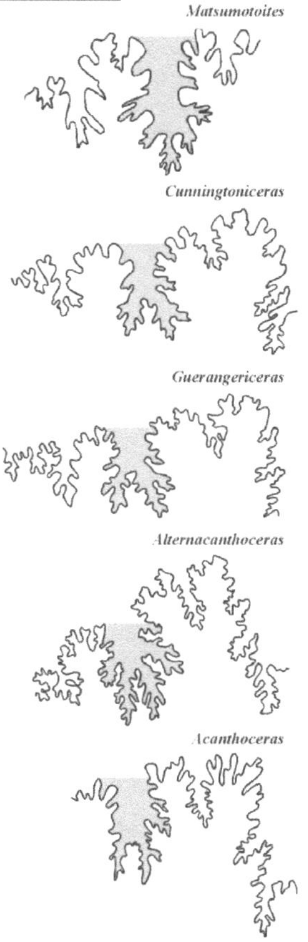

Fig. 346. Acanthoceratid sutures.

more depressed, with sparse weak broad low rectiradiate primary ribs; with growth prominent umbilical tubercles migrate to low lateral position and ventrolaterals coalesce to form exaggerated laterally directed horns; primary ribs connected across venter by finer looped secondaries that carry rounded to clavate siphonal tubercles and are separated by a tuberculate intercalatory so that siphonal tubercles greatly outnumber ventrolaterals; in maturity all ribs and tubercles, save for umbilicals and fused ventrolaterals, may become obsolete; suture like *Acanthoceras*, but with relatively broader E/L and L almost as deep as E; a homoeomorph of *Euomphaloceras*. *M. - U. Cenom.* **Fig. 347N-O** - *C. cunningtoni meridionale* (Kossmat).

***Matsumotoites* gen. nov.** [*Euomphaloceras asura* Matsumoto & Muramoto, 1969, p.277; OD herein]. Like *Cunningtoniceras* but with very depressed coronate whorl section, inclined umbilical wall, sparser more distant primary ribs separated by 3-5 secondaries, umbilicolateral bullae shifting up flank to coalesce with exaggerated ventrolateral horns, early loss of siphonal tubercles, and trifid L to suture. *U. Cenom.* **Fig. 350** - *M. asura* (Matsumoto & Muramoto).

Kennediella COOPER, 1979 [*Pseudotissotia inopinata* Kennedy & Bayliss, 1977; OD]. Early whorls unknown; outer whorl evolute, with rounded subquadratic whorl section, broad tricarinate venter and broad low folds on flanks; a homoeomorph of *Paracompsoceras* and *Pseudotissotia*. *U. Cenom.* **Fig. 347K-L** - *K. inopinata* (Kennedy & Bayliss).

Subfamily Protacanthoceratinae nov.

Diagnosis. Micromorph derivatives of Acanthoceratinae, close to *Acanthoceras*; primitively like the latter but later more inflated, with constrictions and decline of tuberculation. *Upper M. Cenom. - U. Cenom.*

Protacanthoceras SPATH, 1923 [*Ammonites bunburianus* Sharpe, 1853; OD]. Like juvenile *Acanthoceras* but more compressed, with fastigiate to subtabulate venter; tuberculation weakens on adult body chamber of microconchs, when ribs flatten and broaden across venter; suture like *Acanthoceras* but simpler. *Upper M. - U. Cenom.* **Fig. 347C-D** - *P. bunburianum* (Sharpe).

***Askerswellites* gen. nov.** [*Protacanthoceras tegulicum* Wright & Kennedy, 1980; OD herein]. Like *Protacanthoceras* but with reniform whorl section and distant strong primary ribs with only umbilical and inner ventrolateral spines separated by 2-3 weak nontuberculate secondaries; ribs asymmetrical (steepest in front), with deep constriction-like interspaces; suture with squat E/L and non-bifid L/U_2. *Lower U. Cenom.* **Fig. 347G-H** - *A. tegulicus* (Wright & Kennedy).

***Baylissites* gen. nov.** [*Protacanthoceras proteus* Wright & Kennedy, 1980; OD herein]. Like *Protacanthoceras* but smaller, more involute and inflated, with depressed whorls, broadly rounded venter, constricted early whorls, and delicate ventral tuberculation that declines in later growth and is lost on adult body chamber; suture with inner margin of E/L sloping towards broad L. *Lower U. Cenom.* **Fig. 347E-F** - *B. proteus* (Wright & Kennedy).

Provascoceras COOPER, 1979 [*Ammonites diartianus* d'Orbigny, 1850; OD]. Micromorph; globose, moderately evolute, with deep umbilicus, rounded umbilical wall merging with convex flanks, broadly rounded venter and depressed reniform whorl section; prominent slightly prorsiradiate umbilical bullae, becoming pinched on adult body chamber, give rise to bundles of 2-5 faint low ribs, separated by intercalatories, that are weak and faint on outer flank and venter; ribs distinctly prorsiradiate and concave across flanks and recurve to cross venter with a broad convexity where superimposed by weak spiral striations; there are 5 very shallow constrictions on outer whorl and adult aperture is constricted; suture simple, with broad E, broad little incised somewhat rounded asymmetrically bifid E/L, broad asymmetrically trifid L almost as deep as E, and broad low rounded non-bifid L/U_2. *U. Cenom.* **Fig. 351A-B** - *P. diartianum* (d'Orbigny).

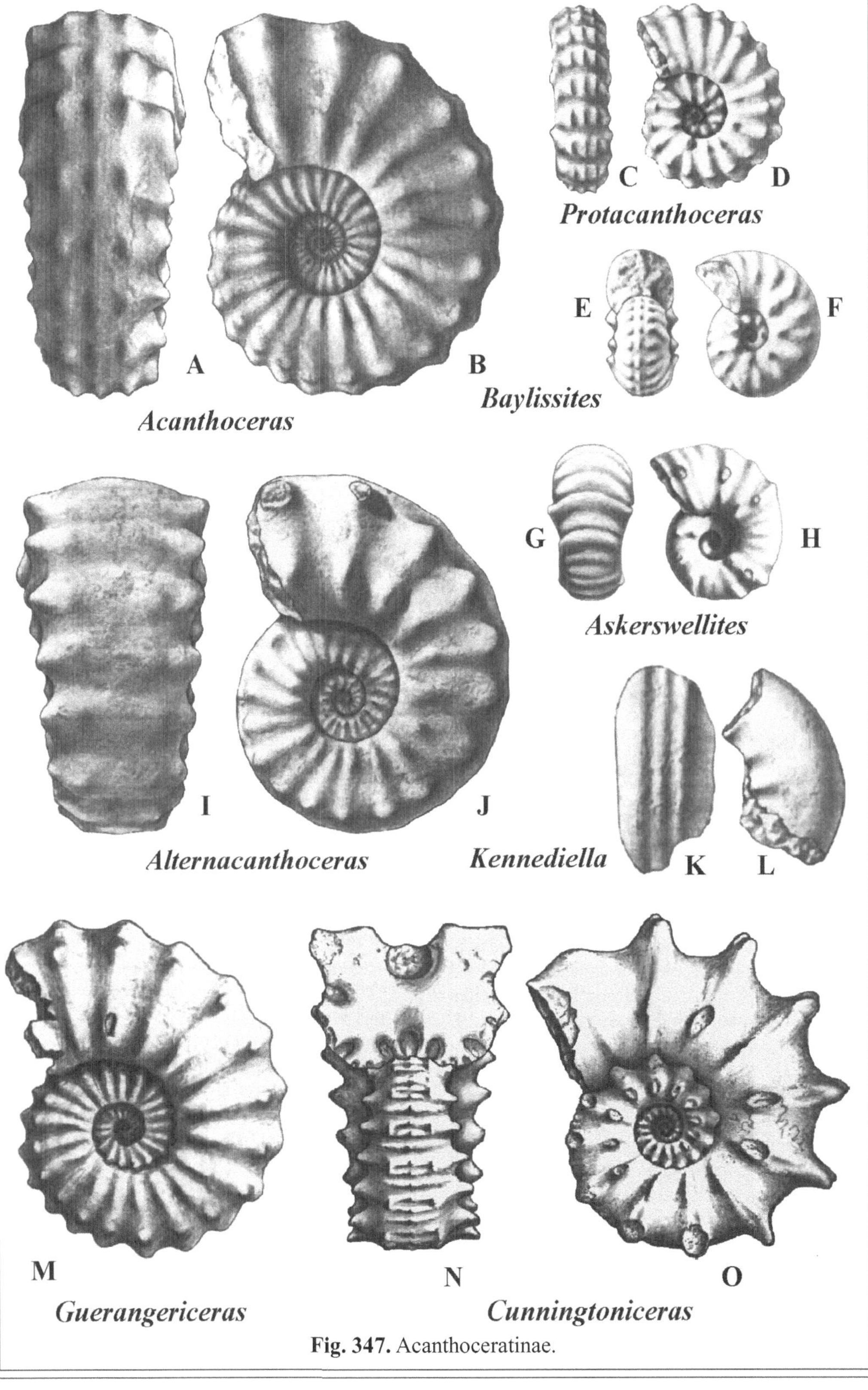

Fig. 347. Acanthoceratinae.

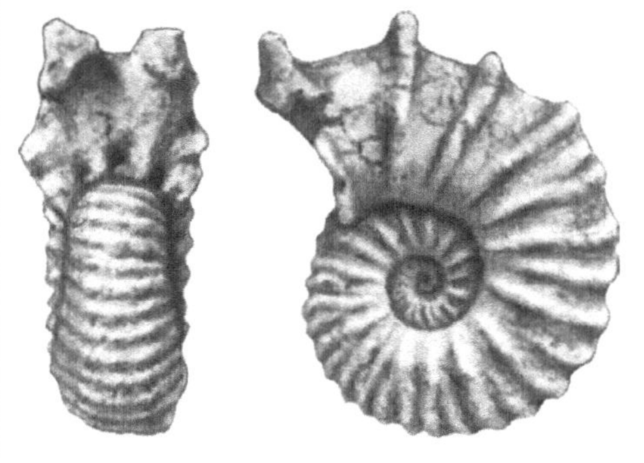

Fig. 348. *Phindaites.*

Family VASCOCERATIDAE Douvillé, 1912

(*nom. correct. et transl.* Spath, 1925 *pro* Vascoceratidés)

Discussion. A problematical group, with many homoeomorphs and, hence, monophyly is uncertain; *Provascoceras* is a derivative of *Baylissites* of Protacanthoceratinae (Cooper, 1979) but its relation to much larger *Vascoceras* is not proven; Zaborski (1998) looked to Eucalycoceratinae for the origin of the group, but this also is not secure. Comprises Vascoceratinae and Neoptychitinae. *Upper U. Cenom. - M. Turon.*

Subfamily Vascoceratinae Douvillé, 1912

(*nom. transl.* Collignon, 1965 *ex* Vascoceratidae)

Discussion. Seemingly wide intraspecific variation among vascoceratines has presented problems of classification. On one hand this has led to splitting

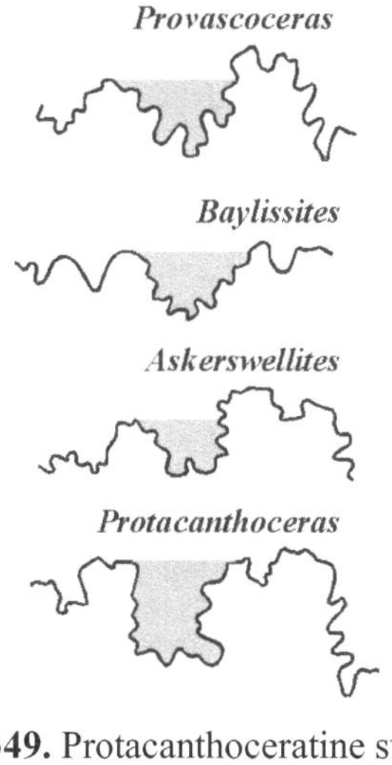

Fig. 349. Protacanthoceratine sutures.

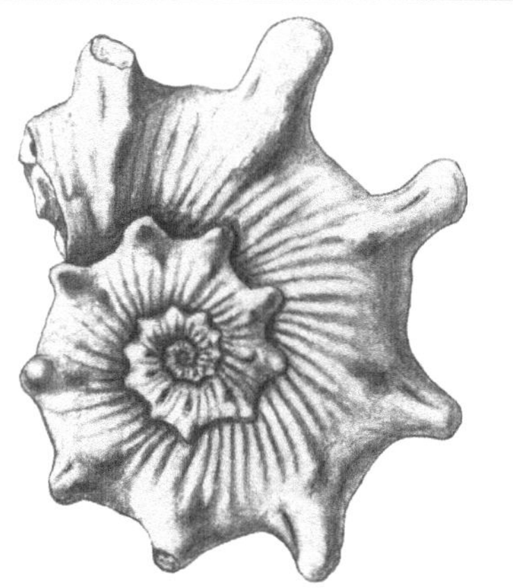

Fig. 350. *Matsumotoites.*

of taxa and, on the other, to wholesale lumping. As currently construed (Wright, 1996), *Vascoceras* includes micromorphs (*Provascoceras*) and macromorphs (*Vascoceras*), unribbed (*Paravascoceras*) and ribbed forms (*Greenhornoceras*), those with prominent umbilical bullae (*Vascoceras*) and those without (*Broggiiceras*), constricted forms (*Nannovascoceras*) and unconstricted forms (*Ezilloella*), inflated forms (*Pachyvascoceras*) and compressed forms (*Paravascoceras*), and almost every imaginable whorl shape. Interestingly, when populations are collected from a limited stratigraphic horizon, not influenced by condensation, e.g. *Greenhornoceras* (Cobban & Scott, 1972), they are not nearly as variable as some (?mixed) samples suggest. Comprises the tribes Vascoceratini and Fagesiini. *U. Cenom. - L. Turon.*

Tribe Vascoceratini Douvillé, 1912

(*nom. transl.* herein *ex* Vascoceratinae)

Type genus: *Vascoceras* Choffat, 1898.

Diagnosis. Inner whorls with or without ribs, umbilical bullae and constrictions; outer whorls generally unornamented. *Upper Cenom. - L. Turon.*

Vascoceras CHOFFAT, 1898 [*V. gamai*; OD]. Large, compressed, evolute; phragmocone with sparse large prominent swollen umbilical tubercles and feeble ribs; ornament weakens and is lost in maturity; smooth adult body chamber has oval to subtrigonal whorl section. *Upper U. Cenom. - L. Turon.* **Fig. 351C-D** - *V. gamai* Choffat.

Nannovascoceras RENZ & ALVAREZ, 1979 [*N. intermedium*; OD]. A progenetic dwarf (D<40 mm); like inner whorls of *Vascoceras* but carrying constrictions to maturity and with strong simple ribs across venter of adult body

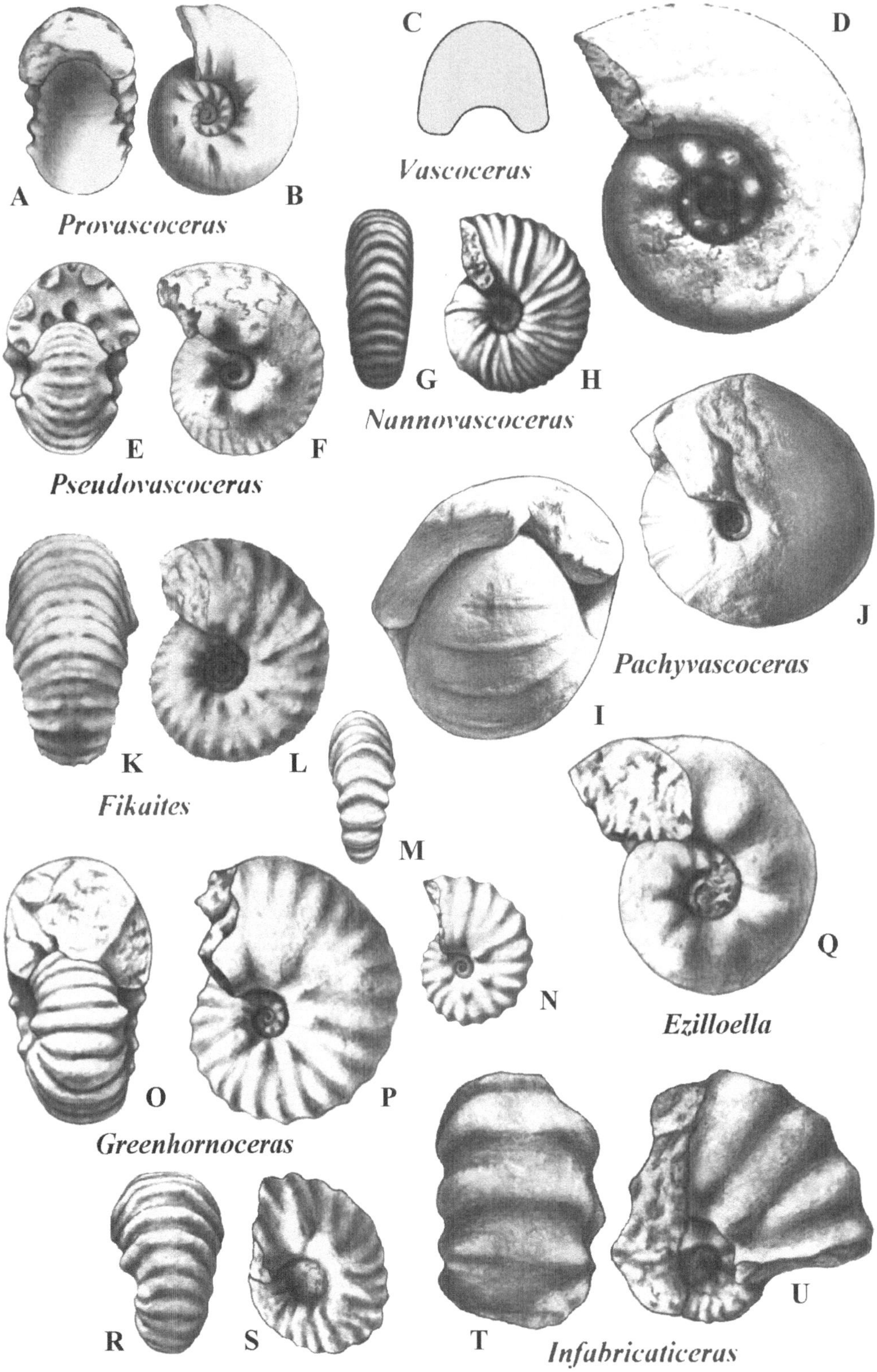

Fig. 351. Vascoceratidae.

chamber. *L. Turon*. **Fig. 351G-H** - *N. intermedium* Renz & Alvarez.

Pachyvascoceras FURON, 1935 [*P. crassum*; OD]. Medium sized, very inflated cadicone in maturity, with deep relatively narrow umbilicus, broad subvertical to overhanging umbilical wall, subangular umbilical shoulder, broadly rounded venter and strongly depressed mature whorls (early whorls tend to be more compressed); innermost whorls with periodic constrictions and weak fold-like ventral ribs; adult body chamber smooth; suture relatively incised with finger-like folioles. *L. Turon*. **Fig. 351I-J** - *P. peroni* (Pervinquière); **Fig. 352** - *P. crassum* (Furon).

Pseudovascoceras ZABORSKI, 1996 [*Vascoceras nigeriense* Woods, 1911; OD]. Variable; moderately involute to fairly evolute, with compressed to depressed whorl section; sparse umbilical tubercles of variable strength weakly connected, if at all, to doubled ventrolateral tubercles joined over venter by fine ribs; in very early growth there are siphonal tubercles; suture simple, with evenly frilled elements; saddles often elongate and rectangular, L fairly broad. *Upper U. Cenom*. **Fig. 351E-F** - *P. nigeriense* (Woods).

Greenhornoceras COBBAN & SCOTT, 1972 (*Vascoceras (G.) birchbyi*; OD]. Inner whorls involute, flat sided, with rounded venter, compressed oval to subquadratic whorl section and periodic collared constrictions; sparse distant weak to prominent umbilical tubercles give rise to strong rounded rectiradiate ribs singly or in pairs, separated by 2-5 intercalatories; ribs may be raised into incipient ventrolateral nodes and cross venter transversely; ornament weakens with growth and becoming obsolete on adult body chamber, when flanks and venter round and whorl section becomes subquadratic; suture fairly simple, with rather narrow E, fairly wide asymmetrically bifid E/L, fairly wide asymmetrically bifid L, and rather narrow weakly bifid L/U_2. *Lower L. Turon*. **Fig. 351M-P** - *G. birchbyi* (Cobban & Scott).

Microdiphasoceras COBBAN, HOOK & KENNEDY, 1989 [*M. novimexicanum*; OD]. Micromorph; involute, compressed, with flattish convergent flanks and fastigiate tritubérculate venter; umbilical bullae give rise to strong sinuous (juv.) to straight (ad.) prorsirasdiate primary ribs separated by 1-2 intercalatories; ribs flex forward on outer flank from incipient to feeble ventrolateral bullae; inner whorls with 6-7 collared constrictions . *U. Cenom*. **Fig. 338L-M** - *M. novimexicanum* Cobban, Hook & Kennedy.

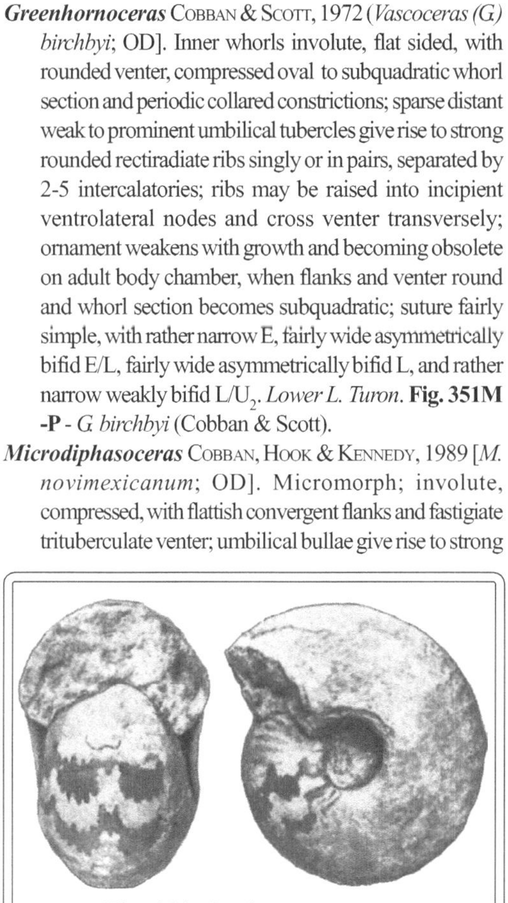

Fig. 352. *Pachyvascoceras*.

Tribe Fagesiini nov.

Type genus: *Fagesia* Pervinquière, 1907.

Diagnosis. Inflated and generally strongly ornamented, with bold ribs arising from persistent umbilical bullae. Derivatives of Vascoceratini, close to *Pachyvascoceras*. *L. Turon*.

Plesiovascoceras SPATH, 1925 [*Ammonites catinus* Mantell, 1822; OD]. Inner whorls moderately evolute with depressed whorl section, with periodic collared constrictions, strongest across venter; prominent nodate umbilical tubercles give rise to 1-2 strong rounded primary ribs, separated by intercalatories, that cross venter transversely or with slight forward arch; in middle growth whorls become much more depressed, broad umbilical wall inclines so that umbilical tubercles shift to midlateral position and whorl section becomes coronate; ribs weaken with growth and evanesce on adult body chamber; suture with deep narrow E, long slightly wider shallowly bifid E/L, narrow bifid L almost as deep as E, and narrow non-bifid L/U_2. *Lower L. Turon*. **Fig. 353A-B** - *P. catinum* (Mantell).

Tolimacococeras KENNEDY, 2017 [*Paramammites colombianus* Etayo-Serna, 1979; OD]. Small, evolute cadicone with reniform whorl section; prominent peg-like umbilical bullae give rise to 2-3 rectiradiate to slightly prorsiradiate ribs, broader than interspaces, that cross venter transversely, or with slight concavity, and are weak or effaced across siphonal line; incipient mid-lateral and prominent ventrolateral bullae; last half whorl with deep constrictions parallel to ribbing. *L. Turon*. **Fig. 355** - *T. colombianum* (Etayo-Serna).

Infabricaticeras COBBAN, HOOK & KENNEDY, 1989 [*I. lunaense*; OD]. Like *Greenhornoceras* but larger, with flattened flanks, subtabulate venter, subquadratic to strongly depressed trapezoidal whorl section, and strong ribbing and tuberculation carried to maturity where robust ribs alternate long and short. *Upper L. Turon*. **Fig. 351R-U** - *I. lunaense* Cobban, Hook & Kennedy.

Parafagesia **gen. nov.** [*Ammonites rudra* Stoliczka, 1865, p.122; OD herein]. Like *Fagesia* but with deep narrow umbilicus, vertical umbilical wall, lack of umbilical bullae, and ribs restricted to outer flank and venter; suture similar to *Plesiovascoceras* but more incised *L. Turon*. **Fig. 353C-D** - *P. rudra* (Stoliczka).

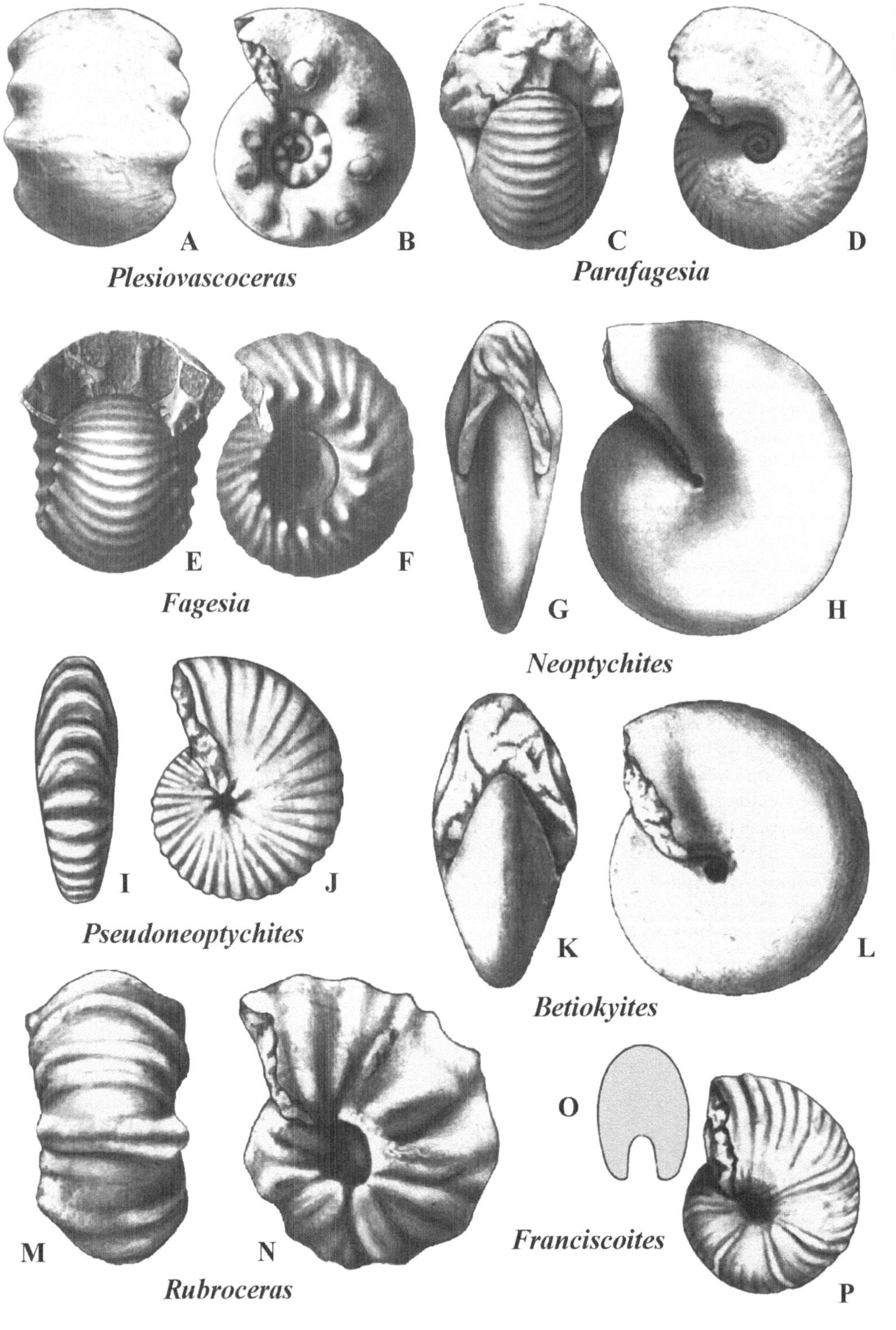

Fig. 353. Vascoceratidae, Pseudotissotiidae.

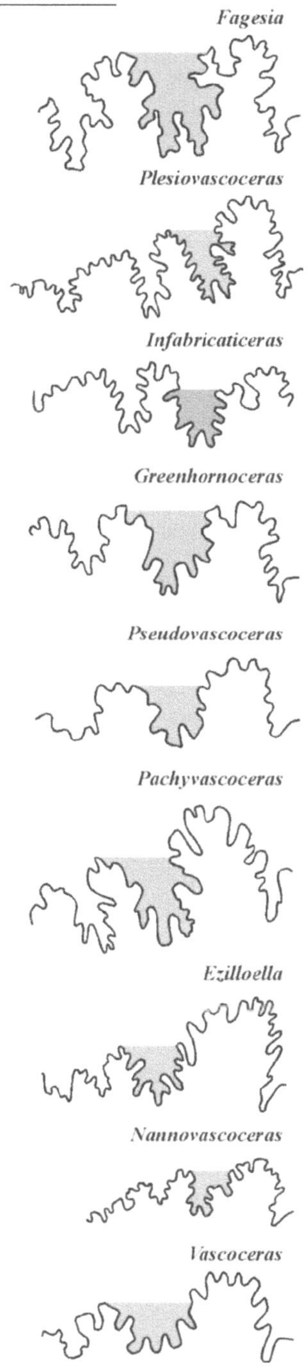

Fig. 354. Vascoceratid sutures.

Fagesia PERVINQUIÈRE, 1907 [*Olcostephanus superstes* Kossmat, 1897; OD]. Like *Plesiovascoceras*, but globose cadicone with deep narrow umbilicus, subvertical umbilical walls, umbilical tubercles that remain on umbilical shoulder and depressed reniform whorl section. *Upper L. Turon.* **Fig. 353E-F** - *F. superstes* (Kossmat).

Subfamily Neoptychitinae Collignon, 1965

Discussion. A monophyletic group, perhaps with an origin close to *Paravascoceras* (Cooper,1979); typically very involute, compressed, with constricted and ribbed early whorls, and contracted adult aperture. *L. - M. Turon.*

Neoptychites KOSSMAT, 1895 [*Ammonites telinga* Stoliczka, 1865, = *A. cephalotus* Courtiller, 1860; SD Solger, 1904]. Inflated, involute; earliest whorls very involute with tiny crater-like umbilicus, flat to gently convex flanks that converge to narrowly rounded venter, compressed or depressed subtrigonal whorl section, and periodic collared constrictions; in middle growth numerous broad low prorsiradiate ribs arise close to umbilical shoulder and are strongest across venter; in later growth all ornament is lost, upper flanks become slightly concave, and whorl section trigonal; adult body chamber may remain compressed and flat sided throughout (?**m**), or may show marked indentation at midflank close to final aperture (?**M**); suture variable, typically with shallow E, rather wide asymmetrically bifid E/L with deep penetrating adventive lobe, narrow asymmetrically bifid L longer than E, and narrow asymmetrically bifid L/U_2. *L. - M. Turon.* **Fig. 353G-H** - *N. cephalotus* (Courtiller).

Franciscoites ETAYO SERNA, 1979 [*Ammonites toroanus* Karsten, 1858; OD]. A progenetic dwarf, like inner whorls of *Neoptychites* with collared constrictions retained to maturity and linguiform across venter; suture relatively simple, with shallow E, relatively broad shallowly bifid E/L and very deep digitate L. *L. Turon.* **Fig. 353O-P** - *N. toroanus* (Karsten).

Betiokytes COLLIGNON, 1965 [*Hemitissotia (B.) besairiei*; OD]. Like *Neoptychites* but with barely incised saddles and a pseudoceratitic suture. *M. Turon.* **Fig. 353K-L** - *N. besairiei* (Collignon).

Pseudneoptychites A. LEANZA, 1967 [*P. andinus*; OD]. A micromorph; like *Neoptychites* but with almost subparallel flanks, a broader subtabulate venter, distinct

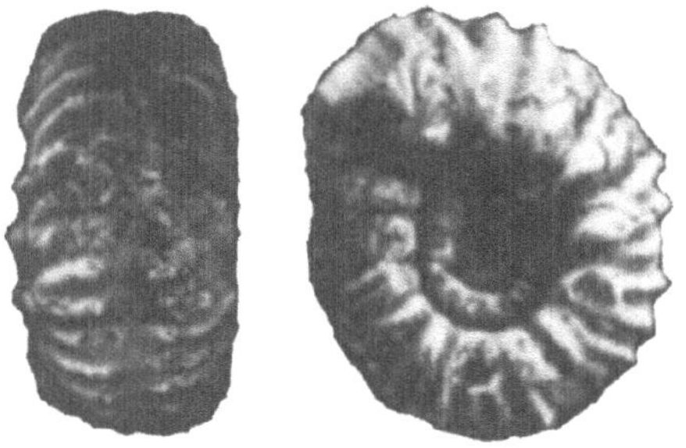

Fig. 355. *Tolimacococeras.*

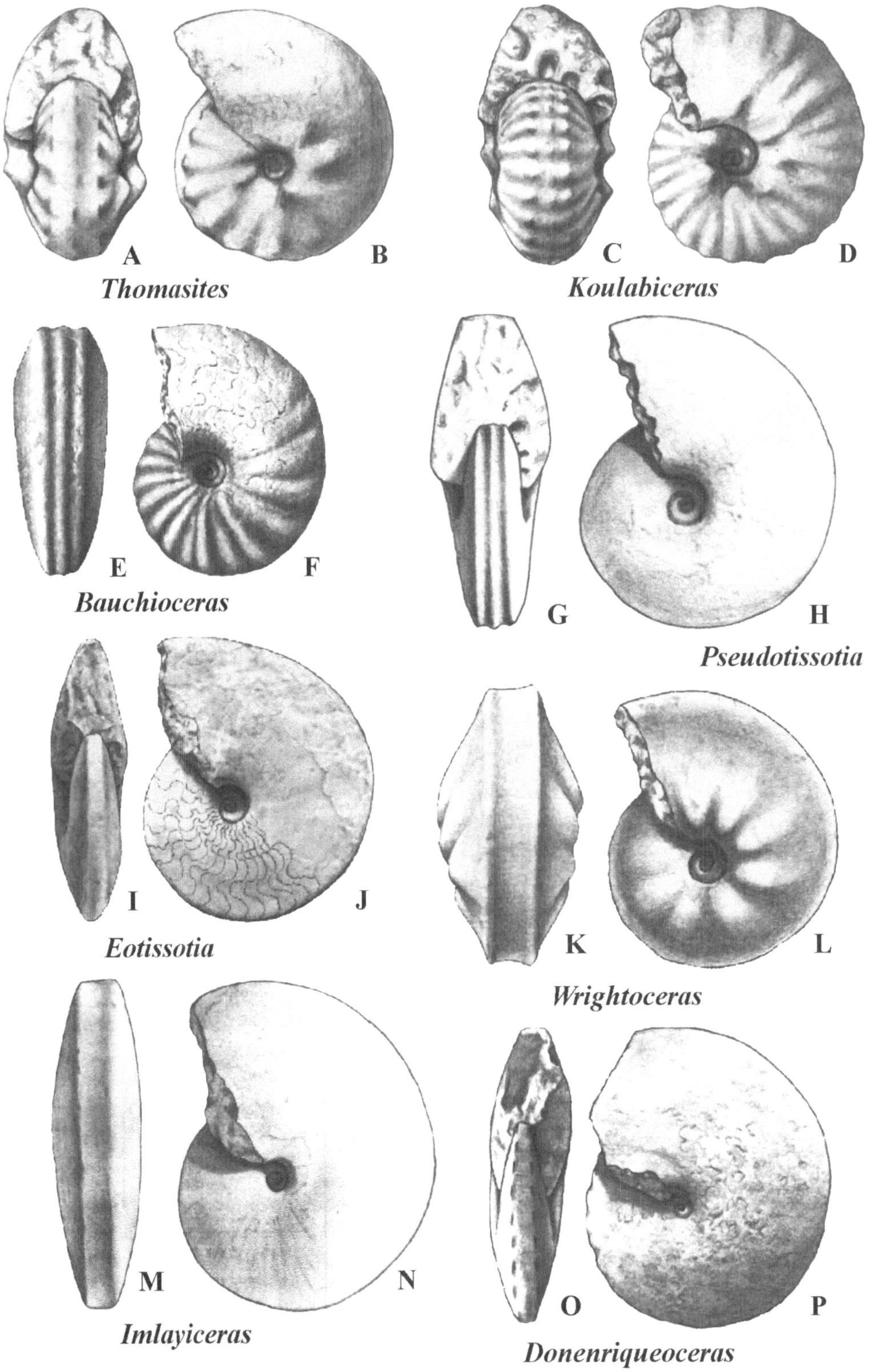

Fig. 356. Pseudotissotiinae.

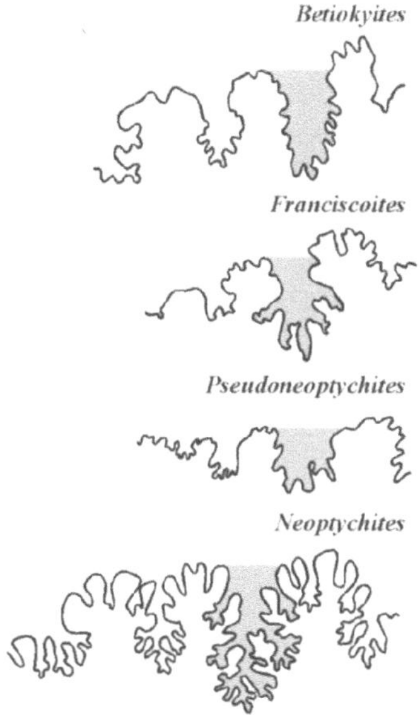

Fig. 357. Neoptychitine sutures.

umbilical bullae to phragmocone, and ribs persisting onto adult body chamber; suture simple, like juvenile *Neoptychites*, but with E as deep as L, broader barely divided quadratic E/L and wider L. *L. Turon.* **Fig. 353I-J** - *P. andinus* A. Leanza.

Superfamily COILOPOCERATOIDEA Hyatt, 1903

(*nom. transl.* herein *ex* Coilopoceratidae)

Discussion. Monophyletic derivatives of Acanthoceratoidea; comprises the families Pseudotissotiidae and Coilopoceratidae. *Upper U. Cenom. - Coni.*

Family PSEUDOTISSOTIIDAE Hyatt, 1903

Discussion. A monophyletic group of uncertain ancestry, either Plesiacanthoceratidae or Acanthoceratidae; comprises the subfamilies Pseudotissotiinae and Hourcqiinae. *Upper U. Cenom. - Coni.*

Subfamily Pseudotissotiinae Hyatt, 1903

(*nom. transl.* Wright, 1952 *ex* Pseudotissotiidae)

Discussion. A monophyletic group of cryptic origin, perhaps Watinoceratinae close to *Paraconlinoceras*; characteristically with trituberculate (primitive) or tricarinate (derived) venter and ornament obsolete in maturity; many homoeomorph Vascoceratidae. *Upper U. Cenom. - M. Turon.*

Gombeoceras REYMENT, 1954 [*Vascoceras gongilense* Woods, 1911; OD]. Involute, compressed to subglobose, with convex to almost flat flanks converging to broadly rounded, subtabulate or subfastigiate venter with depressed reniform to subovate whorl section; 4-5 umbilical tubercles give rise to 2-3 weak broad ribs that terminate in blunt ventrolateral tubercles, but are unconnected to row of siphonal tubercles; ornament weakens with growth and outer whorls are smooth, homoeomorphing *Vascoceras*, but sometimes with low siphonal ridge that may persist to maturity; suture rather simple, shallowly incised, with narrow deep E, broad bifid E/L, wide L and reduced L/U_2. *Upper U. Cenom. - Lower L. Turon.* **Fig. 358M-N** - *G. gongilense* (Woods).

Thomasites PERVINQUIÈRE, 1907 [*Pachydiscus rollandi* Peron, 1889; SD Diener, 1925]. Like *Gombeoceras* but with narrow venter, subtrigonal whorl section, and very different highly incised suture with long narrow saddles and lobes and well developed L/U_2. *Upper L. Turon.* **Fig. 356A-B** - *T. rollandi* (Peron).

Fikaites ZABORSKI, 1993 [*F. varicostatus*; OD]. Like *Gombeoceras* but with stronger more regular ribbing and tuberculation, smaller more numerous umbilical bullae, tendency of ribs to weaken at midflank, frequent short intercalated ribs across venter, siphonal tubercle declining with growth and suture with longer narrower E/L and broader bifid L; a homoeomorph of *Eucalycoceras*. *Upper U. Cenom. - basal Turon.* **Fig. 351K-L** - *F. varicostatus* Zaborski.

Koulabiceras ATABEKIAN, 1966 [*Pseudotissotia koulabica* Kler; 1908; OD]. Like *Gombeoceras* but more strongly ribbed and suture with deeper subrectangular bifid L. *L. Turon.* **Fig. 356C-D** - *T. koulabicum* (Kler).

= *Ferganites* Stankievich & Pojarkova, 1969 (obj.).

Rubroceras COBBAN, HOOK & KENNEDY, 1989 [*R. alatum*; OD]. Inner whorls like *Thomasites*, with rather narrow umbilicus, somewhat flattened flanks, broadly rounded venter, sparse swollen umbilical bullae that give rise to rather coarse rectiradiate ribs which carry weak ventrolateral and siphonal tubercles and cross venter transversely; in middle growth umbilical bullae elongate to form broad robust bar-like ribs that carry ventrolateral tubercles separated across venter by a thin weak intercalatory; adult body chamber moderately evolute (due to scaphitoid uncoiling), with strong distant rectiradiate ribs accentuated on ventrolateral shoulders and crossing venter transversely; suture with deep narrow E, wide subrectangular shallowly bifid E/L, narrow bifid L not quite as deep as E, and moderately narrow to fairly broad non-bifid L/U_2. *Upper U. Cenom.* **Fig. 353M-N** - *R. alatum* Cobban, Hook & Kennedy.

Bauchioceras REYMENT, 1954 [*Hoplitoides nigeriensis* Woods, 1911; OD]. Like *Thomasites* but higher whorled, with broader flatter flanks, flattened venter in early growth that tends to round in maturity, and generally

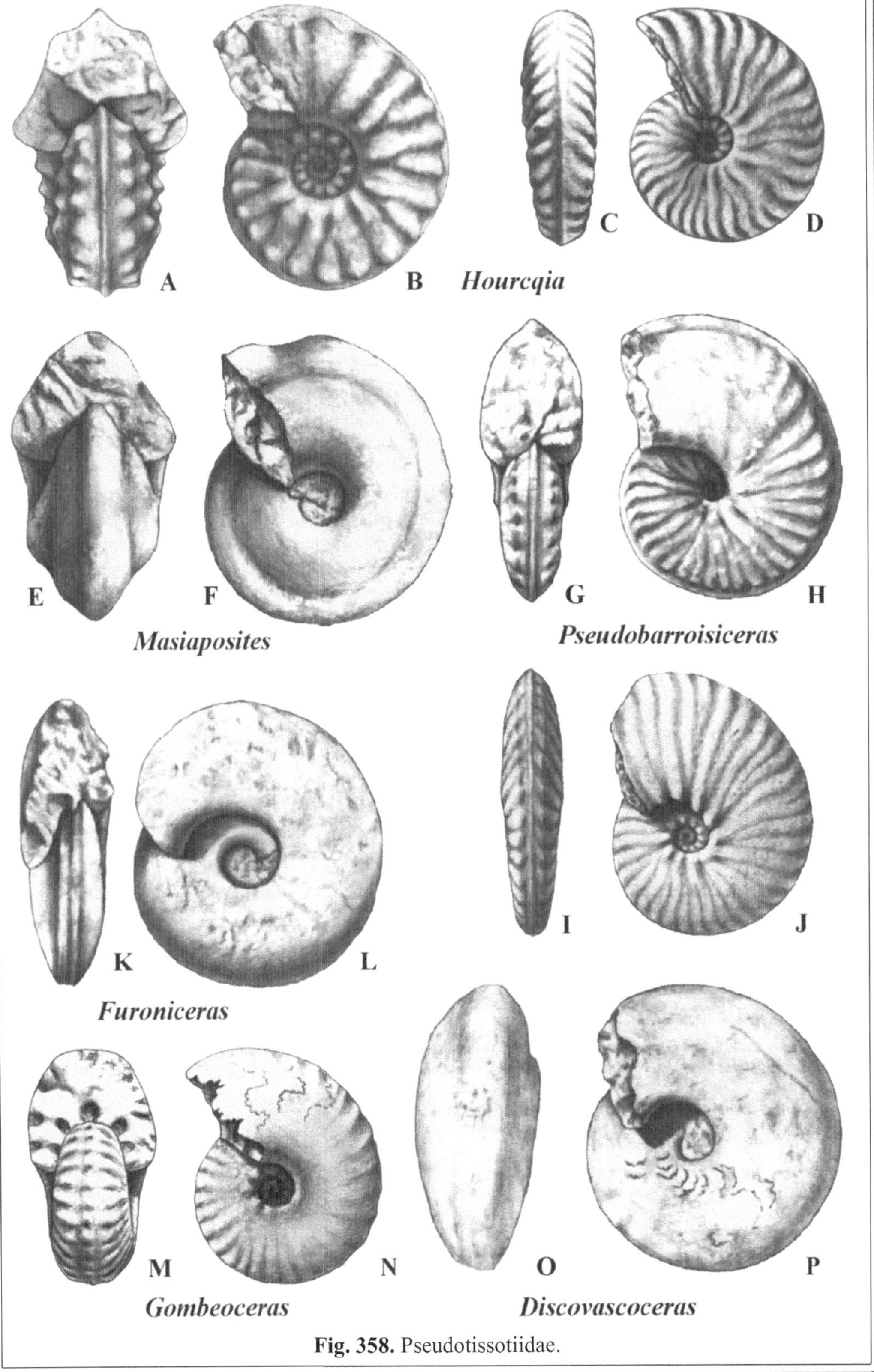

Fig. 358. Pseudotissotiidae.

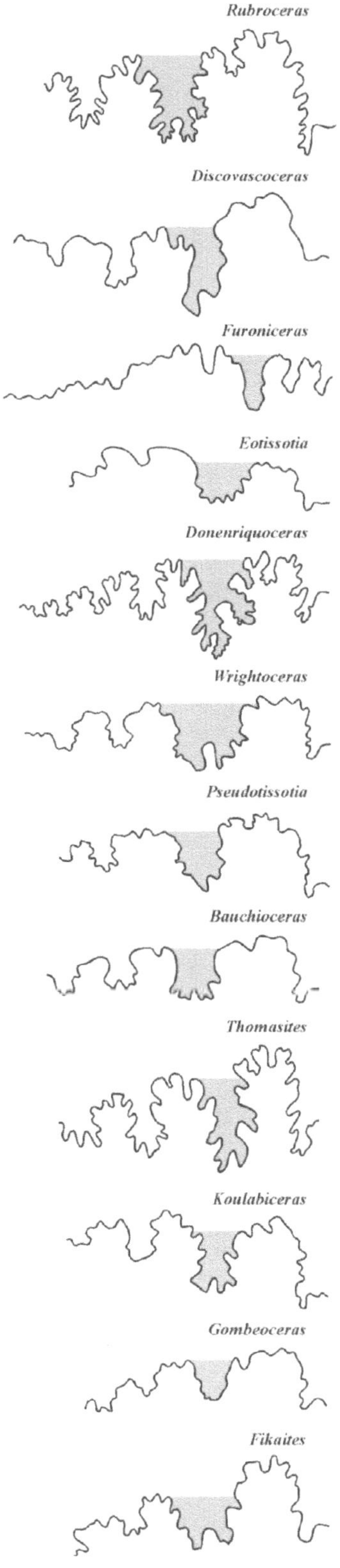

Fig. 359. Pseudotissotiine sutures.

compressed subrectangular to trapezoidal whorl section; inner whorls tricarinate, with nodate or entire keels, of which ventrolaterals weaken first; all ornament lost in maturity; suture similar to early *Thomasites*, little incised, with narrow E, broad shallowly bifid asymmetrical E/L, narrow to moderately wide L that is asymmetrically bifid and as long as E, and broad rounded barely indented L/U_2. *L. Turon.* **Fig. 356E-F** - *B. nigeriense* (Woods).

Discovascoceras COLLIGNON, 1957 [*Vascoceras (D.) tesselitense*; OD]. Like *Bauchioceras* but with deep umbilicus, high vertical umbilical wall, subtrigonal whorl section and suture with very narrow L deeper than E. *L. Turon.* **Fig. 358O-P** - *D. tesselitense* (Collignon).

Furoniceras COLLIGNON, 1957 [*F. trumpyi*; OD]. Like *Discovascoceras* but with complete lack of ornament and different simpler suture with deeply bifid E/L, narrow L and numerous auxiliary elements. *L. Turon.* **Fig. 358K-L** - *F. trumpyi* Collignon.

Pseudotissotia PERON, 1897 [*Ammonites galliennei* d'Orbigny; SD Pervinquière, 1907]. Like earlier *Bauchioceras* but inner whorls smooth and tricarinate from early stage; adult whorls develop sparse low rectiradiate to slightly rursiradiate ribs that arise from weakest of umbilical bullae and broaden ventrally, sometimes reaching ventrolateral keels to produce weak crenulations; suture like *Bauchioceras* but with broader L/U_2. *M. Turon.* **Fig. 356G-H** - *P. galliennei* (d'Orbigny).

Wrightoceras REYMENT, 1954 [*Bauchioceras (W.) wallsi*; OD]. Like *Pseudotissotia* but very inflated, with broadly sulcate venter bounded by sharp ventrolateral keels from early growth; innermost whorls with few broad rectiradiate ribs but these are soon lost and ornament is lacking in later growth when venter becomes tabulate; suture like *Bauchioceras* but with wider bifid L. *L. Turon.* **Fig. 356K-L** - *W. wallsi* (Reyment).

Imlayiceras A. LEANZA, 1967 [*I. washbourni*; OD]. Like *Wrightoceras* but smaller, more compressed, with weak plications around umbilicus and tabulate venter from early growth without bounding ventrolateral keels. *L. Turon.* **Fig. 356M-N** - *I. washbourni* A. Leanza.

Eotissotia BARBER, 1957 [*E. simplex*; OD]. Involute, compressed and high whorled, with initially tabulate venter becoming rounded and finally narrowly arched with compressed oval whorl section; suture like *Bauchioceras* but simpler, with entire L/U_2 and auxiliary saddles. *L. Turon.* **Fig. 356I-J** - *E. simplex* Barber.

Donenriquoceras WIEDMANN, 1960 [*D. forbesiceratiforme*; OD]. Like *Imlayiceras* but with tiny umbilicus, more convex flanks, venter tabulate in early growth and rounded later, and oblique and opposite ventrolateral clavi in early growth; suture like *Thomasites* but with shallower E and deep penetrating adventive saddle to asymmetrical L. *L. Turon.* **Fig. 356O-P** - *D. forbesiceratiforme* Wiedmann.

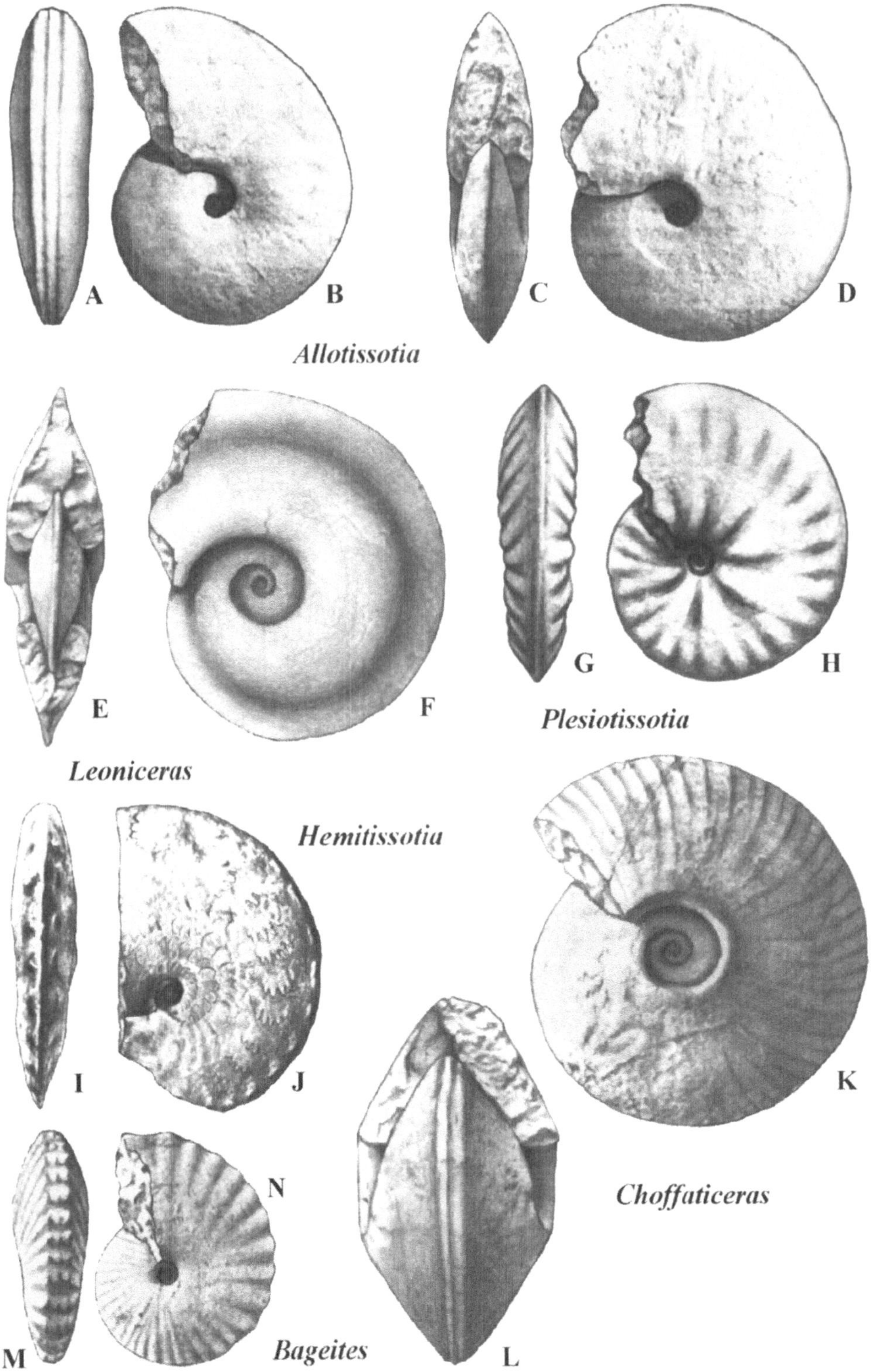

Fig. 360. Pseudotissotiinae, Hemitissotiinae.

Subfamily Hourcqiinae Renz, 1982

(*nom. transl.* Matsumoto & Toshimitsu, 1984 *ex* Hourcqiidae)

Discussion. Monophyletic derivatives of Pseudotissotinae, morphologically closest to *Bauchioceras*, e.g. *B. nigeriense macrocarinatum* (Barber), but suture most like *Donenriquoceras* with deep adventive saddle to L; characteristically with ventrolateral tubercles and blunt siphonal keel, superficially resembling Barroisiceratinae. *U. Turon*.

Hourcqia COLLIGNON, 1965 [*H. mirabilis*; OD]. Macroconchs inflated, rather involute and high whorled; inner whorls with flat convergent flanks and fastigiate venter; strong rounded ribs arise singly or in pairs from robust umbilical bullae and curve forward near venter, terminating in low ventrolateral tubercles separated by shallow smooth groove from distinct siphonal keel; outer whorls inflated, with depressed cordate whorl section and shallow spiral groove on outer flank across which ribs weaken; microconchs with flat parallel sides, subdued umbilical tuberculation, blunt flexuous ribs with weak ventrolateral nodes, and evenly rounded smooth venter with blunt siphonal keel; suture with broad deep E, narrow asymmetrically bifid E/L with narrow inner branch, narrow asymmetrical L not as deep as E and narrow shallowly indented trifid L/U_2. *U. Turon*. **Fig. 358A-D** - *H. mirabilis* Collignon.

Masiaposites COLLIGNON, 1965 [*M. carinatus*; OD]. Inner whorls unknown; outer whorls like *Hourcqia* but more inflated, with subtrigonal whorl section, strongly depressed whorls, more pronounced spiral groove and obsolete ribbing and siphonal keel (at least on outer whorls); suture like *Hourcqia*. *U. Turon*. **Fig. 358E-F** - *M. carinatus* Collignon.

Pseudobarroisiceras SHIMIZU, 1932 [*P. nagaoi*; OD]. Small; inner whorls like compressed high whorled *Hourcquia* microconch, with gently flexuous ribs springing in 2-3's from weak umbilical bullae; in later growth ribs become straight and rigid and, in maturity, ornament declines and adult body chamber is smooth; suture like *Hourcqia* but with broader saddles and narrower L. *U. Turon*. **Fig. 358G-H** - *P. nagaoi* Shimizu; **Fig. 358I-J** - *P. rennense* (de Grossouvre).

= *Pseubarroisiceras* Kennedy, 1995 (*laps. cal.*).

Family COILOPOCERATIDAE Hyatt, 1903

Discussion. A monophyletic group with an origin in Pseudotissotiidae. Populations characteristically vary from smooth and compressed to inflated and relatively strongly ornamented; however, compressed and inflated species are known also. *L. - U. Turon*.

Subfamily Hemitissotiinae Parnes, 1964

Discussion. Monophyletic derivatives of Pseudotissotiinae, close to *Pseudotissotia* (Reyment, 1954), characteristically with oxyconic macroconchs. *L. Turon. - Coni*.

Bageites ZABORSKI, 1998 [*B. bakui*; OD]. Compressed, involute, like inner whorls of *Bauchioceras*, with moderately dense long and short ribs often restricted to outer flank with rounded to clavate ventrolateral and siphonal tubercles arranged transversely across fastigiate venter; ribs tend to decline on flanks with growth; suture with broad shallowly bifid E/L, broad bifid L and frequently excentric E. *Lower L. Turon*. **Fig. 360M-N** - *B. bakui* Zaborski.

Choffaticeras HYATT, 1903 [*Pseudotissotia meslei* Peron, 1897; OD]. Innermost whorls like *Bageites*; later strongly inflated, with flat convergent flanks, narrow fastigiate venter, strongly depressed cordate adult whorl section and sparse distant simple flexuous ribs on outer flank; suture like *Bageites* but simpler, with deeper E, and narrower E/L and L. *L. Turon*. **Fig. 360K-L** - *C. meslei* (Peron).

Allotissotia PARNES, 1964 [*Hemitissotia galeppei* Pervinquière, 1907; OD]. Microconchs unornamented, like *Pseudotissotia*, but more compressed, with marked scaphitoid uncoiling of adult body chamber; macroconchs with steep overhanging umbilical wall with angular edge, acute venter with high sharp keel separated from flanks by shallow spiral depression and deep seated siphuncle (so that breaking off of keel creates narrowly tabulate venter); suture like *Choffaticeras* but saddles and lobes not obviously bifid. *L. Turon*. **Fig. 360A-D** - *A. galeppei* (Pervinquière).

Plesiotissotia PERON, 1897 [*P. michaleti*; OD]. Involute, compressed, high whorled, with fastigiate venter and blunt rounded siphonal keel; 7-8 strong rectiradiate umbilicolateral bullae weakly connected to 2-3 weak low rounded ribs, separated by intercalatories on outer flank; suture very simple, with deeply bifid E/L. *L. Coni*. **Fig. 360G-H** - *P. michaleti* Peron.

Leoniceras DOUVILLÉ, 1911 [*Pseudotissotia (Choffaticeras) luciae* Pervinquière, 1907; OD]. Like *Choffaticeras* but with lanceolate whorl section, convex inner and concave outer flanks, and entire strongly raised siphonal keel from early growth. *L. Turon*. **Fig. 360E-F** - *L. luciae* (Pervinquière).

Hemitissotia PERON, 1897 [*H. cazini*; OD]. Like *Leoniceras* but more compressed, with uniformly convex flanks; inner whorls with sparse, rounded branching ribs; suture

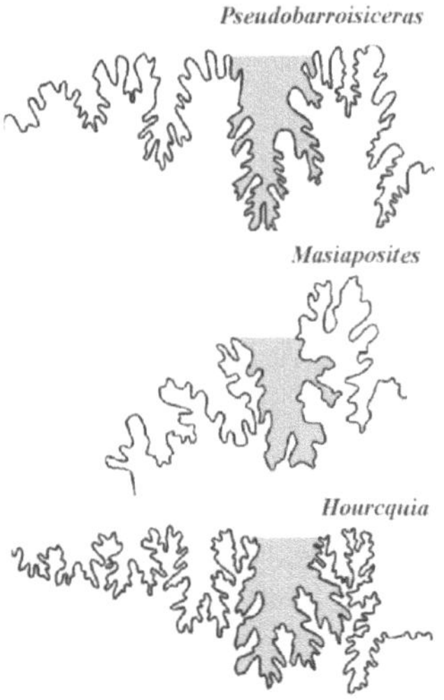

Fig. 361. Hourcquiine sutures.

like *Allotissotia* but even simpler and with broader L. *Coni.* **Fig. 360I-J** - *H. cazini* Peron.

Heterammonites COQUAND, 1880 [*Heterammonites* sp.; M]. Medium sized; very involute, with convex lower flanks, concave upper flanks, narrowly rounded venter and compressed vase-shaped whorl section; ornament lacking on internal mould; peristome seemingly with a subvertical rostrum; suture with broad E, very broad asymmetrically bifid E/L, broad bifid L with elongate dividing lobule and at least 3 entire or almost entire auxiliary saddles. *Coni.* **Fig. 362** - *H.* sp. (Coquand).

A *nomen nudum* because no species was included.

Subfamily Coilopoceratinae Hyatt, 1903

(*nom. transl.* herein *ex* Coilopoceratidae)
= Hoplitoidinae Wright, 1952

Discussion. A monophyletic group with an origin in Hemitissotiinae, close to *Choffaticeras* (Cobban, 1980); characteristically oxycones throughout growth, varying from smooth and compressed to inflated and relatively strongly ornamented. *L. - U. Turon*.

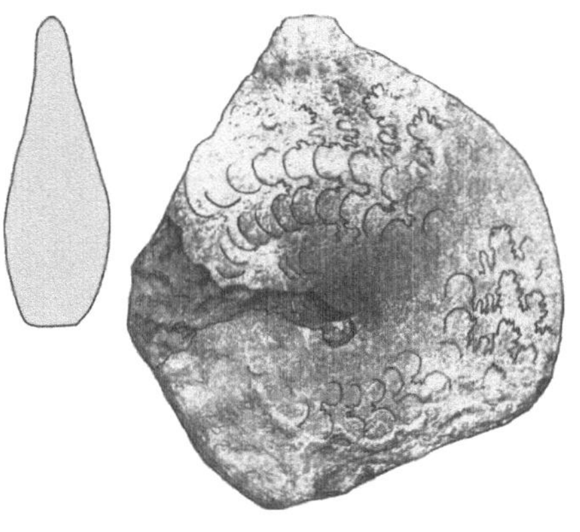

Fig. 362. *Heterammonites*.

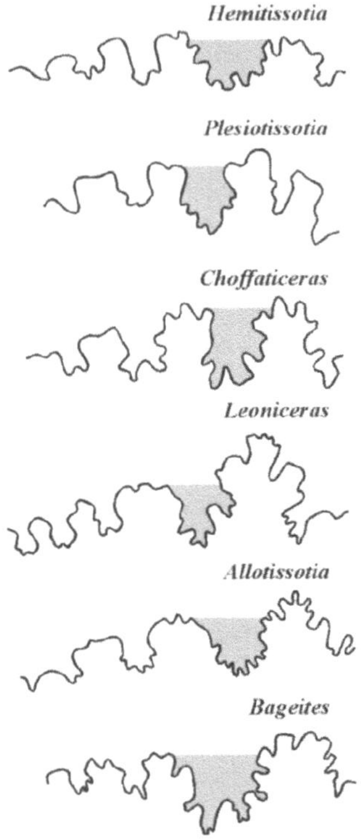

Fig. 363. Hemitissotiine sutures.

Hoplitoides VON KOENEN, 1898 [*H. latesellatus*, = *Neoptychites? ingens* von Koenen, 1897; OD]. Moderately inflated oxycone with convex flanks and very narrow umbilicus (U=4-10%); inner whorls with narrowly tabulate to sulcate venter, later becoming very narrowly rounded; early whorls with weak umbilical bullae giving rise to pairs of weak to moderately strong low sigmoidal ribs that terminate in more or less distinct ventrolateral tubercles; ornament declines with growth and adult whorls are smooth; suture with broad shallow E with diverging arms, broad irregularly dentate E/L, very broad asymmetrically bifid to trifid L with deep penetrating folioles; remaining saddles undivided and cauliflower like. *L. - M. Turon.* **Fig. 366E-F** - *H. ingens* (von Koenen).

Erichsenites MAGALHAES, 1953 (*Hoplitoides mirabilis* Pervinquière, 1907; OD]. Like *Hoplitoides* but smaller, more compressed, with tabulate venter persisting to maturity and simpler suture with less penetrative adventive saddles to L. *L. Turon.* **Fig. 366I-J** - *E. mirabilis* (Pervinquière).

Herrickiceras COBBAN & HOOK, 1980 [*Placenticeras costatum* Herrick & Johnson, 1900; OD]. Like *Erichsenites* but inner whorls with very weak low sickle-

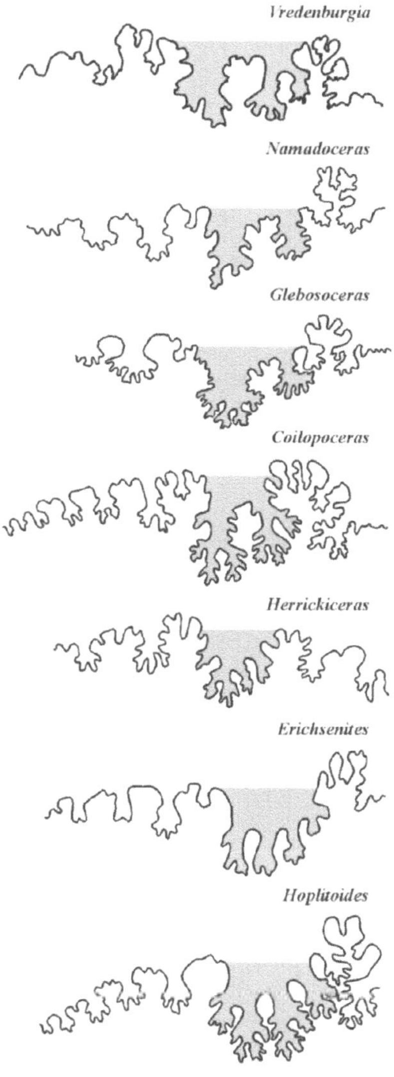

Fig. 364. Coilopoceratine sutures.

shaped rectiradiate ribs joined by 1-2 intercalatories at midflank; all ribs strengthen and broaden on outer flank and form low clavi where they cross ventrolateral keels transversely; ornament obsolete on adult body chamber when venter rounds; suture like *Erichsenites* but with very wide asymmetrically bifid E/L. *M. Turon.* **Fig. 366K-L** - *H. costatum* (Herrick & Johnson).

Coilopoceras HYATT, 1903 [*C. colleti*; OD]. Large, like *Hoplitoides* but with sharp venter from early growth; outer whorls with lanceolate to cordate whorl section; often unornamented, but variable broad low bulges may persist on inner flanks to large size occasionally strengthening to form ventrolateral nodes; very large siphuncle that, when weathered away, leaves a sulcate venter; suture like *Hoplitoides* but asymmetrical L markedly bifid due to enlarged deep penetrating adventive saddle. *M. - U. Turon.* **Fig. 366A-B** - *C. colleti* Hyatt.

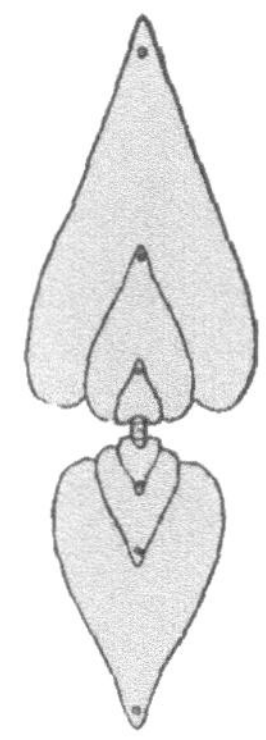

Fig. 365. Whorl section of *Vredenburgia*.

= *Coelopoceras* Bruggen, 1907 (*nom. van.*).

In primitive members of *Coilopoceras* the middle adventive saddle to L is not so enlarged, and the suture thus approaches *Hoplitoides*.

Glebosoceras REYMENT, 1954 [*G. globosum*; OD]. Strongly inflated and discoidal; like *Coilopoceras* but with prominent coarse distant ribs projecting strongly forward on outer part of flank; suture with shallower E, much broader L and entire first auxiliary saddle. **Fig. 366C-D** - *G. glebosum* Reyment.

Namadoceras VREDENBURG, 1907 [*N. scindiae*; OD]. Poorly known; moderately inflated, unornamented oxycone with very narrow umbilicus; suture like *Glebosoceras*, with very broad L, but with narrower asymmetrically bilobed outer lobe to L. **Fig. 366G-H** - *C. scindiae* (Vredenburg).

Vredenburgia CHIPLONKAR & GHARE, 1976 [*V. khadluensis*; OD]. Moderately large, very involute (U=9%), with sharp venter and strongly compressed lanceolate whorl section, convex on lower flank and concave on upper half; hollow keel; very broad, low, fold-like ribs around umbilicus, evanescing on outer flank; suture with E/L trifid. ?*U. Turon. - L. Coni.* **Fig. 365** - *V. khadluensis* Chiplonkar & Ghare.

Although included in the synonymy of *Placenticeras mintoi* (Vredenberg) by Jaitly & Ajani (2013), *Placenticeras* does not have a hollow siphonal keel, and *Vredenburgia* is a coilopoceratid closest to *Namadoceras*. The type species was said to come from the same beds as *Baghiceras* and *Malwiceras* that are Upper Turonian-Lower Coniacian placenticeratids.

Superfamily MAMMITOIDEA Hyatt, 1900

(*nom. transl.* herein *ex* Mammitidae)

Discussion. Monophyletic derivatives of Acanthoceratoidea; comprises the families Plesiacanthoceratidae, Euomphaloceratidae and Mammitidae. *M. Cenom. - L. Coni.*

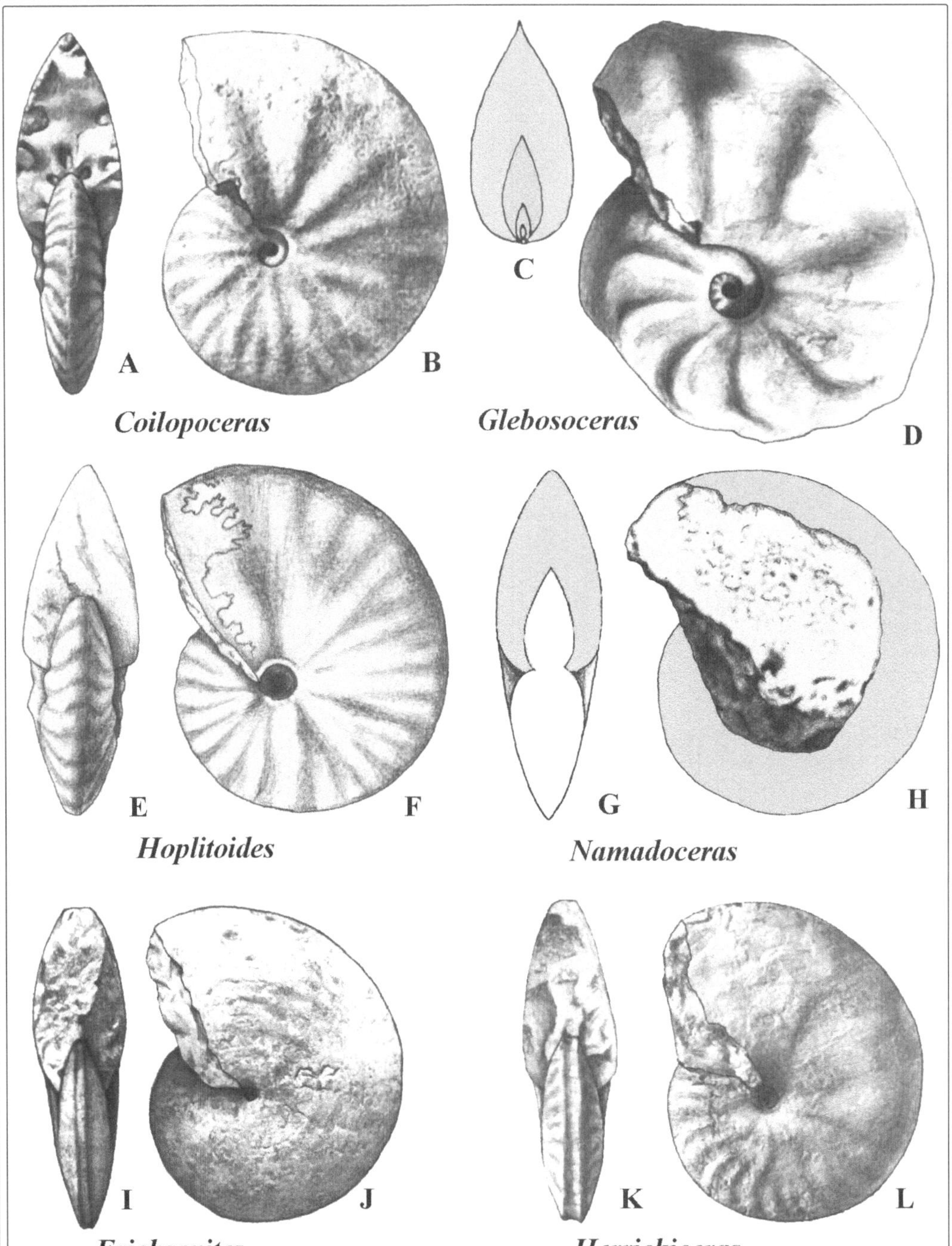

Fig. 366. Coilopoceratidae.

Family PLESIACANTHOCERATIDAE nov.

Diagnosis. Small to medium sized acanthoceratoids, generally with early loss or absence of siphonal tubercles. Monophyletic derivatives of Calycoceratidae comprising the subfamilies Plesiacanthoceratinae and Watinoceratinae. *M. Cenom. - M. Turon*.

Subfamily Plesiacanthoceratinae nov.

Diagnosis. Ventral tubercles asymmetrical, with tendency for inner ventrolateral tubercles to form horns. The most primitive member (*Conlinoceras*) was introduced as a subgenus of *Calycoceras* but some of the horned members were long identified as *Acanthoceras*. A paraphyletic group with an origin in Calycoceratinae, close to *Proeucalycoceras* or *Newboldiceras*; mostly North American, but dispersed at times of high sea level. *M. - U. Cenom.*

Conlinoceras COBBAN & SCOTT, 1972 [*Calycoceras (C.) gilberti*; OD]. Medium sized; inner whorls involute (U=16-21%), compressed, with flattened flanks and venter, generally depressed subquadrate whorl section, strong prorsirad-iate ribs alternating long and short, strong umbilical bullae, conical inner ventrolateral and clavate outer ventrolateral and siphonal tubercles; in middle to later growth ribbing coarsens, umbilicus widens (U=25-31%), umbilical bullae decline and migrate to low-lateral position, inner ventrolaterals and siphonals decline, and outer ventrolaterals strengthen; on adult body chamber ribs very distant, umbilical and inner ventrolaterals and siphonal tubercles obsolete, and ribs cross venter uninter-rupted, with angularity at position of outer ventrolaterals; suture like *Gentoniceras*. *M. Cenom.* **Fig. 368A-B** - *C. gilberti* (Cobban & Scott).

Texacanthoceras COOPER, 1998 [*Acanthoceras amphibolum* Morrow, 1935; OD]. Small (**m**) to moderately large (**M**); inner whorls like *Conlinoceras* but with constr-ictions; in middle growth most ribs long, with additional siphonal tubercles and feebly looped ribs across venter; outer ventrolaterals decline with growth and middle of venter rises into low blunt keel; clavate tubercles characteristically asymmetrical (steepest anteriorly); adult whorls sparsicostate, with prominent umbilicolateral tubercles, laterally protruding (inner) ventrolateral horns, deeply sulcate venter (in costal section) and low siphonal ridge; a homoeomorph of *Cunningtoniceras*. *M. Cenom.* **Fig. 368C-F** - *T. amphibolum* (Morrow).

Paracompsoceras COBBAN, 1971 [*P. landisi*; OD]. Large; inner whorls like *Texacanthoceras*; later moderately involute, high whorled, with flattened flanks, subtabulate to shallowly sulcate venter and rectangular to trapezoidal whorl section; in maturity ornament reduced to broad radial flank bulges arising from umbilical bullae; adult boy chamber smooth with subtabulate to rounded venter; a homoeomorph of *Acompsoceras*. *M. Cenom.* **Fig. 368J-K** - *P. landisi* Cobban.

Plesiacanthoceratoides KENNEDY & COBBAN, 1990 [*Protacanthoceras vetula* Cobban, 1987; OD]. A progenetic dwarf (D<30 mm); like inner whorls of *Texacanthoceras* but with weaker flank ribs and tiny umbilical tubercles; suture very simple; a homoeomorph of *Protacanthoceras*. *Upper M. Cenom.* **Fig. 368G-H** - *P. vetula* (Cobban).

Plesiacanthoceras HAAS, 1964 [*Metoicoceras wyomingense* Reagan, 1924; OD]. Inner whorls like *Texacanthoceras* but more involute and compressed, with early loss of siphonal tubercles; ribs alternate long and short to large diameter; outer whorls compressed with ribbing subordinate to tubercles and with huge blunt ventrolateral horns; suture complex, with rather long elements, and finger-like folioles; moderately wide E, broad almost square asymmetrical bifid E/L, somewhat narrow L,

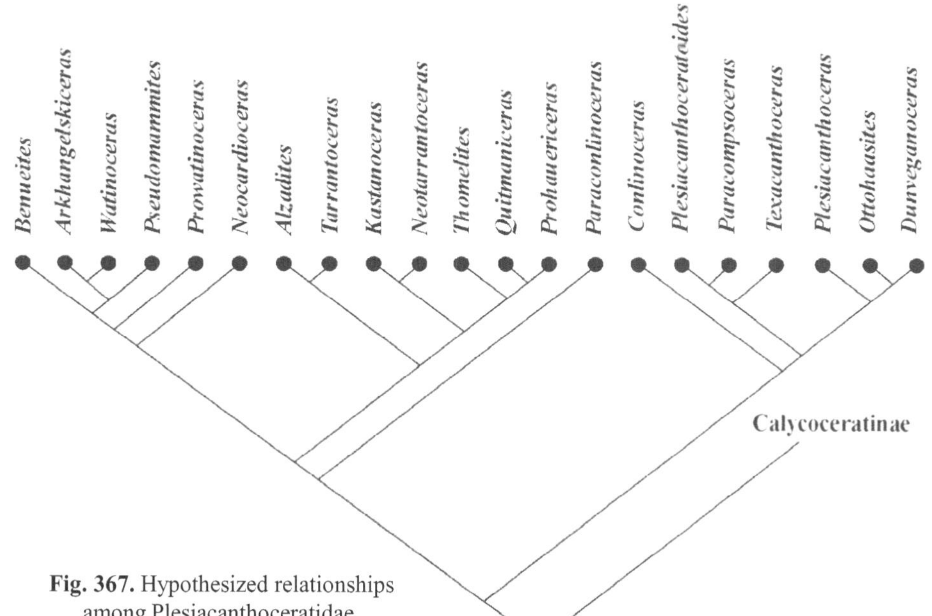

Fig. 367. Hypothesized relationships among Plesiacanthoceratidae.

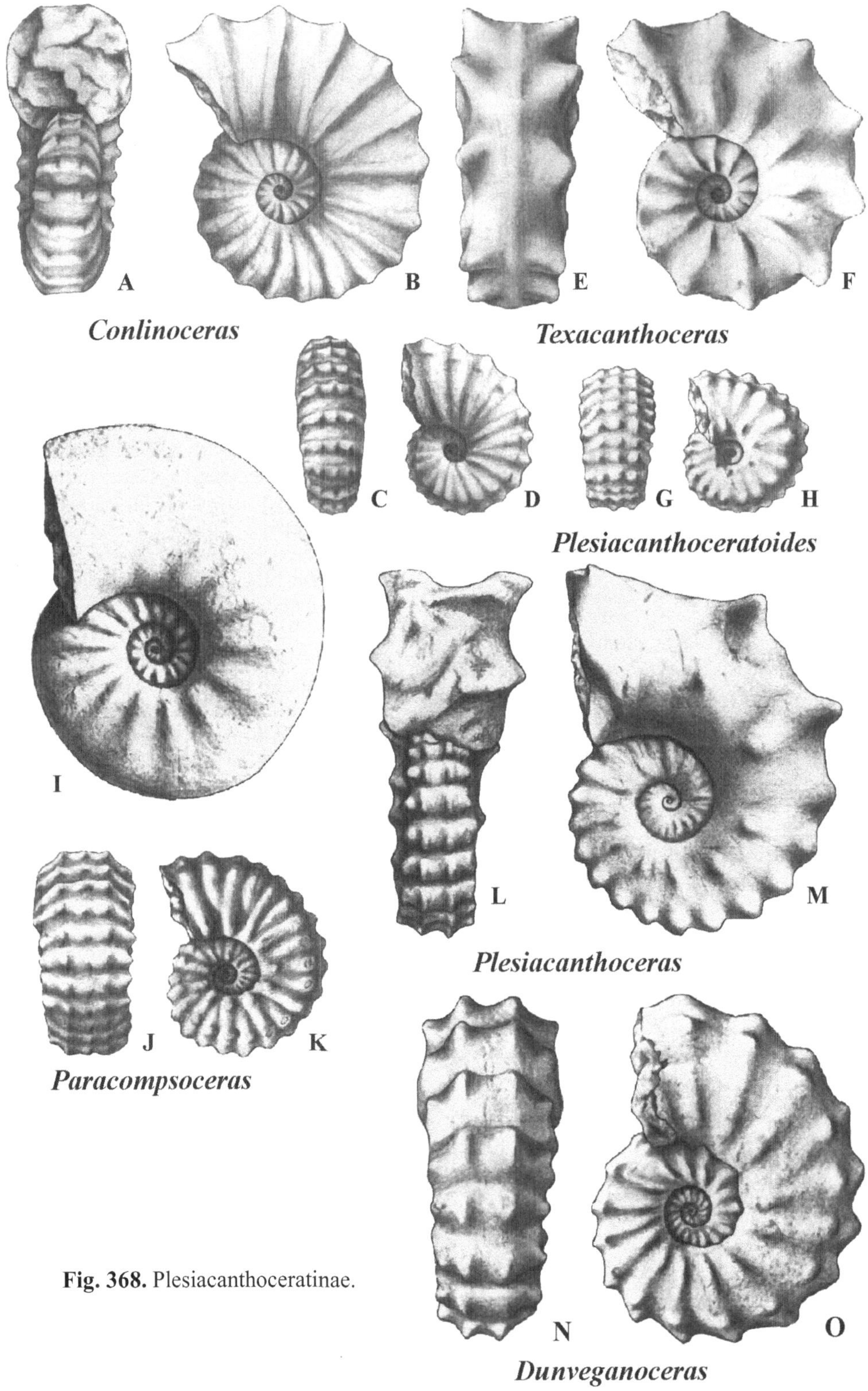

Fig. 368. Plesiacanthoceratinae.

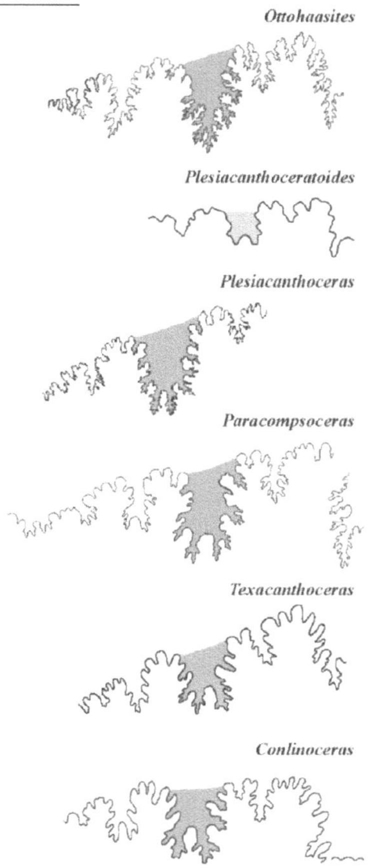

Fig. 369. Plesiacanthoceratine sutures.

and broad digitate bifid L/U_2. *Upper M. - low U. Cenom.* **Fig. 368L-M** - *P. wyomingense* (Reagan).

Ottohaasites COOPER, 1998 [*Dunveganoceras pondi* Haas, 1949; OD]. Large; inner whorls like *Plesiacanthoceras* but with sparse elongate umbilicolateral bullae that, on last whorl of phragmocone, shift to low lateral position and ventrolateral tubercles merge to form blunt horns; on subquadratic adult body chamber ventrolateral horns lost and flared rectiradiate ribs cross venter with distinct depression across siphonal line; suture with broad bifid E/L, deep penetrating, moderately broad, subrect-angular bifid L. *Basal U. Cenom.* **Fig. 382A-B** - *O. pondi* (Haas).

Dunveganoceras WARREN & STELCK, 1940 [*Acanthoceras albertense* Warren, 1930; OD]. Very large; inner whorls like *Ottohaasites*; adult body chamber more compressed, with fastigiate to lanceolate venter and siphonal tubercles; ventrolateral horns lacking. *Middle U. Cenom.* **Fig. 368N-O; Fig. 382I-J** - *D. albertense* (Warren).

Subfamily Watinoceratinae nov.

Diagnosis. Very small to small, evolute, compressed, with umbilical and paired ventrolateral tubercles, and generally strong ribbing; primitively with siphonal tubercles. Monophyletic derivatives of Plesiacanthoceratinae, close to *Conlinoceras*. Comprises the tribes Watinoceratini and Quitmaniceratini. *Upper U. Cenom.- M. Turon.*

Tribe Watinoceratini nov.

Type genus: *Watinoceras* Warren, 1930.

Diagnosis. Mostly densicostate with trend towards loss of siphonal tubercles so that venter becomes sulcate. *U. Cenom. - M. Turon.*

Paraconlinoceras KENNEDY & COBBAN, 1990 [*Eucalycoceras leonense* Adkins; OD]. Small, evolute (U=29-36%); inner whorls with flattened flanks and venter, depressed trapezoidal to poly-gonal whorl section, strong ribs that tend to alternate long short, and prominent umbilical, conical inner ventrolateral and strongly clavate outer ventrolateral and siphonal tubercles; in later growth ventral tuberculation declines, leaving strong ribs crossing venter; suture relatively simple, with deep narrow rectangular E, very broad bifid E/L, shallow narrow bifid L and reduced U_2 on umbilical shoulder; a homoeomorph of *Gentoniceras*. *M. Cenom.* **Fig. 370C-D** - *P. leonense* (Adkins).

Tarrantoceras STEPHENSON, 1955 [*T. rotatile* Stephenson, = *Mantelliceras sellardsi* Adkins, 1928]. Like *Paraconlinoceras* but with subtabulate venter and compressed trapezoidal whorl section to maturity; inner ventrolateral and siphonal tubercles declining in later growth, but outer ventrolaterals remaining strong. *Upper M. - Low U. Cenom.* **Fig. 370A-B** - *T. sellardsi* (Adkins).

Alzadites KENNEDY & COBBAN, 1990 [*A. alzadensis*; OD]. A micromorph (D<20 mm); compressed, involute (U=12-18%), with flat subparallel flanks, rounded venter and constrictions on internal mould; phragmocone with tiny distant umbilical bullae giving rise to prorsiradiate ribs with weak to strong inner ventrolateral tubercles and outer ventrolateral and siphonal clavi; tuberculation declines on adult body chamber when ribs form chevrons over venter; suture simple. *Upper M. - U. Cenom.* **Fig. 370I-J** - *A. alzadensis* Kennedy & Cobban.

***Neotarrantoceras* gen. nov.** [*Acanthoceras cuspidum* Stephenson, 1953, p.202; OD herein]. Like *Tarrantoceras* but smaller still, more strongly tuberculate, with tubercles retained to maturiity and inner ventrolaterals spinose and impressed into umbilical wall of succeeding whorl; outer ventrolateral and siphonal clavi distinctly asymmetrical. *Upper M. - Low U. Cenom.* **Fig. 370E-F** - *N. cuspidum* (Stephenson).

Kastanoceras KENNEDY & COBBAN, 1990 [*K. spiniger*; OD]. A progenetic dwarf (D<12 mm); like *Neotarrantoceras* but with depressed coronate to reniform whorl section, early loss of siphonal and outer ventrolateral tubercles, feeble ribs terminating in large inner ventrolateral spines

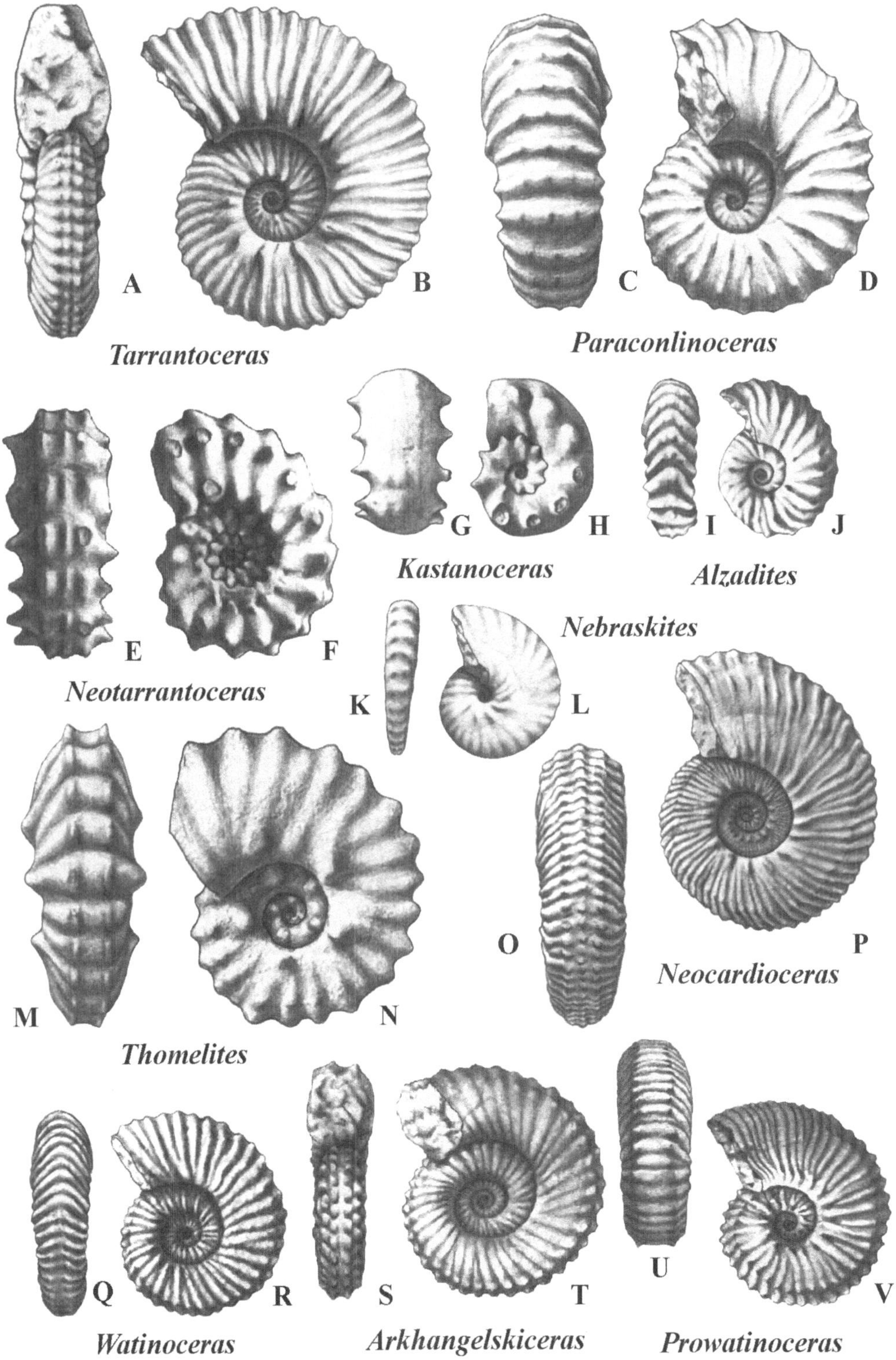

Fig. 370. Watinoceratinae.

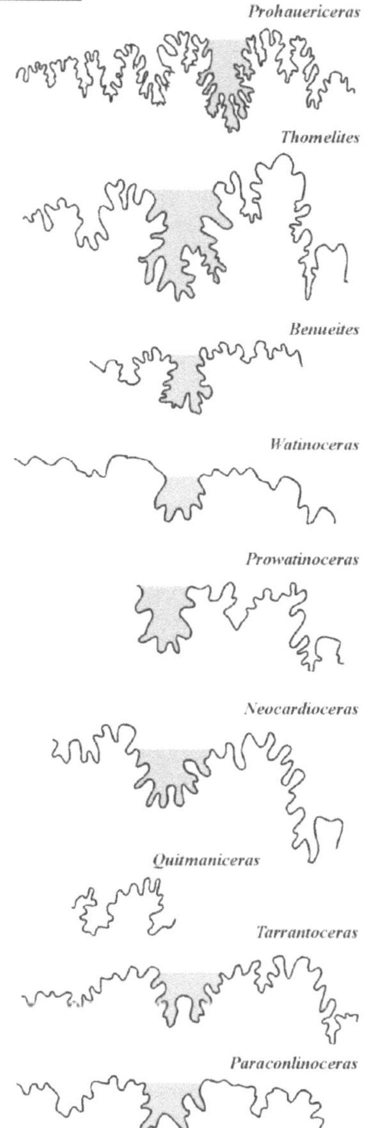

Fig. 371. Watinoceratine sutures.

impressed in umbilical wall, and almost smooth adult body chamber. *Upper M. - Low U. Cenom.* **Fig. 370G-H** - *K. spiniger* Kennedy & Cobban.

Thomelites WRIGHT & KENNEDY, 1973 [*Jeanrogericeras sornayi* Thomel, 1967; OD]. Small; inner whorls like *Neotarrantoceras*; later with flat subparallel to convergent flanks, narrow tabulate venter and compressed to depressed subquadratic whorl section; rectiradiate ribs, varying from broad and flat to rounded, arise in bundles of 1-3 from sparse, sometimes exaggerated, umbilical tubercles, separated by intercalatories; all ribs with weak inner ventrolateral nodes, prominent asymmetrical (steepest anteriorly) outer ventrolateral clavi and siphonal clavi; whereas inner ventrolaterals and siphonal clavi decline with growth, outer ventrolaterals strengthen; suture relatively simple, but may develop finger-like folioles; a *Metoicoceras* homoeomorph. *Lower U. Cenom.* **Fig. 370M-N** - *T. sornayi* (Thomel).

Neocardioceras SPATH, 1926 [*Ammonites juddii* Barrois & de Guerne, 1878; OD]. Small, moderately evolute, compressed, with oval whorl section; densicostate, with numerous rectiradiate to prorsiradiate narrow sharp ribs that flex forward on ventrolateral shoulder and form chevrons across venter; long ribs arise singly or in pairs from umbilical bullae, short ribs higher on flank; all ribs with bullate inner, nodate outer ventrolateral and clavate siphonal tubercles; suture simple, with rather narrow E, broad rounded asymmetrically bifid E/L, moderately wide shallow digitate L not as deep as E, and narrow asymmetrically bifid L/U_2. *Uppermost Cenom.* **Fig. 370O-P** - *N. juddii* (Barrois & de Guerne).

Prowatinoceras **gen. nov.** [*Neocardioceras transiens* Kennedy, Cobban & Landmann, 1996, p.15; OD herein]. Small; like *Neocardioceras* but with irregular vestigial siphonal clavi that are lost in maturity, with outer ventrolateral clavi linked across venter by strong transverse ribs. *Low U. Cenom.* **Fig. 370U-V** - *P. transiens* (Kennedy, Cobban & Landman).

Watinoceras WARREN, 1930 [*W. reesidei*; OD]. Small; like *Prowatinoceras* but more evolute, without siphonal tubercles at any stage; ribs do not cross smooth flat to concave venter; suture little incised, with moderately wide shallow E, very broad weakly bifid E/L, moderately wide asymmetrically bifid L, and broad barely indented L/U_2. *Uppermost Cenom. - M. Turon.* **Fig. 370Q-R** - *W. reesidei* Warren.

Arkhangelskiceras IL'IN, 1957 [*Acanthoceras amudariensis* Arkhangel'skii, 1916; OD]. Small, barely different from *Watinoceras* but with ribs forming chevrons over rounded adult venter. *L. Turon.* **Fig. 370S-T** - *A. amudariense* (Arkhangel'skii).

Pseudomammites **gen. nov.** [*Acanthoceras coloradoense* Henderson, 1908, p.259; OD herein]. Medium sized; inner whorls like *Watinoceras*; in middle to later growth with strong simple rather distant prorsiradiate ribs with prominent inner and outer ventrolateral clavi, the latter rising well above the venter; a *Mammites* homoeomorph. *L. Turon.* **Fig. 372F-G** - *P. coloradoensis* (Henderson).

Nebraskites KENNEDY & COBBAN, 1988 [*N. haresiceratiforme*; OD]. Progenetic dwarf, perhaps of *Watinoceras*; very involute, compressed, flat sided, with narrow tabulate venter, paired prorsiradiate ribs arising from feeble umbilical bullae and with incipient inner and clavate outer ventrolateral tubercles joined across venter by low ribs; on adult body chamber ribbing simplifies, lower ventrolaterals evanesce, and ventral ribbing strengthens; suture like *Watinoceras* but very simple,

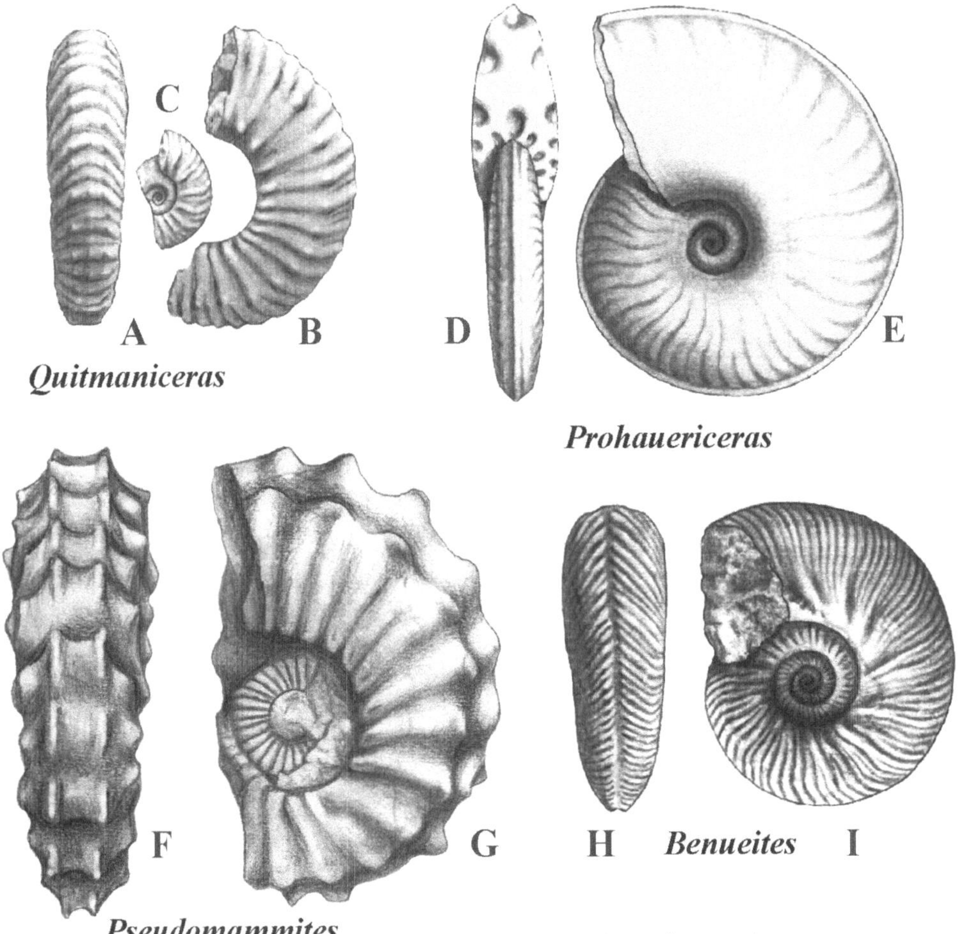

Fig. 372. Watinoceratinae.

barely incised, with entire rounded E/L, rounded entire L, and low rounded entire L/U_2. *M. Turon.* **Fig. 370K-L** - *N. haresiceratiforme* Kennedy & Cobban.

Benuites REYMENT, 1954 [*B. benueensis*; OD]. Like *Watinoceras* but with dense sigmoidal ribs of variable strength, weak umbilical bullae, supressed ventrolateral tuberculation, and narrow ventral sulcus; suture with broad bifid E/L, narrow bifid L, and relatively broad L/U_2. *L. Turon.* **Fig. 372H-I** - *B. benueensis* Reyment.

Tribe Quitmaniceratini nov.

Type genus: *Quitmaniceras* Powell, 1963.

Diagnosis. Typically compressed, with generally weak ribbing and coarsely crenulated to entire siphonal keel. Derivatives of *Neocardioceras*. *L. - M. Turon*.

Quitmaniceras POWELL, 1963 [*Q. reaseri*; OD]. Like *Neocardioceras* but more compressed, with weaker blunter lower broader rather indistinct ribs supplemented by intercalatories across venter; indistinct to absent inner ventrolateral tubercles, clavate outer ventrolaterals, and coarser crenulations to siphonal keel; weakly ornamented forms show only indistinct ribbing and entire siphonal keel; suture simple, little incised. *L. Turon.* **Fig. 372A-C** - *Q. reaseri* Powell.

Prohauericeras NOWAK, 1913 [*Ammonites goupilianus* d'Orbigny, 1841; SD Diener, 1925]. Poorly known and types lost. Involute, compressed, high whorled, with broad gently convex flanks and narrow fastigiate venter with entire siphonal keel; subdued low sinuous ribs bifurcate from umbilical shoulder, separated by intercalatories, and terminate at ventrolateral margin before reaching keel leaving narrow smooth zones; ribbing weakens with growth and evanesces in maturity; suture with fairly broad shallow E, broad square asymmetrically bifid E/L, narrow trifid L deeper than E, and narrow asymmetrically bifid L/U_2; a homoeomorph of *Hauericeras*. *M. Turon.* **Fig. 372D-E** - *P. goupilianum* (d'Orbigny).

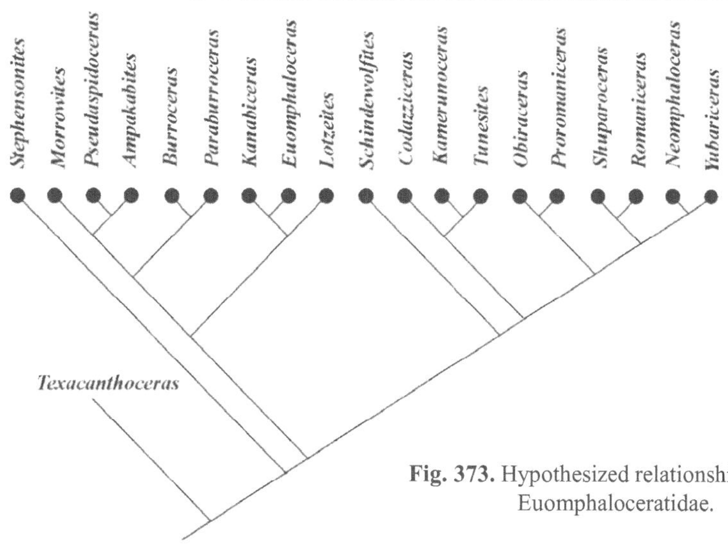

Fig. 373. Hypothesized relationships among Euomphaloceratidae.

Family EUOMPHALOCERATIDAE Cooper, 1978

(*nom. transl.* Cooper, 1997 *ex* Euomphaloceratinae)

Discussion. A monophyletic group with an origin in primitive Mammitidae; comprises the subfamilies Euomphaloceratinae and Romaniceratinae. *U. Cenom. - L. Coni.*

Subfamily Euomphaloceratinae Cooper, 1978

Discussion. A monophyletic group of arguable ancestry; Cooper (1978) looked to *Cunningtoniceras*, whereas Wright & Kennedy (1992) favoured *Lotzeites*. Here an origin in Plesiacanthoceratinae is suggested, close to *Texacanthoceras*, via *Stephensonites*, a *Cunningtoniceras* homoeomorph. Typically with depressed whorls, quinquetuberculate ribs with a multiplicity of ventral tubercles and a suture with broad U. *U. Cenom. - L. Turon.*

Stephensonites **gen. nov.** [*Acanthoceras lonsdalei* Adkins, 1928, p.244; OD herein]. Medium sized, inflated, with depressed subquadratic whorl section; inner whorls involute and constricted, later evolute without constrictions; strong sparser ribs with prominent umbilicolateral bullae and inner ventrolateral horns, the latter joined across venter by looped secondaries separated by an intercalatory so that there is a multiplicity of upper ventrolateral and siphonal tubercles; there is a low siphonal ridge; in maturity the outer ventrolaterals are incorporated into ventrolateral horns joined across the venter by broad bar-like ribs; suture with broad L; a homoeomorph of *Cunningtoniceras*. *Upper M. Cenom.* **Fig. 375** - *S. lonsdalei* (Adkins).

Euomphaloceras SPATH, 1923 [*Ammonites euomphalus* Sharpe, 1855; OD]. Small, like *Stephensonites* but smaller, more evolute, with small spinose umbilical tubercles and rather irregular ribs that do not cross venter in maturity. *Lower U. Cenom.* **Fig. 374L-P** - *E. euomphalum* (Sharpe).

Kanabiceras REESIDE & WEYMOUTH, 1931 [*Scaphites? septemseriatus* Cragin, 1893; OD]. Like *Euomphaloceras* but with obliquely clavate outer ventrolaterals, and siphonal tubercles forming a prominent serrated keel flanked in middle growth by deep grooves. *Upper U. Cenom.* **Fig. 374J-K, 378** - *K. septemseriatum* (Cragin).

Lotzeites WIEDMANN, 1960 [*Acanthoceras aberrans* Kossmat, 1897; OD]. Like *Euomphaloceras* but very evolute, with steep high umbilical wall, flattened flanks, broad gently rounded to almost flat venter, and very depressed subrectangular whorl section that expands rather rapidly in width; on early whorls strong rounded rectiradiate ribs arise from sharp pointed umbilical tubercles and terminate in prominent spinose (inner) ventrolateral tubercles connected across venter by looped ribs with slight forward arch and faint angularity at outer ventrolateral position; on outer whorl looped ribs separated by a non-tuberculate intercalatory arising at ventrolateral shoulder; ventrolateral and siphonal tubercles lacking; marked ventral constrictions present on early whorls; suture with a narrow E/L, and very wide splayed L with plump central foliole. *Lower U. Cenom.* **Fig. 374A-B** - *L. aberrans* (Kossmat).

Wiedmann introduced *Lotzeites* as a subgenus of *Calycoceras*, to which it bears considerable resemblance, but this seems to be a convergence (Wright & Kennedy, 1992).

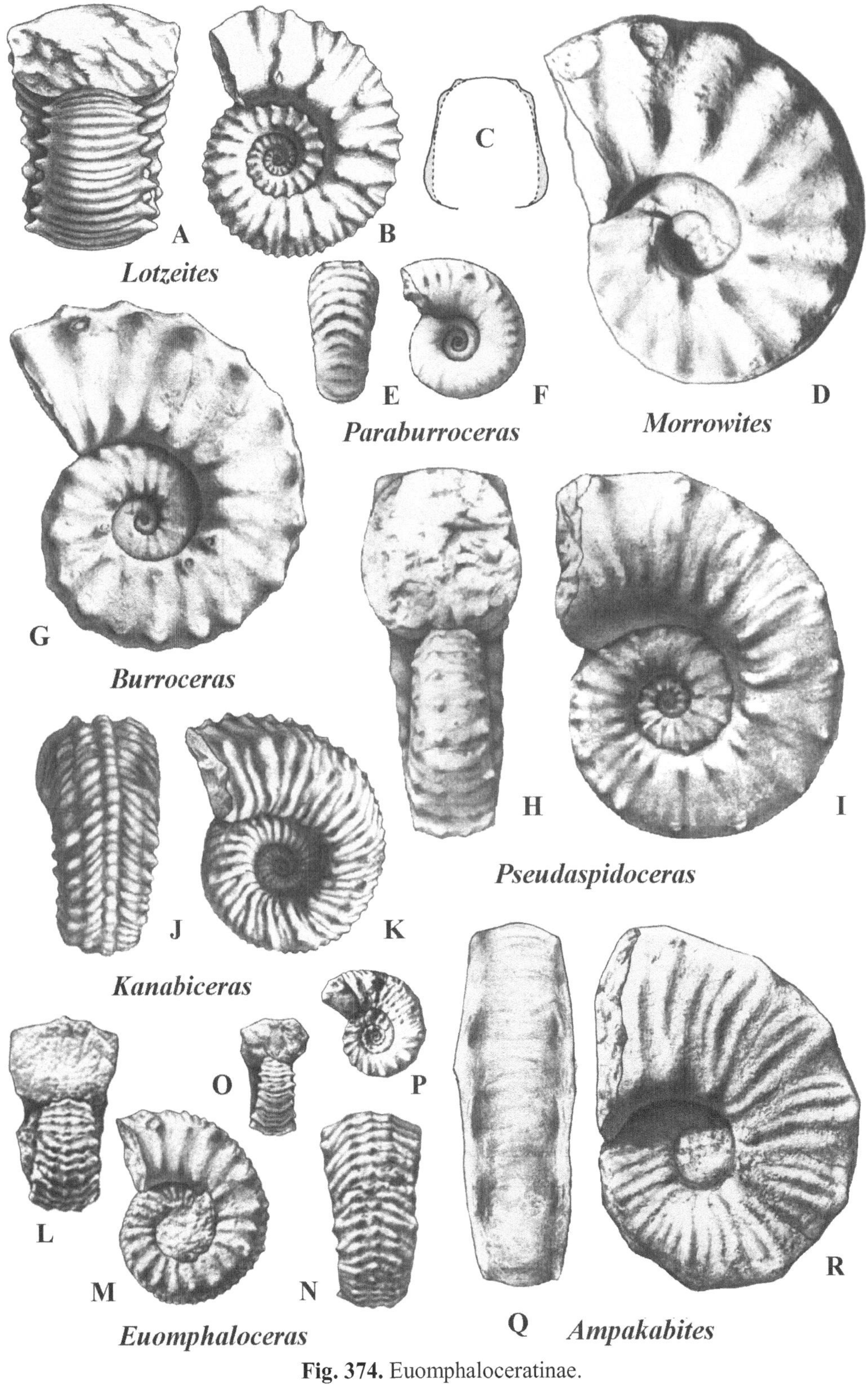

Fig. 374. Euomphaloceratinae.

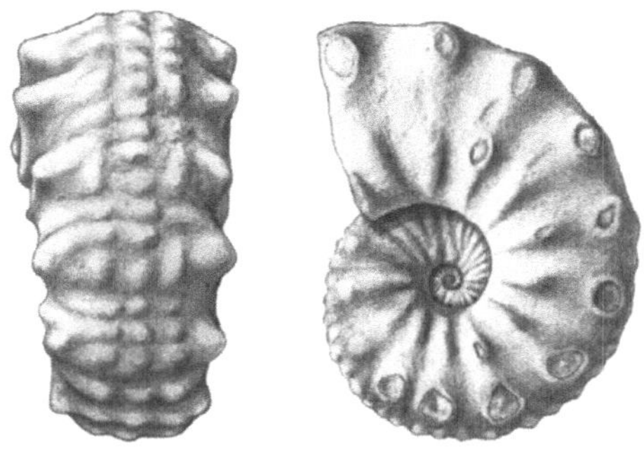

Fig. 375. *Stephensonites*.

Burroceras COBBAN, HOOK & KENNEDY, 1989 [*B. clydense*; OD]. Medium sized; like *Euomphaloceras* but with smooth inner whorls, flatter flanks and irregularly developed siphonal tubercles; inner ventrolateral tubercles larger and fewer than outer ones; suture with deep narrow E, relatively narrow non-bifid E/L with inclined inner side, very broad deeply bifid L, and narrow non-bifid L/U_2. *Uppermost Cenom.* **Fig. 374G** - *B. clydense* Cobban, Hook & Kennedy (see also **Fig. 378**).

Paraburroceras COBBAN, HOOK & KENNEDY, 1989 [*P. minutum*; OD]. A progenetic dwarf, probably of *Burroceras*; flanks and venter broadly rounded with slightly depressed whorl section, most of phragmocone smooth, prorsiradiate ribs appearing on last part only with inner and outer ventrolateral and siphonal tubercles; on body chamber inner ventrolaterals stronger and fewer in number; ribs with asymmetrical profile across venter; suture like *Burroceras* but simpler. *U. Upper Cenom.* **Fig. 374E-F** - *P. minutum* Cobban, Hook & Kennedy.

Pseudaspidoceras HYATT, 1903 [*Ammonites footeanus* Stoliczka, 1864; OD]. Medium sized to moderately large; like *Burroceras* but much larger, with broad flat flanks, broadly rounded venter and square to rectangular whorl section; on inner whorls prominent swollen umbilical tubercles weakly joined, if at all, by low indistinct ribs to sparser swollen inner ventrolateral tubercles that are outnumbered by small nodate outer ventrolaterals which disappear in middle growth to leave venter smooth on outer whorls; inner ventrolaterals may be spinose; suture with long broad stemmed E/L, shallowly bifid L and broad rounded L/U_2. *U. Upper Cenom. - L. Turon.* **Fig. 374H-I** - *P. footeanum* (Stoliczka).

Ampakabites COLLIGNON, 1965 [*Kamerunoceras (A.) auriculatum*; OD]. Moderately large; like *Pseudaspidoceras* but more involute, with strongly compressed rectangular whorls; weak ribs arise in pairs from umbilical bullae, separated by finer irregular intercalatories, become rursiradiate on outer flank, and loop to widely spaced auriculate ventrolateral clavi; suture relatively complex with subphylloid terminations, long narrow stemmed E/L, fairly broad trifid L deeper than E, and narrow L/U_2. *L. Turon.* **Fig. 374Q-R** - *A. auriculatus* (Collignon).

Morrowites COBBAN & HOOK, 1983 [*Mammites wingi* Morrow, 1935; OD]. Inner whorls and suture like *Pseudaspidoceras*; later with nodate inner and weakly clavate outer ventrolateral tubercles equal in number and coalescing to form prominent spinose ventrolateral horns; whorls may be very depressed. *L. Turon.* **Fig. 374C-D** - *M. wingi* (Morrow).

Subfamily Romaniceratinae Cooper, 1997

Discussion. Monophyletic derivatives of Euomphaloceratinae, close to *Burroceras*; typically larger, more evolute and compressed,with 9-11 rows of tubercles; suture with narrow L. *L. Turon. - L. Coni.*

Schindewolfites WIEDMANN, 1960 [*S. inaequicostatus*; OD]. Small; like *Burroceras* but whorls quadratic and flat sided; ribs irregular but relatively strong and rursiradiate; ventrolateral tubercles not spinose; siphonal tubercles slightly clavate sometimes raised on low keel and not markedly outnumbering inner ventrolaterals; suture with fairly broad E, relatively narrow non-bifid E/L, very broad bifid L that is deeper than E (a primitive character), and rather narrow L/U_2. *L. Turon.* **Fig. 376A-B** - *S. inaequicostatus* Wiedmann.

Kamerunoceras REYMENT, 1954 [*Acanthoceras eschii* Solger, 1904; OD]. Like *Schindewolfites* but medium sized, more evolute, with slowly expanding whorls; umbilical tubercles shift to low lateral position early in growth (with or without formation of new weaker umbilical bullae); suture with deep E, broad asymmetrically bifid E/L and narrow asymmetrically bifid L. *L. Turon.* **Fig. 376C-D** - *K. eschii* (Solger).

Tunesites PERVINQUIÈRE, 1907 [*T. salambo*; SD Roman, 1938]. Very small, immature; evolute, compressed; earliest whorls smooth save for weak umbilicolateral bullae and deep collared constrictions; later with strong long and short ribs with umbilical bullae and 5 rows of ventral tubercles; middle and later growth stages unknown. *?Lower U. Cenom.* **Fig. 376I-L** - *T. salambo* Pervinquière.

Tunesites is very similar to the innermost whorls of a number of Romaniceratinae, and the possibility that it has been misplaced stratigraphically cannot be discounted; it is a *nomen dubium*.

Codazziceras ETAYO SERNA, 1979 [*Lyelliceras scheibei* Riedel, 1938; OD]. Like *Kamerunoceras* but serpenticone, with regular ribbing that is strong across venter; tubercles and ribs weaken on adult body chamber; venter lacks siphonal ridge; a homoeomorph of *Lyelliceras*. *L. Coni.* **Fig. 376E-F** - *C. scheibei* (Riedel).

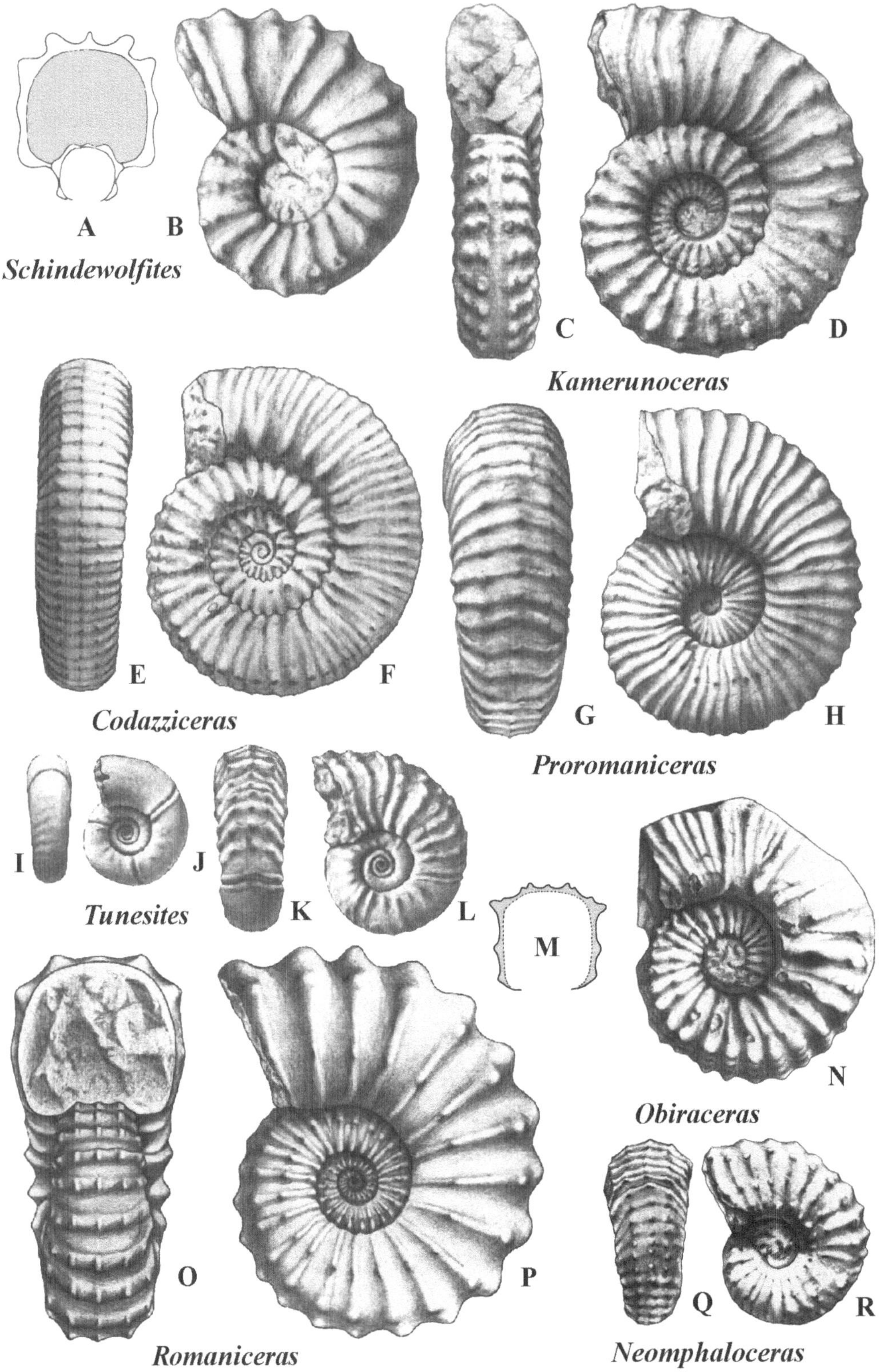

Fig. 376. Romaniceratinae.

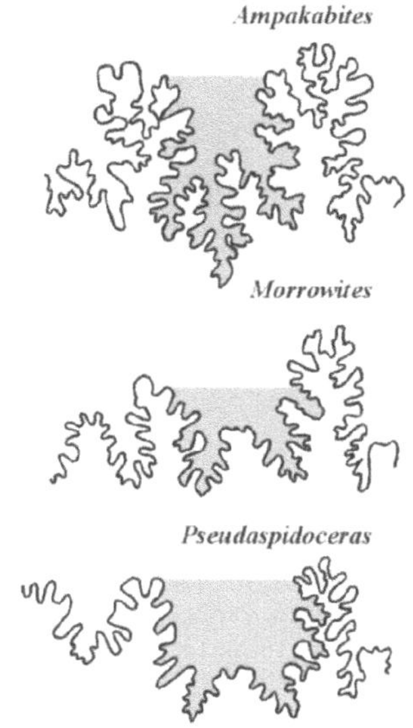

Fig. 377. Euomphaloceratine sutures.

Proromaniceras WIEDMANN, 1960 [*Acanthoceras pseudodeverianum* Jimbo, 1894; OD]. Medium sized; like *Kamerunoceras* but with narrower umbilicus in early growth, stronger, denser, more regular, mostly single ribs that are uninterrupted across venter, and 9 rows of subdued tubercles in which umbilicals and low laterals persist for most of ontogeny. *L. Turon*. **Fig. 376G-H** - *P. pseudodeverianum* (Jimbo).

= *Kossmatia* Yabe, 1927 (*non* Uhlig, 1907).

Obiraceras MATSUMOTO, 1975 [*Romaniceras (Obiraceras) ornatum*; OD]. Like *Proromaniceras* but smaller; in middle to later growth umbilical, lateral and ventrolateral tubercles spinose, the latter doubling on last part of phragmocone; siphonal tubercles weak on adult body chamber; suture with roughly rectangular L and narrow asymmetrically bifid L/U_2. *L. Turon*. **Fig. 376M-N** - *O. ornatum* (Matsumoto).

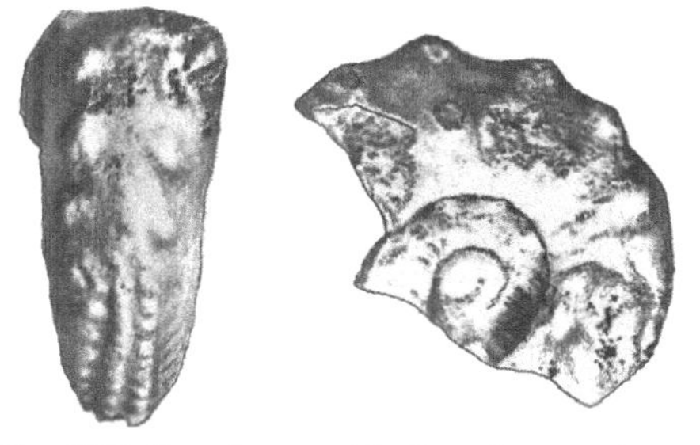

Fig. 378. Specimen with *Kanabiceras* inner whorls and *Burroceras* outer whorl.

Romaniceras SPATH, 1923 [*Ammonites deverianus* d'Orbigny, 1841; OD]. Like *Proromaniceras* but larger, somewhat more evolute, with rapidly expanding whorls; more inflated and coarsely ribbed, with conspicuous tubercles; umbilical tubercles protrude into umbilicus and low lateral tubercles migrates to mid-flank during ontogeny when ribs bifurcate from them; in maturity ribbing coarsens, lateral and inner ventrolateral tubercles weaken and venter rounds. *U. Turon*. **Fig. 376O-P** - *R. medlicottianum* (Stoliczka).

Shuparoceras MATSUMOTO, 1975 [*S. yagii*; OD]. Like *Proromaniceras*; involute and high whorled, with compressed whorl section, overhanging umbilical wall, and dense ribbing that does not simplify in maturity; tuberculation weakens and is lost in middle growth, with ribs becoming broad and flat topped on adult body chamber; constrictions persist throughout; suture like *Kamerunoceras*; a homoeomorph of *Proeucalycoceras*. *U. Turon*. **Fig. 379C-D** - *S. yagii* Matsumoto.

Neomphaloceras MATSUMOTO & OBATA, 1982 [*Yubariceras pseudomphalum* Matsumoto, 1975; OD]. Small; inner whorls like *Romaniceras* but with depressed subquadratic whorl section; intercalated ribs across venter lead to multiplicity of ventral tubercles; outer whorls with very broad venter, distant ribs and short but prominent ventrolateral horns. *Upper L. - M. Turon*. **Fig. 376Q-R** - *N. pseudomphalum* (Matsumoto).

Yubariceras MATSUMOTO, SAITO & FUKADA, 1957 [*Y. yubarense*; OD]. Moderately large; like *Neomphaloceras* but ribs with 11 rows of tubercles. *L. - U. Turon*. **Fig. 379A-B** - *Y. ornatissimum* (Stoliczka).

Family MAMMITIDAE Hyatt, 1903

Discussion. Monophyletic derivatives of Plesiacanthoceratidae; comprises the subfamilies Metoicoceratinae, Mammitinae and Buchiceratinae; typically with rather coarse ribs, paired ventrolateral tubercles that may coalesce to form horns, and a tabulate to sulcate venter without siphonal tubercles (except primitively). *Lower U. Cenom. - L. Coni.*

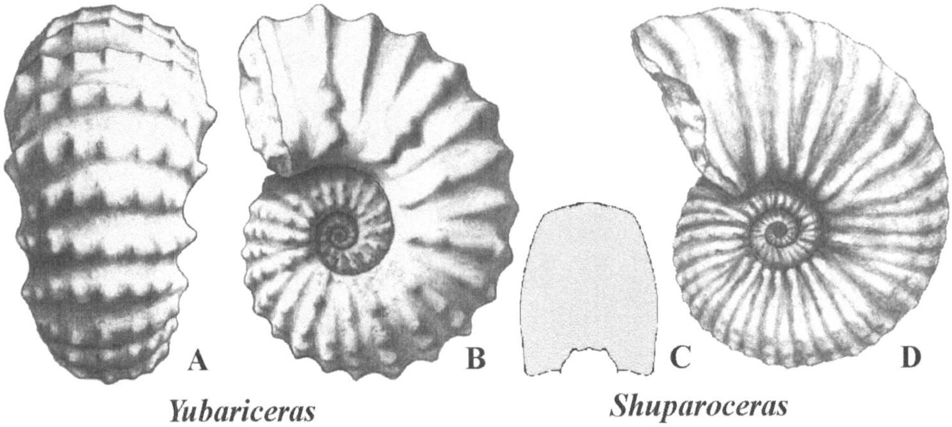

Fig. 379. Romaniceratinae.

Subfamily Metoicoceratinae Hyatt, 1903

Discussion. Monophyletic derivatives of Plesiacanthoceratinae, close to *Ottohaasites*; typically discoidal, compressed and very involute, with subdued ventrolateral tuberculation, and narrow tabulate venter with inner and outer ventrolateral clavi. *U. Cenom.*

Praemetoicoceras COOPER, 1998 [*Metoicoceras whitei praecox* Haas, 1949; OD]. Medium sized; inner whorls like *Ottohaasites* but siphonal tubercles lost early (D<35 mm); outer whorls involute (U = 19-20%), compressed, with narrow venter and weak to absent tuberculation; suture with moderately incised elements. *Lower U. Cenom.* **Fig. 382G-H** - *P. praecox* (Haas).

Cryptometoicoceras KENNEDY & COBBAN, 1990 [*C. mite*; OD]. Dwarf offshoot of *Praemetoicoceras*, with flat flanks, quadratic whorl section; prorsiradiate ribs with conical outer ventrolateral tubercles on body chamber, crossing venter transversely. *Lower U. Cenom.* **Fig. 382K-L** - *C. mite* Kennedy & Cobban.

Metoicoceras HYATT, 1903 [*Ammonites swallovi* Shumard, 1859; OD]. Medium sized to moderately large; like *Praemetoicoceras*, but more involute with siphonal tubercles lacking at all stages; inner whorls with tabulate venter; suture more complex than *Praemetoicoceras*. *U. Cenom.* **Fig. 382M-N** - *M. swallovi* (Shumard).

Nannometoicoceras KENNEDY, 1988 [*Metoicoceras acceleratum* Hyatt, 1903; OD]. Dwarf offshoot of *Metoicoceras*, with flat sides, tabulate venter and very narrow umbilicus; inner whorls with paired ventrolateral tubercles, with or without long and short ribs; adult body chamber with smooth inner flanks and strongly ribbed outer flanks, ribs flexing forward at ventrolateral shoulder and crossing venter transversely. *U. Cenom.* **Fig. 382C-D** - *N. acceleratum* (Hyatt).

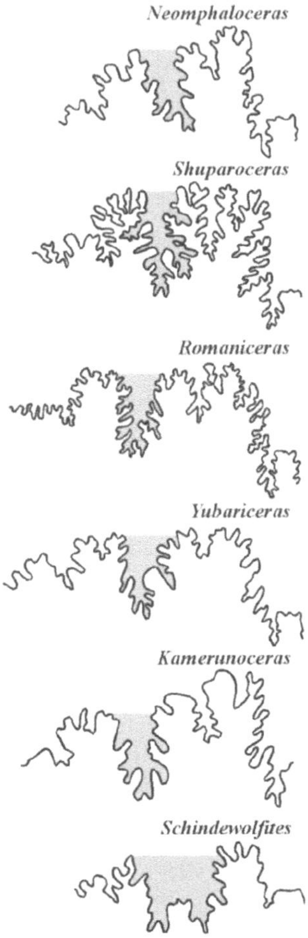

Fig. 380. Romaniceratine sutures.

Buccinammonites KENNEDY & COBBAN, 1990 [*B. minimus*; OD]. Minute, very involute; inner whorls with depressed whorls, sparse coarse ribs ending in strong ventrolateral tubercles; ribs become fine, with minute tubercles; adult

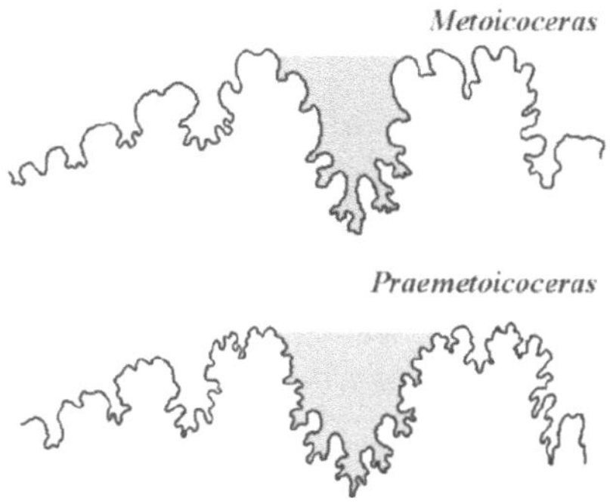

Fig. 381. Metoicoceratine sutures.

body chamber with flared aperture. *U. Cenom.* **Fig 382E-F** - *B. minimus* Kennedy & Cobban.

Subfamily Mammitinae Hyatt, 1903

Discussion. Monophyletic derivatives of Metoicoceratinae, close to *Metoicoceras* (Kennedy *et al.*, 1980); generally with trapezoidal to quadratic whorl section, sparse robust ribbing and prominent umbilical and inner and outer ventrolateral tubercles; ornament tends to decline in later growth becoming obsolete in many. Comprises the tribes Fallotitini, Mammitini and Mitoniini. *Upper U. Cenom. - L. Coni.*

Tribe Fallotitini Wiedmann, 1960

(*nom. transl.* herein *ex* Fallotitinae)

Type genus: *Fallotites* Wiedmann, 1960.

Diagnosis. Characteristically with ventral tuberculation that declines in later growth and a smooth adult body chamber. *Upper U. Cenom. - L. Turon.*

Jeanrogericeras WIEDMANN, 1960 [*Ammonites revelierianus* Courtiller, 1867; OD]. Like *Metoicoceras* but more inflated, with trapezoidal to subquadratic whorl section, broader venter, strong sparser conical umbilical tubercles; highly variable, from compressed and weakly ornamented to inflated and strongly ornamented; outer whorls compressed and high whorled; suture with narrow E, broad quadratic asymmetrically bifid E/L, narrow L as deep as E, and weakly indented L/U_2. *Upper U. Cenom. - L. Turon.* **Fig. 385A-B** - *J. revelierianum* (Courtiller).

Fallotites WIEDMANN, 1960 [*Vascoceras subconciliatum* Choffat, 1898; OD]. Inner whorls like *Jeanrogericeras*; outer whorls remaining broadly trapezoidal and depressed, with venter tending to round and with loss of all ornament except large round umbilical tubercles. *Upper U. Cenom - L. Turon.* **Fig. 385G-H** - *F. subconciliatus* (Choffat).

Spathitoides WIEDMANN, 1960 [*Neoptychites (S.) sulcatus*; OD]. Inner whorls unknown; outer whorls like *Jeanrogericeras* with narrow sulcate venter and rounded undulatory ventrolateral keels but ribs and tubercles obsolete; a homoeomorph of *Neoptychites*. *L. Turon.* **Fig. 385E-F** - *S. sulcatus* (Wiedmann).

Ingridella WIEDMANN, 1960 [*Vascoceras malladae* Fallot, 1931; OD]. Very evolute, inflated, with broadly rounded venter and depressed reniform whorl section; inner whorls with umbilical bullae, very weak inner and outer ventrolateral tubercles and flat venter; ribs lacking at all stages; ventral tubercles lost early and in middle and adult stages venter rounds and only large rounded umbilical tubercles remain; suture simple, with deep E, moderately wide, shallowly incised, trifid E/L, narrow L that is not as deep as E, and narrow weakly trifid L/U_2; yet another homoeomorph of *Vascoceras*. *L. Turon.* **Fig. 385O-P** - *I. malladae* (Fallot).

Reymenticoceras KENNEDY, 2017 [*Mammites nodosoidesappelatus* Etayo-Serna, 1979; OD]. Small, moderately evolute, with quadratic whorl section; prominent umbilical bullae weakly connected, singly and in pairs, to prominent paired ventrolateral clavi. *L. Turon.* **Fig. 386** - *R. nodosoidesappelatus* (Etayo-Serna).

Tribe Mammitini Hyatt, 1903

(*nom. transl.* herein *ex* Mammitinae)

Diagnosis. Like ancestral Fallotitini, but generally larger and retaining strong ornament to maturity, with a tendency for the ventrolateral tubercles to coalesce and form blunt horns. *L. Turon.*

***Powellites* gen. nov.** [*Mammites powelli* Kennedy, Wright & Hancock, 1987, p. 42; OD herein). Inner whorls like *Jeanrogericeras*; in later growth evolute, with slightly depressed subquadratic whorl section, sparse strong distant robust rectiradiate primary ribs separated by 1-2 intercalatories and prominent umbilical and inner and outer ventrolateral tubercles, the ventrolaterals coalescing to form prominent auriculate bullae that almost join across the venter. *Lower L. Turon.* **Fig. 388K-M** - *P. powelli* (Kennedy, Wright & Hancock).

Ramphidoceras KENNEDY & COBBAN, 1990 [*R. saxatile*; OD]. A progenetic dwarf, perhaps of *Powellites*; moderately involute, with flattened flanks, rounded venter and, in maturity, compressed oval whorl section; strong thick rounded slightly flexuous ribs arise singly from pronounced conical umbilical tubercles that become bullate on outer whorl, with intervening intercalatories, and cross venter with slight forward curvature. *L. Turon.* **Fig. 388E-F** - *R. saxatile* Kennedy & Cobban.

Mammites LAUBE & BRUDER, 1887 [*Ammonites nodosoides* Schlüter; SD Roman, 1938]. Like *Powellites* but larger, more compressed and involute, with ribbing effaced from

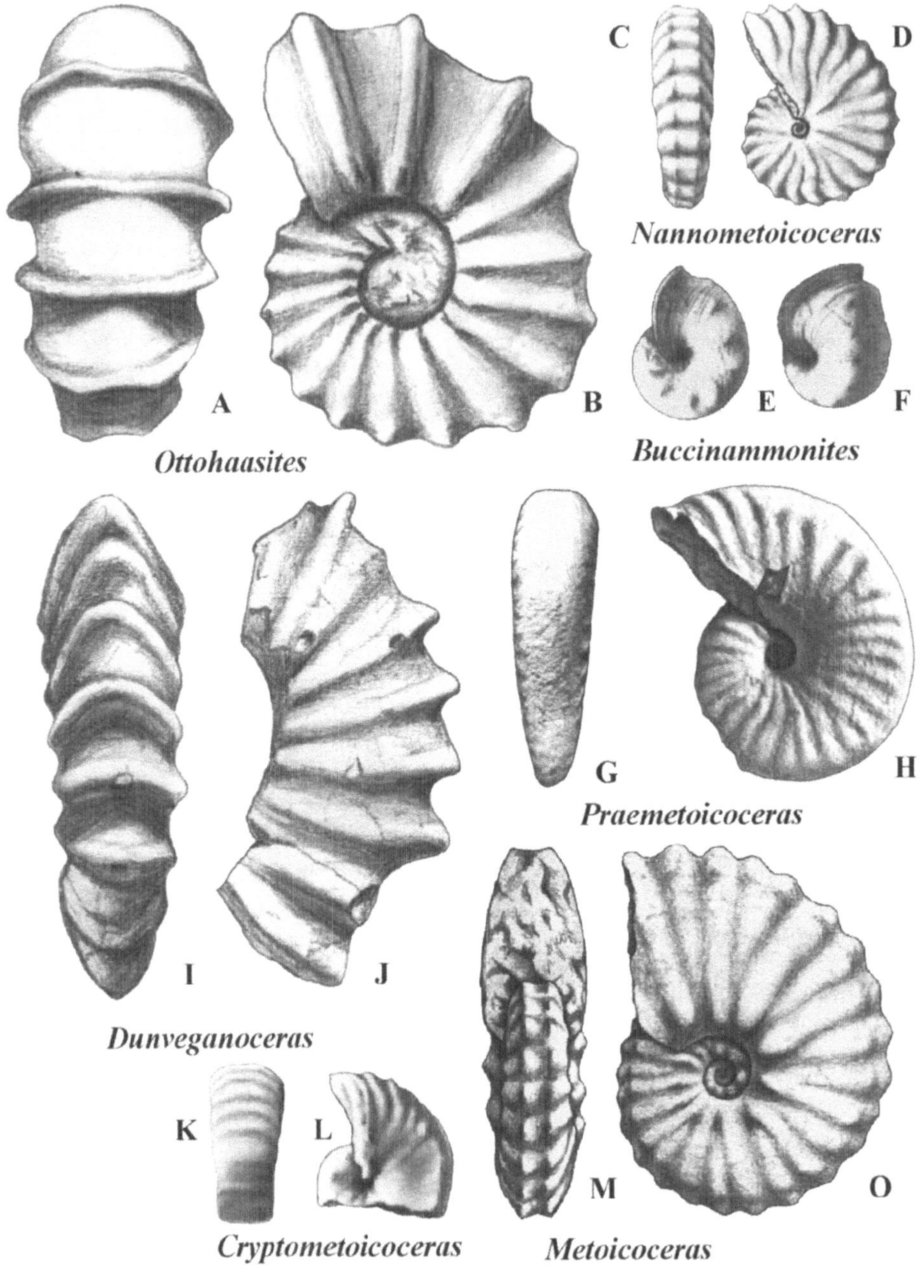

Fig. 382. Metoicoceratinae.

midflank and, in maturity, inner and outer ventrolaterals coalescing to form huge backwardly and outwardly directed horns. *Upper L. Turon*. **Fig. 388A-D** - *M. nodosoides* (Schlüter).

= *Schlueterericeras* Hyatt, 1903 (obj.).

Tribe Mitoniini Renz & Alvarez, 1979

(*nom. correct. et transl.* herein *ex* Mitoniainae)

Discussion. Derived from Mammitini, close to *Powellites*; characteristically very evolute with coronate whorl section, prominent umbilical tubercles in mid-lateral position and only single ventrolateral bullae. *L. Turon.*

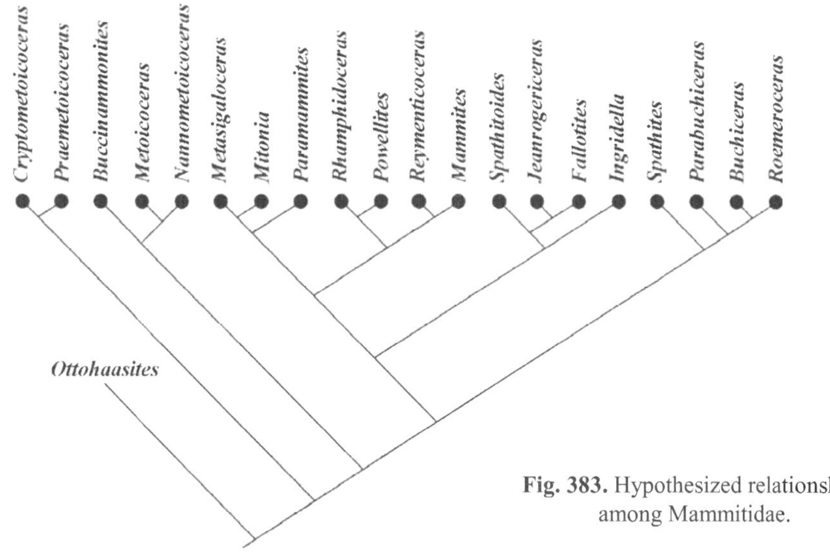

Fig. 383. Hypothesized relationships among Mammitidae.

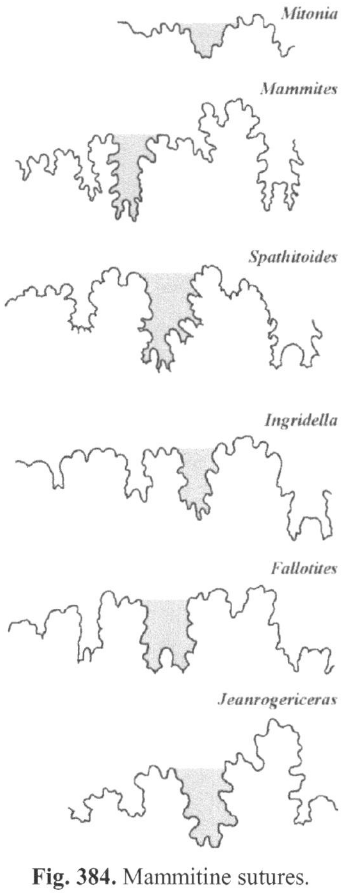

Fig. 384. Mammitine sutures.

Paramammites FURON, 1935 [*Vascoceras polymorphum* Pervinquière, 1907; SD Reyment, 1954]. Like *Powellites* with ribs bifurcating from prominent umbilical tubercles and ventrolateral tubercles declining with growth; suture simple, with narrow E, broad barely incised saddles, fairly broad bifid L almost as deep as E and much reduced U_2. *L. Turon.* **Fig. 388N-O** - *P. polymorphum* (Pervinquière).

Metasigaloceras HYATT, 1903 [*Ammonites rusticus* J. Sowerby; OD]. Very large, evolute, with depressed trapezoidal whorl section; massive sparse umbilical tubercles shifted to midflank, leaving broad sloping unornamented umbilical wall; weak ribs arise in pairs from lateral tubercles and terminate in prominent ventral bullae, leaving siphonal line flat and unornamented. *Upper L. Turon.* **Fig. 388G-H** - *M. rusticum* (J. Sowerby).

Mitonia RENZ & ALVAREZ, 1979 [*M. venezolana*; OD]. A progenetic dwarf; like *Metasigaloceras* but more involute, whorls not so depressed, tubercles weaker, ribs stronger on adult body chamber, and suture simpler. *L. Turon.* **Fig. 388I-J** - *M. venezolana* Renz & Alvarez.

Subfamily Buchiceratinae Hyatt, 1903

(*nom. transl.* herein *ex* Buchiceratidae)

Discussion. Monophyletic derivatives of Mammitinae, close to *Jeanrogericeras* (Fallotitini); typically with subdued ventral tuberculation, simplified suture and, in derived representatives, single ventrolateral tubercles and low blunt siphonal ridge in early growth. *M. Turon. - L. Coni.*

Spathites KUMMEL & DECKER, 1954 [*S. chispaensis*, = *Pseudotissotia? coahuilaensis* Jones, 1938; OD]. Inner whorls like *Jeanrogericeras*; outer whorls virtually smooth, like *Spathitoides*, but with subrounded ventro-

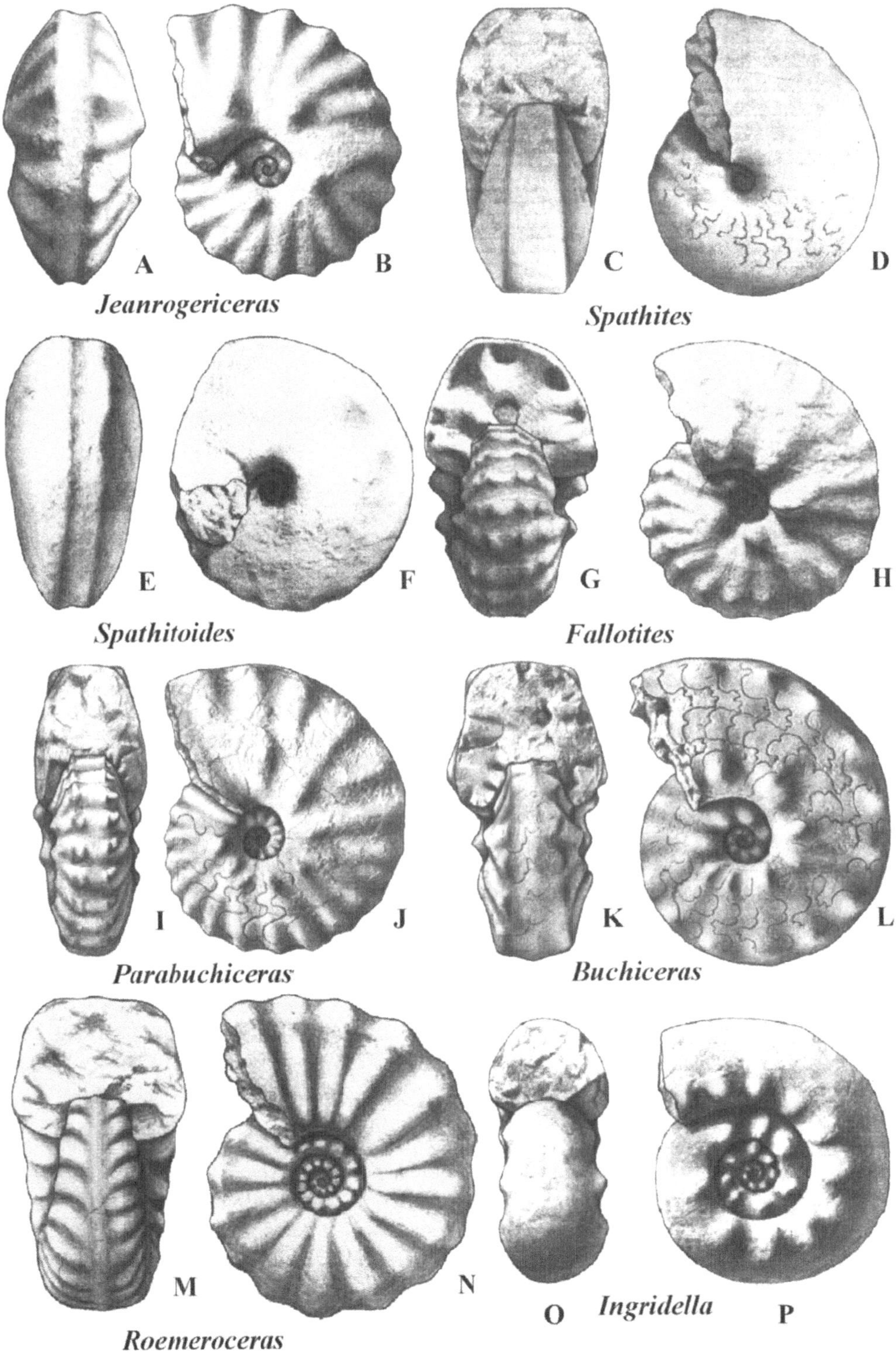

Fig. 385. Mammitinae.

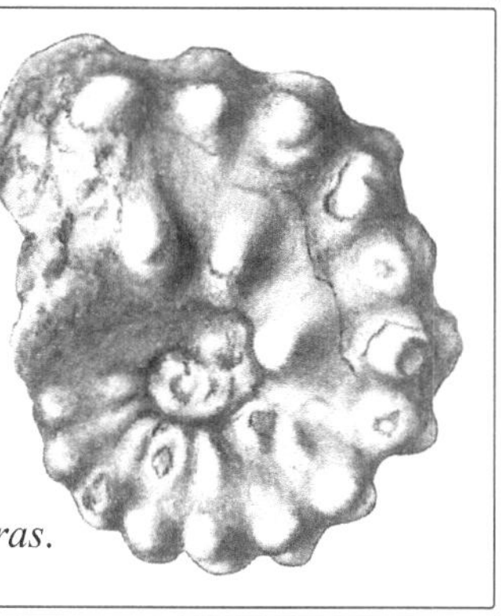

Fig. 386. *Reymenticoceras.*

lateral shoulders and broad tabulate to shallowly concave venter with low transverse undulations; suture with L not as deep as E. *M. Turon.* **Fig. 385C-D** - *S. coahuilaensis* (Jones).

= *Amblydiscus* Adkins, 1933b (*nom. nud.*).

Parabuchiceras COOPER, 1998 [*Buchiceras puercoense* Herrick & Johnston, 1900; OD]. Like *Jeanrogericeras*, but with narrower umbilicus, broader venter, more numerous umbilical bullae, rectiradiate to prorsiradiate ribs and simplified almost pseudoceratitic suture with relatively short E, and broad shallower L with minute denticulations. *U. Turon.* **Fig. 385I-J** - *P. puercoense* (Herrick & Johnston).

Buchiceras HYATT, 1875 [*B. bilobatum*; OD]. Like *Parabuchiceras* but more inflated, with quadratic adult whorls, stronger umbilical tubercles, sparser ribs that are weakly connected to prominent single ventrolateral tubercles that persist to maturity, and broader venter with low siphonal ridge in early to middle growth; suture with narrow L. *Coni.* **Fig. 385K-L** - *B. bilobatum* Hyatt.

Roemeroceras HYATT, 1903 [*R. gabbi*; SD Diener, 1925]. Like *Buchiceras* but more evolute and inflated, with depressed trapezoidal whorl section, weaker more numerous umbilical bullae, simple strong rectiradiate ribs and suture with E deeper than L; at present only the poorly known type species belongs here, other referred species being reassigned to *Buchiceras*. *Coni.* **Fig. 385M-N** - *R. gabbi* Hyatt.

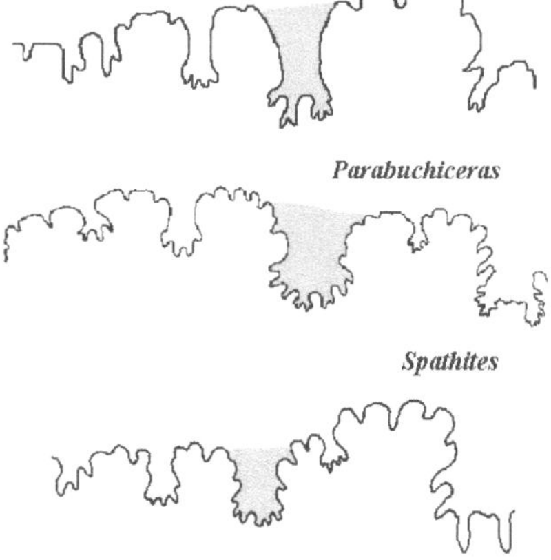

Fig. 387. Buchiceratine sutures.

Hyporder COLLIGNONICERATOIDEI nov.

Diagnosis. Monophyletic derivatives of Acanthoceratoidei; comprises the superfamilies Collignoniceratoidea, Tissotioidea and Sphenodiscoidea. *L. Turon. - U. Maastr.*

Superfamily COLLIGNONICERATOIDEA Wright & Wright, 1951

(*nom. transl.* Cooper, 1991 *ex* Collignoniceratidae)

Discussion. Monophyletic derivatives of Mammitoidea; comprises the families Collignoniceratidae, Peroniceratidae and Texanitidae. *L. Turon. - U. Camp.*

Family COLLIGNONICERATIDAE Wright & Wright, 1951

Discussion. Monophyletic derivatives of Plesiacanthoceratidae, close to *Thomelites* (Watinoceratinae) (Wright & Kennedy, 1990; Wright, 1996); typically relatively large, moderately evolute, with quadrangular whorls, prominent umbilical bullae, strong robust ribs with single or double ventrolateral tubercles that may coalesce to form horns, and siphonal clavi matching ribs. *L. - U. Turon.*

Cibolaites COBBAN & HOOK, 1983 [*C. molenaari*; OD]. Moderately involute, compressed to inflated, with inner whorls like *Thomelites*; flattened flanks converge to strongly fastigiate venter; prominent umbilical tubercles give rise to pairs of rather distant strong robust rectiradiate to prorsiradiate ribs, with occasional intercalatories; umbilical bullae weaken on adult body chamber; all ribs with ventrolateral and siphonal clavi; suture with rather wide E, rectangular asymmetrically bifid E/L and narrow asymmetrically bifid L; a homoeomorph of *Forresteria*. *L. Turon.* **Fig. 389D-E** - *C. molenaari* Cobban & Hook.

Collignoniceras BREISTOFFER, 1947 [*Ammonites woollgari* Mantell, 1822; ICZN Opinion 861]. Large, evolute; inner whorls compressed with flat parallel sides, subrectangular whorl section, strong simple bar-like prorsiradiate ribs, sometimes with intercalatories, that arise from weak umbilical bullae and carry subconical inner and clavate outer ventrolateral tubercles; ventrolaterals more or less matched in number by prominent asymmetrical siphonal

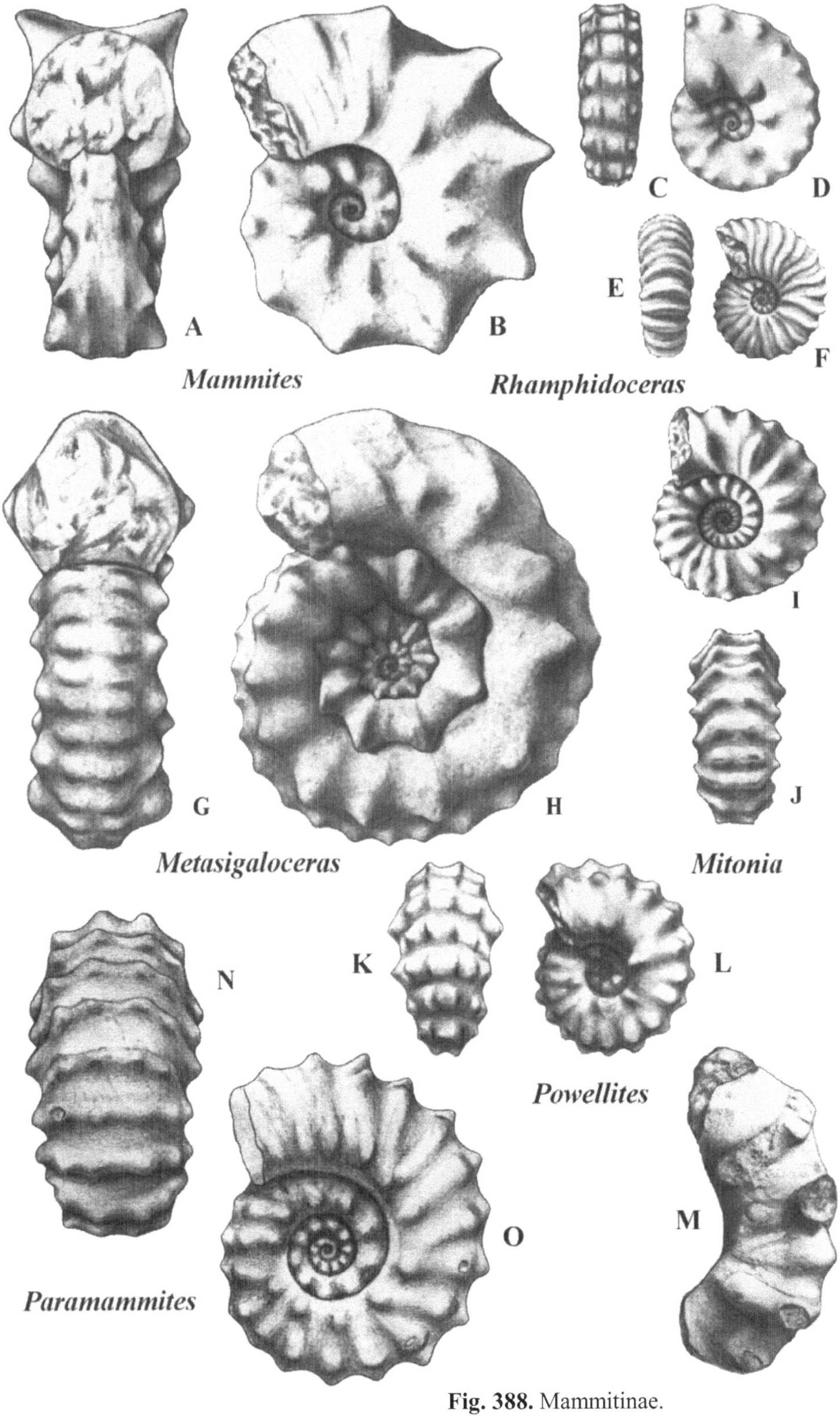

Fig. 388. Mammitinae.

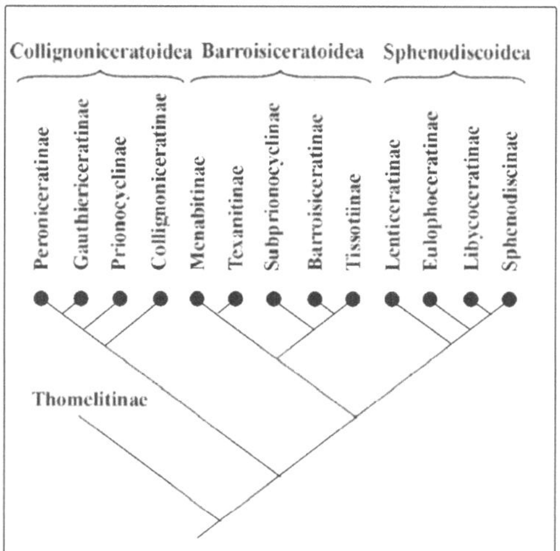

Fig. 388. Hypothesized relationships among Collignoniceratoidei.

clavi; on microconchs tuberculation weakens on body whorl and ribs form chevrons over venter; in macroconchs ribbing coarsens markedly, umbilicals migrate onto flank and ventrolaterals amalgamate to form conspicuous horns that may even absorb umbilical tubercles; suture with square asymmetrically bifid E/L, narrow asymmetrically bifid L and U on umbilical shoulder. *M. Turon.* **Fig. 389A-C** - *C. woollgari* (Mantell).

= *Prionotropis* Meek, 1876 (*non* Fieber, 1853).

Lecointriceras KENNEDY, WRIGHT & HANCOCK, 1980 [*Ammonites fleuriausianus* d'Orbigny, 1841; OD]. Inner whorls much like *Cibolaites* but with prominent sparse conical umbilical bullae giving rise to pairs of low broad ribs with occasional intercalatories; all ribs with outer ventrolateral and siphonal clavi, but appearance and persistence of inner ventrolaterals variable; in middle growth venter flattens, ventrolateral tubercles fuse to form blunt horns, and siphonal clavi may amalgamate into blunt low undulatory keel; close to adult aperture all ornament lost and venter rounds. *L. Turon.* **Fig. 389J-K** - *L. fleuriausianum* (d'Orbigny).

Selwynoceras WARREN & STELCK, 1940 [*Prionotropis borealis* Warren, 1930; OD]. Like *Collignoniceras* but smaller, with distant ribs, persistent umbilical bullae, early development of flared ventrolateral flanges, traces of looped ventral ribs and even ventral tuberculation, without horn development. *M. Turon.* **Fig. 389N-O** - *S. boreale* (Warren)

Sornayella **gen. nov.** [*Prionotropis turoniense* Sornay, 1951; OD herein]. Like *Collignoniceras* but more involute, coarsely ribbed inner whorls with prominent umbilical bullae; in maturity generally with depressed whorl section, convergent flanks, fastigiate venter, prominent umbilicolateral bullae, inner and outer ventrolaterals coalescing to form prominent ventrolateral clavi joined across venter by looped ribs all of which bear rounded siphonal tubercles. *M. Turon.* **Fig. 385F-G** - *S. turoniensis* (Sornay); **Fig. 389H-I** - *S. papalis* (d'Orbigny).

Collignonicerites **gen. nov.** [*Ammonites percarinatus* Hall & Meek, 1856, p.396; OD herein]. Very small (D<35 mm), compressed, evolute (U=32-45%), with flat subparallel flanks, narrow fastigiate venter and dense flexuous sharp periodically strengthened ribs that project forward on venter; all ribs with variably developed inner (nodate) and outer (clavate) ventrolateral tubercles and strengthened ribs with weak umbilical bullae; siphonal keel with clavi corresponding to ribs; suture with broad bifid E/L and L/U_2, and narrow poorly divided L. *M. Turon.* **Fig. 389L-M** - *C. percarinatus* (Hall & Meek).

Prionocyclites KENNEDY, 1988 [*P. mite*; OD]. Micromorph; strongly prorsiradiate ribs arising singly or in pairs from umbilical bullae, terminating in blunt oblique ventrolateral clavi; coarsely undulatory siphonal keel with fewer notches than ribs; last part of body chamber almost smooth, with distant umbilical bullae, weak ribs backed by shallow constrictions, and deep apertural constriction; suture simple with little incised E/L, narrower bifid L which is not as deep as E, and broad irregularly rounded L/U_2. *U. Turon.* **Fig. 392H-I** - *P. mite* Kennedy.

Family PERONICERATIDAE Hyatt, 1900

Discussion. Monophyletic derivatives of Collignoniceratidae with finely serrated to entire siphonal keel; comprises the subfamilies Prionocyclinae, Gauthiericeratinae and Peroniceratinae. *U. Turon. - Coni.*

Subfamily Prionocyclinae Breistroffer, 1947

(*nom. transl.* Cooper, 1991 *ex* Prionocyclidae)

Discussion. Monophyletic derivatives of Collignoniceratinae, close to *Collignoniceras*; typically evolute, with quadratic whorls, subtabulate venter, irregular ribs with single or double ventrolateral tubercles with tendency to form spines, and finely serrated siphonal keel. A paraphyletic taxon. *M.Turon. - U. Coni.*

Prionocyclus MEEK, 1871 [*Ammonites serratocarinatus* Meek (*non* Stoliczka, 1865), = *Prionocyclus wyomingensis* Meek, 1876; SD Meek, 1876]. Very large; moderately evolute (U=30-37%), compressed, with flat flanks, subrectangular whorl section and finely serrated siphonal keel; ribbing irregular, conspicuously differentiated into strong occasionally flared ribs with elongate umbilical bullae that may split to form low lateral tubercles, sometimes bundled ribbing; nodate inner and obliquely clavate to bullate outer ventrolateral tubercles but latter often lost early; ornament may coarsen

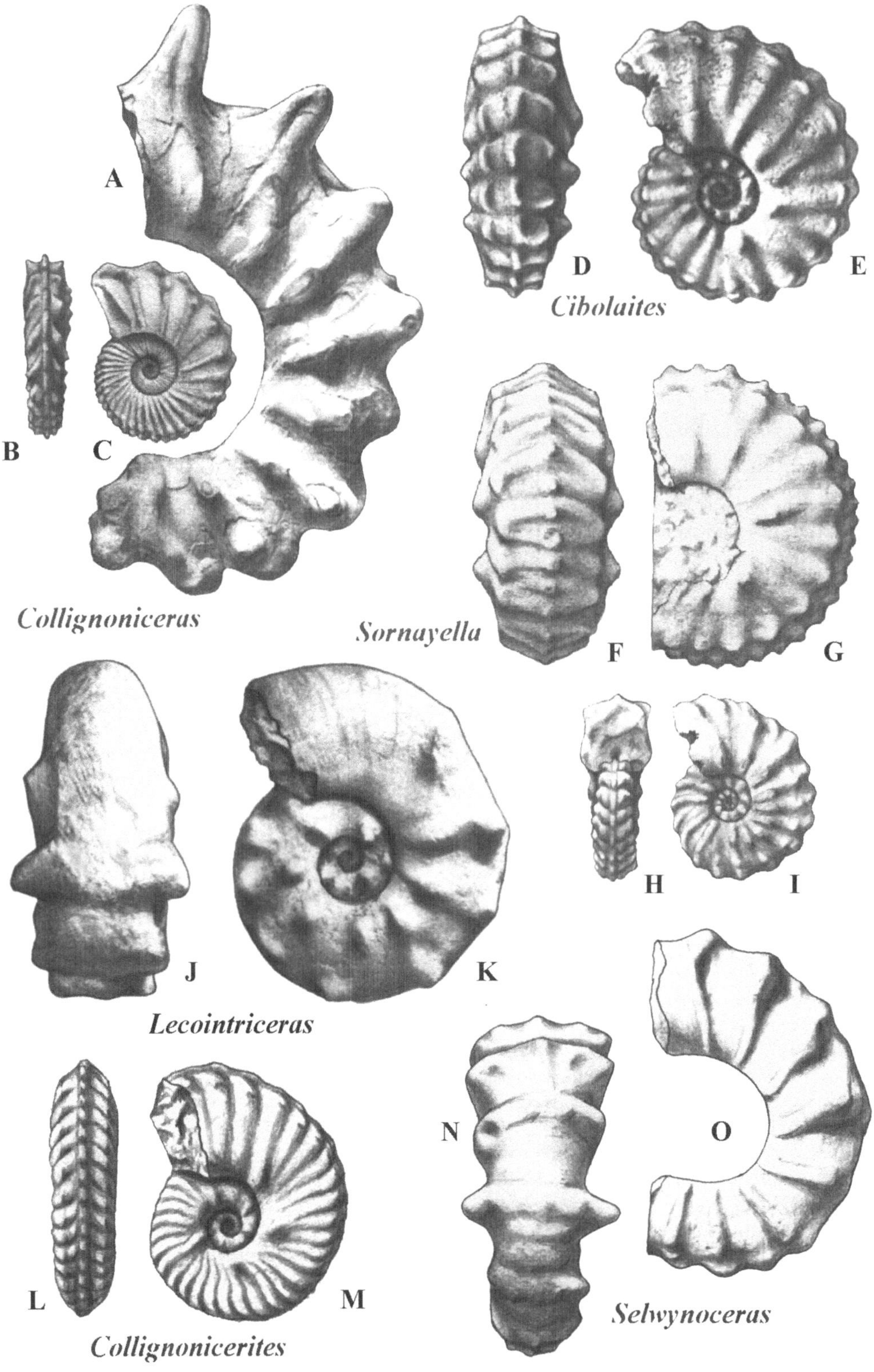

Fig. 389. Collignoniceratinae.

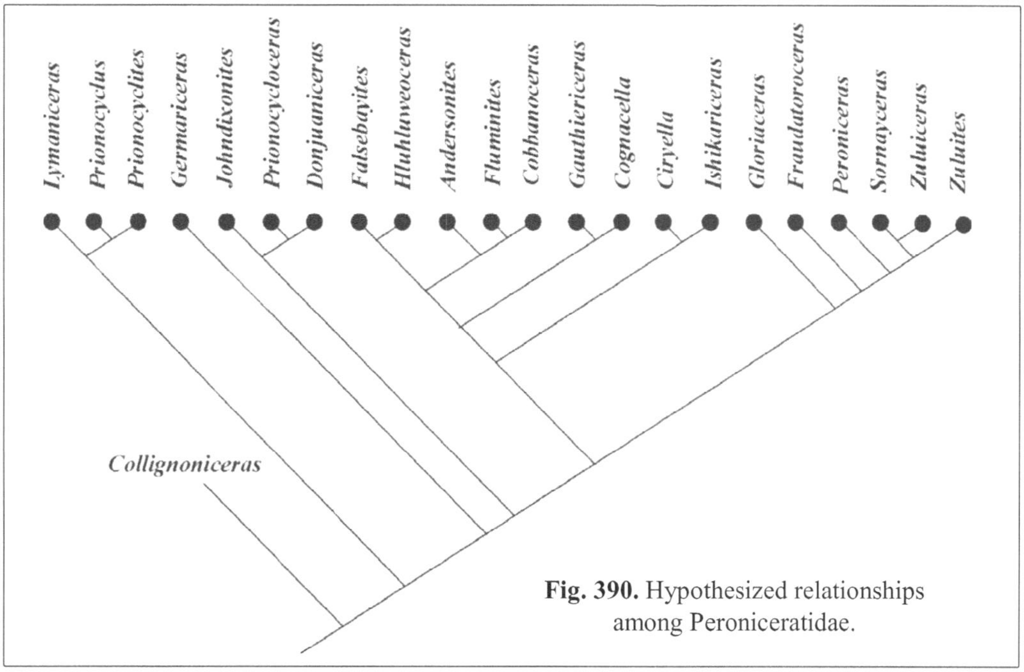

Fig. 390. Hypothesized relationships among Peroniceratidae.

markedly in later growth when some develop flange-like horns; suture with deep E and narrow bifid E/L much longer than E. *M. - U. Turon.* **Fig. 392A-B** - *P. wyomingensis* Meek; **Fig. 392C-D** - *P. macombi* (Meek) (gracile form); **Fig. 395** - *P. hyatti* (Stanton).

= *Spathinella* Shimizu, 1935 (*nom. nud.*).

Lymaniceras MATSUMOTO, 1965 [*L. planulatum*; OD]. Very small; like *Prionocyclus* but with more regular ribs bifurcating from prominent umbilical bullae, single weak ventrolateral clavi; almost smooth adult body chamber with weak flattened ribs, rounded venter and suture with squarer E/L and broader L/U_2. *U. Turon.* **Fig. 392P-Q** - *L. planulatum* Matsumoto.

Germariceras BREISTROFFER, 1947 [*Ammonites germari* Reuss, 1845; OD]. Medium sized; like *Prionocyclus* but with undifferentiated finer sharper more regular generally single ribs to inner whorls, early (D<15 mm) loss of outer ventrolateral tubercles, and loss of umbilical tubercles in later growth; low lateral tubercles never develop; ribs project strongly forward on venter. *Upper U. Turon.* **Fig. 392E-G** - *G. germari* (Reuss).

Prionocycloceras SPATH, 1926 [*Prionocyclus guayabanus* Gerhardt, 1897; OD]. Large; like *Germariceras* but with distant straight to weakly flexuous ribbing of irregular strength from early growth, and tendency for umbilical and inner ventrolateral tubercles to form spinose horns; outer ventrolateral clavi becoming obsolete in later growth. *L. - M. Coni.* **Fig. 392L-M** - *P. guayabanum* (Gerhardt).

Donjuaniceras BASSE, 1950 [*Peroniceras (D.) longispinata*; OD]. Medium sized; like *Prionocycloceras* but more evolute; inner whorls with numerous small umbilical bullae giving rise to fine strongly prorsiradiate ribs, both long and short; flank costae recurve in later growth with oblique ventrolateral clavi, and narrow L to suture. *L. Coni.* **Fig. 392J-K** - *D. longispinatum* (Basse).

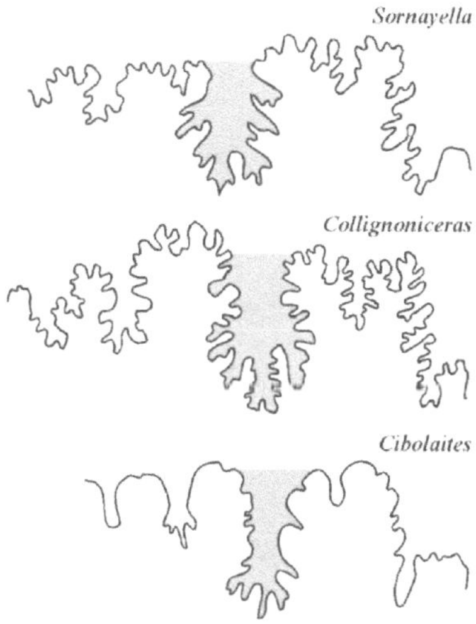

Fig. 391. Collignoniceratine sutures.

Johndixonites **gen. nov.** [*Prionocycloceras multicostatum* Collignon, 1965, p.45; OD herein]. Moderately large, very evolute, with a subquadratic whorl section; inner whorls with simple rectiradiate ribs, narrower than interspaces, with umbilical bullae and ventrolateral spines; on outer whorl ventrolateral spines reduced to small tubercles and finally obsolete when ribs on venter project forward to connect with thick cordate siphonal keel. *M.- U. Coni.* **Fig. 392N-O** - *J. multicostatus* (Collignon).

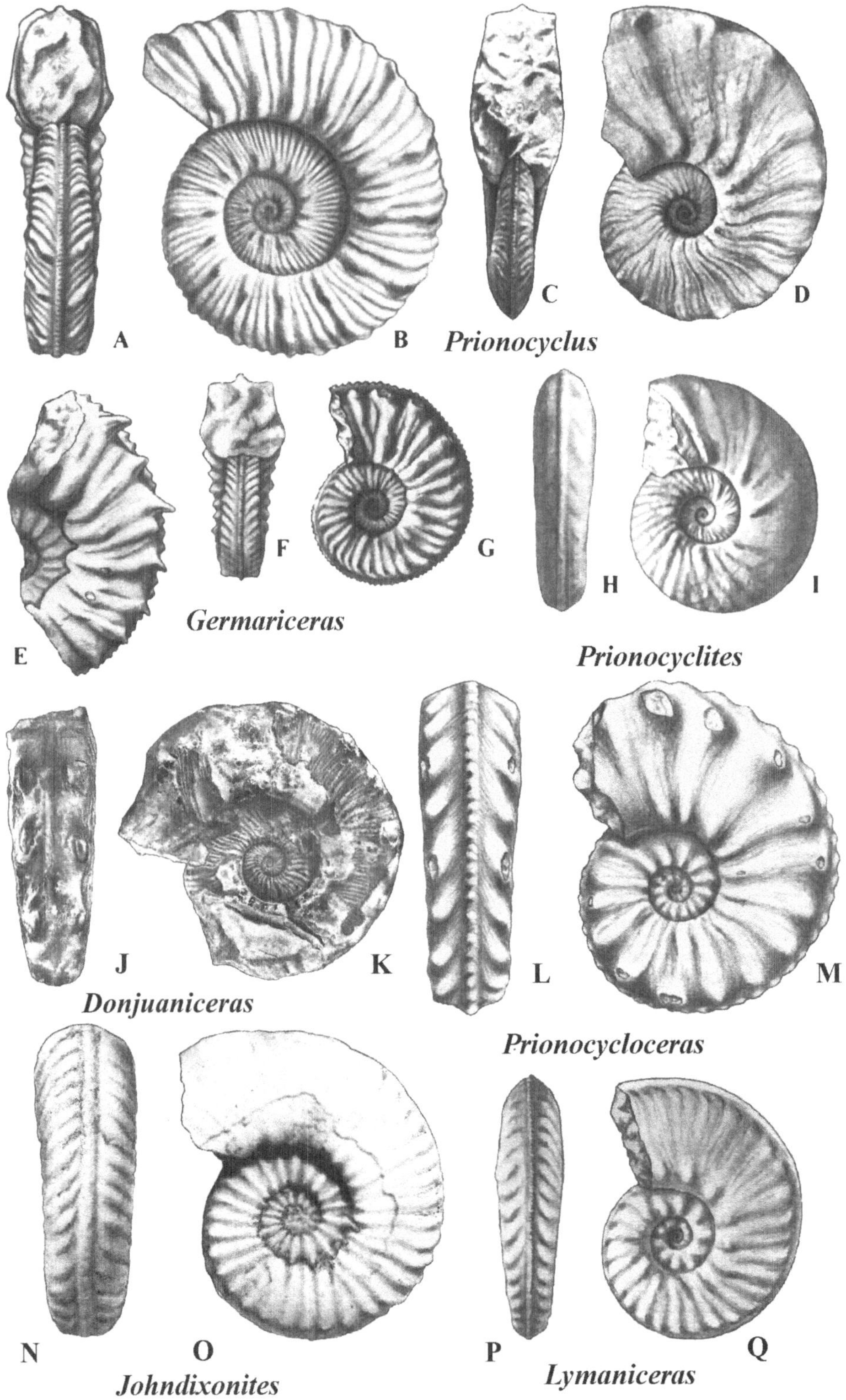

Fig. 392. Prionocyclinae.

Subfamily Gauthiericeratinae van Hoepen, 1955

(*nom. transl.* van Hoepen, 1965 *ex* Gauthiericeratidae)

Discussion. Monophyletic derivatives of Prionocyclinae, close to *Prionocyclus* (Matsumoto, 1965); generally with rather coarse distant regular bituberculate ribs that tend to weaken across the flanks and a finely crenulate to entire siphonal keel; suture with U much smaller than L and slightly retracted to non-retracted suspensive lobe. *L. - U. Coni.*

Ciryella WIEDMANN, 1960 [*Gauthiericeras (C.) vascogoticum*; OD]. Medium sized; compressed, evolute (U=40%), with arched unicarinate venter, umbilical bullae remaining on umbilical shoulder and giving rise to low blunt ribs that frequently bifurcate and terminate in low ventrolateral clavi; suture relatively simple with narrow tapering L, and U_2 situated well above umbilical shoulder. *L. Coni.* **Fig. 398Q** - *C. vascogotica* (Wiedmann).

Ishikariceras MATSUMOTO, 1965 (*I. binodosum*; OD]. Medium sized; like *Ciryella* but more evolute, with subquadratic inner whorls, subelliptical outer ones, straight mostly simple ribs with prominent umbilical and ventrolateral tubercles, loss of ornament in maturity, and suture with broader L and U_2 positioned on umbilical shoulder; a homoeomorph of *Yabeiceras* (Barroisiceratinae). *Coni.* **Fig. 394M-N** - *I. schneebeli* (Boulé, Lemoine & Thevenin).

Gauthiericeras DE GROSSOUVRE, 1894 [*Ammonites margae* Schlüter, 1867; OD]. Moderately large; compressed and moderately evolute (U=29-30%), with trapezoidal whorls increasing rapidly in height; broad subfastigiate venter with finely serrated siphonal keel; sparse prorsiradiate ribs with small umbilical and larger low-lateral bullae in middle to later growth from which ribs may branch and terminate in prominent ventrolateral clavi; suture relatively simple, with deep narrow E, broad shallowly bifid rectangular E/L and narrow bifid L. *U. Coni.* **Fig. 394O-P** - *G. margae* (Schlüter).

Hluhluweoceras VAN HOEPEN, 1965 [*H. fugitivum*; OD]. Moderately large; like *Gauthiericeras* but more evolute (U=36%), with quadratic whorl section, blunt round simple rectiradiate ribs, low arched venter, slightly crenulated siphonal keel; no lateral tubercles; suture with narrow E, broad weakly bifid subquadratic E/L and narrow symmetrically trifid L deeper than E. *Upper Coni.* **Fig. 394K-L, 399** - *H. fugitivum* van Hoepen.

Andersonites VAN HOEPEN, 1965 [*A. listeri*; OD]. Moderately large; like *Hluhweoceras* but with slightly concave parallel flanks, depressed quadratic whorl section, stronger more distant rectiradiate ribs, subspinose ventrolateral tubercles, and tricarinate venter; low-lateral bullae lacking; suture with very broad E/L, trifid L and oblique U_2 on umbilical shoulder. *Upper Coni.* **Fig. 394D-F** - *A. listeri* van Hoepen.

For those who regard *Andersonites*, *Hluhluweoceras* and *Falsebayites* as synonyms (*cf.* Klinger & Kennedy, 1984), *Andersonites* is selected to have priority.

Fig. 393. *Fluminites.*

Falsebayites VAN HOEPEN, 1965 [*F. peregrinus*; OD]. Moderately large; like *Hluhluweoceras* but with rounded fastigiate venter crossed by broad low ribs forming chevrons, vestigial siphonal keel and suture with wider L with finger-like folioles and non-oblique U_2. *Upper Coni.* **Fig. 394A-C** - *F. peregrinus* van Hoepen.

Fluminites VAN HOEPEN, 1965 [*F. albus*; OD]. Like *Andersonites* but with compressed inner whorls, weaker tuberculation, convex forward ribs and suture with broader rectangular L. *Upper Coni.* **Fig. 393, 394I-J** - *F. albus* van Hoepen.

Cobbanoceras MATSUMOTO, 1965 [*C. tanakai*; OD]. Moderately large; like *Fluminites* but with 3 well developed ventral keels as in *Andersonites*. *?L. Sant.* **Fig. 394G-H** - *C. tanakai* Matsumoto.

Cognacella **gen. nov.** [*Schloenbachia boreaui* de Grossouvre, 1894, p.111; OD herein]. Small; compressed, involute (U=22%), with high whorls, broad flat subparallel flanks and broadly rounded venter with high siphonal keel flanked by shallow grooves; sparse distant weak umbilical bullae give rise to 1-2 low flexuous prorsiradiate ribs separated by up to 3 intercalatories of different length; ribs project forward on venter, with incipient oblique ventrolateral bullae; suture very simple, with broad, little incised elements. Very similar to *Pseudobarroisiceras* (Hourqiinae). *M. - U. Coni.* **Fig. 394Q-R** - *C. boreaui* (de Grossouvre).

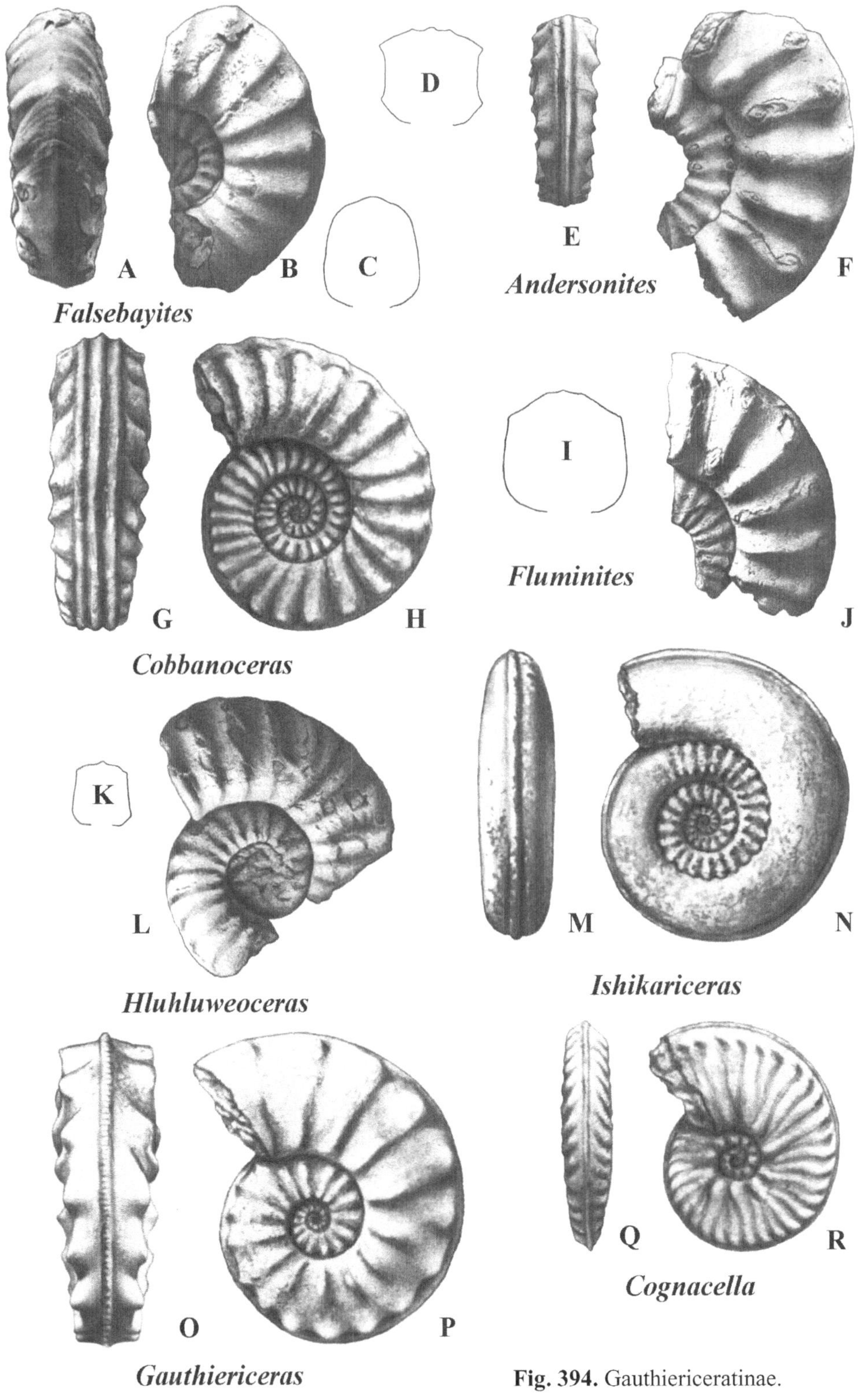

Fig. 394. Gauthiericeratinae.

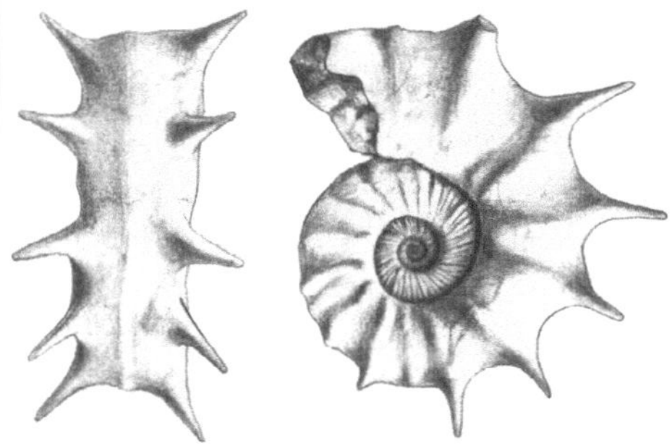

Fig. 395. *Prionocyclus hyatti.*

Subfamily Peroniceratinae Hyatt, 1900

(*nom. transl.* Wright, 1957 *ex* Peroniceratidae)

Discussion. Monophyletic derivatives of Gauthiericeratinae, close to *Ciryella* (Matsumoto, 1965); very evolute to serpenticone, with strong regular ribs, single nail-like ventrolateral clavi and unicarinate to tricarinate (and perhaps even pentacarinate) venter with finely serrated to entire siphonal keel; primitively with suture like Gauthiericeratinae

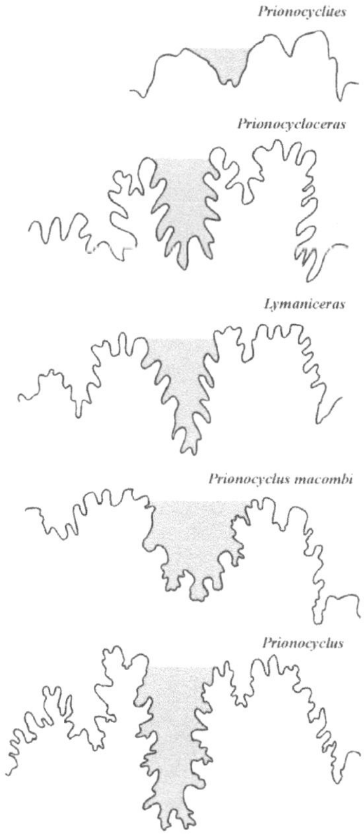

Fig. 396. Prionocycline sutures.

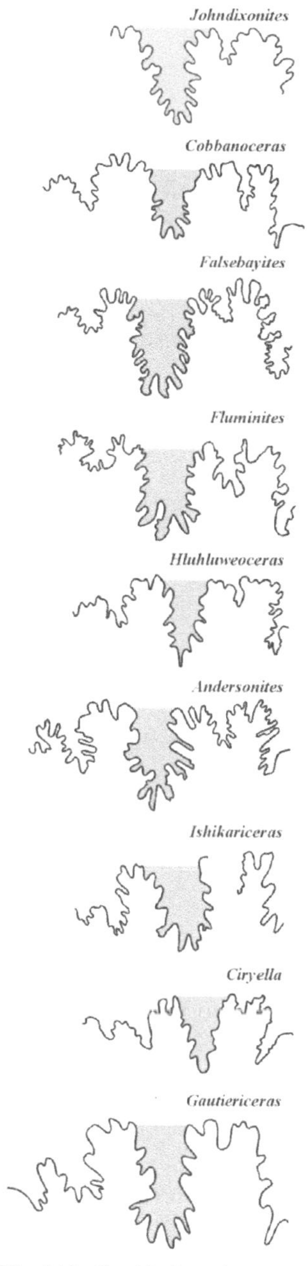

Fig. 397. Gauthiericeratine sutures.

but derived members with highly incised suture with strongly retracted suspensive lobe so that first auxiliary saddle forms an oblique branch off the base of L/U_2. *M. - U. Coni.*

Gloriaceras ETAYO SERNA, 1979 [*G. correai*; OD]. Based on a small phragmocone; compressed, evolute (U=36-41%) with fastigiate bisulcate-tricarinate venter and siphonal keel raised above undulating lateral keels; mostly strong simple concave forward ribs, occasionally bifurcating from umbilical bullae in early to middle

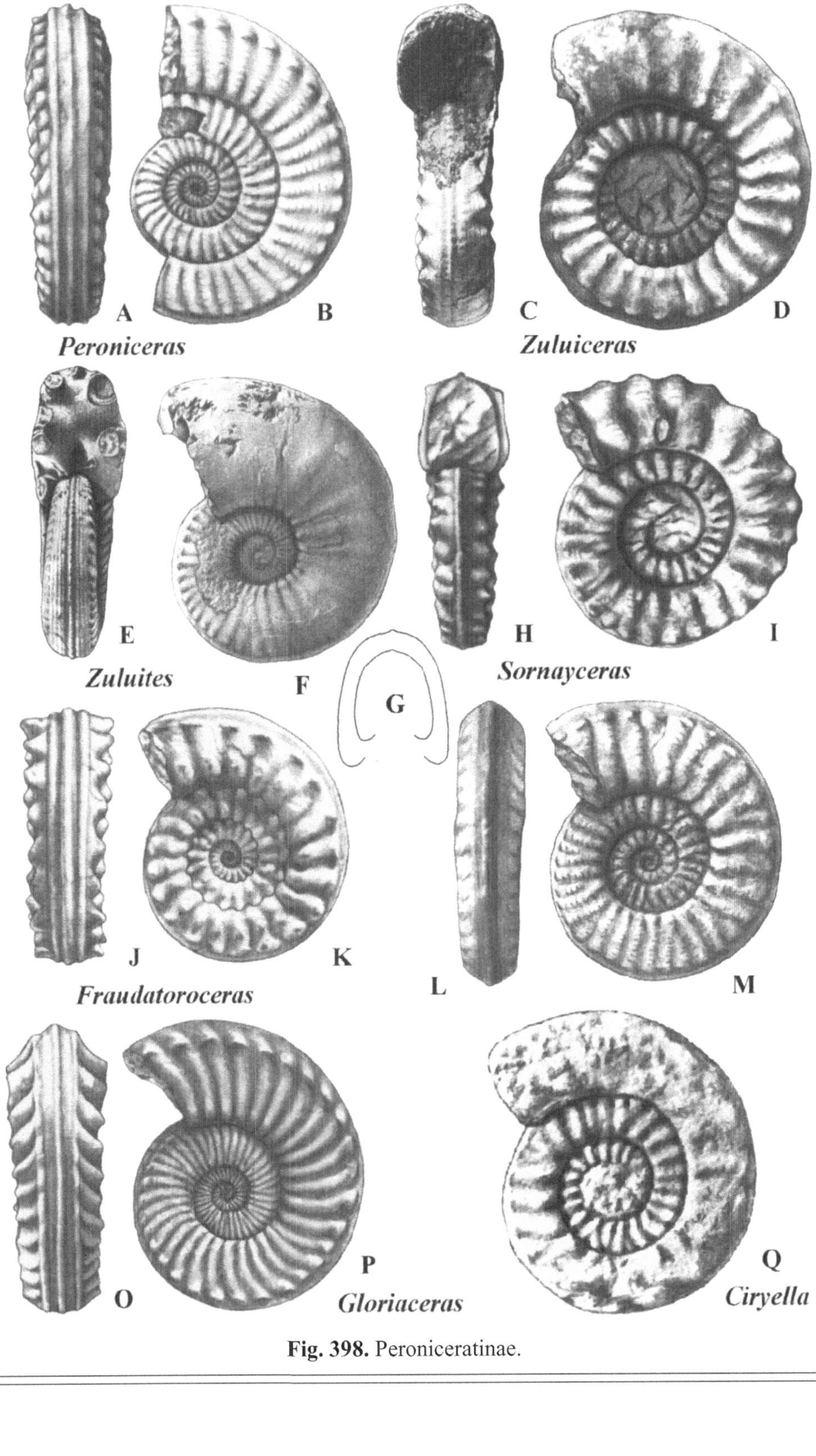

Fig. 398. Peroniceratinae.

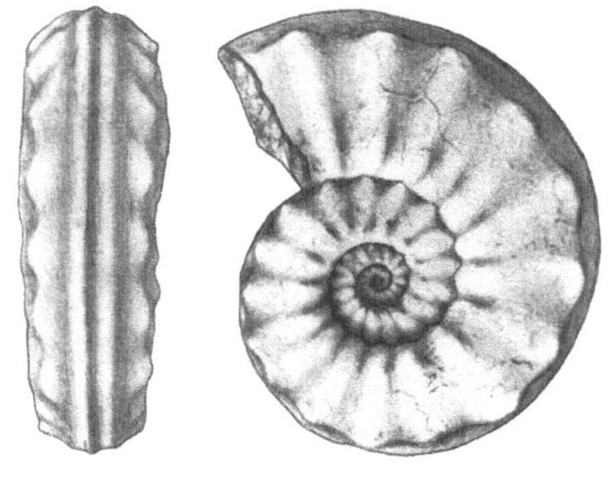

Fig. 399. *Hluhluweoceras*.

growth, with nail-like ventrolateral clavi; ribs project foward on venter to meet lateral keels and, in early growth, form chevrons over venter; umbilical bullae obsolete in later growth; suture with narrow E, sub-rectangular E/L, wide bifid L deeper than E, very narrow L/U_2 and U_2 parallel to L. *M. Coni.* **Fig. 398O-P** - *G correai* Etayo Serna.

Fraudatoroceras VAN HOEPEN, 1965 [*F. besairiei*; OD]. Moderately large; like *Gloriaceras* but more evolute (U = 57%), with lower rounder venter, sparser coarser ribs and swollen umbilicolateral bullae; suture with very narrow lobes. *U. Coni.* **Fig. 398J-K** - *F. besairiei* van Hoepen.

Peroniceras DE GROSSOUVRE, 1894 [*P. moureti*; OD]. Large; like *Gloriaceras* but serpenticone (U=50-56%), with elongate umbilicolateral bullae, broadly rounded low venter and frequently conspicuous spiral ornament; suture with deep narrow E, asymmetrically bifid E/L with finger-like folioles, narrow bifid L as deep as E, long narrow stemmed L/U_2 and strongly retracted susp-ensive lobe with first auxiliary saddle oblique to L/U_2. *M. Coni.* **Fig. 398A-B** - *P. moureti* de Grossouvre; **Fig. 400** - *P. dravidicum* (Kossmat).

Zuluiceras VAN HOEPEN, 1965 [*Peroniceras (Z.) zulu*; OD]. Large; like *Peroniceras* but with arched unicarinate-bisulcate venter, frequently finely notched siphonal keel, whorls increasing more rapidly in height, and lower broader ribs often bifurcating from small umbilical bullae and with vestigial ventrolateral tubercles; rounded venter in later growth. *M. Coni.* **Fig. 398C-D** - *Z. zulu* (van Hoepen).

Sornayceras MATSUMOTO, 1965 [*S. proteus*; OD]. Medium sized; like *Zuluiceras* but with distant robust single rectiradiate ribs and occasional shorter intercalatories, strong umbilicolateral bullae and pronounced rounded to weakly clavate ventrolateral tubercles; spiral ornament present. *Coni.* **Fig. 398H-I** - *S. proteus* Matsumoto.

Introduced simultaneously with *Zuluiceras*, which has priority. The poorly preserved type species of *Sornayceras* is very similar to *Z. regulare* (van Hoepen) and referred species, e.g. *S. omorii* Matsumoto (**Fig. 398L-M**), have finer ribs frequently branching from subdued umbilical bullae and hence approach *Zuluiceras*.

Zuluites VAN HOEPEN, 1965 [*Peroniceras (Z.) modestum*; OD]. Large; like *Zuluiceras* but relatively involute and high whorled, with overhanging umbilical wall, compressed oval whorl section and finer denser ribs declining in later growth so that adult body chamber is smooth. *U. Coni.* **Fig. 398E-G** - *Z. modestum* (van Hoepen).

Superfamily TISSOTIOIDEA Hyatt, 1900

(*nom. transl.* herein *ex* Tissotiidae)

Discussion. Monophyletic derivatives of Collignon-iceratoidea (Wright, 1957), comprising the families Barroisiceratidae, Tissotiidae and Texanitidae. *U. Turon. - U. Camp.*

Family BARROISICERATIDAE Basse, 1947

(*nom. transl.* herein *ex* Barroisiceratinae)

Discussion. Monophyletic derivatives of Collignon-iceratidae (Wright, 1957) comprising the sub-families Subprionocyclinae and Barroisiceratinae. *U. Turon. - U. Coni.*

Subfamily Subprionocyclinae nov.

Diagnosis. Typically relatively small, compressed, involute, with double (primitive) or single (derived) ventrolateral tubercles and crenulated (primitive) or entire (derived) siphonal keel. Monophyletic derivatives of Collignoniceratinae, close to *Collig-noniceras* (Matsumoto, 1959). *U. Turon.- U. Coni.*

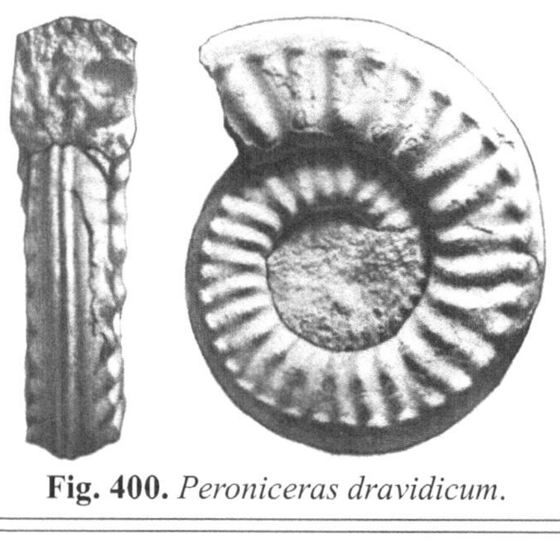

Fig. 400. *Peroniceras dravidicum*.

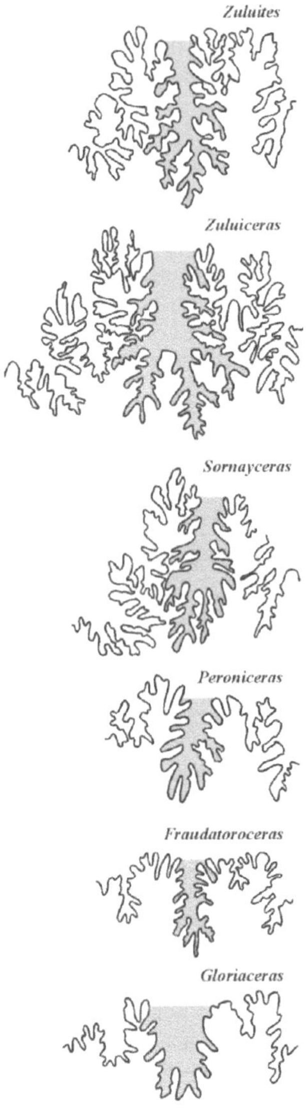

Fig. 401. Peroniceratine sutures.

Subprionocyclus SHIMIZU, 1932 [*Prionocyclus hitchinensis* Billinghurst, 1927; OD]. Small; compressed, high whorled and rather involute (U=21-26%), with steep to undercut umbilical wall, broad flat convergent flanks and narrow fastigiate venter; fine flexuous ribs single at first but later arising in 2-3's from weak umbilical bullae separated by intercalatories; ribs project forward and flatten on outer flank where they carry vestigial inner and clavate outer ventrolateral tubercles; venter normally smooth but for persistent continuous siphonal keel that is notched according to rib density; ribs flatten and simplify in maturity, becoming all long or alternating long and short; suture with E showing "oblique trifidity" and relatively narrow saddles and lobes. *M. - U. Turon.* **Fig. 403A** - *S. hitchinensis* (Billinghurst).

The best known *Subprionocyclus*, *S. neptuni* (Geintiz) (**Fig. 403B-C**) differs from the type species in being more evolute (U=32-40%) and inflated, with subquadratic whorls, coarser less flexuous ribs and prominent inner ventrolateral clavi. It is thus transitional to *Brendaites* of Texanitinae. Macroconchs can be medium sized.

Ledoceras BASSE, 1962 [*L. massoni*; OD]. Small; like *Subprionocyclus* but more involute and high whorled, with umbilical bullae that strengthen with growth, and fastigiate venter on which ribs are projected strongly forwards; inner ventrolateral tubercles present only for short period in middle growth; suture with narrow saddles and lobes and very asymmetrical E/L with reduced outer branch. *U. Turon.* **Fig. 403S-T** - *L. massoni* Basse.

Oregoniceras ANDERSON, 1958 [*Schloenbachia oregonensis* Anderson, 1902; OD]. Small [**m**] to moderately large [**M**], strongly compressed, moderately evolute, with fastigiate venter; siphonal keel entire in early growth, finely serrated later and finally a bevelled ridge; umbilical bullae give rise to pairs strong ribs with paired ventrolateral clavi, projecting forward on venter, becoming obsolete in maturity when ribs form chevrons over venter. *U. Turon.* **Fig. 403N-R** - *O. oregonense* (Anderson).

= *Oregoniceras* Anderson, 1941 (*nom. nud.*).

Subprionotropis BASSE, 1950 [*S. colombianus*; OD]. Very small; like *Subprionocyclus* but in middle to later growth with strong distant almost rectiradiate ribs and prominent inner and vestigial outer ventrolateral clavi; siphonal keel declines on adult body chamber when ribs form chevrons over rounded venter (**m**), or ornament becomes obsolete (**M**); suture with narrow trifid L. *U. Turon. - L. Coni.* **Fig. 403D-G** - *S. colombianus* Basse.

Reesidites WRIGHT & MATSUMOTO, 1954 [*Barroisiceras minimus* Hayasaka & Fukada; OD]. Medium sized; like *Ledoceras* but more compressed, with a narrower umbilicus (U=15-20%), higher whorls, denser costation and only one row of ventrolateral tubercles; suture with broad shallow E, broad asymmetrically bifid E/L that slopes outwards and fairly wide digitate L. *Upper U. Turon.* **Fig. 403H-I** - *R. minimus* (Hayasaka & Fukada).

= *Reesidites* Matsumoto, 1942 (*nom. nud.*).

Vocontiiceras DIEBOLD, BENGTSON, BAUDOUIN, LEES & BERT, 2018 [*V. vocontiense*; OD]. Small to medium sized; involute, compressed, with vertical umbilical wall, flat to weakly convex flanks, tabulate venter and subtrapezoidal to rectangular whorl section; strong ribs arise single or in pairs from umbilical bullae, separated by 1-2 intercalatories, sometimes bifurcating at mid-flank, and terminating in ventrolateral clavi; an entire siphonal keel persists throughout ontogeny, even after rounding of venter and evanescence of ornament on adult body chamber. Similar to *Alstadenites* (Barroisiceratinae). *Uppermost Turon.* **Fig. 407** - *V. vocontiense* Diebold, Bengtson, Baudouin, Lees & Bert.

= *Neoprionocyclus* Diebold, 2012 (*nom. nud.*).

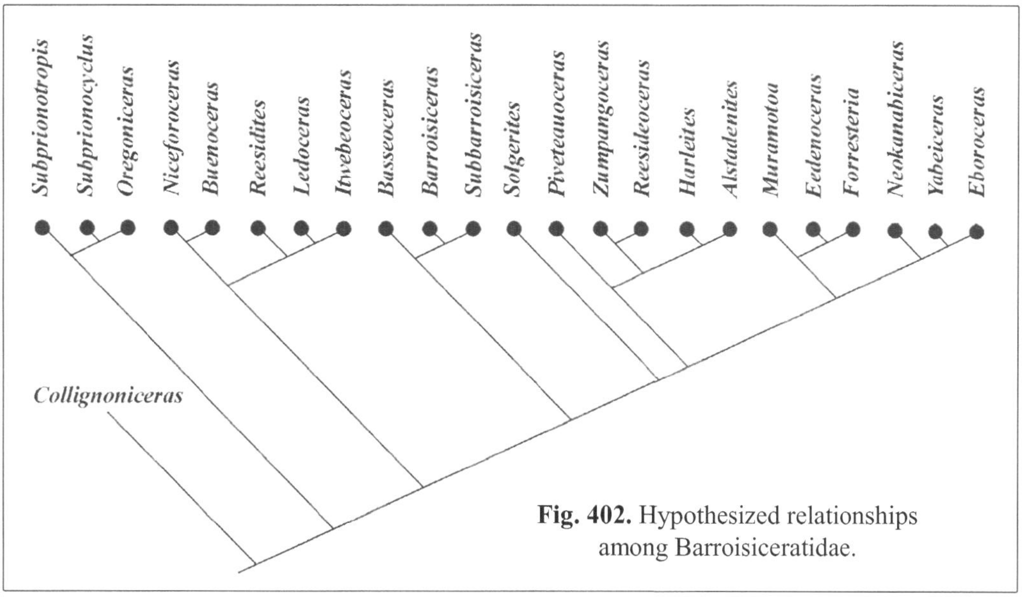

Fig. 402. Hypothesized relationships among Barroisiceratidae.

Itwebeoceras van Hoepen, 1968 [*I. lornae*; OD]. Medium sized; like *Reesidites* but larger, more inflated and evolute, with ribs typically arising in pairs from prominent umbilical bullae that strengthen with growth; initially with entire siphonal keel, later developing low crenulations; suture like *Buenoceras* but with broader E/L. *U. Coni.* **Fig. 403U-V** - *I. lornae* van Hoepen.

Niceforoceras Basse, 1948 [*N. columbianum*; OD]. Very small; like *Reesidites* but more involute (U=10%), with vertical umbilical wall, denser weaker ribbing, and siphonal clavi coalescing to form weakly crenulated keel; a homoeomorph of *Paradolphia* (Forbesiceratidae). *Coni.* **Fig. 403J-K** - *N. columbianum* Basse.

Buenoceras Etayo Serna, 1979 [*B. loboi*; OD]. Small; like *Reesidites* but more evolute, with flat ribbon-like biconcave ribs that are weak across flank, with oblique ventrolateral clavi; suture with long rectangular almost symmetrically bifid E/L and L much deeper than E. *L. Coni.* **Fig. 403L-M** - *B. loboi* Etayo Serna.

Subfamily Barroisiceratinae Basse, 1947

Discussion. Monophyletic derivatives of Subprionocyclinae (Wright, 1957), close to *Reesidites*; typically rather inflated and coarsely ornamented with tendency to lose ornament in later growth; derived members very evolute, with entire siphonal keel. *U. Turon.- U. Coni.*

Barroisiceras de Grossouvre, 1894 [*Ammonites haberfellneri* von Hauer, 1866; SD Solger, 1906]. Small; involute (U=12%), compressed, high whorled, with fastigiate venter with ventrolateral and siphonal clavi, sparse strong, frequently hypernodose umbilical bullae that may protrude into umbilicus, broad coarse sparse flank ribs that are rectiradiate and convex forward, and almost smooth adult body chamber; suture rather simple, with shallow fairly broad E, broad asymmetrically bifid E/L, rather narrow asymmetrically bifid L deeper than E, and broad weakly indented L/U_2. *U. Turon.* **Fig. 405C-D** - *B. haberfellneri* (von Hauer); **Fig. 406** - *B. subtuberculatum* (Gerhardt).

= *Barroisia* de Grossouvre, 1894 (*non* Munier-Chalmas, 1884).

Subbarroisiceras Basse, 1947 [*S. mahafalense*; OD]. Small to medium sized; like *Barroisiceras* but more inflated and evolute (U=22-24%), with deep umbilicus, vertical umbilical wall, strongly convergent flanks and coarser more robust ribbing that is retained to maturity; siphonal tubercles outnumber ventrolaterals. *L. Coni.* **Fig. 405A-B** - *S. mahafalense* Basse.

Ensorites van Hoepen, 1968 [*E. laxus*; OD]. Small, like *Subbarroisiceras* but more compressed and high whorled, with broader flatter flanks, inclined umbilical wall, weaker umbilical bullae and entire siphonal keel; low straight slightly prorsiradiate ribs generally alternate long and short and terminate in blunt ventrolateral tubercles. *M. Coni.* **Fig. 405E** - *E. laxus* van Hoepen.

Basseoceras Collignon, 1965 [*B. colcanapi*; OD]. Like *Barroisiceras* but with earlier weakening of ornament; adult body chamber moderately evolute (U=25-26%), with flattened flanks, broadly rounded venter, subrectangular whorl section, strong distant umbilicolateral bullae, faint to obsolete ribs and siphonal clavi reduced to a blunt keel before evanescing; suture with subtrifid E/L and L with well spaced short folioles; differs from *Solgerites* in narrower mature umbilicus and lateral

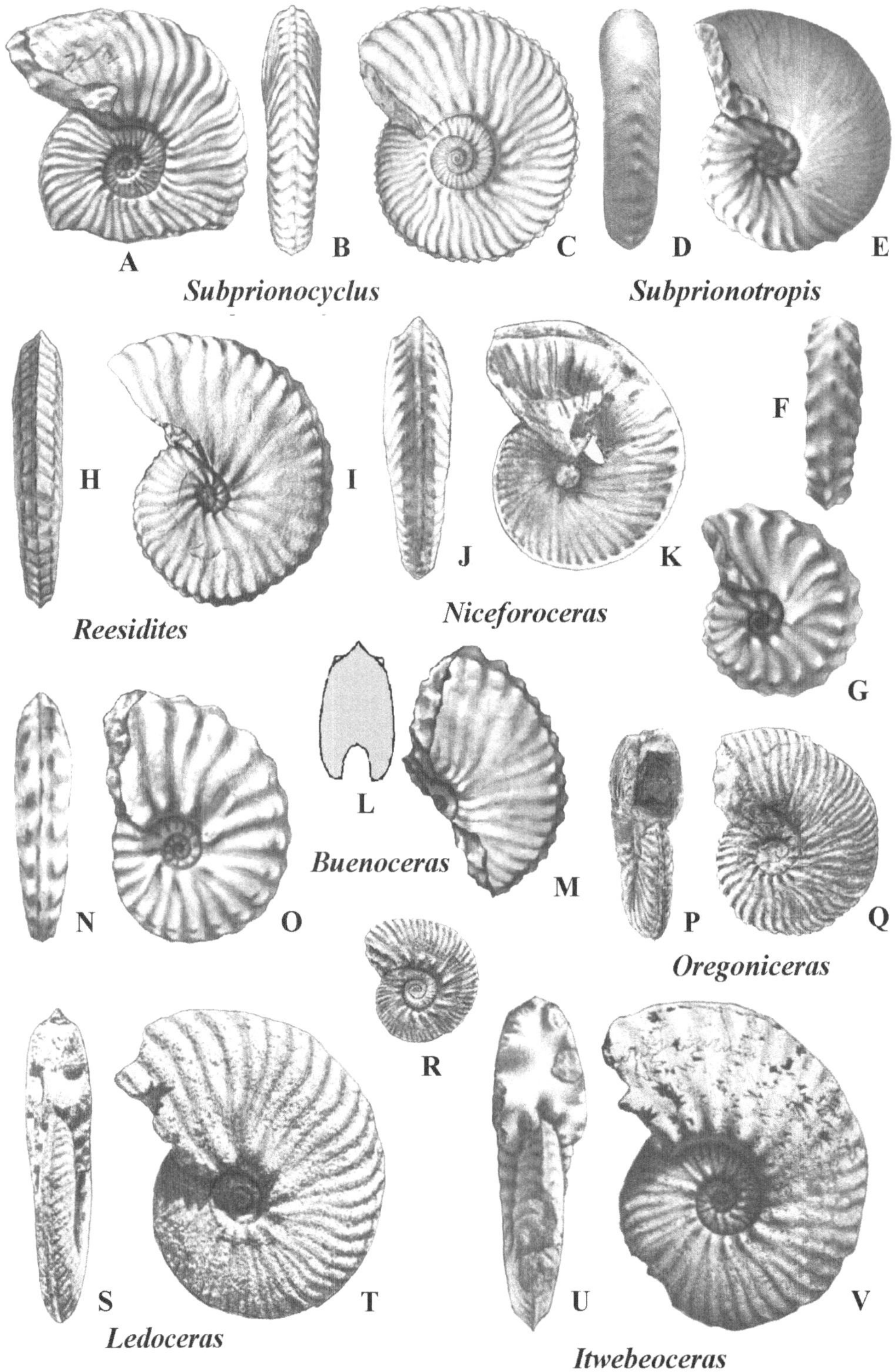

Fig. 403. Subprionocyclinae, Barroisiceratinae.

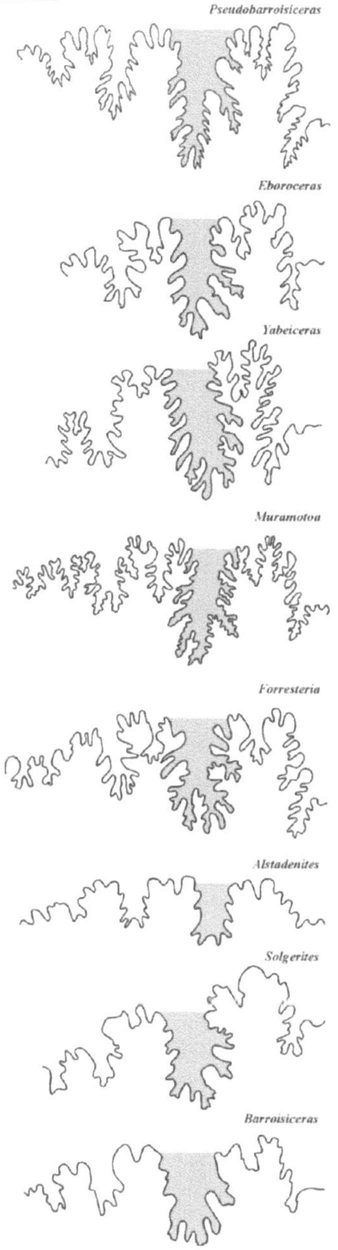

Fig. 404. Barroisiceratine sutures.

tubercles in early to middle growth. *U. Coni.* **Fig. 411A-B** - *B. colcanapi* (Collignon).

Solgerites REESIDE, 1932 [*Barroisiceras brancoi* Solger, 1904; OD]. Medium sized; inner whorls like *Barroisiceras* but with flatter flanks and tiny conical umbilicus; broad low round straight prorsiradiate ribs arise in pairs from small umbilical bullae and connect to blunt ventrolateral clavi before projecting forwards on venter but not connecting with siphonal clavi; feeble lateral tubercles present on innermost whorls (D<15 mm); ribbing declines in middle and later growth, replaced by low undulations and growth striae; adult body chamber evolute, with distant narrow ribs arising on inner flank and terminating in prominent ventrolateral horns, with siphonal clavi replaced by low blunt siphonal ridge; suture florid, with shallow E, bifid E/L, deep penetrating asymmetrically bifid L and small U_2. *Upper U. Turon.* **Fig. 405H-I** - *F. brancoi* (Solger).

Piveteauoceras BASSE, 1947 [*P. eboroense*; OD]. Medium sized; like *Solgerites* but smaller, with adult body chamber remaining involute (U=17%), without ribs or ventrolateral horns. *Coni.* **Fig. 405L-M** - *P. eboroense* Basse.

Harleites REESIDE, 1932 [*Barroisia haberfellneri* var. *harlei* de Grossouvre, 1894, = *Ammonites petrocoriensis* Coquand, 1859; OD]. Small; like inner whorls of *Solgerites* but more compressed, with vertical umbilical wall, and early loss of ornament; siphonal keel entire from early growth. *L. Coni.* **Fig. 411M-P** - *H. petrocoriensis* (Coquand).

Alstadenites REESIDE, 1932 [*Ammonites alstadenensis* Schlüter, 1876; OD]. Small; like *Harleites* but more evolute, with stronger ribs, lateral tubercles persisting to middle growth at least and ventrolateral clavi retained to maturity; suture relatively simple, with symmetrical rectangular L only slightly deeper than E. *L. Coni.* **Fig. 411I-J** - *A. alstadenensis* (Schlüter).

Reesideoceras BASSE, 1947 [*R. gallicum*; OD]. Small; like *Alstadenites* but with umbilical and lateral tubercles fusing in middle growth to form prominent umbilico-lateral bullae, siphonal tubercles becoming obsolete and venter narrowly sulcate. *L. Coni.* **Fig. 411E-F** - *R. gallicum* Basse.

Zumpangoceras BASSE, 1947 [*Z. burkhardti*; OD]. Small; ribs biconcave or straight, with periodically strengthened biconcave striae that are linguoid over venter; umbilical bullae vestigial; lateral tubercles strengthen with growth and migrate up flank to form lateral horns; siphonal clavi alternate with obliquely clavate ventrolaterals; venter initially fastigiate, later tabulate to shallowly sulcate. *L. Coni.* **Fig. 408I-J** - *Z. ospinai* Etayo Serna.

Forresteria REESIDE, 1932 [*Barroisiceras (F.) forresteri*, = *Prionocyclus pitalensis* Steinmann, 1897; OD]. Small to medium sized; like *Alstadenites* but more evolute (U=26-33%) and inflated, with depressed polygonal whorl section, prominent ribs, spinose tuberculation and more incised suture with deeply bifid L and U. *L. - M. Coni.* **Fig. 405J-K** - *F. pitalensis* (Steinmann).

= *Basseoceras* van Hoepen, 1968 (*non* Collignon, 1965) (**Fig. 408A-B**).

Neokanabiceras COLLIGNON, 1965 [*N. madagascariense*; OD]. Small; like *Forresteria* but more inflated, with vertical umbilical wall, broadly rounded venter, vestigial umbilical tubercles and siphonal tubercles coalescing to form a nodate keel. *M. Coni.* **Fig. 411C-D** - *N. madagascariense* Collignon.

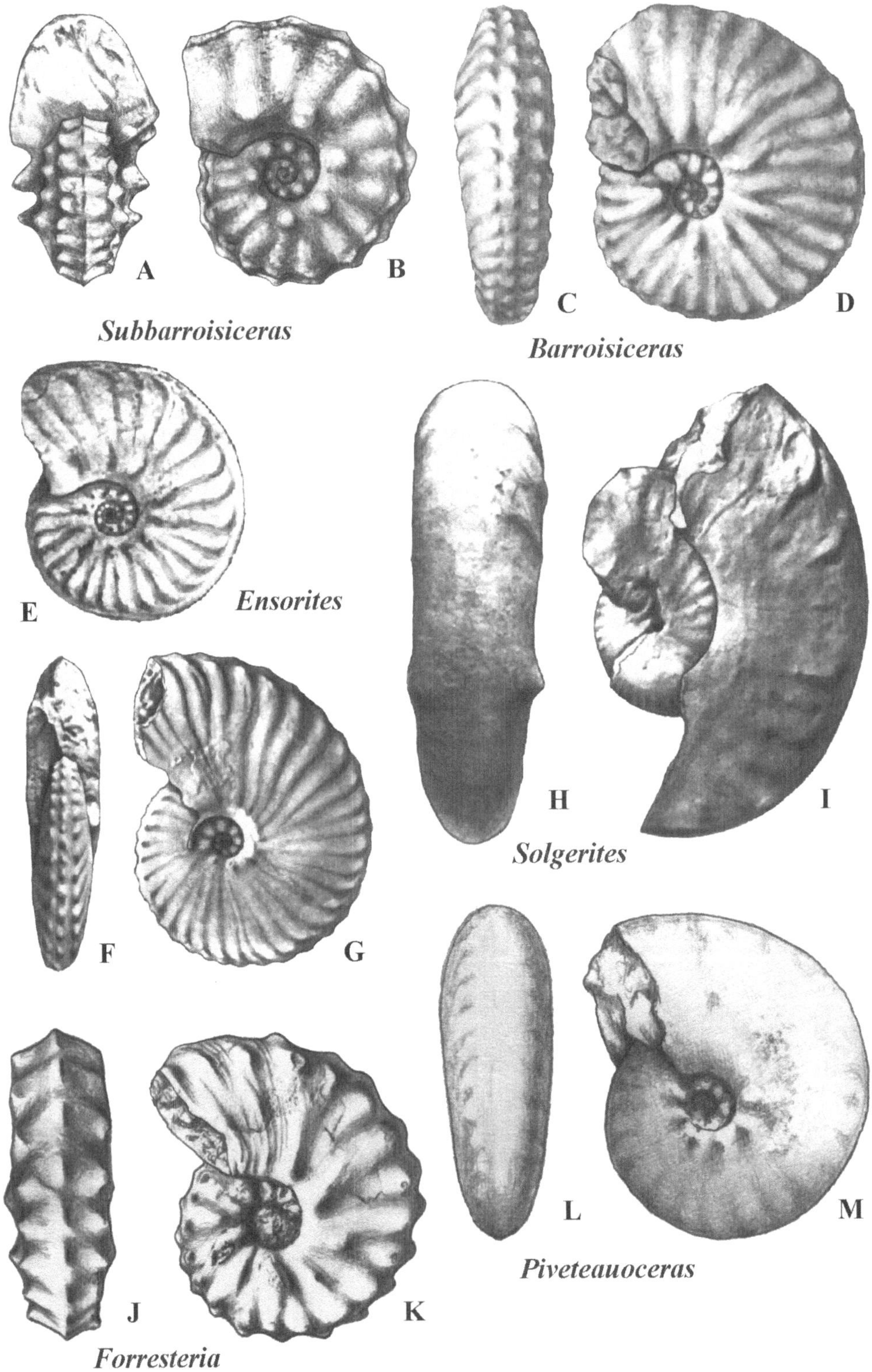

Fig. 405. Barroisiceratinae.

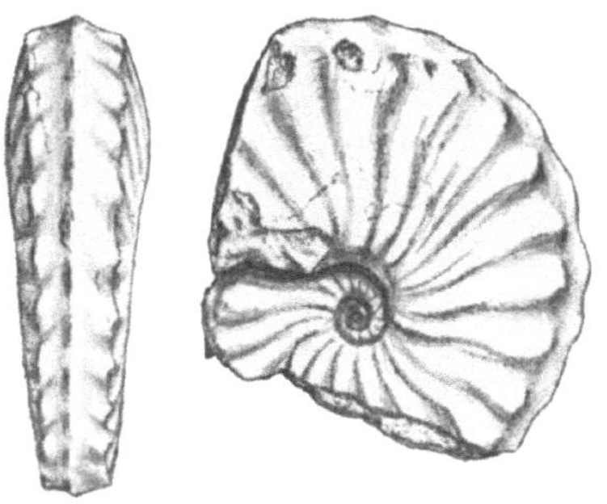

Fig. 406. *Barroisiceras subtuberculatum* (Gerhardt).

Collignonella VAN HOEPEN, 1957 [*C. hammersleyi*; OD]. [**M**]. Proposed in ignorance of *Forresteria* and placed in synonymy of the latter by its author; the large type species is said to fall within the intraspecific variation of *F. alluaudi* (Boulé, Lemoine & Thevenin), suggesting it is a macroconch. *L. Coni.* **Fig. 408K-L** - *C. hammersleyi* (van Hoepen).

= *Collignoniceras* van Hoepen, 1955 (*non* Breistroffer, 1947).

Eedenoceras VAN HOEPEN, 1968 [*E. multicostatum*; OD] [**M**]. Moderately large; like *Forresteria* but in later growth lateral tubercles obsolete, ventrolateral tubercles vestigial and ribbing remains strong bifurcating from umbilical shoulder or higher on flank and with intercalatories. *L. Coni.* **Fig. 408G-H** - *E. multicostatum* van Hoepen.

Muramotoa MATSUMOTO, 1969 [*Forresteria (M.) yezoensis*; OD]. Small; inner whorls like compressed *Forresteria*; beyond 50 mm diameter tubercles and ribs weaken so that adult whorls are smooth. *L. Coni.* **Fig. 408C-F** - *M. yezoensis* (Matsumoto).

Yabeiceras TOKUNAGA & SHIMIZU, 1928 [*Y. orientale*; OD]. Moderately large; compressed, evolute; inner whorls like *Neokanabiceras*; in middle growth (c.90 mm diameter) whorls become compressed with flattened convergent flanks, gently convex to subtabulate venter, and tubercles and ribs obsolete leaving only declining siphonal keel that eventually evanesces; adult body chamber smooth; suture with long narrow bifid E/L and narrow rectangular L as deep as E. *M. Coni.* **Fig. 411K-L** - *Y. orientale* Tokunaga & Shimizu.

Fig. 407. *Vocontiiceras.*

Eboroceras BASSE, 1947 [*E. magnumbilicatum* Basse; OD] Medium sized; almost identical to *Yabeiceras* but more inflated and suture with a broader E/L. *M. Coni.* **Fig. 411G-H** - *E. magnumbilicatum* (Basse).

Family TISSOTIIDAE Hyatt, 1900

Discussion. Monophyletic derivatives of Barroisiceratidae, close to *Barroisiceras* (Wright, 1996). Mostly compressed, involute and high whorled with ventrolateral clavi; venter with siphonal clavi (primitive) or entire keel (derived); suture ceratitic but in derived forms with E/L split by adventive lobe. *L. Coni. - L. Sant.*

Tissotioides REYMENT, 1958 [*Ammonites haplophyllus* Redtenbacher, 1873; OD]. Small, involute, compressed, high whorled, with narrow fastigiate venter; strong sparse umbilical bullae give rise to 1-2 broad low prorsiradiate ribs separated by 1-2 intercalatories; prominent ventrolateral and siphonal clavi; suture with entire saddles and finely denticulate lobes. *L. Coni.* **Fig. 413G-H** - *T. haplophyllus* (Redtenbacher).

Reymentoceras WIEDMANN, 1960 [*Tissotioides (R.) hispanicus*; OD]. Medium sized; like *Tissotioides* but more inflated, with wider umbilicus (U=22-25%), subtrigonal whorl section, narrowly rounded venter, rapid loss of ribs and tubercles leaving adult body chamber with large conical umbilical tubercles only; suture with bifid E/L. *L. Coni.* **Fig. 413M-N** - *R. hispanicum* (Wiedmann).

Dordiella REYMENT, 1958 [*D. bakunda*; OD]. Medium sized; like *Tissotioides* but more inflated, with broader whorls, gently convex flanks, lower venter that rounds in maturity, and different suture with asymmetrically bifid E/L with bilobed inner branch, oblique narrow quadrifid L, and weakly indented L/U_2 with outer margin sloping towards L. *L. Coni.* **Fig. 413K-L** - *D. bakunda* Reyment.

Metatissotia HYATT, 1903 [*Buchiceras fourneli* Bayle, 1878; SD Roman, 1938]. Moderately large; like *Dordiella* but more involute, with slightly concave convergent flanks, high sharply carinate fastigiate venter and swollen rounded umbilical tubercles; suture with inner branch of bilobed E/L swollen and entire, L very broad and finely denticulate, and L/U_2 broad and entire with narrow stem. *Coni.* **Fig. 413E-F** - *M. fourneli* (Bayle).

Tissotia DOUVILLÉ, 1890 [*Buchiceras tissoti* Bayle, 1878; OD]. Moderately large; very involute, moderately inflated, with subtrigonal whorl section and narrowly rounded venter; tricarinate in early to middle growth,

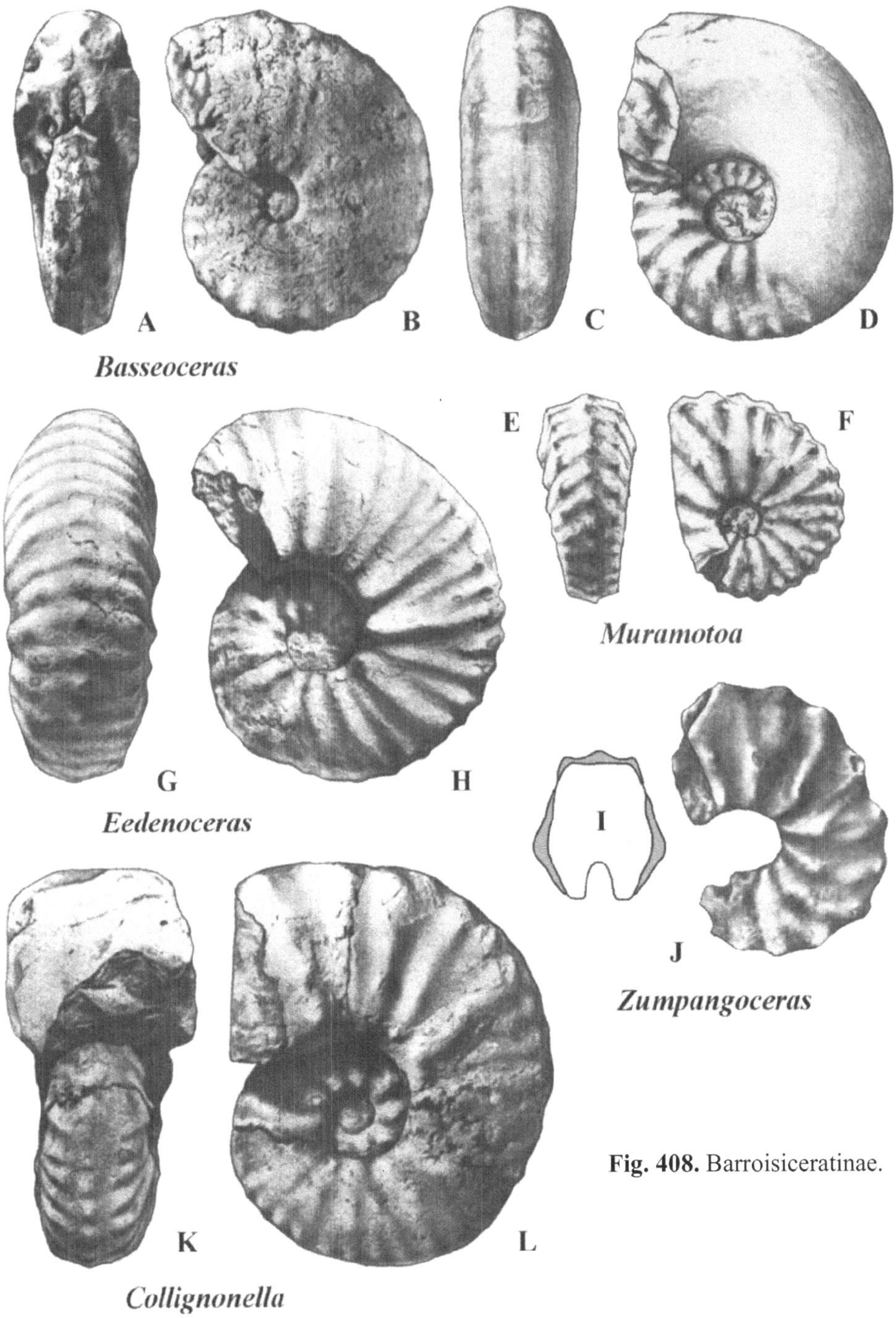

Fig. 408. Barroisiceratinae.

but sulci disappear later; flanks with sparse weak radial folds; suture like *Metatissotia*. *Coni.* - *L. Sant.* **Fig. 412** - *T. tissoti* (Bayle).

Subtissotia HYATT, 1903 [*T. tissoti* var. *inflata* Peron, 1897; SD Roman, 1938]. Small; like *Tissotia* but smaller globose cadicone with fastigiate venter and well spaced feeble ventrolateral tubercles; adult whorls rounded and smooth, resembling *Pachyvascoceras*. *Coni.* **Fig. 413C-D** - *S. intermedia* (Peron).

Paratissotia HYATT, 1903 [*P. regularis*; OD]. Moderately large; earliest whorls like *Tissotia*, but quickly becoming compressed, with acute venter and lanceolate whorl section; tubercles lacking but there may be weak ribs. *Coni.* **Fig. 413O-P** - *P. regularis* Hyatt.

Heterotissotia PERON, 1897 [*H. neoceratites* Peron; OD]. Small to medium sized; involute, compressed, with convex convergent flanks and narrow subtabulate tricarinate venter; 6-8 strong blunt umbilical bullae give rise to 2-3 rounded flexuous ribs, with occasional intercalatories, that terminate in ventrolateral clavi; umbilical tubercles may become more pronounced with growth, forming broad low lateral bulges, or may be lost early, while ventrolaterals coalesce to form undulatory keels; suture like *Metatissotia* but with shallowly incised outer branch to E/L and broader L. *Coni.* **Fig. 413A-B** - *H. neoceratites* Peron.

Family TEXANITIDAE Collignon, 1948

Discussion. Monophyletic derivatives of Barroisiceratidae (Wright, 1979); typically evolute, compressed, flat sided and multituberculate, with more or less entire siphonal keel. Comprises the subfamilies Texanitinae and Menabitinae. *U. Turon. - M. Camp.*

Subfamily Texanitinae Collignon, 1948

(*nom. transl.* Wright, 1957 *ex* Texanitidae)

Discussion. Monophyletic derivatives of Subprionocyclinae (Wright,1979); typically compressed, with regular ribs and 3-6 rows of tubercles, as well as an entire siphonal keel (except in most primitive representatives). Comprises the tribes Protexanitini, Texanitini and Reginaitini. *U. Turon. - M. Camp.*

Tribe Protexanitini nov.

Type genus: *Protexanites* Matsumoto, 1955.

Diagnosis. Typically with 3 rows of tubercles (umbilical and doubled ventrolaterals); primitively with serrated or undulose siphonal keel, later entire. A monophyletic taxon derived from Subprionocyclinae, close to *Subprionocyclus* (Wright, 1979); *U. Turon.- L. Camp.*

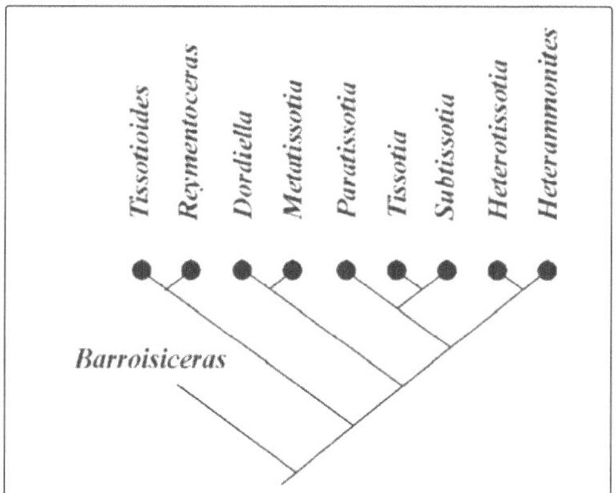

Fig. 409. Hypothesized relationships among Tissotiidae.

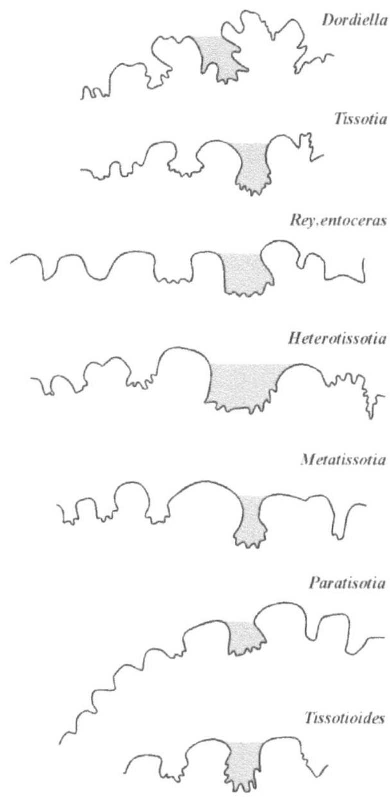

Fig. 410. Tissotiid sutures.

Brendaites **gen. nov.** [*Prionotropis branneri* Anderson, 1902, p.125; OD herein]. Very small; compressed, evolute (U=34-37%), with flat parallel flanks, compressed subquadratic whorl section, strong simple flexuous rather distant generally rectiradiate ribs with inner and outer ventrolateral clavi carried to maturity, and sharp high coarsely crenulate siphonal keel; suture relatively simple, with broad fairly deep E, subquadratic asymmetrically bifid E/L, narrow L, and U_2 on umbilical shoulder. *U. Turon.* **Fig. 415E-F** - *B. branneri* (Anderson).

Protexanites MATSUMOTO, 1955 [*Ammonites bourgeoisianus* d'Orbigny, 1850; OD]. Medium sized; like *Brendaites* but larger, more coarsely ribbed, with round inner ventrolateral tubercles and outer ventrolateral clavi separated from low, entire to undulatory siphonal keel by very shallow grooves; suture with narrow E, very broad E/L, and L longer than E. *L. Coni. - M. Sant.* **Fig. 415C-D** - *P. bourgeoisianus* (d'Orbigny).

Aneuretoceras KENNEDY & COBBAN, 1991 [*A. variable*; OD]. Micromorph; inner whorls smooth with blunt siphonal ridge; outer whorls with flexuous lirae or ribs

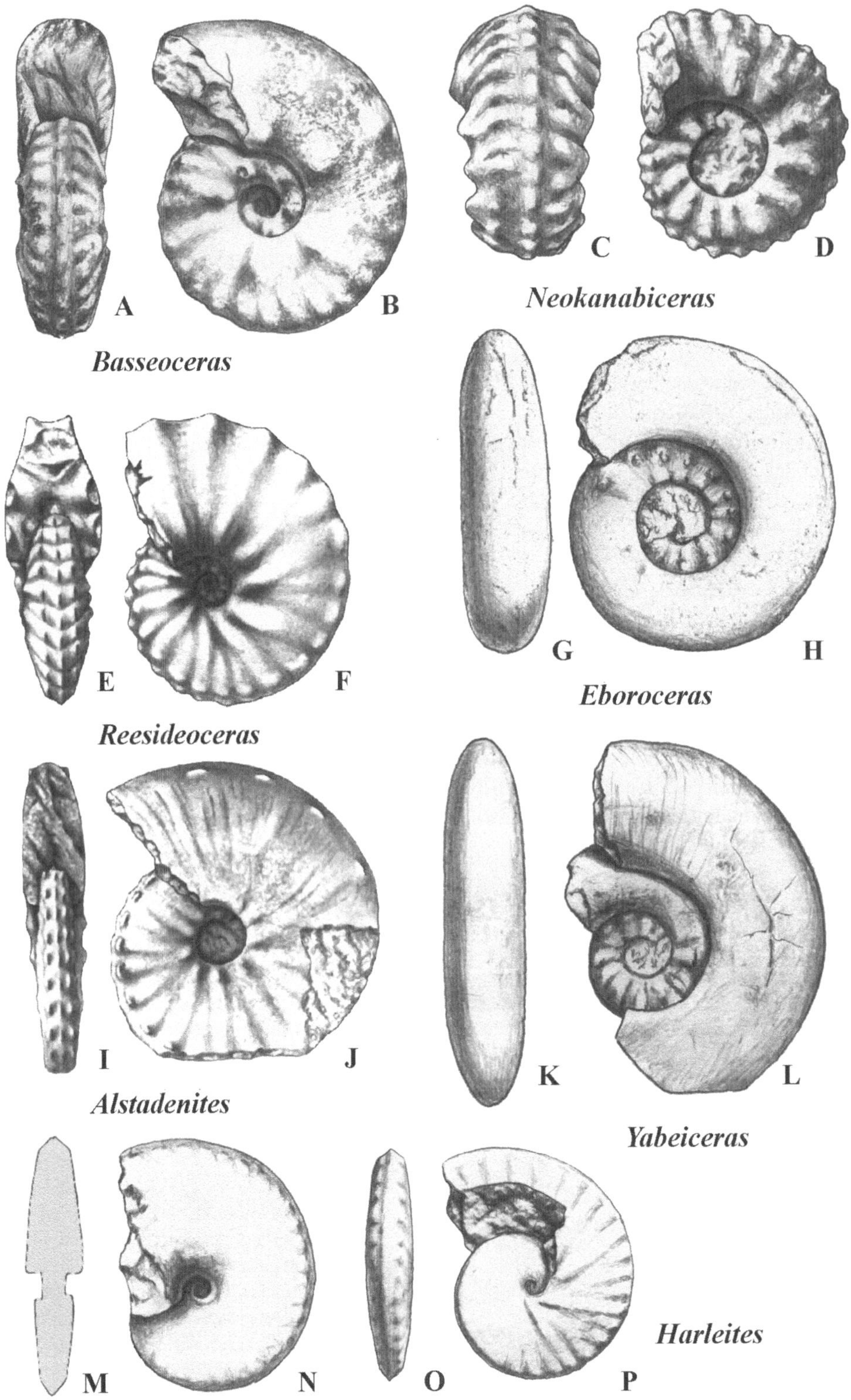

Fig. 411. Barroisiceratinae.

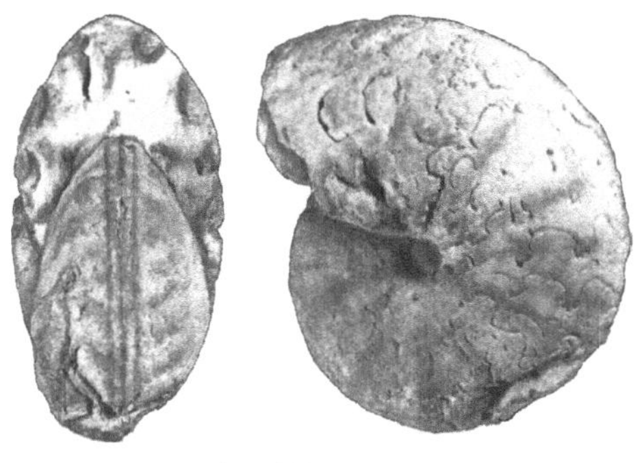

Fig. 412. *Tissotia.*

forming chevrons over venter. *U. Coni.* **Fig. 415O-P** - *A. variable* Kennedy & Cobban.

Hyphaloceras KENNEDY & COBBAN, 1991 [*H. howlettae*; OD]. Very small; evolute, with compressed oval intercostal whorl section; strong distant ribs with umbilical bullae and conical ventrolateral tubercles weakening over broadly rounded venter; suture simple with deep E, broad little incised E/L and narrow L. *L.-M. Sant.* **Fig. 415Q-R** - *H. howlettae* Kennedy & Cobban.

Miotexanites MATSUMOTO, 1970 [*Protexanites (M.) minimus*; OD]. Small; like *Protexanites* but smaller, more densely ribbed, with ribs tending to alternate long and short, long ribs with umbilical bullae; all ribs weak on flank but strengthening ventrally, with weak ventrolateral clavi; arched venter with entire siphonal keel bordered by shallow grooves; indistinct periodic constrictions; suture simple, with moderately wide E/L and L not as deep as E. *U. Coni. - L. Sant.* **Fig. 415G-H** - *M. minimus* (Matsumoto).

Anatexanites MATSUMOTO, 1970 [*Mortoniceras fukazawai* Yabe & Shimizu, 1925; OD]. Large; like *Protexanites* but adding lateral bullae in middle growth. *Sant.* **Fig. 415K** - *A. fukazawai* (Yabe & Shimizu).

Haboroceras TOSHIMITSU, 1988 [*H. haboroense*; OD]. Micromorph; adult body chamber completely smooth, with terminal constriction; suture simple. *U. Sant. - L. Camp.* **Fig. 415I-J** - *H. haboroense* Toshimitsu.

Defordiceras YOUNG, 1963 [*D. hazzardi*; OD]. Large, evolute; in middle growth with single and frequent bifurcating ribs; in later growth with single distant robust ribs with lateral, ventrolateral and ventral tubercles, and narrow sulcate venter; siphonal keel lacking. *Sant.* **Fig. 415A-B** - *D. hazzardi* Young.

Wright (1996) suggested the type species might be teratological.

Tribe Texanitini Collignon, 1948

(*nom. transl.* herein *ex* Texanitinae)

Type genus: *Texanites* Spath, 1932.

Diagnosis. Typically with 4-5 rows of tubercles at least in middle to later growth. Monophyletic derivatives of Protexanitini, close to *Protexanites* (Matsumoto, 1955). *Uppermost Coni.- L. Camp.*

Texanites SPATH, 1932 [*Ammonites texanus* Roemer, 1852; OD]. Moderately large to large; like *Paratexanites* but more compressed, with lateral tubercle added early to become quinquetuberculate in early growth. *U. Coni.- L. Camp.* **Fig. 416** - *T. texanus* (Roemer); **Fig. 418A-B** - *T. soutoni* (Baily).

Plesiotexanites MATSUMOTO, 1970 [*Mortoniceras kawasakii* Kawada, 1929; OD]. Large, very evolute; in early growth like *Protexanites* with only umbilical and doubled ventrolateral tubercles; in middle growth ventrolaterals split to achieve quadrituberculate condition and more or less simultaneously lateral tubercle added to become quinquetuberculate; suture like *Protexanites*. *L. Sant. - L. Camp.* **Fig. 418I** - *P. kawasakii* (Kawada).

Eutexanites KLINGER & KENNEDY, 1980 [*Plesiotexanites (E.) sextuberculatus*; OD]. Moderately large; like *Plesiotexanites* but with 6 rows of tubercles, due to division of umbilicals to create low laterals. *L. Sant.* **Fig. 418J-K** - *E. sextuberculatus* (Klinger & Kennedy).

Neoselwynoceras COLLIGNON, 1966 [*N. paradoxum*; OD]. Based on pathologically deformed individual of *Texanites*, with 4 rows of tubercles on one flank and 5 on the other, the displaced row forming siphonal clavi to mimic *Selwynoceras* (Summesberger, 1985).

Submortoniceras SPATH, 1926 [*Mortoniceras woodsi* Spath, 1921; OD]. Large; like *Texanites* but more involute and high whorled, with overhanging umbilical wall so that umbilical bullae penetrate umbilicus; mostly paired ribs with prominent low lateral tubercles from which they may bifurcate; submarginal and marginal tubercles tend to be clavate; in middle growth ribs typically weaken markedly at midflank. *U. Sant. - L. Camp.* **Fig. 418C-D** - *S. woodsi* (Spath).

Butticeras ANDERSON, 1958 [*B. buttense*; OD]. Medium sized; like *Submortoniceras* but relatively involute in maturity (U=25%), with lateral and ventral tuberculation, as well as declining siphonal keel; outer whorl with only umbilical bullae giving rise to 3-4 thread-like costae that curve forward on outer flank, with narrowly rounded smooth venter; suture simple. *L. Camp.* **Fig. 417C-D** - *B. buttense* Anderson.

Jimenites CARRASCO, 1967 [*J. cantusi*; OD]. Enormous; like *Submortoniceras* but with ribbing weak or effaced from flanks from early growth, regenerating in maturity with strong, simple, distant, weakly tuberculate ribs. *Upper L. Camp.* **Fig. 417A-B** - *J. cantusi* Carrasco.

= *Jimenensites* Klinger & Kennedy, 1980 (*nom. van.*)

= *Jimenesites* Klinger & Kennedy, 1980 (*nom. van.*)

Antsirasirella COLLIGNON, 1969 [*Submortoniceras (A.) stellata*; OD]. Medium sized; like *Submortoniceras* but strongly compressed and involute, almost oxyconic, with

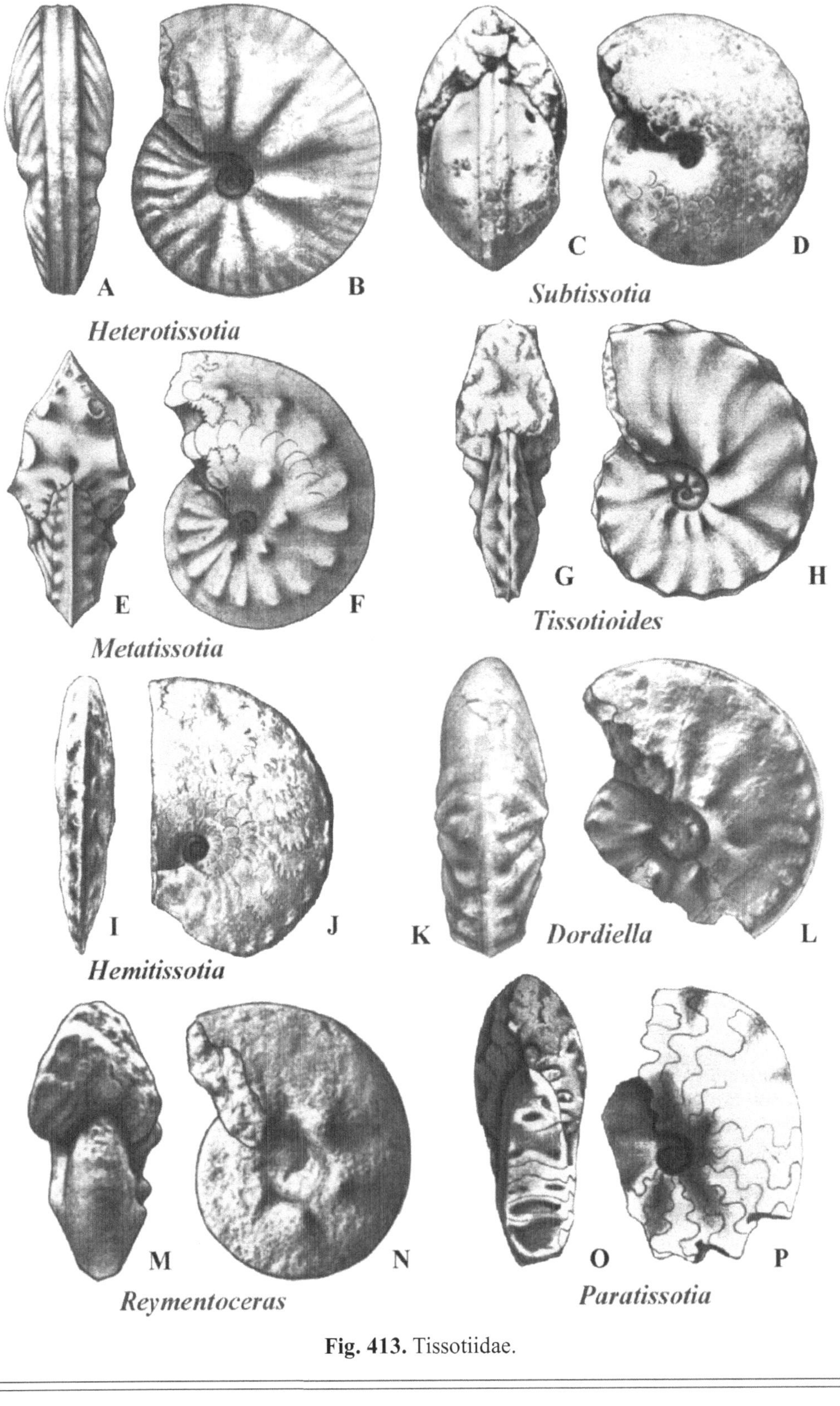

Fig. 413. Tissotiidae.

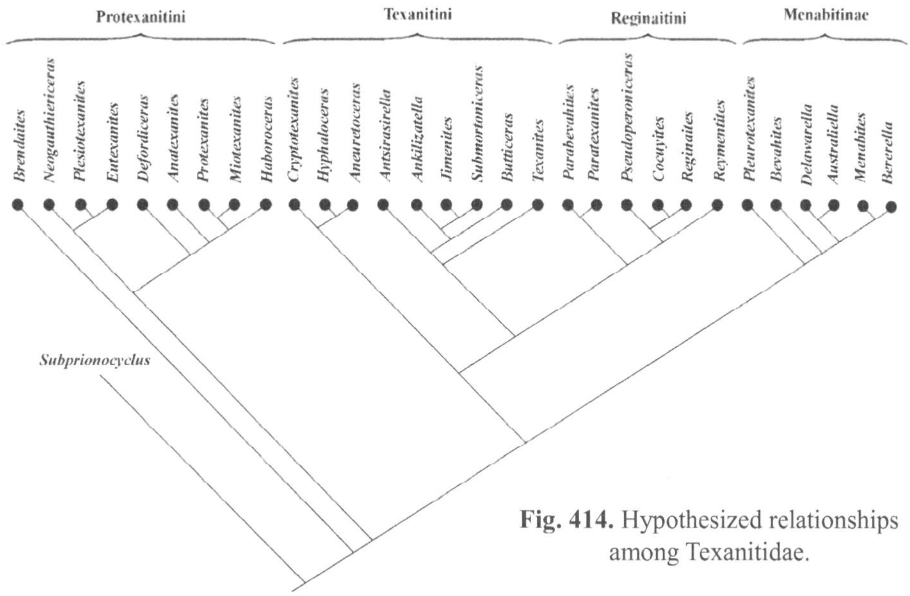

Fig. 414. Hypothesized relationships among Texanitidae.

very broad gently convex flanks and narrowly rounded venter with blunt siphonal keel; weak umbilical bullae give rise to low indistinct rectiradiate ribs that increase by bifurcation or intercalation and terminate in low ventrolateral tubercles; adult body chamber tabulate to weakly sulcate, without siphonal keel. *L. Camp.* **Fig. 418E-F** - *A. stellata* (Collignon).

Ankilizatella COLLIGNON, 1971 [*Texanites (A.) ankilizatensis*, = *Submortoniceras mariscalense* Young, 1963; OD]. Medium sized; like *Submortoniceras* but strongly compressed and very evolute, with rectangular whorl section and ribbing effaced from flanks at early stage leaving only umbilical and paired ventrolateral tubercles. *M. Camp.* **Fig. 418G-H** - *A. mariscalensis* (Young).

Cryptotexanites KENNEDY & COBBAN, 1993 [*C. paedomorphicus*; OD]. Very small; evolute, compressed, with flat parallel flanks, broadly rounded venter and subelliptical whorl section; inner whorls with distinct siphonal keel that becomes obsolete in maturity; indistinct ornament of flexuous prorsiradiate riblets and striae with, on adult body chamber, occasional constrictions; suture with broad little indented E/L and narrow L. *Upper L. Camp.* **Fig. 415L-N** - *C. paedomorphicus* Kennedy & Cobban.

Tribe Reginaitini nov.

Type genus: *Reginaites* Reyment, 1957.

Diagnosis. Primitively with 4 rows of tubercles and siphonal keel, but derived members with 2-3 rows of tubercles and tricarinate venter; includes homoeomorphs of Peroniceratinae. Monophyletic derivatives of Texanitini, close to *Texanites*. *U. Coni. - L. Camp.*

Paratexanites COLLIGNON, 1948 [*Mortoniceras zeilleri* de Grossouvre, 1894; OD]. Moderately large to large; like *Protexanites* but more inflated, with quadrituberculate ribs achieved early by splitting of inner ventrolateral tubercles to produce submarginal and marginal rows, the submarginal row conical and subspinose. *U. Coni.* **Fig. 415S-T** - *P. zeilleri* (de Grossouvre).

Parabevahites COLLIGNON, 1948 [*Ammonites serratomarginatus* Redtenbacher, 1873; OD]. Small to medium sized; like *Paratexanites* but more compressed, with crowded flexuous ribbing to inner whorls and closely spaced clavate submarginal and marginal tubercles; ribbing becomes more distant with growth. *U. Coni.* **Fig. 415U-V** - *P. serratomarginatus* (Redtenbacher).

Pseudoperoniceras **gen. nov.** [*Paratexanites australis* Klinger & Kennedy, 1986, p.44; OD herein]. Medium sized; like *Parabevahites* but with incipiently to strongly tricarinate venter throughout, the central keel strongest; suture with broad asymmetrically bifid E/L with broad outer branch, narrow L as deep as E, and reduced U_2. *U. Coni.* **Fig. 419A-B** - *P. australe* (Klinger & Kennedy).

Reginaites REYMENT, 1957 [*Peroniceras (R.) quadrituberculatum*; OD]. Moderately large; like *Pseudoperoniceras* but more compressed, with entire keels and tricarinate venter throughout; high lateral clavi in later growth; suture with narrow outer branch to E/L and L deeper than E. *Sant. - L. Camp.* **Fig. 419G-H** - *R. quadrituberculatus* (Reyment).

Reymentites MATSUMOTO, 1965 [*R. hataii*; OD]. Medium sized; like *Reginaites* but with fastigiate venter and tubercle spacing equidistant; suture with broad outer branch to E/L and L not so deep penetrating. *Sant.* **Fig. 419E-F** - *R. hataii* Matsumoto.

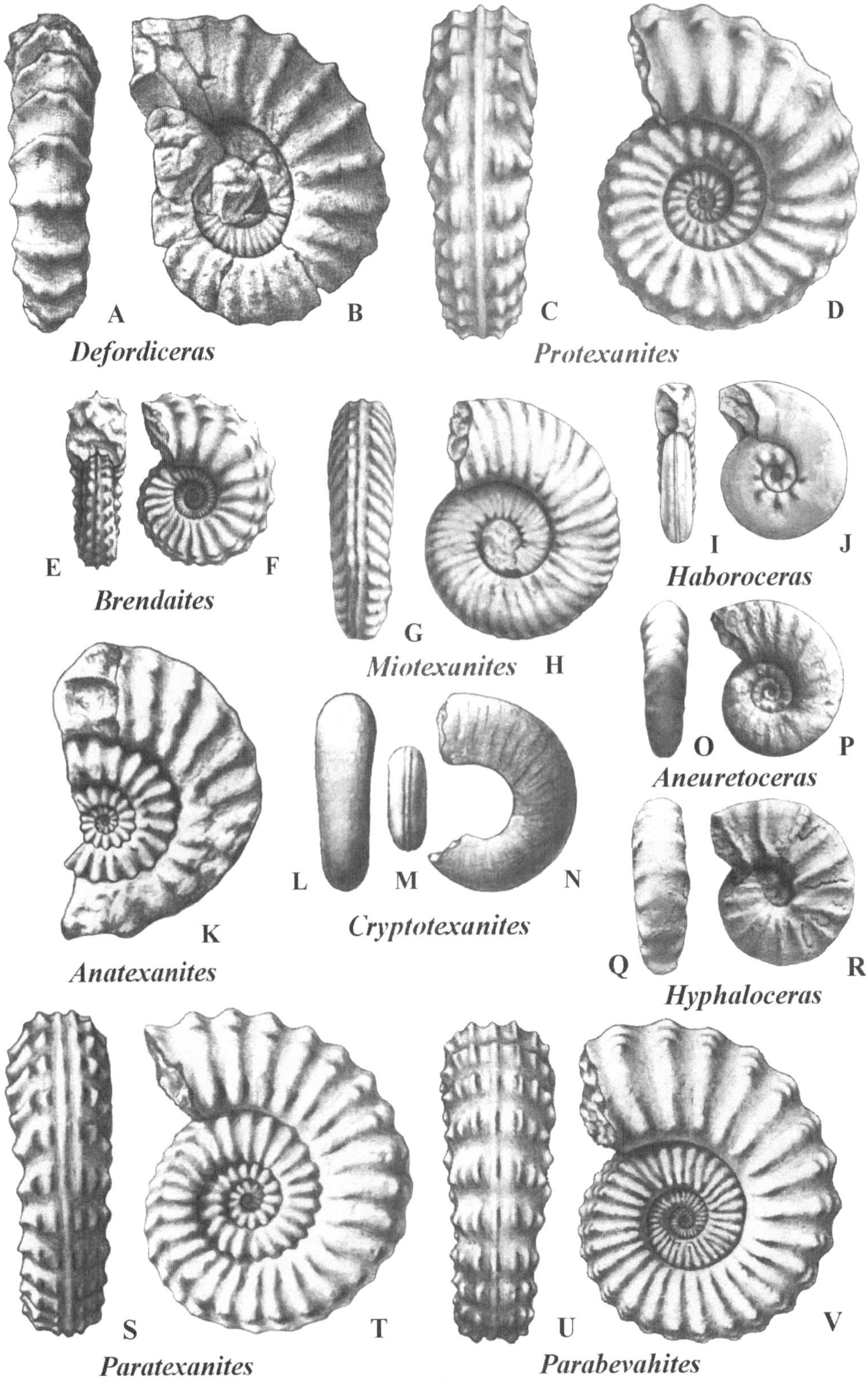

Fig. 415. Texanitinae.

Fig. 416. *Texanites texanus.*

Cocuyites ETAYO SERNA, 1985 [*C. cocuyensis*; OD]. Small; like *Reginaites* but more compressed, with higher whorls, narrower umbilicus, compressed rectangular whorl section, denser ribs with frequent bifurcation and intercalation, and approximated inner and outer ventrolateral clavi from early growth. *Sant.* **Fig. 419C-D** - *C. cocuyensis* Etayo Serna.

Neogauthiericeras COLLIGNON, 1969 [*N. zafimahovai*; OD]. Medium sized; very evolute, with fastigiate unicarinate venter; inner whorls with rounded umbilical tubercles weakly joined to prominent ventrolateral clavi, connecting ribs becoming stronger in middle to later growth when they project forward onto margins of venter; adult body chamber smooth. *Upper U. Sant. - L. Camp.* **Fig. 419I-J** - *N. zafimahovai* Collignon.

Subfamily Menabitinae Cooper, 1991

Discussion. Monophyletic derivatives of Texanitinae, close to *Protexanites* (Matsumoto, 1970), evolving in parallel with texanitines; typically with pillar-like ribs on the lower flank and a multiplicity of tubercles to the ventral rows. *U. Coni. - M. Camp.*

Pleurotexanites MATSUMOTO, 1970 [*Protexanites superbus* Collignon, 1966; OD]. Medium sized (**m**) to moderately large (**M**); like *Protexanites* but with sparser ribbing, outer ventrolateral clavi greatly outnumbering spinose inner ventrolaterals and with undulatory siphonal keel; venter becomes raised and narrow with growth, so that inner ventrolateral tubercles shift to high lateral position; macroconchs with almost smooth adult body chamber and rounded venter; suture with broad asymmetrically bifid E/L, narrower L/U_2 and narrow bifid L. *Upper Coni.* **Fig. 419K (M), 421A-B (m)** - *P. superbus* (Collignon).

Australiella COLLIGNON, 1948 [*Mortoniceras australe* Besairie, 1930; OD]. Large; like *Pleurotexanites* but strongly inflated with broadly rounded venter, flat diverging flanks and very depressed coronate whorl section; tiny umbilical tubercles weakly connected by broad low indistinct rectiradiate ribs to massive blunt inner ventrolateral horns from which secondaries bifurcate, frequently separated by intercalatory, and project forward to meet swollen oblique ventral clavi separated from thick low feebly crenulated siphonal keel by distinct grooves; with growth whorls become rectangular, primary ribs strengthen and become bar like, ventrolateral horns smaller and narrower finally splitting to form inner and outer ventrolateral tubercles, with entire siphonal keel; at large size venter rounds and midlateral tubercles develop; finally ribs strengthen, tending to absorb tuberculation. *L. - M. Camp.* **Fig. 421C-F** - *A. australis* (Besairie).

= *Austinites* Adkins, 1933 (*nom. nud.*)

= *Australiella* Collignon, 1939 (*nom. nud.*).

Collignon (1948) cited the date of introduction of this genus, and the other menabite genera introduced by him, as 1939. However, this reference is not listed in his bibliography, nor is it mentioned by subsequent workers on the group.

Delawarella COLLIGNON, 1948 [*Ammonites delawarensis* Morton, 1830; OD]. Very large; inner whorls like

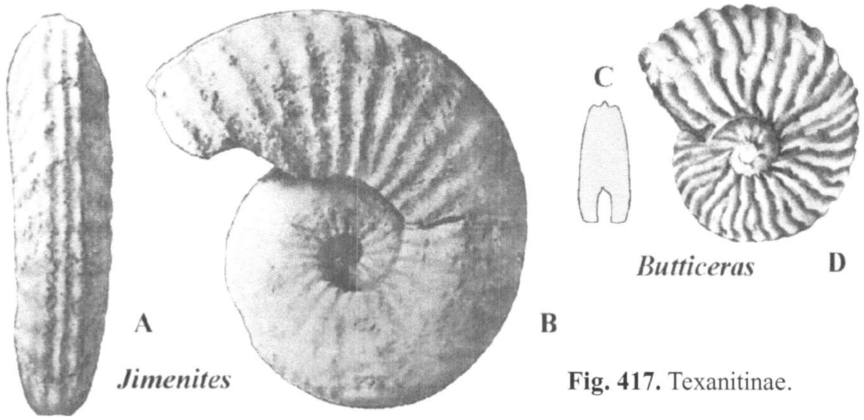

Fig. 417. Texanitinae.

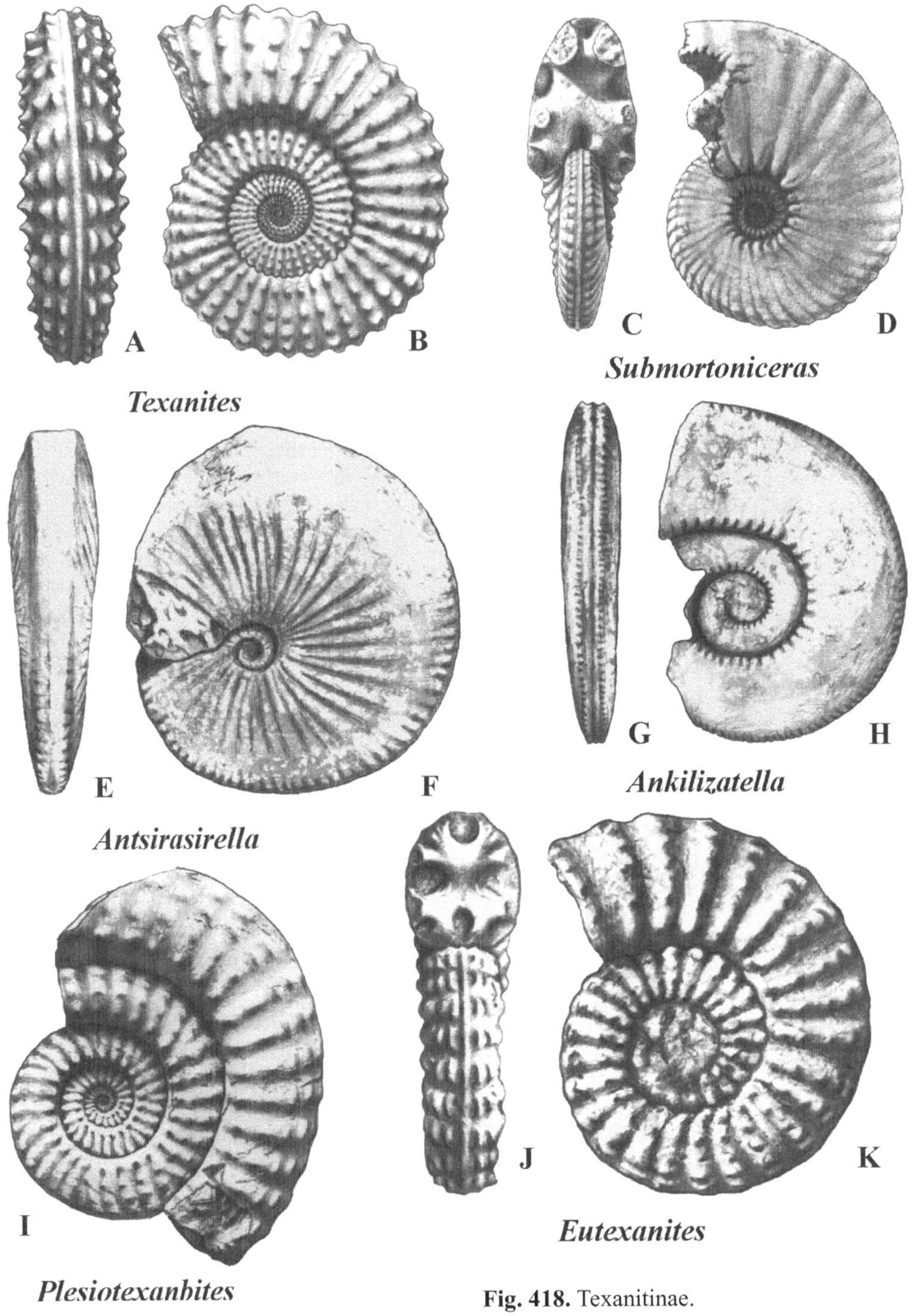

Fig. 418. Texanitinae.

Australiella but whorls soon become higher than wide and with narrower umbilicus (U=21-22%); subspinose umbilical tubercles project into umbilicus and give rise to 1-2 weak low ribs, all with weak lateral bullae, weak conical submarginal tubercles and strong rounded to slightly clavate marginal tubercles; external clavi twice as numerous as tubercles in other rows and separated from entire siphonal keel by shallow grooves; submarginal tubercles become obsolete in maturity; suture with deep penetrating E, rather narrow rectangular asymm-

etrically bifid E/L and broad asymmetrically bifid L. *L. - M. Camp.* **Fig. 421K-L** - *D. delawarensis* (Morton).
= *Delawarella* Collignon, 1939 (*nom. nud.*).

Bererella COLLIGNON, 1948 [*Menabites (B.) bererensis*; OD]. Moderately large; like *Delawarella* but more evolute (U=38-42%) with stronger sparser costation, prominent tuberculation, stout simple bar-like primary ribs and massive submarginal tubercles; suture with small short E with two long fine branches and very large wide L; some species homoeomorph *Submortoniceras*. *L. - M. Camp.* **Fig. 421M-N** - *B. bererensis* (Collignon).
= *Bererella* Collignon, 1939 (*nom. nud.*).

Menabites COLLIGNON, 1948 [*M. menabensis*; OD]. Like *Australiella* but with pentatuberculate stage attained early so that for much of growth shell has flat flanks, quadratic whorl section, and simple strong pillar-like main ribs. *L. - M. Camp.* **Fig. 421G-H** - *M. menabensis* Collignon.
= *Menabites* Collignon, 1939 (*nom. nud.*).

Bevahites COLLIGNON, 1948 [*B. quadratus*; OD]. Like *Menabites* but pentatuberculate condition attained earlier, tending to be more compressed and higher whorled, with frequent bifurcation and intercalation of ribs and not developing strong pillar-like main ribs. *L. Sant. - M. Camp.* **Fig. 421I-J** - *B. quadratus* Collignon.
= *Bevahites* Collignon, 1939 (*nom. nud.*)

Superfamily SPHENODISCOIDEA Hyatt, 1900

(*nom. transl.* Cooper, 1991 *ex* Sphenodiscidae)

Discussion. Monophyletic derivatives of Barroisiceratoidea (Wright, 1957); comprises discoidal oxycones assigned to the the families Lenticeratidae and Sphenodiscidae. *U. Coni.- U. Maastr.*

Family LENTICERATIDAE Hyatt, 1903

Discussion. Monophyletic derivatives of Barroisiceratidae (Wright, 1957); discoidal oxycones with 4 auxiliary lobes to the suture; comprises the subfamilies Lenticeratinae and Eulophoceratinae. *U. Coni. -L. Camp.*

Subfamily Lenticeratinae Hyatt, 1900

(*nom. transl.* Wright, 1952 *ex* Lenticeratidae)

Discussion. A monophyletic taxon with an origin in Subprionocyclinae, close to *Niceforoceras* (Wright, 1957), although *Buenoceras* has the weak biconcave ribs of *Paralenticeras;* the suture of *Reesidites* is like that of Lenticeratinae with an E/L that slopes outwardly. Typically very involute weakly ornamented oxycones with highly incised suture with deep E, broad obliquely trifid E/L the branches of which increase in size outwardly, and 4 auxiliary elements. *U. Coni. - L. Sant.*

Paralenticeras HYATT, 1900 [*Amaltheus sieversi* Gerhardt, 1897; OD]. Medium sized; compressed, very involute (U=3-15%), with broad convergent flanks and fastigiate narrowly rounded venter; inner whorls with weak umbilical bullae giving rise to prorsiradiate biconcave ribs separated by 2-3 intercalatories, all with feeble ventrolateral bullae; in later growth ribbing declines, becoming irregular and eventually reduces to low folds and growth striae, while venter rounds; suture highly incised with finger-like folioles, extremely broad obliquely trifid E/L, long narrow trifid L, broad digitate L/U_2 and 4 auxiliary lobes. *U. Coni. - L. Sant.* **Fig. 425C-F** - *P. sieversi* (Gerhardt).

Lenticeras GERHARDT, 1897 [*Ammonites andii* Gabb, 1877; OD]. Moderately large; like *Paralenticeras* but strongly inflated, with steep and broad umbilical wall, convex flanks and more or less cordate whorl section; inner whorls with low rounded ribs branching from umbilical bullae but fading before reaching venter; outer whorls smooth; suture like *Paralenticeras*, but less incised, without finger-like folioles and with narrow E/L. *U. Coni. - L. Sant.* **Fig. 425A-B** - *L. andii* (Gabb).

Subfamily Eulophoceratinae Hyatt, 1903

(*nom. transl.* herein *ex* Eulophoceratidae)

Discussion. A monophyletic group descended from Lenticeratinae (Wright, 1957); suture with very broad bifid E/L. Comprises the tribes Eulophoceratini and Diaziceratini. *U. Coni.- L. Camp.*

Tribe Eulophoceratini Hyatt, 1903

(*nom. transl.* herein *ex* Eulophoceratinae)

Type genus: *Eulophoceras* Hyatt, 1903.

Diagnosis. Typically strongly compressed oxycones, often with feeble umbilical bullae and low crescents on the outer flank; suture with very broad deeply bifid E/L. *U. Coni. - L. Camp.*

Praelibycoceras DOUVILLÉ, 1912 [*Lenticeras jullieni* Pervinquière, 1907; OD]. Small; involute compressed oxycone with sharp venter becoming subrounded in later growth; strong umbilical folds give rise to low ribs that broaden and decline ventrally; suture relatively simple, with very broad deeply bifid E/L, narrow L, broad shallowly bifid L/U_2, and few shallowly incised auxiliary elements. *U. Coni.* **Fig. 425O-P** - *P. jullieni* (Pervinquière).
= *Prelibycoceras* Spath, 1922 (*laps. cal.*).

Eulophoceras HYATT, 1903 [*E. natalense*; OD]. Medium sized; like *Praelibycoceras* but with almost occluded

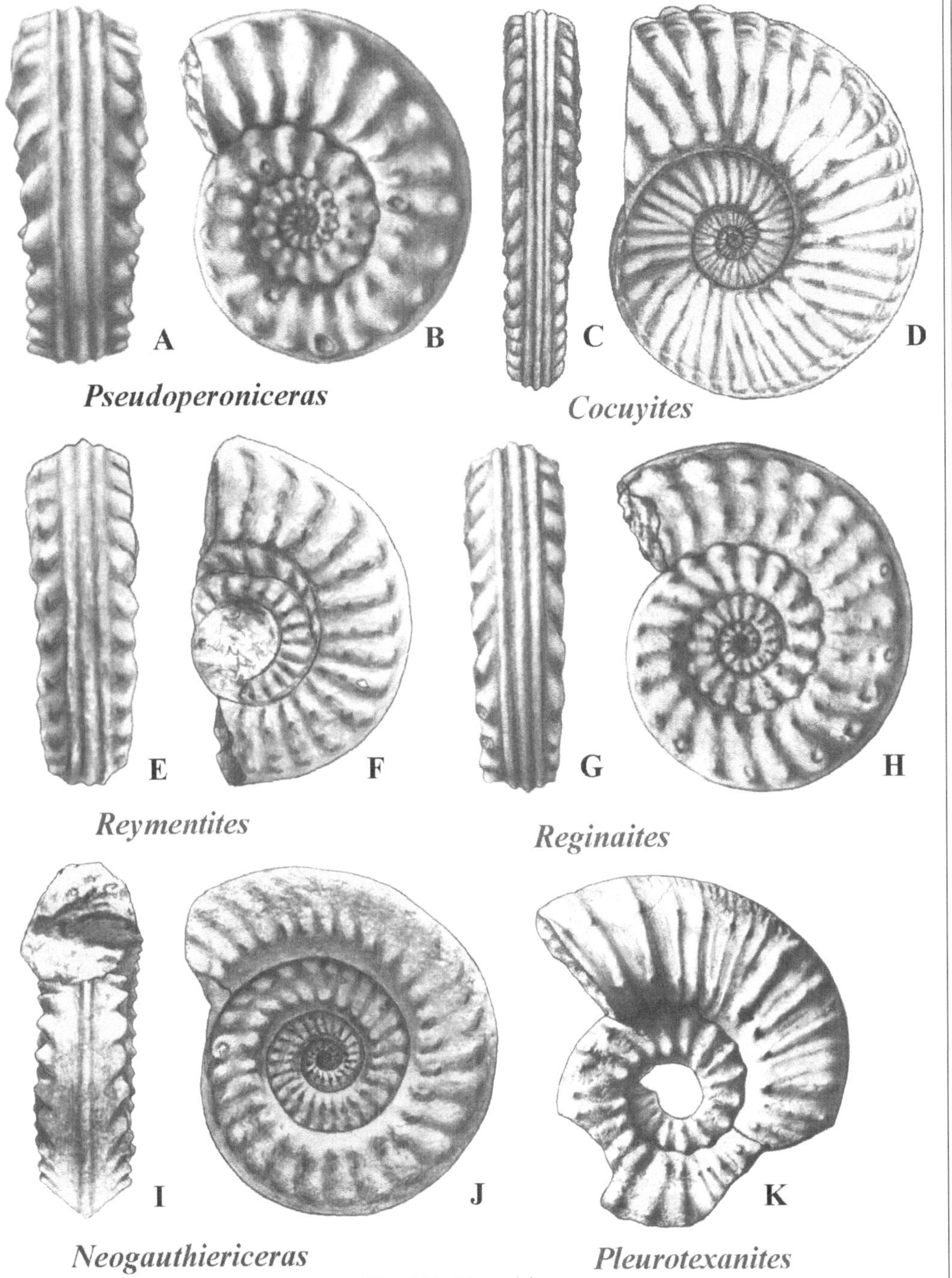

Fig. 419. Texanitinae.

umbilicus, sharper venter, very faint ribs on outer flank and suture with finger-like folioles to saddles. *U. Sant. - L. Camp.* **Fig. 423** - *E. natalense* Hyatt.

Spheniscoceras SPATH, 1921 [*S. africanum* (Crick MS); OD]. Like *Eulophoceras* but with slight differences in suture; E/L markedly asymmetrical and L multifid. *L. Camp.* **Fig. 429C-D** - *S. africanum* Spath.

= *Sphenodiscoceras* Kennedy *et al.*, 1995 (*nom. van.*).

Pelecodiscus VAN HOEPEN, 1921 [*P. umzambiensis*; OD]. Medium sized; like *Eulophoceras* but with simpler suture, less deeply bifid E/L without finger-like folioles and well developed 3rd lateral saddle. *L. Camp*. **Fig. 425M-N** - *P. umzambiensis* van Hoepen.

= *Peleodiscus* Roman, 1938 (*nom. van.*).

= *Pteleodiscus* Roman, 1938 (*laps. cal.*).

***Elninoceras* gen. nov.** [*Eulophoceras jacobi* Hourcq, 1949, p.95; OD herein]. Medium sized; like *Eulophoceras* but strongly ornamented with blunt sinusoidal ribs and both umbilical and ventrolateral tubercles. *L. Camp*. **Fig. 425Q-R** - *E. jacobi* (Collignon).

Daradiceras SORNAY & TESSIER, 1949 [*D. gignouxi*; OD]. Small; inner whorls much like *Elninoceras* with coarse strong flexuous ribs alternating long and short or arising in pairs from feeble umbilical bullae; adult body chamber inflated, with strong rectiradiate ribs terminating in flared ventrolateral bullae, broad tabulate venter with persistent siphonal keel, and umbilical bulge overlapping preceding whorl; suture with relatively narrow shallowly bifid E/L. *Maastr*. **Fig. 428E-H** - *D. gignouxi* Sornay & Tessier.

Tribe Diaziceratini Basse, 1947

(*nom. transl.* herein *ex* Diaziceratinae)

Diagnosis. Typically inflated, with strongly convergent flanks and fastigiate venter; 4-8 prminent umbilical tubercles per whorl weakly connected to straight prorsiradiate ribs that terminate in ventrolateral tubercles. Suture with prominent adventive lobe to E/L, small L much shorter than E, and subphylloid folioles.

The suture of *Diaziceras* is remarkably similar to *Lenticeras*, but the large morphological and stratigraphical gap separating the two favours placement here, with an origin close to *Eulophoceras*. *U. Sant*. - *L. Camp*.

Skoumalia SUMMERSBERGER, 1979 [*S. austriaca*; OD]. Small; like *Eulophoceras* but with high vertical umbilical wall and subangular umbilical edge; macroconchs with 4-8 distinct umbilical tubercles, initially rounded but strengthening with growth and becoming radially bullate on adult body chamber; umbilical tubercles give rise to 2-3 broad low prorsiradiate ribs that, together with intercalatories, terminate in small ventrolateral nodes; flanks with prorsiradiate biconcave growth striae; venter rounds in maturity; microconchs lack umbilical bullae; suture relatively simple, with narrow L. *U. Sant*. **Fig. 425I-J** - *S. austriaca* Summesberger.

Diaziceras SPATH, 1921 [*D. tissotiaeforme*; OD]. Small; like inner whorls of *Skoumalia* but more inflated, with steep umbilical wall, subangular umbilical edge, broad slightly concave flanks that converge to strongly fastigiate venter with high sharp siphonal keel; 5 strong prosocline umbilical bullae give rise to pairs of strong rectiradiate ribs that terminate in ventrolateral clavi projecting forwards on venter; superficially similar to *Metatissotia*. *L. Camp*. **Fig. 427** - *D. tissotiaeforme* Spath.

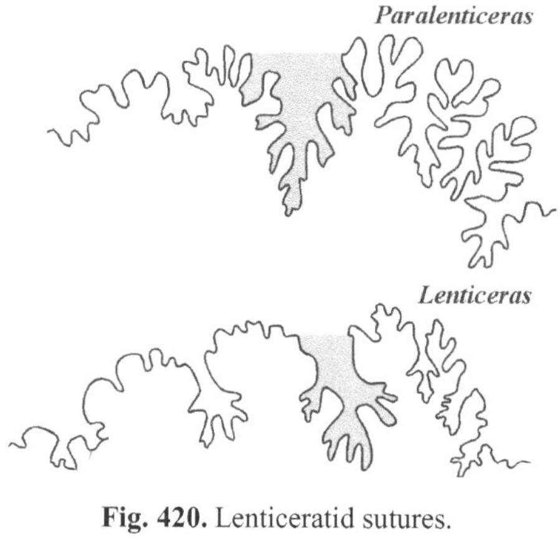

Fig. 420. Lenticeratid sutures.

Family SPHENODISCIDAE Hyatt, 1900

Discussion. Compressed oxycone derivatives of Lenticeratidae with very broad deeply incised E/L and numerous, weakly incised, mushroom-shaped auxiliary saddles. *Upper L. Camp*. - *U. Maastr.*

Subfamily Libycoceratinae Michailov, 1956

(*nom. transl.* Zaborski, 1982 *ex* Libycoceratidae)

= Libycoceratinae Zaborski, 1982

Discussion. Monophyletic derivatives of Eulophoceratinae, close to *Eulophoceras* (Hyatt, 1900); suture typically with one adventive lobe, bifid or trifid E/L and more or less entire saddles to suspensive lobe. *L. Camp*. - *U. Maastr*.

Praemanambolites COLLIGNON, 1969 [*P. hourcqui*; OD]. Moderately large to large; like *Eulophoceras* but with inflated adult body chamber, convex flanks and contracted aperture; suture on arched baseline, with L/U_2 and squat accessory saddles mushroom shaped and entire, and very broad bifid E/L with bilobed branches without finger-like folioles. *Upper L. Camp*. **Fig. 425K-L** - *P. hourcqui* Collignon.

Manambolites HOURCQ, 1949 [*M. piveteaui*; OD]. Medium sized; like *Praemanambolites* but with maximum breadth close to ventrolateral shoulder; adult body chamber with flat parallel sides, rounded venter and non-contracted aperture; suture with very wide shallow E, deeply bifid E/L with narrow elongate branches and moderately narrow asymmetrically bifid L; accessory saddles to suspensive lobe relatively elongate, separated by bifid lobes. *Upper L. Camp*. **Fig. 430C-D** - *M. piveteaui* Hourcq (strongly ornamented individual).

Mzezzemceras BASSE, 1954 [*Coahuilites (M.) pervinquieri* Basse; OD]. Medium sized; like *Manambolites* but non-tuberculate and with feeble folds on outer flank; suture

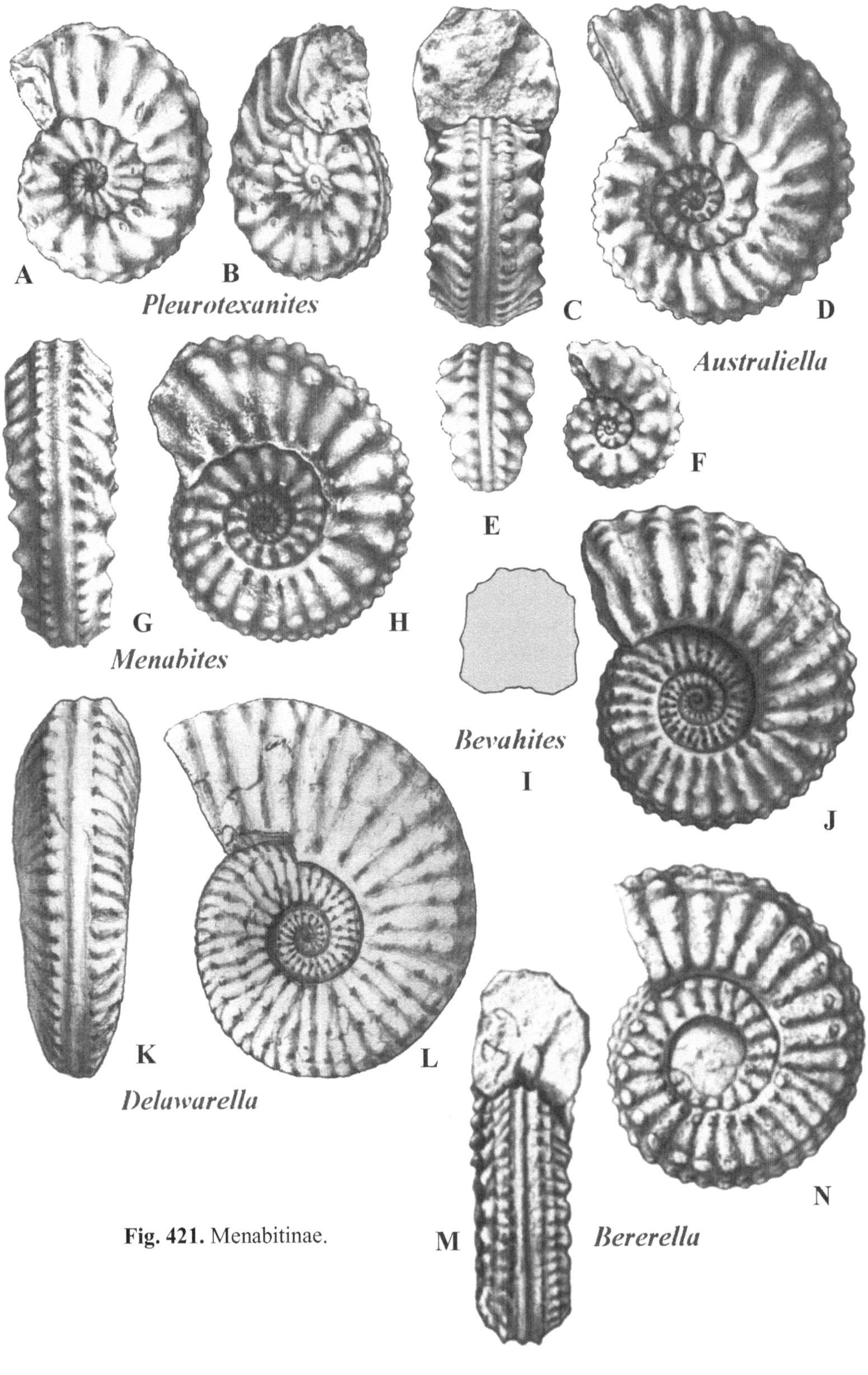

Fig. 421. Menabitinae.

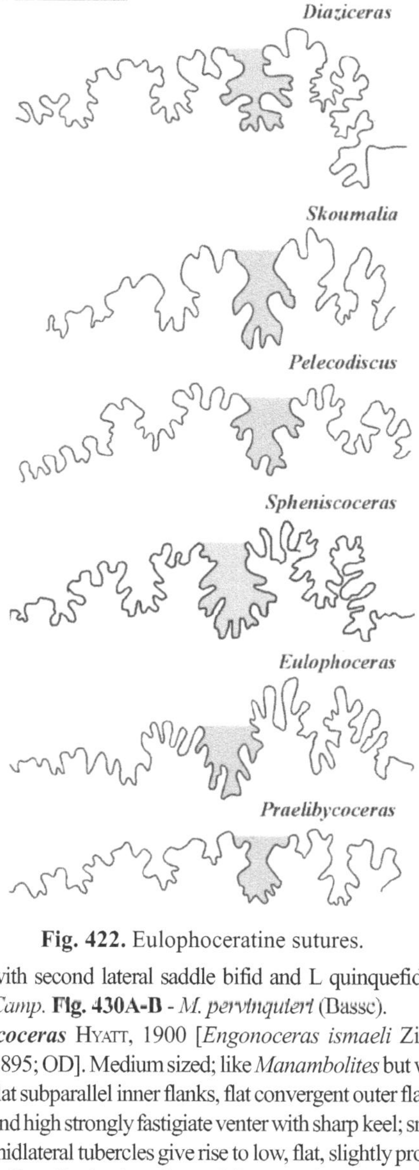

Fig. 422. Eulophoceratine sutures.

with second lateral saddle bifid and L quinquefid. *L. Camp.* **Fig. 430A-B** - *M. pervinquieri* (Basse).

Libycoceras HYATT, 1900 [*Engonoceras ismaeli* Zittel, 1895; OD]. Medium sized; like *Manambolites* but with flat subparallel inner flanks, flat convergent outer flanks and high strongly fastigiate venter with sharp keel; small midlateral tubercles give rise to low, flat, slightly prorsiradiate ribs that broaden rapidly to terminate in enlarged ventrolateral clavi; suture with narrow stemmed entire branches to E/L, deep penetrating rather wide digitate L, and narrow stemmed entire L/U_2 with descending series of at least 6 entire narrow-necked auxiliary saddles. *U. Camp. - L. Maastr.* **Fig. 428C-D** - *L. ismaeli* (Zittel).

Indoceras NOETLING, 1897 [*I. baluchistanense*; OD]. Medium sized; inner whorls with acutely fastigiate venter becoming narrowly rounded in later growth and finally, on adult body chamber, feebly convex with narrowly rounded ventrolateral shoulders; prorsiradiate falcate growth striae and indistinct ribs, with low weak regularly spaced crescents on outer flank; suture as in *Libycoceras* but with saddle-shaped E and E/L with flat-topped accessory saddles. *U. Maastr.* **Fig. 428A-B** - *I. baluchistanense* Noetling.

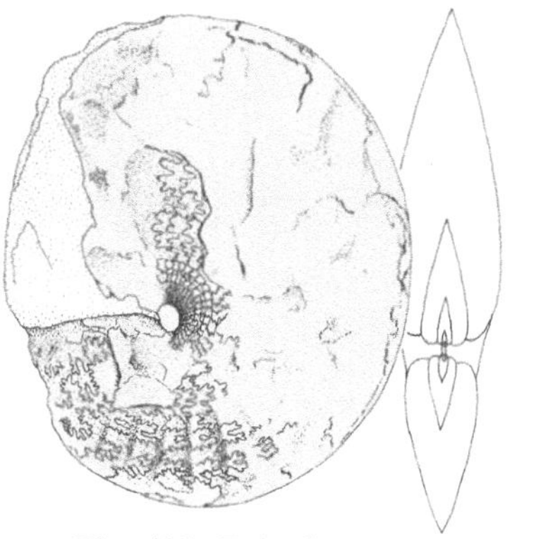

Fig. 423. *Eulophoceras.*

Subfamily Sphenodiscinae Hyatt, 1900

(*nom. transl.* Wright, 1996 *ex* Sphenodiscidae)

Discussion. Seemingly monophyletic derivatives of Libycoceratinae, precise ancestor uncertain, perhaps *Libycoceras*; suture typically with trifid E/L and a multiplicity of auxiliary elements, the outer few saddles of which are digitate. *U. Camp. - Maastr.*

Paciceras OLSSON, 1944 [*P. pacificum*; OD]. Medium sized; like *Libycoceras* with hint of mid-lateral swellings; suture with trifid E/L to suture; lateral and ventrolateral tubercles sometimes elongated into rib-like structures; tuberculation declines in later growth when venter rounds. *U. Camp. - L. Maastr.* **Fig. 428K-L** - *P. pacificum* Olsson.

Sphenodiscus HYATT, 1900 [*Ammonites lenticularis* Owen (*non* Phillips), 1852, = *Ammonites lobatus* Tuomey, 1856; OD]. Very large compressed oxycone with broad convex flanks, sharp venter and lanceolate whorl section; venter narrowly rounded in maturity; generally unornamented save for weak radial bulges in early growth; suture with deeply incised trifid E/L with finger-like

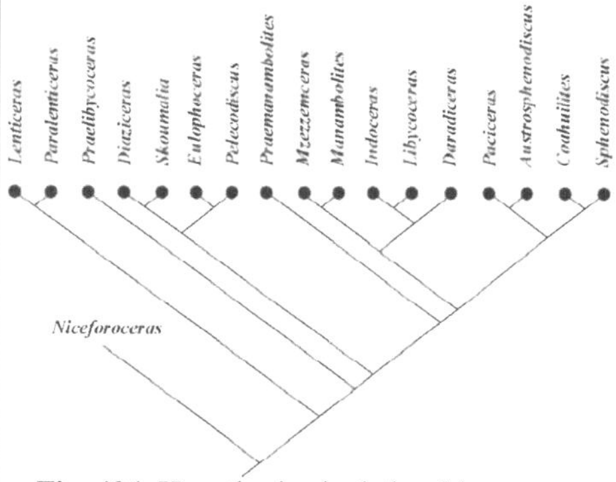

Fig. 424. Hypothesized relationships among Sphenodiscidae.

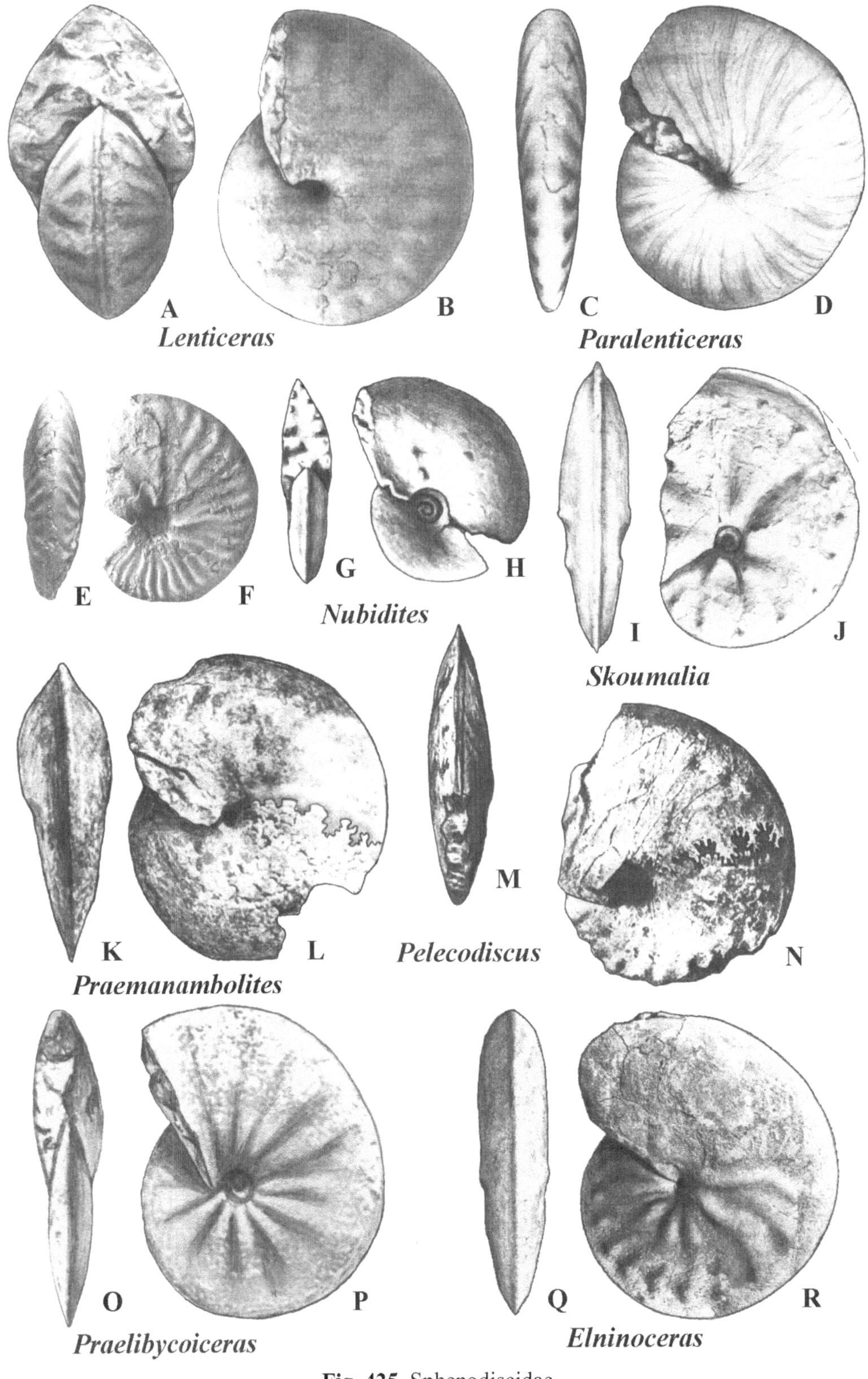

Fig. 425. Sphenodiscidae.

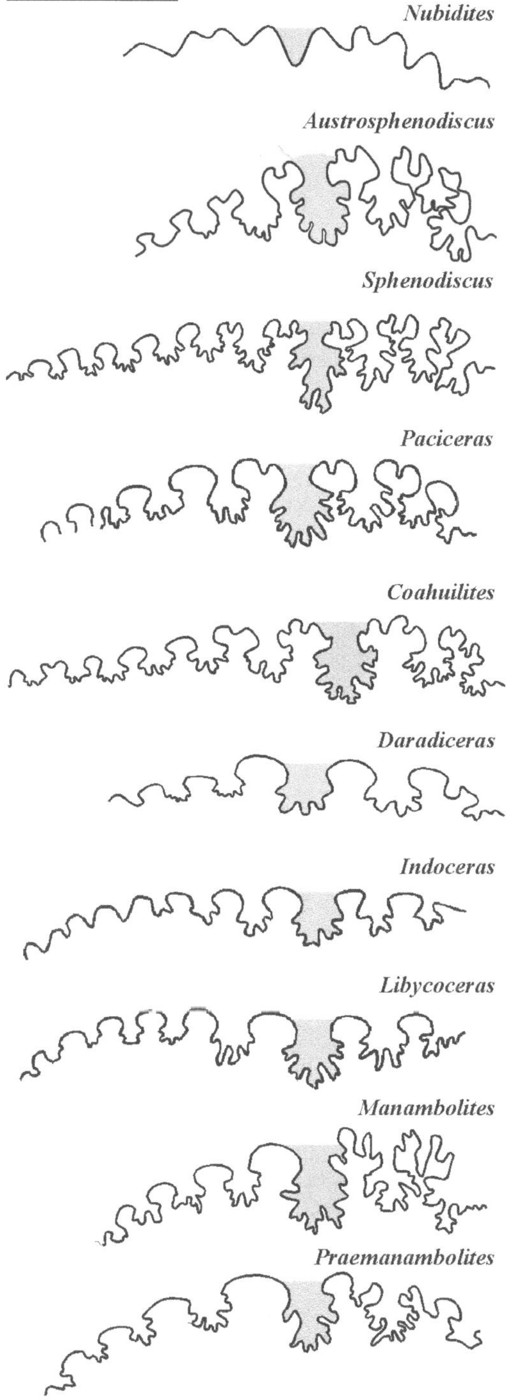

Fig. 426. Sphenodiscid sutures.

folioles and up to 10 narrow necked auxiliary saddles to suspensive lobe, outer two of which are bifid. *Maastr.* **Fig. 428I-J** - *S. lobatus* (Tuomey).

Austrosphenodiscus OLSSON, 1944 [*S. pleurisepta* var. *peruviana* Gerth, 1928; OD]. Medium sized; like *Sphenodiscus* but with two rows of lateral tubercles from early growth, an obtusely fastigiate venter, and simpler suture without finger-like folioles; in later growth venter rounds and upper row of lateral tubercles replaced by arcuate folds. *Maastr.* **Fig. 430A-B** - *A. peruviana* Gerth.

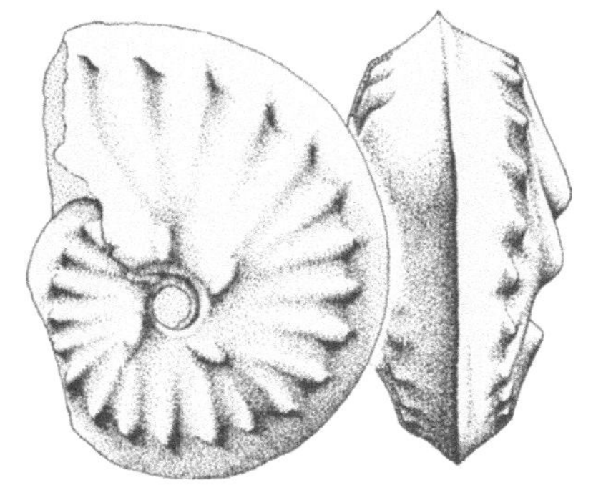

Fig. 427. *Diaziceras*.

Coahuilites BÖSE, 1928 [*C. sheltoni*; OD]. Medium sized to moderately large; like *Austrosphenodiscus* but with tubercles high lateral and ventrolateral in position; ornament declines on adult body chamber that has broad subtabulate venter; suture with wide asymmetrically bifid E/L with narrow moderately incised outer branch and deep penetrating accessory lobe. *Maastr.* **Fig. 428M-N** - *C. sheltoni* Böse.

Nubidites WIEDMANN & KULLMAN, 1979 [*N. amarai*; OD]. Very small; of uncertain stratigraphic position; like *Libycoceras* but with goniatitic suture. *?U. Cret.* **Fig. 425G-H** - *N. amarai* Wiedmann & Kullman.

Order ANCYLOCERATIDA Wiedmann, 1962

(*nom. transl.* herein *ex* Ancyloceratina)

Discussion. Monophyletic derivatives of Ammonitida, comprising the suborders Ataxioceratidina, Ancyloceratidina and Turrilitidina; primitively ammoniticones but later heteromorphs and secondarily recoiled ammoniticones with a quadrilobate suture in later growth. *L. Tithon. - U. Maastr.*

Suborder ATAXIOCERATIDINA nov.

Diagnosis. Monophyletic descendents of Perisphinctidina (Arkell, 1957); comprises the hyporders Ataxioceratoidei and Neocomitoidei. Microconchs commonly with biplicate ribs usually interrupted across venter by smooth band or siphonal groove; macroconchs generally more involute, with biplicate to polyschizotomous and virgatotome ribbing, and frequent intercalatories. *M. Oxford. - Barrem.*

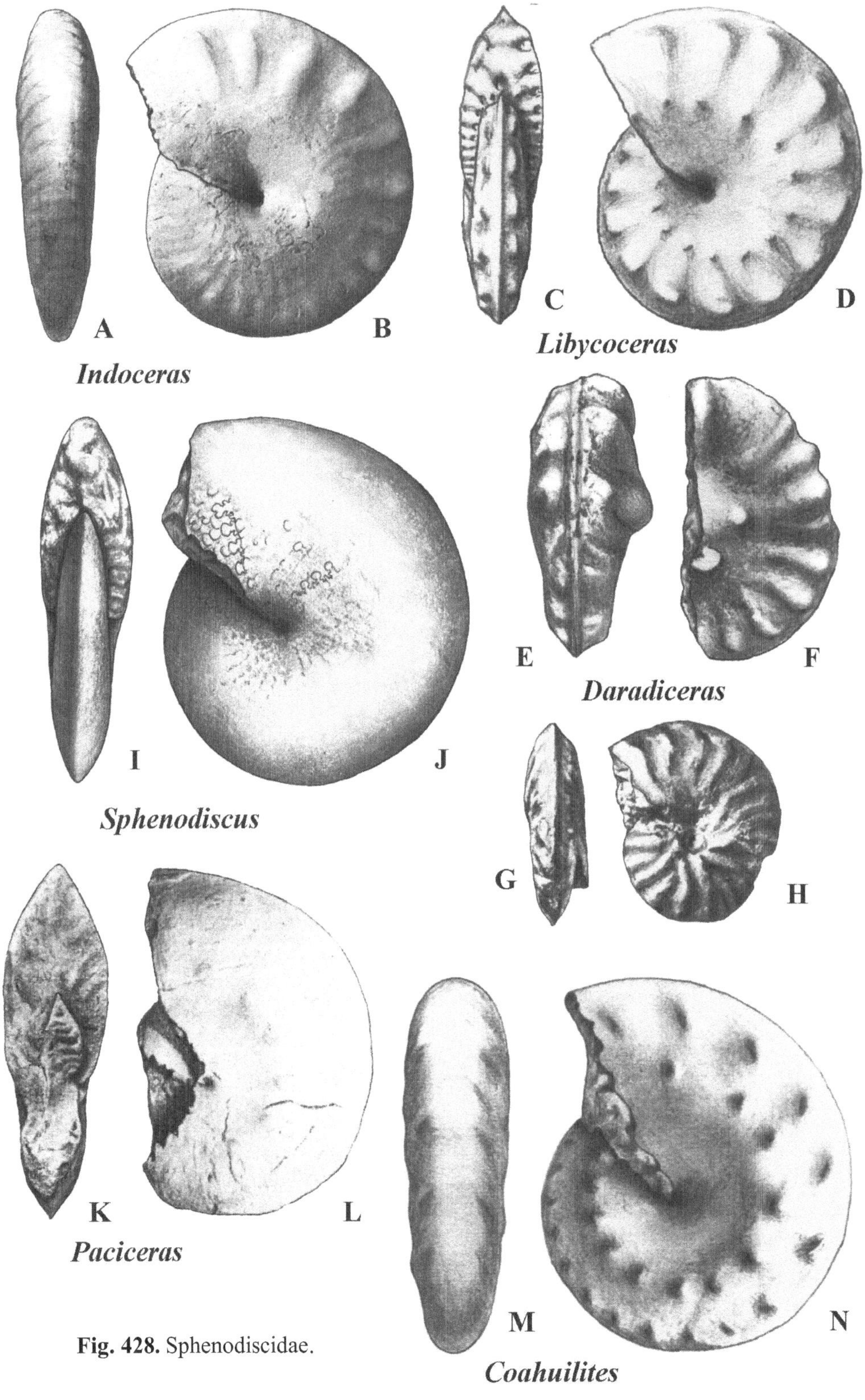

Fig. 428. Sphenodiscidae.

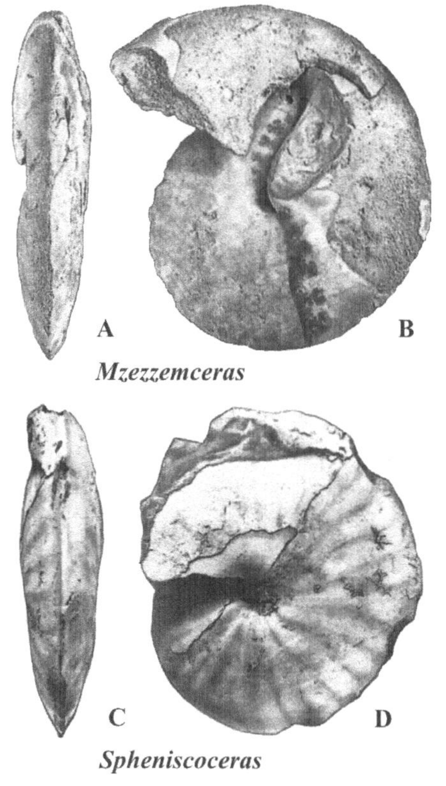

Fig. 429. Sphenodiscidae.

Hyporder ATAXIOCERATOIDEI Buckmann, 1921

(*nom. transl.* herein *ex* Ataxioceratidae)

Discussion. A seemingly monophyletic group with an origin in Perisphictidina, comprising the superfamilies Ataxioceratoidea and Himalayitoidea. *M. Oxford. - U. Tithon.*

Superfamily ATAXIOCERATOIDEA Buckmann, 1921

(*nom. transl.* herein *ex* Ataxioceratidae)

Discussion. Monophyletic derivatives of Idoceratoidea (Arkell, 1950; Melendez *et al.*, 2009); comprises the predominantly Boreal and Tethyan families Ataxioceratidae, Lithacoceratidae, Sublithacoceratidae and Virgatitidae. Dimorphism is conspicuous; microconchs small, commonly with biplicate ribbing and lateral lappets, macroconchs large, with subpolyplocoid to virgatotome and fascipartite ribbing in maturity. *M. Oxford. - U. Tithon.*

Family ATAXIOCERATIDAE Buckmann, 1921

Discussion. A seemingly monophyletic taxon with an origin in Idoceratidae (Meléndez, 1984). The group is commonest in the Submeditterranean Province of the Tethyan Realm. Comprises the subfamilies Ataxioceratinae and Paraboliceratinae. *M. Oxford. - U. Tithon.*

Subfamily Ataxioceratinae Buckmann, 1921

(*nom. transl.* Arkell *et al.*, 1957 *ex* Ataxioceratidae)

Discussion. A seemingly monophyletic taxon with an origin in Passendorferiinae close to *Passendorferia* (Urresti, 1996; Meléndez *et al.*, 2009). Macroconchs large to very large, typically compressed, evolute, with rounded venter, subelliptical whorl section and non-tuberculate primary ribs giving rise to fine secondaries in biplicate to poly-

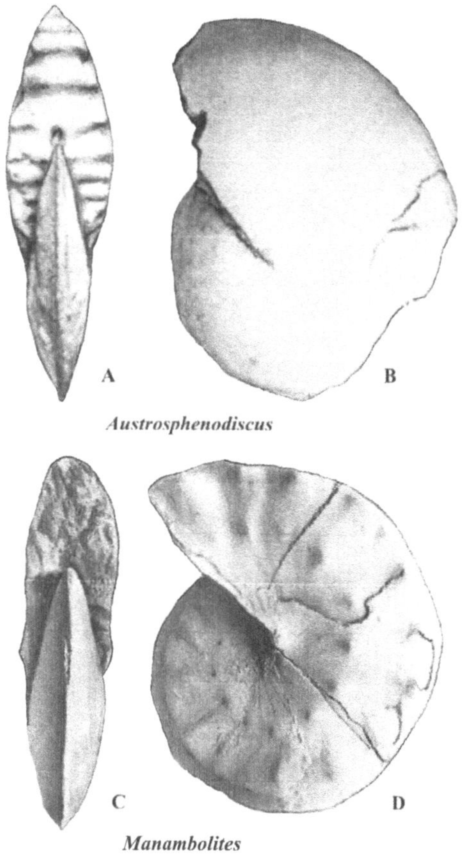

Fig. 430. Sphenodiscidae.

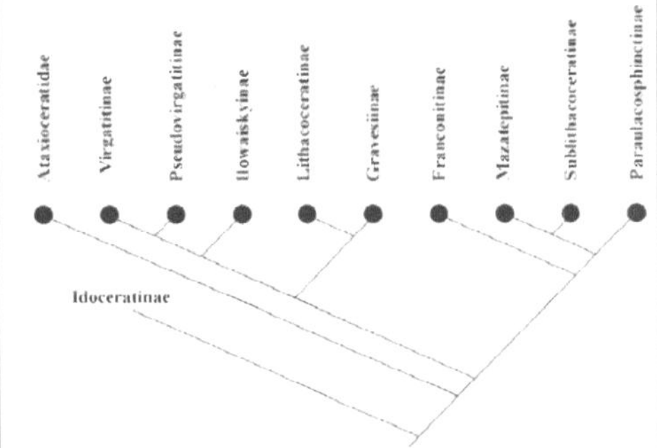

Fig. 431. Hypothesized relationships among Ataxioceratoidea.

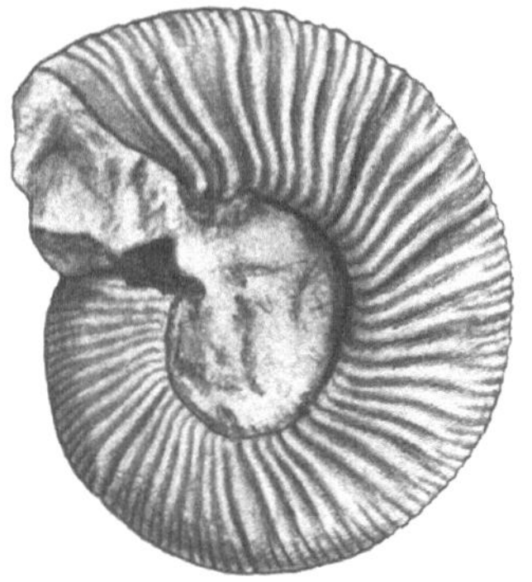

Fig. 433. *Geyericeras.*

plocoidal bundles, with or without parabolae; microconchs small to medium sized, with simple biplicate ribs and lateral lappets. *M. Oxford. - M. Kimmer.*

Eoataxioceras GYGI, 2001 [*E. gansingense*; OD] [**m**]. Small; evolute, compressed, with gently convex flanks converging to narrow rounded venter; well spaced prorsiradiate primary ribs weaken ventrally and connect weakly to indistinct polyplocal bundles of 3-4 fine secondary ribs. *M. Oxford.* **Fig. 436B** - *E. gansingense* Gygi.

Praeataxioceras ATROPS, 1982 [*Perisphinctes laufenensis* Siemiradzki, 1899; OD][**m**]. Small; like *Eoataxioceras* but with more primaries that give rise to only 2-3 secondaries. *Upper U. Oxford. - Lower L. Kimmer.* **Fig. 436A** - *P. laufenense* (Siemiradzki).

Orthosphinctes SCHINDEWOLF, 1925 [*Ammonites tiziani* Oppel, 1863; OD] [**m**]. Small; evolute, with numerous rectiradiate biplicate ribs branching high on flank, and secondaries crossing venter transversely with occasional intercalatories; periodic strong constrictions, slightly oblique to ribbing; adult aperture with lateral lappets. *U. Oxford.* **Fig. 436-G** - *O. tiziani* (Oppel).

= *Biplices* Siemiradzki, 1898 (obj.).

Pseudorthosphinctes ÉNAY, 1966 [*Orthosphinctes (P.) alternans*; OD] [**M**]. Moderately large, with high oval whorl section and numerous rectiradiate ribs branching high on flank, mainly biplicate in early to middle growth, but triplicate later; constrictions tend to parallel ribbing. *U. Oxford. - L. Kimmer.* **Fig. 436C** - *P. alternans* (Énay).

Planites DE HAAN, 1825 [*Ammonites polygyratus* Rienecke, 1818; OD][**m**]. Small; like *Orthosphinctes* but more involute and inflated, with depressed whorl section and fine sharp biplicate and triplicate ribs. *Upper U. Oxford.- Lower L. Kimmer.* **Fig. 436J-K** - *P. polygyratus* (Rienecke).

Ataxioceras FONTANNES, 1879 [*Perisphinctes hypselocyclus* Fontannes, 1879; OD] [**M**]. Moderately large; fairly involute, compressed, high whorled, with flattened subparallel flanks, rounded venter and elliptical whorl section (**m**), or compressed oval whorl section with convergent convex flanks (**M**); strong prorsiradiate primary ribs split above midflank into 2-4 fine secondaries (true polyplocoid ribbing), with some intercalatories, that cross venter transversely; parabolae present. *U. Oxford. - L. Kimmer.* **Fig. 436D-E** - *A. hypselocyclus* (Fontannes).

With 46 nominal species/subspecies (Klein, 2005), a number with query, this is a *nomen commodum.*

Lithacosphinctes OLÓRIZ, 1978 [*Ammonites lictor evolutus* Quenstedt, 1888, = *L. siemiradzkii* Zeiss, 1992; OD] [**M**]. Medium sized (**m**) to very large (**M**); like *Pseudorthosphinctes* but with slightly prorsiradiate primary ribs splitting irregularly into 2-3 secondaries with numerous intercalatories; in later growth primary ribs distant and robust, secondaries obsolete and venter smooth; two deep oblique parabolae per whorl. *Upper U. Oxford. - Lower L. Kimmer.* **Fig. 436H** - *L. evolutus* (Quenstedt).

A macroconch of *Orthosphinctes* (*fide* Atrops, 1982; Molinar & Olóriz, 2010).

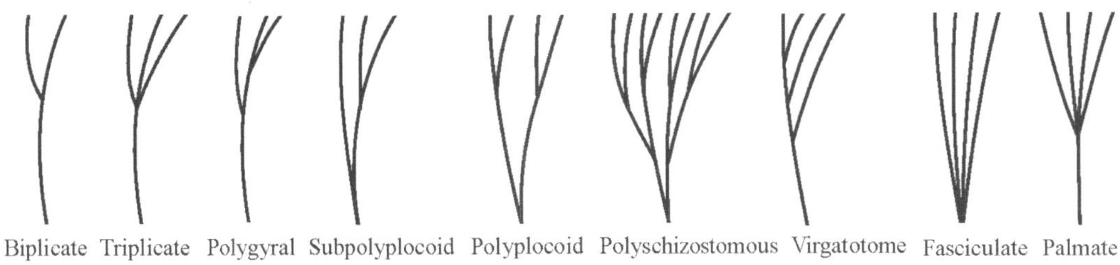

Fig. 432. Ataxioceratoid rib patterns.

Parataxioceras SCHINDEWOLF, 1925 [*Ammonites lothari* Oppel, 1863; OD] [**m**]. Small (**m**) to moderately large (**M**); like *Geyericeras* but larger and more evolute, with polyplocoid ribbing; primary ribs become more distant with growth; parabolae frequent. *Upper U. Oxford. - Lower L. Kimmer.* **Fig. 436L** - *P. lothari* (Oppel).

Ardescia ATROPS, 1982 [*Ataxioceras desmoides* Wegele, 1929; OD] [**m**]. Moderately large (**m**) to very large (**M**); like *Lithacosphinctes* but with narrower umbilicus, less-massive whorls, prorsiradiate primary ribs with more intercalated secondaries, more delicate ribs in early growth, and periodic parabolae. *U. Oxford. - L. Kimmer.* **Fig. 436I** - *A. desmoides* (Wegele).

Geyericeras MOLINER & OLÓRIZ, 2010 [*G. aragoniense*; OD]. Small (**m**) to moderately large (**M**); like *Ardescia* but with finer and tighter ribbing and frequent sub-polyplocoid ribs. *Middle L. Kimmer.* **Fig. 433** - *G. aragoniense* Moliner & Olóriz.

Alienisphinctes ÉNAY & MANGOLD, 2021 [*A. cavelieri*; OD]. Dimorphic; moderately evolute; macroconch with prorsiradiate biplicate primary ribs with frequent intercalatories; microconch with more or less quadratic whorl section, flat venter and secondary ribs more projected forwards on ventrolateral shoulder. *?L. Kimmer.* **Fig. 435** - *A. cavelieri* Énay & Mangold.

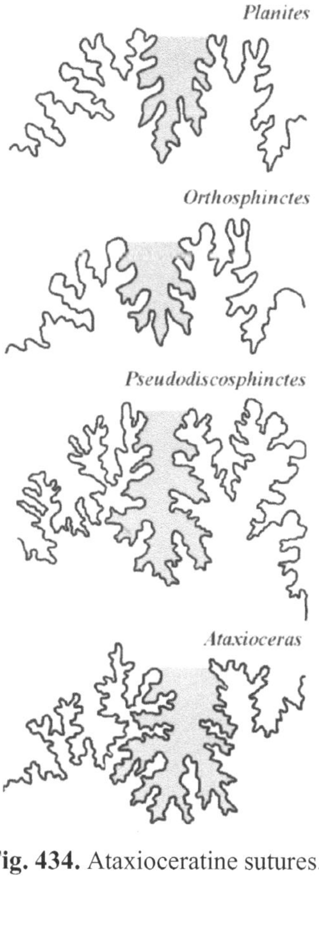

Fig. 434. Ataxioceratine sutures.

Fig. 435. *Alienisphinctes.*

Schneidia ATROPS, 1982 [*Ataxioceras (S.) collignoni*; OD] [**m**]. Medium sized (**m**) to large (**M**); like *Ardescia* but more involute, with oval to subrectangular whorl section; inner whorls densely and finely ribbed, with subpolyplocoid ribbing particularly in later growth; rib curvature decreases after 50 mm diameter; constrictions numerous and well developed; lateral lappets relatively short to obsolete. *Lower L. Kimmer.* **Fig. 437D** - *S. collignoni* (Atrops).

Crussoliceras ÉNAY, 1959 [*Ammonites crusoliensis* Fontannes, 1876; OD] [**M**]. Medium sized (**m**) to very large (**M**); evolute, compressed; inner whorls with convex flanks, low rounded venter and depressed subquadratic whorl section; in early to middle growth with regular rectiradiate biplicate ribs that, following a constriction, become very coarse and distant in maturity; microconchs with variocostate ribbing. *Upper L. - Lower U. Kimm.* **Fig. 438C-D** - *C. crussoliense* (Fontannes).

"Badenia" ÉNAY, 1959 [*B. wegelei*; OD] [m]. Small, with maximum width close to umbilical shoulder and trapezoidal whorl section; strong biplicate ribs becoming triplicate on adult body chamber; deep constrictions bordered anteriorly by simple ribs; relatively simple suture with retracted suspensive lobe. *L. Kimmer.* **Fig. 438I-J** - *B. wegelei* Énay.

Based on a microconch of *Crussoliceras*.

[*non* Finlay, 1930, Gastropoda].

Garnierisphinctes ÉNAY, 1959 [*Ammonites (Perisphinctes) garnieri* Fontannes, 1876; OD] [**m**]. Small (**m**) to moderately large (**M**); like *Crussoliceras* but more evolute, with flat flanks, strongly rounded venter, subquadratic whorl section and weaker denser ribbing. *Upper L. Kimmer.* **Fig. 438F-H** - *G. garnieri* (Fontannes).

Progeronia ARKELL, 1953 [*Perisphinctes progeron* Ammon, 1875; OD] [**M**]. Moderately large; evolute, compressed, with thin long curved primaries that split into finer paired secondaries, branching higher in later growth; in maturity ribs triplicate, separated by an intercalatory. *Upper L. Kimmer.* **Fig. 437A** - *P. progeron* (Ammon).

= *Ammonia* Ilovaisky, 1941 (*non* Brünnich, 1772, *nec* Koch, 1842).

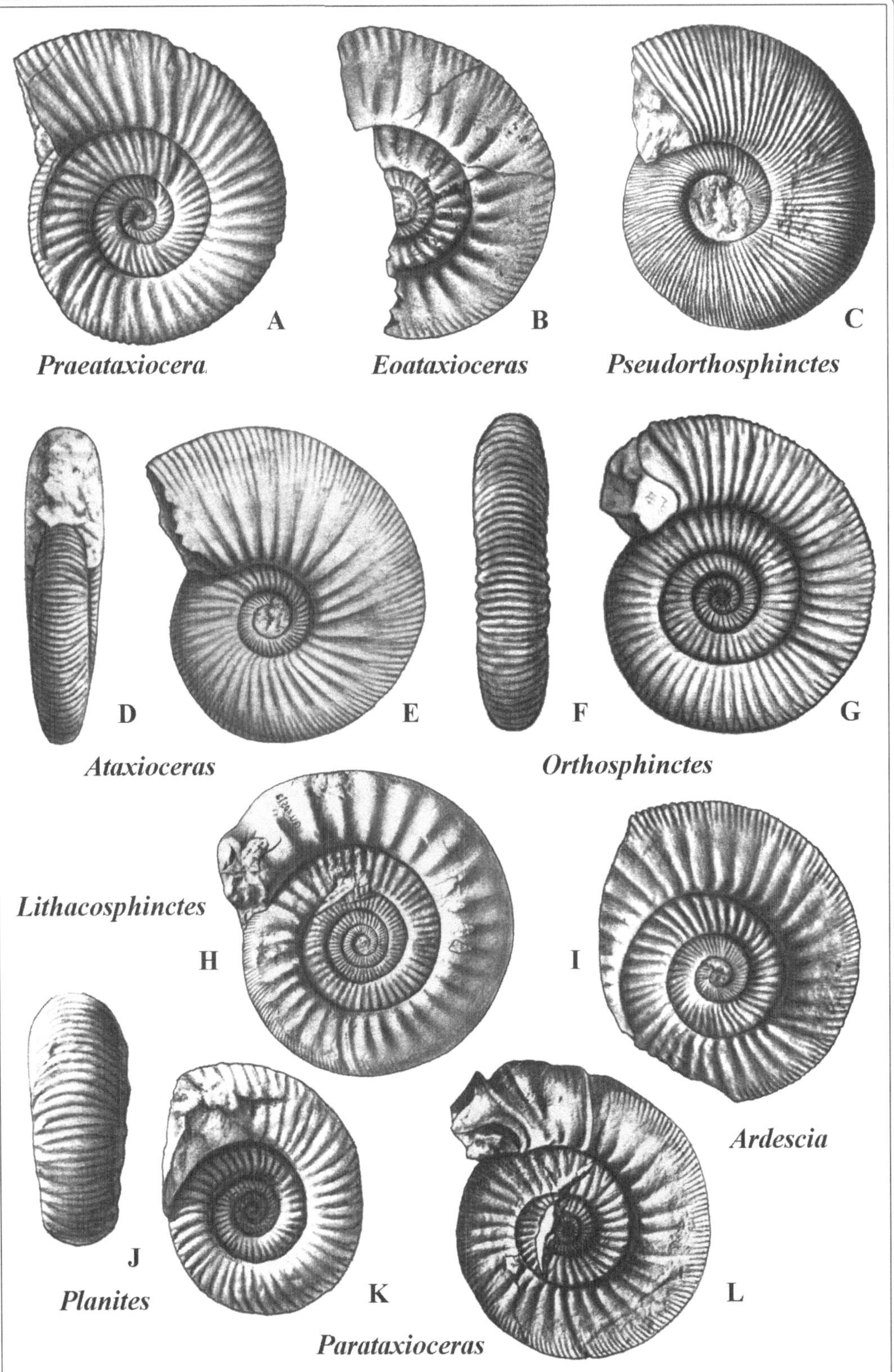

Fig. 436. Ataxioceratinae.

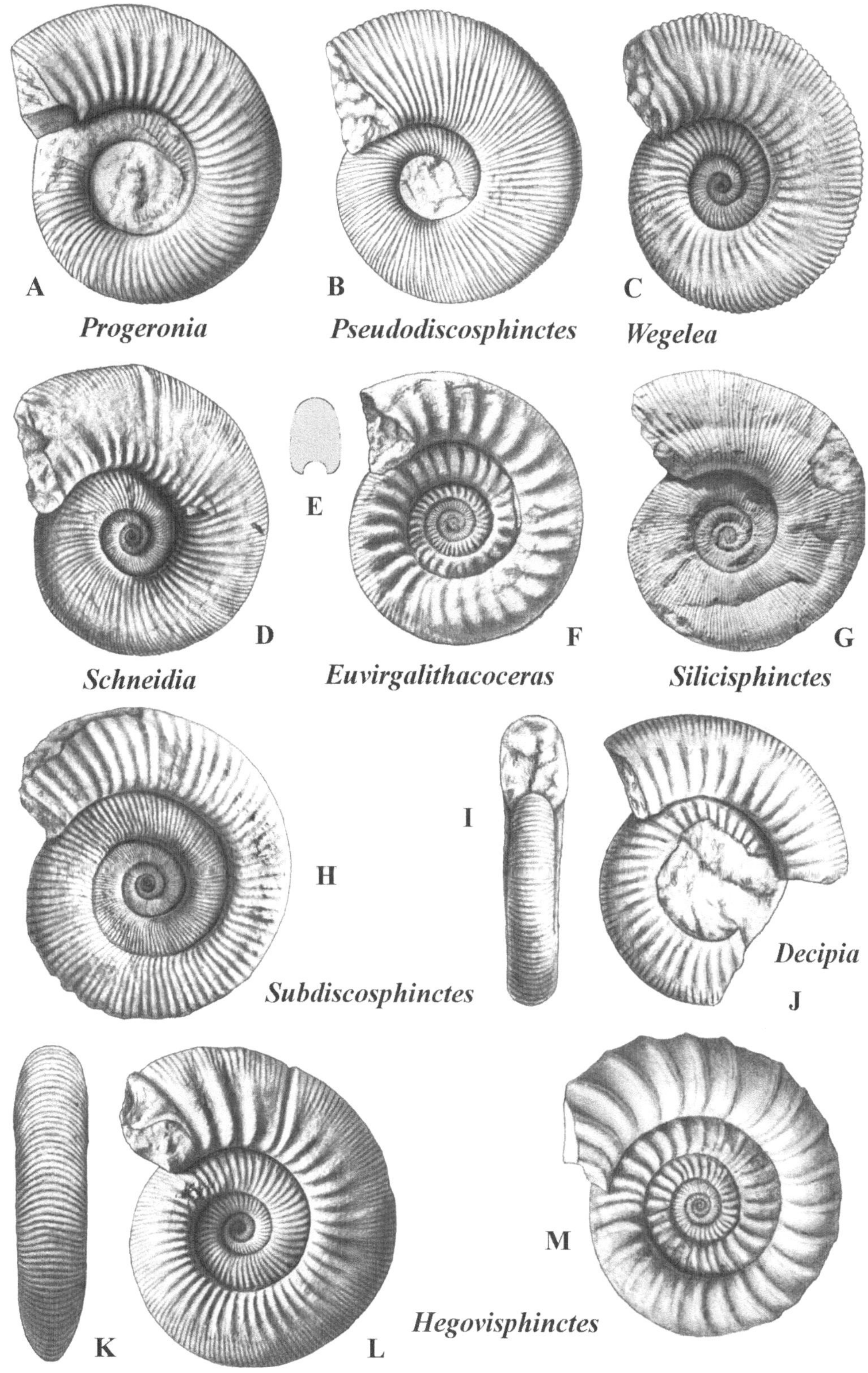

Fig. 437. Ataxioceratinae.

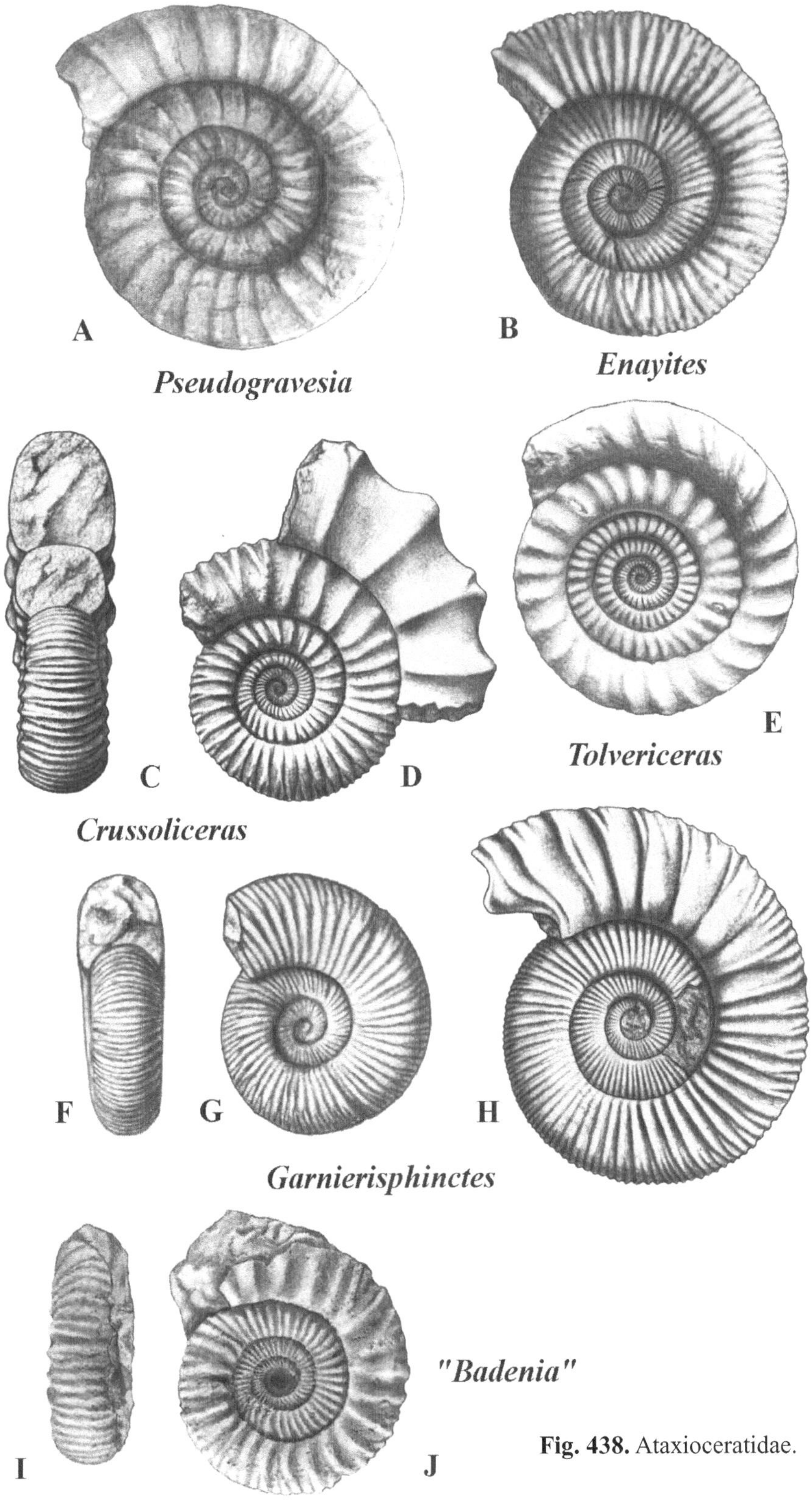

Fig. 438. Ataxioceratidae.

Hugueninsphinctes ATROPS, 1982 [*Ammonites breviceps* Quenstedt, 1882; OD][**m**]. Medium sized; like *Progeronia* but smaller, more compressed, with primary ribs branching lower on flank and with short lateral lappets to adult aperture. *Lower L. Kimmer.* **Fig. 437K-L** - *H. breviceps* (Quenstedt).

Perhaps a microconch of *Progeronia*.

Pseudogravesia HANTZPERGUE, 1989 [*Tolvericeras (P.) hahni*; OD][**m**]. Moderately large (**m**) to very large (**M**); like *Tolvericeras* but with depressed semicircular to subquadratic whorl section; in early growth rectiradiate ribs fine, bifurcating or trifurcating, effaced from venter, later strong, simple, distant. *U. Kimmer. - L. Tithon.* **Fig. 438A** - *P. hahni* (Hantzpergue).

Tolvericeras HANTZPERGUE, 1989 [*T. tolverense*; OD] [**M**]. Medium sized serpenticone; microconch like *Progeronia* but with polygyrate ribbing. *Lower L. Tithon.* **Fig. 438E** - *T. tolverense* Hantzpergue.

Subfamily Paraboliceratinae Spath, 1922

= Richterellinae Sapunov, 1977b.

Discussion. A seemingly monophyletic group with an origin in Ataxioceratinae, close to *Orthosphinctes*. Although Zeiss (2001) looked to Sublithacoceratinae for the ancestry of the group, and Cantú-Chapa (2009) favoured Virgatosphinctinae, close to *Aulacosphinctoides*, both are younger than *Sulaites*. Mostly moderately evolute, densicostate, with fine biplicate to virgatotome prorsiradiate ribs that project forward on the outer flank and are linguoid or form chevrons over the venter; ribs may or may not be interrupted by a narrow siphonal groove, with or without parabolic nodes; suture with broad E/L, narrow trifid L, very narrow L/U_2 and strongly retracted suspensive lobe. Predominantly Indo-Pacific. *Upper Oxford. - U. Tithon.*

Sulaites OLÓRIZ & WESTERMANN, 1998 [*Perisphinctes sularis* Boehm, 1907; OD]. Small serpenticone with rounded depressed (juv.) to compressed ovate (ad.) whorl section; fine mainly dichotomous ribs, ventrally projected forming chevrons (ad.) or narrowly interrupted (juv.) over venter; parabolic ribs and constrictions on inner whorls, and incipient to variable ventral parabolae on outer whorls; microconchs with lappets, macroconch unknown. *U. Oxf. - L. Kimmer.* **Fig. 439C-D** - *S. sularis* (Boehm).

= *Pseudoparaboliceras* Gerth, 1965 (*nom. nud.*)

Paraboliceras UHLIG, 1910 [*Ammonites jubar* Blanford, 1865; SD Roman, 1938]. Small to medium sized; like *Kossmatia* but more inflated, with oval whorl section, subtabulate venter and regularly spaced parabolic nodes that increase in number with growth and to which flank costellae are looped; suture with trifid L much deeper than E. *L. - U. Tithon.* **Fig. 441I-J** - *P. himalayanum* (Uhlig).

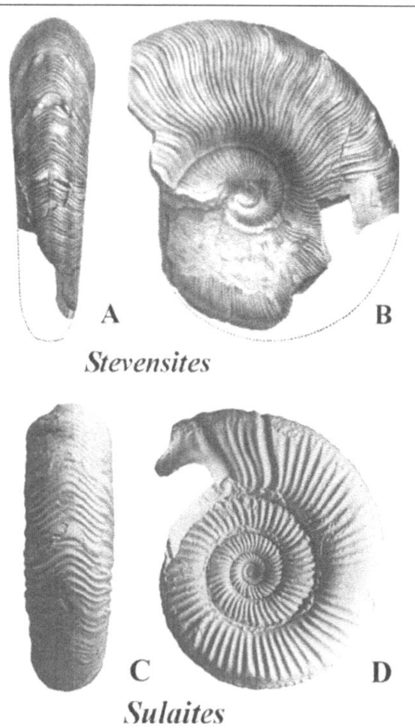

Fig. 439. Paraboliceratinae.

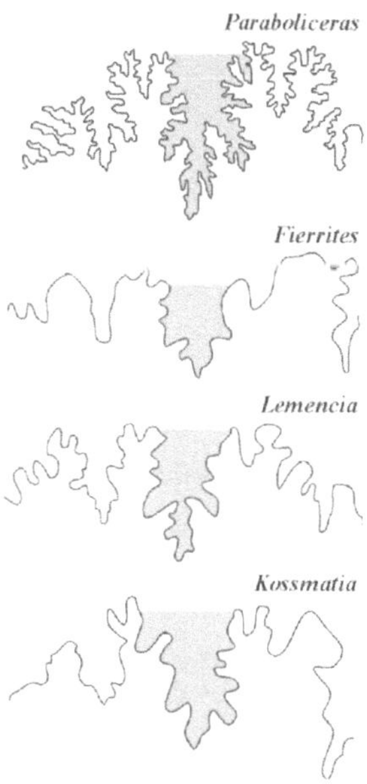

Fig. 440 Paraboliceratine sutures.

Paraboliceratoides SPATH, 1925 [*Ammonites mutilus* Oppel, 1865;OD] [M] Like *Paraboliceras* but much

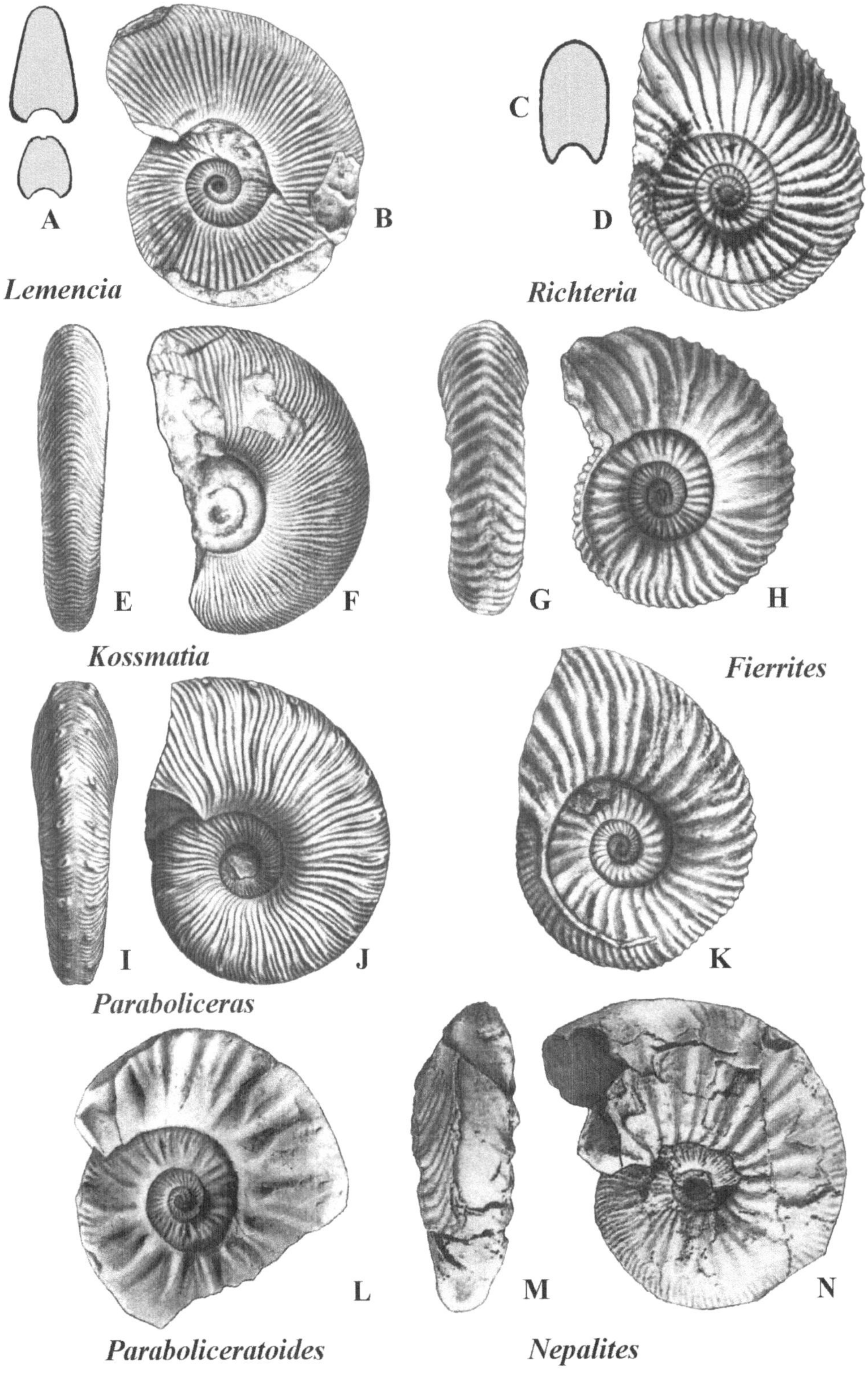

Fig. 441. Paraboliceratidae.

more coarsely ornamented, and without parabolic nodes. *U. Tithon.* **Fig. 441L** - *P. mutilum* (Oppel).

Lemencia DONZE & ÉNAY, 1961 [*L. pseudorichteri*; OD] [**M**]. Medium sized; compressed, with flat to weakly convex flanks converging to narrowly rounded venter and oval to subtrigonal whorls increasing rapidly in height; fine, flexuous, prorsiradiate primary ribs bifurcate above midflank, interrupted in early to middle growth across venter by narrow siphonal groove; ribbing virgatotome in later growth when groove fades. *Upper L. Tithon.* **Fig. 441A-B** - *L. pseudorichteri* Donze & Énay.

= *Richterella* Avram, 1974 *non* Hessler, 1965 (Trilobita).

= *Richteria* Olóriz, 1978 *non* Jones, 1874. (Ostracoda).

Ammonites richteri Oppel (1865) (**Fig. 441C-D**) is the type species of both ***Richteria*** and ***Richterella***; it is small, with subparallel flanks, subelliptical whorl section, and relatively coarse biplicate ribs projecting strongly forward on outer flank and uninterrupted over venter. It is regarded a microconch of *Lemencia*.

Fierrites CANTÚ-CHAPA, 1993 [*Perisphinctes alamitosensis* Aguilera, 1895; OD]. Like *"Richteria"* but more evolute with coarser, more distant ribs branching lower on flank; suture with L as deep as E. *U. Tithon.* **Fig. 441G-H, K** - *F. alamitosensis* (Aguilera).

Kossmatia UHLIG, 1907 [*Ammonites tenuistriatus* Gray, 1832; SD Roman, 1938]. Medium sized; like *Lemencia* but with finer biplicate to polyplocoidal ribs branching lower on flanks, crossing venter with an adoral convexity, with or without interruption by siphonal groove; suture with narrower E, L shorter than E and smaller U_2. *U. Tithon.* **Fig. 441E-F** - *K. tenuistriata* (Gray).

Nepalites ÉNAY, 2009 [*N. chhokorensis*; OD] [**m**] Medium sized; involute, compressed, with rounded venter and subovate whorl section; fine sharp prorsiradiate primary ribs split into 2's and 3's on outer flank, secondaries projecting forwards to cross venter with strong adoral sinus; ornament coarsens on adult body chamber. *U. Tithon.* **Fig. 441M-N** - *N. chhokorensis* Énay.

Stevensites ÉNAY, 2018 [*Kossmatia desmidoptycha* Uhlig, 1910; OD]. Similar to *Nepalites* but moderately large, with finer ribbing effaced from the siphonal line, and parabolic nodes; suture with asymmetrically bifid E/L, trifid L much deeper than E, and retracted suspensive lobe. *U. Tithon.* **Fig. 439A-B** - *S. desmidoptycha* (Uhlig).

= *Stevensia* Énay 2009, *non* Cameron, 1932 (Coleoptera).

Family LITHACOCERATIDAE Zeiss, 1968

Discussion. A seemingly monophyletic taxon of uncertain ancestry; Atrops (1982) regarded *Lithaco-sphinctes* (Ataxioceratinae) as the likely progenitor, whereas Gygi (2003) favoured *Liosphinctes/Ampthillia* (Prososphinctinae). Comprises the subfamilies Lithacoceratinae and Gravesiinae. *U. Oxford. - U. Tithon.*

Subfamily Lithacoceratinae Zeiss, 1968

Discussion. Typically compressed, evolute to moderately involute, with dense fine biplicate ribbing; microconchs with convergent flanks, narrow venter and lateral lappets; jaws praestriaptychi. *U. Oxford. - U. Tithon.*

Discosphinctes DAQUÉ, 1930 [*D. arussiorum*; SD Spath, 1931] [**m**]. Medium sized; evolute, compressed, with broad weakly convex flanks converging to narrow subtabulate venter and subtrapezoidal whorl section; dense fine prorsiradiate biplicate ribs, rarely simple, branching high on flank; in later growth primary ribs strengthen and give rise to secondaries in triplicates. *U. Oxford.* **Fig. 443C-D** - *D. arussiorum* Dacqué.

Wegelea GYGI, 2000 [*Perisphinctes gredingensis* Wegele, 1929; OD] [**M**]. Moderately evolute, flat sided, with weak blunt ribs that attenuate on flank; peristome simple. *U. Oxford.* **Fig. 437C** - *W. gredingensis* (Wegele).

Decipia ARKELL, 1937 [*Ammonites decipiens* J. Sowerby, 1821; OD] [**M**]. Medium sized; very evolute, with flattened flanks, rounded venter and subelliptical whorl section; numerous rectiradiate primary ribs branch above midflank and give rise to fasciculate bundles of secondaries that cross venter transversely. *U. Oxford.* **Fig. 437I-J** - *D. decipiens* (J. Sowerby).

Subdiscosphinctes MALINOWSKA, 1972 [*Perisphinctes kreutzi* Siemiradzki, 1891; OD] [**M**]. Very large; inner whorls like *Discosphinctes*; later more evolute, with ribs

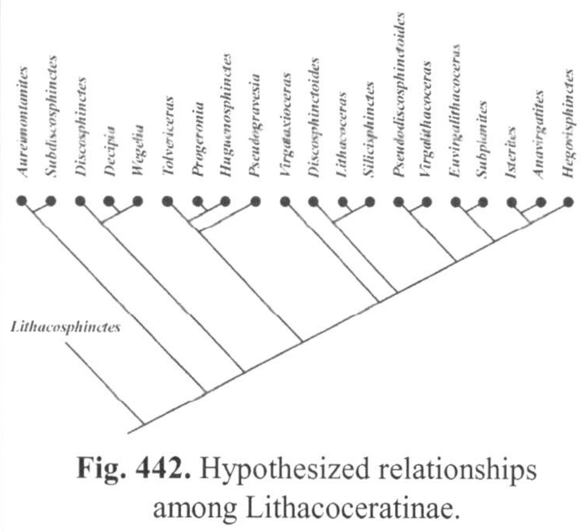

Fig. 442. Hypothesized relationships among Lithacoceratinae.

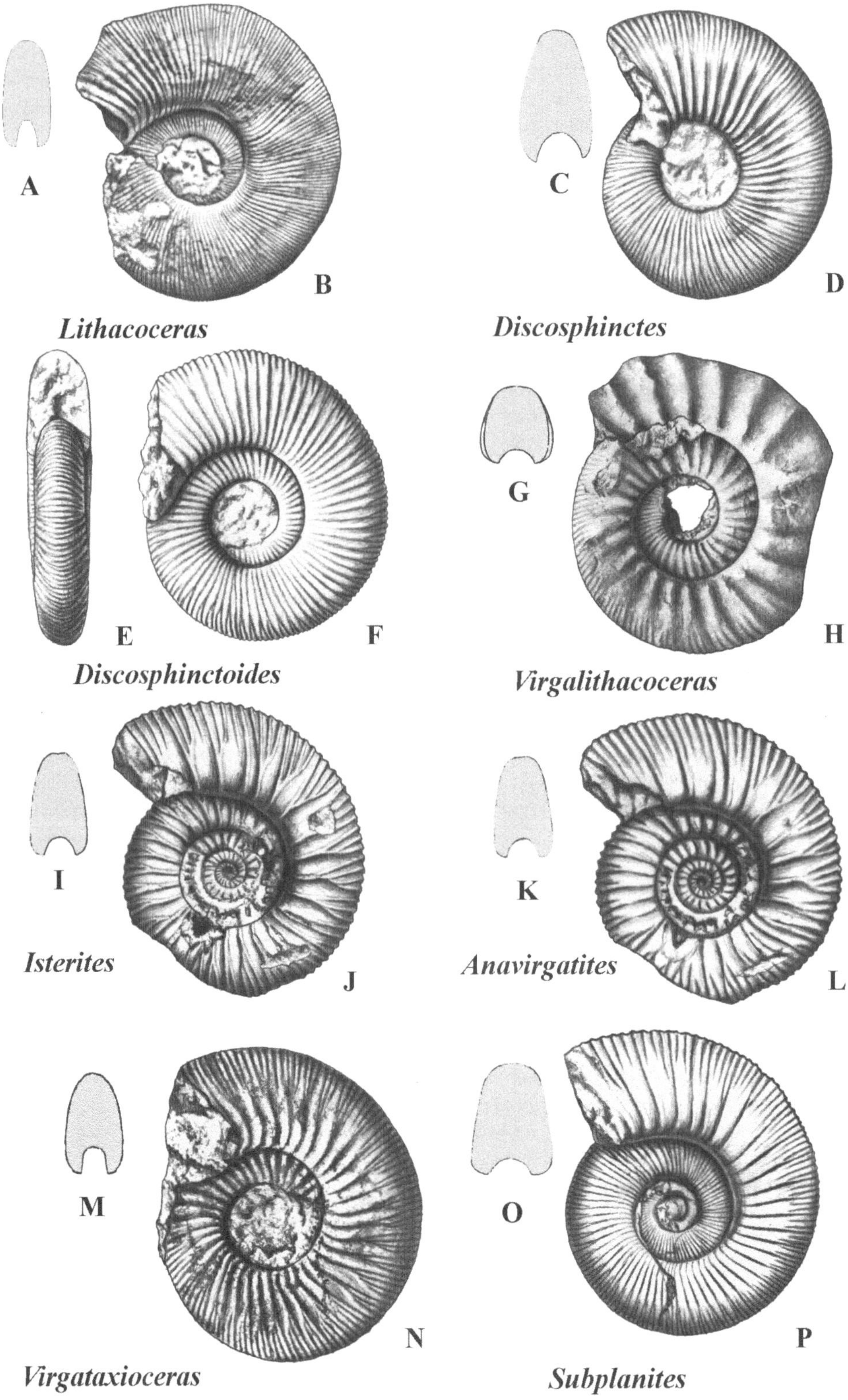

Fig. 443. Lithacoceratinae.

coarser and more distant and commonly in triplicates; secondary ribs obsolete on last part of adult body chamber; shallow constrictions present. *M. Oxford.* **Fig. 437H** - *S. kreutzi* (Siemiradzki).

Discosphinctoides OLÓRIZ, 1978 [*Perisphinctes roubyanus* Fontannes, 1879; OD]. Medium sized; like *Discosphinctes* but more evolute, with subparallel flanks, elliptical whorl section and more regular biplicate ribs branching higher on flank. *Lower U. Kimmer.* **Fig. 443E-F** - *D. roubyanus* (Fontannes).

Virgataxioceras ARKELL, 1951 [*Virgatosphinctes setatus* Schneid, 1914; OD]. Small, inner whorl like *Lithacoceras*; in middle to later growth with fine sharp ribbing including polyplocoidal bundles; later ribs in fasciculate and virgatotome bundles; constrictions lacking. *Middle U. Kimmer.* **Fig. 443M-N** - *V. setatum* (Schneid).

Lithacoceras HYATT, 1900 [*Ammonites ulmensis* Oppel, 1858; OD] [**M**]. Moderately large; like *Discosphinctoides* but with finer denser ribbing and frequent constrictions; in later growth fasciculate bundles of up to 8 secondaries that display virgatotome tendency; suture with long narrow bifid E/L, narrow trifid L as deep as E, very narrow bifid L/U_2, U_2 only half as deep as L, and strongly retracted suspensive lobe. *Upper U. Kimmer.* **Fig. 443A-B** - *L. ulmense* (Oppel).

Silicisphinctes SCHWEIGERT & ZEISS, 1999 [*Subplanitoides hoelderi* Sapunov, 1979; OD] [**m**]. Like *Lithacoceras* but smaller and with lateral lappets. *Upper U. Kimmer.* **Fig. 437G** - *S. hoelderi* (Sapunov).

A microconch of *Lithacoceras*.

Anavirgatites SPATH, 1925 [*A. divisiformis*; OD] [**m**]. Medium sized; compressed, evolute, with flat convergent flanks, subtabulate venter and trapezoidal whorl section; in early growth with distant biplicate ribs becoming virgatotome in later growth with branching high on flank; mostly single ribs in maturity. *U. Kimmer. - M. Tithon.* **Fig. 443K-L** - *A. divisiformis* Spath.

Virgalithacoceras OLÓRIZ, 1978 [*Virgatosphinctes (Perisphinctes) riedensis* Schneid, 1915; OD] [**M**]. Like *Lithacoceras* but with more evolute early whorls, virgatotome ribbing at end of phragmocone, and adult body chamber with bundles of secondaries connected at midflank to thick primary ribs. *U. Kimmer. - L. Tithon.* **Fig. 443G-H** - *V. riedense* (Schneid).

Isterites BARTHEL, 1975 [*Perisphinctes (Pseudovirgatites) palmatus* Schneid, 1915; OD] [**m**]. Small; like *Anavirgatites* but with triplicate and virgatotome ribs branching low on flank, separated by long intercalatories on adult body chamber. *Upper L. Tithon. - Lower M. Tithon.* **Fig. 443E-F** - *I. palmatus* (Schneid).

Hegovisphinctes ZEISS, SCHWEIGERT & SCHERZINGER, 1996 [*H. roettgeri*; OD][**M**]. Very large; serpenticone with fine prorsiradiate biplicate ribbing in early growth; strong coarse, distant, simple, concave ribs in middle to later growth. *Lower L. Tithon.* **Fig. 437M** - *H. roettgeri* Zeiss, Schweigert & Scherzinger.

Euvirgalithacoceras ZEISS, SCHWEIGERT & SCHERZINGER, 1996 [*Virgatosphinctes supremus* Sutner, 1915; OD] [**M**]. Enormous; like *Virgatolithacoceras* but more evolute and coarsely ribbed, without fascipartite ribbing. *Lowermost Tithon.* **Fig. 437E-F** - *E. supremum* (Sutner).

Subplanites SPATH, 1925 [*Virgatosphinctes (Perisphinctes) reisi* Schneid, 1914; OD] [**m**]. Small; like *Anavirgatites* but more inflated and with narrower umbilicus, denser ribbing in early growth and ribs becoming more distant later when intercalatories are added across venter. *L. Tithon.* **Fig. 443O-P** - *S. reisi* (Schneid).

Perhaps a microconch of *Euvirgalithacoceras*.

Pseudodiscosphinctes OLÓRIZ, 1978 [*Perisphinctes ardescicus* Fontannes, 1879; OD] [**M**]. Like *Discosphinctoides* but with narrower umbilicus and denser ribbing without umbilical swellings. *Upper L. Tithon.- Lower M. Tithon.* **Fig. 437B** - *P. ardescicus* (Fontannes).

Subfamily Gravesiinae Zeiss, 1987

Discussion. A questionably monophyletic taxon of enigmatic origin and uncertain familial position (Scherzinger *et al.*, 2006), with a unique aptychus. Typically very inflated cadicones with depressed coronate to reniform whorl section, robust primary ribs that may form blunt umbilical bullae and give rise to coarse blunt secondary ribs that may evanesce in later growth; sinuous non-lapetted aperture in both sexes. *M. Kimmer. - Lowest Tithon.*

Praegravesia ZEISS, 1987 [*P. rolkei*; OD]. Large; very evolute, with sloping umbilical wall, angular umbilical edge, broadly rounded venter and depressed coronate whorl section; strong blunt slightly prorsiradiate primary ribs branch at blunt lateral tubercles in 2-3 's and cross venter with slight forward arch. *M. Kimmer.* **Fig. 445C** - *P. rolkei* Zeiss.

According to Schweigert (1999), the type locality is dubious and the types may represent reworked Middle Jurassic material. However Late Kimmeridgian *Gravesia polypleura* Hahn, the most primitive of the *Gravesia*

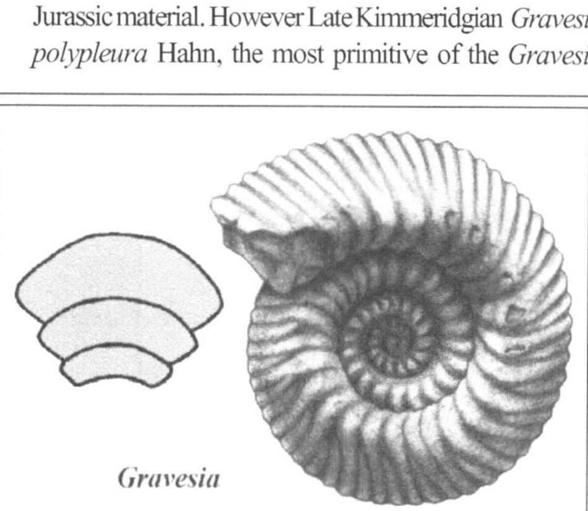

Fig. 444. *Gravesia*.

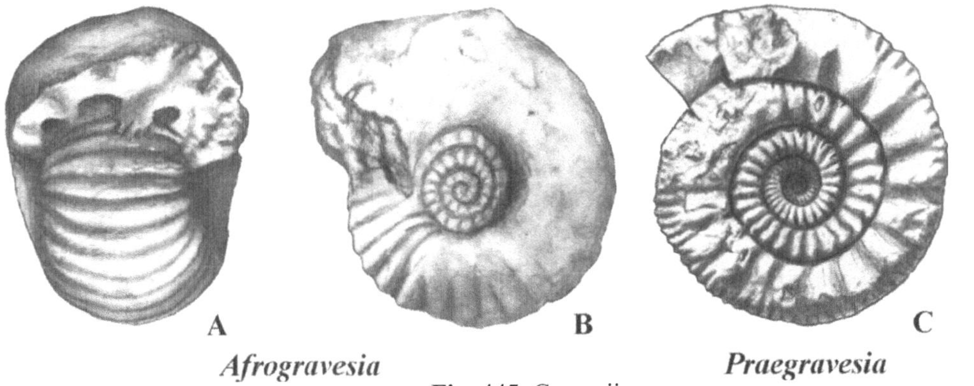

Fig. 445. Gravesiinae.

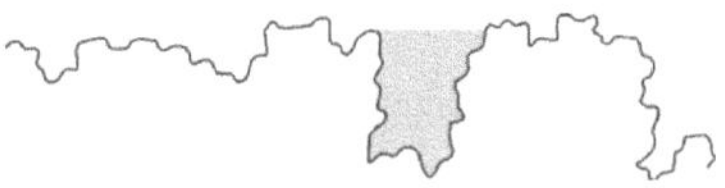

Fig. 446. *Gravesia* suture.

species, shows some similarity to *Praegravesia*. The position of *Praegravesia* is unresolved.

Gravesia SALFELD, 1913 [*Ammonites gravesianus* d'Orbigny, 1847; OD] [**M**]. Moderately large (**m**) to enormous (**M**); like *Praegravesia* but in middle growth more involute, with strongly depressed coronate whorl section, subangular umbilicolateral shoulder, and generally paired prorsiradiate secondaries one branch of which zigzags across venter; ribbing tends to fade on high oval adult body chamber. *Upper U. Kimmer. - Lower L. Tithon.* **Fig. 444** - *G. gravesiana* (d'Orbigny).

Afrogravesia SCHWEIGERT, ZEISS & WESTERMANN, 2012 [*Gravesia loupekinei* Verma & Westermann, 1984; M). Medium sized, inflated, superficially like *Gravesia*; moderately involute; in early to middle growth with strongly depressed reniform whorl section, becoming compressed ovate in maturity; phragmocone with coarse, blunt, slightly prorsiradiate ribs arising singly and in pairs, crossing venter transversely; ornament evanesces in later growth and adult body chamber is smooth with simple peristome; suture with broad bipartide E/L. *U. Kimmer.* **Fig. 445A-B** - *A. loupekinei* (Verma & Westermann).

Family SUBLITHACOCERATIDAE Zeiss, 1968

(*nom. transl.* herein *ex* Sublithacoceratinae)

Discussion. A seemingly monophyletic group with an origin in Lithococeratidae (Zeiss, 2001). Comprises the subfamilies Franconitinae, Sublithacoceratinae, Paraulacosphinctinae and Mazatepitinae. *Middle L. Tithon. - U. Tithon.*

Subfamily Franconitinae Zeiss, 1968

(*nom. transl.* Callomon *et al.*, 1981 *ex* Franconitini)

Discussion. A monophyletic taxon with an origin in Lithacoceratinae, close to *Lithacoceras* (Zeiss, 1968); strongly dimorphic; very evolute, compressed, with numerous long primaries giving rise to finer bundled secondaries; macroconchs with coarse distant subbullate primary ribs in later growth. *Middle L. - M. Tithon.*

Subplanitoides ZEISS, 1968 [*Usseliceras (S.) waltheri*; OD] [**m**]. Small to medium sized; very evolute, compressed, with flattened flanks, subtabulate venter and subrectangular whorl section; numerous dense fine prorsiradiate primaries split into 2-3 secondaries above midflank and cross venter transversely; aperture with lateral lappets. *Middle L. Tithon.* **Fig. 450M** - *S. waltheri* (Zeiss).

Usseliceras ZEISS, 1968 [*U. franconicum*; OD][**M**]. Moderately large; in early to middle growth like *Subplanitoides*, later with sparse distant strong primaries that become obsolete on outer flank leaving venter smooth.

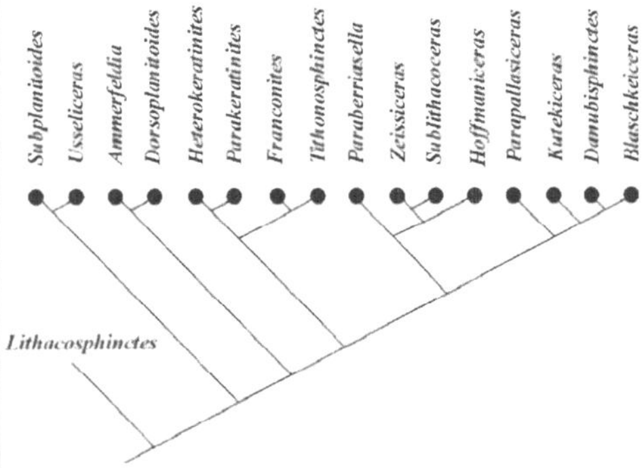

Fig. 447. Hypothesized relationships among Franconitinae.

Middle L. Tithon. **Fig. 450A-B** - *U. franconicum* Zeiss.
A macroconch of *Subplanitoides*.

Dorsoplanitoides ZEISS, 1968 [*D. bavaricus*; OD] [**M**]. Moderately large; like *Usseliceras* but with subquadratic whorl section, more coarsely ribbed inner whorls and, in later growth, with subbullate primaries branching low on flank and giving rise to 2-3 coarse secondaries. *Upper L. Tithon.* **Fig. 450K-L** - *D. bavaricus* Zeiss.

Ammerfeldia ZEISS, 1968 [*Dorsoplanitoides (A.) ammerfeldensis*; OD] [**m**]. Small; in early growth like *Dorsoplanitoides* with long fine slightly prorsiradiate primaries bifurcating high on flank and secondaries flexing foward over venter. *L. Tithon.* **Fig. 450E** - *A. ammerfeldensis* (Zeiss).
A microconch of *Dorsoplanitoides*.

Franconites ZEISS, 1968 [*Perisphinctes (Virgatosphinctes) vimineus* Schneid, 1915; OD] [**M**]. Medium sized; evolute, compressed, with rounded venter and oval whorl section; in early to middle growth with numerous fine primary ribs splitting above midflank into bundles of numerous fine secondaries; in later growth primaries become distant and subbullate. *Middle L. Tithon.* **Fig. 450C-D** - *F. vimineus* (Schneid).

Tithonosphinctes ZEISS, 1968 [*Franconites (T.) stephanovi*; OD]. Moderately large; like inner whorls of *Franconites* but later with somewhat irregular ribbing that remains fine to maturity. *Upper L. Tithon.* **Fig. 450N** - *T. stephanovi* (Zeiss).

Parakeratinites ZEISS, 1968 [*Franconites (P.) communis*; OD] [**m**]. Small; like *Franconites* but more evolute, with narrowly arched venter, trigonally ovate whorl section, and very dense fine biplicate ribs (less commonly single). *Upper L. Tithon.* **Fig. 450H-J** - *P. communis* (Zeiss).

Heterokeratinites ZEISS, 1968 [*Franconites (H.) procurvus*; OD]. Small; like *Parakeratinites* but with flattened convergent flanks, rounded venter, subtrapezoidal whorl section, prominent constrictions and sporadic oblique ventrolateral clavi in later growth. *Upper L. Tithon.* **Fig. 450F-G** - *H. procurvus* (Zeiss).

Subfamily Sublithacoceratinae Zeiss, 1968

(*nom. transl.* Kutek & Zeiss, 1974 *ex* Sublithacoceratini)

Discussion. A monophyletic taxon with an origin in Franconitinae (Zeiss, 2001), close to *Franconites*; typically compressed, evolute, with rounded venter and dense fine biplicate ribbing; in macroconchs primary ribs coarsen in later growth; suture with trifid L deeper than E, and L/U_2 protruding beyond E/L. *Upper L. Tithon. - U. Tithon.*

Sublithacoceras SPATH, 1925 [*Perisphinctes (Aulacosphinctes) penicillatus* Schneid, 1915; OD]. Large; relatively involute, compressed, with flat convergent flanks, narrowly rounded venter and subelliptical to subtrigonal whorl section; inner whorls densely and finely ribbed, with biplicate ribs branching at midflank; periodic constrictions; suture with deep very narrow L. *L. Tithon.* **Fig. 453A-B** - *S. penicillatum* (Schneid).

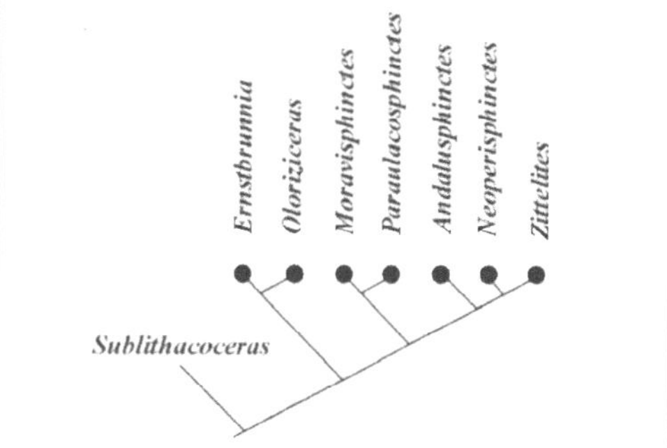

Fig. 448. Hypothesized relationships among Sublithacoceratinae.

Zeissiceras VIGH, 1984 [*Sublithacoceras (Z.) incertaeforme*; OD]. Medium sized; like *Sublithacoceras* but more evolute, with narrowly arched venter, subtrigonal whorl section and denser finer ribbing that branches lower on flank and persists to maturity. *Upper M. Tithon.* **Fig. 453N** - *Z. incertaeforme* (Vigh).

Hofmanniceras VIGH, 1984 [*Sublithacoceras (H.) obliquestriatum*; OD]. Moderately large; like *Sublithacoceras* with compressed elliptical whorl section but with finer ribbing in fasciculate bundles; constrictions lacking. *Upper M. Tithon.* **Fig. 453M** - *H. obliquestriatum* (Vigh).

Paraberriasella DONZE, 1948 [*P. blondeti*; OD][**m**]. Small; compressed, evolute, with broad gently convex flanks, evenly rounded venter and elliptical whorl section; numerous fine prorsiradiate primaries split into finer secondaries high on flank, with occasional intercalatories, crossing venter transversely. *Upper L. Tithon.* **Fig. 453K-L** - *P. blondeti* Donze.

Parapallasiceras SPATH, 1925 [*Berriasella (Aulacosphinctes?) praecox* Schneid, 1915; OD] [**m**]. Small; very evolute, compressed, with flattened flanks, rounded venter and compressed oval whorl section; numerous slightly prorsiradiate biplicate ribs branch on outer flank, with start of furcation displayed in umbilicus; ribs cross venter transversely but in early to middle growth interrupted by shallow siphonal groove that evanesces later;

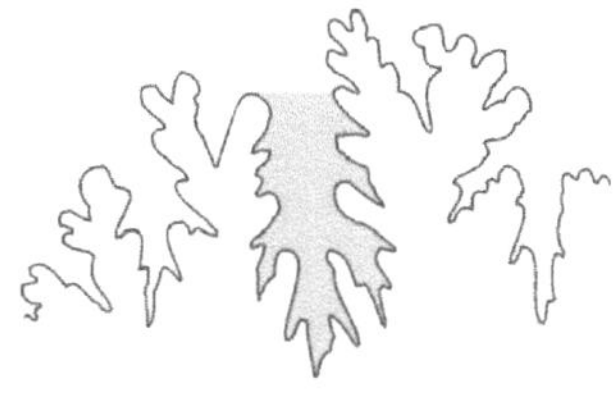

Fig. 449. *Paraberriasella* suture.

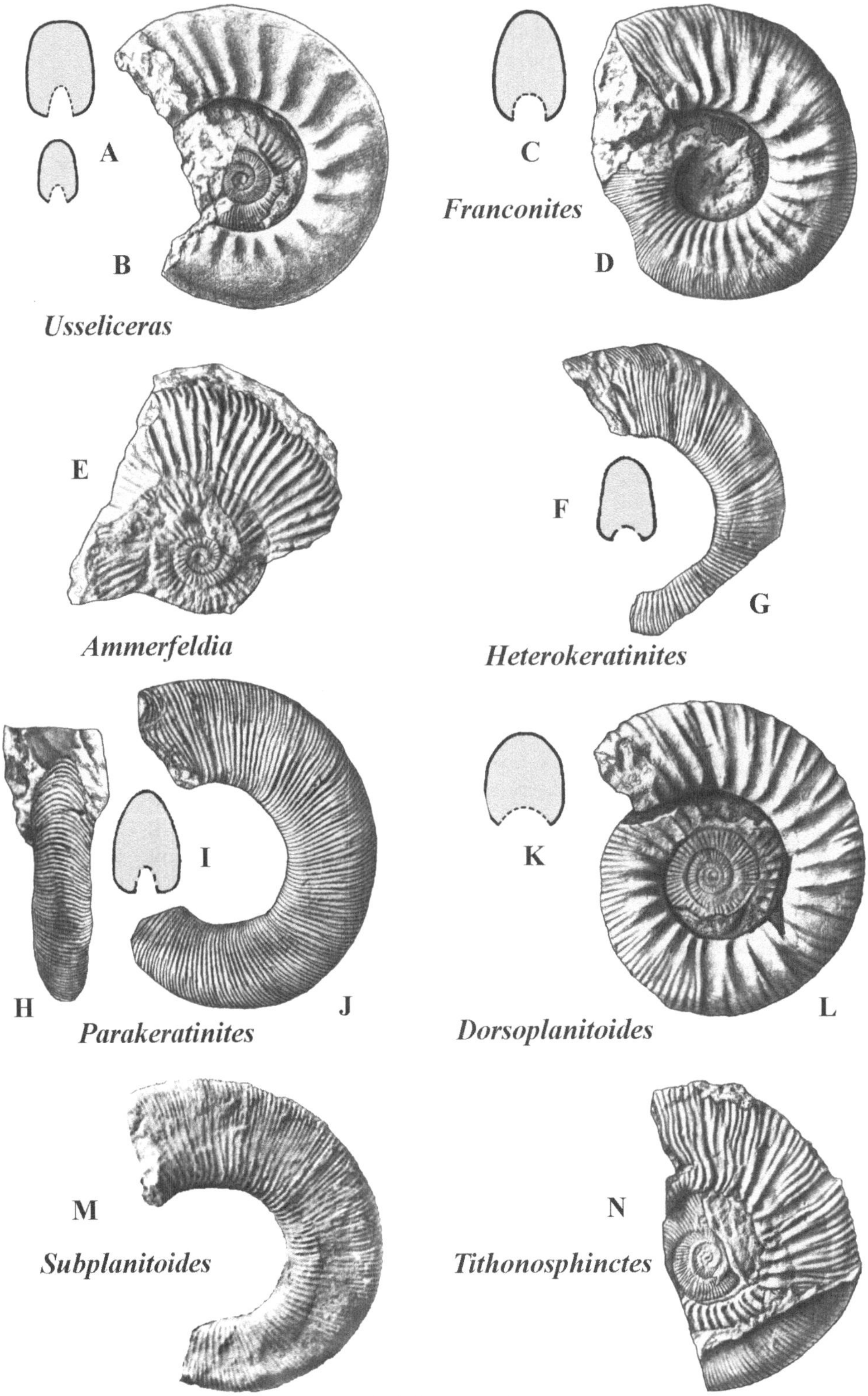

Fig. 450. Franconitinae.

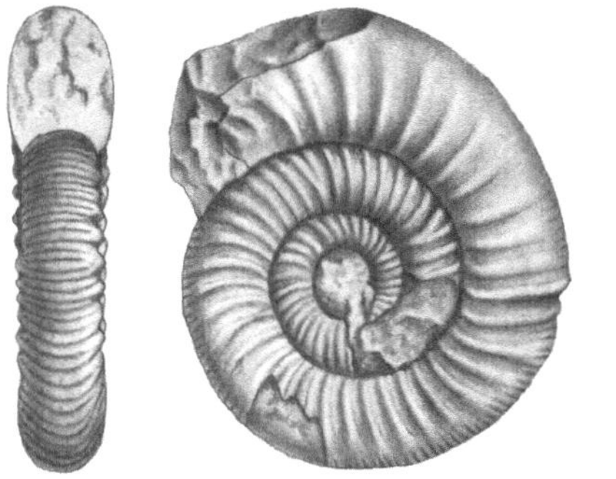

Fig. 451. *Dorsomorphites.*

ribs not modified on adult body chamber; constrictions present; suture with narrow E, elongate bifid E/L and narrow trifid L as long as E; superficially similar to *Torquatisphinctes*. *Upper L. Tithon.* **Fig. 453C-D** - *P. praecox* (Schneid).

Danubisphinctes ZEISS, 1968 [*Parapallasiceras (D.) palatinum*; OD] [**M**]. Medium sized to moderately large; inner whorls like *Parapallasiceras*; in later growth with sublanceolate whorl section and strong primary ribs giving rise to fasciculate bundles of fine secondaries. *Upper L. - Middle Tithon.* **Fig. 453E-F** - *D. palatinum* (Zeiss).

A macroconch of *Parapallasiceras*.

Dorsomorphites SARTI, 2017 [*Ammonites exornatum* Catullo, 1853; OD] [**M**]. Small; like *Danubisphinctes* but with convex forward primary ribs and biplicate secondaries weakening and becoming obsolete on adult body chamber. *L. Kimmer.* **Fig. 451** - *D. selectus* (Neumayr).

= *Pseudokatroliceras* OLÓRIZ, 1978 (*nom. nud.*)

Kutekiceras ZEISS, 2001 [*Perisphinctes pseudocolumbinus* Kilian, 1895; OD] [**m**]. Small; like *Parapallasiceras* but more inflated, with depressed subquadratic to rounded ovate whorl section, denser ribbing that branches lower on flank and periodic constrictions; siphonal groove not always present. *M. - U. Tithon.* **Fig. 453I-J** - *K. pseudocolombinum* (Kilian).

Blaschkeiceras ZEISS, 2001 [*Perisphinctes (Aulacosphinctes) schöpflini* Blaschke, 1911; OD] [**M**]. Moderately large; like *Danubisphinctes* but more evolute, with convex convergent flanks, rounded venter and oval whorl section. *M. - U. Tithon.* **Fig. 453I-J** - *B. schoepflini* (Blaschke).

Subfamily Mazatepitinae Cantú-Chapa, 2009

Discussion. A monotypic taxon, presumably descended from Sublithacoceratinae (Cantú-Chapa, 2009), but precise ancestor uncertain; typically evolute with subrounded venter and sharp rectiradiate distant primary ribs in early to middle growth, in later growth subdividing on outer flank. *Lower U. Tithon.*

Mazatepites CANTÚ-CHAPA, 1967 [*M. arredondensis*; OD]. Medium sized; very evolute (juv.) to evolute (ad.), with convex flanks, rounded venter and oval whorl section; in early growth with sharp distant rectiradiate primary ribs; later primaries split into palmate bundles of secondaries crossing venter transversely; constrictions present; suture with narrow E, narrow surectangular bifid E/L, narrow trifid L longer than E, narrow stemmed widely splayed bifid L/U_2 protruding beyond E/L and strongly retracted suspensive lobe. *Lower U. Tithon.* **Fig. 452** - *M. arredondensis* Cantú-Chapa.

Tentatively included in Torquatisphinctinae by Parent *et al.* (2011).

Subfamily Paraulacosphinctinae Tavera, 1985

Discussion. A doubtfully monophyletic group with an origin in Sublithacoceratinae, close to *Sublithacoceras* (Zeiss, 2001), although Tavera (1985) had looked to *Pseudodiscosphinctes*, and implicated the group in the polyphyletic origin of many berrias-

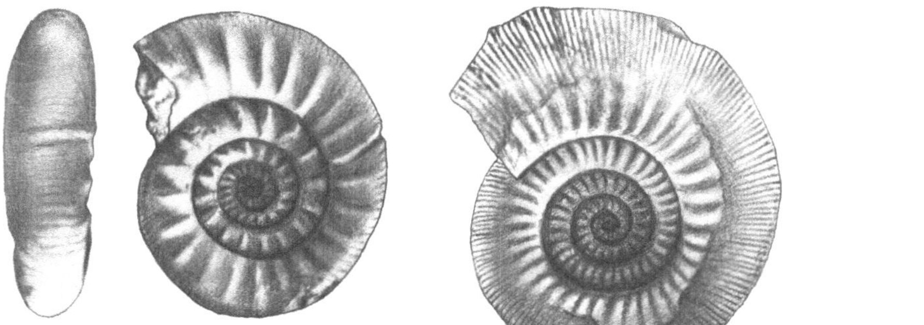

Fig. 452. *Mazatepites.*

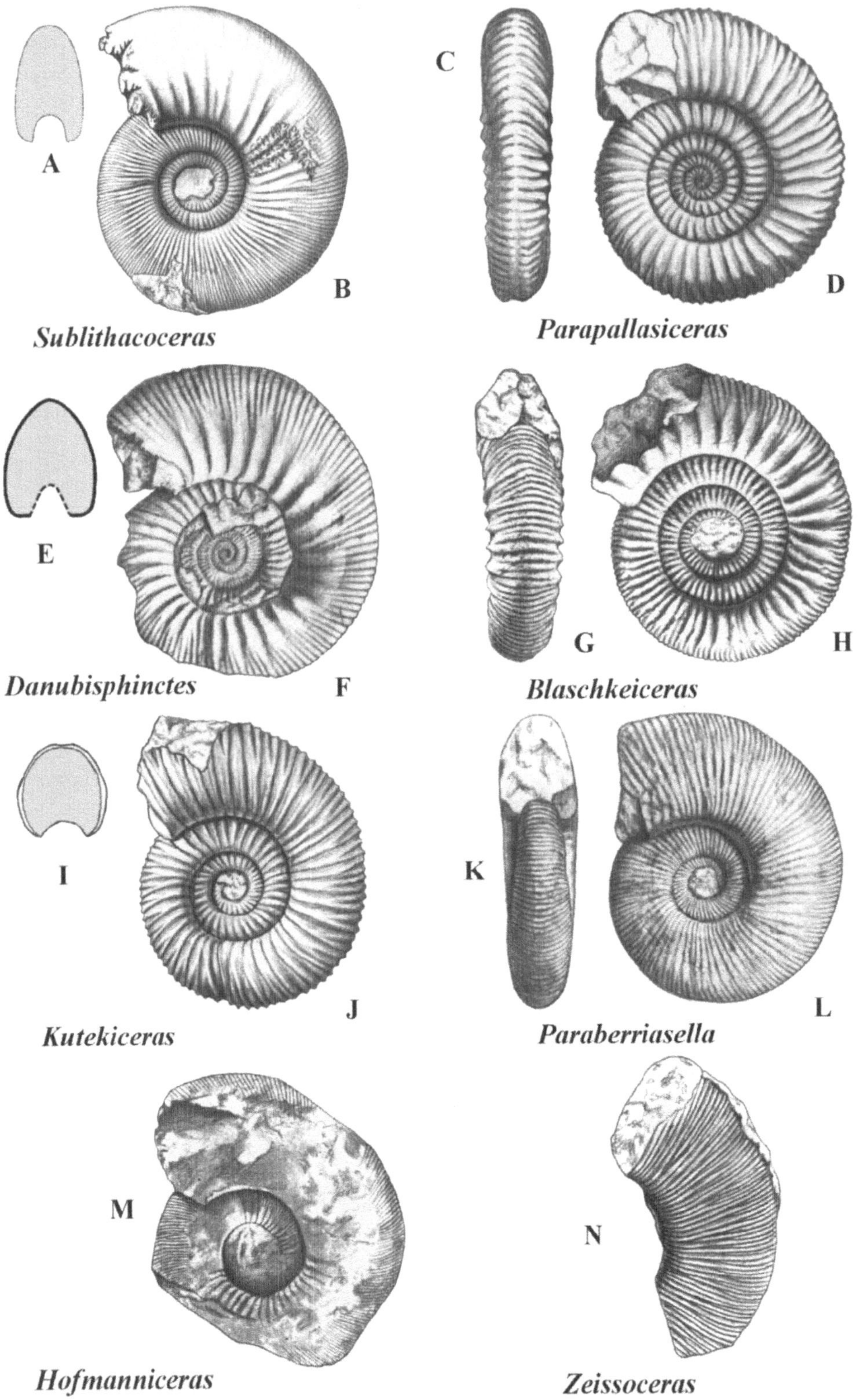

Fig. 453. Sublithacoceratinae.

ellids. Strongly dimorphic; typically compressed, evolute, with flattened flanks, rounded to subtabulate venter, subrectangular whorl section and dense fine biplicate ribbing that may be interrupted by a smooth siphonal band; macroconchs finely ribbed, becoming smooth on adult body chamber. *U. Tithon.*

Oloriziceras TAVERA, 1985 [*O. salarense*; OD] [**m**]. Small; compressed, evolute, with broad flattish flanks, rounded venter and subelliptical to subrectangular whorl section; numerous fine long rectiradiate primary ribs split into paired fine secondaries high on flank and cross venter transversely. *U. Tithon.* **Fig. 455D** - *O. salarense* Tavera.

Ernstbrunnia ZEISS, 2001 [*Perisphinctes (Aulacosphinctes) schoepflini* Blascke, 1911; OD] [**M**]. Large; inner whorls like *Oloriziceras* but with denser finer ribbing; in later growth with flat convergent flanks, flattened venter, trapezoidal whorl section and very distant primary ribs with short feeble secondaries. *Lower U. Tithon.* **Fig. 455C** - *E. bachmayeri* Zeiss.

Perhaps a macroconch of *Oloriziceras*.

Paraulacosphinctes SCHINDEWOLF, 1925 [*Ammonites senex* Oppel, 1865; OD] [**M**]. Moderately large to large; evolute, compressed with flattened subparallel flanks, rounded venter and elliptical whorl section; inner whorls with fine dense prorsiradiate biplicate ribbing; outer whorls smooth. *U. Tithon.* **Fig. 455A-B** - *P. senex* (Oppel).

Moravisphinctes TAVERA, 1985 [*Ammonites moravicus* Oppel, 1868; OD] [**m**]. Small; like *Oloriziceras* but more involute, with fewer coarser primary ribs, smooth body chamber and simple peristome. *Lower U. Tithon.* **Fig. 455E-F** - *M. moravicus* (Oppel).

Perhaps the microconch of *Paraulacosphinctes* (Parent, 2003).

Andalusphinctes TAVERA, 1985 [*A. lemenciformis*; OD] [**m**]. Small; like *Oloriziceras* but inner whorls rounded, equidimensional, later compressed, with fewer coarser primary ribs and biplicate secondaries interrupted across venter by a sulcus. *Middle U. Tithon.* **Fig. 455G-H** - *A. lemenciformis* Tavera.

Neoperisphinctes TAVERA, 1985 [*Perisphinctes falloti* Kilian, 1889; OD] [**M**]. Medium sized; convex flanks, subtabulate venter and subrectangular to subtrapezoidal whorl section; biplicate ribs becoming coarser and sinuous in later growth, with feeble umbilical bullae. *Upper U. Tithon.* **Fig. 455K-L** - *N. falloti* (Kilian).

Zittelites ÉNAY & HOWARTH, 2018 [*Ammonites eudichotomis* Zittel, 1868; OD] [**M**]. Small to medium sized; flanks flat and whorl section subquadratic, with convex venter; ribs biplicate, less commonly simple and polygyral, interrupted across venter by smooth siphonal band; ventrolateral nodes develop at point of rib bifurcation in later growth; periodic constrictions. *U. Tithon.* **Fig. 455I-J** - *T. eudichotomis* (Zittel).

= *Zittelia* Tavera, 1985 *non* Gemmarello, 1869 (Gastropoda).

Family VIRGATITIDAE Spath, 1923

Discussion. A seemingly monophyletic group of uncertain origin. Zeiss (1968) looked to Lithacoceratidae and Djik (1994) suggested Pectinatitinae. Salazar (2012) treat *Virgatites* as a subjective junior synonym of *Virgatosphinctes* but homoeomorphy seems to be involved. It comprises the subfamilies Ilowaiskyinae, Pseudovirgatitinae and Virgatitinae. Endemic to the Subboreal Realm of Eastern Europe. *U. Kimmer. - U. Tithon.*

Subfamily Ilowaiskyinae Zeiss, 1968

(*nom. transl.* Callomon *et al.*, 1981 *ex* Ilowaiskyini)

Discussion. A monophyletic taxon with an origin in Lithacoceratinae (Zeiss, 1968), close to *Discosphinctoides* (Kutek & Zeiss, 1997); typically evolute, compressed, with biplicate to virgatataxioceratoid ribbing, frequent constrictions and a suture with retracted suspensive lobe. *U. Kimmer. - L. Tithon.*

Samartisphinctes KUTEK & ZEISS, 1997 [*Divisosphinctes fallax* Ilovaisky, 1941; OD]. Medium sized; inner whorls with biplicate ribbing replaced, at variable growth stages, by virgatataxioceratoid ribbing; constrictions well developed but variable in number; outer whorls of macroconchs with biplicate and triplicate ribbing; microconchs with lateral lappets. *Uppermost Kimmer.* **Fig. 457B** - *S. fallax* (Ilovaisky).

Ilowaiskya VIALOV, 1940 [*I. sokolovi*; OD]. Small (**m**) to large (**M**); like *Samartisphinctes* but variable, with polygyrate and virgatotome ribbing supplemented by intercalatories; microconchs without lateral lappets. *L. Tithon.* **Fig. 457A** - *I. sokolovi* Vialov.

= *Sokolovia* Ilovaisky, 1934 (*non* Böhm 1933).

= *Ilovaiskya* Janschin, 1941 (obj.).

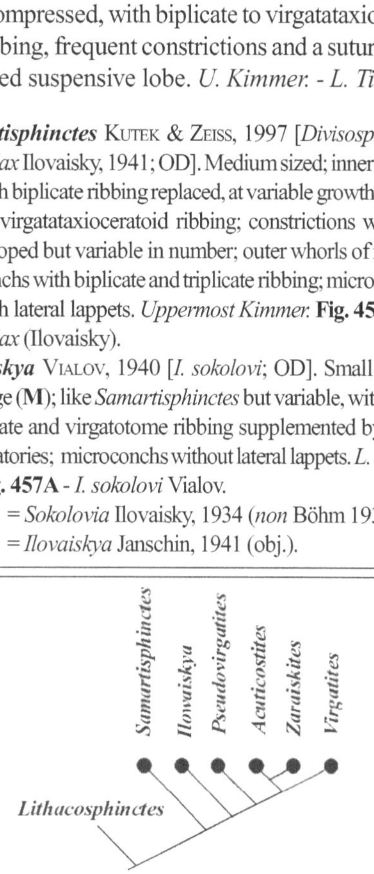

Fig. 454. Hypothesized relationships among Virgatitidae.

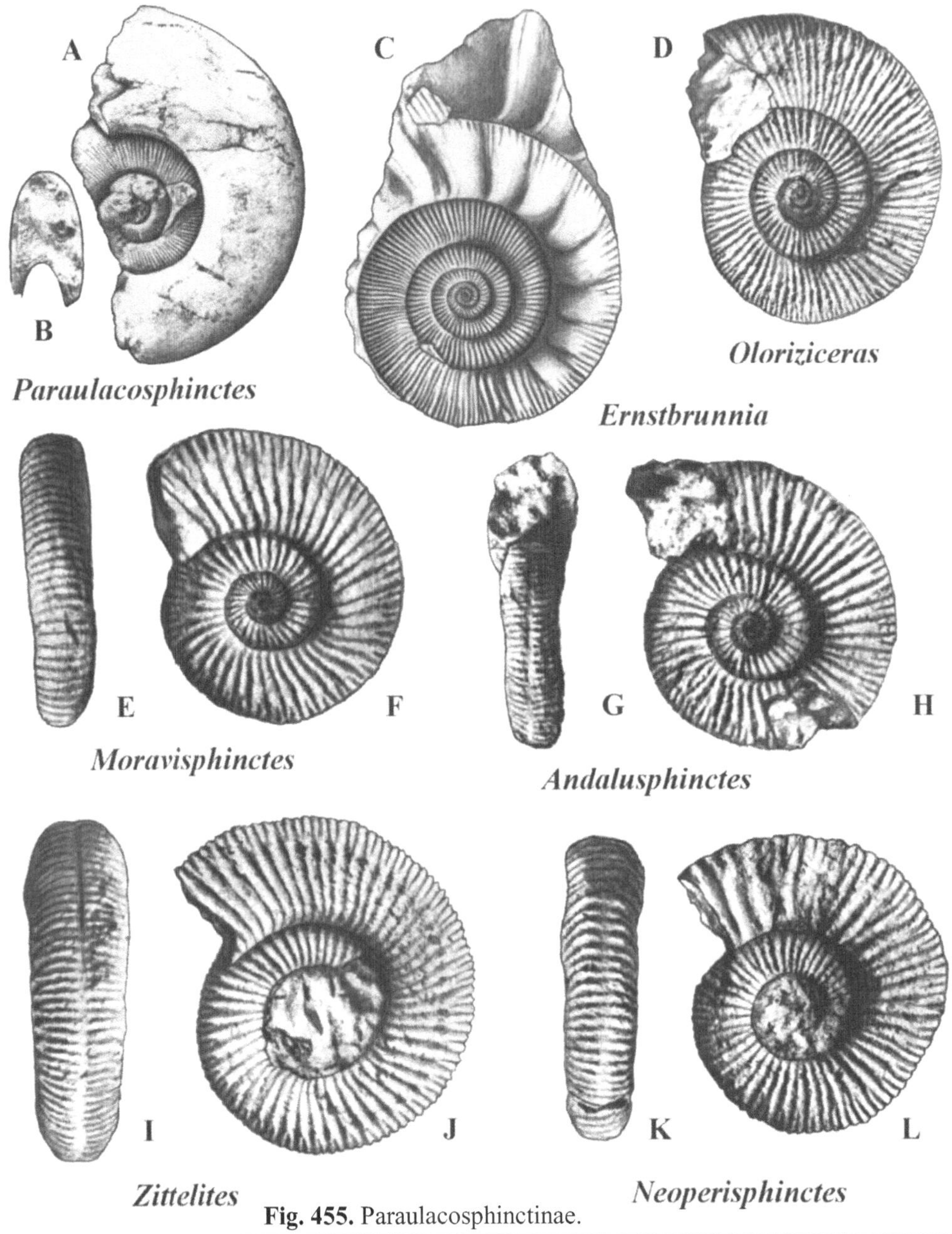

Fig. 455. Paraulacosphinctinae.

Subfamily Pseudovirgatitinae Spath, 1924

Discussion. A monotypic taxon with an origin in Ilowaiskyinae, close to *Ilowaiskya* (Kutek & Zeiss, 1997). Primary ribs typically give rise to virgatome bundles of fine secondaries. *Upper L. - Lower M. Volg.*

Pseudovirgatites VETTERS, 1905 [*Ammonites scruposus* Oppel, 1868; OD] [**M**]. Large; compressed, evolute, with broad flat flanks, narrowly rounded venter and oval whorl section; sinuous primary ribs give rise to virgatome bundles of fine secondaries; inner whorls with constrictions. *Upper L. Volg.* **Fig. 457F-G** - *P. scruposus* (Oppel).

Subfamily Virgatitinae Spath, 1923

Discussion. A monophyletic taxon with an origin in Pseudovirgatitinae, close to *Pseudovirgatites* (Kutek & Zeiss, 1988); typically with dense biplicate ribbing in early growth followed by polygyrate or virgatotome ribbing; microconchs without apertural lappets; suture with broad E/L and enlarged L/U_2 protruding beyond E/L. Confined to the Boreal and Subboreal realms. *M. Volg.*

Zaraiskites SEMENOV, 1898 [*Perisphinctes zarajskensis* Michalski, 1890; SD Basse, 1952]. Small; evolute, compressed, with gently convex flanks converging to rounded venter and oval whorl section; inner whorls with biplicate and triplicate ribs branching low on flank; outer whorls with strong distant primary ribs and finer virgatotome secondaries branching at midflank. *Lower M. Volg.* **Fig. 457C-D** - *Z. zarajskensis* (Michalski).
= *Provirgatites* Lewinski, 1923 (obj.).

Virgatites PAVLOW, 1892 [*Ammonites virgatus* von Buch, 1832; SD Douvillé, 1910]. Small to medium sized; like *Zaraiskites* but more involute and higher whorled, with virgatotome ribbing from early growth and branching occuring close to umbilical edge; ribbing sometimes reverts to simple or biplicate on adult body chamber. *Middle M. Volg.* **Fig. 457E** - *V. virgatus* (von Buch).
= *Euvirgatites* Lewinski, 1923 (obj.).

Acuticostites SEMENOV, 1898 [*Olcostephanus acuticostatus* Michalski, 1980; OD]. Small; evolute, compressed, with coronate inner and oval outer whorls; sharp distant biplicate and simple ribs. *Lower M. Volg.* **Fig. 457H-I** - *A. acuticostatus* (Michalski).
= *Holcostephanoides* Spath, 1924 (obj.).
= *Paravirgatites* Ilovaisky, 1924 (obj.) (*non* Buckmann, 1922).
= *Oxypleurites* Ilovaisky, 1941 (obj.) (*non* Nalepa, 1891, Eriophyidae).

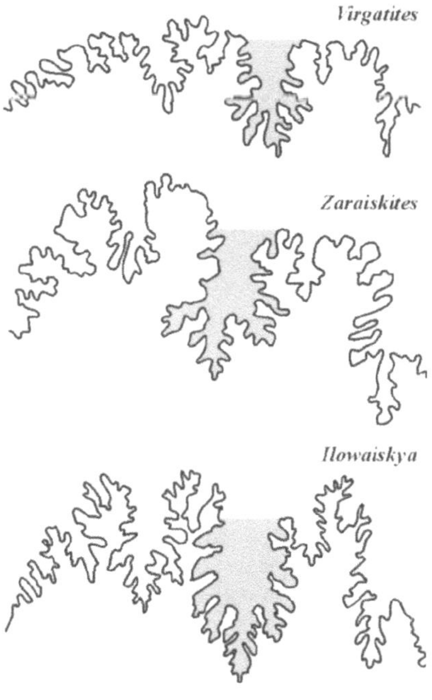

Fig. 456. Virgatitine sutures.

Superfamily HIMALAYITOIDEA Spath, 1925

(*nom. transl.* herein *ex* Himalayitidae)

Discussion. Questionably monophyletic derivatives of Ataxioceratoidea (Donovan *et al.*, 1981), with an origin in Lithacoceratidae (Parent *et al.*, 2015), although Parent *et al.* (2015) noted the very close resemblance of some species of *Choicensisphinctes* to *Malagasites multifasciatus* (Uhlig), suggesting a possible origin independent of *Lithacoceras*. Comprises the families Windhauseniceratidae, Torquatisphinctidae, Himalayitidae, Rias-anitidae, Dalmasiceratidae and Suaritidae. *L. Tithon. - L. Valang.*

Family WINDHAUSENICERATIDAE Zeiss & H. Leanza, 2010

(*nom. transl.* herein *ex* Windhauseniceratinae)

Discussion. Seemingly monophyletic derivatives of Lithococeratidae, virtually confined to the Andean Province (Parent *et al.*, 2017). Variously considered the last ataxioceratids (Zeiss & Leanza, 2010; Parent *et al.*, 2017) or early himalayitids (Parent *et al.*, 2021). Comprises the subfamilies Zapaliinae, Catutosphinctinae and Windhauseniceratinae. *L. Tithon. - L. Berrias.*

Subfamily Zapaliinae Parent, Schweigert, Scherzinger & Garrido, 2017

Discussion. Questionably monophyletic derivatives of Lithacoceratinae, since *Krantziceras* may have a independent origin in Ataxioceratidae; endemic to the Andean Province (Parent *et al.*, 2015, 2017). Evolute platycones and more or less involute suboxycones; ribbing with frequent short virgatotome stage; polyschizotome ribs or branching as fascipartite secondaries, typically with small nodes at point of rib bifurcation; adult ornament variocostate. Strongly dimorphic, microconchs lappetted. *L. Tithon. - L. Berrias.*

Indansites VENNARI, 2016 [*Subplanites malarguensis* Spath, 1931; OD]. Large, evolute, with steep umbilical wall, slightly convergent flanks, broad low subtabulate venter and subtrapezoidal whorl section about as high as wide; phragmocone with fine, straight, wiry, prorsiradiate ribs bifurcating on upper flank and crossing venter transversely; about 4 collared constrictions per whorl, parallel

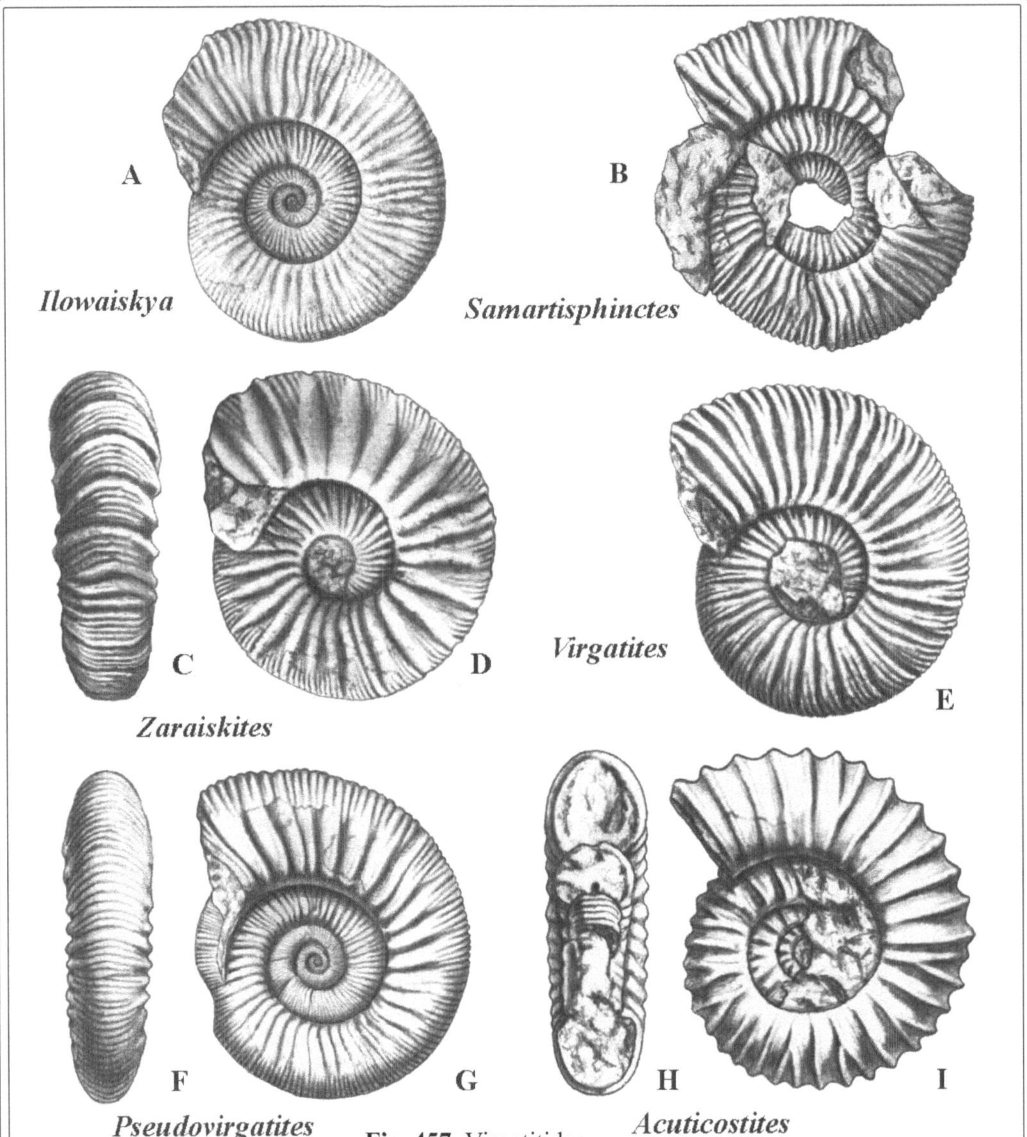

Fig. 457. Virgatitidae.

to ribbing; body chamber with bold slightly prorsiradiate primary ribs, narrower than interspaces, bifurcating or trifurcating on outer flank where separated by 1-3 intercalatories, crossing venter with slight forward arch. *L. Tithon.* **Fig. 459F-G** - *I. malarguensis* (Spath).

= *Indanistes* Parent *et al.*, 2017 (*laps. cal.*)

Choicensisphinctes H. LEANZA, 1980 [*Perisphinctes choicensis* Burckhardt, 1903; OD][**M**]. Moderately evolute, compressed, with depressed subtrapezoidal whorl section; early whorls with regularly biplicate ribbing; outer whorls with strong, prominent and projected primary ribs arising in bundles of 4-10 from umbilical bullae and crossing venter uninterrupted; constrictions present; suture with trifid L as deep as E, E/L wider and higher than L/U_2 and two simple umbilical lobes. *L. - U. Tithon.* **Fig. 459A-C** - *C. choicensis* (Burckhardt).

Zapalia H. LEANZA & ZEISS, 1990 [*Z. fascicostatus*; OD] [**M**]. Giant to medium sized, evolute; inner whorls with dense or coarse, simple or bifurcate ribs tending to be sigmoidal; on penultimate whorl ribs trifurcate or fascipartite; secondary ribs of fascipartite bundles either very dense and equally spaced or branching bundles well separated and split into as many as 5 branches; adult body chamber with coarse distant primary ribs that may divide into fasciculate bundles of secondaries. *Upper L. Tithon.* **Fig. 458B** - *Z. fascicostatus* H. Leanza & Zeiss.

Parazapalia ZEISS & H. LEANZA, 1992 [*P. hillebrandti*; OD] [**m**]. Small; holotype crushed and whorl section unknown; inner whorls like *Zapalia*; adult body chamber

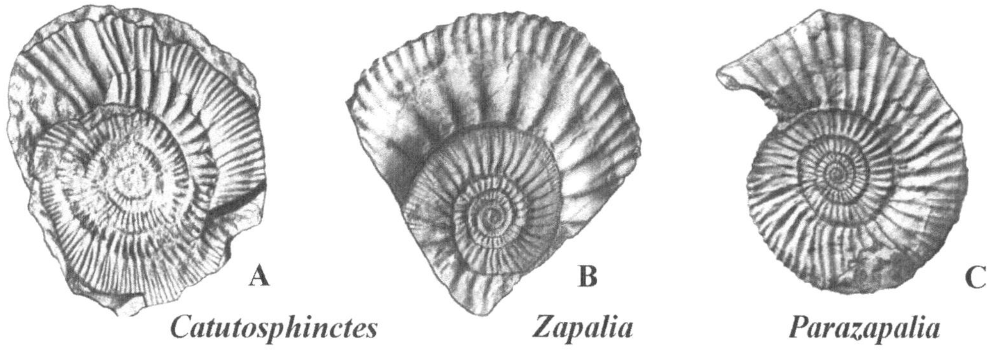

Fig. 458. Catutosphinctinae.

with accentuated single ribs and large lateral lappets. *M. Tithon.* **Fig. 458C** - *P. hillebrandti* H. Leanza & Zeiss. Perhaps a microconch of *Zapalia*.

Platydiscus PARENT, SCHERZINGER & SCHWEIGERT, 2011 [*P. beresii*; OD]. Macroconch with innermost whorls rounded; ribbing prorsiradiate, strong on flanks, faint over venter; middle whorls rounded, moderately involute, densely ribbed; adult bodychamber three quarters to one whorl long, compressed, umbilicus wide and shallow; densely ribbed by acute flexuous primaries bifurcating at midflank and with some simple intercalatories, mostly grouped in pairs, all interrupted over venter by persistent siphonal groove that fades near peristome. Microconch with adult body chamber half a whorl long, compressed, unevenly ribbed, with short lappets. *L. - U. Tithon.* **Fig. 459H-I** - *P. beresii* Parent, Scherzinger & Schweigert.

Kruntziceras PARENT, SCHERZINGER & SCHWEIGERT, 2011 [*K. compressum*; OD][**M**]. Compressed and evolute, with more or less flat flanks and rounded to flattened venter; ribbing wiry, dense, narrowly biplicate on upper flank, with occasional simple ribs; narrow siphonal groove may occur on phragmocone; body chamber isocostate to slightly variocostate. *U. Tithon. - L. Berrias.* **Fig. 459D-E** - *K. compressum* Parent, Scherzinger & Schweigert.

Subfamily Catutosphinctinae Zeiss & H. Leanza, 2010

(*nom. transl.* herein *ex* Catutosphinctini

Discussion. Monophyletic derivatives of Zapaliinae, close to *Choicenisphinctes* (Énay & Cariou, 1997; Parent *et al.*, 2011, 2017); comprises Andean homoeomorphs of Indo-Malgash Torquatisphinctinae that evolved in parallel with Franconitinae and Sublithacoceratinae. Comprises evolute platycones to more or less involute and inflated suboxycones; ribbing with frequent more or less short virgatotome stages, polyschizotome ribs, or branching as fascipartite secondaries. *L. - U. Tithon.*

Catutosphinctes H. LEANZA & ZEISS, 1992 [*C. rafaeli*; OD]. Small (**m**) to very large (**M**); in early growth serpenticone with subrectangular to subrounded whorl section and coarse widely splayed biplicate ribs with incipient to distinct lateral bullae at point of rib furcation; in middle growth ribbing with triplicates and intercalatories, as well as constrictions; in later growth with distant biplicate ribs branching below midflank, coarser and more projected on adult body chamber; microconchs mainly with biplicate and single ribs, but also triplicates, and lateral lappets. *U. Tithon.* **Fig. 458A** - *C. rafaeli* H. Leanza & Zeiss.

Mapuchesphinctes ZEISS & LEANZA, 2010 [*M. garatei*; OD] [**M**]. Gigantic; inner whorls with subquadratic whorl section, later elliptical; strong, regularly spaced biplicate ribs, occasionally trifurcating in later growth, anterior branch projected forward over venter; distant simple ribs on adult body chamber. *M. Tithon.* **Fig. 463G-H** - *M. garatei* Zeiss & Leanza.

Subfamily Windhauseniceratinae Zeiss & H. Leanza, 2010

= Windhauseniceratinae Leanza & Zeiss, 1992 (*nom. nud.*).

Discussion. Monophyletic descendants of Catutosphinctinae, close to *Catutosphinctes* (Parent *et al.*, 2015), although Zeiss & Leanza (2010) had sought an origin in Virgatosphinctinae. Typically evolute, with broadly rounded to subtabulate venter, coronate whorl section in early growth, strong primary ribs terminating in high-lateral nodes, and 2-3 secondary ribs crossing venter transversely; outer whorls with relatively simple ornament; suture with narrow trifid L and retracted suspensive lobe. *U. Tithon., ?Berrias.*

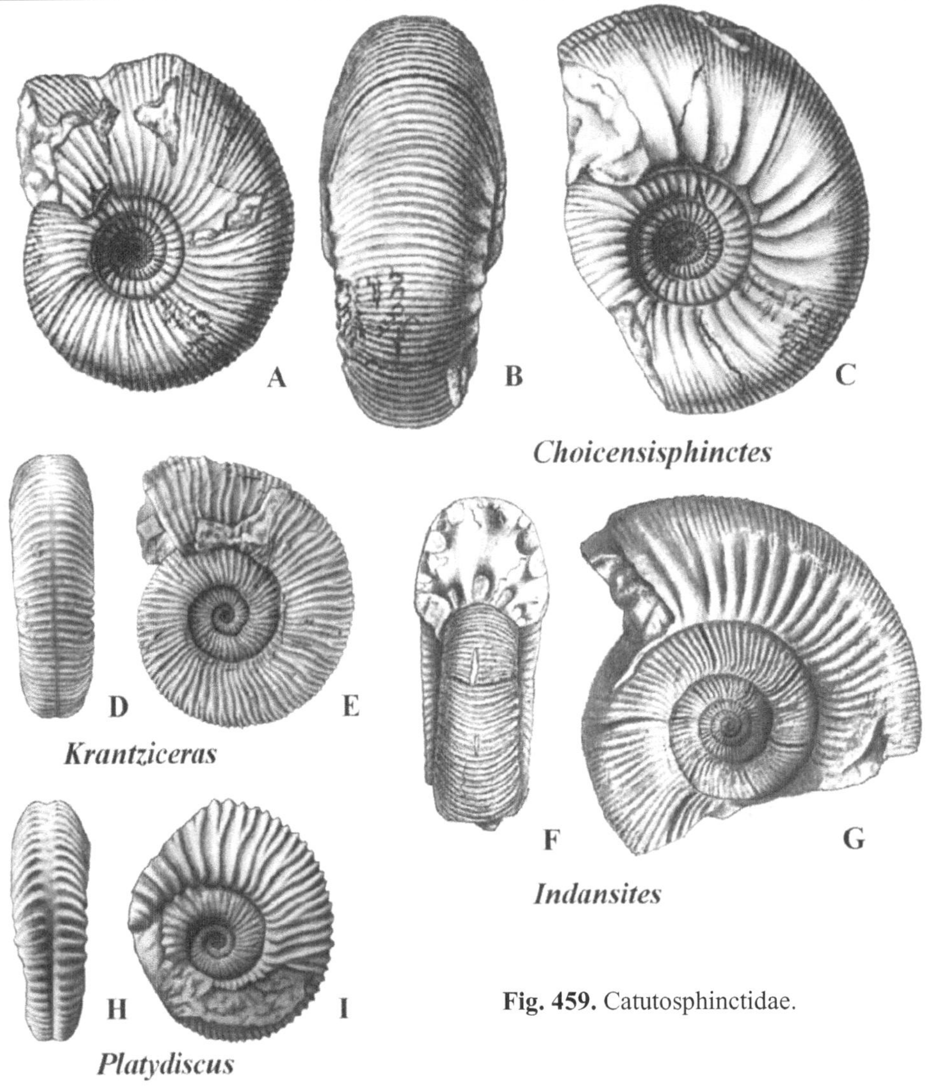

Fig. 459. Catutosphinctidae.

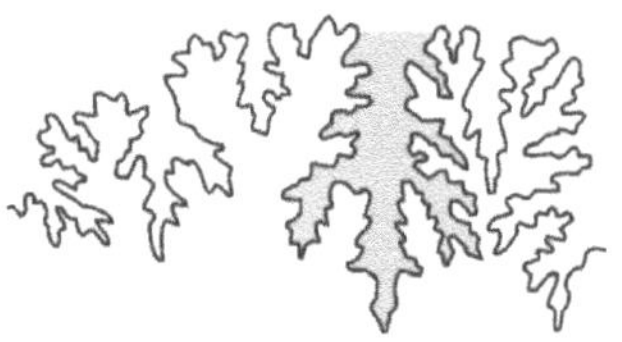

Fig. 460. *Hemispiticeras* suture.

Windhauseniceras A. LEANZA, 1945 [*Perisphinctes internispinosus* Krantz, 1928; SD Parent 2003]. Medium sized (**m**) to moderately large (**M**); inner whorls very evolute with depressed coronate whorl section, strong distant rectiradiate primary ribs terminating in high lateral spines from which 2-3 finer secondaries arise and cross somewhat flattened venter without interruption; outer whorls with rounded venter, compressed oval whorl section and dense non-tuberculate biplicate ribbing; constrictions occasionally present; macroconchs highly variable from coarsely ribbed serpenticones to compressed, more involute, densely ribbed platycones. *Lowermost U. Tithon.* **Fig. 463A-D** - *W. internispinosum* (Krantz).

Hemispiticeras SPATH, 1925 [*Reineckeia steinmanni* Steuer, 1897; SD Parent *et al.*, 2021][**M**]. Moderately large; depressed subcoronate whorl section and broadly rounded to somewhat flattened venter; numerous strong slightly prorsiradiate primary ribs terminate in high lateral tubercles and split into 2-3 finer secondaries over venter; in early to middle growth ribs terminate in weak ventral bullae on either side of shallow siphonal groove that evanesces on adult body chamber; suture with narrow long E, bifid saddles and narrow trifid L as deep as E. *Upper U. Tithon., ?Berrias.* **Fig. 463E-F** - *H. steinmanni* (Steuer).

Dickersonia IMLAY, 1942 [*D. subanillensis*; SD Arkell, 1957]. Small; inner whorls like *Hemispiticeras*; outer whorls with dense fine biplicate ribs uninterruped across venter, retaining vestigial bullae at the point of rib furcation that is above midflank; periodic collared constrictions. *U. Tithon.* **Fig. 463I-J** - *D. subanillensis* Imlay.

Family TORQUATOSPHINCTIDAE Tavera, 1985

Discussion. A seemingly monophyletic taxon with an origin in Ataxioceratidae (Olóriz, 1978; Parent, 2003), the sister taxon of Windhauseniceratidae. Comprises the subfamilies Torquatosphinctinae, Virgatosphinctinae and Nothostephaninae; except for the last, typically evolute with depressed whorls, prorsiradiate biplicate ribs and frequent constrictions. Characteristic of the Indo-Malgash Province. *U. Kimmer. - U. Tithon.*

Subfamily Torquatosphinctinae Olóriz & Tavera, 1985

Discussion. A predominantly Tithonian taxon with an origin in Ataxioceratinae, close to *Orthosphinctes* (Olóriz, 1978), by an increase in size and number of single ribs, lower point of branching of secondary ribs, and occurrence of triplicate ribs on body chamber of microconchs (Pandey *et al.*, 2010). *U. Kimmer. - U. Tithon.*

Biplisphinctes OLÓRIZ, 1978 [*Perisphinctes cimbricus* Neumayr, 1875; OD]. Small, very evolute, with depressed subquadratic whorls and low rounded venter; strong rectiradiate biplicate ribs bifurcating on ventrolateral shoulder and transverse across venter, interrupted by smooth siphonal band. *U. Kimmer.* **Fig. 464A-B** - *B. cimbricus* (Neumayr).

Torquatosphinctes SPATH, 1923 [*Ammonites torquatus* J. de C. Sowerby, 1840; OD]. Small; very evolute, with

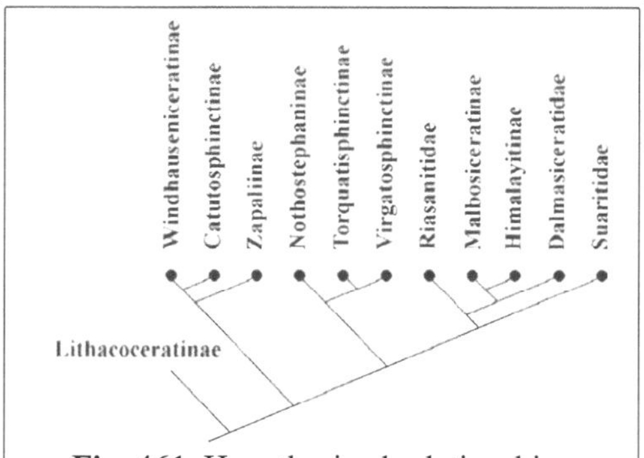

Fig. 461. Hypothesized relationships among Himalayitoidea.

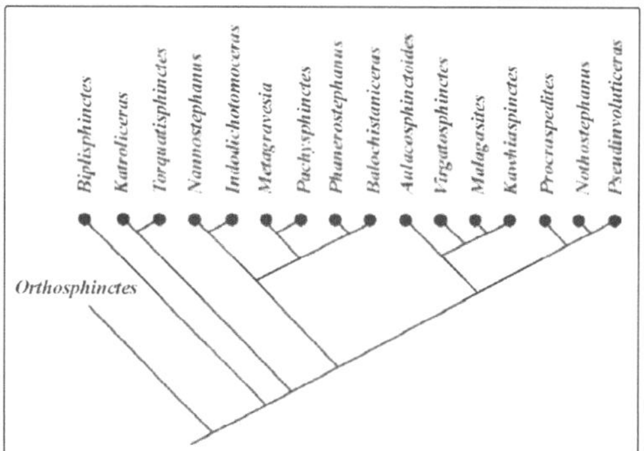

Fig. 462. Hypothesized relationships among Torquatisphinctidae.

rounded depressed subquadratic inner whorls that become compressed in maturity; ribbing comprises dense sharp biplicate ribs that bifurcate on upper third of flank, virgatitid-like triplicates at constrictions and numerous simple ribs; ribbing does not coarsen on adult body chamber. *U. Kimmer.* **Fig. 466A-B** - *T. torquatus* (J. de C. Sowerby).

Katroliceras SPATH, 1924 [*Ammonites pottingeri* J. de C. Sowerby, 1840; OD] [**M**]. Medium sized; like *Pachysphinctes* but with coarser ribbing and adult body chamber with distant very prominent subbullate wedge-shaped primary ribs giving rise to 4-5 secondaries. *Upper U. Kimmer. - L. Tithon.* **Fig. 466C-D** - *K. pottingeri* (J. de C. Sowerby).

Pachysphinctes DIETRICH, 1925 [*Perisphinctes (P.) africogermanus*; OD][**M**]. Large; very evolute, with strongly depressed rounded inner whorls and deep umbilicus; very prominent biplicate ribs with long primaries; outer whorls with sporadic virgatotome triplicate ribs. *U. Kimmer.* **Fig. 466G-H** - *P. africogermanus* (Dietrich).
Perhaps a macroconch of *Torquatisphinctes*.

Balochistaniceras FATMI & ZEISS, 1999 [*B. baragense*; OD]. Large (**m**) to very large (**M**); serpenticone with low steeply inclined umbilical wall and subquadratic to depressed oval whorl section; in early to middle growth coarse long primaries split into bundles 4-5 fine secondaries with some intercalatories near ventrolateral shoulder; in maturity distant pillar-like primaries fade out towards slightly raised smooth venter. *U. Kimmer.* **Fig. 466K-L** - *B. baragense* Fatmi & Zeiss.

Metagravesia SPATH, 1931 [*M. decipiens*; OD][**M**]. Large; moderately involute, highly inflated, with wide overhanging umbilical wall , convex flanks passing imperceptibly into broadly rounded venter and depressed reniform whorl section; inner whorls like *Pachysphinctes*; in later growth with coarse blunt primary ribs with weak umbilical bullae; suture complex. *U. Kimmer.* **Fig. 464C-D** - *M. decipiens* Spath.

Indodichotomoceras KRISHNA & PATHAK, 1993 [*Subdichotomoceras inversum* Spath, 1931; OD][**m**]. Large serpent-

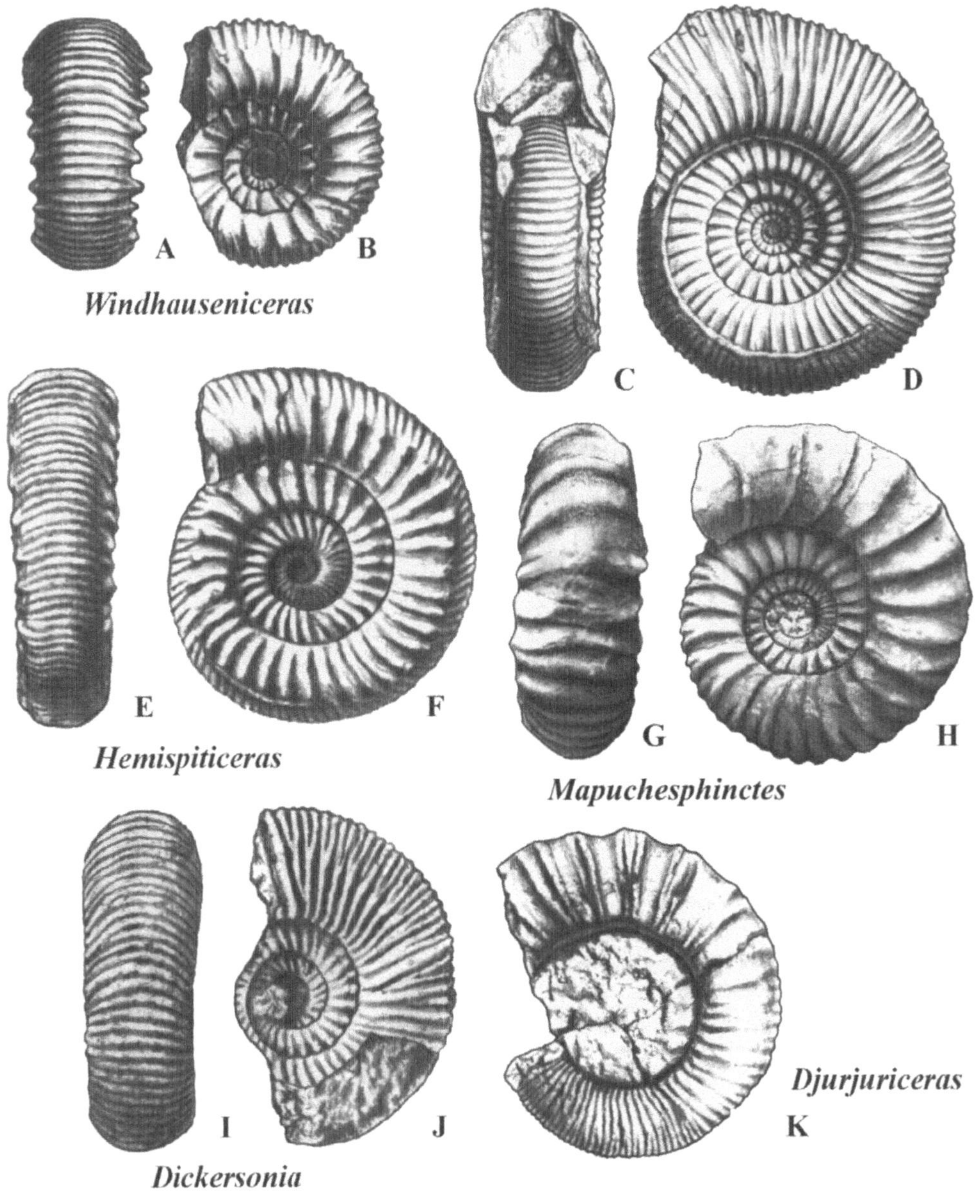

Fig. 463. Windhauseniceratidae, Himalayitidae.

icone with rounded umbilical wall, semicircular whorl section and broadly rounded venter; coarse robust biplicate ribs, some single, projecting forward across venter, with very deep interspaces; constrictions occur throughout but more distant in early growth. *U. Kimmer.* **Fig. 466O-P** - *I. inversum* (Spath).

= *Indodichotomoceras* Krishna & Pathak, 1990 (*nom. nud.*).

Nannostephanus SPATH, 1950 [*N. subcornutus*; OD]. Very small, like *Indodichotomoceras* but with flattened flanks, broad flat to concave venter and quadratic whorl section; inner whorls finely ribbed; in middle to later growth with irregular single and biplicate ribs, the latter branching at ventrolateral tubercles. *U. Tithon.* **Fig. 466M-N** - *N. subcornutus* Spath.

Phanerostephanus SPATH, 1950 [*P. subsenex*; OD]. Medium sized; inner whorls like *Pachysphinctes* but more finely ribbed; in later growth ribs evanesce on flanks leaving umbilical bullae; aperture with ventral lappet; irregularly spaced shallow constrictions; suture with broad E/L and trifid L as deep as E. *L. - U. Tithon.* **Fig. 466E-F** - *P. subsenex* Spath.

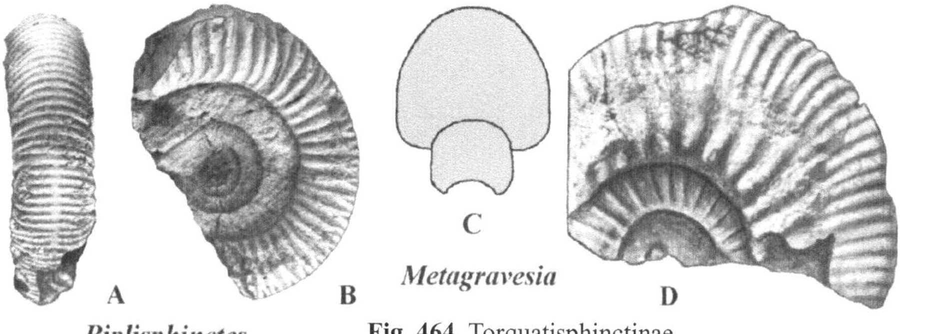

Fig. 464. Torquatisphinctinae.

Subfamily Virgatosphinctinae Spath, 1923

Discussion. A monophyletic taxon with an origin in Torquatisphinctinae, close to *Pachysphinctes* (Kutek & Zeiss, 1974) or *Torquatisphinctes* (Pandey *et al.*, 2010), although Parent *et al.* (2015), like Énay& Cariou (1997) before, also contemplated an origin in Andean *Choicensisphinctes*. Typically evolute, with rounded to slightly compressed whorls, prominent primary ribs giving rise to bundles of biplicate, triplicate or virgatotome secondaries that cross rounded venter uninterrupted, with periodic constrictions. A predominantly Kimmeridgian group largely restricted to the Indo-Malgash Province with some incursions into the Submediterranean Province (Énay, 1972; Pandey *et al.*, 2010). *Upper L. Kimmer. - Upper U. Tithon.*

Virgatosphinctes UHLIG, 1910 [*Perisphinctes (V.) broilii*; SD Douvillé, 1910] [**M**]. Medium sized; evolute, with convex flanks, rounded venter and semicircular whorl section; numerous primary ribs arise at umbilical seam and are slightly prorsiradiate across lower flank, branching above midflank so that only primary ribs visible in umbilicus; ribs initially biplicate, gradually becoming triplicate, virgatotome and finally fasciculate in later growth; periodic parabolae, particularly on inner whorls. *Upper L. - Uppermost Tithon.* **Fig. 468A-B** - *V. broilii* (Uhlig).

Aulacosphinctoides SPATH, 1923 [*Aulacosphinctes infundibulus* Uhlig, 1910; OD] [**m**]. Small; like *Virgatosphinctes* but smaller, with depressed elliptical to ovate whorl section, numerous relatively coarse prorsiradiate sigmoidal ribs bifurcating above midflank and crossing venter transversely; primaries become stiffer in later growth when trifurcation frequently occurs; aperture with lateral lappets. *U. Kimmer. - L. Tithon.* **Fig. 468E-F** - *A. infundibulus* (Uhlig); **Fig. 468K-L** - *A. retrocostatus* Collignon.

= *Subaulacosphinctoides* Pathak, 1989 (*nom. nud.*).

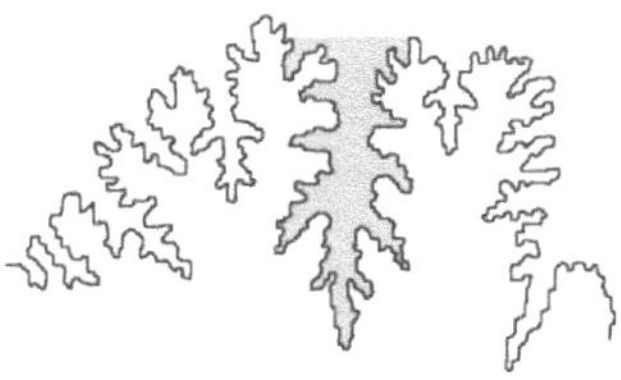

Fig. 465. *Virgatosphinctes* suture.

Malagasites ÉNAY, 2009 [*Perisphinctes (Virgatosphinctes) haydeni* Uhlig, 1910; OD][**m**]. Medium sized to moderately large, evolute, with rounded umbilical shoulder, gently convex flanks converging towards rounded venter and somewhat compressed oval whorl section; in early to middle growth long biplicate primary ribs dense, fine, prorsiradiate, recurving at mid-flank, with an occasional intercalatory, crossing venter transversely; about 5 parabolae per whorl, parallel to ribbing; adult body chamber with coarser more distant polygyrate ribbing in bundles of 3-5; peristome with lateral lappets. *Tithon.* **Fig. 468M-N** - *M. haydeni* (Uhlig).

Kawhiasphinctes STEVENS, 1997 [*K. antipodus*; OD] [**M**]. Very large; evolute, compressed, flat sided in early growth, later with gently convex flanks and subquadratic whorl section slightly higher than wide; strong sharp fine prorsiradiate ribs, single and biplicate, triplicate on body chamber, branching low on flank and recurving on outer flank to cross venter transversely. *M. Tithon.* **Fig. 468R-S** - *K. antipodus* Stevens.

Parent *et al.* (2015) noted the close similarity of the inner whorls of this genus to some Andean specimens questionably assigned to *Choicensisphinctes*.

Subfamily Nothostephaninae nov.

Diagnosis. Typically with umbilical bullae, smooth flanks in later growth, irregular constrictions and evolutionary trend towards increased compression and involution. Monophyletic derivatives of Torquatisphinctinae, close to *Pachysphinctes* (Parent, 2003). *Middle U. Kimmer. - L. Tithon.*

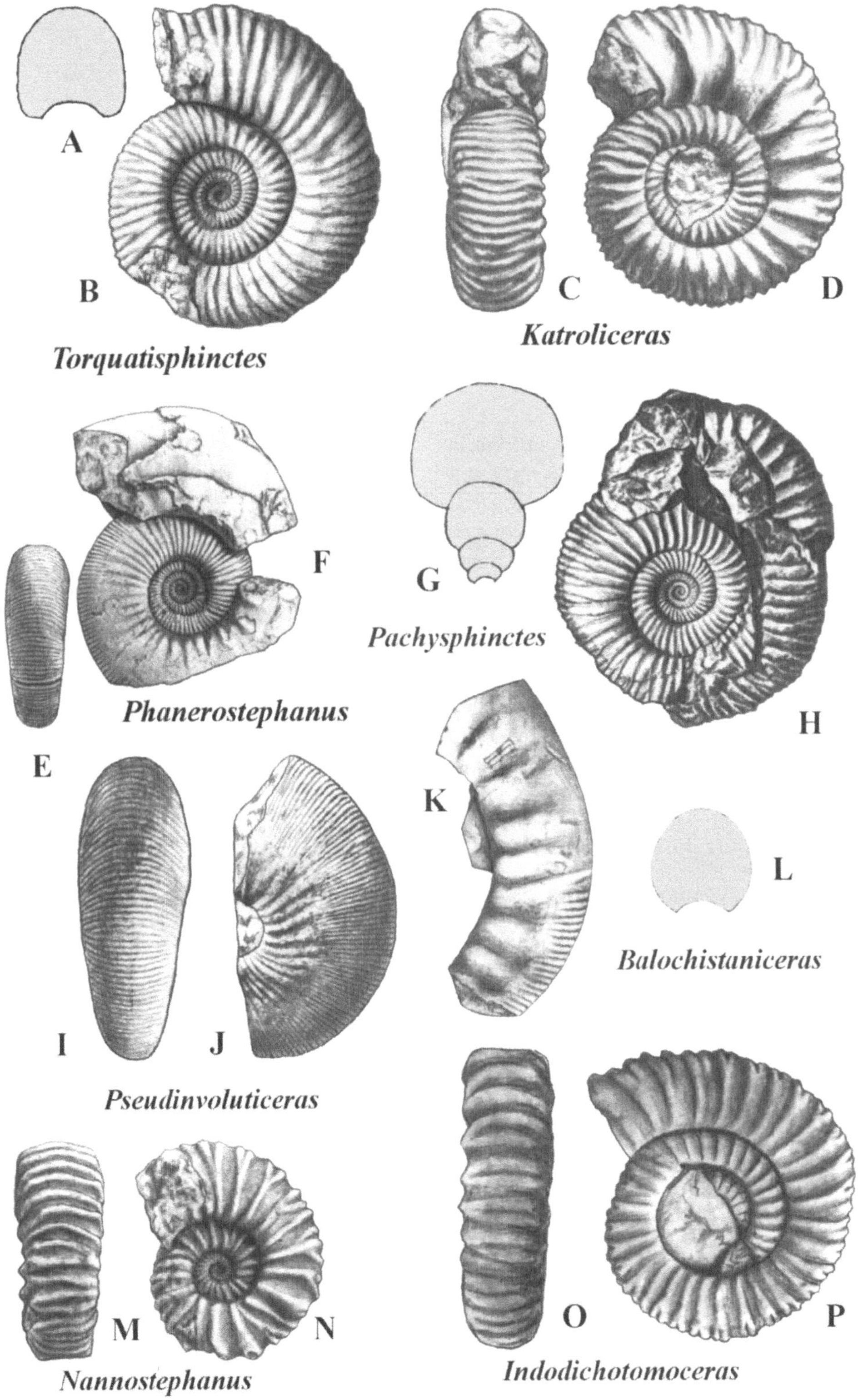

Fig. 466. Torquatisphinctidae.

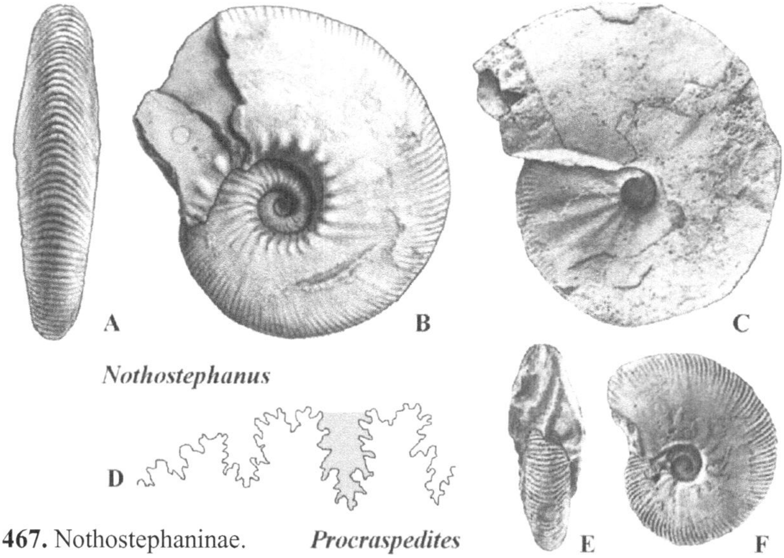

Fig. 467. Nothostephaninae.

Procraspedites SPATH, 1930 [*Craspedites praecursor* Burckhardt, 1906; OD]. Medium sized; involute, compressed, with flattened flanks converging to narrowly rounded venter and long broad primary ribs branching on upper third of flank; in later growth primaries obsolete leaving flanks almost smooth with only short secondaries across venter. *Middle U. Kimmer.* **Fig. 467C-D** - *P. praecursor* (Burckhardt).

Nothostephanus SPATH, 1950 [*N. kurdestanensis*; OD]. Moderately large; like *Phanerostephanus* but compressed and involute, with convergent flanks, high whorls, narrowly rounded venter, subtrigonal whorl section and ribs across venter in later growth. *L. Tithon.* **Fig. 467A-B** - *N. kurdestanensis* Spath.

Pseudinvoluticeras SPATH, 1925 [*P. somalicum*; OD]. Medium sized; like *Nothostephanus* but more inflated, with longer more numerous primary ribs and finer denser secondaries. *L. Tithon.* **Fig. 466I-J** - *P. somalicum* Spath.
= *Pseudoinvoluticeras* Roman, 1938 (*nom. van.*).

Family HIMALAYITIDAE Spath, 1925

Discussion. A questionably monophyletic grouping apparently with an origin in Torquatisphinctidae (Parent, 2003), but complicated by dimorphism, the matching of dimorphs, and much intraspecific variation (Shome & Bardhan, 2007). Typically evolute, with rounded whorls, sharp frequently differentiated ribbing, lateral spines, ventrolateral bullae in many, and a shallow groove or smooth siphonal band in early to middle growth at least. Comprises the subfamilies Himalayitinae, Mazenoticeratinae and Dalmasiceratinae. *U. Tithon. - L. Valang.*

Subfamily Himalayitinae Spath, 1925

(*nom. transl.* Arkell, 1957 *ex* Himalayitidae)

Discussion. Seemingly monophyletic descendents of Torquatisphinctinae (Donovan *et al.*, 1981; Parent, 2001); comprises the tribes Himalayitini and Simplisphinctini. *Upper L. Tithon. - U.Tithon.,?L. Berrias.*

Tribe Himalayitini Spath, 1925

(*nom. transl.* herein ex Himalayitinae)

Diagnosis. Microconchs with biplicate to triplicate and polygyrate ribs, lateral tubercles, prominent ventral furrow and lateral lappets, with or without ventral tubercles; macroconchs larger, more coarsely ribbed, with ventral furrow obsolete in later growth. *Upper L. Tithon. - U. Tithon., ?Berrias.*

***Prohimalayites* gen. nov.** [*Himalayites (Micracanthoceras) peroni* Roman, 1936; OD][**m**]. Small; very evolute, with depressed subcircular whorl section; strong rectiradiate ribs branching from incipient lateral bullae at midflank, with occasional intercalatories, triplicate and polygyrate ribs; ribs interrupted over venter by shallow siphonal groove; periodic constrictions parallel ribbing; adult peristome with lateral lappets. *Upper L. Tithon.* **Fig. 471I-J** - *P. peroni* (Roman).
= *Burckhardticeras* Olóriz, 1978 *non* Erben (1956).

Aulacosphinctes UHLIG, 1910 [*Ammonites moerikeanus* Oppel, 1863; SD Spath, 1924] [**m**]. Small serpenticone with compressed subelliptical to subquadratic whorls,

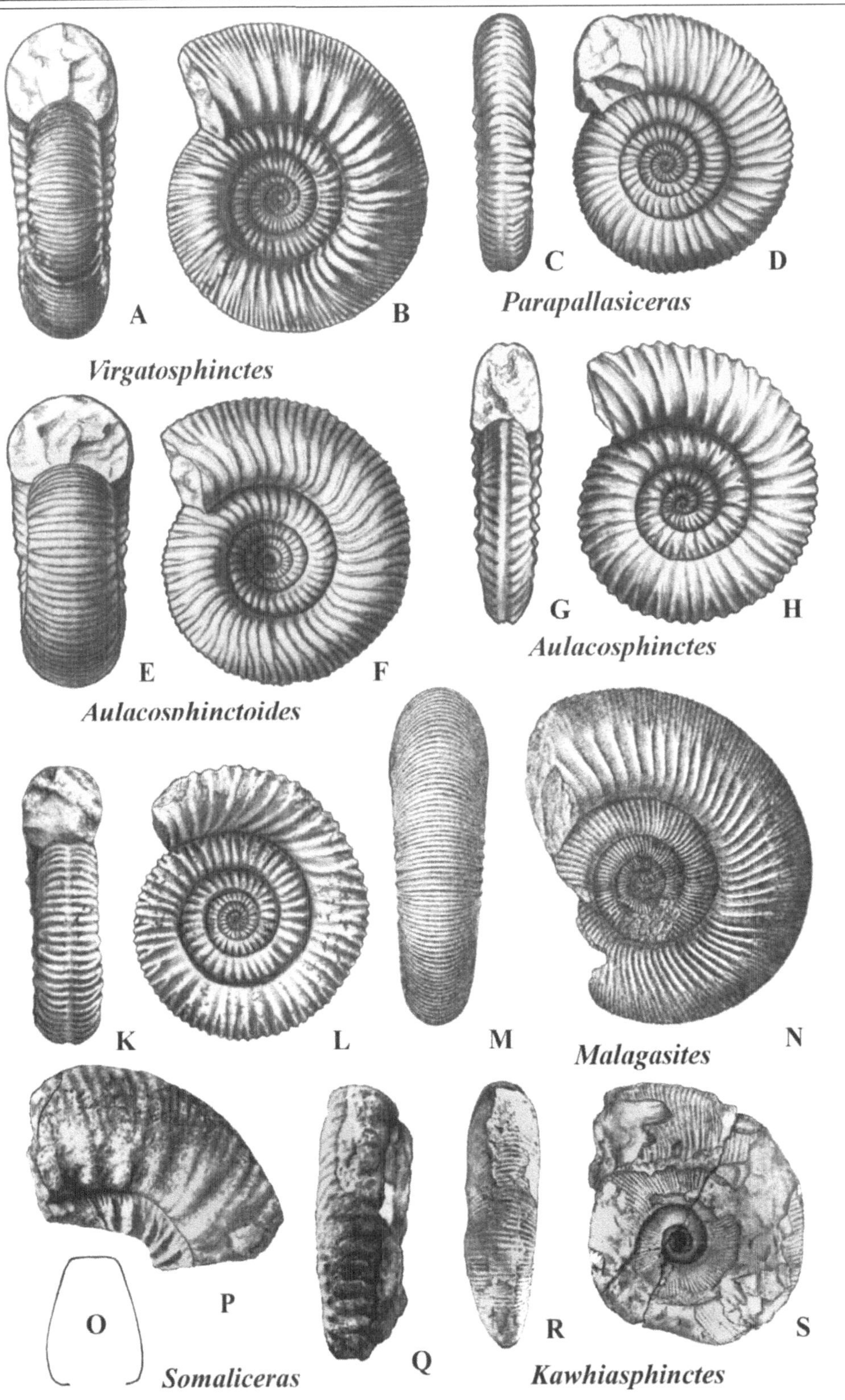

Fig. 468. Virgatosphinctinae, Himalayitinae.

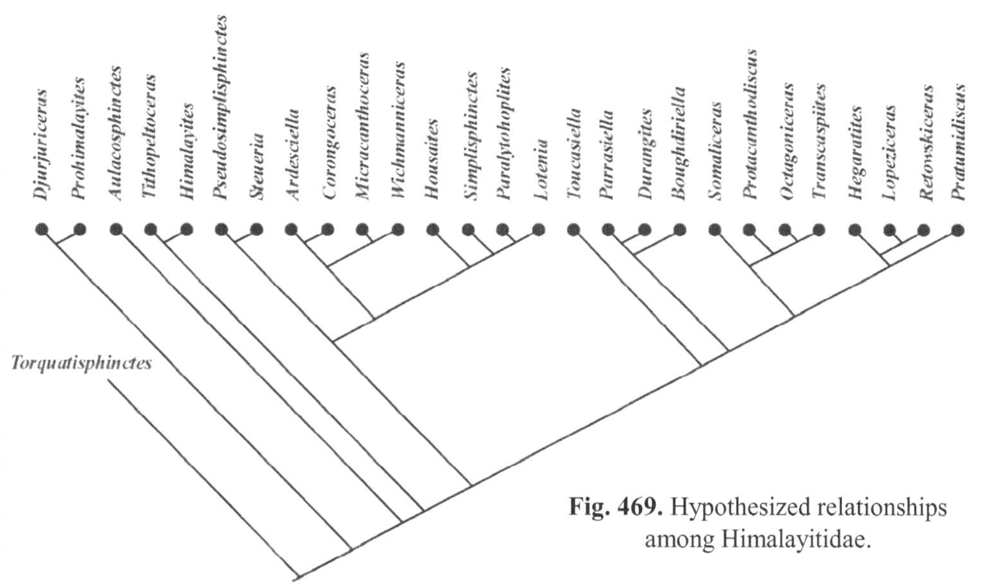

Fig. 469. Hypothesized relationships among Himalayitidae.

flattened flanks and narrowly arched venter; strong distant straight prorsiradiate biplicate ribs, occasionally simple, interrupted across venter by persistent siphonal groove; some ribs bifurcate twice and, rarely, from umbilical shoulder; inner whorls with lateral tubercles; occasional shallow constrictions parallel ribs. *Tithon., ?L. Berrias.* **Fig. 468G-H** - *A. moerikeanus* (Oppel).

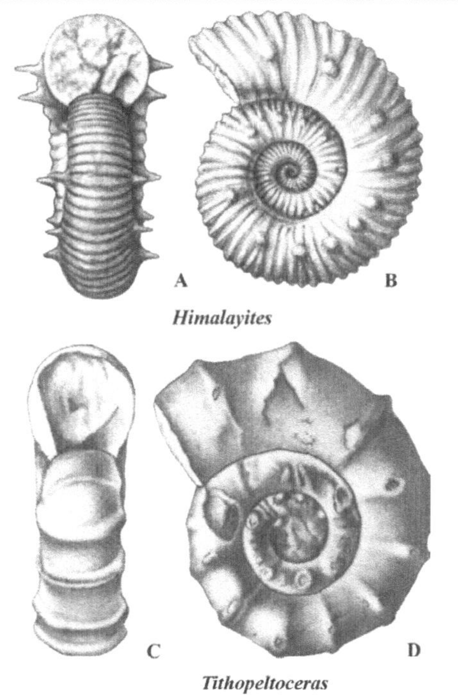

Fig. 470. Himalayitinae.

Micracanthoceras SPATH, 1925 [*Ammonites microcanthus* Oppel, 1868; OD] [**M**]. Medium sized; like *Himalayites* but with denser ribbing in which simple ribs alternate with biplicates or triplicates with tiny lateral tubercles at point of branching; ribs terminate in bullae on either side of smooth siphonal band. *Lowermost U. Tithon.* **Fig. 471M-N** - *M. microcanthum* (Oppel).

Considered the macroconch of *Aulacosphinctes* (Donovan *et al.*, 1981).

Wichmanniceras A. LEANZA, 1945 [*W. mirum*; OD]. Small; strongly compressed serpenticone with subquadratic whorl section and grooved venter in early to middle growth; simple rectiradiate ribs irregularly looped to ventrolateral tubercles; on adult body chamber tubercles obsolete and ribs cross rounded venter uninterrupted. *Upper M. Tithon.* **Fig. 487B-C** - *W. mirum* A. Leanza.

Assigned to Suaritinae by Cantú-Chapa (1967), but considered a possible synonym of *Micracanthoceras* by Parent *et al.* (2011).

Corongoceras SPATH, 1925 [*C. lotenoense*; OD]. Small; inner whorls like *Micracanthoceras*; in middle to later growth with strong distant bituberculate ribs with lateral and ventrolateral tubercles to all ribs with some secondaries zigzagging or looped between them; ventrolateral tubercles may be spinose. *Upper M. - U. Tithon.* **Fig. 471E-F** - *C. lotenoense* Spath; **Fig. 471G-H** - *C. alternans* (Gerth).

With 38 nominal species/subspecies (Klein, 2005), a number with query, this is a *nomen commodum*.

Himalayites UHLIG, 1904 [*H. treubi*; SD Douvillé, 1912] [**M**]. Dimorphic, small (**m**) to medium sized (**M**); inner whorls like *Aulacosphinctes* but becoming rounded and depressed in middle growth when ribbing differentiates, with periodic strengthened primaries terminating in large

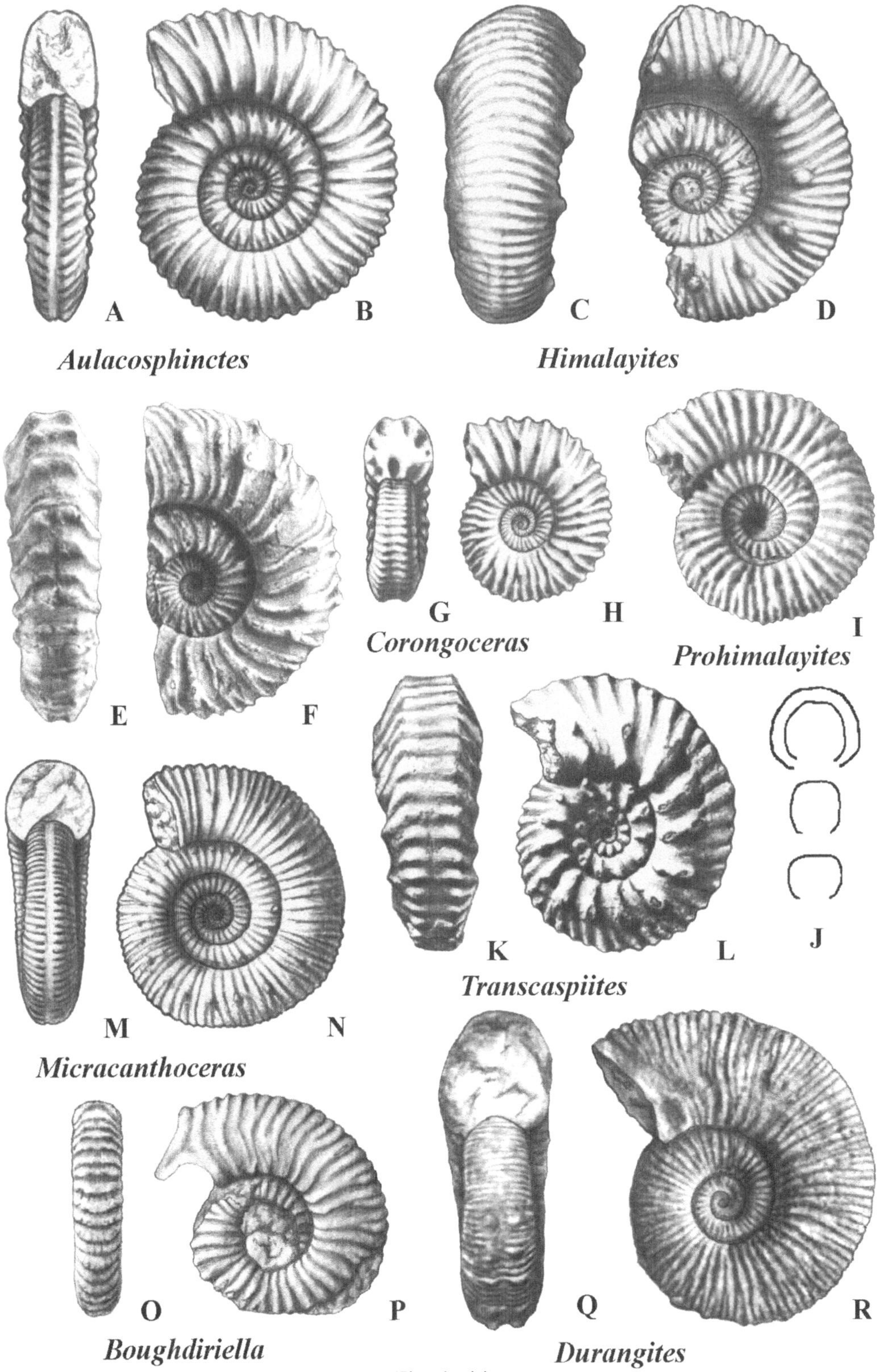

Fig. 471. Himalayitinae.

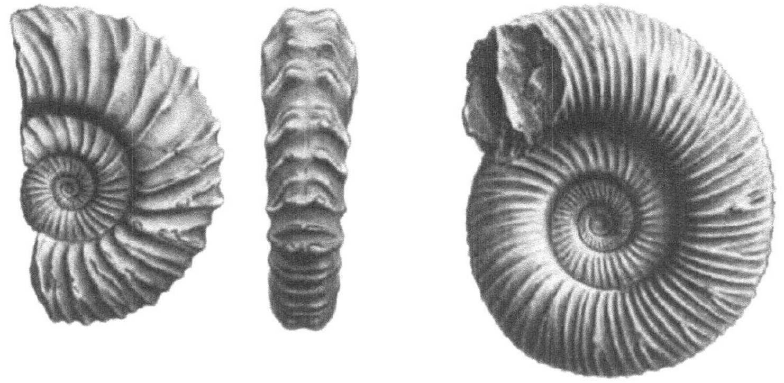

Fig. 472. *Steueria*, microconch and macroconch.

septispinate mid-lateral spines and giving rise to 2-4 secondaries separated by 2-4 non-tuberculate intercalatories; siphonal groove obsolete in maturity. *U. Tithon., ?L. Berrias.* **Fig. 470A-B, 471C-D** - *H. treubi* Uhlig.

Tithopeltoceras ARKELL, 1953 [*Aspidoceras moriconii* Meneghini, 1885; OD] [**M**]. Small to moderately large, very evolute; in early growth with ribs bifurcating from small ventrolateral nodes; in middle growth with depressed coronate whorl section, ribs arising in dublets and triplicates from prominent lateral spines and more or less smooth venter; in maturity with subcircular whorl section and blunt, distant, simple ribs with lateral bullae crossing venter uninterrupted; constrictions present. *U. Tithon.* **Fig. 470C-D** - *T. moriconii* (Meneghini).

Boughdiriella FRAU, BULOT & WIMBLEDON, 2015 [*B. chouetensis*; OD][**m**]. Very small, evolute, with compressed subcircular whorl section in early growth becoming subovate later; early whorls with straight, rigid, simple ribs, in middle to later growth becoming strong and flexuous bifurcating on outer flank, with occasional intercalatories and triplicates; venter with shallow groove; adult aperture with short lateral lappets. *U. Tithon.* **Fig. 471O-P** - *B. chouetensis* Frau, Bulot & Wimbledon.

Ardesciella BULOT, FRAU & WIMBLEDON, 2014 [*Himalayites (?Corongoceras) rhodanicus* Mazenot, 1939; OD]. Small serpenticone; whorl section rounded in early growth, becoming subrectangular to subquadratic later with low vertical umbilical wall, rounded shoulders and strongly convex flanks; prominent siphonal furrow attenuating in later growth; straight, rigid, slightly prorsiradiate, single or bifurcating ribs, the latter branching from tiny tubercles on outer third of flank; in maturity ribs slightly flexuous and rursiradiate on lower flank, most with ventrolateral bullae. *U. Tithon.* **Fig. 474M-O** - *A. rhodanicus* (Mazenot).

Steueria PARENT, SCHERZINGER & SCHWEIGERT, 2011 [*Berriasella alternans* Gerth, 1921; OD]. Medium sized [**m**] to moderately large [**M**]; very evolute, compressed; in middle growth with rounded venter and suboval (intercostal) to subtrapezoidal (costal) whorl section; strong prorsiradiate primary ribs, narrower than interspaces, with more or less prominent lateral bullae, from which secondaries bifurcate, all terminating in septispinate ventrolateral bullae separated by shallow siphonal groove; suture with narrow trifid L deeper than E. *U. Tithon.* **Fig. 472** - *S. alternans* (Gerth).

Pseudosimplisphinctes TAVERA, 1985 [*P. jimenezi*; OD] [**m**]. Very small, very evolute; in early to middle growth with flat flanks converging towards subtabulate venter and trapezoidal whorl section as wide as high, with strong primary ribs bifurcating from ventrolateral tubercles

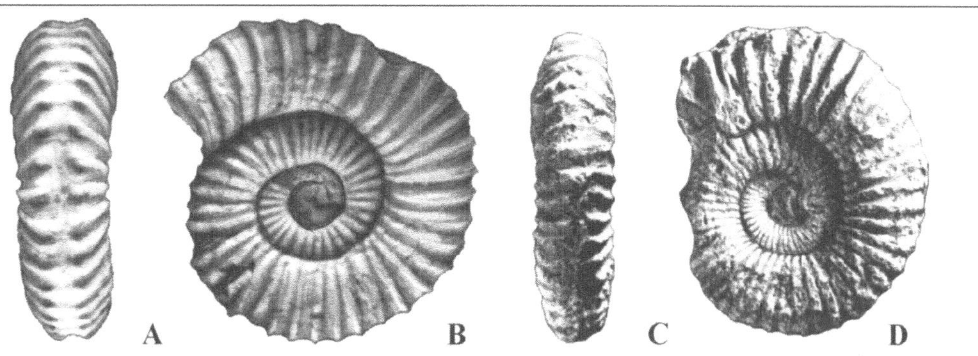

Fig. 473. Simplisphinctini.

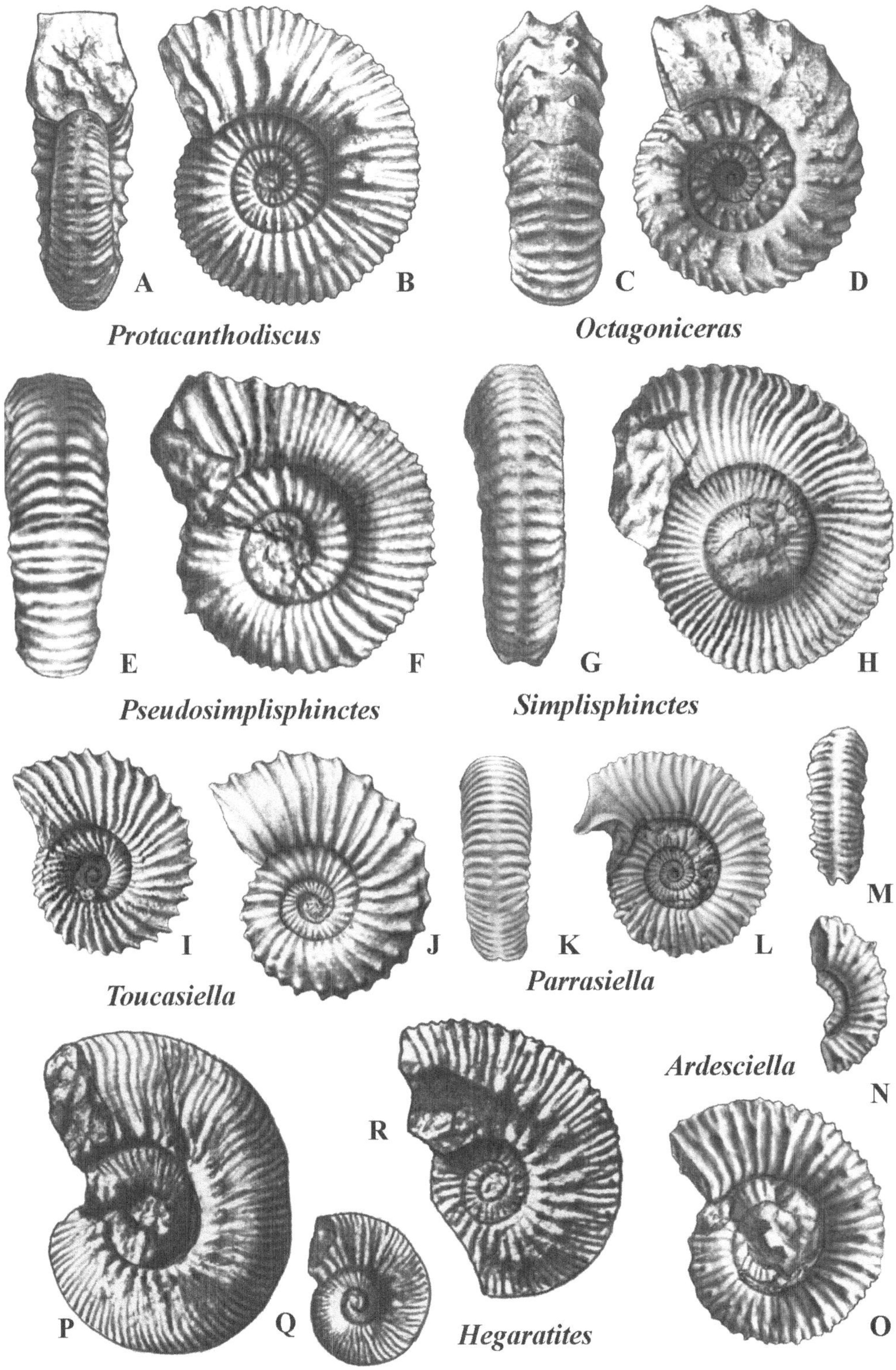

Fig. 474. Himalayitinae.

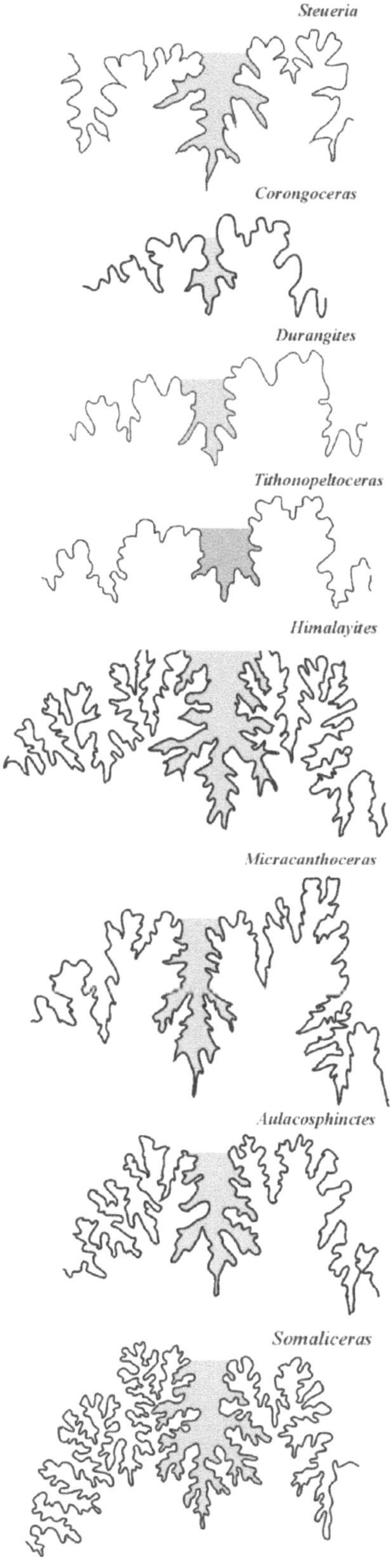

Fig. 475. Himalayitoid sutures.

separated by 2-3 non-tuberculate secondaries, all weakening or interrupted across siphonal line, with periodic constrictions; in maturity whorl section becomes oval or subcircular and ribs single with slight ventrolateral thickening; suture simple, with bifid saddles and trifid lobes. *U. Tithon.* **Fig. 474E-F** - *P. jimenezi* Tavera.

Djurjuriceras ROMAN, 1936 [*D. djurjurense*; OD]. Small to large, evolute with broad venter and subquadratic to subrounded whorl section; in early to middle growth with fine dense flexuous biplicate ribs that cross venter uninterrupted, with irregularly developed ventrolateral bullae where ribs interrupted across venter; lateral tubercles may develop at point of rib furcation; on adult body chamber ribs usually arranged in pairs separated by wide interspaces. *L. - M. Tithon.* **Fig. 463K** - *D. djurjurense* Roman.

Toucasiella ÉNAY, BOUCHDIRI & LE HEGARAT, 1998 [*T. gerardi*: OD]. Very small; very evolute, compressed, with flattened flanks and venter, and compressed rectangular whorl section; straight, simple, prorsiradiate ribs with ventrolateral tubercles bordering siphonal groove; periodic ribs bifurcate from high lateral tubercles; ribs become denser and flexuous near adult peristome of microconchs; on last part of body chamber of macroconchs lateral tubercles evanesce, venter becomes tabulate and ribs cross uninterrupted, with feeble development of umbilical bullae. *U. Tithon.* **Fig. 474I-J** - *T. gerardi* (Énay, Bouchdiri & Le Hegarat).

With 20 nominal species/subspecies assigned here (Klein, 2005), a number with query, this is a *nomen commodum*.

Parrasiella FRAU, BULOT & WIMBLEDON 2015 [*Durangites astillerensis* Imlay, 1939; OD][**m**]. Very small; evolute, with convex flanks and subovate whorl section; in middle growth with strong simple and biplicate ribs periodically with parabolic nodes; posterior branches of biplicate ribs with ventrolateral bullae delimiting shallow ventral groove; body chamber with simple prorsiradiate flexuous biplicate and simple ribs that cross siphonal line weakly; short lateral lappets. *U. Tithon.* **Fig. 474K-L** - *P. astillerensis* (Imlay).

Protacanthodiscus SPATH, 1923 [*Hoplites andreae* Kilian, 1889; OD] [**M**]. Moderately large, evolute; outer whorls with flattened flanks, broad subtabulate venter and subtrapezoidal whorl section; rather irregular ribbing, primary ribs with prominent lateral and ventrolateral bullae separated by 2-3 secondaries with incipient ventral bullae. *L. Berrias.* **Fig. 474A-B** - *P. andreae* (Kilian).

With 23 nominal species/subspecies assigned here (Klein, 2005), some with a query, this is a *nomen commodum*. Perhaps includes macroconchs of *Durangites*.

Durangites BURCKHARDT, 1912 [*D. vulgaris*; SD Cantú-Chapa, 1968][**m**]. Very small (**m**) to small (**M**); evolute, compressed, with convex flanks, flattened venter and

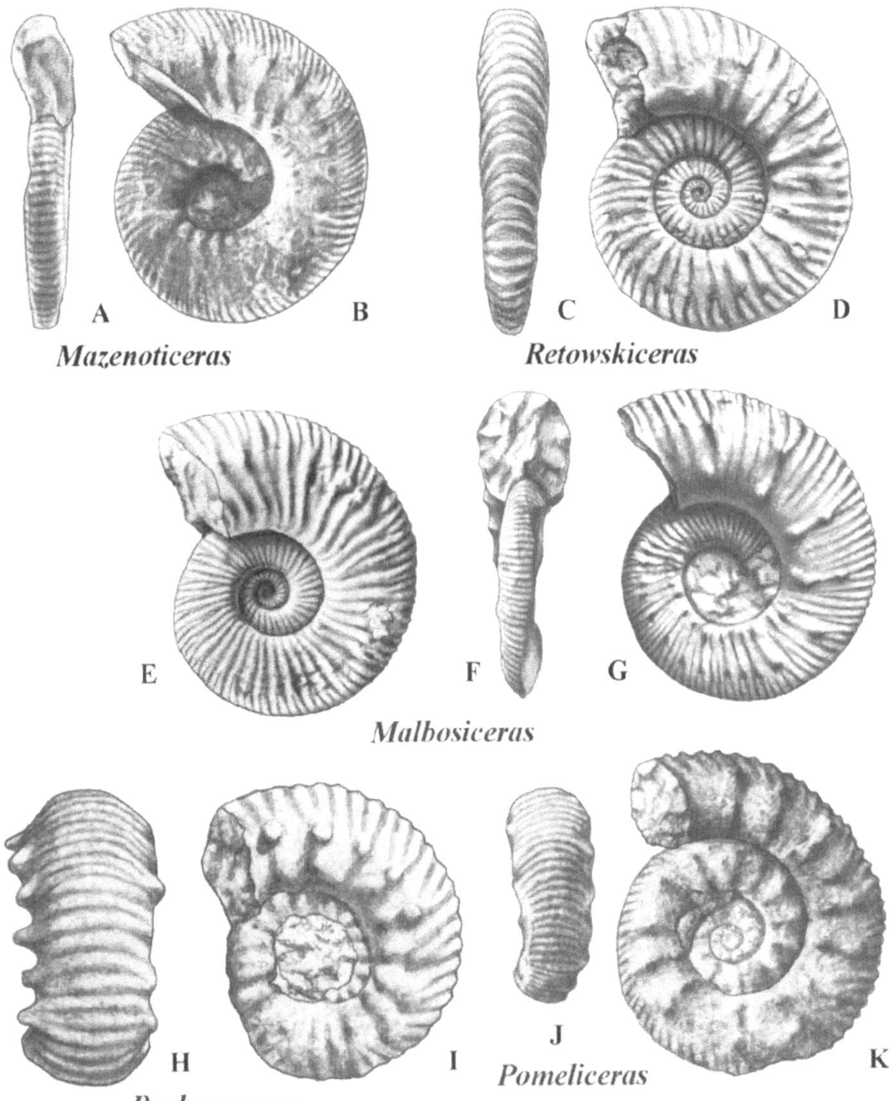

Fig. 476. Mazenoticeratinae.

subrectangular whorl section; inner whorls with fine sharp simple and biplicate rectiradiate ribs with feeble lateral tubercles at point of branching, terminating in small ventrolateral tubercles to leave smooth siphonal furrow; in middle growth lateral tubercles evanesce, periodic ribs with strong ventrolateral bullae; in later growth ventrolateral bullae evanesce, ribs mostly biplicate and single uninterrupted over flattened venter; macroconchs show scaphitoid uncoiling of adult body chamber; suture with narrow E, broad square shallowly bifid E/L, narrow trifid L. *U. Tithon.* **Fig. 471Q-R** - *D. vulgaris* Burckhardt.

Somaliceras SPATH, 1925b [*S. isariforme*; OD] [**M**]. Small, with tabulate venter and compressed trapezoidal whorl section; strong irregular simple prorsiradiate ribs exposed in umbilicus; on outer whorl elongate umbilicolateral bullae give rise to ribs in triplets that are transverse over venter, with distinct ventrolateral bullae; suture with narrow E, trifid L as deep as E, and strongly retracted suspensive lobe with 4 elements. *?Tithon.* **Fig. 468O-Q** - *S. isariforme* Spath.

Although included in the synonymy of *Spiticeras* (Wright, 1996) or assigned to Spiticeratinae (Klein, 2005), the presence of ventrolateral bullae suggest it is a himalayitid (*cf.* Spath, 1925).

Octagoniceras SPATH, 1924 [*Ammonites octagonus* Blanford, 1864; M]. Medium sized; like *Protacanthodiscus* but with octagonal whorl section, sparse prominent primary ribs with umbilical and lateral tubercles, the latter giving rise to 2-3 prorsiradiate secondaries that form prominent ventrolateral bullae and thin slightly across

Fig. 477. *Pratumidiscus.*

siphonal line; adult body chamber with prominent bullate ventrolateral tubercles; suture with deep E and asymmetrically bifid L shallower than E. *Berrias.* **Fig. 474C-D** - *O. octagonum* (Blanford).

Transcaspiites LUPPOV, 1985 [*Protacanthodiscus transcaspius* Luppov, Bodylevsky & Glazunova, 1949; OD]. Small; like *Protacanthodiscus* but with coarser more regular ribbing, strong umbilical bullae, very prominent lateral tubercles and ribs crossing tabulate venter strongly with weak uniform ventrolateral bullae. *Berrias.* **Fig. 471K-L** - *T. transcaspius* (Luppov, Bodylevsky & Glazunova).

Pratumidiscus BULOT, FRAU & WIMBLEDON, 2014 [*P. elsae*; OD] [**m**]. Small, compressed, with wide shallow umbilicus, weakly convex flanks and tabulate venter; ribbing complex and irregular, single, bifurcating, intercalated and simplified virgatotome; all ribs with ventrolateral bullae bounding siphonal groove, with a faint thickening of ribs at point of rib furcation in maturity. *U. Tithon.* **Fig. 477** - *P. elisae* Bulot, Frau & Wimbledon.

Hegaratites BOUGHDIRI, ÉNAY, LE HEGARAT & MEMMI, 1999 [*H. rheouisensis*; OD]. Very small (**m**) to small (**M**); like *Protacanthodiscus* but with subrectangular whorl section; ribs fine, rectiradiate and generally single in juveniles, arising from feeble umbilical bullae; mature microconchs relatively high whorled, with somewhat flexuous prorsiradiate ribs frequently bifurcating at midflank and curving forward ventrally; macroconchs similar but mostly with bifurcating or trifurcating ribs, frequent intercalatories, strong mid-lateral tubercles and sporadic strong ventrolateral tubercles; in some species, flexuous ribbing effaced from flanks of adult macroconchs, lateral tubercles lost, and weak constrictions frequent. *U. Tithon.* **Fig. 474P-R** - *H. rheouisensis* Boughdiri, Énay, Le Hegarat & Memmi.

Lopeziceras FRAU, BULOT, WIMBLEDON & IFRAM, 2016 [*Ammonites chaperi* Pictet, 1868; OD]. Small to medium size, dimorphic; evolute, with flattened flanks and compressed subrectangular whorl section; venter deeply grooved in early to middle growth, with flat smooth band in maturity; in later growth rectiradiate to slightly prorsiradiate primary ribs with feeble umbilical bullae and high lateral tubercles from which secondaries bi- and trifurcate terminating in oblique ventrolateral clavi; on adult body chamber of macroconchs primary ribs distant and prorsiradiate. Superficially similar to *Malbosiceras*. *L. Berrias.* **Fig. 478A-B** - *L. chaperi* (Pictet).

= *Chapericeras* Hoedemaeker, 1981 (*nom. nud.*).

Retowskiceras NIKOLOV, 1966 [*Perisphinctes andrussowi* Retowski, 1893; OD]. Small; like *Lopeziceras* but more evolute (U=43-51%) and coarsely ribbed, with rounded ungrooved venter and well spaced ribs in early to middle growth; umbilical and lateral tubercles develop on last whorl where ribs branch lower on flank. *L. Berrias.* **Fig. 476C-D** - *R. andrussowi* (Retowski).

Kilianites ÉNAY, BOUGHIDIRI & LE HÉGARAT, 1998 [*K. canavarii*; OD][**M**]. Small, evolute, with narrow, shallowly sulcate venter; in early growth primary ribs branch from mid-lateral tubercles, separated by a non-tuberculate secondary; in middle growth primaries develop umbilical and ventrolateral tubercles, and whorl section becomes octagonal; in maturity all ribs simple with prominent ventrolateral clavi. *Lowermost Berrias.* **Fig. 482L-M** - *K. bergeroni* (Kilian).

Tribe Simplisphinctini nov.

Type genus: *Simplisphinctes* Tavera, 1985.

Diagnosis. Small, very evolute, compressed, with mainly simple ribs, occasionally biplicate; venter with inner and outer ventrolateral tubercles, or non-tuberculate. Monophyletic derivatives of Himalayitini, close to *Micracanthoceras* (Tavera, 1985). *L. - U. Tithon.*

Housaites VILLASEÑOR & OLÓRIZ, 2009 [*Paradontoceras butti* Imlay, 1942; OD]. Small to medium sized; in early to middle growth evolute, but becoming more involute later, with flattened flanks and subovate whorl section; phragmocone with dense thin, straight, wiry prorsiradiate ribs, mainly single rarely bifid and branching below mid-flank, crossing venter uninterrupted; adult body chamber with single well spaced ribs, narrower than interspaces. *Upper L. - Lower U. Tithon.* **Fig. 482K** - *H. butti* (Imlay).

= *Butticeras* Houša & De la Nuez, 1975 (*non* Anderson, 1958, Texanitidae).

Paralytohoplites MYCZYŃSKI, 1996 [*Lytohoplites carribeanus* Imlay, 1942; OD]. Small; in early growth with convex flanks, very broadly rounded venter, depressed coronate to subovate whorl section and low simple slightly prorsiradiate ribs much narrower than interspaces; in middle to later growth flanks flatten, whorls become compressed and ribs recurving on outer flank with slight convexity across venter, rarely bifurcating from umbilical shoulder; ribs with triangular profiles, weakening across siphonal line. *U. Tithon.* **Fig. 473C-D** - *P. carribeanus* (Imlay).

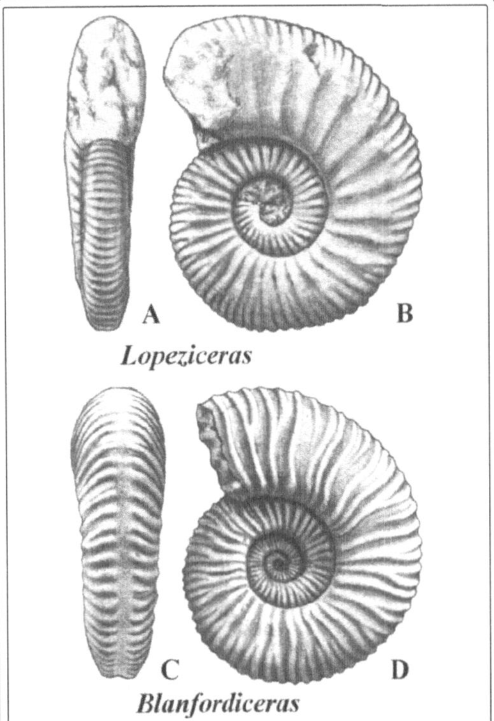

Fig. 478. Mazenoticeratinae.

Simplisphinctes TAVERA, 1985 [*Himalayites? abnormis* Roman, 1936; OD] [**m**]. Small, very evolute, with rounded umbilical rim, slightly convex flanks converging to shallowly sulcate venter and trapezoidal whorl section about as high as wide; ribs simple, fine, rectiradiate to prorsiradiate, becoming sinuous and rursiradiate on adult body chamber; primary ribs with high lateral and spinose ventrolateral tubercles, secondary ribs only with latter, weakening or evanescing over shallow siphonal groove; adult aperture lappetted; suture with shallow symmetrically bifid E/L and narrow trifid L. *U. Tithon.* **Fig. 474G-H** - *S. abnormis* (Roman).

Lotenia ZEISS & H. LEANZA, 2008 [*Simplisphinctes (L.) neuquensis*; OD]. Small; like *Paralytohoplites* but with ventrolateral tubercles. *U. Tithon.* **Fig. 473A-B** - *L. neuquensis* (Zeiss & H. Leanza).

Subfamily Mazenoticeratinae Nikolov, 1966

= Neocosmoceratinae Kvantaliani, 1989
= Pomeliceratinae Kvantaliani, 1989

Discussion. Typically Boreal in distribution; seemingly monophyletic descendents of Himalayitinae (Protacanthodiscini), close to *Kilianites* (Énay *et al.*, 1998); microconchs characteristically evolute, with compressed subrectangular whorls, irregular ribbing and umbilical, lateral and ventrolateral tubercles, although umbilical and lateral tubercles not always present. According to Énay *et al.* (1998), *Mazenoticeras* is based upon macroconchs of *Neocosmoceras*. *U. Tithon. - Berrias.*

Fig. 479. Hypothesized relationships among Mazenoticeratinae.

Blanfordiceras COSSMAN, 1907 [*Ammonites wallachi* Gray, 1832; OD]. Small to medium sized; evolute, with subquadratic whorl section, coarse sparse flexuous ribs with incipient bullae at point of bifurcation; in later growth with many long and short ribs that become stronger and more distant on adult body chamber and may develop feeble umbilical bullae; shallow constrictions follow ribs, strongest on inner whorls; suture with narrow trifid L as deep as E. *U. Tithon.* **Fig. 478C-D** - *B. wallachi* (Gray).

= *Blanfordia* Uhlig, 1905 *non* Adams, 1863 (Gastropoda).

= *Blanfordiceras* Spath, 1923 (obj.).

Malbosiceras GRIGOR'EVA, 1938 [*Ammonites malbosi* Pictet, 1867; OD]. Small (**m**) to large (**M**); like *Lopeziceras* but larger, more inflated, with broad low venter and subrectangular whorl section; bituberculate primary ribs appear early in growth, separated by 2-3 long intercalatories; in later growth up to 5 long intercalatories separate main ribs; last part of adult body chamber with ventrolateral clavi; suture with deep narrow E, broad-

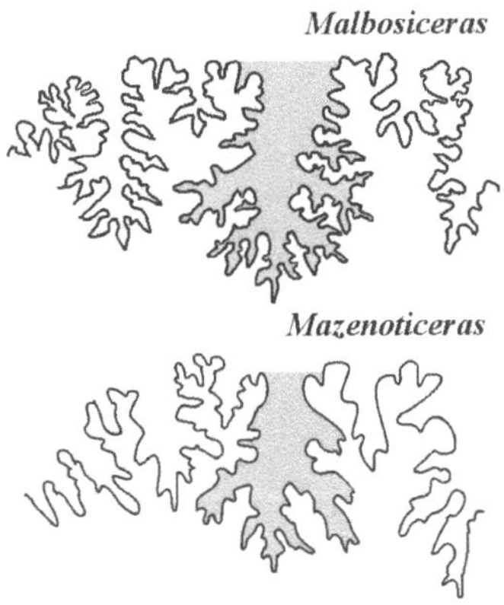

Fig. 480. Mazenoticeratine sutures.

based asymmetrically bifid E/L, and splayed trifid L deeper than E. *L. Berrias.- Lower U. Berrias.* **Fig. 476E** - *M. asper* (Mazenot) [**m**]; **Fig. 476F-G** - *M. malbosi* (Pictet) (**M**).

Mazenoticeras NIKOLOV, 1966 [*Berriasella broussi* Mazenot, 1939; OD] [**M**]. Large; like *Lopeziceras* but more involute and compressed with oval whorl section; inner whorls with fewer stronger umbilical bullae indistinctly connected to finer prorsiradiate secondaries; in maturity bituberculate ribs separated across venter by 3-4 fine secondaries; siphonal groove of early growth passes into smooth band which is lost in maturity. *Upper U. Tithon.- Lower U. Berrias.* **Fig. 476A-B** - *M. broussi* (Mazenot).

Pseudacanthodiscus BARABOSHKIN, 1997 [*P. crymicus*; OD]. Moderately large; like *Mazenoticeras* but inner whorls constricted, with simple and coarse biplicate ribs arising from blunt umbilical bullae; ribs branch on upper third of flank from large blunt lateral bullae from which finer secondary ribs project strongly forwards terminating in low ventrolateral bullae on either side of smooth siphonal band; in later growth high lateral tubercles obsolete, simple and branched ribs alternating. *Lower L. Valang.* **Fig. 482C-D** - *P. crymicus* Baraboshkin.

Pomeliceras GRIGOR'EVA, 1938 [*Ammonites breveti* Pomel, 1889; OD] [**M**]. Large; evolute, with coarse sparse bituberculate ribs from early growth; paired secondaries restricted to outer flank and venter, separated by 3-5 intercalatories; suture with E almost as deep as L, broad trifid E/L and L/U_2, narrow trifid L, and retracted suspensive lobe. *Berrias.* **Fig. 476J-K** - *P. breveti* (Pomel).

Boehmiceras GRIGOR'EVA, 1938 [*Blanfordiceras (B.) caucasicum*; OD] [**M**]. Moderately large to large; like *Pomeliceras* but with broader venter, depressed sub-coronate whorl section, more numerous primary ribs, exaggerated lateral tubercles and coarser secondary ribs. *Berrias.* **Fig. 476H-I** - *B. caucasicum* (Grigor'eva).

Neocosmoceras BLANCHET, 1922 [*Hoplites sayni* Simionescu, 1899; SD Roman, 1938] [**m**]. Small; evolute, with subrectangular whorl section, venter that changes from slightly concave to slightly convex during growth, and coarse straight distant ribs interrupted across venter; 3 rows of strong tubercles, umbilicals and high laterals conical to bullate and ventrolaterals clavate forming recurved spines; aperture with lateral lappets. *Berrias.* **Fig. 482G-H, N** - *N. sayni* (Simionescu).

Euthymiceras GRIGOR'EVA, 1938 [*Ammonites euthymi* Pictet, 1867; OD] [**m**]. Small; like *Neocosmoceras* but with less regular ribbing, frequently bifurcating or alternating long and short, and ventrolateral clavi that do not form recurved spines. *Berrias.* **Fig. 482E-F** - *E. euthymi* (Pictet).

Renngarteniceras GRIGOREVA, 1938 [*Protacanthodiscus (R.) renngarteni*; OD] [**m**]. Small; like *Euthymiceras* but with fine sharp flank costae looped to ventrolateral clavi; umbilical and lateral tubercles lacking. *Berrias.* **Fig. 482A-B** - *R. renngarteni* (Grigor'eva).

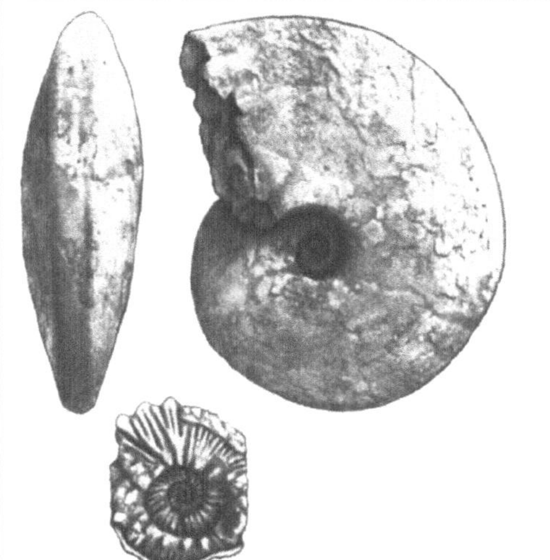

Fig. 481. *Banikoceras.*

Banikoceras HOWARTH, 1992 [*B. involutum*; OD]. Medium sized; inner whorls like *Renngarteniceras*, with ribs looping to ventrolateral tubercles; outer whorls very involute, high-whorled, smooth, with convergent flanks and deeply undercut umbilical wall. *U. Berrias.* **Fig 481** - *B. involutum* Howarth.

Delanoyiceras **gen. nov.** [*Neocosmoceras flabelliforme* Le Hégarat, 1973; OD herein] [**m**]. Small, moderately evolute, with strongly compressed pentagonal whorl section; fine sharp prorsiradiate flexuous flank costae arise from distinct umbilical bullae in fasciculate bundles, many with feeble lateral bullae that move up the flank and become coarser and more distant with growth; ribs terminate in prominent ventrolateral clavi and interrupted across siphonal line. *U. Berrias.* **Fig. 482I-J** - *A. flabelliforme* (Le Hégarat).

Subfamily Dalmasiceratinae Frau, Bulot, Wimbledon & Ifram, 2016

Discussion. Apparently monophyletic derivatives of Himalayitinae close to *Protacanthodiscus* (Frau *et al.*, 2015) but some species of *Praedalmasiceras* are very similar to *Limaites* (Sarasinellinae). Very small to large; in early growth with depressed whorls, distinct siphonal groove, and prorsiradiate ribs with small tubercles at furcation point, and adoral branch of bifid ribs with pinched ventrolateral tubercles; in middle growth umbilical bullae stronger and flexuous primary ribs branching at mid-flank; on adult body chamber ribs fade at mid-flank or evanesce

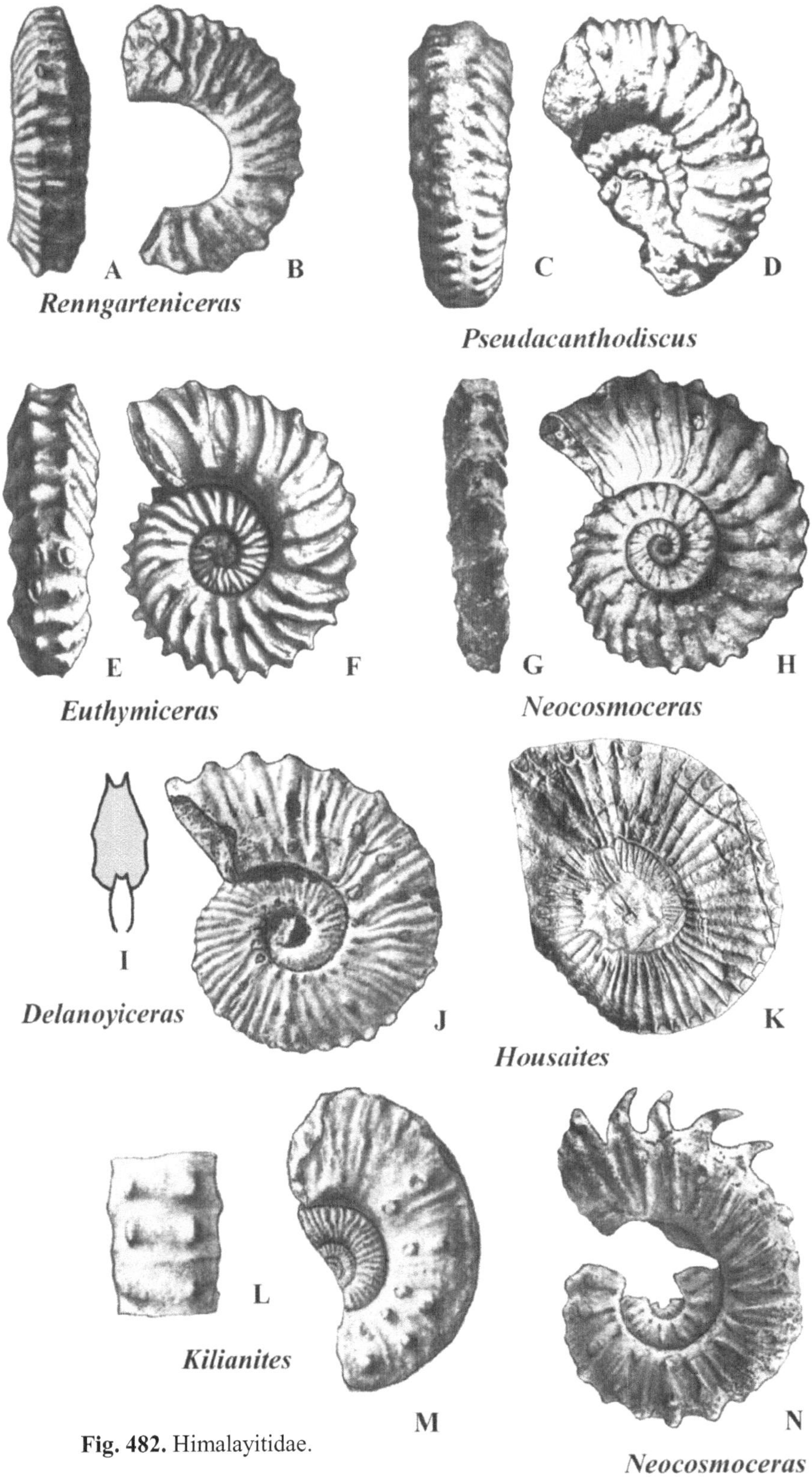

Fig. 482. Himalayitidae.

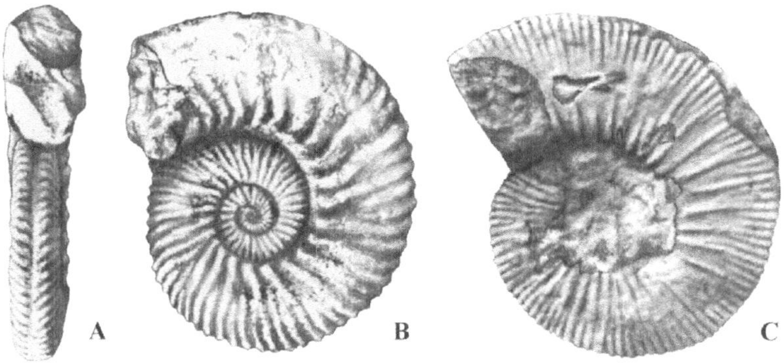

Fig. 483. Dalmaniceratinae.

completely, as does smooth siphonal band. *U. Tithon. - Berrias.*

Praedalmasiceras FRAU, BULOT, WIMBLEDON & IFRAM, 2016 [*Dalmasiceras spiticeratoides* Djanélidze, 1922; OD]. Small to very small, dimorphic; evolute, compressed, with flat flanks, rounded venter and compressed elliptical whorl section; strong rectiradiate primary ribs, single and biplicate, branching above midflank in middle growth from lateral tubercles and terminating in pinched ventrolateral bullae; in later growth secondary ribs in fasciculate bundles; siphonal band of early growth becomes deeply grooved in middle growth, and umbilical bullae strengthen with age; suture with deep asymmetrically trifid L. *L. Berrias.* **Fig. 483C** - *P. spiticeratoides* (Djanélidze).

Some species referred to *Praedalmasiceras*, e.g. *D. aristidis* Mazenot (1949) (**Fig. 562N**), are herein assigned to *Limaites* (Sarasinellinae).

Dalmasiceras DJANÉLIDZÉ, 1922 [*Ammonites dalmasi* Pictet, 1867; SD Roman, 1938]. Small, with compressed elliptical whorl section, siphonal groove that fades in later growth and coarser ribs effaced from midflank leaving strong prorsiradiate ribs on outer flank terminating in weak to distinct ventrolateral bullae. *U. Tithon. - Berrias.* **Fig. 483A-B** - *D. dalmasi* (Pictet); **Fig. 484** - *D. tauricum* Bogdanova & Arkadiev.

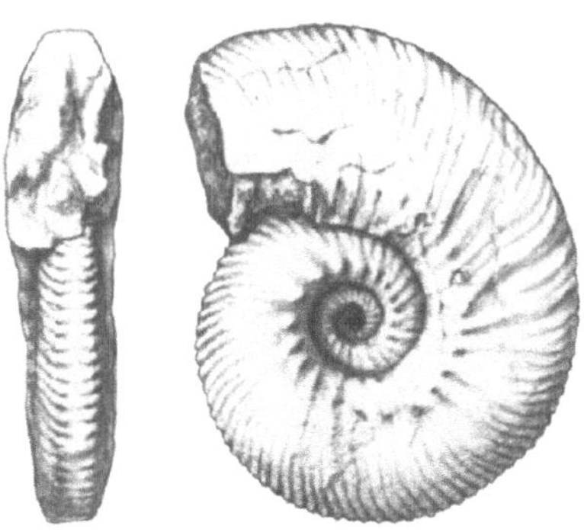

Fig. 484. *Dalmasiceras tauricum.*

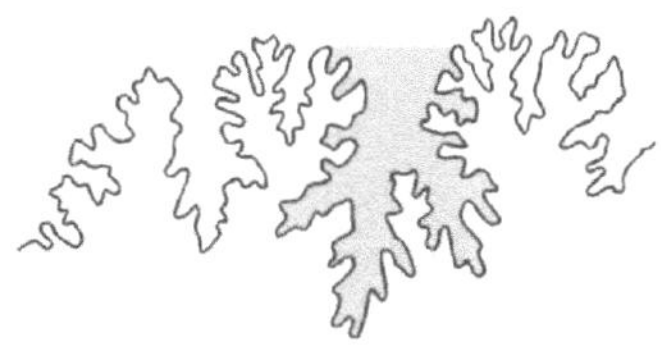

Fig. 485. *Dalmasiceras* suture.

Family RIASANITIDAE Frau, Wimbledon, Ifram, Bulot & Phol, 2020

Discussion. Derivatives of Himalayitinae, close to *Toucasiella* (Mitta, 1988, 2008, 2011). Serpenticone, with subtabulate venter and coarse distant biplicate ribs; microconchs of primitive species with periodic trituberculate ribs. Some homoeomorhs of *Berriasella*. *L. Berrias.*

Mittaites FRAU, WIMBLEDON, IFRAM, BULOT & PHOL, 2020 [*Mazenoticeras ceccai* Mitta, 2011; OD]. Medium sized (**m**) to large (**M**); microconchs with simple ribs in early growth, becoming more distant with rib furcation, rare trifurcation and intercalation in middle growth; in maturity whorl section compressed subhexagonal with biplicate to triplicate ribs, with rounded lateral tubercles and ventrolateral bullae; venter with narrow sulcus across which ribs weaken; macroconchs with simple distant ribs in maturity, thickening over venter. *L. Berrias.* **Fig. 486N-O** - *M. ceccai* (Mitta).

Riasanella MITTA, 2011 [*R. rausingi*; OD][**M**]. Small; evolute, compressed, with flattened convergent flanks, narrow subtabulate venter and trapezoidal whorl section; flexuous single and bifurcating ribs weakening across venter; ribs irregularly form incipient bullae at point of rib furcation; ribs simple and intercalated on last part of

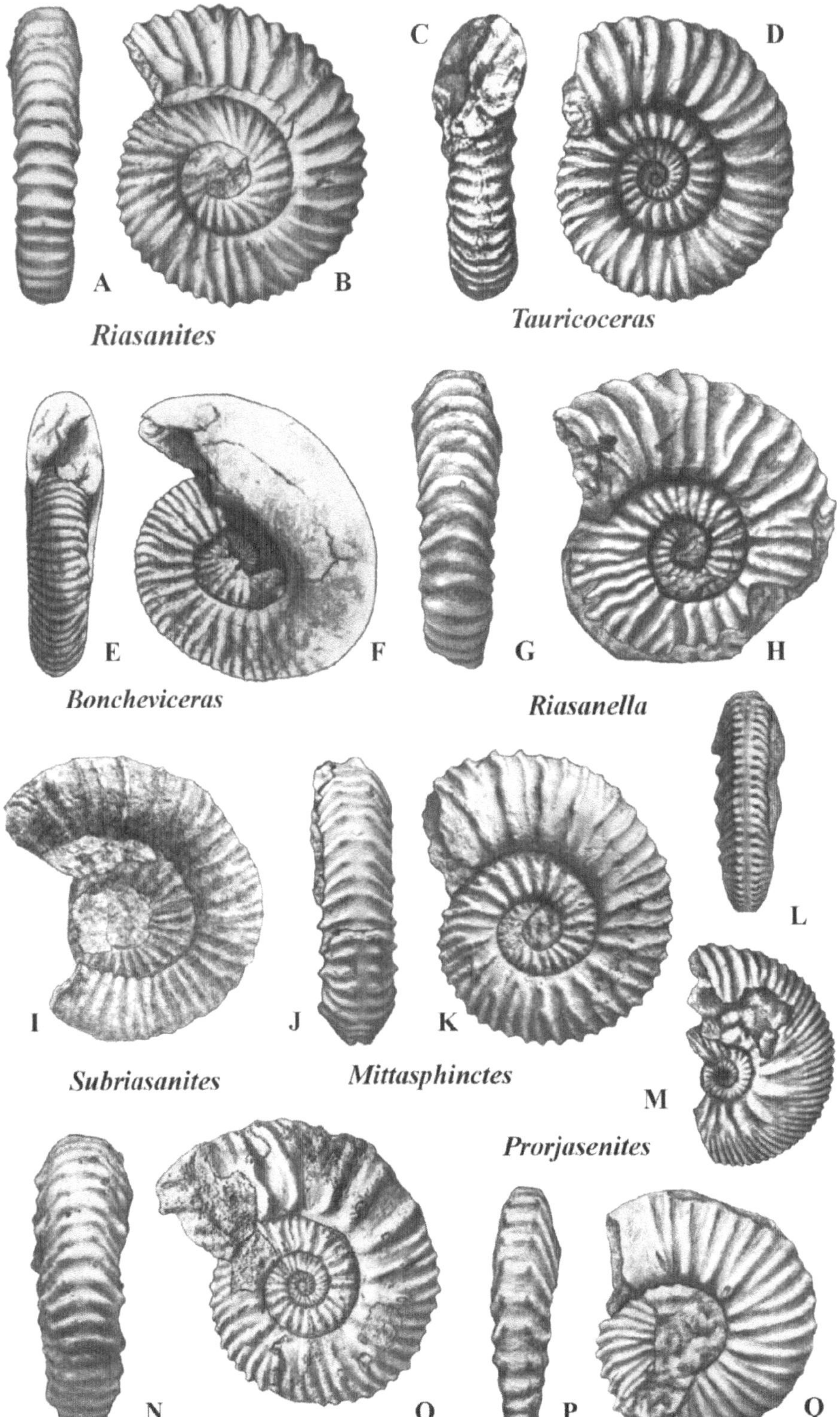

Fig. 486. Riasanitidae.

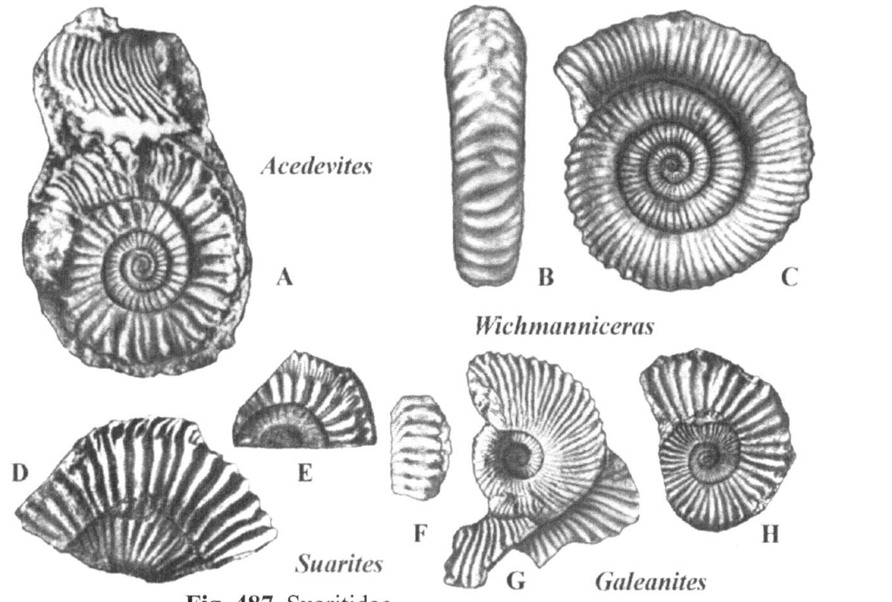

Fig. 487. Suaritidae.

Fig. 488. *Karasyazites.*

adult body chamber; microconch with lateral lappets. *L. Berrias.* **Fig. 486G-H** - *R. rausingi* Mitta.

Riasanites SPATH, 1923 [*Ammonites rjasanensis* Nikitin, 1888; OD; = *rjasensis* Lahusen, 1883, *nom. nud.*]. Medium sized; very evolute in later growth (U>50%), with subtabulate venter and coarse sparse rigid prorsiradiate biplicate ribs that weaken over siphonal line in early to middle growth, but may cross venter uninterrupted in maturity; suture with narrow E, broad bifid E/L and moderately narrow aymmetrically trifid L as deep as E. *L. Berrias.* **Fig. 486A-B** - *R. rjasanensis* (Lahusen).

Some primitive species (*R. swistowianus*) with well developed lateral tubercles in early growth (Mitta, 2011).

Subriasanites SASANOVA, 1991 [*R. rjasensis* var. *maikopensis* Grigor'eva, 1938; OD][**M**]. Like *Riasanites* but with rectiradiate to slightly rursiradiate ribs, single and concave forward on last part of adult body chamber. *L. Berrias.* **Fig. 486I** - *S. maikopensis* (Grigor'eva).

Gechiceras SAKHAROV, 1982 [*G. kistense*; OD]. Small, evolute, strongly compressed, with broad gently convex flanks converging to narrow tabulate venter, with compressed subtrapezoidal whorl section; strong, bold, slightly prorsiradiate straight ribs arise singly or in pairs from feeble umbilical bullae, connecting to ventrolateral clavi and crossing venter transversely; on adult body chamber ribs single, more distant, alternating long and short. *U. Berrias.* **Fig. 486P-Q** - *G. kistense* Sakharov.

Boncheviceras NIKOLOV, 1966 [*Berriasella ardescensis* Mazenot, 1939; OD]. Inner whorls like *Subriasanites* but with narrower umbilicus; coarse robust rigid prorsiradiate ribs arise singly or in pairs from umbilical shoulder and cross venter transversely; ornament declines in later growth and adult body chamber smooth. *L. Berrias.* **Fig. 486E-F** - *B. ardescense* (Mazenot).

Tauricoceras KVANTALIANI & LYSENKO, 1979 [*T. crassicostatum*; OD] [**M**]. Like *Riasanites* but with coarser ribbing that becomes long and short in later growth, sometimes bifurcating from umbilical edge. *L. Berrias.* **Fig. 486C-D** - *T. crassicostatum* Kvantaliani & Lysenko.

Prorjasenites SASONOVA, 1977 [*P. plumatus*; OD]. Very small; convergent flanks, narrow subtabulate venter and compressed subtrapezoidal whorl section; enlarged umbilicolateral bullae give rise to bundles fine, straight prorsiradiate secondaries separated by 2-3 intercalatories; all ribs terminate in ventrolateral bullae leaving smooth

Fig. 489. *Riasanella* suture.

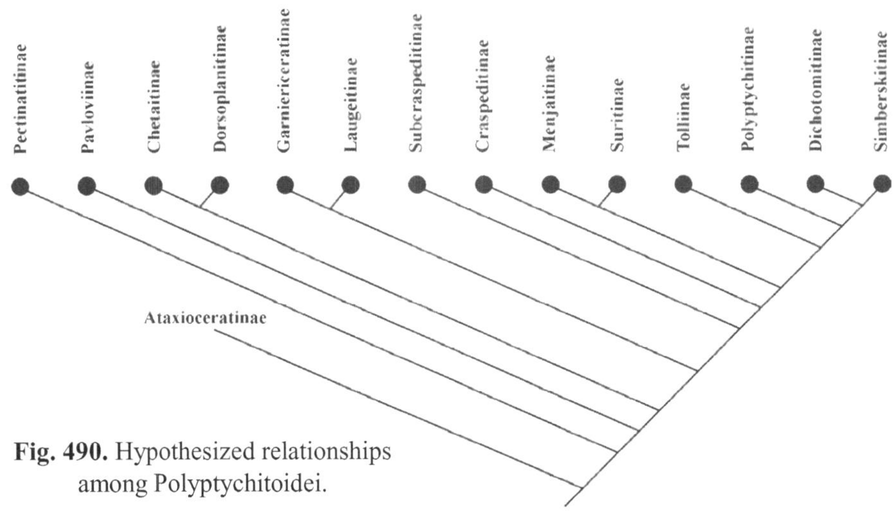

Fig. 490. Hypothesized relationships among Polyptychitoidei.

siphonal band. *L. Berrias.* **Fig. 486L-M** - *P. plumatus* Sasonova.

Regarded a juvenile *Riasanites* (Mitta, 2008).

Karasyazites MITTA, 2018 [*Subalpinites bajarunasi* Luppov, 1988; OD]. Small (**m**) to medium sized (**M**); microconch very evolute, compressed, with flattened flanks, rounded venter and compressed elliptical whorl section; bold, simple or biplicate, strongly prorsiradiate ribs, as wide as interspaces on phragmocone becoming well spaced on adult body chamber; all ribs cross venter transversely, weakening or interrupted by shallow siphonal groove, forming weak ventrolateral bullae; macroconchs more involute; strongly prorsiradiate primary ribs arising from umbilical bullae, separated on outer flank and venter by 1-2 intercalatories that form weak ventrolateral bullae and are interrupted across siphonal line by smooth band. *L. Berrias*. **Fig. 488** - *K. bajarunasi* (Luppov).

Family Suaritidae Cantú-Chapa, 1998

(*nom. transl.* herein *ex* Suaritinae)

Discussion. A family of uncertain ancestry, one species of *Galeanites* having been introduced as a *Corongoceras* (Imlay, 1942) but, more recently. an origin close to *Paralytohoplites* (Himalayitinae) has been suggested (Zeiss & Leanza, 2010). Characteristically very evolute, with flattened flanks and venter, and ribs looping to ventrolateral tubercles, with or without lateral tubercles. *U. Tithon.*

Acedevites CANTÚ-CHAPA, 1968 [*A. acevedensis*; OD]. Small; very evolute, compressed, with subtabulate venter and compressed subrectangular whorl section; phragmocone with strong simple, slightly prorsiradiate ribs and occasional intercalatories, with inner and outer ventrolateral nodes; adult body chamber with fine, wiry, concave forward riblets looped to prominent ventrolateral clavi. *U. Tithon.* **Fig. 487A** - *A. acevedensis* Cantú-Chapa.

Suarites CANTÚ-CHAPA, 1967 [*S. bituberculatus*; OD]. Small; discoidal, evolute, flanks and venter flattened to slightly convex, ribs regularly or irregularly joined in pairs to ventrolateral tubercles, either simple or bifurcating from lateral tubercle; ribs interrupted or attenuated by smooth siphonal band; with or without constrictions to inner whorls. *Lower U. Tithon.* **Fig. 487D-F** - *S. bituberculatus* Cantú-Chapa.

Galeanites CANTÚ-CHAPA, 1998 [*G. garcensis*; OD]. Very small; like *Suarites* but with narrower umbilicus, tabulate venter, and rapidly expanding whorls; flexuous prorsiradiate ribs single or looping to ventrolateral bullae; constrictions and lateral tubercles may be present. *Upper U. Tithon.* **Fig. 487G-H** - *G. garcensis* Cantú-Chapa.

Hyporder POLYPTYCHITOIDEI Spath, 1931

Discussion. A monophyletic taxon descended from Ataxioceratoidei. Comprises the superfamilies Pavlovioidea and Polyptychitoidea. *L.Kimmer. - Lower Barrem.*

Superfamily PAVLOVIOIDEA Spath, 1931

(*nom. transl.* herein *ex* Pavloviidae)

Discussion. Paraphyletic derivatives of Ataxioceratoidea; comprises the families Pavloviidae and Dorsoplanitidae, both of which are Boreal to Suboreal. Members often highly provincialized due to low sea levels and the isolation of sedimentary basins across the Jurassic-Cretaceous boundary. *L. Kimmer.* - *Lower L. Barrem.*

Family PAVLOVIIDAE Spath, 1931

Discussion. A paraphyletic taxon with an origin in Ataxioceratidae (Énay *et al.*, 2014); comprises the Kimmeridgian Pectinatitinae and predominantly Tithonian Pavloviinae. *L. Kimmer.* - *U. Tithon.*

Subfamily Pectinatitinae Zeiss, 1968

(*nom. transl.* Kutek & Zeiss, 1974 *ex* Pectinatitini)

Discussion. A monophyletic taxon with an origin in Ataxioceratinae, close to *Progeronia* (Énay *et al.*, 2014); typically evolute with polygyrate ribbing; except primitively, microconchs with oblique peristomal rostrum. Comprises the tribes Subdicotomoceratini and Pectinatitini. *L. - U. Kimmer.*

Tribe Subdichotomoceratini nov.

Type genus: *Subdichotomoceras* Spath, 1925.

Diagnosis. Very evolute, compressed, with broad low venter and depressed subquadratic whorl section; strong well spaced uniform biplicate ribs uninterrupted over venter; periodic deep collared constrictions; microconch with lateral lappets. *L. - U. Kimmer.*

Virgatosphinctoides NEAVERSON, 1925 [*V. wheatleyensis*; OD] [**M**]. Medium sized to large; very evolute, with rounded venter and oval whorl section; inner whorls unconstricted, with biplicate ribs; on adult body chamber ribs mostly triplicate, occasionally biplicate or simple, with branching above midflank; adult body chamber constricted. *Lower U. Kimmer.* **Fig. 496H** - *V. wheatleyensis* Neaverson.

Allovirgatites NEAVERSON, 1925 [*A. woodwardi*; OD]. Large; like *Virgatosphinctoides* but umbilicus widening with growth; whorl section rounded subquadratic, with shallow depression along siphonal line in middle growth; rectiradiate ribs mainly triplicate, occasionally biplicate, coarsening in later growth; horn development supressed; suture with retracted suspensive lobe. *Lower U. Kimmer.* **Fig. 493E-F** - *A. woodwardi* Neaverson.

Subdichotomoceras SPATH, 1925 [*S. lamplughi*; OD] [**m**]. Medium sized; very evolute, compressed, with broad low venter and depressed subquadratic whorl section; strong well spaced uniform biplicate ribs uninterrupted across venter; periodic deep varices; macroconch aperture trumpet shaped. *L. - M. Kimmer., ?U. Kimmer.* **Fig. 493A-B** - *S. lamplughi* Spath.

Sphinctoceras NEAVERSON, 1925 [*S. crassum*; OD] [**M**]. Large; inflated, evolute, with depressed whorls and deep umbilicus with vertical wall; coarse massive biplicate ribbing from early growth, in some becoming distant in later growth; frequent constrictions parallel ribs; suture with long narrow lobes, trifid L as deep as E, and retracted suspensive lobe. *Lower U. Kimmer.* **Fig. 493K-L** - *S. crassum* Neaverson.

Perhaps a macroconch of *Subdichotomoceras*.

Hoelderia OHMERT & ZEISS, 1980 [*H. schreineri*; OD] [**m**]. Like *Subdichotomoceras* but smaller, with denser costation and ribs branching lower on flank. *Upper U. Kimmer.* **Fig. 496K** - *H. schreineri* Ohmert & Zeiss.

Episphinctoceras NEAVERSON, 1925 [*E. inflatum*; OD] [**M**]. Moderately large; like *Pavlovia* but becoming rapidly inflated with growth, constrictions oblique to ribbing and with simpler suture. *Upper U. Kimmer.* **Fig. 496F** - *E. inflatum* Neaverson.

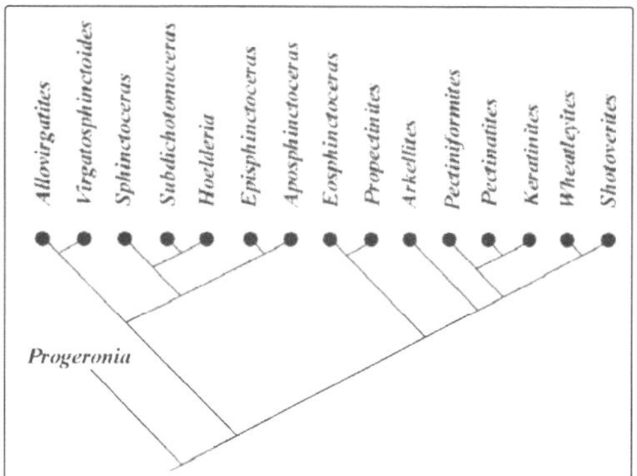

Fig. 491. Hypothesized relationships among Pectinatitinae.

Tribe Pectinatitini Kutek & Zeiss, 1974

Type genus: *Pectinatites* Buckmann, 1922.

Diagnosis. Generally finely ribbed with horn to adult peristome. Descendents of Subdichotomoceratini, close to *Subdichotomoceras*. *U. Kimmer.*

Propectinatites COPE, 1968 [*P. websteri*; OD] [**M**]. Medium sized to moderately large; like *Subdichotomoceras* but with narrower umbilicus and denser, finer, more prorsiradiate ribbing that is slightly reflexed at mid-flank; macroconch peristome simple, microconchs with lateral lappets and ribs almost rectiradiate in later growth. *Lower U. Kimmer.* **Fig. 493D** - *P. websteri* Cope.

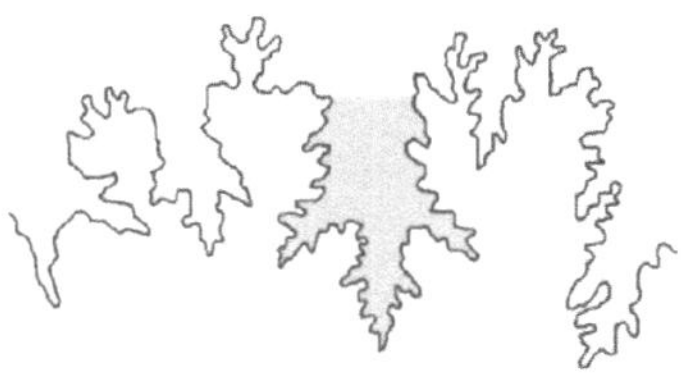

Fig. 492. *Episphinctoceras* suture.

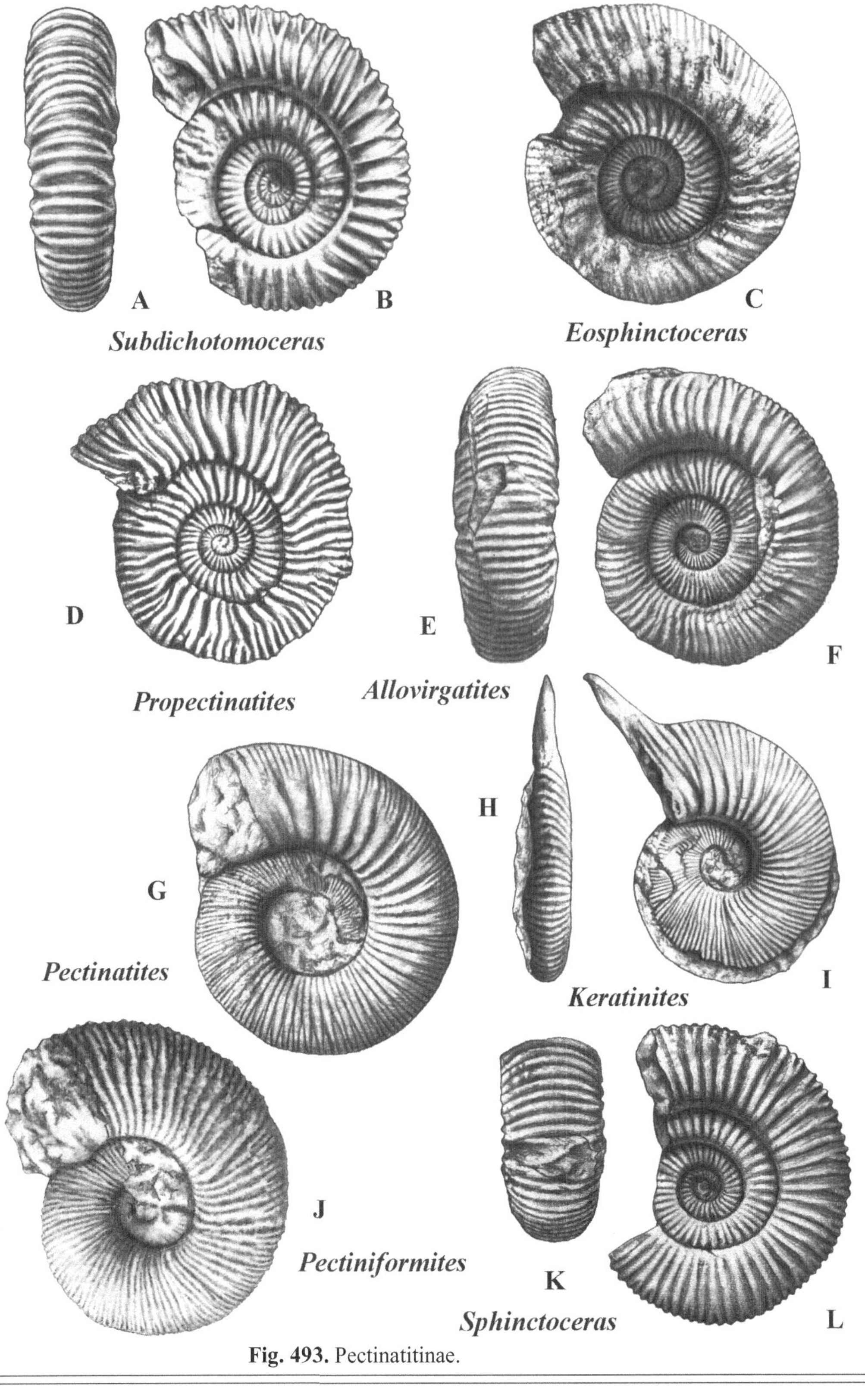

Fig. 493. Pectinatitinae.

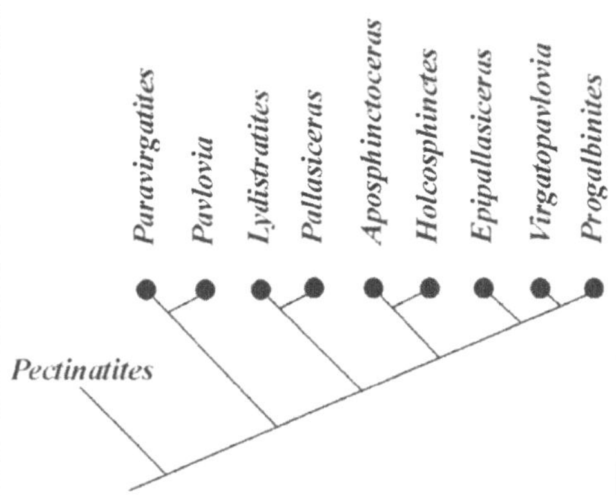

Fig. 494. Hypothesized relationships among Pavloviini.

Eosphinctoceras MEZESHNIKOV, 1969 [*E. magnum*; OD]. Medium sized; very evolute, with rounded subquadratic whorl section; numerous prorsiradiate primary ribs give rise to biplicate, polyplocoidal and virgatome bundles. *Lower U. Kimmer.* **Fig. 493C** - *E. magnum* Mesezhnikov.

Arkellites COPE, 1967 [*Pectinatites (A.) hudlestoni*; OD]. Medium sized (**m**) to large (**M**); like *Subdichotomoceras* but with narrower umbilicus and denser finer ribbing that remains uniform to maturity; apertural horn of microconch sometimes merely an incipient swelling. *Lower U. Kimmer.* **Fig. 496G** - *A. hudlestoni* (Cope).

Pectinatites BUCKMAN, 1922 [*Ammonites pectinatus* Phillips, 1871; OD] [**M**]. Medium sized; like *Arkellites* but more finely ribbed, with branching at all levels on flank, as well as some intercalatories; ribs cross venter with forward convexity, sometimes with impersistent siphonal groove; microconch with ventral peristomal inflation. *Middle U. Kimmer.* **Fig. 493G** - *P. pectinatus* (Buckman).

Keratinites BUCKMAN, 1925 [*K. keratophorus*; OD] [**m**]. Small; like *Pectinatites* but with long rostral horn; macroconchs relatively coarsely ribbed in later growth. *Middle U. Kimmer.* **Fig. 493H-I** - *K. nasutus* Buckman.

Fig. 495. *Pavlovia*.

Pectiniformites BUCKMAN, 1925 [*P. bivius*; OD]. Small; like *Pectinatites* but venter runcinate to 60 mm diameter; strengthened primary ribs lacking in later growth, and ribs remaining biplicate on adult body chamber; periodic shallow constrictions. *U. Kimmer.* **Fig. 493J** - *P. bivius* Buckman.

Wheatleyites BUCKMAN, 1923 [*W. tricostulatus*; OD] [**M**]. Very large; inner whorls finely ribbed like *Pectinatites*; outer whorls with coarse blunt distant primary ribs, secondary ribs gradually declining on adult body chamber; suture with broad stemmed L not markedly trifid. *Middle U. Kimmer.* **Fig. 496A** - *W. tricostulatus* Buckman.

Shotoverites BUCKMAN, 1925 [*S. pringlei*; OD] [**M**]. Large; like *Wheatleyites* but with more densely ribbed early whorls, triplicate ribs in later growth and suture with shorter lobes and narrow stemmed trifid L. *U. Kimmer.* **Fig. 496D-E** - *S. pringlei* Buckman.

Considered by its author a synonym of *Wheatleyites*.

Subfamily Pavloviinae Spath, 1931

Discussion. A monophyletic group with an origin in Pectinatitinae, close to *Pectinatites* (Cope, 1968); typically very evolute, with rounded venter and mostly biplicate ribs with occasional constrictions preceded by a compound rib and succeeded by a single rib; microconchs without terminal rostrum; suture with trifid L; often gigantic (upto 1.37 m diameter) and including the "*Portlandian giants*". Restricted to the Subboreal Province and flourishing in the Anglo-Portlandian basin. The large number of genera is due to taxonomic splitting by Buckland, and many are synonyms (*cf.* Arkell *et al.*, 1957). Comprise the tribes Pavloviini, Glaucolithitini and Titanitini. *Middle U. Kimmer. - U. Tithon.*

Tribe Pavloviini Spath, 1931

(*nom. transl.* herein *ex* Pavloviinae)

Type genus: *Pavlovia* Ilovaisky, 1917.

Diagnosis. Mostly small to medium sized, occasionally very large (D=375 mm), evolute, with strong biplicate ribs crossing the rounded venter transversely and uninterrupted. Jaws Praestriaptychi. *Middle U. Kimmer. - L. Tithon.*

Pavlovia ILOVAISKY, 1917 [*P. iatrensis* var. *primaria*; SD Spath, 1931]. Small to medium sized; evolute, inflated, with broadly rounded venter and depressed subquadratic whorl section; slightly prorsiradiate biplicate and single ribs cross venter uninterrupted; suture with deep narrow E, asymmetrically bifid E/L, narrow trifid L shorter than E, and retracted suspensive lobe. *Middle U. Kimmer. - L. Tithon.* **Figs 495, 496L-M** - *P. iatriensis* Ilovaisky.

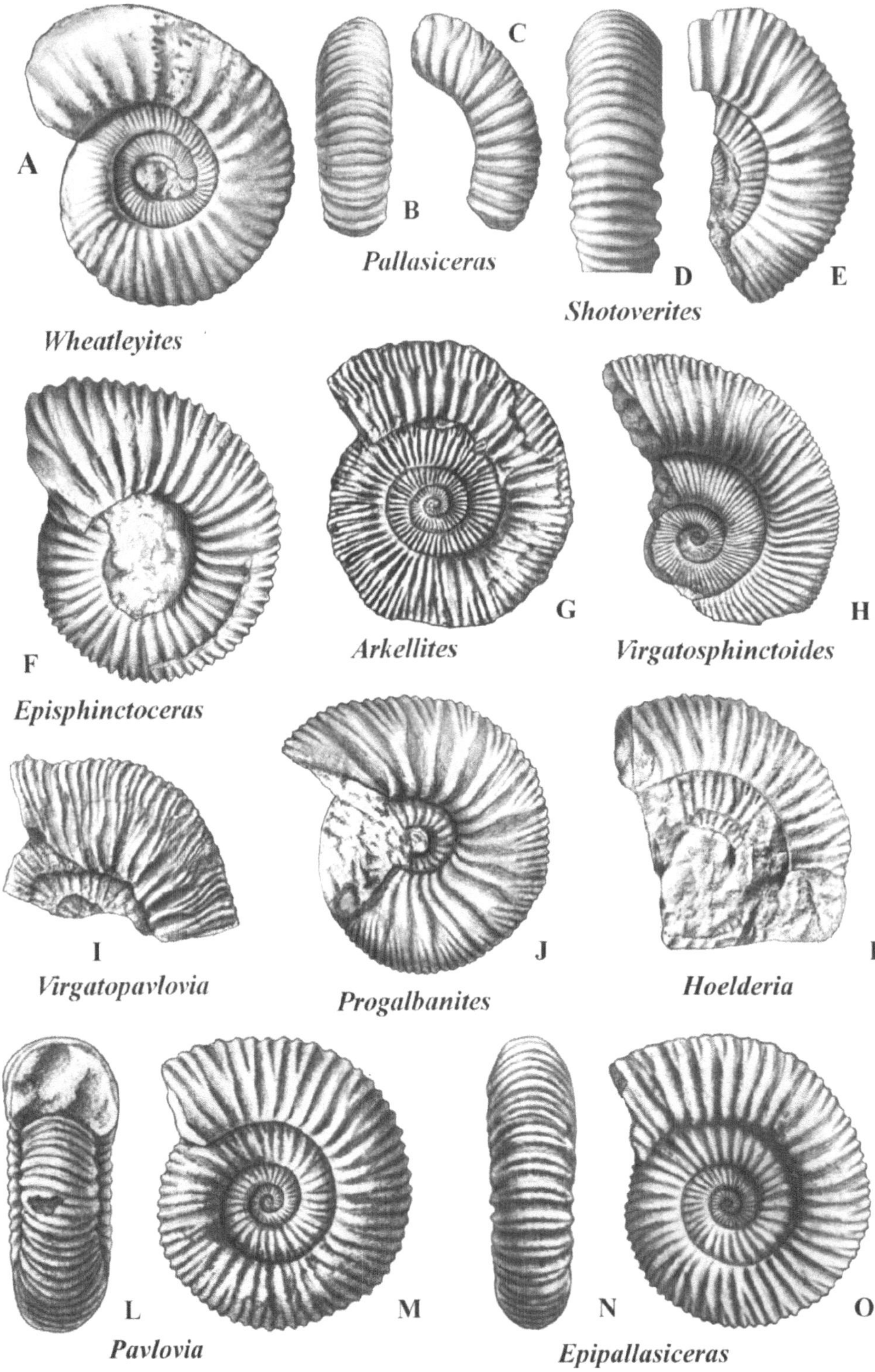

Fig. 496. Pectinatitinae, Pavloviinae.

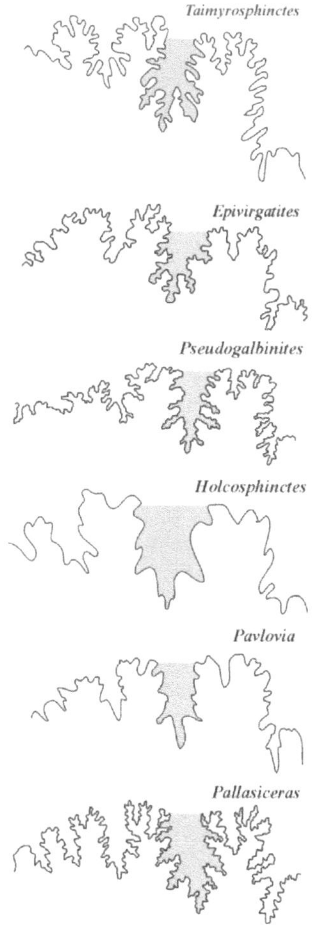

Fig. 497. Pavloviine sutures.

= *Pavlowia* Roman, 1938 (*nom. van.*).
= *Pavlovella* Ilovaisky & Florensky, 1941 (obj.).

Paravirgatites BUCKMANN, 1922 [*P. paravirgatus*; OD]. Moderately large; like *Pavlovia* but with rounded venter and rib branching much lower on flank; ribbing somewhat irregular, varying from simple to triplicate. *Middle U. Kimmer.* **Fig. 498E-F** - *P. paravirgatus* Buckmann.

Pallasiceras SPATH, 1923 [*Ammonites rotundus* J. Sowerby, 1821; OD] [**m**]. Small [**m**] to large [**M**], based on a waterworn body chamber; like *Pavlovia* but very evolute (D=52%) with subtabulate venter, finely ribbed early whorls, and coarse biplicate ribs in later growth with occasional deep constrictions; simpler suture with shorter lobes. *Upper U. Kimmer.* **Fig. 496B-C** - *P. rotundum* (J. Sowerby).

Lydistratites BUCKMAN, 1922 [*L. lyditicus*; OD] [**M**]. Large; inner whorls like *Pavlovia*; in middle growth with frequent trifurcating ribs; in later growth shell very evolute and ribs mostly biplicate; suture with relatively long lobes. *L. Port.* **Fig. 498A-D** - *L. lyditicus* Buckmann.

Holcosphinctes NEAVERSON, 1925 [*H. pallasioides*; OD]. Small, with crowded and delicate ribs on early whorls; later like *Lydistratites* but more coarsely ribbed, with shorter primary ribs, secondaries that commence lower on flank, several polygyrate ribs in later growth, and more lobate suture with E shorter than L. *Upper U. Kimmer.* **Fig. 498H** - *H. pallasioides* Neaverson.

Aposphinctoceras NEAVERSON, 1924 [*A. decipiens*; OD]. Small to medium sized; like *Holcosphinctes* with which it occurs but more inflated, with deeper umbilicus, slightly coarser ribs, periodic single ribs and occasional constrictions preceeded by a polygyrate rib. *Upper U. Kimmer.* **Fig. 498G** - *A. variabile* Neaverson.

Epipallasiceras SPATH, 1936 [*Pavlovia (E.) pseudaperta*; OD] [**m**]. Small to medium sized; like *Pavlovia* but with flatter flanks and secondary ribs grouped into closely spaced pairs; suture simple, with E as deep as L and rather broad saddles. *Upper M. Volg.* **Fig. 496N-O** - *E. pseudapertum* (Spath).

Virgatopavlovia COPE, 1978 [*V. fittoni*; OD]. Inner whorls like *Epipallasiceras* with closely paired biplicate ribs with high furcation point; later with irregular frequently virgatotome ribbing. *L. Tithon.* **Fig. 496I** - *V. fittoni* Cope.

Progalbinites SPATH, 1936 [*P. albani* Arkell, 1935; OD]. Much like *Virgatopavlovia* but with narrower umbilicus and coarser mostly palmate ribs with occasional biplicates. *L. Tithon.* **Fig. 496J** - *P. albani* (Arkell).

Tribe Glaucolithitini nov.

Type genus: *Glaucolithites* Buckman, 1922

Diagnosis. Monophyletic derivatives of Pavloviini. Large to gigantic; evolute, with round whorls, sharp biplicate ribs of uniform density, irregularly branching in maturity, interspersed with frequent simple ribs and constrictions. *L. Tithon. - L. Berrias.*

Crendonites BUCKMANN, 1922 [*C. leptolobatus*; OD] [**m**]. Small to medium sized; like *Pavlovia* but larger serpenticone with rounded whorls, regular biplicate ribs, collared aperture and simpler suture. *L.- M. Port.* **Fig. 498K** - *C. leptolobatus* Buckmann.

Neopavlovia CASEY & MESEZHNIKOV, 1986 [*Crendonites (N.) felix*; OD] [**m**]. Small, evolute (U=35-40%), laterally crushed, with coarse slightly prorsiradiate ribs, mostly biplicate, occasionally single and triplicate, branching at midflank; constrictions lacking. *Lower Berrias.* **Fig. 500N** - *N. felix* (Casey & Mesezhnikov).

Perhaps a microconch of *Crendonites*.

Glaucolithites BUCKMANN, 1922 [*G. glaucolithus*; OD] [**M**]. Very large; very evolute with broadly rounded venter and depressed rounded to subquadratic whorl section; dense sharp prorsiradiate biplicate ribs, about as wide as interspaces, until adult body chamber when simple ribs appear, branching above midflank and crossing venter transversely; frequent conspicuous constrictions at all growth stages. *L. Port.* **Fig. 500A-B** - *G. glaucolithus* Buckmann.

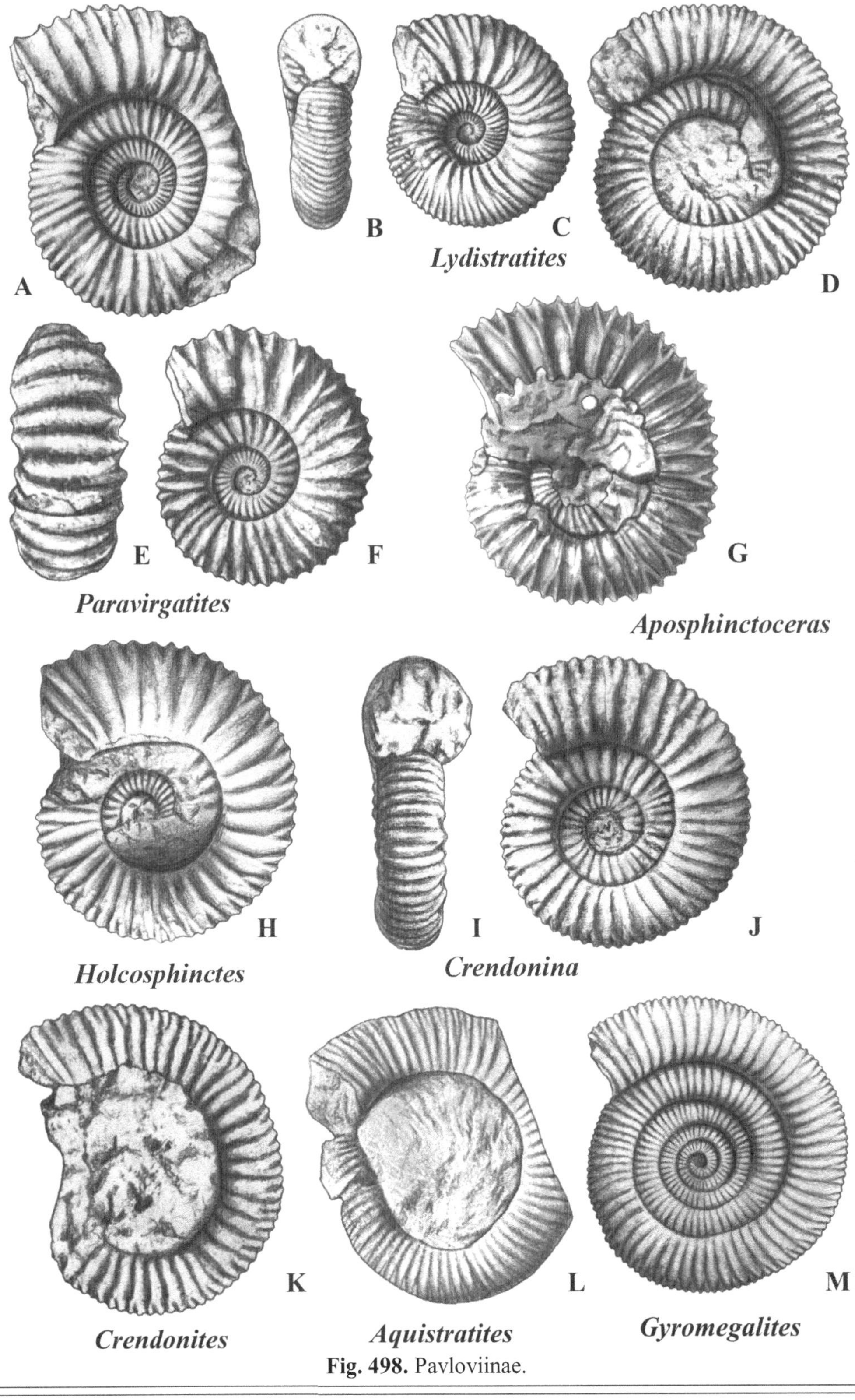

Fig. 498. Pavloviinae.

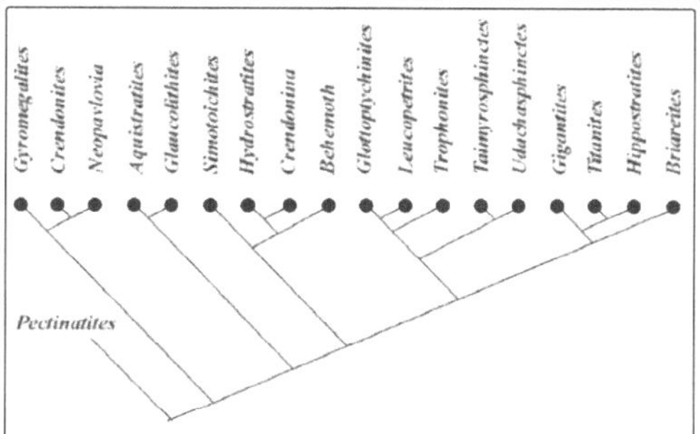

Fig. 499. Hypothesized relationships among Glaucolithitini and Titanitini.

Gyromegalites BUCKMANN, 1925 [*G. polygyralis*; OD]. Like *Glaucolithites* but serpenticone, with fine ribs bifurcating close to venter. *L. Port.* **Fig. 498M** - *G. polygyralis* Buckmann.

Aquistratites BUCKMANN, 1924 [*A. aguator*; OD]. Large; like *Glaucolithites* but serpenticone, with finer ribs branching lower on flank. *L. Port.* **Fig. 498L** - *A. aguator* Buckmann.

Crendonina BUCKMANN, 1925 [*C. subrotundata*; OD]. Like *Glaucolithites* but smaller, more coarsely ribbed and with rounded whorl section. *L. Port.* **Fig. 498I-J** - *C. subrotundata* Buckmann.

Hydrostratites BUCKMANN, 1926 [*H. bifurcus*; OD]. Very large; like *Crendonina* but ribs branching lower on flanks; constrictions lacking. *L. Port.* **Fig. 500H** - *H. bifurcus* Buckmann.

Behemoth BUCKMANN, 1922 [*B. megasthenes*; OD] [**M**]. Very large to enormous; like *Glaucolithites* but with compressed oval whorl section, blunt rectiradiate primary ribs, secondaries that project forwards, and numerous single ribs in later growth. *L. Port.* **Fig. 503G-H** - *B. megasthenes* Buckmann.

Simotoichites BUCKMANN, 1923 [*S. simus*; OD]. Small to medium sized; like *Pavlovia* but flat sided, with subtabulate venter, rectangular whorl section and very long primary ribs that branch near ventrolateral shoulder. *U. Tithon.* **Fig. 500F-G** - *S. simus* Buckmann.

"Fittonia" WIMBLEDON, 1974 (*nom. nud.*) - Unpublished thesis name. **Fig. 504B.**

Tribe Titanitini nov.

Type genus: *Titanites* Buckman, 1921

Diagnosis. Monophyletic derivatives of Glaucolithitini; medium sized to enormous; prorsiradiate biplicate ribs branching above midflank and frequently convex forward across venter; simple and intercalated ribs frequent in maturity; constrictions absent or rare. *Tithon.*

Taimyrosphinctes MESEZHNIKOV, 1972 [*T. excentricus*; OD] [**M**]. Very large, evolute, with broad somewhat flattened venter and trapezoidal whorl section; coarse slightly prorsiradiate biplicate ribs in early to middle growth (D<70 mm), with some intercalatories later; suture with narrow E much deeper than trifid L. *Upper M. Volg.* **Fig. 506N-O** - *T. excentricus* Mesezhnikov.

Udachasphinctes MESEZHNOKOV, 1972 [*Laugeites? udachensis* Shul'gina, 1963; OD] [**m**]. Small to medium sized; like *Taimyrosphinctes* but with subtabulate venter. *Upper M. Volg.* **Fig. 506I-K** - *U. udachensis* (Shul'gina).

Perhaps the microconch of *Taimyrosphinctes*.

Trophonites BUCKMANN, 1922 [*T. trophon*; OD] [**M**]. Large to very large; like *Taimyrosphinctes* but with oval whorl section as wide as high; relatively coarsely ribbed, ribs mostly biplicate but, in later growth, with triplicates. *L. Port.* **Fig. 503K-L** - *T. trophon* Buckmann.

Leucopetrites BUCKMANN, 1922 [*L. leucas*; OD] [**M**]. Very large; like *Trophonites* but with coarser more distant ribbing that is convex forward over venter. *L. Tithon.* **Fig. 500E** - *L. leucas* Buckmann.

Glottoptychinites BUCKMANN, 1923 [*G. glottodes*; OD] [**M**]. Medium sized to moderately large; coarsely ribbed, mainly biplicate but also single ribs, convex forward over venter. *M. Port.* **Fig. 503O-P** - *G. glottodes* Buckmann.

Titanites BUCKMANN, 1921 [*T. titan*; OD] [**M**]. Enormous to gigantic; very evolute (D=47-52%) with depressed rounded whorl section; prorsiradiate biplicate ribs with intercalatories in early growth and simple ribs later, ribbing becoming increasingly crowded with growth. *M. - U. Port.* **Fig. 503A-B** - *T. titan* Buckmann.

Gigantites BUCKMANN, 1921 [*A. giganteus* J. Sowerby, 1826 OD] [**M**]. Enormous; like *Titanites* but more inflated, with rounded whorl section and strong prorsiradiate ribbing retained to maturity, with frequent single ribs. *M. Port.* **Fig. 503C-D** - *G. giganteus* (J. Sowerby).

Hippostratites BUCKMANN, 1924 [*H. hippocephaliticus*; OD] [**M**]. Very large, with compressed oval whorl section; inner whorls cadicone, almost tuberculate; suture with long rectangular bifid saddles and wide asymmetrically bifid L. *M. Port.* **Fig. 500J-K** - *H. hippocephaliticus* Buckmann.

Briareites BUCKMANN, 1921 [*B. polymeles*; OD] [**M**]. Enormous; like *Titanites* but more inflated, with subelliptical whorl section, narrower umbilicus, and finer distinctly flexuous prorsiradiate ribs, mostly biplicates but with some triplicates, simplifying in later growth. *L.-M. Port.* **Fig. 503E-F** - *B. polymeles* Buckmann.

Polymegalites BUCKMANN, 1925 [*P. polypreon*; OD] [**M**]. Enormous; like *Titanites* but with convergent flanks, narrowly rounded venter, oval whorl section and finer ribbing that is concave forward in later growth. *L. - M. Port.* **Fig. 500C-D** - *P. polypreon* Buckmann.

"Ophiolithites" WIMBLEDON, 1974 (*nom. nud.*). Unpublished thesis name. *Uppermost Port.* **Fig. 504C.**

"Portlandia" WIMBLEDON, 1974 (*nom. nud.*). Unpublished thesis name. *Uppermost Port.* **Fig. 504D.**

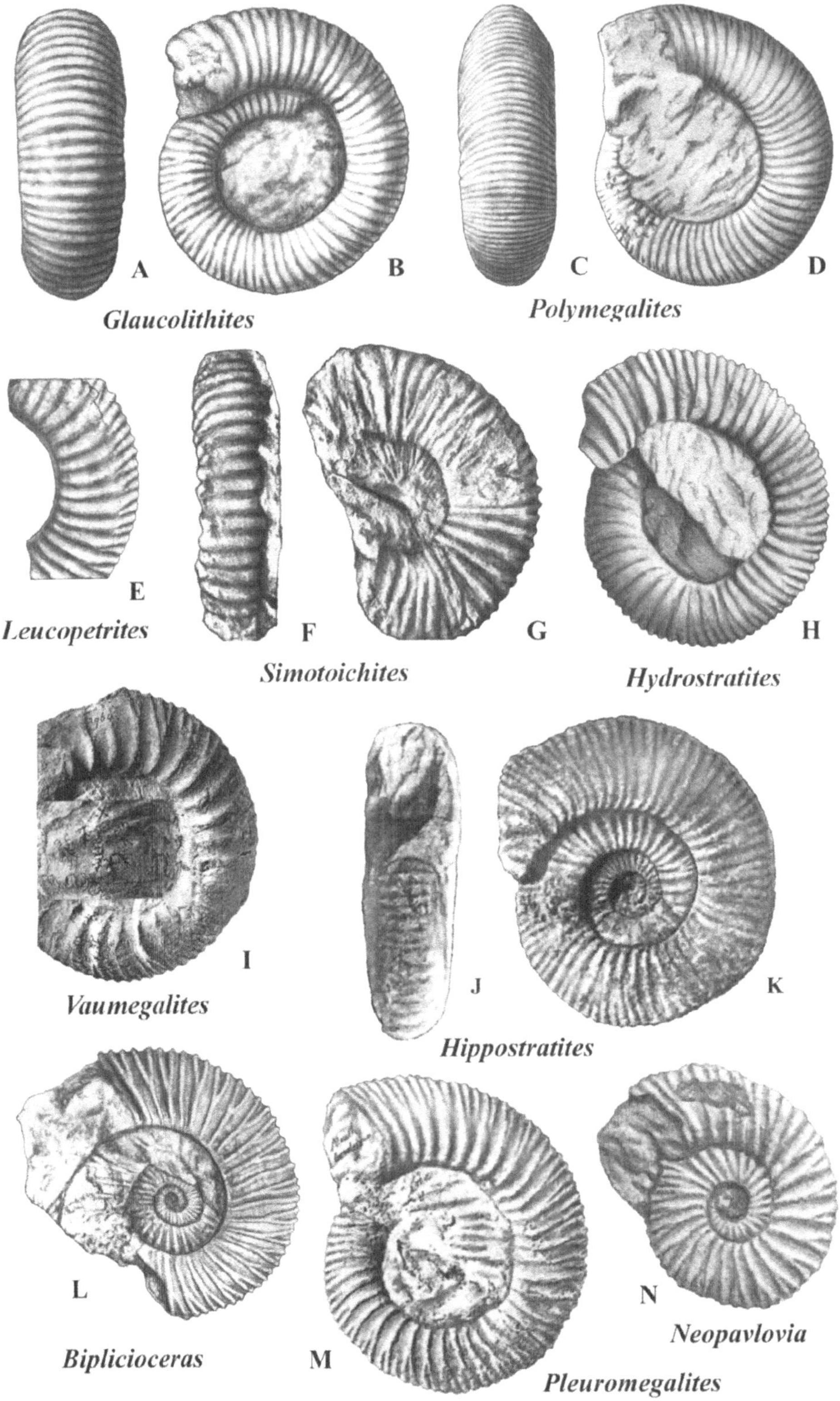

Fig. 500. Pavloviinae.

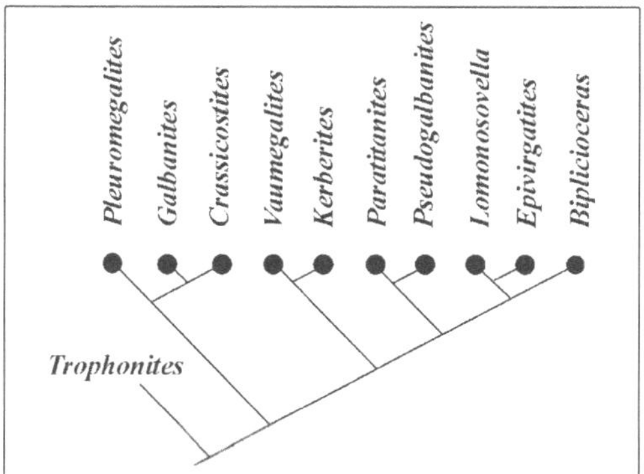

Fig. 501. Hypothesized relationships among Galbanitini.

Tribe Galbanitini nov.

Type genus: *Galbanites* Buckman, 1922

Diagnosis. Large to enormous; relatively coarsely ribbed with triplicates, often crossing venter with forward arch. This taxon could include microconchs of some of the other groups. Monophyletic derivatives of Titanitini. *M. Port.*

Vaumegalites BUCKMANN, 1924 [*V. vau*; OD]. Large, very evolute; coarsely ribbed with well spaced rectiradiate subbullate polygyrate ribs. *L. Port.* **Fig. 500I** - *V. vau* Buckmann.

Pleuromegalites BUCKMANN, 1924 [*P. forticosta*; OD] [**M**]. Very large; like *Galbanites* but with well separated biplicate ribs concave forwards on flanks. *L.-M. Port.* **Fig. 500M** - *P. forticosta* Buckmann.

Galbanites BUCKMANN, 1922 [*G. galbanus*; OD] [**M**]. Moderately large (D_{max}=350 mm); like *Trophonites* but more compressed with ribs branching lower on flanks; suture with trifid L as deep as E. *M. Port.* **Fig. 503I-J** - *G galbanus* Buckmann.

"*Crassicostites*" WIMBLEDON, 1974 (*nom. nud.*) [**M**]. Unpublished thesis name. *L.-M. Port.* **Fig. 504A**.

Kerberites BUCKMANN, 1924 [*K. kerberus*; OD] [**m**]. Very large with depressed ovate whorl section; in early growth densicostate but later with coarse triplicate ribbing; suture with broad bifid E/L and narrow trifid L. *L.-M. Port.* **Fig. 503M-N** - *K. kerberus* Buckmann.

Pseudogalbanites KISELEV, 2018 [*Titanites (P.) triceps*; OD]. [**M**]. Enormous, with equidimensional to slightly depressed ovate whorl section; inner whorls like *Galbanites* with coarse, widely spaces triplicate ribs that persist to later growth; adult body chamber with simple and biplicate ribs. *M. Volg.* **Fig. 504E** - *P. triceps* (Kiselev).

Lomonossovella ILOVAISKY, 1937 [*Olcostephanus lomonossovi* Viaschniakoff, 1882; OD] [**m**]. Moderately large; like inner whorls of *Pseudogalbanites* but with narrower venter, compressed oval whorl section, and ribs mainly triplicate with primaries and secondaries not well differentiated. *Middle M. Volg.* **Fig. 506P-Q** - *L. lomonossovi* (Viaschniakoff).

Epivirgatites SPATH, 1923 [*Perisphinctes nikitini* Michalski, 1890; SD Spath, 1924] [**m**]. Large; like *Lomonosovella* but primary ribs markedly prorsiradiate, not flaring in later growth, and with prominent constrictions; suture with broad E/L, L not as deep as E and retracted suspensive lobe. *Upper M. Volg. - Lower U. Volg.* **Fig. 506L-M** - *E. nikitini* (Michalski).

= *Nikitinella* Ilovaisky, 1941 (obj.).

Biplicioceras IVANOV, 1986 [*Perisphinctes bipliciformis* Nikitin, 1881; OD] [**m**]. Small; like *Epivirgatites* but with thin wiry biplicate and occasional single ribs; no constrictions. *Upper M. Volg.* **Fig. 500L** - *B. bipliciforme* (Nikitin).

Paratitanites KISELEV, 2018 [*Titanites (P.) manipulocostatus*: OD] [**M**]. Enormous, inner whorls moderately evolute with trapezodal whorl section, later evolute with rounded section; innermost whorls with biplicate ribs; in middle growth with biplicate and intercalated ribs, becoming single on adult body chamber; constrictions absent. *M. Volg.* **Fig. 504F** - *P. manipulocostatus* (Kiselev).

Family DORSOPLANITIDAE Arkell, 1950

Discussion. A monophyletic Boreal group with an origin in Pavloviidae (Spath, 1936), but the taxonomy is "*unstable*" (Kiselev, 2017); comprises the subfamilies Dorsoplanitinae and Chetaitinae. *U. Kimmer. - M. Berrias.*

Subfamily Dorsoplanitinae Arkell, 1950

= Polytosphinctinae Schindewolf, 1925

Discussion. Derivatives of Pectinatitinae close to *Sphinctoceras* and *Subdichotomoceras* (Cope, 1968), with Mitta (1994) deriving them polyphyletically from Pectinatitinae. Typically evolute, with

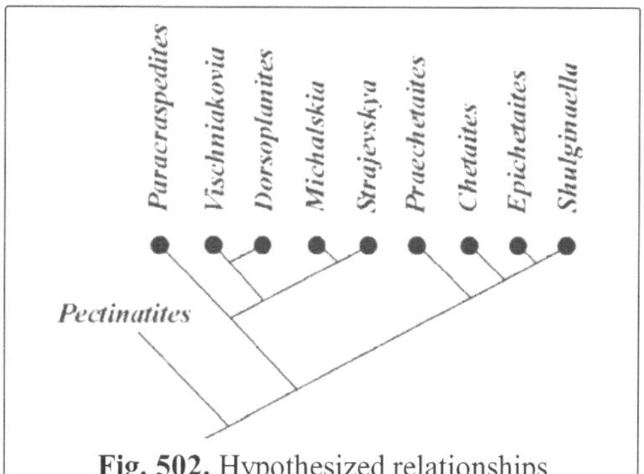

Fig. 502. Hypothesized relationships among Dorsoplanitidae.

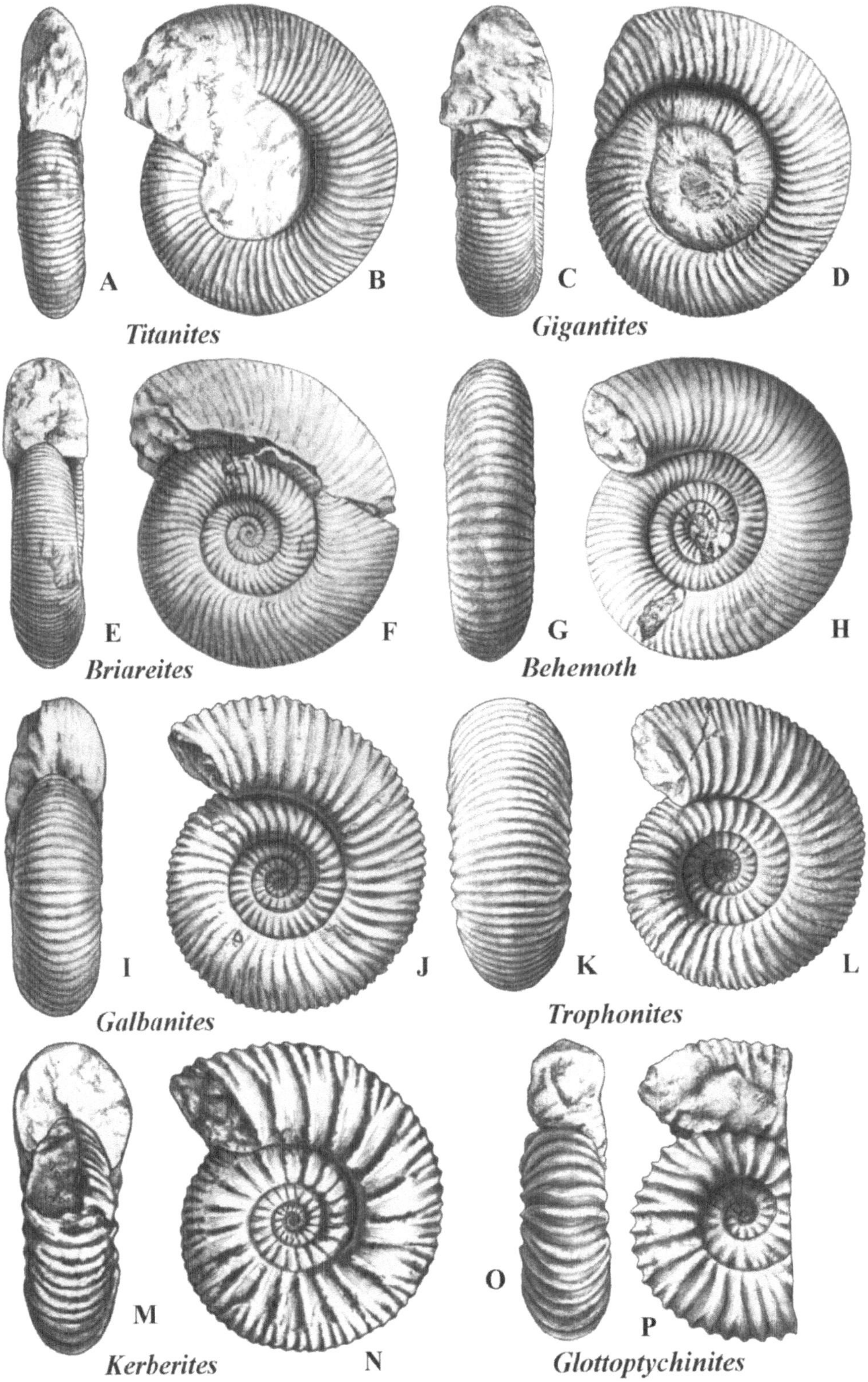

Fig. 503. Dorsoplanitidae.

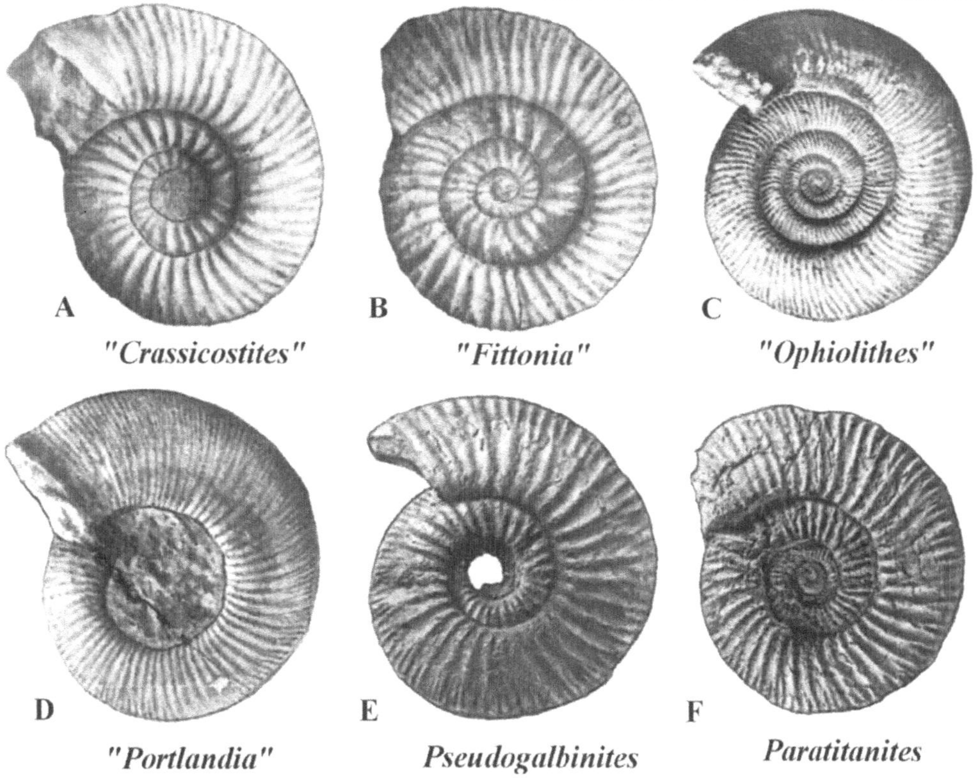

Fig. 504. Pavloviinae.

Fig. 505. *Paracraspedites.*

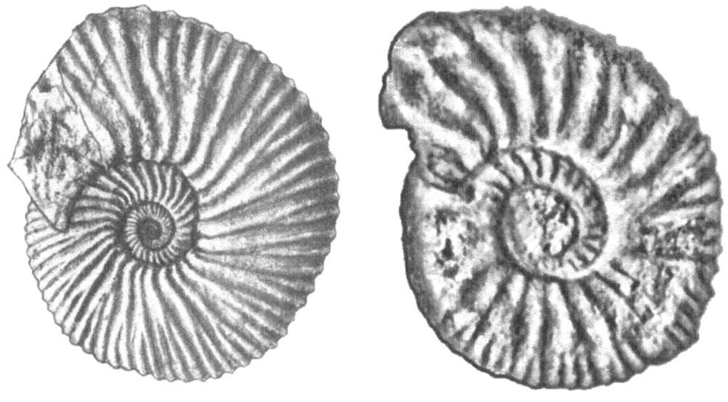

broadly rounded venter and compressed subquadratic to trapezoidal and depressed reniform whorl section. Strong prorsiradiate biplicate and triplicate ribs with intercalatories cross the venter transversely, and may evanesce in maturity. The suture is simple with an asymmetrically trifid L and retracted suspensive lobe. *U. Kimmer. - U. Tithon.*

Dorsoplanites SEMENOV, 1898 [*Ammonites dorsoplanus* Vischniakoff, 1882; SD Roman, 1938] [**M**]. Medium sized; very evolute, inflated; inner whorls flat sided with rounded venter and subquadratic whorl section; later whorls strongly depressed with reniform whorl section; strong slightly prorsiradiate primary ribs split above midflank, giving rise to 2-3 secondaries that cross venter transversely; in maturity venter smooths and primary ribs flare; suture with narrow E, long narrow trifid E/L and narrow trifid L as deep as E. *U. Kimmer. - M. Volg.* **Fig. 506A-B** - *D. dorsoplanus* (Vischniakoff).

= *Polytosphinctes* Schindewolf, 1925 (obj.).

Vischniakovia GERASIMOV, 1978 [*Dorsoplanites (V.) serus*; OD] [**M**]. Large; like *Dorsoplanites* but more comp-

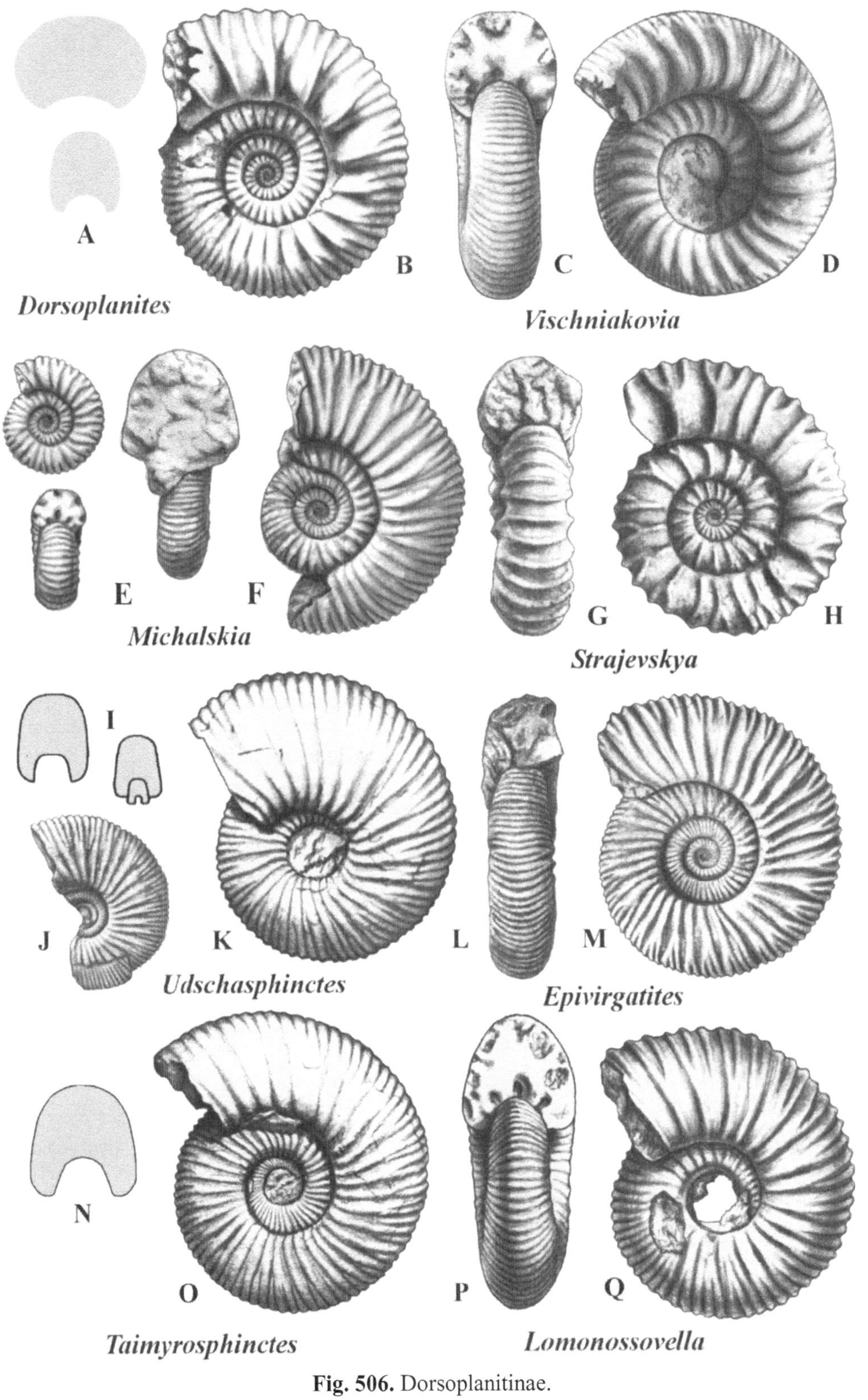

Fig. 506. Dorsoplanitinae.

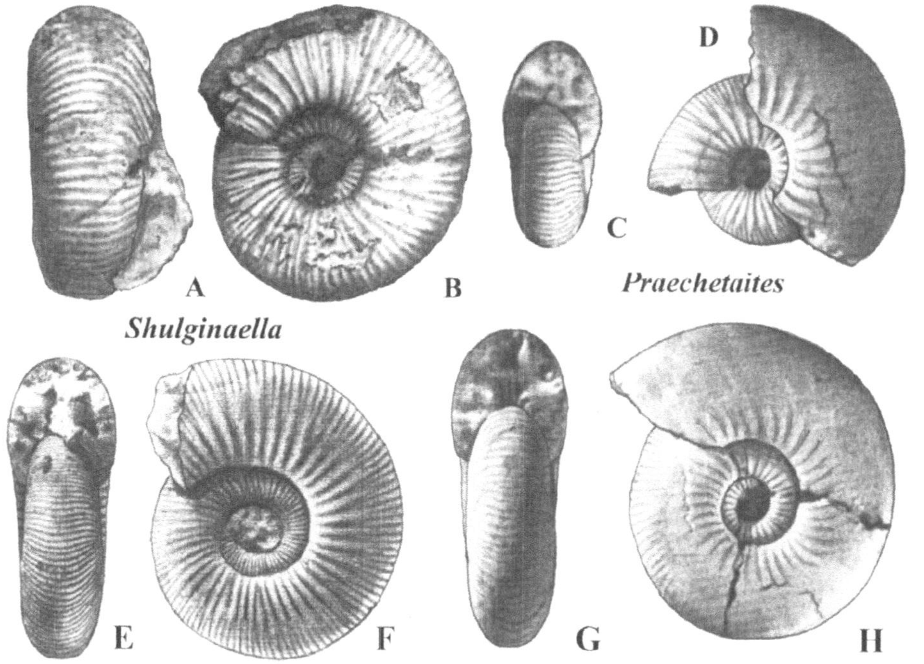

Fig. 507. Dorsoplanitinae.

ressed, with rounded quadratic whorl section, and more numerous concave primaries. *U. Kimmer. - Volg.* **Fig. 506C-D** - *V. sera* (Gerasimov).

Michalskia ILOVAISKY, 1941 [*Perisphinctes miatschkoviensis* Michalski, 1890; OD]. Small; like inner whorls of *Dorsoplanites* but with prominent constrictions. *Lower M. Volg.* **Fig. 506E-F** - *M. miatschkoviensis* (Michalski).

Strajevskya MICHAILOV, 1962 [*Pavlovia strajevski* Ilovaisky, 1917; OD]. Small (**m**) to large (**M**); very evolute, inflated, with depressed subcoronate whorl section, flattened venter in later growth; strong distant widely splayed biplicate and triplicate ribs, occasiona quadrifid, crossing venter transversely. *L. Tithon.* **Fig. 506G-H** - *S. ponomarevi* (Ilovaisky).

Paracraspedites SWINNERTON, 1935 [*P. stenomphaloides*; OD] [**M**]. Enormous; moderately involute, compressed, with flattened flanks, broadly convex venter and rounded rectangular whorl section; only strong rigid rectiradiate ribs exposed in umbilicus; on outer whorl ribs bifurcate at or just above midflank and posterior rib bifurcates again; ribs sharp and strong across venter, sometimes zigzagging; ribbing coarsens and becomes more distant in maturity; suture like *Epivirgatites*. *Lower U. Volg.* **Fig. 505** - *P. stenomphaloides* Swinnerton.

Subfamily Chetaitinae nov.

Diagnosis. Moderately involute, compressed, with numerous umbilicolateral bullae giving rise to 2-3 fine ribs, with intercalatories, crossing venter with slight foward arch; suture with L much shorter than E. Monophyletic derivatives of Dorsoplanitinae (Cope, 1968). *U. Volg. - M. Berrias.*

Praechetaites SASONOVA & SASONOV, 1979 [*Virgatosphinctes exoticus* Shul'gina, 1967; OD]. Dimorphic, small [**m**] to large [**M**]; inner whorls with convergent flanks, narrowly rounded venter, oval whorl section and sharp slightly prorsiradiate ribs in pairs and triplicates separated by intercalatories; in middle growth (D>70 mm) rib branching becomes irregular, virgatome bundles appear and secondary ribs of different length common; constrictions rare; on adult bodty chamber ribs weaken, particularly on flanks, and may evanesce completely (**M**); in maturity ribs cross broadly rounded venter with slight forward inflexion; suture with L considerably shorter than E, and strongly retracted suspensive lobe. *Lower U. Volg.* **Fig. 507C-D** - *P. exoticus* (Shul'gina).

Chetaites SHUL'GINA, 1962 [*C. chetae*; OD]. Moderately large; like *Praechetaites* but more evolute, with flat subparallel flanks, compressed subelliptical whorl section, and secondary ribs persisting to much later stage in growth; ribs cross venter transversely; suture with E much deeper than L. *U. Volg.* **Fig. 507E-H** - *C. chetae* (Shul'gina).

Externiceras SASONOVA, 1971 [*Perisphinctes solowaticus* Bogoslovskii, 1897; OD]. Small; very evolute, with

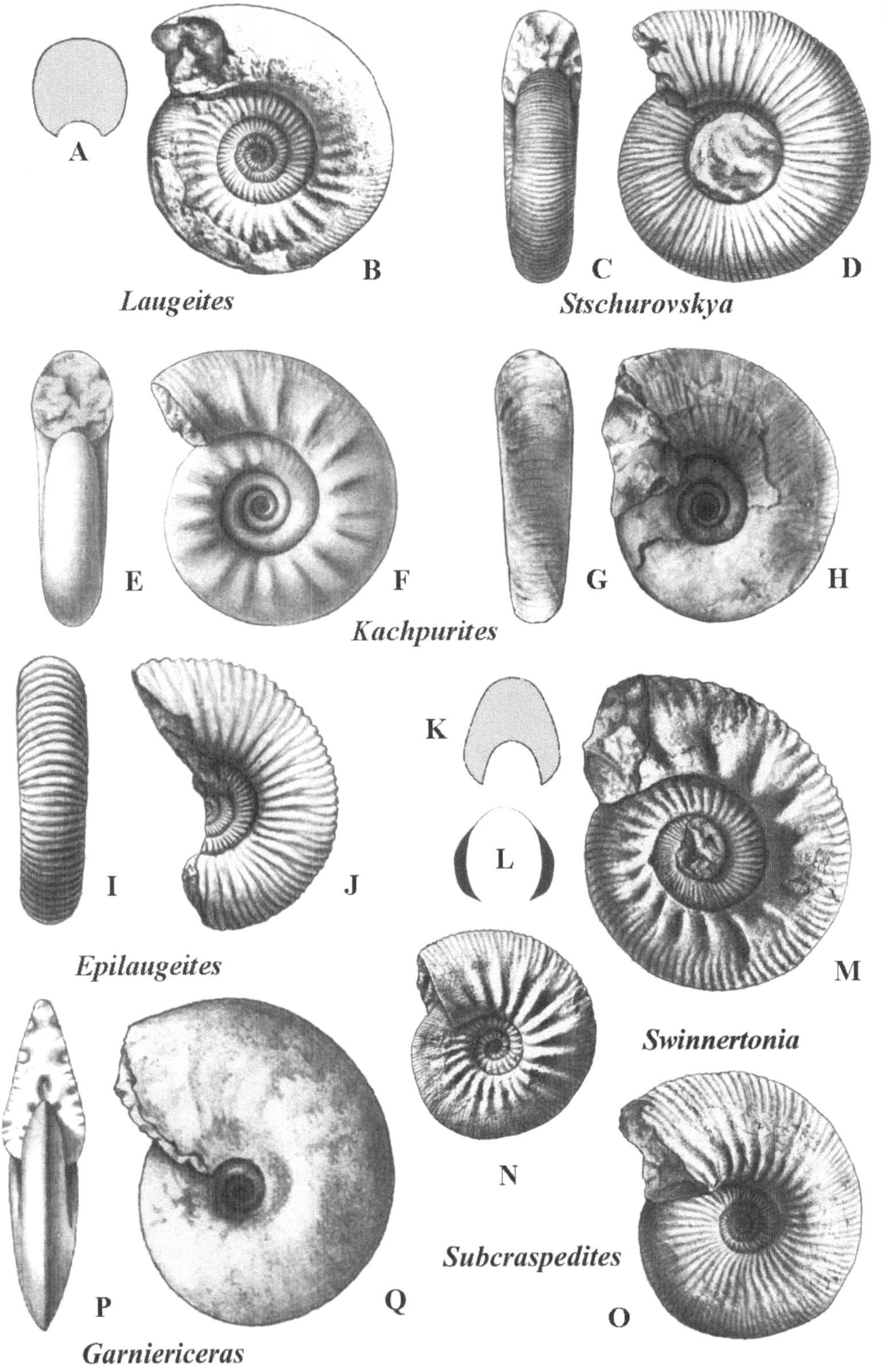

Fig. 508. Subcraspeditinae.

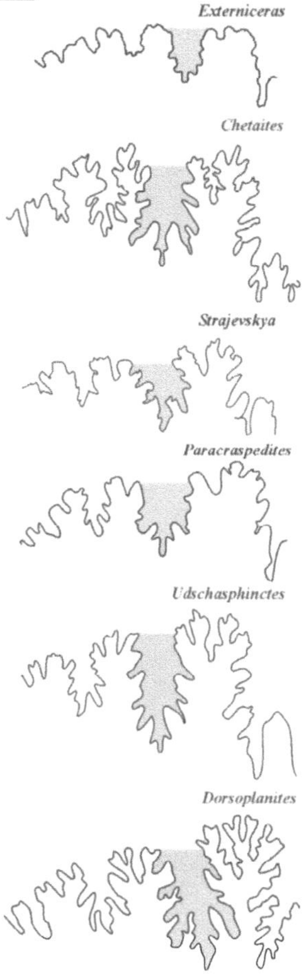

Fig. 509. Dorsoplanitine sutures.

compressed elliptical whorl section and broad well rounded venter; in early growth (D<40 mm) with triplicate and polyplocoid ribs, later with slightly prorsiradiate biplicated ribs that cross venter transversely. *M. Berrias*. **Fig. 511P-Q** - *E. solowaticum* (Bogoslovskii).

***Shulginaella* gen. nov.** [*Olcostephanus? mostjae* Bogoslovskii, 1897; OD herein]. Very small; like *Externiceras* but more inflated with narrower umbilicus, subquadratic whorl section and finer denser prorsiradiate biplicate ribs, closely splayed, that sometimes arise in pairs near umbilicus and bifurcate at midflank, crossing venter transversely. *M. Berrias*. **Fig. 507A-B** - *S. mostjae* (Bogoslovskii).

= *Gerassimovia* Shul'gina, 1985 *(non* Licharew 1956, Brachiopoda).

Family GARNIERICERATIDAE Spath, 1952

Discussion. Monophyletic derivative of Dorsoplanitidae (Spath, 1936); the similarity to *Hectoroceras* is convergent. Primitively evolute finely ribbed ammoniticones giving rise to weakly ornamented oxycones with sharp venter.Typically Boreal. Comprises the subfamilies Laugeitinae and Garniericeratinae. *Upper U. Volg. - L. Berrias*.

Subfamily Laugeitinae Lominadze & Kvantaliani, 1985

Discussion. Monophyletic derivatives of Dorsoplanitinae (Spath, 1936), close to *Dorsoplanites* (Mitta, 1994), although Neale (1961) suggested "*close affinities*" with *Paracraspedites*; typically involute to moderately evolute, with depressed oval to slightly compressed elliptical whorl section with rounded venter; strong rectiradiate to slightly prorsiradiate primary ribs split into finer secondaries on the outer flank, with intercalatories crossing the venter transversely; dimorphic, small (**m**) to large (**M**); suture with narrow trifid L as deep as E. *M. - U. Volg*.

Laugeites SPATH, 1936 [*Kochina groenlandica*; OD]. Medium sized to moderately large, more evolute than *Praechetaites*; compressed, with oval (juv.) to rounded subquadratic (ad.) whorl section; in early growth with dense fine curved slightly prorsiradiate primary ribs splitting into 2-3 finer secondaries high on flank; ornament declines in later growth and adult whorls smooth; suture with 2 auxiliary lobes. *Upper M. Volg*. **Fig. 508A-B** - *L. groenlandica* (Spath).

= *Kochina* Spath, 1936 (*non* Resser, 1935).

= *Prokachpurites* Breistroffer, 1947 (obj.).

Stschurovskya ILOVAISKY, 1941 [*Perisphinctes stschurowaki* Nikitin, 1881; OD]. Large; like *Laugeites* but with rounded venter and compressed oval whorl section. *Upper M. Volg*. **Fig. 508C-D** - *S. stschurowaki* (Nikitin).

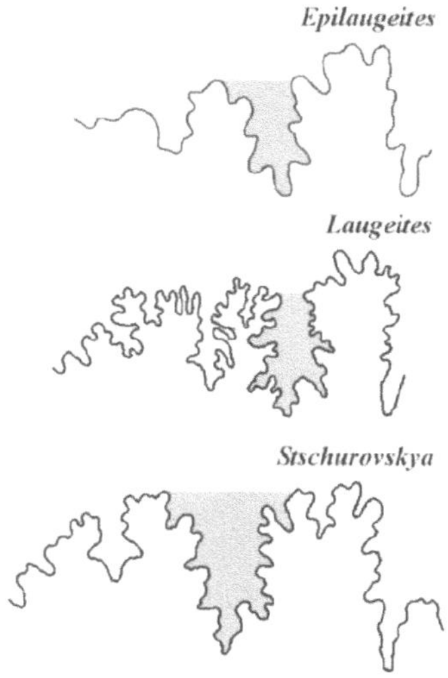

Fig. 510. Laugeitine sutures.

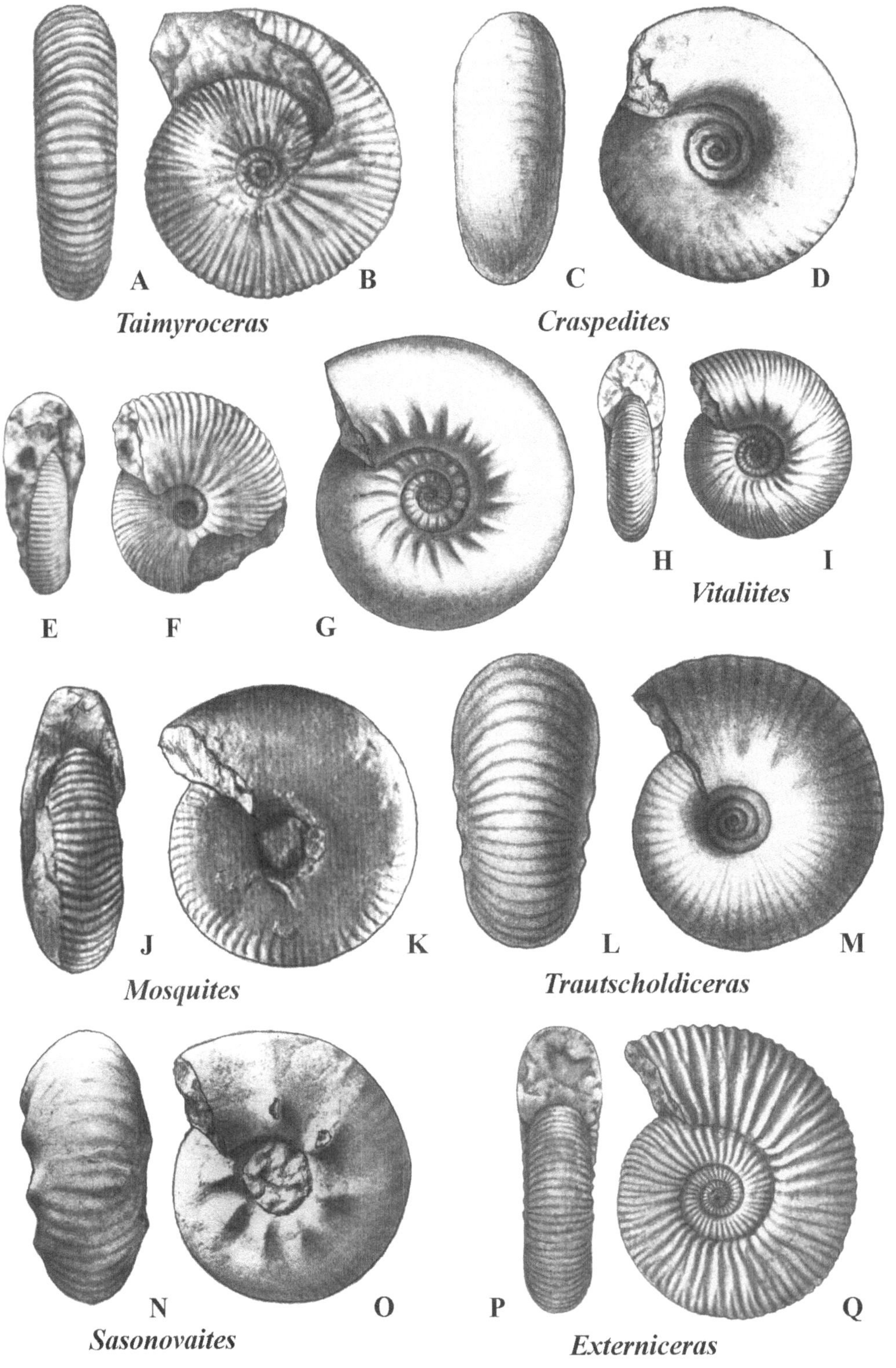

Fig. 511. Craspeditidae.

Epilaugeites MESEZHNIKOV, 1974 [*Pavlovia vogulica* Ilovaisky, 1921; OD]. Like *Laugeites* but retaining strong ribbing to maturity and with simpler suture. *Upper U. Volg*. **Fig. 508I-J** - *E. vogulicus* (Ilovaisky).

Subfamily Garniericeratinae Spath, 1952

Discussion. Monophyletic derivative of Laugeitinae, close to *Laugeites* (Rogov, 2010); the similarity to *Hectoroceras* is convergent. Weakly ornamented oxycones with sharp venter and numerous auxiliary elements to the suture arranged in a curved ascendant series. Typically Boreal. *Lower U. Volg. - L. Berrias*.

Kachpurites SPATH, 1923 [*Ammonites fulgens* Trautschold, 1861; SD Spath, 1924]. Small; like *Laugeites* but with convex flanks converging to narrowly rounded venter and compressed oval whorl section; inner whorls with very fine dense flexuous ribs that thicken ventrally; in middle to later growth strong distant slightly prorsiradiate primary ribs sometimes weakly connected to and separated by 5-8 finer secondaries. *Lowest U. Volg*. **Fig. 508E-F** - *K. f. fulgens* (Trautschold); **Fig. 508G-H** - *K. f. cheremkhensis* Mitta, Michailova & Sumin.
= *Kaschpurites* Luppov *et al*., 1958 (*nom. van*.).

Garniericeras SPATH, 1923 [*Ammonites catenulatus* Fischer von Waldheim, 1837; SD Spath, 1924]. Moderately large; fairly involute, compressed oxycone with sloping umbilical wall, broad convex convergent flanks, sharp to keeled venter and subtrigonal whorl section; in early growth with faint radial primary ribs but these quickly decline and evanesce, leaving unornamented outer whorls; suture relatively simple with curved ascendent suspensive lobe with numerous auxiliary elements. *Upper U. Volg - L. Berrias*. **Fig. 508P-Q** - *G. catenulatum* (Fischer von Waldheim).

Superfamily POLYPTYCHITOIDEA Wedekind, 1918

(*nom. transl.* herein *ex* Polyptychitidae)

Discussion. Monophyletic derivatives of Dorsoplanitoidea, differing from the latter in polyschizostomous ribbing. Comprises the families Craspeditidae, Tolliidae and Polyptychitidae, all of which are predominantly Boreal but with incursions into northern Tethys. Members often highly provincialized. *U. Tithon. - Lower L. Barrem*.

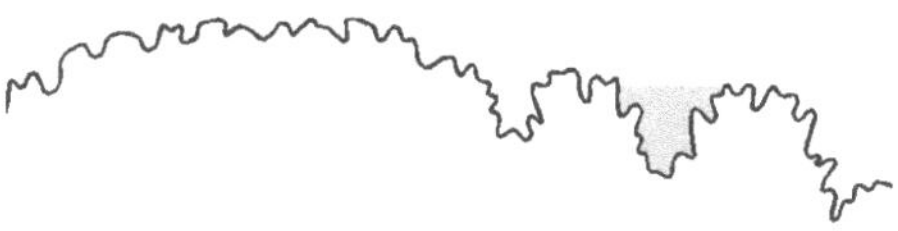

Fig. 512. *Garniericeras* suture.

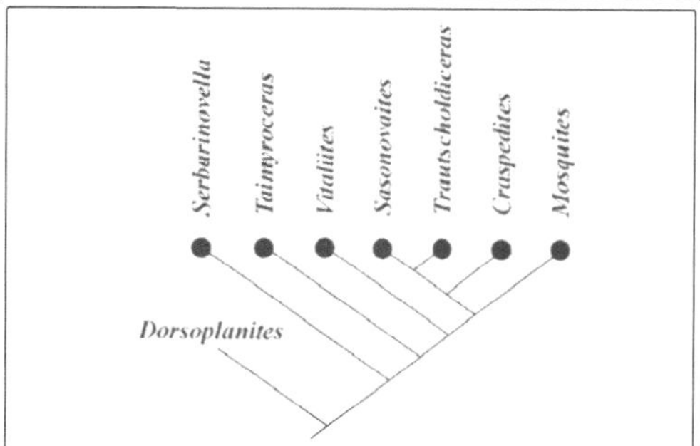

Fig. 513. Hypothesized relationships among Crapeditinae.

Family CRASPEDITIDAE Spath, 1924

Discussion. Monophyletic derivatives of Dorsoplanitidae (Mitta, 1994), although Jelezky (1965) favoured Garniericeratidae, close to *Kachpurites*; a major component of the faunas in the Boreal and Subboreal realms where evolution occurred independently in different basins. Generally small to medium sized, involute and rather inflated with finely ribbed inner whorls; ornament generally declines in later growth and outer whorls often smooth although umbilical bullae may persist; dimorphism conspicuous; relatively simple ascendent suture with 4-6 auxiliary elements to suspensive lobe; jaws *Praestriaptychi*. Comprises the subfamilies Craspeditinae and Menjaitinae. *M. - U. Volg*.

Subfamily Craspeditinae Spath, 1924

(*nom. transl.* Spath, 1931 *ex* Craspeditidae)

Discussion. Monophyletic derivatives of Dorsoplanitinae (Mitta, 1994); generally small to medium sized, involute and rather inflated with finely ribbed inner whorls; ornament generally declines in later

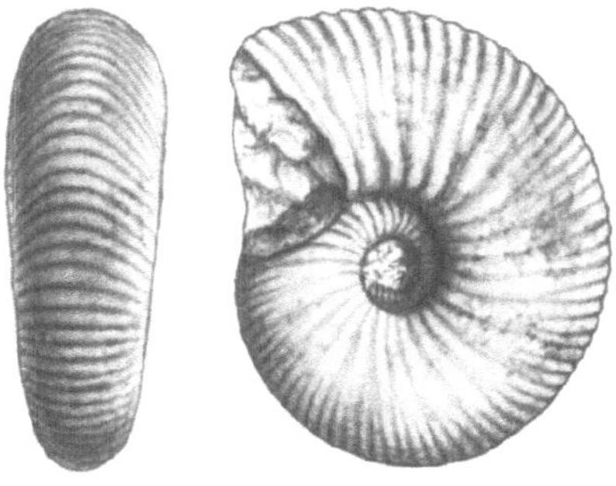

Fig. 514. *Serbarinovella*.

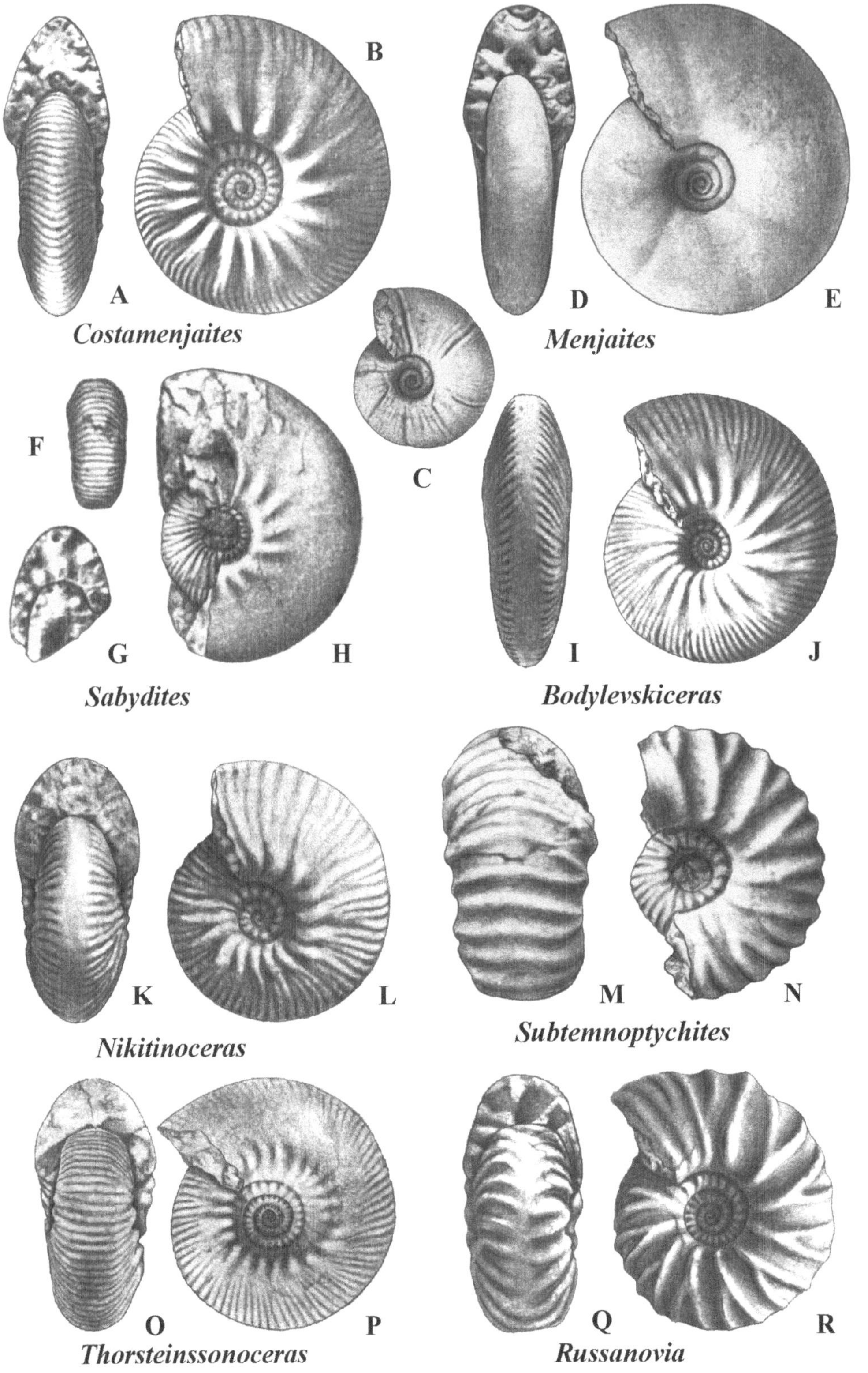

Fig. 515. Menjaitinae.

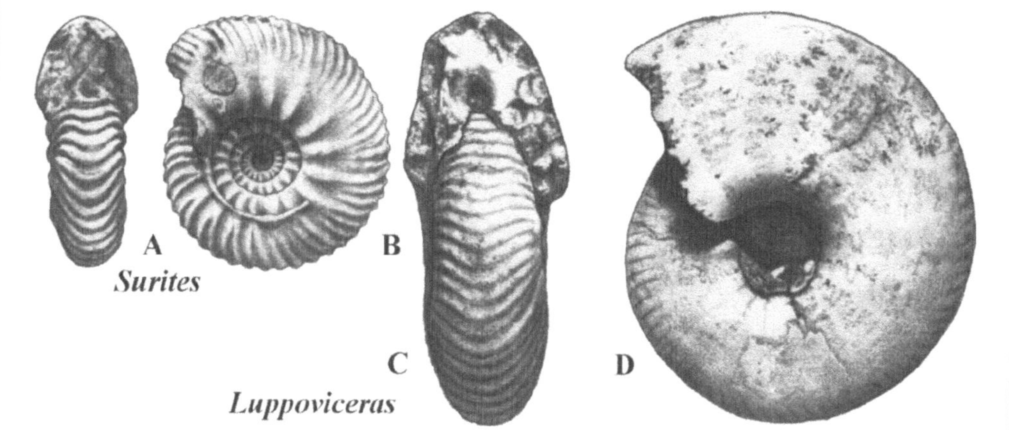

Fig. 516. Menjaitinae, Suritinae.

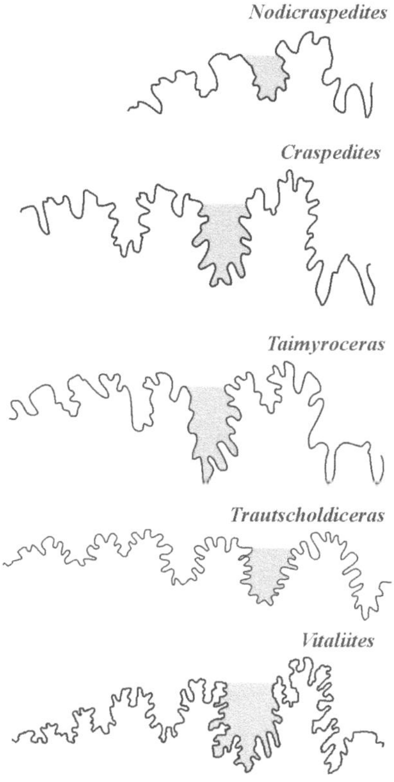

Fig. 517. Craspeditid sutures.

growth and outer whorls often smooth; suture with 4-6 auxiliary elements to suspensive lobe. Typical of the Upper Volgian of the Boreal Realm. *M. Volg. - L. Berrias*.

Serbarinovella MITTA, 1988 [*S. serbarinovi*; OD]. Small; involute, compressed, with rounded venter and oval whorl section; numerous prorsiradiate biplicate and less commonly triplicate primary ribs branch just above midflank, separated by long intercalatory, and crossing venter transversely. *Middle M. Volg*. **Fig. 514** - *S. serbarinovi* Mitta.

Vitaliites SHUL'GINA, 1985 [*Ammonites subditus* Trautschold, 1877; OD]. Small; like *Craspedites* but with convex umbilical wall, numerous umbilicolateral bullae retained to maturity, giving rise to fine prorsiradiate primary ribs that trifurcate above midflank and form weak chevrons over venter; suture with long narrow E/L. *Middle M. - Lower U. Volg*. **Fig. 511E-F** - *V. pseudofragilis* (Gerasimova); **Fig. 511G-I** - *V. subditus* (Trautschold).

Craspedites PAVLOW, 1892 [*Ammonites okensis* d'Orbigny, 1845; SD Douvillé, 1911]. Medium sized; compressed (juv.) to moderately inflated (ad.) and involute, with steep umbilical wall; innermost whorls with prorsiradiate primaries splitting into finer secondaries on outer flanks; ornament weakens early (D~25 mm) and later whorls (D>60 mm) with smooth umbilical wall and inner flank, and ribbing almost effaced from siphonal line. *Lower U. Volg*. **Fig. 511C-D** - *C. okensis* (d'Orbigny).

non *Craspedites* Allman, 1872 (Hydrozoa) - a *nomen oblitum* (Rogov, 2012).

Trautscholdiceras SASONOVA, 1977 [*Ammonites kaschpuricus* Trautschold, 1866; OD]. Small; inflated cadicone with deep narrow umbilicus, broadly rounded venter and depressed reniform whorl section; weak blunt rectiradiate ribs arise from umbilical shoulder and cross venter transversely; distant weak umbilical bullae appear in middle growth; adult body chamber smooth. *U. Volg*. **Fig. 511L-M** - *T. kaschpuricum* (Trautschold).

Sasonovaites **gen. nov.** [*Craspedites parakaschpuricus* Gerassimova, 1969; OD herein]. Small; like *Craspedites* but with subtrigonal whorl section, about 8 strong umbilicolateral bullae feebly connected to 2-3 fine fine secondaries separated by 3-4 intercalatories on outer flank; adult macroconchs with sloping umbilical wall, flat convergent flanks, evenly arched venter and whorls higher than wide; suture like *Craspedites* but with 3

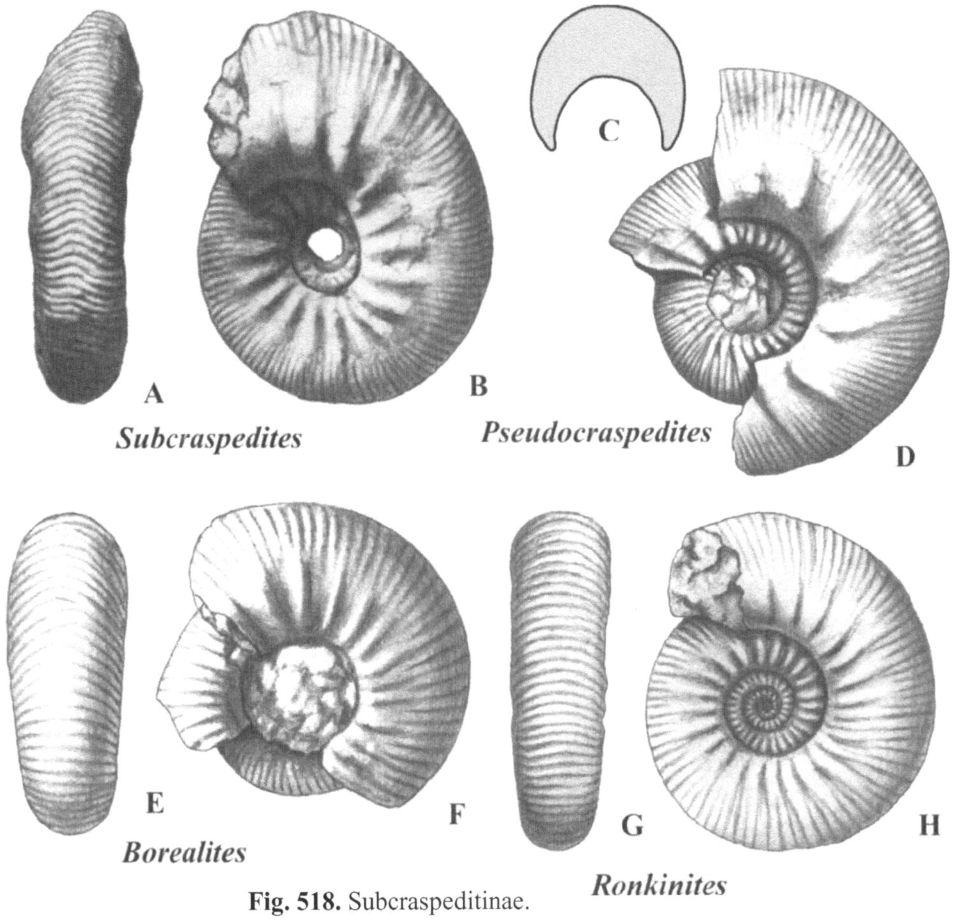

Fig. 518. Subcraspeditinae.

auxiliary elements. *Upper M. Volg. - Upper U. Volgian.* **Fig. 511N-O** - *S. parakaschpuricus* (Gerassimova).

Mosquites SASONOVA, 1971 [*Craspedites mosquensis* Gerassimov, 1960; OD]. Small; like *Craspedites* but more involute and compressed, with flatter convergent flanks, subtabulate venter, trapezoidal whorl section, lack of umbilical bullae and ribbing effaced from lower flank. *Upper U. Volg.* **Fig. 511J-K** - *M. mosquensis* (Gerassimov).

Subfamily Menjaitinae Sasonova, 1971

Discussion. Monophyletic derivatives of Craspeditinae, close to *Craspedites* (Sasonova, 1975); typically compressed, moderately involute, with ribs weakening over the siphonal line; most macroconchs with a smooth adult body chamber; suture like *Craspedites*. A provincial development, largely confined to the central Russian platform. *L. Valang.*

Menjaites SASONOVA, 1971 [*M. imperceptus*; OD]. Small to large (D=250 mm); involute, compressed, with narrowly rounded venter, convergent flanks and subtrigonal whorl section; inner whorls with 4-5 prominent prorsiradiate constrictions, weakening in later growth, and with feeble primary ribs splitting into finer secondaries on outer flank; in middle to later growth ribs reduced to thread-like striae; suture straight, relatively simple with 3-4 elements to suspensive love. *L. Valang.* **Fig. 515C-E** - *M. imperceptus* Sasonova.

Costamenjaites SASONOVA, 1971 [*C. juncundus*; OD]. Small; like *Menjaites* but more evolute, with strongly convergent flanks, very narrowly rounded venter and trigonal whorl section; strong subbullate prorsiradiate primary ribs split into 3-6 secondaries, with intervening intecalatories that are linguoid over venter; ribs weaken across siphonal line and decline with growth so that outer whorls virtually smooth. *L. Valang.* **Fig. 515A-B** - *C. juncundus* Sasonova.

Luppoviceras SASONOVA, 1977 [*L. validum*; OD] [**M**]. Moderately large; like *Costamenjaites* but much larger, more inflated and involute, losing ornament in later growth. *L. Valang.* **Fig. 516C-D** - *L. validum* Sasonova.

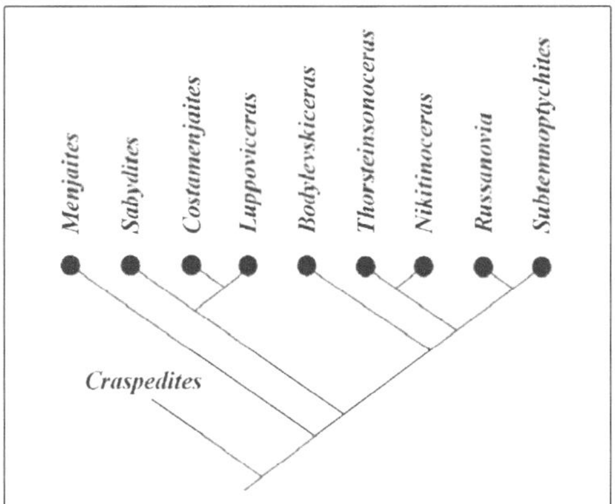

Fig. 519. Hypothesized relationships among Menjaitinae.

Sabydites BOGOMOLOV & BURDYKINA, 1989 [*S. menjaiteformis*; OD]. Small to medium sized; inner whorls like *Costamenjaites* but with more numerous strongly inclined primaries that bifurcate and trifurcate at about midflank; ribs cross venter transversely without interruption; in middle to later growth ribs effaced from flanks leaving umbilicolateral bullae and ventrolateral riblets; adult body chamber smooth. *L. Valang.* **Fig. 515F-H** - *S. menjaiteformis* Bogomolov & Burdykina.

Bodylevskiceras SASONOVA, 1977 [*Temnoptychites elegans* Bodylevskii, 1949; OD]. Small; like *Costamenjaites* but more involute and compressed, with ornament persisting to maturity; numerous (13-18) short primaries give rise to well defined bundles of 5-6 secondary ribs from early growth that are interrupted across venter; suture arched, with narrow lobes and 4 auxiliary elements; a homoeomorph of *Tollia*. *L. Valang.* **Fig. 515I-J** - *B. elegans* (Bodylevsky).

= *Bodylevskiceras* Sasonova, 1971 (*nom. nud.*).

Nikitinoceras SOKOLOV, 1913 [*Olcostephanus hoplitoides* Nikitin, 1888; SD Sazonova, 1961]. Small; like *Bodylevskiceras* but inflated, subglobose, with broadly rounded venter and depressed whorl section; distinct umbilical bullae in middle and later growth; ribs weakening across venter in middle growth; adult body chamber smooth. *L. Valang.* **Fig. 515K-L** - *N. hoplitoides* (Nikitin).

= *Temnoptychites* Pavlow, 1914 (obj.).

Russanovia BODYLEVSKII, 1967 [*Ammonites diptychus* Keyserling, 1845; OD] [**m**]. Small; like *Nikitinoceras* but with broader flatter venter and coarse distant ribs. *L. Valang.* **Fig. 515Q-R** - *R. diptychus* (Keyserling).

= *Keyserlingyceras* Sazonova, 1961 (*nom. nud.*).

Subtemnoptychites KLIMOVA, 1978 [*Temnoptychites (S.) prodigialis*; OD]. Very small; like *Russanovia* but with broader flatter venter, depressed subquadratic whorl section and ribs not interrupted across venter. *L. Valang.* **Fig. 515M-N** - *S. prodigialis* (Klimova).

Thorsteinssonoceras JELETZKY, 1965 [*T. ellesmerense*; OD]. Small; like *Nikitinoceras* but more inflated, cadicone, with sloping umbilical wall, depressed subcoronate whorl section, prominent umbilical bullae, numerous primary ribs and finer secondary ribs that are transverse over venter; constrictions usually present; suture with very wide second auxiliary saddle; a homoeomorph of *Euryptychites*. *L. Valang.* **Fig. 515O-P** - *T. ellesmerense* Jeletzky.

= *Thorsteinssonoceras* Jeletzky, 1964 (*nom. nud.*).

Family TOLLIIDAE Spath, 1952

(*nom. correct.* Arkell, 1957 *ex* Tollinae, *nom. transl.* Sasonova, 1971)

Discussion. Monophyletic derivatives of Craspeditidae (Sasonova, 1971), with ascendent sutures and 4-5 auxiliary lobes; comprises the paraphyletic subfamilies Subcraspeditinae, Suritinae and Tolliinae. A Boreal group. *L. Berrias.* - *L. Haut.*

Subfamily Subcraspeditinae Rogov, 2014

= Peregrinoceratinae Sasonova, 1979 - an untraceable reference and *nomen nudum*.

Discussion. Seemingly monophyletic derivatives of Laugeitinae, close to *Laugeites* (Rogov, 2014, 2020); typically evolute, compressed, with rounded oval to subtrigonal whorl section, strong frequently flared primary ribs and numerous fine secondaries on the outer flank crossing the venter transversely or with a slight forward arch. *M.* - *U. Volg.*

Subcraspedites SPATH, 1923 [*Ammonites plicomphalus* J. de C. Sowerby, 1823 (*non* J. Sowerby 1822), = *S. sowerbyi* Spath; SD Spath, 1924] [**M**]. Medium sized with subtrigonal whorl section; about 8 blunt umbilical bullae weakly connected to bundles of 4-6 fine secondaries (not preserved on internal mould) separated by intercalatories on outer flank; ribs transverse over venter (forming chevrons on distorted type); body chamber uncoiling slightly; suture with trifid L deeper than E and non-retracted suspensive lobe with 4 auxiliary lobes. *U. Tithon.* - *L. Berrias.* **Fig. 508O** - *S. parakaschpuricus* (Gerasimova); **Fig. 518A-B** - *S. sowerbyi* Spath.

Swinnertonia SHUL'GINA, 1972 [*Subcraspedites cristatus* Swinnerton, 1935; OD] [**M**]. Like *Subcraspedites* but more evolute, with more numerous primary ribs in early to middle growth and fewer coarser secondaries that remain strong across venter; primary ribs flared, distant and subalate in maturity, giving rise to generally 3 secondaries separated by an intercalatory; macroconchs with ribs convex forward over venter; suture not markedly ascendent, with broad E/L and L shallower than E. *Up-*

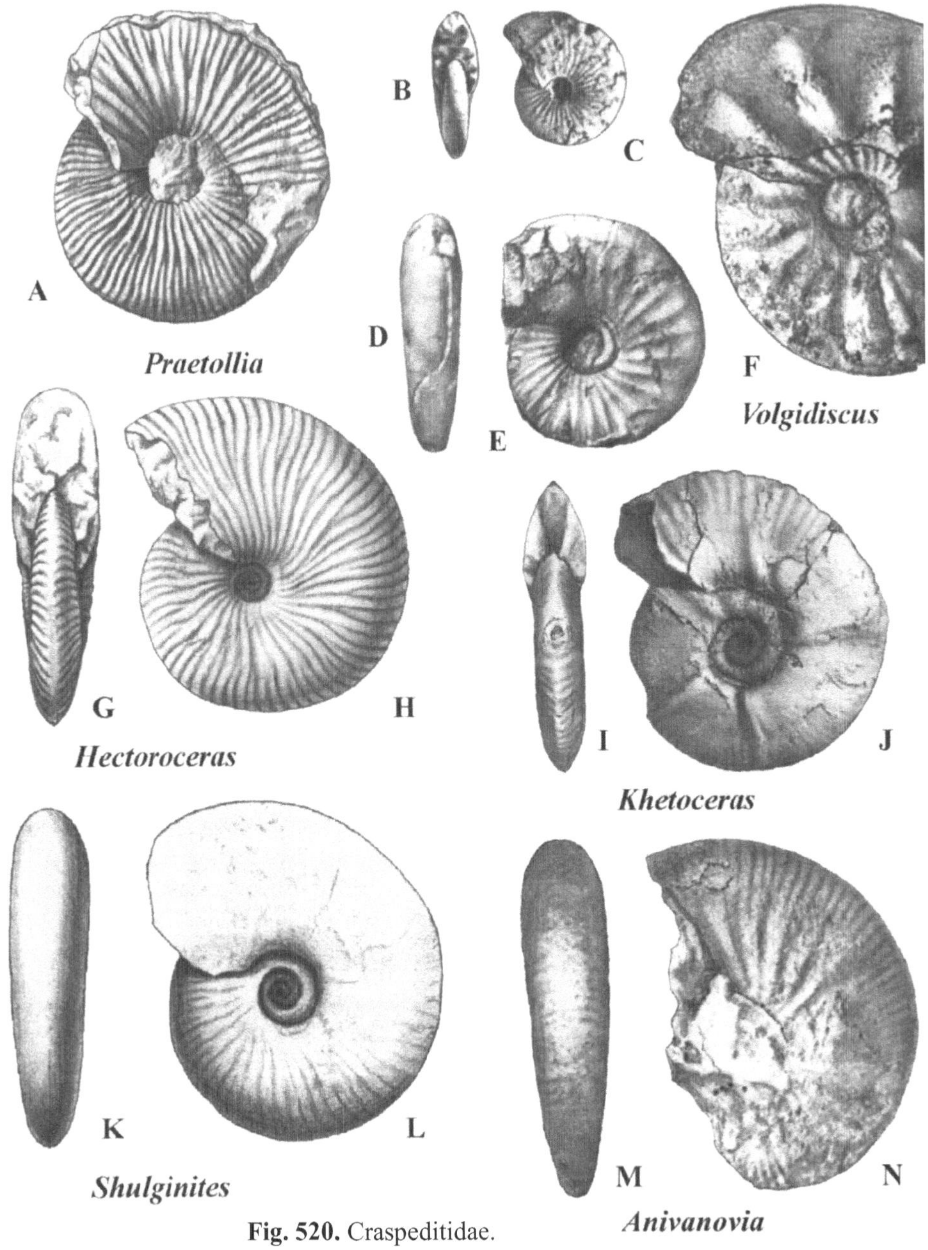

Fig. 520. Craspeditidae.

per M. - *Lower U. Volg.* **Fig. 508K-M** - *S. cristata* (Swinnerton); **Fig. 508N** - *S. primitiva* Swinnerton.

Borealites KLIMOVA, 2003[*B. federovi*; OD] [**M**]. Small [**m**] to moderately large [**M**]; compressed, moderately evolute, with oval whorl section and narrowly arched venter (rounded in macroconchs); in early growth constricted with strong biplicate ribs; in middle growth strong primary ribs bifurcate or trifurcate above mid-flank, separated by short intercalatories, and cross venter transversely; in later growth ribs evanesce below branching point; suture with long narrow saddles and fairly broad trifid L as deep as E. *Berrias.* **Fig. 518E-F** - *B. federovi* Klimova.

Pseudocraspedites CASEY, MESEZHNIKOV & SHUL'GINA, 1977 (*Subcraspedites anglicus* Shul'gina, 1972; OD] [**M**]. Like *Borealites* but large, finely ribbed, with outer whorls more evolute and almost smooth save for distant umbilicolateral bullae; superficially similar to *Subcraspedites*. *Lowermost Berrias.* **Fig. 518C-D** - *P. anglicus* (Shul'gina).

Ronkinites SHUL'GINA, 1972 [*Subcraspedites (R.) rossicus*; OD] [**m**]. Small; like inner whorls of *Pseudocraspedites*

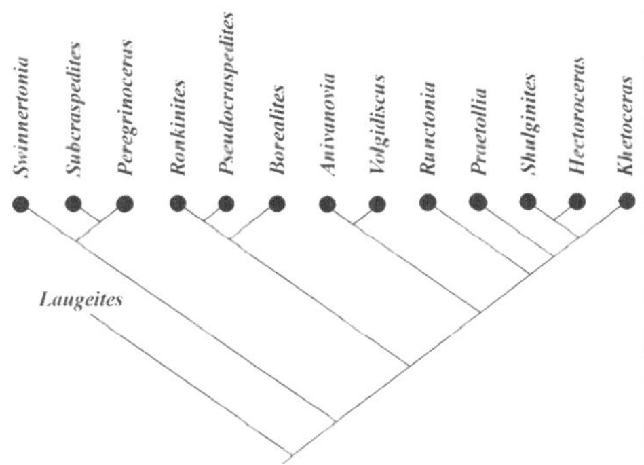

Fig. 521. Hypothesized relationships among Subcraspeditinae.

but with flatter flanks, subquadratic whorl section and scaphitoid uncoiling of body chamber; in early growth ribs biplicate, later triplicate with an intercalatory; 3-4 weak constrictions per whorl typical. *Berrias*. **Fig. 518G-H** - *R. rossicus* (Shul'gina).

Volgidiscus CASEY, 1973 [*Subcraspedites lamplughi* Spath, 1936; OD]. Involute, compressed, with broad convex flanks converging to narrowly rounded venter, and sublanceolate whorl section; prominent closely spaced prorsiradiate primary ribs split on outer flank but do not cross smooth venter; on outer whorls primary ribs extend to near venter without splitting into secondaries; ornament feebly impressed on internal mould; suture with rectangular E/L, bifid L not as deep as E, and strongly ascending series of auxiliary elements. *Lower L. Berrias*. **Fig. 520B-F** - *V. lamplughi* (Spath).

Anivanovia KISELEV, 2003 [*A. mola*; OD] [**M**]. Moderately large (**m**) to large (**M**); involute, compressed, discoidal with narrowly rounded venter; prorsiradiate primaries splitting into 2-3 secondaries on outer flank separated by an intercalatory. *U. Volg*. **Fig. 520M-N** - *A. mola* Kiselev.

This was considered a species of *Kachpurites* by Mitta (2010), but a macroconch of *Volgidiscus* by Rogov (2014).

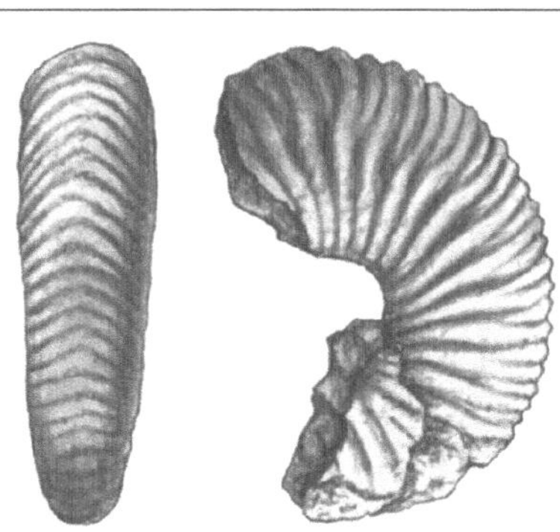

Fig. 522. *Runctonia*.

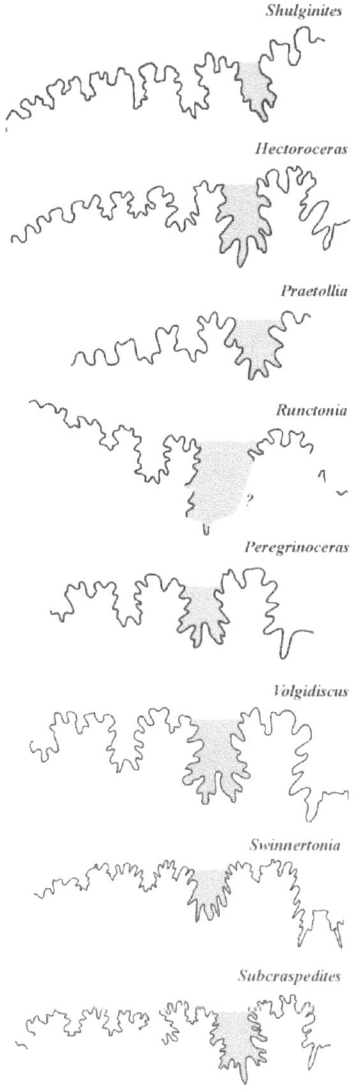

Fig. 523. Subcraspeditine sutures.

Runctonia CASEY, 1973 [*R. runctoni*; OD] [**M**]. Like *Volgidiscus* but with poorly defined umbilical wall; whorl section initially subelliptical with narrowly arched to subacute venter, later becoming subrectangular with flattened or feeblv convex venter; narrow prorsiradiate primary ribs bifurcate at midflank giving rise to falcate secondaries; in early growth venter smooth but later crossed by ribs; in maturity ribs effaced from flanks of macroconchs, leaving only umbilical and ventrolateral ornament; microconchs with subfastigiate venter and ribs forming chevrons over it with tendency to weaken across siphonal line; suture relatively simple, with broad L much deeper than E. *Lower L. Berrias*. **Fig. 522** - *R. runctoni* Casey.

Praetollia SPATH. 1952 [*P. maynci*; OD]. Like *Runctonia* with subquadratic whorl section but more involute, compressed and high whorled, with primaries splitting

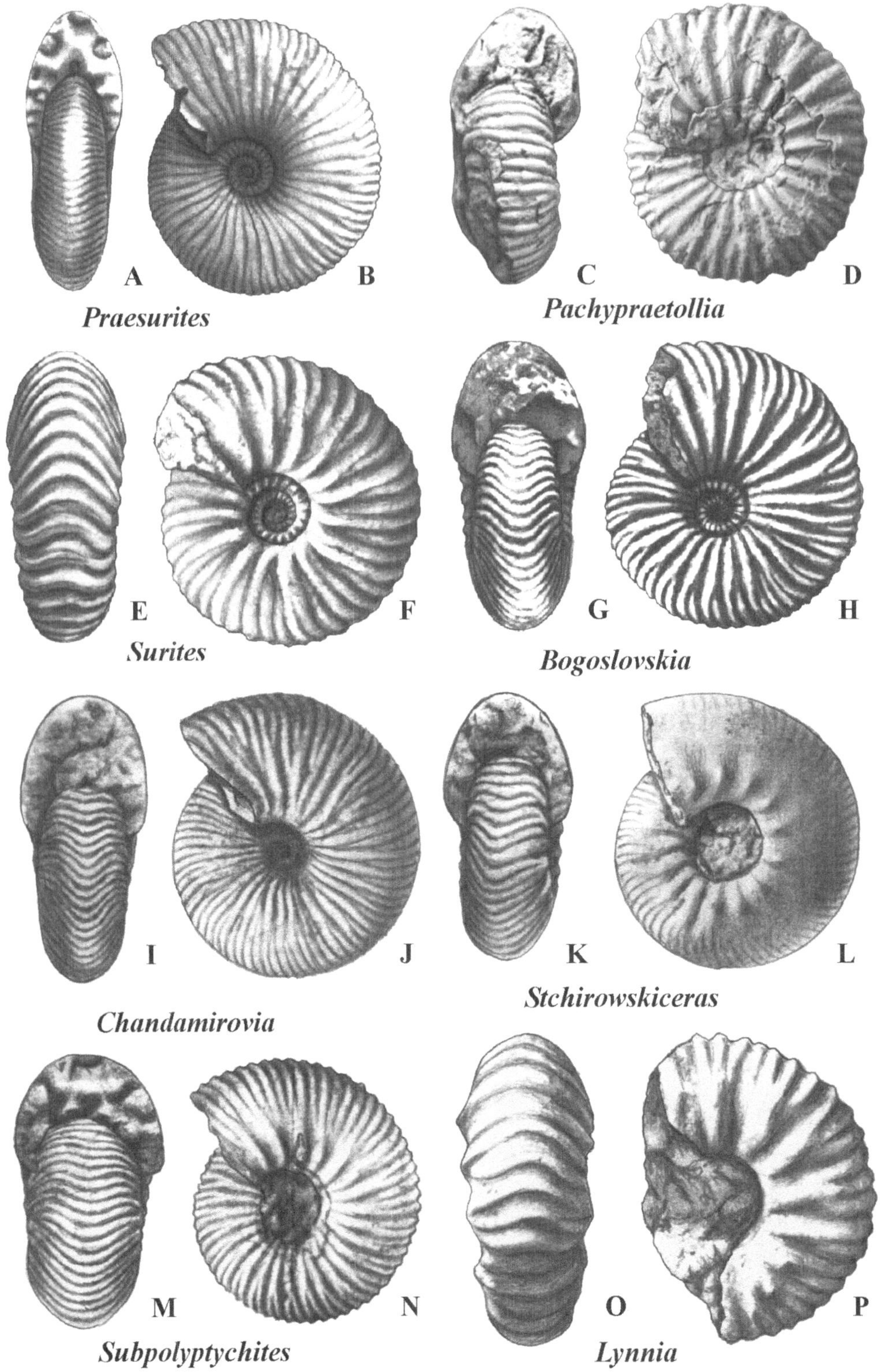

Fig. 524. Suritinae.

into sharp secondaries lower on flank, later intercalatories are added or primaries trifurcate; suture with straight suspensive lobe. *Upper L. Berrias*. **Fig. 520A** - *P. maynci* Spath.

Hectoroceras SPATH, 2014 [*H. kochi*; OD] [**M**]. Moderately large; discoidal oxycone with tiny funnel-shaped umbilicus, acute venter, sickle-shaped ribs and suture with E almost as deep as trifid L and at least 6 auxiliary elements arranged in slightly curved series. *Lower U. Berrias*. **Fig. 520G-H** - *H. kochi* Spath.

Shulginites CASEY, 1973 [*Oxynoticeras tolijense* Nikitin, 1884; OD] [**m**]. Inner whorls like *Hectoroceras* but with weaker prorsiradiate primaries bifurcating or trifurcating just above midflank; ribs become obsolete in later growth when venter rounds; suture with long narrower saddles. *Lower U. Berrias*. **Fig. 520K-L** - *S. tolijensis* (Nikitin).

Considered a microconch of *Hectoroceras* (Mitta, 2019).

= *Toljaiceras* Shul'gina, 1974 (obj.).

= *Schulginites* Mitta, 2005 (*laps. cal.*).

Khetoceras ROGOV, 2014 [*Garniericeras margaritae* Schul'gina, 1969; OD]. Moderately large (**m**) to large (**M**); discoidal, involute (juv.) to moderately evolute (ad.), with acute venter rounding in maturity (**m**), gently sloping umbilical wall and compressed oval whorl section; in early growth (D<50 mm) with prominent constrictions separated by riblets on upper flank; constrictions persist to maturity on microconchs, but ornament weakens in macroconchs. *Lower U. Berrias*. **Fig. 520I-J** - *K. margaritae* (Schul'gina).

Peregrinoceras SASONOVA, 1971 [*Olcostephanus pressulus* Bogoslovskii, 1897; OD]. Small, involute, compressed, with weakly convex flanks converging to evenly rounded venter, oval (juv.) to subrectangular (ad.) whorl section; strong subbullate primary ribs weakly connected to 3-6 fine secondaries separated by 1-2 intercalatories crossing venter with slight forward arch; ribs effacing from adult whorls of macroconchs; suture with L much shorter than E. *Upper U. Berrias*. **Fig. 527C-D** - *P. bellum* Sasonova; **Fig. 527G-H** - *P.?* cf. *albidum* Casey; **Fig. 527I** - *P. pressulum* (Bogoslovskii).

Praeperegrinoceras ALEKSEEV, KRAVED & KUZHINA, 1979. In condensed Berriasian deposits. No figure, description or diagnosis. A *nomen nudum*.

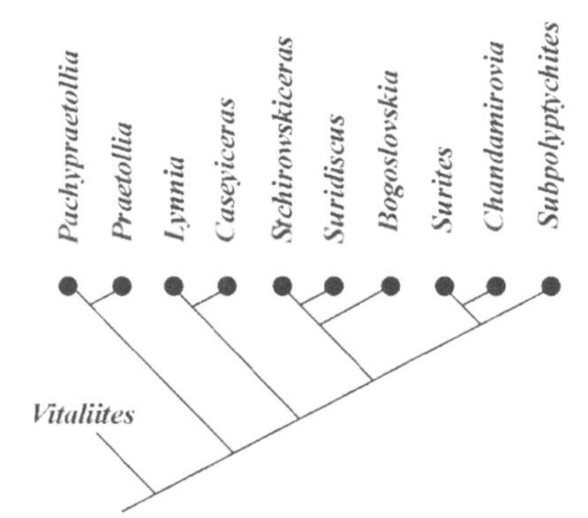

Fig. 525. Hypothesized relationships among Suritinae.

Subfamily Suritinae Sasonov, 1971

Discussion. Monophyletic derivatives of Subcraspeditinae, close to *Subcraspedites* (Sasonova, 1975); small to medium sized, involute with subtrigonal to oval whorl section; relatively widely spaced biplicate and triplicate ribs branching at midflank and linguoid over venter; suture with at least 4 elements to slightly retracted suspensive lobe. *L. Berrias. - L. Valang.*

Praesurites MESEZHNIKOV & ALEKSEEV, 1983 [*Surites (P.) elegans*; OD]. Small, involute, compressed and high whorled, with strongly convergent flanks, narrowly rounded venter and subtrigonal whorl section; numerous flexuous biplicate ribs branch above midflank and are linguoid over venter. *L. Berrias*. **Fig. 524A-B** - *P. elegans* (Mesezhnikov & Alekseev).

Pachypraetollia ALEKSEEV, 1984 [*Praetollia (P.) crassus*; OD]. Small; like *Praesurites* but more inflated and more coarsely ribbed, with straight rectiradiate primary ribs. *L. Berrias*. **Fig. 524C-D** - *P. crassus* (Alekseev).

Surites SASONOVA, 1951 [*S. pechorensis*; OD]. Small to medium sized; like *Praesurites* but more inflated, with evenly rounded venter and stronger coarser more distant bullate primary ribs; suture relatively simple, not much incised, with deep narrow E, subrectangular trifid E/L, quadrifid L and curved slightly retracted series of at least 4 auxiliary elements. *U. Berrias. - L. Valang*. **Fig. 516A-B** - *S. clementinus* Sasonova; **Fig. 524E-F** - *S. pechorensis* Sasonova;

Lynnia CASEY, 1973 [*Surites (L.) icenii*; OD]. Small; like *Surites* but with subquadratic whorl section, flared umbilicolateral bullae that give rise to bundles of 3 coarse

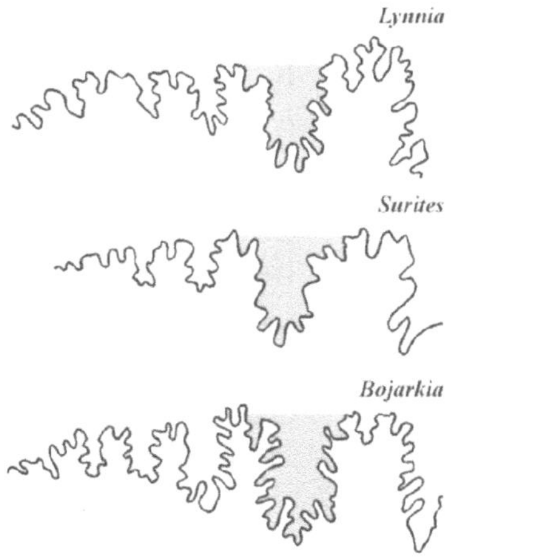

Fig. 526. Suritine sutures.

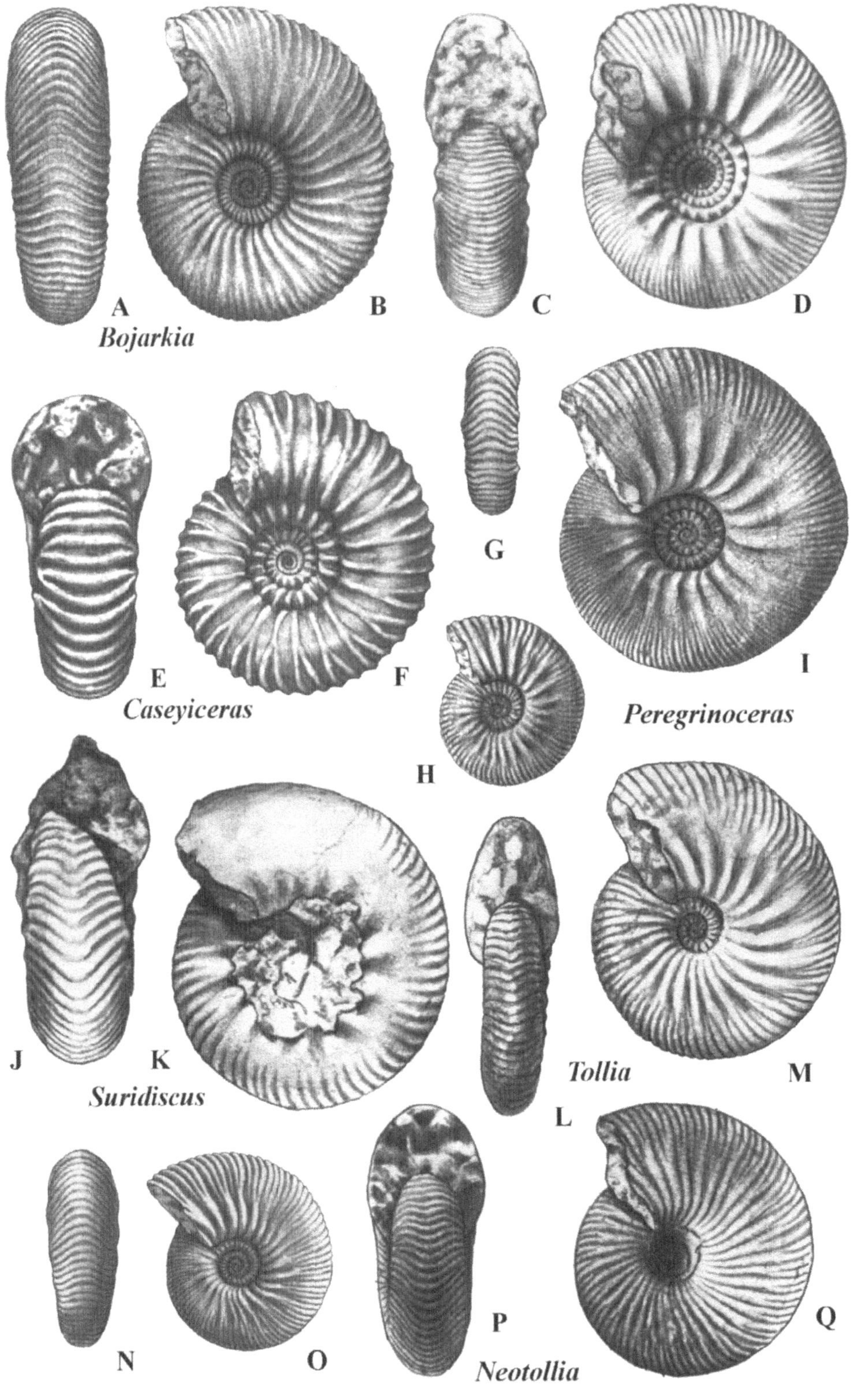

Fig. 527. Suritidae.

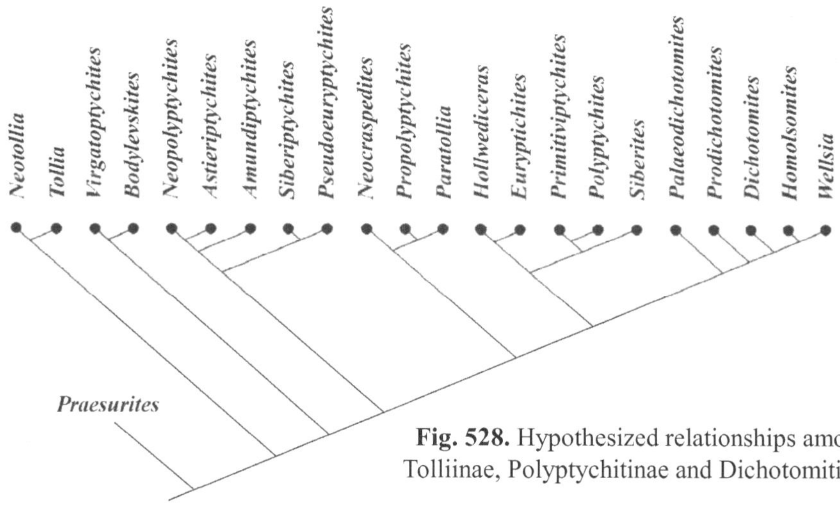

Fig. 528. Hypothesized relationships among Tolliinae, Polyptychitinae and Dichotomitinae.

secondaries; suture like *Surites* but simpler. *Lower U. Berrias.* **Fig. 524O-P** - *L. icenii* (Casey).

Caseyiceras SASONOVA, 1971 [*C. caseyi*; OD]. Small; like *Surites* but more evolute and inflated, with depressed whorls and coarse distant ribbing that is weakly convex over venter; umbilicolateral bullae insignificant. *Berrias.* **Fig. 527E-F** - *C. caseyi* Sasonova.

Chandamirovia SASONOV, 1951 [*C. ilekensis*; OD]. Very small (juv.); like *Surites* but more involute and inflated, with finer denser ribbing that frequently trifurcates. *Upper U. Berrias. - Lower L. Valang.* **Fig. 524I-J** - *C. pechorkensis* Sasonova.

Sasonova (1975) allied it to *Nikitinoceras* (= *Temnoptychites*).

Bogoslovskia SASONOVA, 1965 [*B. pseudostenomphala* Sasonova, 1961; OD]. Very small to small; like *Surites* but slightly more inflated, with narrower umbilicus, denser longer primaries, and secondary ribs markedly linguoid over venter. *Upper U. Berrias. - Lower L. Valang.* **Fig. 524G-H** - *B. pseudostenomphala* Sasonova. = *Bogoslovskia* Sazonova, 1961 (*nom. nud.*).

Stchirowskiceras SAZONOVA, 1971 [*S. principale*; OD]. Small to medium sized; inner whorls like *Bogoslovskia*; later ribbing declines on flanks but strong umbilical bullae persist to maturity; ribs form chevrons over venter. *Lower L. Valang.* **Fig. 524K-L** - *S. principale* Sasonova.

Fig. 529. *Pronjaites.*

Suridiscus SASONOVA, 1977 [*Stchirowskiceras (S.) subprincipale*; OD] [**M**]. Like *Stchirowskiceras* but larger, more evolute, with umbilical bullae obsolete in maturity. *Lowermost Valang.* **Fig. 527J-K** - *S. subprincipalis* (Sasonova).

Subpolyptychites SASONOVA, 1971 [*S. distinctus*; OD]. Very small; like *Suridiscus* but more inflated, with broadly rounded venter and depressed reniform whorl section; ribs frequently trifurcate in later growth. *Lower L. Valang.* **Fig. 524M-N** - *S. orbicularis* Sasonova; **Fig. 530I-J** - *S. distinctus* Sasonova.

Subfamily Tolliinae Spath, 1952

(*nom. correct.* Arkell, 1957 *ex* Tollinae)

Discussion. Monophyletic derivatives of Suritinae (Casey, 1973; Sasonova, 1975), close to *Praesurites*; typically rather involute and high whorled with generally fine dense bundled ribbing that is strong across the venter; macroconchs with smooth flanks in later growth. *M. Berrias. - L. Valang.*

Pronjaites SASONOVA, 1971 [*Olcostephanus bidevexus* Bogoslovskii, 1897; OD]. Small to medium sized; involute (juv.) to evolute (ad.), with steep umbilical wall, convergent flanks, subacute (juv.) to narrowly rounded (ad.) venter, sublanceolate (juv.) to compressed oval (ad.) whorl section; strong subradial primary ribs split into triplicates, later with an occasional intercalatory, that are linguoid over venter; suture with narrow E deeper than narrow bifid L, and with rectracted suspensive lobe of 4 elements. *M. Berrias.* **Fig. 529** - *P. bidevexus* (Bogoslovskii).

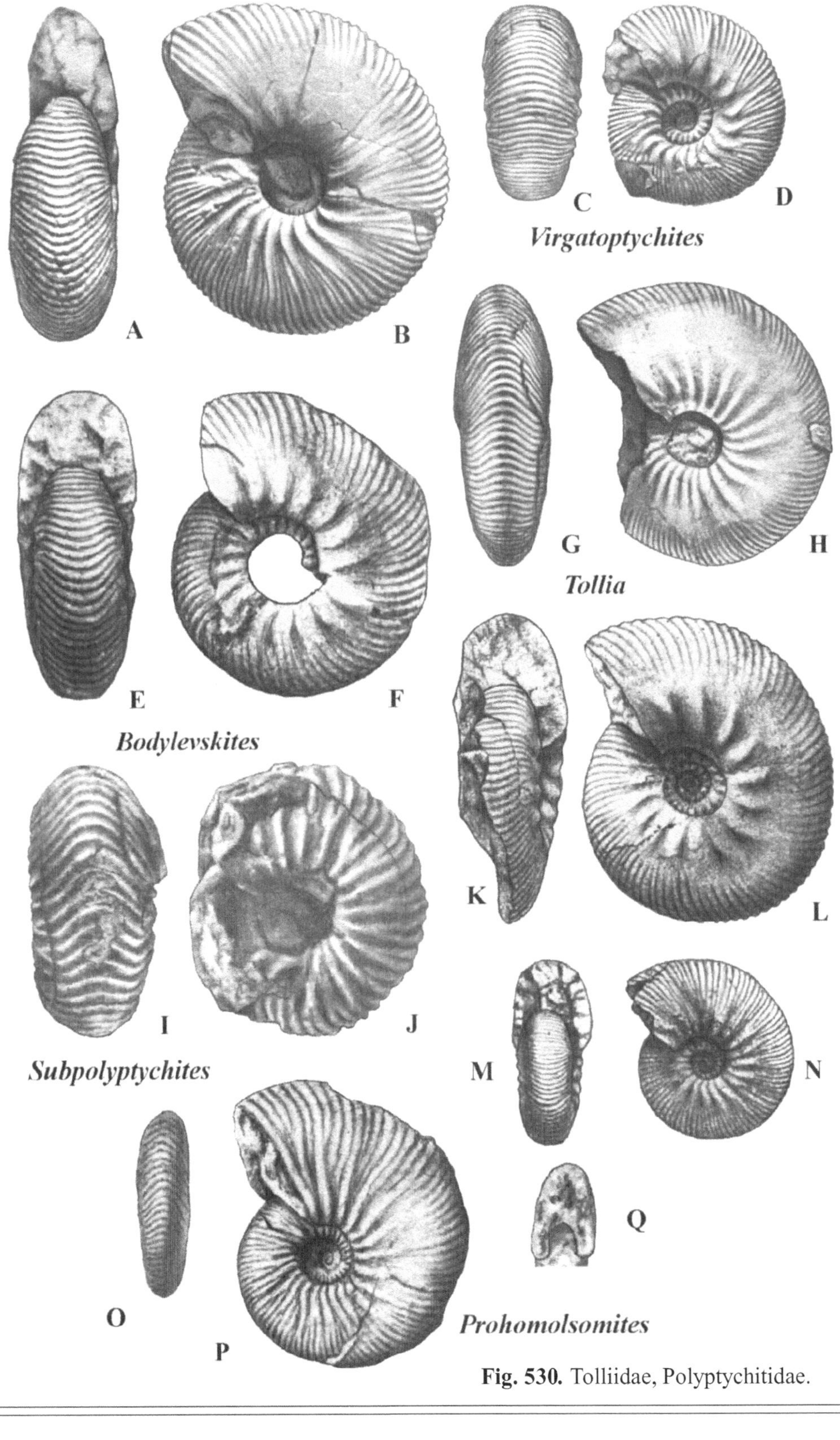

Fig. 530. Tolliidae, Polyptychitidae.

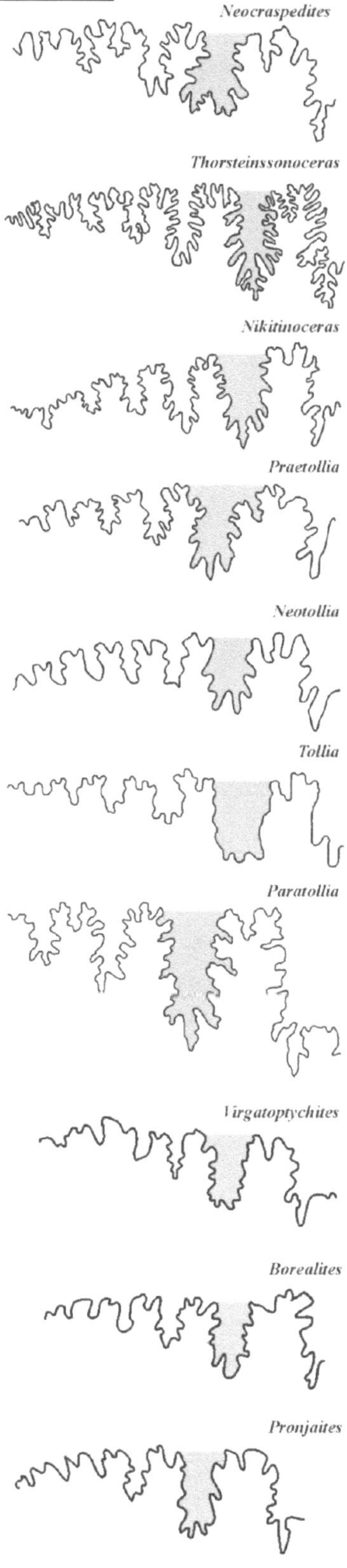

Fig. 531. Tolliine sutures.

Bojarkia SHUL'GINA, 1969 [*B. mesezhnikowi*; OD] [**M**]. Like *Tollia* but more evolute and inflated, with flat subparallel flanks, broadly rounded venter and subelliptical whorl section; ribs predominantly triplicate in juveniles, strong across flank in middle growth, with slight forward convexity over venter; relatively early loss of coarse biplicate ribs. *Upper U. Berrias*. **Fig. 527A-B** - *B. mesezhnikowi* Shul'gina.

Virgatoptychites VORONETS, 1958 [*V. changalassensis*; OD]. Medium sized; like *Tollia* but more inflated, with oval whorl section and regularly virgatotome ribs on outer whorls; suspensive lobe with 5 elements. *Uppermost Berrias. - L. Valang*. **Fig. 530A-D** - *V. pachsaensis* Voronets.

Tollia PAVLOW, 1914 [*T. tolli*; SD Krimgol'ts, 1953][**M**]. Medium sized to moderately larg; compressed, with flattened flanks and compressed oval whorl section; inner whorls with narrowly rounded to acute venter, sharp fine ribs mostly trifurcating at midflank and flexing strongly forward near venter; constrictions common on internal mould of juvenile whorls; during growth venter rounds, primary ribs become blunt, prominent and concave forward, and secondary ribs arise commonly in 2-3's but in some species in 4-5's; on largest septate whorls many ribs intercalated and smooth area may develop between primary and secondary ribs; all ribbing fades on last part of phragmocone and adult body chamber may be smooth; suture with L shorter than E and 5 auxiliary lobes that descend slightly. *L. Valang*. **Fig. 527L-M** - *T. tolli* Pavlow; **Fig. 530G-H** - *T. emeljanvezi* Voronetz.

Neotollia SHUL'GINA, 1969 [*Tollia klimakovskiensis* Krimgol'ts, 1953; OD]. Medium sized; like *Tollia* but very involute, more inflated and densicostate, with oval (juv.) to depressed subtrapezoidal (ad.) whorl section and broader venter; early whorls with mostly biplicate ribs with weak forward flexure and less differentiated primaries, triplicates appearing in middle growth (D>50 mm); suture more denticulate, with narrower and longer lobes and 5-6 auxiliary elements. *L. Valang*. **Fig. 527N-Q** - *N. klimakovskiensis* (Krimgol'ts).

Prohomolsomites ALEKSEEV & SHUL'GINA, 1990 [*Olcostephanus petschorensis* Bogoslovsky, 1902; OD]. Medium sized; involute, compressed, high whorled, with flat flanks, rounded venter and compressed elliptical whorl section; weak umbilical bullae give rise to prorsiradiate bundles of biplicate, polyplocoidal and virgatotome ribs that are linguoid over venter. *U. Valang*. **Fig. 530M-Q** - *P. petschorensis* (Bogoslovsky).

Homolsomites CRICKMAY, 1930 [*H. poecilotomus*, = *Holcodiscus? stantoni* McLellan, 1927; OD]. Very small (**m**) to medium sized (**M**), type a nucleus (D=15 mm); like *Tollia* but with strong constrictions in middle and later growth; pinched primary ribs in middle growth lost in maturity when ornament reduced to fine ribs across venter; suture strongly ascendent, with conspicuous

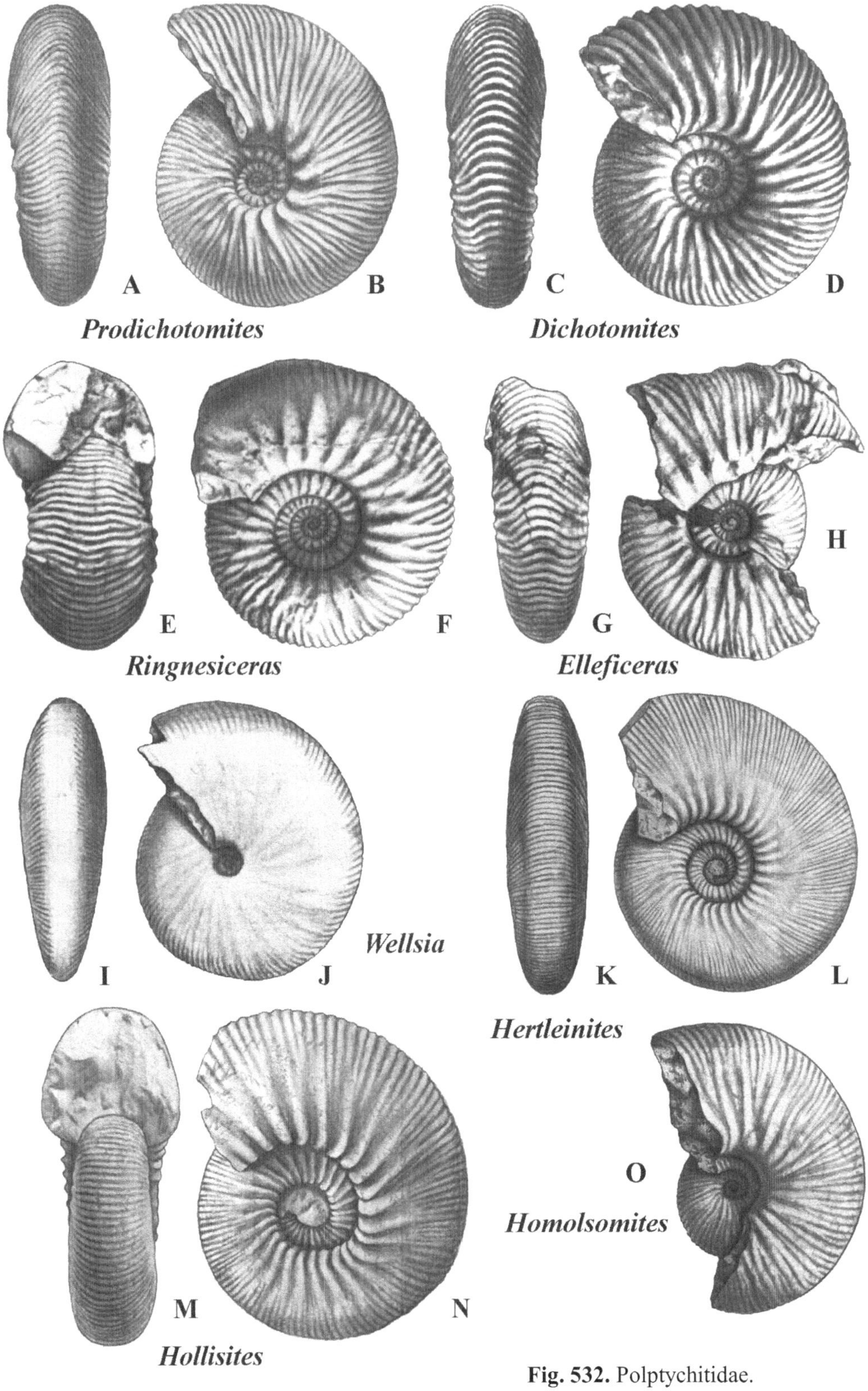

Fig. 532. Polptychitidae.

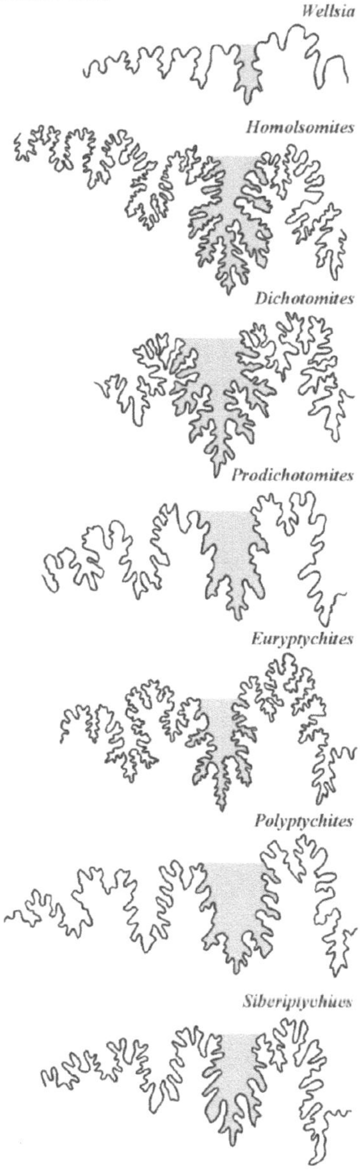

Fig. 533. Polyptychitine sutures.

adventive saddle to E/L and large trifid L deeper than E. *U. Valang. - L. Haut.* **Fig. 532O** - *H. stantoni* (McLellan).

Wellsia IMLAY, 1957 [*Dichotomites oregonensis* Anderson, 1938; OD][**M**]. Small; like *Homolsomites* but discoidal, with flattened flanks converging to narrowly rounded venter; whorl section subtrapezoidal in early growth, later subtrigonal; ornament very weak on flanks, comprises flexuous prorsiradiate primary ribs that bifurcate on middle to outer flank, separated by intercalatories and weakened across siphonal line; ribs decline with growth leaving only weak striae on adult body chamber; suture with broad bifid E/L with prominent adventive saddle and very narrow trifid L as deep as E, and 3 auxiliary lobes. *L. Haut.* **Fig. 532I-J** - *W. oregonensis* (Anderson).

Family POLYPTYCHITIDAE Wedekind, 1918

(*nom. correct.* Spath, 1924 *pro* Polyptichitidae)

Discussion. A monophyletic taxon with an origin in Tolliidae (Wright, 1996); typically with polyptychitine to polyschizotomous and virgatotome ribbing in later growth arising from prominent umbilical bullae; suture radial to ascendent with relatively long elements and 2-5 retracted auxiliary elements. Predominantly Boreal, but with some incursions into northern Tethys. Comprises the subfamilies Polyptychitinae, Dichotomitinae and Simbirskitinae. *U. Volg. - Lower L. Barrem.*

Subfamily Polyptychinae Wedekind, 1918

Discussion. A monophyletic group with an origin in Tolliinae, close to *Virgatoptychites* (Wright, 1996); typically rather inflated and involute, often cadicone, with ribs crossing the venter transversely; suture with slightly retracted suspensive lobe. Comprises the tribes Polyptychitini and Euryptychitini. *L. Valang. - L. Haut.*

Tribe Polyptychitini Wedekind, 1918

(*nom. transl.* herein *ex* Polyptychitinae)

Type genus: *Polyptychites* Pavlow, 1902.

Discussion. Involute, with steep umbilical wall, polyptychid ribbing and 2-3 auxiliary elements to suspensive lobe. *L. Valang.*

Bodylevskites KLIMOVA, 1978 [*B. harabylensis*; OD]. Small, compressed, involute, with fairly shallow step-like umbilicus, smooth vertical umbilical wall, broad almost flat convergent flanks, subtabulate venter and subrectangular to subtrapezoidal whorl section; inner whorls with bidichotomous or tridichotomous ribs arising from umbilical shoulder, with occasional single ribs and frequent intercalatories that may bifurcate; with growth elongated umbilicolateral bullae appear and rib connections weaken to leave smooth zone on inner flank; ribs cross venter with forward convexity; constrictions frequent; length of body chamber exceeds one whorl; suture with only 2 auxiliary elements and strongly retracted suspensive lobe. *Lower L. Valang.* **Fig. 530E-F** - *B. harabylensis* Klimova.

Neopolyptychites SHUL'GINA, 1983 [*N. arcticus*; OD]. Small to medium sized; like *Bodylevskites* but with rounded venter and coarser ribbing that frequently bifurcates on outer flank. *L. Valang.* **Fig. 536C-D** - *N. arcticus* Shul'gina.

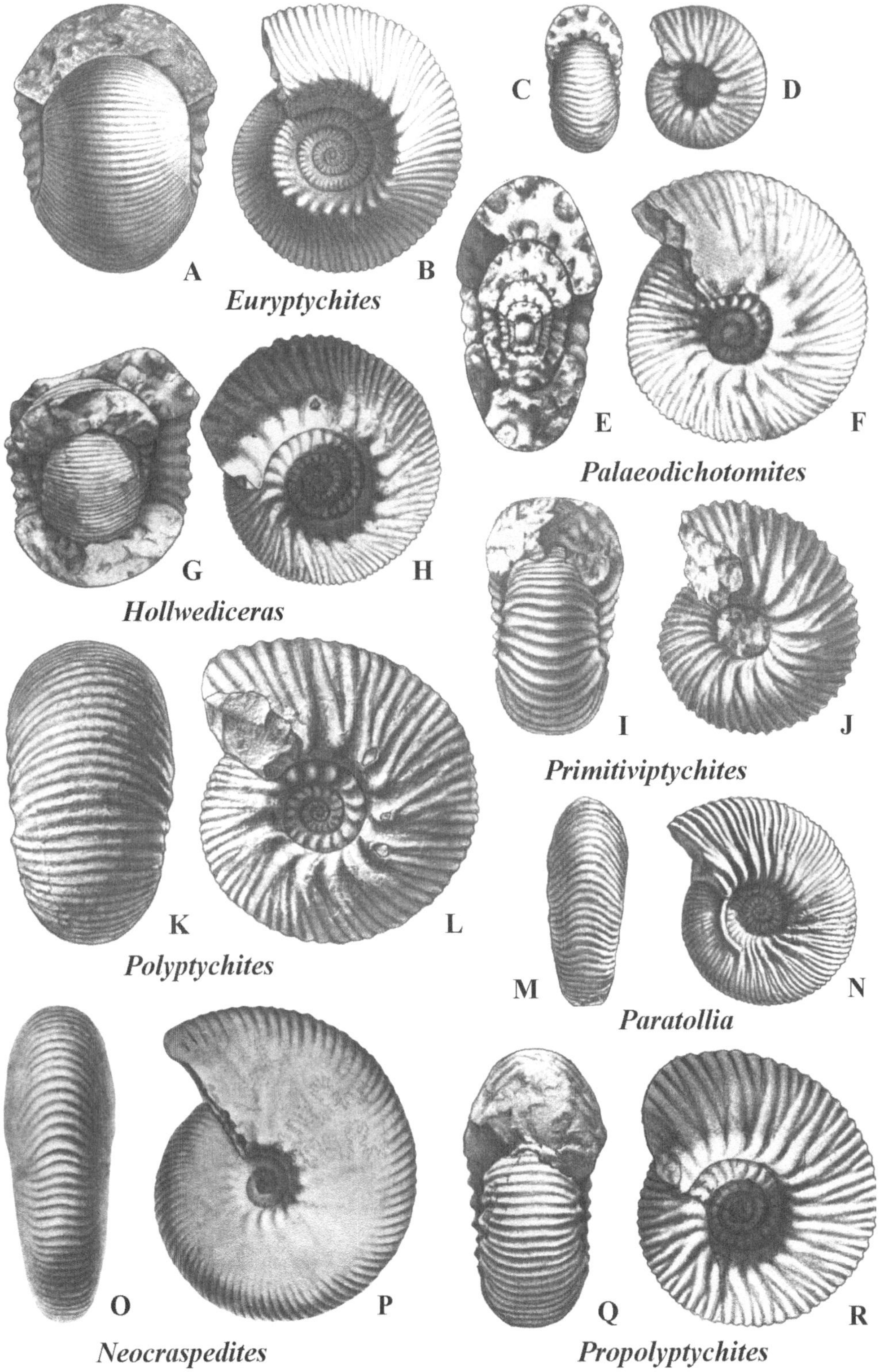

Fig. 534. Polyptychitidae.

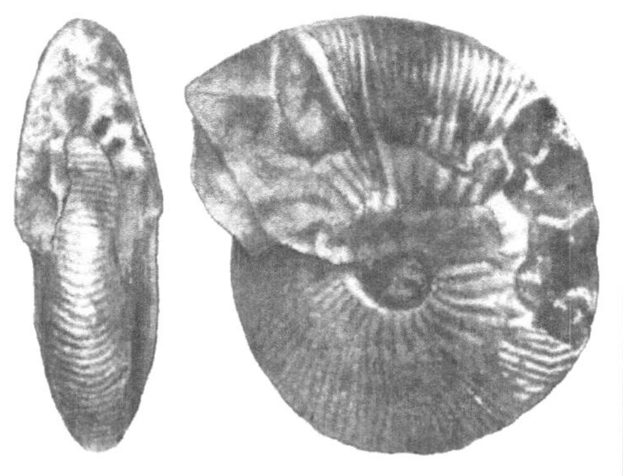

Fig. 535. *Selandites.*

Paratollia CASEY, 1973 [*P. kemperi*; OD]. Small; involute, compressed, with convergent flanks and well rounded narrow venter; in early growth (D<50 mm) about 24 primary ribs per whorl give rise to 3-4 flexuous secondaries; at first branching occurs on lower third of flank (at line of involution), later position of branching varies, one rib of bundle springing from mid-flank; bidichotomy commences at c.40 mm diameter; all ribs thin, sharply elevated, crossing venter with foward convexity; constrictions lacking; suture like *Propolyptychites* but with rounded folioles. *Lower L. Valang.* **Fig. 534M-N** - *P. kemperi* Casey.

Neocraspedites SPATH, 1924 [*Craspedites semilaevis* von Koenen, 1902; OD]. Small; like *Propolyptychites* but more involute and compressed, with subparallel to convex flanks, weaker umbilical bullae and ribs virtually effaced from flanks, leaving only fine prorsiradiate ribs on outermost flank crossing venter with slight forward convexity; suture markedly ascendent throughout, with 3 auxiliary lobes. *Lower L. Valang.* **Fig. 534O-P** - *N. semilaevis* (von Koenen).

Selandites ALEKSEEV & SHUL'GINA, 1990 [*Neocraspedites kotschetkovi* Bodylevskii, 1968; OD]. Small to large, involute with depressed subtrigonal whorl section in early growth, later compressed with narrowly rounded venter; flanks with long prorsiradiate primary ribs splitting into bifurcating and intercalated secondaries, evanescing later; suture relatively simple, with bifid E/L and asymmetrically trifid L/U_2. *L. Valang.* **Fig. 535** - *S. kotschetkovi* (Bodylevski).

Siberites KLIMOVA, 1983 [*S. savitzskii*; OD]. Medium sized; like *Propolyptychites* but with numerous feeble umbilical bullae giving rise to paired fine ribs, posterior one of which commonly bifurcates, separated by intercalatories. *L. Valang.* **Fig. 536L** - *S. savitzskii* Klimova.

Polyptychites PAVLOW, 1902 [*Ammonites polyptychus* Keyserling, 1846; SD Roman, 1938] [**M**]. Medium sized; like *Propolyptychites* but more evolute and inflated, with stronger umbilical bullae and coarser polyptychitine ribs; suture with 2 (primitive) to 3 (advanced) auxiliary elements. *L. Valang.* **Fig. 534K-L** - *P. polyptychus* (Keyserling).

Primitiviptychites KLIMOVA, 1983 [*Olcostephanus rectangulatus* Bogoslovsky, 1902; OD] [**m**]. Small; like *Polyptychites* but with narrower umbilicus, depressed subrectangular whorls, and prominent bullate primaries giving rise to 2-3 strong sharp secondaries that cross venter with slight forward convexity. *L. Valang.* **Fig. 534I-J** - *P. rectangulatus* (Bogoslovsky).

Palaeodichotomites BOGOMOLOV, 1989 [*Olcostephanus (Polyptychites) triplodiptychus* Pavlow, 1892; OD]. Medium sized; in early to middle growth an inflated cadicone with depressed semilunate whorls; later with convergent flanks, rounded venter and subtrigonal whorl section; rursiradiate primary ribs on umbilical wall terminate in prominent umbilical bullae and give rise to 2-3 prorsiradiate ribs that bifurcate or are joined by intercalatories on outer flank. *L. Valang.* **Fig. 534E-F** - *P. triplodiptychus* (Pavlow).

Tribe Euryptychitini nov.

Type genus: *Euryptychites* Pavlow, 1914.

Diagnosis. Inflated cadicones with funnel shaped umbilicus, depressed whorls, prominent umbilical bullae giving rise to bundles of prorsiradiate ribs that vary from polyptychitoid to fasciculate, with frequent constrictions. Seemingly monophyletic derivatives of Polyptychitini, close to *Bodylevskites* (Jeletzky & Kemper, 1988). *L. Valang.*

Astieriptychites BODYLEVSKII, 1960 [*A. astieriptychus*; OD]. Medium sized; inflated cadicone with moderately wide funnel shaped umbilicus and depressed reniform whorl section; inner whorls with polyptychitoid ribs arising from umbilical shoulder; later whorls with prorsiradiate ribs arising in fasciculate bundles (5-6) from small strong comma-shaped umbilical bullae; of secondary ribs, only one (usually anterior rib) firmly attached to umbilical node, and intercalatories may occur; ribs cross venter strongly; constrictions frequent; suture with 3-4 auxiliary lobes, suspensive lobe not retracted; in later growth an *Olcostephanus* homoeomorph. *L. Valang.* **Fig. 536J-K** - *A. astieriptychus* Bodylevskii.

= *Astieriptychites* Bodylevskii, 1950 (*nom. nud.*).

= *Astieriptychites* Bodylevskii, 1958 (*nom. nud.*).

Siberiptychites KEMPER & JELETZKY, 1977 [*Ammonites stubendorffi* Schmidt, 1872; OD]. Medium sized; like *Astieriptychites* but with steeper umbilical wall, more numerous weaker umbilical bullae, finer secondaries in bundles of 3-4 branching ribs with intercalatories, and prominent constrictions; suture with narrow rectangular E, long trifid saddles and narrow trifid lobes. *L. Valang.* **Fig. 536A-B** - *S. stubendorffi* (Schmidt).

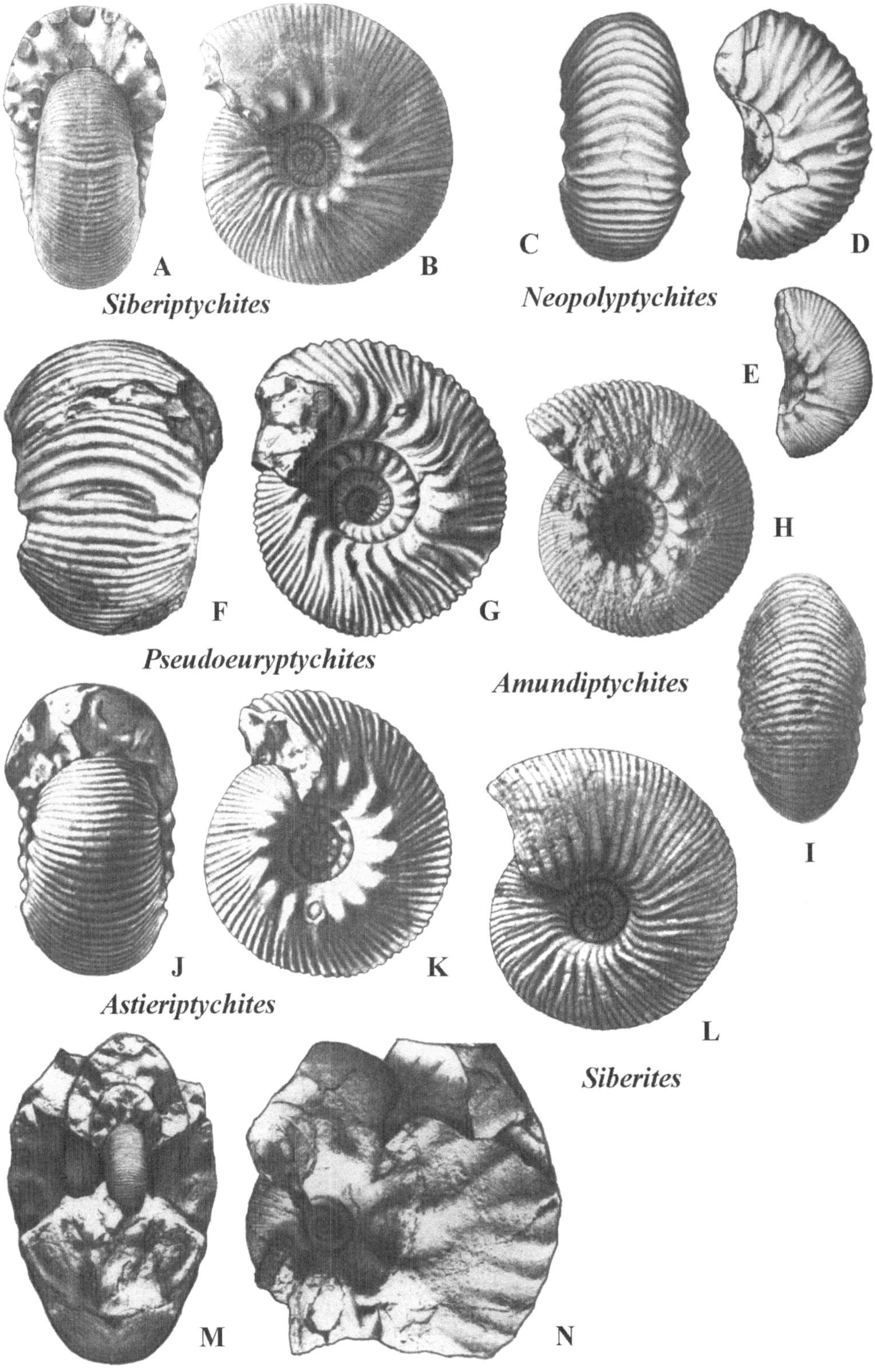

Fig. 536. Polyptychitinae.

Fig. 537. *Dichotomites* (**M**).

Amundiptychites KEMPER & JELETZKY, 1979 [*A. svedrupi*; OD] [**M**]. Moderately large; evolute with funnel shaped umbilicus; inner and middle whorls like *Astieriptychites* but with subcircular whorl section to middle growth (70-90 mm diameter) when ribs weaken on lower flank; adult ornament robust, with prominent oval umbilical bullae and coarse widely spaced bundles of ribs in which 2 or more branches bifurcate at different levels; constrictions lacking; suture with 3-5 auxiliary lobes. *L. Valang*. **Fig. 536H-I** - *A. fasciatus* Kemper & Jeletzky; **Fig. 536M-N** - *A. svedrupi* Kemper & Jeletzky.

Hollwediceras JELETZKY & KEMPER, 1988 [*Polyptychites sphaericus* von Koenen, 1902; OD]. Moderately large; like *Astieriptychites* but more inflated, with strongly depressed whorls, smaller more numerous umbilical bullae and rib bundles that include at least 2 dichotomous elements, that bifurcate at approximately same level rather high on flanks; an additional solitary secondary rib is present in bundles of intermediate growth stages, and may occupy different position in different bundles; suture with long slender denticulate lobes and broader saddles; a *Euryptychites* homoeomorph. *Upper L. - Low U. Valang*. **Fig. 534G-H** - *H. sphaericum* (von Koenen).

Propolyptychites KEMPER, 1964 [*Polyptychites quadrifidus* von Koenen, 1902; OD]. Small to medium sized; moderately stout, compressed, involute, with steep umbilical wall, subrounded umbilical edge, broad gently convex flanks converging to evenly rounded venter and broadly ovate whorl section; in early to middle growth weak umbilical bullae give rise to pairs of biplicate ribs, but later ribbing may become polyschizostomous; suture ascendent, with broad bifid E/L and narrow trifid L as deep as E, and only 2 auxiliary lobes. *Lower L. Valang*. **Fig. 534Q-R** - *P. quadrifidus* (von Koenen).

Pseudoeuryptychites JELETZKY, 1986 [*Euryptychites pavolvi* Voronets, 1962; OD]. Small; like *Siberiptychites* but much larger, more inflated, with broad flattened venter, fewer umbilical bullae that give rise to pairs of dichotomous secondaries of which the posterior is finer, and a more complex suture with all 3 auxiliary lobes on umbilical wall; in later growth inner flanks become smooth and adult body chamber unornamented. *L. Valang*. **Fig. 536F-G** - *P. pavolvi* (Voronets).

Euryptychites PAVLOW, 1914 [*Olcostephanus latissimus* Neumayr & Uhlig, 1888; SD Wright, 1957]. Medium sized; like *Astieriptychites* but more inflated, with strongly depressed whorls, sloping umbilical wall, funnel-shaped umbilicus, sharp umbilical edge and smaller more numerous umbilical bullae; posterior rib of each rib bundle bifurcating; suture with prominent median saddle to E, fairly broad E/L, narrow parallel sided trifid L, bifid L/U_2 and two retracted auxiliary lobes. *Lower L. Valang*. **Fig. 534A-B** - *E. latissimus* (Neumayr & Uhlig).

Subfamily Dichotomitinae nov.

Diagnosis. Typically compressed with bundles of fine virgatotome ribs arising from umbilical bullae and flexing forward across venter; suture subradial, with 3-5 auxiliary elements. Monophyletic derivatives of Polyptychitinae, close to *Polyptychites* (Jeletzky & Kemper, 1988). *U. Valang.*

Prodichotomites KEMPER, 1971 [*Polyptychites polytomus* von Koenen, 1902; OD]. Small to medium sized; rather involute, compressed and high whorled, with convex sides, narrowly rounded venter and prominent umbilical bullae giving rise to bundles of prorsiradiate virgatotome ribs; ribs flex forward on outer flank and cross venter with forward convexity; suture with long narrow bifid E/L, narrow trifid L almost as deep as E and 4-5 auxiliary elements. *U. Valang*. **Fig. 532A-B** - *P. polytomus* (von Koenen)

Elleficeras KEMPER & JELETZKY, 1979 [*Ringnesiceras (E.) ellefense*; OD]. Small; like *Prodichotomites* but more evolute, with radial primary ribs extending onto lower flank where they split into relatively coarse paired secondaries of which both or only posterior one bifurcates high on flank; bifurcating intercalatories appear in later growth; secondaries linguoid over venter; primary ribs become subbullate in maturity; suture with 4 auxiliary elements. *U. Valang*. **Fig. 532G-H** - *E. ellefense* (Kemper & Jeletzky).

Dichotomites VON KOENEN, 1909 [*Ammonites bidichotomus* d'Orbigny, 1841; M]. Small (**m**) to large (**M**); like *Prodichotomites* but with broad flattish flanks converging to well rounded venter; numerous umbilical bullae give rise to pairs prorsiradiate ribs that bifurcate high on flank, separated by up to 2 intercalatories some of which may also bifurcate; ribs cross venter with slight forward convexity; outer whorls relatively evolute; suture with retracted suspensive lobe. *U. Valang*. **Fig. 532C-D** - *D. bidichotomus* (d'Orbigny) (**m**); **Fig. 537** - *D. grotriani* Neumayr & Uhlig (**M**).

= *Bidichotomites* Sasonova, 1951 (obj.).

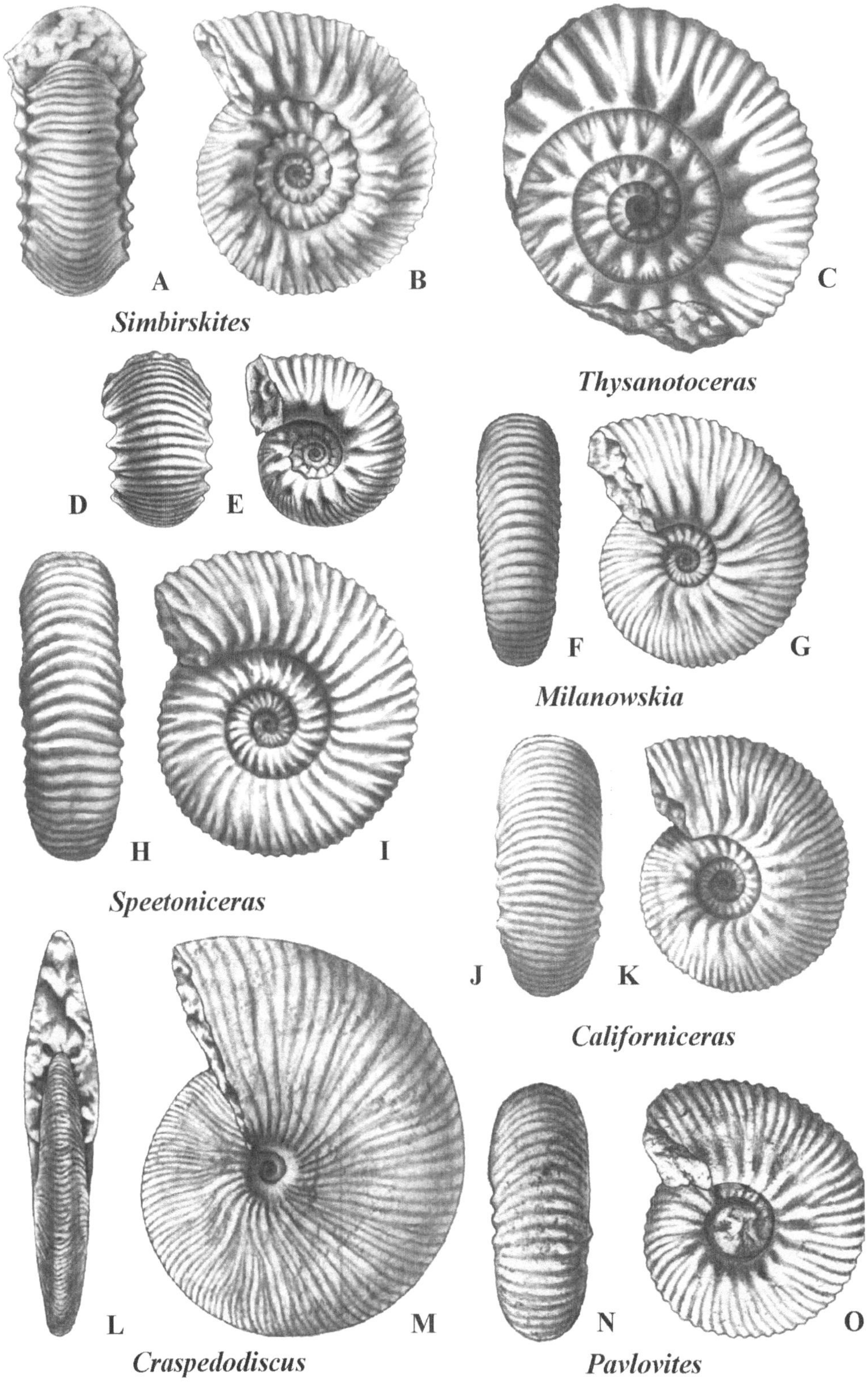

Fig. 538. Simbirskitinae.

Ringnesiceras KEMPER & JELETZKY, 1979 [*R. amudense*; OD]. Medium sized to moderately large; early whorls inflated with almost semicircular whorl section; later moderately evolute with shallow step-like umbilicus, rounded umbilical edge, convex flanks converging to narrowly rounded venter and oval whorl section; in middle growth 23-26 short primary ribs terminate on lower flank in weak bullae that give rise to fine paired secondaries, the posterior one of which branches high on flank (sometimes there is an intercalated rib in bundles); in early and later growth secondary ribs arise from umbilical bullae; ribs flex forward near venter and cross siphonal line with shallow V; slightly ascendent suture relatively simple, with weakly retracted suspensive lobe of 2-3 auxiliary lobes. *Lower U. Valang*. **Fig. 532E-F** - *R. amudense* Kemper & Jeletzky.

Hertleinites IMLAY, 1958 [*Neocraspedites aguila* Anderson, 1938; OD]. Medium sized to moderately large; fairly involute, compressed, with subquadratic early whorls, short primary ribs that branch well below midflank and give rise to finer secondaries in polyptychitoid bundles with no sign of lateral tubercles; in later growth umbilical seam egresses and point of rib furcation becomes indistinct and moves above midflank; suture descendent with narrow E/L with prominent adventive saddle, narrow trifid L not as deep as E, and somewhat retracted suspensive lobe with 2 auxiliary lobes. *Upper U. Haut*. **Fig. 532K-L** - *H. aguila* (Anderson).

= *Hertleinia* Imlay, 1957 *non* Marks (1949).

Subfamily Simbirskitinae Spath, 1924

(*nom. transl.* Spath, 1931 *ex* Simbirskitidae)

Discussion. Monophyletic derivatives of Dichotomitinae, close to *Ringnesiceras* (Kemper & Jeletzky, 1979); primitively evolute and inflated with rounded venter throughout and secondary ribs branching from lateral tubercles (at least in early growth) and crossing venter with forward arch; derived members becoming more involute and compressed, with loss of lateral tubercle and ascendent suture with L/U_2 projecting beyond E/L and only 2-3 auxiliary elements. An important group in the Late Hauterivian of the Boreal Realm. *Upper L. Haut. - Lower L. Barrem.*

Hollisites IMLAY, 1957 [*H. lucasi*; OD]. Large; moderately evolute and rather inflated with vertical umbilical wall, abruptly rounded umbilical edge and gently convex flanks; in early growth with rounded venter and oval whorl section, later with subtabulate venter and trapezoidal whorl section; in early growth prominent prorsiradiate primary ribs terminate at midflank in blunt tubercles from which finer secondaries project forward in pairs, crossing venter with forward convexity; in middle to later growth lateral tubercles obsolete, primary ribs branching into polyptychitoid bundles of 3-5 finer secondaries on lower flank; only primaries exposed in umbilicus; shallow constrictions common; suture with moderately deep E, broad asymmetrically bifid E/L, wide bifid or trifid L as deep as E, narrow asymmetrically bifid narrow stemmed L/U_2 and retracted suspensive lobe with 3 elements. *Upper L. Haut*. **Fig. 532M-N** - *H. lucasi* Imlay.

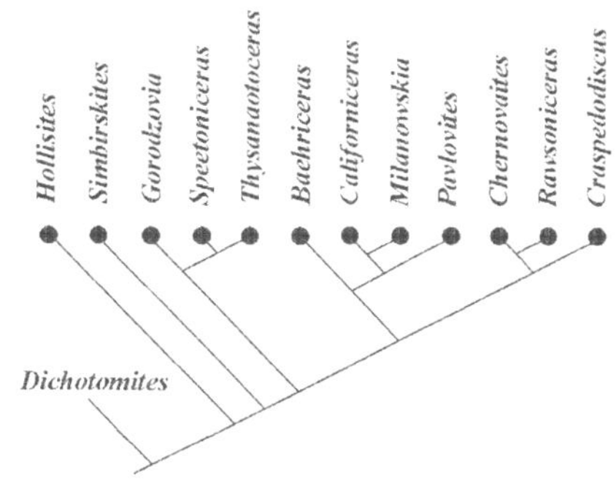

Fig. 539. Hypothesized relationships among Simbirskitinae.

Speetoniceras SPATH, 1924 [*S. subbipliciforme*; OD][**m**]. Small; compressed, evolute, with subcircular to coronate whorl section, prominent prorsiradiate primary ribs that regularly bifurcate at subdued lateral tubercle just below or at midflank and flex forward on outer flank; in later growth lateral tubercles obsolete and flanks flattened; constrictions lacking; suture relatively simple, with deep narrow E, broad bifid E/L and asymmetrically trifid L not as deep as E. *Upper L. Haut*. **Fig. 538H-I** - *S. subbipliciforme* Spath.

Subspeetoniceras IVANOV & ARISTOV, 1969 [*Speetoniceras inversioides* Aristov, 1967; OD][**M**]. Large; like *Speetoniceras* but with coarser more distant ribs that bifurcate lower on flank. *U. Haut*. **Fig. 541G-H** - *S. inversioides* (Aristov).

= *Subspeetoniceras* Aristov, 1968 (*nom. nud.*).

Simbirskites PAVLOW, 1892 [*Ammonites decheni* Roemer, 1841; SD Roman, 1938][**m**]. Small; like *Speetoniceras*

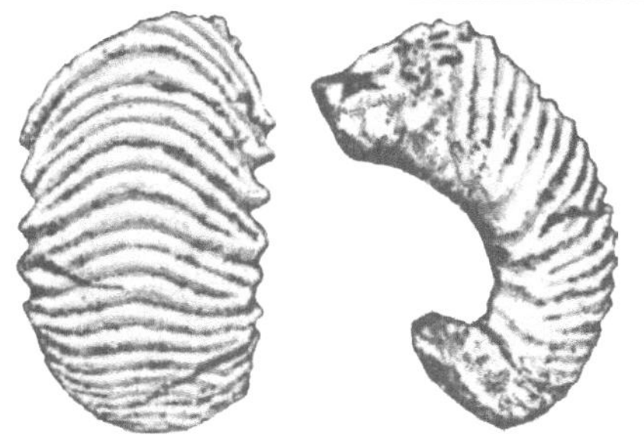

Fig. 540. *Progorodzovia*.

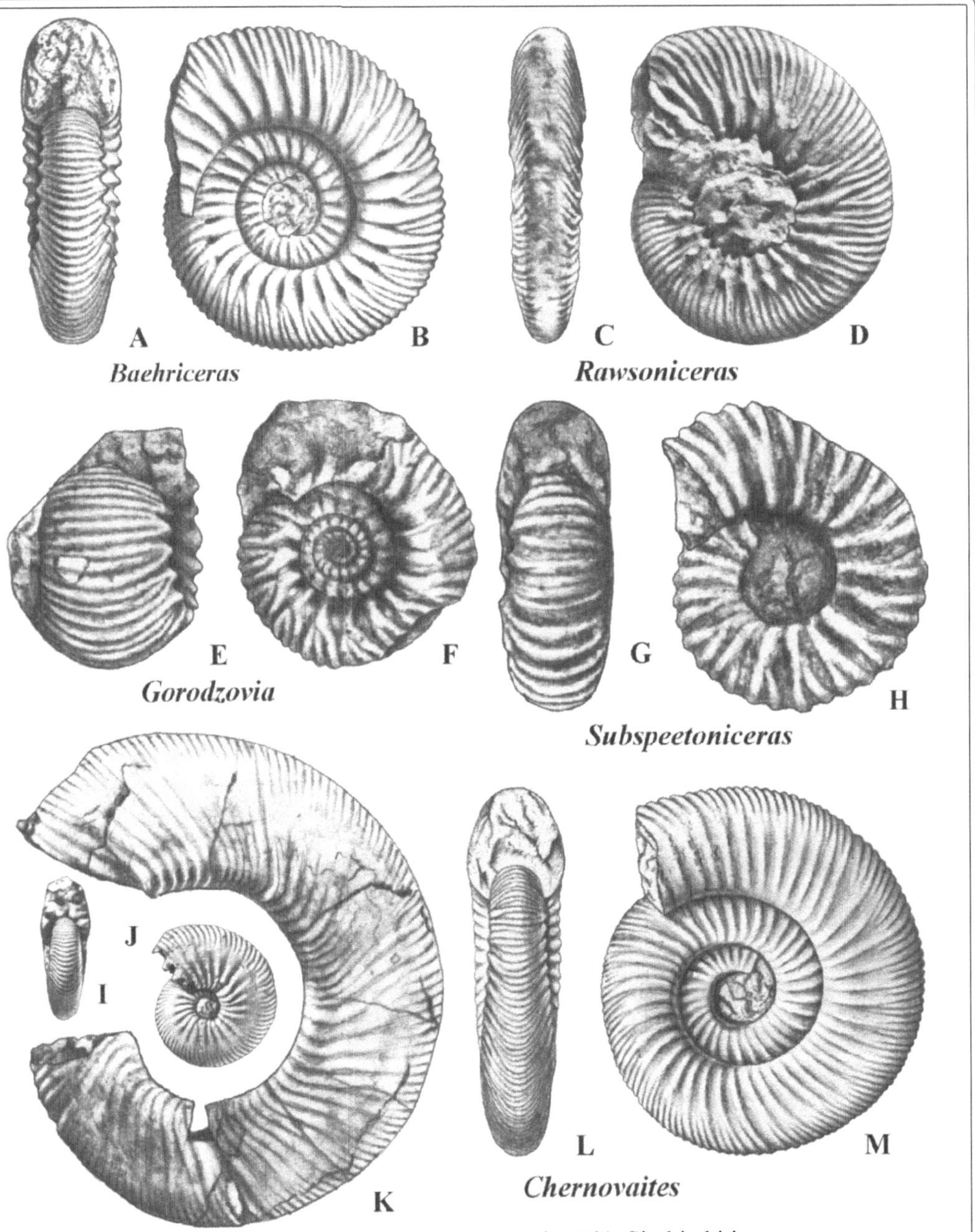

Fig. 541. Simbirskitinae.

but more inflated, with depressed coronate whorl section, sparser coarser shorter primaries, more prominent lateral tubercles, frequent rib trifurcation leading to more numerous secondaries; microconchs with biplicate ribbing to adult body chamber; suture with narrow E/L with prominent adventive saddle and trifid L. *U. Haut.* **Fig. 538A-B** - *S. decheni* (Roemer).

Thysanotoceras WHITEHOUSE, 1927 [*Ammonites picteti* Weerth, 1884; OD] [**M**]. Medium sized; like *Simbirskites* but more evolute, with steeper umbilical wall, coarser ribbing and secondary ribs well exposed in umbilicus in middle to later growth. *U. Haut.* **Fig. 538C-E** - *T. picteti* (Weerth).

Gorodzovia IVANOV & ARISTOV, 1969 [*G. mosquitini*; OD]. Small; like *Simbirskites* but an inflated cadicone with steeper umbilical wall, broadly rounded venter, depressed reniform whorl section and coarser ribbing. *U. Haut.* **Fig. 541E-F** - *G. mosquitini* Ivanov & Aristov.

Progorodzovia ARISTOV, 1984 [*Gorodzovia (P.) jaroslavlensis*; OD]. Like *Gorodzovia* but less inflated with ribs

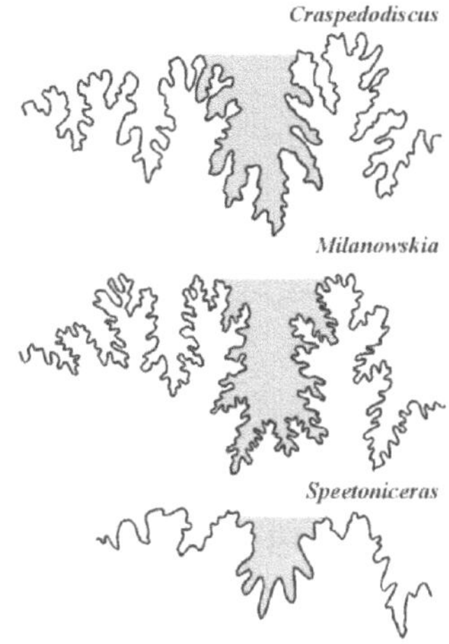

Fig. 542. Simbirskitine sutures.

zigzagging across venter with forward arch. *U. Valang.* **Fig. 540** - *P. jaroslavlensis* (Aristov).

Californiceras SHIMIZU, 1931 [*Ammonites traski* Gabb, 1864; OD]. Small; moderately involute, with subvertical umbilical wall, slightly flattened flanks, broadly rounded venter and slightly depressed subcircular whorl section; prominent short primary ribs terminate in small lateral tubercles below midflank and give rise to bundles of 3-5 prorsiradiate secondaries that occasionally bifurcate and cross venter with slight forward convexity; secondary ribs exposed in umbilicus and lateral tubercles becoming obsolete in later growth. *U. Haut.* **Fig. 538J-K** - *C. lecontei* (Anderson).

Gabb's holotype of *A. traski* Gabb was destroyed in the San Francisco fire of 1906 but, according to Imlay (1960), is possibly identical with *Polyptychites lecontei* Anderson (a macroconch).

Milanowskia CHERNOVA, 1952 [*Ammonites spetonensis* Young & Bird, 1828; OD]. Small; barely separable from *Californiceras* but slightly more involute and compressed, with higher whorls and narrower venter. *Uppermost L. - Lower U. Haut.* **Fig. 538F-G** - *M. spetonensis* (Young & Bird).

Pavlovites IVANOV & ARISTOV, 1969 [*Speetoniceras polyptychoides* Aristov, 1967; OD][**M**]. Medium sized to moderately large; like *Californiceras* but with slightly coarser more rigid ribs. *U. Haut.* **Fig. 538N-O** - *P. polyptychoides* (Aristov).

Baehriceras **gen. nov.** [*Olcostephanus virgifer* Neumayr & Uhlig, 1888, p.160; OD herein] [**M**]. Like *Milanowskia* but more evolute with compressed oval whorl section and shorter primaries that generally give rise to 3 secondaries. *U. Haut.* **Fig. 541A-B** - *B. virgifer* (Neumayr & Uhlig).

= *Hannoverites* Baehr, 1964 (*nom. nud.*)

= *Hannoverites* Kemper, 1968 (*nom. nud.*).

Chernovaites **gen. nov.** [*Craspedites gottschei* von Koenen, 1904, p.32; OD herein]. Moderately large; inner whorls involute, compressed, with narrowly arched venter and numerous slightly prorsiradiate primary ribs that extend to beyond midflank; 2-4 secondaries associated with each primary but with no definite point of branching, ribbing tends to weaken at midflank and to cross venter with a slight forward arch (juv.), later transversely (ad.); in middle growth secondaries may branch and long intercalatories appear; outer whorls evolute, relatively inflated, with numerous strongly prorsiradiate primary ribs that branch high up on flank with 4-6 shallow constrictions. *Middle U. Haut.* **Fig. 541I-K** - *C. gottschei* (von Koenen); **Fig. 541L-M** - *C. kayseri* (Neumayr & Uhlig).

Craspedodiscus SPATH, 1924 [*C. clypeiformis*, = *Ammonites discofalcatus* Lahusen, 1874; OD]. Large; inner whorls like those of *Chernovaites*; later a compressed oxycone with tiny umbilicus, broad flattened flanks, very narrowly rounded venter and high very compressed subtrigonal whorl section; densicostate, with primary ribs splitting into sheaves of many fine secondaries on lower flank; umbilical tubercles weak to absent; ribbing declines with growth and, on outer whorls, is reduced to weak folds and striae; suture like *Simbirskites* but more indented, with narrow E/L and L deeper than E. *U. Haut. - Lower L. Barrem.* **Fig. 538L-M** - *C. discofalcatus* (Lahusen).

Rawsoniceras **gen. nov.** [*Simbirskites (Craspedodiscus) variabilis* Rawson, 1968, p.57; OD herein]. Small; compressed, involute (juv.) to evolute (ad.), with elliptical whorl section; short primary ribs arise from feeble umbilical bullae and terminate in lateral tubercles below midflank from which arise 2-3 finer secondaries some of which bifurcate high on flank; intercalatories more common in later growth; ribs cross venter with forward arch. *Upper U. Haut., ?Lower L. Barrem.* **Fig. 541C-D** - *R. variabile* (Rawson).

Volgaia BAEHR, 1964 (*nom. nud.*).

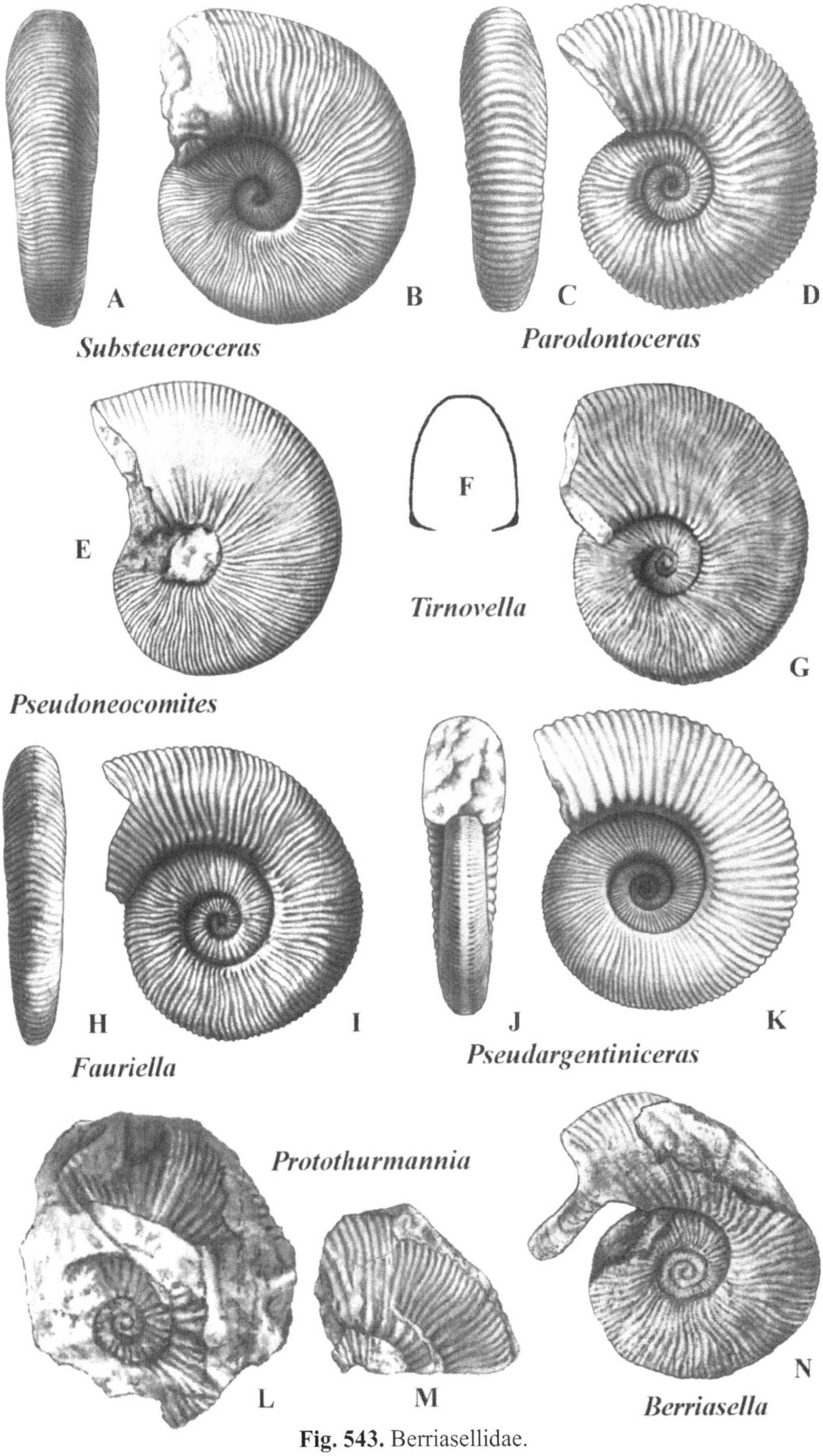

Fig. 543. Berriasellidae.

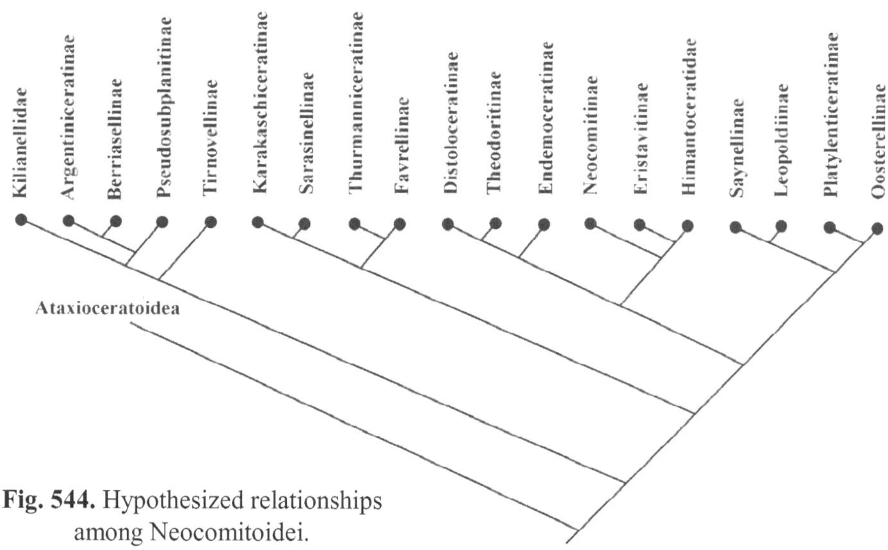

Fig. 544. Hypothesized relationships among Neocomitoidei.

Hyporder NEOCOMITOIDEI Salfeld, 1921

(*nom. transl.* herein *ex* Neocomitoidea)

Discussion. A monophyletic group with an origin in Ataxioceratoidei (Zeiss, 1968); comprises the superfamilies Berriaselloidea, Thurmanniceratoidea, Leopoldioidea and Endemoceratoidea. *U. Tithon. - U. Apt.*

Superfamily BERRIASELLOIDEA Spath, 1922

(*nom. transl.* Luppov *et al.*, 1957 *ex* Berriasellidae)

Discussion. A questionably monophyletic group with multiple suggested origins in Ataxioceratoidea (Zeiss, 1968) and relationships not fully resolved. Typically evolute with prorsiradiate biplicate ribs interrupted by a siphonal groove or smooth band, and suture with a retracted suspensive lobe. Comprises the families Berriasellidae and Kilianellidae. *Upper U. Tithon. - Lower U. Valang.*

Family BERRIASELLIDAE Spath, 1922

Discussion. A questionably monophyletic group of uncertain origin. Zeiss (1968) and Nikolov (1982) looked to Sublithacoceratidae for the antecedent, Le Hégarat (1979) to Ataxioceratidae and Tavera (1985) to Torquatisphinctidae; *Kossmatia* (Parodonticeratidae) also has been assigned to Berriasellinae (Verma & Westermann, 1973). Typically evolute, compressed, with biplicate ribs that in early to middle growth are interrupted by a siphonal groove or smooth band. Dimorphic, microconchs with lateral lappets. Comprises the subfamilies Tirnovellinae, Berriasellinae, Pseudosubplanitinae and Argentiniceratinae. *U. Tithon. - L. Valang.*

Subfamily Tirnovellinae nov.

Diagnosis. Involute to evolute, compressed, whorls mostly increasing rapidly in height, with grooved venter in early growth, later a smooth band and finally ribs crossing venter uninterrupted; fine sinuous prorsiradiate ribs arising in bundles from tiny umbilical bullae and bifurcating at various levels on flank. Shallow constrictions may parallel ribbing. A paraphyletic taxon of uncertain origin, perhaps Sublithococeratinae, close to *Sublithococeras* (Le Hégarat, 1979) or *Paraulacosphinctes* (Tavera, 1985) but, also, with a close similarity to *Kossmatia*. *U. Tithon. - L. Valang.*

Substeueroceras SPATH, 1923 [*Odontoceras koeneni* Steuer, 1897; OD]. Medium sized; moderately evolute (U= 30%), compressed, with gently convex flanks, broad subtabulate venter and subtrapezoidal whorl section increasing rapidly in height; in early to middle growth with fine dense biconcave prorsiradiate riblets arising singly or in subpolyplocoidal and polyplocoidal bundles flexing forward on outer flank to cross venter with slight forward convexity; ribs coarsen and become irregular in later growth, with development of feeble umbilical bullae; ventral groove lost early (D<50 mm); suture with trifid L deeper than E, and retracted suspensive lobe. *U. Tithon. - L. Berrias.* **Fig. 543A-B** - *S. koeneni* (Steuer).

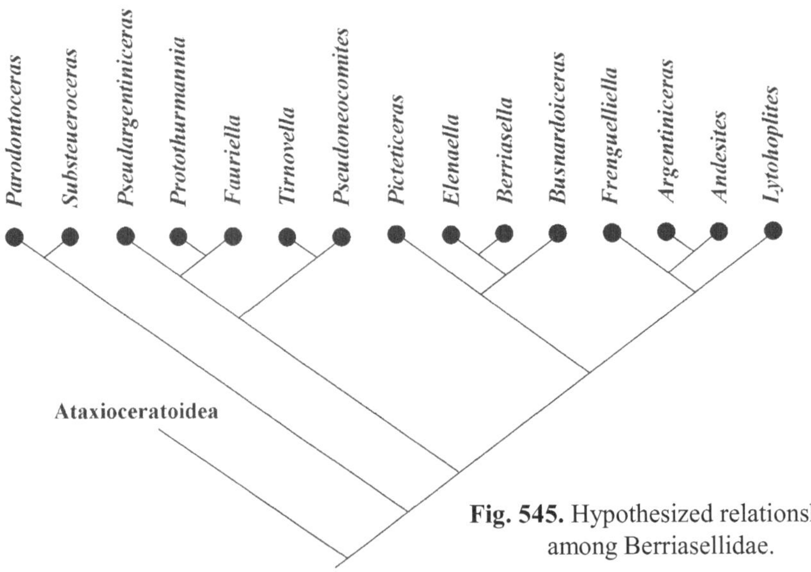

Fig. 545. Hypothesized relationships among Berriasellidae.

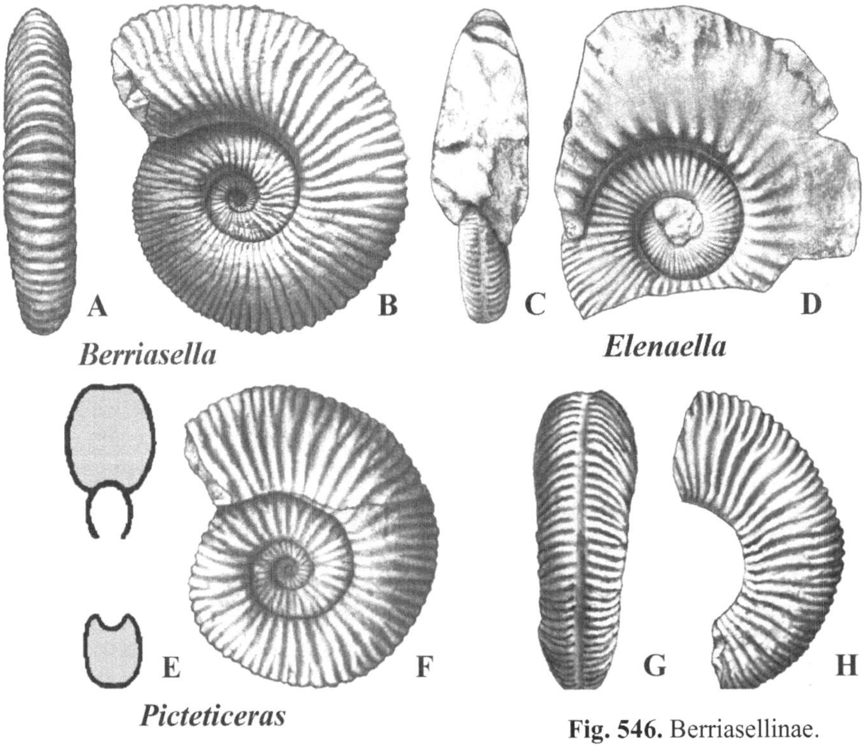

Fig. 546. Berriasellinae.

Parodontoceras SPATH, 1923 [*Hoplites calistoides* Behrendsen, 1891; OD]. Small to medium sized; like *Substeueroceras* but with evenly rounded venter and compressed oval whorls increasing rapidly in height; numerous fine flexuous biplicate ribs, occasionally single or triplicate, branching just above midflank (lower later) and crossing venter transversely; suture with L much deeper than E, and strongly retracted suspensive lobe. *U. Tithon.* **Fig. 543C-D** - *P. calistoides* (Behrendsen). = *Paradontoceras* Howarth, 1998 (*nom. van.*).

Tirnovella NIKOLOV, 1966 [*Berriasella alpillensis* Mazenot, 1939; OD]. Small (**m**) to medium sized (**M**); compressed, initially moderately involute (U=32%) but showing scaphitoid uncoiling with whorls increasing rapidly in

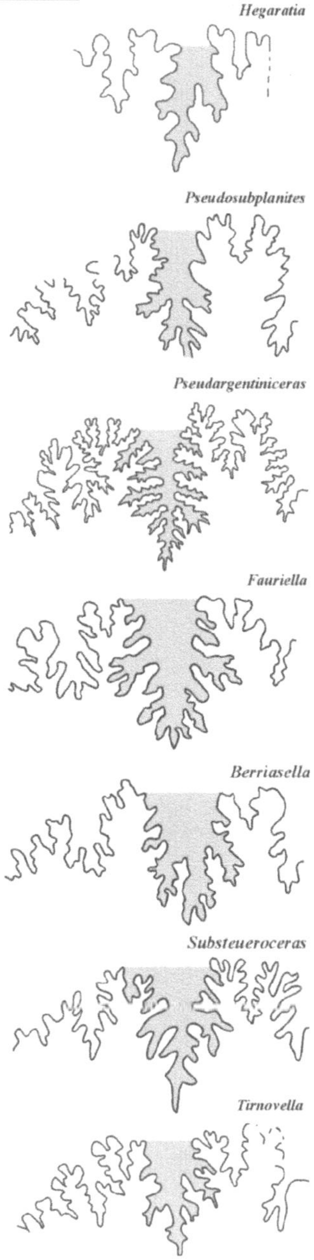

Fig. 547. Berriasellid sutures.

height; umbilical wall vertical; in early growth (D<25 mm) venter grooved; in middle growth with thin smooth siphonal band that fades at D=50-60 mm; densicostate, with 1-3 fine prorsiradiate sinusoidal riblets arising from small umbilical bullae in subpolyplocoidal bundles with frequent intercalatories of various length; single ribs and one branch of bundles generally bifurcates below midflank; ribs cross venter transversely, with slight thinning over siphonal line, and weakening across flanks in maturity. *U. Tithon.* - *U. Berrias.* **Fig. 543F-G** - *T. alpillensis* (Mazenot).

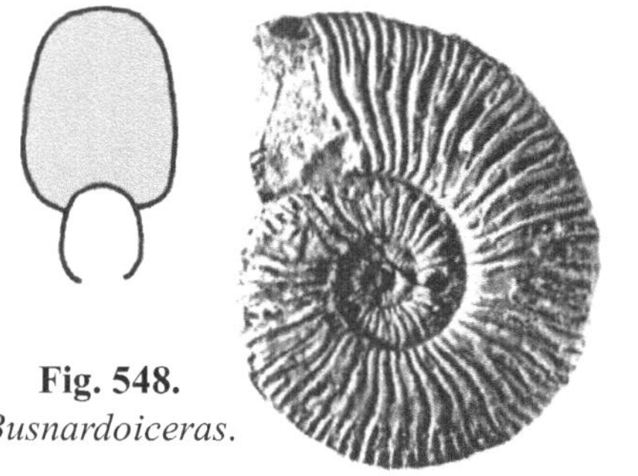

Fig. 548. *Busnardoiceras.*

Pseudoneocomites HOEDEMAKER, 1982 [*Hoplites retowskyi* Sarasin & Schöndelmayer, 1893; OD]. Small, involute and high whorled, with steep to overhanging umbilical wall, rounded venter, bundles 2-3 fine flexuous ribs originating from umbilical edge, and smooth adult body chamber. *L. Berrias.* **Fig. 543E** - *P. retowskyi* (Sarasin & Schöndelmayer).

Protothurmannia CRICKMAY, 1932 [*P. rezanoffiana*; OD]. Large, evolute; early whorls with compressed subovate whorls and subtabulate venter; slightly flexed biplicate ribs interrupted on venter; in later growth ribs flexuous, branching close to umbilical edge, and finally massive, with umbilical bullae and loss of ventral sulcus at c.80 mm diameter; suture complex, with exceptionally long narrow lobes and narrow trifid L. *U. Tithon.* **Fig. 543L-M** - *P. rezanoffiana* Crickmay.

Fauriella NIKOLOV, 1966 [*Berriasella gallica* Mazenot, 1939; OD]. Small (**m**) to medium sized (**M**); like *Tirnovella* but with ribs arising singly and in pairs from tiny umbilical bullae, with intercalatories, bifurcating just above midflank, then flexing forward to cross venter uninterrupted with slight forward arch; microconch lappetted. *Upper U. Tithon.* - *Lower L. Valang.* **Fig. 543H-I** - *F. gallica* (Mazenot).

Pseudargentiniceras SPATH, 1925 [*Ammonites abscissus* Oppel, 1868; OD]. Moderately large, evolute, compressed, with slightly convergent flanks, broad subtabulate venter and compressed elliptical whorl section; ribs fine, dense, flexuous, prorsiradiate, mostly bifurcating above midflank with frequent single ribs and persistent smooth siphonal band; in later growth uniform ribs arise mostly in pairs from umbilical bullae; suture with narrow trifid L deeper than E and retracted suspensive lobe. *L. Berrias.* **Fig. 543J-K** - *P. abscissum* (Oppel).

Subfamily Berriasellinae Spath, 1922

(*nom. transl.* Roman, 1938 *ex* Berriasellidae)

Discussion. A taxon of questionable ancestry. Some have looked to Virgatosphinctinae (Zeiss, 1968; Nikolov, 1982) close to *Aulacostephanoides*,

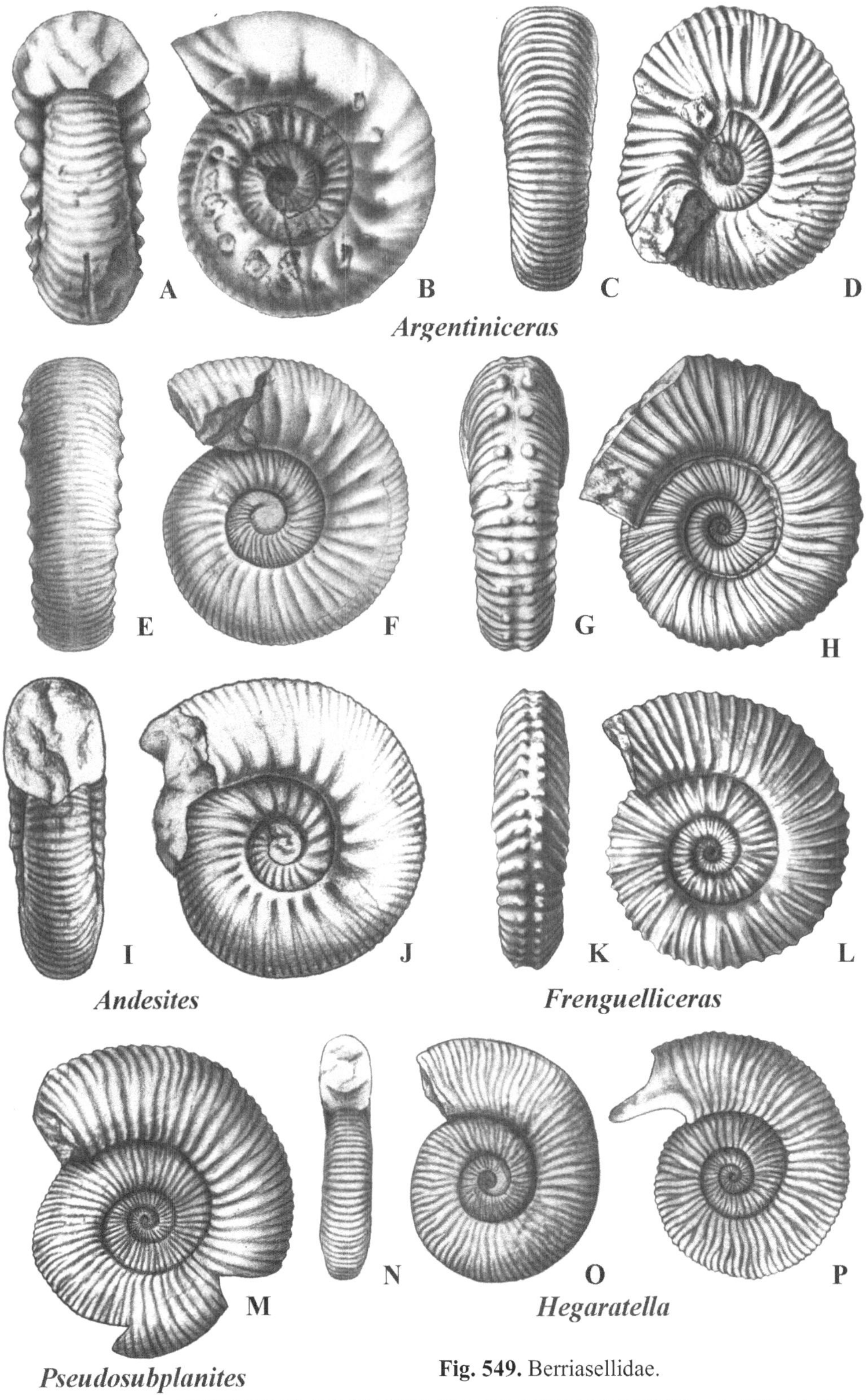

Fig. 549. Berriasellidae.

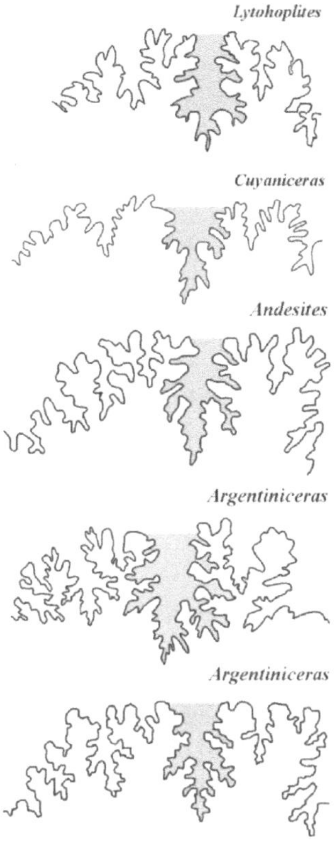

Fig. 550. Argentiniceratine sutures.

whereas Tavera (1985) favoured Andalusisphinctinae close to *Neoperisphinctes* and, by implication, Verma & Westerman (1973) suggested Paraboliceratidae, close to *Kossmatia*. An origin in Tirnovellinae is favoured here. Mostly moderately evolute, compressed, with ovate to subelliptical whorl section and subrounded venter with siphonal groove that is lost in maturity. Strong biplicate ribs fork high on the flank and cross the venter transversely, coarsening in maturity (**M**) and sometimes developing umbilical and/or ventrolateral nodes; microconchs lappetted. Suture with narrow E, trifid L and retracted suspensive lobe. *U. Tithon. - L. Valang.*

Busnardoiceras TAVERA, 1985 [*Parapallasiceras busnardoi* Le Hégarat, 1973; OD]. Small, evolute, high whorled, with compressed subelliptical whorl section in maturity; phragmocone with rectiradiate to slightly prorsiradiate biplicate ribs, branching occurring on upper third of flank; body chamber with irregular intercalation of slightly sinuous simple and virgatotome ribs, interrupted across venter by shallow groove. *U. Tithon. - L. Berrias.* **Fig. 548** - *B. busnardoi* (Le Hégarat).

Berriasella UHLIG, 1905 [*Ammonites privasensis* Pictet, 1867; SD Roman, 1938] [**m**]. Small; like *Parodontoceras* but more evolute (U=37-40%), with subelliptical whorl section, whorls not increasing in height so rapidly and with coarser flexuous ribs branching higher on flank in later growth; inner whorls with siphonal groove that evanesces later when transverse ribs cross venter uninterrupted; ribbing coarsens on body chamber of macroconchs, microconchs with lateral lappets; suture with narrow bifid E/L, large trifid L deeper than E, and retracted suspensive lobe. *Uppermost Tithon. - Berrias.* **Fig. 543N** - *B. jauberti* Mazenot (**m**), **Fig. 546A-B** - *B. privasensis* (Pictet); **Fig. 546G-H** - *"B". besairiei* Collignon.

Frau *et al.* (2016) suggested that *Berriasiella* is restricted to the Tethyan Realm and that records from the Indo-Pacific are endemic berriasellids.

= *Beriasella* Gerth, 1931 (*nom. van.*).

Elenaella NIKOLOV, 1966 [*Berriasella curalensis* Mazenot, 1939; OD] [**M**]. Small, in early to middle growth like *Berriasella*; later with broad flat convergent flanks, narrowly arched venter, small pinched umbilical bullae to primaries, and finer ribs that bifurcate higher on flank and tend to weaken at midflank; macroconchs with smooth adult body chamber, superficially similar to *Dalmasiceras*. *U. Berrias.* **Fig. 546C-D** - *E. cularensis* (Mazenot).

Picteticeras LE HÉGARAT, 1971 [*Hoplites picteti* Jacob, 1912; SD Sapunov, 1979] [**m**]. Small; like *Berriasella* but more evolute (U=47-50%), with broadly rounded to subtabulate venter, subrectangular whorl section and coarser straighter ribs. *U. Tithon. - Berrias.* **Fig. 546E-F** - *P. picteti* (Jacob).

Subfamily Argentiniceratinae Nikolov, 1979

Discussion. Seemingly monophyletic derivatives of Berriasellinae (Nikolov, 1982), with an origin close to *Picteticeras*; medium sized to large, evolute, with broadly rounded to subtabulate venter, subquadratic to rounded rectangular whorl section in maturity, with or without siphonal groove or smooth band. Prominent umbilical bullae give rise to fasciculate bundles of ribs that cross venter transversely, and may develop lateral or ventrolateral nodes. Suture with E as deep as L, broad stemmed rectangular E/L and small U_2. *?U. Tithon., L. Berrias.*

Argentiniceras SPATH, 1924 [*Odontoceras malarguense* Steuer, 1897, = *Reineckeia argentina* Steuer, 1897; OD] [**M**]. Moderately large; in early to middle growth (D<100 mm) fairly involute (U=31%), compressed, with flat subparallel flanks, rounded venter and slightly depressed subquadratic to subrounded whorl section; weak umbilical bullae give rise to pairs slightly prorsiradiate sinuous ribs, rarely single, crossing venter transversely; occasional rare constrictions; adults evolute (U=40-46%),

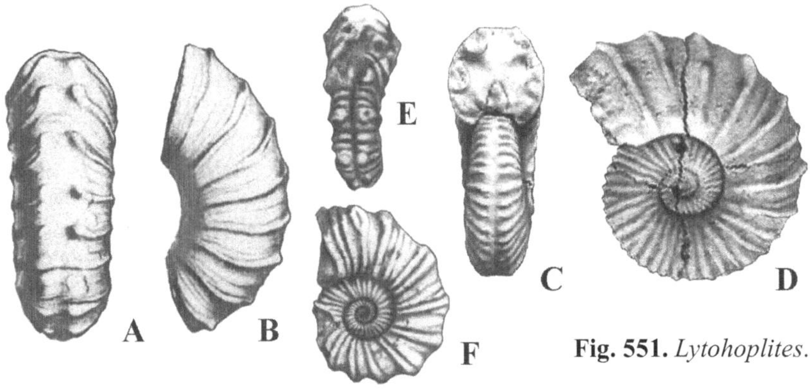

Fig. 551. *Lytohoplites.*

with convex flanks and venter, and subrounded whorl section; ribs bifurcate on lower flank in early growth, furcation rising to midflank during ontogeny, with strong tubercles at point of branching in maturity; 2-3 constrictions per whorl; suture with narrow E almost as deep as or deeper than L. *U. Tithon. - L. Berrias.* **Fig. 549A-F** - *A. argentinum* (Steuer).

= *Argenticeras* Weaver, 1931 (*nom. van.*).

Berriasella steinmanni Krantz (1928) (**Fig. 549G-H**) is said to fall within the intraspecific variation of *A. fasciculatum*; in later growth it has wiry ribs looped to swollen rounded ventrolateral tubercles separated by 1-2 ribs with or without small tubercles; suture with narrow trifid L.

Andesites GERTH, 1925 [*Perisphinctes loncochensis* Steuer, 1897; OD]. Like *Argentiniceras* but with distinct umbilical bullae giving rise to fasciculate bundles of 3-4 prorsiradiate ribs that cross venter transversely; suture with very narrow E, subquadratic E/L, narrow trifid L, broad L/U_2 and retracted suspensive lobe. *L. Berrias.* **Fig. 549I-J** - *A. loncochensis* (Steuer).

Frenguelliceras A. LEANZA, 1945 [*F. magister*; OD]. Small to medium sized; very evolute, with slightly convex flanks, rounded venter and compressed whorl section; umbilical bullae give rise to 2-3 gently sinuous, rectiradiate to slightly prorsiradiate splayed ribs, with frequent singles, all with small rounded ventrolateral tubercles; 2-3 constrictions per whorl that parallel ribbing on adult body chamber. *L. Berrias.* **Fig. 549K-L** - *F. magister* A. Leanza.

Lytohoplites SPATH, 1925 [*Hoplites burckhardti* Mayer-Eymar, 1900; OD]. Small, evolute, moderately inflated, with convergent convex flanks, trapezoidal to quadratic whorl section and narrow subtabulate venter; strong distant sinuous prorsi-radiate ribs bifurcate at midflank, posterior branch thicker than anterior one; ribs may form weak lateral bullae at point of branching and form prominent ventrolateral bullae, weakening or interrupted across venter by a siphonal groove; in maturity all ribs single and flanks strongly convex. *U. Tithon., ?L. Berrias.* **Fig. 551A-D** - *L. burckhardti* (Mayer-Eymar); **Fig. 551E-F** - *L. besairei* Collignon.

Here interpreted as a progenetic offshoot of *Argentiniceras.*

Subfamily Pseudosubplanitinae Nikolov & Sapunov, 1977

Discussion. A monophyletic taxon with an origin in Berriasellinae, close to *Busnardoiceras* (Frau *et al.*, 2014). Typically strongly compressed, evolute, densicostate, with ribs uninterrupted or with slight thinning across siphonal line. *Upper U. Tithon. - Lower L. Berrias.*

Pseudosubplanites LE HÉGARAT, 1973 [*P. berriasensis*; OD] [**M**]. Medium sized to moderately large; evolute (U=30-32%), compressed, with broad weakly convex flanks converging to rounded venter and strongly compressed oval to subrectangular whorl section; ribs dense, fine, sharp, weakly flexuous, crossing venter transversely without interruption; in early growth ribs mostly biplicate, branching at various levels but mostly above midflank; in later growth variocostate with ribs trifurcating, polygyrate, fasciculate or polyplocoidal; suture with deep narrow E, bifid E/L and asymmetrically trifid L. *Upper U. Tithon. - Lower L. Berrias.* **Fig. 549M** - *P. berriasensis* (Le Hégarat).

Hegaratella NIKOLOV & SAPUNOVA, 1977 [*Berriasella paramacilenta* Mazenot, 1939; OD] [**m**]. Small; like *Pseudosubplanites* but with flexuous biplicate ribs (occasionally single) that zigzag across venter with slight weakening over siphonal line in maturity; in early growth ribs bifurcate high on flank, but later lower down; adult aperture with lateral lappets; suture with very narrow asymmetrically bifid L. *Upper U. Tithon. - Lower L. Berrias.* **Fig. 549N-P** - *H. jauberti* (Mazenot).

Regarded the microconch of *Pseudosubplanites*, but in Bulgaria they occur at different stratigraphical levels (Nikolov, 1982).

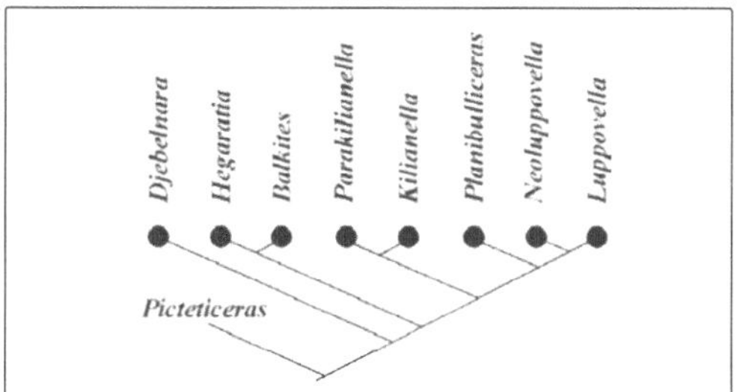

Fig. 552. Hypothesized relationships among Kilianellidae.

Family KILIANELLIDAE nov.

Diagnosis. Typically small, evolute with narrow subtabulate venter, coarse flexuous ribs arising from umbilical bullae, ventrolateral tubercles, frequent deep constrictions particularly in early growth and smooth siphonal band; lateral tubercles commonly develop at rib furcation. Monophyletic derivatives of Berriasellidae (Uhlig, 1905; Mazenot, 1939), close to *Picteticeras* (Company, 1987). *L. Berrias. - Lower U. Valang.*

***Djebelnara* gen. nov.** [*Berriasella constricta* Arnould-Saget, 1951, p.57; OD herein]. Very small; evolute, with convex flanks, tabulate venter, vertical umbilical wall and rounded quadratic whorl section; strong rectiradiate non-tuberculate primary ribs bifurcate or trifurcate on outer flank, with 6-7 constrictions on outer whorl generally fronted by single rib; ribs interrupted by shallow siphonal groove that evanesces on body chamber when ribs become sigmoidal. *L. Berrias.* **Fig. 554P-Q** - *D. constricta* (Arnould-Saget).

Hegaratia PATRULIUS & AVRAM, 1976 [*Corongoceras? (H.) busnardoi*; OD] [**m**]. Very small; like *Djebelnara* but with simple and biplicate ribs terminating in ventrolateral tubercles separated by shallow siphonal groove; lateral tubercles at point of rib furcation and ventrolateral clavi in maturity. *L. Berrias.* **Fig. 554H-I** - *H. busnardoi* (Patrulius & Avram).

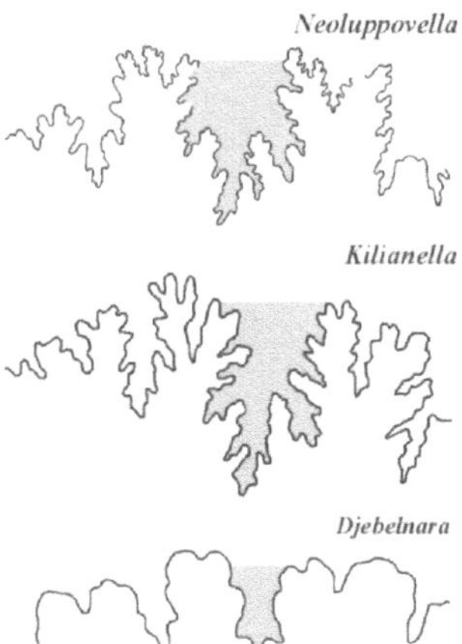

Fig. 553. Kilianellid sutures.

Balkites BOGDANOVA & KVANTALIANI, 1983 [*B. balkensis*; OD]. Small; like *Kilianella* but more inflated, with coarser ribs that terminate perpendicular to siphonal line without forming significant tubercles. *U. Berrias.* **Fig. 554F-G** - *B. balkensis* Bogdanova & Kvantaliani.

Kilianella UHLIG, 1905 [*Hoplites pexiptychus* Uhlig, 1882; SD Roman, 1938, = *Ammonites roubaudianus* d'Orbigny, 1850] [**m**]. Small; compressed, evolute, with tabulate venter and rectangular whorl section; coarse sinuous single and biplicate ribs terminate in oblique ventrolateral clavi separated by smooth siphonal groove; ribs may show incipient tubercles at point of branching; frequent deep flexuous constrictions; suture with inner branch of E/L larger than outer branch, asymmetrically trifid L deeper than E, narrow L/U_2 and retracted suspensive lobe. *U. Berrias. - L. Valang.* **Fig. 554A-B** - *K. pexiptycha* (Uhlig).

Parakilianella SATO, 1961 [*P. umazawensis*; M] [**M**]. Moderately large to large; inner whorls like *Kilianella*; in later growth with compressed quadrangular whorl section and simple flexuous ribs that are periodically strengthened and loop to ventrolateral clavi. *U. Berrias.* **Fig. 554E** - *P. umazawensis* Sato.

Perhaps a macroconch of *Kilianella* (Wright, 1997).

Planibulliceras BARABOSHKIN & MIKHAILOVA, 1994 [*Luppovella (P.) kachensis*; OD]. Small; like *Balkites* but with 2-3 simple ribs between branched ones, prominent lateral tubercles at point of rib furcation, and unequal ventrolateral tubercles. *L. Valang.* **Fig. 554N-O** - *P. kachense* (Baraboshkin & Mikhailova).

Neoluppovella BARABOSHKIN & MIKHAILOVA, 2000 [*Luppovella (N.) spinosa*; OD]. Very small; like *Planibulliceras* but with uniform equal-sized ventrolateral tubercles. *Upper L. Valang.* **Fig. 554J-M** - *N. spinosa* (Baraboshkin & Mikhailova).

Luppovella NIKOLOV, 1966 [*Thurmannia (Kilianella) superba* Sayn, 1907; OD] [**m**]. Small; like *Neoluppovella* but with robust bar-like primary ribs with prominent rounded high-lateral tubercles from which paired prorsiradiate secondaries arise and terminate in strong ventrolateral bullae; some ribs lack lateral tubercles. *Upper L. - Lower U. Valang.* **Fig. 554C-D** - *L. superba* (Kilian).

Superfamily THURMANNICERATOIDEA nov.

Diagnosis. Primitively with ventrolateral bullae, later also lateral tubercles. A monophyletic taxon with an origin in Berriaselloidea comprising the families Thurmanniceratidae and Sarasinellidae. *U. Tithon. - L. Haut.*

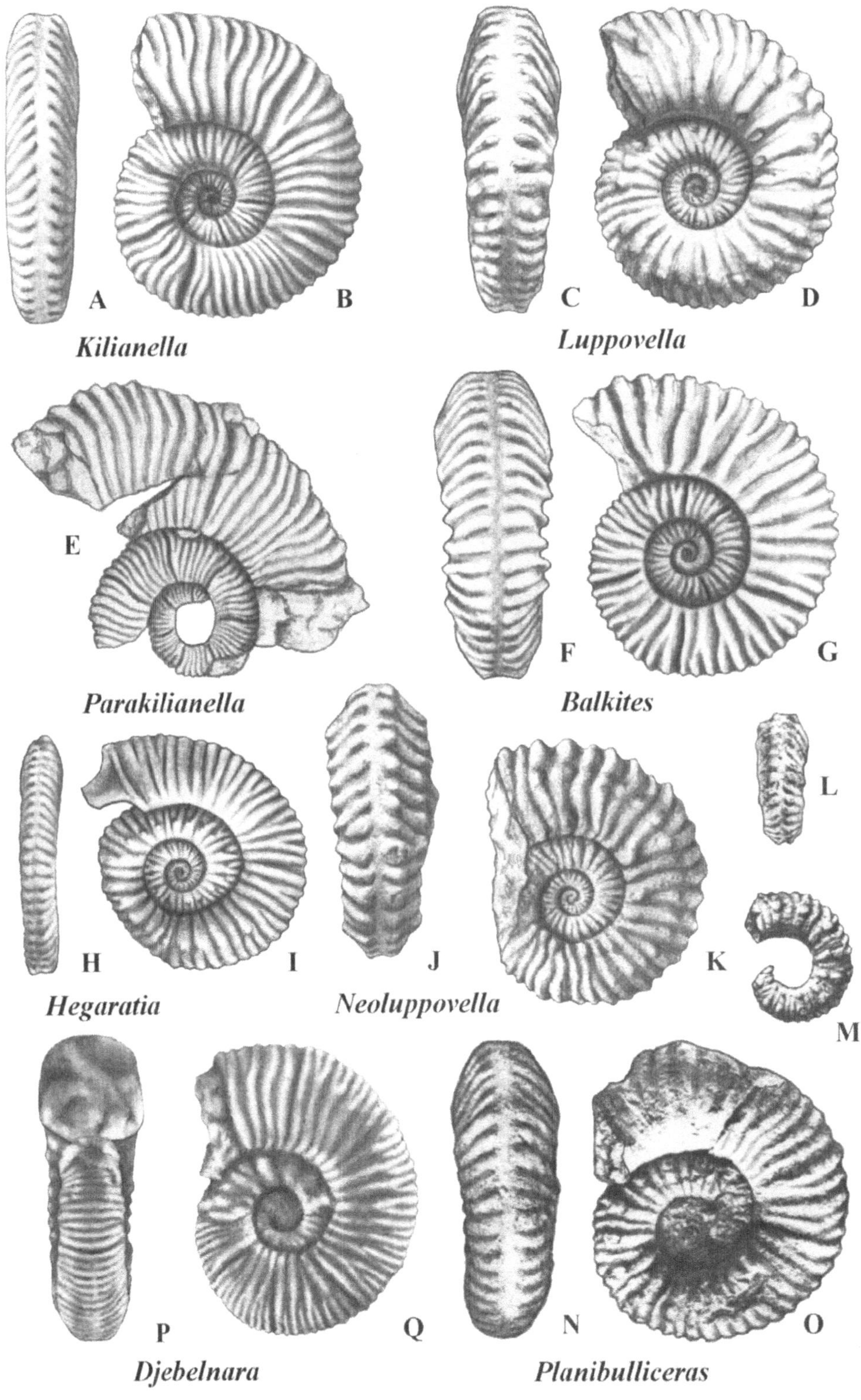

Fig. 554. Kilianellidae.

Family THURMANNICERATIDAE nov.

Discussion. Typically moderately evolute, compressed and high whorled, with narrow tabulate shallowly grooved to smooth venter; sinusoidal biplicate to polyplocoidal prorsiradiate ribs arise from incipient to small umbilical bullae, terminating in ventrolateral bullae. Suture with trifid L and retracted suspensive lobe. Monophyletic derivatives of Berriasellidae (Sayn, 1907; Mazenot,1939; Baraboshkin & Mikhailova, 2000) comprising the subfamilies Thurmanniceratinae and Favrellinae. *U. Tithon. - L. Haut.*

Subfamily Thurmanniceratinae nov.

Diagnosis. Umbilical bullae feeble. Seemingly monophyletic derivatives of Berriasellinae close to *Berriasella* (Le Hégarat, 1973). *U. Tithon. - L. Valang.*

Delphinella LE HÉGARAT, 1971 [*Hoplites delphinensis* Kilian, 1889; OD] [**m**]. Small to medium sized; fairly evolute (U=36-37%), compressed, whorls increasing rapidly in height, with broad gently convex flanks, narrow subtabulate venter with subangular shoulders and subtrapezoidal whorl section; venter with siphonal groove in early growth, later smooth band; fine prorsiradiate biplicate ribs branching on outer third of flank with ventrolateral bullae; ribs show incipient thickening at midflank; on adult body chamber (**M**) ribs weak or effaced from flank; microconchs with lateral lappets. *U. Tithon. - M. Berrias.* **Fig. 557A-B** - *D. delphinensis* (Kilian); **Fig. 557I-J** - *D. storssi* (Stanton) (**M**).
= *Stenoceras* Uhlig, 1911 (*non* d'Orbigny, 1849).

Strambergella NIKOLOV, 1966 [*Ammonites carpathicus* Zittel, 1868; OD]. Medium sized to large; like *Delphinella* but with narrower umbilicus (U=32%), umbilical seam egressing in maturity; adult body chamber with triplicate and polyplocoidal ribs arising from small umbilical bullae. *U. Tithon. - Berrias.* **Fig. 557C-D** - *S. carpathica* (Zittel).

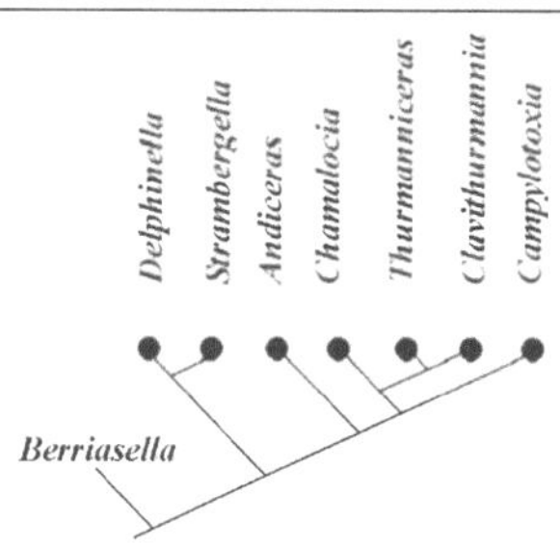

Fig. 555. Hypothesized relationships among Thurmanniceratinae.

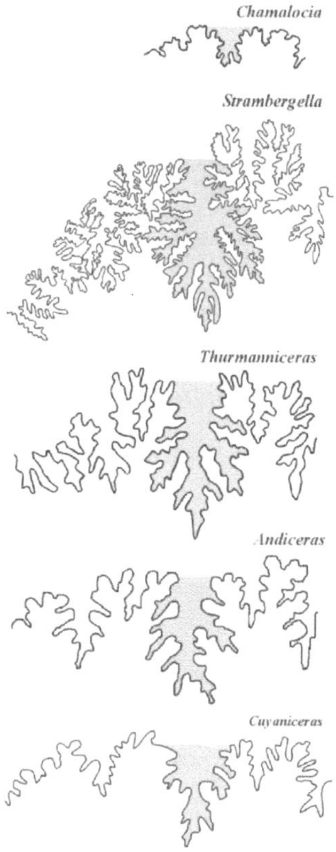

Fig. 556. Thurmanniceratine sutures.

Andiceras KRANTZ, 1928 [*A. trigonostomum*; SD Arkell, 1957]. Medium sized; very evolute (U=46%), compressed, with convergent flanks, subtrigonal whorl section and wiry biplicate ribs that, in later growth, frequently branch below midflank and with many single ribs; ribs thicken on ventrolateral shoulders and cross ventral sulcus. *Uppermost Tithon. - Lowermost Berrias.* **Fig. 557G-H** - *A. trigonostomum* Krantz.

Thurmanniceras COSSMAN, 1901 [*Ammonites thurmanni* Pictet & Campiche, 1860; SD Bauberger, 1906]. Small (**m**) to moderately large (**M**); in early to middle growth moderately evolute (U=29-33%), strongly compressed, with broad gently convex flanks, narrow tabulate venter, subrectangular whorl section, prorsiradiate paired ribs and frequent constrictions; in later growth ribs sinusoidal, arising mostly in pairs from incipient umbilical bullae, but also singly, with one or both branches bifurcating above midflank; all ribs terminate in ventrolateral bullae leaving smooth narrow siphonal band; suture with asymmetrically bifid E/L and trifid L much deeper than E. *L. Valang.* **Fig. 557E-F** - *T. thurmanni* (Pictet & Campiche).
= *Thurmannia* Hyatt, 1900 (*non* Heer, 1852).
= *Thurmannites* Kilian & Reboul, 1914 (obj.).

Clavithurmannia THIEULOY, 1979 [*Thurmanniceras (C.) foraticostatum*; OD] [**M**]. Moderately large; inner and middle whorls like *Thurmanniceras*; last part of phrag-

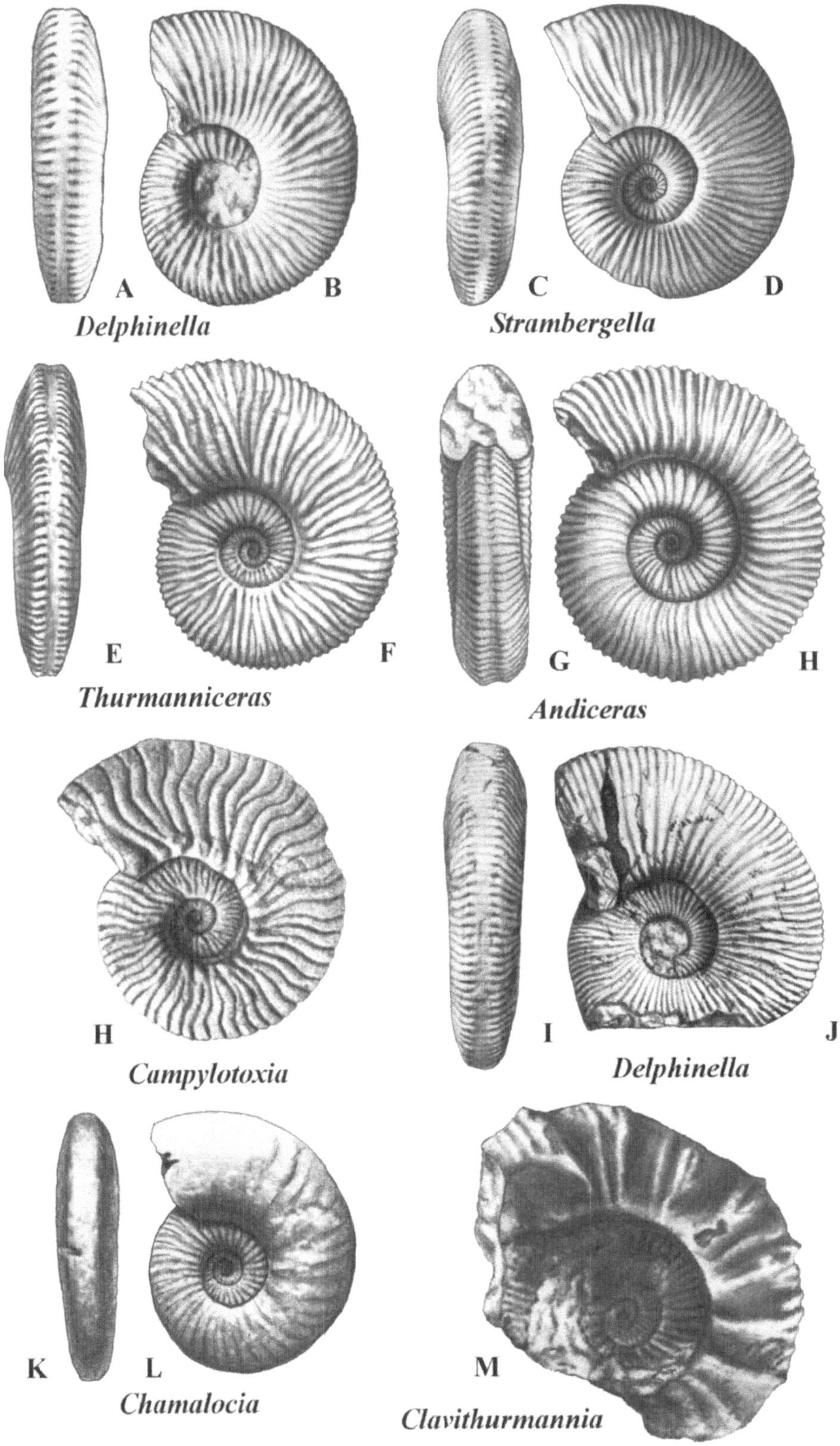

Fig. 557. Thurmanniceratinae.

mocone and adult body chamber with robust distant trituberculate main ribs, each with umbilical and inner and outer ventrolateral tubercles, separated by fine non-tuberculate secondary. *L. Valang.* **Fig. 557M** - *C. foraticostata* (Thieuloy).

Campylotoxia BARABOSHKIN & MIKHAILOVA, 2000 [*Hoplites campylotoxus* Uhlig, 1902; OD]. Medium sized, moderately evolute, with subtabulate venter and compressed trapezoidal whorl section; coarse sickle-shaped, prorsiradiate ribs, broader than interspaces, arise from small umbilical bullae, bifurcate on the flank in later growth, and terminate in oblique ventrolateral clavi that become bullate in maturity, separate by a smooth siphonal band; suture with trifid L deeper than E. *Upper L. Valang.* **Fig. 557H** - *C. campylotoxa* (Uhlig).

Chamalocia THIEULOY, 1971 [*Leopoldia subaenigmatica* Sayn, 1907; OD]. Very small; like *Thurmanniceras* but much smaller, more involute, with flat flanks, ventrolateral facets, tabulate venter and subrectangular whorl section; early whorls strongly ribbed, with pointed ventrolateral tubercles; in later growth ribs effaced from flanks; suture relatively simple, with very narrow E, very broad shallowly bifid E/L and narrow asymmetrically bifid L as deep as E. *Upper L. - Lower U. Valang.* **Fig. 557K-L** - *C. subaenigmatica* (Sayn).

Subfamily Favrellinae nov.

Diagnosis. Involute to very evolute with feeble to distinct umbilical bullae giving rise to simple and biplicate ribs that project strongly forwards and are linguoid or form chevrons over the venter in maturity. Ribs with ventrolateral tubercles and grooved or smooth siphonal band in early to middle growth, with or without lateral tubercles. A questionably monophyletic Andean group descended from Thurmanniceratinae, close to *Thurmanniceras* (Parent *et al.*, 2015). *U. Tithon. - L. Haut.*

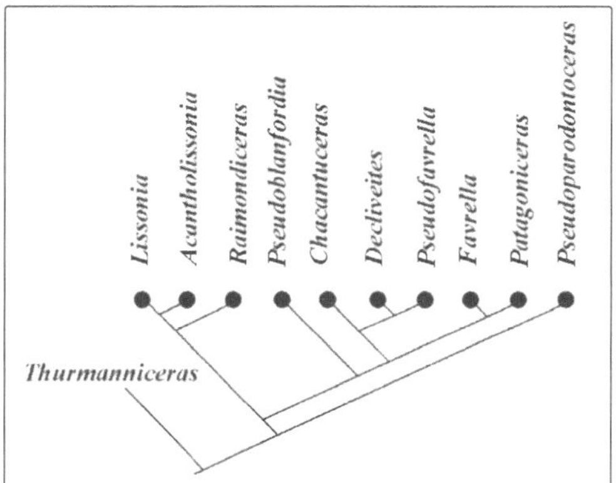

Fig. 558. Hypothesized relationships among Favrellinae.

Fig. 559. *Lissonia.*

Raimondiceras SPATH, 1924 [*Hoplites raimondii* Lissón, 1907; OD]. Small; evolute, compressed, with broad flattened flanks, angular ventrolateral shoulders and compressed subrectangular whorl section; strong prorsiradiate primary ribs with feeble umbilical bullae and prominent high-lateral tubercles give rise to 2-3 secondary ribs that project strongly forward on venter; primary ribs separated by 1-2 non-tuberculate ribs; on phragmocone all ribs terminate in feeble ventral tubercles on either side of smooth siphonal band; adult body chamber with narrowly arched venter crossed by linguoid ribs; suture with subtrifid L. *U. Tithon.* **Fig. 560P-Q** - *R. raimondii* (Lissón).

= *Pfluckeria* Lissón, 1924 (obj.).

Acantholissonia A. LEANZA, 1972 [*Spiticeras gerthi* Weaver, 1931; OD]. Small to medium sized; like *Raimondiceras* but with coarser ribbing and prominent tuberculation; ribs meet deep ventral groove at very acute angle without forming tubercles. *L. Valang.* **Fig. 560N-O** - *A. gerthi* (Weaver).

Chacantuceras AGUIRRE-URRETA & RAWSON, 1999 [*Neocosmoceras ornatum* Oliviero, 1983; OD]. Very large to enormous; compressed, with whorls increasing rapidly in height, inclined umbilical wall, strong coarse distant ribs separated by 1-2 intercalatories, with umbicolateral and lateral tubercles from which ribs project fowards and terminate in oblique ventrolateral clavi; suture with very broad bifid L unlike other members of the subfamily. *U. Valang.* **Fig. 561** - *C. ornatum* (Oliviero).

Pseudofavrella A. LEANZA & H. LEANZA, 1973 [*Hoplites angulatiformis* Behrendsen, 1892; OD]. Small, evolute, compressed with tabulate (juv.) to narrowly rounded (ad.) venter and compressed oval whorl section; primary ribs with incipient to weak umbilical bullae, ending in subdued lateral tubercles; intercalated and bifurcating secondary ribs projecting strongly forward, terminating in distinct ventrolateral tubercles to leave smooth siphonal; band; in maturity ribs cross venter, forming chevrons; suture with wide L deeper than E, and L/U_2 projecting beyond E/L. *Lower L. Valang.* **Fig. 560C-D** - *P. angulatiformis* (Behrendsen).

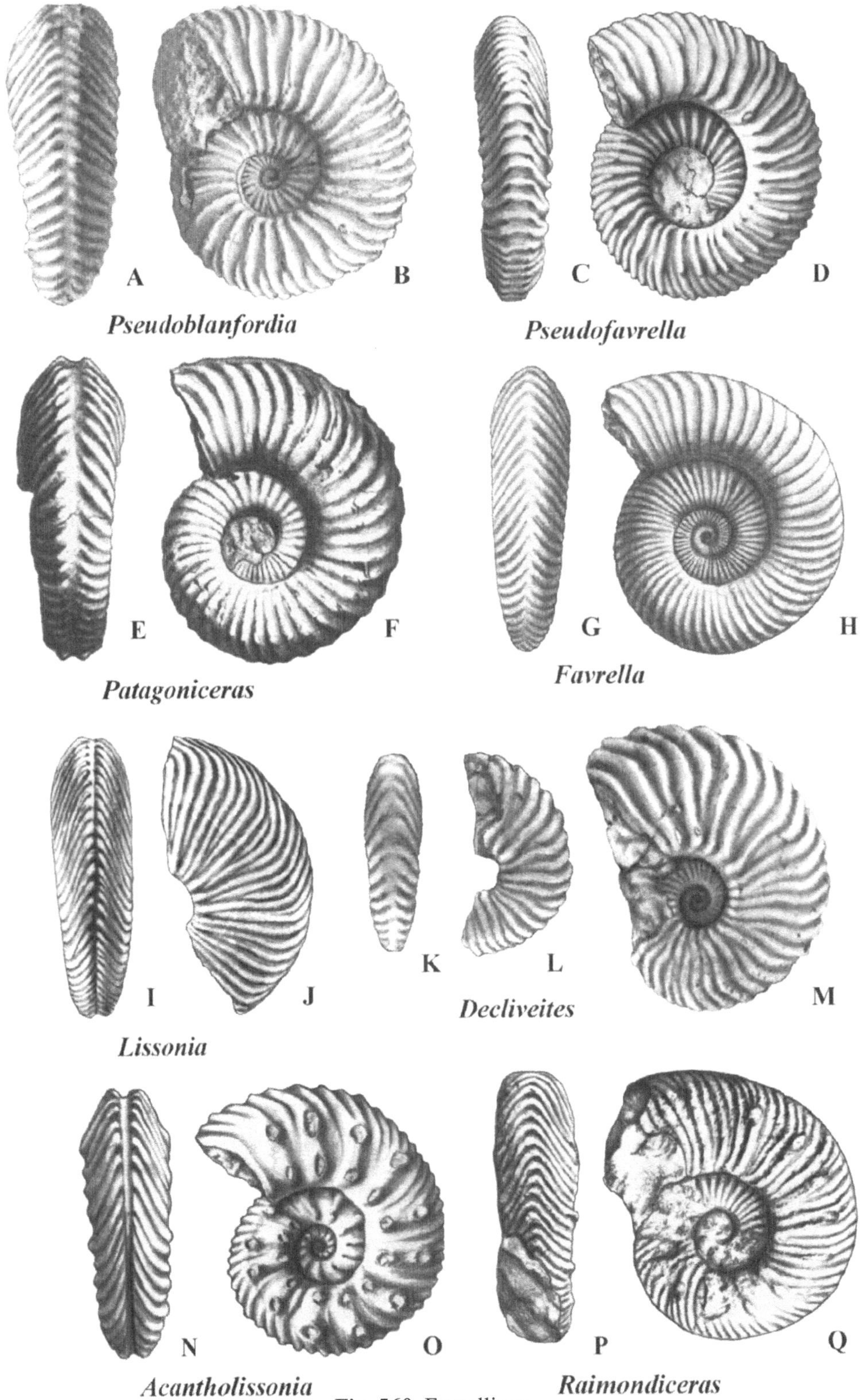

Fig. 560. Favrellinae.

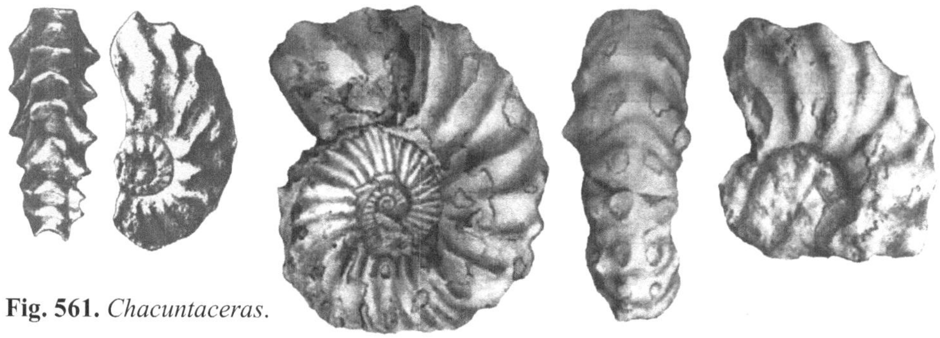

Fig. 561. *Chacuntaceras*.

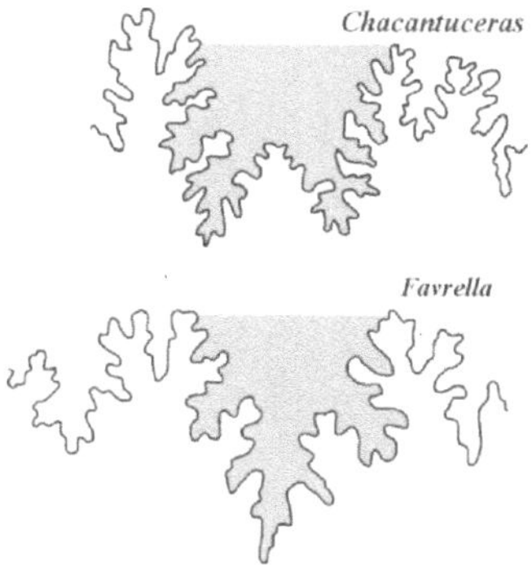

Fig. 562. Favrelline sutures.

Pseudoblanfordia SPATH, 1925 [*Hoplites australis* Burckhardt, 1903; OD]. Small, evolute, compressed, with steep umbilical wall, weakly curved convergent flanks and very narrow smooth venter; strong wiry slightly prorsiradiate primary ribs, narrower than interspaces, arise from feeble umbilical bullae and bifurcate on outer flank, curving forward to terminate in ventrolateral tubercles. *L. Valang.* **Fig. 560A-B** - *P. australis* (Burckhardt).

Lissonia GERTH, 1925 [*Hoplites riveroi* Lissón, 1907; OD]. Small to medium sized; involute (U=20%), with convergent flanks, subtabulate venter, subangular ventrolateral shoulders, dense prorsiradiate flat to slightly rounded falciform ribs mostly bifurcating from umbilical shoulder and projecting strongly forward on outer flank to meet deep ventral groove at acute angle; ribs may reunite on outer flank and some develop incipient umbilical bullae in maturity; suture with relatively broad trifid L much deeper than E. *L. Valang.* **Fig. 559, 560I-J** - *L. riveroi* (Lissón).

Decliveites AGUIRRE-URRETA & RAWSON, 1998 [*D. crassicostatus*; OD]. Medium sized [**m**] to moderately large [**M**]; slightly evolute to moderately involute, compressed, with umbilical wall sloping gently inward at all growth stages; flat flanks converge gently towards venter; prorsiradiate primary ribs arise singly or occasionally in pairs at umbilical shoulder, sometimes from feeble bullae, curving forward on outer flank, terminating in strong oblique ventro-lateral clavi that form chevrons over venter, interrupted by almost smooth siphonal groove; secondary ribs may be intercalated low on flank or on outer third. *Uppermost Valang.* **Fig. 560K-M** - *D. crassicostatus* Aguirre-Urreta & Rawson.

Favrella DOUVILLÉ, 1909 [*Neocomites americanus* Favre, 1908; OD]. Moderately large, serpenticone, with well spaced single rectiradiate ribs from early growth, occasionally branching from umbilical tubercles; ribs flex strongly forwards near venter and interrupted across siphonal line on inner whorls but forming chevrons in later growth; strongly ribbed morphs develop lateral tubercles. *L. Haut.* **Fig. 560G-H** - *F. americana* (Favre).

Patagoniceras WETZEL, 1960 [*Neocomites steinmanni* Favre, 1908; OD]. Small; like *Favrella* but with weak ventrolateral tubercles and ribs terminating in feeble oblique clavi at well developed siphonal groove. *L. Haut.* **Fig. 560E-F** - *P. steinmanni* (Favre).

Pseudoparodotoceras PARENT, GARRIDO, SCHERZINGER, SCHWEIGERT & FOZY, 2015 [*P. dezai*; OD]. Small, evolute, compressed, with oval (juv.) to subrectangular (ad.) whorl section; in early to middle growth (D<80 mm) with slightly prorsiradiate sinusoidal ribs, biplicate and single, branching high on flank so that only single ribs exposed in umbilicus; primary ribs with spiny lateral and ventrolateral tubercles interrupted by broad smooth siphonal band; on adult body chamber lateral tubercles obsolete and ventrolaterals bullate. *U. Tithon.* **Fig. 573C-E** - *P. dezai* Parent, Garrido, Scherzinger, Schweigert & Fozy.

Family SARASINELLIDAE nov.

Diagnosis. Typically evolute, compressed, with shallowly grooved to smooth venter. Umbil-ical bullae give rise, singly and in pairs, to strong ribs with lateral and ventrolateral tubercles; lateral tubercles may evanesce in maturity. Monophyletic derivatives of Thurmanniceratidae comprising the subfamilies Sarasinellinae and Karakaschiceratinae. *Upper U. Tithon. - U. Valang.*

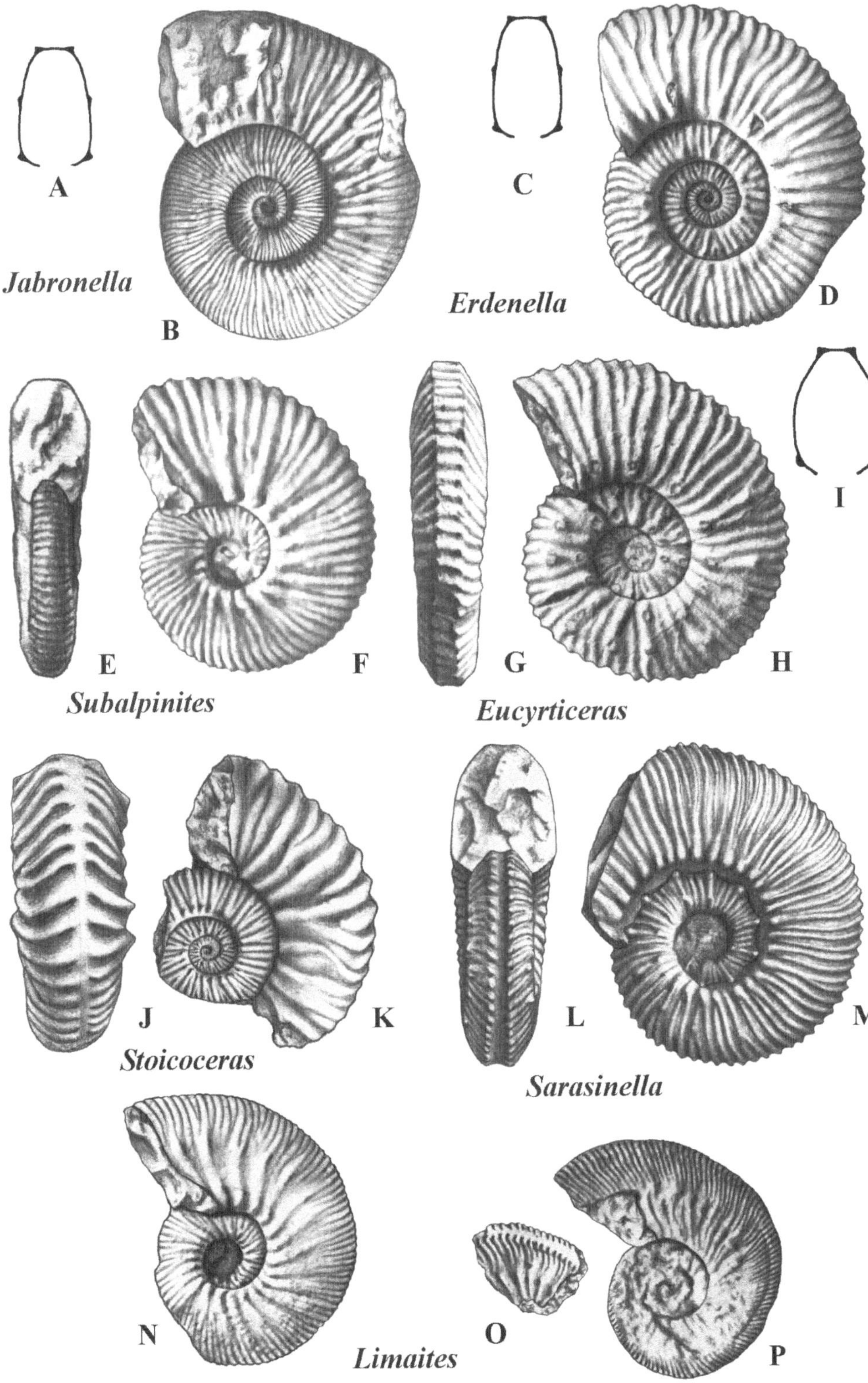

Fig. 563. Sarasinellinae.

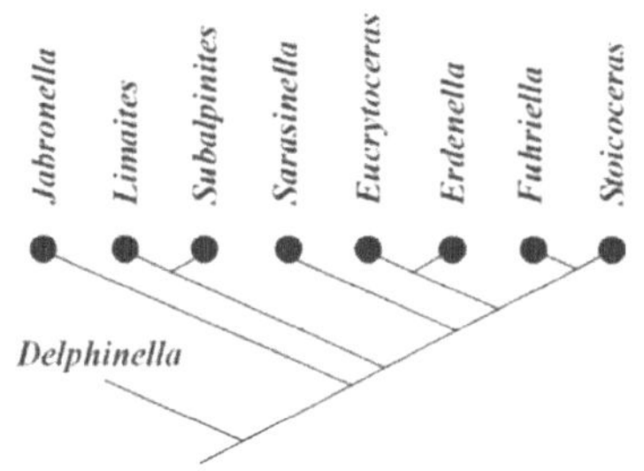

Fig. 564. Hypothesized relationships among Sarasinellinae.

Subfamily Sarasinellinae nov.

Diagnosis. Typically evolute, compressed, with grooved venter that in later growth passes into smooth siphonal band; ribbing differentiated with trituberculate primary ribs appearing late (primitive) or early (derived) in ontogeny. A seemingly monophyletic group with an origin in Thurmanniceratinae, close to *Delphinella*. *Upper U. Tithon. - L. Haut.*

Jabronella NIKOLOV, 1966 [*Berriasella jabronensis* Mazenot, 1939; OD]. Medium sized; very evolute (U=41%), compressed, with elliptical whorl section and siphonal groove that in later growth passes into smooth band; densicostate, with fine rectiradiate to slightly sinusoidal prorsiradiate ribs arising from feeble umbilical bullae in fasciculate bundles separated by intercalatories; ribs bifurcate and trifurcate at midflank from small lateral tubercles, and terminate in weak ventrolateral tubercles; ribbing coarsens on adult body chamber. *Uppermost Tithon. - L. Berrias.* **Fig. 563A-B** - *J. jabronensis* (Mazenot).

Erdenella NIKOLOV, 1979 [*Hoplites paquieri* Simionescu, 1899; OD]. Small to medium sized; like *Jabronella* but more coarsely ribbed, with fewer stronger umbilical bullae that give rise to paired flexuous rectiradiate ribs, one of which frequently bifurcates; in later growth lateral tubercles develop at point of rib furcation. *Uppermost Tithon. - L. Berrias.* **Fig. 563C-D** - *E. paquieri* (Simionescu).

Subalpinites MAZENOT, 1939 [*S. fauriensis*; OD] [**m**]. Small to medium sized; moderately evolute (U=30-34%), with flat convergent flanks, elliptical whorl section and feeble siphonal band; bundled prorsiradiate ribs arise from umbilical bullae, with main ribs branching from weak lateral tubercles. *Upper U. Tithon. - Berrias.* **Fig. 563E-F** - *S. fauriensis* (Mazenot).

Limaites LISSÓN, 1924 [*Hoplites leopoldinus* var. *peruanum*; M] [**m**]. Very small; compressed, evolute, with flat flanks, sharp ventrolateral edges, narrow tabulate venter with smooth siphonal band and rectangular whorl section; prorsiradiate primary ribs arise from umbilical bullae and terminate in sharp lateral tubercles from which prorsiradiate secondaries bifurcate or trifurcate with occasional intercalatories, terminating in small ventrolateral nodes; in later growth ribs effaced from most of flank remaining strong only near venter; aperture with lateral lappets. *Berrias. or L. Valang.* **Fig. 563N** - *L. aristidis* (Kilian), **Fig. 563O-P** - *L. peruanus* (Lissón).

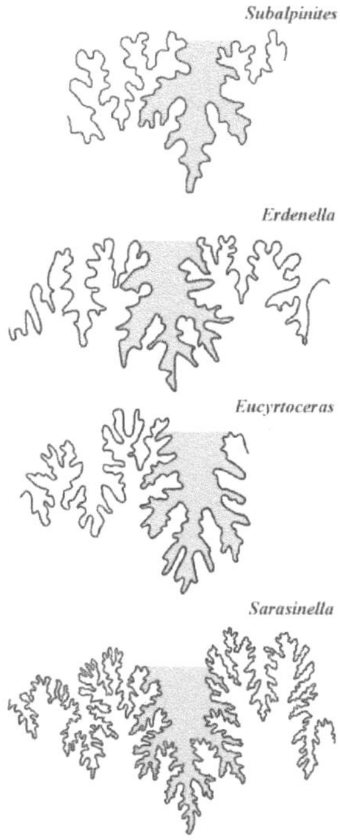

Fig. 565. Sarasinelline sutures.

Sarasinella UHLIG, 1905 [*Hoplites ambiguus*; SD Lemoine 1906]. Medium sized; like *Eucyrticeras* but more inflated, with trapezoidal whorl section that becomes more compressed with growth and with deep smooth siphonal groove; in early to middle growth two fine thread-like non-tuberculate ribs separate strong rectiradiate primary ribs with umbilical and spinose lateral tubercles; on final whorl strong umbilical bullae give rise to pairs uniform thin sinuous ribs that occasionally branch on outer flank and with rare intercalatories; lateral tubercles obsolete in

Fig. 566. *Fuhriella.*

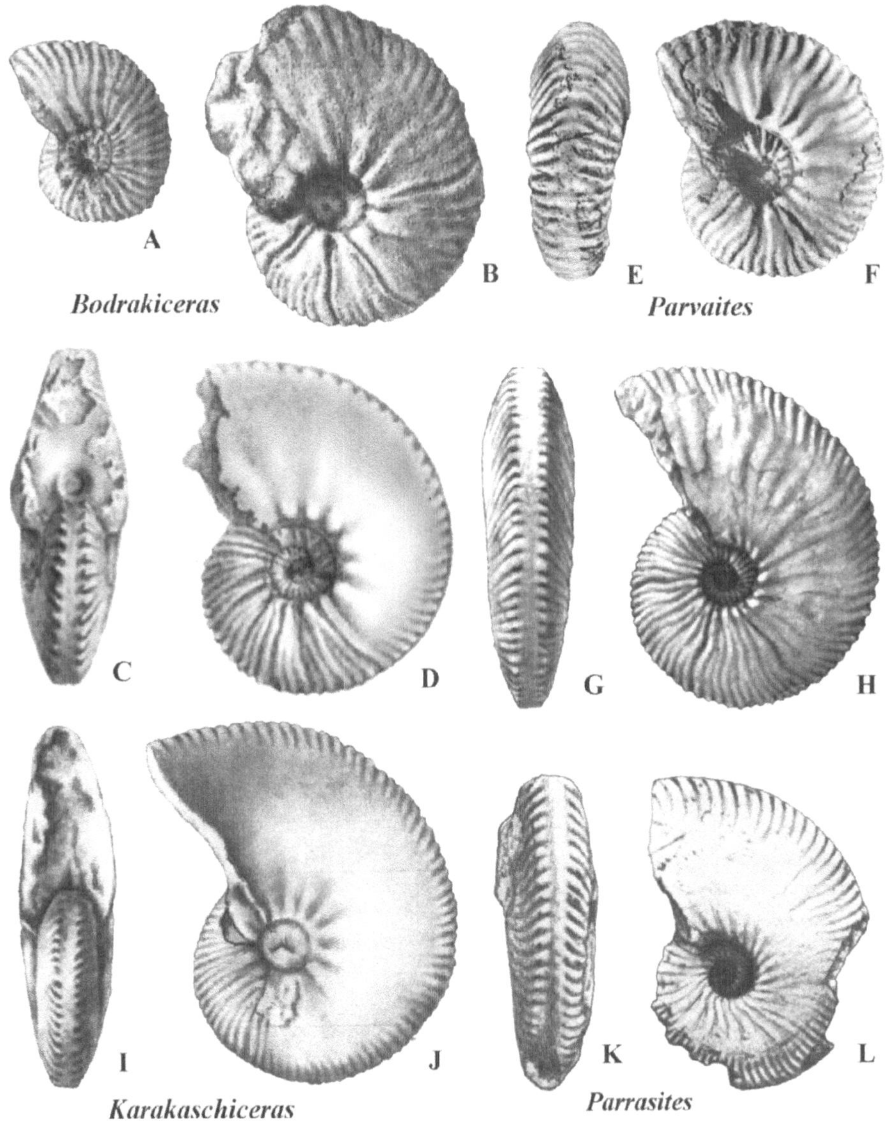

Fig. 567. Karakaschiceratinae.

later growth; few shallow constrictions; suture with narrow E, deeply bifid E/L and trifid L. *L. Valang.* **Fig. 563L-M** - *S. ambigua* (Uhlig).

According to Reboulet (1996) poorly known Upper Valanginian representatives probably belong to a new genusl

***Eucyrticeras* gen. nov.** [*Neocomites eucyrta* Sayn, 1907, p.36; OD herein]. Small; like *Erdenella* but with broad convex flanks, narrow subtabulate venter and polygonal whorl section; in early to middle growth with strong rectiradiate primary ribs separated by 1-2 finer secondaries; primary ribs with prominent umbilical and midlateral tubercles from which secondaries branch and terminate in tiny conical ventrolateral tubercles; on last part of adult body chamber ventrolateral tubercles joined across venter by transverse ribs. *Lower L. Valang.* **Fig. 563G-I** - *E. eucyrtum* (Sayn).

Fuhriella BULOT, 1995 [*Hoplites michaelis* Uhlig, 1902; OD]. Medium sized serpenticone with convergent flanks, rounded venter and compressed oval whorl section; strong distant rectiradiate primary ribs terminate in midlateral tubercles and give rise to pairs of prorsiradiate secondaries separated by 2-4 intermediate ribs; all ribs terminate in ventrolateral bullae leaving smooth siphonal band; lateral tubercles weaken and ribs simplify in maturity. *L. Valang.* **Fig. 566** - *F. michaelis* (Uhlig).

Stoicoceras WHITEHOUSE, 1927 [*Ammonites (Hoplites) teutobergensis* Weerth, 1884; OD]. Small; inner whorls much

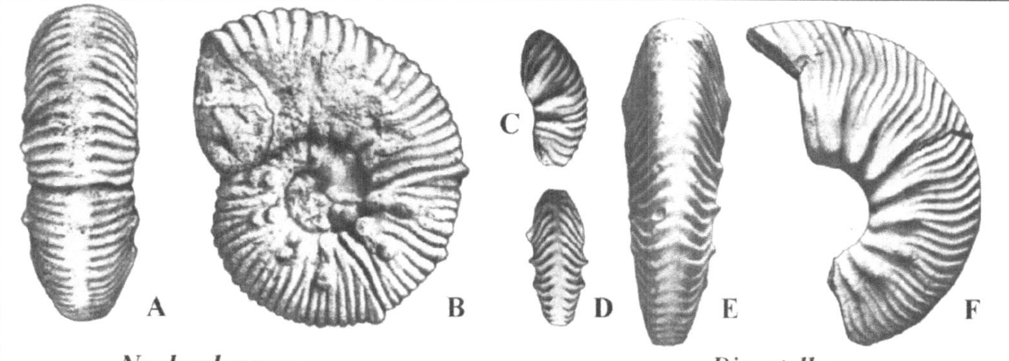

Fig. 568. Karakaschiceratinae.

like *Erdenella* with straight rigid ribs arising singly and in pairs (rarely triplicates) from prominent umbilical bullae; in later growth more inflated, with irregular curved secondary ribs branching from weak midlateral bullae and terminating in feeble ventral bullae on either side of smooth siphonal band; suture with broad bifid E/L and narrow trifid L. *U. Valang.* **Fig. 563J-K** - *S. teutobergense* (Weerth).

Subfamily Karakaschiceratinae nov.

Diagnosis. Medium sized to large, primitively with grooved venter and trituberculate ribs in early growth; in middle to latter growth involute, compressed, high whorled, with subtrigonal whorl section, tabulate venter and ornament weak or effaced from flanks. Seemingly monophyletic derivatives of Sarasinellinae, close to *Jabronella* (Baraboshkin & Mikhailova, 2000) comprising the tribes Karakaschiceratini and Neohoploceratini. *L. - U. Valang.*

Tribe Karakaschiceratini nov.

Type genus: *Karakaschiceras* Thieuloy, 1971.

Diagnosis. Involute, compressed and high whorled, with convergent flanks and narrow grooved to tabulate venter. Ribs with umbilical and rounded ventrolateral tubercles and, primitively, lateral tubercles; constrictions conspicuous in early growth; ribs effaced from siphonal line and from flanks in maturity; suture primitively with very broad L (like *Leopoldia*). *Lowermost Valang. - Lower U. Valang.*

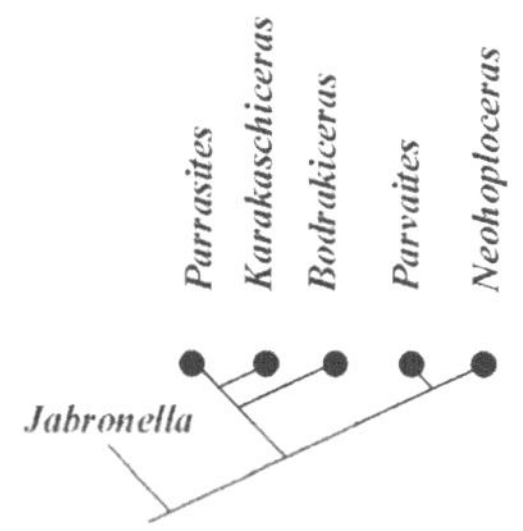

Fig. 569. Hypothesized relationships among Karakaschiceratinae.

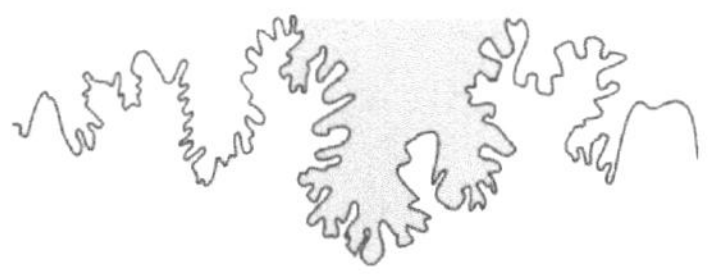

Fig. 570. *Bodrakiceras* suture.

Bodrakiceras BARABOSHKIN & MIKHAILOVA, 2000 [*Hoplites inostranzewi* Karakasch, 1889; OD]. Medium sized to large, discoidal, with moderate narrow umbilicus and abrupt umbilical wall; whorl section hexagonal to compressed trapezoidal, juveniles with ventral groove; in early growth (D<20 mm) periodic narrow constrictions separated by a single rib, one biplicate and one triplicate; in middle growth (D=20-40 mm) a single rib and complex rib; in later growth ribs single, bi- or trifurcate; ribs arise from spinose umbilical tubercle directed into umbilicus, and terminate in ventrolateral tubercle, with strong lateral tubercles at point of branching; ribbing weak to obsolete in maturity. *Lowermost Valang.* **Fig. 567A-B** - *B. constrictum* Baraboshkin & Mikhailova; **Fig. 567C-D** - *B. inostranzewi* (Karakasch); **Fig. 567G-H** - *B. copei* (Fatmi).

Karakaschiceras THIEULOY, 1971 [*Hoplites biassalensis* Karakasch, 1889; OD]. Like *Bodrakiceras* but with coarser ornament, well developed lateral and ventrolateral tubercles and deep constrictions; suture highly incised, with inner branch of E/L smaller than outer branch and L much deeper than E; in maturity superficially much like *Leopoldia*. *L. - U. Valang.* **Fig. 567I-J** - *K. biassalense* (Karakasch).

Parrasites CANTÚ-CHAPA, 2009 [*Leopoldia victoriensis* Imlay, 1938; OD]. Like *Karakaschiceras* but in early growth with mainly single falcoid ribs arising from blunt umbilical bullae, bifurcating or joined by intercalatory at mid-flank; faint constrictions; rib terminations projecting obliquely forwards on venter; suture with narrow

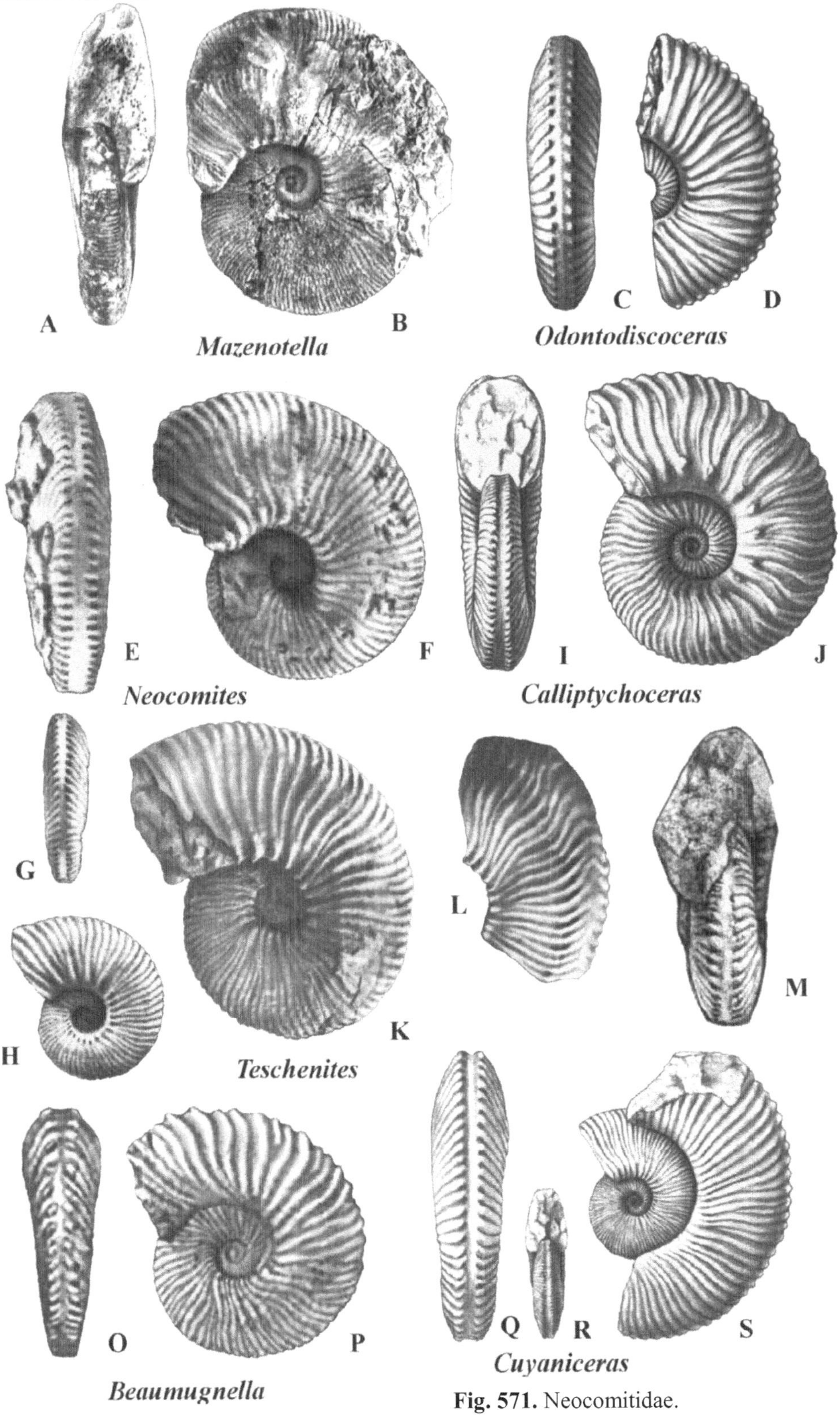

Fig. 571. Neocomitidae.

E/L and L. *U. Valang.* **Fig. 567K-L** - *P. victoriensis* (Imlay).

Dicostella BUSNARDO, 1966 [*D. pitrei*, = *Hoplites (Neocomites) houdardi* var. *tuberculata* Roman, 1933; OD]. Medium sized, with high slender whorls, strong blunt short primary ribs that arise from umbilical bullae and terminate in prominent lateral bullae below midflank from where 2-3 secondary ribs separated by an intercalatory curve strongly forward and interrupted across flattened venter by smooth band; there may be very weak ventrolateral tubercles; hypernodose specimens superficially resemble *Acanthodiscus*. *Upper U. Valang.* **Fig. 568C-F** - *D. tuberculata* (Roman).

=*Arnoldia* Stolley, 1937 (*non* Mayer-Eymar, 1887).

Tribe Neohoploceratini nov.

Type genus: *Neohoploceras* Spath, 1939.

Diagnosis. Small to medium sized, inflated, moderately evolute, with prominent umbilical and lateral tubercles in early to middle growth when constrictions are conspicuous. Monophyletic derivatives of Karakaschiceratini (Reboulet, 1996), close to *Bodrakiceras* (Baraboshkin & Mikhailova, 2000). *U. Valang.*

Neohoploceras SPATH, 1939 [*Ammonites submartini* Mallada, 1887; OD]. Small; inner whorls evolute, inflated, with straight deep constrictions backed by collar that bears prominent ventral bullae; small umbilical bullae give rise to 1-2 primary ribs that connect to enlarged midlateral tubercles which give rise to 2-3 rectiradiate secondaries separated by 1-2 nontuberculate intercalatories; ribs interrupted across venter by shallow smooth siphonal groove; constrictions and lateral tubercles become obsolete on discoidal and weakly ornamented outer whorls. *U. Valang.* **Fig. 568A-B** - *N. submartini* (Mallada).

Parvaites CANTÚ-CHAPA, 2009 [*P. agirrei*; OD]. Inner whorls like *Neohoploceras* but without lateral tubercles and with compressed whorls; feeble ventral bullae restricted to collar ribs backing flexuous and irregular constrictions. *Valang.* **Fig. 567E-F** - *P. aguirrei* Cantú-Chapa.

Superfamily NEOCOMITOIDEA Salfeld, 1921

(*nom. transl.* Vermeulen & Bulot, 2007 *ex* Neocomitidae)

Discussion. A monophyletic group with an origin in Berriaselloidea; comprises the ammoniticone Neocomitidae with narrow umbilicus, bituberculate ribs and narrow tabulate venter, and Himantoceratidae of trituberculate criocones. *U. Tithon. - U. Haut.*

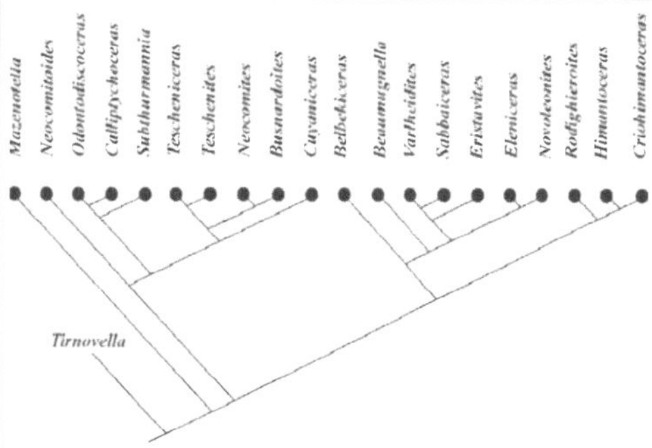

Fig. 572. Hypothesized relationships among Neocomitidae and Himantoceratidae.

Family NEOCOMITIDAE Salfeld, 1921

(*nom. transl.* Spath, 1923 *ex* Neocomitinae)
= Palaeohoplitidae Roman, 1938 (*nom. nud.*).

Discussion. A seemingly monophyletic group with an origin in Berriasellidae (Spath, 1939). Typically small to medium sized, involute, compressed, high whorled with narrow subtabulate venter; flexuous prorsiradiate ribs arise in bundles from distinct umbilical bullae and terminate in oblique ventrolateral clavi or transverse bullae separated by a smooth siphonal band; comprises the paraphyletic subfamilies Neocomitinae and Eristavitinae. *U. Tithon. - L. Haut., ?Barrem.*

Subfamily Neocomitinae Salfeld, 1921

Discussion. A monophyletic group with an origin in Tirnovellinae close to *Tirnovella* (Le Hégarat, 1973; Reboulet, 1996). Typically compressed, with narrow tabulate venter and bundled biplicate to subpolyplocoidal and fasciculate ribs arising from distinct umbilical bullae and terminating in bullate to clavate ventrolateral tubercles; siphonal band or groove in early to middle growth, evanescing later. Suture with narrow asmmetrically trifid L. Dimorphic, microconchs with lateral lappets. Comprises the tribes Neocomitini and Neocomitoidini. *U. Tithon. - U. Valang., ?Barrem.*

Mazenotella **gen. nov.** [*Neocomites suprajurensis* Mazenot, 1939; OD herein] [?**M**]. Medium sized to moderately large; involute, compressed, high whorled, with vertical umbilical wall, broad gently convex flanks, subtabulate venter and compressed trapezoidal whorl section; umbilical seam egressing in maturity; very fine flexuous slightly prorsiradiate ribs arising in bundles from small

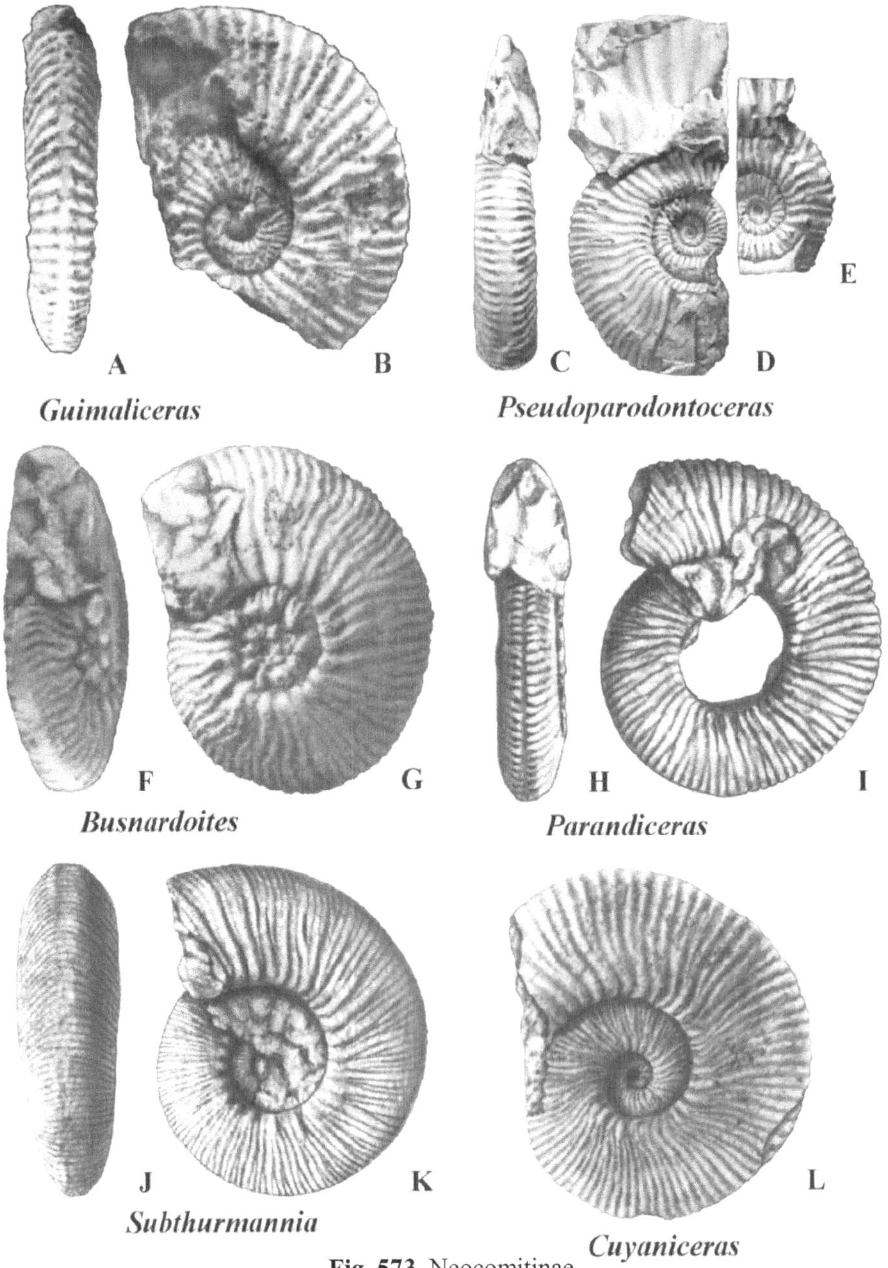

Fig. 573. Neocomitinae.

umbilical bullae, bifurcating at various levels on flanks, with ventrolateral bullae, crossing venter uninterrupted in maturity when ribs coarsen; suture with L as deep as E, and L/U_2 protruding beyond E/L. *U. Tithon.* **Fig. 571A-B** - *M. suprajurensis* (Mazenot).

Subthurmannia SPATH, 1939 [*S. fermori*; OD]. Medium sized to moderately large; in early to middle growth fairly evolute (U=38%) with tabulate to shallowly sulcate venter and elliptical whorl section; rather distant blunt umbilical bullae give rise to polyplocoid ribs separated by non-bullate single and biplicate ribs that terminate in small ventrolateral bullae separate by smooth siphonal band; adult body chamber with irregular primary ribs, with secondaries and intercalatories restricted to outer flank and venter, crossing rounded venter uninterrupted. *?Berrias*. **Fig. 573J-K** - *S. fermori* Spath.

= *Subthurmanniceras* Nikolov, 1960 (obj.).

Guimaliceras LUKENEDER, SUTTNER & BERTEL, 2011 [*G. guimaliense*; OD]. Small, moderately evolute, with steep umbilical wall, flat subparallel flanks, tabulate venter

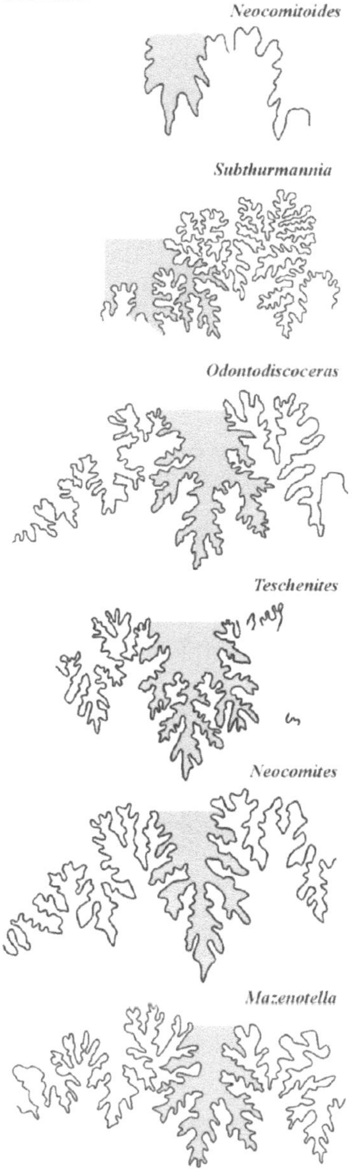

Fig. 574. Neocomitid sutures.

rounding in maturity, and rectangular whorl section; strong prorsiradiate ribs arise singly or in pairs from umbilical bullae, the posterior rib of a pair frequently bifurcating at mid-flank; all ribs terminate in spinose ventrolateral tubercles separated by a very deep siphonal groove or smooth band. *?U. Berrias., L. Valang.* **Fig. 573A-B** - *G. guimaliense* Lukeneder, Suttner & Bertel.

Neocomites Uhlig, 1905 [*Ammonites neocomiensis* d'Orbigny, 1841; SD Sayn, 1906]. Small (**m**) to medium sized (**M**); like *Mazenotella* but with subangular ventrolateral shoulders, narrow tabulate venter and strongly compressed subrectangular whorl section; small umbilical bullae give rise to sinusoidal to falcate prorsiradiate ribs, singly or in pairs, with frequent bifurcation on flank; all ribs terminate ventrolaterally in oblique clavi (juv.) to transverse bullae (ad.) separated by a siphonal groove (juv.) or smooth band (ad.); ribbing weakening on flanks and may become obsolete on adult body chamber of macroconch; suture with L much deeper than E and retracted suspensive lobe. *L. - Lower U. Valang.* **Fig. 571E-H** - *N. neocomiensis* (d'Orbigny).

Calliptychoceras Spath, 1924 [*Neocomites calliptychus* Uhlig, 1910; M]. Small, evolute, compressed, with inclined umbilical wall, convex flanks, rounded ventrolateral shoulders and oval (juv.) to subrectangular (ad.) whorl section; broad venter changing from initially sulcate to tabulate then evenly arched; coarse sinuous single and biplicate ribs with small acute tubercles at point of branching; umbilical tubercles strengthen and ribs become more distant on adult body chamber. *L. Valang.* **Fig. 571I-J** - *C. calliptychus* (Uhlig).

Odontodiscoceras Spath, 1924 [*Neocomites odontodiscus* Uhlig, 1910; M]. Small; like *Calliptychoceras* but more compressed, with blunt siphonal ridge, subfastigiate venter and unusually flat subpolyplocoidal ribs that terminate in small ventrolateral clavi. *L. Valang.* **Fig. 571C-D** - *O. odontodiscus* (Uhlig).

Parandiceras Spath, 1939 [*P. rota*; OD]. Small, moderately evolute, with convergent flanks and narrow tabulate to sulcate venter; straight prorsiradiate ribs arise singly and in pairs from small umblical bullae and branch mostly at midflank, frequently separated by an intercalatory; distinct oblique ventrolateral clavi form two prominent ventral ridges in maturity. *L. Valang.* **Fig. 573H-I** - *P. rota* Spath.

Busnardoites Nikolov, 1966 [*Ammonites desori* Pictet & Campiche, 1860; OD] [**m**]. Small to medium sized, moderately evolute (U=29%), with narrowly rounded venter and compressed subovate whorl section; strong conical umbilical tubercles give rise to 2-3 ribs; on outer whorl umbilical tubercles weaken and usually give rise to 2, occasionally 1 or 3, flexuous ribs, one of which bifurcates at mid-flank; ribs terminate in oblique ventrolateral clavi separated by a smooth band; aperture with lateral lappets. *Lower Valang.* **Fig. 573F-G** - *B. desori* (Pictet & Campiche).

Cuyaniceras A. Leanza, 1945 [*Odontoceras transgrediens* Steuer, 1897; SD Arkell, 1952]. Small to medium sized; involute (juv.) to very evolute (ad.), with compressed subrounded whorl section and sulcate venter; in early to middle growth ribs rectiradiate to prorsiradiate, gently flexuous, mostly bifurcating below midflank but with some singles, and occasional weak umbilical bullae, interrupted across venter; in maturity ribs straighter, prorsiradiate, mostly single; all ribs with outwardly pointing ventrolateral tubercles. *U. Berrias.* **Fig. 571Q-S, 573L** - *C. transgrediens* (Steuer).

Teschenites Thieuloy, 1971 [*Ammonites neocomiensiformis* Uhlig, 1902; OD]. Medium sized; inner whorls like *Neocomites*; outer whorls more evolute, with rounded ven-

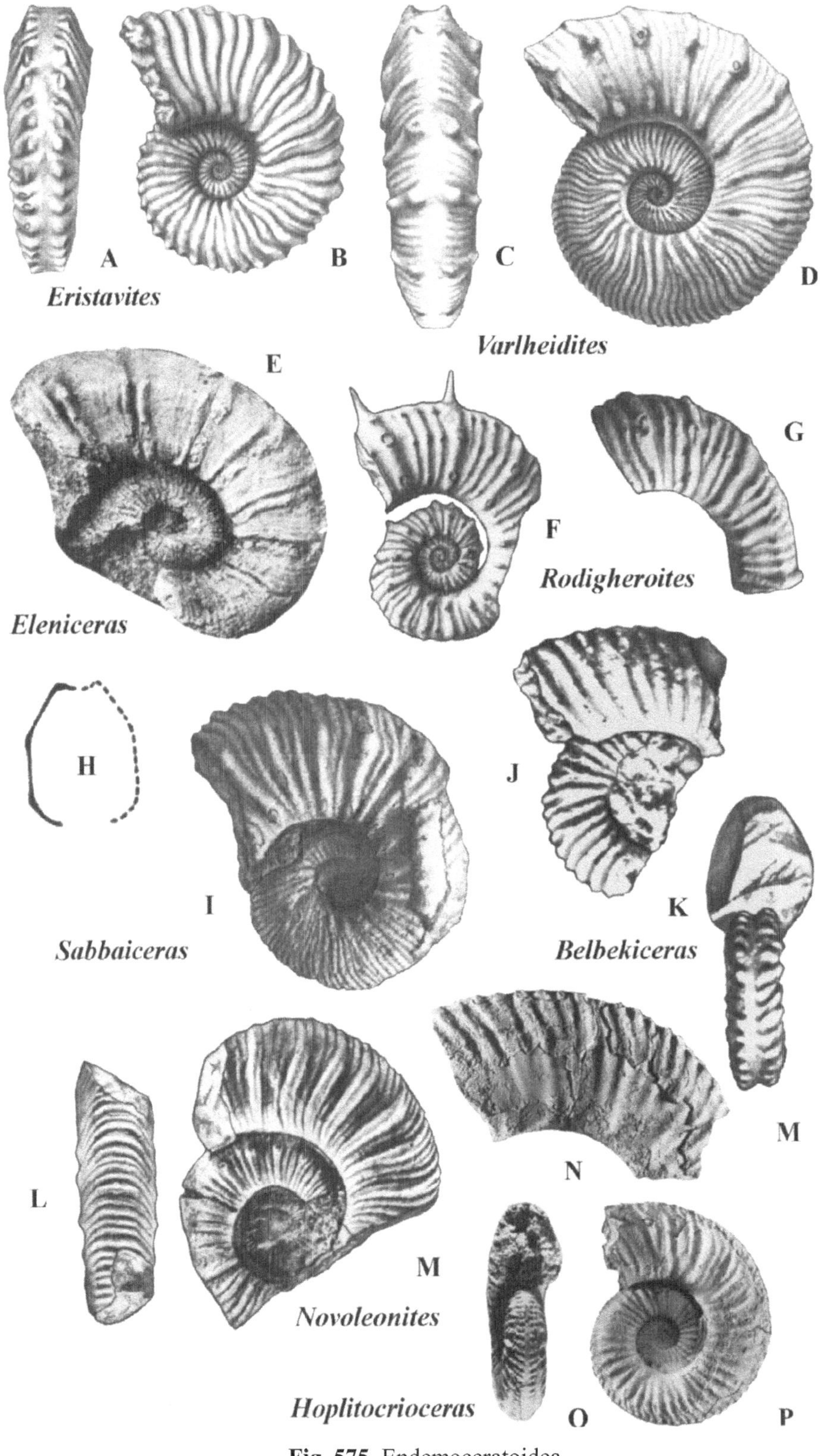

Fig. 575. Endemoceratoidea.

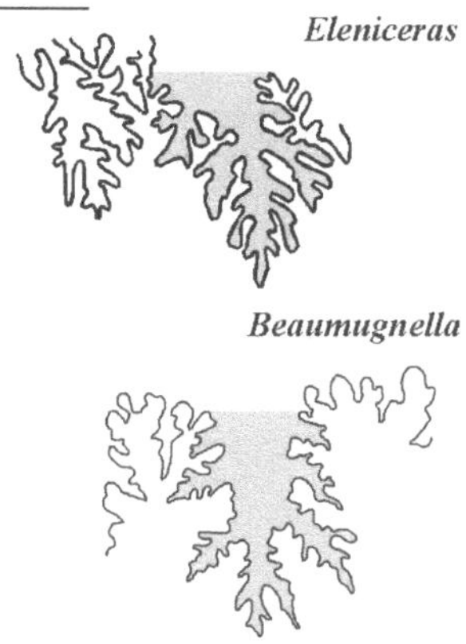

Fig. 576. Eristavitine sutures.

ter, more numerous umbilical bullae that strengthen with growth and irregular ribs with frequent bifurcation and intercalation on outer flank and hint of ventrolateral bullae; ribs tend to weaken at midflank and project forward on outer flank, crossing venter strongly with forward convexity; suture with slightly asymmetrical trifid L. *Uppermost Valang. - L. Haut.* **Fig. 571K** - *T. neocomiensiformis* (Uhlig).

Tescheniceras VAŠÍCEK, 2020 [*Neocomites (Teschenites) flucticulus* Thieuloy, 1977; OD] [**M**]. Small (**m**) to medium sized (**M**); primary ribs arise from umbilical bullae separated by 2 intercalatories appearing on inner third of flanks, rarely bifurcating, terminating in fairly prominent ventrolateral tubercles, larger on primary ribs; venter smooth to peristome (**M**), but microconchs with linguoid ribs over tabulate venter in maturity. *Uppermost Valang. - basal Haut.* **Fig. 571L** - *T. flucticulus* (Thieuloy).

Tribe Neocomitoidini nov.

Type genus: *Neocomitoides* Obata & Matsukawa, 2007.

Diagnosis. As for genus. A monotypic taxon of uncertain origin, superficially similar to *Neocomites* but separated by a substantial stratigraphical gap and with some characters similar to Holcodiscidae. *Barrem.*

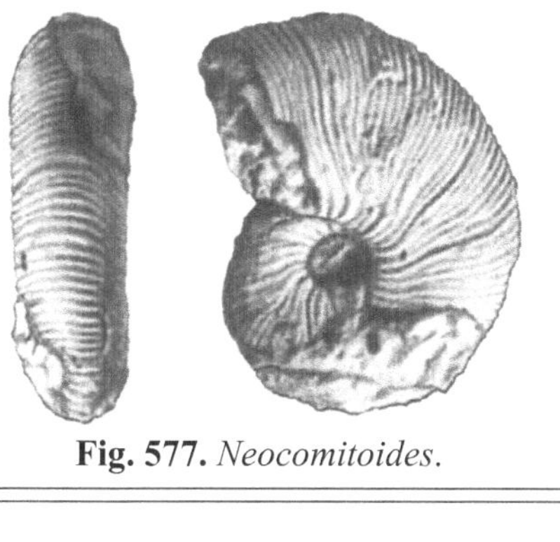

Fig. 577. *Neocomitoides.*

Neocomitoides OBATA & MATSUKAWA, 2007 [*N. minimus*; OD]. Small; compressed, involute with rounded venter and elliptical whorl section increasing rapidly in height; ribs fine, dense, prorsiradiate, biconcave, arising from weak umbilical bullae, splitting in 2-3's on lower flank and swinging forward on outer flank to cross venter transversely and uninterrupted; periodic constrictions parallel ribbing; suture with trifid L as deep as E. *Barrem.* **Fig. 577** - *N. minimus* Obata & Matsukawa.

Subfamily Eristavitinae nov.

Diagnosis. Typically with more or less differentiated ribbing, trituberculate primary ribs and a tendency in some for the body chamber to uncoil so that derived representatives are criocone. Monophyletic derivatives of Neocomitinae, close to *N. teschenensis* (Reboulet, 1996), although Baraboshkin & Mikhailova (2000) sought an origin in Himalayitidae, close to *Euthymiceras*; convergence is involved. *Lower L. Valang.- Lower L. Haut.*

Belbekiceras BARABOSHKIN, 1997 [*B. belbeki*; OD]. Medium sized, evolute, with broad flat flanks and compressed hexagonal whorl section; ribs prorsiradiate, simple or biplicate, rarely triplicate, branching from small sharp high-lateral tubercles that project fowards and terminate in oblique ventrolateral clavi on either side of smooth siphonal groove; some ribs bifurcate from feeble umbilical bullae; frequent weak constrictions. *Lower L. Valang.* **Fig. 575J-K** - *B. belbeki* Baraboshkin.

Beaumugnella **gen. nov.** [*Neocomites beaumugnesis* Sayn, 1907; OD herein]. Very small; initially involute but uncoiling later to become evolute, compressed, with broad gently convex flanks, narrow tabulate venter and compressed trapezoidal whorl section; in early to middle growth with dense fine strongly prorsiradiate ribs arising in fasciculate bundles from small umbilical tubercles and terminating in oblique ventrolateral clavi separated by a smooth siphonal band; in maturity strong bifurcating ribs followed by simple ribs separated by short intercalatory, all terminating in prominent oblique ventrolateral tubercles. *Lower U. Valang.* **Fig. 571O-P** - *B. beaumugnesis* (Sayn).

Sabbaiceras AVRAM & GRĂDINARU, 1993 [*S. stephanescui*; OD]. Early whorls like *Neocomites* but with high-octagonal whorl section; in middle growth trituberculate primary ribs bifurcate from lateral tubercle); later with strong trituberculate primaries separated by 1-2 secondaries of various length terminating in small sharp ventrolateral tubercles alongside a smooth siphonal band; constrictions lacking; suture like *Neocomites*. *Lower U. Valang.* **Fig. 575H-I** - *S. stephanescui* Avram & Grădinaru.

Varlheidites RAWSON & KEMPER, 1978 [*V. peregrinus*; OD] [**M**]. Small (**m**) to large (**M**); inner whorls with strongly

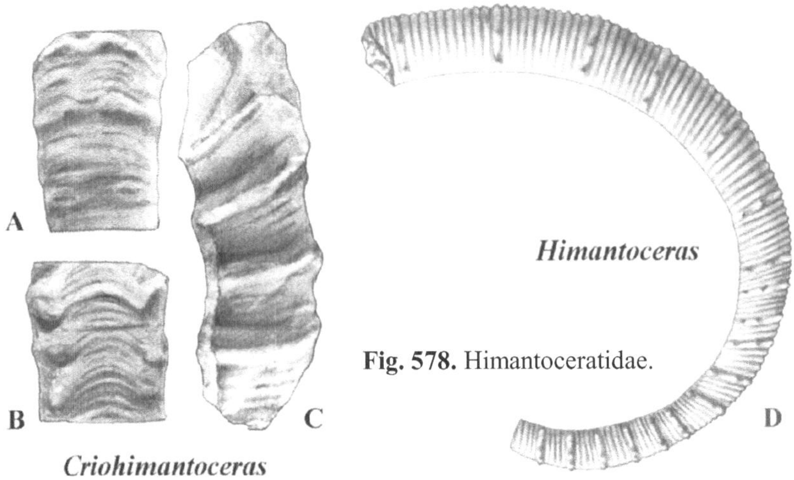

Fig. 578. Himantoceratidae.

flexed ribs; adult body chamber with periodic strengthened trituberculate ribs separated by 3-4 non-tuberculate secindaries. *Lower U. Valang*. **Fig. 575C-D** - *V. peregrinus* Rawson & Kemper.

Eristavites NIKOLOV, 1966 [*Neocomites platycostatus* Sayn, 1907; OD] [**m**]. Small; moderately evolute, compressed, with gently convex flanks and tabulate venter; coarse slightly prorsiradiate ribs arise singly or in pairs from umbilical shoulder and flex strongly forward on outer flank, terminating in rounded ventrolateral tubercles to leave smooth siphonal band; some ribs thicker than others, with incipient tubercles at point of inflexion. *Lower U. Valang*. **Fig. 575A-B** - *E. platycostatus* (Sayn).

Eleniceras BRESKOVSKI, 1967 [*E. stevrecense*; OD] [**M**]. Medium sized to large; like *Teschenites* but with enlarged single trituberculate ribs appearing in early growth, separated by 1-2 bifurcating secondary ribs, and ribbing may weaken across venter; adult body chamber uncoils. *Uppermost Valang. - Lower L. Haut.* **Fig. 575E** - *E. tchechitevi* Breskovski.

Novoleonites CANTÚ-CHAPA, 1992 [*N. tovarensis*; OD]. Moderately large; very evolute, compressed, with flat subparallel flanks and rectangular whorl section; in early growth with tabulate venter and rectiradiate trituberculate main ribs separated by 2-3 nontuberculate secondaries; in later growth with pillar-like nontuberculate main ribs separated by 2-3 finer ribs; primary ribs split into finer secondaries that project forward across venter. *U. Valang.* **Fig. 575L-M** - *N. tovarensis* Cantú-Chapa.

Family HIMANTOCERATIDAE Dimitrova, 1970

Discussion. A monophyletic group with an origin in Neocomitinae (Dimitrova, 1970) close to *Beaumugnella* according to Reboulet (1996) but the phenotypic discontinuity is substantial; open criocones with slowly expanding whorls; strong trituberculate main ribs separated by fine non-tuberculate secondaries. *Uppermost Valang.- basal Haut.*

Rodighieroites COMPANY, 1987 [*R. cardulus*; OD]. Small, with subcircular whorl section in early growth, becoming compressed oval in maturity; secondary ribs with high-lateral and ventrolateral tubercles, and tubercles to primary ribs spinose; smooth siphonal band appears on last part of body chamber. *Uppermost Valang.* **Fig. 575F-G** - *R. cardulus* Company.

Himantoceras THIEULOY, 1964 [*H. trinodosum*; OD]. A low very open helicone with compressed oval to subelliptical whorl section, rounded venter and strong trituberculate prorsiradiate primary ribs separated by 3-7 fine non-tuberculate secondaries; ribs cross venter transversely where occasional secondaries bifurcate or are joined by an intercalatory; suture relatively simple, with narrow E, narrow bifid E/L and wide trifid L much deeper than E. *Uppermost Valang. - Lower L. Haut.* **Fig. 578D** - *H. trinodsum* Thieuloy.

Criohimantoceras THIEULOY & BULOT, 1992 [*C. gigas*; OD]. Known only from fragments; like *Himantoceras* but

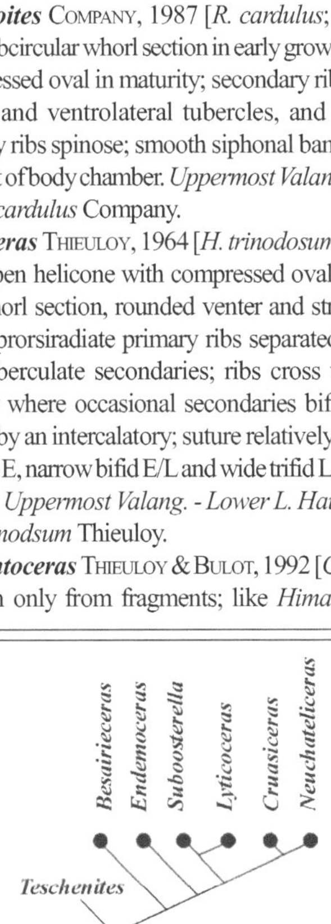

Fig. 579. Hypothesized relationships among Endemoceratidae.

much larger, with subcircular to subquadratic whorl section and robust trituberculate main ribs separated by weak non-tuberculate secondaries of variable strength; primaries split into fine convex forward secondaries across dorsum. *Upper U. Valang.* **Fig. 578A-C** - *C. gigas* Thieuloy & Bulot.

Perhaps a macroconch of *Himantoceras*.

Superfamily ENDEMOCERATOIDEA Schindewolf, 1966

(*nom. transl.* Vermeulen, 1996 *ex* Endemoceratidae)

Discussion. A monophyletic group derived from Neocomitoidea (Thieuloy, 1971). Comprises the families Endemoceratidae and Theodoritidae. *U. Valang. - L. Haut.*

Family ENDEMOCERATIDAE Schindewolf, 1966

Discussion. A monophyletic group with an origin in Neocomitinae close to *Teschenites* (Fatmi, 1972). Typically evolute, compressed, with shallowly depressed to fastigiate venter; umbilical bullae give rise to strong flexuous ribs arising mostly in pairs from umbilical bullae and projecting forwards on outer flank, with small ventrolateral tubercles; ribs either effaced from siphonal line or forming chevrons over venter. *U. Valang. - L. Haut.*

Besairieceras COLLIGNON, 1962 [*B. colcanapi*; OD]. Medium sized to moderately large; evolute, compressed, with wide shallow umbilicus, flattened flanks and subrectangular whorl section; venter sulcate on inner whorls but subtabulate later; umbilical bullae appear at about 50 mm diameter and strengthen with growth, sometimes protruding into umbilicus; flanks with pairs of sinuous prorsiradiate ribs that frequently bifurcate and project forward near venter forming weak oblique ventrolateral clavi and forwardly directed chevrons over venter; ribs weaken across siphonal line in maturity; fairly deep constrictions on inner whorls, declining with growth; typically Indo-Malgash. *U. Valang.* **Fig. 581E-F** - *B. colcanapi* Collignon.

Besaireiceras Aguirre-Urreta & Rawson, 1998 (*laps. cal.*).

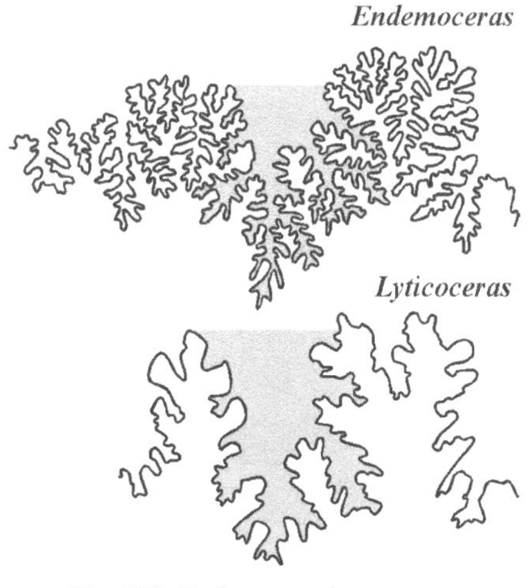

Fig. 580. Endemoceratine sutures.

Endemoceras THIERMANN, 1964 [*Hoplites amblygonius* Neumayr & Uhlig, 1881; OD]. Large; like *Besairieceras* but more compressed and evolute, with narrower fastigiate venter, more numerous subdued umbilical bullae that do not strengthen with growth nor protrude into umbilicus, biconcave ribs that flex strongly forward on outer flank and frequently bifurcate from umbilical shoulder in later growth; ornament declines on outer whorls of macroconchs, when venter rounds; suture with asymmetrical L and U_2; typically Boreal. *Lower L. Haut.* **Fig. 581C-D** - *E. amblygonium* (Neumayr & Uhlig).

Lyticoceras HYATT, 1900 [*Ammonites cryptoceras* d'Orbigny, 1840; OD]. Moderately large; like *Endemoceras* but with rounded venter, small pointed ventrolateral nodes, ribs that cross venter with forward convexity (but do not form chevrons) and considerable weakening across siphonal line, becoming coarser and less flexuous with growth; suture with more symmetrical L; typically Tethyan. *Upper L. Haut.* **Fig. 581A-B** - *L. cryptocerum* (d'Orbigny).

Suboosterella SPATH, 1924 [*Ammonites heliacus* d'Orbigny, 1840; SD Roman, 1938] [**m**]. Small; inner whorls like *Lyticoceras* but with subdued umbilical tubercles throughout, no ventrolateral tubercles, and ribs becoming more distant and simplifying in later growth, at first alternating long and short and then straight and simple. *Upper L. Haut.* **Fig. 581H-I** - *S. heliaca* (d'Orbigny).

Perhaps a microconch of *Lyticoceras*.

Neuchateliceras **gen. nov.** [*Lyticoceras claveli* Busnardo & Thieuloy, 1989, p.137, = *Hoplites (Leopoldia) bargemensis* Kilian, 1910; OD herein]. Large; like *Lyticoceras* but very evolute (U=45%), with slightly convergent flanks, prominent slightly rursiradiate bullate primary ribs splitting at midflank into pairs of thick robust secondaries that curve forward and are separated in later growth by an intercalatory; ventrolateral bullae feeble. *L. Haut.* **Fig. 581J-K** - *N. bargemense* (Kilian).

Cruasiceras BUSNARDO, 1970 [*Ammonites cruasensis* Torcapel, 1884; OD]. Large; like *Neuchateliceras* but more compressed and involute, with narrowly rounded venter and denser finer rectiradiate to slightly rursiradiate biplicate ribs that in later growth are mostly single. *L. Haut.* **Fig. 581L** - *C. cruasense* (Torcapel).

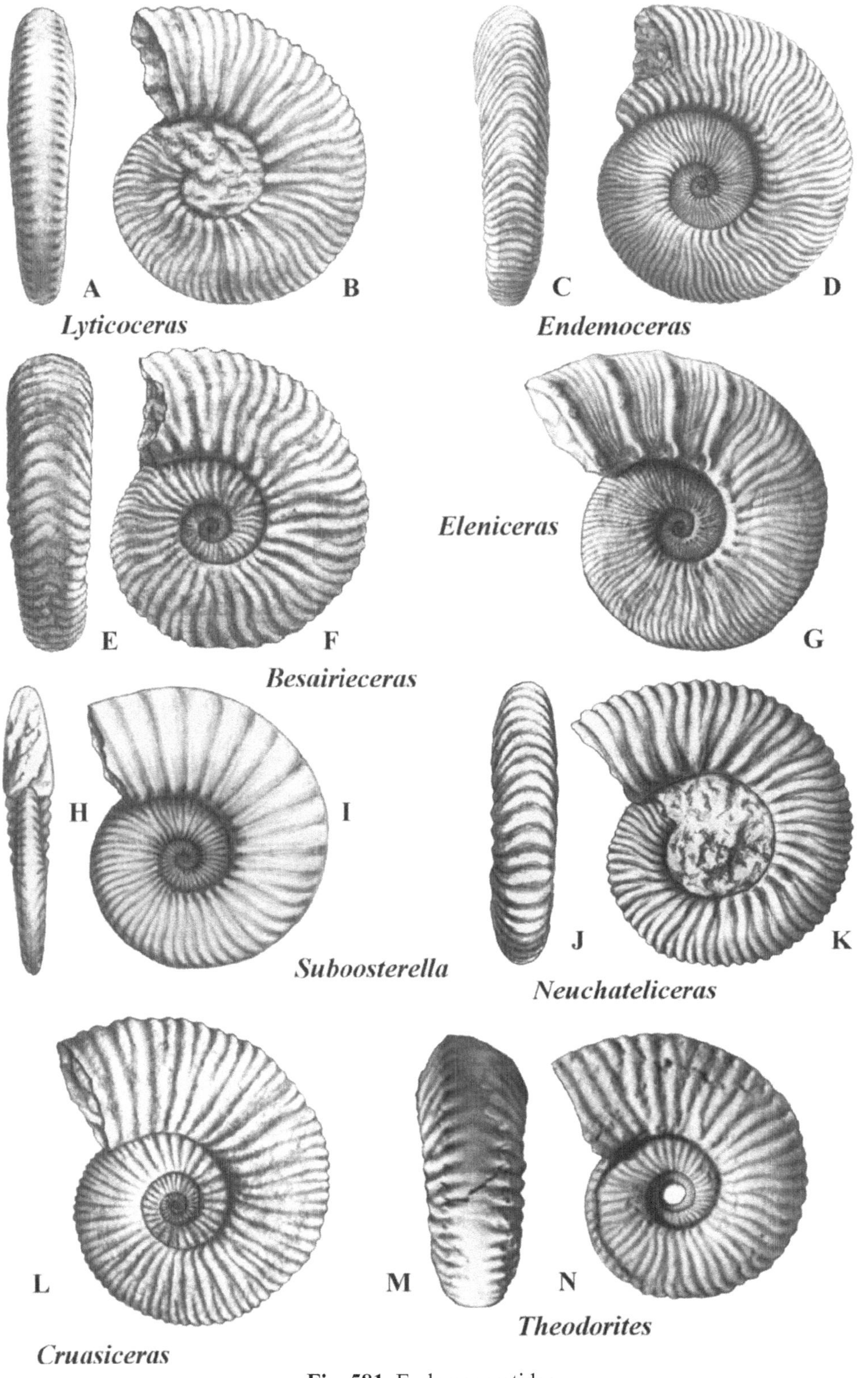

Fig. 581. Endemoceratidae.

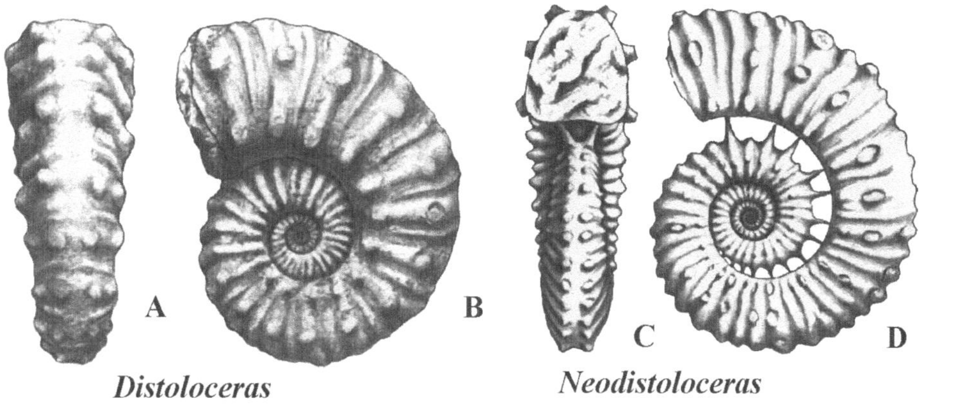

Fig. 582. Distoloceratinae.

Subfamily Distoloceratinae nov.

Diagnosis. Initial planispire followed by more or less uncoiled criocone; typically with trituberculate main ribs often separated by thinner secondaries that may or may not be tuberculate; ribs interrupted over venter. Seemingly monophyletic derivatives of Endemoceratinae but precise antecedent unknown. *Upper L. Haut.*

Distoloceras HYATT, 1900 [*Ammonites hystrix* Phillips, 1829; OD]. Small; very evolute ammoniticone; inner whorls with broad flat flanks, distinct ventrolateral facets and gently rounded to subtabulate venter and strong simple rather distant rectiradiate ribs that become differentiated in later growth by intervention of 1-2 thin non-tuberculate secondaries; main ribs with strong umbilical and inner ventrolateral tubercles that become spinose with growth, also outer ventrolateral clavi which decline in maturity; ribs project forward from inner ventrolaterals and cross venter transversely; suture like *Endemoceras* but simpler and with L not so deep; convergent towards *Rodighieiroites*. *Upper L. Haut.* **Fig. 582A-B** - *D. hystrix* (Phillips).

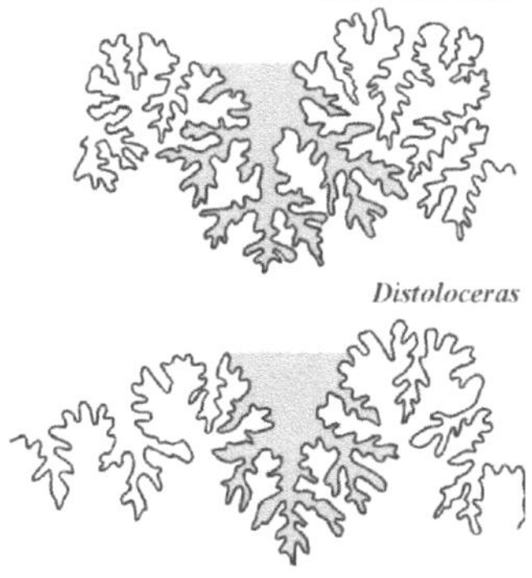

Fig. 583. Distoloceratine sutures.

***Neodistoloceras* gen. nov.** [*Crioceras roemeri* Neumayr & Uhlig, 1881, p.59; OD herein]. Large criocone; like *Distoloceras* but with numerous umbilical tubercles, lateral tubercles positioned lower on flank and present from early growth, robust spinose ventral tubercles, primary ribs separated by 2-3 fine flexuous riblets and growth striae, and more incised suture with widely splayed L. *Upper L. Haut.* **Fig. 582C-D** - *N. roemeri* (Neumayr & Uhlig).

Hoplitocrioceras GIOVINE, 1950 [*H. gentilis*; OD]. Like *Neodistoloceras* but with trigonal to subtrigonal whorl section; in early to middle growth planispiral, with strong slightly prorsiradiate primary ribs terminating in prominent high-lateral tubercles from which secondaries arise in pairs, projecting forwards and terminating in small ventrolateral tubercles, leaving a smooth siphonal band; in later growth whorls uncoil slightly, linked by thick dorsal shell draped over tips of sharp, upwardly pointing ventrolateral spines impressed into dorsum of succeeding whorls; strong trituberculate primary ribs, with mid-lateral spines, separated by 1-4 slightly finer primary or secondary ribs with short ventrolateral spines; suture with shallow E and broad asymmetrically trifid L. *Upper L. Haut.* **Fig. 575N-P** - *H. gentile* Giovine.

Family THEODORITIDAE Baraboshkin & Mikhailova, 2006

Discussion: Monotypic derivatives of Endemoceratidae (Mikhailova & Baraboshkin, 2009), close to *Lyticoceras*. Diagnosis as for genus. *L. Haut.*

Theodorites MIKHAILOVA & BARABOSHKIN, 2006 [*T. theodori*; OD]. Small, compressed, evolute, with perforated umbilicus; whorls compressed, rectangular with broad subtabulate venter; strong simple or bifurcating, slightly prorsiradiate ribs, thicker than interspaces flex slightly

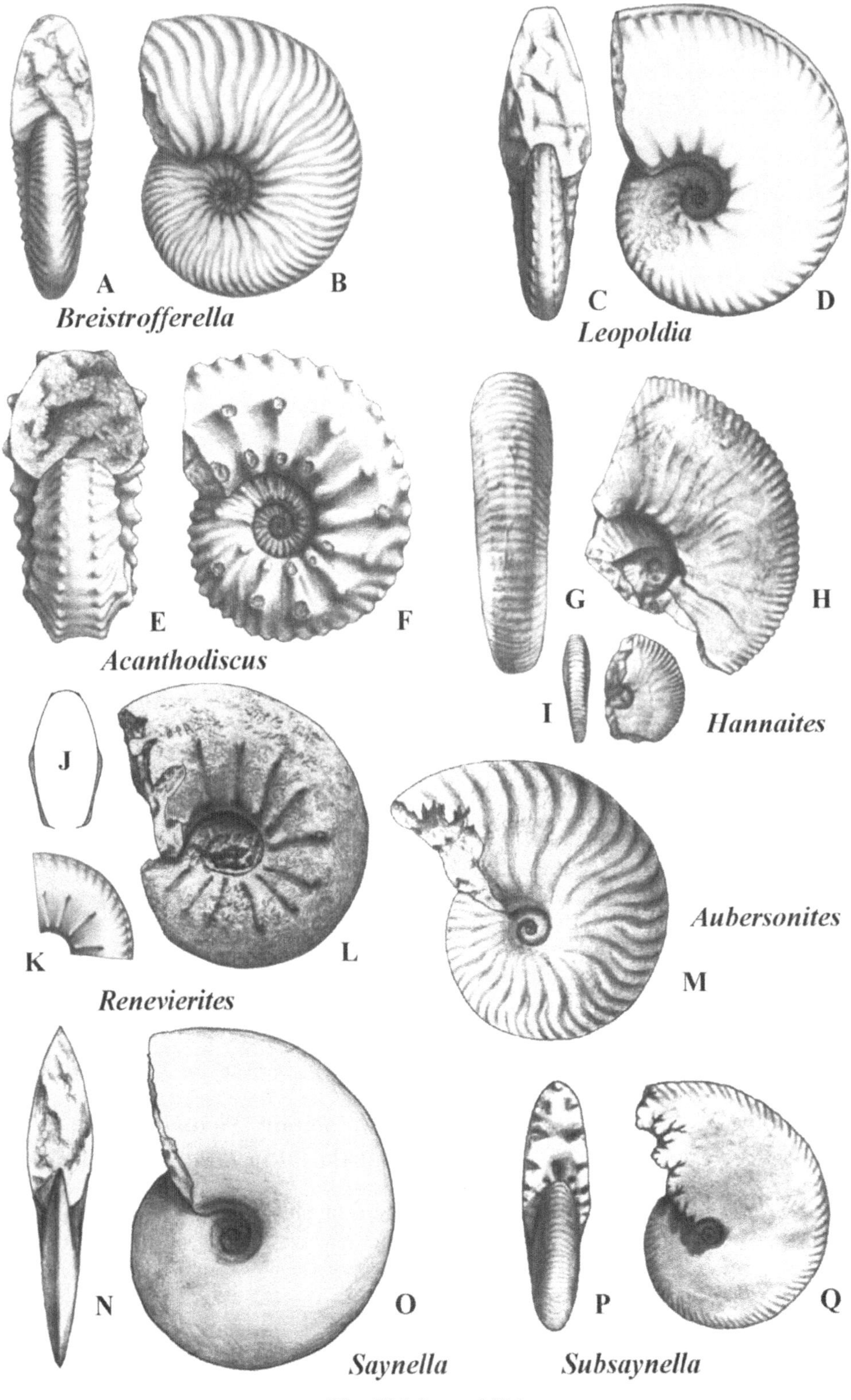

Fig. 584. Leopoldiidae.

forward near venter where an occasional interalatory appears; ribs weak or effaced over slightly convex venter, with small ventrolateral clavi; feeble lateral tubercles in some species; suture complex, narrow E with high median saddle, deeper symmetrically trifid L. *Lower Haut.* **Fig. 581M-N** - *T. drushitsi* Mikhailova & Baraboshkin.

Superfamily LEOPOLDIOIDEA Thieuloy, 1971

(*nom. transl.* herein *ex* Leopoldiinae)

Discussion. A monphyletic taxon with an origin in Neocomitoidea (Thieuloy, 1971). Comprises the families Leopoldiidae and Oosterellidae. *L. Haut. - U. Barrem.*

Family LEOPOLDIIDAE Thieuloy, 1971

(*nom. transl.* herein *ex* Leopoldiinae)

Discussion. Monophyletic derivatives of Neocomitidae (Thieuloy, 1971); compressed, involute, high whorled with narrow tabulate (primitive) to acute (derived) venter and suture with very broad asymmetrically bifid E/L. Comprises the subfamilies Leopoldiinae and Saynellinae. *L. Haut. - L. Barrem.*

Subfamily Leopoldiinae Thieuloy, 1971

Discussion. A monophyletic group, with an origin in Neocomitinae close to *Teschenites* (Thieuloy, 1971); distinct umbilical bullae give rise to prorsiradiate bundles of ribs that terminate in ventrolateral clavi, effacing from flanks in middle to later growth; suture with large asymmetrically bifid L and 3

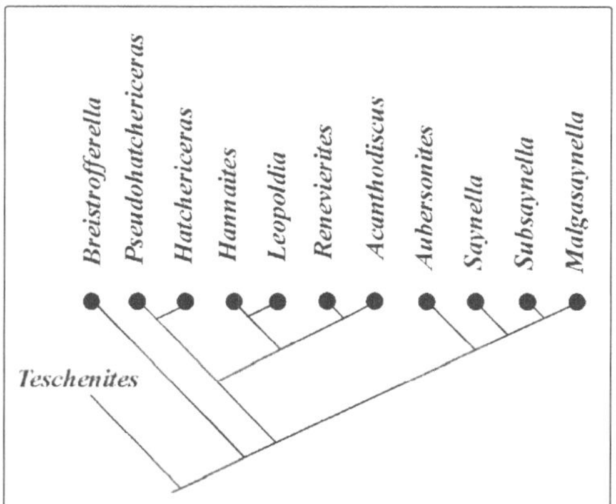

Fig. 585. Hypothesized relationships among Leopoldiidae.

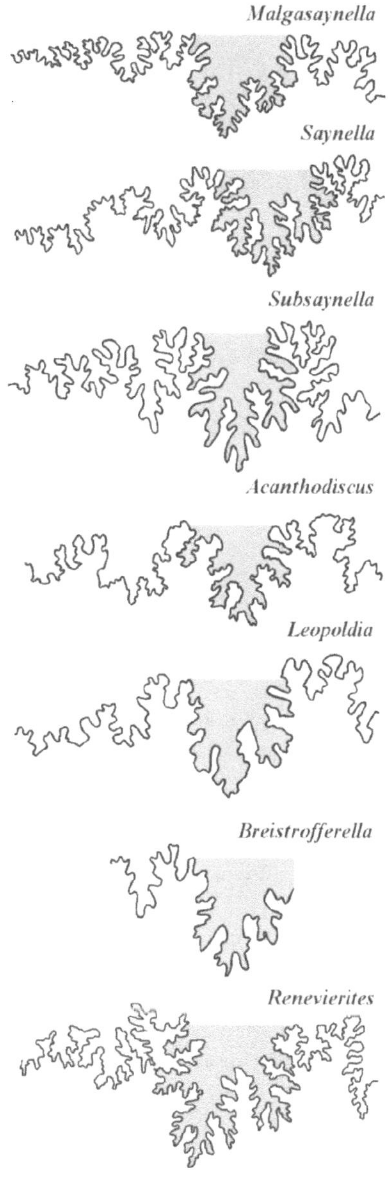

Fig. 586. Leopoldiine sutures.

auxiliary elements. Comprises the tribes Leopoldiini and Hatchericeratini. *L. Haut. - L. Barrem.*

Tribe Leopoldiini Thieuloy, 1971

(*nom. transl.* herein *ex* Leopoldiinae)

Type genus: *Leopoldia* Mayer-Eymar, 1887.

Diagnosis. Typically involute, compressed and high whorled, with narrow tabulate to sulcate venter, distinct umbilical bullae giving rise to prorsiradiate bundles of ribs that terminate in ventrolateral clavi; suture with large asymmetrically bifid L and three auxiliary elements. *L. - Lower U. Haut.*

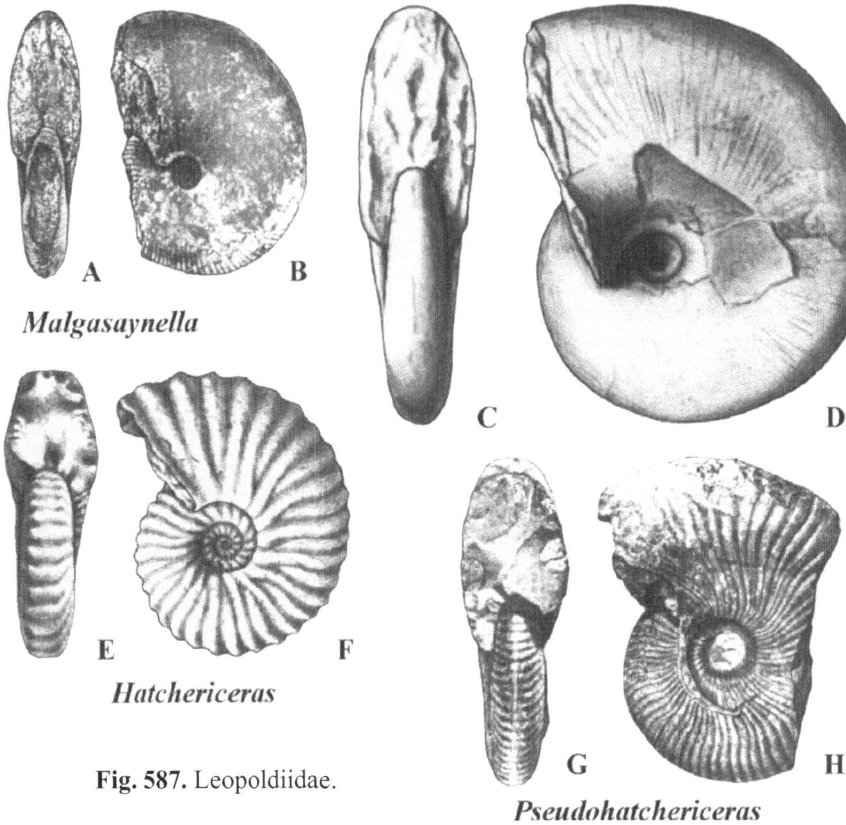

Fig. 587. Leopoldiidae.

Breistrofferella THIEULOY, 1971 [*Ammonites castellanensis* d'Orbigny, 1840; OD]. Small; compressed, involute, high whorled, with broad gently convex flanks, narrowly rounded venter and compressed oval whorl section; strong generally single falcate ribs arise from umbilical bullae and bifurcate, or are joined by 1-2 short intercalatories, on outer flanks; in maturity ribs become more distant, biconcave, and alternate long and short, projecting strongly forward and terminating in ventrolateral clavi to leave venter smooth; suture with broad asymmetrically trifid L. *L. Haut.* **Fig. 584A-B** - *B. castellanensis* (d'Orbigny).

Leopoldia MAYER-EYMAR, 1887 [*Ammonites leopoldinus* d'Orbigny, 1840; T]. Small (**m**) to moderately large (**M**); like *Breistrofferella* but in middle growth with ribs effaced from midflank leaving only umbilical bullae and very short uniform flat prorsiradiate ribs close to venter; adult whorls smooth (**M**); suture with narrow E, asymmetrically bifid E/L and broad asymmetrically bifid L and retracted suspensive lobe. *L. Haut.* **Fig. 584C-D** - *L. leopoldinus* (d'Orbigny).

= *Hoplitides* von Koenen, 1902 (obj.).

= *Solgeria* Uhlig, 1905 (obj.).

Renevierites BUSNARDO & THIEULOY, 1989 [*Leopoldia renevieri* Baumberger, 1905; OD]. Medium sized; like *Leopoldia* but with lozenge-shaped whorl section, maximum width at midflank and rigid slightly prorsiradiate primary ribs on inner flank terminating in lateral bullae; suture with broad L. *L. Haut.* **Fig. 584J-L** - *R. renevieri* (Baumberger).

Acanthodiscus UHLIG, 1905 [*Ammonites radiatus* Bruguière, 1789; SD Sayn, 1907]. Medium sized; inner whorls like *Renevierites* but more evolute and inflated, with flat subparallel inner flanks, flat strongly convergent outer flanks, tabulate to slightly sunken venter, and octagonal whorl section; trituberculate from early growth, with strong rather distant rigid bar-like primariy ribs with prominent umbilical bullae and spinose high-lateral tubercles from which secondaries arise in pairs, separated by a various number of weaker intercalatories; all ribs project forwards and terminate in small rounded

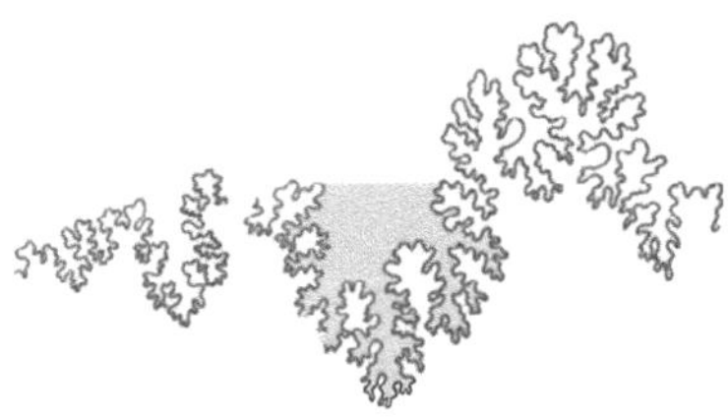

Fig. 588. *Hatchericeras* suture.

ventrolateral nodes at smooth shallowly concave venter; in later growth secondaries become obsolete, but their tubercles remain, leaving strong distant bituberculate bar-like ribs on inner flank; secondary ribs weaken and eventually become obsolete, leaving more numerous ventrolateral tubercles disconnected from lateral tubercles. *Lower U. Haut.* **Fig. 584E-F** - *A. radiatus* (Bruguière).

Hannaites IMLAY, 1957 [*Neocomites riddlensis* Anderson, 1938; OD]. Small; inner whorls like *Leopoldia* but with sharply defined ventrolateral shoulders; later with broader venter, subquadratic whorl section and strong dense fine ribs uninterrupted across venter; suture with narrow L. *L. - M. Haut.* **Fig. 584G-I** - *H. riddlensis* (Anderson).

Tribe Hatchericeratini nov.

Type genus: *Hatchericeras* Stanton, 1901.

Diagnosis. In early to middle growth with coarse flexuous ribs bifurcating from umbilical shoulder, with or without bullae, crossing tabulate venter uninterrupted; venter rounds and ornament lost in maturity; suture incised with L much deeper than E. Like Leopoldiini (Uhlig, 1910) but separated by a substantial stratigraphical break, and Austral in distribution. *L. Barrem.*

Hatchericeras STANTON, 1901 [*H. patagonense*; OD][**M**]. Large, involute, compressed; in early to middle growth with flat subparallel flanks and subtabulate venter; strong umbilical bullae give rise to generally 2 strong rigid prorsiradiate ribs, separated by an intercalatory, crossing venter transversely; in middle to later growth macroconchs smooth, with rounded venter and elliptical whorl section; suture incised, with narrow E, broad deeply bifid asymmtrical E/L, and broad asymmetrically bifid L much deeper than E. *L. Barrem.* **Fig. 587C-F** - *H. patagonense* Stanton.

= *Hatchiceras* Howlett, 1989 (*laps. cal.*).

Pseudohatchericeras A. LEANZA, 1970 [*Hatchericeras argentinense* Stanton, 1901; OD]. Medium sized; like *Hatchericeras* but with dense fine flexuous flat-topped ribs separated by narrow interspaces. *L. Barrem.* **Fig. 587G-H** - *P. argentinense* (Stanton).

Subfamily Saynellinae nov.

Diagnosis. Typically small compressed oxycones with acute venter and 4 auxiliary lobes to suture. Monophyletic derivatives of Leopoldiinae (Kilian, 1910), close to *Renevierites*. *L. Haut. - U. Barrem.*

Aubersonites **gen. nov.** [*Ammonites sueuri* Pictet & Campiche, 1859, p.170; OD herein]. Small compressed oxycone with subvertical umbilical wall, broad weakly convex convergent flanks and narrowly rounded venter; low falcate ribs, faint on inner flank, show weak mid-lateral bullae from which pairs of secondaries project forward and terminate in distinct small ventrolateral clavi; suture like *Leopoldia*, with asymmetrically bifid L. *L. Haut.* **Fig. 584M** - *A. sueuri* (Pictet & Campiche).

Saynella KILIAN, 1920 [*Ammonites clypeiformis* d'Orbigny, 1841; SD Spath 1924]. Small; in early growth like *Aubersonites*; subsequently involute, strongly compressed oxycone with inclined umbilical wall, convex flanks, acute venter, lanceolate whorl section and smooth outer whorls; suture with narrow shallow E, small digitate E/L and very broad subtrifid L. *L. Haut.* **Fig. 584N-O** - *S. clypeiformis* (d'Orbigny).

Subsaynella SPATH, 1923 [*Ammonites sayni* Paquier, 1900; OD]. Very small; like *Saynella* but with oblique umbilical wall, sharply rounded venter, weak prorsiradiate sigmoidal ribs that split into bundles of 2-3 secondaries and strengthen across venter, and suture with narrower lobes; superficially similar to *Aconeceras* (Aconeceratidae). *Lower U. Haut.* **Fig. 584P-Q** - *S. sayni* (Paquier).

= *Sunsaynella* Breskovski 1980 (*laps. cal.*).

Malgasaynella BUSNARDO, 1970 [*Saynella besairiei* Collignon, 1949; OD]. Moderately large; like *Subsaynella* but with very fine prorsiradiate secondaries on outer flank and venter, and suture with broad E/L and L/U_2 and wider L. *U. Barrem.* **Fig. 587A-B** - *M. besairiei* (Collignon).

= *Malgesaynella* Wright, 1996 (*nom. van.*).

Hemisaynella MORET & MAHMOUD, 1953 [*nom. nud.*].

Family OOSTERELLIDAE Breistroffer, 1940

(*nom. transl.* Wright, 1957 *ex* Oosterellinae)

Discussion. Monophyletic derivatives of Leopoldiidae (Wright, 1957); primitively compressed, weakly ornamented with tabulate venter and suture with broad E/L; later strongly ribbed and with siphonal keel. Comprises the subfamilies Platylenticeratinae and Oosterellinae. *L. Valang.- L. Haut.*

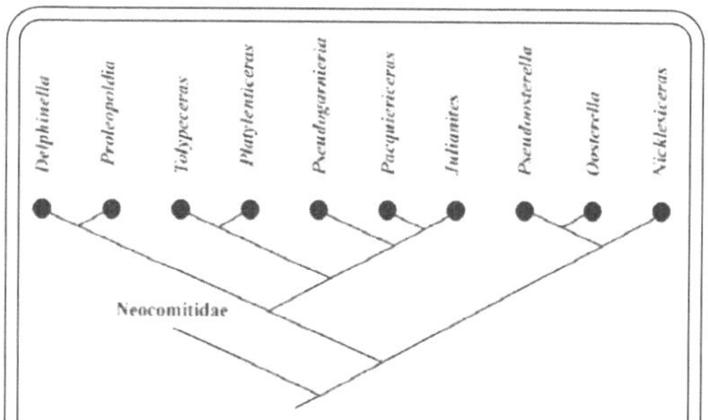

Fig. 589. Hypothesized relationships among Platylenticeratinae.

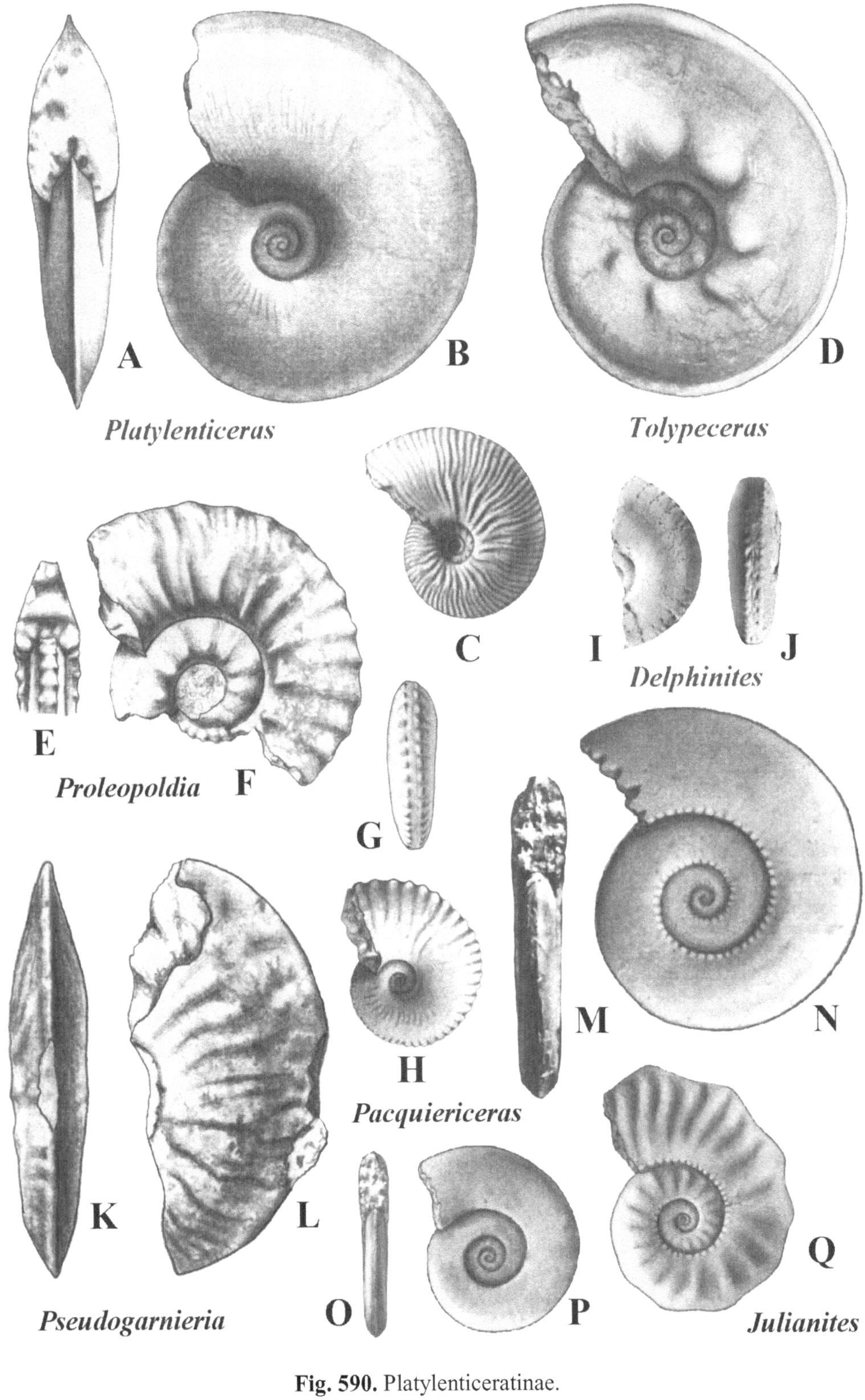

Fig. 590. Platylenticeratinae.

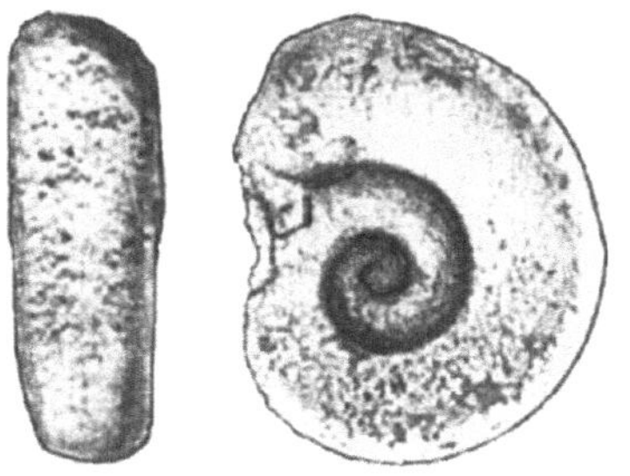

Fig. 591. *Eopaquiericeras.*

Subfamily Platylenticeratinae Casey, 1973

Discussion. Monophyletic descendants of uncertain ancestry, Leopoldiinae close to *Leopoldia* according to Wright (1996), but Jeletzky & Kemper (1988) looked to *Spiticeras*, Bulot (1990) suggested Craspeditinae and Alsen & Rawson (2005) favoured Neocomitinae. These are the "*... degenerate oxycones of the Lower Neocomian*" (Spath, 1947: 17); typically very evolute, weakly ornamented with high keeled venter and broad E/L to suture. Predominantly Boreal, but with some incursions into northern Tethys. *?U. Tithon., L. Valang. - L. Haut.*

Eopaquiericeras Cantú-Chapa, 2006 [*E. peraltai*; OD]. Small, evolute, with steep umbilical wall, parallel flanks, broadly rounded venter and subquadratic whorl section; unornamented; suture simple, with broad irregularly incised L. *U. Tithon*. **Fig. 591** - *E. peraltai* Cantú-Chapa.

The position of this genus is uncertain.

Delphinites Sayn, 1900 [*D. ritteri*; M]. Very small involute platycone with broadly convex flanks, narrow tabulate to slightly concave venter and subrectangular whorl section; inner flanks smooth, outer flanks with weak prorsiradiate ribs terminating in ventrolateral clavi; suture distinctive, with very broad multilobed E/L. *L. Valang*. **Fig. 590I-J** - *D. ritteri* (Sayn).

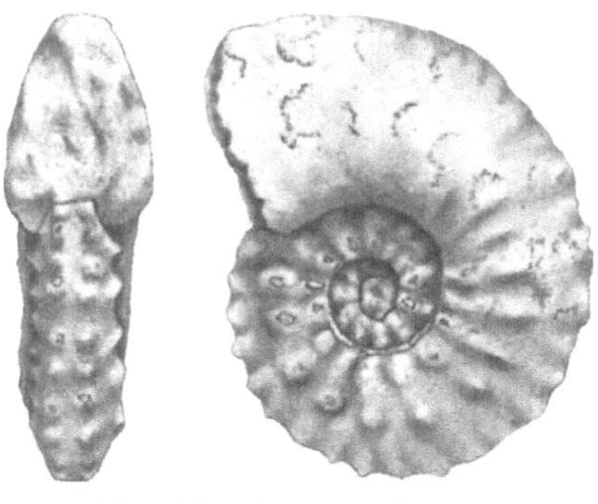

Fig. 592. *Greenlandiceras.*

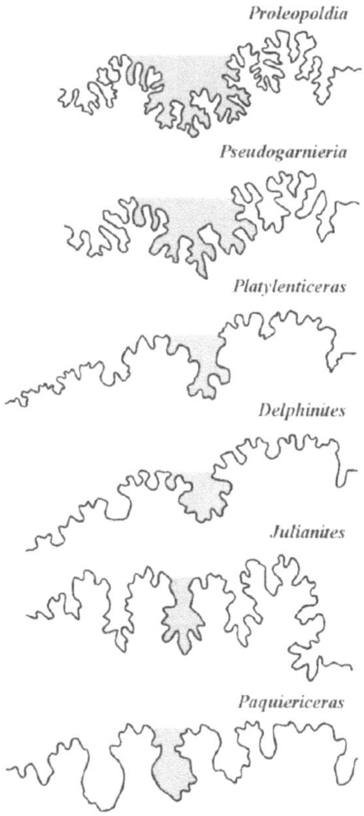

Fig. 593. Oosterellid sutures.

Pseudogarnieria Spath, 1923 [*Oxynoticeras unduloplicatile* Stchirowski, 1894; OD]. Small, like *Delphinites* but more evolute, with low simple flexuous ribs; inner whorls with a corona of small periumbilical tubercles and sporadic bifurcation of some ribs on outer flank; suture relatively complex, with shallow E, broad asymmetrically bifid E/L and deep trifid L. *L. Valang*. **Fig. 590K-L** - *P. unduloplicatilis* (Stchirowski).

Although included in the synonymy of *Delphinites* (Mitta, 2018) the sutures are very different.

Proleopoldia Spath, 1923 [*Hoplites kurmyschensis* Stchirowsky, 1895; OD]. Medium sized; like *Pseudogarnieria* but more evolute, with narrow flattened venter throughout, periodic ventrolateral nodes, and weaker more numerous umbilical bullae that on outer whorls develop into strong primary ribs from which arise bundles of finer secondaries at midflank separated by intercalatories; suture like *Pseudogarnieria*. *L. Valang*. **Fig. 590E-F** - *P. kurmyschensis* (Stchirowsky).

Greenlandiceras **gen. nov.** [*Delphinites (Pseudogarnieria) donovani* Alsen & Rawson 2005, p.208; OD herein]. Small; like *Proleopoldia* but more inflated, with narrow flat venter, subpentagonal whorl section and prominent lateral tubercles in early to middle growth. *L. Valang*. **Fig. 592** - *G. donovani* (Alsen & Rawson).

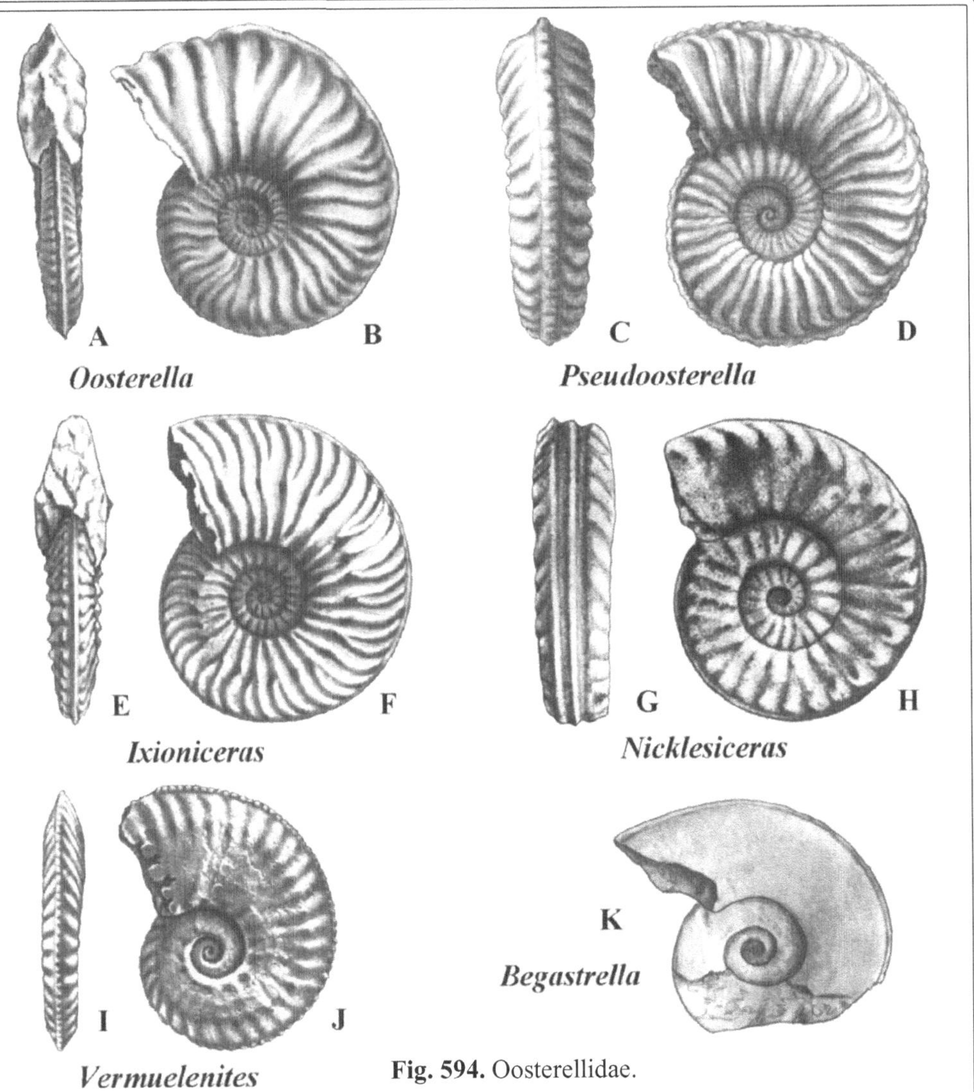

Fig. 594. Oosterellidae.

Tolypeceras HYATT, 1903 [*Ammonites marcousianus* Pictet & Campiche, 1859; OD]. Medium sized; rather evolute compressed oxycone with broad weakly convex flanks converging to narrowly rounded venter, and subtrigonal whorl section; ornament dominated by sparse distant large rounded umbilical tubercles, although inner whorls may be ribbed. *Lower L. Valang*. **Fig. 590D** - *T. denticulatum* (Roemer).

Platylenticeras HYATT, 1900 [*Amaltheus (Oxynoticeras) heteropleurus* Neumayr & Uhlig, 1881; OD]. Small; like *Tolypeceras* but more compressed and involute, with sharp venter, absence of umbilical tubercles and suture often strongly asymmetrical. *Lower L. Valang*. **Fig. 590A-B** - *P. heteropleurum* (Neumayr & Uhlig), **590C** - *P. cardioceroides* (Sayn).

= *Garnieria* Sayn, 1901 (*non* Bourguignat, 1877).

Paquiericeras SAYN, 1901 [*P. paradoxum*; OD]. Very small; compressed, very evolute (U=37-43%), with flat subparallel flanks, narrowly rounded to fastigiate and sometimes keeled venter, and compressed elliptical whorl section; with or without conspicuous bead-like umbilical tubercles and flexuous growth striae; faint spiral lines may be present. *Upper L. - Lower U. Valang*. **Fig. 590G-H** - *P. paradoxum* Sayn; **Fig. 590M-N** - *P. tuberculatum* Vermeulen.

***Vermeulenites* gen. nov.** [*Paquiericeras mourrei* Vermeulen, 1972, p.43; OD herein]. Small; like *Paquiericeras* but more involute (U=28-33%) and compressed, with small umbilical tubercles, conspicuous ribs that carry weak high-lateral tubercles and are strongest on outer flank, finely beaded siphonal keel and more incised suture with less massive L. *U. Valang*. **Fig. 594I-J** - *V. mourrei* (Vermeulen).

Julianites THIEULOY, 1977 [*Paquiericeras (J.) undulatum*; OD]. Very small; like *Vermeulenites* but with broad low distant rectiradiate to rursiradiate ribs on outer flank and siphonal keel with broad gentle undulations in maturity. *Upper L. - Lower U. Valang.* **Fig. 590O-Q** - *J. undulatus* (Thieuloy).

Subfamily Oosterellinae Breistroffer, 1940

Discussion. Descendents of Platylenticeratinae (Vermeulen, 1972), close to *Julianites* (Reboulet, 1996); typically evolute, compressed, with a siphonal keel that may be crenulated and, in all but the most primitive genus, with strong ribs that arise from bullae and may have lateral and/or ventrolateral tubercles. *U. Valang. - L. Haut.*

Begastrella **gen. nov.** [*Oosterella begastrensis* Company, 1987, p.194; OD herein]. Small; relatively evolute (U=33-36%), compressed, with flat subparallel flanks and subtabulate venter with distinct siphonal keel; ornament lacking; suture simple with broad shallowly bifid E/L and narrow L. *Lower U. Valang.* **Fig. 594K** - *B. begastrensis* (Company).

Nicklesiceras **gen. nov.** [*Mortoniceras gaudryi* Nicklès, 1892, p.188; OD herein] [**m**]. Small; very evolute, compressed, with flat parallel flanks, distinct ventrolateral shoulders and rectangular whorl section; broad subtabulate bisulcate-tricarinate venter; strong rectiradiate ribs mostly single, occasionally bifurcating, weakening at midflank and with prominent umbilical tubercles and ventrolateral clavi; a homoeomorph of *Peroniceras* (Collignoniceratoidea). *U. Valang.* **Fig. 594G-H** - *N. gaudryi* (Nicklès).

Oosterella KILIAN, 1911 [*Ammonites cultratus* d'Orbigny, 1841; SD Roman, 1938]. Small to medium sized; like *Begastrella* but with subrectangular whorl section and strong, blunt, irregularly simple and branching ribs that, in some, arise in fasciculate bundles from umbilical bullae; high keel, crenulated in juveniles, and sometimes undulatory later; suture with narrow L. *Upper U. Valang. - L. Haut.* **Fig. 594A-B** - *O. cultrata* (d'Orbigny).

= *Nicklesia* Kilian, 1910 (*non* Hyatt, 1903).

Pseudoosterella SPATH, 1924 [*Mortoniceras fischeri* Nicklès, 1892; SD Roman, 1938]. Very small; like *Oosterella* but very evolute, with strong simple rectiradiate ribs that project forward near venter and robust serrated siphonal keel. *U. Valang.* **Fig. 594C-D** - *P. fischeri* (Nicklès).

= *Pseudooosterella* Roman, 1938 (*nom. van.*).

Ixioniceras **gen. nov.** [*Ammonites ixion* D'ORBIGNY 1841, p.186; OD herein). Like *Oosterella* but very small and more finely ribbed, with tubercles at point of rib furcation; suture with broad trifid L. *L. Haut.* **Fig. 594E-F** - *A. ixion* d'Orbigny.

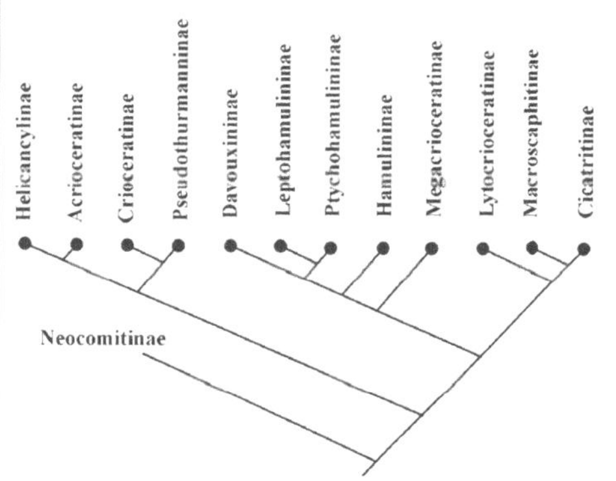

Fig. 595. Hypothesized relationships among Crioceratitoidei.

Hyporder CRIOCERATITOIDEI Gill, 1871

(*nom. transl.* herein *ex* Crioceratitidae)

Discussion. Monophyletic derivatives of Neocomitoidei (Kilian, 1907). Comprises mostly criocones, recoiled criocones and hamulinicones assigned to the superfamilies Crioceratitoidea, Hamulinoidea and Macroscaphitoidea. *Uppermost Valang. - M. Albian.*

Superfamily CRIOCERATITOIDEA Gill, 1871

(*nom. transl.* herein *ex* Crioceratitidae)

Discussion. Monophyletic derivatives of Neocomitoidea (Kilian, 1907), comprising ammoniticones, criocones and aspinocones, convergent towards Emericiceratidae; strongly ribbed, with or without tubercles and constrictions. Comprises the families Crioceratitidae and Helicancylidae. *Upper Valang. - U. Apt.*

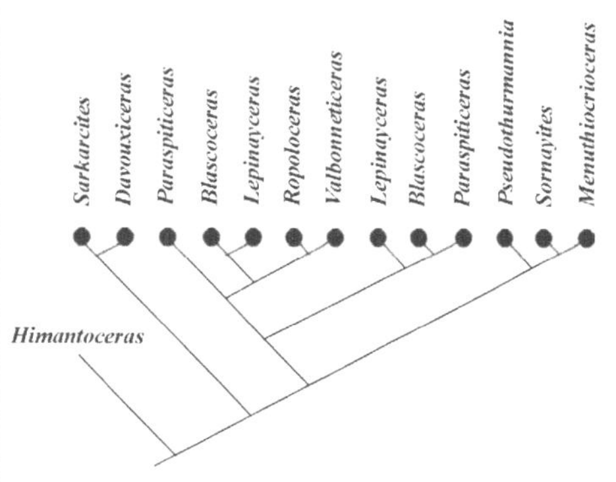

Fig. 596. Hypothesized relationships among Crioceratitinae.

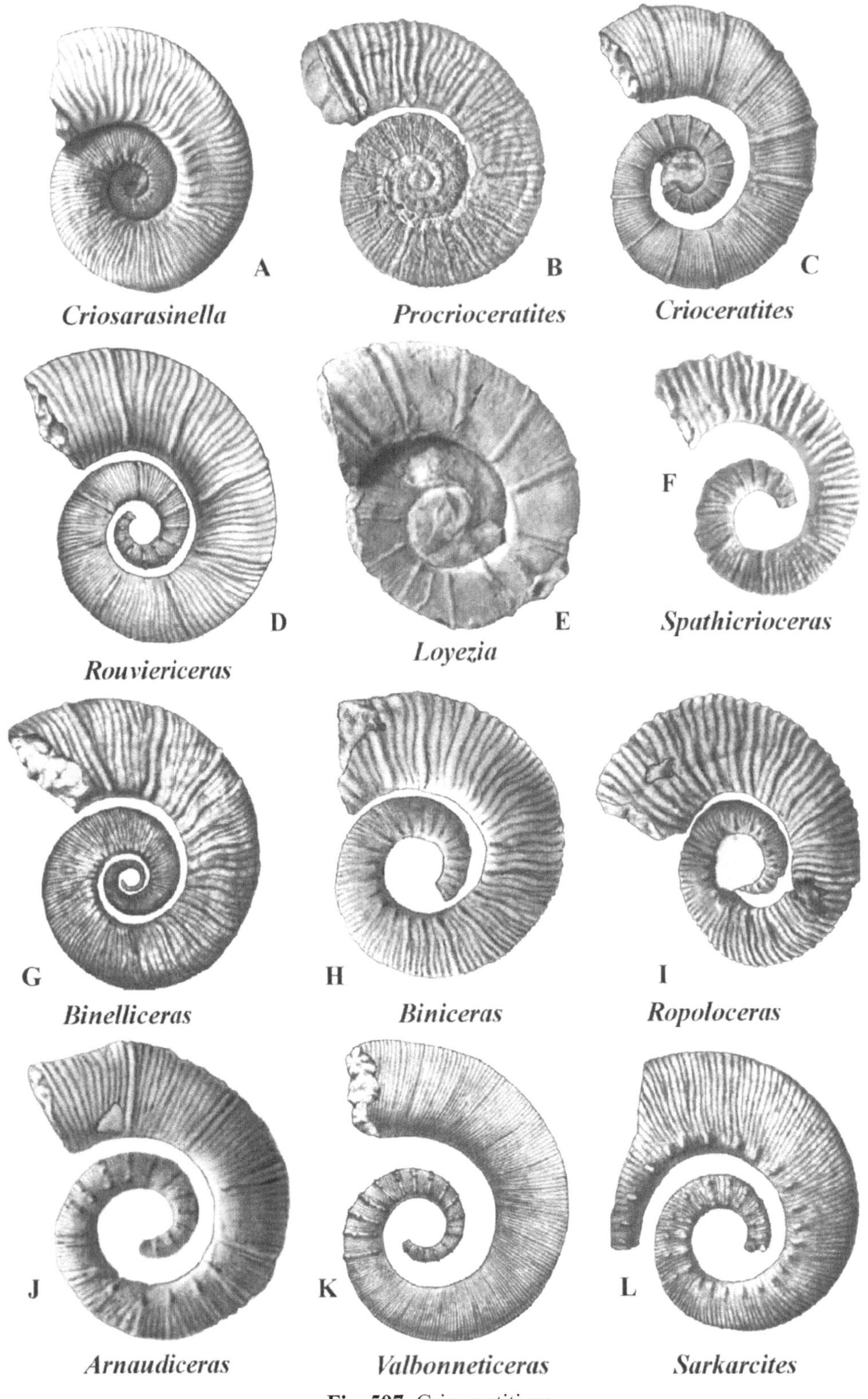

Fig. 597. Crioceratitinae

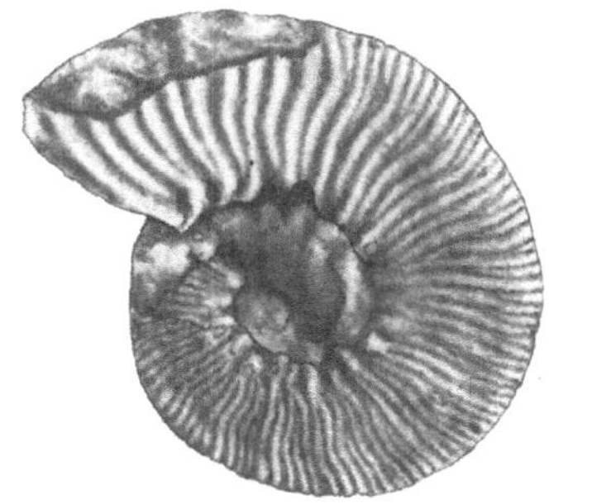

Fig. 598. *Criosarasinella* (**m**).

Family CRIOCERATITIDAE Gill, 1871

(*nom. correct.* Wright, 1952 *pro* Crioceratidae)

Discussion. A monophyletic group of medium sized to large constricted criocones with up to three rows of tubercles, descended from Neocomitidae (Thieuloy, 1977; Baraboshkin & Mikhailova, 2000); comprises the subfamilies Crioceratitinae and Pseudothurmanniinae. *Uppermost Valang. - U. Barrem.*

Subfamily Crioceratitinae Gill, 1871

(*nom. transl.* Wright, 1957 *ex* Crioceratitidae)

= Protacrioceratinae Dimitrova, 1970

= Crioceratinae Reboulet, 1995 (*nom. van.*)

Discussin. A seemingly monophyletic group with an origin in Neocomitinae, close to *Teschenites* (Thieuloy, 1977); the dominant larger heteromorphs of the Hauterivian in the Meditterranean Province. Characteristically fairly large loose criocones with rapidly expanding whorls, broad subparallel flanks, rounded venter and compressed whorl section; trituberculate primary ribs separated by non-tuberculate secondaries, at least in early growth; ornament not generally coarsening markedly in later growth; suture with narrow stemmed bifid saddles, symmetrically trifid L and small trifid U. *Uppermost Valang. - U. Haut.*

Criosarasinella THIEULOY, 1977 [*C. furcillata*; OD]. Medium sized; initially a criocone, later with whorls in contact; dense fine flexuous ribs bifurcating from weak umbilical bullae and then again higher up. *Upper U. Valang.* **Fig. 597A (M), 598 (m)** - *C. furcillata* Thieuloy.

= *Croisarasinella* Reboulet, 1996 (*nom. van.*).

Procrioceratites VERMEULEN, BORRO, LAZARIN, LEPINAY, LEROY & MASCARELLI, 2020 [*P. etiennei*; OD]. Small to medium sized; a tight criocone with flat flanks, rounded venter and compressed elliptical whorl section; strong flexuous slightly prorsiradiate primary ribs with weak dorsolateral bullae separated by 3-6 fine secondaries; ribs coarsening with growth. *Uppermost Valang. - basal Haut.* **Fig. 597B** - *P. etiennei* Vermeulen, Borro, Lazarin, Lepinay, Leroy & Mascarelli.

Crioceratites LÉVEILLÉ, 1837 [*C. duvalii*; SD Diener, 1925]. Medium sized to large criocone with flattened flanks and dorsum, and compressed oval whorl section; in early to middle growth strong simple slightly flexuous primary ribs arise from umbilical bullae, separated by 10-20 fine secondaries some bifurcating; primaries closer in later growth, separated by 5-10 secondaries, and fronted by constriction; suture highly incised, with trifid L. *Uppermost Valang. - U. Haut.* **Fig. 597C** - *C. duvalii* Léveillé.

= *Crioceras* d'Orbigny, 1842 (unjust. emend.).

According to Vermeulen *et al.* (2009), in early growth there may be ventrolateral tubercles and, in maturity, primary ribs may be trituberculate. With 28 nominal species (Klein *et al.*, 2007) this genus comprises more than one phyletic line.

Ropoloceras VERMEULEN, LAZARIN, LÉPINAY, LEROY, MASCARELLI, MEISTER & MENKVELD-GFELLER, 2012 [*Crioceras seitzi* Sarkar, 1955; OD]. Small to medium sized criocone tightening with growth; in early growth primary ribs with dorsolateral spines and lateral and ventrolateral tubercles sometimes separated by 1-2 secondaries; midlateral tubercles evanesce at various stages when ribs become closer, finer and more flexuous, separated by 1-2 intercalatories; in maturity umbilical bullae weaken and ventrolateral tubercles become very tiny, while ribs strengthen, becoming straighter and more distant. *L. Haut.* **Fig. 597I** - *R. seitzi* (Sarkar).

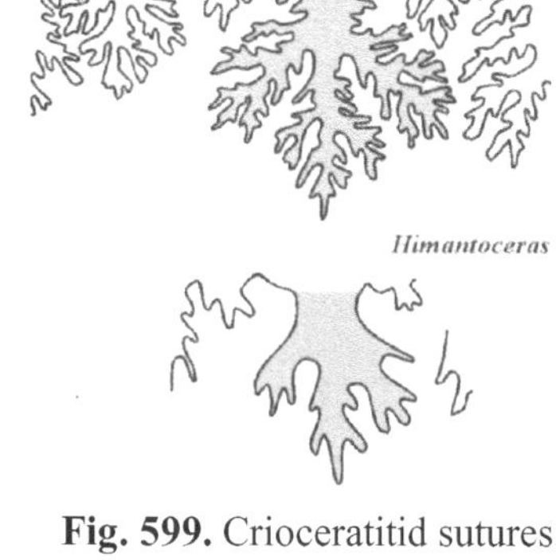

Fig. 599. Crioceratitid sutures.

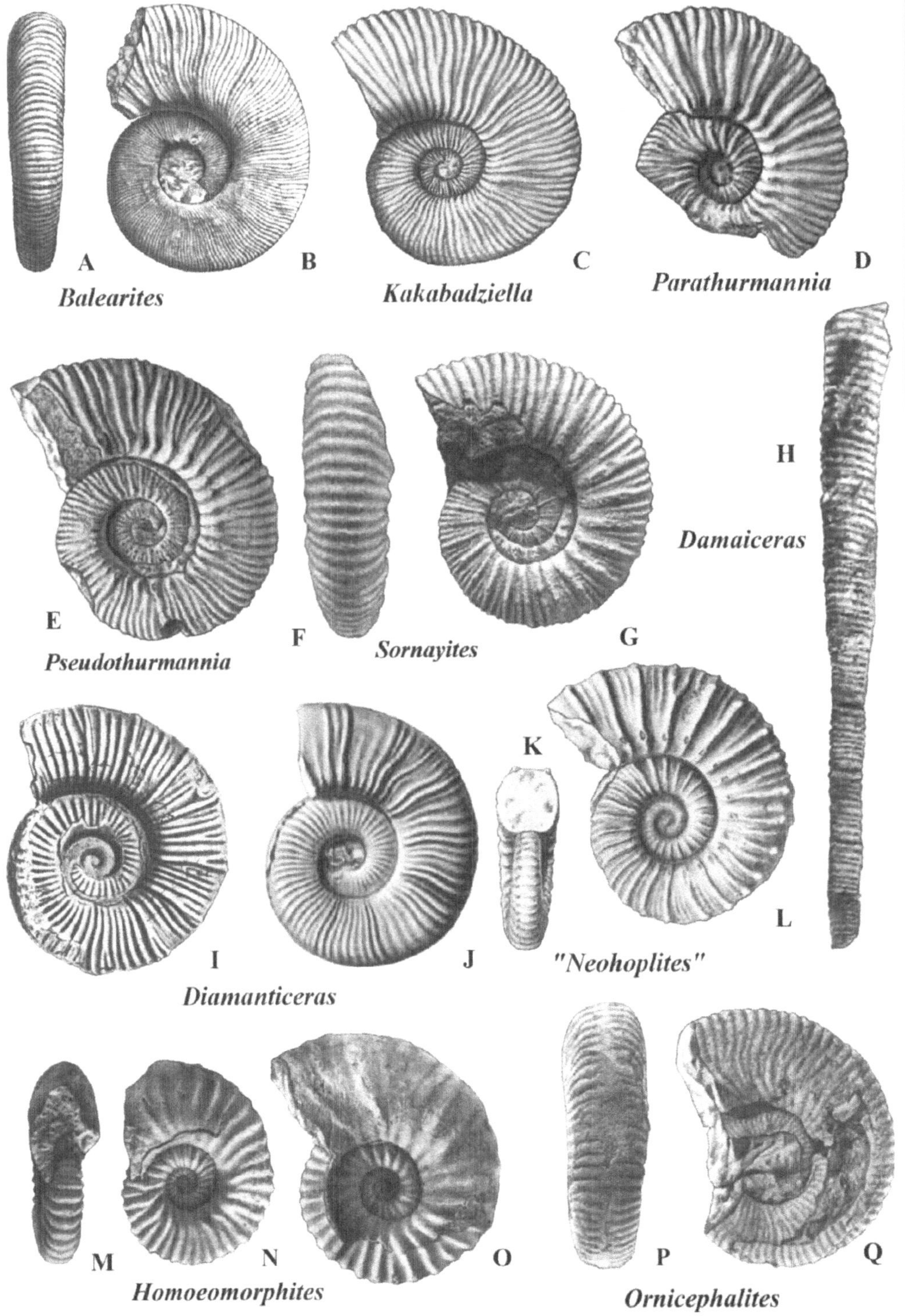

Fig. 600. Pseudothurmanninae, Homoeomorphitidae.

Spathicrioceras SARKAR, 1955 [*S. sornayi*; OD]. Earliest growth stages unknown; in middle growth a loose criocone with flattened flanks; primary ribs bituberculate with only umbilical and midlateral tubercles, separated

Fig. 601. *Pseudomoutoniceras.*

by 2-4 fine non-tuberculate secondaries; in later growth ribbing uniform, undifferentiated, non-tuberculate, bifurcating from umbilical shoulder. *U. Haut.* **Fig. 597F** - *S. sornayi* Sarkar.

= *Sapthicrioceras* Sarkar, 1954 (*nom. nud.*).

Rouviericeras VERMEULEN, LAZARIN, LEROY & MASCELLARI, 2009 [*Crioceras picteti* var. *majoricensis* Nolan, 1894; OD]. Earliest stages (D<36 mm) like *Spathicrioceras*; in middle growth ribs flexuous, lateral tubercles obsolete and secondaries arise in bundles from umbilical shoulder; on adult body chamber ventrolateral tubercles obsolete, primaries fronted by constrictions, and secondaries simple or sometimes bifurcating. *Upper U. Haut.* **Fig. 597D** - *R. majoricense* (Nolan).

Protacrioceras SARKAR, 1955 [*Ancyloceras ornatum* d'Orbigny, 1850; OD]. Small to medium sized; criocone becoming ellipticone in maturity; whorl section oval, with rounded venter; trituberculate primary ribs separated by 2-8 non-tuberculate secondaries that in later growth occasionally bifurcate, persisting to maturity. *Upper U. Haut. - Lower L. Barrem.* **Fig. 605K** - *P. ornatum* (d'Orbigny).

Biniceras VERMEULEN, DUYE, LAZARIN, LEROY & MASCARELLI, 2009 [*Crioceratites andersoni* Sarkar, 1955; OD]. Moderately tight criocone with compressed whorls, trituberculate in early growth; in middle to later growth with numerous slightly flexuous prorsiradiate primary ribs arising from weak umbilical bullae, separated by 2-6 non-tuberculate secondaries. *U. Haut.* **Fig. 597H** - *B. andersoni* (Sarkar).

Arnaudiceras VERMEULEN, DUYE, LAZARIN, LEROY & MASCARELLI, 2009 [*Crioceratites shibaae* Sarkar, 1955; OD]. In early growth like *Crioceratites*; later with prominent umbilical bullae giving rise to 2-3 strong rectiradiate non-tuberculate ribs separated by 4-6 secondaries; in middle growth ventrolateral tubercles evanesce followed later by umbilicals, leaving mostly simple primary ribs in maturity that flex forward near venter, separated by up to 12 fine secondaries that may bifurcate; adult body chamber with collared constriction. *L. Haut.* **Fig. 597J** - *A. shibaae* (Sarkar).

Loyezia LEROY, VERMEULEN, LAZARIN, LÉPINAY & MASCARELLI, 2017 [*L. coluccii*; OD]. A tight criocone with broad flat flanks, rounded venter and compressed elliptical whorl section; in earliest growth umbilical bullae give rise to 1-2 fine ribs, followed by slightly flexuous or straight bullate primary ribs separating long and short secondaries; in maturity primary ribs with prominent dorsolateral bullae flanked by constriction, secondary ribs fading. *U. Haut.* **Fig. 597E** - *L. robertsi* Vermeulen, Borro, Lazarin, Lépinay, Leroy & Mascarelli.

Damaisiceras VERMEULEN, LAZARIN, LÉPINAY, LEROY, MASCARELLI, MEISTER & MENKVELD-GFELLER, 2009 [*D. baguei*; OD]. Large criocone with very small spire; type a long straight slowly expanding pro-versum with strong annular ribs, fairly broad and irregularly spaced, most with low, rounded, subspinose dorsolateral and ventrolateral tubercles; frequent broad shallow constrictions bordered by doubled ventrolateral tubercles between which tubercles may loop; ribs become stronger, wider and more evenly spaced on body chamber when dorsolateral tubercles weaken and ventrolaterals strengthen. *U. Haut.* **Fig. 600H** - *D. baguei* Vermeulen, Lazarin, Lépinay, Leroy, Mascarelli, Meister & Menkveld-Gfeller.

Pseudomoutoniceras AUTRAN, DELANOY & THOMEL, 1986 [*Toxoceras annulare* d'Orbigny, 1842; = *Crioceras honnorati* Léveillé, 1837; OD]. Large open criocone with slowly expanding whorls, oval whorl section, periodic collars with adjacent wider interspace and slightly rursiradiate, uniform, non-tuberculate ribs uninterrupted over venter; suture florid, with large trifid L and small U. *U. Haut. - L. Barrem.* **Fig. 601** - *P. honnorati* (Léveillé).

Sarkarcites **gen. nov.** [*Ancyloceras koechlini* Astier, 1851, p.13; OD herein]. Medium sized; in earliest growth with trituberculate primary ribs separated by 2-8 fine secondaries; later with uniform undifferentiated fine ribs arising in fasciculate bundles from prominent umbilical bullae. *U. Haut. - Lower U. Barrem.* **Fig. 597L** - *S. koechlini* (Astier).

Valbonnetticeras **gen. nov.** [*Crioceratites (Emericiceras) tenuicostatum* Thomel, 1964, p.39; OD herein]. Medium sized; circular to subcircular whorl section and rapidly expanding whorls; initially with trituberculate primary ribs separated by fine non-tuberculate secondaries; in middle growth (D>60 mm) primary ribs vestigial and non-tuberculate, leaving mainly fine rursiradiate secondaries crossing venter with forward sinus. *Lower L. Barrem.* **Fig. 597K** - *V. tenuicostatum* (Thomel).

Subfamily Pseudothurmanniinae nov.

Diagnosis. Tight criocones with whorls barely separated and secondarily recoiled ammoniticones with trituberculate flanks in earliest growth; strong flexuous primary ribs arise from umbilical bullae, separated by 1-5 secondaries; primaries sometimes fronted by constrictions. A Meditterranean group. Monophyletic derivatives of Crioceratitinae (Vermeulen, 2006), close to *Criosarasinella* (Vermeulen *et al.*, 2020), with which they evolved in parallel. *U. Haut.* - *L. Barrem.*

Binelliceras SARKAR, 1977 [*Ancyloceras binelli* Astier, 1851; OD]. Tight criocone with subquadrangular whorls just separated; in early growth (D<15 mm) with trituberculate primaries separated by 4-5 fine non-tuberculate secondaries with ventrolateral tubercles to D=35 mm; later (D>50 mm) with fine flexuous secondary ribs arising in fasciculate bundles from umbilical bullae; in maturity strong flexuous to biconcave non-tuberculate primary ribs bounded by constrictions, separated by 2-10 secondaries commonly branching above midflank. *U. Haut.* - **Fig. 597G** - *B. binelli* (Astier).

Pseudothurmannia SPATH, 1923 [*P. angulicostata* Spath *non* d'Orbigny; = *P. picteti* Sarkar, 1955; SD Company *et al.*, 2002]. Very large, very evolute, with steep umbilical wall, broad weakly convex flanks, subangular ventrolateral shoulders, relatively broad tabulate venter and compressed trapezoidal whorl section; more or less sigmoidal primary ribs separated by 1-4 secondaries of variable length, all with weak umbilical bullae and small ventrolateral clavi; in middle growth (D=65-140 mm) primary ribs with high lateral bullae that are lost in maturity. *L. Barrem.* **Fig. 600E** - *P. picteti* (Sarkar).

= *Prieuriceras* Vermuelen, 2003 (obj.).

Balearites SARKAR, 1954 [*Crioceras baleare* Nolan, 1894; OD]. Like *Pseudothurmannia* but much larger, with evenly rounded venter and dense fine flexuous ribs that bifurcate at various levels on flank; adult body chamber uncoils slightly, with several flexuous constrictions bordered by collars. *U. Haut.* **Fig. 600A-B** - *B. balearis* (Nolan).

Kakabadziella HOEDEMAEKER, 2013 [*Ammonites mortilleti* Pictet & de Loriol, 1858; OD]. Evolute, with low umbilical wall, broad flat flanks and compressed elliptical whorl section; finely ribbed; in early growth feeble umbilical bullae give rise to prorsiradiate primary ribs separated by 1-3 finer secondaries; in middle to later growth sinuous primaries flex forward near venter, separated by 1-2 finer secondaries often intercalated or bifurcating high on flank. *U. Haut.* **Fig. 600C** - *K. mortilleti* (Pictet & de Loriol).

= *Kakabadziella* Hoedemaeker & Herngreen, 2003 [*nom. nud.*].

Parathurmannia BUSNARDO, 2003 [= *Hoplites anguli-costatus* Sarasin & Schoendelmayer (*non* d'Orbigny), = *Pseudothurmannia sarasini* Sarkar, 1955 ; OD]. Medium sized; evolute, with flat flanks and compressed subrectangular whorl section; inner whorls like *Balearites*; in maturity regular prorsiradiate primary ribs separated by 4-6 finer flexuous secondaries, often bifurcating, with ventrolateral nodes and weakening over venter; lateral tubercles absent. *U. Haut.* **Fig. 600D** - *P. sarasini* (Sarkar).

Sornayites WIEDMANN, 1962 [*Emericiceras paronai* Sarkar, 1955; OD]. Like *Pseudothurmannia* a recoiled ammoniticone with whorls in contact; gently convex convergent flanks, subtabulate venter and compressed trapezoidal whorl section; straight slightly prorsiradiate trituberculate primary ribs separated by 0-4 intercalatories of various length crossing venter uninterrupted. *U. Haut.* **Fig. 600F-G** - *S. eduoardsi* (Honnorat-Bastidé).

"Neohoplites" GERTH, 1921 [*N. bederi*; *nom. nud.*]. Much like *Sornayites* but in early to middle growth with simple rectiradiate ribs, narrower than interspaces, with feeble umbilical and lateral tubercles and distinct ventrolateral

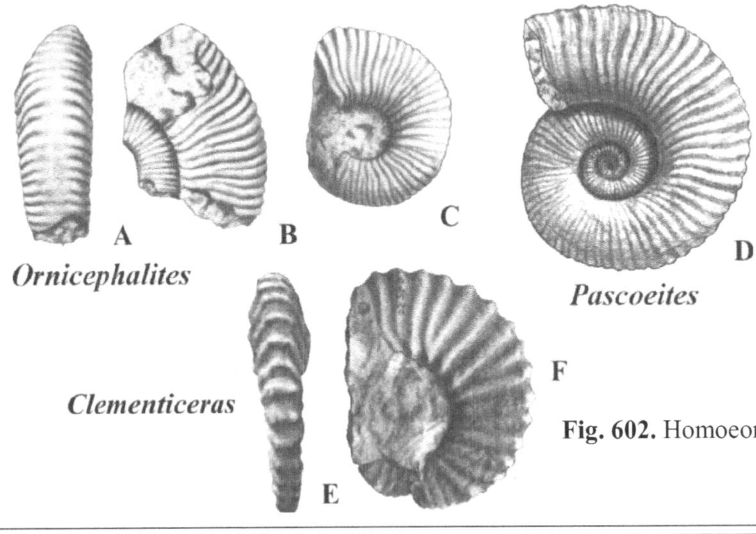

Fig. 602. Homoeomorphitinae.

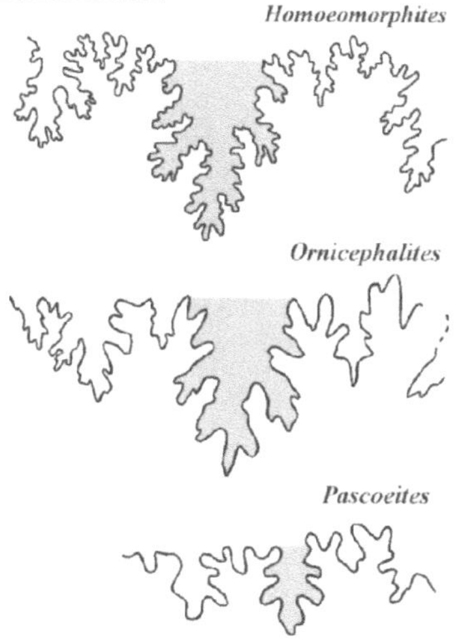

Fig. 603. Homoeomorphitine sutures.

tubercles connected across venter by transverse ribs; in maturity ribs differentiated into strong primaries with distinct umbilical bullae separated by 1-2 fine non-tuberculate intercalatories. **Fig. 600K-L** - *"Neohoplites" bederi* Gerth.

Diamanticeras VERMEULEN, 2004 [*Crioceras diamantense* Gerth, 1926; OD]. Tight criocone with whorls almost in contact, barely impressed on flat dorsum, convex flanks converging to narrowly rounded venter and compressed subovate whorl section; early whorls with strong slightly sinuous rectiradiate non-tuberculate ribs that, beyond D=50 mm, differentiate into stronger frequently bifurcating flexuous primaries separated by 3-7 weaker secondaries; ribs coarsen and become closer in maturity when whorl section is depressed subrectangular; suture with irregular L as deep as E, broad subquadratic trifid E/L and bifid L/U. *U. Haut. - L. Barrem.* **Fig. 600I-J** - *D. diamantense* (Gerth).

Subfamily Homoeomorphitinae Bert, Bersac & Canut, 2020

Discussion. Monophyletic derivatives of Pseudothurmanninae, close to *Pseudothurmannia* (Skwarko & Thieuloy, 1989); evolute, compressed, with flattened venter; non-tuberculate flexuous ribs bifurcating from umbilical edge, primitively weak or effaced from venter. *Upper L. Haut. - U. Apt.*

Ornicephalites SKWARKO & THIEULOY, 1989 [*Pseudothurmannia (O.) indonesiana*; OD]. Medium sized to moderately large; evolute, with flattened venter and subquadratic whorl section; dense finer non-tuberculate flexuous ribs bifurcate from umbilical edge, broaden ventrally and weaken across or efface from siphonal line; in maturity primary ribs separated by 2-4 finer intercalatories; suture with broad asymmetrically bifid L and narrow asymmetrically trifid L much deeper than E. *Upper L. Haut. - L. Barrem.* **Fig. 600P-Q** - *O. ploszkiewiczi* (Riccardi & Aguirre Urreta); **Fig. 602A-C** - *O. indonesiana* (Skwarko & Thieuloy).

Homeomorphites BERT, BERSAC & CANUT, 2020 [*H. aguirreurretae*; OD]. Dimorphic; small (**m**) to moderately large (M**)**; evolute, with rounded [**M**] to flattish flanks [**m**], slightly convex venter and compressed [**M**] to subrectangular [**m**] whorl section; strong usually flexuous ribs, simple or bifurcating from umbilical edge (inner whorls) or just below midflank, crossing venter without weakening; on macroconch ribs disappear progressively, becoming less marked and more distant and irregular; end of phragmocone and body chamber smooth; constrictions and strengthened ribs scarce; suture much like *Ornicephalites* but E/L shallowly bifid. Superficially similar to *Hatchericeras*. *L. Barrem.* **Fig. 600M-O** - *H. aguirreurretae* Bert, Bersac & Canut.

Clementiceras VERMEULEN, LAZARIN, LÉPINAY, LEROY & MASCARELLI, 2009 [*C. jansseni*; OD]. Small, very evolute, with flattened flanks, low rounded venter and compressed trapezoidal whorl section; pinched prosocline umbilical bullae give rise, singly and in pairs, to strong prorsiradiate ribs that quickly recurve to become rectiradiate and cross venter transversely; initially (D<14 mm) with subspinose ventrolateral tubercles that become clavate before evanescing (D>20 mm). *U. Barrem.* **Fig. 602E-F** - *C. jansseni* Vermeulen, Lazarin, Lépinay, Leroy & Mascarelli.

Pascoeites SPATH, 1933 [*P. budavadensis*; OD]. Very small; similar to *Ornicephalites* but ribs mostly single, crossing venter uninterrupted; suture simple, with narrow E, broad asymmetrically bifid E/L, moderately narrow trifid L deeper than E, and fairly broad asymmetrically bifid L/U. *Apt.* **Fig. 602D** - *P. budavadensis* Spath.

Family HELICANCYLIDAE Hyatt, 1894

Discussion. Seemingly monophyletic derivatives of Crioceratitidae (Vermeulen, 2004). Wright (1996)

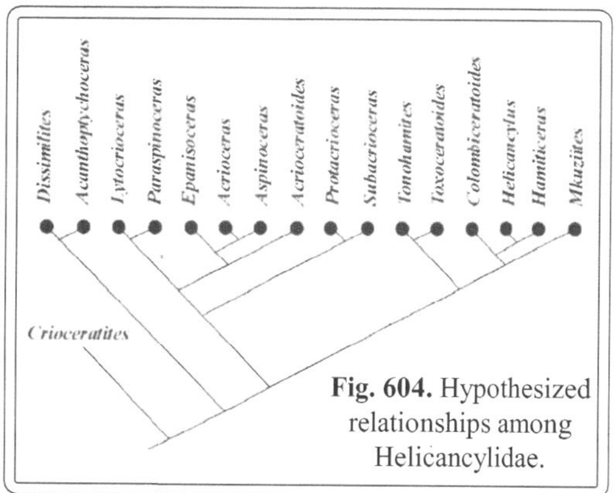

Fig. 604. Hypothesized relationships among Helicancylidae.

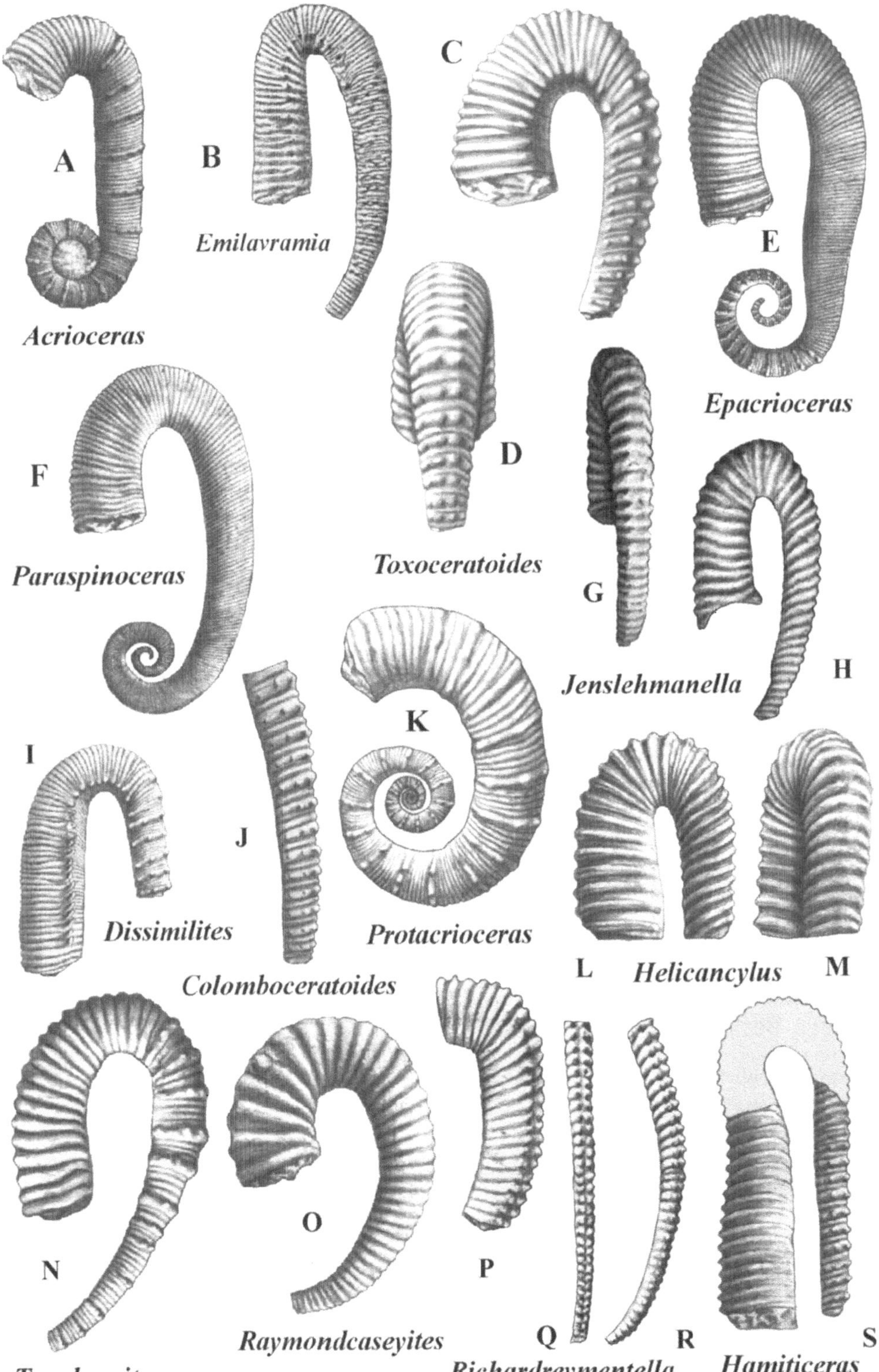

Fig. 605. Helicancylidae.

believed this group to comprise microconchs of Crioceratitidae but this interpretation has been challenged (Vermeulen, 2006, 2008; Bert, 2009) since hellicancylids are common in the Aptian where crioceratitids are absent. Typically small criocones, ancylocones and aspinocones, primitively with trituberculate main ribs separated by non-tuberculate secondaries. This taxon records an important Barremian-Aptian radiation of small heteromorphs displaying marked provincialism. Comprises the subfamilies Monoditinae, Acrioceratinae and Helicancylinae. *Uppermost Haut. - U. Apt.*

The family and generic names have been considered *nomina nuda* (Vermeulen, 2010; Vermeulen *et al.*, 2020) due to the fragmentary nature of the holotype of *Helicancylus*. However Anderson (1938) collected topotype material from Alderson Creek (**Fig. 599M-N**) that he compared closely with the holotype, concluding they were the same. It preserves a length of proversum with strong simple uniform slightly flexuous prorsiradiate ribs, as wide as the interspaces, all with weak ventrolateral bullae, and a suture much like *Toxoceratoides*. Anderson's description clarified the characters of the type species and genus that are not *nomena nuda*, thus validating the family name.

Subfamily Monoditinae Bert, 2009

(*nom. transl.* herein *ex* Monoditidae)

Discussion. A monotypic taxon derived from Crioceratitinae, but precise antecedent unknown. Diagnosis as for genus. *U. Haut.*

Monodites BERT, 2009 [*Hamiticeras crevolai* Vermeulen, Thieuloy & Lapeyre, 1999; OD]. Medium sized; criocone followed by long straight proversum and long retroversum; whorl section to shaft subhexagonal increasing in height with growth; regular strong simple slightly prorsiradiate trituberculate ribs strengthen on flexus; suture with trifid L. *U. Haut.* **Fig. 606D** - *M. crevolai* (Vermeulen, Thieuloy & Lapeyre).

Subfamily Acrioceratinae Vermeulen, 2004

(*nom. transl.* herein *ex* Acrioceratidae)

Discussion. A questionably monophyletic group with an origin in Crioceratitinae close to *Protacrioceras* (Dimitrova, 1970); typically small to medium sized ancylocones with initial shallow helicone followed by long straight to curved proversum, U-shaped flexus and straight retroversum; spire and proversum with trituberculate main ribs separated by non-tuberculate secondaries; flexus and retroversum with simple ribbing sometimes bifurcating from umbilical bullae, weakening or interrupted across siphonal line. Comprises the tribes Acrioceratini and Toxoceratoidini. *L. Barrem. - U. Apt.*

Tribe Acrioceratini nov.

Type genus: *Acrioceras* Hyatt, 1900.

Diagnosis. Trituberculate main ribs to proversum, separated by non-tuberculate secondaries; retroversum with non-tuberculate ribs. Seemingly monophyletic derivatives of Crioceratitinae close to *Protacrioceras* (Vemeulen *et al.*, 2020). *?U. Haut., L. Barrem. - U. Apt.*

Paraspinoceras SARKAR, 1955 [*Ancyloceras pulcherrimum* d'Orbigny, 1842; OD]. Medium sized; like *Acrioceras* but with uniform fine dense prorsiradiate non-tuberculate ribbing throughout, although occasionally 1-2 tubercles may appear on the spire. *?U. Haut., L. Barrem.* **Fig. 605F** - *P. pulcherrimum* (d'Orbigny).

Acrioceras HYATT, 1900 [*Ancyloceras tabarelli* Astier, 1851; OD]. Small; shallow helicone followed by long straight barely expanding proversum and retroversum with compressed oval whorl section; strong trituberculate primary ribs separated by 3-5 fine non-tuberculate secondaries; ribs coarsen and become more uniform on flexus and retroversum, bifurcating from umbilicolateral bullae, without lateral or ventrolateral tubercles; suture with trifid L. *L. Barrem.* **Fig. 605A** - *A. tabarelli* (Astier).
= *Mesocrioceras* Breistroffer, 1951 (*nom. nud.*).

Dissimilites SARKAR, 1954 [*Hamites dissimilis* d'Orbigny, 1842; OD]. Medium sized; like *Acrioceras* but flexus and retroversum with tongue-like dorsolateral clavi from which flat-topped crowded ribs arise, sometimes in pairs, separated by very narrow interspaces. *Lower Barrem.* **Fig. 605I** - *D. dissimilis* (d'Orbigny).

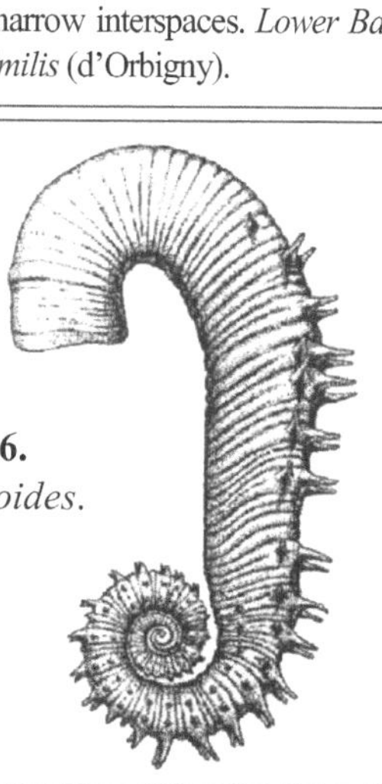

Fig. 606.
Acrioceratoides.

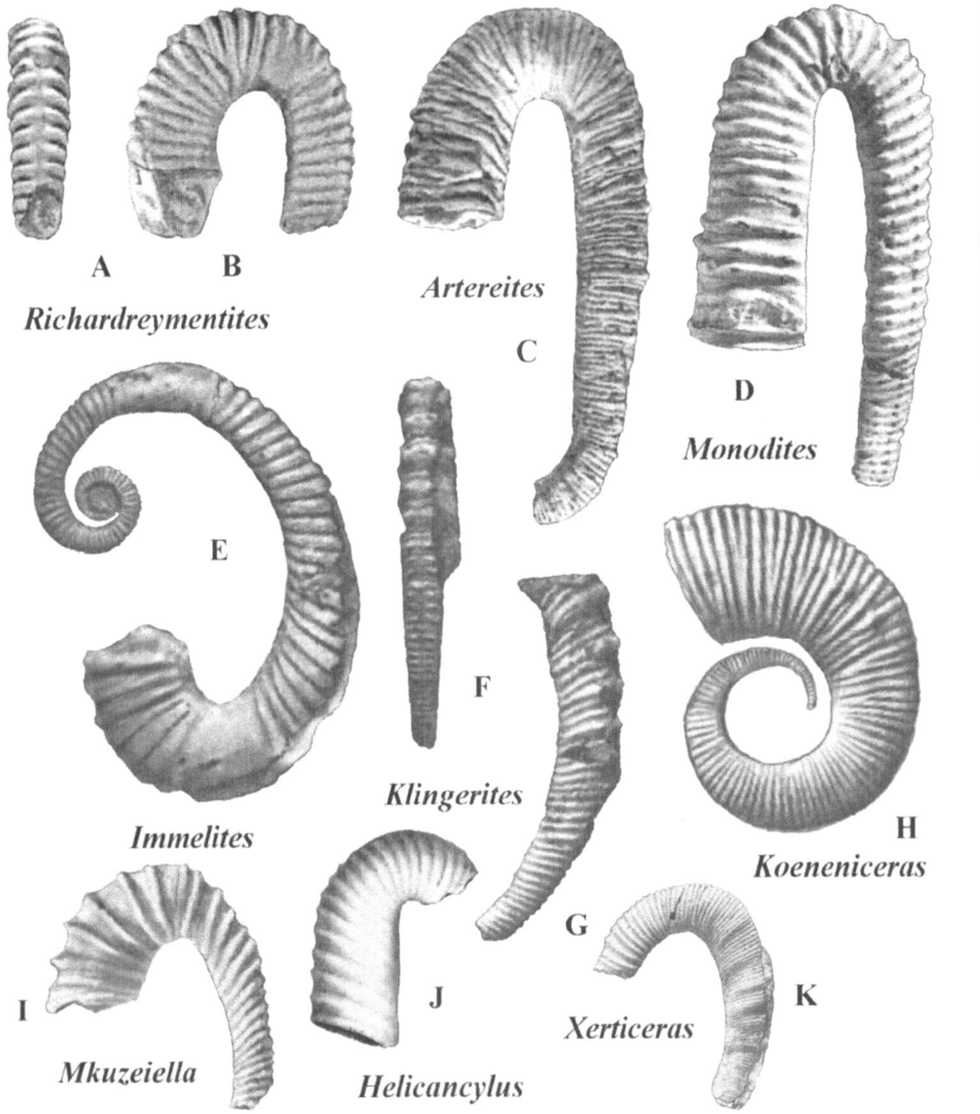

Fig. 607. Helicancylidae.

***Emilavramia* gen. nov.** [*Hamulina subalternata* Breskovski, 1966; OD herein]. Small; like *Dissimilites* but with fine ribs looped between tubercles on proversum; retroversum with fine biplicate rursiradiate to rectiradiate ribs arising from rounded umbilical tubercles. *Uppermost L. Barrem. - lowermost U. Barrem.* **Fig. 605B** - *E. subalternata* (Breskovski).

Artareites BERT, 2009 [*A. landii*; OD]. Medium sized; like *Acrioceras* but with reduced spire and whorls increasing rapidly in height, with relatively long retroversum; primary ribs with feeble umbilical, ventrolateral and ventrolateral tubercles separated by 2-3 non-tuberculate secondaries; ribs frequently looped on shaft and bifurcating on flexus. *U. Barrem.* **Fig. 607C** - *A. landii* Bert.

***Acrioceratoides* gen. nov.** [*Acrioceras zulu* Klinger & Kennedy 1992, p.111; OD herein]. Small; like *Acrioceras* but more coarsely ornamented with numerous recurved prorsiradiate often hypernodose primaries separated by up to 4 fine secondaries; finely ribbed forms (superficially similar to *Paraspinoceras*) have poorly differentiated ribbing, lack umbilical and sometimes inner ventrolateral tubercles to primaries and may show bundled secondaries. *U. Barrem.* **Fig. 606** - *A. zulu* (Klinger & Kennedy).

Koeneniceras MIKHAILOVA & BARABOSHKIN, 2002 [*Ancyloceras tenuiplicatum* van Koenen, 1902; OD]. Small criocone with partially coiled whorls (**m**) or completely uncoiled (**M**); whorl section hexagonal, subrounded to round; simple and occasional bifurcating ribs with small

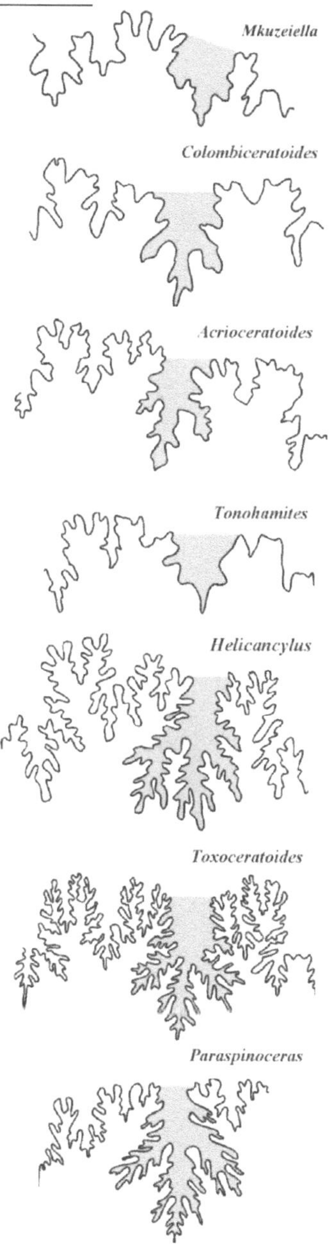

Fig. 608. Helicancyline sutures.

ventrolateral and well developed ventral tubercles on early whorls only; all ribs cross venter uninterrupted; very rare constrictions; suture with narrow E, broader asymmetrically bifid E/L, broad trifid L, small trifid U and larger trifid I. *Lower L. Apt.* **Fig. 607H** - *K. tenuiplicatum* (von Koenen).

Colomboceratoides ETAYO SERNA, 1979 [*Toxoceratoides (C.) renzoni*; OD]. Known only by a gently curved proversum with flattened dorsum, compressed subovate whorl section, tuberculate primary ribs separated by 1-2 non-tuberculate secondaries; ribs with tiny umbilical tubercle, small lateral tubercle to which ribs may be weakly looped and prominent ventrolateral tubercles; ribs attenuated across siphonal line; suture relatively simple with broad bifid E/L and trifid L deeper than E. *U. Apt.* **Fig. 605J** - *C. renzoni* (Etayo Serna).

Tribe Toxoceratoidini nov.

Type genus: *Toxoceratoides* Spath, 1924.

Diagnosis. Rapidly expanding curved proversum generally with only 1-2 non-tuberculate secondaries between trituberculate primaries and coarse non-tuberculate ribs to flexus and retroversum. Monophyletic derivatives of Acrioceratinae close to *Dissimilites* (Vermeulen *et al.*, 2020). *U. Barrem.* - *U. Apt.*

Toxoceratoides SPATH, 1924 [*Toxoceras royerianum* d'Orbigny, 1842; OD]. Ancylocone with rapidily expanding curved proversum, trituberculate primaries separated by 1-2 secondaries that tend to weaken across siphonal line, and flexus and retroversum with undifferentiated rectiradiate ribs arising in 2's and 3's from umbilical tubercles and crossing venter transversely. *U. Barrem.* - *L. Apt.* **Fig. 605C-D** - *T. royerianus* (d'Orbigny).

Hamiticeras ANDERSON, 1938 [*H. pilsburyi*; OD]. Early growth stages unknown; limbs circular in cross section; proversum like *Toxoceratoides* but not expanding rapidly, with alternating trituberculate primaries and fine non-tuberculate secondaries; ribs interrupted across venter by nearly smooth sulcus and long retroversum with coarser non-tuberculate ribs. *M. Apt.* **Fig. 605S** - *H. pilsburyi* Anderson.

Tonohamites SPATH, 1924 [*T. decurrens*; OD]. Like *Toxoceratoides* but with primary ribs almost rectiradiate on curved shaft and separated by 1-4 secondaries; undifferentiated simple ribs on hook and final shaft mainly straight and non-tuberculate. *L.* - *U. Apt.* **Fig. 605N** - *T. decurrens* Spath.

Raymondcaseyites AVRAM, 2001 [*Tonohamites limbatus* Casey, 1960; OD]. Small; like *Tonohamites* but initially with rounded whorl section expanding rapidly and becoming depressed elliptical in maturity; relatively short retroversum; coarse simple prorsiradiate non-tuberculate ribs broaden and strengthen ventrally, flat topped over venter and separated by narrow interspaces; suture with trifid L and narrow deep penetrating U. *L.* - *U. Apt.* **Fig. 605O** - *R. koeneni* (Casey).

Xerticeras DELANOY, MORENO-BEDMAR, RUIZ & TOLÓS LLÁDSER, 2013 [*X, salasi*; OD]. Microconch; criocone phragmocone followed by curved proversum, open flexus and fairly long retroversum; fine prorsiradiate flexuous non-tuberculate riblets superimposed on bulges on proversum, coarsening and simplifying on retro-

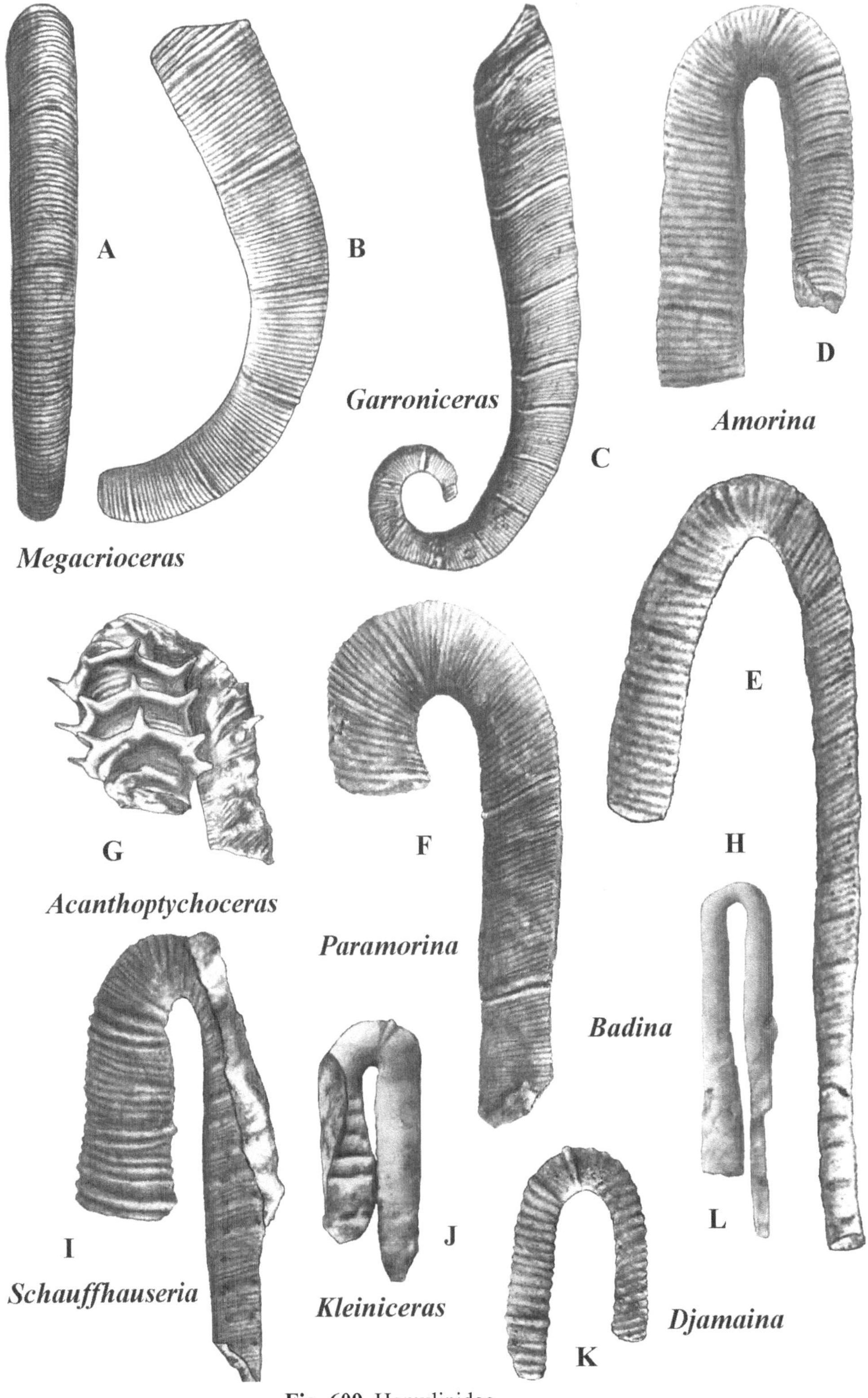

Fig. 609. Hamulinidae.

versum. *L.-M.. Apt.* **Fig. 607K** - *X. salasi* Delanoy, Moreno-Bedmar, Ruiz & Tolós Lládser.

Immelites BULOT, FRAU & PICTET, 2018 [*Tonohamites multituberculatus* Immel & Guoixiong, 2002; OD]. Very small, coiling labeceratid with arcuate proversum and short retroversum; whorl section rounded with nearly flat dorsum; small spire a criocone, with minutely trituberculate ribs; proversum with close, simple, weakly tuberculate, prorsiradiate ribs that become very thick and blunt on flexus, with wide interspaces; aperture simple. *Lower U. Apt.* **Fig. 607E** - *I. multituberculatus* (Immel & Guoixiong).

Richardreymentella BULOT, FRAU & PICTET, 2018 [*Ancyloceras patagonicum* Stolley, 1912; OD]. Small ancylocone with subhexagonal whorl section; uniform undifferentiated ribbing with rounded inner and outer ventrolateral tubercles to all ribs; ribs interrupted across siphonal line by marked ventral sulcus; weak dorsolateral tubercles may or may not be present. *?Upper U. Apt.* **Fig. 605Q-R** - *R. patagonicus* (Stolley).

Subfamily Helicancylinae Hyatt, 1894

(*nom. transl.* Casey, 1961 *ex* Helicancylidae)

Discussion. Seemingly monophyletic derivatives of Acrioceratinae, but precise ancestor unknown. Proversum with simple prorsiradiate ribs with ventrolateral bullae, weakening across the venter. *Upper L. - U. Apt., ?M. Alb.*

Helicancylus GABB, 1869 [*Ptychoceras aequicostatum* Gabb, 1864; M]. Small; the lectotype designated herein is the original of Gabb's (1864) figured specimen, a part of an open flexus and partial retroversum with slightly flattened venter and subcircular whorl section; all ribs strong, simple, rounded, non-tuberculate; flexus with curved rectiradiate ribs and retroversum with rursiradiate ribs, narrower than interspaces, weakening across dorsum (**Fig. 607J**); topotype material shows flexuous prorsiradiate ribs on the straight proversum, all with weak ventrolateral bullae. *Apt.* **Fig. 605L-M** - *H. aequicostatus* (Gabb).

Jenslehmannella BULOT, FRAU & PICTET, 2018 [*J. bangestanensis*; OD]. Small, gracile, with flattened venter for most of gently curved proversum, and moderately long retroversum; strong, simple prorsiradiate ribs to proversum, as wide as interspaces, interrupted by narrow ventral furrow on upper part of proversum, as narrow as interspaces, becoming flexuous and rursiradiate on retroversum; minute ventrolateral tubercles on all ribs at beginning of proversum. *Upper L. Apt.* **Fig. 605G-H** - *J. bangestanensis* Bulot, Frau & Pictet.

Klingerites BULOT, FRAU & PICTET, 2018 [*Toxoceratoides? haughtoni* Klinger & Kennedy, 1977; OD]. Earliest whorls unknown; long curved proversum with strong simple prorsiradiate ribs, all with ventrolateral bullae, interrupted across venter by smooth zone; towards flexus periodically strengthened ribs appear, eventually becoming bold, inflated and trituberculate on inflated flexus and retroversum (like a miniature *Audouliceras*); suture with broad bifid saddles, narrow trifid L and reduced U. *U. Apt.* **Fig. 607F-G** - *K. haughtoni* (Klinger & Kennedy).

Mkuzeiella KLINGER & KENNEDY, 2008 [*M. andersoni*; OD], Small; earliest growth stage unknown, perhaps an open criocone; proversum curved with open flexus and relatively long proversum; phragmocone with sharp prorsiradiate ribs with incipient ventrolateral tubercules; on internal mould ribs appear broad and flat; towards body chamber ribs become more distant and prominently flared with angular ventrolateral edges; aperture constricted; suture with asymmetrically trifid L. *M. Alb.* **Fig. 607I** - *M. andersoni* Klinger & Kennedy.

The long time gap between this genus and the last helicancylid makes assignment here questionable; it shows some similarity to *Rencurelites* (Anisoceratidae).

Superfamily HAMULINOIDEA Gill, 1871

(*nom. transl.* Vermeulen, Lazarin, Lépinay, Leroy & Mascarelli, 2014 *ex* Hamulinidae)

Discussion. Monophyletic derivatives of Crioceratitoidea (Vermeulen *et al.,* 2014), comprising the families Hamulinidae, Leptohamulinidae and Macroscaphitidae (Vermeulen *et al.,* 2014). Very small to very large hamulinicones, criocones, aspinocones and ammoniticones, with or without tubercles; varices and constrictions commom. *U. Haut. - U. Alb.*

Family HAMULINIDAE Gill, 1871

Discussion. Apparently monophyletic derivatives of Crioceratitidae (Vermeulen *et al.,* 2014). Mostly hamulinicones with trituberculate primary ribs separated by non-tuberculate secondaries. Comprises the subfamilies Megacrioceratinae and Hamulininae (Vermeulen *et al.,* 2010). *U. Haut. - L. Barrem.*

Subfamily Megacrioceratinae Vermeulen, 2006

(*nom. transl.* Vermeulen *et al.*, 2007 *ex* Megacrioceratidae)

Discussion. Seemingly monophyletic derivatives of Crioceratitinae close to *Pseudomoutoniceras*. Large;

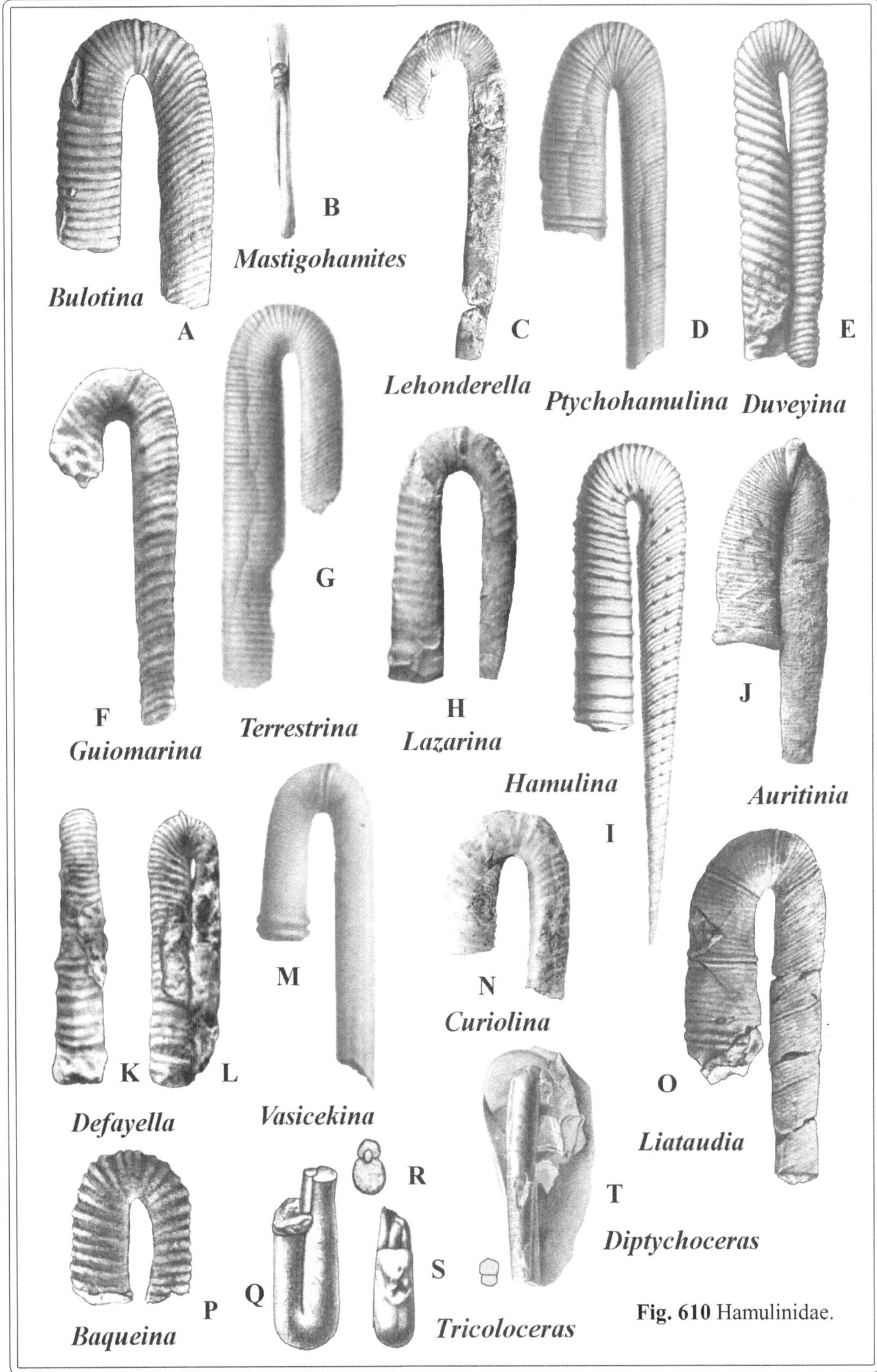

Fig. 610 Hamulinidae.

criocone spire followed by long, straight, finely ribbed, non-tuberculate proversum with periodic constrictions,well separated from retroversum. *U. Haut. - L. Barrem.*

Megacrioceras DELANOY, AUTRAN & THOMEL, 1987 [*Ancyloceras doubieri* Jaubert, 1854; OD]. Large, with initial tight criocone followed by very long almost straight proversum; compressed elliptical whorl section with rounded venter; fine, dense, simple prorsiradiate riblets, with periodic straight collared constrictions. *U. Haut.* **Fig. 609A-B** - *M. doubieri* (Jaubert).

Amorina VERMEULEN, 2005 [*Hamulina cincta* d'Orbigny, 1850; OD]. Medium sized, with close subparallel shafts and long retroversum well separated from, and parallel to, proversum; proversum with thin, regularly spaced, slightly prorsiradiate ribs, some stronger and some interspaces deeper; beginning and apex of flexus with deep collared constrictions; retroversum with coarser, regularly spaced, slightly rursiradiate, simple ribs with deep constriction shortly before peristome. *U. Haut. - L. Barrem.* **Fig. 609D** - *A. cincta* (d'Orbigny).

The genus is linked to *Megacrioceras* by *A. intermedia* Vermeulen, Lazarin, Lépinay, Leroy, Mascarelli, Meister & Menkveld-Gfeller (2012) (**Fig. 609E**) that has fine dense ribbing and deep collared constrictions to the proversum. Vermeulen *et al.* (2015) recognized three species groups within the genus based on the number and position of constrictions.

Garroniceras VERMEULEN, 1990 [*Ancyloceras seringei* Astier, 1851; OD]. Like *Megacrioceras* but criocone stage smaller; proversum expanding rather rapidly, with dense fine ribbing with bifurcation and intercalation, and more frequent slightly flexuous constrictions. *U. Haut.* **Fig. 609C** - *G. seringei* (Astıer).

Bastelia VERMEULEN, LAZARIN, LÉPINAY, LEROY & MASCARELLI, MEISTER & MENKVELD-GFELLER, 2012 [*B. schloegli*; OD]. Very large hamulinicone with very close proversum and retroversum; fine non-tuberculate ribs, narrower than interspaces, prosiradiate on proversum, slightly rursiradiate on retroversum; prominent constriction on flexus bordered by varix in front; apertural constriction bordered on both sides by varices. *U. Haut.* **Fig. 612B** - *B. schloegli* Vermeulen, Lazarin, Lépinay, Leroy & Mascarelli, Meister & Menkveld-Gfeller.

Liautaudia VERMEULEN, LAZARIN, LÉPINAY, LEROY, MASCARELLI, MEISTER & MENKVELD-GFELLER, 2012 [*L. fumisuginiformis*; OD]. Very large ptychocone with straight tapering proversum and tight flexus; fine simple prorsiradiate ribs and periodic collared varices; ribs coarsening on flexus, becoming very coarse simple and rursiradiate to rectiradiate on retroversum. *U. Haut.* **Fig. 610O** - *L. fumisuginiformis* Vermeulen, Lazarin, Lépinay, Leroy & Mascarelli, Meister & Menkveld-Gfeller.

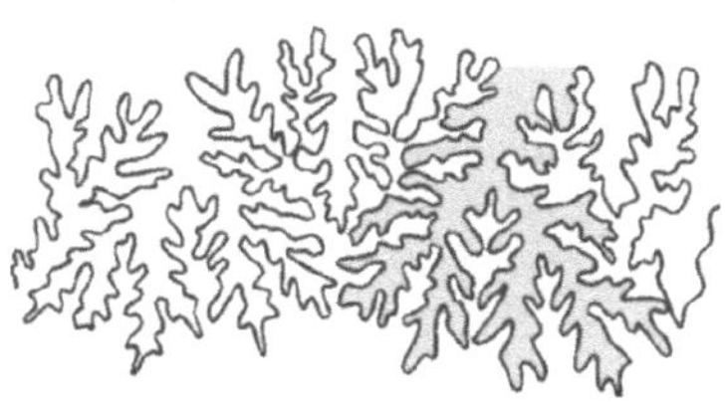

Fig. 611. *Hamulina* suture.

Subfamily Hamulininae Gill, 1871

(*nom. trans.* Vermeulen, Cecca & Kruta, 2006 *ex* Hamulinidae)

Discussion. Moderately large to large ptychocones and hamulinicones with trituberculate primary ribs separated by finer secondaries; constrictions present only in early growth, with or without collars (Vermeulen *et al.,* 2015). *Uppermost Haut. - L. Barrem.*

Hamulina D'ORBIGNY, 1850 [*H. astieriana*; SD Roman, 1938]. Very large ptychocone (55 cm long); earliest stages unknown; straight, strongly expanding proversum followed by tight flexus and long retroversum of constant dimensions closely parallel to proversum; strong prorsiradiate trituberculate main ribs separated by 2-4 fine non-tuberculate secondaries; ribbing coarsens on flexus and becomes uniform, distant and non-tuberculate on retroversum; suture highly indented, with oblique E/L, widely splayed asymmetrically bifid L deeper than E, and large bifid U deeper than I. *Upper U. Haut. – M. Barrem.* **Fig. 610I** - *H. astieriana* d'Orbigny.

Schauffhauseria VERMEULEN, LAZARIN, LÉPINAY, MASCARELLI, MEISTER & MENKVELF-GFELLER, 2012 [*S. veveysensis*; OD]. Medium sized hamulinicone with shaft and retroversum slightly convergent to divergent; proversum with bi- or trituberculate primary ribs separated by finer intercalatories; at approach of hook main trituberculate ribs evanesce and ribs become uniform and of similar length, some bifurcating; constrictions bordered by two varices; one or two varices before or after middle of flexus; retroversum with prominent ribs, sometimes irregular in strength and width; suture with very asymmetrically bifid L. *Uppermost Haut.* **Fig. 609I** - *S. veveysensis* Vermeulen, Lazarin, Lépinay, Mascarelli, Meister & Menkvelf-Gfeller.

Lehonderella VERMEULEN, 2005 [*L. gouvenauxae*; OD]. Very large, like *Schauffhauseria* in its tuberculate shaft and non-tuberculate retroversum, but with thinner more distant primary ribs, trituberculation more fleeting, absent or very rare constrictions on retroversum and narrower less prominent varices. *Uppermost Haut.* **Fig. 610C** - *L. gouvenauxae* Vermeulen.

Acanthoptychoceras MANOLOV, 1962 [*A. spinatocostatum*; OD]. Medium sized; two straight, parallel, slowly-expanding shafts not in contact, with depressed elliptical whorl section; trituberculate bar-like primary ribs

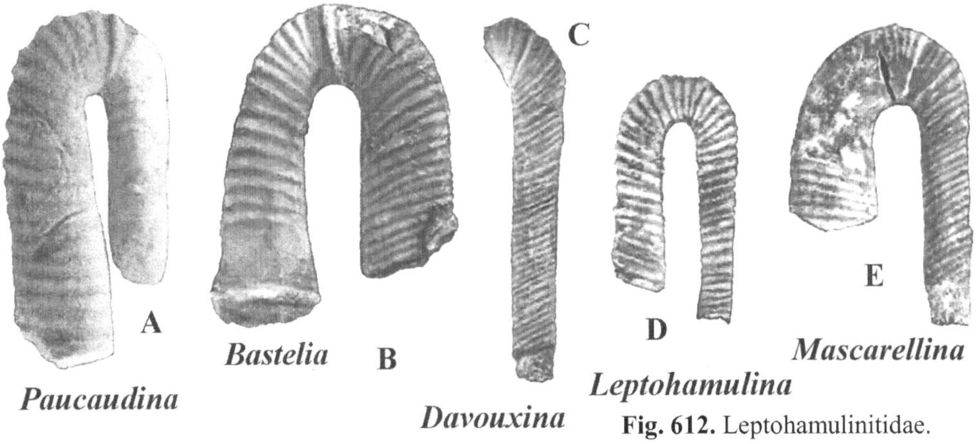

Fig. 612. Leptohamulinitidae.

(tubercles dorsolateral, midlateral and ventrolateral in position) separated by 3-4 low broad rounded non-tuberculate secondaries. *L. Barrem.* **Fig. 609G** - *A. spinatocostatum* Manolov.

Family LEPTOHAMULINIDAE Vermeulen, 2010

(*nom. transl.* Vermeulen, Borro, Lazarin, Lépinay, Leroy & Mascarelli, 2020).

Discussion. Seemingly monophyletic derivative of Hamulininidae (Vermeulen, 2006). Small to very small, feebly tuberculate to non-tuberculate, without varices or well marked constriction on the flexus and retroversum. Comprises the subfamilies Leptohamulininae, Ptychohamulininae and Davouxininae (Vermeulen *et al.*, 2014). *U. Haut. - U. Alb.*

Subamily Leptohamulininae Vermeulen, 2010

Discussion: Medium sized to very small hamulinicones with slowly expanding non-tuberculate, shafts; proversum finely ribbed; flexus with or without constrictions; retroversum with simple ribs. *L. Barrem. - U. Apt.*

Leptohamulina VERMEULEN, 2006 [*L. distans* Vašíček, 1972; OD]. Medium sized (L>85 mm), with low expansion rate and parallel to divergent shafts; proversum with strong prorsiradiate riblets thicker than interspaces, recurving slightly on outer flank and crossing venter transversely; ribs coarsening on flexus and coarser still and prorsiradiate on long retroversum. *L. - U. Barrem.* **Fig. 612D** - *L. distans* Vašíček.

Some referred species have collared constriction on flexus and/or retroversum.

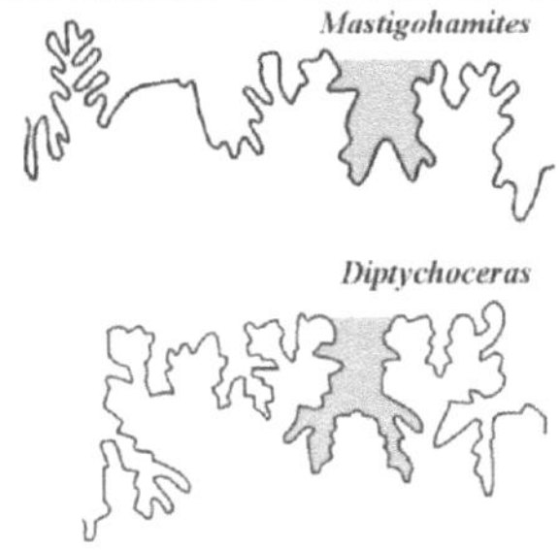

Fig. 613. Ptychohamulinine sutures.

Mascarellina VERMEULEN, 2009 [*M. batangae*; OD]. Small, with fairly close parallel shafts expanding slowly; ribs on proversum regular, fine, spaced, low, becoming stronger, more widely spaced and rectiradiate; two intercostal furrows deeper than others on first half of flexus; retroversum with strong relatively thin rursiradiate ribs. *Barrem.* **Fig. 612E** - *M. batangae* Vermeulen.

Vasicekina VERMEULEN, 2005 [*Hamulina paxillosa* Uhlig, 1883; OD]. Like *Leptohamulina* but flexus with two varices and very long retroversum with uniform, rounded, rectiradiate ribs. *Barrem.* **Fig. 610M** - *V. papillosa* (Uhlig).

Badina VERMEULEN & VAŠÍČEK, 2011 [*B. lazarini*; OD]. Very small, with slender subparallel shafts with a low to medium expansion rate; proversum unornamented or with barely perceptible prorsiradiate ribs, rarely strong; flexus generally smooth, but sometimes with collared constrictions on adapical half; retroversum smooth or commonly with weak rursiradiate to rectiradiate ribs generally with a prominent wide collared constriction; suture with massive bifid E/L and asymmetrically trifid L. *U. Barrem. - L. Apt.* **Fig. 609L** - *B. bedouliensis* Vermeulen, Lazarin, Lépinay, Leroy & Mascarelli.

Tricoloceras WHITEHOUSE, 1928 [*Ptychoceras? closteroides* Etheridge, 1904; OD]. Like *Mastigohamites* but with first shaft encapsulated by penultimate and final shafts. *U. Apt.* **Fig. 610Q-S** - *P. closteroides* (Etheridge).

Subfamily Ptychohamulininae Vermeulen, 2010

Discussion. Very small derivatives of Leptohamulininae (Vermeulen *et al.*, 2010), with very close to adpressed shafts, with or without collared constrictions on flexus and retroversum; proversum may be partially or completely unornamented and retroversum may also be smooth. *L. Barrem. - U. Alb.*

Defayella VERMEULEN, LAZARIN, LÉPINAY, LEROY & MASCARELLI, 2013 [*D. dubalae*; OD]. Ptychoceratid, with long retroversum adpressed to proversum with small teardrop opening at flexus; strong biplicate ribs on flexus and first part of retroversum, becoming simple and rursiradiate on collared retroversum; collar on flexus. *L. Barrem.* **Fig. 610K-L** - *D. dubalae* Vermeulen, Lazarin, Lépinay, Leroy & Mascarelli.

Djamaina VERMEULEN, LAZARIN, LÉPINAY, LEROY & MASCARELLI, 2010 [*D. manangai*; OD]. Very small, with weakly divergent slowly expanding shafts with slightly compressed whorl section; proversum with strong, rounded, straight prorsiradiate ribs about as wide as interspaces; ribs fine on flexus with prominent varix at apex, and another at begining of long straight retroversum and shortly before adult peristome. *L. Barrem.* **Fig. 609K** - *D. manangai* Vermeulen, Lazarin, Lépinay, Leroy & Mascarelli.

Ptychohamulina VERMEULEN, 2005 [*Hamulina ptychoceroides* Hohenegger *in* Uhlig, 1883; OD]. Very small; proversum and retroversum very close, parallel to weakly divergent, with low expansion rate; early part of proversum may be unornamented, later generally with very fine uniform prorsiradiate riblets; flexus with large rib, very high and flared; retroversum with rursiradiate ribs of variable strength; constrictions lacking. *U. Barrem.* **Fig. 610D** - *P. ptychoceratoides* (Hohenegger).

Kleiniceras VERMEULEN, LAZARIN, LÉPINAY, LEROY & MASCARELLI, 2014 [*K. compsensiforme*; OD]. Very small, with shafts adpressed at peristome leaving a keyhole aperture at flexus; proversum smooth or with fine prorsiradiate ribs in some species; flexus with circular whorl section or with flattened dorsum, with prominent collared constriction just before apex; retroversum smooth or with fine more or less spaced ribs. *U. Barrem.* **Fig. 609J** - *K. compsensiforme* Vermeulen, Lazarin, Lépinay, Leroy & Mascarelli.

In some species the flexus and retroversum are not in contact, and may be ornamented with fine riblets.

Diptychoceras GABB, 1869 [*D. laevis*, = *Ptychoceras gabbi* Pervinquière, 1907; OD]. Small (L=75 mm) with 3 shafts; initial shaft subelliptical with slightly flattened dorsum; proversum subcircular; retroversum extending to mid-length of shell; proversum with extremely faint undulations; retroversum with feeble constrictions not apparent on external mould; suture simple with bifid saddles and lobes. *Alb.* **Fig. 610T** - *P. gabbi* (Pervinquière).

Mastigohamites BREISTROFFER, 1947 [*Hamites adpressus* J. Sowerby, 1814; OD]. A progenetic dwarf with 4 smooth adpressed shafts; initial subcircular whorl section later becoming depressed with conspicuous dorsal furrow; proversum perpendicular to earlier two shafts, and retroversum short; constrictions lacking; suture simple, at first goniatitic later with bifid or trifid L. *Lower U. Alb.* **Fig. 610B** - *M. adpressus* (J. Sowerby).

= *Mastigoceras* Böhm, 1925 (*non* Handschin, 1924, Collembola).

Subfamily Davouxininae Vermeulen, Borro, Lazarin, Lépinay, Leroy & Mascarelli, 2020

(*nom. correct.* herein *pro* Davouxinae)
= Pacaudininae Vermeulen *et al.*, 2013.

Discussion. Derivatives of Leptohamulininae (Vermeulen *et al.*, 2020). The subfamily Pacaudininae was introduced by Vermeulen *et al.* (2013), without diagnosis, and thus a *nomen nudum*. Subsequently it was included in the synonymy of Davouxininae (Vermeulen *et al.*, 2020). However the large number of genera now represented in Davouxininae comprise more than one phyletic line and more work is needed. Medium sized to very small, with non-tuberculate to feebly tuberculate ribs; close parallel to adpressed shafts, often with a keyhole opening below the flexus; flexus with varices or constrictions; retroversum with simple rursiradiate ribs; aperture sometimes constricted. *Uppermost Haut. - Lower L. Apt.*

Terrestrina VERMEULEN, 2006 [*Hamites (Hamulina) quenstedti* Uhlig, 1883; OD]. Medium sized, with straight proversum separated from very long parallel retroversum by U-shaped flexus; proversum with thin prorsiradiate ribs wider than interspaces, coarsening on flexus; retroversum with thin straight rursiradiate ribs narrower than interspaces; constrictions and varices lacking. *Uppermost Haut. - basal Barrem.* **Fig. 610G** - *T. quenstedti* Vermeulen.

Davouxina VERMEULEN, 2009 [*D. rouvieri*; OD]. Small, with slowly expanding, subparallel proversum and retroversum, U-shaped flexus and rounded to elliptical whorl section; most of proversum with fine prorsiradiate ribs, regularly spaced and as wide as interspaces; flexus with strong, distant rectiradiate ribs with a few intercalatories and prominent varix at apex; retroversum with well spaced rursiradiate ribs with frequent intercalatories and

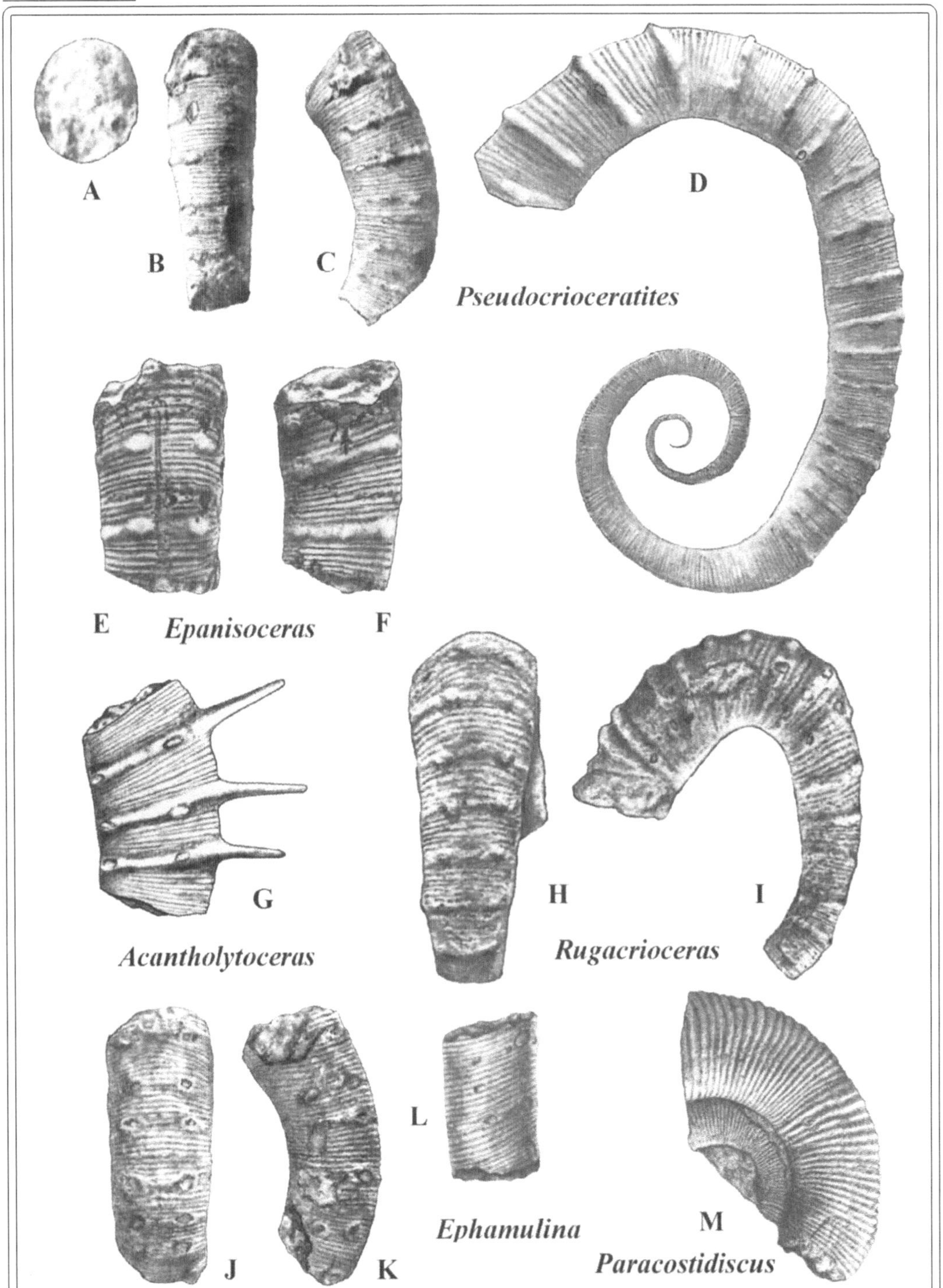

Fig. 614. Macroscaphitidae.

a collared peristome. *L. Barrem.* **Fig. 612C** - *D. rouvieri* Vermeulen.

Duyeina VERMEULEN, 2005 [*Anahamulina glemmbachensis* Immel, 1987; OD]. Shaft and retroversum in contact save for teardrop opening at flexus, slowly expanding; ribs strong, robust, even, rounded, rectiradiate to slightly prorsiradiate, of variable spacing on proversum, finer on flexus, and strong distant and rursiradiate on retroversum, becoming rectiradiate as aperture is approached; constrictions and flares lacking; suture with bifid L with

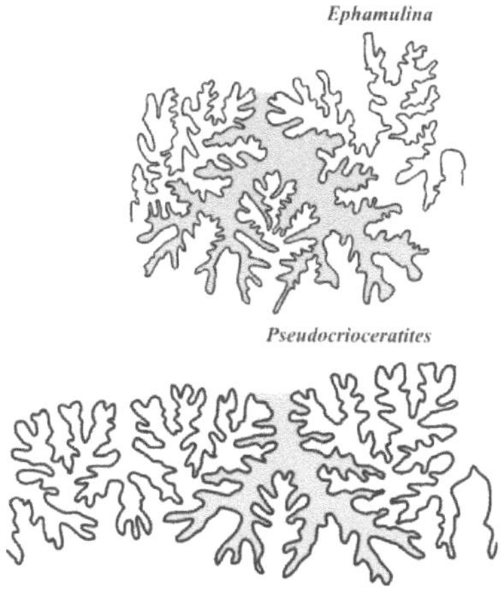

Fig. 615. *Pseudocrioceratites* suture.

bifid terminal lobules. *U. Barrem. - basal Apt.* **Fig. 610E** - *D. glemmbachensis* (Immel).

Bulotina VERMEULEN, 2009 [*B. tronchettii*; OD]. Small, with U-shaped flexus and subparallel shafts; primary ribs to proversum and flexus, later with collared constriction at apex; retroversum with slightly rursiradiate ribs. *Barrem.* **Fig. 610A** - *B. tronchettii* Vermeulen.

Curiolina VERMEULEN, 2010 [*C. rosae*; OD]. Small, with weakly divergent shafts; most of proversum with rounded prorsiradiate ribs, either regular or with intercalatories; flexus commences with prominent collared constriction (additional varices present in some species) followed by more widely spaced ribs; ribs of retroversum variable, rursiradiate to subradial, regular or feebly differentiated. *Barrem.* **Fig. 610N** - *C. rosae* Vermeulen.

Lazarina VERMEULEN, 2009 [*L. lepinayi*; OD]. Small with close, slowly expanding parallel shafts, and long retroversum; prorsiradiate ribs on proversum bold, blunt, broader than interspaces, becoming rectiradiate on flexus and slightly rursiradiate on retroversum; prominent collared constriction at middle of flexus and at adult aperture. *L. Barrem.* **Fig. 610H** - *L. lepinayi* Vermeulen.

Auritina EGOIAN, 1989 [*A. aurita*; OD]. Small, with shafts closely adpressed; proversum with dense, fine, flexuous, prorsiradiate riblets; prominent varix at apex of flexus; retroversum with dense, fine, straight, slightly rursiradiate riblets and collared aperture. *L. Barrem.* **Fig. 610J** - *A. aurita* Egoian.

Pacaudina VERMEULEN, LAZARIN, LÉPINAY, LEROY & MASCARELLI, 2010 [*P. enteronensis*; OD]. Small, with convergent shafts leaving keyhole opening; proversum unornamented; apex of flexus with prominent varix; adoral part of flexus and retroversum with blunt rounded slightly rursiradiate ribs; suture very simple, with barely indented trifid lobes and bifid saddles. *L. Barrem.* **Fig. 612A** - *P. enteronensis* Vermeulen, Lazarin, Lépinay, Leroy & Mascarelli.

Referred species with parallel shafts and U-shaped flexus, frequently with a large varix preceded by a smooth zone.

Baqueina VERMEULEN, LAZARIN, LÉPINAY, LEROY & MASCARELLI, 2010 [*B. guieui*; OD]. Very small to small with convergent shafts almost meeting; ribs simple, blunt, wider than interspaces, slightly prorsiradiate on proversum and slightly rursiradiate on retroversum, of more or less uniform strength. *L. Barrem.* **Fig. 610P** - *B. guieui* Vermeulen, Lazarin, Lépinay, Leroy & Mascarelli.

Guiomarina VERMEULEN, CECCA & KRUTA, 2007 [*G. guiomarae*; OD]. Small, with U-shaped flexus and short retroversum slightly inclined to proversum; shafts expanding slowly with strong feebly bituberculate primary ribs separated by weaker finer secondaries; flexus with two shallow constrictions, anterior one weaker; adoral part of flexus with fine strongly rursiradiate riblets. *Barrem.* **Fig. 610F** - *G. guiomarae* Vermeulen, Cecca & Kruta.

Superfamily MACROSCAPHITOIDEA Hyatt, 1900

(*nom. transl.* Vermeulen, 2006 *ex* Macroscaphitidae)

Discussion. A monophyletic group derived from Crioceratitoidea (Vermeulen *et al.*, 2015); comprises the families Macroscaphitidae and Cicatritidae. *L. Barrem. - U. Alb.*

Family MACROSCAPHITIDAE Hyatt, 1900

Discussion. A seemingly monophyletic group with an origin in Crioceratitidae (Wright, 1996; Vermeulen *et al.*, 2015). Medium sized compressed serpenticones, in some with ancylocone uncoiling of the adult body chamber; fine dense, straight, rectiradiate to rursiradate ribs, sometimes with umbilical and ventrolateral tubercles, and periodic collared constrictions. Comprises the subfamilies Lytocrioceratinae and Macroscaphitinae. *L.Barrem - Lowermost M. Alb.*

Subfamily Lytocrioceratinae Vermuelen, 2000

(*nom. transl.* herein *ex* Lytocrioceratidae)

Discussion. Seemingly monophyletic derivatives of Hamulininae, close to *Hamulina* (Ebbo *et al.*, 1999). Small to very large ancylocones with differentiated ribbing, with or without tubercles; suture with bifid L. *L. Barrem. - M. Alb.*

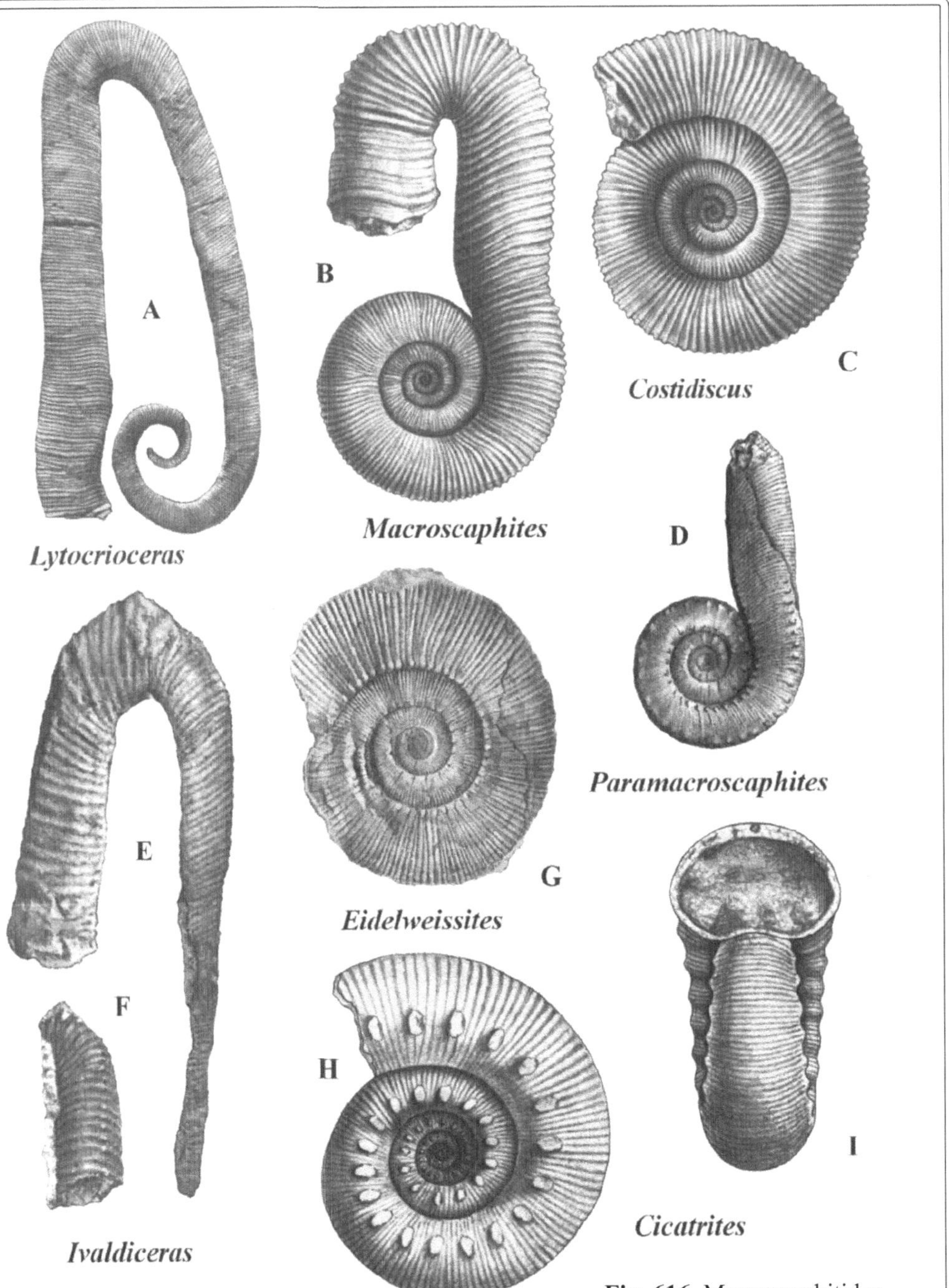

Fig. 616. Macroscaphitidae.

Lytocrioceras SPATH, 1924 [*Ancyloceras jauberti* Astier, 1851; OD]. Moderately large; initially a relatively large loose criocone followed by a long gently curved proversum, an acutely rounded flexus and a very long retroversum that may extend beyond spire; ribs simple, dense, fine, slightly prorsiradiate on proversum and slightly rursiradiate on retroversum, more widely spaced on flexus; suture with asymmetrically bifid L, external branch better developed than internal one. *L. Barrem.* **Fig. 616A** - *L. jauberti* (Astier).

Pseudocrioceratites EGOIAN, 1969 [*P. pseudoelegans*; OD]. Very large slowly expanding ancylocone; open spire with dense fine non-tuberculate riblets; tubercles appear towards beginning of proversum which, together with retroversum, is ornamented by prominent trituberculate primary ribs separated by 6-8 fine non-tuberculate

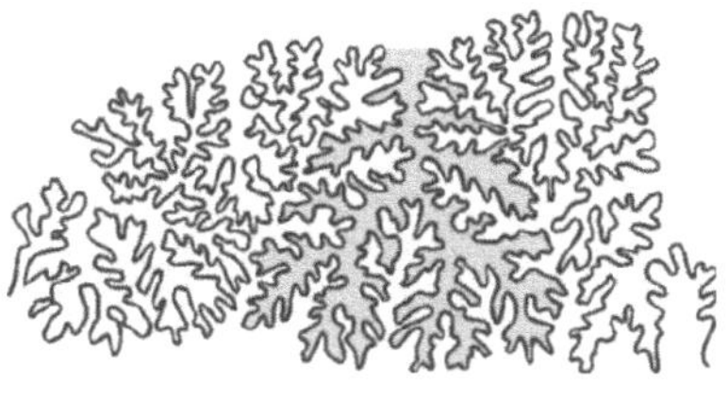

Fig. 617. *Costidiscus* suture.

secondaries; suture with very large bifid L and large, tending to cruciform, internal lobe. *L. Barrem. - L. Apt.* **Fig. 614A-C** - *P. pseudoelegans* Egoian; **Fig. 614D** - *P. vocontiense* (Ebbo, Poupon, Delanoy & Gonnet).

Perhaps a macroconch of *Lytocrioceras*.

Acantholytoceras SPATH, 1923 [*Hamites (Pictetia) longispinus* Uhlig, 1883; OD]. Medium sized loose criocone with oval whorl section showing low expansion rate; strong quadrituberculate primary ribs with long dorsolateral, high lateral and ventrolateral spines separated by upto 8 fine non-tuberculate secondaries; suture with bifid L. *Barrem.* **Fig. 614G** - *A. longispinum* (Uhlig).

Epanisoceras COLLIGNON, 1962 [*E. raulinianiforme*; OD]. Based on a fragment but at least moderately large; trituberculate primary ribs separated by up to a dozen fine non-tuberculate secondaries; suture with L and U trifid. *U. Apt.* **Fig. 614E-F** - *E. raulinianiforme* (Collignon).

The fragment is almost indistinguishable from older *Acantholytoceras*.

Rugacrioceras VERMEULEN, 2002 [*Ancyloceras martinsii* Reynès, 1876; OD]. Small ancylocone (estimated height 80-85 mm), the spire not preserved; proversum gently curved, with strongly depressed oval whorl section, open flexus, and moderately long, expanding, divergent retroversum; proversum with tuberculate (1-3) primary ribs separated by 6-9 fine non-tuberculate secondaries, ribs coarsening and becoming approximated on flexus (3-5), and crossing dorsum with slight forward convexity; retroversum with prominent rursiradiate primary ribs only, with attenuated tubercles. *U. Barrem.* **Fig. 614H-I** - *R. martinsii* (Reynès).

= *Rugacrioceras* Vermeulen, 1990 (*nom. nud.*).

Referred material is said to have a subcircular whorl section, moderate sized open spire and straight or curved, frequently short proversum and suture with bifid L. The lectotype was assigned to the genus *Acrioceras* by Delanoy & Bulot (1990).

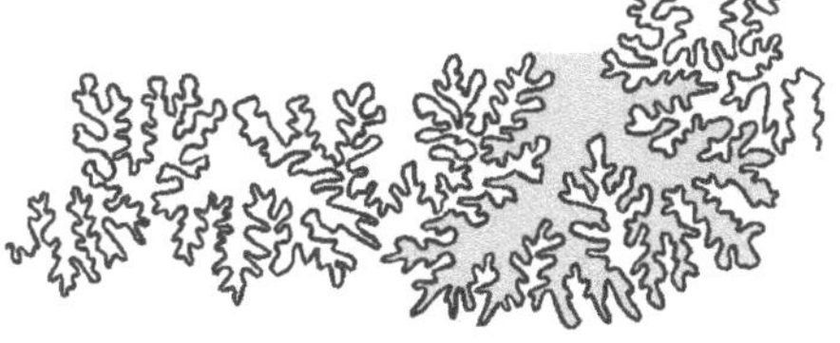

Fig. 618. *Cicatrites* suture.

Ivaldiceras DELANOY, BAUDOUIN, BOSELLI & PENAGE, 2016 [*I. beratteroi*; OD]. Medium sized hamulinicone; proversum straight, with fine single or bifurcating ribs with small tubercles arranged more or less regularly on the ribs or between them; flexus with thickened ribs that broaden towards conical ventral tubercles; simple relatively thick ribs of retroversum with small dorsolateral and ventrolateral tubercles; dorsum with very fine lirae. *L. Apt.* **Fig. 616E-F** - *I. beratteroi* Delanoy, Baudouin, Bosseli & Penage.

Epacrioceras EGOIAN, 1974 [*E. rarum*; OD]. Like small *Pseudocrioceratites* but body chamber expanding in height more rapidly; trituberculation restricted to criocone stage; long proversum and flexus non-tuberculate, with fine dense prorsiradiate ribs that coarsen slightly on flexus; long retroversum almost reaching phragmocone; suture with bifid L. *U. Apt.* **Fig. 605E** - *E. rarum* Egoian.

Ephamulina COLLIGNON, 1963 [*Anisoceras? trituberculatum* Collignon, 1949; OD]. Large; known only from short fragments suggesting loose criocone coiling, subcircular whorl section and trituberculate primary ribs separated by 2-10 finer non-tuberculate secondaries the number of which increases with growth; ribs split into finer pairs across dorsum; suture highly incised, with short E, oblique E/L and widely splayed bifid L covering almost entire flank. *U. Apt. - Lowermost M. Alb.* **Fig. 614J-L** - *E. trituberculata* (Collignon).

Subfamily Macroscaphitinae Hyatt, 1900

(*nom. transl.* herein *ex* Macroscaphitidae)

Discussion. A monophyletic group descended from Lytocrioceratinae, close to *Lytocrioceras* (Delanoy & Poupon, 1992); typically serpenticone, with uniform dense wiry ribs and periodic constrictions. The claim that the small scaphitoid forms (*Macroscaphites*) may be the microconchs of the larger serpenticones is not accepted by all workers, but remains a possibility. *Barrem. - L. Apt.*

Paracostidiscus BUSNARDO, CHAROLLAIS, WIEDMANN & CLAVEL, 2003 [*P. radians*; OD]. Medium sized serpenticone, whorls barely in contact, with an oval whorl section; simple sharp fine rectiradiate non-tuberculate ribs dense on inner whorls more spaced later, crossing venter transversely (slight flexuosity perhaps due to deformation); two barely discernible constrictions at start of body chamber. *Uppermost Haut.* **Fig. 614M** - *P. radians* Busnardo, Charollais, Wiedmann & Clavel.

Paramacroscaphites VERMEULEN, 2009 [*P. pineli*; OD]. Ancylocone with very evolute spire and subcircular to elliptical whorl section; ribs fine to very fine, sometimes filiform, prorsiradiate on proversum and subradial on retroversum; ribs fibulate, looped between umbilical and

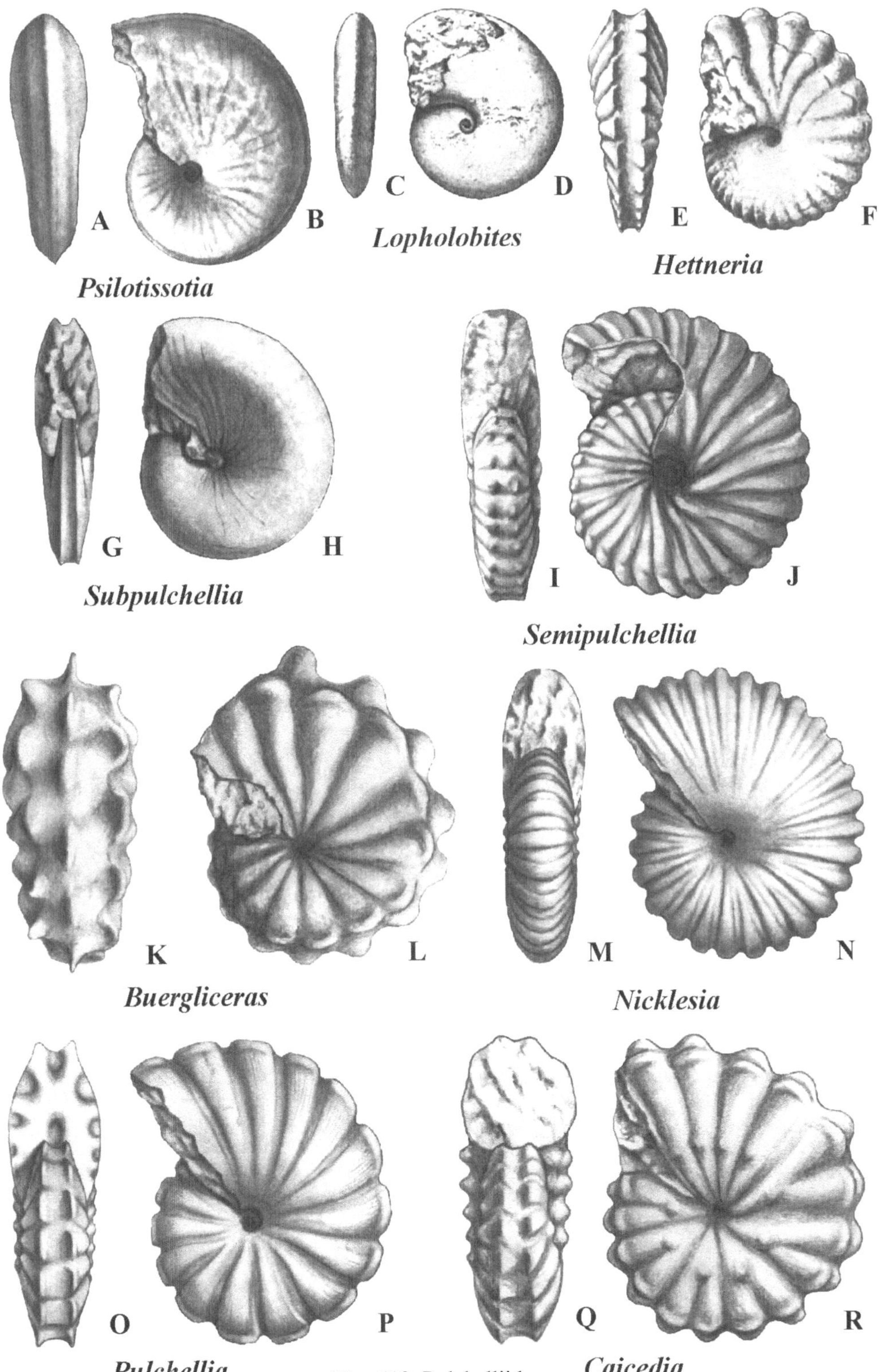

Fig. 619. Pulchelliidae.

ventrolateral tubercles; flexus and retroversum non-tuberculate, with irregular ribbing. *Barrem.* **Fig. 616D** - *P. binodosum* (Uhlig).

Macroscaphites MEEK, 1876 [*Scaphites yvani* Puzos, 1832; SD Roman, 1938]. Phragmocone serpenticone, with rounded flanks and venter, and depressed oval whorl section; body chamber uncoiled to form scaphitoid hook with collared aperture; dense strong straight slightly prorsiradiate non-tuberculate ribs and periodic collared constrictions; suture with deep penetrating very narrow E, straight narrow stemmed bifid E/L and very narrow widely splayed bifid L. *Barrem.* - *L. Apt.* **Fig. 616B** - *M. yvani* (Puzos).

Considered by many the microconch of *Costidiscus* (Wright, 1996).

Costidiscus UHLIG, 1882 [*Ammonites recticostatus* d'Orbigny, 1841; OD]. Like *Macroscaphites* but serpenticone throughout, without scaphitoid hook; suture highly florid, with deep narrow E, narrow stemmed bifid E/L, very narrow widely splayed bifid L, well developed U and small I. *Barrem.* - *L. Apt.* **Fig. 616C** - *C. recticostatus* (d'Orbigny).

Eidelweissites **gen. nov.** [*Costidiscus nodosostriatus* Uhlig, 1883, p.197; OD herein]. Like *Costidiscus* but with small umbilical bullae that give rise to dense bundles of 2-3 fine rectiradiate ribs by bifurcation and intercalation; constrictions not obvious. *L. Apt.* **Fig. 616G** - *E. nodosostriatae* (Uhlig).

Mascarelliceras VERMEULEN *et al.* (2013). No description or figure. A *nomen nudum*.

Family CICATRITIDAE Spath, 1927

Discussion. A monotypic taxon of cryptic origin, Koloboceratinae according to Wright (1996) but the sutures are vastly different. Vermeulen (1999) looked to *Costidiscus*, and Busnardo *et al.* (2003) raised the possibility of Simbirskitinae; differs from *Costidiscus* in very shallow E, oblique E/L and non-retracted suspensive lobe. *Apt.*

Cicatrites ANTHULA, 1899 [*C. abichi*; OD]. Serpenticone with very strongly convex flanks, broadly rounded venter and depressed reniform whorl section; dense fine simple rectiradiate ribs cross venter transversely; periodic large septispinate umbilicolateral bullae cover 3 ribs and are separated by 2-5 nontuberculate ribs; suture with very short E, small oblique E/L, very large widely splayed asymmetrically bifid L and well developed trifid U and I. *L.* - *U. Apt.* **Fig. 616H-I** - *C. abichi* (Anthula).

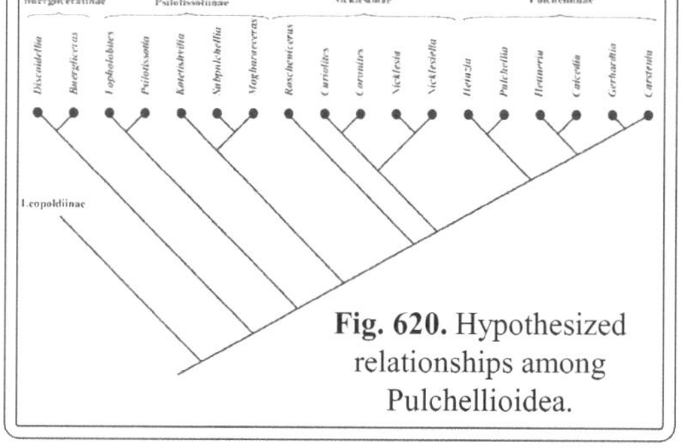

Fig. 620. Hypothesized relationships among Pulchellioidea.

Hyporder PULCHELLIOIDEI nov.

Discussion. A questionably monophyletic group with an origin in Neocomitoidei (Vermeulen, 1995); comprises the superfamilies Pulchellioidea and Engonoceratoidea. *U. Haut.* - *M. Alb.*

Superfamily PULCHELLIOIDEA Douvillé, 1890

(*nom. transl.* Drushchits, 1958 *ex* Pulchelliidae)

Discussion. A monophyletic group with an origin, in Leopoldioidea (Vermeulen, 1995, 2002); comprises only the family Pulchelliidae. *Lower U. Haut.* - *U. Barrem.*

Family PULCHELLIIDAE Douvillé, 1890

(*nom. correct.* Hyatt, 1903 *pro* Pulchellidés)

Discussion. A monophyletic group with an origin in Leopoldiidae (Vermeulen 1995, 2002); mostly very small to medium sized, involute, compressed with strong flat topped ribs, simple suture, variably developed tubercles and keeled, tabulate or sulcate venter. Comprises the subfamilies Buergliceratinae, Psilotissotiinae and Pulchelliinae. Suture primitively with narrow bifid L. Typically Tethyan. *U. Haut.* - *U. Barrem.*

Subfamily Buergliceratinae Vermeulen, 1995

Discussion. A monophyletic group with an origin in Saynellinae (Vermeulen, 1995), close to *Aubersonites*; typically small, strongly compressed, very involute, with sloping umbilical wall and circumumbilical depression; primitively with narrowly arched keeled venter, later adding ventrolateral clavi, *Lower U. Haut.* - *L. Barrem.*

Discoidellia VERMEULEN, 1995 [*D. couratieri*; OD]. Small, strongly compressed, very involute, with sloping umbilical wall, funnel shaped umbilicus, and convex flanks converging to narrowly arched venter with tiny siphonal clavi in some; feeble rectiradiate ribs, broader than interspaces, separated by up to 4 long intercalatories. *Lower U. Haut.* **Fig. 623C-D** - *D. couratieri* Vermeulen.

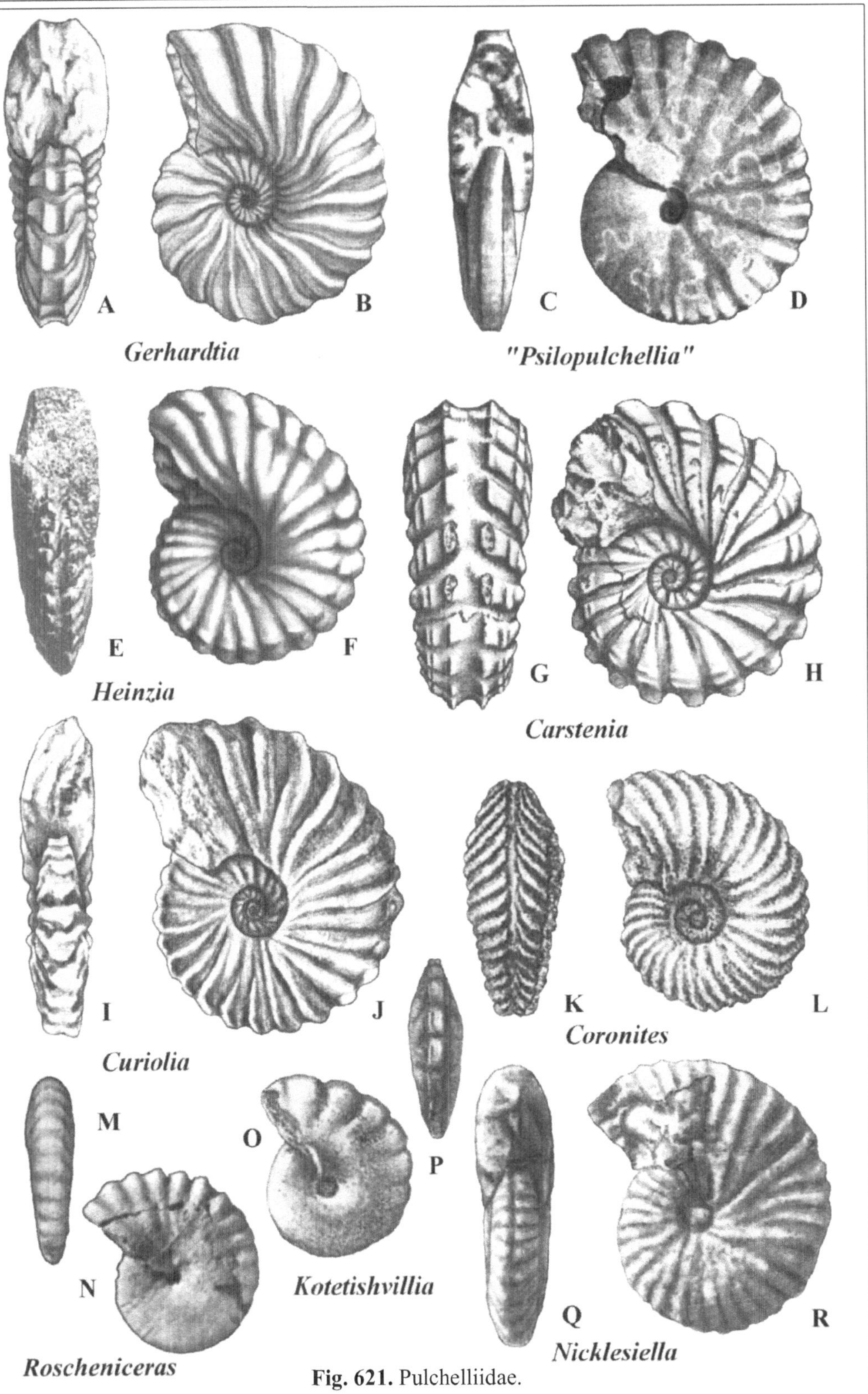

Fig. 621. Pulchelliidae.

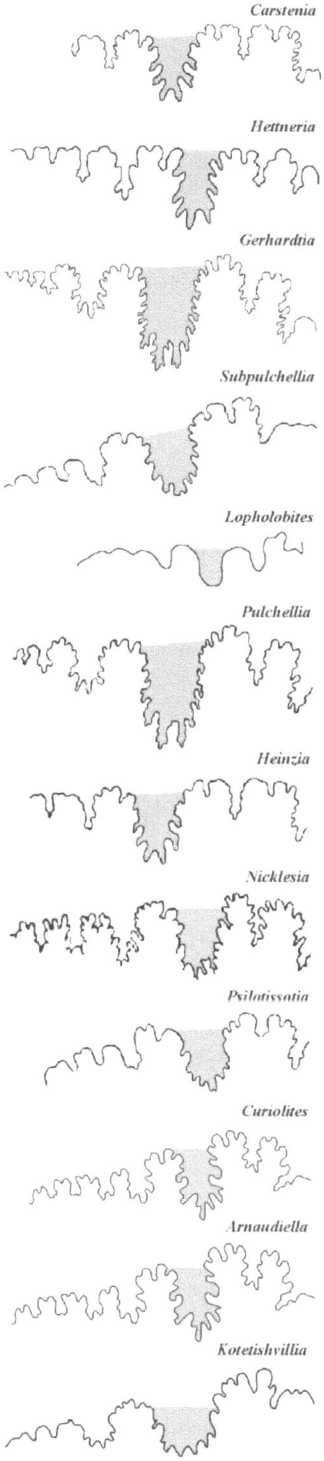

Fig. 622. Pulchelliid sutures.

Some species with feeble mid-lateral bullae to primary ribs in later growth. *Discoidellia gabyi* Vermeulen, with ventrolateral and siphonal clavi provides the link to *Buergliceras*.

Buergliceras ETAYO SERNA, 1968 [*B. buerglii*; OD]. Small; very involute, with sloping umbilical wall, maximum width high on flanks, broad subtabulate venter, sparse rectiradiate ribs broadening markedly towards venter, with exaggerated ventrolateral and siphonal clavi. *Barrem.* **Fig. 619K-L** - *B. buergli* Etayo Serna.

Subfamily Psilotissotiinae Vermeulen, 1995

Discussion. A paraphyletic taxon with an origin in Buergliceratinae, close to *Discoidellia* (Vermeulen 1995, 2002); typically small, very involute and compressed; primitively with dense fine prorsiradiate growth striae and lirae and fastigiate keeled venter; later with sulcate venter bordered by ventrolateral clavi. *U. Haut.* - *L. Barrem.*

Psilotissotia HYATT, 1900 [*Pulchellia chalmasi* Nicklès, 1890; OD]. Very involute compressed oxycone with low siphonal keel that is denticulate in some species; smooth in early growth but later with broad low slightly prorsiradiate folds and striae on flanks; suture simple with shallowly bifid E/L, broad digitate L deeper than E and at least 4 auxiliary lobes; convergent towards *Vergoliceras. U. Haut.* - *L. Barrem.* **Fig. 619A-B** - *P. chalmasi* (Nicklès).

Lopholobites HYATT, 1900 [*Neolobites? cotteaui* Nicklès, 1894; OD]. Micromorph; like *Psilotissotia* but more involute, unornamented, with subacute venter, scaphitoid uncoiling of adult body chamber and very simple asymmetrical suture with deeply bifid E/L and remaining saddles entire. *Barrem.* **Fig. 619C-D** - *L. cotteaui* (Nicklès).

Kotetishvilia VERMEULEN, 1995 [*Pulchellia changarnieri* SAYN, 1890; OD]. Very small to small; like *Subpulchellia* but with strong flexuous ribs arising in pairs or alternating long and short; ribs broaden markedly ventrally with deep interspaces and elongate ventrolateral clavi. *Upper L. Barrem.* **Fig. 621O-P** - *K. changarnieri* (Sayn).

Roscheniceras **nom. nov.** [*Arnaudiella anglasense* Vermeulen, 1994; OD]. Very small; compressed, very involute with smooth inner whorls; later with strong rounded ribs that cross venter uninterrupted. *L. Barrem.* **Fig. 621M-N** - *R. malladae* (Nicklès).

= *Arnaudiella* Vermeulen, 1995 (*non* Douvillé 1907, Foraminiferida).

Subpulchellia HYATT, 1903 [*S. castellanensis*, = *Pulchella oehlerti* Nicklès, 1894; OD]. Very small; compressed, involute, unornamented, with sulcate venter bordered by sharp keels; siphonal line may be slightly raised on internal mould; suture with broad shallow E, deep penetrating L and auxiliary saddles rounded and barely indented. *L. Barrem.* **Fig. 619G-H** - *S. oehlerti* (Nicklès).

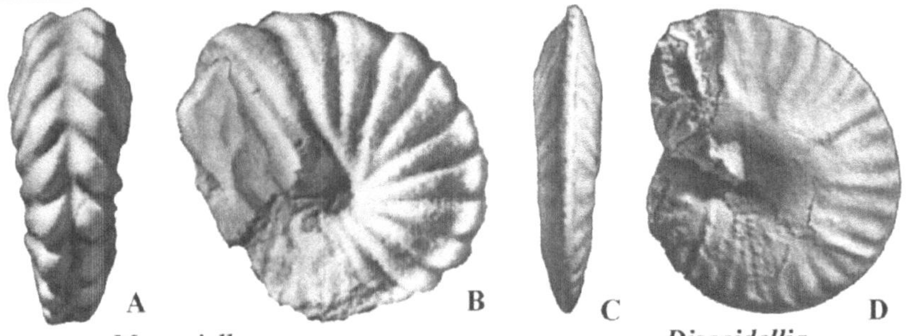

Fig. 623. Pulchelliidae.

Subfamily Pulchelliinae Douvillé, 1890

(*nom. transl.* Vermeulen, 2002 *ex* Pulchelliidae)

Discussion. Monophyletic derivatives of Psilotissotiinae, close to *Subpulchellia* (Hyatt, 1903); typically very involute, more or less compressed, with single or paired ribs broadening markedly ventrally and terminating in ventrolateral clavi; venter moderately to deeply sulcate. *Upper L. - U. Barrem.*

Tribe Pulchelliini Douvillé, 1890

(*nom. transl.* herein *ex* Pulchelliinae)

Diagnosis. Typically with simple rectiradiate ribs broadening ventrally and terminating in ventrolateral clavi. *Upper L. - U. Barrem.*

Nicklesia HYATT, 1903 [*Ammonites dumasianus* d'Orbigny, 1842, = *Ammonites pulchella* d'Orbigny, 1842; OD]. Small; involute, compressed, with circum-umbilical depression, no umbilical wall or umbilical shoulder, evenly-rounded venter and subelliptical whorl section; early whorls smooth, then with flat-topped ribs which broaden ventrally where broader than interspaces; ribs cross venter transversely and uninterrupted; suture like *Psilotissotia* but more complex, with more deeply bifid E/L. *Upper L. Barrem.* **Fig. 619M-N** - *N. pulchella* (d'Orbigny).

Nicklesiella BÜRGL, 1956 [*Pulchellia (N.) leivaesensis*; OD]. Like *Nicklesia* but with flat or shallowly sulcate venter to inner whorls. *Barrem.* **Fig. 621Q-R** - *N. leivaesensis* (Bürgl).

Semipulchellia BÜRGL, 1956 [*Pulchellia (S.) communis*; OD]. Small; involute, compressed, with broad weakly convex flanks, subtabulate venter and subelliptical whorl section; strong fine flexuous ribs bifurcate from feeble umbilical bullae in early growth and alternate long and short later on; ribs cross shallowly sulcate venter strongly with swollen rounded ventrolateral clavi. *Upper L. Barrem.* **Fig. 619I-J** - *S. communis* (Bürgl).

Caicedia BÜRGL, 1956 [*Ammonites caicedi* Karsten, 1858; OD]. Small; very involute, moderately inflated; strong simple rectiradiate ribs terminate in small lateral tubercles from which secondary ribs arise in pairs and broaden rapidly towards venter, terminating in prominent ventrolateral clavi; ribs cross deeply sulcae venter. *Lower U. Barrem.* **Fig. 619Q-R** - *C. caicedi* (Karsten).

Pulchellia UHLIG, 1883 [*Ammonites galeatus* von Buch, 1839; SD Gignoux 1921]. Small; like *Caicedia* but more compressed, with coarser broader slightly flexuous frequently single ribs with elongate ventrolateral clavi; lasteral tubercles obsolete; suture with shallower E and deeper penetrating L. *U. Barrem.* **Fig. 619O-P** - *P. orbignyi* Bürgl.

The type species is poorly known and interpretation of the genus has been based upon allied forms such as *P. orbignyi* Bürgl (Wright, 1996). The former has narrower more numerous ribs that almost certainly branch close to the umbilicus, and hence with smaller less prominent ventrolateral clavi; lateral tubercles appear to be absent.

Psilopulchellia HYATT, 1903 (*nom. nud.*). Roman (1938) designated *Pulchellia nicklesi* Hyatt (**Fig. 621C-D**) as the type species, but no species were cited or described in the original reference and the designation was invalid.

Hettneria BÜRGL, 1956 [*Pulchellia selecta* Gerhardt, 1897; OD]. Small; like *Pulchellia* but with smooth inner flanks to early whorls, flexuous ribs, broad ventral groove and suture with shallower E, broader E/L and narrower L. *U. Barrem.* **Fig. 619E-F** - *H. selecta* (Gerhardt).

Gerhardtia HYATT, 1903 [*Ammonites galeatoides* Karsten, 1858, = *Ammonites sartousianus* d'Orbigny, 1842; OD]. Small; with generally single prorsiradiate flexuous ribs in later growth that broaden ventrally and carry very elongate ventrolateral clavi. *U. Barrem.* **Fig. 621A-B** - *G. sartousiana* (Karsten).

Carstenia HYATT, 1903 [*Ammonites lindigii* Karsten, 1858; OD]. Small; like *Gerhardtia* but more inflated and ribs with prominent lateral and inner and outer ventrolateral clavi; lateral and inner ventrolateral tubercles become obsolete in maturity; suture like *Heinzia* but with broader E/L and shallower L. *U. Barrem.* **Fig. 621G-H** - *C. lindigi* (Karsten).

= *Karstenia* Sayn, 1904 (*nom. van.*).

Tribe Heinziini Hyatt, 1903

(*nom. transl.* herein *ex* Heinziidae)

Diagnosis. Venter shallowly grooved, ribs with lateral tubercles and ventrolateral clavi. Derivatives of Pulchellini, close to *Caicedia* (Vermeulen, 1996). *Upper L. Barrem. - Lower U. Barrem.*

Coronites HYATT, 1900 [*Heinzia coronatoides* Sayn, 1891; OD]. Small; moderately involute, with convergen convex flanks, tabulate venter with siphonal groove from early growth, and compressed coronate whorl section; uniform rectiradiate to slightly prorsiradiate ribs, as wide as interspaces, bifurcating or single with associated intercalatory, curving strongly fowards on outer flank and projecting on venter; in early growth with lateral tubercles; suture simple, with narrow E, very broad bifid or trifid E/L, narrow bifid or trifid L and 3 auxiliary elements. *Lower U. Barrem.* **Fig. 621K-L** - *C. coronatoides* (Sayn).

Curiolia VERMEULEN, 1999 [*Pulchellia heinzi* Sayn, 1890; OD.]. Small; like *Coronites* but with coarser ornament and vigorous blunt oblique ventrolateral tubercles. *Uppermost L. Barrem.* **Fig. 621I-J** - *C. heinzi* (Sayn).
= *nom. correct. pro Curiolites* Vermeulen, 1995.

Memmiella VERMEULEN, 1999 [*M. crevolai*; OD]. Like *Curiolia* but with fewer broader ribs, more massive, obliquely clavate ventrolateral tubercles in early to middle growth, and with narrow sulcate venter. *Upper L. Barrem.* **Fig. 623A-B** - *M. crevolai* Vermeulen.

Heinzia SAYN, 1891 [*Ammonites provincialis* d'Orbigny, 1850; SD Roman, 1938]. Small; relatively evolute (U=25%), compressed, with steep umbilical wall, subrectangular whorl section, flattened prorsiradiate ribs arising in pairs and terminating in prominent ventrolateral clavi; venter shallowly grooved; suture like *Nicklesia* but simpler, with very broad E/L. *Barrem.* **Fig. 621E-F** - *H. heinzi* (Sayn). *Lower U. Barrem.*

Superfamily ENGONOCERATOIDEA Hyatt, 1900

(*nom. transl.* Basse, 1952 *ex* Engonoceratidae)

Discussion. Sufficient unifying characters suggest this group is monophyletic, although substantial differences indicate that a number of primitive species remain to be discovered; workers have looked to Parahoplitidae (Wright, 1996) or Pulchelliidae (Robert & Bulot, 2004) for the origin of the group but uncertainty remains. Characteristically discoidal and involute, with variably developed ribbing and tuberculation, and a distinctive suture with numerous adventive and auxiliary elements. *L. Alb. - U. Cenom.*

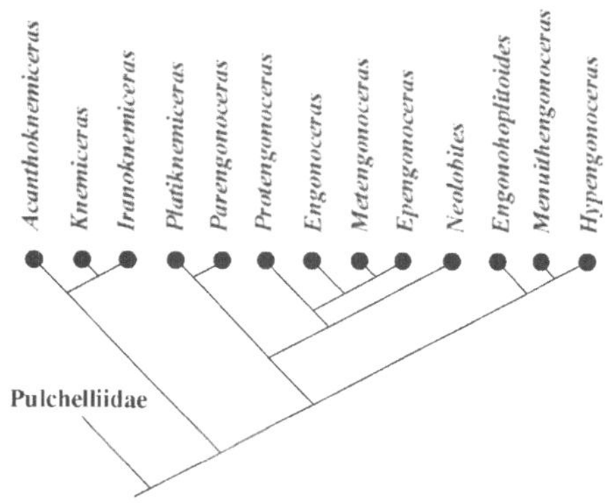

Fig. 624. Hypothesized relationships among Engonoceratoidea.

Family ENGONOCERATIDAE Hyatt, 1900

Discussion. A monophyletic group of cryptic origin, comprising homoeomorphs of Placenticeratinae; mostly oxycones with a pseudoceratitic suture. Typically Tethyan in distribution, but with some incursions into the Boreal Realm. Comprises the subfamilies Knemiceratinae, Engonoceratinae and Neolobitinae. *L. Alb. - U. Cenom.*

Subfamily Knemiceratinae Hyatt, 1903

Discussion. A seemingly monophyletic taxon of uncertain ancestry; suture typically with very broad deeply divided L and highly frilled saddles and lobes. *L. - U. Alb.*

Knemiceras BÖHM, 1898 [*Ammonites syriacus* von Buch, 1850; OD][**M**]. Moderately large, fairly inflated, involute, with steep umbilical wall, rounded umbilical shoulder, gently convex flanks that converge to tabulate to shallowly sulcate venter, and trapezoidal whorl section (quadrangular in early growth); sparse swollen umbilical tubercles give rise to pairs of low robust rectiradiate ribs with large ventrolateral nodes that are weakly connected and opposite across venter; suture on straight baseline with very shallow E, massive weakly indented asymmetrically bifid E/L that slopes towards L, very broad deeply bifid L and narrow necked frilled adventive lobes and bifid saddles with entire terminations. *L. - U. Alb.* **Fig. 625A-B** - *K. syriacum* (von Buch).
= *Cnemioceras* Haug, 1900 (obj.).
= *Cnemidoceras* Haug, 1900 (*laps. cal.*).

Omaimaiceras MAHMOUD, 1955 [*Knemiceras rittmani* Mahmoud, 1952; OD; Mahmoud had over 200 specimens and, although he introduced the genus and species three times, did not designate a holotype; here the original

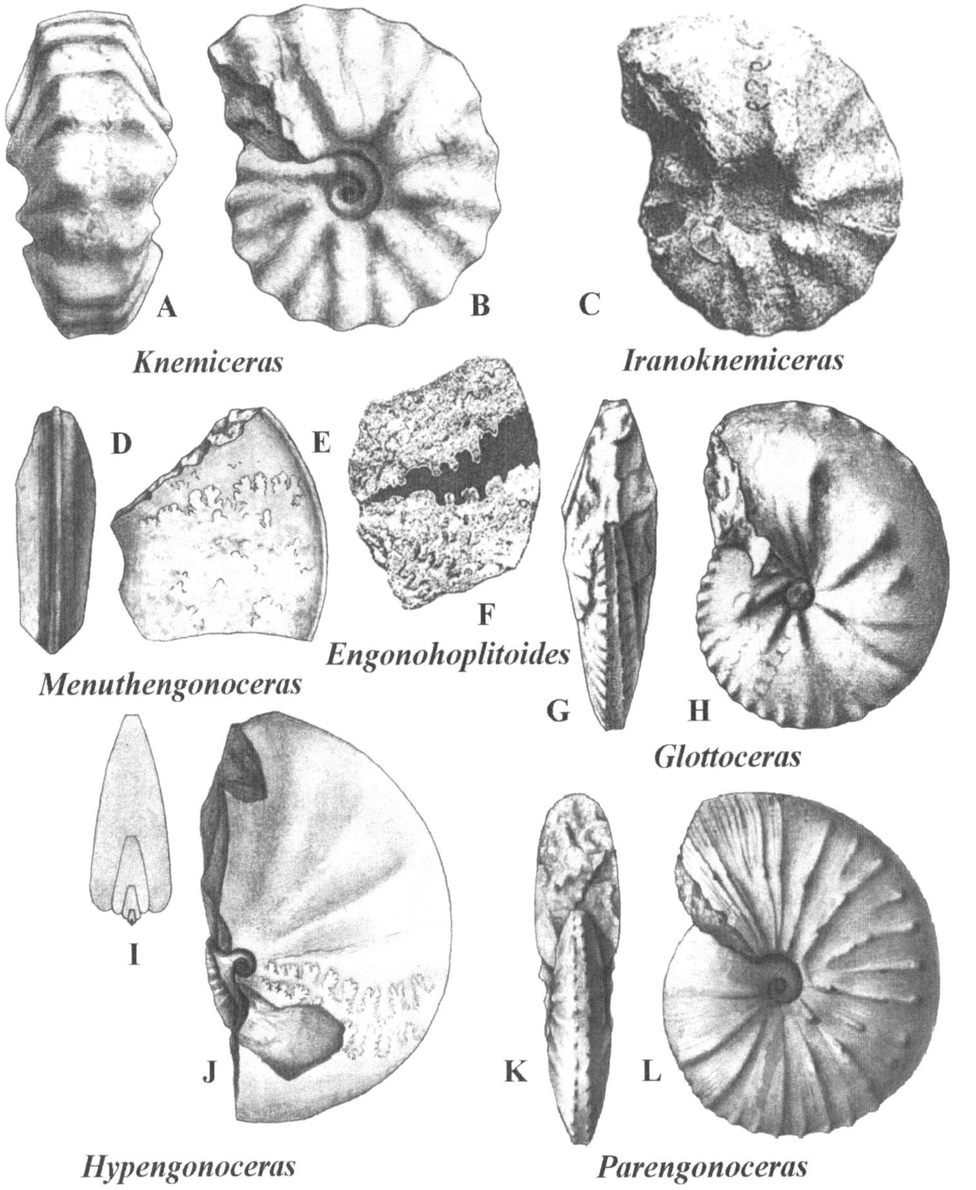

Fig. 625. Engonoceratidae.

of his 1955, plate 3, fig. 2, is selected as lectotype]. Like *Knemiceras* but small, relatively evolute, with 5-7 prominent conical umbilical bullae, and 20 ribs per whorl joined across venter by broad, strong, low transverse ribs; suture with entire elements. *L. Alb.* **Fig. 627** - *O. rittmani* (Mahmoud).

Engonohoplitoides Basse, 1940 [*E. khenchelaensis*; OD]. Type highly weathered; involute, compressed, with feeble, distant umbilicolateral bullae, distinguished mainly by sutural characteristics; although similar to *Glotticeras* its suture has a subdivided L like *Knemiceras*. *L. Alb.* **Fig. 625F** - *E. khenchelaensis* Basse.

Glottoceras Hyatt, 1875 [*Buchiceras attenuatum*; OD]. Small to medium sized; like *Knemiceras* but more compressed, involute and high whorled, with convergent flanks, narrow tabulate venter and subtrigonal whorl section; sparse distant umbilicolateral bullae give rise to low falciform ribs that may bifurcate from tubercles above midflank; lateral tubercles outnumber umbilicals, and are outnumbered by ventrolateral clavi that alternate across venter; ribbing declines with growth; suture like *Knemiceras* but simpler, with little incised E/L and bulbous phylloid terminations to remaining saddles. *M. Alb.* **Fig. 625G-H** - *G. attenuatum* (Hyatt).

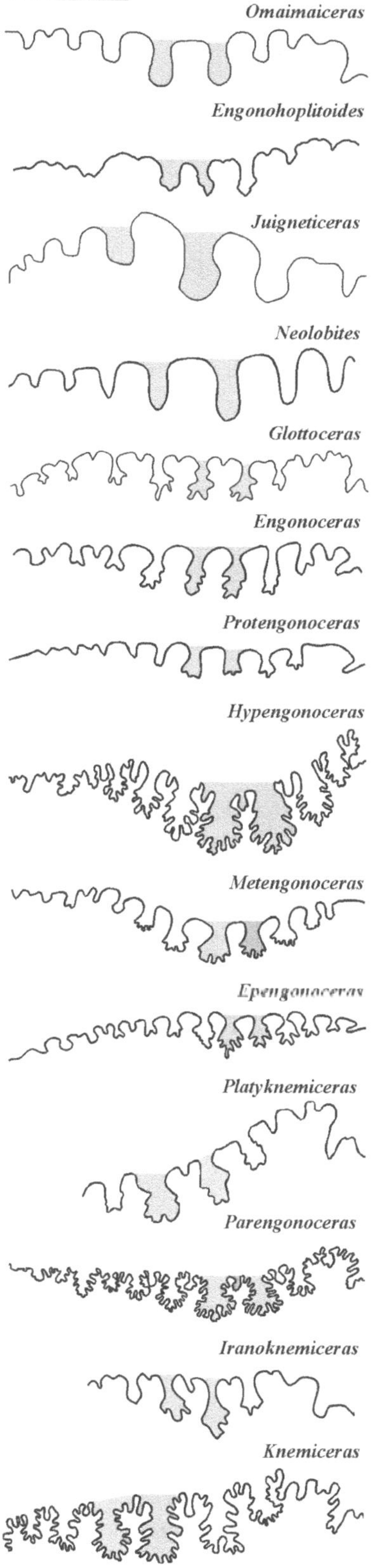

Fig. 626. Engonoceratid sutures.

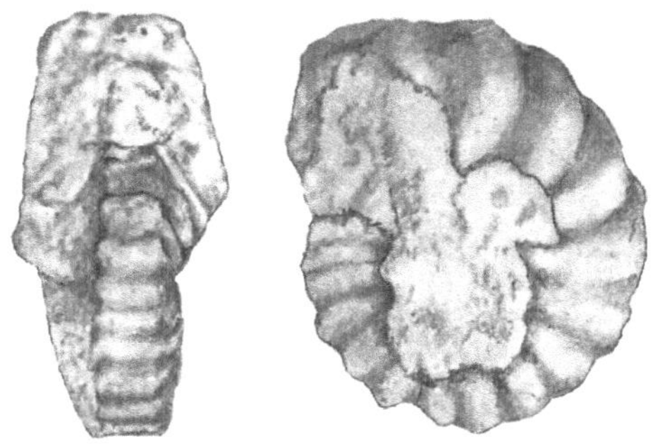

Fig. 627. *Oumaimaiceras.*

= *Glossoceras* Douvillé, 1907 (*non* Barrande, 1867).

Platiknemiceras BATALLER, 1954 [*Knemiceras (P.) bassei*; OD]. Small; compressed, involute, with broad convergent flanks and narrow sulcate venter with sharp ventrolateral edges, becoming tabulate in later growth; tubercles lacking and ornament of flexuous biconcave growth striae only; suture with little incised elements, shallow E, very broad E/L sloping towards L subdivided by two adventitious lobes and with broad outer adventitious saddle. *L.- M. Alb.* **Fig. 628A-B** - *P. bassei* (Bataller).

= *Platyknemiceras* Bataller, 1957 (*nom. null.).*

Iranoknemiceras COLLIGNON, 1981 [*Knemiceras uhligi* var. *douvillei* Basse, 1940; OD]. Small; like *Knemiceras* but more coarsely ornamented and with simpler suture. *L. Alb.* **Fig. 625C-D** - *I. douvillei* (Basse).

Parengonoceras SPATH, 1924 [*Ammonites ebrayi* de Loriol, 1882; OD]. Moderately large; inner whorls strongly compressed, involute, with broad gently convex flanks converging to narrow tabulate venter; indistinct ribs terminate in ventrolateral clavi which alternate across venter; adult body chamber moderately inflated with rounded venter, compressed oval whorl section, and distant strong narrow convex forward ribs separated by 1-2 shorter intercalatories; ribs terminate in rounded ventrolateral nodes and carry irregularly developed umbilicolateral tubercles and 1-2 rows of lateral tubercles; highly indented suture on curved baseline, with deeply tripartite E/L sloping towards L and fingerlike folioles. *L. - M. Alb.* **Fig. 625K-L** - *P. ebrayi* (de Loriol).

Subfamily Engonoceratinae Hyatt, 1900

Discussion. A monophyletic group with an origin in Knemiceratinae, close to *Knemiceras*; suture typically with entire saddles and indented lobes, and more elements than in Knemiceratinae. *L. Alb. - U. Cenom.*

Engonoceras NEUMAYR & UHLIG, 1881 [*Ammonites pierdenalis* von Buch, 1850; SD Pervinquière, 1907]. Small;

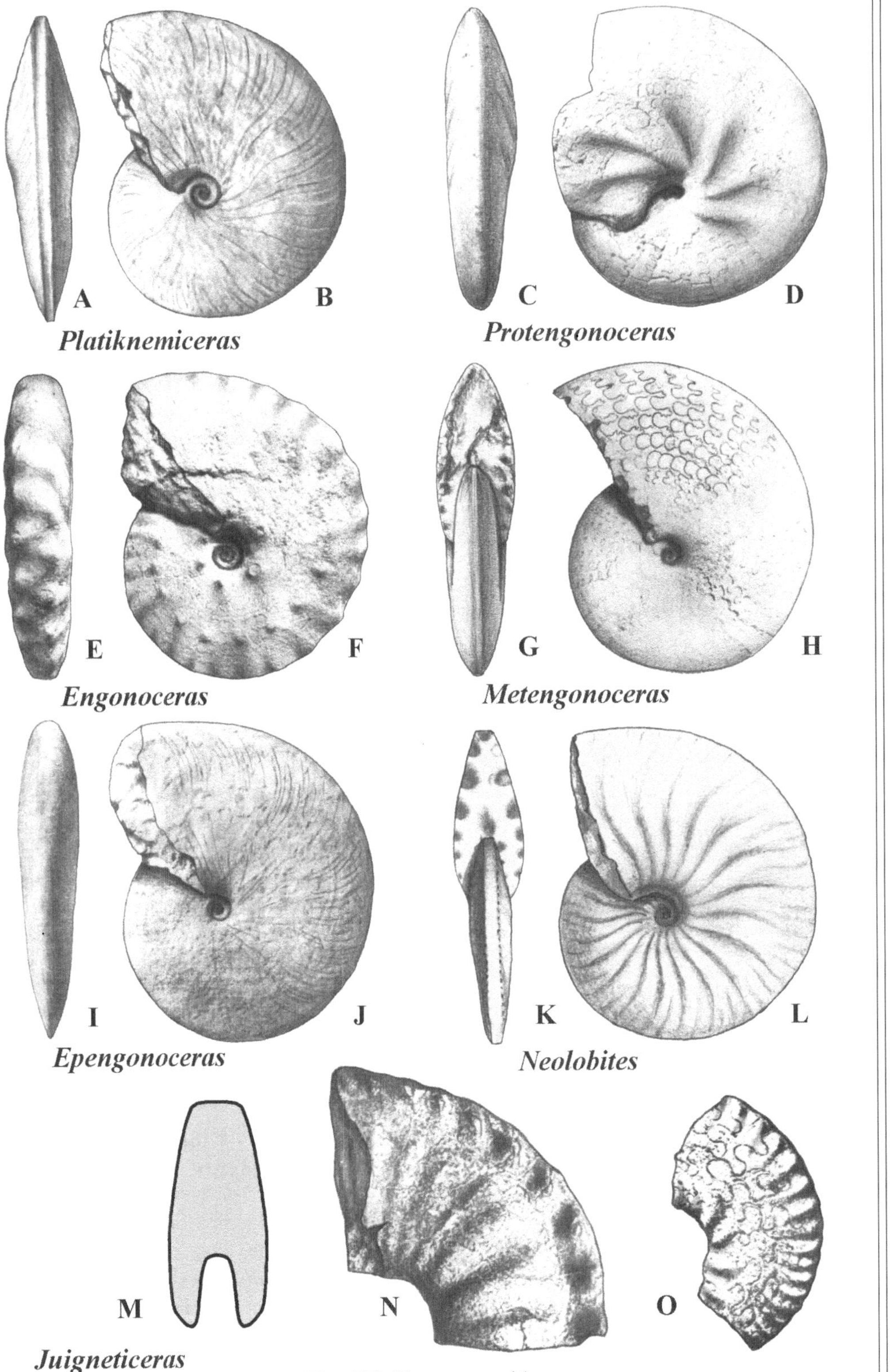

Fig. 628. Engonoceratidae.

like *Protengonoceras* but ribs with weak high-lateral tubercles and ending in small ventrolateral clavi that alternate across venter, sometimes joined by zigzagging ribs; suture with entire saddles except outermost one that is bifid. *L. Alb. – M. Cenom.* **Fig. 628E-F** - *E. stolleyi* (Böhm).

The type species of the genus, *E. pierdenale* (von Buch), is not well known, with discrepant descriptions; material is assigned to the genus mainly on sutural characteristics.

Protengonoceras HYATT, 1903 [*Engonoceras gabbi* Böhm, 1898; OD]. Small to medium sized; in early to middle growth like *Platiknemiceras* but, in maturity, flanks become more convex, venter rounds andthen flattens on adult body chamber; low blunt concave forward prorsiradiate ribs appear on lower to middle flanks; suture on nearly straight baseline, with splayed E, barely incised outer adventitious saddle to E/L, weakly indented lobes and entire auxiliary elements. *M. Alb.* **Fig. 628C-D** - *P. gabbi* (Böhm).

Metengonoceras HYATT, 1903 [*M. inscriptum*; SD Roman, 1938]. Medium sized; in early growth like *Protengonoceras* with shallowly concave venter becoming acute in maturity and rounding on final part of adult body chamber; umbilical margin with weak folds; growth striae falcoid, sometimes strengthening to form arcuate ribs on outer flank; suture on sinusoidal baseline, with short narrow splayed "bow-tie" E, very broad E/L with 4 narrow entire to trifid adventitious lobes and 5 entire adventitious saddles, a narrow trifid L and up to 7 mostly entire auxiliary saddles separated by very narrow bifid lobes that become single close to umbilical margin. *U. Alb. - U. Cenom.* **Fig. 628G-H** - *M. inscriptum* (Hyatt).

Epengonoceras SPATH, 1924 [*Engonoceras dumbli* Cragin, 1893; OD]. Medium sized; like *Metengonoceras* but generally more compressed and involute, with only falcoid growth striae; suture on slightly sinuous to almost straight baseline, with mushroom-shaped adventitious saddles to E/L separated by bifid adventitious lobes, a narrow bifid L and up to 11 auxiliary saddles. *?L. Cenom., M. - U. Cenom.* **Fig. 628I-J** - *M. dumbli* (Cragin).

Subfamily Neolobitinae Luppov & Mikhailova, 1958

Discussion. A monophyletic group with an origin in Engonoceratinae (Luppov & Mikhailova, 1958) close to *Metengonoceras*; suture with entire saddles and lobes. *M. - U. Cenom.*

***Juigneticeras* gen. nov.** [*Neolobites fortaui* Pervinquière, 1907, p.209; OD herein]. Medium sized, involute (U=17%), with flat slightly convergent flanks, strongly compressed whorl section and tabulate to slightly sulcate venter with subangular edges; strong blunt umbilical bullae, in later growth connected to rectiradiate ribs that may develop lateral bullae, poorly connected to prominent ventrolateral clavi that are joined across the venter by low ribs; suture with inner auxilliary saddle of E/L broadly rectangular and very shallowly bifid. *Lower M. Cenom.* **Fig. 628M-O** - *J. fortaui* (Pervinquière).

Neolobites FISCHER, 1882 [*Ammonites vibrayeanus* d'Orbigny, 1841; M]. Like *Juigneticeras* but very involute (U=8%), not so compressed; convex flanks with convex forward ribs and striae, weak or effaced from outer flank (juv.), with minute ventrolateral clavi, forming fasciculate bundles reaching venter in maturity; suture with entire leaf-like saddles and lobes. *U. Cenom.* **Fig. 628K-L** - *N. vibrayeanus* (d'Orbigny).

Subfamily Hypengonoceratinae Chiplonkar & Ghare, 1976

Discussion. A monotypic taxon with an origin in Engonoceratinae close to *Parengonoceras* (Klinger & Kennedy, 1989). The distinctive suture has large pincer-like bifid folioles. *?M. Alb., U. Alb.*

Hypengonoceras SPATH, 1922 [*Placenticeras warthi* Kossmat, 1895; OD]. Very large; inner whorls with broad low prorsiradiate falcoid ribs that terminate in ventrolateral clavi which alternate across tabulate venter; in middle growth ribs become single and are lost in maturity when venter becomes subacute; suture with relatively narrow E/L, bifid saddles with large pincer-like frequently bifid folioles and much incised lobes wider than saddles. *U. Alb.* **Fig. 625I-J** - *H. warthi* (Kossmat).

Menuthengonoceras COLLIGNON, 1963 [*M. komihevitraense*; OD]. Medium sized, known only from a fragment; like *Hypengonoceras* but with strong siphonal keel and acute venter; suture florid. *M. Alb.* **Fig. 625D-E** - *M. komihevitraense* Collignon.

Wright (1996) included this genus in the synonymy of *Oxytropidoceras* but the sutures are different. There is some resemblance to much younger *Tropitoides*.

Suborder ANCYLOCERATIDINA Wiedmann, 1962 (1966)

(*nom. correct.* herein *pro* Ancyloceratina)

Discussion. A monophyletic taxon with an origin in Ataxioceratidina (Zeiss, 1968). A major problem with classification of the group is compelling evidence for sexual dimorphism that involves both size and morphology (Delanoy *et al.*, 1995). While this seems clear in some cases it is not easily proven in others, with microconchs and macroconchs assigned to different subfamilies/families. The result is much speculation, and taxa that are difficult to define.

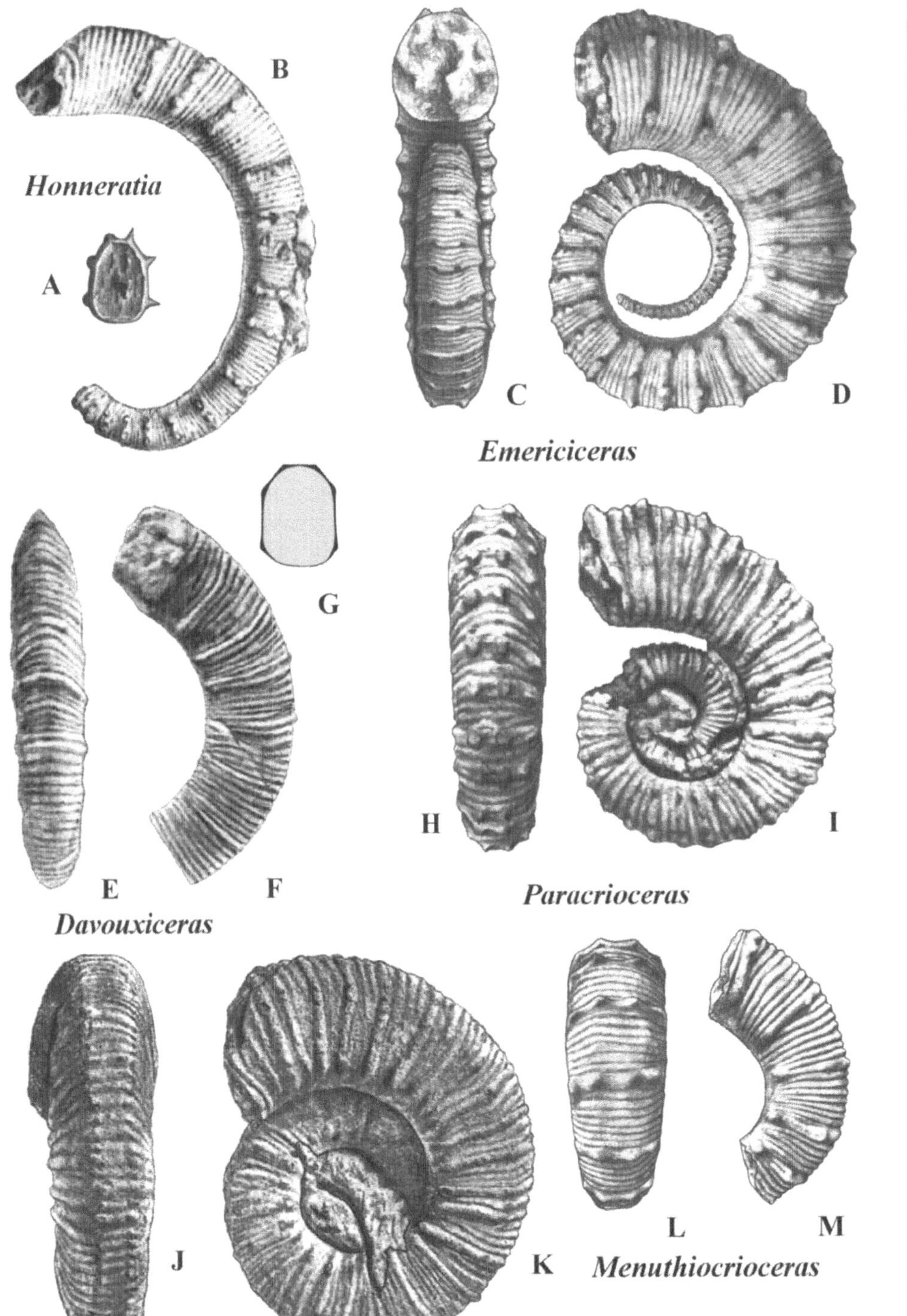

Fig. 629. Pedioceratoidea.

Moreover the taxonomy is in a status of flux due to convergence among different lineages. Comprises the hyporders Ancyloceratoidei and Douvilleiceratoidei (Wiedmann, 1966). *U. Haut. - M. Alb.*

Hyporder ANCYLOCERATOIDEI nov.

Discussion. A monophyletic taxon with an origin in Neocomitoidei (Thieuloy, 1977) comprising the

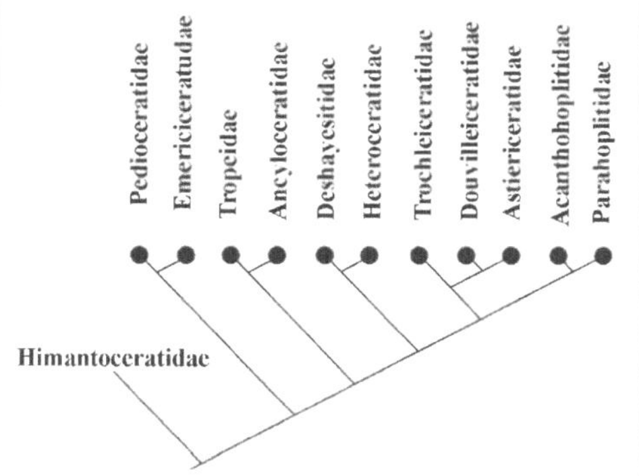

Fig. 630. Hypothesized relationships among Ancyloceratina.

superfamilies Pedioceratoidea and Ancyloceratoidea. *U. Haut. - L. Alb.*

Superfamily PEDIOCERATOIDEA Hyatt, 1900

(*nom. transl.* herein *ex* Pedioceratidae)

Discussion. A monophyletic group with an origin in Neocomitoidea (Thieuloy, 1977). Typically moderately large criocones and ellipticones with recoiling in several lineages leading to secondary ammoniticones; strong ribbing, mostly trituberculate, may coarsen in later growth and the complex suture has a very reduced U. Comprises the families Emericiceratidae, Pedioceratidae and Hemihoplitidae, but relationships are complicated by a convergent similarity to Crioceratitoidea, and the rapid rate at which new taxa are being introduced. *L. Haut. - M. Alb.*

Family EMERICICERATIDAE Vermeulen, 2004

= Paracrioceratidae Bert & Bersac, 2017

Discussion. Comprises the subfamilies Emericeratinae and Paraspiticeratinae, with an origin in Himantoceratidae (Vermeulen *et al.*, 2020). Bert & Bersac (2017) introduced the family Paracrioceratidae, without diagnosis or definition, for the hypothetical phyletic line *Fissicostaticeras - Paracrioceras - Parancyloceras* but this is an artificial assemblage. Loose criocones to recoiled ammoniticones commonly with flattened venter and compressed trapezoidal to subcircular whorl section; at least in early growth with trituberculate primary ribs separated by non-tuberculate secondaries. A predominantly Hauterivian radiation, but including recoiled ammoniticones in the Barremian. *L. Haut. - L. Barrem.*

Subfamily Emericiceratinae Vermeulen, 2004

Discussion. A seemingly monophyletic group with an origin in Himantoceratidae (Vermeulen *et al.*, 2020a). Moderately large to large tight criocones and recoiled ammoniticones with flattened venter and compressed subrectangular whorl section; trituberculate primary ribs with umbilical and inner and outer ventrolateral tubercles, separated by non-tuberculate secondaries; ribbing may simplify in later growth, with loss of tubercles. *L. Haut. - L. Barrem.*

Davouxiceras VERMEULEN, 2004 [*Crioceras nolani* Kilian, 1907; OD]. Large to very large, compressed and high whorled criocone with subparallel flanks and compressed elliptical intercostal whorl section in later growth; primary ribs weakly trituberculate, with rounded ventrolateral tubercles, separated by 2-7 non-tuberculate secondaries the number increasing with growth; in later growth intercalatories appear at various levels on flank; mature ornament comprises mainly primary ribs with occasional short intercalatories; suture with narrow E, asymmetrically bifid saddles and widely splayed trifid lobes. *L. Haut.* **Fig. 629E-G** - *D. nolani* (Kilian).

Honnoratia BUSNARDO, 2003 [*Toxoceras honneratianum* d'Orbigny, 1842; OD]. Moderately large loosely coiled criocone with a slightly compressed oval whorl section; conspicuously trituberculate primary ribs separated by generally 3 non-tuberculate secondaries in early growth, increasing to 6-8 in middle growth and up to 10 later, occasionally looped; tubercles rounded. *Upper U. Haut. - Lower L. Barrem.* **Fig. 629A-B** - *H. honneratiana* (d'Orbigny).

Emericiceras SARKAR, 1954 [*Crioceratites emerici* Léveillé, 1837; OD]. Large to enormous (up to 1 m diameter)

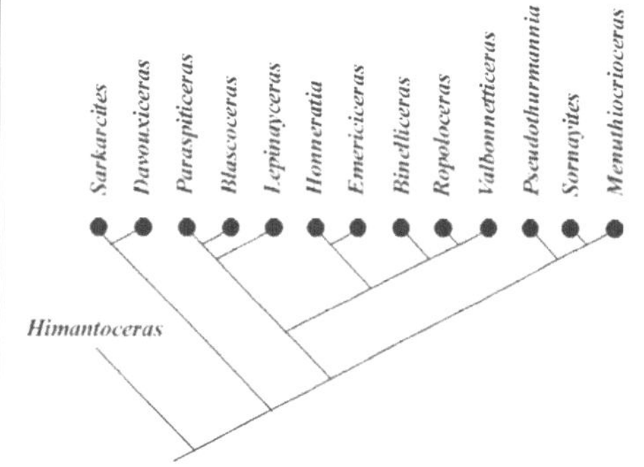

Fig. 631. Hypothesized relationships among Emericiceratidae.

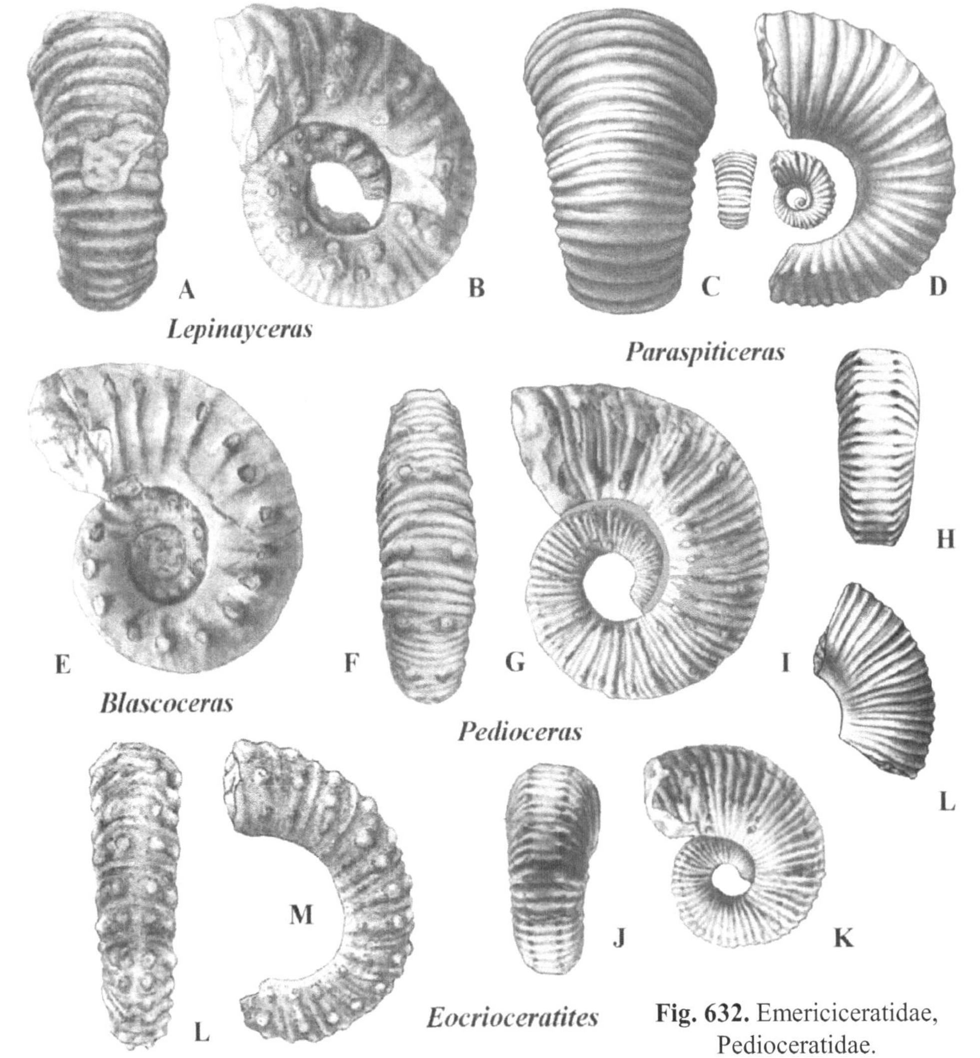

Fig. 632. Emericiceratidae, Pedioceratidae.

loose criocone with subcircular whorl section in early growth; in middle growth dorsum and venter flattened, flanks slightly convex and convergent and whorl section trapezoidal, becoming compressed and oval in later growth; slightly curved primary ribs with umbilical, high lateral and ventrolateral septispinate tubercles, the latter clavate, separated by 1-6 fine nontuberculate secondaries the number increasing with growth; ribs become uniform on the adult body chamber where only umbilical bullae remain and ribs cross venter with forward arch. *U. Haut. - Lower U. Barrem.* **Fig. 629C-D** - *E. emerici* (Léveillé).

Borroceras VERMEULEN, LAZARIN, LEPINAY, LEROY & MASCARELLI, 2018 [*B. borroi*; OD]. Small criocone with compressed subelliptical whorl section, venter flattening with age; in early growth trituberculate primary rib with prominent rounded lateral and ventrolateral tubercles separated by 1-2 fine non-tuberculate secondaries; in middle growth umbilical bullae prominent, lateral tubercles weak to obsolete, 1-4 secondaries many looped

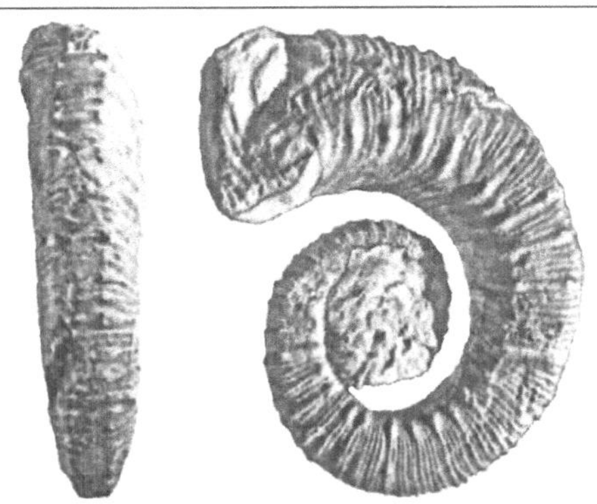

Fig. 633. *Borroceras*.

to additional ventrolateral tubercles; later ribs increasingly flexuous, ventrolateral tubercles feeble and discrimination of primaries and secondaries no longer obvious; primary ribs strengthen in maturity. *L. Barrem.* **Fig. 633** - *B. borroi* Vermeulen, Lazarin, Lepinay, Leroy & Mascarelli.

Paracrioceras SPATH, 1924 [*Ammonites (Crioceras) occultus* Seeley, 1865; OD]. Medium sized criocone with whorls almost in contact, flat flanks and polygonal whorl section; straight prorsiradiate trituberculate primary ribs separated by 2-4 non-tuberculate secondaries uninterrupted across venter that they cross transversely; lateral tubercles evanesce in middle to later growth; there may be sutural differences from *Emericiceras* (Sarkar, 1977). *Upper L. - U. Barrem.* **Fig. 629H-I** - *P. occultum* (Seeley).

Subfamily Paraspiticeratinae Vermeulen, 2009

Discussion. Monophyletic descendents of *Emericiceras* by recoiling of the shell (Vermeulen, 2006; Baraboshkin & Mikhailova, 2020). Serpenticones with inflated depressed whorls generally in contact; primitively with prominent umbilical and lateral tubercles; some with perforated umbilicus. *Uppermost Haut. - L. Barrem.*

Lepinayceras VERMEULEN, LAZARIN, LEROY & MASCARELLI, 2020 [*L. sandovali*; OD]. Medium sized, very evolute, inflated, with convex flanks and depressed subelliptical whorl section; about 6 prominent, rounded mid-lateral spine bases per half whorl; in middle growth umbilical tubercles appear and loop to lateral ones, from which they bi- or trifurcate and cross venter transversely with feeble ventrolateral bullae, separated by 1-2 non-tuberculate rectiradiate to rursiradiate secondaries; in later growth (whorl height >25 mm) bifurcating and trifurcating ribs evanesce. *U. Haut.* **Fig. 632A-B** - *L. sandovali* Vermeulen, Lazarin, Leroy & Mascarelli.

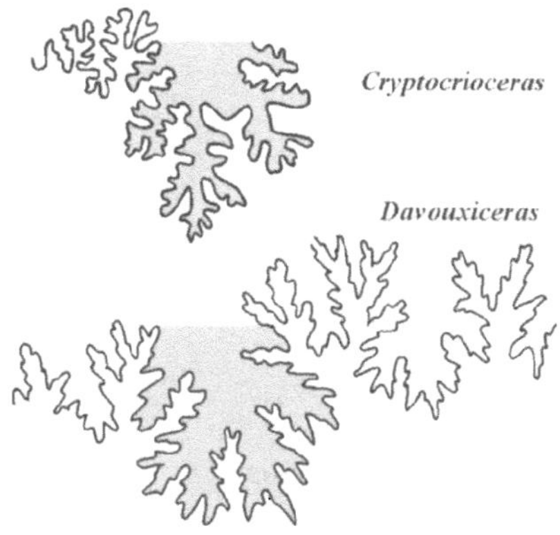

Fig. 634. Pedioceratoid sutures.

Blascoceras VERMEULEN, LAZARIN, LEROY & MASCARELLI, 2012 [*Ammonites nodulosus* Catulo, 1848; OD]. Like *Lepinayceras* but umbilical tubercles lacking, ribs feeble and weak secondaries sparse; in maturity flank tubercles become bullate and finally are lost leaving only distant simple ribs. *U. Haut.* **Fig. 632E** - *B. nodulosum* (Catulo).

Paraspiticeras KILIAN, 1910 [*Aspidoceras percevali* Uhlig, 1883; SD Spath, 1921]. Medium sized; perforate umbilicus followed by tight serpenticone coiling; rapidly enlarging whorls with strongly convex flanks, broadly rounded venter and depressed reniform whorl section; strong rounded slightly rursiradiate ribs cross venter with slight forward inflexion and carry prominent umbilical and midlateral tubercles on inner whorls; suture with deep rectangular E, long narrow bifid E/L, wide bifid L much shallower than E, bifid L/U and narrow U not as deep as I. *L. Barrem.* **Fig. 632C-D** - *P. percevali* (Uhlig).

Family PEDIOCERATIDAE Hyatt, 1900

Discussion. Barremian derivatives of Emericiceratidae (Delanoy, 1992; Vermeulen, 2000) close to *Paracrioceras*, although Company *et al.* (2008) and Bert & Bersac (2017) looked to Gassendiceratinae for the origin of the group. Serpenticone to tight criocone; whorl section polygonal with ventrolateral facets and subtabulate venter; ribs mostly simple, trituberculate (except in earliest growth), strengthened primary ribs with umbilical and rounded inner and outer ventrolateral tubercles. *Barrem.*

Pedioceras GERHARDT, 1897 [*Ammonites cundinamarcae* Gerhardt, 1897, = *caquesensis* Karsten, 1858; SD Hyatt, 1903]. Medium sized (?adults undescribed); serpenticone with shallowly impressed dorsum and whorls just in contact; broad flattish flanks, facetted ventrolateral shoulders, subtabulate venter and polygonal to oval whorl section; earliest whorls with straight simple ribs with lateral and ventral tubercles, the former rapidly shifting to an outer ventrolateral position; ribs weakly joined across venter by transverse ribs; in middle to later

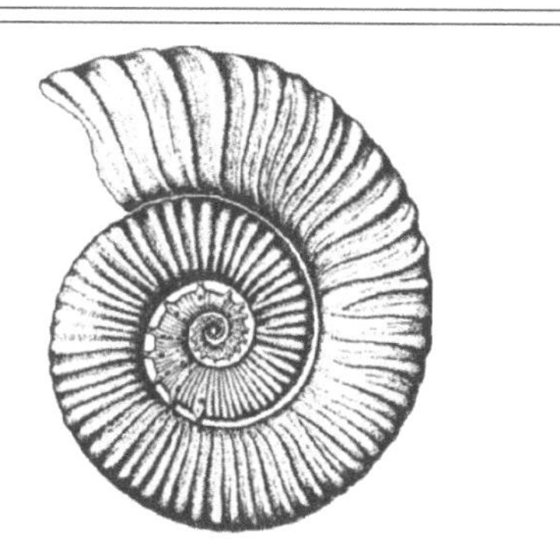

Fig. 635. *Cryptocrioceras.*

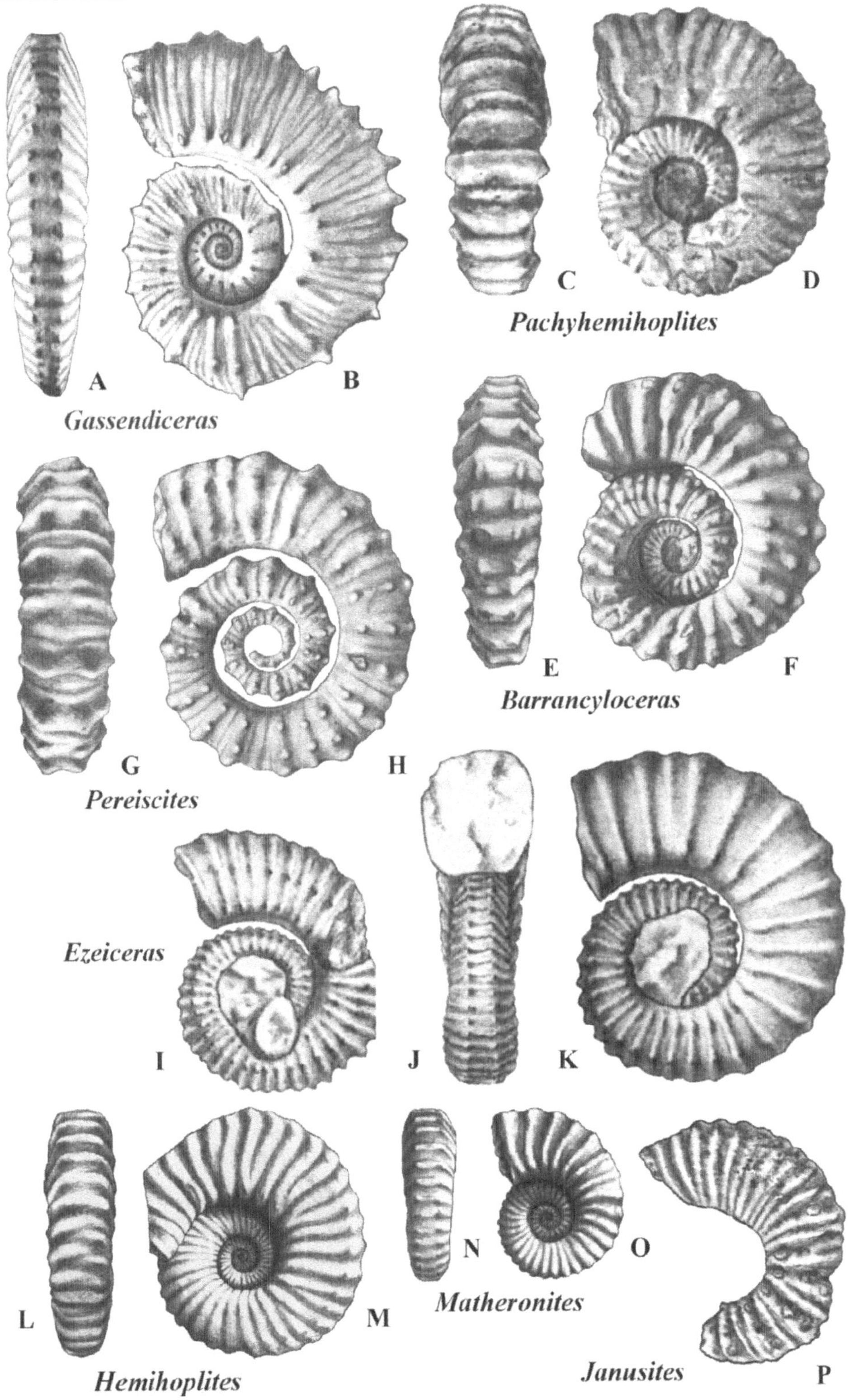

Fig. 636. Hemihoplitidae.

growth strong primary ribs develop with prominent umbilical bullae to become trituberculate, separated by 2-3 finer secondaries; ventrolateral tubercles rounded. *L? Barrem.* **Fig. 632F-K** - *P. caquesense* (Karsten).

Eocrioceratites WIEDMANN, 1973 [*Protancyloceras rebillyi* Collignon, 1962; OD]. Known only from curved fragment with subcircular whorl section; regular strong concave-forward primary ribs separated by fine generally non-tuberculate secondaries although some with small lateral tubercle; primary ribs with small dorsolateral tubercle, large rounded lateral tubercle slightly above midflank and rounded ventrolateral tubercle; ribs form strong chevrons across venter where interrupted by smooth zone; suture with large trifid L. *L. Barrem.* **Fig. 632L-M** - *E. rebillyi* (Collignon).

Menuthiocrioceras COLLIGNON, 1949 [*Crioceras (M.) lenoblei*; OD]. Like *Pedioceras* but with vertical umbilical wall, flat subparallel flanks and compressed rectangular (juv.) to trapezoidal (ad.) whorl section; in early to middle growth with strong sigmoidal primary ribs with feeble to distinct umbilical bullae and spinose inner and outer ventrolateral tubercles, separated by 1-3 non-tuberculate secondaries that cross venter transversely; in later growth intercalated ribs almost as strong as primaries. *L. Barrem.* **Fig. 629J-M** - *M. lenoblei* (Collignon).

Cryptocrioceras AGUIRRE URRETA & RAMOS, 1981 [*Mantelliceras yrigoyeni* A. Leanza, 1971; OD]. Moderately large; a tight criocone with barely impressed dorsum, subtabulate venter and trapezoidal (juv.) to quadratic (ad.) whorl section; in early growth strong trituberculate primary ribs separated by 1-3 finer secondaries with or without ventral tubercles; >50 mm diameter with low, rounded, uniform, flexuous ribs with weak umbilicolateral bullae and slight thickening on ventrolateral shoulders; ribs weak across siphonal line; suture with shallow E, broad bifid E/L, large trifid L much deeper than E and small U. *U. Barrem.* **Fig. 635** - *C. yrigoyeni* (Leanza).

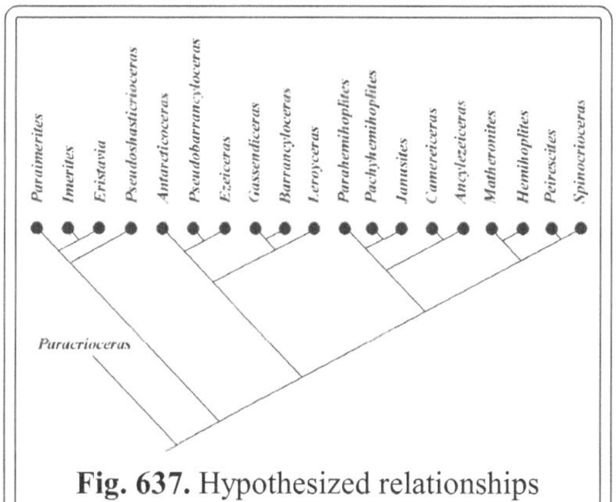

Fig. 637. Hypothesized relationships among Hemihoplitidae.

Family HEMIHOPLITIDAE Spath, 1924

Discussion. Monophyletic secondarily recoiled derivatives of Emericiceratidae (Wiedmann, 1962; Vermeulen & Lazarin, 2007); characteristically ammoniticone with more or less uniform bi- or trituberculate ribbing; suture with trifid L. Comprises the subfamilies Gassendiceratinae, Hemihoplitinae and Peirescitinae. *U. Barrem.*

Subfamily Gassendiceratinae Bert, Delanoy & Bersac, 2006

= Barrancyloceratinae Bert, Delanoy & Bersac, 2006

Discussion. A monophyletic group with an origin in Emericiceratinae, close to *Honneratia* (Vermeulen & Lazarin (2007) or *Paracrioceras* (Bert & Bersac, 2014). Evolute ammoniticone in early to middle growth, uncoiling slightly later; whorls compressed, with trituberculate primary rib (umbilical, high lateral, ventrolateral) crossing venter transversely, with or without intervening non-tuberculate secondaries. *U. Barrem.*

Gassendiceras BERT, DELANOY & BERSAC, 2006 [*G. quelquejeui*; OD]. Medium sized to large criocone with whorls touching in early growth, but uncoiling later, with flat dorsum, broad gently convex convergent flanks, narrow subtabulate venter and compressed subovate whorl section; in middle growth trituberculate primaries separated by non-tuberculate secondaries, the ventrolaterals clavate; in later growth with dense rursiradiate ribs separated by 1-2 intercalatories. *U. Barrem.* **Fig. 636A-B** - *G. quelquejeui* Bert, Delanoy & Bersac.

Barrancyloceras VERMEULEN, 2000 [*Crioceras barremense* Kilian, 1895; OD]. Tight criocone with compressed elliptical whorl section; robust main ribs with umbilical and high lateral tubercles, and ventrolateral clavi, normally separated by a finer non-tuberculate secondary but with up to 3; ribs split into 2-3 finer secondaries across flattened dorsum. *U. Barrem.* **Fig. 636E-F** - *B. barremense* (Kilian).

According to Bert *et al.* (2013) the type is a "*doubtful species*" that they regard a *nomen dubium* (but see rebutal by Vermeulen, 2013); they assign Vermeulen's material (but not the type) to *Gassendiceras alpinum* (d'Orbigny).

= *Barrancyloceras* Vermeulen & Bert, 1998 (*nom. nud.*).

Leroyceras VERMEULEN, 2006b [*Barrancyloceras mascarellii* Vermeulen, 2006a; OD]. Large ancylocone; spire with flat to concave venter and subpolygonal whorl section increasing rapidly in height; spire with strong rectiradiate trituberculate primary ribs, with umbilical and inner and

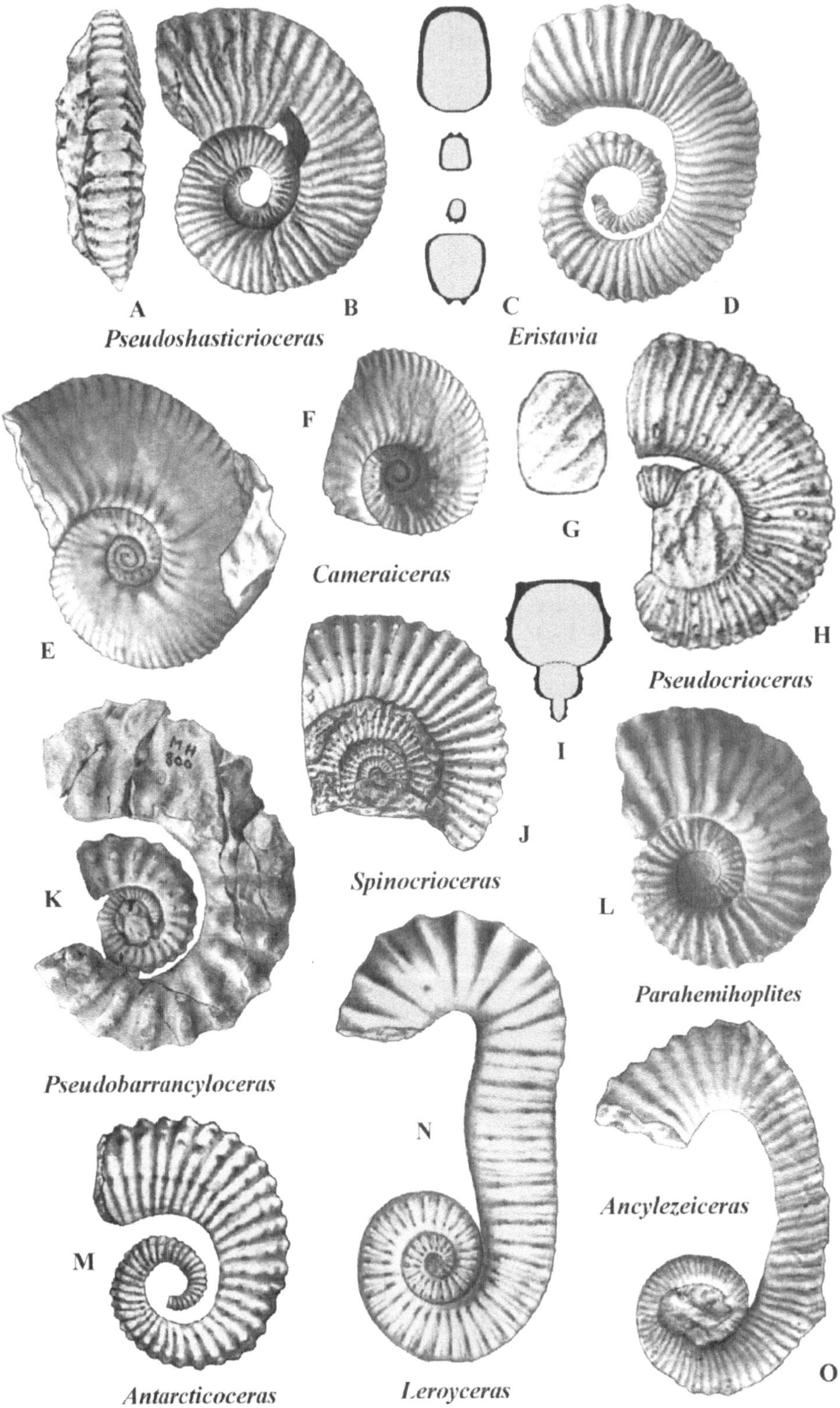

Fig. 638. Hemihoplitidae.

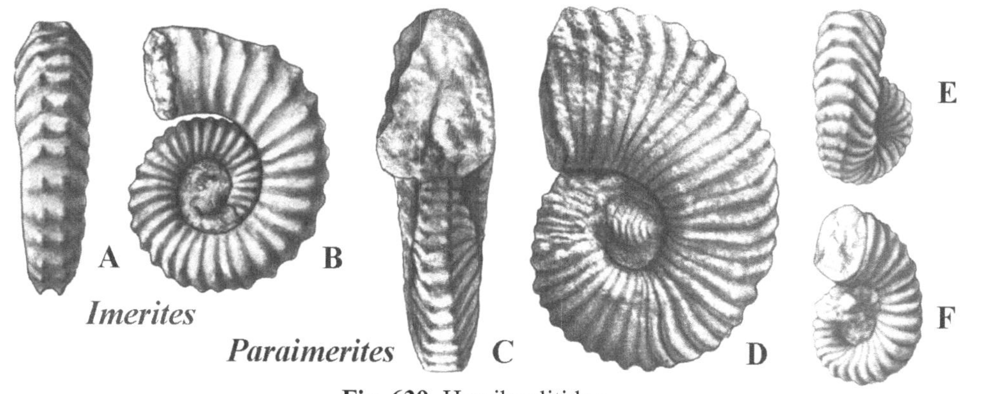

Fig. 639. Hemihoplitidae.

outer ventrolateral tubercles separated by 2-4 fine secondaries; ribs coarsen on proversum with only 1-2 secondaries, becoming bold, distant and feebly tuberculate on flexus and short retroversum; suture with oblique bifid E/L, narrow trifid L much deeper than E and L/U. *U. Barrem.* **Fig. 638N** - *L. hoheneggeri* (Uhlig).

Pseudoshasticrioceras DELANOY, 1999 [*Emericiceras magnini* Delanoy, 1992; OD]. Moderately large to large criocone increasing rather rapidly in height, with subquadratic whorl section in early to middle growth, later slightly compressed and subtrapezoidal; in early growth with bi- and trituberculate ribs, the ventrolaterals clavate; in middle growth non-tuberculate secondaries appear; in maturity ribs only with ventrolateral clavi, ornament weakening on adult body chamber. *U. Barrem.* **Fig. 638A-B** - *P. magnini* (Delanoy).

Imerites ROUCHADZÉ, 1933 [*Heteroceras giraudi* Kilian, 1889; ? = *Crioceras cristatum* d'Orbigny, 1842; OD] [**m**]. Very small; like *Pseudoshasticrioceras* but body chamber not expanding so rapidly and with coarse simple distant ribs with inner and outer ventrolateral tubercles; a homoeomorph of some Heteroceratinae. *Upper U. Barrem.* **Fig. 639A-B** - *I. giraudi* (Kilian).

Crioceras cristatus d'Orbigny, regarded by some a senior synonym of *I. giraudi*, is treated by Bert *et al.* (2009) as a *nomen dubium*.

= *Atopoceras* Jaubert, 1889 (*nom. nud.*).

= *Escragnolleites* Sarkar, 1954 (obj.).

Eristavia KAKABADZE, 1971 [*Colchidites (Imerites) dichotomus* Eristavi, 1955; OD] [**M**]. Like *Imerites* but larger, more loosely coiled; early whorls (D<45 mm) with fibulate ribbing and doubled ventrolateral tubercles at some growth stage. *Upper U. Barrem.* **Fig. 638C-D** - *E. dichotomum* (Eristavi).

= *Eristavia* Kakabadze, 1967 (*nom. nud.*).

Paraimerites KAKABADZE, 1967 [*Imerites densecostatus* Renngarten, 1926; OD]. Initial helicone tightly embraced by rapidly expanding planispire increasing rapidly in height; ribs strong simple, slightly rursiradiate, straight to flexuous, curving forward near venter, with ventrolateral tubercles, crossing venter transversely. *Upper U. Barrem.* **Fig. 639C-F** - *P. densecostatus* (Renngarten).

Subfamily Hemihoplitinae Spath, 1924

(*nom. transl.* Bert, Delanoy & Bersac, 2006 *ex* Hemihoplitidae)
= Mathoceratinae Casey, 1964.

Discussion. Monophyletic derivatives of Emericiceratinae, close to *Pseudothurmannia* (Wiedmann, 1962); medium sized ancylocones and ammoniticones, primitively with trituberculate primaries separated by non-tuberculate secondaries; later with uniform flexuous ribs, simple or bifurcating, with ventrolateral tubercles. *U. Haut. - U. Barrem.*

Camereiceras DELANOY, 1990 [*Matheronites limentinus* Thieuloy, 1979; OD]. Moderately large; like *Hemihoplites* but early whorls trituberculate, with strong clavate umbilical and inner and outer ventrolateral tubercles; inner ventrolaterals evanesce in middle growth when umbilicals become bullate; adult whorls with

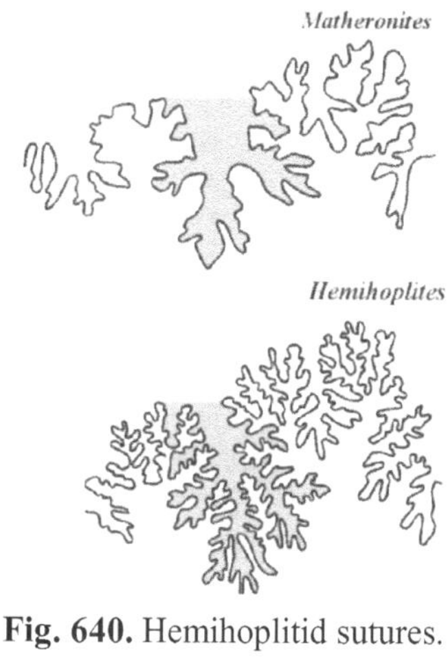

Fig. 640. Hemihoplitid sutures.

strong non-tuberculate ribs. *Lower U. Barrem.* **Fig. 638E-F** - *C. limentinum* (Thieuloy).

Ancylezeiceras VERMEULEN, 2006 [*Acrioceras breistrofferi* Sarkar, 1955; OD] [**m**]. Small ancylocone with planispiral early whorls; proversum long, weakly curved; long retroversum reaching spire; strong straight trituberculate main ribs separated by occasional very fine secondaries that appear at base of proversum; tuberculation weakens on flexus; on retroversum umbilical bullae give rise to paired non-tuberculate ribs that cross venter uninterrupted. *Lower U. Barrem.* **Fig. 638O** - *A. breistrofferi* (Sarkar).

Perhaps a microconch of *Camereiceras*.

Pseudobarrancyloceras VERMEULEN, LAZARIN, LÉPINAY, LEROY & MASCARELLI, 2013 [*P. companyi* Vermeulen & Lazarin, 2007; OD]. Moderately large; a tight criocone with whorls barely separated; strong, simple, slightly curved rectiradiate to slightly rursiradiate ribs, narrower than interspaces, with prominent umbilicolateral and lateral tubercles splitting into finer secondaries across dorsum, separated by fine non-tuberculate secondaries, with prominent ventrolateral clavi. *U. Barrem.* **Fig. 638K** - *P. companyi* (Vermeulen & Lazarin).

Ezeiceras VERMEULEN & BERT, 1998 [*Crioceras herberti* Fallot, 1884; OD]. Tight criocone with barely separated subquadratic whorls with distinct ventrolateral facets; ribs single, narrower than or equal to interspaces, with umbilical, high-lateral and ventrolateral tubercles; transverse ribs cross venter uninterrupted. *U. Barrem.* **Fig. 636I-K** - *E. herberti* (Fallot).

Antarcticoceras THOMPSON, 1924 [*A. antarcticum*; OD]. Small; in early growth like *Ezeiceras* with strong simple rursiradiate trituberculate ribs, but with whorls increasing rapidly in height; in later growth umbilicals shift to midflank before evanescing. *U. Barrem., ?L. Apt.* **Fig. 638M** - *A. antarcticum* Thompson.

Hemihoplites SPATH, 1924 [*Ammonites feraudianus* d'Orbigny, 1841; OD][**M**]. Large; evolute, compressed, with flat to gently convex subparallel flanks, subtabulate venter and subrectangular whorl section; in early to middle growth umbilical bullae give rise singly or in pairs to almost straight to slightly flexuous ribs that form weak ventrolateral bullae and cross venter with slight forward convexity; ribbing and tuberculation decline on adult body chamber of macroconchs that becomes almost smooth; suture with narrow E, broad rather narrow stemmed asymmetrically bifid E/L, narrow asymmetrically bifid L deeper than E, and fairly broad asymmetrically bifid L/U. *U. Barrem.* **Fig. 636L-M** - *H. feraudianus* (d'Orbigny), juv.

Matheronites RENNGARTEN, 1926 [*Ammonites soulieri* Matheron, 1879; OD][**m**]. Small, with perforated umbilicus, simple uniform flexuous slightly rursiradiate ribs narrower than interspaces, and strong ventrolateral tubercles that become clavate in maturity; suture like *Hemihoplites* but simpler. *U. Barrem. - Lower L. Apt.* **Fig. 636N-O** - *M. soulieri* (Matheron).

Said to be microconchs of *Hemihoplites*.

Pachyhemihoplites DELANOY, 1992 [*P. thomeli*; OD]. Small, evolute, moderately inflated, with flattened venter and slightly compressed trapezoidal whorl section; in early to middle growth with strong simple rectiradiate ribs and occasional intercalatories, with small umbilical, mid-lateral and ventrolateral tubercles; in maturity umbilical bullae strengthen, ribs coarsen and lateral tubercles evanesce. *U. Barrem.* **Fig. 636C-D** - *P. thomeli* Delanoy.

A *nomen nudum* fide Vermeulen *et al.* (2009).

Janusites BERT & DELANOY, 2000 [*Crioceratites janus* Thieuloy, 1979; OD] [**m**]. Small tight criocone with whorls almost in contact, sometimes slightly uncoiling in maturity; early whorls with suboctogonal to suboval whorl section, and simple trituberculate ribs; in middle growth tuberculate main ribs strengthen and simple non-tuberculate secondary ribs appear progressively, sometimes bifurcating at the umbilical shoulder; adult body chamber with almost circular whorl section and simple, closely spaced non-tuberculate ribs except for occasional pinched umbilical bullae. *Barrem.* **Fig. 636P** - *J. janus* (Thieuloy).

Perhaps a microconch of *Camereiceras* (Bert *et al.*, 2006) or *Pachyhemihoplites* (Bert & Bersac, 2012).

Parahemihoplites VERMEULEN, LAZARIN, LÉPINAY, LEROY & MASCARELLI, 2009 [*P. gerthiformis*; OD]. Small, evolute, compressed planispire with steep convex umbilical wall, weakly convex venter and compressed trapezoidal to quadratic whorl section; robust mainly single trituberculate ribs with spiny umbilical and clavate lateral and ventrolateral tubercles. *U. Barrem.* **Fig. 638L** - *P. gerthiformis* Vermeulen, Lazarin, Lépinay, Leroy & Mascarelli.

Subfamily Peirescitinae Bert, Delanoy & Bersac, 2006

Discussion. A seemingly monophyletic group with an origin in Gassendiceratinae, close to *Gassendiceras* (Bert *et al.*, 2006). According to Vermeulen *et al.* (2013) *Peirescites* is a subjective junior synonym of *Spinocrioceras* that they assign to Douvilleiceratidae. However *Peirescites* is a criocone with

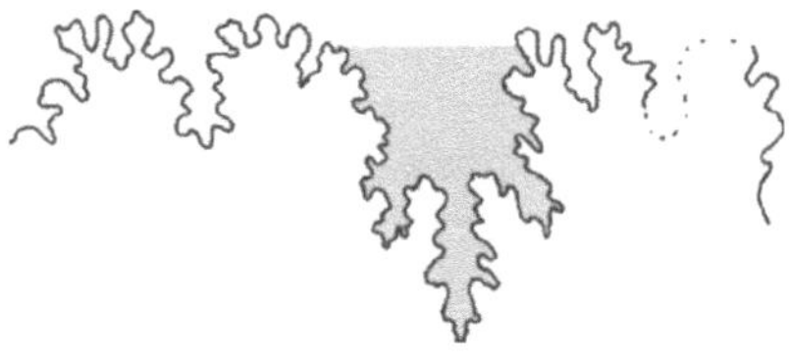

Fig. 641. *Spinocrioceras* suture.

umbilical and doubled ventrolateral tubercles of more or less equal size, and *Spinocrioceras* has a narrow trifid L to the suture, unlike the very broad bifid L typical of Douvilleiceratidae; typically very evolute criocones with coarse trituberculate primary ribs separated by 1-2 strong non-tuberculate intercalatories; tubercles umbilical, high lateral and ventrolateral in position, latter bullate. *U. Barrem. - L. Apt.*

Peirescites BERT, DELANOY & BERSAC, 2006 [*P. gygii;* OD]. Small criocone with whorls almost in contact and a subcircular whorl section; in early to middle growth with strong rectiradiate trituberculate primary ribs separated by 1-2 fine non-tuberculate secondaries; prominent incurved ventrolateral spines (subtriangular in cross section) in contact with succeeding whorl; in later growth secondary ribs obsolete and primaries slightly rursiradiate. *U. Barrem.* **Fig. 636G-H** - *P. gygii* Bert, Delanoy & Bersac.

Spinocrioceras KEMPER, 1973 [*S. polyspinosum*; OD]. Medium sized, very evolute, with whorls barely overlapping; broad flat venter and slightly depressed quadratic whorl section; after a short bituberculate stage all ribs strong, simple, thick and trituberculate, crossing venter uninterrupted, with tiny umbilical, lateral and ventrolateral tubercles; intercalatories rare. *Uppermost Barrem.* **Fig. 638I-J** - *S. polyspinosum* Kemper.

Superfamily ANCYLOCERATOIDEA Gill, 1871

(*nom. transl.* Klein *et al.*, 2007 *ex* Ancyloceratidae)

Discussion. A seemingly monophyletic group with an origin in Pedioceratoidea (Thomel, 1964). These are the "*ponderous heteromorphs*" of the Barremian and Aptian, comprising the families Ancyloceratidae, Tropaeidae and Lenicostitidae. *L. Barrem. - U. Apt.*

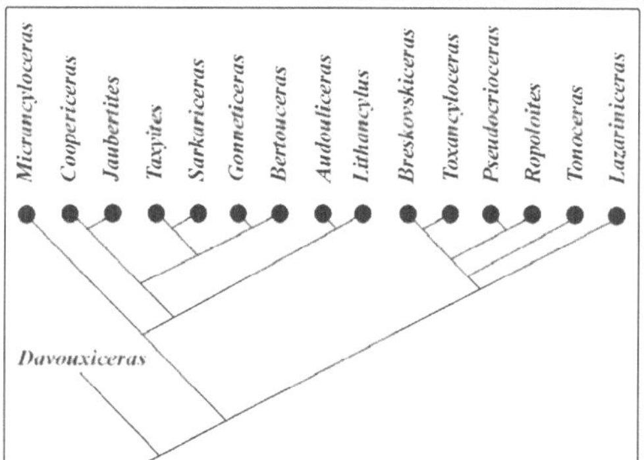

Fig. 642. Hypothesized relationships among Ancyloceratinae.

Family ANCYLOCERATIDAE Gill, 1871

Discussion. Monophyletic derivatives of Emericiceratidae (Thomel, 1964); typically very large to gigantic heteromorphs with rapidly expanding whorls; primitively ancylocones but later secondarily recoiled criocones and ammoniticones with very coarsely ribbed and tuberculate last part (crozier). The large number of genera assigned here indicate more than one phyletic line, and a need for taxonomic subdivision. *U. Barrem. - U. Apt.*

Subfamily Ancyloceratinae Gill, 1871

(*nom. transl.* Wright, 1957 *ex* Ancyloceratidae)

Discussion. A seemingly monophyletic group with an origin in Hauterivian Emericiceratinae (Thomel, 1964), close to *Davouxiceras*; characteristically large with initial small trituberculate criocone followed by J-shaped adult body chamber with coarsely ribbed and tuberculate flexus and retroversum without fine secondary ribs; complex suture with narrow trifid L. The number of genera assigned here indicates that more than one phyletic line is represented *L. U. Barrem. - U. Apt.*

Ancyloceras D'ORBIGNY, 1842 [*A. matheronianum*; SD Haug 1889]. Large; initial loose criocone followed by enlarged J-shaped body chamber; whorls expanding rapidly in height with subcircular to suboctagonal whorl section; in early growth strong prorsiradiate primary ribs with umbilical and inner and outer ventrolateral tubercles separated by 2-5 fine non-tuberculate secondaries (number increasing with growth); primary ribs may bifurcate across dorsum; long almost straight proversum with flat subparallel flanks, somewhat flattened dorsum and subquadratic whorl section; on flexus and retroversum secondary ribbing obsolete, umbilicals strongly bullate, and inner ventrolateral tubercles weaken and become obsolete before peristome, leaving distant single ribs; suture with asymmetrically bifid saddles, trifid L deeper than E and reduced trifid U smaller than trifid I. *U. Barrem. - L. Apt.* **Fig. 643A** - *A. matheronianum* d'Orbigny.

= *Hoploancyloceras* Nowak, 1913 (*nom. nud.*).

Sarkariceras VERMEULEN, 2006 [*S. thomeli*; OD]. Large ancylocone with small loose criocone followed by

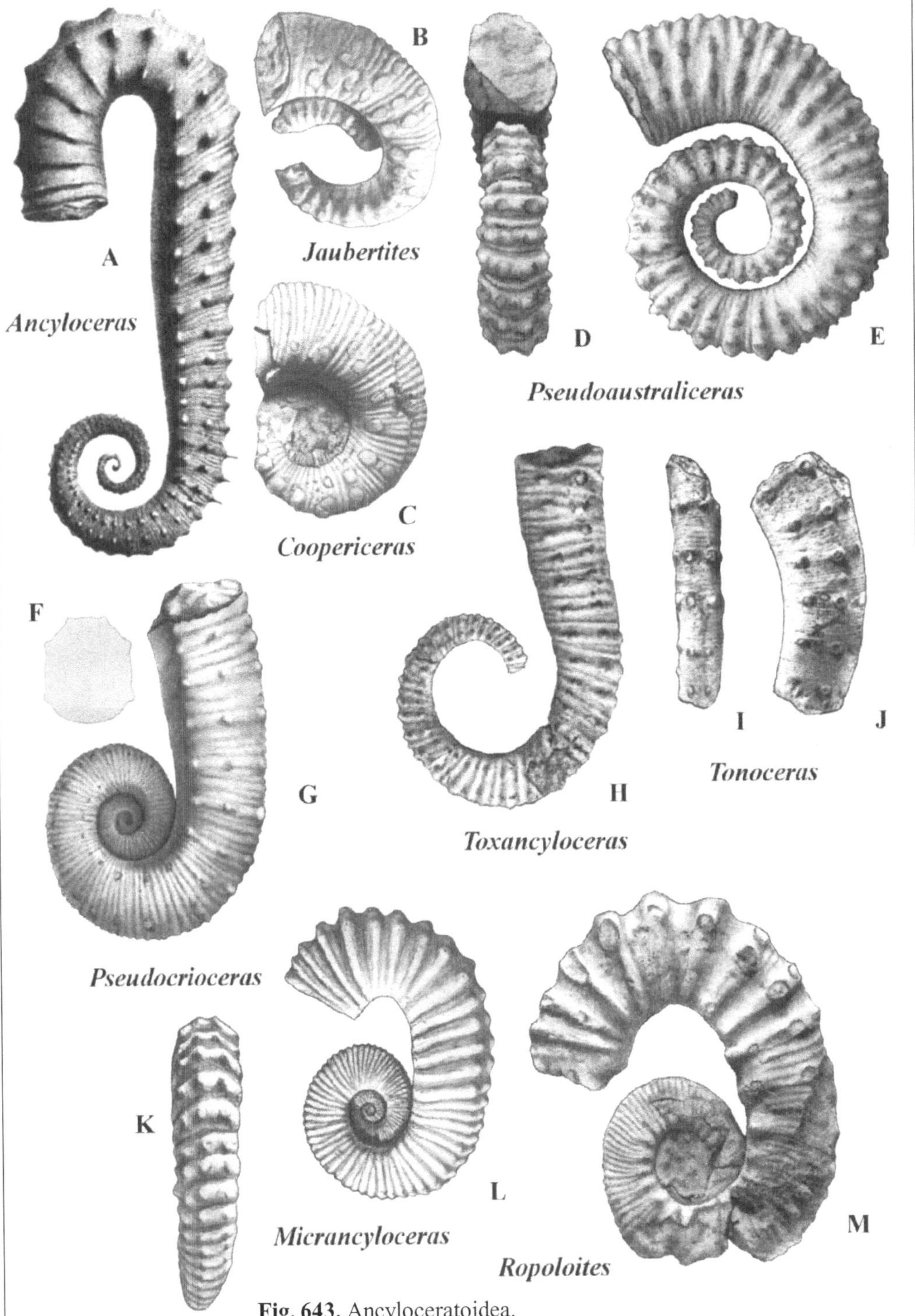

Fig. 643. Ancyloceratoidea.

rapidly expanding proversum and retroversum; spire trituberculate, tubercles dorsolateral, lateral and ventrolateral in position, separated by 2-3 fine non-tuberculate secondaries; on proversum rounded dorsolateral tubercles shift to lateral position, fine secondaries increase in number and loop between lateral and ventrolateral tubercles; retroversum short with coarse simple trituberculate ribs. *U. Barrem.* **Fig. 644H** - *S. thomeli* Vermeulen.

Jaubertites SARKAR, 1955 [*J. dubius*; OD]. Large ancylocone; spire a tight criocone with rapidly expanding subcircular to elliptical whorls; rursiradiate to rectiradiate trituberculate primary ribs with 1-3 ribs looped between large umbilical, lateral and ventrolateral septispinate tubercles, separated by fine non-tuberculate secondaries; ribs cross venter with adoral convexity; last part of spire and rapidly expanding proversum with fine slightly rursiradiate to slightly prorsiradiate fine ribs, as wide as interspaces; adoral part of proversum and flexus hypernodose with simple distant trituberculate ribs; retroversum short, with fine trituberculate and non-tuberculate ribs. *Upper U. Barrem.* **Fig. 643B** - *J. dubius* Sarkar.

= *Joubertites* Kakabadze, 1978 (*nom. van.*)

Berthouceras VERMEULEN, LÉPINAY & MASCARELLI, 2011 [*B. leroyi*; OD]. Medium sized criocone; whorls barely separate, with convex dorsum, flattened flanks and compressed elliptical whorl section; ribs fine, rectiradiate, periodically strengthened, some looped to ventrolateral tubercles; in later growth with rounded dorsolateral nodes and ribs weakening or effaced from venter; with growth ribs fine, and bituberculate primaries are separated by single intercalatories with variably developed ventrolateral nodes and ribs effaced from venter. *U. Barrem.* **Fig. 644E** - *B. leroyi* Vermeulen, Lépinay & Mascarelli.

Coopericeras VERMEULEN & LAZARIN, 2007 [*Ancyloceras (Adouliceras) cooperi* Klinger & Kennedy, 1977; OD]. Medium sized; planispirally coiled phragmocone with flattened but not impressed dorsum and subcircular whorl section; in early growth single low, broad, rectiradiate to slightly rursiradiate ribs with three rows of equal sized, large, septispinate spine bases separated by 1-2 thin non-tuberculate secondaries that cross venter transversely; with growth primary ribs widen and lateral and ventrolateral tubercles evanesce leaving bundles of 2-4 ribs arising from weak umbilical bullae. *Uppermost Barrem.?, L. Apt.* **Fig. 643C** - *C. cooperi* (Klinger & Kennedy).

Audouliceras THOMEL, 1964 [*Ancyloceras audouli* Astier, 1851; OD]. Very large ancylocone; spire with dense fine undifferentiated ribs which, periodically, loop to variably developed tubercles; proversum rapidly expanding towards highly inflated flexus, with fine prorsiradiate non-tuberculate ribs; strong distant prorsiradiate flank bulges with large midlateral tubercles and huge blunt ventrolateral horns develop around flexus; suture with large trifid L and U much smaller than I. *Upper U. Barrem. - L. Apt.* **Fig. 646A-C** - *A. audouli* (Astier).

= *Adouliceras* Klinger & Kennedy, 1977 (*nom. van.*).

Lithancylus CASEY, 1960 [*Hamites grandis* J. de C. Sowerby, 1828; OD]. Very large, apparently hamuliniform, a spire not having been recorded and, if present, very small; straight to slightly curved proversum and open flexus; proversum with non-tuberculate fine prorsiradiate ribs of irregular strength, replaced by coarse distant trituberculate ribs on flexus; suture highly incised, with very narrow cruciform trifid lobes. *U. Barrem. - Lower L. Apt.* **Fig. 644B** - *L. bucket* (Thieuloy); **Fig. 648A** - *L. grandis* (J. de C. Sowerby).

= *Lythancylus* Delanoy *et al.* (2008) (*laps. cal.*).

Taxyites DELANOY, MADOU, BARATERRO, PENAGÉ, E. & LHAUMET, G. 2008 [*T. agostinii*; OD]. Very large (D up to 500 mm); initial spire unknown; whorl section initially compressed oval becoming subcircular on proversum and flexus; low swollen trituberculate primary ribs separated by finer non-tuberculate secondaries on proversum; tubercles dorsolateral, lateral and ventrolateral in position; primaries split into 4-6 fine secondaries across dorsum; on flexus swollen primary ribs with exaggerated tubercles separated by 1-3 short intercalatories over venter. *Lower L. Apt.* **Fig. 644C** - *T. agostinii* Delanoy, Madou, Baraterro, Penagé & Lhaumet.

Micrancyloceras BUSNARDO, 2006 [*Ancyloceras brevis* d'Orbigny, 1842; OD]. Medium sized ancylocone, in early to middle growth very evolute with somewhat compressed whorl section; straight, uniform, regularly spaced rectiradiate ribs cross venter transversely without any signs of tuberculation. *Uppermost Barrem.* **Fig. 643K-L** - *M. breve* (d'Orbigny).

Toxancyloceras DELANOY, 2003 [*Ancyloceras vandenheckei* Astier, 1851; OD]. Initial loose criocone followed by long straight proversum with subcircular whorl section; strong generally looped trituberculate ribs separated by 1-4 thin non-tuberculate secondaries. *U. Barrem.* **Fig. 643H** - *T. vandenheckei* (Astier).

Breskovskiceras VERMEULEN & LAZARIN, 2007 [*B. anglesensis*; OD]. Very large, spire unknown but probably small criocone; proversum long gently curved, slowly expanding, with fine straight initially slightly rursiradiate ribs becoming slightly rectiradiate, bifurcating across dorsum; periodic shallow constrictions. *U. Barrem.* **Fig. 644D** - *B. anglesense* Vermeulen & Lazarin.

Gonneticeras VERMEULEN & LAZARIN, 2007 [*Ancyloceras renauxianus* d'Orbigny, 1842; OD]. Very large ancylocone; spire a loose criocone, long straight slowly expanding proversum with open flexus with compressed elliptical whorl section intercostally, octagonal costally, and short retroversum; spire and proversum finely ribbed initially trituberculate with 0-3 non-tuberculate secondaries; flexus with trituberculate primary ribs, tubercles evanescing on retroversum. *Middle L. Apt.* **Fig. 644A** - *G. renauxianum* (d'Orbigny).

Tonoceras HYATT, 1900 [*Ancyloceras duvalianum* d'Orbigny, 1842; OD]. Known by a criocone fragment with broad flanks and compressed ellliptical whorl section; primary ribs with low lateral and larger rounded high lateral tubercles projected forwards to subrounded ventrolateral tubercles, separated by 7-10 fine non-

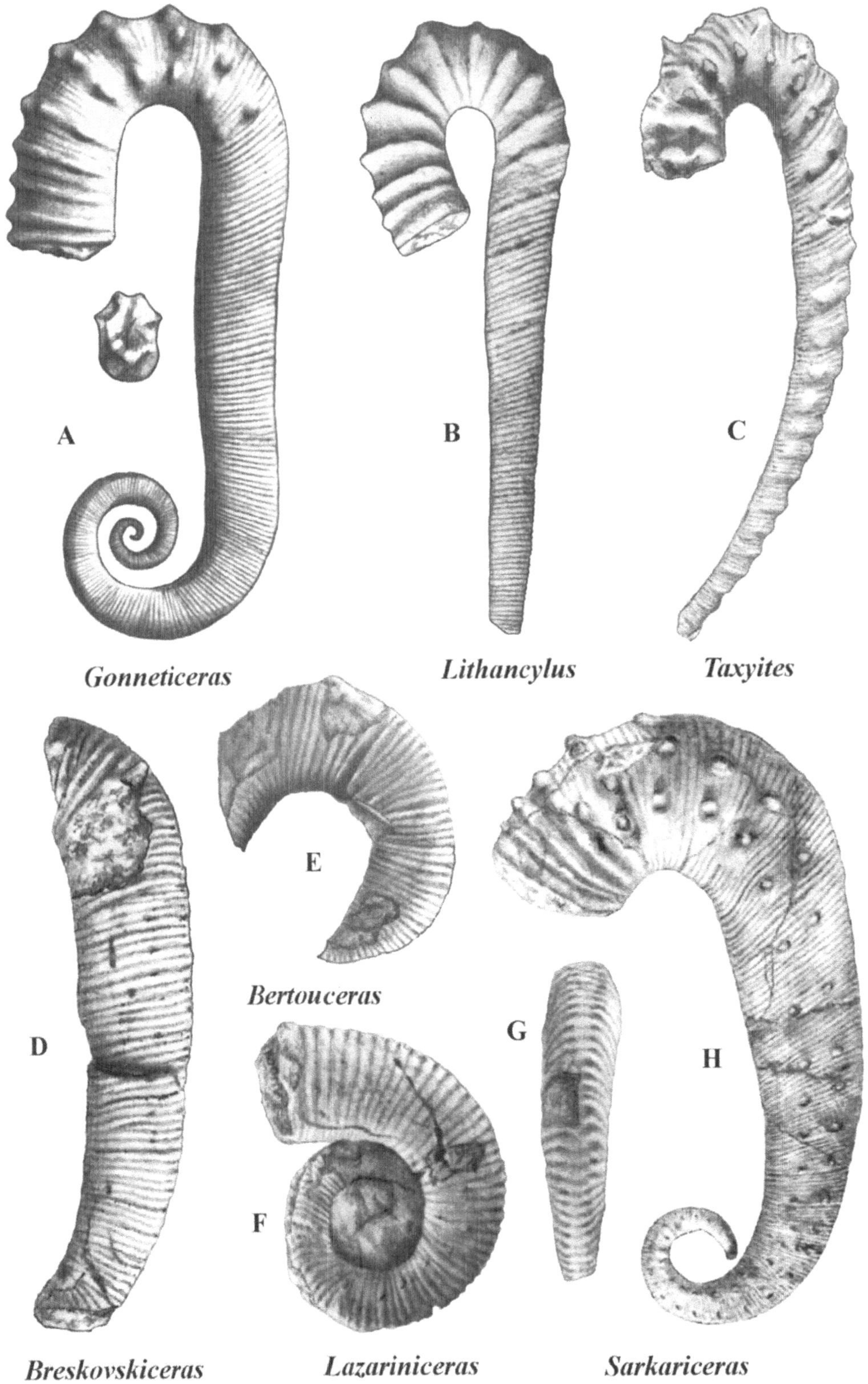

Fig. 644. Ancyloceratidae.

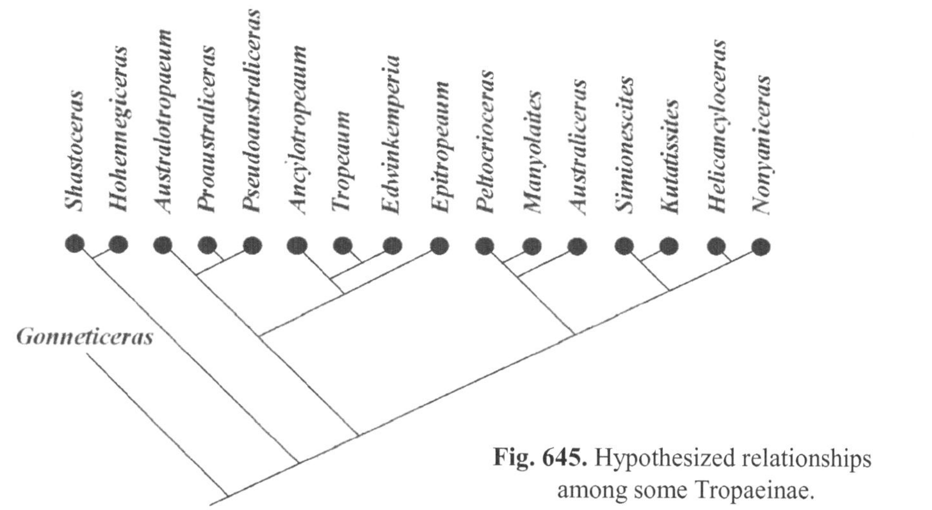

Fig. 645. Hypothesized relationships among some Tropaeinae.

tuberculate secondaries that cross dorsum. *L. Apt.* **Fig. 643I-J** - *T. duvalianum* (d'Orbigny).

Kilian & Reboul (1915) regarded it an Hauterivian *Crioceratites*.

Pseudocrioceras SPATH, 1924 [*Scaphites abichii* Simonovich & Batsevich, 1873; OD]. Large; inner whorls a very evolute ammoniticone with uniform, slightly prorsiradiate ribs, simple or bifurcating, periodically with small umbilical and inner and outer ventrolateral tubercles; very slightly curved proversum with convex dorsum, flat parallel flanks, distinct ventrolateral facets, subtabulate venter and slightly compressed subhexagonal whorl section; ribs coarsen on proversum when inner ventrolaterals shift to midflank, separated by 2-3 secondaries; suture with large trifid L deeper than E. *Upper U. Barrem. - Lower L. Apt.* **Fig. 643F-G** - *P. abichii* (Simonovich & Batsevich).

Ropoloites **gen. nov.** [*Ammonites fascicularis* d'Orbigny, 1840, p.117; OD herein]. Moderately large ancylocone with flattened dorsum and venter and compressed trapezoidal whorl section; phragmocone with prominent umbilical bullae giving rise to bundles of flexuous rursiradiate ribs; last part of flexus and short retroversum with bold, simple ribs with prominent inner and outer ventrolateral tubercles; weakening umbilical bullae become obsolete on retroversum. *Barrem.* **Fig. 643M** - *R. fascicularis* (d'Orbigny).

Lazariniceras VERMEULEN, LÉPINAY & MASCARELLI, 2011 [*L. rouvieri*; OD]. Medium sized criocone with flattened dorsum and subrectangular whorl section; mostly single rectiradiate to slightly rursiradiate and concave forward ribs, as wide as interspaces, strengthening towards venter; main ribs trituberculate, barely stronger than intercalatories that have ventrolateral tubercles; primary ribs with feeble dorsolateral bullae, weak, rounded and subspinose lateral tubercles and strong ventrolateral tubercles that are rounded and subspinose at the beginning of the last whorl but weaken aperturally; main ribs disappear at start of straight proversum that has fine uniform ribs with feeble ventrolateral nodes; flexus and retroversum unknown. *U. Barrem.* **Fig. 644F-G** - *L. rouvieri* Vermeulen, Lépinay & Mascarelli.

Family TROPAEIDAE nov.

Discussion. Monophyletic derivatives of Ancyloceratidae (Casey, 1960; Thomel, 1974). Many workers (Spath, 1931; Klinger & Kennedy, 1977) have commented on the difficulty of distinguishing fragments of *Australiceras* and *Tropaeum* between which, alleged, there are transitions, and McKenzie *et al.* (2014) synonymized the two. However the latter adopted a typological approach, recognizing that typical *Australiceras* was part of a lineage that evolved in parallel with European *Tropaeum* during the Late Aptian. Comprises the subfamilies Tropaeinae and Ammonoceratitinae. *Upper U. Barrem. – U. Apt.*

Subfamily Tropaeinae nov.

Diagnosis. Generally very large; initial toxicone or helicone followed by planispire; body chamber ancylocone (primitive) or criocone (derived); trituberculate in earliest growth, generally non-tuberculate in middle to later growth; adult body chamber a scaphitoid hook but, in some, adpressed to the phragmocone, with coarse distant robust single tuberculate ribs. Monophyletic derivatives of Ancyloceratinae close to *Gonneticeras* (Thomel, 1974). Although the subfamily seems to fall into distinctive groups, only the tribe Kutassitini is

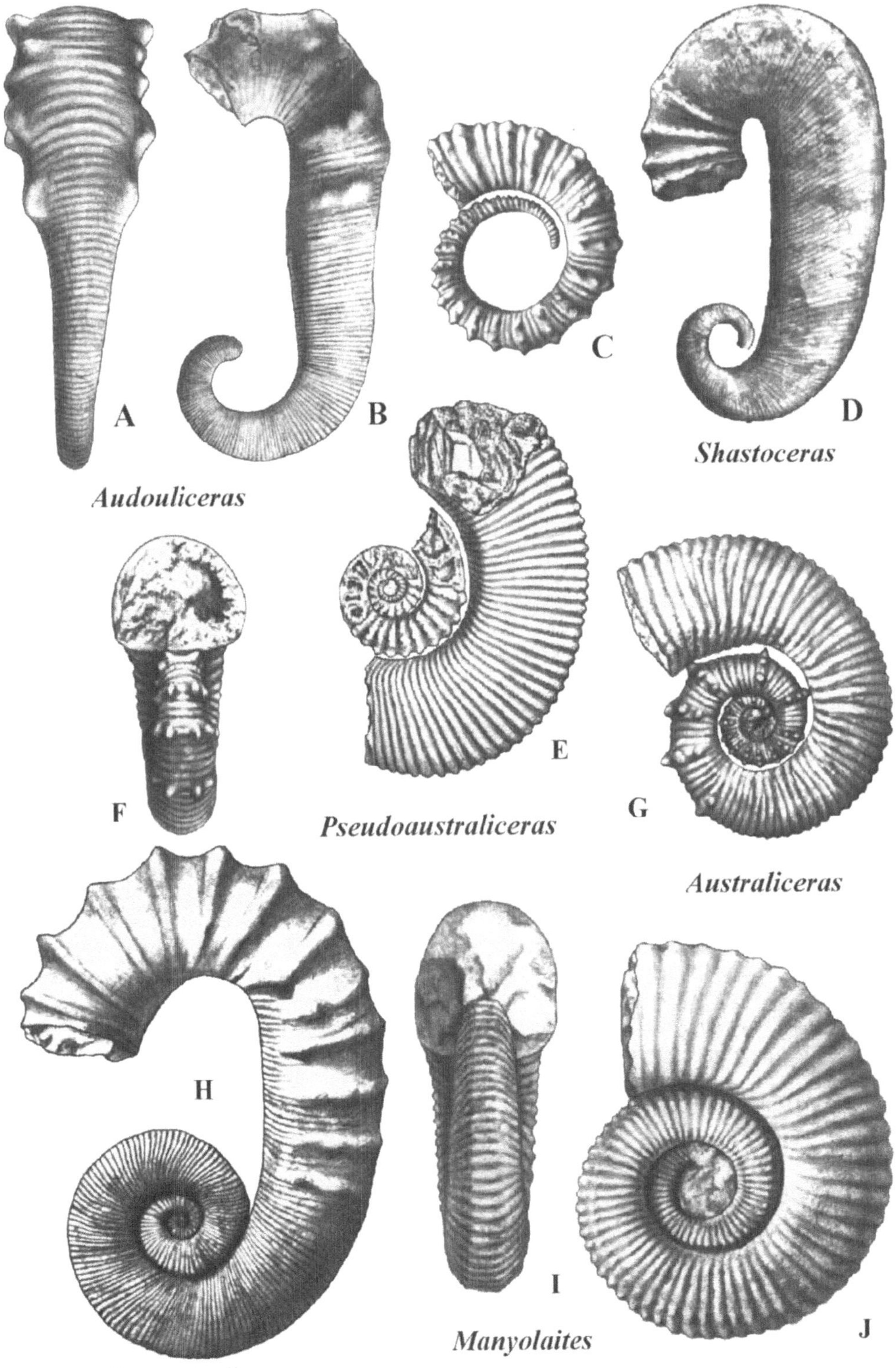

Fig. 646. Ancyloceratidae.

separated here, remaining subdivision must await testing of the hypothesized relationships which are complex. *Upper U. Barrem. - U. Apt.*

Tribe Tropeaini nov.

Type genus: *Tropeaum* J. de C. Soowerby, 1837.

Diagnosis. Very large to gigantic criocones, ancylocones and serpenticones with trituberculate earliest whorls and coarse blunt ribs to adult body chamber. *U. Barrem. - U. Apt.*

Hoheneggericeras DELANOY, BAUDOUIN, GONNET & BERT, 2008 [*Crioceras fallauxi* Uhlig, 1883; OD]. Moderately large; a very tight criocone increasing rapidly in height; in earliest growth (D<30 mm) trituberculate primary ribs separated by 1-2 non-tuberculate secondaries; primaries soon evanesce to leave most of spire and proversum with fine single non-tuberculate ribs, occasionally bifurcating on outer flank; trituberculate primaries reappear on last part of proversum, separated by fine non-tuberculate secondaries; flexus with very broad distant trituberculate ribs with large conical to clavate tubercles; short retroversum with simple, distant, non-tuberculate ribs. *Uppermost Barrem. - Lower L. Apt.* **Fig. 647, 648C** - *H. fallauxi* (Uhlig).

Shastoceras ANDERSON, 1938 [*S. californicum*; OD]. Poorly known; small criocone with slightly flattened dorsum and elliptical whorl section, becoming subcircular on flexus and short retroversum; curved proversum increases rapidly in height, with tight U-shaped flexus; fine dense prorsiradiate ribs weaken across venter and on early whorls carry small ventrolateral bullae that soon evanesce; flexus and retroversum mostly smooth, with few strong distant rounded primary ribs near final aperture. *L. Apt.* **Fig. 646D** - *S. californicum* Anderson.

Australotropaeum AGUIERRE-URRETA, 1985 [*Tropaeum (A.) magnum*; OD]. Very large (D=495 mm); in early to middle growth like *Tropaeum*, but uncoiling and expanding rapidly to become an aspinocone with depressed oval whorl section, shallowly concave dorsum and narrowly rounded venter, becoming slightly compressed in middle to later growth; in early growth trituberculate, but later with dense fine rectiradiate to slightly rursiradiate ribs, sometimes bifurcating from umbilical shoulder, crossing venter transversely; adult body chamber with 18 coarse simple ribs, narrower than interspaces, that split into striae on dorsum. *L. Apt.* **Fig. 651G-H** - *A. magnum* Aguierre-Urreta.

Fig. 647. *Hohennegiceras*.

Tropaeum J. DE C. SOWERBY, 1837 [*T. bowerbanki*; M]. Gigantic (up to 900 mm diameter), with tightly coiled phragmocone, inflated whorls, generally coarser ribbing, incipient umbilical bullae in middle growth, short curved body chamber barely separated from, or in contact with, phragmocone, and contracted adult aperture; suture highly incised with subrectangular E, narrow deep trifid L and much shorter U. *Upper L. - U. Apt.* **Fig. 650, 651F** - *T. bowerbanki* J. de C. Sowerby; **Fig. 651A** - *T. hillsi* (J. de C. Sowerby); **Fig. 654A-B** - *T. subsimbirskense compressum* (Sinzow).

***Edwinkemperia* gen. nov.** [*Tropaeum drewi spinosum* Kemper, 1964, p.36; OD herein]. Like *Tropaeum* but serpenticone with slowly expanding whorls barely in contact, rounded flanks and depressed reniform whorl section; primary ribs with prominent septispinate ventrolateral tubercles in early growth that shift to lateral position in middle growth, separated by non-tuberculate secondaries; adult body chamber slightly more than half whorl, remaining in contact with phragmocone, with bold simple uniformly spaced primary ribs. *U. Apt.* **Fig. 651D-E** - *E. spinosum* (Kemper).

Epitropaeum KAKABADZE, 1977 [*Tropaeum subarcticum* Casey, 1960; OD]. Very large serpenticone; inner whorls with flat sides, subtabulate venter and quadratic whorl section to about 120 mm diameter; in middle and later growth venter rounds and whorl section becomes depressed reniform; relatively coarse ribbing on phragmocone with ventrolateral tubercles to 55 mm diameter; adult body chamber with uniform ribbing tending to efface across venter, with much closer ribbing than in *Tropaeum*. *Upper U. Apt.* **Fig. 651B-C** - *E. subarcticum* (Casey).

Proaustraliceras KAKABADZE, 1977 [*Hamites gigas* J. de C. Sowerby, 1828; OD]. Large ancylocone; phragmocone a planispire with whorls in contact; body chamber uncoiling to form crozier with moderately short curved proversum; in middle growth spire with trituberculate primaries separated by finer non-tuberculate secondaries; last whorl of spire and initial proversum non-tuberculate, with fine flexuous rursiradiare ribs; strong trituberculate ribs appear on proversum alternating with thinner non-tuberculate ribs, with strongly ribbed dorsum; flexus and retroversum with strong single trituberculate ribs. *L. Apt.* **Fig. 646H** - *P. gigas* (J. de C. Sowerby).

Pseudoaustraliceras KAKABADZE, 198 [*Crioceras ramososeptatum* Anthula, 1899; OD]. Like *Proaustraliceras*

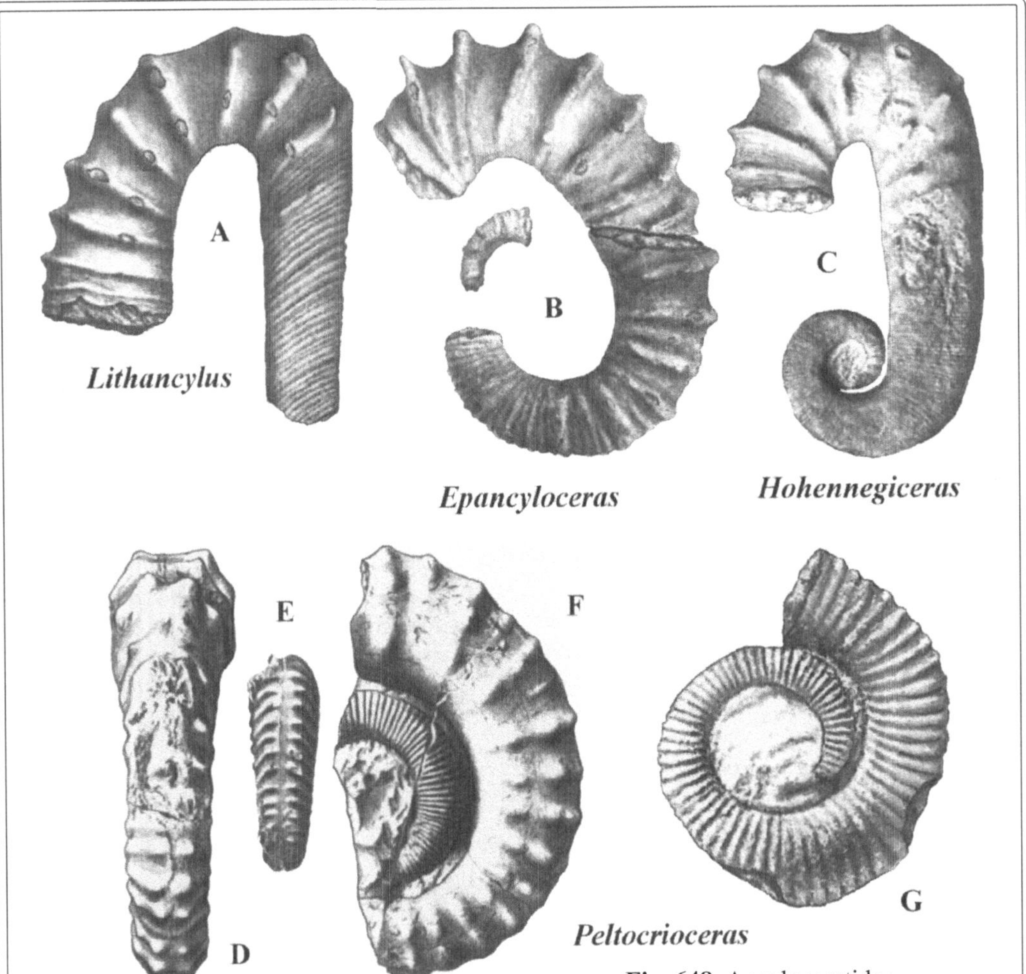

Fig. 648. Ancyloceratidae.

but with very short straight proversum with flattened dorsum and subtrigonal whorl section; spire with prominent strongly trituberculate ribs separated by thread-like rib at umbilical shoulder; body chamber with uniform fine sometimes bifurcating ribs; suture highly frilled, with very narrow trifid L longer than E and long cruciform I. *Lower U. Apt.* **Fig. 643D-E, 646E** - *P. ramososeptatum* (Anthula).

Ancylotropaeum CASEY, 1980 [*Tropaeum (A.) baylissi*; OD]. Very large ancylocone; phragmocone with flattened dorsum, gently convex convergent flanks and compressed suboval whorl section; dense fine straight slightly rursiradiate ribs, some bifurcating or trifurcating on flank, replaced on massive adult body chamber by distant strongly flared non-tuberculate to feebly trituberculate ribs with wide concave interspaces; suture with very narrow cruciform L. *Lower U. Apt.* **Fig. 651J-K** - *A. baylissi* (Casey).

Australiceras WHITEHOUSE, 1926 [*Crioceras jackii* Etheridge, 1880; OD]. Enormous; a tightly coiled criocone throughout, with whorls almost touching; dorsum shallowly concave to flat, venter narrowly rounded and intercostal whorl section subquadratic (juv.) to subtrigonal (ad.), but octagonal costally; for first 3-4 whorls robust prominently trituberculate primary ribs separated by 2-4 finer non-tuberculate secondaries, ventrolateral tubercles strongest and persisting longest, thereafter all ribs non-tuberculate; body chamber with ribs as wide as interspaces, losing definition; suture with broad stemmed asymmetrically bifid saddles and large trifid L. *Lower U. Apt.* **Fig. 646F-G** - *A. jackii* (Etheridge).

Australiceras lampros (Etheridge), which has been included in the synonymy of *A. jackii* (McKenzie *et al.*, 2009), is based on an adult body chamber with distant single trituberculate ribs.

***Manyolaites* gen. nov.** [*Australiceras wandalina* (Boshoff MS) Klinger & Kennedy, 1977, p.297; OD herein]. Like *Epitropaeum* but whorl section intially compressed suboctagon becoming subrectangular later; undifferentiated simple rectiradiate ribs, all with umbilical, high-

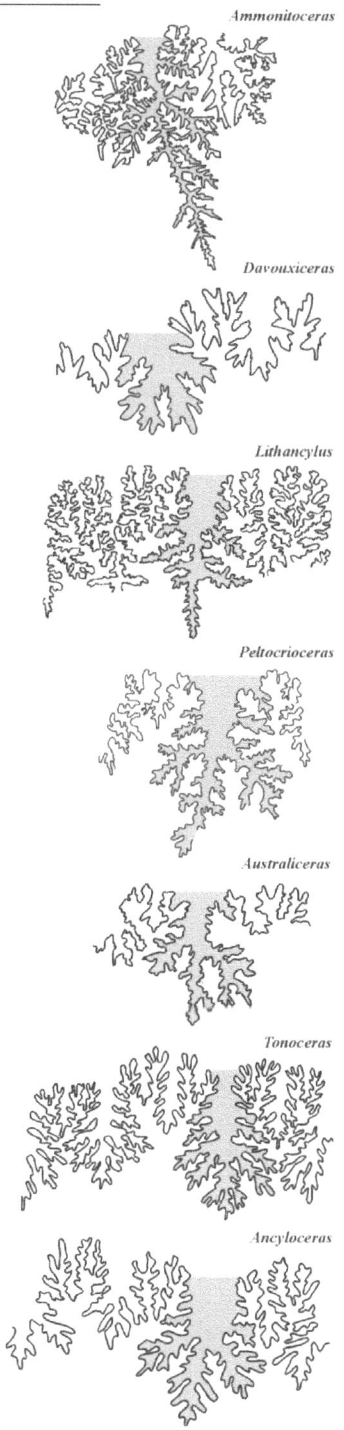

Fig. 649. Ancyloceratoid sutures.

lateral and ventral bullae persist to large size; umbilical tubercles evanesce first, followed by laterals but ventral tubercles persist; suture with very large asymmetrically bifid L. *U. Apt.* **Fig. 646I-J** - *M. wandalina* (Klinger & Kennedy).

Fig. 650. *Tropaeum.*

Peltocrioceras SPATH, 1924 [*Crioceras deeckei* Favre, 1908; OD]. Gigantic (D>1 m); early whorls a loose criocone with whorl section intially compressed suboctagonal becoming subrectangular later; undifferentiated simple rectiradiate ribs, all with umbilical, high-lateral and ventrolateral bullae persisting to large size, interrupted across siphonal line; adult body chamber with coarse distant trituberculate ribs with rounded tubercles; suture with narrow asymmetrically bifid L. *U. Apt.* **Fig. 648D-G** - *P. deeckei* (Favre).

Tribe Kutatissitini nov.

Type genus: *Kutatissites* Kakabadze, 1970.

Diagnosis. Initial helix followed by trituberculate criocone or ancylocone; convergent towards Heteroceratinae but with typical ancyloceratid sculpture and shape of whorls in later growth. A seemingly monophyletic group of uncertain ancestry, perhaps close to *Ancyloceras*. *L. - U. Apt.*

Kutatissites KAKABADZE, 1970 [*K. bifurcatus*; OD]. Medium sized; initial helicone enclosed by planispiral whorl; adult growth stages of type species unknown but some referred species with uncoiled shaft and hook; strong trituberculate main ribs separated by non-tuberculate secondaries; with growth umbilical tubercles strengthen, whereas lateral and ventrolateral tubercles weaken and may evanesce; ribs interrupted between ventrolateral tubercles and thin over dorsum; suture with very deep wide L and narrow E/L. *Lower L. Apt.* **Fig. 652** - *K. creutzbergi* Kakabadze & Hoedemaeker; **Fig. 654K-L** - *K. bifurcatus* Kakabadze.

There is substantial variation in the shell shape of species of *Kutatissites* and, although generic separation of ancylocones from criocones has been suggested (Stoykova, 1992), this is made difficult by the uncertain adult characters of the type species.

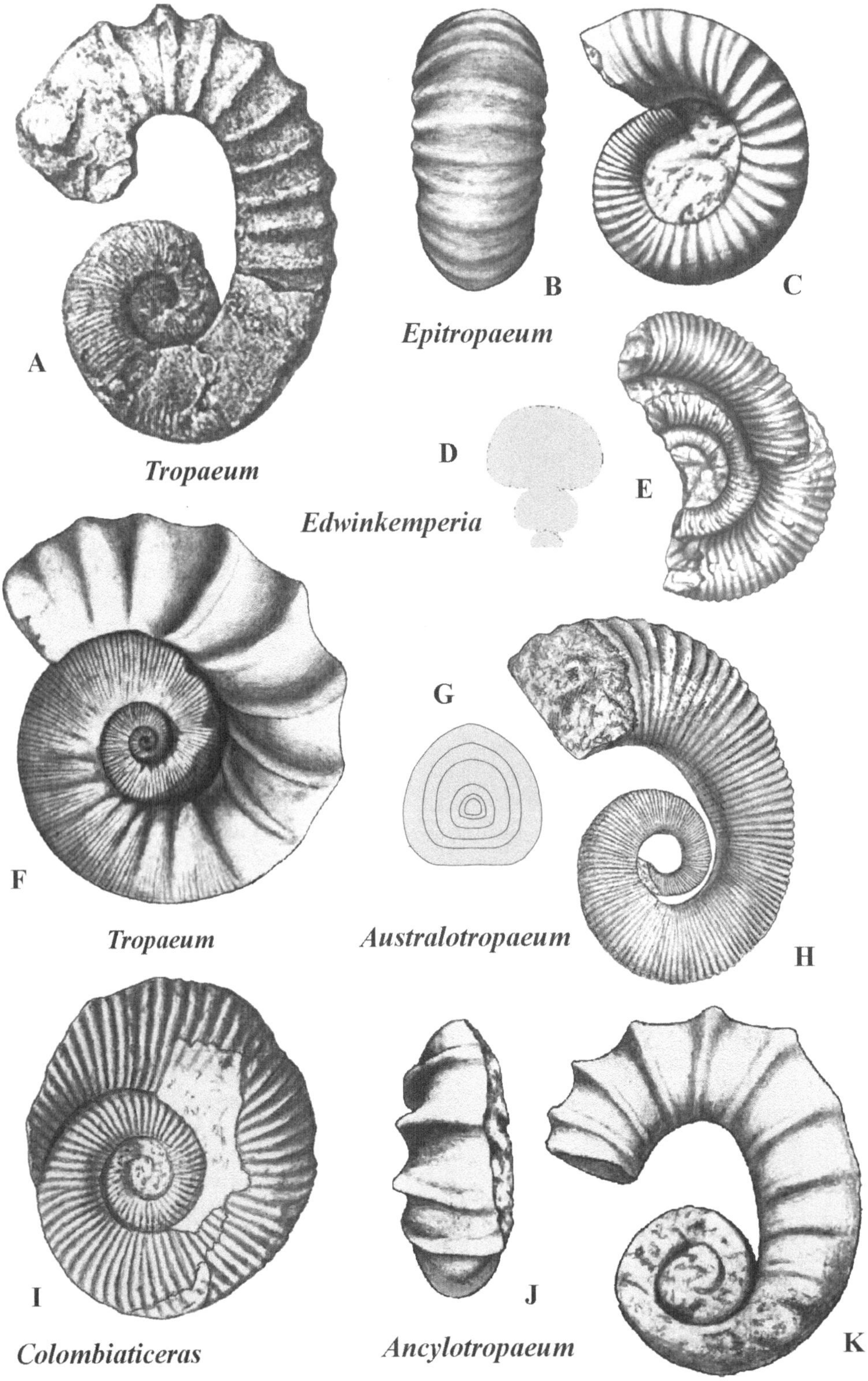

Fig. 651. Ancyloceratidae.

Fig. 652. *Kutatussites.*

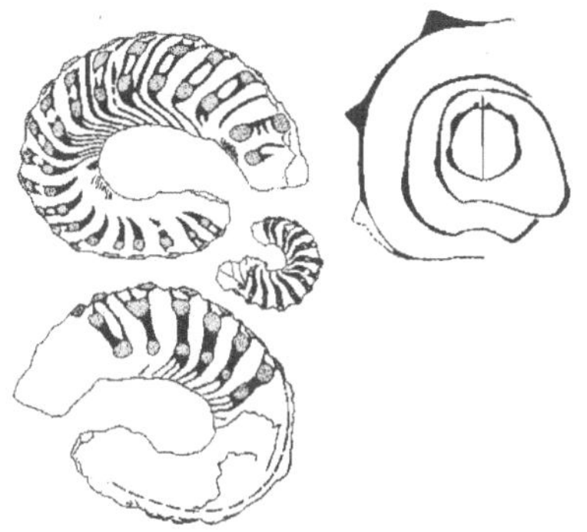

Fig. 653. *Simionescites* rib pattern and whorl section (after Avram, 1976).

Simionescites AVRAM, 1976 [*S. simionescui*; OD]. Very large (D=260 mm); intial helicone unknown; in middle growth a loose ellipticone with rapidly expanding whorl enclosed by planispiral body chamber; in middle growth whorl section subcircular, becomes slightly depressed, subquadratic in maturity with feebly concave venter; prominent fine rectiradiate ribs looped between auriculate umbilical bullae, lateral tubercles and ventrolateral clavi, crossing venter obliquely where interruped by a shallow siphonal furrow; on outer whorl lateral tubercles shift to dorsolateral position and ventrolaterals to lateral position. *Lower L. Apt.* **Fig. 653, 654J** - *S. simionescui* Avram.

Perhaps a macroconch of *Kutatissites*.

Nonyaniceras KLINGER & KENNEDY, 1977 [*Helicancyloceras (N.) nonyani*; OD]. Relatively small; early low helical spire of 1-2 whorls followed by planispiral coiling; whorl section initially subcircular, later oval; spire with trituberculate main ribs (umbilical tubercles may be very weak) separated by weaker non-tuberculate secondaries; ribs oblique across venter and somewhat broadened and flattened over siphonal line. *U. Apt.* **Fig. 654H-I** - *N. nonyani* (Klinger & Kennedy).

Helicancyloceras KLINGER & KENNEDY, 1977 [*Heteroceras (Argvethites?) vohimaranitraensis* Collignon, 1962; OD]. Like *Nonyaniceras* but larger; spire with strong simple rounded annular ribs, broadening ventrally; ribs weak or interrupted across siphonal line with small ventrolateral tubercles on either side; on planispiral whorls tuberculation obsolete and ribbing becomes finer, rursiradiate and crosses venter transversely; suture like *Australiceras*, with large trifid L. *U. Apt.* **Fig. 654E-G** - *H. vohimaranitraense* (Collignon).

Perhaps a macroconch of *Nonyaniceras*. For those who regard them as synonyms *Helicancyloceras* is chosen as having priority.

Subfamily Ammonitoceratinae nov.

Diagnosis. Large to very large criocones, body chamber sometimes hooked; initially trituberculate but most of phragmocone bituberculate; primary ribs with umbilical and lateral tubercles separated by non-tuberculate secondaries. Seemingly monophyletic derivatives of uncertain origin, perhaps *Lithancylus* (Casey 1960) evolving in parallel with Tropaeinae. *L. - U. Apt.*

Caseyites DELANOY, BAUDOUIN, PICTET, DELATTRE & PENAGE, 2017 [*C. esteronensis*; OD]. Very large ancylocone with straight to slightly curved proversum, open flexus and short retroversum; spire bituberculate with dorsolateral bullae giving rise to 2-3 secondaries separated by 0-2 intercalatories; spire with subcircular whorl section; ribs strengthen as flexus is approached but intercalatories disappear; ribs on retroversum non-tuberculate. *Upper L. Apt.* **Fig. 654C** - *C. esteronensis* Delanoy, Baudouin, Pictet, Delattre & Penage.

Perhaps a microconch of *Ammonitoceras* (Bersac & Bert, 2019).

Epancyloceras SPATH, 1930 [*E. hythense*; OD]. According to Casey (1961) the holotype is teratological; earliest whorls (D<50 mm) with strongly raised trituberculate primary ribs separated by 3 non-tuberculate secondaries; strong ventrolateral tubercles evanescing beyond 60 mm diameter; in middle growth lateral bullae give rise to paired secondaries; strongly arched proversum, flexus and very short retroversum with trituberculate ribs separated by deeply concave interspaces that appear in early growth. *Upper L. Apt.* **Fig. 648B** - *E. hythense* Spath.

Ammonitoceras DUMAS, 1876 [*A. ucetiae*; OD]. Very large with phragmocone whorls almost in contact and body chamber forming part of loose criocone; whorl section depressed and subcoronate; inner whorls with broad bold primary ribs connecting umbilical and high lateral tubercles, separated by 1-2 fine non-tuberculate intercalatories; high-lateral tubercles give rise to 3 fine secondaries across venter; earliest whorls may have

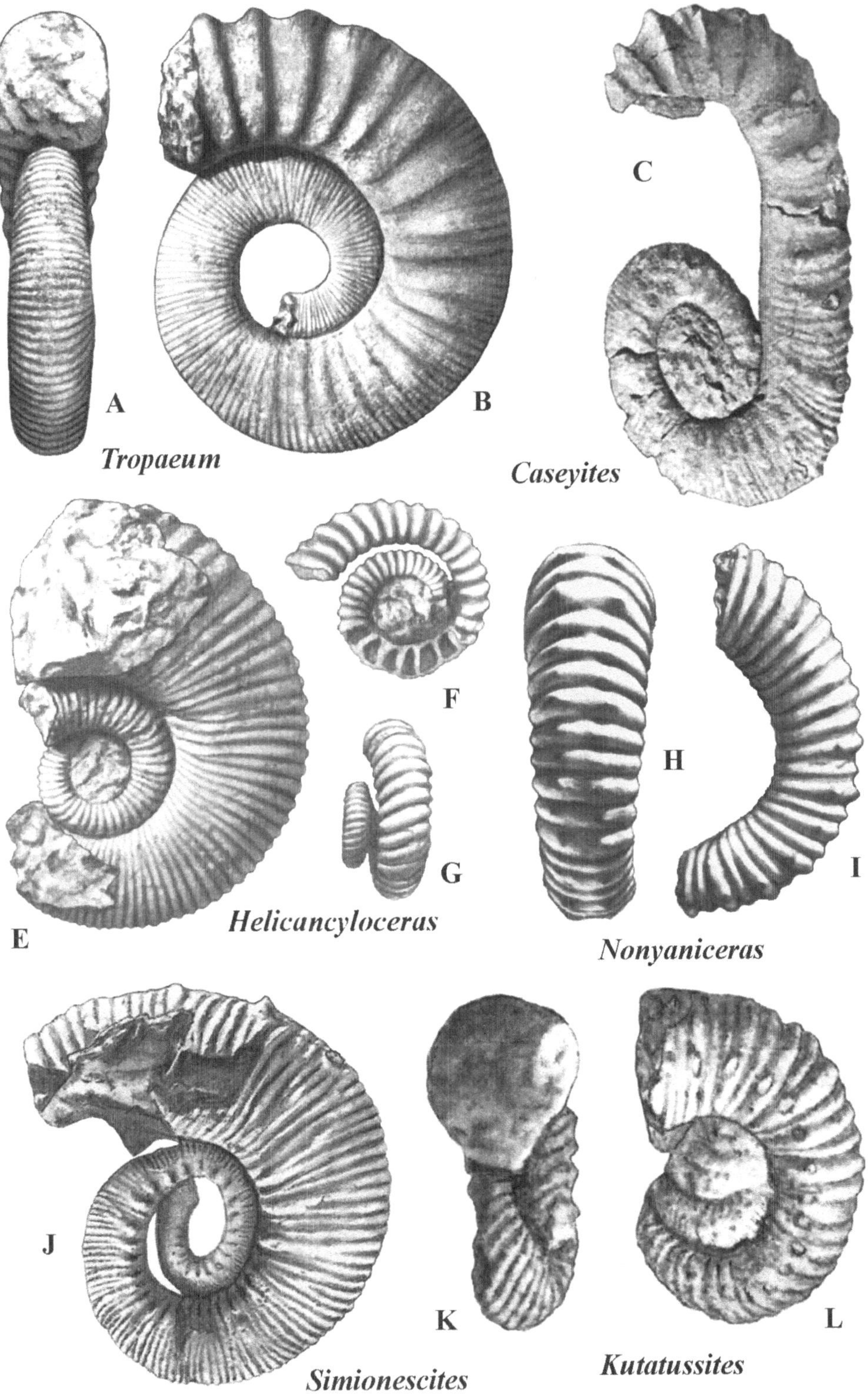

Fig. 654. Ancyloceratidae, Tropaeidae.

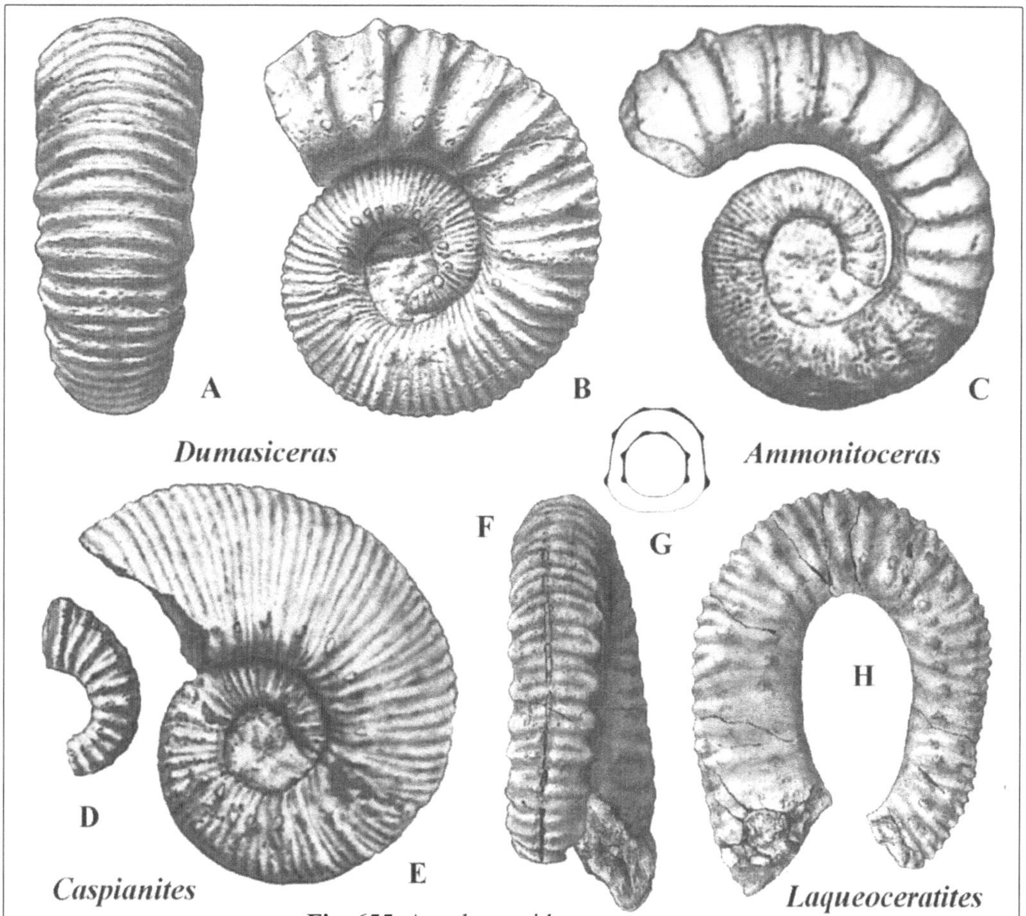

Fig. 655. Ancyloceratidae.

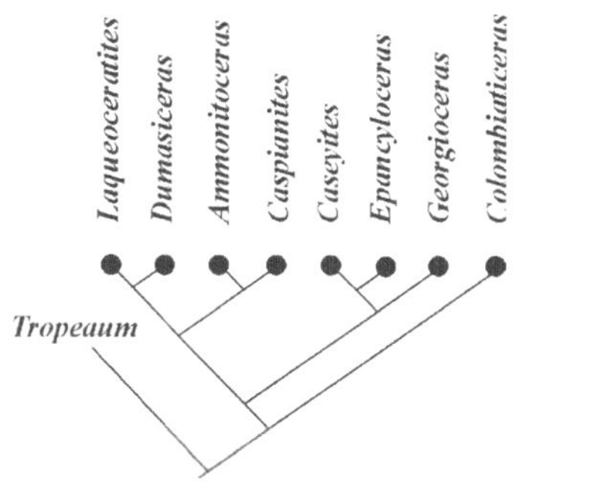

Fig. 656. Hypothesized relationships among Ammonitoceratinae.

feeble ventrolateral tubercles; all secondaries and intercalatories lost on massively expanded adult body chamber when flared primaries become simple and distant. *Lower U. Apt*. **Fig. 655C** - *A. ucetiae* Dumas.

Caspianites CASEY, 1961 [*Crioceras (Ammonitoceras) wassiliewskyi* Renngarten, 1926; OD]. Earliest whorls trituberculate, with umbilical perforation; ribs mostly branching from umbilical bullae, but first lateral, then ventrolateral and finally umbilical tubercles evanesce; ribs fine and uniform in maturity. *U. Apt*. **Fig. 655D-E** - *C. wassiliewskyi* (Renngarten).

***Dumasiceras* gen. nov.** [*Crioceras lahuseni* Sinzow, 1906, p.192; OD herein]. Like *Ammonitoceras* but a recoiled ammoniticone; earliest whorls unknown; phragmocone with prominent umbilical bullae giving rise to pairs fine slightly rursiradiate (juv.) to slightly prorsiradiate (ad.) ribs separated by 2-3 fine secondaries; ribs coarsen in maturity when secondaries evanesce leaving strong distant primary ribs bifurcating from weak ventrolateral bullae; suture with very long narrow cruciform L. *U. Apt*. **Fig. 655A-B** - *D. lahuseni* (Sinzow).

***Laqueoceratites* nom. nov.** [*Laqueoceras laqueus* Kakabadze & Hoedemaeker, 2004; OD herein]. Large, early growth stages unknown; adult whorl a low helicoid ellipticone with whorl section changing from subcircular to subquadratic and subtrapezoidal; on adapical half distant main ribs with umbilical and lateral tubercles, the latter giving rise to paired secondaries looped across broad low venter, occas-

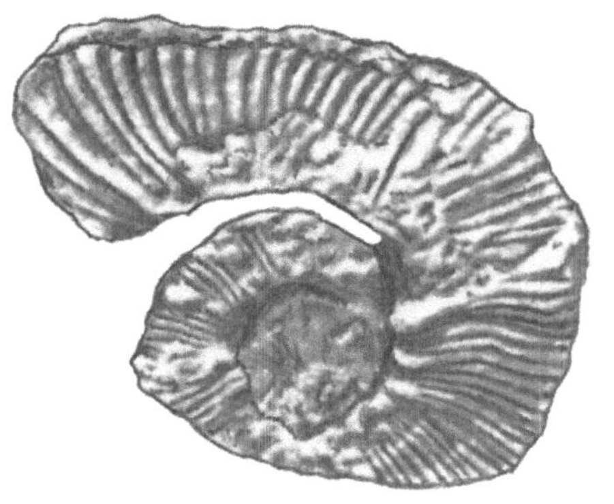

Fig. 657. *Georgioceras.*

ionally separated by a thin, non-tuberculate intercalatory; on adoral half (body chamber) only bifurcating, bituberculate, distant main ribs. *?Lower U. Apt.* **Fig. 655F-H** - *L. laqueus* (Kakabadze & Hoedemaeker).

= *Laqueoceras* Kakabadze & Hoedemaeker, 2004, *non* Lange, 1925 (Psiloceratidae).

Georgioceras WILKENS, 1947 [*G. kohllarseni*; OD]. Medium sized criocone with open umbilicus; following a smooth earliest stage (D>10 mm) straight thin ribs that curve backwards appear with umbilical and lateral nodes, sometimes on a broad rib, looping in 2-4 between nodes with occasional bifurcation, separated by simple or bifurcating intercalatories; in middle growth (D>35 mm) lateral nodes evanesce and umbilicals elongate; tuberculate primaries separated by secondaries, followed by a stage in which fine bundled ribs spring from weak umbilical bullae; in later growth with strong single prorsiradiate flank costae that recurve on outer flank. *U. Apt.* **Fig. 657** - *G. kohllarseni* Wilckens.

Colombiaticeras ROYO Y GOMEZ, 1945 [*C. bolivari*; OD]. Moderately large; very tight criocone with whorls almost in contact, rounded umbilical shoulder, broad gently convex flanks, rounded venter and slightly compressed coronate to turbinate whorl section; ribs simple, fine, rectiradiate to slightly prorsiradiate, about as wide as interspaces, with rare long intercalatories; in early growth with tiny umbilical and ventrolateral tubercles on alternate ribs. *U. Barrem.* **Fig. 651I** - *C. bolivari* Royo y Gomez.

Family LENICOSTITIDAE Bert, Bersac & Canut, 2020

Discussion. A monotypic taxon of uncertain ancestry, the perforated umbilicus suggesting assignment to the Ancyloceratoidea, but there is a strong resemblance to Hemihoplitidae. Diagnosis as for genus. *Lower U. Barrem.*

Lenicostites BERT, BERSAC & CANUT, 2020 [*Hemihoplites rusticus* Vermeulen, 1996; OD]. Small (**m**) to medium sized (**M**) planispire with steep umbilical wall, and wide shallow umbilicus, perforated in earliest growth; flanks and venter convex, subcircular whorl section tending to heighten with growth; thick undifferentiated ribs mostly simple, rectiradiate to slightly rursiradiate, sometimes a little sinuous with thin intercalatory ribs present on robust morphotypes; in early growth ribs of uniform width but in maturity thicken on outer flank and venter; ribs show incipient umbilical and ventrolateral thickening without forming distinct bullae. *Lower U. Barrem.* **Fig. 658** - *L. rusticum* (Vermeulen).

Hyporder DOUVILLEICERATOIDEI nov.

Discussion. A monophyletic taxon of contentious ancestry. Wiedmann (1962) and Vašíček & Wiedmann (1994) favoured an origin in Ancyloceratoidei. However Kakabadze (1967) looked to Protancyloceratidina for the origin of the group with Dimitrova (1970) suggesting *Aegocrioceras* as the likely progenitor, whereas Vermeulen (2006) favoured *Moutoniceras* (Toxoceratinae) as the antecedent. Although prevailing opinion seems to favour an origin in Protancyloceratidina this is dependent on the position of *Toxoceras* that is referred to theo Ancyloceratidae by Delanoy & Magnum (1991) and Ebbo & Delanoy (1999). Except primitively the group comprises recoiled heteromorphs belonging to the superfamilies Deshayesitoidea, Douvilleiceratoidea and Parahoplitoidea. *L. Barrem. - M. Alb.*

Superfamily DESHAYESITOIDEA Stoyanow, 1949

(*nom. transl.* Wiedmann, 1966 *ex* Deshayesitidae)

Discussion. A seemingly monophyletic group with an origin in Toxoceratoidea, comprising the

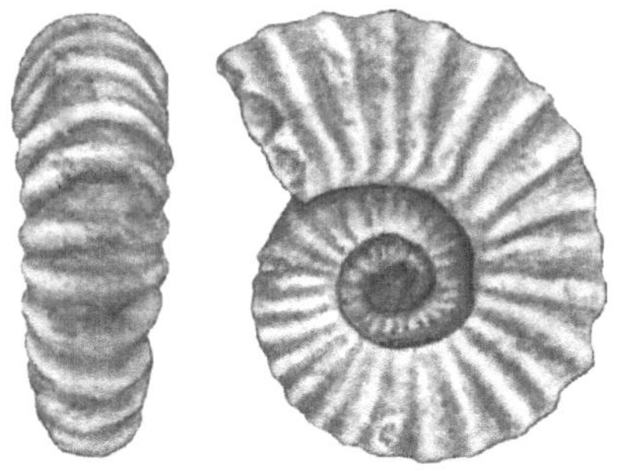

Fig. 658. *Lenicostites.*

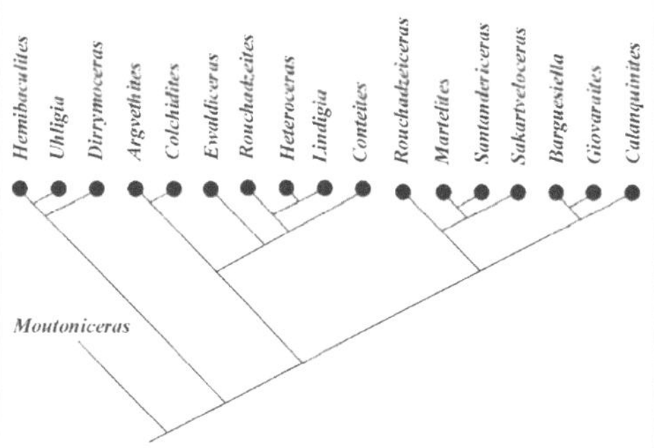

Fig. 659. Hypothesized relationships among Heteroceratidae.

primitive heteromorph family Heteroceratidae and the secondarily recoiled ammoniticone Deshayesitidae. *Upper L. Barrem. - Uppermost Apt.*

Family HETEROCERATIDAE Spath, 1922

Discussion. A monophyletic group perhaps with an origin in Toxoceratidae (Kakabadze, 1967; Dimitrova, 1970; Vermeulen, 2006); comprises only the family Heteroceratinae but evolutionary relationships are complex and several phyletic lines are involved (Delanoy, 1997). Typically with initial helix followed by straight or curved proversum and hook in different plane, often at right angles. Dimorphism conspicuous but not well documented; macroconchs allegedly with more numerous flat sided planispiral whorls increasing rapidly in height. *Upper L. Barrem. - L. Apt.*

Subfamily Heteroceratinae Spath, 1922

(*nom. transl.* Kakabadze, 1967 *ex* Heteroceratidae)
= Colchiditinae Kakabadze, 1967
= Epicolchiditinae Egoian, 1976 (*nom. nud.*)

Discussion. A seemingly monophyletic group with an origin in Toxoceratidae, close to *Moutoniceras* (Sarkar, 1955; Kakabadze, 1967; Vermeulen, 2006), although Vašíček & Wiedmann (1994) looked to Leptoceratoidinae close to *Eoheteroceras* and Wright (1996) favoured some unidentified helicancyline. Kakabadze (1967) separated Heteroceratinae, without planispiral whorls, and Colchiditinae with planispiral whorls, but later abandoned this notion (Kakabadze, 2004). Subsequently (Kakabadze, 2015) he distinguished 5 fundamental modes of coiling that may have phylogenetic significance - heterocone (*H. elegans*), longiheterocone (*H. baylei*), heterotrianglicone (*H. moulladei*), colchicone (*C. colchidus*) and colchicriocone (*Egoianiceras mikhaili*). Comprises the tribes Heteroceratini and Martelitini. *U. Barrem.*

Tribe Heteroceratini Spath, 1922

(*nom. transl.* herein ex Heteroceratinae)
Type genus: *Heteroceras* d'Orbigny, 1849.

Diagnosis. Medium sized; typically with initial helix, that was very small in primitive members, followed by long straight to curved proversum and J-shaped hook. The intraspecific variation and possibilities of dimorphism were discussed at length by Kakabadze (2004, 2015). *Upper L. - Uppermost Barrem.*

Ewaldiceras VERMEULEN, 2003 [*E. eigenheeri*; OD]. Enormous (D=500-600 mm); uncoiled spire with long proversum not expanding rapidly; whorl section compressed, oval; in early growth bituberculate main ribs separated by non-tuberculate secondaries; in middle growth ribs uniform, thickened and nodate on upper flank; proversum with strong primary ribs with ventrolateral clavi and rare lateral tubercles separated by 0-3 simple non-tuberculate secondaries; flexus with irregular strong ribs with 1-3 tubercles, alternate or not, with intercalatories. *Upper L. Barrem.* **Fig. 660N** - *E. eigenheeri* Vermeulen.

Heteroceras D'ORBIGNY, 1849 [*Ammonites astierianus* d'Orbigny; SD d'Orbigny, 1851]. Initial high helix of 4-8 whorls with axis oblique to proversum that increases rapidly in height; U-shaped flexus and relatively long subparallel retroversum; simple annular ribs in early growth; fine flexuous ribs on proversum and flexus commonly bifurcate on upper flank and coarsening on flexus; retroversum with simple rursiradiate to rectiradiate ribs. *U. Barrem.* **Fig. 660A** - *H. astierianum* (d'Orbigny).
= *Atopoceras* Jaubert, 1889 (*nom. nud.*).
= *Epiheteroceras* Egoian, 1976 (*nom. nud.*)

Dirrymoceras HYATT, 1900 [*Ancyloceras simplex* d'Orbigny, 1842, OD]. Earliest stages unknown; later like *Heteroceras* but with compressed oval (?distorted) whorl section and coarser straighter ribs that frequently bifurcate low on flank, do not coarsen on retroversum, and are convex forward and looped across dorsum. *U. Barrem.* **Fig. 660L-M** - *D. simplex* (d'Orbigny).

Rouchadzeites DELANOY & BERT, 2006 [*R. doli*; OD]. Like *Heteroceras* but with coarser ribs that rarely bifurcate on the proversum, and long curved body chamber not expanding so markedly and wrapping loosely around initial helix. *U. Barrem.* **Fig. 662J-K** - *R. doli* Delanoy & Bert.

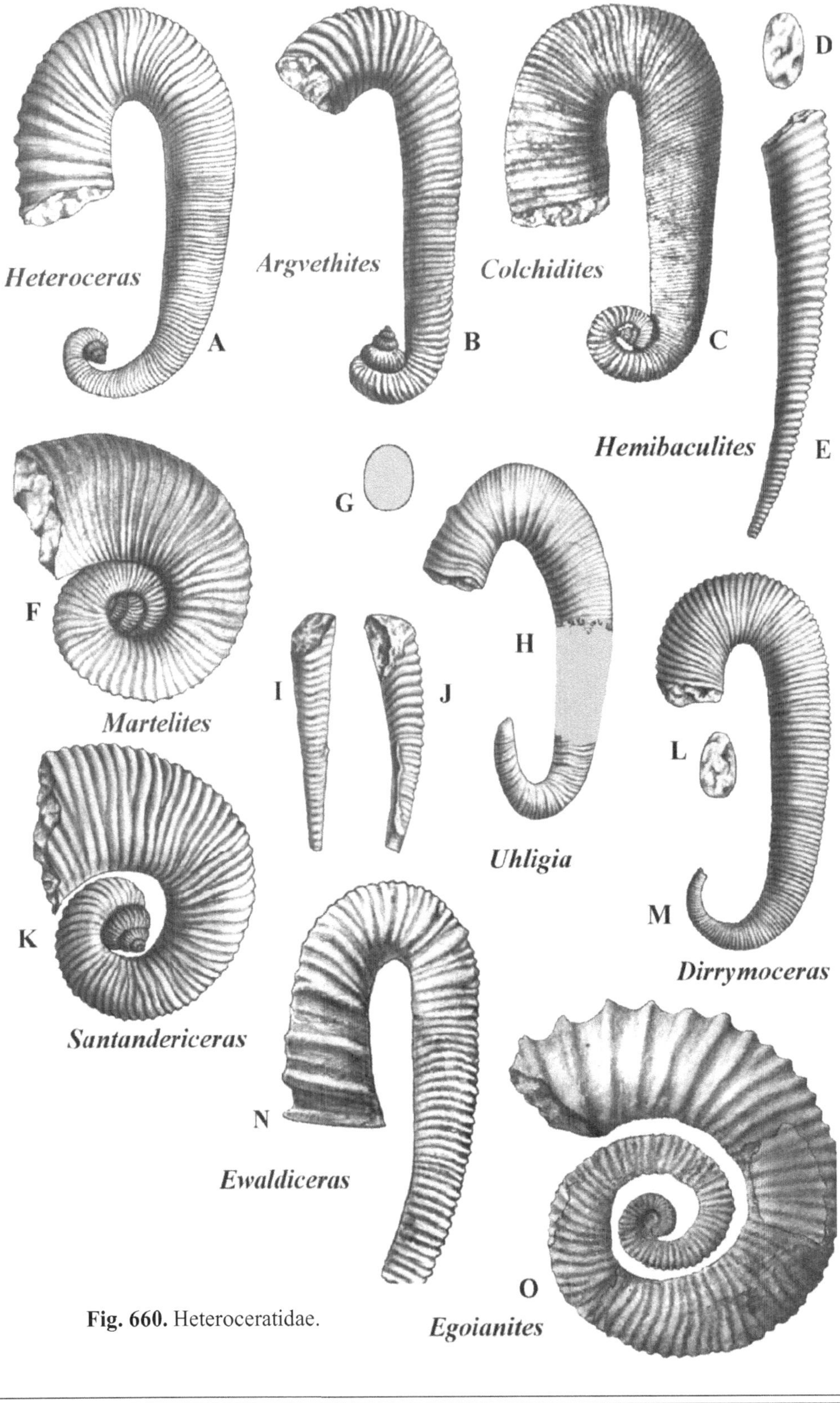

Fig. 660. Heteroceratidae.

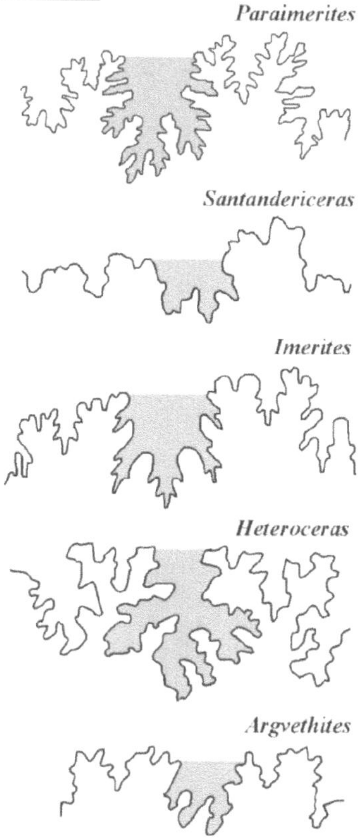

Fig. 661. Heteroceratid sutures.

Lindigia KARSTEN, 1858 [*L. helicoceroides*; OD]. Small, like *Heteroceras* but with plane of proversum almost perpendicular to axis of helix; long straight slender proversum expanding rapidly; ribs form chevrons over venter. *U. Barrem.* **Fig. 662H-I** - *L. helicoceroides* Karsten.

Questionably a microconch of *Heteroceras* (Aguirre Urreta & Klinger, 1986).

Hemibaculites HYATT, 1900 [*Toxoceras obliquatus* d'Orbigny, 1842; OD]. Long gently curved torticone showing rapid expansion in height, and compressed elliptical whorl section; regular simple strong slightly prorsiradiate annular ribs weaken across dorsum; intercalated ribs lacking; no evidence of initial helix. *Barrem.* **Fig. 660D-E** - *H. obliquatus* (d'Orbigny).

The type is lost, the suture unknown and its stratigraphical level imprecise, and hence impossible to assign either to Heteroceratidae or Ancyloceratidae (Autran *et al.*, 1986); a *nomen dubium*.

Uhligia VON KOENEN, 1904 [*Crioceras minutum* Neumayr & Uhlig, 1888; OD]. Based on tiny weakly curved shaft, much of which is body chamber; like *Hemibaculites* but with ribbing differentiated into stronger primaries and weaker secondaries; ribs recurve on outer flank to cross venter transversely where they are much broader than interspaces, with every third or fourth seeming to swell across siphonal line; in maturity with elliptical coiling, J-shaped body chamber and fine differentiated ribbing initially showing bifurcating ribs separated by an intercalatory; number of intercalatories increases around flexus and coarsens on retroversum. *L. Barrem.* **Fig. 660G-J** - *U. minuta* (Neumayr & Uhlig).

Conteites DELANOY, FRAU, BUSNARDO & TENDIL, 2018 [*Heteroceras fuhrae* Delanoy, 1997; OD]. Medium sized, with relatively large helix with compressed subrectangular whorl section; helix with sinuous oblique ribs over venter, weakening across shallow siphonal groove; proversum with simple prorsiradiate ribs, narrower than interspaces; flexus with bifurcating ribs; retroversum short. *Uppermost Barrem.* **Fig. 662L-M** - *C. fuhrae* (Delanoy).

Colchidites DJANÉLIDZÉ, 1926 [*C. colchicus*; OD]. Like *Kakabadzia* but with helical spire of 4-7 whorls (sometimes more), followed by 1-2 planispiral whorls; rapidly expanding proversum normal to axis of helix, with dense fine prorsiradiate ribs, irregularly bifurcating; ribs coarsening and regularly bifurcating on flexus, simplifying on relatively long retroversum. *Upper U. Barrem.* **Fig. 660C** - *C. colchicus* Djanélidzé.

Argvethites ROUCHADZÉ, 1933 [*Heteroceras (A.) lashense*; OD]. Long straight proversum, not expanding so rapidly, perpendicular to relatively large spire; ribs relatively coarse, rarely bifurcating on proversum, with ventrolateral tubercles; crozier frequently with siphonal depression. *U. Barrem.* **Fig. 660B** - *A. lashensis* (Rouchadzé).

Tribe Martelitini nov.

Type genus: *Martelites* Conte, 1989.

Diagnosis. Medium sized; typically with final whorl barely or not separated from helix. Derivatives of Heteroceratini (Rouchadzé, 1933) close to *H. coulleti* Delanoy (Delanoy, 1997), thus evolving in parallel with Heteroceratini. *Upper Barrem.*

Egoianites DELANOY & BERT, 2006 [*E. mikhaili*; OD]. Moderately large (D=210 mm); small helix with axis at 50° to slowly expanding criocone of two whorls with compressed oval whorl section; simple rursiradiate ribs, rarely bifurcating, coarsening markedly on adult body chamber, with ventrolateral clavi. *U. Barrem.* **Fig. 660O** - *E. mikhaili* Delanoy & Bert.

A *nomen nudum* fide Vermeulen *et al.* (2014).

Sakartveloceras DELANOY, FRAU, BUSNARDO & TENDIL, 2018 [*Ammonites securiformis* Simonovich, Batesevich & Sorokin, 1875; OD]. Medium sized, with relatively small helix of 2-3 whorls tightly enclosed by two elliptical whorls of planispire; ribs mostly single and bifurcating, with some intercalatories, interrupted across siphonal line in youth but crossing venter in maturity; planispiral

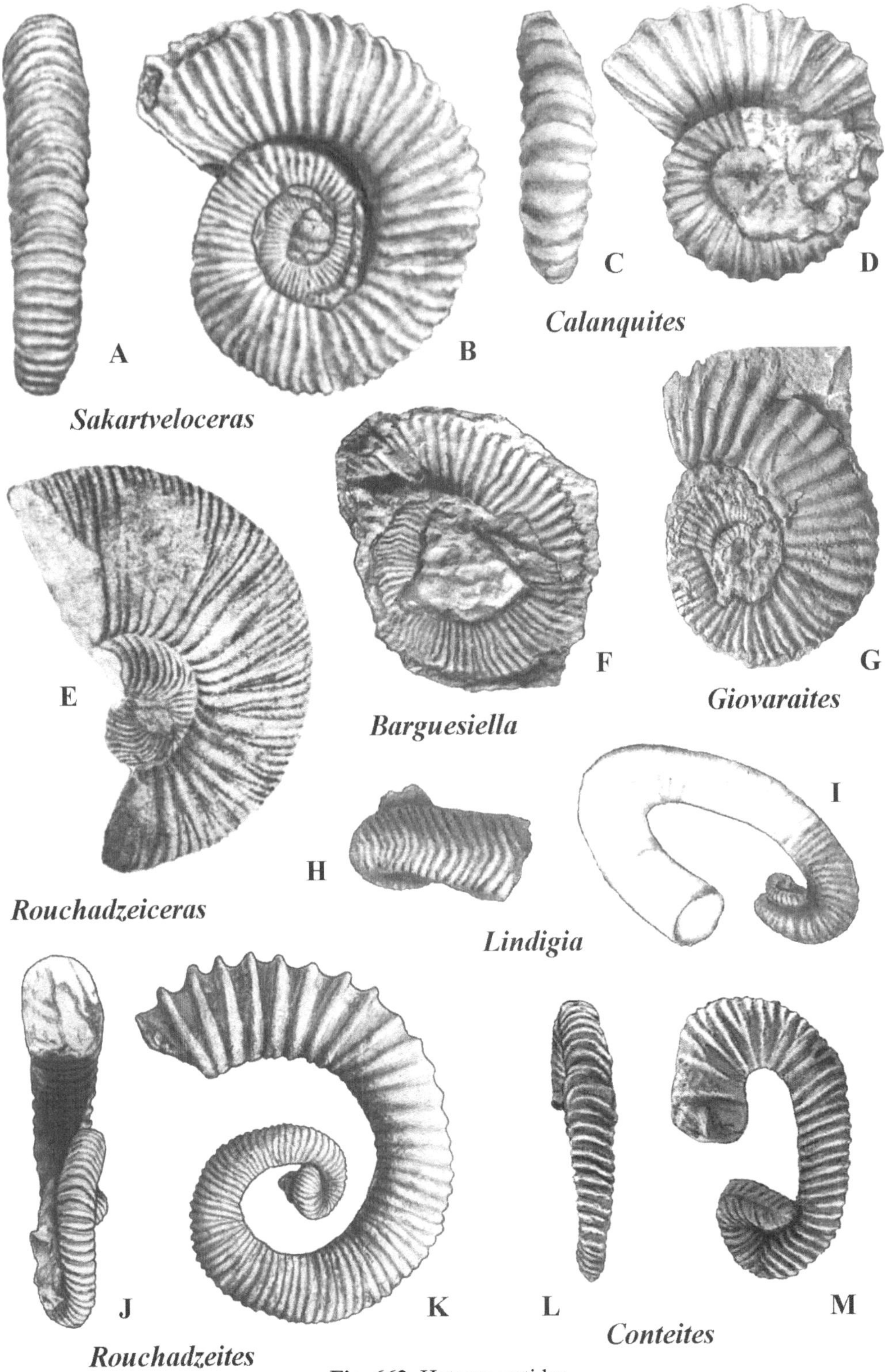

Fig. 662. Heteroceratidae.

whorls seem to show incipient mid-lateral bullae. *Uppermost Barrem.* **Fig. 662A-B** - *S. securiforme* (Simonovich, Batesevich & Sorokin).

Santandericeras ROYO Y GOMEZ, 1949 [*Heteroceras (S.) apolinari*; OD]. Initial helix of 4-5 whorls just separated from rapidly expanding elliptically coiled body chamber; axis of helix almost perpendicular to planispire; ribs mostly single, annular on helix but slightly flexuous with some bifurcation and trifurcation on body chamber where ribs do not coarsen markedly. *Upper U. Barrem.* **Fig. 660K** - *S. apolinari* (Royo y Gomez).

Martelites CONTE, 1989 [*M. marteli*; OD]. Like *Santandericeras* but larger, with helix of 2-7 whorls (sometimes more), axis of helix in more or less same plane as planispire; 1-2 tightly coiled planispiral whorls in contact with helix. *Uppermost Barrem.* **Fig. 660F** - *M. marteli* Conte.

= *Epicolchidites* Egoian, 1976 (*nom. nud.*).

***Rouchadzeiceras* gen. nov.** [*R. rouchadzei* sp. nov., = *Colchidites* sp. indet., Rouchadzé 1933: 183, pl. 4, fig. 4; OD herein]. Like *Martelites* but helix with simple, fine, flexuous ribs; body chamber with fine ribs in biplicate, fasciculate and virgatome bundles. *U. Barrem.* **Fig. 662E** - *R. rouchadzei* sp. nov.

Calanquites FRAU, DELANOY, MASSE, LANTAUME & TENDIL, 2016 [*Imerites katsharavai* Djanélidzé, 1933; OD]. Small, strongly compressed, whorls slowly expanding; whorl section compressed subrectangular on phragmocone, subelliptical on body chamber; small helix of 4-6 whorls enclosed by 1-2 planispiral whorls; helix with dense, sinuous simple ribs developing small ventrolateral tubercles on last whorl of helix; ribs of adult whorl thickening on ventrolateral shoulders, and terminal part of body chamber tends to be tightly uncoiled. *Uppermost Barrem.* **Fig. 662C-D** - *C. katsharavai* (Djanélidzé).

Barguesiella FRAU, DELANOY, MASSE, LANTAUME & TENDIL, 2016 [*B. goudesense*; OD]. Small, with small to medium helix enclosed by one elliptial planispiral whorl; helix with slightly flexuous simple ribs crossing venter uninterrupted; after helix whorl section suboval; body chamber with simple, bifurcating and intercalated ribs, straight or slightly flexuous, thickening markedly. *Uppermost Barrem.* **Fig. 662F** - *B. goudesensis* Frau, Delanoy, Masse, Lantaume & Tendil.

Family DESHAYESITIDAE Stoyanow, 1949

(*nom. transl.* Wright, 1955 *ex* Deshayesitinae)

Discussion. A monophyletic group with an origin in Heteroceratidae (Tovbina, 1963); comprises the subfamilies Deshayesitinae and Dufrenoyiinae. Suture with lobe regeneration by subdivision of saddle U_2/I (Wiedmann, 1970). *U. Barrem.- U. Apt.*

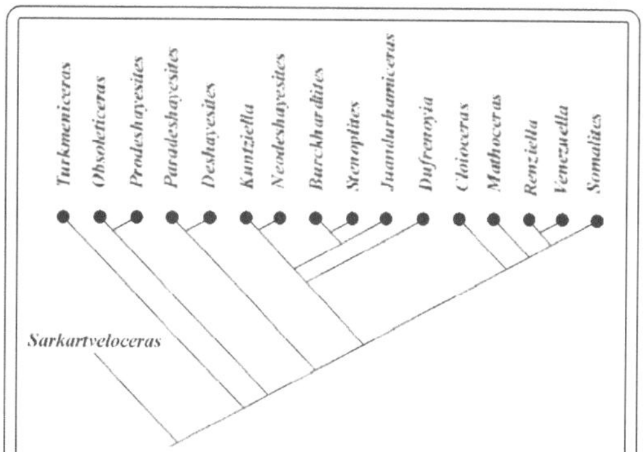

Fig. 663. Hypothesized relationships among Deshayesitidae.

Subfamily Deshayesitinae Stoyanow, 1949

Diagnosis. Typically evolute, compressed, with strong flexuous ribs that may weaken across subtabulate venter. Dimorphism conspicuous. Restricted to the Boreal and Tethyan realms, comprising monophyletic derivatives of Heteroceratinae, close to *Sakartveloceras* (Delanoy *et al.*, 2018). *U. Barrem. - L. Alb.*

Turkmeniceras TOVBINA, 1962 [*T. turkmenicum*; OD]. Medium sized with perforate umbilicus; compressed, evolute, with steep umbilical wall, broad flat subparallel flanks, evenly rounded venter and subelliptical whorl section; strong sinuous primary ribs flex forward near venter where bifurcation and intercalation occurs; suture with broad shallow lobes. *U. Barrem.- basal Apt.* **Fig. 664A-B** - *T. turkmenicum* Tovbina.

Prodeshayesites CASEY, 1961 [*Ammonites fissicostatus* Phillips, 1829; OD]. Like *Turkmeniceras* but umbilicus imperforate; umbilical bullae give rise to ribs that bifurcate low on flank, with intercalatories, and form chevrons over venter, weakening across siphonal line; suture with broad E/L and wide asymmetrically trifid L. *Lower L. Apt.* **Fig. 664C-D** - *P. fissicostatus* (Phillips).

Obsoleticeras BOGDANOVA & MIKHAILOVA, 1999 [*Prodeshayesites obsoletus* Casey, 1964; OD] [**M**]. Small (**m**) to large (D_{max} = 250 mm), like *Prodeshayesites* but with high rapidly expanding whorls with rounded venter and oval sections; in early to middle growth weak but distinct prosocline umbilical bullae give rise to feeble flexuous flat topped single and bifurcating ribs; beyond 100 mm diameter ribbing obsolete, reappearing on adult body chamber with distant strong straight ribs; suture with deeply incised bifid E/L and very broad asymmetrically trifid L much deeper than E. *Lower L. Apt.* **Fig. 665** - *O. obsoletum* (Casey).

Paradeshayesites KEMPER, 1967 [*Hoplitides laeviusculus* von Koenen, 1902; OD]. Like *Prodeshayesites* but more

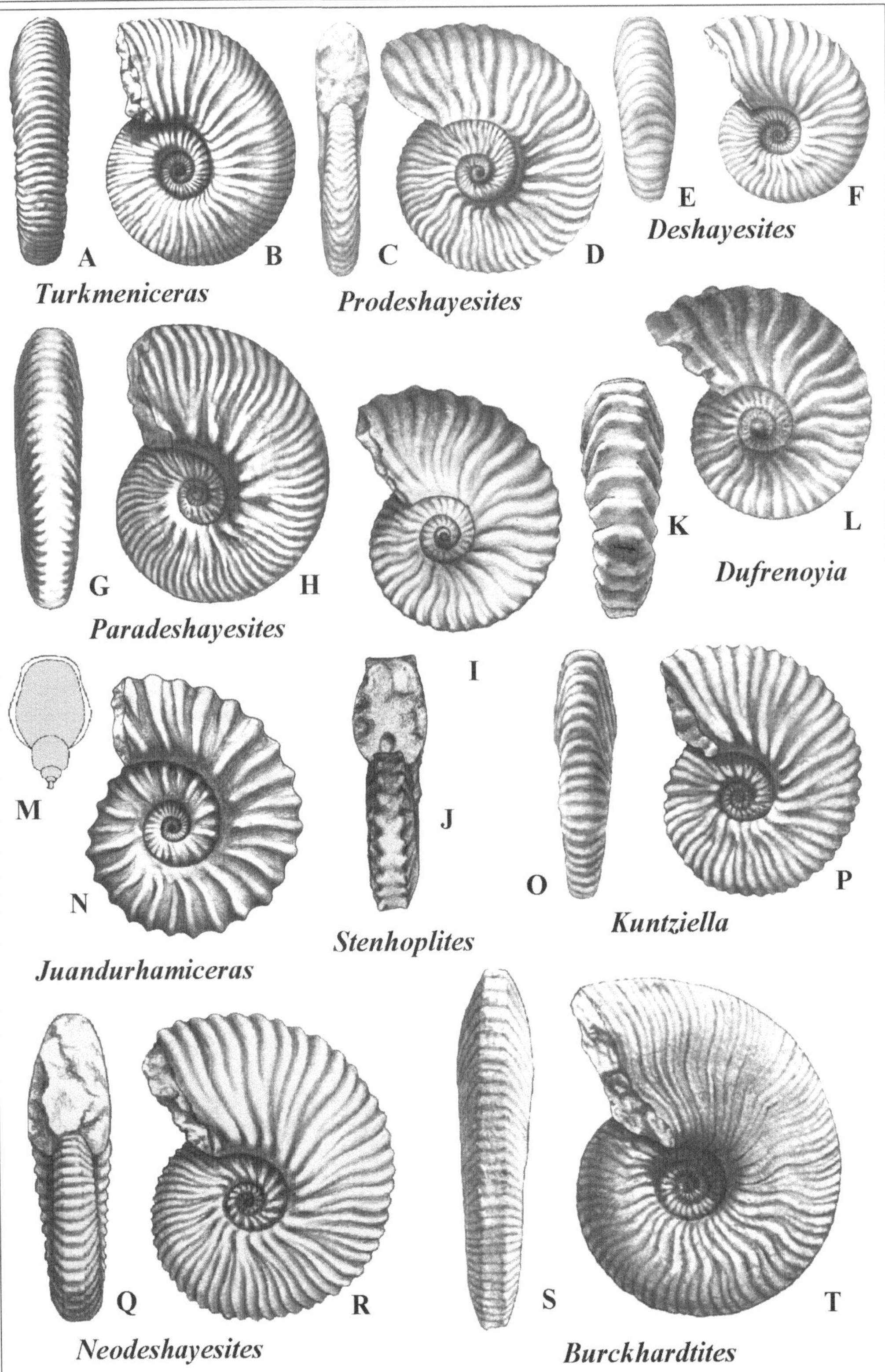

Fig. 664. Deshayesitidae, Acanthohoplitidae

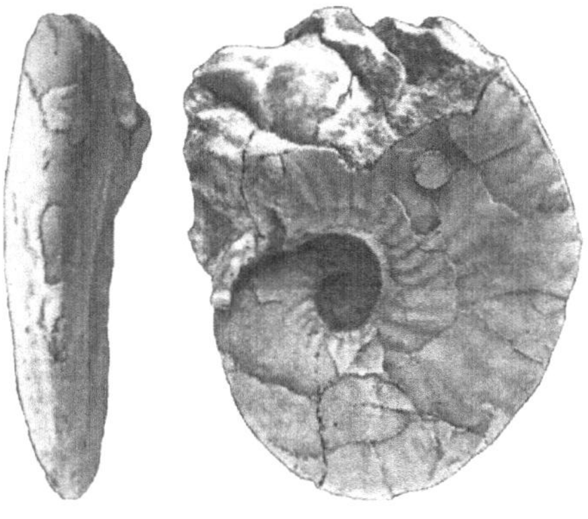

Fig. 665. *Obsoleticeras.*

involute, compressed and high whorled, with narrowly rounded venter and subtrigonal whorl section; fasciculate bundles of fine ribs arise from distinct umbilical bullae and weaken markedly across venter; suture with 4 inner lobes. *Lower L. Apt.* **Fig. 664G-H** - *P. laeviusculus* (von Koenen),.

Deshayesites KAZANSKY, 1914 [*Ammonites deshayesi* d'Orbigny, 1840; OD]. Like *Prodeshayesites* but with narrower umbilicus, more inflated whorls, flattened venter across which ribs remain strong, and absence of umbilical bullae; suture with shallowly bifid E/L and relatively narrow trifid L; dimorphic, **M** up to 400 mm diameter. *L. Apt.* **Fig. 664E-F** - *D. deshayesi* (d'Orbigny).

= *Parahoplitoides* Spath, 1922 (obj.).

Subfamily Dufrenoyiinae nov.

Diagnosis. Characteristically compressed, relatively evolute, flat sided with subtabulate venter, blunt sinuous ribs and ventrolateral clavi that evanesce in maturity. A monophyletic group with an origin in Deshayesitinae, close to *Deshayesites*. *Upper L. Apt. - L. Alb.*

Dufrenoyia KILIAN & REBOUL, 1915 [*Ammonites furcatus* J. de C. Sowerby, 1836; OD]. Evolute, compressed, with convergent flanks, subtabulate venter and compressed subtrapezoidal whorl section; strong robust distant sigmoidal ribs alternate long and short, with distinct umbilicolateral bullae and ventrolateral clavi; ribs continuous across flattened venter except in earliest growth; ventrolateral clavi lost at D>70 mm; suture like *Deshayesites*. *Upper L. Apt. - Lower U. Apt.* **Fig. 664K-L** - *D. furcata* (J. de C. Sowerby).

= *Dufrenoya* Kilian & Reboul, 1915 (*nom. van.*).

Stenhoplites SPATH, 1922 [*Ammonites dufrenoyi* d'Orbigny, 1840; OD]. Like *Dufrenoyia* but more compressed and

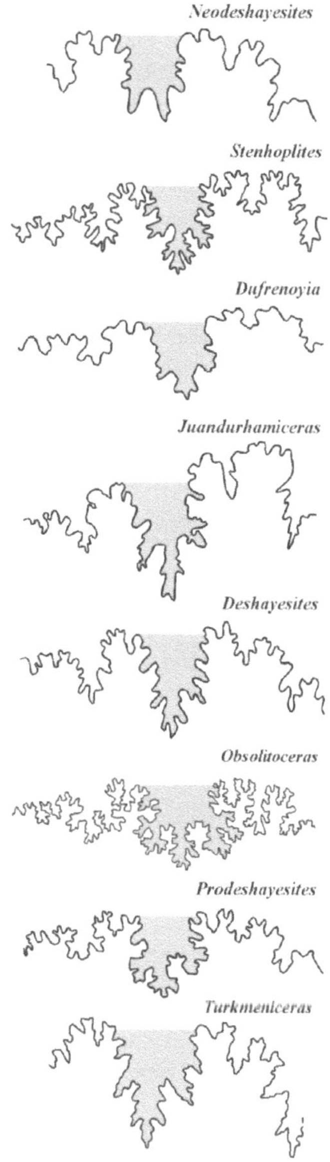

Fig. 666. Deshayesitid sutures.

high whorled, with finer ribs that flatten on outer flank; up to 50 mm diameter ribs effaced from venter, after which feeble ribs cross venter transversely. *Upper L. Apt.* **Fig. 664I-J** - *S. dufrenoyi* (d'Orbigny).

Burckhardtites HUMPHREY, 1949 [*Neocomites nazasensis* Burckhardt, 1925; OD]. Like *Stenhoplites* but involute and compressed, with whorls increasing rapidly in height, narrow tabulate smooth venter, tiny ventrolateral tubercles persisting to maturity, and biconcave ribs that become fine and irregular on adult body chamber. *M. Apt.* **Fig. 664S-T** - *B. nazasensis* (Burckhardt).

Juandurhamiceras ETAYO SERNA, 1979 [*J. juandurhami*; OD]. Like *Stenhoplites* but with rounder venter, in middle to later growth with distant almost straight ribs which recurve slightly on outer flank and tend to depress across siphonal line; in maturity ribs become flat topped; suture with asymmetrically bifid E/L and deep-penetrating trifid L. *U. Apt.* **Fig. 664M-N** - *J. juandurhami* Etayo Serna.

Kuntziella COLLIGNON, 1962 [*Deshayesites (K.) kuntzi*; OD]. Like *Deshayesites* but more compressed and flat sided, with oblique umbilical wall, flatter venter, less flexuous and more inclined primary ribs that frequently bifurcate from umbilical bullae that strengthen with growth and, later, with 1-2 short intercalatories on outer flank. ?*L. Apt.* **Fig. 664O-P** - *K. kuntzi* (Collignon).

The type material was found 80 m above a Lower Barremian fauna with *Menuthiocrioceras* and 80 m belowAlbian marls; its age was given as Lower Aptian but Wright (1996) assigned it to the Upper Aptian, allying it to "early" *Dufrenoyia*.

Collignon (1962) makes no mention of ventrolateral tubercles and these are not evident in his poor illustrations but, in all other respects, *Kuntziella* is very similar to *Neodeshayesites*.

Neodeshayesites CASEY, 1964 [*Deshayesites stutzeri* Riedel; OD]. Like inner whorls of *Burckhardtites* but with less-flexuous ribs; in later growth with rounded venter and coarse asymmetrical ribs (steepest anteriorly) that, in maturity, broaden into wide flattened bands separated by very narrow interspaces; suture relatively simple, with broad bifid E/L and generally narrow asymmetrically-trifid L deeper than E. *L. Alb.* **Fig. 664Q-R** - *N. stutzeri* (Riedel).

Records of *Neodeshayesites* from the Upper Aptian seem to be in error, the material figured by Renz (1982) is very close to *Penaceras*. Both Casey (1964) and Wright (*in* Guillaume *et al.* 1972, 1996) considered *Neodeshayesites* transitional between *Deshayesites* and *Dufrenoyia*, but its Early Albian age (Etayo Serna, 1981) indicates convergence. It is very similar to *Immunitoceras* (Ovando-Figueroa *et al.*, 2017).

Superfamily DOUVILLEICERATOIDEA Parona & Bonarelli, 1897

(*nom. transl.* Luppov, 1958 *ex* Douvilleiceratidae)

Discussion. Monophyletic recoiled derivatives of Pedioceratoidea with an unstable 5-lobed primary suture (Mikhailova, 1980), new elements appearing by division of umbilical (U) and inner lateral (I) lobes (Bogdanova & Mikhailova, 2016); comprises the family Douvilleiceratidae, Astiericeratidae and questionably Trochleiceratidae. *Lower Apt. - Lower M. Alb.*

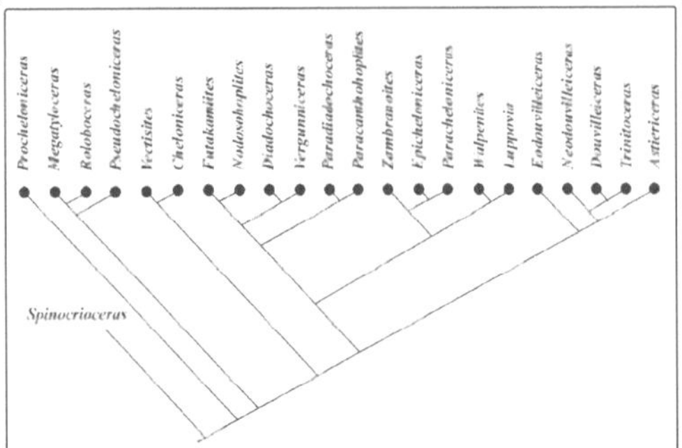

Fig. 667. Hypothesized relationships among Douvilleiceratoidea.

Family DOUVILLEICERATIDAE Parona & Bonarelli, 1897

Discussion. A monophyletic group seemingly with an origin in Hemihoplitidae (Wiedmann, 1966; Bogdanova & Mikhailova, 2016). However, in view of the substantial time gap between the two Bogdanova & Mikhailova (2016) suggested a possible origin in Crioceratidae but this seems less likely. Comprises the subfamilies Cheloniceratinae, Walpenitinae, Roloboceratinae, Diadochoceratinae and Douvilleiceratinae. *Lower Apt. - Lower M. Alb.*

Subfamily Cheloniceratinae Spath, 1923

(*nom. transl.* Breistroffer, 1953 *ex* Cheloniceratidae)

Discussion. A paraphyletic taxon with an origin in Peirescitinae, close to *Spinocrioceras* (Vermeulen, 2006). Mostly inflated, with depressed whorls, flattened venter, small umbilical tubercles, prominent septispinate lateral tubercles from which ribs frequently bifurcate, and mostly a broad asymmetrically bifid L to suture; derived members with

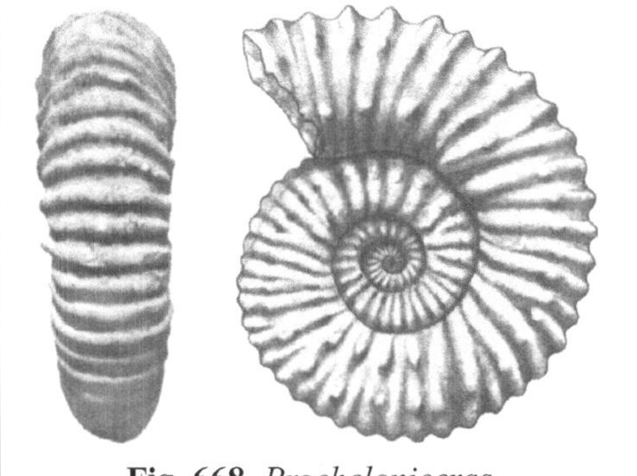

Fig. 668. *Procheloniceras*.

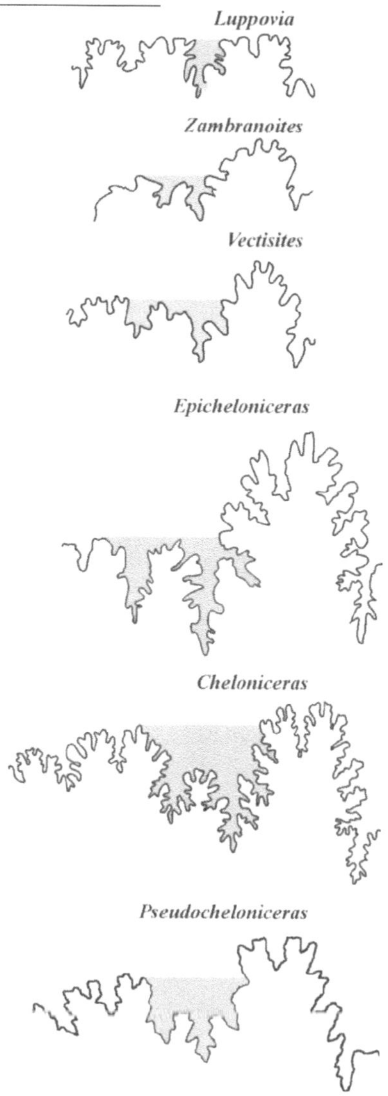

Fig. 669. Cheloniceratid sutures.

ventrolateral bullae. Sexual dimorphism conspicuous, large macroconchs losing tubercles and becoming finely ribbed in maturity. *Apt.*

Procheloniceras SPATH, 1923 [*Ammonites stobieckii* d'Orbigny, 1850; OD]. Moderately large; very evolute, compressed, with steep umbilical wall, slightly convex flanks and subcircular to oval whorl section; strong rursiradiate to rectiradiate ribs arise singly (with intercalatory) or in pairs from small umbilical bullae and pass strongly over flanks, crossing venter transversely; a midlateral tubercle is present in early growth, from which ribs sometimes bifurcate, migrating up flank to inner ventrolateral position in later growth; tuberculation lost on adult body chamber; suture with a deep narrow E, narrow multifid E/L, very broad bifid L, and broad bifid L/U. *Lower L. Apt.* **Fig. 668** - *P. stobieckii* (d'Orbigny).

Cheloniceras HYATT, 1903 [*Ammonites cornuelianus* d'Orbigny, 1841; ICZN Opinion 428]. Small to medium sized, inflated, with broad tabulate venter, depressed coronate to reniform whorl section, narrow umbilicus, persistent septispinate midlateral tubercle from which ribs branch (anterior branch generally stronger), separated by 1-3 intercalated ribs; in middle growth venter flattens, ventrolateral facets develop, whorl section becomes octagonal and ribbing uniform; on outer whorls of macroconchs ribbing simplifies, lateral tubercle are lost and point of rib bifurcation moves to umbilical edge; suture with deep E, broad asymmetrically bifid E/L, deep narrow L as deep as E, and small asymmetrically bifid L/U_2. *L. Apt.* **Fig. 670A-B** - *C. cornuelianum* (d'Orbigny); **Fig. 672I** - *C.* sp. (**M**).

As shown by Pictet (2012) the genus *Cheloniceras* is an evolutionary complex group that requires phylogenetic subdivision.

Pseudocheloniceras SHARIKADZE, 2015 [*P. accomodatum*; OD]. Small, evolute, with subtabulate venter and depressed subtrapezoidal whorl section; coarse ribs single or bifurcating from feeble umbilical bullae and periodic flat-topped ribs with rounded mid-lateral tubercles; all ribs with ventrolateral angulations. *L. Apt.* **Fig. 672A-B** - *P. accomodatum* Sharikadze.

= *Eocheloniceras* Egoian, 1977 (*nom. nud.*).

Vectisites CASEY, 1962 [*V. caprotinus*; OD]. Very small; inner whorls very depressed with primary ribs bearing midlateral tubercles separated by 3-4 non-tuberculate secondaries; later with subcircular whorl section and simple, uniform, non-tuberculate ribs alternating long and short, broadening across venter. *M. Apt.* **Fig. 670C-D** - *V. caprotinus* Casey.

Epicheloniceras CASEY, 1954 [*Douvilleiceras tschernyschewi* Sinzow, 1906; OD]. Moderately large; like *Cheloniceras* but with narrower umbilicus, infrequent branching of ribs from midlateral tubercle (of which posterior branch tends to be stronger), distinct ventrolateral bullae to main ribs and weakening of ribs across siphonal line in middle and adult growth to form shallow sulcus; as in *Cheloniceras* tuberculation declines on body chamber of macroconchs. *M. Apt.* **Fig. 670E-F** - *E. tschernyschewi* (Sinzow).

Paracheloniceras COLLIGNON, 1962 [*Epicheloniceras (P.) wrighti*, = *E. (P.) rerati* Collignon); OD]. Small; like *Epicheloniceras* but with large auriculate outer ventrolateral tubercles and flat-topped ribs. *U. Apt.* **Fig. 670G** - *P. rerati* (Collignon).

Zambranoites ETAYO SERNA, 1979 [*Vectisites (Z.) zambranoi*; OD]. Very small; like *Vectisites* but more evolute, with depressed subquadratic whorl section, flattened venter, sharper more distant ribs that form low ventrolateral bullae and are depressed across the siphonal line; suture with shallower E and broader E/L; perhaps a paedomorphic offshoot of *Epicheloniceras*. *U. Apt.* **Fig. 670J-K** - *Z. zambranoi* (Etayo Serna).

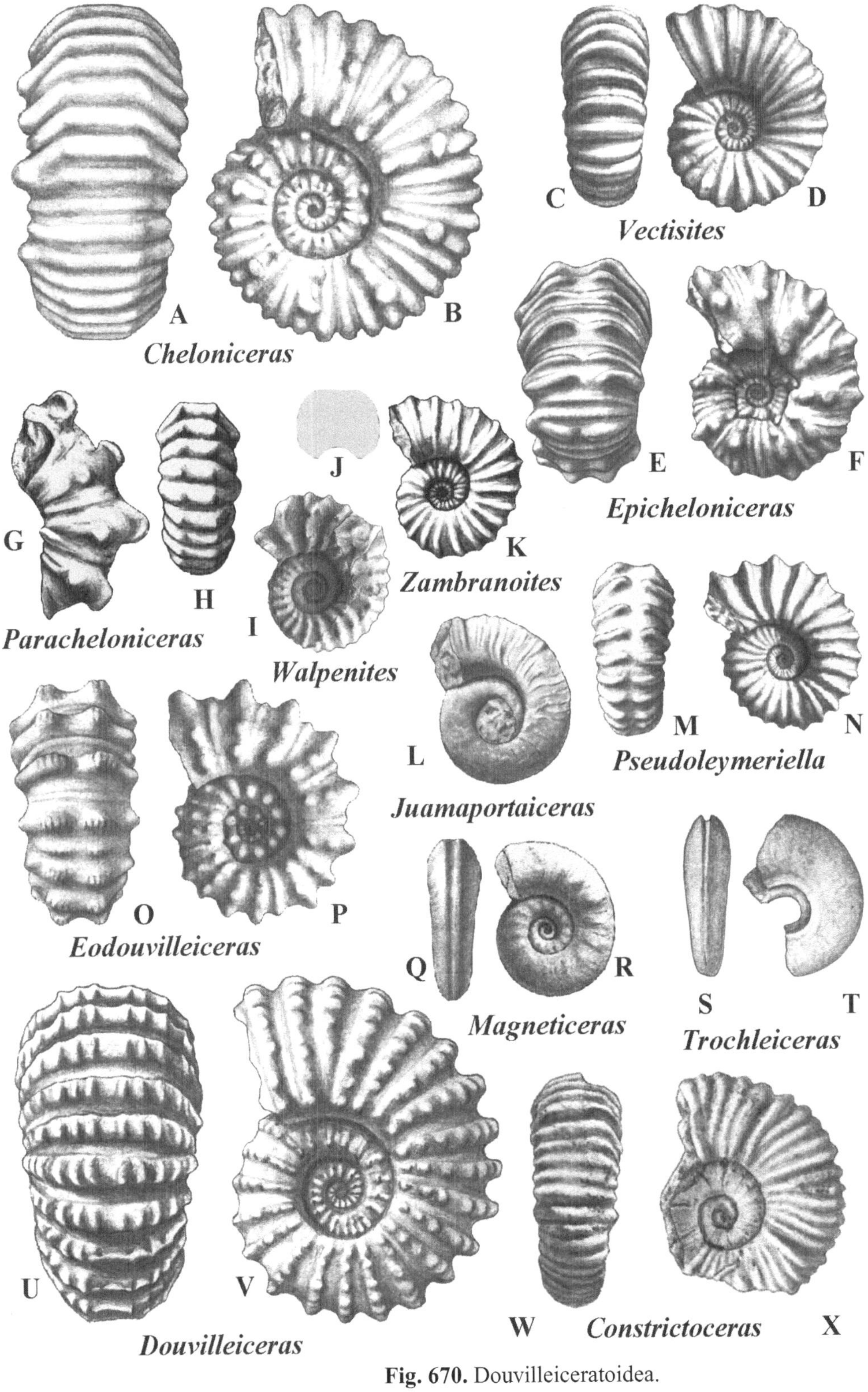

Fig. 670. Douvilleiceratoidea.

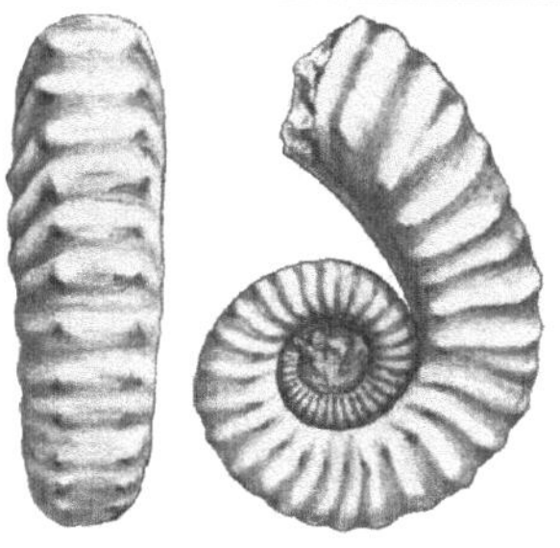

Fig. 671. *Luppovia.*

Subfamily Walpenitinae nov.

Diagnosis. Very small ancylocones with perforated umbilicus, trituberculate ribs and narrow L to suture; analogous to Astiericeratidae. Dwarf derivatives of Cheloniceratinae close to *Epicheloniceras*. *M. Apt.*

Walpenites CASEY, 1962 [*W. tardespinatus*; OD]. Very small, with uncoiling body chamber; latter depressed with suboctagonal whorl section, flattened venter and sparse coarse ribs with spinose inner and outer ventrolateral clavi, the former shifting to a lateral position with growth. *Lower U. Apt.* **Fig. 670I** - *W. tardespinatus* Casey.

Luppovia BOGDANOVA, KAKABADZE & MIKHAILOVA, 1978 [*L. dostshanensis*; OD]. Like *Walpenites* with very evolute phragmocone with perforated umbilicus, followed by uncoiled straight shaft with subquadratic whorl section; coarse rather distant trituberculate ribs (umbilical, lateral, ventrolateral) occasionally bifurcating at umbilical or lateral tubercles; ribs seem to be looped between ventrolateral clavi; suture with broad bifid E/L and narrow L. *M. Apt.* **Fig. 671** - *L. dostshanensis* Bogdanova, Kakabadze & Mikhailova.

Subfamily Roloboceratinae Casey, 1961

Discussion. Probable descendents of Paraspiticeratinae, close to *Blascoceras*, Bogdanova & Mikhailova (2016) noting that the external sutures of the two correspond "*almost completely*"; typically inflated, very evolute, with umbilical and lateral tubercles, the latter becoming exaggerated in later growth. *Middle L. - Lower U. Apt.*

Roloboceras CASEY, 1954 [*Ammonites hambrovi* Forbes, 1845; OD]. Medium sized, evolute, inflated; innermost whorls coronate with periodic collared constrictions with prominent lateral tubercle separated by weaker non-tuberculate ribs; thereafter strongly inflated, with deep umbilicus, broad subvertical umbilical wall, gently convex flanks, broadly rounded venter and depressed reniform whorl section; in middle growth 2-3 thick blunt round rectiradiate ribs, separated by 2-3 intercalatories of equal strength, arise from sparse large blunt umbilical bullae and cross venter uninterrupted; in maturity umbilical tubercles degenerate into radial bulges that may become very large, connected to 1-2 ribs; suture with very broad little incised multifid E/L, very broad asymmetrically bifid L, and small broad bifid L/U. *Middle L. Apt.* **Fig. 672E-F** - *R. hambrovi* (Forbes), **672G-H** - *R. regale* Casey.

Megatyloceras HUMPHREY, 1949 [*Douvilleiceras coronatum* Rouchadzé, 1933; ICZN Opinion 428]. Large; like *Roloboceras* but inner whorls very depressed and coronate throughout, with broad sloping umbilical wall and exaggerated umbilicolateral tubercles; suture with long narrow saddles. *Middle L. Apt.* **Fig. 672C-D** - *M. coronatum* (Rouchadzé).

Subfamily Diadochoceratinae Kvantaliani, 1978

Discussion. Seemingly monophyletic derivatives of Cheloniceratinae close to *Epicheloniceras*, but with narrow L to suture. Sharikadze (2015) sought an origin in Acanthohoplitinae, close to *Protacanthohoplites*. *Upper U. Apt.*

Paradiadochoceras SHARIKADZE, 2015 [*P. fragosum*; OD]. Small, inflated, with octagonal whorl section, trituberculate primaries separated by 2-3 non-tuberculate secondaries, and shallowly sulcate venter; suture with narrow E and broad trifid L. *U. Apt.* **Fig. 673F-G** - *P. fragosum* Sharikadze.

Diadochoceras HYATT, 1900 [*Ammonites nodosocostatus* d'Orbigny, 1841; OD]. Small, inflated, with octagonal whorl section, trituberculate primary ribs separated by 2-3 non-tuberculate secondaries and shallowly sulcate venter; convergent towards *Epicheloniceras* but suture with narrow E as deep as trifid L. *Uppermost Apt.* **Fig. 673H-I** - *D. nodosocostatum* (d'Orbigny).

Vergunniceras THOMEL, 1980 [*Ammonites pretiosus* d'Orbigny, 1841; OD]. Very small; evolute, inflated, with depressed subreniform whorl section; simple straight, rectiradiate primary ribs with prominent umbilical bullae, large round high lateral and ventrolateral nodes (septispinate spine bases) widely separated by 1-2 fine non-tuberculate secondaries; suture relatively simple, with wide trifid L. *Uppermost Apt.* **Fig. 673D-E** - *V. pretiosum* (d'Orbigny).

Nodosohoplites EGOIAN, 1965 [*N. subplanatus*; OD]. Very small; evolute, with flattened flanks and polygonal costal whorl section; weak umbilical bullae give rise to 1-2 ribs with high lateral and ventrolateral clavi, separated by up to 4 non-tuberculate secondaries; on adult body chamber venter rounds, rib simplify and tubercles

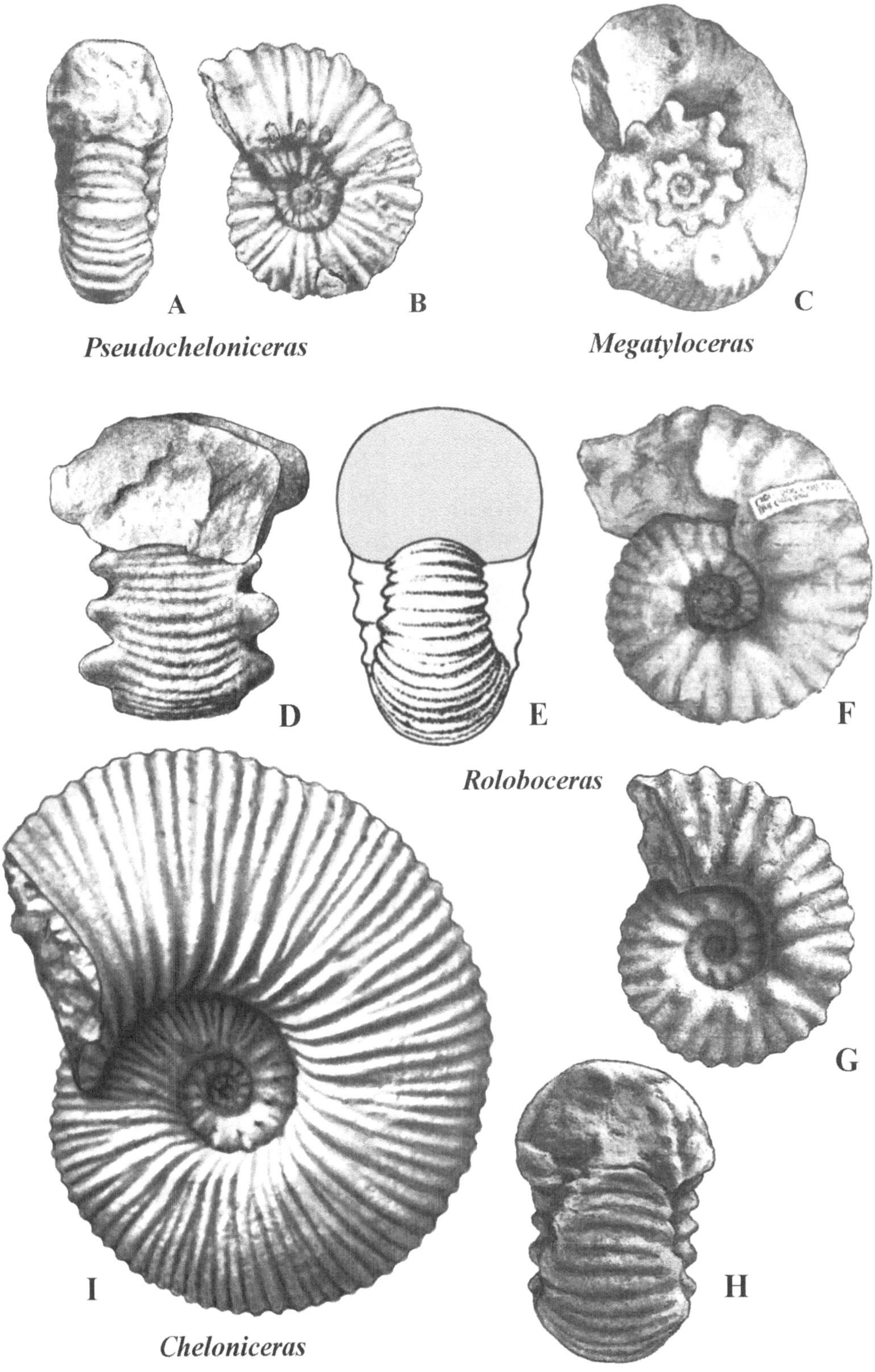

Fig. 672. Douvilleiceratidae.

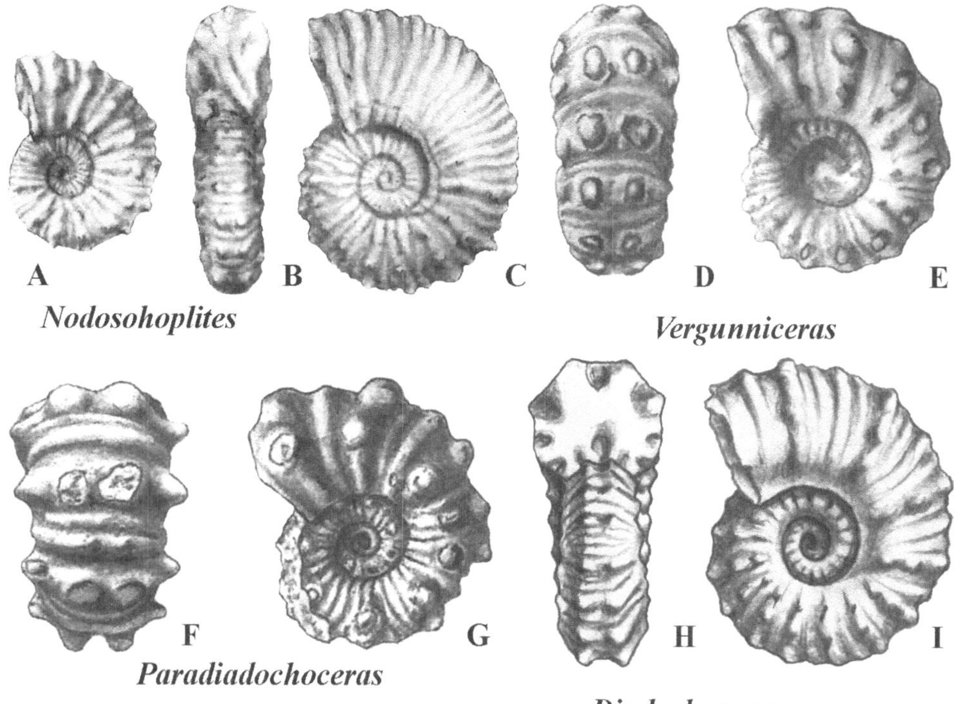

Fig. 673. Diadochoceratinae.

evanece; suture with wide trifid L as broad as E/L. *U. Apt.* **Fig. 673A-C** - *N. subplanatus* Egoian.

***Futakamiites* gen. nov.** [*Ammonites subcornuerianus* Shimizu, 1931, p.32; OD herein]. Barely separable from *Nodosohoplites*; compresse quadratic whorl section; phragmocone with flexuous slightly prorsiradiate primary ribs separated by 1-2 non-tuberculate secondaries; tuberulation evanesces on adult body chamber leaving only flexuous prorsiradiate ribs with feeble umbilical bullae separated by 1-2 intercalatories of different length. *Uppermost Apt.* **Fig. 674** - *F. subcornuelianae* (Shimizu).

Paracanthohoplites STOYANOW, 1949 [*P. meridionalis*; OD]. Medium sized, compressed, moderately involute, with flattened flanks, tabulate venter, angular ventrolateral shoulders and polygonal whorl section; strong slightly prorsiradiate main ribs with spinose midlateral tubercles separated by 1-3 non-tuberculate secondaries, cross venter transversely; in later growth lateral tubercles obsolete, flexed primary ribs attenuate at midflank and fewer secondaries; in maturity venter rounds and whorl section becomes oval; primary and secondary ribs alternate long and short; suture with bifid saddles and broad symmetrically trifid L. *U. Apt.* **Fig. 687H-I** - *P. meridionalis* Stoyanow.

= *Paracanthoplites* Wright, 1996 (*nom. van.*).

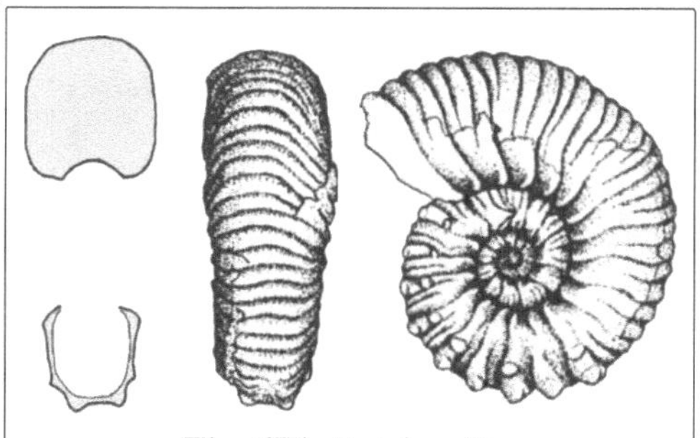

Fig. 674. *Futakamiites*.

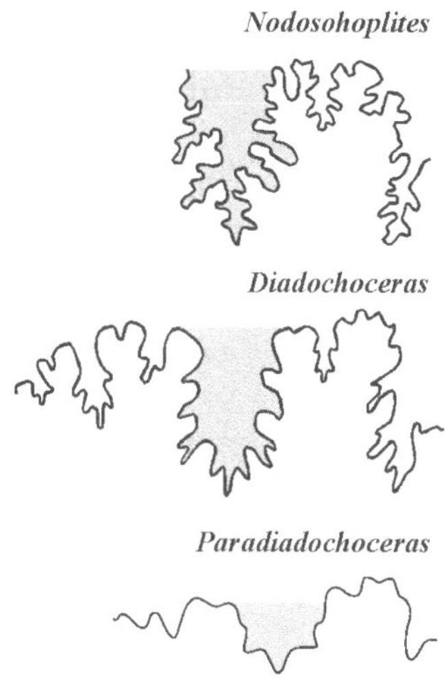

Fig. 675. Diadochoceratine sutures.

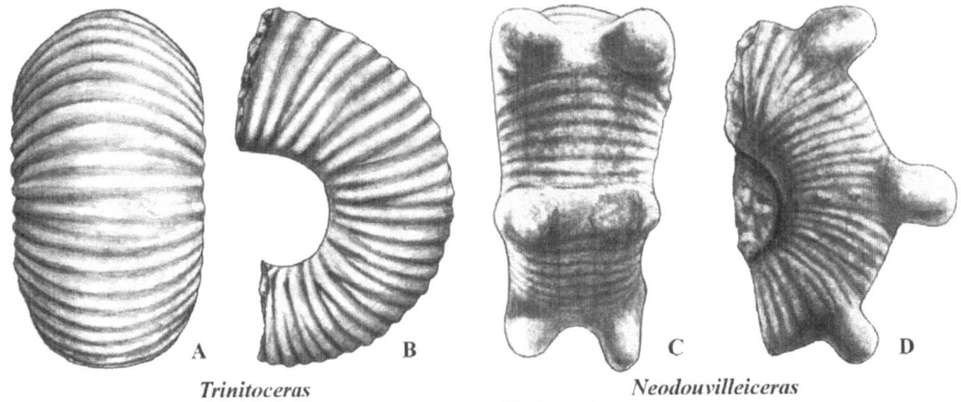

Fig. 676. Douvilleiceratinae.

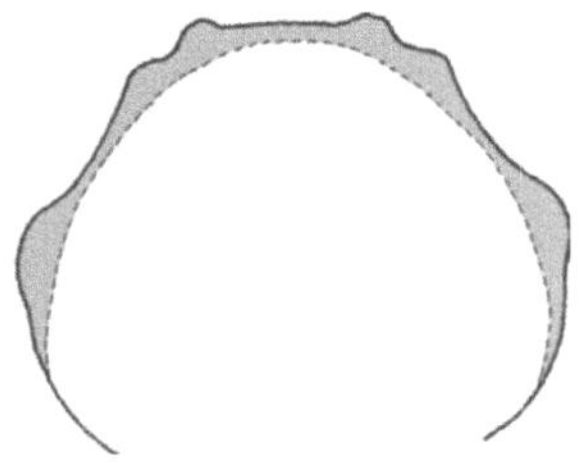

Fig. 677. Whorl section of *Trinitoceras* at 220 mm diameter.

Subfamily Douvilleiceratinae Parona & Bonarelli, 1897

(*nom. transl.* Spath, 1922 *ex* Douvilleiceratidae)

Discussion. Monophyletic derivatives of Cheloniceratinae, close to *Epicheloniceras*; typically with pronounced notching to ventral bullae, grooved siphonal line and, in derived members, multituberculate ribs; suture with very broad L. Sexual dimorphism conspicuous, macroconchs like those of Cheloniceratinae, with densicostate non-tuberculate adult whorls. *Uppermost Apt. - Lowermost M. Alb.*

Eodouvilleiceras CASEY, 1961 [*Douvilleiceras horridum* Riedel, 1938; OD]. Small; like *Epicheloniceras* but with sparse distant simple robust rectiradiate main ribs separted by thin intercalatories; main ribs with prominent conical umbilical and lateral tubercles and elevated ventral bullae which may or may not be subdivided by 2-3 notches. *Uppermost Apt. - Lower L. Alb.* **Fig. 670O-P** - *E. horridum* (Riedel).

Douvilleiceras DE GROSSOUVRE, 1894 [*Ammonites mammillatus* von Schlotheim, 1813; ICZN Opinion 422]. Small (**m**) to large (**M**); like *Eodouvilleiceras* but

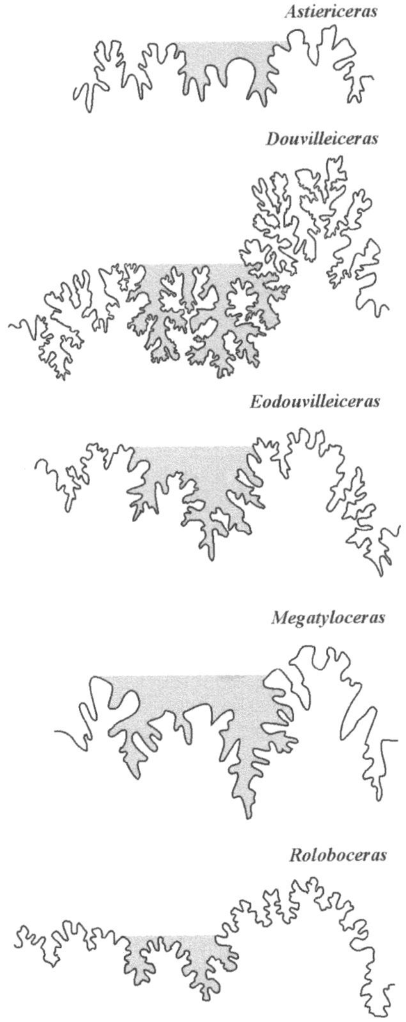

Fig. 678. Douvilleiceratid sutures.

with multituberculate ribs (up to 8 rows) in later growth; outer whorls of macroconchs like those of *Cheloniceras*, ultimately becoming smooth. *L. - Lowermost M. Alb.* **Fig. 670U-V** - *D. mammillatum* (von Schlotheim).

Trinitoceras SCOTT, 1940 [*T. rex*; OD] [**M**]. Large; like *Douvilleiceras* but with strongly convergent flanks and trapezoidal whorl section in middle growth (c. 200 mm diameter); primary ribs with prominent lateral bullae and doubled ventrolateral tubercles. *L. Alb.* **Fig. 676A-B, 677** - *T. rex* Scott.

***Neodouvilleiceras* gen. nov.** [*Douvilleiceras benonae* Besairie, 1936, p.164; OD herein]. Medium sized; inner whorls unknown but probably like *Douvilleiceras;* in later growth with fine flexuous non-tuberculate ribs with periodic large ventrolateral knobs joined across venter by broad rib; analogue of *Parachelonicera*s. *Upper L. - Lower M. Alb.* **Fig. 676C-D** - *N. benonae* (Besairie).

Includes also *N. magnodosum* (Casey) and *N. orbignyi* (Quenstedt).

Family ASTIERICERATIDAE Breistroffer, 1953

Discussion. A monotypic family with an origin in Douvilleiceratinae, close to *Douvilleiceras* (Spath, 1926); a small ancylocone with coronate whorl section, prominent umbilicolateral tubercles and suture with narrow U. A homoeomorph of *Yezoites* (Otoscaphitidae). Sexual dimorphism conspicuous. *Lower M. Alb.*

Astiericeras PARONA & BONARELLI, 1897 [*Scaphites astierianus* d'Orbigny, 1842; OD]. Very small to small; serpenticone phragmocone with deep conical umbilicus with strongly sloping wall, narrowly rounded edge, broadly rounded venter and depressed coronate whorl section; fine primary ribs on umbilical wall loop in groups of 3 to prominent septispinate tubercles at midflank that, in turn, connect over venter with similar number of looped ribs; ribs coarsen on uncoiled body chamber, those of microconchs scale-like and asymmetrical (steepest posteriorly); suture relatively simple, like *Eodouvilleiceras*, with short squat E/L. *Lower M. Alb.* **Fig. 679** - *A. astierianum* (d'Orbigny).

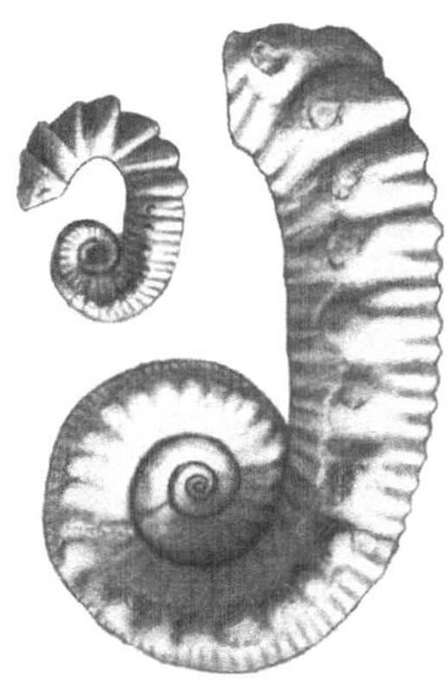

Fig. 679. *Astiericeras.*

Family TROCHLEICERATIDAE Breistroffer, 1951

Discussion. A micromorph group of uncertain affinity, originally placed with Pulchelliidae (Wiedmann, 1962), but later Wiedmann (1969) suggested an origin in *Paraspiticeras* that led to linking with Cheloniceratinae. It was assigned to Douvilleiceratoidea by Wright (1996), and *Pseudoleymeriella* seems close to *Zambranoites*, but Bogdanova & Mikhailova (2016) considered the evidence weak; there is also some resemblance to Silesitoidinae (Vermeulen, 2006) and its taxonomic position is unresolved. Typically evolute, with sulcate venter, early loss of lateral tubercles and simple quadrilobate suture with wide L. *U. Apt. - L. Alb.*

Pseudoleymeriella CASEY, 1957 [*Hoplites haidaquensis* Whiteaves, 1893; OD]. Very small, evolute, with rounded inner whorls; simple flexuous rursiradiate ribs interrupted over venter by narrow siphonal groove, distinct ventrolateral bullae and suture with narrow trifid L. *U. Apt.* **Fig. 670M-N** - *P. haidaquensis* (Whiteaves).

Trochleiceras FALLOT & TERMIER, 1923 [*Waagenia? balearensis* Fallot, 1920; OD]. Micromorph; evolute, compressed, with shallow umbilicus, low umbilical wall, gently convex flanks and narrowly rounded venter with flat to sulcate siphonal line; inner whorls with growth striae and indistinct riblets only that may persist to maturity; larger specimens (?**M**) with simple rectiradiate ribs alternating long and short, long ribs with weak umbilical bullae; suture almost goniatitic, with barely incised saddles and lobes, and very wide L. *U. Apt.* **Fig. 670S-T** - *T. balearense* (Fallot).

Magneticeras COLLIGNON, 1950 [*M. magneti*; OD]. Micromorph; like *Trochleiceras* but more evolute and inflated, whorls about as wide as high, with pronounced siphonal groove bordered by low keels; phragmocone unornamented but body chamber with low concave ribs; suture with deep narrow E, subrectangular E/L and relatively narrow L. *L. Alb.* **Fig. 670Q-R** - *M. magneti* Collignon.

Juamaportaiceras ETAYO SERNA, 1979 [*J. hoffstetteri*; OD]. Very small; like *Magneticeras* but with siphonal groove weakly developed on inner whorls and obsolete in maturity; low rounded irregular ribs on adult body chamber with ventrolateral bullae producing subquadratic costal whorl section; suture with broad weakly incised E/L and narrow L. *L. Alb.* **Fig. 670L** - *J.*

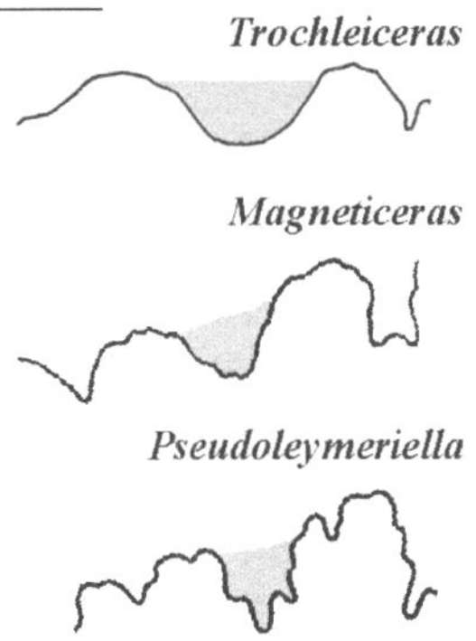

Fig. 680. Trochleiceratid sutures.

hoffstetteri Etayo Serna.
= *Juamportoceras* Wright, 1996 (*nom. van.*).

Superfamily PARAHOPLITOIDEA Spath, 1922

(*nom. transl.* Mikhailova & Baraboshkin, 2009 *ex* Parahoplitidae)

Discussion. A monotypic taxon of controversial affinity, but probably Douvilleiceratoidea (Sharikadze, 2015), although Mikhailova & Baraboshkin (2009) looked to Hemihoplitidae. The superficially similar sutures of Deshayesitoidea, that share an unstable 5-lobed primary suture, are fundamentally different in the appearance of new elements (Bogdanova & Mikhailova, 2016). In view of its uncertain ancestry the group is maintained here at the superfamily level. It comprises the families Acanthohoplitidae and Parahoplitidae. *L. Apt?., M. Apt. - L. Alb*

Family ACANTHOHOPLITIDAE Stoyanow, 1949

Discussion. Comprises the subfamilies Colombiceratinae and Acanthohoplitinae, perhaps with an origin in Cheloniceratidae (Sharikadze, 2015). Lateral tubercles prominent in early growth but obsolete later. *L. Apt?., M. Apt. - L. Alb.*

Subfamily Colombiceratinae Tovbina, 1979

Discussion. Seemingly monophyletic derivatives of Cheloniceratinae, close to *Pseudocheloniceras* (Sharikadze, 2015). Typically with somewhat flattened to grooved venter; ribs arise singly or in pairs from the umbilical shoulder and at some growth stage are broad and flat topped; suture with trifid L. *L. - U. Apt.*

Procolombiceras Sharikadze, 1979 [*P. atum*; OD]. Small, moderately involute, compressed, with steep umbilical wall, subparallel convex flanks, broadly rounded venter and subquadratic whorl section; strong flexuous rectiradiate ribs, broader than interspaces, arise singly or in pairs from feeble umbilical bullae and cross venter subtransversely; suture with deep E, shallowly bifid E/L, shallow very broad trifid L. *L. Apt.* **Fig. 682A-B** - *P. atum* Sharikadze.

Colombiceras Spath, 1923 [*Ammonites crassicostatus* d'Orbigny, 1840; OD]. Very evolute, compressed, with subtabulate venter, ventrolateral angulations and subrectangular whorl section; early whorls with flat-

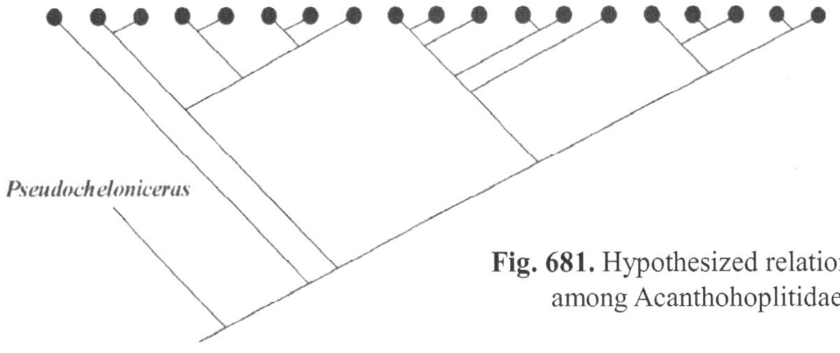

Fig. 681. Hypothesized relationships among Acanthohoplitidae.

topped ribs arising singly or in pairs from umbilical shoulder, with midlateral tubercle; in later growth venter rounds, midlateral tubercles evanesce and ribs round; suture with narrow rectangular E, broad quadratic bifid saddles, trifid L and simplified auxiliaries. *M. Apt.* **Fig. 682C-D** - *C. crassi-costatum* (d'Orbigny).

Gargasiceras CASEY, 1954 [*Ammonites gargasensis* d'Orbigny, 1841; OD]. Like *Colombiceras* but more densicostate, with flexuous ribs separated by 1-2 intercalatories that are broad, flat and slightly depressed over venter; on inner whorls major ribs raised into thin flange above which ribs branch at minute lateral tubercle. *M. Apt.* **Fig. 682I-J** - *G gargasense* (d'Orbigny).

Penaceras CANTU-CHAPA, 1963 [*Hypacanthoplites? rursiradiatus* Humphrey, 1949; OD]. Like *Colombiceras* but with primary ribs rursiradiate on outer flank, thickening on ventrolateral shoulders and flattening over venter. *Upper L. - U. Apt.* **Fig. 682K-L** - *P. rursiradiatum* (Humphrey).

= *Pegnaceras* Etayo Serna, 1979 (*nom. van.*).

Egoianiceras AVRAM, 1974 [*Colombiceras crassicost-atum angulatum* Egoian, 1969; OD]. Like *Colombiceras* but with flatter venter and without lateral tubercles and flat-topped band-like ribs. *U. Apt.* **Fig. 682E-F** - *E. angulatum* (Egoian).

Riedelites ETAYO SERNA, 1979 [*R. esthersernae*; OD]. Like *Egoianiceras* but with broadly rounded venter and suboval to subquadratic whorl section; ribs prorsiradiate, flexuous, broadening ventrally, flat topped on inner whorls only; in later growth ribs rather distant, tending to alternate long and short, becoming well rounded over venter; suture with sloping inner margin to E/L and L deeper than E. *U. Apt.* **Fig. 687F-G** - *R. esthersernae* Etayo Serna.

Fig. 683. *Hokodziceras.*

Pseudogargasiceras COLLIGNON, 1962 [*Gargasiceras (P.) enigmaticum*; OD]. Like *Gargasiceras* but small, very compressed, with fine non-tuberculate ribs on inner whorls coarsening later and forming weak chevrons over venter where ribs may weaken or efface in maturity. *U. Apt.* **Fig. 682G-H** - *P. enigmaticum* (Collignon).

Hokodziceras SHARIKADZE, 2015 [*H. druschitsi*; OD]. Small, very evolute, compressed, with compressed oval whorl section, simple flexuous ribs broader than interspaces, occasionally bifurcating on outer flank, and much weakened over venter. *U. Apt.* **Fig. 683** - *H. druschitsi* Sharikadze.

An obvious synonym of *Pseudogargasiceras*.

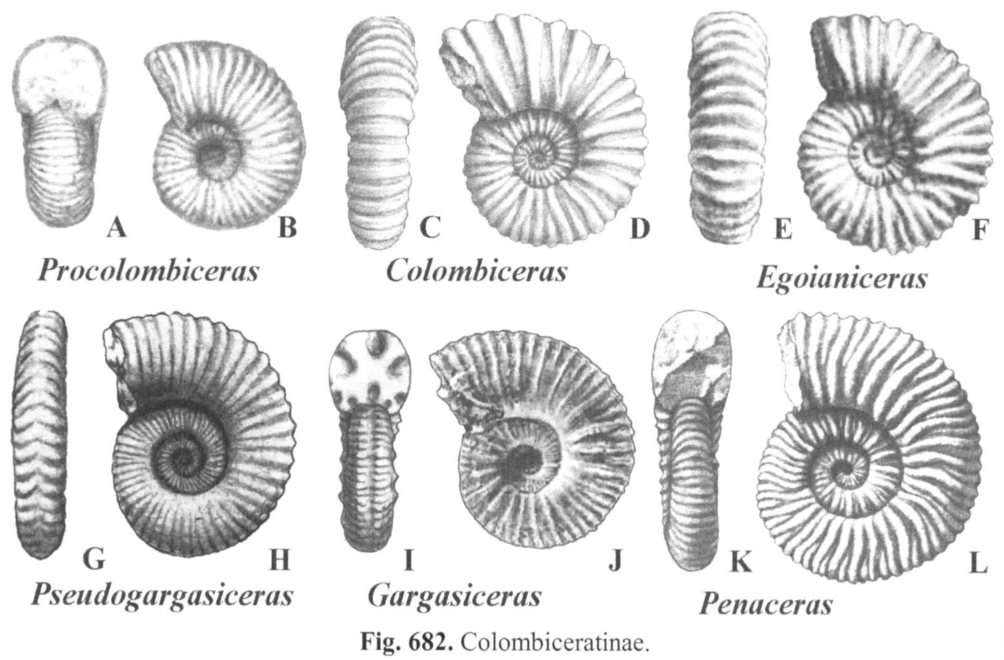

Fig. 682. Colombiceratinae.

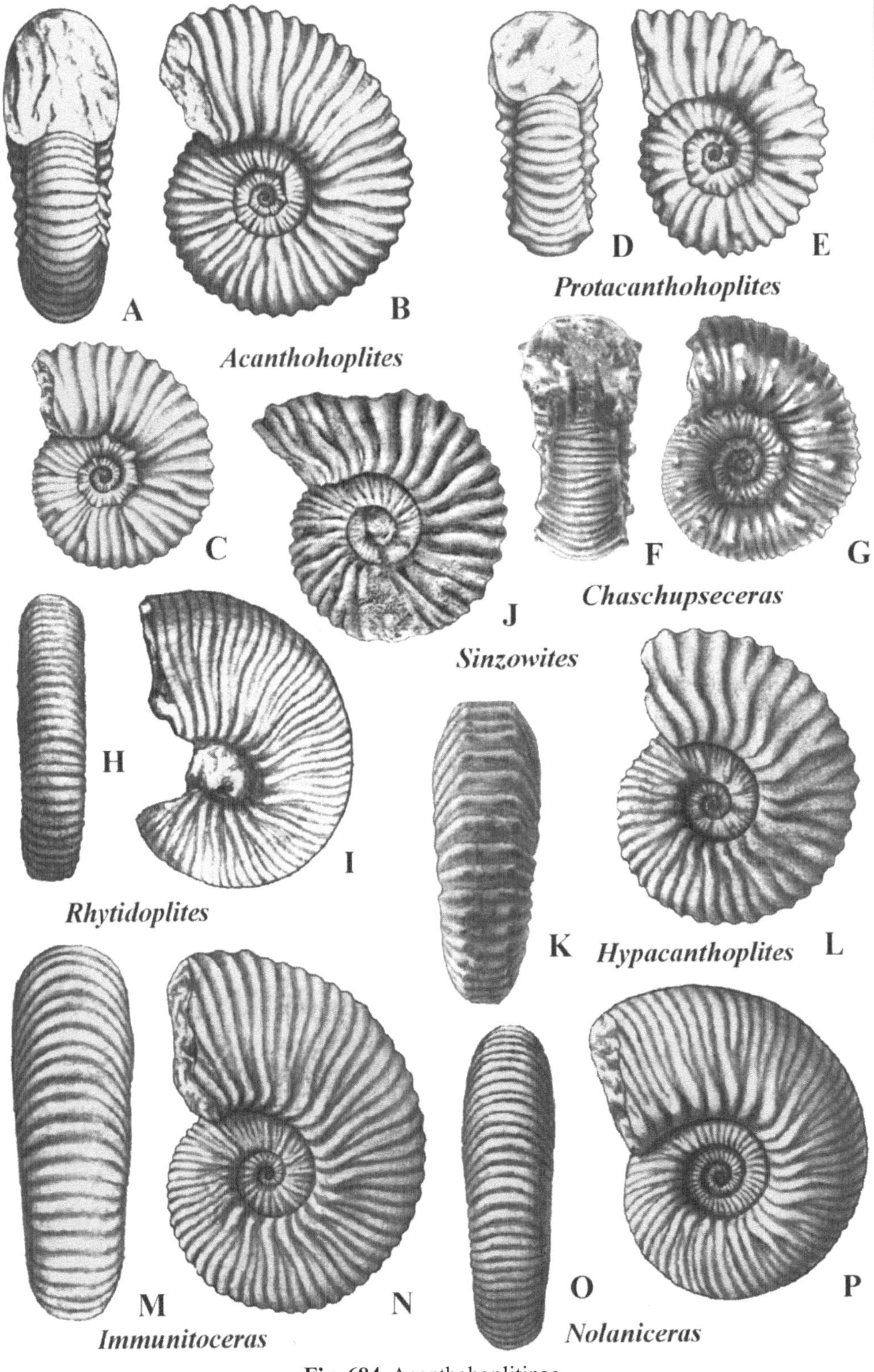

Fig. 684. Acanthohoplitinae.

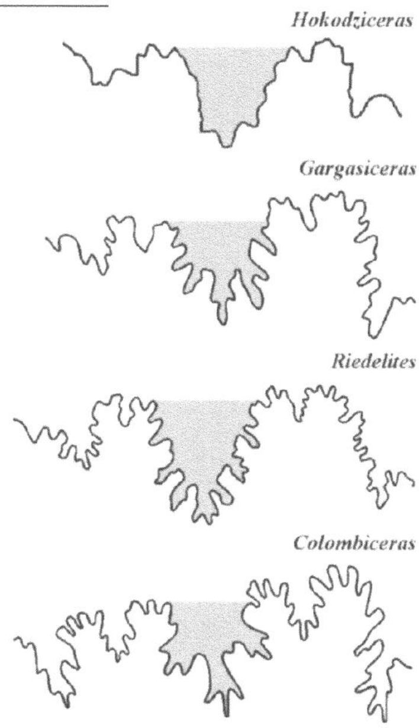

Fig. 685. Colombiceratine sutures.

Subfamily Acanthohoplitinae Stoyanow, 1949

Discussion. A monophyletic taxon with an origin in Colombiceratinae, close to *Colombiceras*; typically with umbilical and midlateral tubercles from which ribs branch; lateral tubercles persist into middle growth and, in derived members, are joined by ventrolateral tubercles. Comprises the tribes Acanthohoplitini and Constricticeratini nov. *M. Apt. L. Alb.*

Tribe Acanthohoplitini Stoyanow, 1949.

Type genus: *Acanthohoplites* Stoyanow, 1949.

Diagnosis. Medium sized to large, strongly ribbed with lateral tubercles in early growth. *M. Apt. - L. Alb.*

Acanthohoplites SINZOW, 1907 [*Parahoplites aschiltaensis* Anthula, 1899; SD Roman, 1938]. Medium sized; moderately evolute with steep umbilical wall, rounded venter and oval whorl section that becomes increasingly compressed with growth; in early growth primary ribs with incipient to feeble umbilical bullae; ribs bifurcate at midflank from prominent lateral tubercle, separated by 3-4 finer non-tuberculate secondaries, with periodic constrictions; in later growth lateral tubercles obsolete, umbilical bullae and intercalatories strengthen and ribbing simplifies, with uniform, slightly prorsiradiate flexuous ribs, about as wide as interspaces, single or alternating long and short, crossing venter uninterrupted; suture with deep narrow rectangular E, broad shallowly bifid subrectangular E/L, and narrow trifid L not as deep as E. *U6 Apt.* **Fig. 684A-B** - *A. aschiltaensis* (Anthula); **Fig. 686** - *A. sarasini* (Collet).

= *Acanthoplites* Douvillé 1912 (*nom. van.*).

Chaschupseceras KVANTALIANI & SHARIKADZE, 1968 [*C. abchasicum*; OD]. Like *Acanthohoplites* but inflated, with depressed coronate whorl section, denser finer costation and persistent bituberculate primary ribs separated by non-tuberculate secondaries. *U. Apt.* **Fig. 684F-G** - *C. abchasicum* Kvantaliani & Sharikadze.

Nolaniceras CASEY, 1961 [*Hoplites nolani* Seunes, 1887; OD]. Small to medium sized, moderately evolute and strongly compressed with compressed elliptical whorl section (somewhat flattened in earliest growth); dense prorsiradiate ribs, single with intercalatory or branching low on flank, arise from prominent umbilical bullae and cross venter uninterrupted, with minute lateral tubercles in earliest growth; ribs tend to become more prorsiradiate as aperture is approached. *Lower U. Apt.* **Fig. 684O-P** - *N. nolani* (Seunes).

= *Noloniceras* Rey *et al.*, 1986 (*laps. cal.*).

Rhytidoplites SCOTT, 1940 [*R. robertsi*; OD]. Like *Nolaniceras* but more compressed and involute, with relatively distant umbilical bullae giving rise to single slightly prorsiradiate primary ribs that recurve slightly on outer flank where separated by 2 intercalatories one of which invariably branches. *M. Apt.* **Fig. 684H-I** - *R. robertsi* Scott.

= *Rhytidohoplites* Scott, 1940 (*laps. cal.*).

= *Rhyditoplites* Young, 1974 (*laps. cal.*).

Immunitoceras STOYANOW, 1949 [*I. immunitum*; OD]. Like *Nolaniceras* but larger, with whorls increasing more rapidly in height, with biconcave flank costae well differentiated into primary and secondary ribs, becoming uniform later, with early loss of lateral and ventrolateral tubercles. *U. Apt. - Lower L. Alb.* **Fig. 684M-N** - *I. immunitum* (Stoyanow).

Protacanthoplites TOVBINA, 1970 [*Hoplites abichi* Anthula, 1899; OD]. Medium sized, very evolute, with steep umbilical wall, flat slightly divergent flanks and depressed polygonal whorl section; strong rectiradiate primary ribs split into secondaries at prominent midlateral

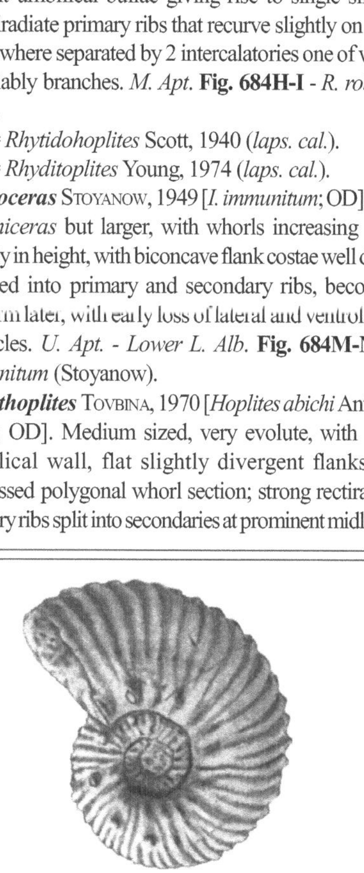

Fig. 686. *Acanthohoplites.*

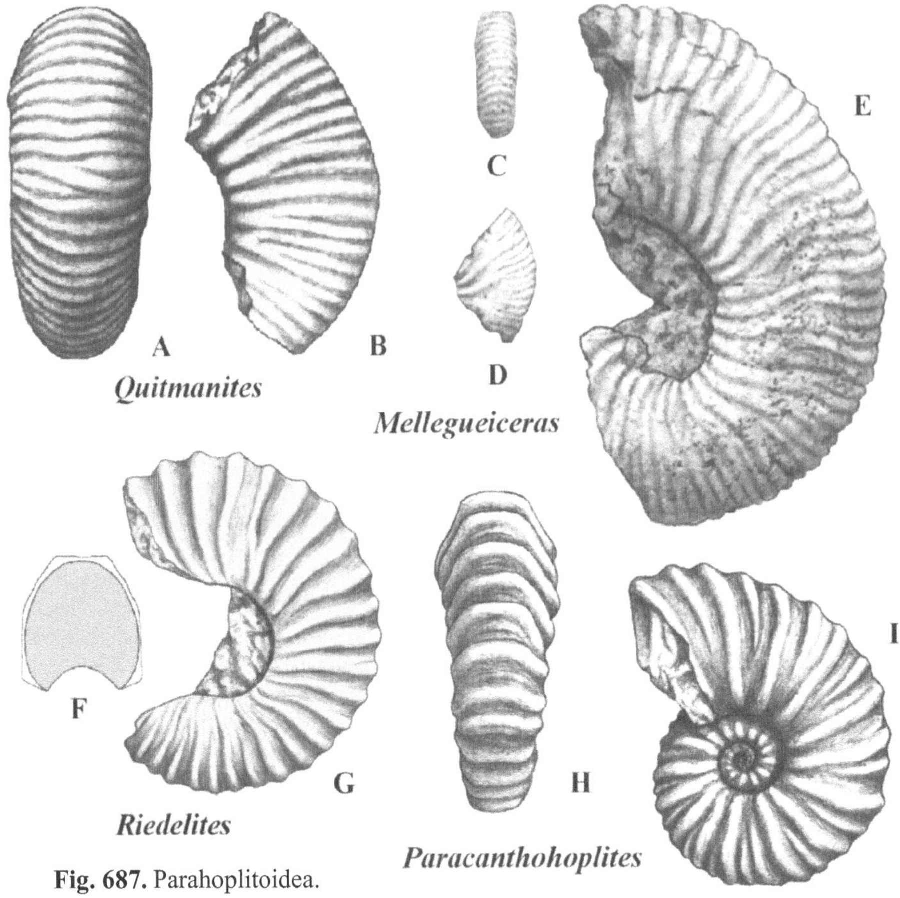

Fig. 687. Parahoplitoidea.

bullae, separated by 1-4 nontuberculate secondaries, crossing venter transversely; suture with L deeper than E. *U. Apt.* **Fig. 684D-E** - *P. abichi* (Anthula).

Hypacanthoplites SPATH, 1923 [*Acanthoceras milletianus* var. *plesiotypica* Fritel, 1906; OD]. Very large; evolute, with convex flanks and subtabulate venter; sharp umbilical bullae give rise to pairs of ribs separated by 1-2 intercalatories all with ventrolateral bullae; in later growth umbilical bullae decline and coarse ribs alternate long and short; midlateral tubercles retained to later growth; suture like *Acanthohoplites* but with narrower deeply bifid E/L and large trifid L deeper than E. *Upper U. Apt. - L. Alb.* **Fig. 684K-L** - *H. plesiotypicus* (Fritel), **Fig. 688** - *H. elegans* (Fritel).

Sinzowites GLAZUNOVA, 1949 [*Parahoplites jacobi* Collet, 1907, 1906; OD]. Like *Hypacanthoplites* but more compressed, with simpler ribs in early growth that frequently bifurcate from mid-lateral tubercles, earlier loss of mid-lateral and ventrolateral tubercles and rounder venter. *Upper U. Apt.* **Fig. 684J** - *S. jacobi* (Collet).

Cuchillites SCOTT, 1940 [*C. evolutus*; OD]. Holotype crushed; moderately large, evolute, inflated, with rounded venter and depressed subcircular whorl section;

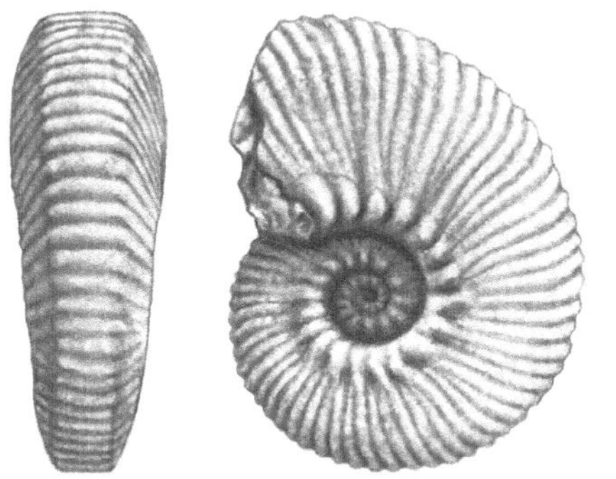

Fig. 688. *Hypacanthoplites*.

strong flexuous prorsiradiate ribs with asymmetrical profiles (steepest anteriorly), broader than interspaces, cross venter uninterrupted without forward inflexion. *U. Apt.* **Fig. 691C-D** - *Q. evolutus* Scott.

Mellegueiceras LATIL, 2011 *[Parahoplites ouenzaensis* Breistroffer, 1953; OD]. Like *Hypacanthoplites* but with irregular ribbing and lack of tubercles at all growth stages. *L. Alb.* **Fig. 687C-E** - *M. ouenzaense* (Breistroffer).

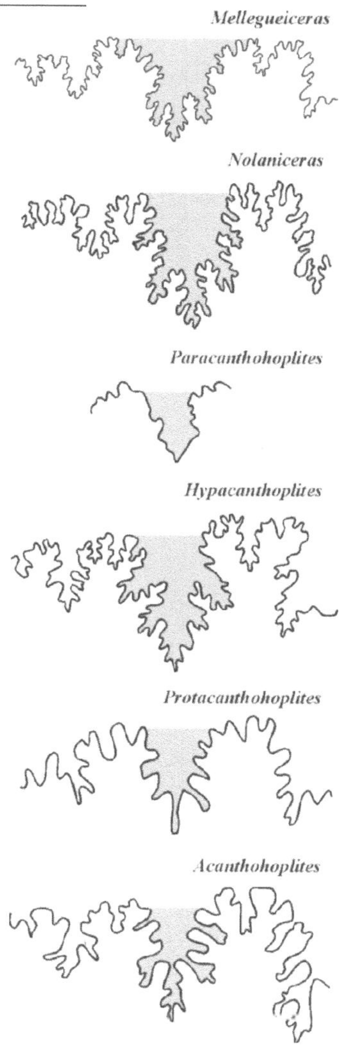

Fig. 689. Acanthohoplitine sutures.

Tribe Constrictoceratini nov.

Type genus: *Constrictoceras* Szives, 2008.

Diagnosis. Medium sized, moderately involute with subcircular whorl section; numerous deep straight constrictions per whorls separated by up to 3 broad ribs with small pointed midlateral tubercle. Assigned to Acanthohoplitinae by Szives (2008). *M? - U. Apt.*

Constrictoceras SZIVES, 2008 [*C. steineri*; OD]. Small, evolute, with rounded whorl section becoming subcircular and slightly depressed with growth; 8-9 deep collared constrictions per whorl; strong, narrow, simple prorsiradiate ribs, in early to middle growth with periodic mid-lateral tubercles, lost later; in later growth 3-5 coarse flexuous rectiradiate to slightly rursiradiate ribs between constrictions, with a short intercalatorybehind each collar, crossing venter uninterrupted; umbilical bullae develop on collars in maturity, sometimes also on adjacent rib. *M? Apt.* **Fig. 670W-X** - *C. steineri* Szives.

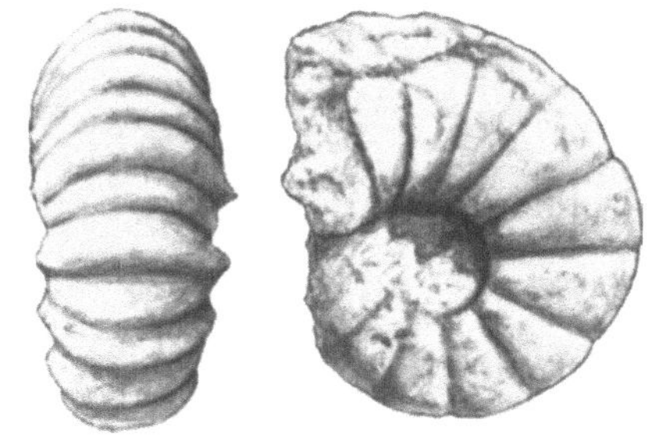

Fig. 690. *Cincticeras.*

***Cincticeras* gen. nov.** [*Cheloniceras? percrassum* Collignon, 1962; OD herein]. Medium sized, moderately involute, with subvertical umbilical wall, gently convex flanks, rounded venter and subcircular whorl section; 13 straight deep rigid prorsiradiate constrictions per whorl crossing venter uninterrupted, separated by extremely wide ribs with small pointed tubercle at mid-flank from which a shallow groove divides the rib into unequal parts. *U. Apt.* **Fig. 690** - *C. percrassum* (Collignon).

Family PARAHOPLITIDAE Spath, 1922

Discussion. Most workers have looked to Deshayesitidae (Casey 1965; Wright 1996) for the origin of the family but the sutural development is different (Mikhailova 1976, 1979, 1982) and an origin in Acanthohoplitidae seems most likely (Bogdanova & Mikhailova, 2016). Characterist-ically evolute, compressed, with oval to subrectangular whorl section, strong ribs and suture with trifid L; microscopic lateral tubercles on first few whorls. *U. Apt.*

Subfamily Parahoplitinae Spath, 1922

(*nom. transl.* Roman, 1938 *ex* Parahoplitidae)

Discussion. Monophyletic derivatives of Colombiceratinae, close to *Riedelites*; typically rather involute and compressed, with oval to subrectangular whorl section and strong rounded ribs with feeble umbilical bullae but no ventrolateral tubercles; dimorphic, macroconchs large (D_{max} = 300 mm), with almost smooth adult whorls; suture with trifid L. *U. Apt., ?L. Alb.*

Parahoplites ANTHULA, 1899 [*P. melchioris*; OD]. Moderately involute and high whorled, with flattish flanks, somewhat flattened venter and compressed sub-

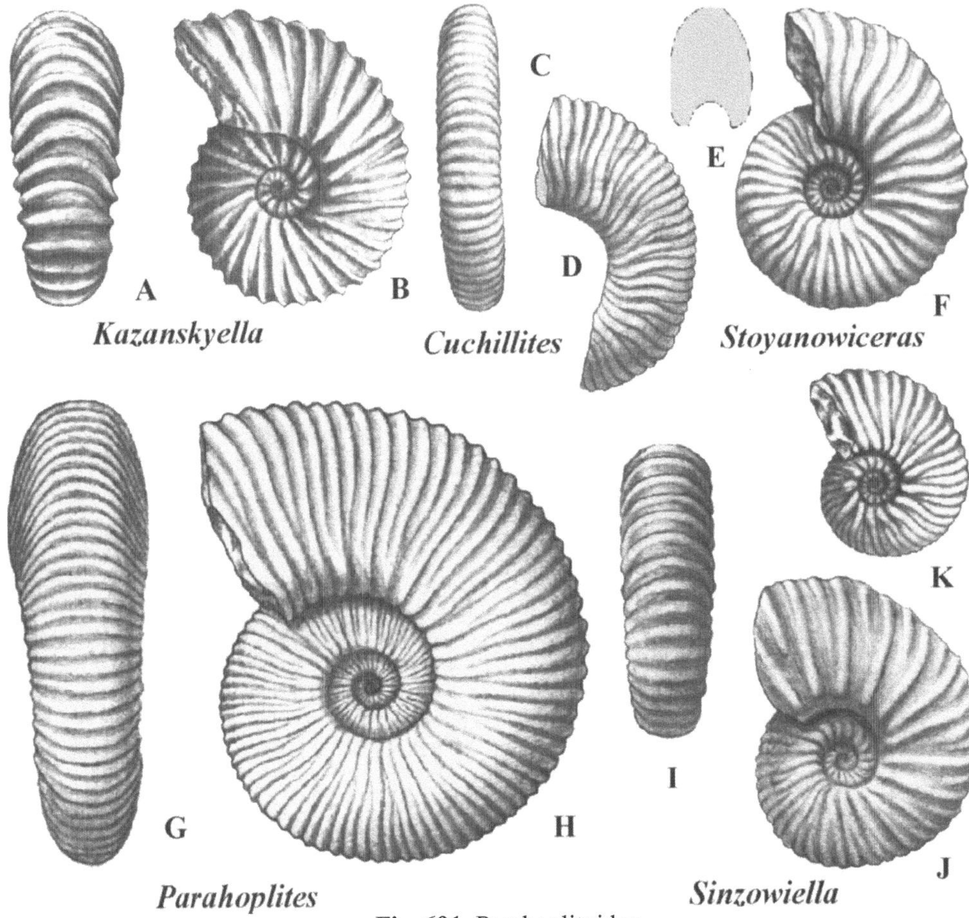

Fig. 691. Parahoplitoidea.

rectangular whorls; sinuous prorsiradiate ribs bifurcate from umbilical bullae with occasional intercalatories, or tend to alternate long and short, crossing venter with slight forward convexity; lateral and ventrolateral tubercles lacking; suture with deep rectangular E, large shallowly bifid E/L and narrow asymmetrically trifid L not as deep as E. *U. Apt.* **Fig. 691G-H** - *P. melchioris* (Anthula).

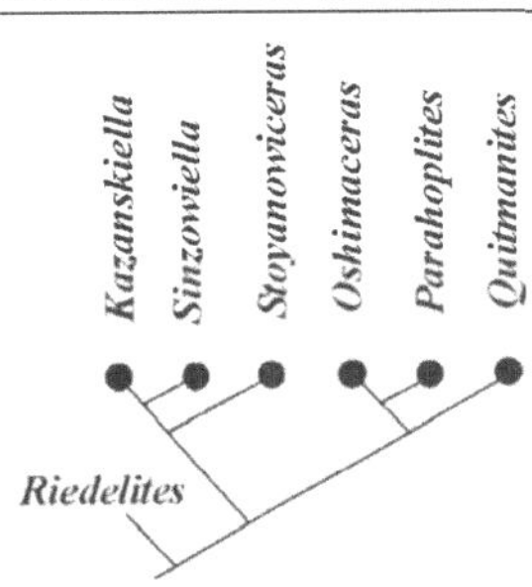

Fig. 692. Hypothesized relationships among Parahoplitidae.

Oshimaceras OBATA & FUTAKAMI, 2006 [*O. kanazawai*; OD]. Large, moderately evolute (U=28%), with compressed oval whorl section; rectiradiate to slightly prorsiradiate primary ribs arise singly incipient umbilical bullae, some bifurcating at midflank and crossing venter transversely, separated by 1-2 intercalatories; suture complex, with E deeply divided by narrow median lobule, narrow stemmed bifid E/L and L/U, asymmetrically trifid L as deep as E, and bifid L. *U. Apt.* **Fig. 694** - *O. kanazawai* Obata & Futakami.

Stoyanowiceras ETAYO SERNA, 1979 [*Ammonites treffryanus* Karsten, 1858; OD]. Like *Parahoplites* but more compressed, with rounder venter and oval adult whorl section; strong biconcave prorsiradiate ribs with incipient umbilical bullae in early growth tend to alternate long and short and broaden ventrally, in *Colombiceras* fashion, but not flat topped; suture with subrectangular E/L and broad trifid L deeper than E. *U. Apt.* **Fig. 691E-F** - *S. treffryanum* (Karsten).

= *Stoyawiceras* Etayo Serna, 1979 (*laps. cal.*).

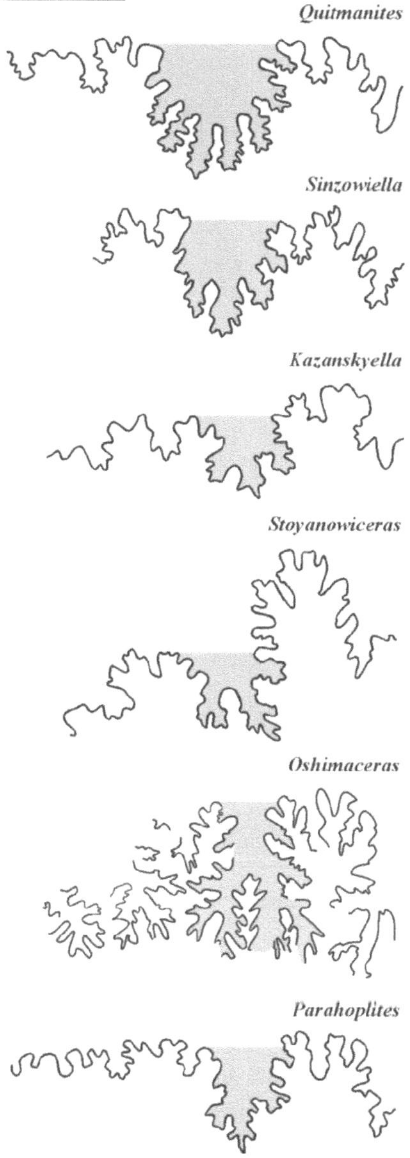

Fig. 693. Parahoplitid sutures.

Kazanskyella STOYANOW, 1949 [*K. arizonica*, = *Sonneratia minima* Scott, 1940; OD]. Like *Parahoplites* but with narrower umbilicus (U=22-29%), maximum width at midflank, oval to elliptical whorl section, straighter more rigid prorsiradiate ribs that alternate long and short for much of growth, and flex forward near venter; suture with broad E with oblique sides, broad asymmetrically bifid E/L and wide trifid L. *Upper U. Apt.* **Fig. 691A-B** - *K. minima* (Scott).

= *Kasanskyella* Young, 1974 (*nom. van.*).

Sinzowiella STOYANOW, 1949 [*S. spathi*; OD]. Like *Kasanskyella* but slightly more inflated, with denser ribbing in early growth (D<30 mm) and more delicate sculpture; later ribs finer and less oblique; suture like

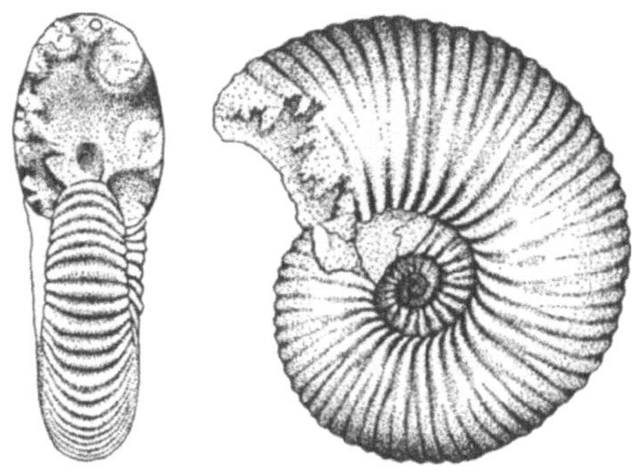

Fig. 694. *Oshimaceras*.

Parahoplites but with oblique sides to E, symmetrically bifid E/L and broad short obscurely trifid L. *Upper U. Apt.* **Fig. 691I-K** - *K. spathi* (Stoyanow).

Quitmanites SCOTT, 1940 (*Q. ceratitosus*; OD]. Like *Parahoplites* but more evolute and inflated, with convex flanks, broadly rounded venter and compressed subcircular whorl section; low, broadly rounded to flattish, dense, strong, rectiradiate to slightly rursiradiate ribs arising in pairs or triplets from long low umbilical bullae, with occasional intercalatories, crossing venter transversely; suture with broad asymmetrically bifid E/L and very broad digitate L deeper than E. *?Upper U. Apt.* **Fig. 687A-B** - *Q. ceratitosus* Scott.

= *Quitmanensis* Young, 1972 (*laps. cal.*)

Suborder TURRILITIDINA Besnosov & Mikhailova, 1984

(*nom. correct.* herein *pro* Turrilitina)

Discussion. Monophyletic derivatives of Ataxioceratidina. Heteromorphs characteristically with quadrilobate suture throughout and, in most, bifid L and U; primitively hamulinicones and ancylocones, later helicones, scaphiticones and baculicones. Comprises the superfamilies Anisoceratoidea, Hamitoidea, Turrilitoidea, Scaphitoidea and Baculitoidea. *U. Apt. - U. Maastr.*

Superfamily ANISOCERATOIDEA Hyatt, 1900

(*nom. transl.* herein *ex* Anisoceratidae)

Discussion. A seemingly monophyletic taxon with an origin in Macroscaphitoidea (Vermeulen, 2006). Mostly small to medium sized ancylocones, helicones and recoiled ammoniticones commonly with 2-3 rows of tubercles; suture with bifid L. Comprises the families Anisoceratidae and Labeceratidae. *U. Apt. - U. Cenom.*

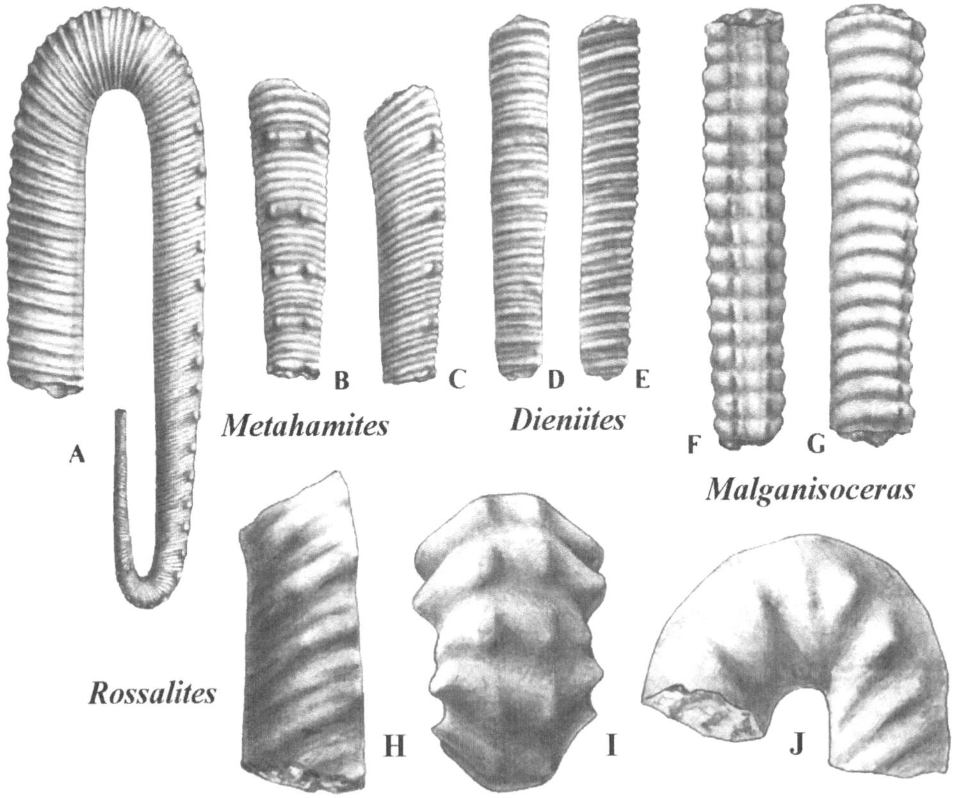

Fig. 695. Anisoceratidae.

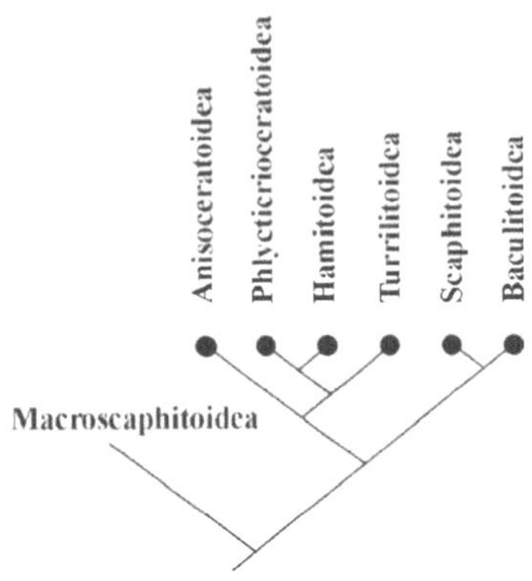

Fig. 696. Hypothesized relationships among Turrilitina.

Family ANISOCERATIDAE Hyatt, 1900

Discussion. A seemingly monophyletic group with an origin in Macroscaphitidae (Vermeulen, 2006). Mostly helicones and ancylocones. These are the common tuberculate heteromorphs of the Albian and Cenomanian. Comprises the subfamilies Protanisoceratinae and Anisoceratinae. *U. Apt. - U. Cenom.*

Subfamily Protanisoceratinae nov.

Diagnosis. Primitively hamulinicones with 3 well separated subparallel shafts and strong bituberculate ribs; suture with narrow E, bifid saddles, bifid L deeper than E, and very reduced trifid U much smaller than I. A monophyletic group with an origin in Lytocrioceratinae close to *Pseudocrioceratites* comprising the tribes Metahamitini and Protanisoceratini. *U. Apt. – M. Alb.*

Tribe Metahamitini nov.

Type genus: *Metahamites* Spath, 1930.

Diagnosis. Hamulinicones with 3 well separated subparallel shafts; fine oblique ribs single or looped to periodic rounded septispinate ventrolateral spines joined across the venter by looped ribs; tuberculation is obsolete on the final shaft when ribbing coarsens; suture with narrow stemmed E/L. *U. Apt. - M. Alb.*

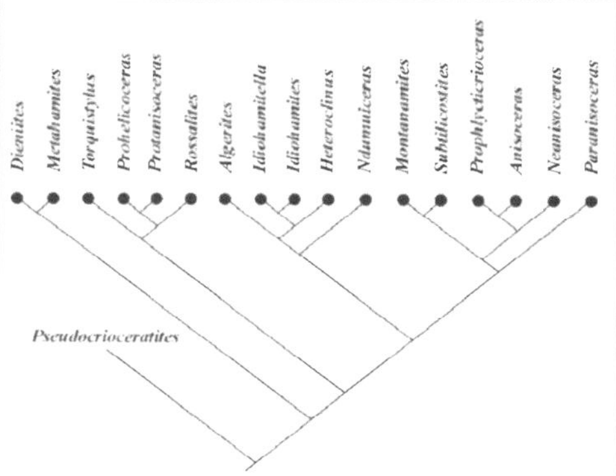

Fig. 697. Hypothesized relationships among Anisoceratidae.

Metahamites SPATH, 1930 [*Hamites sablieri* d'Orbigny, 1842; OD]. Small ptychocone with 3 well separated expanding parallel shafts and long retroversum; whorl section subcircular, asymmetrical, with adapical face vertical and adapertural face inclined; fine prorsiradiate flank costae loop to periodic ventrolateral clavi, separated by 2-4 intermediaries, splitting into fine concave bundles across somewhat flattened dorsum; ribs strengthen and coarsen on flexus and retroversum; suture complex with widely splayed bifid E/L. *U. Apt. - M. Alb.* **Fig. 695A** - *M. elegans* (d'Orbigny); **Fig. 695B-C** - *M. parcetuberculatus* (Collignon).

***Dieniites* gen. nov.** [*Hamites (Metahamites) dalpiazi* Wiedmann & Dieni, 1968, p.60; OD herein]. Known only by straight shaft that has differentiated ribbing with strong non-tuberculate primaries separated by 1-2 weaker finer secondaries crossing venter slightly obliquely; some deepened interspaces may be feeble constrictions; incised suture with asymmetrically bifid L and deep U. *U. Alb.* **Fig. 695D-E** - *D. dalpiazi* (Wiedmann & Dieni).

Tribe Protanisoceratini nov.

Type genus: *Protanisoceras* Spath, 1923.

Diagnosis. Small ancylocones with low loose helix followed by J-shaped body chamber; simple uni- or bituberculate primary ribs commonly separated by non-tuberculate secondaries; suture usually with broad stemmed E/L. *U. Apt. - M. Alb.*

Protanisoceras SPATH, 1923 [*Hamites raulinianus* d'Orbigny, 1842; OD]. Whorl section subcircular (intercostally) to subhexagonal (costally); strong blunt slightly prorsiradiate to rectiradiate primary ribs with midlateral and strong conical ventrolateral tubercles, separated by 1-3 finer non-tuberculate secondaries; ribbing weak or effaced from dorsum; suture with non-splayed asymmetrically bifid L. *U. Apt. - M. Alb.* **Fig. 699A-B** - *P. raulinianum* (d'Orbigny).

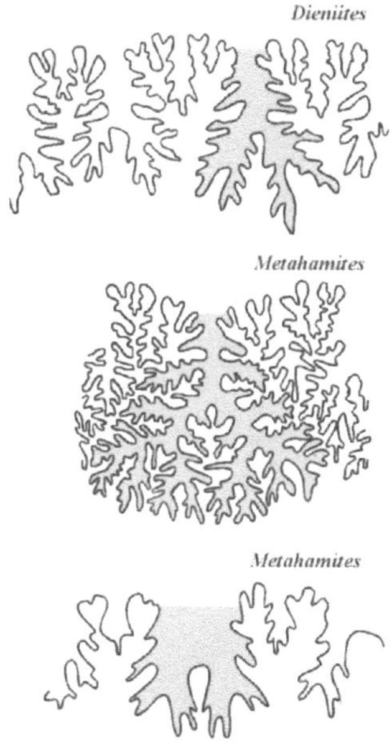

Fig. 698. Metahamitine sutures.

Rossalites CASEY, 1961 [*Protanisoceras? superbus* Collignon, 1949; OD]. Moderately large; known only from penultimate shaft and hook; proversum like *Protanisoceras* but with coarse blunt strongly prorsiradiate ribs; on flexus lateral tubercles become swollen and exaggerated, and ventrolaterals clavate; suture highly incised, with splayed almost symmetrical L. *Upper L. Alb.* **Fig. 695H-J** - *R. superbus* (Collignon).

***Malganisoceras* gen. nov.** [*Protanisoceras convexum* Collignon, 1963, p.33; OD herein]. Medium sized, known only from proversum with tabulate venter, rounded dorsum and compressed subelliptical whorl section; uniform convex forward ribs broader than interspaces, all with prominent ventrolateral clavi; ribs split into fine weak riblets across dorsum. *M. Alb.* **Fig. 695F-G** - *M. convexum* (Collignon).

Torquistylus CASEY, 1961 [*Prohelicoceras anglicum* Spath, 1939; OD]. Small, earliest known stage loose planispire followed by straight shaft and then twisted body chamber; whorl section subcircular to slightly depressed; strong blunt annular ribs with lateral and ventrolateral tubercles becoming oblique on adult body chamber; suture like *Protanisoceras* but simpler, with symmetrically bifid L. *Alb.* **Fig. 699V-W** - *T. anglicus* (Spath).

Prohelicoceras SPATH, 1925 [*Helicoceras thurmanni* Pictet & Campiche, 1861; OD]. Small; like *Protanisoceras* but with loose helical coiling throughout; sharp oblique annular ribs differentiated into thickened primaries with midlateral and ventrolateral tubercles irregularly joined across venter by looped ribs, separated by 1-2 finer non-tuberculate secondaries; suture simple. *L. - M. Alb.* **Fig. 699G-H** - *P. thurmanni* (Pictet & Campiche).

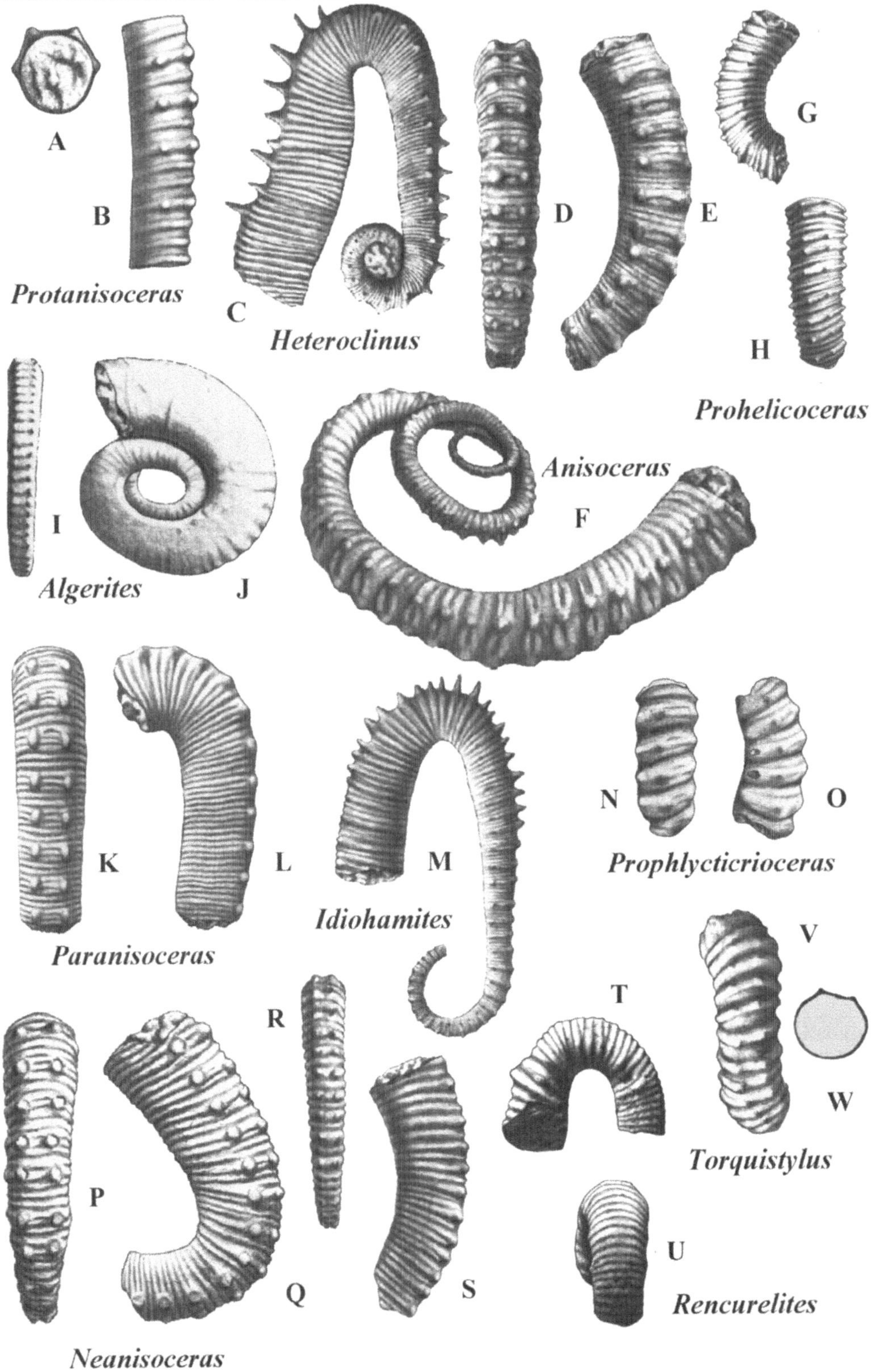

Fig. 699. Anisoceratidae.

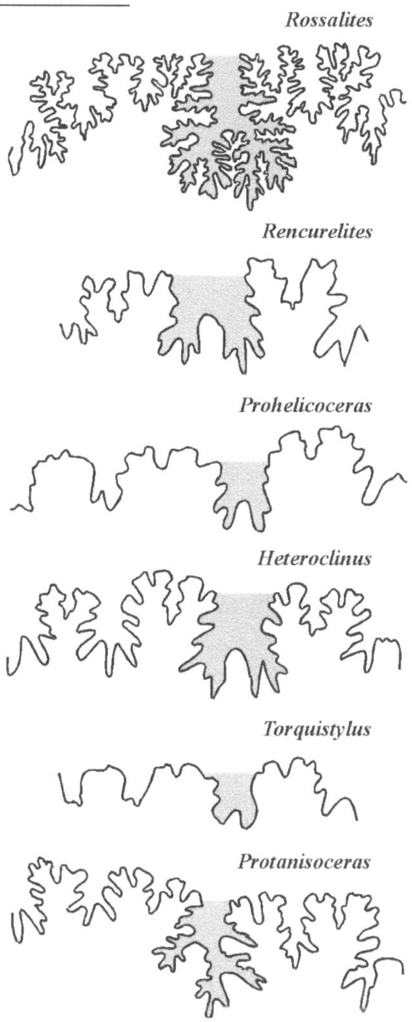

Fig. 700. Anisoceratid sutures.

Rencurelites THIEULOY, 1964 [*Protanisoceras (R.) ambiguum*; OD]. Poorly known; like *Protanisoceras* but small, with non-tuberculate annular ribs to last portion of proversum; ribs to flexus branch either from prominent dorsolateral tubercles or higher up flank; aperture with collared constriction; simple suture. *L. Alb.* **Fig. 699T-U** - *R. ambiguus* (Thieuloy).

Subfamily Anisoceratinae Hyatt, 1900

(*nom. transl.* herrein *ex* Anisoceratidae)

Discussion. A monophyletic group descended from Protanisoceratinae, typically with ribbing looped between tubercles and suture with narrow stemmed elements, large well developed bifid U and small trifid I. Turonian records are based upon early exiteloceratines. *U. Alb. - U. Cenom.*

Tribe Anisoceratini Hyatt, 1900

(*nom. transl.* herrein *ex* Anisoceratidae)
Type genus: *Anisoceras* Pictet, 1854.

Diagnosis. Mostly loosely coiled helicones with an obtuse spiral angle, with strong quadritubercul-ate primary ribs. Seemingly monophyletic descend-ents of Protanisoceratini, close to *Heteroclinus* with relatively large U. *U. Alb. - U. Cenom.*

Anisoceras PICTET, 1854 [*Hamites saussureanus* Pictet, 1847; OD]. Moderately large, with initial loose helix followed by straight or sinuous and sometimes twisted shaft and final hook; ribs conspicuously looped between prominent rounded dorsolateral and ventrolateral tubercles, and again across venter, with or without intervening non-tuberculate secondaries, splitting into finer ribs across dorsum; ribbing simplifies on adult body chamber; suture highly incised, with E/L substantially smaller than L/U, large deep almost symmetrically bifid L, well developed widely splayed U and short narrow trifid I. *U. Alb. - U. Cenom.* **Fig. 699D-E** - *A. saussureanum* (Pictet); **Fig. 699F** - *A. perarmatum* Pictet & Campiche.

***Subtilicostites* gen. nov.** [*Anisoceras haasi* Cooper & Kennedy, 1979, p.210; OD herein]. Small to medium sized; a very low helicone with subcircular to elliptical whorl section and straight body chamber; 2-4 fine ribs on broad low rursiradiate swellings loop between dorsolateral and septispinate ventrolateral spines, and also across the venter, separated by 2-9 fine non-tuberculate secondaries; with growth dorsolateral spines shift to mid-lateral position and ventrolaterals become clavate. *Uppermost Alb. - M. Cenom.* **Fig. 201F** - *S. haasi* (Cooper & Kennedy).

***Paranisoceras* gen. nov.** [*Anisoceras leckenbyi* Wright & Kennedy, 2002, p.305; OD herein]. Like *Anisoceras* but coiling planispiral in middle and adult growth; whorls flat sided, compressed with elliptical whorl section; ribs fine, slightly prorsiradiate, looped in 2's and 3's to subrounded to clavate ventrolateral tubercles; dorsolateral tubercles absent to vestigial; a homoeomorph of *Metahamites*. *L. Cenom.* **Fig. 699K-L** - *P. leckenbyi* (Wright & Kennedy).

Prophlycticrioceras CLARK, 1965 [*Hamites tanima* Adkins & Winton, 1920; OD]. Known only from short slightly curved fragments; whorl section hexagonal to laterally compressed; very broad blunt rounded ribs, oblique over venter and thinning across dorsum, with ventrolateral and siphonal tubercles; ribs seem to show incipient looping between tubercles and encompass a fine non-tuberculate secondary rib on their posterior slope. *Low U. Alb.* **Fig. 699N-O** - *P. tanima* (Adkins & Winton).

Perhaps a malformed *Anisoceras* (Wright, 1996).

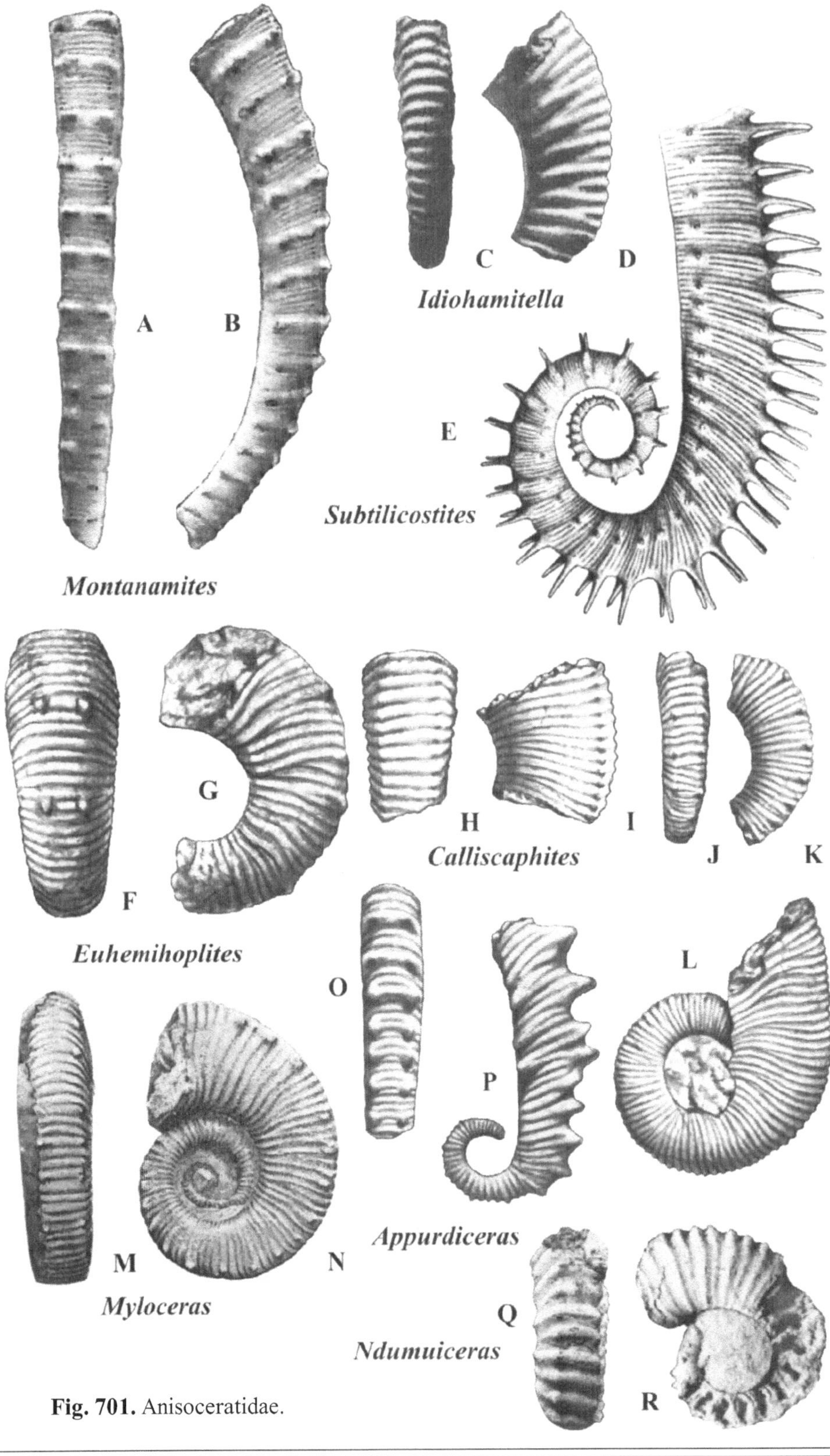

Fig. 701. Anisoceratidae.

Neanisoceras **gen. nov.** [*Anisoceras coloradoense* Cobban, Hook & Kennedy, 1989, p.59; OD herein]. Like *Anisoceras*, but with compressed subrectangular whorl section, uniform ribbing without swollen primaries, and tubercles high-lateral and ventrolateral in position; suture with broad stemmed rectangular saddles, narrow bifid L, L/U smaller than E/L and penetrative I. *Upper U. Cenom.* **Fig. 699P-Q** - *N. coloradoense* (Cobban, Hook & Kennedy).

Montanamites **gen. nov.** [*Idiohamites pulchellus* Kennedy & Cobban, 1990, p.416; OD herein]. Small, very loosely coiled criocone, expanding rather rapidly; bold prorsiradiate primary ribs with rounded dorsolateral and bullate ventrolateral tubercles, separated by 3-4 fine non-tuberculate secondaries, without looping. *U. Cenom.* **Fig. 701A-B** - *M. pulchellus* (Kennedy & Cobban).

Tribe Algeritini Spath, 1925

(*nom. transl.* herein *ex* Algeritidae)

Type genus: *Algerites* Spath, 1925.

Diagnosis. Ancylocones, criocones and secondarily recoiled ammoniticones; flat sided with subrectangular whorl section and bituberculate venter. A seemingly monophyletic group of uncertain origin, perhaps Metanisoceratini. *M. - U. Alb.*

Ndumuiceras KENNEDY & KLINGER, 2009 [*N. variabile*; OD]. Small criocone to aspinocone with whorls barely separated; whorl section subtrapezoidal with feebly convex dorsum and venter; phragmocone with coarse straight rectiradiate to prorsiradiate primary ribs with feeble high-lateral and spinose ventrolateral clavi linked across venter by broad ribs; body chamber with flexuous primary ribs separated by a finer long intercalatory on which lateral tubercle is obsolete; suture with non-splayed bifid L. *Upper L. - Lower M. Alb.* **Fig. 701Q-R** - *N. variabile* Kennedy & Klinger.

Heteroclinus CASEY, 1961 [*Hamites nodosus* J. Sowerby, 1818; OD]. Medium sized aspinicone; spire followed by long slightly oblique proversum and long rapidly expanding retroversum; mature whorls flat sided, with subrectangular whorl section; fine oblique annular ribs periodically quadrituberculate, with tubercles joined by looped ribs; in later growth tubercles high lateral and ventrolateral in position, the former lost on retroversum; suture with relatively large U. *Lower M. Alb.* **Fig. 699C** - *H. nodosus* (J. Sowerby).

Idiohamites SPATH, 1925 [*Hamites tuberculatus* J. Sowerby, 1818; OD]. Small, known mostly by septate fragments of loose planispire with compressed elliptical whorl section; simple rectiradiate primary ribs with ventrolateral tubercles separated by 2-3 finer non-tuberculate secondaries, without looping across venter; suture with bifid L/U same size as E/L, and small trifid I.

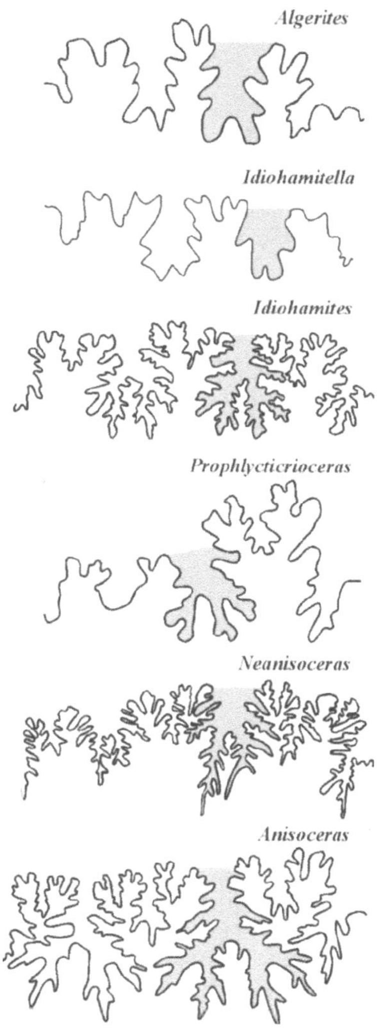

Fig. 702. Anisoceratid sutures.

The species *I. spiniger* (J. Sowerby), based upon a complete ancylocone, has long been used to interpret the genus (Wright, 1957, 1996). It shows a rapidly expanding proversum and retroversum with flattened flanks, subtabulate venter and compressed subrectangular whorl section, much like *Heteroclinus*, but with low-lateral tubercles and ventrolateral spines on flexus and retroversum. *Lower U. Alb. - M. Cenom.* **Fig. 699M** - *I. spiniger* (J. Sowerby).

Idiohamitella **gen. nov.** [*Idiohamites ellipticoides* Spath, 1934, p.594; OD herein]. Small; early whorls a loose planispire, body chamber straightening but not known to form a hook; whorl section compressed subelliptical, with rounded dorsum and venter, and flattened sides; strong flexuous non-tuberculate ribs bifurcate at midflank and may be separated by an intercalatory; ribs strong over venter but effaced from dorsum; suture simple, with small saddles and bifid lobes. *Lower U. Alb.* **Fig. 701C-D** - *I. ellipticoides* (Spath).

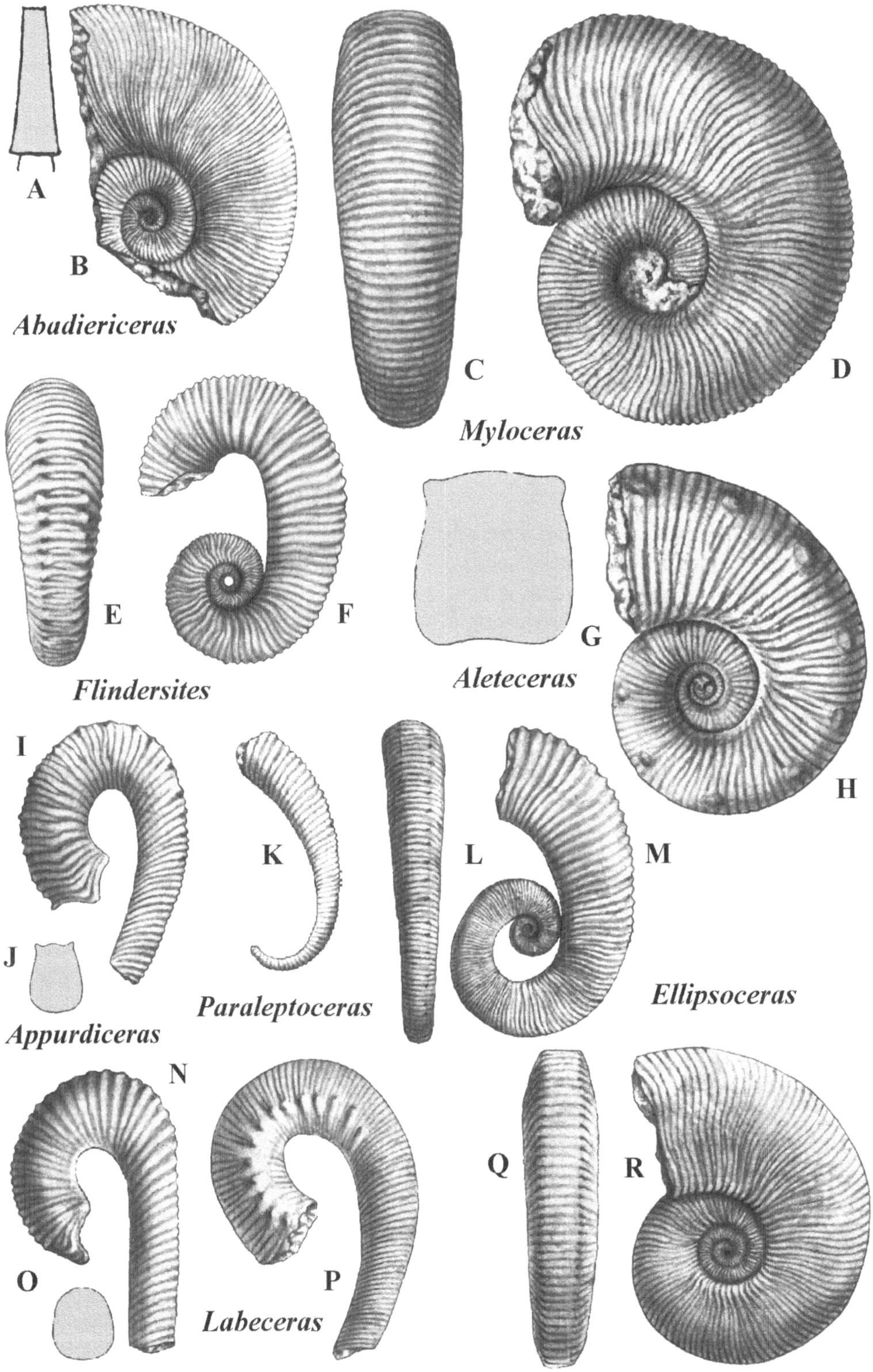

Fig. 703. Labeceratidae.

Algerites PERVINQUIÈRE, 1910 [*A. sayni*; OD]. Like *Idiohamites* but subelliptical serpenticone in middle growth apparently followed by straight body shaft with terminal hook; very strongly compressed rectangular whorls, very weak blunt straight to slightly curved ribs arise singly or in pairs from umbilical shoulder with intercalatories and slightly clavate ventrolateral tubercles joined by low transverse rib across narrowly rounded venter; regular weak constrictions to phragmocone; simple suture with very broad E, long narrow saddles, L/U longer than E/L, nonsplayed lobes and bifid I. *L. Cenom.* **Fig. 699I-J** - *A. sayni* Pervinquière.

Family LABECERATIDAE Spath, 1925

= Labeceratidae Whitehouse, 1926

Discussion. A monophyletic group with an origin in Anisoceratidae (Wiedmann, 1962). Large aleteceratines are invariably accompanied by much smaller labeceratines with different coiling, and dimorphism has been claimed (Klinger, 1989) and, according to Wright (1996: 233) "... There is little doubt that *Labeceras* .... comprises microconchs and *Myloceras* macroconchs of the same genus". However Etheridge (1909) figured both macroconchs and microconch of *Flindersites* with the same coiling, and Henderson (1990) found little support for this view. Here the subfamilies Labeceratinae and Aleteceratinae are recognized, recording a fascinating radiation that has been masked by a conservative taxonomy. Restricted to the Austral Realm. *Upper M. - Lower U. Alb., ?L. Cenom.*

Subfamily Labeceratinae Spath, 1925

Discussion. Monophyletic derivatives of Algeritini, close to *Ndumuiceras*; the suture of *Labeceras* is a simplified less incised version of *Protanisoceras*. Typically small ancylocones with strong prorsiradiate ribbing, with or without umbilical and/or ventrolateral tubercles, lappetted adult aperture and simple suture with narrow trifid L and reduced U. *Upper M. - Lower U. Alb.*

Labeceras SPATH, 1925 [*L. bryani* Whitehouse, 1926; ICZN Opinion 556]. Small ancylocone with small crioconic spire, almost straight proversum and incurved retroversum; whorl section subcircular, slightly compressed; rather coarse non-tuberculate ribs prorsiradiate on proversum becoming more distant and rectiradiate to slightly rursiradiate on retroversum; adult aperture with rather long lateral lappets; suture with exceptionally broad shallowly bifid saddles and very narrow trifid lobes; U of similar size to I. *Upper M. - Lower U. Alb.* **Fig. 703N-O** - *L. bryani* Whitehouse; **Fig. 703P** - *L. papulatum* Whitehouse.

= *Labeceras* Whitehouse, 1926 (junior homonym).

Paraleptoceras A. LEANZA, 1970 [*P. singulare*; OD]. Like *Labeceras* but with periodic ventrolateral tubercles to flexus and incised suture with longer narrower deeply bifid E/L and L much deeper than E. *U. Alb.* **Fig. 703K** - *P. singulare* Leanza.

Appurdiceras WHITEHOUSE, 1926 [*Ancyloceras cordycepoides* Etheridge, 1905; OD]. Small, spire and slightly curved proversum in one plane; venter tabulate, dorsum rounded and whorl section compressed oval to subquadratic; strong prorsiradiate ribs coarsening on body chamber where blunt ventrolateral horns cover 1-3 ribs, with umbilical bullae on intermediate ribs. *Lower U. Alb.* **Fig. 701O-P** - *A. cordycepoides* (Etheridge).

Henderson (1990) interpreted the genus to comprise all labeceratids with ventrolateral tubercles, e.g. *L. etheridgei* (Whitehouse) (**Fig. 703I-J**), but this interpretation seems too broad. Inclusion of *Appurdiceras* in the synonymy of *Anisoceras* cannot be supported, since it has a trifid L to the suture.

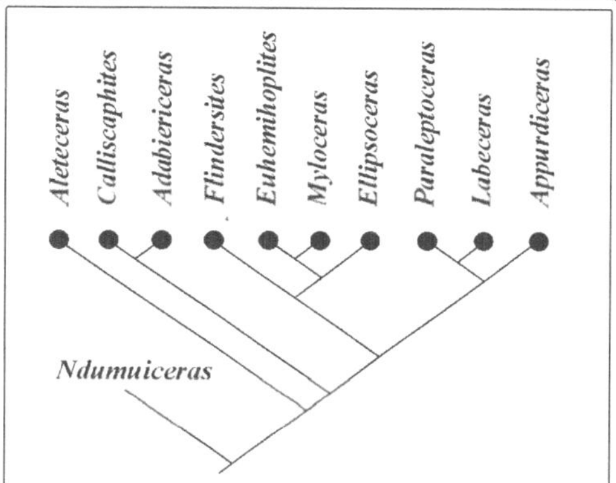

Fig. 704. Hypothesized relationships among Labeceratidae.

Subfamily Aleteceratinae Whitehouse, 1926

(*nom. transl.* herein *ex* Aleteceratidae)
= Myloceratidae Spath, 1939c.

Discussion. The sister taxon of Labeceratinae; small to gigantic with criocone earliest whorls; later evolute with planispiral or scaphitoid coiling and strongly compressed to depressed quadratic whorls; dense fine flexuous ribs commonly looping to ventrolateral tubercles and across venter; tubercles

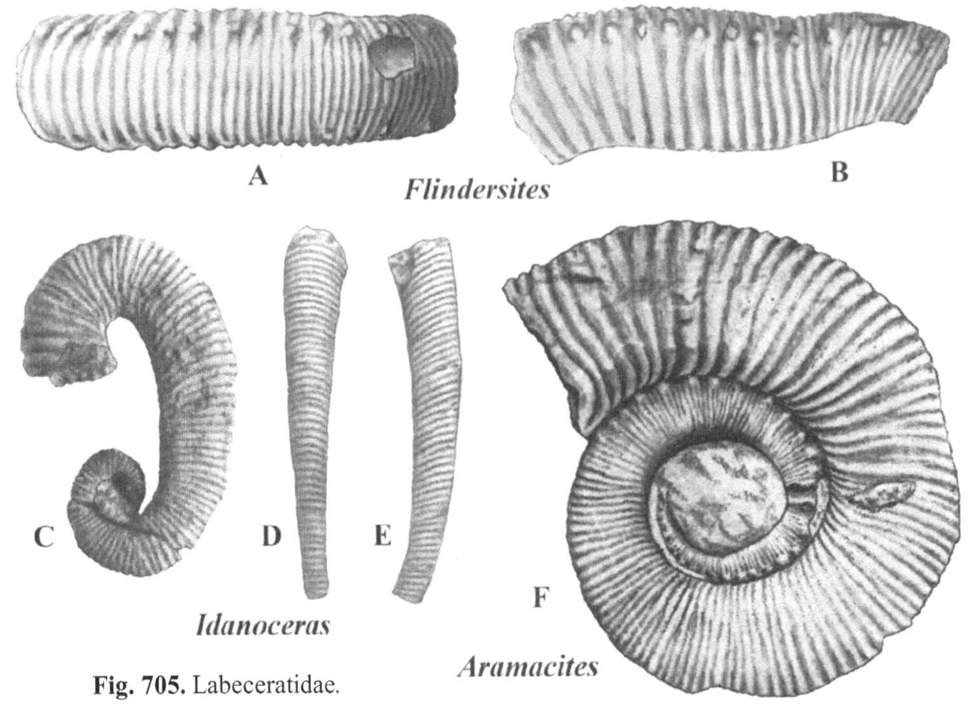

Fig. 705. Labeceratidae.

may be lost in later growth; relatively complex suture with bifid saddles and trifid lobes. *Upper M. - Lower U. Alb.*

Myloceras SPATH, 1925 [*Crioceras ammonoides* Etheridge, 1909; OD]. Small to medium sized; initial loose criocone followed by rapidly expanding planispire; flat sided with subtabulate venter and very compressed rectangular whorl section; ribs fine, flexuous, prorsiradiate, joined singly or in pairs to small elongate ventrolateral spines and mostly single across venter; tuberculation lost on adult body chamber. *Upper M. - Lower U. Alb.* **Fig. 701M-N** - *L. ammonoides* (Etheridge).

Myloceras orbiculus Whiteaves (**Fig. 703C-D**), a questionable macroconch, differs from the type in its larger size and feeble tuberculation in middle to later growth.

= *Myloceras* Whitehouse, 1926 (obj.).

Aleteceras WHITEHOUSE, 1926 [*Crioceras plectoides* Etheridge, 1909; OD]. Like *Myloceras* but with more inflated subcircular to subquadrate whorls and periodic large rounded ventrolateral tubercles that cover up to 3 ribs; suture with L not as deep as E. *Lower U. Alb.* **Fig. 703G-H** - *A. plectoides* (Etheridge) (juv.).

Flindersites WHITEHOUSE, 1926 [*F. baccatus*; OD] [**M**]. Very large ancylocone with body chamber uncoiling to form scaphitoid hook; subquadratic whorl section with low rounded venter; sharp straight ribs looping to rounded ventrolateral tubercles and again across the venter; suture with broad-based E/L. *Lower U. Alb.* **Fig. 705A-B** - *F. baccatus* Whitehouse (proversum).

Some specimens of *F. flindersi* (McCoy) are said to attain "gigantic" proportions (Etheridge, 1909) and, presumably, are macroconchs since the microconch (Etheridge 1909, pl. 40, figs 5-6) is much smaller with strongly clavate ventrolateral tubercles joined across the venter by 3-4 ribs. *Flindersites serotinus* (Spath) (**Fig. 703E-F**) shows the features of a *Flindersites* microconch but giant macroconchs are unknown from Southern Africa.

Ellipsoceras COLLIGNON, 1950 [*E. expansum*; OD]. Like *Flindersites* but more compressed and high whorled, with narrow tabulate venter, elliptically uncoiled body chamber, finer flexuous ribs that bifurcate from weak umbilical bullae and again at various levels on flank, and tiny ventrolateral tubercles to all ribs; suture with fairly narrow bifid saddles, large trifid L and small trifid U and I. *U. Alb.* **Fig. 703L-M, Q-R** - *E. expansum* Collignon.

Adabiericeras COLLIGNON, 1950 [*A. altissimum*; OD]. Like *Ellipsoceras* but very strongly compressed, discoidal, with flat concave flanks, sharp ventrolateral edges, narrow tabulate venter, trapezoidal whorl section, prorsiradiate ribs with frequent long intercalatories, and a body chamber that does not uncoil. *U. Alb.* **Fig. 703A-B** - *A. altissimum* Collignon.

= *Adabieceras* Wright, 1996 (*nom. van.*).

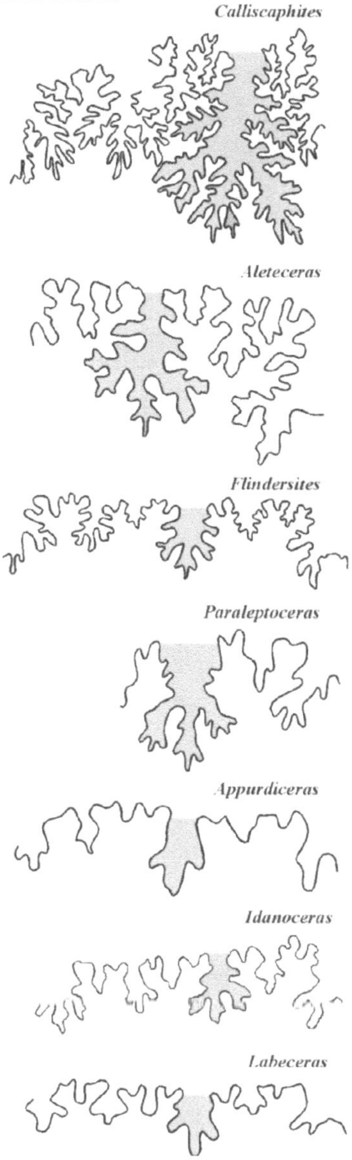

Fig. 706. Labeceratid sutures.

Euhemihoplites COLLIGNON, 1964 [*E. paradoxus*; OD]. Based on single curved fragment; rounded venter and compressed elliptical whorl section; numerous weak umbilicolateral bullae give rise to flexuous rursiradiate ribs that bifurcate at midflank, separated by 1-2 intercalatories; periodically ribs loop to large ventrolateral clavi joined across venter by looped ribs; suture with narrow trifid L deeper than E. *?L. Cenom.* **Fig. 701F-G** - *E. paradoxus* Collignon.

It has been suggested this specimen is based upon a stratigraphically misplaced, perhaps reworked, specimen (Wright, 1996).

Calliscaphites A. LEANZA, 1970 [*C. andinus*; OD]. Like *Aleteceras* but smaller, more compressed, in early growth

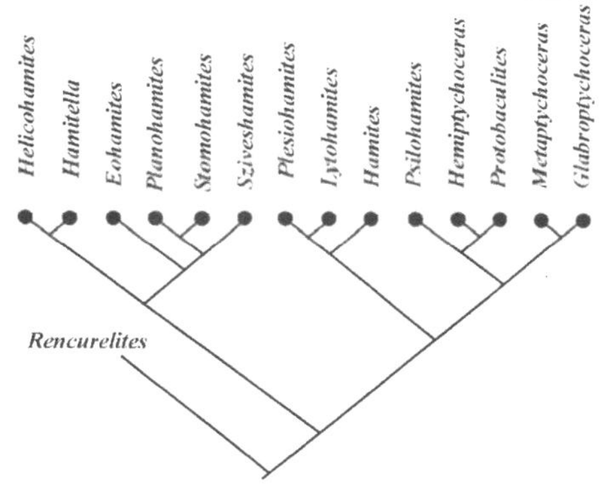

Fig. 707. Hypothesized relationships among Hamitinae and Metaptychoceratinae.

with fine tuberculate primary ribs separated by non-tuberculate secondaries, and a more complex suture with short E, oblique E/L and prominent trifid U. *U. Alb.* **Fig. 701H-L** - *C. andinum* Leanza.

Aramacites MCKENZIE, 1999 [*Crioceras axonoides* Etheridge, 1908; OD]. Medium sized criocone with ribs looped in 2's and 3's to large ventrolateral tubercles in early growth, separated by a non-tuberculate intercalatory; on adult body chamber tubercles joined across venter by paired ribs. *L. - M. Alb.* **Fig. 705F** - *A. axonoides* (Etheridge).

Idanoceras HENDERSON & MCKENZIE, 2002 [*I. klingeri*; OD]. Small; in early growth a very open planispire becoming more open with growth; bilaterally symmetrical, with slender, subcircular, gradually expanding whorls; fine closely spaced prorsiradiate annular ribs broader than interspaces coarsening on retroversum; suture with bifid saddles and trifid L as deep as E. *U. Alb.* **Fig. 705C-E** - *I. klingeri* Henderson & McKenzie.

Superfamily HAMITOIDEA Gill, 1871

(*nom. transl.* Wright & Wright, 1951 *ex* Hamitidae)

Discussion. Monophyletic derivatives of Anisoceratoidea that evolved along two lines constituting the families Hamitidae and Polyptychoceratidae. Mostly non-tuberculate, loosely coiled criocones, ancylocones and ptychocones, with simple annular ribbing, with or without constrictions. The rather simple suture has bilobed saddles and lobes. *L. Alb. - U. Maastr.*

Family HAMITIDAE Gill, 1871

Discussion. Monophyletic derivatives of Anisoceratinae; characteristically small, with non-tuberculate annular ribs and simple suture with

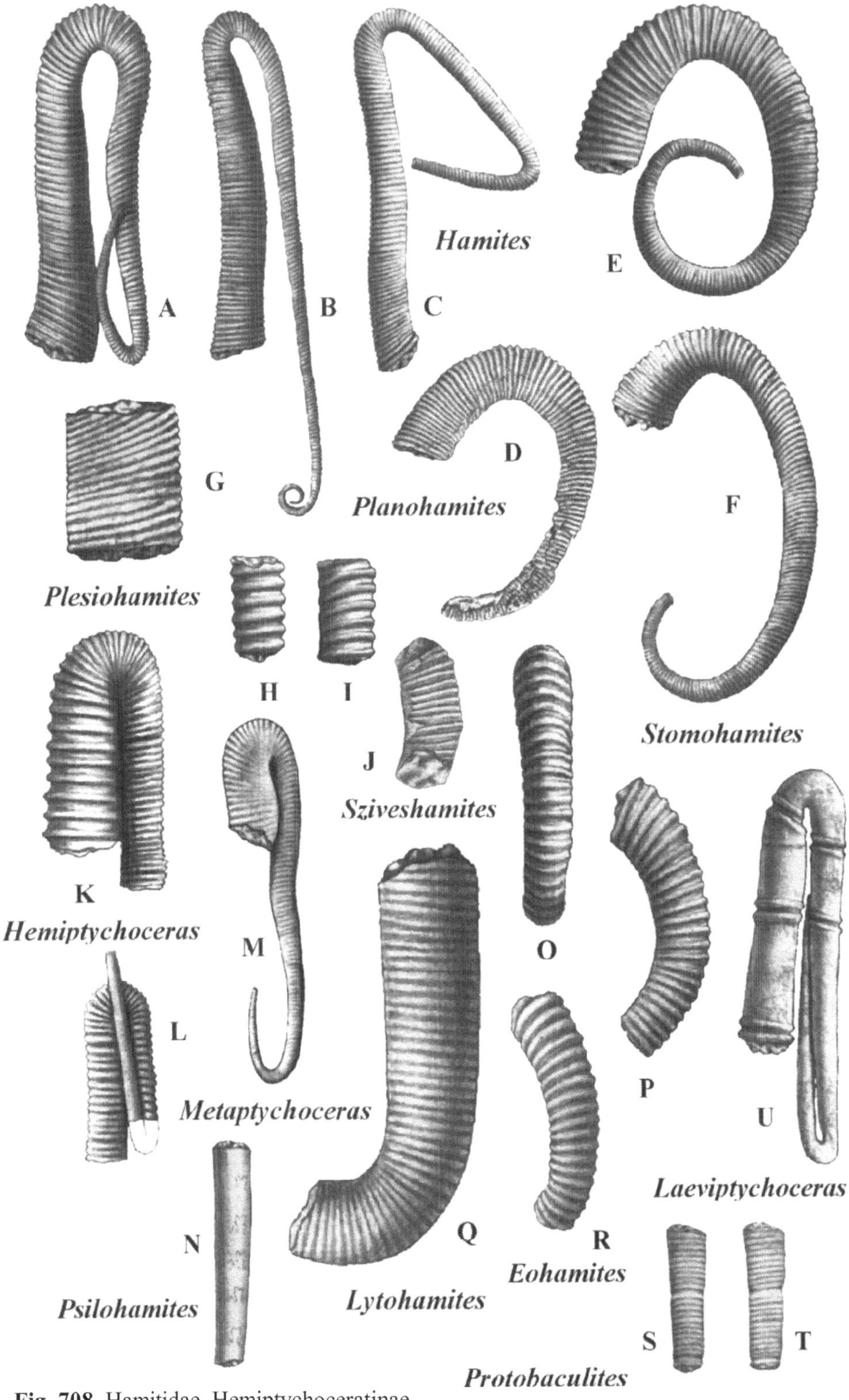

Fig. 708. Hamitidae, Hemiptychoceratinae.

bifid saddles and lobes. These are the common small non-turriliticone heteromorphs of the Albian and Cenomanian, at the base of an important late Cretaceous radiation. Comprises the subfamilies Stomohamitinae, Hamitinae and Metaptychoceratinae; here helicones are referred to the Turrilitidae. *L. Alb. - U. Turon.*

Subfamily Stomohamitinae nov.

Diagnosis. Small criocones, ellipticones and aspinocones, without collars or constrictions. Monophyletic derivatives of Protanisoceratinae, close to *Rencurelites*. *L. Alb. - U. Cenom.*

Eohamites MONKS, 2002 [*Hamites hybridus* Casey, 1961; OD]. Small aspinocone with simple, coarse, blunt annular ribs about as wide as interspaces; suture with large, subtrifid L, smaller trifid U and I with elongate, bifid median lobule. *L. Alb.* **Fig. 708R** - *E. hybridus* (Casey).

Stomohamites BREISTOFFER, 1940 [*Hamites virgulatus* Brongniart, 1822; OD]. Small, loose criocone coiling retained to middle growth followed by a subelliptical adult whorl; whorl section compressed oval to subelliptical; coarse blunt ribbing weak or effaced from dorsum, particularly in derived members, and sometimes weakening across siphonal line also; suture like *Hamites* but with L/U of similar size to E/L and relatively large bifid U in derived representatives. *M. Alb. - L. Cenom.* **Fig. 708E** - *S. virgulatus* (Brongniart); **Fig. 708F** - *S. gardneri* (Spath).

A problem with *Stomohamites* is that the holotype is lost and the selected neotype (Renz, 1968) is from a different locality to the holotype, and is based on such a short fragment its coiling is effectively unknown, as is the suture. Moreover, the whorl section is subcircular unlike the compressed subelliptical whorl section shown by the holotype and generally taken to be typical of the species. A new neotype, better displaying the generally accepted characters of the species, needs to be selected from topotype material.

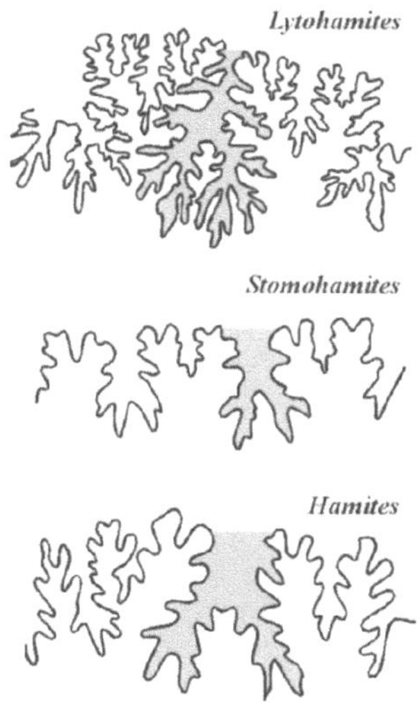

Fig. 709. Hamitid sutures.

Planohamites MONKS, 2002 [*Hamites compressus* J. Sowerby, 1811; OD]. Small aspinocone with strongly compressed elliptical whorls increasing rapidly in height on adult body chamber; dense, sharp annular ribs; suture with large bifid lateral lobes and slightly distorted and much smaller trifid U. *M. - U. Alb.* **Fig. 708D** - *P. compressus* (J. Sowerby).

Sziveshamites MONKS, 2002 [*Hamites nokonis* Adkins & Winton, 1920; OD]. Small aspinocone with compressed elliptical whorl section, simple annular ribs and band-like apertural collar (not shown by the type); suture with large bifid L and trifid U that are smaller and similar in size to trifid I. *Upper U. Alb. - M. Cenom.* **Fig. 708H-I** - *S. subvirgulatus* (Spath); **Fig. 708J** - *S. nokonis* (Adkins & Winton).

Subfamily Hamitinae Gill, 1871

(*nom. transl.* herein *ex* Hamitidae)

Discussion. Derivatives of Stomohamitinae in which the penultimate shaft and body chamber are almost straight, with a collared aperture. *L. Alb. - L. Cenom.*

Hamites PARKINSON, 1811 [*Hamites attenuatus* J. Sowerby, 1814; SD Diener, 1925]. Small to moderately large; initial low helix or criocone followed by 2-3 more or less straight shafts; whorls compressed, sometimes with slight dorsal impression due to whorl contact; sharp annular ribs, strong across dorsum and transverse over venter; apertural constriction in derived members; suture like *Protanisoceras* but simpler, with bifid saddles, large bifid L, L/U protruding beyond E/L, reduced trifid U and trifid I. *M. Alb. - L. Turon.* **Fig. 708A** - *H. attenuatus* J. Sowerby, **Fig. 708B** - *H. maximus* J. Sowerby, **Fig. 708C** - *H. simplex* d'Orbigny.

= *Torneutoceras* Hyatt, 1900 (obj.).

Lytohamites CASEY, 1961 [*L. similis*; OD]. Medium sized; known only from internal moulds of almost straight shafts and tight U-shaped bends; like *Hamites* but much larger, with finer denser blunter ribbing and much more incised suture with large trifid U; a homoeomorph of Upper Cretaceous *Diplomoceras*. *Uppermost Alb.* **Fig. 708Q** - *L. similis* Casey.

Plesiohamites BREISTROFFER, 1947 [*Hamites multicostatus* Brown, 1837; OD]. Known only from single flattened fragment; like *Lytohamites* but dense ribs more oblique, crowded dorsally and becoming more widely spaced and stronger ventrally. *L. Cenom.* **Fig. 708G** - *P. multicostatus* (Brown).

Subfamily Hemiptychoceratinae nov.

Diagnosis. Typically with vestigial to obsolete criocone stage followed by several adpressed more or less straight shafts; ribbed or not, with well developed constrictions and, in some, a collared aperture. Monophyletic derivatives of Stomohamitinae close to *Stomohamites* (Monks, 2002). *U. Alb. - U. Turon.*

Hemiptychoceras SPATH, 1925 [*Ptychoceras gaultianum* Pictet, 1847; OD]. Small; earliest stages of type species unknown; straight circular proversum adpressed into grooved dorsum of retroversum; proversum with simple annular ribs and periodic deep constrictions, ribs coarsening on retroversum; simple suture with narrow bifid L, smaller U and large I. *U. Alb. - L. Turon.* **Fig. 708K** - *H. gaultianum* (Pictet).

Protobaculites COLLIGNON, 1964 [*P. ambiguus*; OD]. Based on short fragment of straight shaft showing strong blunt annular ribs and prominent constriction. *U. Alb.* **Fig 708S-T** - *P. ambiguus* Collignon.

Indistinguishable from proversum of *Hemiptychoceras*.

Psilohamites SPATH, 1941 [*Hamites bouchardianus* d'Orbigny, 1842; OD]. Known only from fragments of very small, almost smooth straight shafts with compressed oval whorl section and well spaced septa; simple suture with subrectangular bifid saddles and well developed bifid L and U of similar size, and trifid I. *U. Alb.* **Fig. 708N** - *P. bouchardianus* (d'Orbigny).

Hemiptychoceras crassum (Kennedy) (**Fig. 708L**) shows an unornamented straight initial shaft which, if found isolated in the Upper Albian, would be identified as *Psilohamites* that Wright (1996) treated as a subgenus of *Hamites*. Monks (2002) regarded them generically distinct, and suggested the similarity of *Psilohamites* to Baculitidae was convergent. However the derived sutural characteristics of *Psilohamites*, particularly its well developed subrectangular bifid U, and shell form make it a plausible antecedent for Baculitidae.

Metaptychoceras SPATH, 1926 [*Ptychoceras smithi* Woods, 1896; OD]. Like *Hemiptychoceras* but smaller, with rapidly expanded retrorsum; fine ribs do not coarsen in maturity; constrictions restricted to proversum; suture relatively simple, with broad L. *?L. Cenom., U. Turon.* **Fig. 708M** - *M. smithi* (Woods).

Laeviptychoceras **gen. nov.** [*Metaptychoceras annulatum* Kennedy, 1988, p.100; OD herein]. Coiling ptychoceratid, comprising 3 straight subparallel shafts; first shaft in contact with second, except at flexus, and retroversum slightly separated from proversum; shell generally smooth but with shallow collared constrictions to last part of proversum and retroversum between which are occasional weak single ribs; a homoeomorph of *Alabushevites* (Polyptychoceratinae). *U. Turon.* **Fig. 708U** - *G. annulatum* (Kennedy).

Family POLYPTYCHOCERATIDAE Matsumoto, 1938

Discussion. A family of questionable ancestry with no derived characters to support an origin in Scalaritinae (Wright & Matsumoto, 1954). The first polyptychoceratines appear in the Late Turonian, contemporaneous with the last metaptychoceratines, and differ from the latter mainly in details of coiling and a suture with U consistently larger than I. Seemingly monophyletic derivatives of Hemiptychoceratinae; typically small to medium sized with up to 5 parallel shafts that may or may not be adpressed; generally with annular ribs and constriction, but a few are unornamented; suture relatively simple, with narrow stemmed bifid saddles and lobes, L generally deeper than E. *U. Turon. - U. Maastr.*

Polyptychoceras YABE, 1927 [*Ptychoceras pseudogaultinum* Yokoyama, 1890; OD]. Medium sized, with 5 straight parallel shafts in one plane separated by U-shaped bends; ribs prorsiriradiate in early growth but rectiradiate later, becoming coarser and more distant on flexus and adult body chamber, with periodic constrictions. *Coni. - U. Camp.* **Fig. 712B-C** - *P. pseudogaultinum* (Yokoyama).

If Hayakawa & Tashiro's (1994) assertion is correct, that *P. pseudogaultinum* is descended from *Subptychoceras* in the Campanian, then all pre-Campanian *Polyptychoceras* require a new generic name.

Rhyoptychoceras MATSUMOTO, 1977 [*R. mikasaense*; OD]. Medium sized; comprises 5 nearly straight shafts coiled in two different planes connected by tight U-shaped

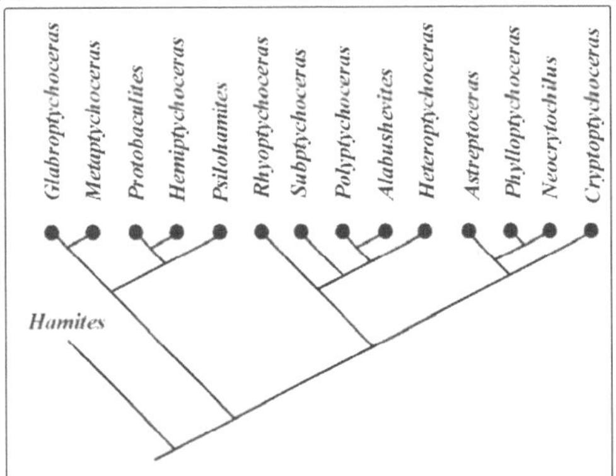

Fig. 710. Hypothesized relationships among Hemiptychoceratinae and Polyptychoceratidae.

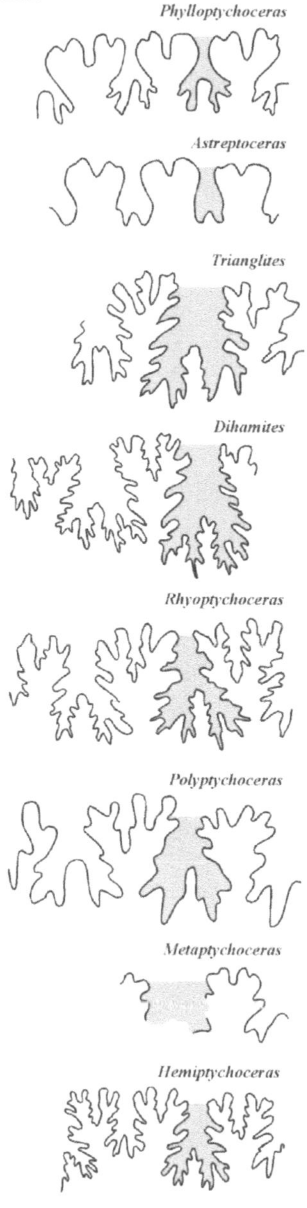

Fig. 711. Polyptychoceratid sutures.

bends; 1st, 4th and 5th (body chamber) shafts in same plane, 2nd and 3rd at right angles; body chamber hooked; initial shaft almost smooth but with fine sharp simple rectiradiate ribs in middle to later growth and periodic constrictions that become obsolete in maturity; suture with narrow stemmed deeply bifid saddles and lobes and small I. *Coni.* **Fig. 712J-K** - *R. mikasaense* Matsumoto.

Heteroptychoceras MATSUMOTO, 1977 [*H. obatai*; OD]. Like *Polyptychoceras* but smaller, with obliquely uncoiled adult body chamber forming wide U. *Sant.* **Fig. 712L** - *H. obatai* Matsumoto.

Alabushevites **gen. nov.** [*Polyptychoceras sakhalinum* Alabushev & Wiedmann, 1997, p.16; OD herein]. Like *Polyptychoceras* but shell smooth, with last two shafts in contact and frequent constrictions accompanied by flares. *L. Sant.* **Fig. 712G-H** - *A. sakhalinus* (Alabushev & Wiedmann).

Subptychoceras SHIMIZU, 1935 [*Hamites (Polyptychoceras) yubarensis* Yabe, 1927; OD]. Like *Polyptychoceras* but last two straight shafts with strong radial folds on which 2-6 riblets are superimposed. *Camp.* **Fig. 712D** - *S. yubarense* (Yabe).

Phylloptychoceras SPATH, 1953 [*Ptychoceras sipho* Forbes, 1846; OD]. Like *Polyptychoceras* but with virguliform opening at first bend; prorsiradiate growth striae only in early growth but later with low, broad, distant, rounded, slightly prorsiradiate folds that weaken over dorsum; suture very simple, with very narrow E with widely splayed branches, broad E/L, narrow L and fairly broad bifid L/U. *Maastr.* **Fig. 712E-F** - *P. sipho* (Forbes).

Neocyrtochilus ANDERSON, 1958 [*N. bryani*; OD]. Small (L=22 mm), with adpressed limbs and long retroversum; like *Phylloptychoceras* but smooth, without folds; growth striae on penultimate shaft strongly prorsiradiate, much less so on retroversum. *U. Camp.* **Fig. 712M** - *N. bryani* Anderson.

Astreptoceras HENDERSON, 1970 [*Ptychoceras zelandicum* Marshall, 1926; OD]. Small; known only from short straight fragments; like *Neocyrtochilus* but with faint folds and strong oblique constrictions. *Camp.* **Fig. 712A** - *A. zelandicum* (Marshall).

Cyrtoptychoceras KENNEDY & HENDERSON, 1992 [*Hamites undulatus* Forbes, 1846; OD]. Small, comprising single endogastrically curved shaft with circular whorl section and broad low blunt prorsiradiate ribs that cross venter transversely; aperture constricted, trumpet like with linguoid ventral lappet. *U. Maastr.* **Fig. 712I** - *C. undulatum* (Forbes).

Superfamily PHLYCTICRIOCERATOIDEA Spath, 1926

(*nom. transl.* Cooper, 1994 *ex* Phlycticrioceratidae)

Discussion. Monophyletic derivatives of Hamitoidea, comprising the families Phlycticrioceratidae and Neocrioceratidae; typically relatively large heteromorphs with variable coiling, uniform ribbing and ventrolateral tubercles. *U. Cenom. - U. Maastr.*

Family PHLYCTICRIOCERATIDAE Spath, 1926

Discussion. A monophyletic taxon with an origin in Hamitidae; venter bituberculate or trituberculate.

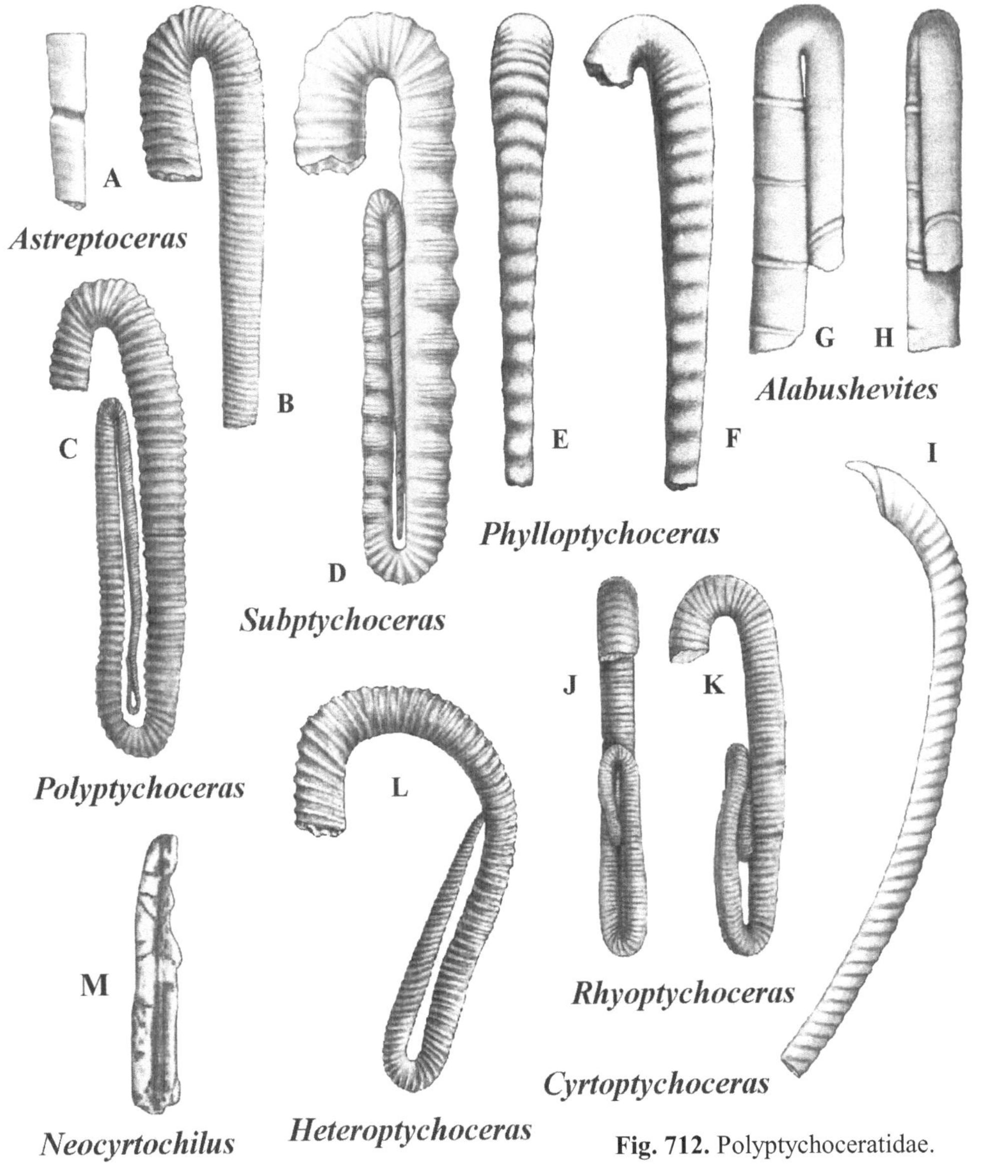

Fig. 712. Polyptychoceratidae.

Comprises the subfamilies Exiteloceratinae and Phlycticrioceratinae. *U. Cenom. - Maastr.*

Subfamily Exiteloceratinae Cooper, 1994

Discussion: Monophyletic derivatives of Hamitinae, close to *Stomohamites*, characteristically with criocone to loose helical coiling, spinose ventrolateral tubercles often joined across the subtabulate venter by looped ribs, thus mimicing *Idiohamites*; suture with well developed U. *Upper U. Cenom.-Maastr.*

***Tuberohamites* gen. nov.** [*Hamites larvatus* Conrad, 1855, p.265; OD herein]. For most of growth a loose criocone, intially with subcircular coiling, later elliptical; whorls subcircular at first but increasing rapidly in height and becoming compressed subrectangular to elliptical in middle to later growth, with flattened flanks and subtabulate venter; initially with slightly prorsiradiate non-tuberculate annular ribs but, in middle growth, small ventrolateral tubercles appear on some or all ribs joined across venter by looped ribs; suture incised, with deeply bifid saddles, narrow asymmetrically bifid lobes and L as long as E. *Upper U. Cenom. - U. Turon.* **Fig. 714E** - *T. larvatus* (Conrad).

Allocrioceras SPATH, 1926 [*Crioceras ellipticum* Woods, 1896, = *Hamites angustus* J. de C. Sowerby, 1850; OD]. Like *Tuberohamites* but, after initial ellipticone coiled in more or less regular loose helix with axial plane of whorls inclined to spiral axis; oblique ribs across venter except on straight adult body chamber, ventrolateral tubercles sometimes becoming clavate in maturity, constrictions present in early growth in some, and simpler less incised suture with broad square E/L, L much longer than E, and broader U/I. *Upper U. Cenom. - L. Coni.* **Fig. 714A-B** - *A. angustum* (J. de C. Sowerby); **Fig. 714C-D** - *A. annulatum* (Shumard).

Neancyloceras SPATH, 1926 [*Ancyloceras bipunctatum* Schlüter, 1872; OD]. Lile *Tuberohamites* but much larger, with very low expansion rate and simple distant uniform annular ribs with wide interspaces and tiny ventrolateral tubercles; suture with narrow stemmed E/L. *U. Camp.* **Fig. 714J** - *N. bipunctatum* (Schlüter).

Paralectotypes show loose elliptical coiling and periodic flared ribs; not all Schlüter's material came from the same level and there is every possibility that more than one species (or genus) is involved.

Exiteloceras HYATT, 1894 [*Ancyloceras jenneyi* Whitfield, 1880; SD Diener, 1925]. Like *Tuberohamites* but larger and coiled in a loose planispire throughout; early whorls with two or more open shafts connected by elbows followed by elliptical to criocone coiling; dorsum and flanks broadly rounded, venter narrowly rounded, whorl section oval to elliptical; mostly single weakly flexed rectiradiate to slightly rursiradiate ribs all with spinose ventrolateral tubercles connected across venter by broad flat or looped ribs; suture like *Tuberohamites* but more incised. *U. Camp.* **Fig. 714K-L** - *E. jenneyi* (Whitfield).

Axonoceras STEPHENSON, 1941 [*A. compressum*; OD]. Like *Exiteloceras* but small; coiling an initial open planispire followed by loose to subtrigonal criocone; whorl section initially circular, later compressed and subrectangular with flattened dorsum and subtabulate venter; strong sharp oblique ribs, weak or effaced from dorsum, mostly carry small rounded ventrolateral tubercles; aperture constricted and collared; suture very simple. *U. Camp. - L. Maastr.* **Fig.714F-G** - *A. compressum* Stephenson.

Lewyites MATSUMOTO & MIYAUCHI, 1984 [*Idiohamites? oronensis* Lewy, 1969; OD]. Like *Exiteloceras* but with loose subelliptical coiling, a subcircular to oval whorl section, flattened dorsum and flanks converging to broadly rounded to subtabulate venter; low, blunt, narrow, rectiradiate to weakly rursiradiate ribs link in 2's and 3's to rounded opposite ventrolateral spines joined across venter by delicate looped ribs on low broad transverse swellings; tuberculate ribs separated by 1-2 non-tuberculate ribs; suture like *Exiteloceras* but less incised, with splayed symmetrical L and short I. *U. Camp.* **Fig. 714M-N** - *L. oronensis* (Lewy).

Exicrioceras ANDERSON, 1958 [*E. ortigalitense*; OD]. Large criocone with whorls separating progressively with growth; dorsum and flanks flattened, venter narrowly rounded, with strongly compressed ovate whorl section; strong fine sinuous annular ribs with small ventrolateral nodes; constrictions oblique to ribbing prominent in early growth, declining later. *?Maastr.* **Fig. 714H-I** - *E. ortigalitense* Anderson.

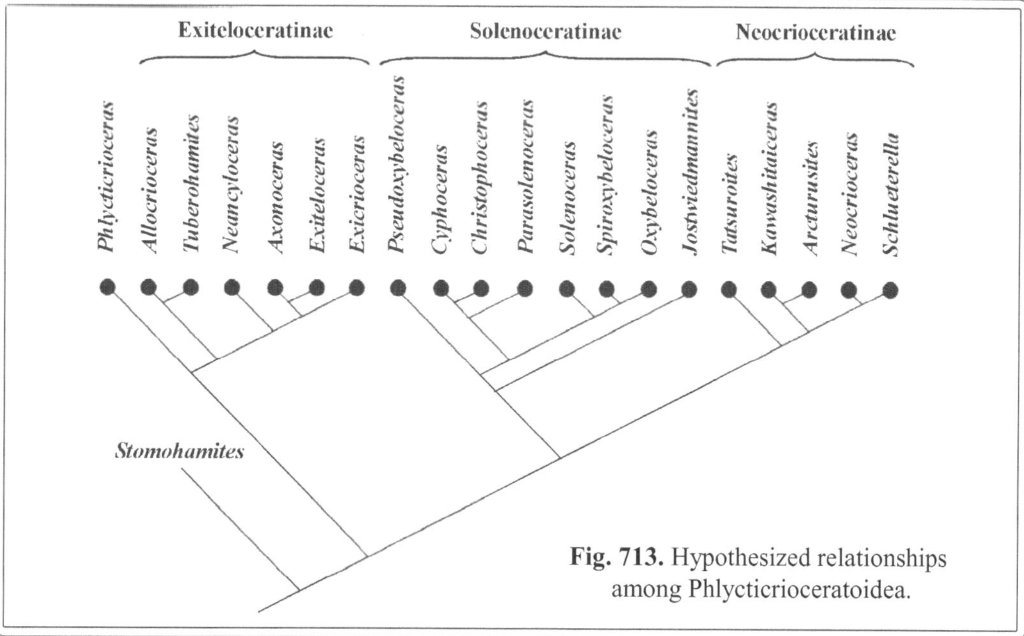

Fig. 713. Hypothesized relationships among Phlycticrioceratoidea.

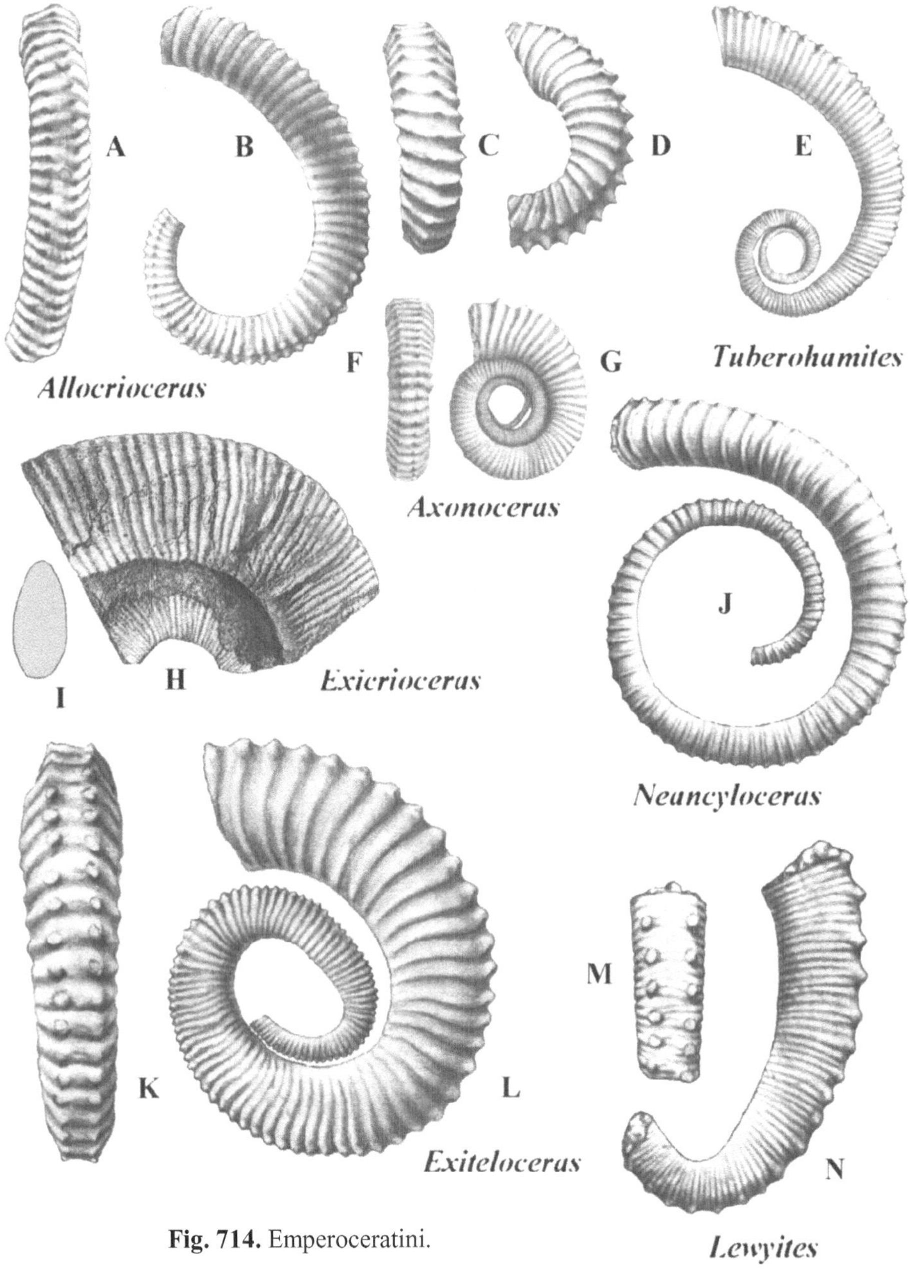

Fig. 714. Emperoceratini.

Subfamily Phlycticrioceratinae Spath, 1926

(*nom. transl.* herein *ex* Phlycticrioceratidae)

Discussion. Monotypic derivative of Exiteloceratinae, close to *Tuberohamites*; diagnosis as for genus. *U. Coni. - U. Sant.*

Phlycticrioceras SPATH, 1926 [*Ancyloceras? douvillei* de Grossouvre, 1894, =*Hamites trinodosus* Geinitz, 1850; OD]. Medium sized; early criocone followed by loose ellipticone with flat convergent flanks, acutely fastigiate venter, and teardrop-shaped costal whorl section; strong annular ribs, weak across dorsum, carry ventrolateral and siphonal clavi, the latter joining to form a serrated keel; suture with bifid saddles and lobes, L much deeper than E. *U. Coni. - U. Sant.* **Fig. 719C-D** - *P. trinodosum* (Geinitz).

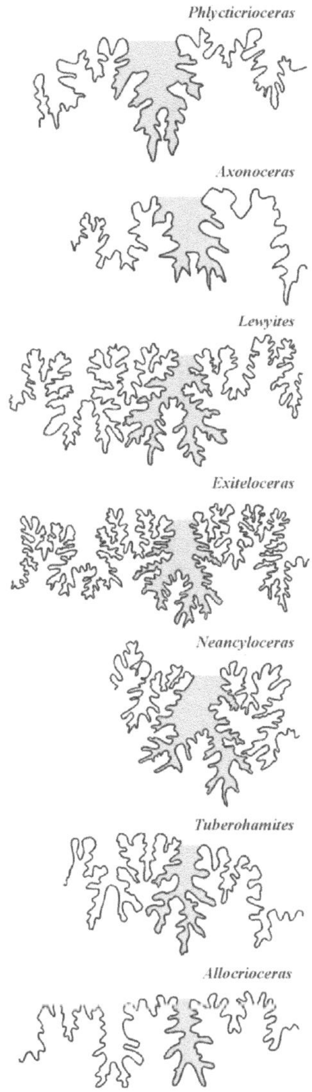

Fig. 715. Exiteloceratine sutures.

Family NEOCRIOCERATIDAE Spath, 1953

(*nom. transl.* Cooper, 1994 *ex* Neocrioceratinae)

Discussion. Monophyletic derivatives of Phlycticrioceratidae; mostly bi- or quadrituberculate ellipticones and ptychocones, but also criocones; comprises the subfamilies Solenoceratinae and Neocrioceratinae. *U. Turon. - U. Camp.*

Subfamily Solenoceratinae Cooper, 1994

Discussion. Monophyletic derivatives of Exiteloceratinae, close to *Tuberohamites*; mostly ellipticones and ptychocones with single or paired ventrolateral tubercles. *U. Turon. - L. Maastr.*

Jostwiedmannites COOPER, 1994 [*Hamites multinodosus* Schlüter, 1872; OD]. Known only from short weakly curved fragments of middle to adult stages with loose elliptical coiling,d low expansion rate and oval whorl section; differentiated ribbing with simple primaries sometimes backed by constriction separated by 4-11 finer secondary ribs; all ribs with 4 rows of tubercles, those of secondaries smaller; ribs may show incipient looping between tubercles. *U. Turon. - M. Coni.* **Fig. 716L-M** - *J. multinodosus* (Schlüter).

Pseudoxybeloceras WRIGHT & MATSUMOTO, 1954 [*Hamites quadrinodosus* Jimbo, 1894; OD]. Moderately large; straight parallel shafts joined by tight U-shaped bends; strong, sharp, slightly prorsiradiate ribs, narrower than interspaces, with ventrolateral tubercles in early growth; later quadrituberculate, with high lateral and ventrolateral rows of tubercles; species discrimination largely on shape of tubercles that vary from club-shaped to bullate, clavate and spinose. *L. Coni. - U. Camp.* **Fig. 716I-J** - *P. quadrinodosum* (Jimbo) with rounded tubercles; **Fig. 716K** - *P. matsumotoi* Collignon with club-shaped tubercles; **Fig. 718** - *P. amapondense* (van Hoepen) with clavate tubercles.

Parasolenoceras COLLIGNON, 1969 [*P. splendens*; OD]. Loose ellipticone with several straight subparallel shafts joined by tight flexi, compressed oval whorl section with tabulate venter, strong fine uniform annular ribs all with small spinose ventrolateral tubercles joined singly across venter. *L. Camp.* **Fig. 716C-D** - *P. splendens* Collignon.

Christophoceras COLLIGNON, 1969 [*C. ramboulai*; OD]. Like *Parasolenoceras* but with periodic quadrituberculate primary ribs with flared high-lateral tubercles that become auriculate on retroversum. *L. Camp.* **Fig. 716A-B** - *C. ramboulai* Collignon.

Cyphoceras WARD & MALLORY, 1977 [*Ancyloceras? lineatum* Gabb, 1869; OD]. Like *Christophoceras* but with criocone coiling persisting to middle growth; periodically enlarged spinose ventrolateral clavi joined by 2-4 flank costae, separated by simple bituberculate ribs; sometimes with broad wedge-shaped rib in front of looped ribs and, in some, fibulate ribs superimposed on broad low bulges. *L. Camp.* **Fig. 716E-F** - *C. lineatum* (Gabb).

Oxybeloceras HYATT, 1900 [*Ptychoceras crassum* Whitfield, 1880; OD]. Earliest whorls perhaps an open criocone followed by a broadly curved limb, then two tightly adpressed straight shafts with flattened venter, impressed dorsum and subcircular whorl section; strong, sharp, oblique ribs with small sharp ventrolateral nodes linked across venter by looped ribs; suture simple, with deep penetrating E, little incised bifid saddles, deeply bifid L and broad asymmetrically bifid U. *M. - U. Camp.* **Fig. 716G-H** - *O. crassum* (Whitfield).

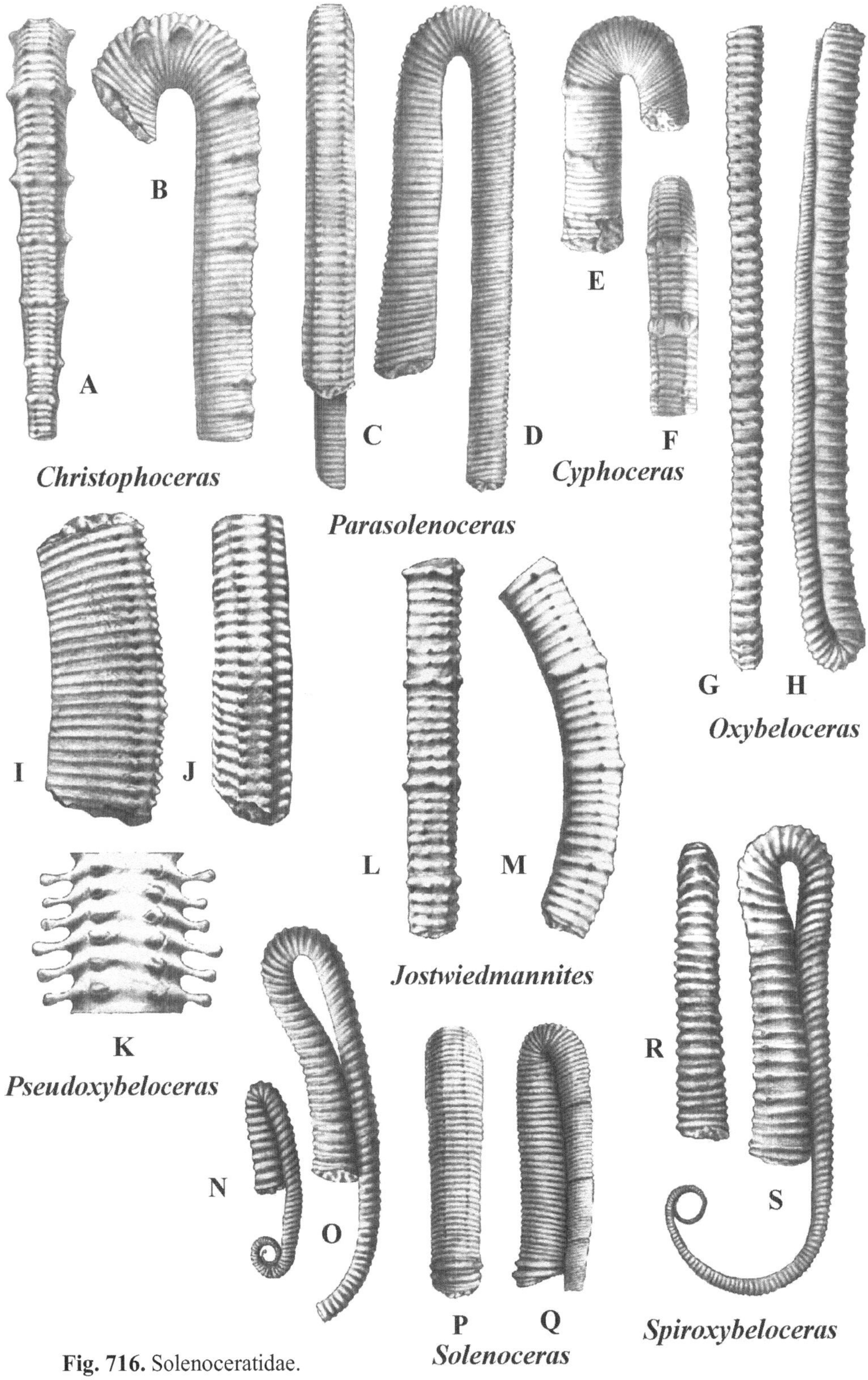

Fig. 716. Solenoceratidae.

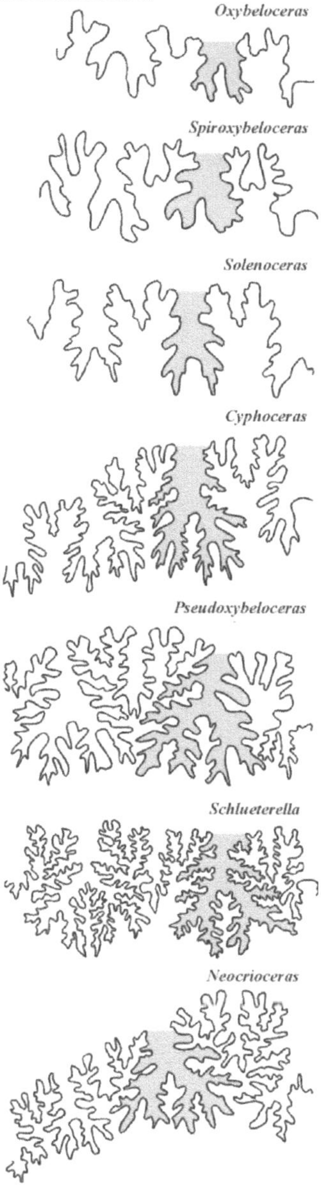

Fig. 717. Solenoceratine and Neocrioceratine sutures.

Spiroxybeloceras KENNEDY & COBBAN, 1999 [*Ptychoceras meekanum* Whitfield, 1880; OD]. Like *Oxybeloceras* but smaller, with early whorls a loose planispire followed by two parallel shafts barely in contact and teardrop-shaped opening to final elbow; constrictions present; suture simple, like *Oxybeloceras* but with shallower E. *U. Camp.* **Fig. 716N-O, R-S** - *S. meekanum* (Whitfield).

Solenoceras CONRAD, 1860 [*Hamites annulifer* Morton, 1842; OD]. Like *Spiroxybeloceras* but even smaller, with initial ammonitella followed by two closely adpressed straight shafts with impressed dorsal furrow; constrictions present or not, but adult aperture with prominent collared constriction; ventrolateral tubercles minute. *U. Camp. - L. Maastr.* **Fig. 716P-Q** - *S. annulifer* (Morton).

Subfamily Neocrioceratinae Spath, 1953

Discussion: A monophyletic group with an origin in Solenoceratinae, close to *Jostwiedmannites*; characteristically with looped quadrituberculate ribs separated by non-tuberculate secondaries. *U. Turon. - U. Camp.*

Tanabeites **gen. nov.** [*Neocrioceras? undulosum* Matsumoto, 1977, p.343; OD herein]. Earliest stages unknown; in middle to later growth an irregular low wide helix with moderate rate of expansion, subcircular whorl section, and body chamber forming part of helix; periodic main ribs with lateral and ventrolateral tubercles, separated by fine weak secondaries that, also, are superimposed on main ribs where they loop between tubercles; suture like *Neocrioceras*. *M. Turon.* **Fig. 719H** - *T. undulosus* (Matsumoto).

Kawashitaiceras MATSUMOTO, 1984 [*Neocrioceras dentatum* Matsumoto & Obata, 1981; OD]. Known only from almost straight slowly expanding body shaft; ornament dominated by very large extremely clavate ventrolateral and ventral tubercles, steepest anteriorly and with crenulated crests, separated by 4-5 simple fine, rather distant, annular, non-tuberculate ribs weakly looped between tubercles. *U. Turon., ?Coni.* **Fig. 719E-F** - *K. dentatum* (Matsumoto & Obata).

Arcturusites **gen. nov.** [*Neocrioceras annelisae* Klinger & Kennedy, 2003, p.313; OD herein]. Like *Kawashitaiceras* but known only from curved septate fragment suggesting loose criocone coiling; fine uniform ribs with lateral and ventrolateral clavi, the latter formed of sets of 3 smaller tubercles that increase in size posteriorly; tubercles of lateral row conical, but those of ventral row becoming increasingly clavate posteriorly. *M. Coni.* **Fig. 719G** - *A. annelisae* (Klinger & Kennedy).

Neocrioceras SPATH, 1921 [*Crioceras spinigerum* Jimbo, 1894; OD]. Criocone with rapidl expanding depressed whorls, maximum width at midflank and broad subtabulate venter; fine dense flexuous uniform ribs loop to

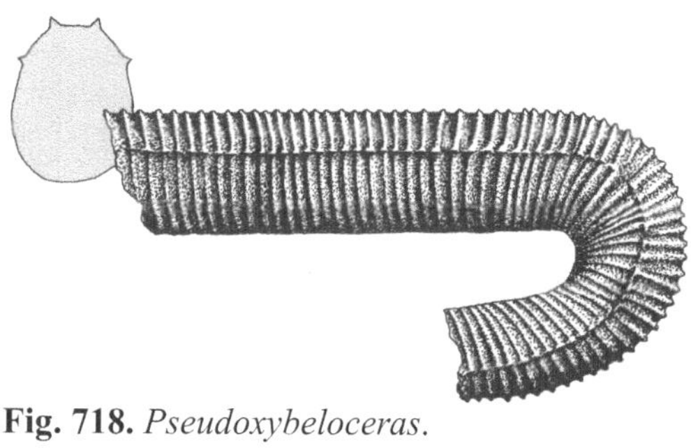

Fig. 718. *Pseudoxybeloceras.*

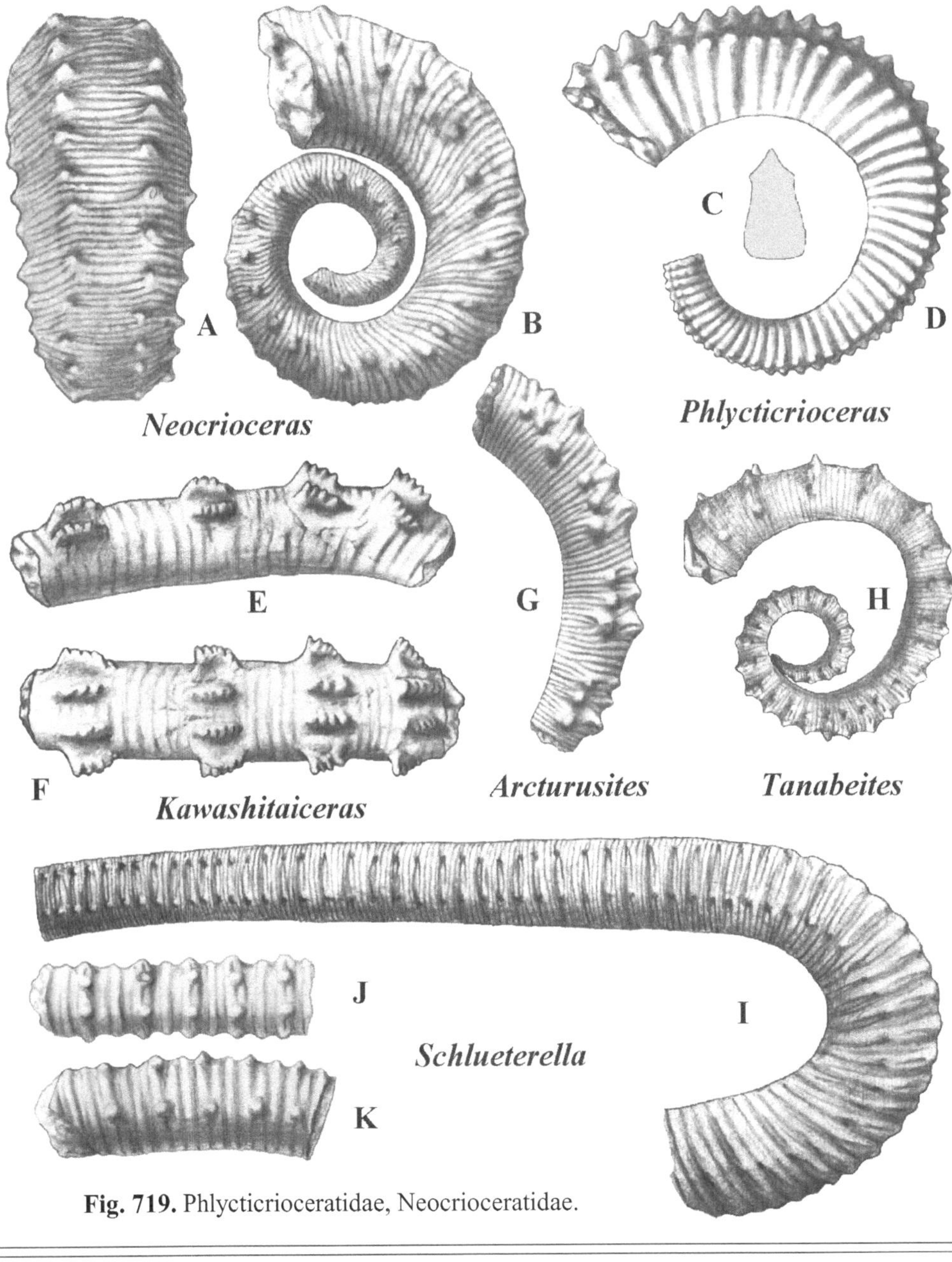

Fig. 719. Phlycticrioceratidae, Neocrioceratidae.

prominent mid-lateral tubercles, separated by 1-4 non-tuberculate secondaries, and then loop again to spinose ventrolateral tubercles that tend to alternate across venter; suture with bifid saddles and lobes, and deep I. *Sant.* **Fig. 719A-B** - *N. spinigerum* (Jimbo).

= *Cyrtohamites* Shimizu, 1933 [obj.].

Schlueterella WIEDMANN, 1962 [*Ancyloceras pseudoarmatum* Schlüter, 1872; OD]. Large; in early growth probably like *Tanabeites*; later with looped quadrituberculate ribs on swollen bases separated by up to 5 secondary ribs of equal strength; tubercles low-lateral and ventrolateral in position and opposite across venter; ribbing coarsens on adult body chamber. *L. - U. Camp.* **Fig. 719I-K** - *S. pseudoarmata* (Schlüter).

Superfamily TURRILITOIDEA Gill, 1871

(*nom. transl.* Wright, 1957 *ex* Turrilitidae)

Discussion. Monophyletic derivatives of Hamitoidea; spirally coiled with asymmetrical quadrilobate suture throughout, generally with bifid lobes.

Comprises the families Turrilitidae, Diplomoceratidae and Nostoceratidae. *M. Alb. - Maastr.*

Family TURRILITIDAE Gill, 1871

Discussion. Questionably monophyletic derivatives of Hamitidae comprising the subfamilies Pseudhelicoceratinae, Turrilitinae and Ostlingoceratinae. These are the common helical heteromorphs of the Middle Albian to Upper Cenomanian. *M. Alb. - U. Cenom.*

Subfamily Pseudhelicoceratinae Breistroffer, 1953

Discussion. Whereas the origin of Turrilitinae in Hamitidae is secure, that of Pseudhelicoceratinae is less certain (*cf.* Wright, 1996); it is possible it was derived independently from Protanisoceratinae, evolving in parallel with Albian Turrilitinae. Comprises small to medium sized helicones whose body chamber may uncoil in cork-screw fashion, initially with two rows of tubercles but later four. *M. Alb. - Lower U. Alb.*

Parahelicoceras BREISTROFFER, 1953 [*Turrilites catenatus* d'Orbigny, 1842; OD]. Small; coiled in a fairly tight high spire (apical angle ~25°) with whorls almost in contact; siphuncle at midflank; weak ribs loop between 2 rows of tubercles exposed on outer face, sometimes with intervening non-tuberculate ribs; twice as many ribs on upper whorl face as on lower; on lowest part of flank and base of spire, single ribs strengthen into radially elongate bullae; suture with bifid saddles, L/U smaller than E/L, bifid L much deeper than E, smaller bifid U (trifid in some species) and very small trifid I. *Upper M. Alb.* **Fig. 724H** - *P. catenatum* (d'Orbigny).

= *Fibuloceras* Chiriac, 1981 (obj.).

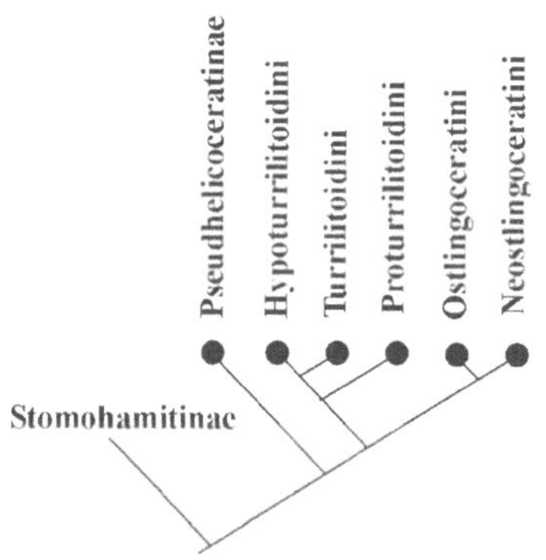

Fig. 720. Hypothesized relationships among Turrilitidae.

Subhelicoceras BREISTROFFER, 1953 [*Turrilites bituberculatus* d'Orbigny, 1842; OD]. Small; like *Parahelicoceras* but with larger spiral angle, bigger tubercles, weaker finer ribs that are more numerous on upper whorl face, almost smooth outer face and suture with narrower deeper U and I. *M. Alb.* **Fig. 724P-Q** - *S. bituberculatum* (d'Orbigny).

Hemiturrilites BREISTROFFER, 1953 [*Turrilites elegans* d'Orbigny, 1842; OD]. Small; like *Parahelicoceras*, but with simple uniform oblique quadrituberculate ribs, without intervening non-tuberculate riblets. *Lower U. Alb.* **Fig. 724E-F** - *H. elegans* (d'Orbigny).

Pseudhelicoceras SPATH, 1922 [*Turrilites robertianus* d'Orbigny, 1842; OD]. Medium sized helicone with very acute spiral angle; quadrituberculate main ribs separated by non-tuberculate secondaries, and body chamber uncoiling in cork-screw fashion; suture very incised, with widely splayed L. *U. Alb.* **Fig. 724I** - *P. robertianum* (d'Orbigny).

= *Spiroceras* Meek, 1876 (*non* Quenstedt, 1858).

= *Pseudohelicoceras* Matsumoto, 1977 (*nom. van.*).

Subfamily Turrilitinae Gill, 1871

(*nom. transl.* Cooper 1990 *ex* Turrilitidae)

Discussion. Questionably monophyletic derivatives of Hamitidae, typically coiled in a more or less tight helix with dorsolaterally positioned siphuncle and asymmetrical suture; comprises the tribes Proturrilitoidini, Turrilitini and Hypoturrilitini. *M. Alb. - L. Coni.*

Tribe Proturrilitoidini nov.

Type genus: *Proturrilitoides* Breistroffer, 1940.

Diagnosis. Partially to completely helically coiled, typically non-tuberculate, with oblique strong annular ribs, with or without a collared aperture; asymmetrical suture and siphuncle at middle of exposed whorl face. Monophyletic derivatives of Stomohamitinae, close to *Stomohamites*; the common turrilites of the Middle Albian. *M. Alb. - L. Coni.*

Helicohamites MONKS, 2002 [*Baculites parkinsoni* Fleming, 1814; OD]. Small, known only by fragments; a helicone with simple ribs and collared aperture; suture with large, bifid L and U, and smaller trifid I; third lateral saddle similar in breadth to first and second. *U. Alb. - U. Cenom.* **Fig. 721H-I** - *H. parkinsoni* (Fleming); **Fig. 721J-K** - *H. duplicatus* (Pictet & Campiche).

Hamitella BREISTOFFER, 1947 [*Helicoceras annulatum* d'Orbigny, 1842; OD]. Small; a loose low helix, perhaps with almost straight adult body chamber; whorl section circular to subcircular; annular ribs oblique across venter and dorsum; suture like *Hamites*, with subrectangular

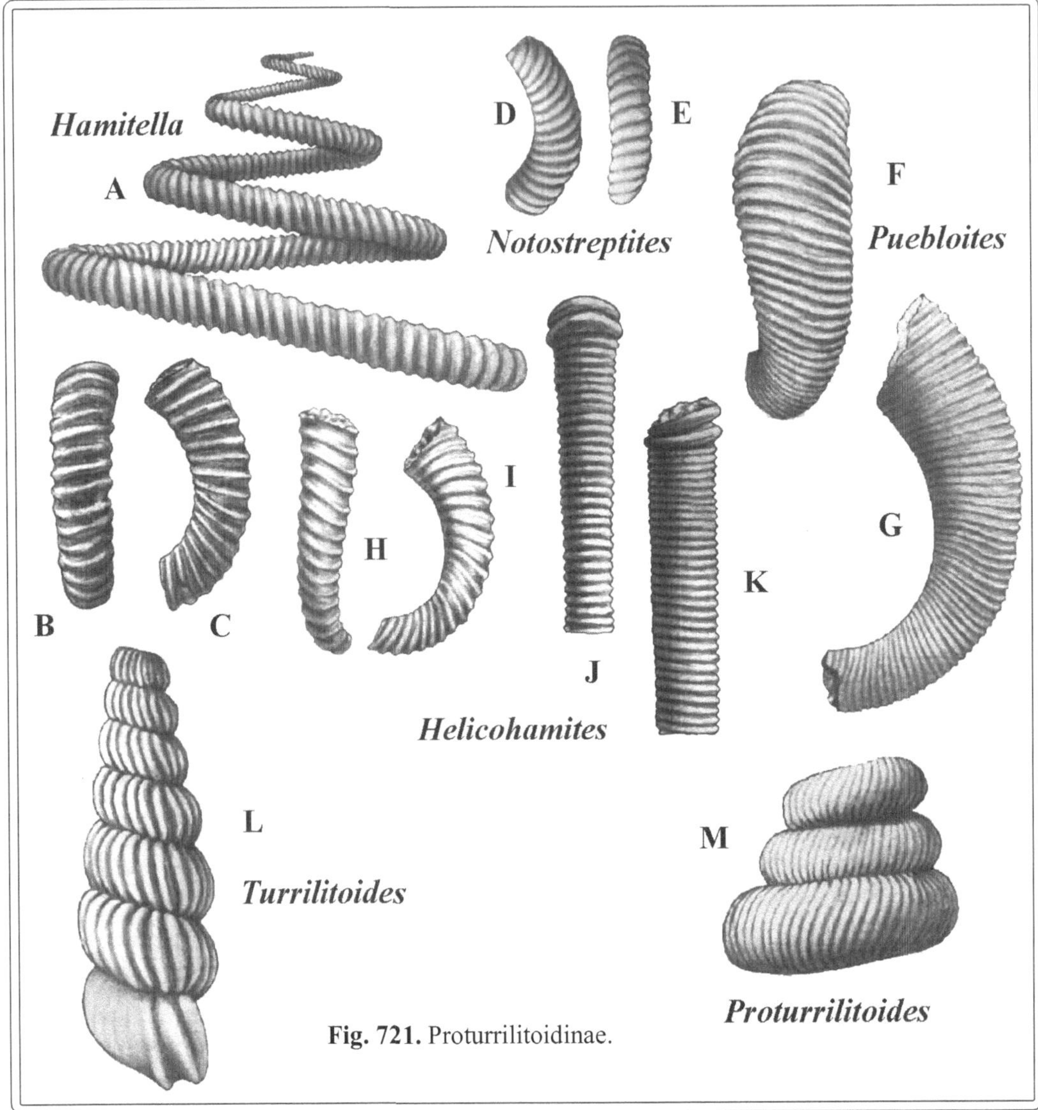

Fig. 721. Proturrilitoidinae.

bifid saddles and lobes, L as deep as E, and reduced U. *M. - U. Alb.* **721A-C** - *H. ibex* (Spath); **Fig. 721H-I** - *H. annulata* (d'Orbigny).

= *Helicoceras* d'Orbigny, 1842 (*non* Koenig, 1825).

Proturrilitoides BREISTROFFER, 1940 [*Turrilites astierianus* d'Orbigny, 1842; OD]. Small; like *Hamitella* but more tightly coiled, with whorls almost in contact and body chamber forming part of helix; ribs frequently bifurcating on adoral face; suture with widely splayed L and narrow stemmed saddles. *Lower M. Alb.* **Fig. 721M** - *P. astierianus* (d'Orbigny).

= *Proturrilites* Henderson, 1990 (*laps. cal.*).

Notostreptites HENDERSON, 1990 [*N. exilis*; OD]. Small; known only from slender fragments suggesting low loose helix with little helical translation; subcircular whorl section, sharp annular rursiradiate ribs, and suture with shallow E, oblique E/L, splayed bifid L much deeper than E, and bifid L/U much larger than E/L. *U. Alb.* **Fig. 721D-E** - *N. exilis* Henderson.

Puebloites COBBAN & SCOTT, 1972 [*Helicoceras? corrugatum* Stanton, 1894; OD]. Medium sized; like *Hamitella* but larger, with narrower apical angle and rapidly expanding whorls; compressed oval to elliptical whorl section, finer ribs that coarsen on adult body chamber, and more incised suture with narrower stemmed deeply bifid saddles and deeper E. *U. Cenom. - L. Coni.* **Fig. 721F-G** - *P. corrugatus* (Stanton).

Tribe Turrilitini Gill, 1871

(*nom. transl.* herein *ex* Turrilitinae)

Type genus: *Turrilites* Lamarck, 1801.

Diagnosis. Small to medium sized, primitively non-tuberculate, later with 2-4 rows of generally rounded

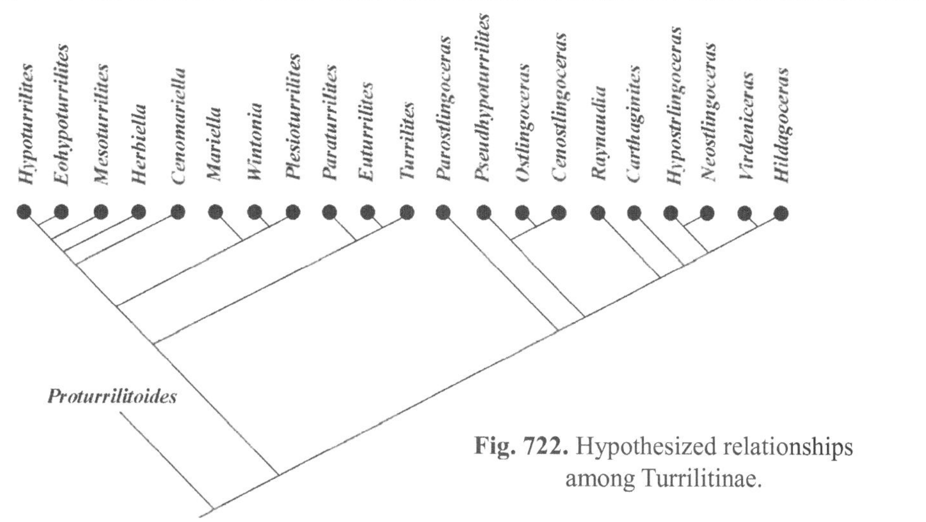

Fig. 722. Hypothesized relationships among Turrilitinae.

to bullate tubercles of similar size. Seemingly monophyletic derivatives of Proturrilitoidini, close to *Proturrilitoides*. *M. Alb. - U. Cenom.*

Turrilitoides SPATH, 1923 [*Turrilites hugardianus* d'Orbigny, 1842; OD]. Small to medium sized; like *Notostreptites* but more tightly coiled with whorls in contact, flatter flanks, oval to angulated whorl section and acute spiral angle; microconch with almost smooth adult body chamber, with collared and constricted aperture; suture with L frequently bifid. *Upper M. - Upper U. Alb.* **Fig. 721L** - *T. hugardianus* (d'Orbigny).

Paraturrilites BREISTROFFER, 1947 [*Turrilites gresslyi* Pictet & Campiche, 1861; OD]. Medium sized helicone with only 3 rows of tubercles, those on umbilical shoulder obsolete or vestigial; ribbing weak to strong, weakly tuberculate variants closely approach *Turrilitoides* but with angulated whorls. *Upper U. Alb.* **Fig. 724L** - *P. gresslyi* (Pictet & Campiche); **Fig. 726A** - *P. circumtaeniatus* (Stoliczka).

Wintonia ADKINS, 1928 [*W. graysonensis*, = *Turrilites bosquensis* Adkins, 1920; OD]. Like *Hemiturrilites* but with upper part of outer whorl face smooth and only 3 rows of tubercles exposed on outer whorl face; upper two rows of tubercles very prominent, followed by deep spiral groove with ridge-like lower margin carrying small clavi; 4th row of tubercles hidden on outer part of lower whorl face that has strong radial ribs; whorls remain in contact throughout and, in one species (*rhacioformis*), the adult aperture curves upwards. *Upper U. Alb.* **Fig. 724M** - *W. bosquensis* (Adkins).

Wintonia was created for small pyritic specimens with a straight smooth grooved initial shaft and only 3 rows of tubercles. Subsequent studies have shown the material is juvenile, the straight shaft is pathological (it is developed also by gastropods in the same deposit) and that a 4th row of tiny tubercles was overlooked.

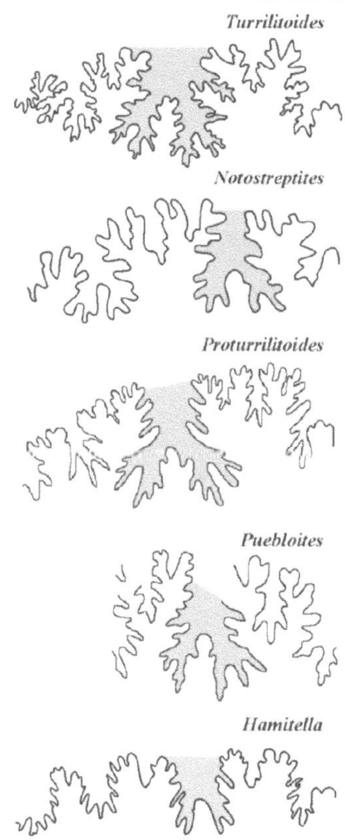

Fig. 723. Turrilitid sutures.

Mariella NOWAK, 1916 [*Turrilites bergeri* Brongniart, 1822; OD]. Like *Wintonia*, with 4 rows of tubercles exposed on outer face of whorls, but no spiral groove and tubercles of uniform size. *Upper U. Alb.* **Fig. 724K** - *M. bergeri* (Brongniart).

= *Bergericeras* Wiedmann, 1962 (obj.).

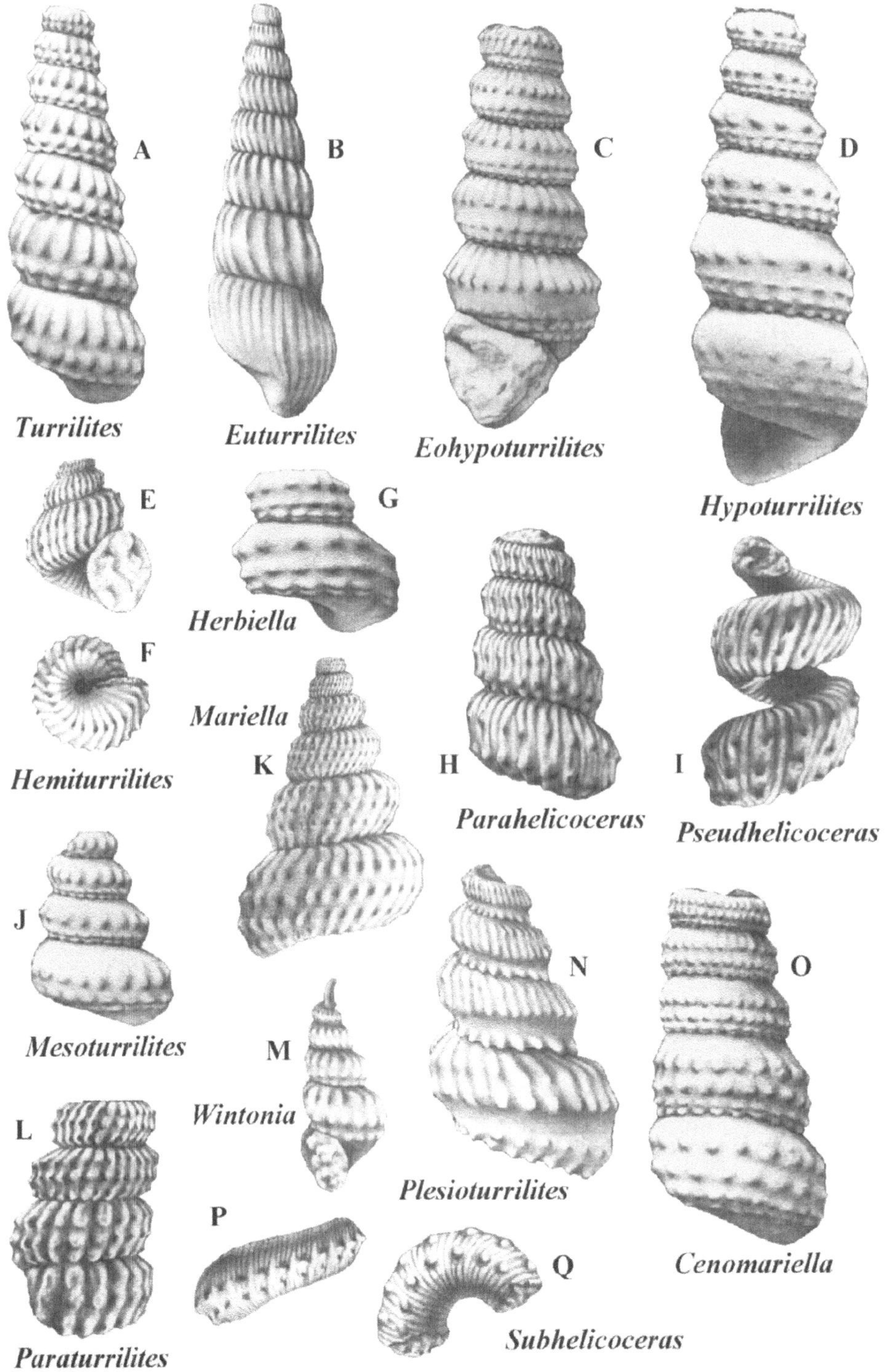

Fig. 724. Turrilitinae.

Plesioturrilites BREISTROFFER, 1953 [*Turrilites brazoensis* Roemer, 1852; OD]. Like *Wintonia* but much larger, with different whorl profile, strong ribs to upper part of exposed whorl face, conical tubercles to 3rd row and detached pendant body chamber. *Upper U. Alb.* **Fig. 722N** - *P. brazoensis* (Roemer).

Turrilites LAMARCK, 1801 [*Turrilites costatus*; OD]. Like *Paraturrilites* but with stronger ribs and only 3 rows of tubercles; two rows of sharply bullate to pointed tubercles on outer face separated by smooth spiral band, with third row of smaller markedly clavate tubercles sometimes merged into spiral ridge, hidden by spiral suture; suture with broad asymmetrical E on upper whorl face, somewhat narrow E/L, very broad L and narrower trifid L/U. *L. - M. Cenom.* **Fig. 724A** - *T. costatus* Lamarck; **Fig. 726B** - *T. acutus* Passy.

Euturrilites BREISTROFFER, 1953 [*Turrilites scheuchzerianus* Bosc, 1801; OD]. Like *Turrilites* but with flattened outer whorl face, strong radial ribs that weaken or are interrupted by a spiral groove at midflank, bordered by weak bullae in early growth; spiral groove becomes obsolete in later growth when ribs cross outer flank strongly; upper and lower whorl faces smooth; adult aperture hooded; suture with L much deeper than E. *L. - U. Cenom.* **Fig. 724B** - *E. scheuchzerianus* (Bosc).

= *Eoturrilites* Scholz, 1979 (*laps. cal.*).

Tribe Hypoturrilitini nov.

Type genus: *Hypoturrilites* Dubourdieu, 1953.

Diagnosis. Ttypically with 4 rows of tubercles, the lower 2-3 rows spirally elongate; in derived members the upper row forms enlarged septispinate spines. Monophyletic derivatives of Turrilitini, close to *Mariella*. *U. Alb. - L. Cenom.*

Cenomariella COOPER, 1999 [*Turrilites oehlerti* Pervinquière, 1910; OD]. Medium sized; like *Mariella*, with

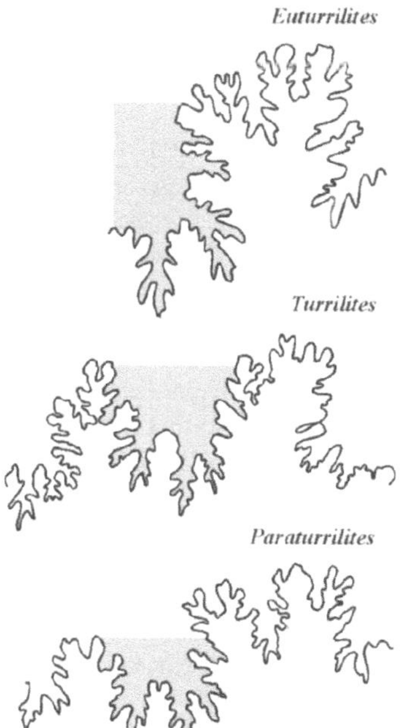

Fig. 725. Turrilitid sutures.

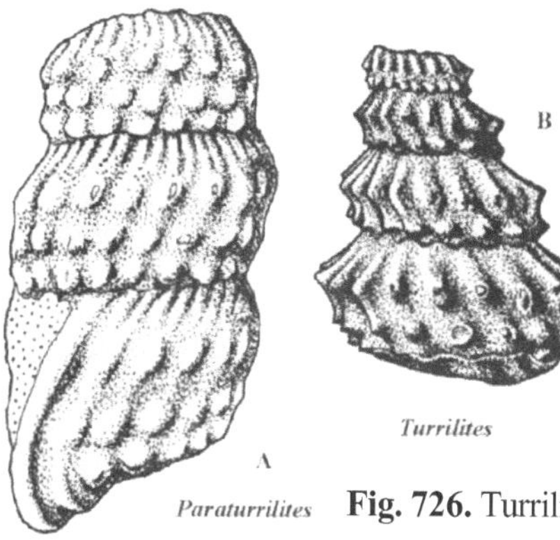

Fig. 726. Turrilitidae.

strong oblique quadrituberculate ribs, but lower (adoral) two rows of tubercles approximated and spirally elongated. *Upper U. Alb. - L. Cenom.* **Fig. 724O** - *C. oehlerti* (Pervinquière).

Herbiella COOPER, 2012 [*Turrilites bicarinatus* Kner, 1852; OD]. Like *Cenomariella*, but with lower spire, tubercles dominant and lower three rows of tubercles spirally elongated. *L. Cenom.* **Fig. 724G** - *H. bicarinata* (Kner).

= *Klingerella* Cooper, 1999 (*non* Dingle 1982, Ostracoda).

Mesoturrilites BREISTROFFER, 1953 [*Turrilites aumalensis* Coquand, 1862; OD]. Small; like *Herbiella* but with upper part of outer whorl face smooth; second and third rows of tubercles joined by spiral ridges and separated by spiral groove; 4th row of tubercles covered by succeeding whorl; base of whorl with radial ribs; rib and tubercle density increases with growth. *L. Cenom.* **Fig. 724J** - *M. aumalensis* (Coquand).

Eohypoturrilites SCHOLZ, 1973 [*Turrilites mantelli* Sharpe, 1857; OD]. Like *Herbiella* but larger, high spired, with much narrower (28°) spiral angle, convex outer flanks, deeply impressed spiral suture, and 3 rows of small, strongly clavate tubercles on lower part of outer whorl that outnumber pointed tubercles of upper row. *Upper U. Alb. - L. Cenom.* **Fig. 724C** - *E. mantelli* (Sharpe).

Hypoturrilites DUBOURDIEU, 1953 [*Turrilites gravesianus* d'Orbigny, 1842; OD]. Large, like *Eohypoturrilites* but with somewhat larger apical angle, narrower smooth band at midflank, far fewer much enlarged mid-lateral bosses (septate bases to prominent spines) and weak to obsolete ribs on adapical shoulder. *L. Cenom.* **Fig. 724D** - *H. gravesianus* (d'Orbigny).

= *Hypoturrilites* Shimizu, 1935 (*nom. nud.*).

Subfamily Ostlingoceratinae Cooper, 1999

Discussion. Monophyletic derivatives of uncertain ancestry, perhaps *Turrilitoides* (Turrilitinae) but Pseudhelicoceratinae close to *Parahelicoceras* remains a possibility; characteristically with small

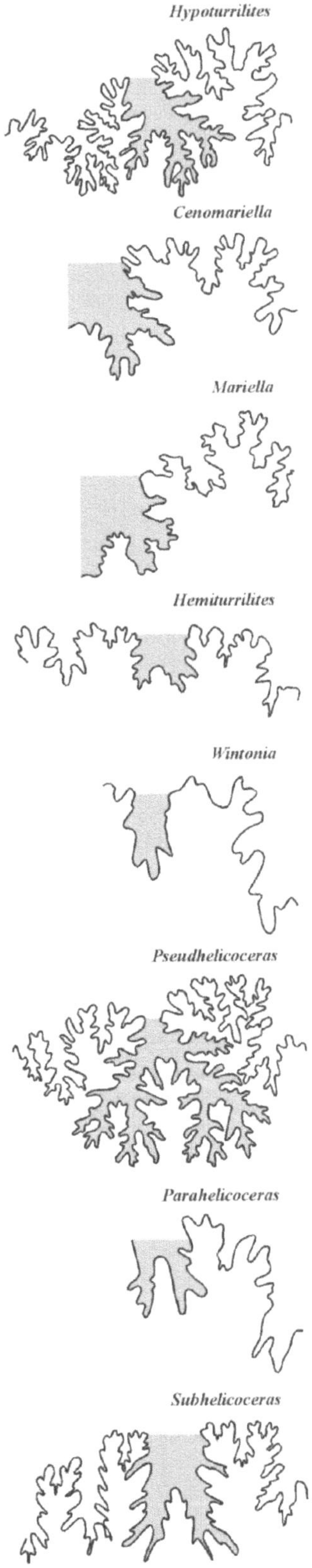

Fig. 727. Turrilitid sutures.

apical angle, flat sided whorls and 2-3 rows of tubercles concentrated on adoral shoulder, but also developing a row at midflank in derived members. Comprises the tribes Ostlingoceratini and Neostlingoceratini. *M. Alb. - U. Cenom.*

Tribe Ostlingoceratini Cooper, 1999

(*nom. transl.* herein *ex* Ostlingoceratinae)
Type genus: *Ostlingoceras* Hyatt, 1900.

Diagnosis. Whorls with flattish to rounded outer face and strong simple ribs with two rows of tubercles on adoral shoulder. Monophyletic derivatives of Turrilitoidini, close to *Turrilitoides*. *M. Alb. - Lower Cenom.*

Parostlingoceras BREISTROFFER, 1953 [*Turrilites moutonianus* d'Orbigny, 1842; OD]. Helically coiled, but with whorls barely in contact; phragmocone with rounded to flat outer face, becoming flat or even slightly concave on adult body chamber; strong annular ribs interrupted by spiral zone of weakening at midflank; every 2-4 ribs terminate above spiral zone of weakening in feeble but distinct bullae; strong ribs on anterior face terminate on lower flank in elongate bullae that are sometimes weakly binodate; siphuncle position unknown. *M. Alb.* **Fig. 729A** - *P. moutonianum* (d'Orbigny).

Ostlingoceras HYATT, 1900 [*Turrilites puzosianus* d'Orbigny, 1842; OD]. Like *Parostlingoceras* but with tighter coiling, 3 rows of tubercles to distinctly facetted anterior shoulder, poorly defined spiral zone of weakening, and far more rows of tubercles on anterior shoulder than ribs on posterior edge; some species show weak tendency to develop lateral bullae just above midflank. *Upper U. Alb. - L. Cenom.* **Fig. 729B** - *O. puzosianum* (d'Orbigny).

Cenostlingoceras COOPER, 1999 [*Turrilites bechii* Sharpe, 1857; OD]. Like *Ostlingoceras* but with rounded outer face with dense prorsiradiate slightly flexuous ribs that terminate in bullae at midflank and are separated by weakly ornamented to smooth spiral band from 3 rows of spirally elongate tubercles on lower flank, those of the upper two rows arranged on spiral ribs in maturity; base of whorl ornamented with slightly prorsiradiate ribs. *Lower L. Cenom.* **Fig. 729D** - *C. bechii* (Sharpe).

Pseudhypoturrilites COOPER, 1999 [*Hypoturrilites sharpei* Wright & Kennedy, 1996; OD]. Like *Ostlingoceras* but with stronger coarser ribs carrying 3 rows (?4, there is a suggestion the adoral row may be bipartite) of bullae, the upper row above midflank and separated from lower two rows by spiral smooth zone. *Lower L. Cenom.* **Fig. 729K** - *P. sharpei* (Wright & Kennedy).

Tribe Neostlingoceratini nov.

Type genus: *Neostlingoceras* Klinger & Kennedy, 1978.

Diagnosis. Typically small, with flat to shallowly concave and even grooved flank, and feeble ribbing if at all. Primitively with prominent mid-lateral tubercles outnumbered by 1-2 rows of smaller tubercles on the adoral shoulder, but tubercles obsolete in derived members. Monophyletic derivatives of Ostlingoceratinae close to *Pseudhypoturrilites*.The dominant turrilites in the Upper Cenomanian. *Upper U. Alb. - U. Cenom.*

Raynaudia DUBOURDIEU, 1953 [*Turrilites (Carthaginites) raynaudianus* Collignon; OD]. Very small helicone with acute spiral angle, flat outer surface to smooth unornamented whorls and pair of spiral ridges defining oblique facet on lower flank; siphuncle at upper shoulder; suture very simple. *Upper U. Alb.* **Fig. 729F** - *R. raynaudianus* (Collignon).

Carthaginites PERVINQUIÈRE, 1907 [*Turrilites (C.) kerimensis* Pervinquière; OD]. Like *Raynaudia* but with subparallel flanks with narrowly rounded shoulders, spiral groove at midflank and siphuncle at inner upper angle of whorl; some species with row of small tubercles on adoral shoulder; suture barely indented, with very broad weakly bifid E/L and shallow L. *U. Cenom.* **Fig. 729H** - *C. kerimensis* (Pervinquière).

Connected to *Hildagoceras* by *C. aquillonius* Kennedy & Cobban (**Fig. 729G**).

Neostlingoceras KLINGER & KENNEDY, 1978 [*Turrilites carcitanensis* Matheron, 1842; OD]. Small, like *Carthaginites* but with slightly concave flanks with prominent midlateral tubercles greater outnumbered by 2-3 rows small tubercles on adoral shoulder and more complex suture. *L. Cenom.* **Fig. 729C** - *N. carcitanense* (Matheron).

Hypostlingoceras MATSUMOTO & TAKAHASHI, 2000 [*H. japonicum*; OD]. In early growth like *Neostlingoceras*, later like *Cenostlingoceras* with fine oblique ribs lacking lateral tubercles. *L. Cenom.* **Fig. 729E** - *H. japonicum* Matsumoto & Takahashi.

Hildagoceras COOPER, 1999 [*Neostlingoceras virdenense* Cobban, Hook & Kennedy; OD]. Superficially like *Neostlingoceras* but with external lobe (E) at midflank, a single row of clavate tubercles on adoral shoulder, and distinct spiral groove to whorls. *U. Cenom.* **Fig. 729J** - *H. virdenense* (Cobban, Hook & Kennedy).

***Virdeniceras* gen. nov.** [*Neostlingoceras procerum* Cobban, Hook & Kennedy, 1988, p.60; OD herein]. Small, like *Neostlingoceras* but without mid-lateral tubercles (except vaguely on adult body chamber) and with strong spiral groove near base of whorl. *U. Cenom.* **Fig. 729I** - *V. procerum* (Cobban, Hook & Kennedy).

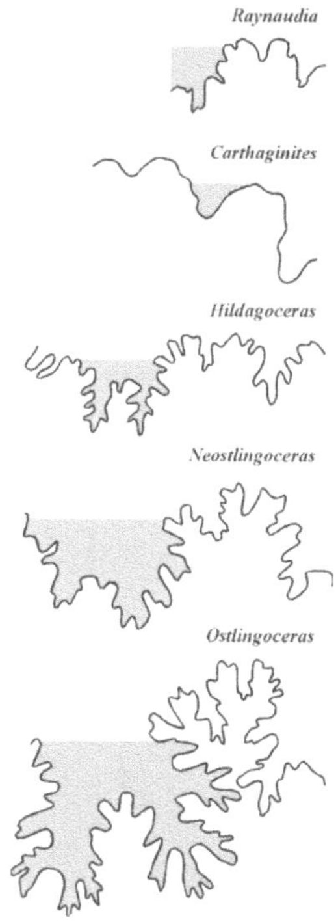

Fig. 728. Ostlingoceratine sutures.

Family DIPLOMOCERATIDAE Spath, 1926

Discussion. Monophyletic derivatives of Turrilitidae; mostly relatively large with variable coiling, simple non-tuberculate annular ribs, periodic constrictions and a suture with L/U commonly larger than E/L. These are the "*neohamites*" of the Upper Cretaceous. *Sant. - U. Maastr.*

Subfamily Diplomoceratinae Spath, 1926

Discussion. Monophyletic derivatives of Proturrilitoidinae, close to *Puebloites*. Coiling variable, but mostly non-tuberculate, finely ribbed loose criocones, helicones and ellipticones, rarely turriliticones, with constrictions generally confined to early growth; sutures relatively complex. Comprises the tribes Diplomoceratini and Smedaliceratini. *Sant. - U. Maastr.*

Tribe Diplomoceratini Spath, 1926

Type genus: *Diplomoceras* Hyatt, 1900.

Diagnosis. Primitively small, becoming medium sized to large with time; venter rounded, without tubercles. *Sant. - U. Maastr.*

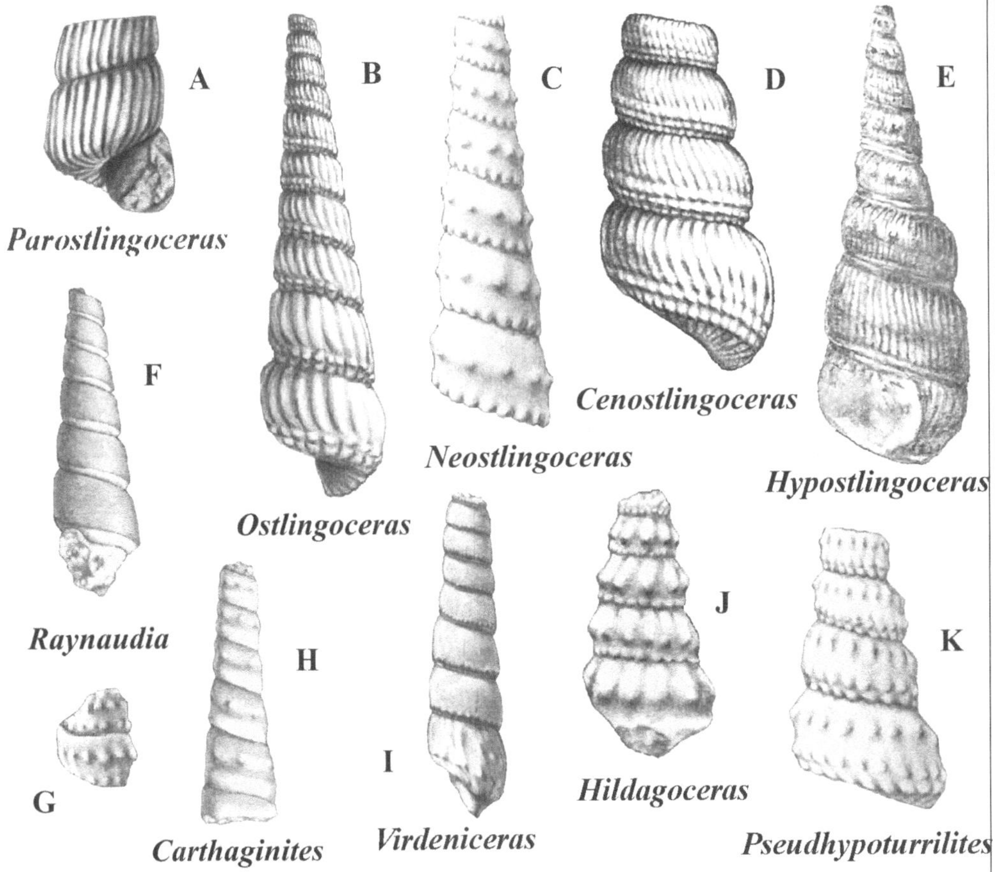

Fig. 729. Ostlingoceratinae.

Dihamites MATSUMOTO, 1977 [*D. obiraensis*; OD]. Small, earliest whorls unknown; first shaft long and gently curved, recurving on itself to form oblique long gently curved penultimate shaft and hooked adult body chamber; simple rather distant annular ribs coarsen and become scale-like in maturity; periodic weak irregular constrictions. *Sant.* **Fig. 733A** - *D. obiraensis* Matsumoto.

Ryugasella WRIGHT & MATSUMOTO, 1954 [*R. ryugasensis*; OD]. Small; comprises straight or gently curved fragments, seemingly coiled in one plane; sharp narrow uniform annular ribs slightly prorsiradiate on flanks, crossing venter transversely and with forward arch on dorsum; occasional deep uncollared constrictions in early growth; suture with narrow stemmed elements. *U. Sant. - Camp.* **Fig. 733P-Q** - *R. ryugasensis* Wright & Matsumoto.

= *Ryugasella* Matsumoto 1942 (*nom. nud.*).

Glyptoxoceras SPATH, 1925 [*Hamites rugatus* Forbes, 1846; OD]. Medium sized; early whorls an open criocone followed by triangular later whorls with almost straight shafts joined by angular bends; whorl section subcircular to distinctly compressed; sharp simple annular ribs cross venter and dorsum transversely; adult body chamber straight with collared and constricted peristome; suture with rather narrow stemmed bifid saddles and lobes, E long and narrow, L as deep as E, and L/U larger than E/L. *U. Sant.- U. Maastr.* **Fig. 733N-O** - *G. rugatum* (Forbes).

Neohamites BRUNNSCHWEILER, 1966 [*N. giraliensis*; OD]. Like *Glyptoxoceras* but with compressed oval whorl section, narrower venter and frequent shallow constrictions bordered anteriorly by weak collar; suture with shallowly bifid E/L, narrow bifid L as deep as E, very broad deeply bifid L/U and prominent bifid U/I; the claim that coiling was polyptychoceratoid has not been supported by subsequent studies. *U. Maastr.* **Fig. 733K-M** - *N. giraliensis* Brunnschweiler.

Morewites SHIGETA, 2014 [*M. sakakibarai*; OD]. Small with initial straight shaft followed by tiny helicone of two whorls, then by a slowly expanding ellipticone whose axis is perpendicular to that of the spire; simple transverse annular ribs, narrower than interspaces, increase in density towards the aperture; suture simple, with

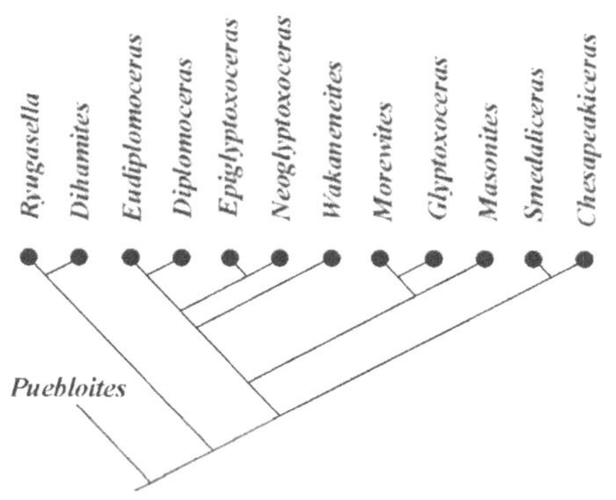

Fig. 730. Hypothesized relationships among Diplomoceratidae.

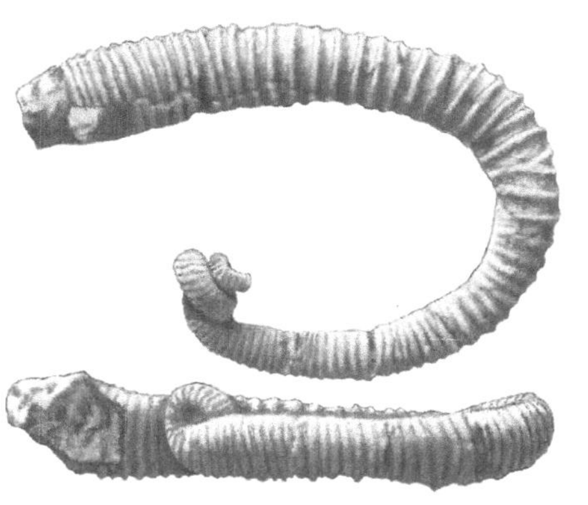

Fig. 731. *Morewites.*

asymmetrically bifid saddles, narrow bifid lobes and shallow I. *L. Camp*. **Fig. 731** - *M. sakakibarai* Shigeta.

Morewites aquisgranensis (Schlüter) (**Fig. 733F**) is larger (?macroconch), with an initial smooth straight shaft followed by a helix of about 5 whorls and then a loose planispire of at least two whorls; flexuous ribs become sharp, more distant and rursiradiate on body chamber; suture with rectangular little incised bifid lobes and saddles.

Masonites HENDERSON, 1970 [*M. biannulatus*; OD]. Initial straight shaft followed by open subelliptical whorl; dense fine rectiradiate ribs superimposed on broad low radial bulges; periodic collared constrictions; suture very simple. *Camp*. **Fig. 733B** - *M. biannulatus* Henderson.

***Wakaneneites* gen. nov.** [*Glyptoxoceras wakanenei* Henderson, 1970, p.25; OD herein]. Earliest whorls unknown; later like *Glyptoxoceras* but shell bilaterally symmetrical for most of growth, coiled in open planispire with rapidly expanding whorls and tuba-shaped adult aperture; suture highly incised, with widely splayed narrow stemmed saddles and lobes. *Sant. - L. Camp*. **Fig. 733C** - *W. wakanenei* (Henderson).

Neoglyptoxoceras COLLIGNON, 1979 [*N. magnificum*; OD]. Early stages unknown; later like *Morewites* but a much larger low helicone with obtuse spiral angle and subcircular whorl section; sharp distant ribs weaken across dorsum, separated by wide concave interspaces over venter; suture like *Wakaneneites* with very broad deeply incised L/U; in some cases widely spaced septa do not interlock. *L. - M. Camp*. **Fig. 733G** - *N. magnificum* Collignon.

Epiglyptoxoceras COLLIGNON, 1969 [*E. abnorme*; OD]. Like *Neoglyptoxoceras* but with different suture, L/U constricted at base and with reduced median incision. **Fig. 736E-F** - *E. annulatum* Collignon.

Diplomoceras HYATT, 1900 [*Baculites cylindracea* Defrance, 1816; OD]. Very large, with polyptychoceratid coiling from very early growth comprising gradually expanding subparallel straight shafts joined by tight bends; whorl section oval to subcircular; dense narrow fine sharp annular ribs weaken slightly across dorsum and cross venter transversely; internal moulds almost smooth; suture florid, with deeply incised bifid elements except for I which is small and trifid. *U. Camp. - U. Maastr.* **Fig. 733J** - *D. cylindraceum* (Defrance).

Eudiplomoceras BRUNNSCHWEILER, 1966 [*E. raggatti*; OD]. Like *Diplomoceras* but with circular whorl section and smooth internal mould to phragmocone; suture with short narrow E, oblique bifid E/L, widely splayed L much deeper than E, very large L/U bigger than E/L and very small I. *Maastr*. **Fig. 733H-I** - *E. raggatti* Brunnschweiler.

Tribe Smedaliceratini nov.

Type genus: *Smedaliceras* Young, 1963.

Diagnosis. Some or all ribs with siphonal tubercles. Monophyletic derivatives of Diplomoceratini, close to *Morewites*. *L. Camp*.

Chesapeakiceras KENNEDY & COBBAN, 1998 [*C. nodatum*; OD]. Coiling probably criocone or low open helix, with broadly rounded dorsum, narrowly rounded venter and compressed oval whorl section; distant coarse ribs strengthen markedly across venter where they carry blunt siphonal bullae, but weaken across dorsum; suture with very widely separated septa. *L. Camp*. **Fig. 736C-D** - *C. nodatum* Kennedy & Cobban.

= *Chesapeakella* Kennedy & Cobban, 1993 (*non* Campbell, 1993).

Smedaliceras YOUNG, 1963 [*S. durhami*; OD]. Moderately large, based on fragments; like *Chesapeakiceras* but with sharp dense slightly rursiradiate ribs with mid-lateral

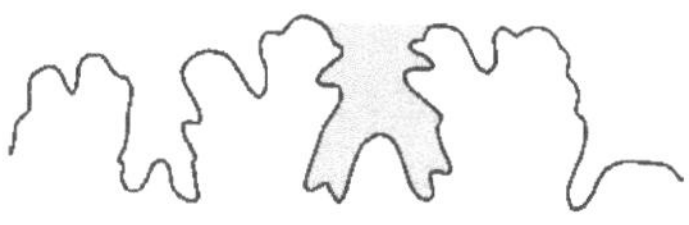

Fig. 732. *Morewites* suture.

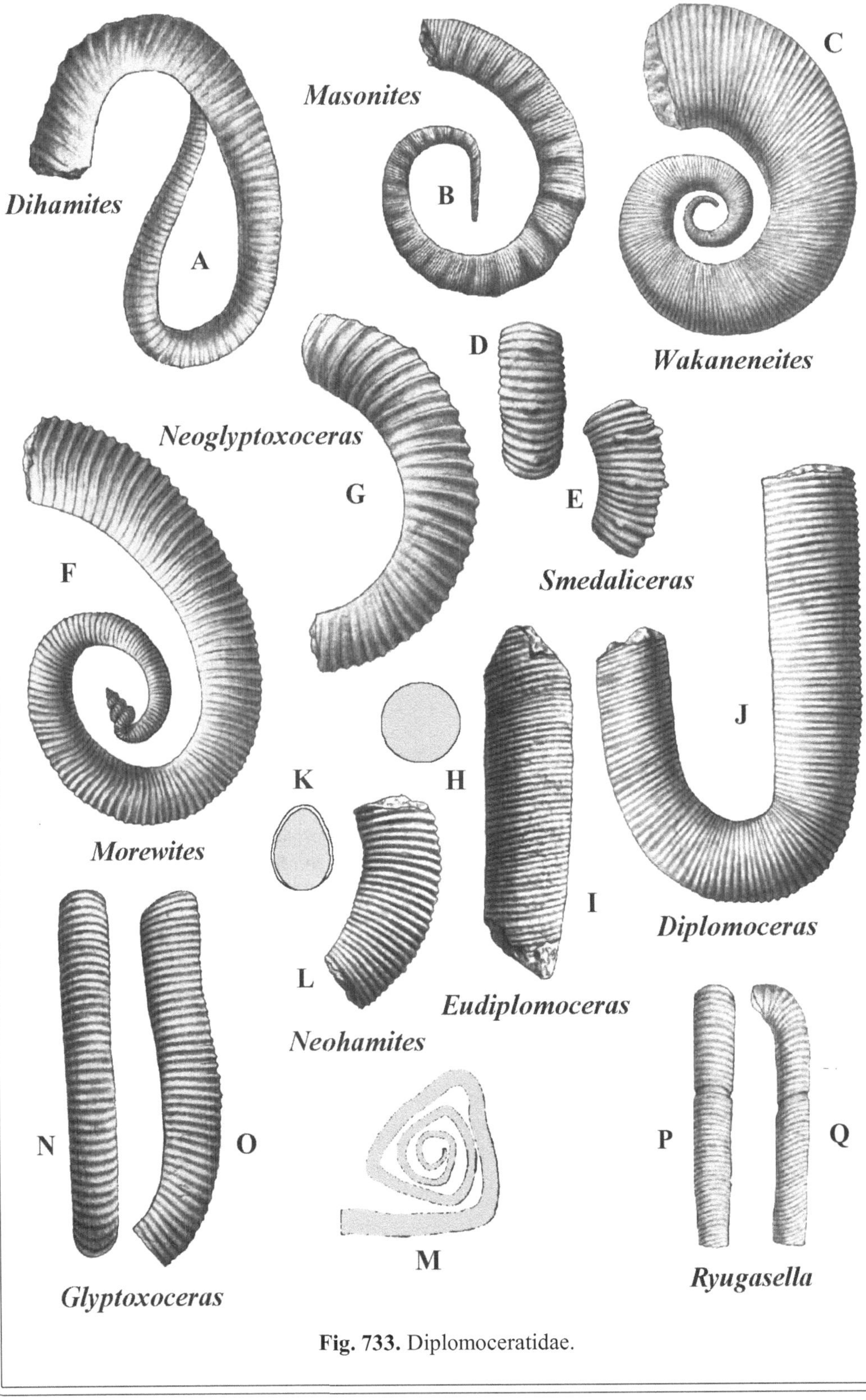

Fig. 733. Diplomoceratidae.

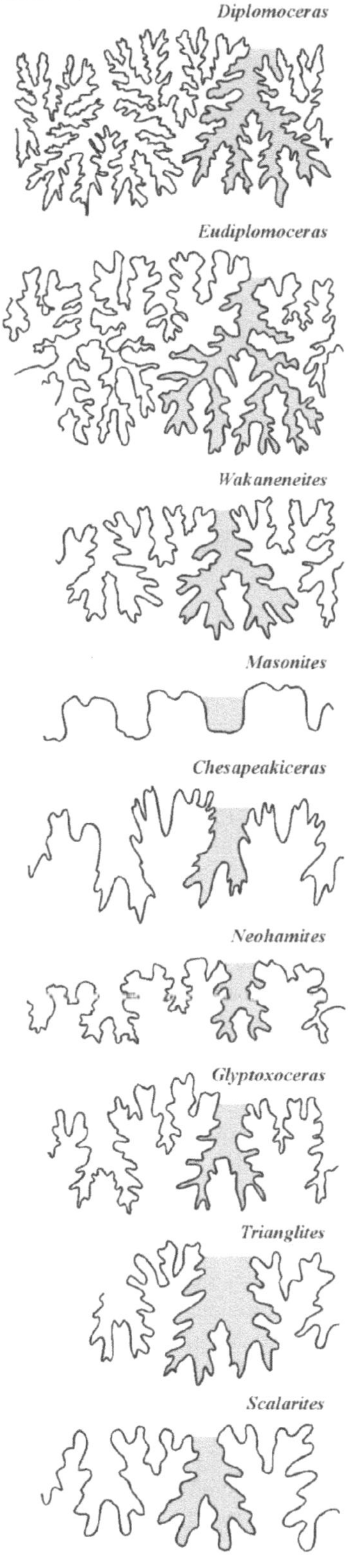

Fig. 734. Diplomoceratid sutures.

bullae and siphonal nodes, not necessarily on same rib. *L. Camp*. **Fig. 733D-E, 736A-B** - *S. durhami* Young.

Family NOSTOCERATIDAE Hyatt, 1894

(*nom. transl.* herein *ex* Nostoceratidae)

Discussion: Monophyletic derivatives of Turrilitidae, mostly larger and with uncoiled adult body chamber; comprises the subfamilies Bostrychoceratinae, Nostoceratinae and Hyphantoceratinae. Helical coiling is both sinistral and dextral, with the siphuncle at midflank. Size dimorphic. These are the common larger heteromorphs of the Upper Cretaceous. *M. Turon. - U. Maastr.*

Subfamily Bostrychoceratinae nov.

Diagnosis. Typically with loose (primitive) to tight (derived) helical coiling, a U-shaped adult body chamber and, when present, two rows of tubercles; constrictions occur in some members. Monophyletic derivatives of Turrilitinae, thus evolving in parallel with Diplomoceratidae. Wright (1996) looked to *Turrilitoides* for the origin of the group but there is no evidence that genus survived the Albian-Cenomanian boundary. Cenomanian records of *Eubostrychoceras*, e.g. *E. thomasi* (Pervinquière), claiming to link the two, seem to be based on stratigraphically misplaced Turonian material. One species of *Wintonia* shows an upturned aperture to the body chamber, like *Eubostrychoceras*, and the body chamber of *Plesioturrilites* is a pendant hook as in *Bostrychoceras*. *M. Turon. - L. Maastr.*

Eubostrychoceras MATSUMOTO, 1967 [*E. indopacificum*; OD]. Medium sized; a tight helicone with moderate spiral angle and whorls in contact; initial straight shaft forms axis about which helical coiling occurred; whorl section varies from round to subquadratic; 2-3 regular oblique collared constrictions per whorl; fine regular sharp to rounded single or branching ribs, slightly narrower than interspaces; adult body chamber curves upward, without uncoiling, to hug penultimate whorl; final aperture collared and constricted; suture with rather florid bifid elements. *M. Turon. - Coni.* **Fig. 738A-B** - *E. indopacificum* Matsumoto.
= *Eurostrychoceras* Matsumoto, 1977 (*laps. cal.*).

Nipponites YABE, 1904 [*N. mirabilis*; OD]. Medium sized; initially like *Eubostrychoceras*, but helix followed by tangled succession of 8 or more U-shaped segments; adult body chamber forms last U or descends to form

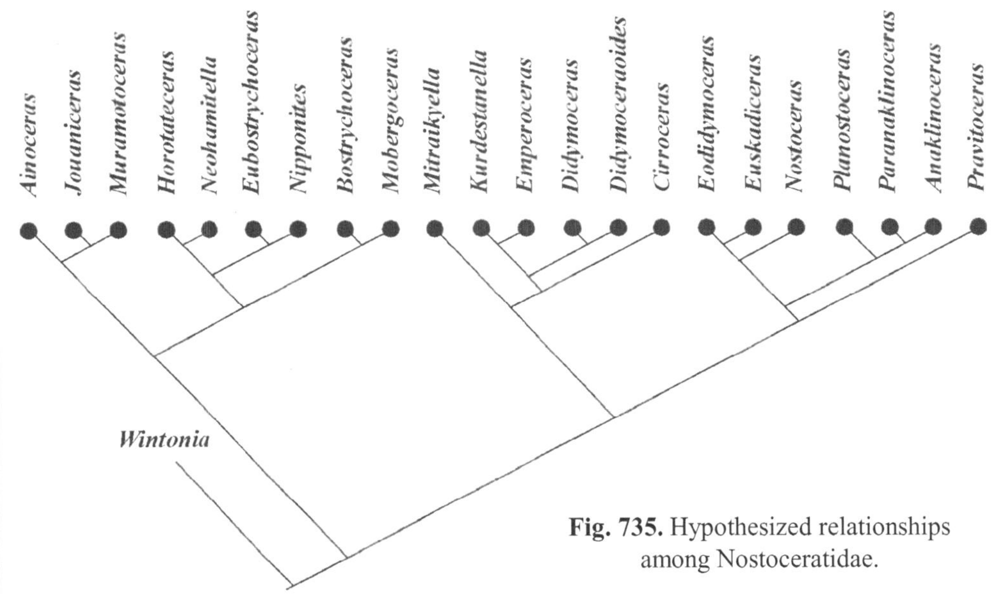

Fig. 735. Hypothesized relationships among Nostoceratidae.

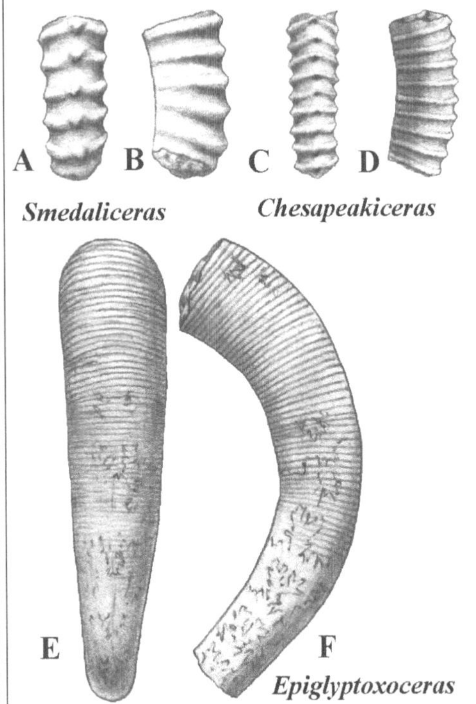

Fig. 736. Diplomoceratidae.

retroversal hook; whorl section subcircular to subelliptical; constrictions lacking. *M. Turon. - Coni.* **Fig. 738C-D** - *N. mirabilis* Yabe.

Neohamitella **gen. nov.** [*Bostrychoceras protractum* Collignon, 1969, p.31; OD herein]. Moderately tall loosely coiled helix with small spiral angle; whorl section subcircular; strong simple oblique non-tuberculate ribs weakening across dorsum, with or without periodic constrictions. *?Turon., L. Coni. - L. Camp.* **Fig. 738E** - *N. condamyi* (Collignon).

Horotateceras HAYAKAWA, 1998 [*H. tatsuyai*; OD]. Medium sized; only known from last part of phragmocone and body chamber; like *Neohamitella* but with flattened outer flank, compressed elliptical whorl section and bluntly rounded spiral keel high on outer flank that is crenulated by the ribbing; occasional flares and shallow constrictions; ribs coarsen on adult body chamber. *M. Turon.* **Fig. 738G** - *H. tatsuyai* Hayakawa.

Bostrychoceras HYATT, 1900 [*Turrilites polyplocus* Roemer, 1841; OD]. Large; like *Eubostrychoceras* but with more-inflated whorls, higher spire, finer more flexuous ribs, occasional constrictions, irregular tubercles to helical whorls, and pendant body chamber with apertural constriction. *M. Camp. - L. Maastr.* **Fig. 738H** - *B. elongatum* (Whiteaves), **738I** - *B. polyplocum* (Roemer).

Mobergoceras SCHMID & ERNST, 1975 [*Turrilites junior* Moberg, 1885; OD]. Small, with very acute apical angle (12°) and low expansion rate; whorls non-tuberculate, with 15-17 prominent sharp ribs per whorl, narrower than interspaces, with a varix almost every 180°; body chamber unknown. *U. Camp.* **Fig. 741I** - *M. junior* (Moberg).

Subfamily Nostoceratinae Hyatt, 1994

(*nom. transl.* herein *ex* Nostoceratidae)

Discussion. Seemingly monophyletic derivatives of

uncertain ancestry; an origin in Bostrychoceratinae has generally been assumed, but fragments of *Mitraikyella* are indistinguishable from *Allocrioceras* (Klinger & Kennedy, 2003) and an origin in Exiteloceratinae seems more likely, suggesting the similarity to Bostrychoceratinae is convergent. Comprises the tribes Emperoceratini, Nostoceratini and Pravitoceratini. *L. Camp. - L. Maastr.*

Tribe Emperoceratini Spath, 1953

(*nom. transl.* herein *ex* Emperoceratinae)
Type genus: *Emperoceras* Hyatt, 1894.

Diagnosis. Typically relatively large bituberculate nostoceratids with loose helical coiling and body chamber forming a more or less uncoiled pendant hook. *L. Camp. - L. Maastr.*

***Mitraikyella* gen. nov.** [*M. collignoni* sp. nov., = *Nostoceras hyatti* var. *mitraikyensis* Collignon, 1970, p.67; OD herein]. Moderately large; earliest stages unknown; a loose helicone with small spiral angle, narrow umbilicus and numerous whorls; strong coarse annular ribs with every second carrying two prominent tubercles on lower half of outer face. *L. Camp., ?U. Camp.* **Fig. 738F** - *M. collignoni* sp. nov.

Didymoceratoides KENNEDY & COBBAN, 1993 [*D. binodosum*; OD]. Moderately large, earliest whorls a loose criocone, followed by 2-3 helical whorls almost in contact and, finally, a body chamber that barely uncoils and is weakly upturned; ribs distant, annular, uniformly bituberculate, with constrictions. *U. Camp.* **Fig. 738K** - *D. binodosum* Kennedy & Cobban.

Criocone early whorls make it uncertain whether this genus is convergent towards *Didymoceras* in maturity or if, like *Mitraikyella*, its early whorls merely homoeomorph the exiteloceratine condition.

Cirroceras CONRAD, 1868 [*Ammonceratites conradi* Morton, 1843; OD]. Moderately large, earliest stages unknown; later like *Didymoceratoides* with barely uncoiled adult body chamber; spiral whorls with flattened outer face bounded by strong tubercles weakly joined by looped ribs. *M. - U. Camp.* **Fig. 738L** - *C. conradi* (Morton).

Didymoceras HYATT, 1894 [*Ancyloceras? nebraskense* Meek & Hayden, 1857; OD]. Like *Didymoceratoides* but with initial straight shaft followed by loosely coiled helix and pendant J-shaped body chamber; whorls initially elliptical, later subcircular with rounded outer face; numerous fine, sharp, slightly oblique ribs carry two rows of irregularly developed nodes on lower half of outer whorl face, mostly paired but sometimes alternating, separated by 2-5 non-tuberculate ribs; ribs may loop between tubercles that frequently evanesce on last part of phragmocone and first part of body chamber; ornament coarsens on adult body chamber when tubercles become especially prominent; frequent constrictions; suture with deep narrow E, oblique narrow stemmed bifid E/L, deep splayed bifid L much deeper than E, and broad asymmetrically bifid L/U. *Camp. - Maastr.* **Fig. 738J** - *D. nebraskense* (Meek & Hayden).

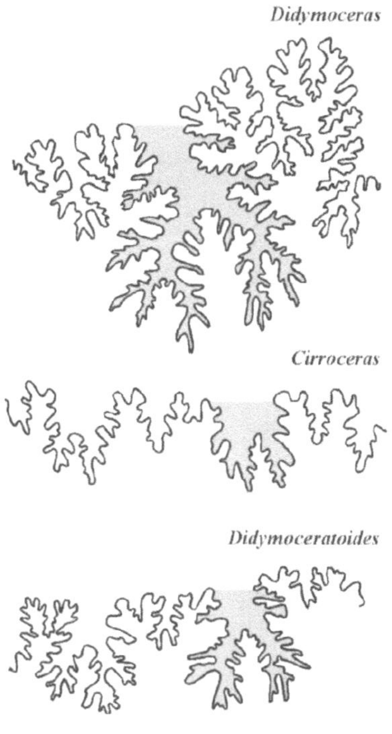

Fig. 737. Nostoceratine sutures.

Emperoceras HYATT, 1894 [*E. beecheri* Meek & Hayden, 1857; OD]. Like *Didymoceras* but early whorls an ellipticone in more or less one plane, with finer ribbing and loss of tubercles in later growth leaving only simple flexuous ribs. *U. Camp.* **Fig. 741G-H** - *E. beecheri* (Meek & Hayden).

***Kurdestanella* gen. nov.** [*Nostoceras ellipticum* Kennedy, 2000, p.465; OD herein]. Small, early whorls elliptically coiled as in *Emperoceras*, apparently in low helix; constrictions lacking but with periodic flares; fine dense ribs all with minute ventrolateral tubercles; body chamber with sparse fine distant bituberculate ribs. *U. Camp.* **Fig. 741D** - *K. ellipticus* (Kennedy).

Tribe Nostoceratini Hyatt, 1994

(*nom. transl.* herein *ex* Nostoceratinae)
Type genus: *Nostoceras* Hyatt, 1894.

Discussion. Probably monophyletic derivatives of Bostrychoceratini, close to *Bostrychoceras*; characterized by small to medium size, tightly coiled spire, uncoiled pendant body chamber (except in derived representatives) and bituberculate ribs with or without constrictions. *U. Camp. - Maastr.*

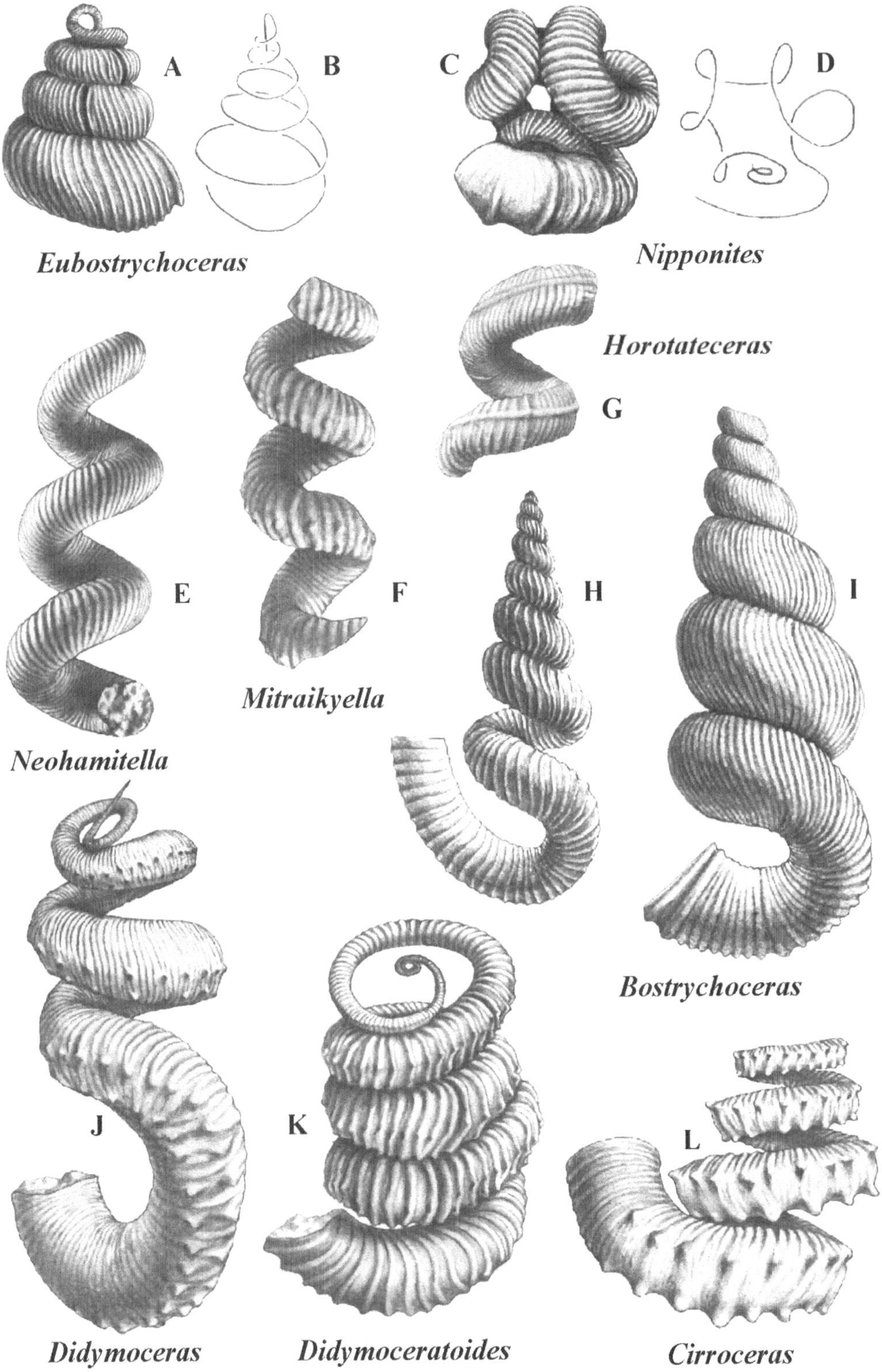

Fig. 738. Nostoceratidae.

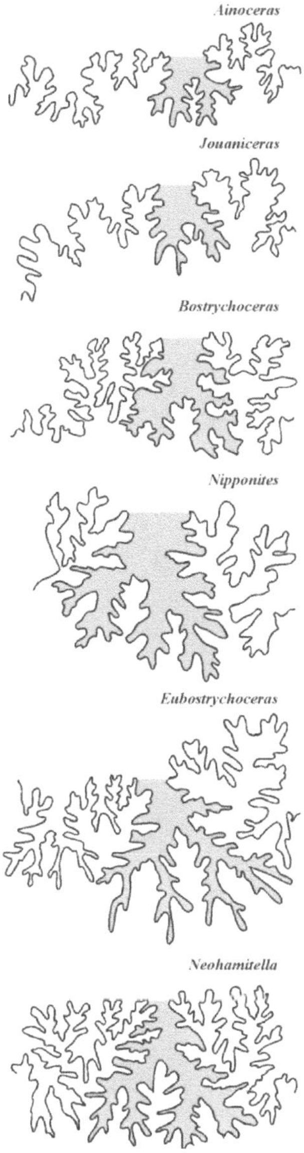

Fig. 739. Nostoceratine sutures.

Nostoceras HYATT, 1894 [*N. stantoni retrorsum*, = *Ancyloceras approximans* Conrad, 1856; OD]. Medium sized; low tight helical spire (apical angle 55-70°) followed by uncoiled U-shaped retroversal hook with aperture terminating close to base of spire; whorl section subcircular with flattened outer face; 2-3 broad constrictions per whorl irregularly developed; strong, fine, prorsiradiate ribs, crowded and somewhat concave on outer face, frequently bifurcating on lower whorl face; tuberculation to spire irregular, lacking in some, irregularly developed in others, with upper row slightly below midflank and lower row at spiral suture or concealed by younger whorl; ribs coarsen on bituberculate body chamber but weaken across dorsum, looping, zigzagging or intercalated between tubercles that alternate across venter, and simplify near aperture. *U. Camp. - Maastr.* **Fig. 741A** - *N. approximans* (Conrad).

Euskadiceras KÜCHLER, 2000 [*Nostoceras (E.) euskadiense*; OD]. Like *Nostoceras* but with body chamber an open hook well separated from helix; coarsely ornamented helix with two rows of prominent rounded nodes exposed on outer face, the upper row above midflank linked to 2 ribs on the adoral surface, separated by a single rib. *Uppermost Camp.* **Fig. 740** - *E. euskadiense* (Küchler).

The type species is very similar to *E. donezianum* (Mikhailov), the earliest whorls of which are an open helix with simple bituberculate ribs.

Eodidymoceras KLINGER & KENNEDY, 2003 [*Nostoceras hyatti var. mitraikense* Collignon, 1970, = *N. hyatti*; OD]. Like *Euskadiceras* but with last part of body chamber curving upwards in *Eubostrychoceras* fashion. *Upper U. Camp.* **Fig. 741J** - *E. hyatti* (Stephenson).

Eodidymoceras Klinger & Kennedy (2003) was created for a Malgash specimen introduced as a variety of *N. hyatti*; since the varietal name is without nomenclatural standing (Article 16, ICZN code), the type species of *Eodidymoceras* is *N. hyatti*.

***Paranaklinoceras* gen. nov.** [*Nostoceras liratum* Kennedy, 1992; OD herein]. Small; like *Nostoceras* but with tight coiling, fine bituberculate ribs and body chamber curving upwards, remaining in contact with helix. *Camp.* **Fig. 743** - *P. liratum* (Kennedy).

Anaklinoceras STEPHENSON, 1941 [*A. reflexum*; OD]. Like *Eodidymoceras* but smaller, with body chamber wrapping around spire in plane of axis; collared constrictions regularly developed; ribbing somewhat distant, with both bifurcation and intercalation; two rows of tubercles on lower half of outer whorl, often separated by 1-2 non-tuberculate ribs, ribs coarse and convex on lower whorl face, joining in pairs close to umbilicus; suture relatively simple, with narrow asymmetrically bifid E/L, broad bifid L, very broad asymmentrically bifid L/U and narrow bifid U. *U. Camp.* **Fig. 741C** - *A. reflexum* Stephenson.

***Navarroites* gen. nov.** [*Turrilites saundersorum* Stephenson, 1941, p.416; OD herein]. Small; like *Nostoceras* but with tall, tightly coiled helix and acute spiral angle (23-

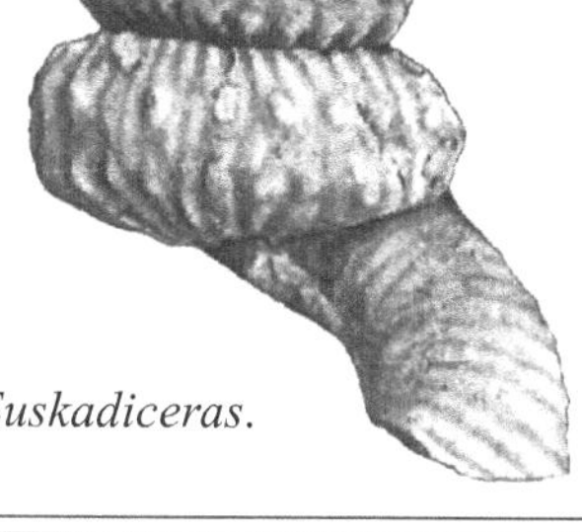

Fig. 740. *Euskadiceras*.

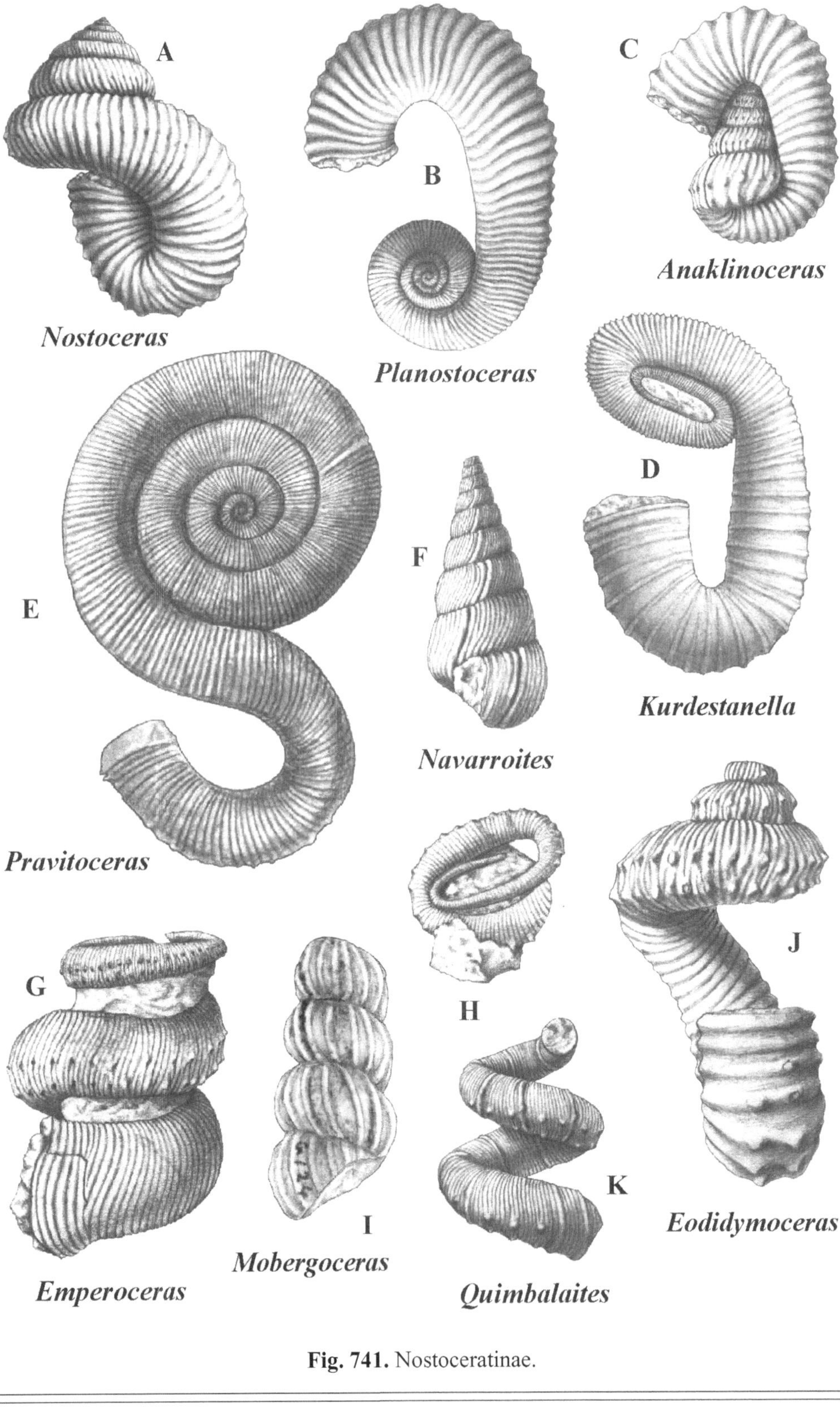

Fig. 741. Nostoceratinae.

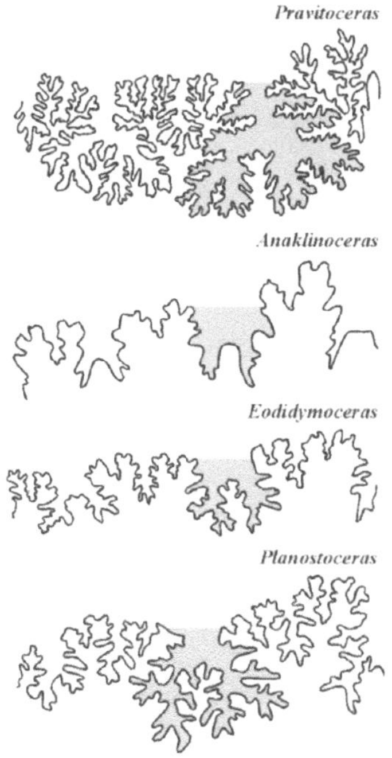

Fig. 742. Nostoceratine sutures.

37°); outer whorl face somewhat flattened; regular collared constrictions occur throughout, at 180° intervals; ribbing dense, fine, regular, rounded, slightly sinuous and oblique, with occasional bifurcation; 1-2 rows of weak indistinct nodes on lower flank in later growth; body chamber does not uncoil; aperture collared and constricted; a homoeomorph of Albian *Turrilitoides*. *L. Maastr.* **Fig. 741F** - *N. saundersorum* (Stephenson).

Planostoceras LEWY, 1967 [*Nostoceras (P.) rehmanni*; OD]. Medium sized; like *Nostoceras* but with low spire, obtuse spiral angle and hooked body chamber uncoiling horizontally, in plane of last helical whorl. *U. Camp.* **Fig. 741B** - *P. rehmanni* (Lewy).

Tribe Pravitoceratini Spath, 1953

(*nom. transl.* herein *ex* Pravitoceratinae, *nom. correct.* Wright, 1996 *ex* Proavitoceratinae)
Type genus: *Pravitoceras* Yabe, 1902.

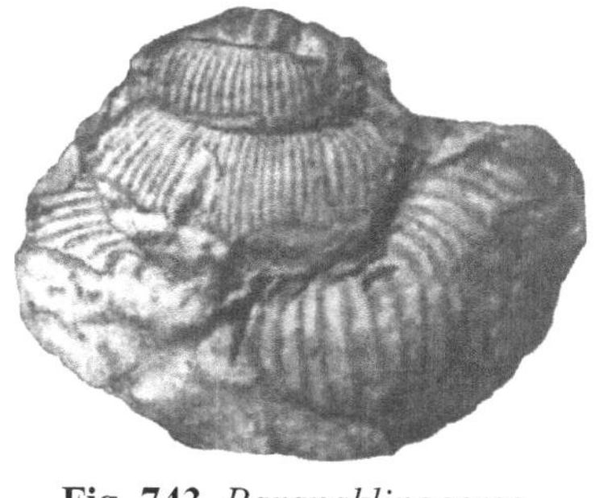

Fig. 743. *Paranaklinoceras*.

Diagnosis. A monotypic taxon with an origin in Nostoceratini; the change to serpenticone coiling in middle growth, normal to the spiral axis, and the S-shaped adult body chamber are evolutionary novelties that warrant taxonomic recognition. *U. Camp.*

Pravitoceras YABE, 1902 [*P. sigmoidale*; OD]. Large, with helix restricted to earliest growth, followed by serpenticone with slowly expanding whorls just touching and, finally, S-shaped body chamber that uncoils in plane of planispire; whorl section subcircular; early whorls with distant bituberculate rectiradiate ribs, with or without intervening non-tuberculate secondaries; later whorls with dense slightly flexuous ribs, some tuberculate; irregular shallow constrictions present; suture highly incised, with narrow stemmed saddles and lobes, oblique E/L and widely splayed L much deeper than E. *Upper U. Camp.* **Fig. 741E** - *P. sigmoidale* Yabe.

Subfamily Jouaniceratinae Wright, 1952

Discussion. A monophyletic group with an origin in Bostrychoceratinae, close to *Eubostrychoceras*; small helicone with strong simple non-tuberculate ribs, coarsening markedly and becoming distant on recoiled body chamber that may envelop the spire. *M. Sant. - L. Camp.*

Jouaniceras BASSE, 1939 [*Heteroceras? sicardi* de Grossouvre, 1894; OD]. Small; earliest growth stages a small helix of 3 whorls barely in contact, with coarse distant transverse ribs effaced from outer whorl face; following prominent constriction, two adpressed planispiral whorls wrap around helix in axis of coiling and carry coarse distant feebly prorsiradiate sharp ribs (flat topped on internal moulds); adult body chamber slightly depressed, with periodic non-tuberculate flares separated by up to 3 ribs; suture with deep narrow E, broad narrow stemmed asymmetrically bifid E/L, rather broad asymmetrically bifid L deeper than E, and very narrow asymmetrically bifid L/U. *M. - U. Sant.* **Fig. 745A** - *J. sicardi* (de Grossouvre).

Muramotoceras MATSUMOTO, 1977 [*M. yezoense*; OD]. Initially gently sinuous, nearly straight or arcuate shaft forms axis for loose helical coiling of middle growth (like *Ryuella*); adult body chamber uncoils oblique to helix; whorl section subcircular; ribbing fine and simple in early growth, later with distant flares, and weak umbilicolateral and septispinate ventrolateral tubercles separated by several faint riblets. *M. Turon.* **Fig. 744A-C** - *M. yezoense* Matsumoto.

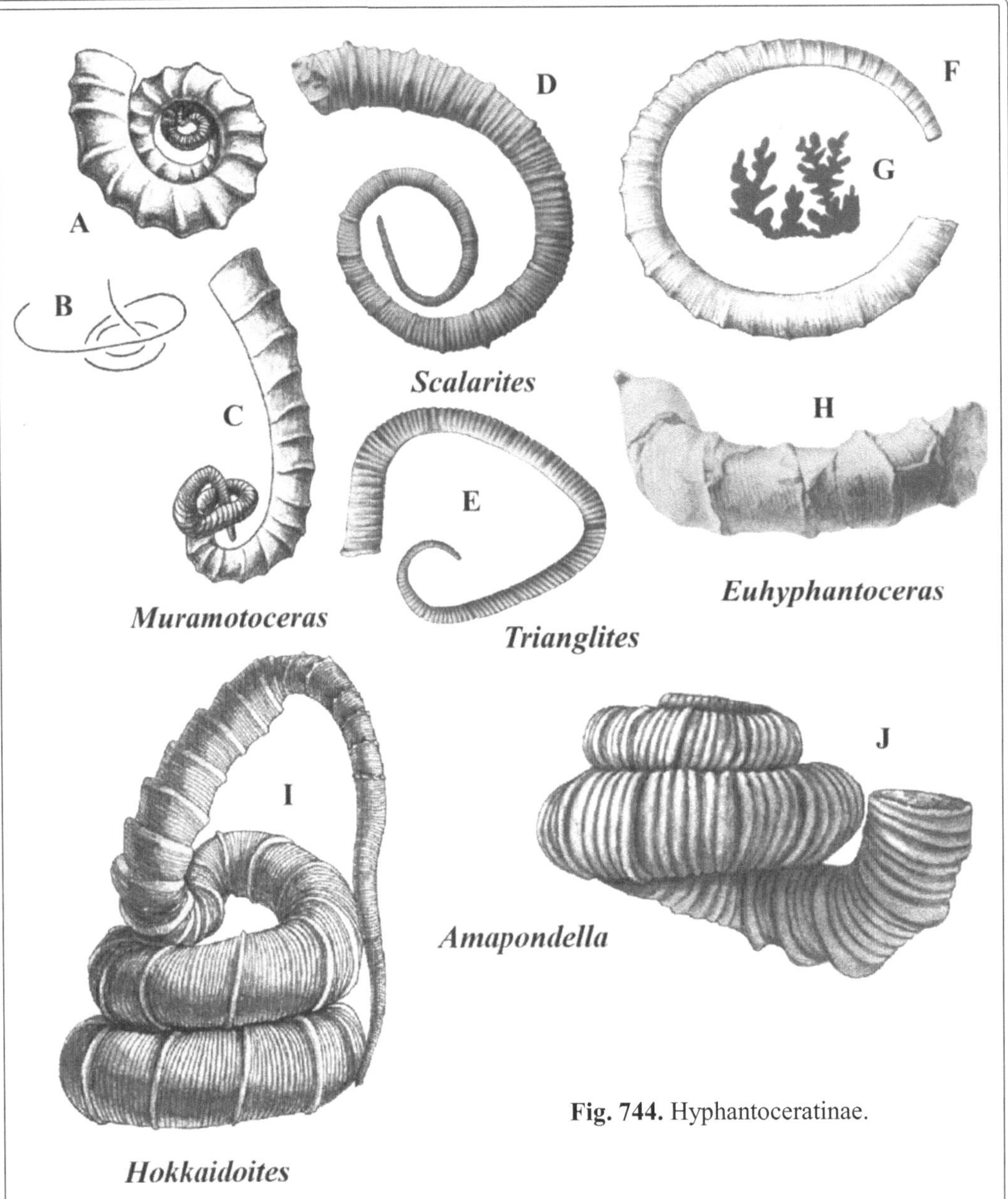

Fig. 744. Hyphantoceratinae.

Ainoceras MATSUMOTO, 1967 [*A. kamuy*; OD]. Medium sized; like *Jouaniceras* but with final body chamber partially uncoiled and not enveloping spire. *?L. Camp.* **Fig.745B-C** - *A. kamuy* Matsumoto.

Subfamily Hyphantoceratinae Spath, 1953

Discussion. A monophyletic group with an origin in Bostrychoceratinae, close to *Neohamitella*; characteristically with initial helix followed by loose helicone with tuberculate flares separated by fine non-tuberculate secondaries. *L. Coni. - M. Camp.*

***Ebisuites* gen. nov.** [*Heteroceras japonicum* Yabe, 1904, p.17; OD herein]. Medium sized; elliptically coiled earliest whorls become torticone in middle to later growth, with coarse blunt annular ribs; flares frequent in early growth and on adult body chamber where joined by periodic constrictions; middle growth stages indistinguishable from *Neohamitella*; suture highly incised, with narrow stemmed bifid saddles and lobes. *M. Turon.* **Fig. 746F** - *E. japonicus* (Yabe).

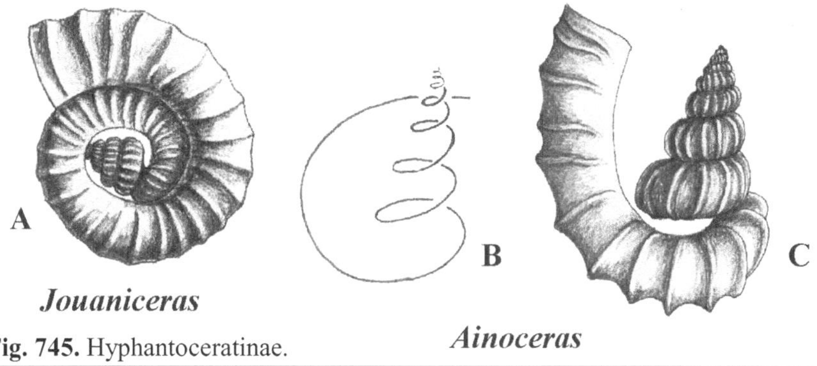

Fig. 745. Hyphantoceratinae.

Hyphantoceras HYATT, 1900 [*Heteroceras reussianum* d'Orbigny, 1842; OD]. Medium sized; initial small turriliticone followed by loose helical coiling and pendant U-shaped body chamber; periodic flares separated by 2-5 fine ribs, spacing between flares tending to increase with growth; on adult body chamber flared ribs normally with 4 rows of tubercles but, in earlier growth, there may be only two or even none; suture with shallow E, narrow stemmed oblique E/L and widely splayed symmetrical L, well developed U and small I. *U. Turon. - Sant.* **Fig. 746A-D** - *H. reussianum* (d'Orbigny).

Euhyphantoceras SHIMIZU, 1935 [*E. maestrichtiense*, = *Heteroceras venustum* Yabe, 1904; OD]. Earliest stages unknown; later like *Hyphantoceras* but with looser more open coiling from earlier growth, finer secondary ribs between bituberculate flares; ribs and flares weaken on umbilical side where there are numerous intercalated stria; suture simpler with deeper E and non-splayed L. *U. Coni. - Sant.* **Fig. 744F-H** - *E. venustum* (Yabe).

Orientoceras SHIMIZU, 1935 [*Heteroceras? orientale* Yabe, 1904; OD]. Like *Hyphantoceras* but with very acute spiral angle; earliest part of shell straight and pointing away from helix; flared ribs lacking; body chamber barely upturned, essentially forming part of helix; simple fine strong gently sigmoidal quadrituberculate primary ribs with indistinct elevation on upper shoulder, two bullate tubercles on lower half of external face, and rounded tubercle at lower shoulder, separated by 1-2 slightly weaker non-tuberculate secondaries; suture like *Hyphantoceras* but with deeper E and smaller markedly asymmetrical U. *U. Sant. - L. Camp.* **Fig. 746E** - *O. orientale* (Yabe).

Tridenticeras WIEDMANN, 1962 [*Turrilites tridens* Schlüter, 1876; OD]. Medium size; like early stages of *Hyphantoceras*, but carrying tight helical coiling to maturity; periodic strong main ribs, narrower than interspaces, with 3 rows of bullae on outer face separated by 1-3 fine non-tuberculate secondaries; there may be 4th vestigial tubercle to primary ribs at umbilical edge; ribs may loop between flank tubercles; tubercles evanesce on adult body chamber leaving only strong primary ribs; a homoeomorph of *Turrilites*. *U. Turon. - Coni.* **Fig. 746H** - *T. tridens* (Schlüter).

Yezoceras MATSUMOTO, 1977 [*Y. nodosum*; OD]. Like *Tridenticeras* but with siphuncle at base of whorl and body chamber retroversed or not; periodic main ribs with up to 4 rows of bullae on adult body chamber; suture like *Hyphantoceras* but with asymmetrical L. *L. - U. Coni.* **Fig. 746K** - *Y. nodosum* Matsumoto.

Ankinatsytes COLLIGNON, 1965 [*A. yabei*; OD]. Known fragment like *Hyphantoceras*, but flares with only 3 rows of tubercles, and suture simpler with less splayed elements similar to *Allocrioceras*. *L. Coni.* **Fig. 746J** - *A. yabei* Collignon.

Madagascarites COLLIGNON, 1966 [*M. andimakensis*; OD]. Moderately large; early coiling like *Ankinatsytes* but embraced by U-shaped body chamber; midlateral and ventrolateral tubercles joined by looped ribs. *Sant.* **Fig. 746G** - *M. andimakensis* Collignon.

Quimbalaites **gen. nov.** [*Didymoceras subtuberculatum* Howarth, 1965, p.374; OD herein]. Medium sized loose helicone with variable apical angle; slowly expanding whorls with fine ribs, 3-4 tuberculate flares per whorl precede weak slight constrictions, with tubercles positioned low on whorls; body chamber unknown. *M. Camp.* **Fig. 741K** - *Q. subtuberculatus* (Howarth).

Ryuella KLINGER & KENNEDY, 1997 [*Madagascarites ryu* Matsumoto & Muramoto, 1967; OD]. Like *Hyphantoceras*, but with different coiling; initial straight shaft enveloped by 2 helical whorls, after which coiling becomes irregular and *Nipponites*-like. *U. Turon.* **Fig. 746I** - *R. ryu* (Matsumoto & Muramoto).

Amapondella KLINGER & KENNEDY, 1997 [*Heteroceras amapondense* van Hoepen, 1921; OD]. Early stages reputedly helical; later very low helix with whorls almost in contact, wide umbilicus and adult body chamber an upturned limb that extends up to or slightly beyond apex of helix; initially only simple ribs but later interspersed (every 3-8 ribs) with periodic flares. *U. Sant.- L. Camp.* **Fig.744J** - *A. amapondensis* (van Hoepen).

Hokkaidoites **gen. nov.** [*Hyphantoceras? heteromorphum* Matsumoto, 1977, p.314; OD herein]. Like *Amapond-*

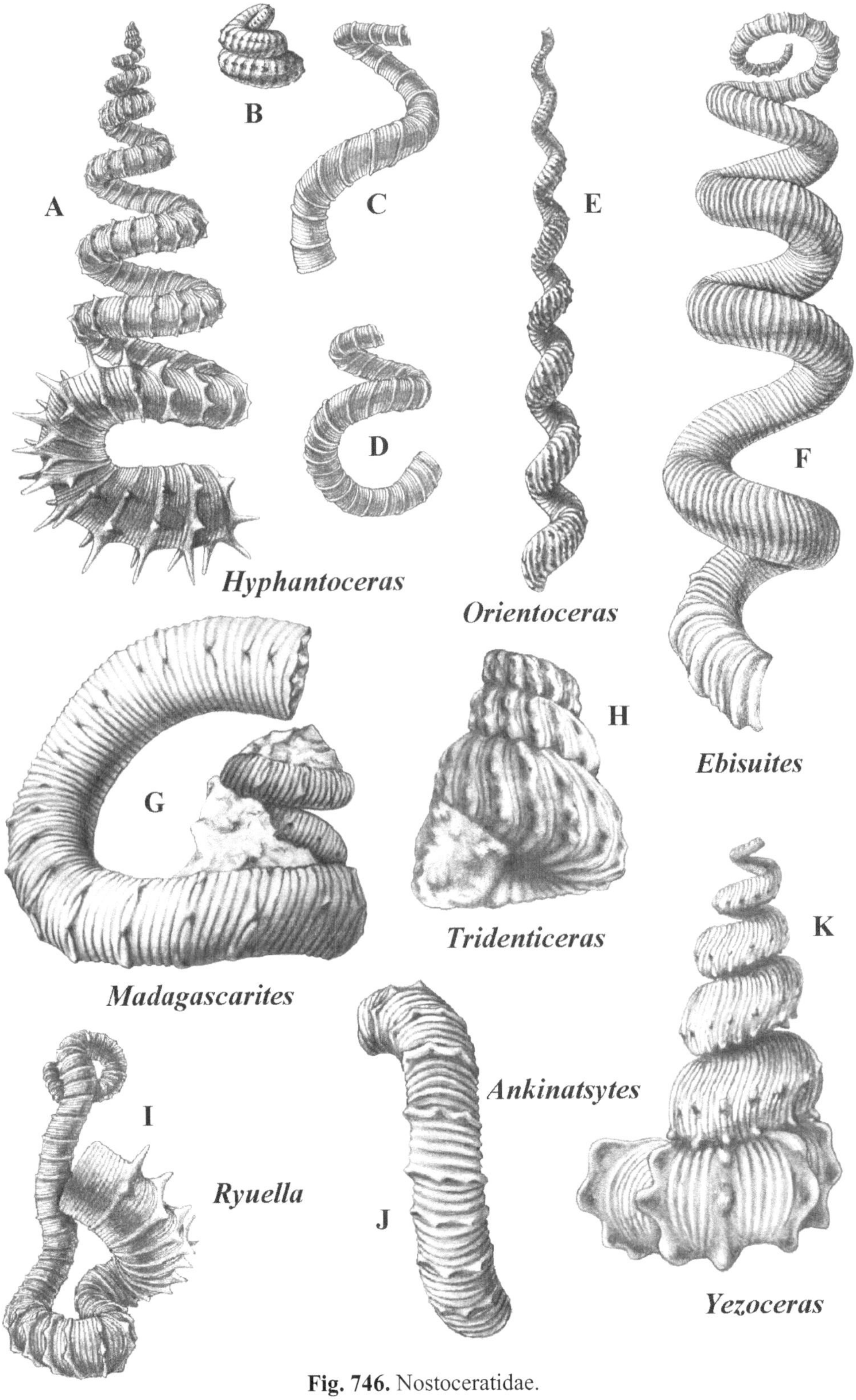

Fig. 746. Nostoceratidae.

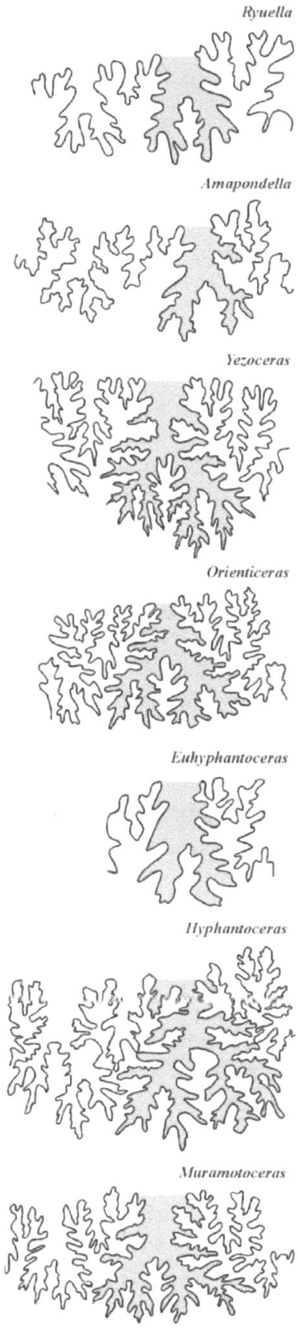

Fig. 747. Hyphantoceratine sutures.

ella but initially a long sinuous tapering shaft that after initial bend forms loose low helix; initial shaft with simple annular ribbing but after first bend with regular flares separated by 10-22 fine ribs; adult body chamber unknown. *Turon.* **Fig. 744I** - *H. heteromorphus* (Matsumoto).

Neoturrilites SHIMIZU, 1935 [*N. karafutoensis*; OD]. The unfigured type species was characterized by having oblique ribbing with 3 rows of weak tubercles.

According to Wright & Matsumoto (1954), this description "... is quite insufficient for recognition", and the name is a *nomen nudum*.

Subfamily Scalaritinae Ward, 1976

Discussion. Monophyletic derivatives of Hyphantoceratinae, close to *Tanabeites* (Matsumoto, 1977). Medium sized, with initial constricted smooth straight shaft followed by variable coiling; in middle to later growth with simple, annular, non-tuberculate ribs and frequent flares. *M. Turon. - L. Camp.*

Scalarites WRIGHT & MATSUMOTO, 1954 [*Helicoceras scalare* Yabe, 1904; OD]. Medium sized; initial straight smooth shaft with well marked constrictions followed by low loose helix with subelliptical coiling; whorl section subcircular in early growth, somewhat flat sided and compressed later; uniform simple annular ribs in middle growth with frequent regular non-tuberculate flares, backed by constrictions; flares irregular in later growth when ribbing becomes more distant; suture with narrow-stemmed deeply bifid saddles and lobes of more or less same size, and L as deep as E. *M. Turon. - L. Coni.* **Fig. 744D** - *S. scalare* (Yabe).

= *Scalarites* Matsumoto, 1938 (*nom. nud.*).

Trianglites MATSUMOTO, 1977 [*T. antiquus* (=*anitqus*); OD]. Like *Scalarites* but smaller, with long almost straight shafts and subtrigonal coiling in later growth; constrictions prominent but flares lacking; adult aperture collared. *Turon.* **Fig. 744E** - *T. antiquus* Matsumoto.

Superfamily SCAPHITOIDEA Gill, 1871

(*nom. transl.* Wright & Wright 1951 *ex* Scaphitidae)

Discussion. Monophyletic derivatives of Hamit-

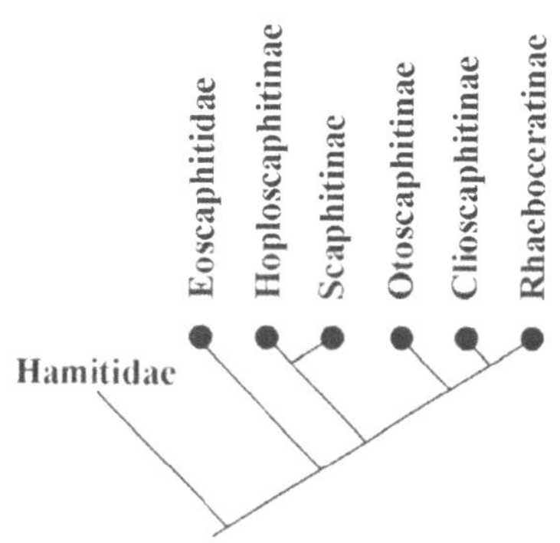

Fig. 748. Hypothesized relationships among Scaphitoidea.

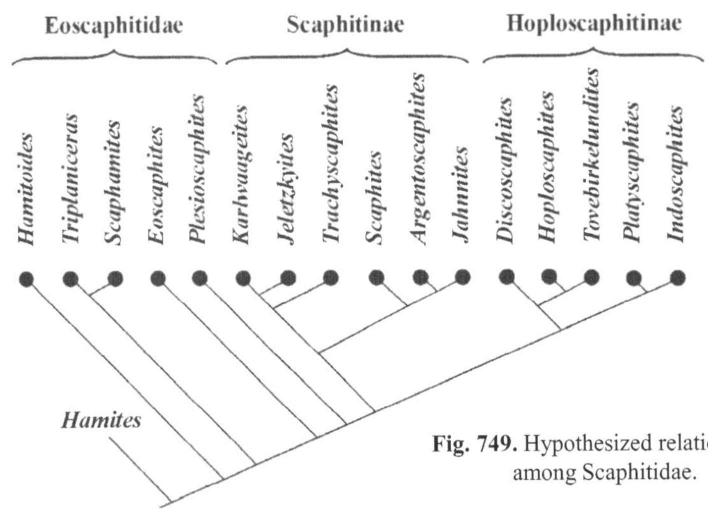

Fig. 749. Hypothesized relationships among Scaphitidae.

oidea (Spath, 1921), independent of Turrilitoidea; typically with bifurcating ribs and final body chamber uncoiling to form a hook, although recoiling occurs in several lineages. Comprises the families Eoscaphitidae, Scaphitidae and Otoscaphitidae. *U. Alb. - U. Maastr.*

Family EOSCAPHITIDAE Cooper, 1994

Discussion: A paraphyletic taxon with an origin in Hamitidae (Spath, 1921); primitively with early whorls a very shallow loose helix or criocone with perforate umbilicus; adult body chamber a J-shaped hook; fine ribs increasing by bifurcation and intercalation across the venter; suture with long narrow E, bifid saddles, bifid L, L/U without pseudolobes, and trifid I deeper than U. Comprises the subfamilies Triplaniceratinae and Eoscaphitinae. *Lower U. Alb. - L. Cenom.*

Subfamily Triplaniceratinae Bert, 2008

Discussion. Monophyletic derivatives of Hamitinae, close to *Hamites* (Bert, 2008); generally small, with rapidly expanding whorls, initially irregular hamitid coiling, followed by open hook with short final shaft; ribs simple, coarsening in maturity. *M. - U. Alb.*

Scaphamites WIEDMANN & MARCINOWSKI, 1985 [*S. passendorferi*; OD]. Initial low helix, with rapidly expanding whorls, followed by twisted scaphitoid body chamber; whorl section depressed reniform to subcircular with flattened dorsum; simple annular ribs transverse across venter, with incipient biplication on adult body chamber; suture with subrectangular bifid elements, L as wide as E/L and U trifid. *Albian* (precise level uncertain). **Fig. 750A** - *S. passendorferi* Wiedmann & Marcinowski.

Triplaniceras BERT, 2008 [*T. delanoyi*; OD]. Like *Scaphamites* but very small, with shorter shaft, more open hook and ribs weakening across venter. *Upper M. Alb.* **Fig. 750G-I** - *T. delanoyi* Bert.

Subfamily Eoscaphitinae Cooper, 1994

(*nom. transl.* herein *ex* Eoscaphitidae)

Discussion. Paraphyletic derivatives of Triplaniceratinae, close to *Triplaniceras* (Bert, 2008); early whorls planispirally coiled, initially with perforate umbilicus, later closed; ribs generally bifurcating and intercalated across the non-tuberculate venter. *U. Alb. - M. Cenom.*

Hamitoides SPATH, 1925 [*Hamites studerianus* Pictet, 1847; OD]. Phragmocone unknown; body chamber uncoiled to form a J-shaped hook; whorl section subcircular to compressed elliptical; fine prorsiradiate ribs bifurcating or trifurcating across venter, with point of branching marked by weak bulla; where bifurcation occurs, second bifurcation may occur higher on flank; dorsum smooth; suture apparently like *Hamites*, with bifid saddles and trifid lobes. *M. - U. Alb.* **Fig. 750B** - *H. studerianus* (Pictet).

The derived characters of *Hamitoides* ally it to *Eoscaphites*, and it is possible that some records of *Hamitoides* actually represent body chambers of *Eoscaphites*.

Eoscaphites BREISTROFFER, 1947 [*Ammonites? circularis* J. de C. Sowerby, 1836; OD]. Initial criocone with wide

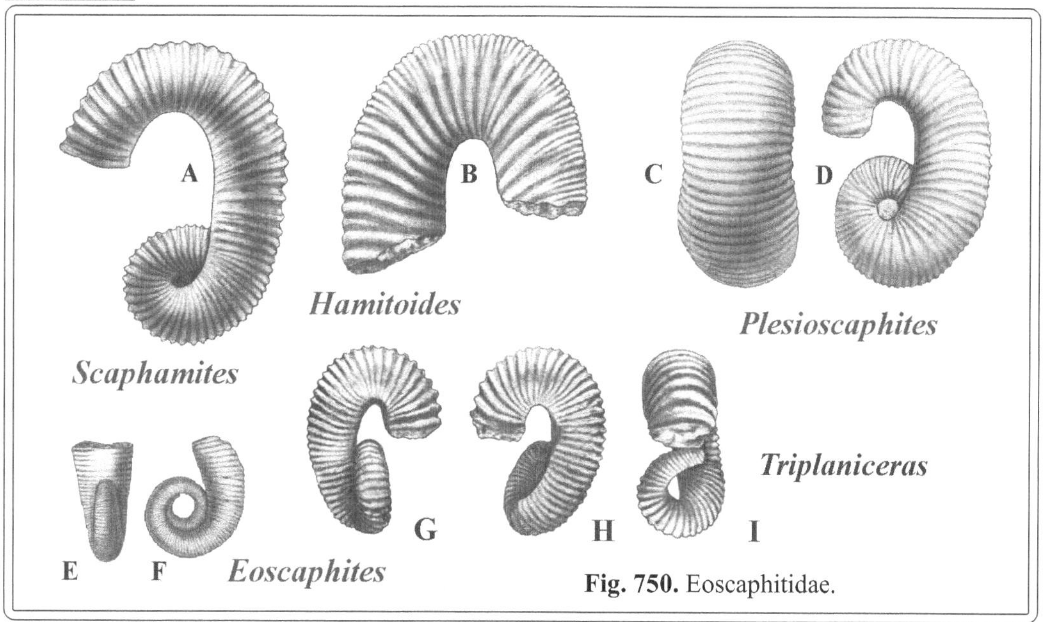

Fig. 750. Eoscaphitidae.

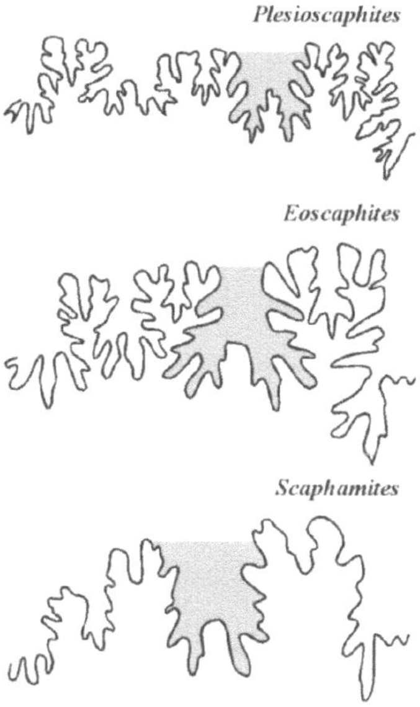

Fig. 751. Eoscaphitid sutures.

perforate umbilicus followed by rapidly expanding planispire with depressed elliptical whorls barely in contact and shallowly impressed dorsum; adult body chamber presumed to be a scaphitoid hook; fine dense rursiradiate ribs bifurcate across venter and weaken or efface from dorsum, with incipient tubercles at point of branching; suture like *Scaphamites* but more incised. *Lower U. Alb.* **Fig. 750E-F** - *E. circularis* (J. de C. Sowerby).

***Plesioscaphites* gen. nov.** [*Eoscaphites chardensis* Wright & Kennedy, 1996, p.386; OD herein]. Like *Eoscaphites* but phragmocone tightly coiled, rather involute (U=27-36%), with impressed dorsum; whorls inflated, with depressed reniform whorl section; tubercles lacking; adult body shaft with umbilical bulge (?**M**); ribs strong, single or branching at midflank; adult aperture obtuse to shaft. *U. Alb. - M. Cenom.* **Fig. 750C-D** - *P. chardensis* (Wright & Kennedy).

Comprises the phyletic line *subcircularis-tenuicostatus - chardensis*. In the most primitive species, *P. subcircularis* (Spath), the umbilicus is wider, the whorl section subcoronate and the ribbing finer than in *P. chardensis*.

Family SCAPHITIDAE Gill, 1871

Discussion: A monophyletic taxon whose origin lies in Eoscaphitidae (Spath, 1921); characteristically with imperforate umbilicus and J-shaped body chamber. Comprises the subfamilies Scaphitinae and Hoploscaphitinae. *U. Alb. - U. Maastr.*

Subfamily Scaphitinae Gill, 1871

Discussion. A monophyletic taxon whose ancestry lies in Eoscaphitinae, close to *Plesioscaphites*; typically with one or more rows of tubercles, and a suture with very wide L/U subdivided by pseudolobes and with U forming part of the internal suture. Macroconchs with a bulge at the base of the adult body chamber that may occlude the umbilicus. *U. Alb. – Maastr.*

Scaphites PARKINSON, 1811 [*Scaphites equalis* J. Sowerby, 1813; SD Meek, 1876]. Relatively small (<40 mm long);

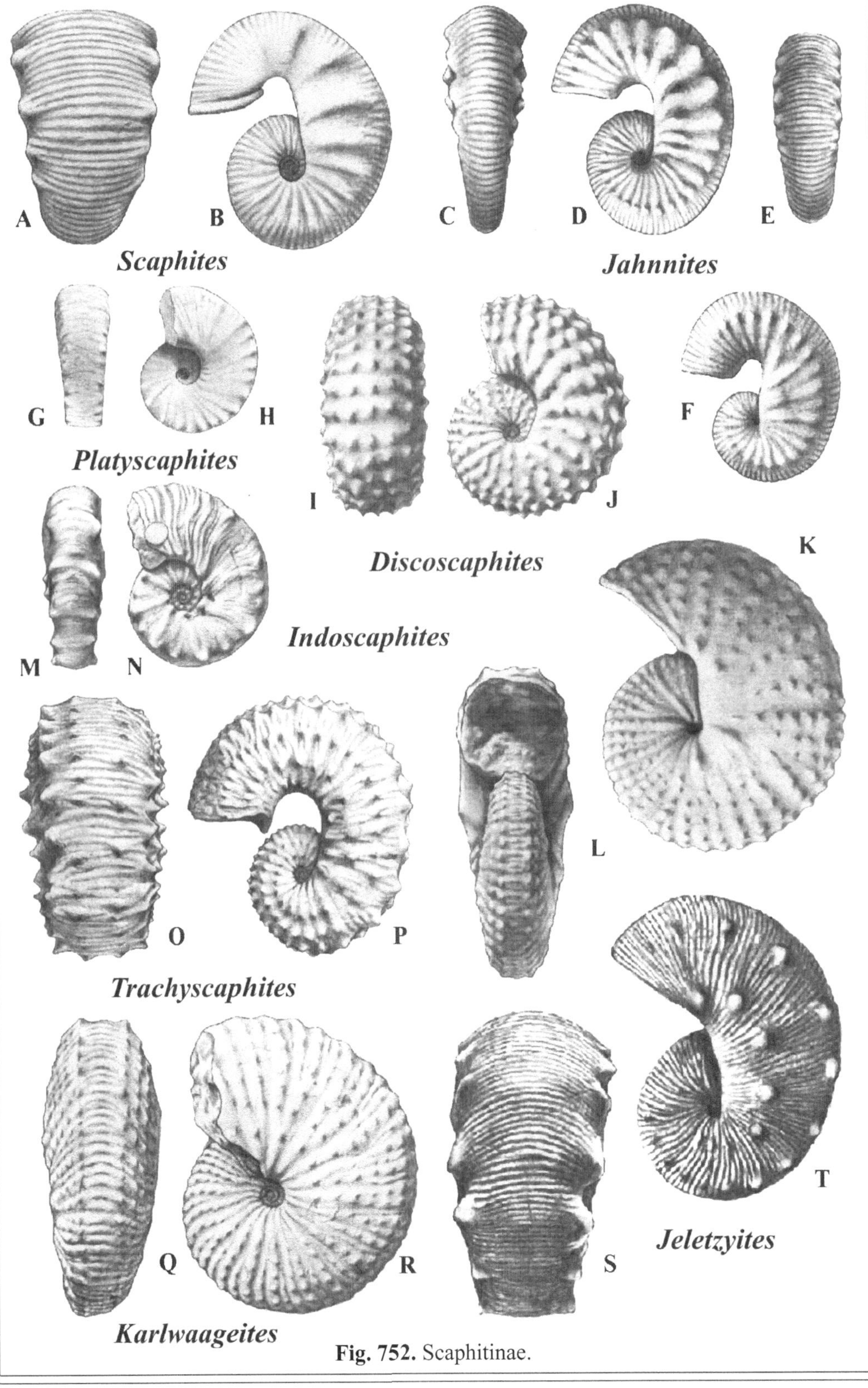

Fig. 752. Scaphitinae.

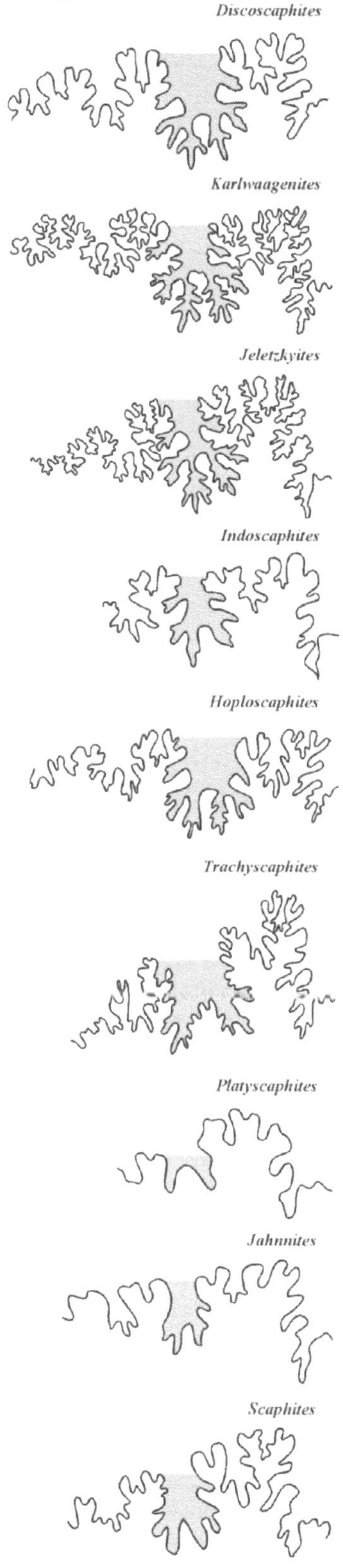

Fig. 753. Scaphitid sutures.

early whorls involute, very inflated to slightly compressed, with prominent deep dorsal impression to whorls; early whorls with numerous fine primary ribs that bifurcate or trifurcate across venter, separated by 1-2 shorter intercalatories; on adult body chamber primaries become strong, distant and bullate, sometimes developing ventrolateral tubercles, and giving rise to bundles of up to 6 fine dense secondaries; adult aperture constricted, forming acute angle to shaft; suture with moderately wide E, wide asymmetrically bifid E/L, shallow moderately wide L, and L/U with single pseudolobe. *U. Alb. – U. Cenom.* **Fig. 752A-B** - *S. equalis* J. Sowerby; **Fig. 756** - *S? reesidei* Collignon [**m**].

= *Holcoscaphites* Nowak, 1911 (obj.).

Trachyscaphites COBBAN & SCOTT, 1964 [*T. redbirdensis*; OD]. Like *Jahnnites* but larger, inflated, more robust, with multinodose body chamber with ribs looped between 3-5 rows of tubercles and (in derived species) zigzagging across venter; ribbing tends to weaken in later growth when tubercles dominate; suture with small trifid U. *L. - M. Camp.* **Fig. 752O-P** - *T. redbirdensis* Cobban & Scott.

Jeletzkyites RICCARDI, 1983 [*Scaphites (Ammonites?) nodosus* Owen, 1852; OD]. Like *Jahnnites* but larger and more inflated, with generally depressed whorl section, strong fine flexuous bifurcating riblets, low-lateral and ventrolateral tubercles in middle and adult growth, and undifferentiated ornament on adult body chamber; suture moderately complex. *L. Camp. - Maastr.* **Fig. 752S-T** - *J. nodosus* (Owen).

Karlwaageites COOPER, 1994 [*Scaphites nebrascensis* Owen, 1852; OD]. Like *Jeletzkyites* but more compressed, with more acute apertural angle, coarser ribs all of which carry ventrolateral tubercles; multinodose phragmocone with 2-7 rows of lateral tubercles, tubercles commonly obsolete on adult body chamber; microconchs with looser coiling; convergent towards *Discoscaphites*. *Maastr.* **Fig. 752Q-R** - *K. nebrascensis* (Owen).

Jahnnites HYATT, 1900 [*S. geinitzii* var. *binodosus* Jahn (*non* Roemer), = *S. g. intermedius* Scupin, 1907; OD]. Like *Scaphites* but more compressed, with flatter flanks (particularly in microconchs) and numerous primaries to shaft and hook with strong rounded or bullate ventrolateral tubercles that appear on phragmocone; incipient to strong umbilical bullae on shaft, particularly in more derived species; suture with small L shorter than E and two pseudolobes to L/U. *U. Turon. - Coni.* **Fig. 752C-D** - *J. kieslingswaldensis* (Langenhan & Grundy); **Fig. 752E-F** - *J. geinitzii* (d'Orbigny).

= *Jahnites* Reeside, 1927 (*nom. van.*).

Argentoscaphites BLASCO DE NULLO, NULLO & PROSERPIO, 1980 [*A. mutans* (*nom. correct. pro mutandibus*, Wright 1996); OD]. Small; phragmocone with primary ribs bifurcating from lateral bullae, separated by secondary

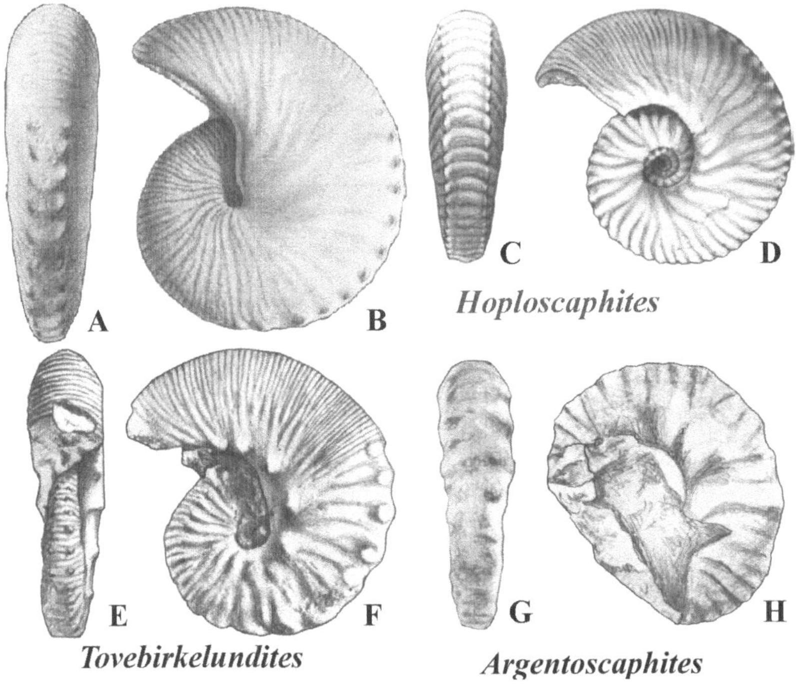

Fig. 754. Hoploscaphitinae.

Fig. 755. *Tovebirkelundites* (**M**).

and intercalated ribs, all with small oblique ventrolateral clavi offset across venter; body chamber (**m**) narrow, with coarse Y-shaped ribs with transverse ventral clavi that persist onto adapical part of proversum before evanescing on the flexus; body chamber (**M**) high whorled, with strong Y-shaped ribs with variable bullae at point of branching on proversum. *Sant. - L. Camp.* **Fig. 754G-H** - *A. mutans* Blasco de Nullo, Nullo & Proserpio.

Subfamily Hoploscaphitinae Cooper, 1994

Discussion. Monophyletic derivatives of Scaphitinae, close to *Jahnnites*; characteristically with a compressed adult body chamber with subtabulate venter and ventrolateral tubercles; in derived members body chamber recoils. *Sant. - U. Maastr.*

Platyscaphites COBBAN, KENNEDY & LANDMAN, 1999 [*P. fremontensis*; OD]. Small, compressed with flat flanks, tabulate venter and subrectangular whorl section; adult shaft short, so that body chamber remains in contact with phragmocone (although uncoiled in microconchs of primitive species); flanks with prorsiradiate striae and generally weak but variably developed riblets, and strong conical ventrolateral tubercles joined across venter by 2-3 weak looped ribs; suture simple, with deep rather broad E, broad irregular E/L, short narrow oblique L, and very narrow almost entire L/U; superficially resembles *Indoscaphites*. *L. Camp.* **Fig. 752G-H** - *P. fremontensis* Cobban, Kennedy & Landman.

Hoploscaphites NOWAK, 1911 [*Scaphites constrictus* J. Sowerby, 1817; OD]. Like *Platyscaphites* but strongly compressed; inner whorls with narrowly rounded venter and non-tuberculate phragmocone; uncoiled adult body chamber with short shaft, weakly recurved hook and obtuse apertural angle; straight to slightly flexuous

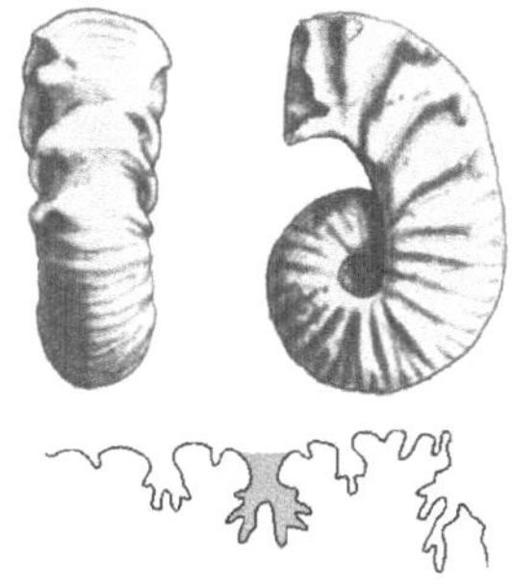

Fig. 756. *Scaphites? reesidei.*

prorsiradiate ribs bifurcate at various levels on flank; clavate ventrolateral tubercles appear on last part of phragmocone but evanece on finely ribbed terminal stage; U of suture with 3 pseudolobes. *U. Camp. - Maastr.* **Fig. 754A-B** - *H. constrictus* (J. Sowerby) (**M**); **Fig. 754C-D** - *H. comprimus* (Owen) [**m**].

= *Mesoscaphites* Atabekian, 1979 (*nom. nud.*).

Tovebirkelundites COOPER, 1994 [*Acanthoscaphites schmidi* Birkelund, 1982; OD]. Like *Hoploscaphites* but microconchs with inner and outer ventrolateral and siphonal tubercles, and strong umbilical bullae to adult body chamber; macroconchs with umbilical bullae replaced by tiny low-lateral tubercles; superficially similar to *Acanthoscaphites*. *Maastr.* **Fig. 754E-F** - *T. schmidi* (Birkelund) [**m**], **Fig. 755** [**M**].

Discoscaphites MEEK, 1871 [*Ammonites conradi* Morton, 1834; OD]. Like *Hoploscaphites* but multinodose, with umbilical, 1-3 lateral, and ventrolateral tubercles from early growth, as well as siphonal tubercles. *U. Maastr.* **Fig. 752I-L** - *D. conradi* (Morton).

Indoscaphites SPATH, 1953 [*Ammonites cunliffei* Forbes, 1846; OD]. Like *Hoploscaphites* but small, with weak distant umbilical bullae connected by low rounded straight to feebly concave prorsiradiate ribs that broaden ventrally and connect with conical ventrolateral tubercles linked across broadly rounded venter by broad low ribs or ribs separated by intercalatories; on adult body chamber umbilical bullae strongly prorsiradiate, ribbing finer sharper flexuous and more irregular, linking in groups to ventrolateral tubercles that become obsolete on final part of body chamber; prominent umbilical and ventrolateral tubercles; suture rather simple, with narrow E, broad asymmetrically bifid E/L, narrow asymmetrically trifid L almost as deep as E, and narrow asymmetrically bifid narrow stemmed pseudolobe. *U. Maastr.* **Fig. 752M-N** - *I. cunliffei* (Forbes).

Family OTOSCAPHITIDAE Wright, 1953

(*nom. transl.* Wiedmann, 1962 *ex* Otoscaphitinae)

Discussion. A monophyletic group of uncertain origin, but probably Eoscaphitidae; comprises the subfamilies Worthoceratinae, Otoscaphitinae, Clioscaphitinae and Rhaeboceratinae. *U. Alb. - U. Maastr.*

Subfamily Worthoceratinae nov.

Diagnosis. Typically comprises small, weakly ornamented ammonites with a scaphitoid shell, serpenticone phragmocone, long body shaft in microconchs, and without the sutural development, i.e. subdivision of U, typical of true scaphitids. A monophyletic taxon, probably with an origin in Eoscaphitidae, close to *Eoscaphites*. Claims of an

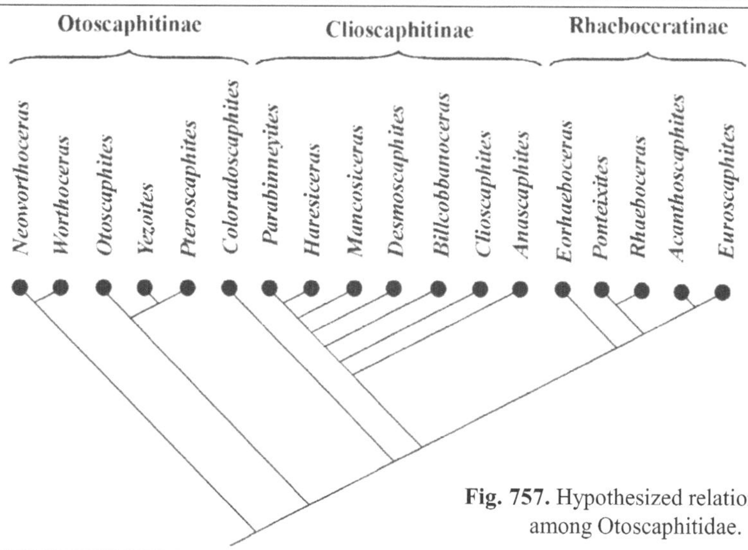

Fig. 757. Hypothesized relationships among Otoscaphitidae.

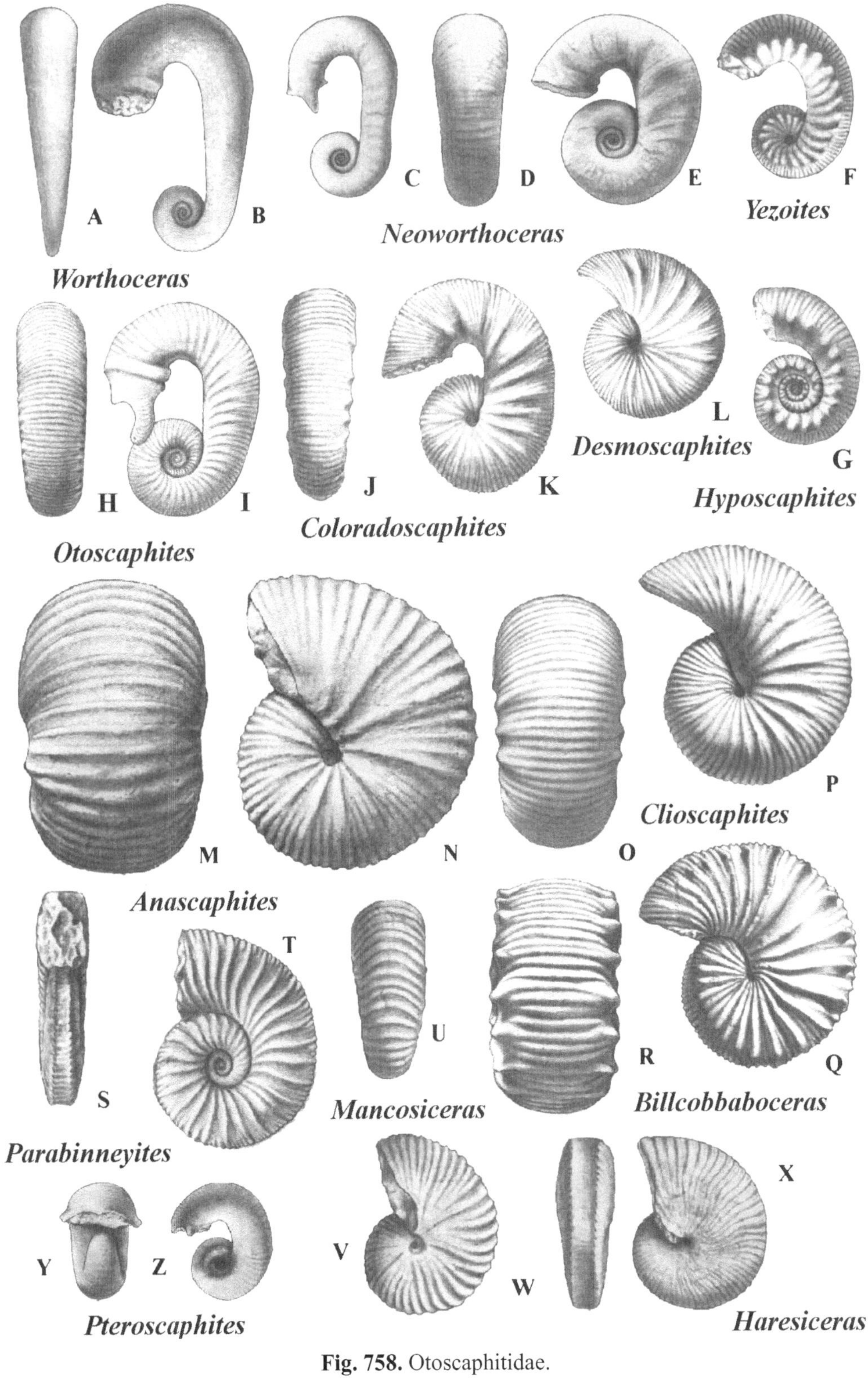

Fig. 758. Otoscaphitidae.

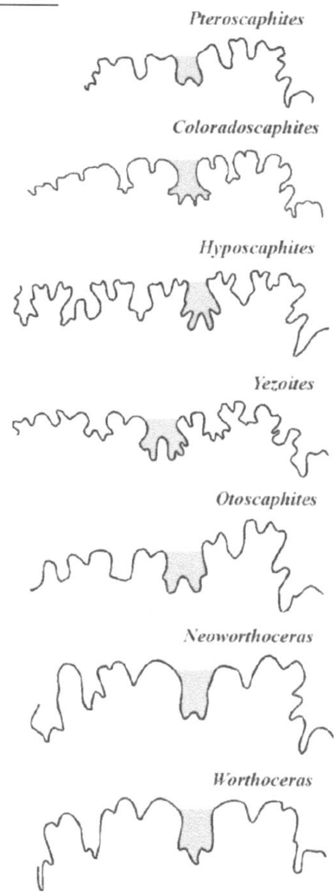

Fig. 759. Otoscaphitid sutures.

origin in Ptychoceratidae (Wiedmann, 1965; Monks, 2002) are not supported by sutural characteristics, the broad stemmed wide E/L and L/U of *Worthoceras* differ markedly from *Ptycho ceras* and, in this respect, there is a closer similarity to *Scaphamites*. Moreover, as shown by Wiedmann *et al.* (1990: 378), whether L is bifid or trifid is of "*very limited*" use in classification, and may vary within a species. As interpreted here, Worthoceratinae are the sister taxon of Otoscaphitinae. *U. Alb. - U. Turon.*

Worthoceras ADKINS, 1928 [*Macroscaphites platydorsus* Scott, 1924; OD]. Small; serpenticone phragmocone followed by long straight proversum and terminal hook with flattened dorsum; ornament lacking although some species have radial ribs; microconchs with tubular shaft of uniform width and lateral lappets to aperture; macroconchs with rapidly expanding (in width) shaft and simple aperture; suture simple, with deep rectangular E, bifid subrectangular E/L and L/U, narrow asymmetrically bifid L, narrow vermiform U and narrow I deeper than E. *U. Alb. - L. Cenom.* **Fig. 758A-B** - *W. platydorsum* (Scott).

***Neoworthoceras* gen. nov.** [*Scaphites vermiculus* Shumard, 1860, p.419; OD herein]. Like *Worthoceras* but larger, with relatively large planispire with impressed dorsum, shorter curved proversum with flat sides and constant width in both sexes, and more incised suture with broader E/L and bifid L; weak broad low flank ribs may be present in some species. *U. Cenom. - U. Turon.* **Fig. 758C-E** - *N. vermiculum* (Shumard).

Subfamily Otoscaphitinae Wright, 1953

Discussion: A monophyletic group with an origin in Worthoceratinae (Wright, 1979) close to *Neoworthoceras*. Dimorphism is ubiquitous; although there is size overlap, macroconchs have an umbilical bulge to base of adult body chamber and microconchs with lateral lappets; suture with symmetrically bifid L. *U. Alb. - Coni.*

Otoscaphites WRIGHT, 1953 [*Ammonites? bladenensis* Schlüter, 1871; OD]. Small, with very evolute spire with gently convex flanks, narrowly rounded venter and moderately compressed whorl section; phragmocone with strong bifurcating sigmoidal ribs that bend forward on outer flank; ribbing weak to fairly strong on shaft, with fine prorsiradiate primaries recurving on inner flank to become radial, breaking into 2-3 lirae separated by intercalatories across venter; aperture with collared constriction and lateral lappets. *U. Alb. - Coni.* **Fig. 758H-I** - *O. bladenensis* (Schlüter).

Yezoites YABE, 1910 [*Scaphites perrini* Anderson; SD Diener, 1925]. Very small; very involute with broad flat inclined umbilical wall, subangular ventrolateral shoulders, depressed coronate whorl section; long primary ribs on umbilical wall terminate in ventrolateral tubercles; suture with very broad E/L. *U. Cenom. - Coni.* **Fig. 758F** - *Y. seabeensis* (Cobban & Gryc).

Hyposcaphites WIEDMANN, 1965 [*Scaphites stephanoceroides* Yabe, 1909; OD]. Like *Yezoites* but spire evolute, with angular lateral keels situated at midflank, and strong round umbilicolateral tubercles giving rise to coarse ribs in triplets. *U. Turon.* **Fig. 758G** - *H. stephanoceroides* (Yabe).

Coloradoscaphites COOPER, 1994 [*Scaphites whitfieldi* Cobban, 1952; OD]. Innermost whorls like *Yezoites* with coronate whorl section and angulated ventrolateral shoulders; later with round inflated whorls with ribs splitting into 2's and 3's at midflank; hook with strongly prorsiradiate primary ribs separated by 4-8 fine secondaries; suture with 2 pseudolobes to U. *U. Turon.* **Fig. 758J-K** - *C. whitfieldi* (Cobban).

Subfamily Clioscaphitinae Cooper, 1994

Discussion: Monophyletic derivatives of Otoscaphitinae, close to *Coloradoscaphites*, that evolved in parallel with Scaphitinae; typically with adult body chamber remaining in contact with phragmocone and suture with asymmetrically bifid to trifid L. With one questionable exception (*Parabinneyites*), endemic to the Western Interior of North America. *Coni. - Camp.*

Anascaphites HYATT, 1900 [*Scaphites ventricosus* Meek & Hayden, 1863; OD]. Like *Coloradoscaphites* but medium sized, more inflated, with depressed whorl section, strongly rounded flanks, swollen curved body chamber that uncoils only slightly (at least in macroconchs), and coarser ornament with fewer primary ribs to adult body chamber. *Coni.* **Fig. 758M-N** - *A. ventricosus* (Meek & Hayden).

Clioscaphites COBBAN, 1952 [*C. montanensis*; OD]. Like *Anascaphites* but body chamber tightly adpressed to phragmocone, with finer ribbing, more primary ribs and secondary ribs intercalated higher on flank. *L. Sant.* **Fig. 758O-P** - *C. montanensis* (Cobban).

Billcobbanoceras COOPER, 1994 [*Scaphites vermiformis* Meek & Hayden, 1876; OD]. Like *Clioscaphites* but with broadly rounded to flattened venter, more or less prominent ventrolateral tubercles and intercalated ribs virtually restricted to venter; suture usually with trifid L. *M. - U. Sant.* **Fig. 758Q-R** - *B. vermiforme* (Meek & Hayden).

Desmoscaphites REESIDE, 1927 [*Desmoscaphites bassleri*; OD]. Like *Billcobbanoceras* but smaller, with constricted inner whorls, coarse forwardly arching ventral ribs that become fine in maturity, compressed adult body chamber and suture with trifid L. *U. Sant. - L. Camp.* **Fig. 758L** - *D. bassleri* Reeside.

Mancosiceras COBBAN, 1964 [*Puzosia (Latidorsella) mancosensis* Reeside, 1927; OD]. Like *Desmoscaphites* but smaller, more involute, with more compressed outer whorls. *U. Sant.* **Fig. 758U-V** - *M. mancoense* (Reeside).

Haresiceras REESIDE, 1927 [*H. placentiforme*; OD]. Like *Mancosiceras* but still smaller, weakly ornamented; suture with numerous auxilliary elements. *L. Camp.* **Fig. 758W-X** - *H. placentiforme* Reeside.

Parabinneyites A. LEANZA, 1964 [*Leopoldia paynensis* Favre, 1908; OD]. Like *Haresiceras*, but more evolute and with rather distant sinuous flank costae separated by 1-4 short intercalatories, all with tiny ventrolateral nodes connected across venter by fine striae. *L. Camp.* **Fig. 758S-T** - *P. paynensis* (Favre).

= *Patagoniceras* Leanza, 1963 (*non* Wetzel, 1960).

Although *Parabinneyites* resembles *Haresiceras*, the sutures are very different and it is probably a homoeomorph but, at present, there seem few other places to assign it.

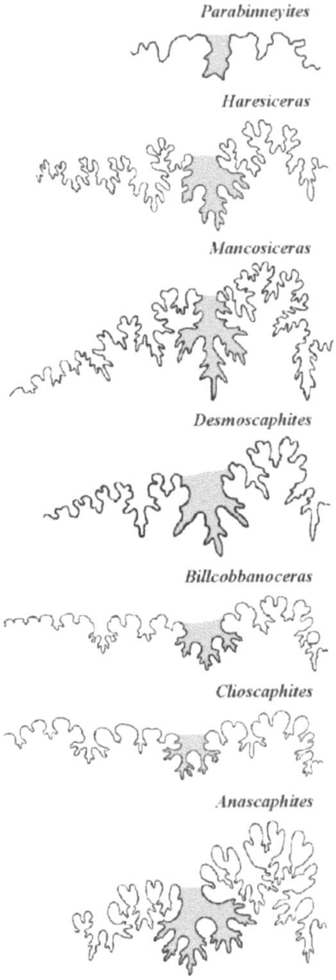

Fig. 760. Clioscaphitine sutures.

Pteroscaphites WRIGHT, 1953 [*Scaphites auriculatus* Cobban; OD]. Weakly ornamented micromorphs with sparse subdued ribs, sometimes with weak nodes at angular to subangular ventrolateral shoulder; hollow lateral lappets to adult aperture and simple suture. *Coni.* **Fig. 758Y-Z** - *P. auriculatus* (Cobban).

Landman (1987) recognized a succession of *Pteroscaphites* species, allegedly derived from different contemporaneous clioscaphites. If confirmed, these are all homoeomorphs of *Pteroscaphites*, for which new generic names are required.

Subfamily Rhaeboceratinae Cooper, 1994

Discussion: A cryptogenic taxon but probably derived from Clioscaphitinae, close to *Clioscaphites*. Typically large with tightly coiled body chamber and bundles of fine branching ribs arising from weak umbilical bullae; derived species are multinodose. *U. Sant. - Maastr.*

Eorhaeboceras ALABUSHEV, 1989 [*E. derivatum*; OD]. Small, somewhat inflated, evolute, with rapidly expanding whorls; inner whorls with wide shallow umbilicus, rounded umbilical wall and shoulder, convex flanks, broadly rounded venter and subcircular whorl section; inner flanks with numerous convex forward narrow folds that evanesce before reaching venter but give rise to fasciculate bundles of fine secondaries separated across venter by 1-3 intercalatories; adult body chamber does not uncoil, and has compressed-oval whorl section with overhanging umbilical wall; suture with deep narrow E, moderately wide deeply bifid E/L, moderately wide bifid L as deep as E, and narrow asymmetrically trifid L/U$_2$. *Upper U. Sant. - L. Camp.* **Fig. 762H-I** - *E. derivatum* Alabushev.

Rhaeboceras MEEK, 1876 [*Phylloceras? halli*; OD]. Like *Eorhaeboceras* but much larger, more involute and compressed, without plications on inner flank but with distinct umbilicolateral bullae giving rise to bundles of bifurcating and trifurcating prorsiradiate riblets; suture complex to very complex, with very narrow asymmetrically bifid to trifid L and narrow deeply incised E/L and L/U$_2$. *Maastr.* **Fig. 762D-E** - *R. halli* (Meek).
= *Amopalus* Owen, 1852 (*nom. oblit.*).

Ponteixites WARREN, 1934 [*P. robustus*; OD]. Micromorph, with peculiar sparse slightly flexuous rounded fold-like ribs that are conspicuous across venter; suture relatively simple, with very wide subquadratic E/L and wide subsymmetrically bifid L. *L. Maastr.* **Fig. 762F-G** - *P. robustus* Warren.
Perhaps a microconch of *Rhaeboceras*.

Acanthoscaphites NOWAK, 1911 [*Scaphites tridens* Kner, 1850; SD Diener, 1925]. Moderately large; microconchs with small deep umbilicus throughout, macroconchs with umbilicus partly occluded by initial part of body chamber; phragmocone with dense fine prorsiradiate ribs arising singly or in pairs from numerous weak umbilical bullae with frequent branching and intercalatories of different length; in maturity umbilical bullae shift to umbilicolateral position (**m**) or are lost (**M**); flank ribs loop in 2's and 3's to ventrolateral clavi joined across venter by looped ribs; variably developed periodic large blunt siphonal tubercles, sometimes clavate, develop in maturity; suture complex. *L. Maastr.* **Fig. 762B-C** - *A. tridens* (Kner), macroconch and microconch.

Euroscaphites JAGT, KENNEDY & MACHALSKI, 1999 [*Scaphites varians* Lopuski, 1911; OD]. Like *Acanthoscaphites* but with broad straight flank ribs carrying 3 rows of lateral tubercles, lowest row strongest and most persistent sometimes continuing onto adult body chamber; ribs flat topped and striate in maturity. *Upper L. - U. Maastr.* **Fig. 762A** - *E. varians* (Lopuski).

Superfamily BACULITOIDEA Gill, 1871

(*nom. transl.* Cooper, 1994 *ex* Baculitidae)

Discussion. Monophyletic derivatives of Polyptychoceratoidea with which they evolved in parallel throughout their existence and hence are given equal taxonomic rank. Small to large straight and curved baculicones, either smooth or with fine ribbing, dorsolateral crescents, tiny ventrolateral tubercles and/or a siphonal keel. *U. Alb. - U. Maastr.*

Family BACULITIDAE Gill, 1871

Discussion: Monophyletic derivatives of Polyptychoceratidae; initial ammoniticone followed by long tapering shaft with subcircular to compressed oval and elliptical whorl section; shaft generally straight but occasional curved to crioconic; ornament varying from smooth to strong annular ribbing, commonly with crescentic bullae on the lower flank;

Fig. 761. Rhaeboceratine sutures.

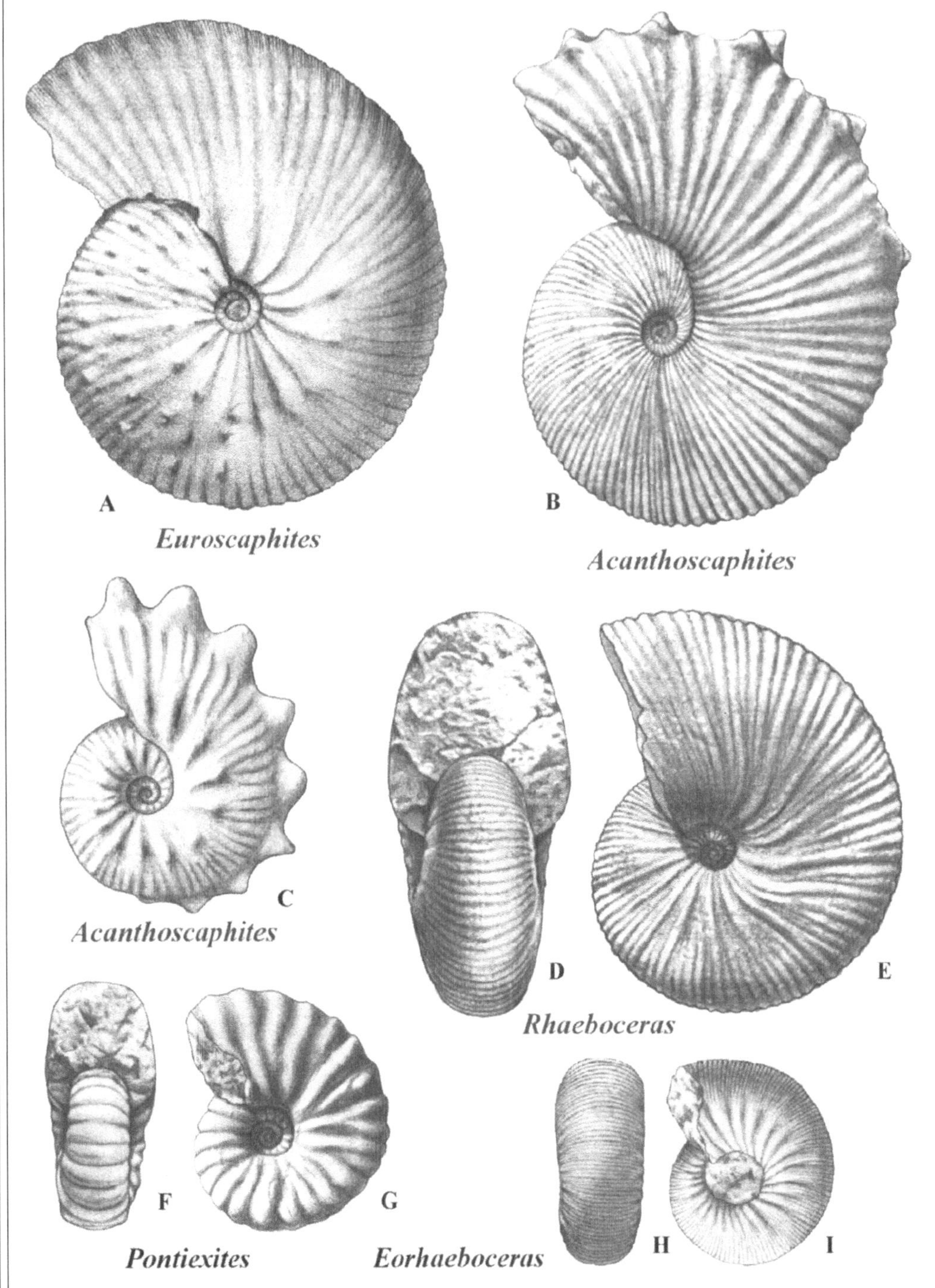

Fig. 762. Rhaeboceratinae.

constrictions and a siphonal keel are present in some. Comprises the subfamilies Baculitinae and Eubaculitinae. *U. Alb. - U. Maastr.*

Subfamily Baculitinae Gill, 1871

(*nom. transl.* Brunnschweiler, 1966 *ex* Baculitidae)

Discussion. Monophyletic derivatives of Metaptychoceratinae close to *Psilohamites* since the early

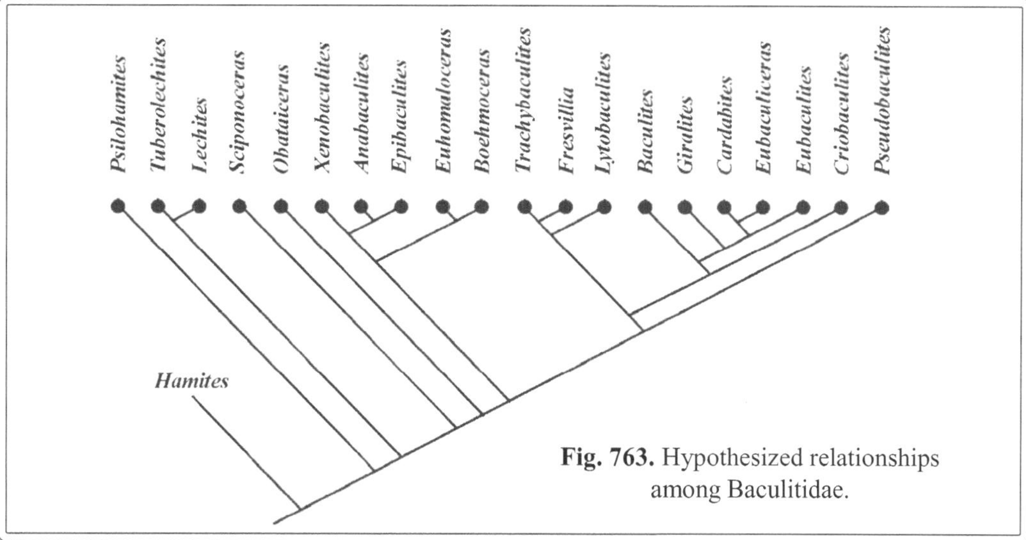

Fig. 763. Hypothesized relationships among Baculitidae.

growth stage of *Hemiptychoceras crassum* (Kennedy) (**Fig. 708L**) is a smooth straight shaft that, in isolation, would probably be identified as *Psilohamites*. Sexually dimorphic, microconchs commonly with down turned apertures whereas that of macroconchs is trumpet-like. The group underwent noteworthy radiation in the Late Cretaceous and, with over 80 accepted species (Klinger & Kennedy, 2001), the diversity of the baculitids records rapid evolution and a complex evolutionary history that has not been reflected in prevailing taxonomies. Particularly noteworthy is the evolution of the group in the Western Interior of North America, where isolation saw the rise of endemic species and genera. *U. Alb. - U. Maastr.*

Lechites NOWAK, 1908 [*Baculites gaudini* Pictet & Campiche, 1861; OD]. Medium sized; long, straight, slowly expanding shaft with subcircular whorl section; low robust prorsiradiate ribs, generally narrower than interspaces, cross venter transversely and weaken over dorsum; suture like *Psilohamites* but with reduced trifid I. *U. Alb* **Fig. 765A-B** - *L. gaudini* (Pictet & Campiche).

Tuberolechites COOPER & KENNEDY, 1977 [*Lechites (T.) regifex*; OD]. Like *Lechites* but smaller, with tiny ventrolateral nodes. *Upper U. Alb.* **Fig. 765C-D** - *T. regifex* (Cooper & Kennedy).

Sciponoceras HYATT, 1894 [*Hamites baculoides* Mantell, 1822; OD]. Like *Lechites* but with periodic oblique constrictions to phragmocone deepening over venter, generally weaker ribbing often with intercalatories, adult body chamber with coarse to fine sometimes scale-like ribs that strengthen markedly towards aperture, and suture with E almost as deep as L. *U. Alb. - U. Cenom.* **Fig. 765E-F** - *S. baculoides* (Mantell).

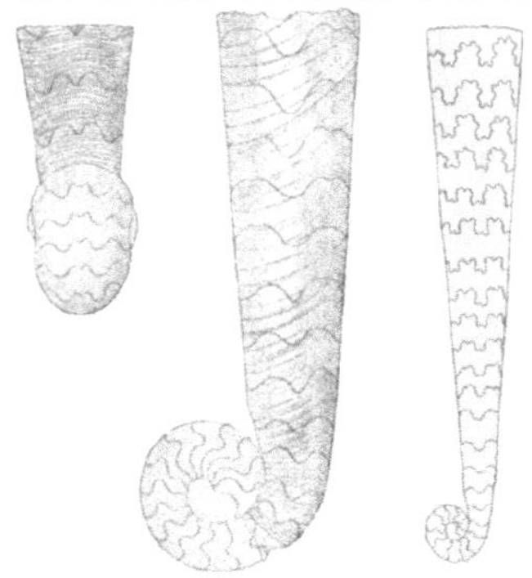

Fig. 764. Baculitid protoconch (after Smith, 1901).

= *Cyrtochilus* Meek, 1876 (*non* Jakowlew, 1875).
= *Cyrtochilella* Strand, 1929 (obj.).

Tuberosciponoceras MEDINA & RICCARDI, 2007[*T. insolitum*; OD]. Like *Sciponoceras* but with ventrolateral tubercles and prominent constrictions; suture with a deep trifid I. *Lower U. Alb.* **Fig. 767A-B** - *T. insolitum* Medina & Riccardi.

***Obataiceras* gen. nov.** [*Baculites bohemicum* Fritsch, 1875, p.49; OD herein]. Like *Sciponoceras* but slowly expanding, whorl section elliptical and ribs forming crescents on lower flank, with or without development of nodes; strong constrictions still present on phragmocone; adult aperture unknown. *U. Turon.* **Fig. 765O-P** - *O. bohemicum* (Fritsch).

Assignment of the type species to *Sciponoceras* on the basis of constrictions emphasizes primitive characters; crescentic ribs and presence of nodes are derived characters typical of later baculites.

Euhomaloceras SPATH, 1926 [*Baculites incurvatus* Dujardin, 1837; OD]. Like *Obataiceras* but shaft

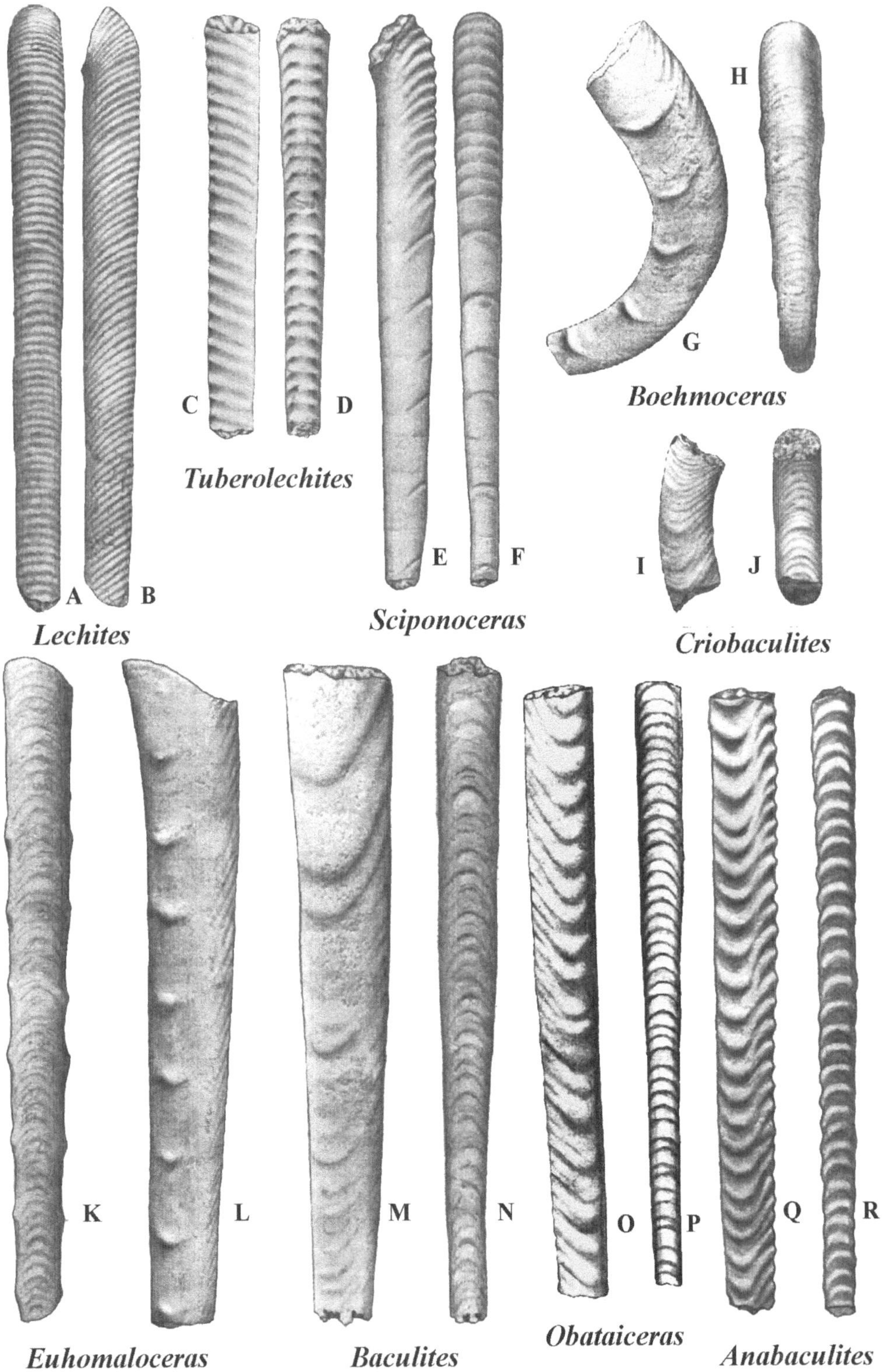

Fig. 765. Baculitidae.

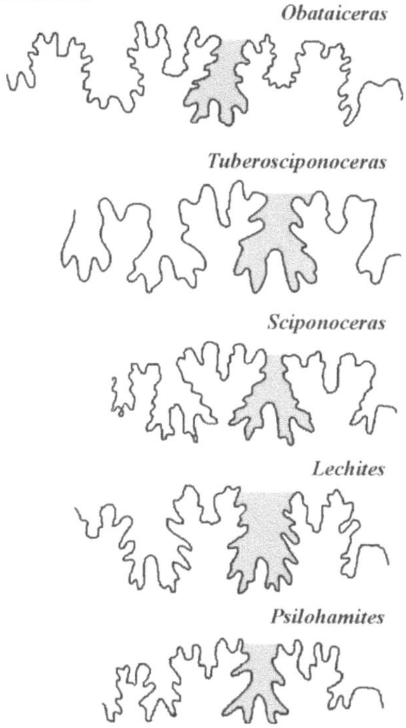

Fig. 766. Baculitid sutures.

expanding in height more rapidly and with somewhat flattened dorsum and obsolete constrictions; adult aperture gently downturned but opening forward, with short ventral rostrum and dorsal lappet; suture with subquadratic bifid elements, E/L generally wider than high, subrectangular L, well developed bifid U and reduced trifid I. *U. Turon. - U. Camp.* **Fig. 765K-L, 771A-B** - *E. incurvatum* (Dujardin).

Pseudoboehmoceras **gen. nov.** [*Ammonites arculus* Morton, 1834; OD herein]. Like *Euhomaloceras* but with distinctly curved shaft; well spaced dorsolateral crescents give rise to a narrow rib that effaces across upper flank; interspaces with numerous minor ribs best developed over rounded venter, giving it a feebly crenu-

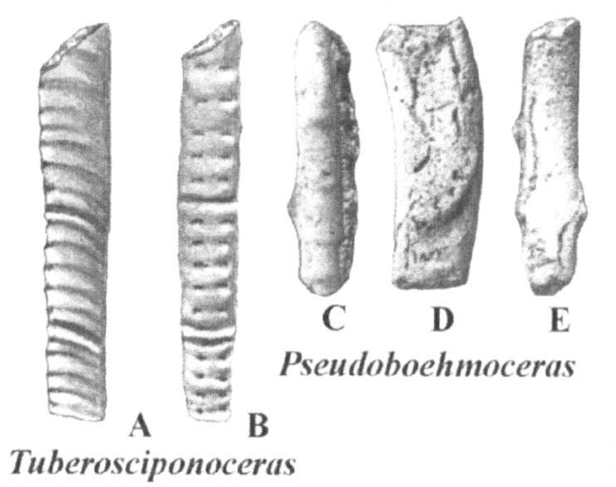

Fig. 767. Baculitidae.

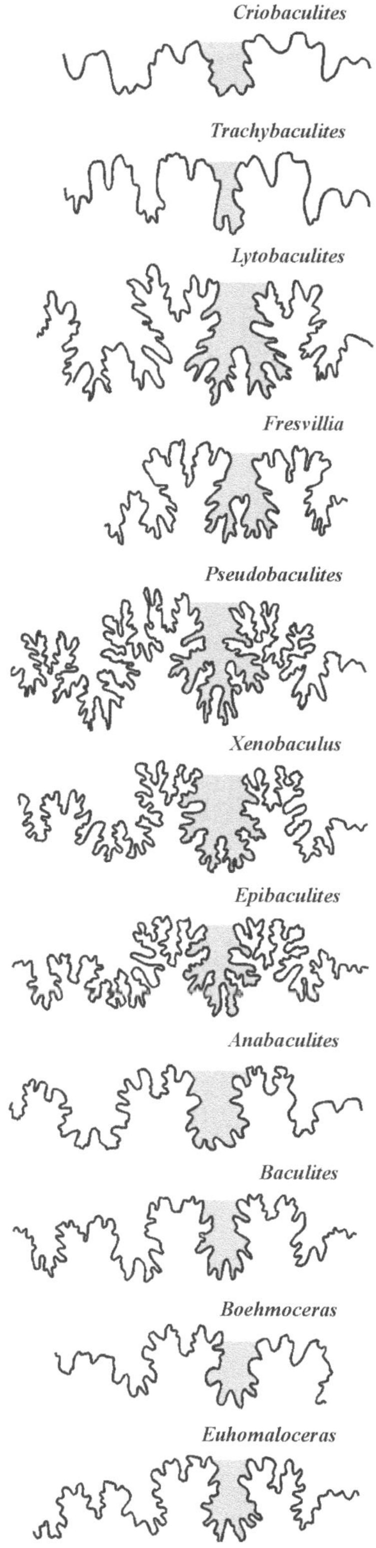

Fig. 768. Baculitid sutures.

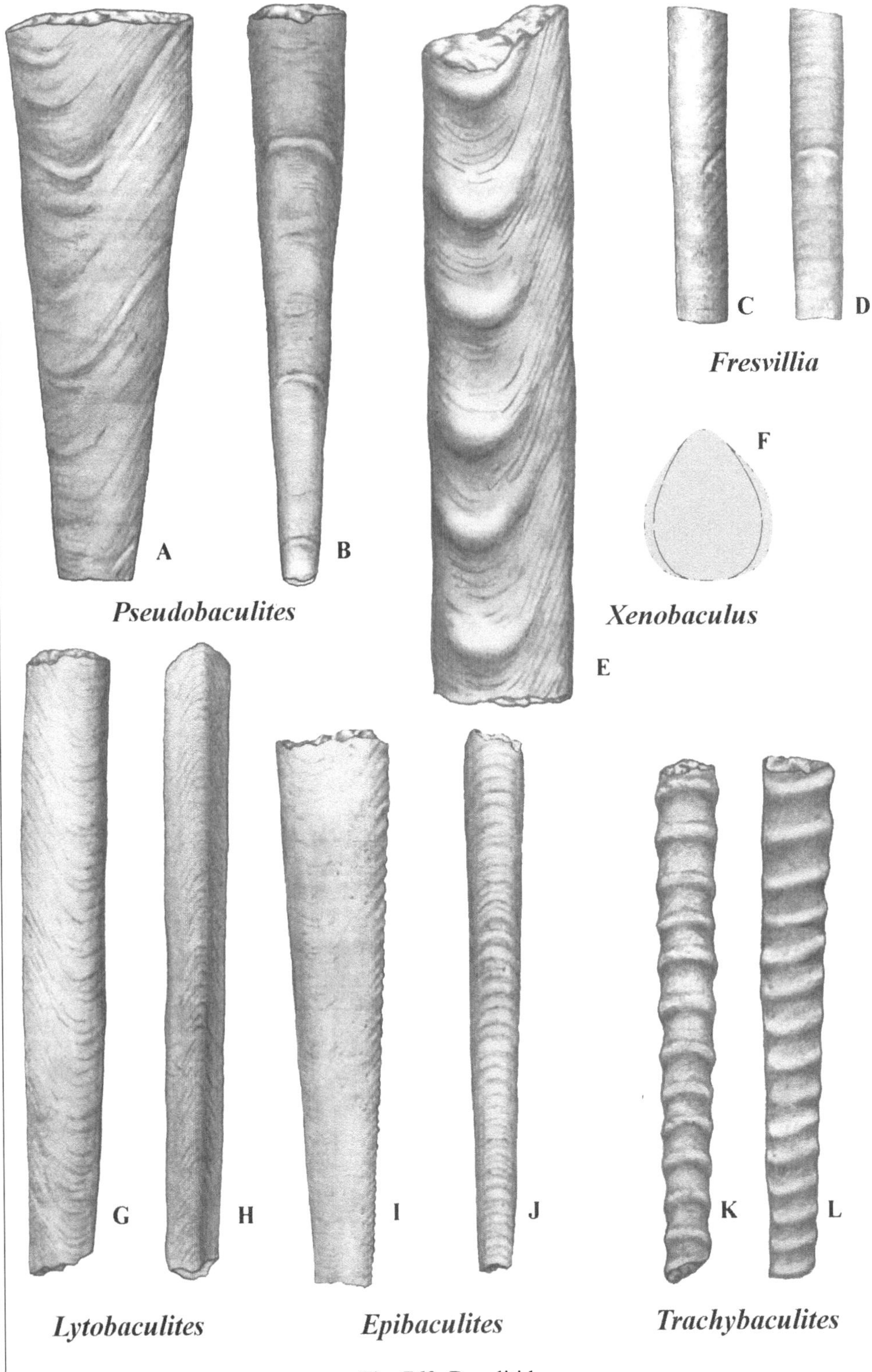

Fig. 769. Baculitidae.

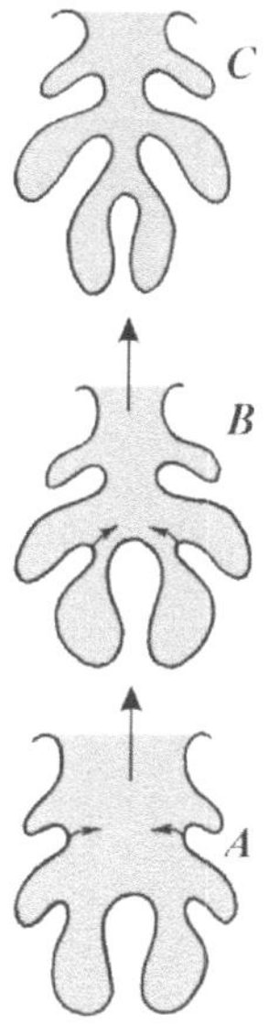

Fig. 770. Evolutionary trend within the shape of the lateral lobe (L) in some species of *Anabaculites*. A: *A. perplexus*, B: *A. gregoriensis*, C: *A. eliasi*. Adapted from Gill & Cobban (1973).

lated appearance. *U. Sant.* **Fig. 767C-E** - *P. arculum* (Morton).

Boehmoceras RIEDEL, 1931 [*Ancyloceras krenkeleri* Wegner, 1905; SD Wright, 1957]. Small, strongly compressed tight criocone with whorls increasing rapidly in height; flanks with crescentic lateral bulges, sometimes flared, splitting into finer ribs on outer flank; venter with prominent coarsely serrated keel; periodic deepened interspaces may be constrictions; suture like *Euhomaloceras* but with broader shorter U/I. *U. Sant.* **Fig. 765G-H, 771C** - *B. krenkeleri* (Wegner).

Pseudobaculites COBBAN, 1952 [*P. nodosus*; OD]. Like *Euhomaloceras* but larger and more compressed, with shaft having high degree of taper and expanding rapidly in height; suture complex, with deep penetrating folioles to L and U, and deeply bifid U/I; endemic to Western Interior of USA. *U. Coni. - L. Maastr.* **Fig. 769A-B** - *P. nodosus* Cobban.

Anabaculites **gen. nov.** [*Baculites thomi* Reeside, 1927, p.13; OD herein]. Like *Euhomaloceras* but larger, slowly expanding, with strong lateral crescents outnumbered by well developed ventral ribs that corrugate venter in middle and later growth; suture like *Euhomaloceras* but more incised and with reduced U/I; largely endemic to Western Interior of USA. *U. Coni. - L. Camp.* **Fig. 765Q-R** - *A. thomi* (Reeside).

Epibaculites **gen. nov.** [*Baculites cuneatus* Cobban, 1962, p.127; OD herein]. Like *Anabaculites* but generally with weaker ornament mostly comprising faint prorsiradiate ribs across venter; suture like *Anabaculites* but more incised, with terminal branches to L with constricted bases; endemic to the Western Interior of USA. *M. Camp. - L. Maastr.* **Fig. 769I-J** - *E. cuneatus* (Cobban).

Xenobaculus **gen. nov.** [*Baculites clinolobatus* Elias, 1933, p.310; OD herein]. Moderately large, with oval whorl section, large subdued crescents that cover entire flank,

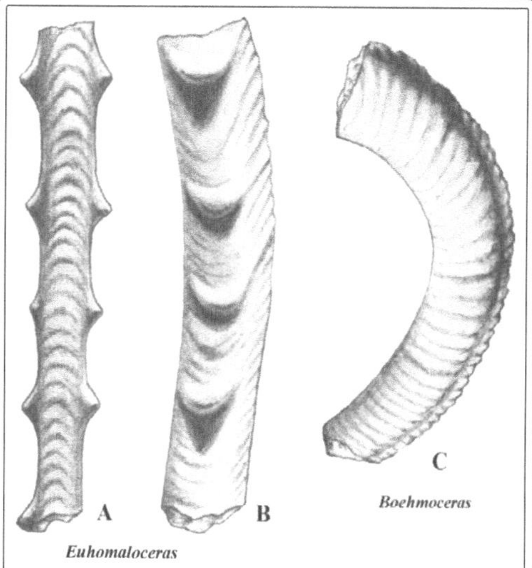

Fig. 771. Baculitinae.

fine weak riblets and growth striae over venter; suture like *Epibaculites* but with broader lobes; endemic to Western Interior of USA. *L. Maastr.* **Fig. 769E-F** - *X. clinolobatus* (Elias).

Baculites LAMARCK, 1799 [*Baculites vertebralis* Lamarck; SD Meek, 1876]. Slowly expanding shaft with elliptical to compressed ovate whorl section; both venter and dorsum uniformly rounded; ornament lacking; suture like *Euhomaloceras*. *Maastr.* **Fig. 772** - *B. vertebralis* (Lamarck).

In *B. anceps* Lamarck (**Fig. 765M-N**) the venter is narrowly rounded to fastigiate and keeled, and there are weak crescents on the lower flank, periodic strong broad shallow constrictions that are arcuate in early growth but biconcave and strongly projected on venter in maturity; adult aperture with long ventral rostrum.

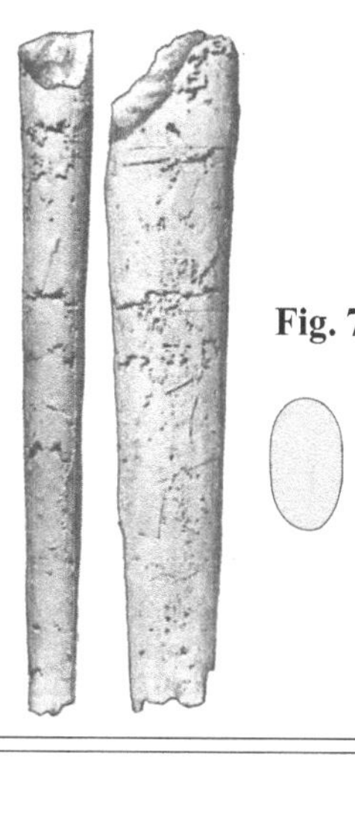

Fig. 772. *Baculites.*

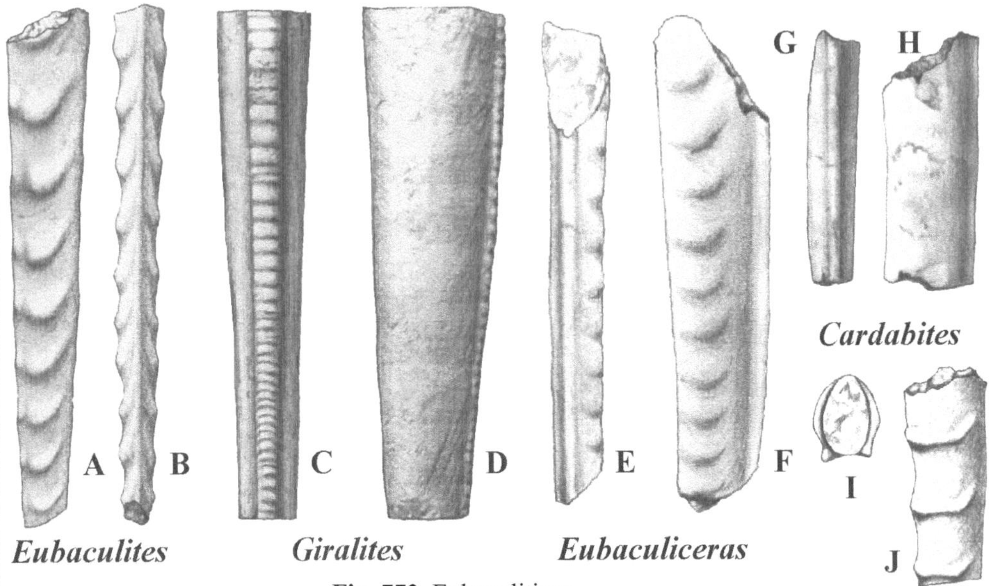

Fig. 773. Eubaculitinae.

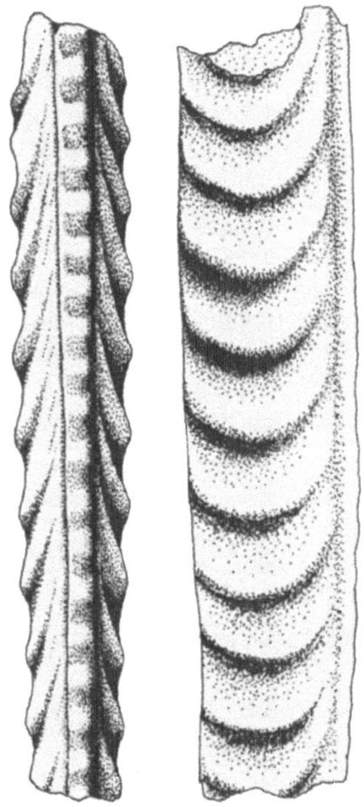

Fig. 774. *Eubaculiceras*

Lytobaculites **gen. nov.** [*Baculites chicoensis* Trask, 1856, p.92; OD herein]. Like *Baculites* but expanding slowly; suture distinctive, with narrow stemmed saddles and splayed bifid L. *L. - U. Camp., ?Maastr.* **Fig. 769G-H** - *L. chicoensis* (Trask).

Fresvillia KENNEDY, 1986 [*F. constricta*; OD]. Small; like *Lytobaculites* but smaller, expanding slowly, with circular whorl section; periodic constrictions strongest across venter; strongly prorsiradiate growth lines and feeble broad low prorsiradiate riblets markedly concave across flanks but effaced from dorsum; suture like *Lytobaculites* but simpler; superficially resembles shafts of *Phylloptychoceras*. *U. Maastr.* **Fig. 769C-D** - *F. constricta* Kennedy.

Trachybaculites COBBAN & KENNEDY, 1995 [*Baculites columna* Morton, 1834; OD]. Like *Euhomaloceras* but smaller, slowly expanding, with coarse, distant, blunt, slightly prorsiradiate to nearly rectiradiate ribs that do not follow growth lines and cross venter and dorsum with broad convexity, weakening across dorsum; suture with narrow E. *U. Maastr.* **Fig. 769K-L** - *T. columna* (Morton).

Criobaculites KLINGER & KENNEDY, 1997 [*Baculites trabeatus* Morton, 1834; OD]. Small, compressed, gently curved, with broadly rounded dorsum and flanks, somewhat narrower venter, and ovoid whorl section; delicate riblets of various strength arcuate on lower flank, projected forward on outer flank, and crossing venter (strongly) and dorsum (weakly) with broad convexity; suture very simple, with little incised bifid elements. *Maastr.* **Fig. 765I-J** - *C. trabeatus* (Morton).

Subfamily Eubaculitinae Brunnschweiler, 1966

Discussion. Monophyletic derivatives of Baculitinae, close to *Baculites*, with which they evolved in

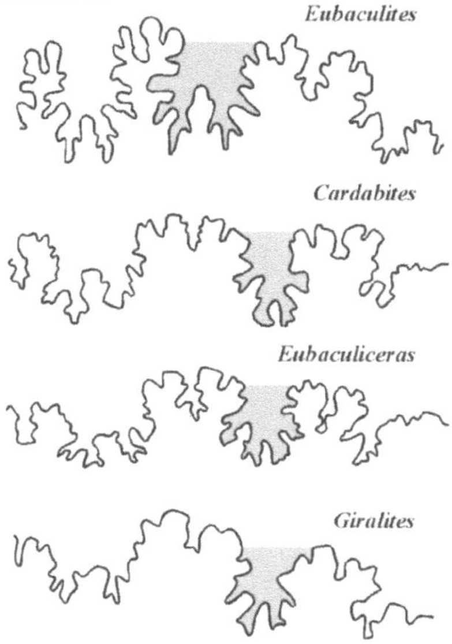

Fig. 775. Eubaculitine sutures.

parallel for much of the Maastrichtian; Kennedy (1986) distinguished 3 species groups. Sometimes very large baculicones, typically with a fastigiate to carinate or tabulate venter, flattened dorsum, with or without flank ornament; suture with very reduced I. *Upper L. - U. Maastr.*

Eubaculites SPATH, 1926 [*Baculites vagina* var. *ootacodensis* Stoliczka, 1866; = *labyrinthicus* Morton, 1834; OD]. Baculicone with flattened and convergent upper flanks, fastigiate venter with blunt siphonal keel, and cuneiform whorl section; strong dorsolateral clavi connected by broad low ribs to similarly elongate slightly oblique midlateral tubercles; forwardly projected fine riblets and striae on outer flank form chevrons over crenulated venter; suture with deeper E, broader E/L, broader L not as deep as E, much narrower L/U that protrudes well beyond E/L, narrower U and larger I. *U. Maastr.* **Fig. 773A-B** - *E. labyrinthicus* (Morton), **I-J** - *E. ootacodensis* (Stoliczka).

As noted by Klinger & Kennedy (1993: 199), binodose flanks and fastigiate venter distinguish this species (genus) from all others assigned to *Eubaculites*.

Giralites BRUNNSCHWEILER, 1966 [*G latecarinatus*; OD]. In early growth like *Eubaculites*; in middle to later growth with greater expansion rate, upstanding broad tabulate siphonal keel bounded by shallow sulci, flattened to concave dorsum with subangular dorsolateral shoulders, smooth flanks subparallel on lower half and convex on upper half, and suture with broad E, L/U larger than E/L, very wide U and tiny I. *Upper L. Maastr.* **Fig. 773C-D** - *G latecarinatus* Brunnschweiler.

Eubaculiceras BRUNNSCHWEILER, 1966 [*E. compressum*, = *Baculites vagina* var. *simplex* Kossmat, 1895; OD]. Like *Giralites* but compressed, with lower rate of taper; convex flanks, teardrop-shaped to pyriform whorl section; fastigiate to narrowly tabulate notched siphonal keel, without bounding grooves; smooth or with prominent crescents to flank. *Low U. Maastr.* **Fig. 773E-F** - *E. simplex* (Kossmat); **Fig. 774** - *E. carinatus* (Morton).

Cardabites BRUNNSCHWEILER, 1966 [*C. tabulatus*; OD]. Like *Eubaculiceras* but unornamented, with narrow tabulate venter and larger deeper penetrating I to suture. *Low U. Maastr.* **Fig. 773G-H** - *C. tabulatus* Brunnschweiler.

The type species is said to fall within the intraspecific variation of *E. simplex* (Kossmat); for those who regard them as synonymous, *Eubaculiceras* is selected here as the name having priority.

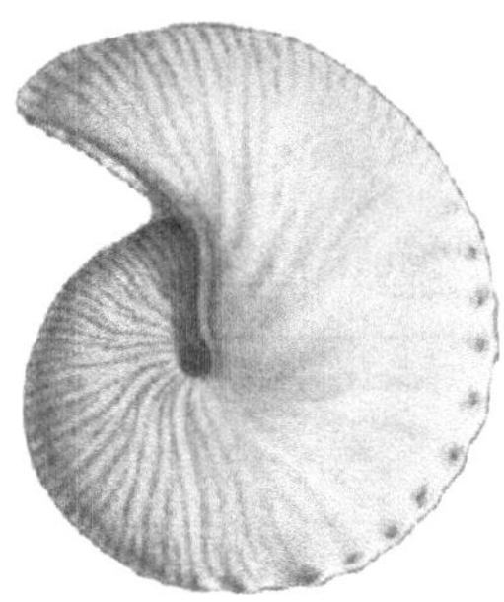

References

Adkins, W.S. 1920. The Weno and Pawpaw Formations of the Texas Comanchean. *University of Texas Bulletin* **1856**: 1-172.

——— 1928. Handbook of Texas Cretaceous fossils. *University of Texas Bulletin* **2838**: 1-385.

——— 1933a. The Mesozoic systems in Texas. *In*, Sellards, E.H., Adkins, W.S. & Plummer, H.J., The geology of Texas, 1, Stratigraphy. *University of Texas Bulletin* **3232**: 239-518.

——— 1933b. New ammonite fauna from Mexican Turonian [abstract]. *Pan-American Geologist* **59**: 238.

——— & Winton, W.M. 1920. Paleontological correlation of the Fredericksburg and Washita Formations in North Texas. *University of Texas Bulletin* **1945**: 1-128.

Agrawal, S.K. 1956. Contribution a l'etude stratigraphique et paléontologique du Jurassique du Kutch (Inde). *Annales du Centre d'etudes et de documentation paléontologiques* **19**: 1-188.

Aguirre Urreta, M.B. 1985. Ancyloceratidos (Ammonoidea) Aptianos de la Cordillera Patagonica Austral provincia de Santa Cruz, Argentina. *Boletin de la Academia nacional de Ciencias, Cordoba* **56**: 135-257.

——— 1986. Aptian ammonites from the Argentinian Austral Basin. The subfamily Helicancylinae Hyatt, 1894. *Annals of the South African Museum* **96** (7): 271-314.

——— 1993. Neocomian ammonite biostratigraphy of the Andean basins of Argentina and Chile. *Revista Española de Paleontología* **8**: 57-74.

——— 1995. *Spitidiscus riccardii* Leanza y Wiedmann (Ammonoidea) en el Hauteriviano del Neuquén. *Ameghiniana* **32**: 407-410.

——— 1998. The ammonites *Karakaschiceras* and *Neohoploceras* (Valanginian Neocomitidae) from the Neuquén Basin, west-central Argentina. *Journal of Paleontology* **72** (1): 39-59.

——— 2002. Hemihoplitid ammonoids from the Austral Basin of Argentina and Chile. *Abhandlungen der Geologischen Bundesanstalt, Abh. Geol. B.-A*, **57**: 491-500.

——— & Alvarev, P.P. 1999. The Berriasian genus *Groebericeras* in Argentina and the problem of its age. *Scripta Geologica, Special Issue* **3**: 15-29.

———, Concheyro, A., Lorenzo, M., Ottone, E.G & Rawson, P.F. 1999. Advances in the biostratigraphy of the Agrio Formation (Lower Cretaceous) of the Neuquén basin, Argentina: ammonites, palynomorphs and calcareous nannofossils. *Palaeogeography, Palaeoclimatology, Palaeoecology* **150**: 33-47.

———, Lazo, D.G, Griffin, M., Vennari, V., Parras, A.M., Cataldo, C., Garberoglio, R. & Luci. L. 2011. Megainvertebrados del Cretacico y su importancia bioestratigraphica. *Relatorio del XVIII Congreso Geologico Argentino, Neuquen* **2011**: 465-488.

———, Mourgues, F.A., Rawson, P.F., Bulot, L.G & Jaillard, E. 2007. The Lower Cretaceous Chañarcillo and Neuquén Andean basins: ammonoid biostratigraphy and correlations. *Geological Journal* **42**: 143-173.

——— & Ramos, V.A. 1981. Estratigrafia y paleontologia de la Alta Cuenca del Rio Roble Cordierra Patagonica, Provincia de Santa Cruz. *Congress geológico Argentino* **VIII**: 101-133.

——— & Rawson, P.F. 1997. The ammonite sequence in the Agrio Formation (Lower Cretaceous), Neuquén basin, Argentina. *Geological Magazine* **134**: 449-458.

——— & ——— 2001. Lower Cretaceous ammonites from the Neuquén basin, Argentina: a Hauterivian *Olcostephanus* fauna. *Cretaceous Research* **22**: 763-778.

——— & ——— 2003. Lower Cretaceous ammonites from the Neuquén basin, Argentina: the Hauterivian genus *Holcoptychites*. *Cretaceous Research* **24**: 589-613.

——— & Alvarez, P.P. 1999. The Berriasian genus *Groebericeras* in Argentina and the problem of its age. *Scripta Geologica, Special Issue* **3**: 15-29.

———, Amaro Mourges, F., Rawson, P.F., Bulot, L.G & Jaillard, E. 2007. The Lower Cretaceous Chañarcillo and Neuquén Andean basins: ammonoid biostratigraphy and correlations. *Geological Journal* **42**: 143-173.

——— & Klinger, H.C. 1986. Upper Barremian Heteroceratinae (Cephalopoda, Ammonoidea) from Patagonia and Zululand, with comments on the systematics of the subfamily. *Annals of the South African Museum* **96** (8): 315-358.

——— & Rawson, P.F. 1999. The Early Cretaceous (Valanginian) ammonite *Chacantuceras* (gen. nov.), a link between the Neuquén and Austral basins. *Revista de la Asociación Geológica Argentina* **53** (for 1998): 354-364.

——— 2001. Lower Cretaceous ammonites from the Neuquén Basin, Argentina: the Hauterivian neocomitid genus *Hoplitocrioceras* (Giovine, 1950). *Cretaceous Research* **22**: 201-218.

——— & ——— 2010. Lower Cretaceous ammonites from the Neuquén Basin, Argentina: the neocomitids of the *Pseudofavrella angulatiformis* Zone (Upper Valanginian. *Cretaceous Research* **31**: 321-343.

———, Suárez, M., Cruz, R. de la & Ramos, V.A. 2007. Ammonoids (Crioceratitinae, Hauterivian) from the Austral Basin. *Ameghiniana* **44** (2): online ISSN 1851-8044.

Aiba, D. 2018. A possible phylogenetic relationship of two species of *Hyphantoceras* (Ammonoidea: Nostoceratidae) in the Cretaceous Yezo Group, northern Japan. *Paleontological Research* doi:10.2517/2018 PR010

Aitken, W.G 1959. Geology and palaeontology of the Jurassic and Cretaceous beds of southern Tanganyika. Unpublished Ph.D. thesis, University of Glasgow.

Alabushev, A.I. 1988. *Rapidoplacenticeras* - a new genus of Cretaceous ammonite. *Paleontological Journal* **1988** (1): 110-113.

——— 1989a. Scaphitidae and certain other Late Cretaceous ammonites of the Far East. *Akademii nauk SSSR Dal'nevostokoe Otdelenie Severo-Vostochnii Komitetskii Nauchno-Issledovatel'skii Institut, Magadan*: 1-52 [in Russian].

——— 1989b. *Morphogenesis of Albian and Early Cenomanian ammonitids of the north-east of the USSR.* Magadan [in Russian].

——— 1995. Ammonite faunas and biostratigraphy of the Albian to Middle Cenomanian (Cretaceous) in western Korjak-Kamchatka, NE Russia. *Neues Jahrbuch fur Geologie und Paläontologie* **196**: 109-139.

——— & Alabusheva, A.V. 1988a. Albian and Lower Cenomanian ammonites from northeastern USSR. *Akademii nauk SSSR Dal'nevostokoe Otdelenie Severo-Vostochnii Komitetskii Nauchno-Issledovatel'skii Institut, Magadan*: 1-41 [in Russian].

———, ——— 1988b. Dimorphism of the late Cretaceous ammonite *Yokoyamoceras jimboi* Matsumoto. Magadan: Svknii Dvo An SSR: 1-17.

——— & Wiedmann, J. 1997. Upper Cretaceous ammonites from southern Sakhalin and northwestern Kamchatka (north-east Russia). *Palaeontographica, A*, Band 244, Lieferung 1-3: 1-36.

Alekseev, S.N. 1982. Development of septal sutures in the ontogeny of some late Jurassic and early Cretaceous Craspeditidae and Polyptychitidae. *Stratigraphy of Triassic and Jurassic deposits of oil-and-gas bearing basins of the USSR*: 115-128. Leningrad: VNIGRI [in Russian].

ALEKSEEV, S.N. & VAVILOV, M.N. 1983. On principles of the development and terminology of septal suture of Mesozoic ammonoids. *Annual All-Union Paleontological Society* **26**: 93-104 [in Russian].

ALIEV, M.M. & ALIEV, R.A. 1959a. Ammonites from the Campanian and Maastrichtian of central Kopet-Dagh. *Transactions of the Institute of Geology, Academy of Sciences, Turkmenian SSR* **2**: 199-211 [in Russian].

———, ——— 1959b. Ammonites from the Cenomanian of central Kopet-Dagh. *Transactions of the Institute of Geology, Academy of Sciences, Turkmenian SSR* **2**: 212-221 [in Russian].

ALIFIROV, A.S. 2009. *Craspedites schulginae* sp. nov., a new ammonite species from the Volgian Stage. *Paleontological Journal* **43** (6): 612-614.

ALKAYA, F. 1982. Taxonomic revision of the Lower Jurassic (Liassic) phylloceratids from northern Turkey (part I). *Bulletin of the geological Society of Turkey* **25**: 31-40 [in Turkish].

——— 1983. Taxonomic revision of the Lower Jurassic (Liassic) phylloceratids from northern Turkey (part II). *Bulletin of the geological Society of Turkey* **26**: 65-71 [in Turkish].

ALLMAN, G.J. 1871-72. A monograph of the gymnoblastic or tubularian hydroids. I. The hydroids in general (1871). II. The genera and species of the Gymnoblastea (1872). *Ray Society Monographs* **46** (1871): i-xxii, 1-154; **47** (1872): xxiii-xxiv, 155-450.

ALSEN, P. & RAWSON, P.F. 2005. The early Valanginian (early Cretaceous) ammonite *Delphinites (Pseudogarnieria)* from north-east Greenland. *Bulletin of the geological Society of Denmark* **52**: 201-212.

ALY, M.F. 2006. Aptian cephalopods from Gabal Aru Ruqum, north Sinai, Egypt. *Egyptian Journal of Paleontology* **6**: 347-385.

——— & ABDEL GAWAD, G.L. 2001. Early Cretaceous ammonites of Gebel Lagama, north Sinai, Egypt. *Palaeontographica*, Abteilung A, **262**: 25-52.

AMANNIYAZOV, K.N. 1962. *Stratigraphy and ammonites of the Upper Jurassic of Tuarkyr*. Turkyr Academy of Science Publishers: Ashkhabad [in Russian].

AMEDRO, F. 1976. Une nouvelle espèce d'ammonite: *Neophlycticeras (Eotropitoides) destombesi* nov. sp. (Lyelliceratidae) de l'Albien de Wissant (Boulonnais). *Annales de la Societe géologique du Nord* **96** (2): 107-112.

——— 1992. L'Albien du bassin anglo-parisien: ammonites, zonation phyletique, sequences. *Bulletin des Centres de Recherche Exploration-Production, Elf Aquitaine* **16**: 187-233.

———, MATRION B., TOMMASON, R., MAGNIEZ-JANNIN, F. & COLLETÉ, C. 2003. The Upper Albian of Vallentigny in the stratotypical region (Aube, F.): new data and revision of the ammonite *Mortoniceras (M.) inflatum* (J. Sowerby, 1818). *Bulletin de la Société géologique de Normandie et des Amis du Muséum du Havre* **90** (2) (for 2002): 5-28.

ANDELKOVIC, M.Ž. 1966. Amoniti iz slojeva sa *Aspidoceras acanthicum* stare planine (istocna Srbija). *Palaeontologia Jugoslavica* **6**: 1-136.

ANDERSON, F.M. 1902. Cretaceous deposits of the Pacific Coast. *Proceedings of the California Academy of Sciences* (3, Geology) **2**: 1-154.

——— 1938. Lower Cretaceous deposits in California and Oregon. *Geological Society of America, Special Papers* **16**: x, 1-340.

——— 1943. Synopsis of the later Mesozoic in California. *California, Division of Mines and Geology Bulletin* **118**: 183-186.

——— 1958. Upper Cretaceous of the Pacific Coast. *Geological Society of America Memoir* **71**: 1-378.

ANDERT, H. 1934. Die Kreideablagerungen zwischen Elbe und Jeschken. Teil III: Die Fauna der obersten Kreide in Sachsen, Böhmen und Schlesien. *Abhandlungen der Preußischen Geologischen Landesanstalt*, N.F., Heft **159**: 1-477.

ANTHULA, D.J. 1899. Über die Kreidefossilien des Kaukasus mit einem allegemeinen Ueberblick über die Entwicklung der Sedimentärbildungen des Kaukasus. *Beiträge zur Paläontologie und Geologie Oesterreich-Ungarns und des Orients* **12**: 55-160 (1-106).

ARISTOV, V.N. 1967. The boreal Lower Hauterivian and its fauna in the USSR. *Geologiia i Geofizika, Novosibirsk* **1967** (9): 6-22 [in Russian].

——— 1968. On a polyptychitan-simbirskitan ammonite fauna from the Yaroslavl Oblast. *In*, *Essay on the history of geologico-geographical knowledge*: 165-178. Pedagogicheskii Institut: Yaroslavl [in Russian].

——— 1974. On ammonites of the genus *Homolsomites* from the Lower Cretaceous of the Russian Platform. Biostratigraphy of the Boreal Mesozoics. *Transactions of the Institute of Geology & Geophysics, Siberian Branch, Academy of Science USSR* **136**: 149-154.

——— 1979. On the questions of the stratigraphic significance of the Yaroslavl assemblage of Hauterivian ammonites. *In*, IVANOV, A.N., Late Mesozoic cephalopods of the Upper Volga area. *Intercollege collection of scientific papers, Issue* **183**: 55-60. Yaroslavl: Pedagogical University [in Russian].

——— 1984. About the fauna and age of the *Homolsomites* beds of the Yaroslavl Volga area. Problems of geology of the european north of the USSR. *Collection of scientific papers.* V.I.: 87-93 Lenin Pedagogical Institute: Moscow [in Russian].

ARKADIEV, V.V. 2002. A new genus *Leiophylloceras* (Phylloceratida, Ammonoidea) from the Berriasian of the mountainous Crimea. *Palaeontological Journal* **2002** (6): 41-45 [in Russian].

——— 2003a. Zone *Berriasiella jacobi - Pseudosubplanites grandis* of the Berriasian of mountainous Crimea. *Bulletin of Moscow Society of Naturalists*, Geological Series, **78** (6): 29-35 [in Russian].

——— 2003b. Morphogenesis of the genus *Pseudosubplanites* (Perisphinctidae, Ammonoidea) from the Berriasian of the mountainous Crimea. *Bulletin of the St Petersburg University* (7) **2** (15): 67-93 [in Russian].

——— 2007a. Some Upper Berriasian ammonites of the Crimean Mountains. *Questions of stratigraphy, palaeontology and paleogeography (in memoir of 100 anniversary of G.Ya. Krimholz)*: 103-117. Saint-Petersburg [in Russian].

——— 2007b. Ammonite *Fauriella boissieri* (Pictet), the index species of the Berriasian upper zone from the Crimean Mountains. *Stratigraphy and geological Correlation* **15** (2): 185-192 [in Russian].

——— 2011a. Representatives of the genera *Fauriella*, *Tirnovella* and *Jabronella* (Neocomitidae, Ammonoidea) from the Berriasian of the mountainous Crimea and their stratigraphic significance. *Bulletin of the Saint-Petersburg State University*, (7, Geology, Geography) **2**: 3-20 [in Russian].

——— 2011b. New data on ammonoids of the genus *Paraulacosphinctes* from the Upper Tithonian of the mountainous Crimea. *Stratigraphy and Geological Correlation* **19** (2): 238-242.

——— 2014. Representatives of the *Substeuroceras* genus (Neocomitidae, Ammonoidea) from the Berriasian of the mountainous Crimea. *Bulletin of the Saint-Petersburg State University*, (7, Geology, Geography), **4**: 79-87 [in Russian].

———, ATABEKIAN, A.A., BARABOSHKIN, E.Y. & BOGDANOVA, T.N. 2000. Stratigraphy and ammonites of Cretaceous deposits of south-west Crimea. *Palaeontographica, Abt. B*, **255** (4-6): 85-128.

——— & BOGDANOVA, T.N. (eds) 1997. *Atlas of the Cretaceous fauna of south-west Crimea*. Mining Institute: Saint-Petersburg [in Russian].

———, ——— 2001. Revision of the genus *Ptychophylloceras* (Phylloceratida, Ammonoidea) from the Berriasian of the Crimean Peninsula. *Palaeontological Journal* **2001** (5): 27-34 [in Russian].

———, ——— 2004. The genus *Berriasella* (Ammonoidea) and zonal subdivision of the Berriasian of the Crimean Mountains. *Stratigraphy and Geological Correlation* **12** (4): 54-67 [in Russian]

———, ——— 2005. Genus *Delphinella* (Ammonoidea) from the Berriasian of the Crimean Mountains. *Palaeontological Journal* **39** (5): 487-497.

———, ——— 2009. Representatives of the *Neocosmoceras* (Neocomitidae, Ammonoidea) genus from the Berriasian of the Crimean Mountains and their stratigraphic significance. *Stratigraphy and Geological Correlation* **17** (4): 415-429.

———, ——— & Lysenko, N.I. 2007. Representatives of genus *Malbosiceras* and *Pomeliceras* (Neocomitidae, Ammonoidea) from the Berriasian of the Crimean Mountains. *Stratigraphy and Geological Correlation* **15** (3): 277-296.

———, Rogov, M.A. & Perminov, V. A. 2011. New occurrences of heteromorph ammonites in the Berriasian-Valanginian of the Crimean Mountains. *Paleontological Journal* **45** (4): 390-396.

Arkell, W.J. 1933. *The Jurassic System in Great Britain.* Clarendon Press: Oxford.

——— 1935. Seven new genera of Jurassic ammonites. *Geological Magazine* **90**: 36-40.

——— 1950. A classification of the Jurassic ammonites. *Journal of Paleontology* **24**: 354-364.

——— 1951. A Middle Bathonian ammonite fauna from Schwandorf, northern Bavaria. *Schweizerische palaeontologische Abhandlungen* **69**: 1-18.

——— 1952. Jurassic Ammonoidea. *Journal of Paleont-ology* **26**: 860-861.

——— 1957. [Jurassic Ammonoidea]. *In*, Moore, R.C. (ed.), *Treatise on Invertebrate Paleontology.* Part L, Mollusca 4, Cephalopoda, Ammonoidea. Geological Society of America: New York; University of Kansas Press: Lawrence.

———, Furnish, W.M., Kummel, B., Miller, A.K., Moore, R.C., Schindewolf, O.H., Sylvester-Bradley, P.C. & Wright, C.W. 1957. Part L, Mollusca 4, Cephalopoda, Ammonoidea. *In*, Moore, R.C. (ed.), *Treatise on Invertebrate Paleontology.* Geological Society of America: New York; University of Kansas Press: Lawrence.

Arkhangelsky, A.D. 1916. Upper Cretaceous molluscs of Turkestan. *Trudy geologicheskago Komiteta, S.-Peterburg*, n.s., **152**: iv, 1-57 [in Russian].

Arnould-Saget, S. 1951. Les ammonites pyriteuses du Tithonique supérieur et du Berriasien de Tunisie centrale. *Annales Mineralogie et Géologie* **10**: 1-132.

——— 1956. *Contribution a` l'étude des Engonoceratidae (les couches a* Knemiceras *du sud tunisien).* Imprimerie La Rapide: Tunis.

Astier, J.E. 1851. Catalogue descriptif des *Ancyloceras* appartenant à l'étage Néocomien d'Escragnolles et des Basses-Alpes. *Annales de la Societe nationale d'Agriculture, d'Histoire naturelle et des Arts utiles de Lyon* **3** (2): 435-456 (1-27).

Atabekian, A.A. 1960. Upper Cretaceous stratigraphy of the western flank of the Kopet Dag. *Trudy Vsesoiuznogo Nauchno-Issledovatel'skogo geologicheskogo Instituta (VSEGEI) Leningrad* (n.s.) **42**: 182-210 [in Russian].

——— 1966. New genus *Koulabiceras* gen. nov. from the Turonian of the eastern part of central Asia. *Izvestiia Akademii Nauk Armianskoi SSR* (Earth Sciences) **19** (4): 75-78 [in Russian].

——— 1979. Correlation of the Campanian Stage in Kopetdag and Western Europe. *Aspekte der Kreide Europas. International Union of Geological Sciences* (A) **6**: 511-526.

——— 1985. *Turrilitids of the Late Albian and Cenomanian of the southern part of USSR*. Nauka: Leningrad [in Russian].

——— 1987. *Turrilitidae de l'Albien supérieur et du Cénomanien du sud de l' URSS*. Association Géologique: Auboise.

——— 1972. Late Cretaceous ammonites of the Armenian SSR (Collignoceratinae, Peroniceratinae). *Proceedings of the Academy of Science Armenian SSR*, Earth Sciences, **2**: 3-12 [in Russian].

——— & Akopyan, V.T. 1969. Late Cretaceous ammonites of the Armenian SSR (Pachydiscidae). *Izvestiya AN Armyanskoj SSR, nauki o zemle* **22** (6): 3-20 [in Russian].

——— & Khakimov, F.H. 1976. *Campanian and Maastrichtian ammonites of Middle Asia*. Dushanbe: Donish [in Russian].

——— & Mikhailova I.A. 1976. Peculiarities of the development of turrilitids in an example of *Hypoturrilites gravesianus* Orb. *Doklady, Academy of Sciences USSR* **231** (5): 1230-1233 [in Russian].

Atrops, F. 1982. La sous-famillie Ataxioceratinae (Ammonitina) dans le Kimméridgian Inferieur du sud-est de la France. Systématique, évolution, chronostratigraphie des genres *Orthosphinctes* et *Ataxioceras*. *Documentaire Laboratoire géologique de Lyon* **83**: 1-463.

——— & Meléndez, G. 1988. Palaeobiogeography and evolutionary trends in Lower Kimmeridgian ataxioceratids from northern Spain. *In*, Wiedmann, J. & Kullmann, J. (eds), *Cephalopods - present and past*: 437-446. Schweizerbart'sche Verlagsbuchhandlung: Stuttgart.

Autran, G., Delanoy, G. & Thomel, G. 1986. Discussion critique des genres d'ammonites déroulées: *Toxoceras* d'Orbigny, 1840, *Hemibaculites* Hyatt, 1900 et *Moutoniceras* Sarkar, 1954. Proposition d'un nouveau genre *Pseudomoutoniceras* pour l'espèce *T. annulare* d'Orbigny. *Comptes Rendus des Seances de l'Academie des Sciences* (2) **303**: 1059-1062.

Avram, E. 1973. Position et valuer taxonomique du groupe *"Berriasella" richteri* (Oppel). *Dari de Seama ala sedintelor* (3, Paléontologie) **60** (1972-1973): 10-22 (1-12).

——— 1974. *Egoianiceras* nouveau sous-genre du genre *Colombiceras* Spath, 1923 (Ammonitina). *Dari de Seama ale Sedintelor* (3, Paleontologie) **60** (1972-1973): 3-10.

——— 1976a. Nouvelles ammonites hétéromorphes Bédouliennes du Couloir de la Dîmbovicioara. Contributions à la paléontologie du Jurassique terminal et Crétacé des Carpates, Bucarest. *Institut de Geologie et de Geophysique, Memoire* **24**: 75-82.

——— 1976b. Fossils from the Eocretaceous flysch and Tithonian limestones in basins of the Doftana and Tirlung vallies (east Carpathians). *Institut de Géologie et Géophysique Bucarest, Mémoire* **24**: 1-69 [in French].

——— 1978a. Observations sur les espèces d'ammonites de la region de Svinita (Banat) décrites par Tietze (1872) et Uhlig (1883). *Dari de seama ale sedintelor* (3, Paleont-ologie) **64** (3) (1976-1977): 9-25.

——— 1978b. Une nouvelle espece de céphalopode dans le Barrémien supérieur de la Roumanie: *Costidiscus tardus* n. sp. *Dari de seama ale sedintelor* (3, Paleontologie) **64** (3) (1976-1977): 5-8.

——— 1984. Correspondent species of the genera *Macroscaphites* Meek and *Costidiscus* Uhlig. *75 Years of the Laboratory of Paleontology, University of Bucharest, Special Volume*: 67-80.

——— 1990a. *Patruliusiceras*, a new genus of the family Silesitidae Hyatt, 1900 (Ammonitina). *Dari de Seama ale Sedintelor* (3, Paleontologie) **74** (for 1987): 69-86.

——— 1990b. *Haplobrancoceras* n.g., ammonite Barremiéne a ligne cloisonnaire simplifiée. *Dari de seama ale Sedintelor* **74/3** (1987): 27-32.

——— 1997. *Barremitites*, new genus of the family Eodesmoceratidae Wright, 1955 (Ammonitina). *Acta palaeontologica Romaniae* **24**: 176-179.

——— 1999. Some new species of the 'subfamily' Leptoceratoidinae (Ancyloceratina, Ammonoidea) in uppermost Hauterivian and Lower Barremian deposits of Rumania. *Scripta Geologia, Special Issue* **3**: 31-43.

——— 2001. Representatives in Romania of the genera *Dissimilites* Sarkar, 1954 and *Toxoceratoides* Spath, 1924 (Ancyloceratina, Ammonoidea. *Acta paleontologica Romaniae* **3**: 23-30.

——— 2002. The taxonomic position and biostratigraphic value of the genus *Pseudocrioceratites* EGOIAN, 1969 (Lytoceratina, Lower Cretaceous). *In*, SUMMESBERGER, H., HISTON, K. & DAURER, A. (eds), Cephalopods - present and past. *Abhandlungen geologischen Bundesanstalt* **57**: 257-264.

———, BUCUR, I.I. & POPESCU, O. 1987. Considérations sur quelques faunes d'ammonites éocrétacées de la Zone de Resita (SW de la Roumanie). *Dari de Seama ale Sedintelor* (3, Paleontologie) **72-73** (3): 21-35.

——— & GRADINARU, E. 1993. A peculiar Upper Valanginian cephalopod fauna from the Carpathian bend (Codlea town area, Romania): paleostratigraphic and paleobiogeographic implications. *Jahrbuch der Geologischen Bundesanstalt* **136** (Heft 4): 665-700.

AZARYAN, N.R. 1982. *Jurassic ammonites of the Armenian SSR.* Academy of Sciences of the Armenian SSR: Erevan [in Russian].

BAILY, W.H. 1855. Description of some Cretaceous fossils from South Africa; collected by Capt. Garden of the 45th Regiment. *Quarterly Journal of the geological Society of London* **11**: 450-465.

BACHNOU, A. 1992. *Stratigraphie et faunes d'ammonites du Malm-Berriasien basal dans la région du moyen Ouerrha (Prérif interne, Maroc).* Thesis, Université Claude Bernard, Sciences de la Terre: Lyon.

BALAN, T.M. 1979. Materials for systematics of family Gaudryceratidae Spath. *In*, *Biostratigraphical investigations of Mesozoic-Cenozoic deposits of the southern margins of the Russian Platform*: 28-43. Schtiinca: Kishinev [in Russian].

BARABOSHKIN, E.YU. 1992. The Lower Albian of central parts of the Russian Platform. *In*, SHIK, S.M. (ed.), *The Phanerozoic stratigraphy of the central part of the East - European Platform*: 20-36. Tzentrgeologia Press: Moscow [in Russian].

——— 1999. Berriasian-Valanginian (Early Cretaceous) seaways of the Russian Platform basin and the problem of Boreal/Tethyan correlation. *Geologica carpathica* **50** (1): 5-20.

——— & FOKIN, P. A. 2019. Cephalopods from Santonian/Campanian (Upper Cretaceous) boundary interval of Aksudere section (Mountain Crimea). *Bulletin de le Societe impériale des Naturalistes, Moscou* **94** (4): 77-84 [in Russian].

——— & MIKHAILOVA, I.A. 1994. Revision of Neocomian ammonites of the Crimea: the genera *Neohoploceras* Spath and *Luppovella* Nikolov. *Paleontological Journal* **1994** (4): 41-53 [in Russian].

———, ——— 2000. New and poorly-known Valanginian ammonites from south-west Crimea. *Bulletin d'Institut Royal des Sciences naturelle de Belgique* (Science de la Terre) **79**: 80-120.

———, ——— 2002. New stratigraphic scheme for the Lower Aptian in the Volga River middle courses. *Stratigraphy and Geological Correlation* **10**: 603-626.

———, ——— 2005. Genus *Arcthoplites* Spath, 1925: distribution and taxonomic position. *In*: ARKADIEV, V.V. & PROZOROVSKY, V.A. (eds), Cretaceous System of Russia: problems of stratigraphy and paleogeography. *Transactions of Second All-Russian Conference, Saint-Petersburg University*: 68-90 [in Russian].

———, ——— 2006. *Theodorites* - a new genus of Lower Hauterivian heteromorph ammonites from the mountainous Crimea. *In*, LEONOVA, T.B. (ed.), *Modern problems of studying cephalopod molluscs: morphology, taxonomy, evolution, biostratigraphy.* Materials of All-Russia meeting, Moscow, November 8-10: 19-21. Palaeontological Institute of Russian Academy of Science: Moscow [in Russian].

———, ——— 2020. Revision of the early Cretaceous genus *Paraspiticeras* Kilian, 1910 (Ancyloceratoidea, Ammonoidea). *Proceedings of the Geological Association*, doi 10.1016/j.pgeola. 2019.06.006.

BARBER, W. 1957. Lower Turonian ammonites from north-eastern Nigeria. *Geological Survey of Nigeria, Bulletin* **26**: 1-88.

BARRANDE, J. 1867. *Système Silurien du centre de la Bohême, première partie, Recherches paléontologiques, Classe des Mollusques, Ordre des Céphalopodes.* J. Barrande: Prague, Paris: .

BARROIS, C. & DE GUERNE, J. 1878. Description de quelques espèces nouvelles de la Craie de l'est du bassin de Paris. *Annales de la Société géologique du Nord* **5**: 42-64.

BARROSO-BARCENILLA, F. 2007. Revision and new data of the ammonite family Acanthoceratidae de Grossouvre, 1894, from the lower Turonian of the Iberian Trough, Spain. *Palaeontographica*, Abt. A, Bd. **280**: 123-163.

——— & GOY, A. 2010. The ammonite genus *Vascoceras* Choffat, 1898 (familiy Vascoceratidae Douvillé, 1912) in the Iberian Trough, Spain. *Palaeontographica*, Abt. A, Bd. 290, Lief 4-6: 199-235.

———, JOSÉ, M., CALLAPEZ, P.M. & SANTOS, V.F. 2015. An unexpected discovery in the storage room of the National Museum of Natural History and Science, Lisbon, Portugal: The missing holotypes of the Upper Cretaceous ammonoids *Vascoceras gamai* Choffat, 1898, and *Vascoceras barcoicense* (Choffat, 1898). *Jornadas de Paleontología, Sociedad Española de Paleontología* **31**: 67-68.

BARTHEL, K.W. 1959. Die Cephalopoden des Korallenkalks aus dem oberen Malm von Laisacher bei Neuburg a.d. Donau. I. *Gravesia, Sutneria, Hybonoticeras. Neues Jahrbuch fur Geologie und Paläontologie, Abhandlungen* **108** (1): 47-74.

——— 1962. Zur Ammonitenfauna und Stratigraphie der Neuberger Bankkalke. *Bayerische Akademie der Wissenschaften Mathematisch-naturwissenschaftliche Klasse, Abhandlungen,* n.f., Heft **115**: 1-29.

——— 1975. *Isterites* gen. nov. (Ammonitina). *Journal of Paleontology* **49**: 426.

——— & GEYSSANT, J.R. 1973. Additional Tethydian ammonites from the lower Neuburg Formation (Middle Tithonian, Bavaria). *Neues Jahrbuch fur Geologie und Paläontologie*, Heft **1**: 18-36.

BASOV, V.A., ZAKHAROV, V.A., IVANOVA, E.F., SAKS, V.N., SHUL'GINA, N.I. & YUDOVNY, E.G. 1970. Zonal division of the Upper Jurassic and Lower Cretaceous deposits on Cape Urdyuk-Khaya (Pakhsa Peninsula, Anabar Bay). *Uchenye Zapiski Nauchno-Issledovatel'skogo Instituta Geologii* (Paleontologiia, Biostratigrafiia) **29**: 14-31 [in Russian].

BASSE, É. 1939. Sur quelques mollusques Crétacés des Corbières méridionales. *Bulletin de la Société géologique de France* (5) **9**: 35-58.

——— 1940. Les céphalopodes Crétacés des massifs côtiers syriens. Deuxième partie. Notes et memoires Haut-Commissariat de la Republique Francaise en Syrie et Liban. *Service des Travaux Publics-Section d'Etudes géologique* **3**: 411-472.

——— 1947a. Sur deux ammonites nouvelles du Coniacien du sud-ouest de Madagascar: *Subbarroisiceras* n.g. *mahafalense* n.sp. et *Eboroceras* n.g. *magnumbilicatum* n.sp. *Bulletin de la Société géologique de France* (5) **16** (1946): 71-76.

——— 1947b. Paléontologie de Madagascar. N. XXVI. Les peuplements malgaches de *Barroisiceras*. Révision du genre *Barroisiceras* de Gross. *Annales de Paléontologie* **33**: 97-178 (1-82).

——— 1948. Quelques ammonites nouvelles du Crétacé de Colombie (Am. Sud). *Bulletin de la Société géologique de France* (5) **18**: 691-698.

——— 1950. Quelques mollusques du Crétacé de Colombie. *Bulletin de la Société géologique de France* (5) **20**: 245-256.

——— 1952a. Sous-classe des Ammonoidea: caractères généraux. *In*, Piveteau, J. (ed.), *Traite de Paléontologie* **2**: 522-555. Masson: Paris.

——— 1952b. Ammonoidea *s. str. In*, Piveteau, J. (ed.) , *Traite de Paléontologie* **2**: 581-688. Masson: Paris.

——— 1954. Sur la présence du genre *Coahuilites* Böse1927 dans le sud tunisien et du genre *Eulophoceras* Hyatt 1903 dans le Proche-Orient. *Bulletin de la Société géologique de France* (6) **3** (1953): 865-871.

Basse de Ménorval, É. 1963. Quelques ammonites nouvelles du Crétacé Supérieur d'Angola. *Bulletin de la Société géologique de France* (7) **4** (1962): 871-876.

Bataller, J.R. 1950. Sinopsis de la especies neuvas de Cretácico de España. Part VIII. Mollusca. 3, Cephalopoda. *Memorias de la Real Academia de Ciencias y Artes de Barcelona, No. 586,* **28** (12): 62-141.

——— 1954. Los engonoceratidos en España. *Homenage postuma Dr. Vaquer. Facultad de Ciencias Universidad Barcelona*: 173-178.

——— 1959. Primer suplemento a la "Sinopsis de las especies nuevas del Cretáceo de España". *Boletin del Instituto Geologico y Minero de Espana* **70**: 1-77.

Baudouin, C., Boselli, P. & Bert, D. 2011. The Oppeliidae of the Acanthicum Zone (Upper Kimmeridgian) from Mount Crussol (Ardèche, France): ontogeny, variability and dimorphism of the genera *Taramelliceras* and *Streblites* (Ammonoidea). *Revue de Paléobiologie* **30** (2): 619-684.

———, Delanoy, G., Boselli, P., Bert, D. & Boselli, M. 2012. Les faunes d'ammonites de la sous-zone à Sarasini (Barrémien supérieur) dans les Baronnies (Drôme, France). *Revue de Paléobiologie* **31**: 601-677.

———, ———, Ifram, C. & Moreno-Bedmar, J.A. 2018. The ammonoid fauna of the *Prionocyclus germari* Zone (Upper Turonian, Upper Cretaceous) from Rochefort-en-Valdaine (Drôme, France). *Carnets de géologie (Notebooks on geology)* **18** (14): 313-351.

Baumberger, E. 1906. Fauna der untern Kreide im westschweizerischen Jura. Zweiter Teil. Die Ammonitiden der untern Kreide im westschweizerischen Jura. *Abhandlungen der Schweizerischen paläontologischen Gesellschaft* **23** (1905): 1-80.

Bayle, E. 1878. Fossiles principaux des Terrains. *Explication de la Carte géologique de la France* 4 (1) (atlas). Imprimerie Nationale: Paris.

——— & Coquand, H. 1851. Mémoire sur les fossiles secondaires recueillis dans le Chili par Ignace Domeyko, et sur les terrains auxquels ils appartiennent. *Mémoire de la Société géologique de France* **2** (4-1): 1-47.

Behrendsen, O. 1891. Zur Geologie des Ostabhanges der argentinischen Cordillere, Theil 1. *Zeitschrift der deutschen geologischen Gesellschaft* **43**: 369-420.

——— 1892. Zur Geologie des Ostabhanges der argentinischen Cordillere, Theil 2. *Zeitschrift der deutschen geologischen Gesellschaft* **44**: 1-42.

Benavides-Cáceres, V.E. 1956. Cretaceous System in northern Peru. *Bulletin of the American Museum of natural History* **108**: 353-494.

Benett, E. 1831. *A catalogue of the organic remains of the county of Wilts*. E. Benett: Warminster.

Benetti, A. & Pezzoni, N. 1983. I generi *Pseudhimalayites* Spath, 1925 (Aspidoceratidae) e *Hemispiticeras* Spath, 1925 (Berriasellidae) nei Monti Lessini (Verona) (Mollusca, Ammonitina). *Bollettino del Museo civico di storia naturale di Verona* (Geologia paleontologia preistoria) **12**: 501-507.

Bengtson, P. & Nordlund, U. 1987. The ammonite fauna and genesis of a mid-Cretaceous siliceous oolite from the Alagoas Basin, Brazil. *Cretaceous Research* **8** (4): 305-333.

———, Zucon, M.H. & Sobral, A. d'C.S. 2018. Cretaceous ammonite zonation of the Sergipe Basin, northeastern Brazil. *Cretaceous Research* **88**: 111-122.

Benzaggagh, M. & Atrops, F. 1997. Stratigraphie et associations de faude d'ammonites des zones du Kimméridgien, Tithonien et Berriasien basal dans le Prérif interne (Rif, Maroc). *Newsletters on Stratigraphy* **35** (3): 127-163.

Bentz, A. 1928. Über Strenoceraten und Garantianen insbesondere aus dem mittleren Dogger von Bielefeld. *Jahrbuch der koeniglich preussischen geologischen Landesanstalt* **49** (1): 138-206.

Bert, D. 2008. Description de deux nouvelles ammonites hétéromorphé albiennes de genres *Astiericeras* Parona & Bonarelli, 1897 et *Triplaniceras* nov. gen. (Crétacé Inférieur, sud-est de la France) - implications évolutives et taxinomique. *Revue du Paléobiologie* **27** (1): 73-79.

——— 2009. Description de *Artereites landii* nov. (Ammonoidea) du Barrémiene Supérieur de Majastre (sud-est de la France) et discussion sur les Helicancylidae Hyatt, 1894. *Annales de Paléontologie* **95**: 139-163.

——— 2012a. Les ammonites du Niveau Paquier (Albien basal, Crétacé inférieur) du sud-est de la France. *Fossilies* **12**: 6-27.

——— 2012b. Phylogenetic relationships among the Hemihoplitidae Spath, 1924 (Ammonoidea, Upper Barremian). *In*, Bert, D. & Bersac, S. (eds), First meeting of the research group for paleobiology and biostratigraphy of the ammonites - Communications. *Boletin del Instituto de Fisiografia y Geologia, Rosario* **82**: 37-38.

——— 2014. Factors of intraspecific variability in ammonites, the example of *Gassendiceras alpinum* (d'Orbigny, 1850) (Hemihoplitidae, Upper Barremian). Facteurs de la variabilité intraspécifique chez les ammonites, l'exemple de *Gassendiceras alpinum* (d'Orbigny, 1850) (Hemihoplitidae, Barremien supérieur). *Annales de Paléontologie* **100** (3): 217-236.

——— & Bersac, S. 2006. Descriptions de représentants ou peu connus de la familie des Hemihoplitidae Spath, 1924 (Barrémien Supérieur, sud-est de la France): conséquences taxinomiques et phylétiques. *Annales de Muséum d'Histoire naturelle de Nice* **21**: 179-253.

———, ——— 2012. Rediscovery of the type specimen and status of *Ezeiceras heberti* Fallot, 1884 (Ammonoidea, Hemihoplitidae, Barremian). *In*, Bert, D. & Bersac, S. (eds), First meeting of the research group for paleobiology and biostratigraphy of the ammonites - Communications. *Boletin del Instituto de Fisiografia y Geologia, Rosario* **82**: 42-44.

———, ——— 2014. Origin of the Tethyan Hemihoplitidae tested with cladistics (Ancyloceratina, Ammonoidea, Early Cretaceous): an immigration event? *Carnets de Geologie* **14** (13): 255-272.

———, ——— & Canut, L. 2020. Implications of the 'hemihoplitid-like' ammonites iterative morphology in the context of the late Tethyan Barremian (Early Cretaceous). *Cretaceous Research* **106**: 104239.

———, ———, Delanoy, G & Canut, L. 2013. Paleontology, taxonomic revision and variability of some species of the genus *Gassendiceras* Bert *et al*., 2006 (Ammonitina, Upper Barremian) from southeastern France. *Acta geologica Polonica* **63** (3): 355-397.

———, Busnardo, R., Delanoy, G & Bersac, S. 2010. Problems in the identity of *"Crioceras" barremense* Kilian, 1895

(Ancyloceratida, Late Barremian), and their proposed resolution. *Carnets de Géologie/Notebooks on Geology* **2010/01** (CG2010_A01).

——— & Delanoy, G. 2000. Considérations nouvelles sur quelque représentants Barrémiens des Pulchelliidae Douvillé, 1890 et des Hemihoplitidae Spath, 1924. *Annales de Muséum d'Histoire naturelle de Nice* **15**: 63-89.

——— & ——— 2009. *Pseudoshasticrioceras bersaci* sp. nov. (Ammonoidea, Gassendiceratinae), and new ammonite biohorizon for the Upper Barremian of southeast France. *Carnets de Géologie* **2009/02**: 1-9.

———, ——— & Bersac, S. 2006. Descriptions de représentants nouveaux ou peu connus de la famille des Hemihoplitidae Spath, 1924 (Barrémien supérieur, sud-est de la France): conséquences taxinomiques et phylétiques. *Annales de Muséum d'Histoire naturelle de Nice* **21**: 179-253.

———, ———, ——— 2008. Nouveaux biohorizons et propositions pour le découpage biozonal ammonitique du Barrémien supérieur du sud-est de la France. *Carnets de Géologie/Notebooks on Geology* - Article 2008/03 (CG2008-A03).

———, ——— & Canut, L. 2009. L'origine des *Imerites* Rouchadzé, 1933: résultat d'un innovation chez les Gassendiceratinae Bert, Delanoy et Bersac, 2006 (Ammonoidea, Ancyloceratina). *Annales de Paléontologie* **95**: 21-35.

———, ———, Juarez-Ruiz, J. & Hughes, Z. 2018. Size reduction and ornamental oscillation within a Barremian lineage of giant heteromorphic ammonites (Early Cretaceous, northwestern Tethyan margin). *Cretaceous Research* **88**: 173-186.

———, Peres, V. & Marchand, D. 2006. Nouvelles données stratigraphiques sur le Turonien moyen (Crétacé supérieur) du versant septentrional du Seuil du Poitou (centre-ouest de la France): description de *Pseudotissotia faustinleybachae* sp. nov., et évolution des Pseudotissotinae Hyatt, 1903 (Ammonoidea) ouest-européens. *Annales de Muséum d'Histoire naturelle de Nice* **21**: 297-317.

Berthou, P.Y., Brower, J.C. & Reyment, R.A. 1976. Morphometrical study of Choffat's vascoceratids from Portugal. *Bulletin of the geological Institutions of the University of Uppsala*, n.s., **6**: 73-83.

Besairie, H. 1930. Les rapports du Crétacé malgache avec le Crétacé de l'Afrique australe. *Bulletin de la Société géologique de France* (4) **30**: 613-643.

——— 1936. Recherches géologiques à Madagascar. 1. La géologie du nord-ouest. *Mémoires de l'Academie Malgache* **21**: 1-259.

Beurlen, K. 1926. Zur Systematik der Perisphincten. *Centralblatt für Mineralogie, Geologie und Paläontologie, Abteilung B*: 78-95.

Beznosov, N.V. 1956. New and little known genera of Jurassic Lytoceratina. *Biulleten' Moskovskogo Obshchestva Ispytatelei Prirody, Geologiia* **31** (4): 109-110 [in Russian].

——— 1957. New morphological and systematic data on the Phylloceratаceae. *Biulleten' Moskovskogo Obshchestva Ispytatelei Prirody, Geologiia* **32** (2): 166 [in Russian].

——— 1958a. Development of septal suture in suborder Phylloceratina. *Proceedings of Higher Educational Establishments, Geology and exploration* **11**: 34-40 [in Russian].

——— 1958b. Jurassic ammonites of the northern Caucasus and Crimea. Phylloceratina and Lytoceratina. *Leningrad: State science-technology publ. oil and mining-fuel literature*: 1-118 [in Russian].

——— 1958c. Material on systematics of families Lytoceratidae Neumayr and Nannolytoceratidae Spath. *Bulletin of the Moscow State University* **1**: 109-117 [in Russian].

——— 1960. Taxonomy of the Jurassic Ammonitida. *Palaeontological Journal* **2** (1): 29-44.

——— 1982. On systematics of perisphinctids. *Paleontological Journal* **1**: 54-64 [in Russian].

——— & Kutuzova, V.V. 1990. Systematics of the Middle Jurassic heteromorphic ammonites. *Paleontological Journal* **24** (4): 19-27.

——— & Mikhailova, I.A. 1983. The evolution of the Jura-ssic-Cretaceous ammonoids. *Doklady Akademii nauk SSSR* **269** (3): 733-737.

———, ——— 1984a. Higher taxa of the Jurassic and Cretaceous Phylloceratida. *Bulletin de le Societe impériale des Naturalistes, Moscou* (Geology) **59** (3): 82-91 [in Russian].

———, ——— 1984b. Higher taxa of the Jurassic and Cretaceous Lytoceratida. *Bulletin de le Societe impériale des Naturalistes, Moscou* (Geology) **60** (3): 100-113.

———, ——— 1991. Higher taxa of the Jurassic and Cretaceous Ammonitida. *Paleontological Journal* **4**: 3-18.

——— & Mitta, V.V. 1995. Polymorphism in the Jurassic ammonoids. *Paleontological Journal* **29** (2): 46-57.

———, ——— 1995. *Upper Jurassic ammonites and black shales of Central Asia.* VNIGNI: Moscow [in Russian].

———, ——— 2000. Jurassic geology and ammonites of the Great Balkhan (western Turkmenistan). *Bulletin of CF VNIGNI* **5**: 1-115 [in Russian].

Billinghurst, S.A. 1927. On some new Ammonoidea from the Chalk Rock. *Geological Magazine* **64**: 511-518.

Binkhorst, J.T. 1861. *Monographie des gastéropodes et des céphalopodes de la craie supérieure du Limbourg.* Muquardt: Brussels.

Birkelund, T. 1965. Ammonites from the Upper Cretaceous of west Greenland. *Meddelelser om Grønland* **179** (7): 1-192.

——— & Callomon, J.H. 1985. The Kimmeridgian ammonite faunas of Milne Land, central east Greenland. *Bulletin Grønland Geologiske Undersøgelse* **153**: 5-56.

———, ———, Clausen, C.K., Nøhr Hansen, H. & Salinas, I. 1983. The Lower Kimmeridge Clay at Westbury, Wiltshire, England. *Proceedings of the Geological Association* **94** (4): 289-309.

———, Hancock, J.M., Hart, M.B., Rawson, P.F., Remane, J., Robaszynski, F., Schmidt, F. & Surlyk, F. 1984. Cretaceous stage boundaries - proposals. *Bulletin of the geological Society of Denmark* **33** (1-2): 3-20.

Biró-Bagóczky, L. 1980. Algunos ammonites nuevos en la Formación Lo Valdés, Titoniano-Neocomiano, Provincia de Santiago (33° 50' lat. Sur.), Chile. *Actas II Congreso Argentino Paleontología Bioestratigrafia I Congreso Lationoamericano Paleontología, Buenos Aires 1978*, **1**: 223-242.

Blanchet, F. 1922. Sur un groupe d'ammonites éocrétacées dérivées des *Cosmoceras. Compte Rendu Sommaire des Seances de la Société géologique de France* **13**: 158-160.

Blanford, H.F. 1863. On Dr. Gerard's collection of fossils from the Spiti valley, in the Asiatic Society's Museum. *Journal of the Asiatic Society of Bengal, Calcutta* **32**: 124-138.

Blasco de Nullo, G, Nullo, F. & Proserpio, C. 1980. Santoniano-Campaniano: estratigrafia y contenido ammontifero. Cuenca austral. *Revista de la Asociacion geologica Argentina* **35**: 467-493.

———, Nullo F. E. & Ploszkiewicz, J. 1980. El genero *Colchidites* Djanelidze, 1926 y la posicion estratigrafia del genero *Hatchericeras* Stanton, 1901 en le estancia Tucu-Tucu, provincial de Santa Cruz. *Revista Asociacion Geologica Argentina* **35** (1): 41-58.

Blaszkiewicz, A. 1980. Campanian and Maastrichtian of the Middle Vistula river, Poland: a stratigraphic-paleontological study. *Prace Instytutu Geologiczniego* **92**: 1-63.

Blau, J. & Meister, C. 1991. Liassic (Pliensbachian) ammonites from the Leinz Dolomites (eastern Tyrol, Austria). *Jahrbuch der Geologischen Bundesanstalt* **134** (2): 171-204.

Bodylevski, V.I. 1936. Fauna of the upper Volgian Stage of Novaya Zemlya. *Trudy Arkticheskogo Nauchno-Issledovatel'skogo Instituta, Leningrad* **49**: 113-136 [in Russian].

——— 1956. Superfamily Perisphinctacea Wedekind, 1917, family Craspeditidae Spath, 1924, genus *Taimyroceras* Bodylevski gen. nov. *In*, KIPARISOVA, L.D., MARKOWSKI, B.I. & RADCHENKO, G.P. (eds), Material on palaeontology. New families and genera. *Vsesoiuznyi Nauchno-Issledovatel'skii geologicheskii Institut (VSEGEI)*, n.s., Paleontologiia Vypusk **12**: 82-84 [in Russian].

——— 1960. New late Valanginian ammonites from northern Siberia. *In*, MARKOVSKII, B.P. (ed.), New species of prehistoric plants and invertebrates of the USSR, part 2. *Vsesoiuznyi Nauchno-Issledovatel'skii, Otdel geologicheskogo Instituta, Moscow (VSEGEI)*: 172-176 [in Russian].

——— 1967. Jurassic and Cretaceous faunas of Novaya Zemlya. *Zapiski Leningradskogo Gornogo Instituta* **53** (2): 99-112 [in Russian].

——— & SHUL'GINA, N.I. 1958. Jurassic and Cretaceous faunas of the lower course of Yenisei. *Transactions of the Institute of Arctic Geology* **93**: 1-196 [in Russian].

BOEHM, G. 1904. Beiträge zur Geologie von Niederländish-Indien. I Abt. Die Südküsten der Sula-Inseln Taliabu und Mangoli. 1 Abs. Grenzschichten zwischen Jura und Kreide. *Palaeontographica*, Supplement 4, Lief 1: 1-46.

BOGDANOVA, T.N. 1971. New Barremian ammonites from western Turkmenia. *Paleontological Journal* **3**: 334-344 [in Russian].

——— 1979. Ammonites from the Deshayesitidae family in Turkmenistan. *Plankton and pelagic organic life in the Earth's evolution.* Transactions XIX Session of Paleontological Society: 152-169. Nauka: Leningrad [in Russian].

——— 1983. *Deshayesites tuarkyricus* Zone - the lower zone of the Aptian in Turkmenia. *Annual of the All-Union Paleontological Society* **26**: 128-147.

——— 1991. New ammonite species from the Lower Aptian of Turkmenia. *Annual of the All-Union Paleontological Society* **34**: 77-98 [in Russian].

——— 1999. The Lower Aptian of the Mangyshlak Mountains. *Stratigraphy and Geological Correlation* **7** (4): 343-355.

——— & ARKADIEV, V.V. 1999. The genus *Dalmasiceras* (Ammonoidea) from the Berriasian of the mountainous Crimea. *Paleontological Journal* **1999** (4): 20-26 [in Russian].

———, ——— 2005. Revision of species of the ammonite genus *Pseudosubplanites* from the Berriasian of the Crimean Mountains. *Cretaceous Research* **26**: 488-506.

———, ——— 2011. Representatives of genus *Retowskiceras* (Neocomitidae, Ammonoidea) from the Berriasian of the mountainous Crimea. *Bulletin of the Saint-Petersburg University*, Series 7 (Geography, Geology) **3**: 19-25.

——— & HOEDEMAKER, P.J. 2004. Barremian-early Albian Deshayesitidae, Desmoceratidae and Silesitidae of Colombia. *Scripta Geologica* **128**: 183-312.

——— & KAKABADZE, M.V. 1976. On the systematics of the genus *Ammonitoceras* Dumas. *Bulletin of the Academy of Sciences, Georgian SSR* **82** (1): 125-128.

———, KALACHEVA E.D. & SEY, I.I. 1999. On the presence of the *Tirnovella occitanica* Zone (Lower Cretaceous, Berriasian) in the Feodosia section, eastern Crimea. *Regional geology and metallogeny* **9**: 27-32 [in Russian].

——— & KVANTALIANI, I.V. 1983. New Berriasian ammonites of the Crimea. *Bulletin de le Societe impériale des Naturalistes, Moscou*, Ser. Geol., **58** (3): 70-83 [in Russian].

———, ——— & SCHARIKADZE, M.Z. 1979. Some early Aptian Deshayesitidae from central Dagestan. *Geologica Balcanica* **9** (3): 3-12.

———, LUPPOV, N.P. & MIKHAILOVA, I.A. 1985. Genus *Transcaspiites* Luppov, gen. nov. and its systematic position. *In*, MENNER, V.V. (ed.), *Unknown cephalopodous molluscs*: 149-155. Nauka: Moscow [in Russian].

——— & MIKHAILOVA, I.A. 1999. Origin and evolution of the family Deshayesitidae (Ammonoidea). *Paleontological Journal* **5**: 522-529

———, ——— 2004. Origin, evolution and stratigraphic significance of the superfamily Deshayesitaceae. *Bulletin de l'Institut Royal des Sciences naturelles de Belgique* (Sciences de la Terre) **74**: 189-243.

BOGOMOLOV, Y.I. 1989. The polyptychitids (ammonites) and biostratigraphy of the boreal Valanginian. *Trudy Instituta Geologii i Geofizika ANSSSR, Sibirskoe Otdelenie Novosibirsk* **696**: 1-200 [in Russian].

——— & BURDYKINA, M.D. 1989. A new craspeditid genus from the Valanginian of northern Siberia. *Geologiia i Geofizika* **1989** (5): 17-22 [in Russian].

BOGOSLOVSKII, N.A. 1897. The Ryazan horizon: its fauna, stratigraphical relations and probable age. *Materialy dlia Geologii Rossii* **18**: 1-158 [in Russian].

——— 1902. Der unterkretacischen Ammonitenfauna von Central- und Nored-Russland. *Trudy geologicheskogo Komiteta*, n.s., **2**: 1-110 [in Russian], 111-161 [in German].

BOGOSLOVSKYA, M.F., MIKHAILOVA, M.I. & SHEVYREV, A.A. 1990. Ammonoid systematics. *In*, *Systematics of phylogeny of invertbrates: criteria for identifying higher taxa*. Nauk: Moscow.

BØGVAD, R. & ROSENKRANTZ, A. 1934. Beiträge zur Kenntnis der unteren Kreide Ostgrönlands. *Meddelelser om Grønland* **93** (1): 1-28.

BÖHM, J. 1895. Review of "Recherches sur la craie supérieure. 2ième partie: Paléontologie. Les ammonites de la craie supérieure" by A. de Grossouvre. *Neues Jahrbuch für Mineralogie, Geologie und Paläontologie* **1895** (Band 2): 360-366.

——— 1898. Ueber *Ammonites Pedernalis*, v. Buch. *Zeitschrift der deutschen geologischen Gesellschaft* **50**: 183-201.

——— 1904. Beiträge zur Geologie von Niederländish-Indien. I Abt. Die Südküsten der Sula-Inseln Taliabu und Mangoli. 1 Abs. Grenzschichten zwischen Jura und Kreide. *Palaeontographica*, Suppl. 4, Lief. 1: 1-46.

——— 1910. Review of "Études de paléontologie tunésienne," by L. Pervinquière. *Neues Jahrbuch fur Mineralogie, Geologie und Paläontologie* (Band 2, Section 2, Mineralogie) **1910**: 149-155.

——— 1926. Zur Fauna des Oberen Alb mit *Mastigoceras adpressum* Sow. sp., bei Ootmarsum (Holland). *Zeitschrift der deutschen geologischen Gesellschaft* **77** (1925): 198-205

——— & RIEDEL, L. 1933. Über eine als *Placenticeras* beschriebene *Oppelia* (*Bornhardticeras* n. gen.) aus dem Neocom von Deutsch-Ostafrika. *Jahrbuch der preussischen geologischen Landesanstalt* **53**: 112-124.

BONARELLI, G. 1894. *Hecticoceras,* novum genus Ammonidarum. *Bolletino della Societa malacolica Italiana* **18** (for 1893): 73-104.

——— & NÁGERA, J.J. 1921. Observaciones geológicas en las immediaciones del Lago San Martín (territorio de Santa Cruz). *Boletin de Ministerio de Agricultura, Buenos Aires* (B, Geologia) **27**: 1-39.

BHOSALE, S. SURAJ, D.K., FÜRSICH, F.T., ALBERTI, M., CHASKAR, K. & THAKKAR, M.G. 2021. Palaeobiogeography of the late Jurassic ammonite genus *Paraboliceras* in the Indo-Pacific realm and newly collected material from Spiti valley, India. *Neues Jahrbuch für Geologie und Paläontologie-Abhandlungen* **302** (1): 63-87

BOSC, L.A.G. 1801. *In*, ROISSY, F., *Histoire naturelle générale et particulière des Mollusques, Animaux sans vertèbres et à sangblanc. Oeuvrage faisant suite aux oeuvres de Leclerc de Buffon, et parti du cours complet d'histoire naturelle, rédigé par C. S. Sonnini, membre de plusieurs sociétés savants. Continué par F. de Roissy.* Deterville: Paris.

BÖSE, E., 1923. Algunas faunas cretacicas de Zacatecas, Durango y Guerrero. *Boletín del Instituto Géologica de México* **42**: 1-219.

———1928. Cretaceous ammonites from Texas and northern Mexico. *University of Texas Bulletin* **2748**: 143-312.

BOUGHDIRI, M. 1994. *Les genres d'ammonites* Durangites *et* Protacanthodiscus *(Tithonien superiéur) dans la Tethys occidentale (SE Espagne, SE France, Algerie et Tunisie). Stratigraphie, Paléontologie et Biogeographie.* These, Universite Claude-Bernard-Lyon I.

———, ÉNAY, R., HEGARAT, G LE & MEMMI, L. 1999. *Hegaratites* gen. nov. (Ammonitina): Himalayitidae nouveau du Tithonien Supérieur de la coupe du Jebel Rhéouis (Axe nord-sud, Tunisie centrale). Précisions stratigraphiques, approches phylétiques et signification biogéographique. *Revue de Paléobiologie* **18** (1): 105-121.

BOULE, M. 1937. Types du prodrome de paléontologie stratigraphique de d'Orbigny. *Annales de Paléontologie* **26** (1-2): 17-48.

———, LEMOINE, P. & THEVENIN, A. 1906. Paléontologie de Madagascar. III.Céphalopodes Crétacés des environs de Diego-Suarez. *Annales de Paleontologie* **1**: 173-192 (1-20).

——— 1907. Paléontologie de Madagascar. III. Céphalopodes Crétacés des environs de Diego-Suarez. *Annales de Paléontologie* **2**: 1-56 (21-76).

BOURGUIGNAT, J. R. 1877. Histoire des clausilies de France vivantes et fossiles. *Annales des Sciences naturelles, Zoologie, Paris* (6) **5** (4): 1-50.

BOURILLOT, R., NEIGE, P., PIERRE, A. & DURLET, C. 2008. Early-Middle Jurassic lytoceratid ammonites with constrictions from Morocco: palaeobiogeographical and evolutionary implications. *Palaeontology* **51** (3): 597-609.

BOWDITCH, T. E. 1822. *Elements of conchology, including the fossil genera of the animals. Part 1: Univalves.* J. Smith: Paris.

BREISTROFFER, M. 1936. Les subdivisions du Vraconien dans le sud-est de la France. *Bulletin de la Société géologique de France* (5) **6**: 63-68.

——— 1940. Révision des ammonites du Vraconien de Salazac (Gard) et considérations générales sur ce sous-étage albien. *Travaux du Laboratoire de Géologie de la Faculte des Sciences de l'Universite de Grenoble* **22** (1938-1939): 71-171 (1-101).

——— 1946. Sur l'âge exact des grès verts de Cambridge (Angleterre). *Compte Rendu Sommaire des Seances de la Société géologique de France* **1946**: 309-312.

——— 1947a. Notes de nomenclature paléozoologique. I. Ammonites jurassiques et crétacées. *Proces-verbaux Mensuels de la Société scientifique du Dauphine* **26** (195): 5 unnumbered pages.

——— 1947b. Sur les zones d'ammonites dans l'Albien de France et d'Angleterre. *Travaux du Laboratoire de Géologie de la Faculte des Sciences de l'Universite de Grenoble* **26** (1946–1947): 17-104 (1-88).

——— 1951. Sur quelques ammonites de l'Albien inférieur de Madagascar. *Compte Rendu Sommaire des Seances de la Société géologique de France* **1951**: 266-268.

——— 1952a. À propos des genres *Jacobella* et *Hoplites* (auct.). *Compte Rendu Sommaire des Seances de la Société géologique de France* **1952**: 152-154.

——— 1952b. Sur la découverte de Knemiceratinae (ammonites albiennes) en Équateur, en Colombie et au Vénézuela. *Comptes Rendus hebdomadaires des Seances de l'Academie des Sciences* **234**: 2633-2635.

——— 1952c. Sur la position systématique du genre *Ptychoceras* d'Orb. *Travaux du Laboratoire de Géologie de la Faculte des Sciences de l'Universite de Grenoble* **29** (1951): 47-54.

——— 1953a. L'évolution des turrilitidés albiens et cénomaniens. *Comptes Rendus hebdomadaires des Seances de l'Academie des Sciences* **237**: 1349-1351.

——— 1953b. Commentaires taxonomiques. *In*, BREISTOFFER, M. & DE VILLOUTREYS, O. Les ammonites albiennes de Peille (Alpes-Maritimes). *Travaux du Laboratoire de Géologie de la Faculte des Sciences de l'Universite de Grenoble* **30**: 69-74.

BRESKOVSKI, S. 1965. Sur la valeur stratigraphique de quelques représentants du genre *Neocomites* Uhlig, 1906. *CBGA, VII Congress Sofia, Reports*, part II, 1: 207-209.

——— 1966. Biostratigraphie du Barrémien au sud du village de Brestak, dans la région de Varna. *Travaux sur la Géologie de Bulgarie*, Série Paléontologie, **8**: 71-121.

——— 1967. *Eleniceras* - genre nouveau d'ammonites Hautériviens. *Bulgarian Academy of Sciences, Bulletin of the geological Institute* (Paleontology) **16**: 47-52.

——— 1975. Les zones et sous-zones ammonitiques dans l'étage Barrémien en Bulgarie du nord-est. *Geologica Balcanica* **5** (2): 47-66.

——— 1977a. Sur la classification de la famille Desmoceratidae Zittel, 1895 (Ammonoidea, Crétacé). *Comptes Rendus de l'Academie Bulgare des Sciences* **30**: 891-894.

——— 1977b. Genres nouveaux du Crétacé inférieur de la famille Desmoceratidae Zittel, 1895 (Ammonoidea). *Comptes Rendus de l'Academie Bulgare des Sciences* **30** (6): 891-894.

——— 1980. Des genres nouveaus du Crétacé inférieur de la famille Desmoceratidae Zittel,1895 (Ammonoidea). *Comptes Rendus de l'Academie Bulgare des Sciences* **33** (2): 245-248.

BRINKMANN, R. 1937. Biostratigraphie des Leymeriellen-stammes nebst Bemerkungen zür Paläogeographie des nordwestdeutschen Alb. *Mitteilungen aus dem geologischen Staatsinstitut in Hamburg* **16**: 1-18.

BROCHWICZ-LEWINSKI, W.B. 1973. Some remarks on the origin of the subfamily Idoceratinae Spath, 1924 (Peri-sphinctidae, Ammonoidea). *Acta geologica Polonica* **18** (3): 299-320.

——— 1975. On the Oxfordian genus *Subdiscosphinctes* Malinowska, 1972, and subgenus *S. (Aureimontanites)* nov. (Perisphinctidae, Ammonoidea). *Acta geologica Polonica* **20** (1): 87-96.

——— & RÓZAK, Z. 1976. Oxfordian idoceratids (Ammonoidea) and their relation to *Perisphinctes* proper. *Acta geologica Polonica* **21** (4): 299-320.

BROWN, T. 1837-1849. *Illustrations of the fossil conchology of Great Britain and Ireland.* Smith, Elder, Maclachlan & Stewart: London, Edinburgh.

BRÜGGEN, H. 1910. Die Fauna des unteren Senons von Nord-Perú. *Neues Jahrbuch für Mineralogie, Geologie und Paläontologie*, Beilage-Band **30**: 717-788.

BRUGUIÈRE, J.G 1789. *Histoire naturelle des Vers et des mollusques. Encyclopédie méthodique.* Part 1. Panckoucke: Paris.

BRUNNSCHWEILER, R.O. 1959. New Aconeceratinae (Ammon-oidea) from the Albian and Aptian of Australia. *Bureau of Mineral Resources, Geology and Geophysics, Australia, Bulletin* **54**: 1-19.

——— 1966. Upper Cretaceous ammonites from the Carnarvon Basin of Western Australia. I. The heteromorph Lytoceratina. *Bureau of Mineral Resources, Geology and Geophysics, Australia, Bulletin* **58**: 1-58.

BUCH, L. VON. 1839. *Petrifications recueillis en Amérique par A. de H. Humboldt et par C. Degenhardt.* Imprimerie de l'Académie royale des sciences: Berlin.

——— 1850. Über Ceratiten. *Physikalische Mathematische Abhandlungen der koeniglichen Akademie der Wissenschaften zu Berlin* **1848**: 1-33.

BUCK, W. 1958. *Die Gattung Aspidoceras in Europäischen, insbesondere im Malm.* Unpublished Doctoral Dissertation, University of Tübingen.

BUCKMAN, S.S. 1905. On certain genera and species of Lytoceratidae. *Quarterly Journal of the geological Society of London* **61**: 142-154.

——— 1909-1930. *Yorkshire type ammonites.* Vol. 1-7. Wesley & Son: London.

BULOT, L. 1992. Les Olcostephaninae Valanginiens et Hauteriviens (Ammonitina, Cephalopoda) du Jura Franco-Suisse: systematique et interet biostratigraphiqie, *Revue de Paléobiologie* **11** (1): 149-166.

——— 2010. Systematic palaeontology of Aptian and Albian ammonites from southwest Iran. *GeoArabia, Special Publication* **4** (1): 167-195.

——— & AUTRAN, G. 1989. Révision des types et figures de la collection Matheron 1. *Olcostephanus stephanophorus* (Matheron, 1878). *Mésogée* **49**: 15-19.

———, COMPANY, M. & THIEULOY, J.-P. 1990. Origine, évol-ution et systématique du genre Valanginien *Saynoceras* (Ammonitina, Olcostephaninae). *Geobios* **23**: 399-413.

———, KENNEDY, W.J., JAILLARD, E. & ROBERT, E. 2005. Late Middle - early late Albian ammonites from Ecuador. *Cretaceous Research* **26**: 450-459.

BURCKHARDT, C. 1903. Beiträge zur Kenntnis der Jura und Kreideformation der Cordillere. *Palaeontographica* **50**: 1-144.

——— 1906. La faune jurassique de Mazapil avec un appendice sur les fossiles du Crétacique supérieur. *Boletim Instituto Geologico de Mexico* **23**: 1-216.

——— 1911. Bemerkungen über die russisch-borealen Typen im Oberjura Mexikos und Südamerikas. *Centralblatt für Mineralogie, Geologie und Paläontologie* **1911**: 477-483.

——— 1912. Faunes Jurassiques et Crétacés de San Pedro del Gallo (Durango, Mexico). *Boletim Instituto Geologico de Mexico* **29**: 1-264.

——— 1919-21. Faunas jurassicas de Symon (Zacatecas). *Boletim Instituto Geologico de Mexico* **33** (1): 1-135 (pls published 1921)

——— 1925. Faunas del Aptiano de Nazas (Durango). *Boletim Instituto Geologico de Mexico* **45**: 1-71.

BURDYKINA, M.D. 1981. Valanginian ammonites of the Great Begichev Island. *Russian Geology and Geophysics* **10**: 49-58 [in Russian].

BÜRGL, H. 1956. Catalogo de las amonitas de Colombia. Parte 1: Pulchelliidae. *Boletín Geológico, Bogotá* **4** (1): 1-119.

BUSNARDO, R. 1965. Le stratotype du Barrémien. Lithologie et macrofaune. *In*, Colloque sur le Crétacé inférieur (Lyon, Septembre 1963). *Mémoire du Bureau de Recherches Géologiques et Minières* **34**: 101-116.

——— 1966a. IV. Paléontologie. *In*, BUSNARDO, R., ÉNAY, R., LATREILLE, G. & ROUQUET, P., Le Crétacé moyen détritique à céphalopodes près de Poncin (Jura méridional). *Travaux des Laboratoires de Géologie de la Faculte des Sciences de Lyon*, n.s., **13**: 219-226.

——— 1966b. II. Paléontologie. *In*, BUSNARDO, R., ÉNAY, R. & PITRE, B., L'Hauterivien inférieur et ses ammonites près de Hauteville (Ain). *Travaux des Laboratoires de Géologie de la Faculte des Sciences de Lyon*, n.s., **13**: 231-242.

——— 1970a. *Torcapella,* nouveau genre d'ammonites du Barremien Inférieur. *Documents des Laboratoires de Géologie de la Faculte des Sciences de Lyon* **37**: 85-130.

——— 1970b. Les *Pseudothurmannia* (Ammonoidea) de l'Hauterivien Supérieur de la Montagne de Lure. *Documents des Laboratoires de Géologie de la Faculte des Sciences de Lyon* **37**: 133-142.

——— 1984. Crétacé inférieur: 1.3.1. Ammonites. *In*, DEBRAND-PASSARD, S. (ed.), Synthèse géologique du sud-est de la France. I: Stratigraphie et paléogéographie. *Mémoires du Bureau de Recherches géologiques et Minières* **125**: 292-294.

———, CHAROLLAIS, J., WIEDMANN, M. & CLAVEL, B. 2003. Le Crétacé Inférieur de la Veveyse de Châtel (Ultra-helvétique des Préalpes externed; canton de Fribourg, Suisse). *Revue de Paléobiologie* **22** (1): 1-174.

——— & DAVID, L. 1957. Contribution à l'étude des faunes d'ammonoïdes de Medjez Sfa (est Constantinois). *Publications du Service de la Carte géologique de l'Algerie, Bulletin*, n.s., **13** (1956): 67-123.

——— & SORNAY, J. 1965. Au sujet de *Delphinites,* genre d'ammonites valanginiennes. *Travaux des Laboratoires de Géologie de la Faculte des Sciences de Lyon*, n.s., **12**: 137-141.

——— & THIEULOY, J.-P. 1989. Les ammonites de l'Hauterivien jurassien: révision des faunes de la region du stratotype historique de l'étage Hauterivien. *Mémoires de la Société Neuchateloise des Sciences Naturelles* **11**: 101-147.

BÜRGL, H. 1956. Catálogo de los amonitas de Colombia. Parte I. Pulchelllidae. *Boletin Geologico* **4** (1): 1-119.

BUSSON, G. & ALBANESI, C. 1967. Le Crétacé inférieur et le Jurassique terminal de l'extrême-sud tunisien. *Rivista italiana di paleontologia e stratigrafia* **73** (2): 591-634.

BUTJOR, L. 2010. Systematics, phylogeny and homoeomorphy of the Engonoceratidae Hyatt, 1900 (Ammonoidea, Cretaceous) and revision of *Engonoceras duboisi* Latil, 1989. *Carnéts de Géologie* **2010/08**: 1-31.

CALLOMON, J.H. & BIRKELUND, T. 1982. The ammonite zones of the Boreal Volgian (Upper Jurassic) in East Greenland. *In*, EMBRY, A.F. & BALKWILL, H.R. (eds), Arctic geology and geophysics. *Memoirs of the Canadian Society of Petroleum Geologists, Memoir* **8**: 349-369.

CANAVARI, M. 1900. Hopliti titoniani dell'Apennino Centrale. *Atti della Societá Toscana di Scienze Naturali, Memorie* **17**: 95-103.

CANTÚ-CHAPA, A. 1963. Étude biostratigraphique des ammonites du centre et de l'est du Mexique (Jurassique Supérieur et Crétacé). *Mémoires de la Société géologique de France* (5) **99**: 1-103.

——— 1966. Se propone una nueva subdivisión de la familia Olcostephanidae (Ammonoidea) del Cretacico Inferieur (Taraisitinae subfam. nov. y *Taraisites* gen. nov.). *Ingenieria Petrolera* **6** (12): 15-17.

——— 1967. El limite Jurásico-Cretácico en Mazatepec, Puebla (México): Mexico, D. F. *Instituto Mexicano del Petróleo (Geologia), Monografia* **1**: 3-24.

——— 1976a. El contacto Jurásico-Cretácico, la estratigr-aphia del Neocomiano, el hiato Hauteriviano Superior-Eoceno Inferior y las ammonites del Pozo Bejuco 6 (centro-este de México). *Boletin de la Sociedad geologica Mexicana* **37**: 60-83.

——— 1976b. Estratigrafia de la formacion La Peña (Aptiano Sup.) en le area de Monterrey, N. L. *Revista del Instituto mexicano del Petroleo* **8**: 7-16.

——— 1983. Suaritinae new subfamily and *Galeanites garcense* new genus and new species of the Upper Jurassic Perisphinctaceae ammonite. *Revista Mexicana de Ciencias Géologicas* **15** (1): 21-24.

——— 1987. Las amonitas del Albiano superior de Tepexi de Rodríguez, Puebla. *Sociedad Mexicana de Paleontología* **1**: 159-160.

——— 1992a. Importancia de la segunda silli lateral en la caracteracion de la ammonita Perisphinctidae *Idoceras* de la Kimeridgiano Inferior. *Revista de la Sociedad Mexicana de Paleontologia* **5** (1): 15-35.

——— 1992b. *Novoleonites tovarensis* gen. nov. sp. nov. (Ammonoideo Neocomitino) del Valangiano Superior del NE de Mexico. *Revista de la Sociedad Mexicana de Paleontologia* **5** (1): 37-43.

——— 2001. The Taraises Formation (Berriasian-Valanginian) in northeastern Mexico: subsurface and outcrop studies. *In*, BARTOLINI, C., BUFFLER, R.T. & CANTÚ-CHAPA, A. (eds). The western Gulf of Mexico Basin: Tectonics, sedimentary basins, and petroleum systems. *American Association of petroleum Geologists Memoir* **75**: 353-370.

——— 2006. New Upper Tithonian (Jurassic) ammonites from the Chinameca Formation in southern Veracruz, eastern Mexico. *Journal of Paleontology* **80** (2): 294-308.

——— 2009. Ammonites of the Cretaceous Taraises and lower Tamaulipas Formations in eastern Mexico. *In*, BARTOLINI, C. & RAMOS, J.R.R. (eds), Petroleum systems in the southern Gulf of Mexico. *American Association of petroleum Geologists Memoir* **90**: 191-216.

CARRASCO, V.B. 1967. *Jimensites,* una nuova ammonita del Campaniano inferior de Mexico. *Ingenieria Petrolero* **7** (2): 15-21.

CARTER, J.G. *et al.* 2011. A synoptical classification of the Bivalvia (Mollusca). *Paleontological Contributions* **4**: 1-47.

CASEY, R. 1952. The ammonite genera *Arcthoplites* Spath and *Tetrahoplites* gen. nov. *Abstracts of the Proceedings of the geological Society of London* **1490**: 134-135.

——— 1954a. New genera and subgenera of Lower Cretaceous ammonites. *Journal of the Washington Academy of Sciences* **44**: 106-115.

——— 1954b. *Falciferella,* a new genus of Gault ammonites, with a review of the family Aconeceratidae in the British Cretaceous. *Proceedings of the Geologists' Association* **65**: 262-277.

——— 1957. The Cretaceous ammonite genus *Leymeriella,* with a systematic account of its British occurrences. *Palaeontology* **1**: 29-59.

——— 1960a. *Hengestites,* a new genus of Gault ammonites. *Palaeontology* **2**: 200-209.

——— 1960b. A monograph of the Ammonoidea of the Lower Greensand. Part 1. *Palaeontographical Society Monograph* **1959**: xxxvi, 1-44.

——— 1961a. A monograph of the Ammonoidea of the Lower Greensand. Part 2. *Palaeontographical Society Monograph* **1960**: 45-118.

——— 1961b. The stratigraphical palaeontology of the Lower Greensand. *Palaeontology* **3**: 487-621.

——— 1961c. The Cretaceous (Albian) ammonite genus *Platiknemiceras* Bataller. *Annals and Magazine of Natural History* (13) **4**: 353-357.

——— 1961d. A monograph of the Ammonoidea of the Lower Greensand. Part 3. *Palaeontographical Society Monograph* **1961**: 119-216.

——— 1962. A monograph of the Ammonoidea of the Lower Greensand. Part 4. *Palaeontographical Society Monograph* **1962**: 217-288.

——— 1964. A monograph of the Ammonoidea of the Lower Greensand. Part 5. *Palaeontographical Society Monograph* **1963**: 289-398.

——— 1965. A monograph of the Ammonoidea of the Lower Greensand. Part 6. *Palaeontographical Society Monograph* **1964**: 399-546.

——— 1966. A monograph of the Ammonoidea of the Lower Greensand. Part 7. *Palaeontographical Society Monograph* **1966**: 547-582.

——— 1973. The ammonite succession at the Jurassic-Cretaceous boundary in eastern England. *In*, CASEY, R. & RAWSON, P.F., The Boreal Lower Cretaceous. *Geological Journal, Special Issue* **5**: 193-266.

——— 1978. A monograph of the Ammonoidea of the Lower Greensand. Part 8. *Palaeontographical Society Monograph* **1978**: 583-632.

——— 1980. A monograph of the Ammonoidea of the Lower Greensand. Part 9. *Palaeontographical Society Monograph* **1979**: 633-660.

———, MESEZHNIKOV, M.S. & SHUL'GINA, N.I. 1977. Correlation of the Jurassic-Cretaceous boundary deposits of England, the Russian Platform, the Subarctic Urals and Siberia. *Izvestiia Akademii Nauk SSSR, Seriia geologicheskaia* **7**: 14-33 [in Russian].

——— & RAWSON, P.F. (eds) 1973. The Boreal Lower Cretaceous. *Geological Journal, Special Issue* **5**: 1- 448.

CATULLO, T. 1847a. *Appendice al catalogo degli ammoniti delle Alpi Venete: accompagnata da una tavola, che sarebbe la 12. del Prodromo (specie Neocomiane).* Sicca, Magio: Padova.

——— 1847b. *Seconda appendice al catalogo degli ammoniti delle Alpi Venete: accompagnata da una tavola, che sarebbe la XIII del Prodromo (specie Neocomiane).* Sicca, Magio: Padova.

CECCA, F. 1990a. Etude des périsphinctidés de la zone à Darwini (Tithonique inférieur) des Apennins des Marches (Italie): la paléontologie et paléobiogéographie. *Atti II Conv. F.E.A. Pergola, 1987. Countato Centenario Raffaele Pieccinini*: 39-55.

——— 1990b. *'Subplanitoides' mediterraneus*, nuova specie di perisphinctide (Ammonitina) della Zona a Semiforme (Titonico inferiore) della provincia mediterranea. *Atti II Conv. F.E.A. Pergola, 1987. Countato Centenario Raffaele Pieccinini*: 57-62.

——— 1990c. Studio paleontologico di alcuni rappresent-anti del genre *Danubisphinctes* Zeiss (Ammonitina) della Zona a Fallauxi, Sottozona a Biruncinatum (Titonico inferiore) di due sezoni dell'Appennino marchigiano (Italia centrale). *Bollettino del Servizio geologico d'Italia* **107**: 21-42.

——— (ed.) 1995. Proceedings of 3rd workshop on Early Cretaceous cephalopods. *Memorie Descrittive della Carta Geologica d'Italia* **51**: 1-218.

———, FARAONI, P., MARINI, A. & PALLINI, G. 1995. Fieldtrip across the representative sections for the Upper Hauterivian - Barremian ammonite biostratigraphy in the Maiolica exposed at Monte Nerone, Monte Petrano and Monte Catria (Umbria-Marche Apennines). *Memorie descrittive della Carta Geologica d'Italia* **51**: 187-211.

——— & LANDRA, G. 1994. Late Barremian-early Aptian ammonites from the Maiolica Formation near Cesana Brianza (Lombardy Basin, northern ltaly). *Rivista ltaliana di Paleontologia e di Stratigrafia* **100**: 395-422.

———, ROPOLO, P. & GONNET, R. 1998. La base de l' Aptien à Cassis-La Bédoule (SE France). *Géologie Méditerranéenne* **25**: 149-157.

———, ——— & ——— 1999. The appearance of the genus *Deshayesites* (Kazansky, 1914, Ammonoidea) in the lowermost Aptian (Lower Cretaceous) of La Bédoule (SE France). *Rivista Italiana di Paleontologia e Stratigrafia* **105**: 267-286.

———, SEYED-EMAMI, K., SCHNYDER J., BENZAGGAGH M., MAJIDIFARD, M.R. & MONFARED, M.M. 2012. Early Berriasian ammonites from Shal, Talesh region (NW Alborz Mountains, Iran). *Cretaceous Research* **33**: 106-115.

CHAO, K. 1976. Jurassic and Cretaceous ammonites from Mount Jolmo Lungma region. *In*, *A report of the scientific expedition in the Mount Jolmo Lungma region.* III. Palaeontology: 504-545. Academia Sinica: Tibetan Scientific Expedition Team; Science Press: Beijing [in Chinese].

CHANCELLOR, G.R. 1982. Cenomanian-Turonian ammonites from Coahuila, Mexico. *Bulletin of the Geological Institute of Uppsala University*, n.s., **9**: 77-129.

———, KENNEDY, W.J. & HANCOCK, J.M. 1994. Turonian ammonite faunas from central Tunisia. *Special Papers in Palaeontology* **50**: 1-118.

CHAPUT, E. 1921. Les desmocératidés du paléocrétacé. Revision du genre *Desmoceras*. *In*, KILIAN, C.C.C.W. (ed.), Contributions à l'étude des céphalopodes paléocrétacés du sud-est de la France. *Mémoires pour Servir a l'Explication de la Carte géologique detaillee de la France* **1920**: 167-188.

CHECA, A. 1984. Phylogenetic relationship among Oxfordian and Kimmeridgian Aspidoceratidae "classical species", deduced from the Subbetic record (south Spain). A proposal. *Géobios* **17**: 21-31.

——— 1985. *Los aspidoceratiformes en Europa (Ammonitina, fam. Aspidoceratidae: subfamilias Aspidoceratinae y Physadoceratinae)*. Tesis Doctorale, Faculdad de Ciencias, Departamento de Paleontologia y Geologia general, University of Granada, 413 pp.

——— & OLÓRIZ, F. 1984. Significative Mediterranean "*Aspidoceras*" in Upper Jurassic biostratigraphy. *International Symposium on Jurassic Stratigraphy, Erlanger, Sept. 1-8, 1984* II: 393-414.

———, ——— 1986. Evolutionary trends in Oxfordian and Kimmeridgian subbetic Aspidoceratinae (southern Spain). A proposition of Null Hypotheses about the evolutionary course in a highly significant group of tethydian Upper Jurassic ammonites. *Bollettino del Servizio geologico d'Italia* **24** (2-3): 145-159.

——— & ROUGET, I. 2006. Anagenetic evolution of the early Tithonian ammonite genus *Semiformiceras* tested with cladistic analysis. *Palaeontology* **49** (5): 1069-1080.

CHERNOVA, E.S. 1951. On the age and division of the Simbirskitid Beds and the Belemnite Series of the Volga region. *Biulleten Moskovskogo Obshchestva Ispytatelei Prirody, Otdel Geologicheskii* **26**: 46-81 [in Russian].

——— 1952. On the systematics of *Simbirskites*. *Biulleten Moskovskogo Obshchestva Ispytatelei Prirody, Otdel Geologicheskii* **27** (6): 45-58 [in Russian].

CHIPLONKAR, G.W., & GHARE, M.A. 1976. Palaeontology of the Bagh beds. Part 7: Ammonoidea. *Bulletin of Earth Sciences* **4 & 5**: 1-10.

——— & ——— 1977. Palaeontology of the Bagh beds. Part 10: Scaphitidae. *Proceedings of the Indian Academy of Sciences* **85B**: 67-76.

——— & ——— 1978. Taxonomic comments on Placenticeratidae with diagnosis of a new genus *Sancarlosia*. *Biovigyanam* **4**: 75-80.

CHOFFAT, P. 1898. Les ammonées du Bellasien, des couches à *Neolobites vibrayeanus*, du Turonien et du Sénonien. Part 2. *In*, Receuil d'études paléontologiques sur la faune Crétacique du Portugal. *Direction des Travaux géologique du Portugal* **1898**: 41-86.

——— 1903. Contributions à la connaissance géologique des colonies portugaises d'Afrique. I. La Crétacique de Conducia. *Commission du Service géologique du Portugal* **1903**: 1-32l.

CHUDEAU, R. 1909. Ammonites du Damergou. *Bulletin de la Societe geologique de France* (4) **9**: 67-71.

——— 1921. Ammonites turoniennes du Soudan. *Bulletin du Museum national d'Histoire naturelle, Paris* **27**: 463-470.

CIESLINSKI, S. 1959. The Albian and Cenomanian in the northern periphery of the Holy Cross Mountains. *Prace Panstwowego Instytutu Geologicznego* **28**: 1-95.

CLARK, D.L. 1965. Heteromorph ammonoids from the Albian and Cenomanian of Texas and adjacent areas. *Geological Society of America Memoir* **95**: viii, 1-99.

COBBAN, W. A. 1952a. Scaphitoid cephalopods of the Colorado Group. *United States geological Survey Professional Papers* **239**: iv, 1-42.

——— 1952b. A new Upper Cretaceous ammonite genus from Wyoming and Utah. *Journal of Paleontology* **26**: 758-760.

——— 1953. Cenomanian ammonite fauna from the Mosby sandstone of central Montana. *United States geological Survey Professional Papers* **243-D**: iv, 45-56.

——— 1961. The ammonite family Binneyitidae Reeside in the Western Interior of the United States. *Journal of Paleontology* **35**: 737-758.

——— 1964. The late Cretaceous cephalopod *Haresiceras* Reeside and its possible origin. *United States geological Survey Professional Paper* **454-I**: iv, 1-21.

——— 1971. New and little-known ammonites from the Upper Cretaceous (Cenomanian and Turonian) of the Western Interior of the United States. *United States geological Survey Professional Papers* **699**: 1-24.

——— 1987a. A new ammonite from the Upper Cretaceous of Kansas. *United States geological Survey Bulletin* **1690**: iii, 1-3.

——— 1987b. Some Middle Cenomanian (Upper Cretaceous) acanthoceratid ammonites from the Western Interior of the United States. *United States geological Survey Professional Papers* **1445**: iv, 1-28.

——— & GRYC, G. 1961. Ammonites from the Seabee Formation (Cretaceous) of northern Alaska. *Journal of Paleontology* **35**: 176-190.

——— & HOOK, S.C. 1980. The Upper Cretaceous (Turonian) ammonite family Coilopoceratidae Hyatt in the Western Interior of the United States. *United States geological Survey Professional Papers* **1192**: iv, 1-28.

——— & ——— 1983. Mid-Cretaceous (Turonian) ammonite fauna from Fence Lake area of west-central New Mexico. *New Mexico Bureau of Mines and Mineral Resources Memoir* **41**: 1-50.

———, ——— & KENNEDY, W.J. 1989. Upper Cretaceous rocks and ammonite faunas of southwestern New Mexico. *New Mexico Bureau of Mines and Mineral Resources Memoir* **45**: iv, 5-137.

——— & KENNEDY, W.J. 1989. *Acompsoceras inconstans* zone, a Lower Cenomanian marker horizon in Trans-Pecos Texas, U.S.A. *Neues Jahrbuch für Geologie und Paläontologie, Abhandlungen* **178**: 133-145.

———, ——— 1991a. Evolution and biogeography of the Cenomanian (Upper Cretaceous) ammonite *Metoicoceras* Hyatt, 1903, with a revision of *Metoicoceras praecox* Haas, 1949. *United States geological Survey Bulletin* **1934**: B1-B11.

———, ——— 1991b. New records of the ammonite subfamily Texanitinae in Campanian (Upper Cretaceous) rocks in the Western Interior of the United States. *United States geological Survey Bulletin* **1985**: B1-B4.

———, ——— 1991c. Some Upper Cretaceous ammonites from the Nacatoch Sand of Hempstead County, Arkansas. *United States geological Survey Bulletin* **1985**: C1-C5.

———, ——— 1991d. Upper Cretaceous (Maastrichtian) ammonites from the *Nostoceras alternatum* zone in southwestern Arkansas. *United States geological Survey Bulletin* **1985**: E1-E6.

———, ——— 1991e. *Pachydiscus* (Ammonoidea) from Campanian (Upper Cretaceous) rocks in the Western Interior of the United States. *United States geological Survey Bulletin* **1985**: F1-F4.

———, ——— 1991f. *Baculites thomi* Reeside, 1927, an Upper Cretaceous ammonite in the Western Interior of the United States. *United States geological Survey Bulletin* **1934**: C1-C8.

———, ——— 1992a. Campanian ammonites from the Upper Cretaceous Gober Chalk of Lamar County, Texas. *Journal of Paleontology* **66**: 440-454.

———, ——— 1992b. Campanian *Trachyscaphites spiniger* ammonite fauna in north-east Texas. *Palaeontology* **35**: 63-93.

———, ——— & LANDMAN, N. H. 1999. *Platyscaphites*, a new ammonite from the Lower Campanian (Upper Cretaceous) of the United States Western Interior. *Bulletin de l'Institut Royal des Sciences Naturelles de Belqique (Sciences de la Terre)* **69**: 47-54.

——— & SCOTT, G.W. 1964. Multinodose scaphitid cephalopods

from the lower part of the Pierre Shale and equivalent rocks in the conterminous United States. *United States geological Survey Professional Papers* **483-E**: iv, 1-14.

———, ——— 1972. Stratigraphy and ammonite fauna of the Graneros Shale and Greenhorn Limestone near Pueblo, Colorado. *United States Geological Survey Professional Papers* **645**: 1-108

Collet, L. W. 1907. Sur quelques espèces de l'Albien inférieur de Vöhrum (Hanovre). *Mémoires de la Société de Physique et d'Histoire naturelle de Geneve* **35**: 519-529.

——— 1924. Sur quelques ammonites du Barrémien de Colombie. *Eclogae Geologicae Helvetiae* **18**: 485-493.

Collignon, M. 1928-29. Paléontologie de Madagascar. XV. Les céphalopodes du Cénomanien pyriteux de Diégo-Suarez. *Annales de Paléontologie* **17** (1928): 137-160; **18** (1929): 1-56.

——— 1931a. Paléontologie de Madagascar. XVI. La faune du Cénomanien à fossiles pyriteux du nord de Madagascar. *Annales de Paléontologie* **20**: 43-104.

——— 1931b. Faunes sénoniennes du nord et de l'ouest de Madagascar. *Annales géologiques du Service des Mines, Madagascar* **1**: 1-64.

——— 1932. Les ammonites pyriteuses de l'Albien supérieur du Mont Raynaud á Madagascar. *Annales géologiques du Service des Mines, Madagascar* **2**: 2-36.

——— 1936. Les *Oxytropidoceras* de l'Albien moyen (niveau supérieur) de la province d'Analalava. *In*, Besairie, H., Recherches géologiques à Madagascar. 1. La géologie du nordouest. *Mémoires de l'Academie Malgache* **21**: 176-190.

——— 1937a. Ammonites cénomaniennes du sudouest de Madagascar. *Annales géologiques du Service des Mines, Madagascar* **8**: 31-69 (7-45).

——— 1937b. Paléontologie de Madagascar. XXII. Les ammonites pyriteuses de l'Aptien d'Antanatanamirafy. *Annales de Paléontologie* **26**: 107-132 (1-28).

——— 1938. Ammonites campaniennes et maestrichtiennes de l'ouest et du sud de Madagascar. *Annales géologiques du Service des Mines, Madagascar* **9**: 53-118.

——— 1939. Fossiles cénomaniens et turoniens du Menabe de Madagascar. *Annales géologiques du Service des Mines, Madagascar* **10**: 1-49.

——— 1948. Ammonites néocrétacées du Menabe (Madagascar). I. Les Texanitidae. *Annales géologiques du Service des Mines, Madagascar* **13**: 47-107 (2-62); **14**: 5-60, 99-101 (63-120).

——— 1949a. Faune néocomienne des couches à criocères de Belohasifaka (Cercle de Sitampiky) (Madagascar). *Annales géologiques du Service des Mines, Madagascar* **15**: 53-86.

——— 1949b. Recherches sur les faunes albiennes de Madagascar. I. L'Albien d'Ambarimaninga (Madagascar). *Annales géologiques du Service des Mines, Madagascar* **16**: 1-128.

——— 1950a. Recherches sur les faunes albiennes de Madagascar. III. L'Albien de Komihevitra (Cercle de Soalala). *Annales géologiques du Service des Mines, Madagascar* **17**: 19-54.

——— 1950b. Recherches sur les faunes Albiennes de Madagascar. IV. L'Albien de Mokaraha (Cercle de Soalala). *Annales géologiques du Service des Mines, Madagascar* **17**: 55-85.

——— 1952. Ammonites néocrétacées du Menabe (Madagascar). II. Les Pachydiscidae. *Travaux de la Bureau géologique de Madagascar* **41**: 1-114.

——— 1954. Ammonites néocrétacées du Menabe (Madagascar). III. Les Kossmaticeratidae. *Travaux de la Bureau géologique de Madagascar* **62**: 1-59.

——— 1955. Ammonites néocrétacées du Menabe (Madagascar). II. Les Pachydiscidae. *Annales géologiques du Service des Mines, Madagascar* **21**: 1-98.

——— 1956. Ammonites néocretacées du Menabe (Madagascar). IV. Les Phylloceratidae. *Annales géologiques du Service des Mines, Madagascar* **23**: 7-38.

——— 1957. Céphalopodes néocrétacés du Tinrhert (Fezzan). *Annales de Paléontologie* **43**: 113-136 (1-24).

——— 1959a. *Atlas des fossiles caractéristiques de Madagascar.* Fascicle **4** (Argovian-Rauracien). Service Géologique: Tananarive.

——— 1959b. *Atlas des fossiles caractéristiques de Madagascar.* Fascicle **5** (Kimmeridgian). Service Géologique: Tananarive.

——— 1960. *Atlas des fossiles caractéristiques de Madagascar.* Fascicle **6** (Tithonique). Service Géologique: Tananarive.

——— 1961. Ammonites néocrétacées du Menabe (Madagascar). VII. Les Desmoceratidae. *Annales géologiques du Service des Mines, Madagascar* **31**: 1-116.

——— 1962a. *Atlas des fossiles caractéristiques de Madagascar (Ammonites).* Fascicle **8** (Berriasien, Valanginien, Hauterivien, Barremien). Service Géologique: Tananarive.

——— 1962b. *Atlas des fossiles caractéristiques de Madagascar (Ammonites).* Fascicle **9** (Aptien). Service Géologique: Tananarive.

——— 1963. *Atlas des fossiles caractéristiques de Madagascar (Ammonites).* Fascicle **10** (Albien). Service Géologique: Tananarive.

——— 1964. *Atlas des fossiles caractéristiques de Madagascar (Ammonites).* Fascicle **11** (Cenomanien). Service Géologique: Tananarive.

——— 1965a. *Atlas des fossiles caractéristiques de Madagascar (Ammonites).* Fascicle **12** (Turonien). Service Géologique: Tananarive.

——— 1965b. *Atlas des fossiles caractéristiques de Madagascar (Ammonites).* Fascicle **13** (Coniacien). Service Géologique: Tananarive.

——— 1966. *Atlas des fossiles caractéristiques de Madagascar (Ammonites).* Fascicle **14** (Santonien). Service Géologique: Tananarive.

——— 1967. Les cephalopodes Crétacés du bassin côtier de Tarfaya. *Notes et Mémoires du Service des Mines et de la Carte géologique du Maroc* **175** (1966): 7-148.

——— 1969. *Atlas des fossiles caractéristiques de Madagascar (Ammonites).* Fascicle **15** (Campanien Inférieur). Service Géologique: Tananarive.

——— 1970. Une remarquable faune à *Stoliczkaia* de la région d'Analalava (Madagasikara). *Comptes Rendus des Semaines Géologiques, Tananarive* **1968**: 27-32.

——— 1971a [1970]. *Atlas des fossiles caractéristiques de Madagascar (Ammonites).* Fascicle **16**. (Campanien moyen, Campanien supérieur). Service Géologique: Tananarive.

——— 1971b. *Atlas des fossiles caractéristiques de Madagascar (Ammonites).* Fascicle **17**. (Maestrichtien). Service Géologique: Tananarive.

——— 1977. Ammonites néocrétacées de la Nouvelle-Calédonie. *Bulletin du Bureau de Recherches géologiques et Minieres, Paris* (2, 4) **1**: 7-36.

——— 1979. Ammonites du Cretacé moyen-supérieur de l'Angola. *In*, Reconhecimento cientifico de Angola: estudos de géologia, de paléontologia e de micologia. *Publicacoes do II Centenario da Academia das Ciencias de Lisboa* **1977**: 1-75.

——— 1981. Faune albo-cénomanienne de la formation des marnes de Kazhdumi, région du Fars-Khuzestan (Iran). *Documents des Laboratoires de Géologia de la Faculte des Sciences du Lyon*, hors serie, **6**: 251-291.

Company, M. 1987. *Los ammonites del Valanginiense del sector oriental de las Cordilleras Béticas (SE de España).* Tesis doctoral, Universidad de Granada.

———, Fozy, I., Sandoval J. & Tavera, J.M. 2006. *Deitanites* n.g. and other related ammonites. Their significance within the family Holcodiscidae (Lower Cretaceous, Mediterranean region). *Neues Jahrbuch für Geologie und Paläontologie, Monatshefte* **2006**: 1-14.

Conrad, T.A. 1855. Remarks on the fossil shells from Chile, collected by Lieut. Gilliss, with descriptions of the species. *In*, Gilliss, J.M., *United States naval astronomical expedition to the southern hemisphere during the years 1849-50-'51-'52. Supplementary Papers*. A.O.P. Nicholson: Washington. U.S. House of Representatives, 33rd Congress, First Session. House Executive Document 121, Serial Set 729: 282-286.

——— 1856. Descriptions of eighteen new Cretaceous and Tertiary fossils, etc. *Proceedings of the Academy of natural Sciences of Philadelphia* **7** (1855): 265-268.

——— 1857. Description of Cretaceous and Tertiary fossils. *Report on U.S. and Mexican Boundary Survey* **1** (2): 141-174.

——— 1860. Descriptions of new species of Cretaceous and Eocene fossils of Mississippi and Alabama. *Journal of the Academy of Natural Sciences of Philadelphia* (2) **4**: 275-298.

——— 1868. Synopsis of invertebrate fossils [Cretaceous and Eocene]. *In*, Cook, G.H., *Geology of New Jersey*: 721-732. New Jersey geological Survey: Newark.

Conte, G 1989. Fossiles du plateau de l'Albion. *Les Alpes de Lumiere* **99**: 1-72.

——— 1995. Une description de *Hamiticeras carcitanense* (Matheron) (Cephalopodes, Ammonites). *Bulletin de la Société d'Etude des Sciences Naturelles de Vaucluse* **1994-1995**: 13-21.

Conte, P.G 1997. "*Parahoplites melchioris* Anthula, variété *transitans* Sinzow" découverte d'une espèce index d'ammonite dans l'Aptien du Gard. *Bulletin dela Société d'Etude des Sciences Naturelles de Nîmes et du Gard* **61**: 67-71.

——— 1999. Eléments de la faune de l'Aptien inférieur (Bédoulien) de Serviers la Baume (Gard). *Bulletin de la Société d'Etude des Sciences Naturelles de Nîmes et du Gard* **62**: 11-15.

Contreras, M.B., Martínez, A. & Gómez, M.E. 1988. Bioestratigrafía y sedimentología del Jurásico Superior en San Pedro del Gallo, Durango, México. *Revista del Instituto Mexicano del Petróleo* **20** (3): 5-49.

Cooper, M.R. 1977. Eustacy during the Cretaceous; its implications and importance. *Palaeogeography, Palaeoclimatology, Palaeoecology* **21**: 165-208.

——— 1978. Uppermost Cenomanian-basal Turonian ammonites from Salinas, Angola. *Annals of the South African Museum* **75**: 51-152.

——— 1979. Ammonite evolution and its bearing on the Cenomanian-Turonian boundary problem. *Paläontologische Zeitschrift* **53**: 120-128.

——— 1981. Revision of the late Valanginian Cephalopoda from the Sundays River Formation of South Africa, with special reference to the genus *Olcostephanus*. *Annals of the South African Museum* **83**: 147-366.

——— 1982. Lower Cretaceous (Middle Albian) ammonites from Dombe Grande, Angola. *Annals of the South African Museum* **89**: 265-314.

——— 1991. Towards a phylogenetic classification of the Cretaceous ammonites. I. Collignoniceratаceae. *Neues Jahrbuch für Geologie und Paläontologie, Abhandlungen* **182** (1): 1-19.

——— 1992. Towards a phylogenetic classification of the Cretaceous ammonites. II. Lyelliceratаceae. *Neues Jahrbuch für Geologie und Paläontologie, Abhand-lungen* **185**: 21-389.

——— 1994a. Towards a phylogenetic classification of the Cretaceous ammonites. III. Scaphitaceae. *Neues Jahrbuch für Geologie und Paläontologie, Abhandlungen* **193** (2): 165-193.

——— 1994b. Towards a phylogenetic classification of the Cretaceous ammonites. IV. Phlycticrioceratаceae. *Neues Jahrbuch für Geologie und Paläontologie, Abhandlungen* **194** (2/3): 361-378.

——— 1997. Towards a phylogenetic classification of the Cretaceous ammonites. V. Euomphaloceratidae. *Neues Jahrbuch für Geologie und Paläontologie, Abhandlungen* **203** (1): 1-21.

——— 1998. Towards a phylogenetic classification of the Cretaceous ammonites. VI. Mammitinae. *Neues Jahrbuch für Geologie und Paläontologie, Abhandlungen* **209** (2): 217-230.

——— 1999. Towards a phylogenetic classification of the Cretaceous ammonites. VIII. Turrilitidae. *Neues Jahrbuch für Geologie und Paläontologie, Abhandlungen* **213** (1): 1-18.

——— 2003. Upper Cretaceous (Turonian-Coniacian) ammonites from São Nicolau, Angola. *Annals of the South African Museum* **110**: 89-146.

——— 2015. On the Pterotrigoniidae (Bivalvia, Trigoniida): their biogeography, evolution, classification and relationships. *Neues Jahrbuch für Geologie und Paläontologie, Abhandlungen* **277** (1): 11-42.

——— 2018. The Arestoceratinae van Hoepen, 1942: a resurrected subfamily of late Albian brancoceratid ammonites, with description of a new genus. *Cretaceous Research* **88**: 302-324.

——— 2019. On the status of *Kaiparaites* Matsumoto, 1955 (Cretaceous; Ammonoidea: Kossmaticeratidae). *Neues Jahrbuch für Geologie und Paläontologie, Abhandlungen* **294/3**: 307-310.

——— & Kennedy, W.J. 1977. A revision of the Baculitidae of the Cambridge Greensand. *Neues Jahrbuch für Geologie und Paläontologie, Monatshefte* **1977**: 641-658.

——— & ——— 1979a. Uppermost Albian (*Stoliczkaia dispar* Zone) ammonites from the Angolan littoral. *Annals of the South African Museum* **77** (10): 175-308.

——— & ——— 1979b. A revision of the Puzosiidae of the Cambridge Greensand. *Neues Jahrbuch für Geologie und Paläontologie, Monatshefte* **1979**: 641-658.

——— & Owen, H.G 2011a. The phylogeny and classification of primitive Placenticeratidae (Cretaceous Hoplitina, Hoplitoidea). *Neues Jahrbuch für Geologie und Paläontologie, Abhandlungen* **260**: 331-342.

———, ——— 2011b. Evolutionary relationships within Hoplitidae (Cretaceous Ammonoidea). *Neues Jahrbuch für Geologie und Paläontologie, Abhandlungen* **261**: 337-351.

———, ——— 2011c. Evolutionary relationships among Schloenbachiidae (Cretaceous Ammonoidea: Hoplitoidea), with a revised classification of the family. *Neues Jahrbuch für Geologie und Paläontologie, Abhandlungen* **262/3**: 289-307.

———, ——— 2012. Sonneratiidae: stem Hoplitoidea and their relationships. *Neues Jahrbuch für Geologie und Paläontologie, Abhandlungen* **267/1**: 9-21.

Cope, J.C.W. 1967. The palaeontology and stratigraphy of the lower part of the Upper Kimmeridge Clay of Dorset. *Bulletin of the British Museum of natural History* (Geology) **15**: 1-79.

——— 1968. *Propectinatites*, a new Lower Kimmeridgian ammonite genus. *Palaeontology* **11** (1): 16-18.

——— 1974. Upper Kimmeridgian ammonite faunas of the Wash area and a subzonal scheme for the lower part of the Upper Kimmeridgian. *Bulletin of the Geological Survey of Great Britain* **47**: 29-37.

Coquand, H. 1859. Synopsis des animaux et des végétaux fossiles observés dans la formation Crétacée du sud-ouest de la France. *Bulletin de la Société géologique de France* (2) **16**: 945-1023.

——— 1862. Géologie et paléontologie de la region sud de la province de Constantine. *Mémoires de la Société d'Emulation de la Provence, Marseille* **2**: 1-320, 321-341 (supplement).

——— 1880. Études supplémentaires sur la paléontologie algérienne. *Bulletin de l'Academie Hippone, Bone* **15**: 1-449.

Cossmann, M. 1899. Rectifications de nomenclature. *Revue critique de Paleozoologie* **3**: 45-46.

——— 1901. Rectifications de nomenclature. *Revue critique de Paleozoologie* **5**: 58.

——— 1907. Rectifications de nomenclature. *Revue critique de Paleozoologie* **11**: 64.

Cottreau, J. 1925. Types du prodrome de paléontologie stratigraphique universelle de d'Orbigny. Vol. II. *Annales de Paléontologie* **14** (4): 133-164.

——— 1927. Types du prodrome de paléontologie stratigraphique universelle de d'Orbigny. Vol. II. *Annales de Paléontologie* **16** (2-3): 101-132.

——— 1932. Types du prodrome de paléontologie stratigraphique universelle de d'Orbigny. Vol. II. *Annales de Paléontologie* **20** (3-4): 165-184.

——— 1934. Types du prodrome de paléontologie stratigraphique universelle de d'Orbigny. *Annales de Paléontologie* **23**: 45-80.

——— 1937. Types du prodrome de paléontologie stratigraphique universelle de d'Orbigny. B. Néocomien Supérieur ou Urgonien. *Annales de Paléontologie* **26**: 17-48.

Courtiller, A. 1860. Description de trois nouvelles espèces d'ammonites du terrain Crétacé des environs de Saumur et des ammonites *carolinus* et *fleuriausianus* à l'état adulte. *Mémoires de la Société Imperiale d'Agriculture, Sciences et Arts d'Angers* **3**: 246-252.

——— 1867. Les ammonites du Tuffeau. *Annales de la Société linneenne de Maine-et-Loire, Angers* **9**: 1-8.

Courville, Ph. 1992. Les Vascoceratinae et les Pseudo-tissotinae (Ammonitina) d'Ashaka (NE Nigeria): relations avec leur environnement biosédimentaire. *Bulletin des centres de recherches exploration-production Elf-Aquitaine* **16** (2): 407-457.

Cragin, F.W. 1893. A contribution to the invertebrate palaeontology of the Texas Cretaceous. *Geological Survey of Texas, Annual Report* **4**: 139-294.

Crick, G.C. 1896. On the aperture of a baculite from the Lower Chalk of Chardstock, Somerset. *Proceedings of the malacological Society of London* **2**: 77-80.

——— 1899. Note on *Ammonites euomphalus*, Sharpe. *Geological Magazine*, n.s., (4) **6**: 251-256.

——— 1910. Note on two cephalopods [*Pachydiscus farmeryi*, n.sp., and *Heteroceras reussianum* (d'Orbigny)] from the Chalk of Lincolnshire. *Geological Magazine*, n.s., (5) **7**: 345-348.

Crickmay, C.H. 1927. On *Beudanticeras breweri* and *Coloboceras stantoni*. *American Journal of Science* (5) **13**: 503-516.

——— 1930. Fossils from Harrison Lake area, British Columbia. *Bulletin of the national Museum of Canada* **63**: 33-66.

——— 1932. A new Jurassic ammonite from the coast ranges of California. *American Midland Naturalist* **13**: 1-11.

Cuvier, G. & Brongniart, A. 1822. *Description géologique des environs de Paris*. New edition. d'Ocagne: Paris.

Dacqué, E. 1905. Beiträge zur Geologie des Somalilandes, II Teil, Oberer Jura. *Beiträge zur Paläontologie und Geologie Österraich-Ungarns Orients*, Bd. 17, Hft. 3-4: 119-160.

——— 1910. Dogger und Malm aus Ostafrica. *Beiträge zur Paläontologie und Geologie Österraich-Ungarns*, Band 23, Heft 1-2: 1-63.

Dan, G. 2010. Idoceratinae (*Presimoceras*, *Trenerites*, and *Lessiniceras* genera) from "Acanthicum Beds" of the Haghimas Mts. (the eastern Carpathians - Romania). Muzeul Olteniei Craiova. *Oltenia. Studii si comunicari, Stiintele Naturii* **26** (2): 287-295.

——— Idoceratinae (*Idoceras* and *Nebrodites* genera) from "Acanthicum Beds" of the Haghimas Mts. (the eastern Carpathians-Romania). Muzeul Olteniei Craiova. *Oltenia. Studii si comunicari, Stiintele Naturii* **26** (2): 303-314.

——— 2018. Oppelid and haploceratid ammonites from the Upper Jurassic deposits of Haghimas Mts. (the eastern Carpathians - Romania). Muzeul Olteniei Craiova. *Oltenia. Studii si comunicari, Stiintele Naturii* **34** (2): 15-22.

Darwin, C. 1859. *On the origin of species by means of natural selection, or the preservation of favoured races in the struggle for life*. John Murray: London.

——— 1871. *The descent of Man, and selection in relation to sex*. John Murray: London.

——— & Wallace, A. 1858. On the tendency of species to form varieties; and on the perpetuation of varieties and species by natural means of selection. *Journal of the Proceedings of the Linnean Society* (Zoology) **3** (9): 45-62.

Dawkins, R. 2004. *The ancestor's tale: a pilrimage to the dawn of evolution*. Weidenfeld & Nicolson: London.

Defrance, M.J.L. 1816. *In*, Defrance, M.J.L., de Blainville, H.M.D., Leach, W.E. *et al.*, *Dictionnaire des sciences naturelles... par plusieurs professeurs du Jardin du Roi, et des principales Écoles de Paris* **1-61**. Strasbourg: Paris.

DeKay, J.E. 1828. Report on several multilocular shells from the state of Delaware, with observations on a second specimen of the new fossil genus *Eurypterus*. *Annals of the Lyceum of natural History of New York* **2**: 273-279.

Delamette, M. & Kennedy, W.J. 1991. Cenomanian ammonites from the condensed deposits of the helvetic domain (western Alps, France and Switzerland). *Journal of Paleontology* **65**: 435-465.

——— & Latil, J.-L. 1989. Révision del'*Ammonites bonnetianus* Pictet, 1847 (Cephalopoda, Ammonoidea), Stoliczkaiinae de l'Albien supérieur de la Plateforme Helvétique (Haute-Savoie, S.E. France). *Revue de Paléobiologie*, special volume **3**: 55-61.

Delanoy, G. 1990a. *Camereiceras* nov. gen. (Ammonoidea, Ancyloceratina) du Barrémien Supérieur du sud-est de la France. *Geobios* **23**: 71-93.

——— 1990b. Données nouvelles sur les espèces-index *Hemihoplites feraudianus* (d'Orb., 1841) (Ammonoidea, Ancyloceratina). *Comptes Rendu Acadamie Sciences, Paris*, Series II, **310**: 661-666.

——— 1992. Les ammonites du Barrémien supérieur de Saint–Laurent de l'Escarène (Alpes-Maritimes, sud-est de la France). *Annales du Muséum d'Histoire Naturelle de Nice* **9**: 1-148.

——— 1994. Nouvelles conceptions génériques au sein de la familie Barrémien des Heteroceratidae Spath, 1922 (Ammonoidea, Ancyloceratina). *Comptes Rendu l'Académie Sciences, Paris*, Series II, **318**: 543-548.

——— 1995a. About some significant ammonites from the Lower Aptian (Bedoulian) of the Angles-Barrême area (south-east France). *Memorie Descrittive della Carta geologica d'Italia* **51**: 65-101.

——— 1995b. Les zones à Feraudianus, Giraudi et Sarasini du Barremien supérieur de la région stratotypique d'Angles-Barrême-Castellane (sud-est de la France). *Géologie Alpine Mémoire*, Hors Série, **20**: 279-319.

——— 1997. Biostratigraphie des faunes d'ammonites à la limite Barremien-Aptien dans la région d'Angles-Barrême-Castellane. Etude particulière de la famille des Heteroceratidae Spath, 1922 (Ancyloceratina, Ammonoidea). *Annales du Muséum d'Histoire Naturelle de Nice* **12**: 1-270.

———, Autran, G. & Thomel, G. 1987. Proposition d'un nouveau genre d'Ammonoidea de l'Hauterivien Supérieur: *Megacrioceras*, après révision d'une espèce méconnue de la littérature paléontologique: *Ancyloceras doublieri* Jaubert et comparaison avec les formes affines. *Comptes Rendus de l'Académie des Sciences, Paris* (2) **305**: 311-315.

———, Baudouin, C., Gonnet, R. & Bert, D. 2008. Sur les faunes d'ammonites (Cretace Inferieur) du niveau glauconieux de la carriere des Trois-Vernes, pres de Crest (Drome, sud-est de la France). *Annales des Museum d'Histoire naturelle de Nice* **23**: 11-65.

——— & Bert, D. 2006. Nouvelles unités taxonomiques de la famille des Heteroceratidae Spath, 1924). *Annales du Muséum d'Histoire naturelle de Nice* **21**: 255-295.

——— & Bulot, L.G. 1990. Révision des types et figures des collections Matheron et Reynès 3. Genres: *Acrioceras, Heteroceras, Kutatissites* et *Pseudocrioceras* (Ancyloceratina, Cephalopoda). *Mésogée* **50**: 15-21.

——— & Busnardo, R. 2007. *Anglesites* gen. nov. (Ammonoidea, Ancyloceratina), une nouveau genre d'ammonites hétéromorphs du Barrémien Supérieur du sud-est de la France. *Geobios* **40** (6): 801-807.

———, ———, Ropolo, P., Gonnet, R., Conte, G., Moullade, M. & Masse, J.-P. 1997. The "*Pseudocrioceras* beds" at La Bédoule (SE France) and the position of the Barremian/Aptian boundary in the historical Lower Aptian stratotype. *Comptes Rendus de l'Académie des Sciences, Paris* **325**: 593-599.

——— & Ebbo, L. 2008. Une nouvelle espèces d'*Heteroceras*: *H. mascarellii* sp. nov. (Ammonoidea, Ancyloceratina), dans le Barrrémien Supérieur du S-E. de la France. *Annales des Museum d'Histoire naturelle de Nice* **15**: 1-17.

——— & Féraud, P. 1995. On the genus *Spinocrioceras*, Kemper, 1973 (Ammonoidea, Ancyloceratina). *Geologische Jahrbuch* **A141**: 201-223.

———, Frau, C., Busnardo, R. & Tendil, A.J.B. 2018. Nouveaux genres d'ammonite Heteroceratidae du Barrémien supérieur de la Téthys méditerranéenne. *Annales des Museum d'Histoire naturelle de Nice* **33**: 33-45.

———, Madou, E., Penagé, E., Baraterro, X. & Lhaumet, G. 2000. *Taxyites* nov. gen. un nouveau genre d'ammonite heteromorphe de l' Aptien Inferieur du sud-est de la France (Ammonoidea, Ancyloceratina). *Annales des Museum d'Histoire naturelle de Nice* **23**: 1-9.

——— & Magnin, A. 1994. Évolution de l'enroulement des ammonites hétéromorphes avec les événements marins. *Comptes Rendu l'Académie Sciences, Paris*, Series II, **318**: 843-848.

———, ———, Sélébran M. & Sélébran J. 1991. *Moutoniceras nodosum* d'Orbigny, 1850 (Ammonoidea, Ancyloceratina): une très grande ammonite hétéromorphe du Barrémien inférieur. *Revue de Paléobiologie* **10** (2): 229-245.

———, Moreno-Bedmar, J.A., Ruiz, J.J. & Tólos Lladser, D. 2013. *Xerticeras* gen. nov., a new genus of micromorphic heteromorph ammonite (Ancyloceratina, Ancyloceratidae) from the Lower Aptian of Spain. *Carnets de Géologie [Notebooks on Geology]* **2013/02**: 89-103 (CG2013_A02).

——— & Poupon, A. 1992. Sur le genre *Lytocrioceras* Spath, 1924. (Ammonoidea, Ancyloceratina). *Geobios* **25** (3): 367-382.

———, Ropolo, P., Gonnet, R. & Ebbo, L. 2000. Le genre *Kutatissites* Kakabadze, 1970 (Ancyloceratina, Ammonoidea) dans le Barrémien supérieur/Aptien basal de la Bédoule (sud-est de la France). *Annales du Muséum d'Histoire naturelle de Nice* **15**: 19-61.

———, ———, Magnin, A., Autran, G., Poupon, A. & Gonnet, R. 1995. Sur le dimorphisme chez les Ancyloceratina (Ammonoidea) du Crétacé Inférieur. *Comptes Rendu l'Académie Sciences, Paris*, Series IIa, **321**: 537-543.

Destombes, P. 1970. Biostratigraphie des ammonites dans l'Albien inférieur et moyen, argileux, du Bassin de Paris. *Comptes Rendus hebdomadaires des Seances de l'Académie des Sciences* (D, Sciences Naturelles) **270**: 2061-2064.

———, Juignet, P. & Rioult, M. 1973. Ammonites de l'Aptien-Albien du Bec de Caux, Normandie (NW France). *Bulletin de la Société géologique de Normandie et des Amis du Museum du Havre* **61** (1971-1972): 49-106.

Diebold, F., Bengtson, P., Baudouin, C., Lees, J.A. & Bert, D. 2018. *Vocontiiceras vocontiense* nov. gen. et sp. (Collignoniceratidae, Ammonoidea) from the Upper Turonian (Cretaceous) *Prionocyclus germari* Zone of the Vocontian Basin, France. *Cretaceous Research* **88**: 375-383.

Diedrich, C. 2000. Faziesabhängige Schalenmorphologie des Grossammoniten *Puzosia dibleyi* (Spath 1922) aus dem Puzosia-Event I (Ober-Cenoman) von Europa. *Senckenbergiana lethaea* **80**: 463-483.

Diener, C. 1926. Ammonoidea neocretacea. *In*, Diener, C. (ed.), *Fossilium catalogus. I. Animalia.* Pars 29. W. Junk: Berlin.

Dimitrova, N. 1967. *Fossils of Bulgaria. IV. Lower Cretaceous Cephalopoda (Nautiloidea & Ammonoidea).* Bulgarian Academy of Sciences: Sofia [in Bulgarian, with French summary].

——— 1970. Phylogénèse des ammonites heteromorphes du Crétacé Inférieur. *Bulletin of the geological Institute, Sofia* (Paleontology) **19**: 71-110.

Dixon, F. 1850. *The geology and fossils of the Tertiary and Cretaceous formations of Sussex.* Longman, Brown, Green, & Longmans: London.

Djanélidzé, A. 1922a. *Dalmasiceras,* un sous-genre nouveau du genre *Hoplites. Bulletin de la Société géologique de France* (4) **21**: 256-274.

——— 1922b. Les *Spiticeras* du sud-est de la France. *Mémoires pour Servir a l'Explication de la Carte géologique detaillee de la France* **1922**: vi, 1-255.

——— 1926. Mélanges géologiques et paléontologiques. *Bulletin de l'Université de Tiflis* **6**: 232-266 [in Georgian with French summary].

Doguzhaeva, L. & Mikhailova, I. 1982. The genus *Luppovia* and the phylogeny of Cretaceous heteromorphic ammonoids. *Lethaia* **15**: 55-65.

——— & Mutvei, H. 1989. *Ptychoceras* - a heteromorphic lytoceratid with truncated shell and modified ultrastructure (Mollusca: Ammonoidea). *Palaeontographica* (Abt. A) **208**: 91-121.

Dommergues, J.-L. 2002. Les premiers Lytoceratoidea du nord-ouest de l'Europe (Ammonoidea, Sinémurien inférieur, France). Exemple de convergence évolutive vers les morphologies capricornes. *Revue de Paléobiologie* **21**: 257-277.

———, David, B. & Marchand, D. 1986. Les relations ontogenèse-phylogenèse: applications paléontologiques. *Geobios* **19** (3): 335-356.

Donovan, D.T. 1953. The Jurassic and Cretaceous stratigraphy and palaeontology of Traill Ø, east Greenland. *Meddelelser om Grønland* **3** (4): 1-150.

——— 1964. Stratigraphy and ammonite fauna of the Volgian and Berriasian rocks of east Greenland. *Meddelelser om Grønland* **154** (4): 1-34.

———, Callomon, J.H. & Howarth, M.K. 1981. Classification of the Jurassic Ammonitina. *In*, House, M.R. (ed.), The Ammonoidea. The evolution, classification, mode of life and geological usefulness of a major fossil group. *The Systematic Association, Special Volume* **18**: 101-155.

Donze, P. 1948. *Paraberriasella*, nouveau genre péri-sphinctidés. *Bulletin de la Société géologique de France* (5) **18** (4-5): 237-239.

——— & Énay, R. 1961. Les céphalopodes du Tithonique Inférieur de la Croix-de-Saint-Course près Chambéry (Savoie). *Travaux du Laboratoire de Géologie de la Faculté des Sciences de Lyon*, n.s., **7**: 1-236.

Douglas, J.A. 1921. Geological sections through the Andes of Peru and Bolivia. III. From the Port of Callao to the River Perene.

Quarterly Journal of the geological Society of London **77**: 246-284.

DOUVILLÉ, H. 1879. Note accompagnant la présentation de l'atlas du t. iv de l'explication de la carte géologique de France de E. Bayle et R. Zeiller. *Bulletin de la Société géologique de France* (3) **7**: 91-92.

——— 1890. Sur la classification des cératites de la Craie. *Bulletin de la Société géologique de France* (3) **18**: 275-292.

——— 1907. Sur les ammonites du Crétacé sud-américain. *Annales de la Societe royale Zoologique et Malacologique de Belgique* **41** (1906): 142-155.

——— 1909a. Sur les ammonites provenant des environs de Lima et sur une nouvelle coupure de la famille des hoplitidés: *Favrella* nov. gen. *Compte Rendu Sommaire des Seances de la Société géologique de France* **17**: 164-166.

——— 1909b. *Ammonites deslongchampsii, Ammonites bajocensis. Palaeontologia Universalis*, Centuria 2, fiche 132-133. Congres géologique International (Laval).

——— 1910a. Céphalopodes argentins. *Mémoires de la Société géologique de France* (Paléontologie, 17) **43**: 1-24.

——— 1910b. Un *Virgatites* du Caucase occidental; origine méditerranéenne de ce genre; *Ataxioceras, Pseudovirgatites* et *Virgatosphinctes. Bulletin de la Société géologique de France*, Séries 4, **18**: 730-739.

——— 1911a. Évolution et classification des pulchelliidés. *Compte Rendu Sommaire des Seances de la Société géologique de France* **8** & **9**: 85-86.

——— 1911b. *Ammonites okensis* d'Orbigny 1845. *Palaeontologia Universalis* (3), fascicle 3, fiche 213.

——— 1912a. Évolution et classification des pulchelliidés. *Bulletin de la Société géologique de France* (4) **11** (1911): 285-320.

——— 1912b. Review of "Untersuchung einiger Ammonitiden aus dem unteren Gault Mangyshlaks und des Kaukasus" by J. Sinzow, 1908. *Revue critique de Paléozoologie* **16**: 260.

——— 1917. Les terrains secondaires dans le massif du Moghara a l'est de l'isthme de Suez, d'après les explorations de M. Couyat-Barthoux. Paléontologie. *Mémoires de l'Academie des Sciences de l'Institut de France* (2) **54** (1916): 1-184.

——— 1931. Contribution à la géologie de l'Angola. Les ammonites de Salinas. *Boletim do Museu de Mineralogico e Geologico da Universidade de Lisboa* (1) **1**: 17-46.

DOYLE, J.C. 1963. A new heteromorph ammonite from the Lower Cretaceous of Yorkshire. *Palaeontology* **6**: 575-578.

DRUSHCHITS, V.V. 1953a. Study of the ontogenetic development of *Biasaloceras subsequens* (Karakasch). *Vestnik Moskovskogo Universiteta* **6**: 157-161 [in Russian].

——— 1953b. Data on the ontogeny of the new ammonite genus *Euphylloceras. Vestnik Moskovskogo Universiteta* **9**: 141-147 [in Russian].

——— 1956a. *Lower Cretaceous ammonites, Crimea and northern Caucasus.* Izdatel'stvo Moskovskogo Universiteta: Moscow [in Russian].

——— 1956b. On systematics of the family Parahoplitidae. *Scientific Reports of Moscow State University* **176**: 105-113 [in Russian].

——— 1960 Ammonites. *In*, DRUSHCHITS, V.V. & KUDRIAVTSEVA, M.P. (eds), *Atlas of the Lower Cretaceous faunas of the northern Caucasus and the Crimea*: 249-355. Vsesoiuznyi Nauchno-Issledovatel'skii Institut Prirodnykh Gasov: Moscow [in Russian].

——— 1962. Scope of the order Lytoceratida. *Bulletin de la Societe imperiale des Naturalistes de Moscow* (Geology) **37** (2.1): 154-155.

———, BOGOSLOVSKAYA, M.F. & DOGUZHAEVA, L.A. 1976. Evolution of septal necks in ammonoids. *Paleontological Journal* **1**: 41-56.

——— & DOGUZHAYEVA, L.A. 1978. The structure of the skeletal elements of early Cretaceous ammonites and their systematic importance. *Malacological Review* **11**: 134-137.

——— & KHIAMI, N. 1970. Structure of septum, protoconch wall and early whorls of some Lower Cretaceous ammonites. *Paleontological Journal* **1**: 35-47 [in Russian].

——— & KNORINA, M.V. 1980. Morphogenesis of boreal representatives of the genus *Phyllopachyceras* and general characteristics of phylloceratids (Ammonitea). *Bulletin of the Moscow University* (Series 4, Geology) **6**: 55-71 [in Russian].

——— & KUDRYAVTSEV, M.P. (eds) 1960. *Atlas of the Lower Cretaceous fauna of northern Caucasus and Crimea.* GONTI: Moscow [in Russian].

———, KVANTALIANI, I.V., KNORINA, M.V. & SHARIKADZE, M.Z. 1982. Morphogenesis of the shell of two Aptian ammonites *Cheloniceras* Hyatt and *Epicheloniceras* Casey (Ammonoidea, Cephalopoda). *Bulletin of the Moscow University* (Series 4, Geology) **3**: 51-66 [in Russian].

——— & MIKHAILOVA, I.A. 1973. On the systematic position of the Tetragonitidae (Lytoceratida, Ammonoidea). *Transactions of the Academy of Sciences, USSR* **209** (4): 953-955.

———, ——— 1974. On the systematics of the Lower Cretaceous ammonites. *Paleontological Journal* **4**: 32-44 [in Russian].

———, ———, KABANOV, G.K. & KNORINA, M.V. 1980. Morphogenesis of the *Simbirskites* group. *Paleontological Journal* **14** (1): 42-57.

———, MURATIV, E.S., & BARANOV, V.N. 1983. Morphogenesis of the shells of Middle Volgian ammonites of the genera *Virgatites, Lomonosovella, Epivirgatites. Bulletin of the Moscow State University* (Geology) **4**: 35-44 [in Russian].

———, ———, ——— 1985. Shell morphogenesis of the late Volgian ammonites *Craspedites* and *Kachpurites* from Yaroslavl Volga area. *In*, MENNER V.V. (ed.), *Fossil cephalopods*: 132-145. Nauka: Moscow [in Russian].

DUBOURDIEU, G. 1953. Ammonites nouvelles des Monts du Mellègue. *Bulletin du Service de la Carte géologique de l'Algerie* (1, Paléontologie) **16**: 1-76.

DUBOWSKI, A.G. 1971. On the ontogeny of *Craspedites subditus* Trautschold. *Materials on geology, geophysics and geochemistry of the Ukraine, Kazakhstan and Trans-Baikalia. Collection of scientific papers* **3**: 177-181 [in Russian].

DUJARDIN, M.J.L. 1837. Mémoire sur les couches du sol en Touraine, et description des coquilles de la craie et des faluns. *Mémoires de la Société géologique de France* (1) **2** (1835): 211-311.

DUMAS, E. 1876. *Statistique géologique, minéralogique, métallurgique et paléontologique du Départment du Gard.* **2**. Nîmes: Paris.

DUMORTIER, E. & FONTANNES, F. 1876. *Description des ammonites de la zone a* Ammonites tenuilobatus *de Crussol (Ardeche) et de quelques autres fossiles jurassiques nouveaux ou peu connus*. F. Savy Libraire: Lyon, Paris.

DURAND DELGA, M. 1954. À propos de *"Bochianites" superstes* Perv.: Remarques sur les ammonites droites du Crétacé Inférieur. *Compte Rendu Sommaire des Seances de la Société géologique de France* **1954**: 134-137.

DUTOUR, Y. 2005. *Biostratigraphie, évolution et renouv-ellements des ammonites de l'Aptien supérieur (Gargasien) du bassin Vocontien (sud-est de la France).* Thèse de 3° cycle, Université Claude Bernard Lyon I.

DZIK, J. 1981. Origin of the Cephalopoda. *Acta palaeontologica Polonica* **26** (2): 161-189.

——— 1994. Sexual dimorphism in the virgatitid ammonites. *Palaeopelagis, Special Publication,* Proceedings of the 3rd Pergola International Symposium **1**: 129-141.

EBBO, L. & DELANOY, G. 1999. Nouvelle acceptation du genre *Toxoceras* d'Orbigny, 1842 (Ammonoidea, Ancyloceratina). *Géologie Alpine* **75**: 81-89.

———, Poupon, A., Delanoy, G. & Gonnet, R. 1999. Nouveau données sur le genre *Lytocrioceras* Spath, 1924 dans le Barrémienne Inférieur du sud-est de la France. *Annales des Museum d'Histoire naturelle de Nice* **14**: 1-25.

Egoian, V.L. 1955. *Upper Cretaceous deposits of the south-west part of the Armenian SSR*. Academy of Sciences, Armenian SSR: Erevan [in Russian].

——— 1961a. Aptian and Albian ammonites of the northern Caucasus. *Transactions of the Geological Institute and Academy of Sciences, Georgian SSR*, Ser. geol., **12** (17): 41-77 [in Russian].

——— 1961b. Einige Cephalopodenarten aus der Unteren Kreide Zentralkarpaten. *Geologické práce, Zprávy* **21**: 81-108 [in Russian].

——— 1965a. Some ammonites of the Clansayesian of the western Caucasus. *Trudy Krasnodarskogo Filiala Vsesoiuznogo Neftegazovogo Nauchno-Issledovatel'skogo Instituta* **16**: 112-160 [in Russian].

——— 1965b. The first find of *Colchidites* in northwestern Caucasus. *Trudy Vsesoyuznogo Nauchno-Issledovatel'skii Geologicheskogo Nefte Gazov Instituta* **44**: 116-126.

——— 1969. Ammonites of the Clansayesian beds of the western Caucasus. *In*, V.L. Egoian (ed.), Geology and oil and gas resources in the western Caucasus and pre-Caucasus. *Trudy Krasnodarskogo Filiala Vsesoiuznogo Neftegazovogo Nauchno-Issledovatel'skogo Instituta* **19**: 126-188 [in Russian].

——— 1974. New family of uncoiled ammonites from the Aptian stage of the western Caucasus. *Doklady Akademii nauk SSSR* **217** (4): 939-942 [in Russian].

Eichwald, E. 1865-1868. *Lethaea rossica ou paléontologie de la Russie. Second volume. Période moyenne*. **2** (1, 1865: xxxvi, 1-832; 2, 1868: 833-1304). E. Schweizbart (E. Koch): Stuttgart.

Eldredge, N. & Gould, S.J. 1972. Punctuated equilibria: an alternative to phyletic gradualism. *In*, Schopf, T.J.M. (ed.), *Models in paleobiology*: 82-115. Freeman Cooper: San Francisco.

Emerson, B.L., Emerson, J.H., Akers, R.E. & Akers, T.J. 1994. Texas Cretaceous ammonites and nautiloids. *Texas Paleontology Series, Publication* **5**: 1-439.

Énay, R. 1959. La faune des couches à *Perisphinctes crusoliensis* (Fontannes) dans le Jura méridional. *Compte Rendu Sommaire des Séances de la Société géologique de France* **8**: 229-231.

——— 1983. Spéciation phylétique dans le genre d'ammonite téthysien *Semiformiceras* Spath, du Tithonique inférieur des chaînes bétiques (Andalousie, Espagne). *In*, Modalités, rythmes, mécanismes de l'évolution biologique. Gradualisme phylétique ou équilibres ponctués? Dijon, 1982. *Colloques Internationaux CNRS, Édition du Centre National de la Recherche Scientifique, Paris* **330** :115-123.

——— 2009. Les faunes d'ammonites de l'Oxfordien au Tithonien et la biostratigraphie des Spiti-shales (Callovien supérieur-Tithonien) de Thakkhola, Népal central. *Documents des Laboratoires de géologie, Lyon* **166**: 1-350.

———, Boughdiri, M. & Le Hegarat, G. 1998. *Durangites*, *Protacanthodiscus* et formes voisines du Tithonien superieur - Berriasien dans la Tethys mediterraneenne (SE France, Espagne, Algerie et Tunisie). *Comptes Rendus de l'Académie des sciences, Paris* **327** (Serie IIa): 425-430.

———, ———, ——— 2008a. *Kilianites* gen. nov., Himalayitidae (Ammonitina) nouveau du Berriasien de la region de Cabra (province de Cordoba, SE Espagne): descendance de *Protacanthodiscus*. *Comptes Rendus de l'Académie des sciences, Paris* **327**: 549-554.

———, ———, ——— 2008b. *Toucasiella* gen. nov., Himalayitidae (Ammonitina) nouveau du Tithonien superieur: origine de *Durangites*. *Comptes Rendus de l'Académie des sciences, Paris* **327**: 471-477.

——— & Cecca, F. 1986. Structure et évolution des populations tithoniques du genre d'ammonite téthysien *Haploceras* Zittel, 1868. *Atti Ist Conv. Fossili evoluzione ambiente. Pergola, 25-28 ottobre 1984*: 37-53.

——— & Debrand-Passard, S. 2005. Présence de *Trenerites* Sarti (Ammonitina, Idoceratinae) dans le Kimméridgien inférieur du Berry (sud du bassin de Paris). *Revue de Paléobiologie* **24** (1): 97-109.

——— & Cariou, E. 1999. Jurassic ammonite faunas from Nepal and their bearing on the palaeobiogeography of the Himalayan belt. *Journal of Asian Earth Sciences* **17**: 829-848.

———, Gallois, R. & Etches, S. 2014. Origin of the Kimmeridgian-Tithonian Boreal perisphinctid faunas: migration and descendants of the Tethyan genera *Crussoliceras* and *Garnierisphinctes*. *Revue de Paléobiologie* **33** (2): 299-377.

——— & Howarth, M.K. 2017. The Upper Oxfordian and Lower Kimmeridgian ammonite genera *Idoceras* Burckhardt, 1906, and *Subnebrodites* Spath, 1925. *Paleontological Contributions* **17**: 1-3.

———, ——— 2018. *Stevensites*, *Zittelites*, and *Gyrophylloceras*, new generic names proposed for *Stevensia* Enay, 2009, *Zittelia* Tavera Benitez, 1985, and *Gyrophyllites* Wiedmann, 1963 (Mollusca: Ammonoidea). *Paleontological Contributions* **19**: 1-2.

———, ——— 2019. Systematic descriptions of the Perisphinctoidea. Part L, revised, vol. 3B, chapter 7. *Treatise Online* **120**: 1-184.

——— & Mangold, C. 2021. Jurassic ammonites from central Saudi Arabia (Jebel Tuwaiq and adjacent areas). *Revue de Paléobiologie* **40** (1): 1-197.

Engeser, T. & Keupp, H. 2002. Phylogeny of the aptychi possessing Neoammonoidea (Aptychophora nov., Cephalopoda). *Lethaia* **34**: 79-96.

Eristavi, M. S. 1955. Lower Cretaceous fauna of Georgia. *Akademiya Nauk Gruzinskoi* SSR **6**: 1-225 [in Georgian].

——— 1957. *Correlation of the Lower Cretaceous deposits of Georgia and Crimea*. Academy of Sciences USSR Publishing House: Moscow [in Georgian].

——— 1961a. Aptian and Albian ammonites of the northern Caucasus. *Transactions of the Geological Institute and Academy of Sciences, Georgian SSR*, Ser. geol., **12** (17): 41-77 [in Georgian]

——— 1961b. Einige Cephalopodenarten aus der Unteren Kreide Zentralkarpaten. *Geologické práce, Zprávy* **21**: 81-108 [in Russian].

——— & Egoian, V.L. 1959. *Lower Cretaceous fauna of the Kafan district of the Armenian SSR*. Academy of Sciences, Armenian SSR: Erevan [in Russian].

Ershova, E.S. 1969. New finds of late Volgian ammonites on western Spitzbergen. *Uchenye Zapiski Nauchno-Issledovatel'skogo Instituta Geologii Arktiki, Paleontologiia i Biostratigrafiia* **26**: 52-69.

——— 1972a. Some Berriasian ammonoids of Spitzbergen Island. *In*, *Mesozoic deposits of Svalbard*: 82-89. NIIGA: Leningrad [in Russian].

——— 1972b. Hauterivian ammonites of Spitsbergen Island. *In*, *Mesozoic deposits of Svalbard*: 90-99. NIIGA: Leningrad [in Russian].

——— 1980. Some Early Valanginian ammonites of Spitsbergen. *In*, *Geology of the sedimentary cover of the Svalbard archipelago*: 70-80. NIIGA: Leningrad [in Russian].

Etayo Serna, F. 1968. Apuntaciones acerca de algunas amonitas interesantes del Hauteriviano y del Barremiano de la region de villa de Leiva (Boyaca, Colombia, S.A.). *Boletin de Géologia, Facultad de Ingenieria de Petroleos, Universidad Industrial de Santander, Bucaramanga* **24**: 51-70.

——— 1969. *Lenticeras baltai* Lissón en Colombia y su probable posicion zonal Santoniana (Noticia preliminar). *Géologia Colombiana* **6**: 17-29.

——— 1979. Zonation of the Cretaceous of central Colombia by ammonites. *Publicaciones geologicas Especiales del Ingeominas, Bogota* **2**: 1-186.

——— 1981. On the supposed Aptian occurrence of the ammonite genus *Neodeshayesites* in Colombia and Venezuela; with an appendix on *Neodeshayesites karsteni* (Marcou). *Geologia Norandina* **3**: 45-51.

——— 1983. The Georgian heteromorph ammonite genera *Kutatissites* and *Pseudoaustraliceras* in northwest South America. *Geologica Norandina* **7**: 3-13.

——— 1985. El límite Jurásico-Cretácico en Colombia. *Publicaciones geologicas Especiales del Ingeominas, Bogota* **16**: 1-46.

ETHERIDGE, R. 1872. Description of the Palaeozoic and Mesozoic fossils of Queensland, Appendix 1. *In*, DAINTREE, R., Notes on the geology of the colony of Queensland. *Quarterly Journal of the geological Society of London* **28**: 317-350.

——— 1880. On a collection of fossils from Bowen River coalfield and the limestone of the Fanning River, N. Queensland. *Proceedings of the Royal physical Society of Edinburgh* **5**: 263-328.

——— 1902. The Cretaceous Mollusca of South Australia and the Northern Territory. *Memoirs of the Royal Society of South Australia* **2**: 1-54.

——— 1904. On the occurrence of the genus *Ptychoceras(?)* and other additional fossils in the Cretaceous beds of the Northern Territory of South Australia. *Records of the Australian Museum* **5**: 108-115.

——— 1905. Contributions to the palaeontology of South Australia. No. 14. Cretaceous fossils from Dalhousie Springs. *Parliamentary Papers of South Australia* **71**: 13-17.

——— 1909. Lower Cretaceous fossils from the sources of the Barcoo, Ward and Nive Rivers, south central Queensland. Part 2: Cephalopoda. *Records of the Australian Museum* **7** (3): 135-165; (4): 235-240.

ETTACHFINI, M. 2004. Les ammonites du Néocomien dans l'Atlas atlantique, Maroc. Biostratigraphie, paléontologie, paléobiogéographie et paléoécologie. *Strata* **44**: 1-225.

FABRE, S. 1940. Le Crétacé Supérieur de la Basse-Provence occidentale. I. Cénomanien et Turonien. *Annales de la Faculte des Sciences de Marseille* (2) **14**: 1-355.

FALLOT, P. 1920. La faune des marnes aptiennes et albiennes de la région d'Andraitx (Majorque). *Trabajos del Museo nacional de Ciencias naturales, Madrid* (Geologica) **26**: 1-68.

——— 1921. Remarques sur le genre *Silesites*. *In*, KILIAN, C.C.C.W. *et al*., Contributions à l'étude des céphalopodes paléocrétacés du sud-est de la France. *Memoires pour Servir a l'Explication de la Carte géologique Detaillee de la France* **1920**: 207-227.

——— 1932. Note préliminaire sur le Crétacé de la region de Soria. *Bulleti de la Institucio Catalana d'Historia natural* (2) **11** (7-9) (1931): 1-6.

FALLOT, P. & TERMIER, H. 1923. Ammonites nouvelles des îles Baléares. *Trabajos del Museo Nacional de Ciencias naturales, Madrid* (Geologica) **32**: 1-85.

FANTINI-SESTINI, N. 1974. Phylloceratina (Ammonoidea) del Pliensbachiano italiano. *Rivista Italiana di Paleontologia e Stratigrafia* **80**: 193-250.

FARAUD, M. 1936. Sur deux ammonites du Turonien des environs de Bagnols-sur-Cèze. *Bulletin de la Société d'Étude des Sciences naturelles de Vaucluse, Avignon* **1**: 9-17.

——— 1940. Le genre *Vascoceras* dans le Turonien du Gard. *Bulletin de la Société d'Étude des Sciences naturelles de Vaucluse, Avignon* **3-4**: 3-24.

——— 1951. La famille des Tissotiidae dans le Turonien inférieur du Gard. *Bulletin de la Société géologique de France* (6e série) **1** (1-3): 147-157.

FATMI, A.N. 1972. Stratigraphy of the Jurassic and Lower Cretaceous rocks and Jurassic ammonites from northern areas of West Pakistan. *Bulletin of the British Museum (Natural History)*, Geology **20** (7): 299-380.

——— & KENNEDY, W.J. 1999. Maastrichtian ammonites from Balochistan, Pakistan. *Journal of Paleontology* **73**: 641-662.

——— & ZEISS, A. 1991. First Upper Jurassic and Lower Cretaceous (Berriasian) ammonites from the Sembar Formation (Belemnite shales), Windar nai Lasbela - Balochistan, Pakistan. *Geological Survey of Pakistan Memoir* **19**: 1-114.

FAVRE, F. 1908. Die Ammoniten der unteren Kreide Patagon-iens. *Neues Jahrbuch für Mineralogie, Geologie und Paläontologie, Beilage-Band* **25**: 601-647.

FIEBER, F.X. 1853. Synopsis der europäischen Orthopteren mit besonderes Rücksicht der Böhmischen Arten. *Lotos* **3**: 90-104, 115-129, 138-154, 168-176, 184-188, 201-207, 232-238, 252-261.

FINKEL, R. 1992. Eine Ammoniten-Fauna aus dem Kimmeridgium des nordöstlichen Keltiberikums (Spanien). *Profil* **3**: 227-297.

FISCHER, J.-C. 1994 (ed.), *Révision critique de la Paléontologie Française d'Alcide d'Orbigny. **I.** Céphalopodes Jurassiques.* Masson: Paris.

———& GAUTHIER, H. 2006. *Révision critique de la Paléontologie Française d'Alcide d'Orbigny. **IV** Céphalopodes Crétacés.* Backhuys Publishers: Leiden.

FISCHER, P. 1880-1887. *Manuel de conchyliologie et de paléontologie conchyliologique*, **1-11**. F. Savy: Paris.

FISCHER, R. & ZEISS, A. 1987. Zwei neue Gravesienfunde (Ammonoidea) aus dem Mittleren Kimmeridge von Wolfsburg, Niedersachse (*Praegravesia rolkei* nov. gen., nov. sp.). *Geologica et Palaeontologica* **21**: 227-235.

FISCHER VON WALDHEIM, G.F. 1830-1837. *Oryctographie du gouvernement de Moscou*. August Semen: Moscow.

FITTON, W.H. 1836. Observations on some of the strata between the Chalk and the Oxford Oolite in the south-east of England. Including Appendix A: Descriptive notes respecting the shells figured in pl. xi–xxiii, by J. de C. Sowerby. *Transactions of the geological Society of London* (2) **4**: 103-390.

FONTANNES, F. 1879. *Des ammonites des calcaires du chateau du Crussol, Ardeche (Zones a* Oppelia tenuiloba *et* Waagenia beckeri*).* Georg: Lyon.

FORBES, E. 1845. Catalogue of Lower Greensand fossils in the museum of the Geological Society, with notices of species new to Britain. *Quarterly Journal of the geological Society of London* **1**: 237-250, 345-355.

——— 1846. Report on the fossil Invertebrata from southern India, collected by Mr. Kaye and Mr. Cunliffe. *Transactions of the geological Society of London* (2) **7**: 97-174.

FÖRSTER, R. 1975. Die geologische Entwicklung von Süd-Mozambique seit der Unterkreide und die Ammoniten-Fauna von Unterkreide und Cenoman. *Geologisches Jahrbuch* **B12**: 1-324.

FŐZY, E. 1988. Tithonian ammonites (Oppeliidae, Haploceratidae and Simoceratidae) from the Transdanubian Central Range, Hungary. *Annales Universitatis Scientiarum Budapestinensis de Rolando Eötvös Nominatatae* **28**: 43-119.

——— 1995. Upper Jurassic ammonites from Seno di Guidaloca (western Sicily). *Hantkeniana* **1** (Gézcy Jubilee Volume): 131-143.

——— 2001. Campanian (late Cretaceous) cephalopods from Sumeg (Transdanubian Central Range, Hungary). *Fragmenta palaeontologica Hungarica* **19**: 25-37.

———, BULOT, L.G & DELANOY, G, 2017. New and poorly known Aptian Acrioceratidae (Ancyloceratoidea, Ammonoidea) from

Cassis (Bouches-du-Rhône). *Neues Jahrbuch für Geologie und Paläontologie Abhandlungen* **238/3**: 335-346.

——— & MELÉNDEZ, G. 1996. Oxfordian ammonites from Hungary. *GeoResearch Forum* **1-2**: 187-194.

———, REHÁKOVÁ, D., WIMBLEDON, W.A. & IFRIM C. 2016. Revision of the ammonite index species *Berriasella jacobi* Mazenot, 1939 and its consequences for the biostratigraphy of the Berriasian Stage. *Cretaceous Research* **66**: 94-114. DOI 10.1016/j.cretres.2016. 05.007.

———, ———, WIMBLEDON, W.A.P. & IFRAM, C. 2016a. Systematic palaeontology of the Perisphinctoidea in the Jurassic/Cretaceous boundary interval at Le Chouet (Drôme, France), and its implications for biostratigraphy. *Acta geologica Polonica* **66** (2): 175-204.

———, ———, ———, ——— 2016b. Upper Tithonian ammonites (Himalayitidae Spath, 1925 and Neocomitidae Salfeld, 1921) from Charens (Drôme, France). *Geologica Carpathica* **66**: 543-559.

———, DELANOY, G., MASSE, J.-P., LANTEAUME, C. & TENDIL, A.J.B. 2016. New Heteroceratidae (Ammonoidea) from the late Barremian deepening succession of Marseille (Bouches-du-Rhône, France). *Acta geologica Polonica* **66** (2): 205-225.

———, SCHERZINGER, A. & SZIVES, O. 2022. Late Jurassic–Early Cretaceous (Kimmeridgian-Barremian) ammonites of the Bakony Mountains (Transdanubian Range, Hungary). *In*, FOZY, I. (ed.), *Fauna, biostratigraphy, facies and paleotectonic evolution of the Late Jurassic-Early Cretaceous formations in the Bakony Mountains (Trans-danubian Range, Hungary)*: 243-360. Institute of Geosciences, University of Szeged, GeoLitera Publishing House.

———, WIMBLEDON, W.A.P., IFRAM, C., BULOT, L. & POHL, A. 2020. Berriasian ammonites of supposed Tethyan origin from the type 'Ryazanian', Russia: a systematic re-interpretation. *Palaeoworld*, 2020.

FRITEL, P. 1906. Sur les variations morphologiques d'*Acanthoceras milletianum,* d'Orbigny sp. *Le Naturaliste, Paris*, 28th year (2) **472**: 245-257.

FRITSCH, A. 1872. *Cephalopoden der böhmischen Kreideformation*. Commission bei Fr. Rivnác: Prague.

——— & KAFKA, J. 1887. *Die Crustaceen der böhmischen Kreideformation.* Selbstverlag: Prague.

FRY, C.H. & KEITH, S. 2004. *The birds of Africa*. *7 - Sparrows to Buntings*. Christopher Helm: London.

FUCINI, A. 1901. Cefalopodi liassici del Monte di Cetona, part 1. *Palaeontographia italica* **7**: 1-89.

——— 1923-1935. Fossili domeriani dei Dintorni di Taormina. *Palaeontographia italica* **26** (1923): 1-42; **27** (1924): 43-64; **29** (1929): 65-102; **31** (1931): 103-160; **35** (1935): 161-176.

FÜLÖP, J. 1964. Unterkreide-Bildungen (Berrias-Apt) des Bakony-Gebirges. *Geologica Hungarica, Institutum Geologicum Hungaricum, Budapest*, ser. Géologie, **13**: 1-194.

FURON, R. 1935. Le Crétacé et le Tertiaire du Sahara soudanais (Soudan, Niger, Tchad). *Archives du Museum d'Histoire naturelle, Paris* (6) **13**: 1-96.

FUTAKAMI, M. 1990. Turonian collignoniceratid ammonites from Hokkaido, Japan - stratigraphy and palaeontology of the Cretaceous in Ishikari province, central Hokkaido, Part 3. *Journal of Kawamura Gakuen Woman's University* **1**: 235-280.

——— 1992. Distribution of Turonian collignoniceratid ammonites in the Ishikari province of central Hokkaido, with special reference to the validity of regional correlations based on the zonal indices. *Journal of Kawamura Gakuen Woman's University* **1**: 217-232 [in Japanese with English abstract].

FUTTERER, K. 1894. Beiträge zur Kenntniss des Jura in Ost-Afrika. *Zeitschrift deutsche geologisches Gesellschaft, Berlin* **46** (1): 1-49.

GABB, W. M. 1864. Description of the Cretaceous fossils. *Geological Survey of California* (Palaeontology) **1**: 55-236.

——— 1869. Palaeontology of California. Cretaceous and Tertiary fossils: Descriptions of new species. *Geological Survey of California* (Palaeontology) **2**: xiv, 1-299.

——— 1877. Description of a collection of fossils, made by Doctor Antonio Raimondi in Peru. *Journal of the Academy of natural Sciences of Philadelphia* (2) **8**: 263-336.

GALE, A.S., BENGTSON, P. & KENNEDY, W.J. 2005. Ammonites at the Cenomanian-Turonian boundary in the Sergipe Basin, Brazil. *Bulletin of the Geological Society of Denmark* **52** (2): 167-191.

GANDOPADHYAY, T.K. & BARDHAN, S. 2000. Dimorphism and a new record of *Barroisiceras* de Grossouvre (Ammonoidea) from the Coniacian of Bagh, central India. *Canadian Journal of Earth Sciences* **37**: 1377-1387.

GARRIDO, A.C., PARENT, H. & BRAMBILLA, L. 2018. Tithonian stratigraphy and ammonite fauna of the Vaca Muerta Formation in Mallín Quemado (Neuquén Basin, Argentina), with remarks on the Andean chronostratigraphy. *Volumina Jurassica* **16**: 1-26.

GAUTHIER, H., BUSNARDO, R., COMBÉMOREL, R., DELANOY, G., FISCHER, J.C., GUÉRIN-FRANIATTE, S., JOLY, B., KENNEDY, W.J., SORNAY J. & TINTANT, H. 2006. Volume IV. Céphalopodes crétacés. *In*, FISCHER, J.-C. (ed.), *Révision critique de la Paléontologie Française d'Alcide d'Orbigny, incluant la réédition de l'original*. Backhuys Publishers: Leiden.

GEBHARD, G. 1983. Taxonomische Konsequenzen aus der Synonymie von *Mojsisovicsia* Steinmann und *Fallot-iceras* Parona & Bonarelli (Ammonoidea, Alb). *Paläontologische Zeitschrift* **57**: 111-119.

GEGENBAUR, C. 1878. *Grundriss der vergleichenden Anatomie. Zweite verbesserte Auflage*. W. Engelmann: Leipzig.

GEINITZ, H.B. 1849-1850. *Das Quadersandsteingebirge oder Kreidegebirge in Deutschland.* Craz & Gerlach: Freiberg.

GEMMELLARO, G.G. 1872. Sopra i Cefalopodi délia zona con *Aspidoceras acanthicum*, Opp. sp. di Burgillamuni, presso Favara, provinciadi Girgenti. *Giornale Scienze naturali ed economichi, Palermo* **8** (1): 137-159.

GÉRARD, C. & CONTAUT, H. 1936. Les ammonites de la zone à *Peltoceras athleta* du centre-ouest de la France. *Mémoires de la Société géologique de France*, n.s., **13** (29): 1-79.

GERASIMOV, P.A. 1960. New late Jurassic ammonites from the Russian Platform. *In*, *New species of ancient plants and invertebrates of the USSR*: 168-172. Gosgeoltekhisdat: Moscow [in Russian].

——— 1969. *The upper substage of the Volgian Stage of the central part of the Russian Platform*. Akademiia nauk SSSR. Ministerstvo Geologii SSSR. Mezhvedomstvennyi Stratigraficheskii Komitet: Moscow [in Russian].

———, MITTA, V.V. & KOCHANOVA, M.D. 1995. *Fossils of the Volgian Stage of central Russia.* VNIGNI: Moscow [in Russian].

GERHARDT, K. 1897a. Beitrag zur Kenntniss der Kreide-formation in Venezuela und Peru. *In*, STEINMANN, G., Beiträge zur Geologie und Paläontologie von Süd-amerika. *Neues Jahrbuch für Mineralogie, Geologie und Paläontologie, Beilage-Band* **11**: 65-117.

——— 1897b. Beitrag zur Kenntniss der Kreideformation in Columbien. *In*, STEINMANN, G., Beiträge zur Geologie und Paläontologie von Südamerika. *Neues Jahrbuch für Mineralogie, Geologie und Paläontologie, Beilage-Band* **11**: 118-208.

Gerth, H. 1921. Fauna und Gliederung des Neocoms in der argentinischen Kordillere. *Zentralblatt für Mineralogie, Geologie und Paläontologie* **1921**: 140-147.

——— 1925. La fauna Neocomiana de la Cordillera Argentina en la parte meridional de la provincia de Mendoza. *Actas de la Academia nacional de Ciencias, Argentina* **9**: 57-132.

——— 1928a. Beiträge zur Paläontologie und Stratigraphie des Lias, Doggers, Tithons und der Unterkreide in den Kordilleren im Süden der Provinz Mendoza (Argentinien). III. Die Fauna des Neokom in der argentinischen Kordillere. *Geologische Rundschau* (Steinmann Festschrift), Sonderband **17A**: 463-494.

——— 1928b. Neue Faunen der oberen Kreide mit Hippuriten aus Nordperu. *Leidsche geologische Mededeelingen* **2**: 231-241.

Geyer, O.F. 1969. The ammonite genus *Sutneria* in the Upper Jurassic of Europe. *Lethaia* **2**: 63-72.

——— 1983. Obertithonische Ammoniten-Faunen von Peru. *Zentralblatt für Geologie und Paläontologie*, Teil I, Heft 3/4: 335-350.

Geyssant, J.R. 1979. Evolution, systematique et dimorphisme d'un nouveau genre d'ammonite: *Baeticoceras* (Ammonitina, Simoceratinae) dans le Tithonique inferieur des Cordilleres betiques (Espagne). *Palaeontographica* **166**, Abt. A: 1-36.

Gignoux, M. 1921. Les pulchelliidés du paléocrétacé. *In*, Kilian, C.C.C.N. *et al*., Contributions à l'étude des céphalopodes paléocrétacés du sud-est de la France. *Mémoires pour servir a l'Explication de la Carte géologique Detaillee de la France* **1920**: 135-165.

Gill, T. 1871. Arrangement of the families of mollusks. *Smithsonian Miscellaneous Collections* **227**: xvi, 1-49.

Giovine, A.T.Y. 1950. Algunos cefalopodos del Hauterivense de Neuquen. *Revista de la Asociacion geologica Argentina* **5**: 35-76.

Glaessner, M.F. 1958. New Cretaceous fossils from New Guinea (with a contribution on a new ammonite genus by R. Casey). *Records of the South Australian Museum* **13**: 199-226.

Glazunova, A.E. 1949a. Subdivision of the Albian of Kopet Dag. *Izvestiia Turkmenskogo Filiala Akademii nauk SSR* **1**: 22-27 [in Russian].

——— 1949b. Some Upper Albian ammonites of Kopet-Dagh. *Annual of the All-Russian Palaeontological Society* **13**: 43-51 [in Russian].

——— 1952. New and little-known Upper Albian ammonites of mountainous Turkmenia. *Transactions VSEGEI. Miscellaneous paleontology and stratigraphy*: 90-106 [in Russian].

——— 1953. *Ammonites of the Aptian and Albian of Kopet-Dagh, Lesser and Great Balkhan and Mangyshlak*. Gosgeolizdat: Moscow [in Russian].

——— 1955. On a study of the Cretaceous ammonites of west-Siberian lowlands. *Materials of VNIGRI* **9**: 178-195 [in Russian].

——— 1960a. On a new Albian genus of ammonite from Transcaspia. *Informatsionnyi Sbornik Vsesoiuznogo geologicheskogo Instituta (VSEGEI)* **35**: 91-94 [in Russian].

——— 1960b. New Cretaceous representatives of the Suborder Lytoceratinae from the Asian part of the USSR. *In*, *New species of ancient plants and animals of the USSR*. Part 2: 161-163. Gosgeoltechizdat: Moscow.

——— 1960c. New Cenomanian desmoceratid from South Sakhalin. *In, New species of ancient plants and animals of the USSR.* Part 2: 176-177. Gosgeoltechizdat: Moscow.

——— 1960. On a new Albian genus of ammonite from Transcaspia. *Informational miscellania of the All-Union Geological Institute (VSEGEI)* **35**: 91-94 [in Russian].

——— 1968. New records of Cretaceous ammonites on the Russian Platform. *Transactions VSEGEI*, n.s., **129** (3): 156-165 [in Russian].

Gol'bert, A.V. & Klimova, I.G. 1983. New Valanginian ammonites from northern Siberia. *In*, Zakharov, V.A. & Nal'nyaeva, T.-I. (eds.), Mesozoic of the Soviet Arctic. *Trudy Instituta Geologii i Geofiziki ANSSSR, Sibirskoe Otdelenie Novosibirsk* **555**: 137-154 [in Russian].

Goldfuss, A. 1826-1833. *Petrefacta Germaniae, tam ea, Quae in Museo Universitatis Regiae Borussicae Fridericiae Wilhelmiae Rhenanea, serventur, quam alia quaecunque in Museis Hoeninghusiano Muensteriano aliisque, extant, iconibus et descriiptionns illustrata*. I. Arnz & Co.: Düsseldorf.

——— & Münster, G. 1841-1844. *Petrefacta Germaniae, tam ea, Quae in Museo Universitatis Regiae Borussicae Fridericiae Wilhelmiae Rhenanea, serventur, quam alia quaecunque in Museis Hoeninghusiano Muensteriano aliisque, extant, iconibus et descriiptionns illustrata. III., Divisio quinta. Molluscorum gasteropodum reliquiae.* Arnz & Co.: Düsseldorf.

González-Arreola, C., Barrágan, R. & Moreno-Bedmar, J. 2014. Olcostephanidae from the Upper Valanginian Taraises Formation (Durango state, Mexico). *Cretaceous Research* **49**: 55-62.

Gorn, N.K. 1969. *Almella almensis* - new ammonite from the Barremian deposits of the Crimea. *Bulletin of the Leningrad State University* **12**: 84-90 [in Russian].

Gottsche, C. 1878. Ueber jurassische Versteinerungen aus der argentinischen Cordillere. *Palaeontographica*, Supplement 3, Lief. 2, Heft 2: 1-50.

——— 1925. Contribuciones a la paleontología de la República Argentina. Sobre fósiles jurásicos de la cordillera Argentina (Paso del Espinacito, provincia de San Juan). Completado por los resultados paleontológicos de la obra "El Dogger en el Paso de Espinacito", por A. Tornquist. *Actas Academia Nacional de Ciencias, Córdoba* **8** (for 1924): 231-296.

Gray, J.E. 1830-32. *Illustrations of Indian zoology.* **1-2**. Treuttel, Wurtz, Treuttel & Richter: London.

——— 1855. *Catalogue of Pulmonata or air-breathing Mollusca in the collection of the British Museum. Part 1.* British Museum: London.

Grigore, D. 2009. Aulacostephanid species (*Sutneria* genus) from "Acanthicum Beds" of Ghilcos Massif (eastern Carpathians - Romania). *Oltenia, Studii si Comunicari, Stiintele Naturii, Craiova* **25**: 366-374.

——— 2011a. Kimmeridgian-Lower Tithonian ammonite assemblages from Ghilcos-Haghimas massif (eastern Carpathians - Romania). *Acta palaeontologica Romaniae* **7**: 177-189.

——— 2011b. Phylloceratids from "Acanthicum Beds" of the Haghimas Mts. (the eastern Carpathians - Romania). *Oltenia, Studii si Comunicari, Stiintele Naturii, Craiova* **27/2**: 191-202.

Gregorio, A. de. 1886. Fossiles tithoniques des Stramberg-Schichten du "Biancone" de "Roverè di Velo" de Verone. *Annales de Géologie et de Paléontologie* **3**: 1-8.

——— 1922. Monografia dei fossili titonici di Casale di sopra (Busambra). *Annales de Géologie et de Paléontologie du Marquis Antoine de Gregoria, Palermo* **36**: 1-28.

Grigorieva, O.K. 1938. A Lower Valanginian ammonite fauna from the basin of the Belaya River on the northern slope of the Caucasus (Maikop area). *In, Azov-Black Sea geological Trust. Materials on the geology and economic minerals* **1**: 83-122. Azchergeolizdat: Rostov on Don.

Grossouvre, A. de. 1894. Recherches sur la Craie Supérieure. Deuxième partie: Paléontologie. Les ammonites de la Craie Supérieure. *Mémoires du Service de la Carte géologique Detaillee de la France* **1893**: 1-264.

——— 1901. Recherches sur la Craie Supérieure. 1. Stratigraphie générale. *Mémoires pour Servir a l'Explication de la Carte géologique Detaillee de la France* **1901**: 1-1013.

GUÉRANGER, E. 1867. *Album paléontologique du département da la Sarthe.* Beauvois et Vallienne: Le Mans.

GYGI, R.A. 2000. Integrated stratigraphy of the Oxfordian and Kimmeridgian (Late Jurassic) in northern Switzerland and adjacent southern Germany. *Mémoires de l'Académie Suisse des Sciences naturelle* **104**: 1-151.

——— 2001. Perisphinctacean ammonites of the type Transversarium Zone (Middle Oxfordian, Late Jurassic) in northern Switzerland. *Schweizerische Paläontologische Abhandlungen* **122**: 1-170.

——— 2003. Perisphinctacean ammonites of the Late Jurassic in northern Switzerland. *Schweizerische Paläontologische Abhandlungen* **123**: 1-232.

HAAS, O. 1942. The Vernay Collection of Cretaceous (Albian) ammonites from Angola. *Bulletin of the American Museum of natural History* **81**: 1-224.

——— 1946. Intraspecific variation in, and ontogeny of, *Prionotropis woollgari* and *Prionocyclus wyomingensis. Bulletin of the American Museum of Natural History* **86**: 141-224.

——— 1963. *Paracanthoceras wyomingense* (Reagan) from the Western Interior of the United States and from Alberta (Ammonoidea). *American Museum Novitates* **2151**: 1-19.

——— 1964. *Plesiacanthoceras,* new name for *Paracanthoceras* Haas, 1963 *non* Furon, 1935. *Journal of Paleontology* **38**: 610.

HAECKEL, E. 1865. Ueber fossile Medusen. *Zeitschrift für wissenschaftliche Zoologie, Leipzig* **15**: 504-514.

HAGGART, J.W. 1989. New and revised ammonites from the Upper Cretaceous Nanaimo Group of British Columbia and Washington State. *Bulletin of the Geological Survey of Canada* **396**: 181-221.

——— & WARD, P.D. 1989. New Nanaimo Group ammonites (Cretaceous, Santonian-Campanian) from British Columbia and Washington State. *Journal of Paleontology* **63**: 218-227.

HAHN, W. 1963. Die Gattung *Gravesia* Salfeld (Ammonoidea) im Oberjura (Mittel- und Nord-Westeuropas. *Palaeontographica* Band **122**, Abt. A., Lieferung 1-3: 90-110.

HALANYCH, K.M., BACHELLER, J.D., AGUINALDO, A.M., LIVA, S.M., HILLIS, D.M. & LAKE, J.A. 1995. Evidence from 18S ribosomal DNA that the lophophorates are protostome animals. *Science, N.Y.* **267**: 1641-1643.

HANDSCHIN, E. 1924. Neue myrmecophile und termitophile Collembolenformen aus Süd-Amerika. *Neue Beiträge zur Systematischen Insektenkunde, Berlin* **3**: 13-19, 21-28.

HANCOCK, J.M. & KENNEDY, W.J. 1993. The high Cretaceous ammonite fauna from Tercis, Landes, France. *Bulletin de l'Institut royal des Sciences naturelles de Belgique, Sciences de la Terre* **63**: 149-209.

HANTZPERGUE, P. 1989. Les ammonites kimméridgiennes du haut-fond d'Europe occidentale. Biochronologie, systématique, evolution, paléobiogéographie. *Cahiers de Paléontologie*: 1-428.

HAUER, F.R. VON. 1854. Beiträge zur Kenntniss der Heterophyllen der österreichischen Alpen. *Sitzungsberichte der kaiserlichen Akademie der Wissenschaften, Mathematisch-Naturwissenschaftliche Klasse, Wien* **12**: 861-910.

——— 1858. Über die Cephalopoden der Gosauschichten. *Beiträge zur Paläontographie von Oesterreich* **1**: 7-14.

——— 1866. Neue Cephalopoden aus den Gosaugebilden der Alpen. *Sitzungsberichte der kaiserlichen Akademie der Wissenschaften in Wien* **53**: 300-308.

HAUG, E. 1889. Beitrag zur Kenntniss der oberneocom Ammonitenfauna der Puezalpe bei Corvara (Südtirol). *Beiträge zur Paläontologie und Geologie Oesterreich - Ungarns und des Orients* **7** (for 1888): 193-231.

—— 1898. Rectification de nomenclature. *Revue critique de Paléozoologie* **2**: 78.

——— 1900. Céphalopodes. *Revue critique de Paléozoologie* **4**: 19-24, 78-87.

——— 1910. Période Crétacée. *In*, HAUG, E., *Traité de géologie.* **2**. Les périodes géologiques (2): 1153-1395. Colin: Paris.

HAYAKAWA, H. 1998. Keeled heteromorph ammonite, *Horotatceras tatsuyai* gen. et sp. nov. from the Upper Cretaceous of Hokkaido, Japan. *Bulletin of the the Mikasa City Museum* **2**: 41-45.

HAYASAKA, I. & FUKADA, A. 1951. On the ontogeny of *Barroisiceras minimum* Yabe from the upper ammonite bed in Hokkaidô. *Journal of the Faculty of Science, Hokkaido University* (4, Geology and Mineralogy) **7**: 324-330.

HE, G. & XIA, J. 1984. Early early Cretaceous ammonites from Bangoin, northern Xizang. *Acta palaeontologica Sinica* **23** (6): 657-671.

HÉBERT, E. & MUNIER-CHALMAS, E.C.P.A. 1875. Description du terrain Crétacé Supérieur du bassin d'Uchaux. Appendice paléontologique. *Annales des Sciences géologiques, Paris* **6**: 113-132.

HEER, O. 1852. Ueber die Lias-Insel im Aargau. *In*, HEER, O. & ESCHER VON DER LINTH, A., *Zwei geologische Vorträge, Gehalten im März* **1852**: 1-15.

HÉGARAT, G. LE. 1964. Deux ammonites nouvelles dans le Berriasien de Berrias (Ardeche). *Travaux du Laboratoire de géologie de la Faculté des Sciences de Lyon,* n. s., **11**: 95-101.

——— 1965. Presentation d'un neotype de *Neocosmoceras (Euthymiceras) euthymi* Pictet, ammonite berriasienne. *Travaux du Laboratoire de géologie de la Faculté des sciences de Lyon*, n.s., **12**: 125-128.

——— 1971. Perisphinctidae et Berriasellidae de la limite Jurassique-Crétacé. Genres nouveaux et révision critique de quelques définitions taxinomiques antérieures. *Comptes Rendus des Seances de l'Academie des Sciences, Paris* 273 (D), **10**: 850-853.

——— 1973. Le Berriasien du sud-est de la France. *Documents des Laboratoires de Géologique, Faculte des Sciences, Lyon* **43** (1971): 1-576.

——— & REMANE, Y. 1968. Tithonique superieur et Berriasien de l'Ardeche et de l'Herault. Correlation des ammonites et des calpionelles. *Geobios* **1**: 7-70.

HENDERSON, R.A. 1970. Ammonoidea from the Mata Series (Santonian-Maastrichtian) of New Zealand. *Special Papers in Palaeontology* **6**: v, 1-82.

——— 1973. Clarence and Raukumara Series (Albian-Santonian) Ammonoidea from New Zealand. *Journal of the Royal Society of New Zealand* **3**: 71-123.

——— & KENNEDY, W.J. 2002. Occurrence of the ammonite *Goodhallites goodhalli* (J. Sowerby) in the Eromanga Basin, Queensland: an index species for the late Albian (Cretaceous). *Alcheringa* **26**: 233-247.

———, ——— & MACNAMARA, K.J. 1992. Maastrichtian heteromorph ammonites from the Carnarvon Basin, Western Australia. *Alcheringa* **16**: 133-170.

——— & MACNAMARA, K.J. 1985. Maastrichtian non-heteromorph ammonites from the Miria Formation, Western Australia. *Palaeontology* **28**: 35-88.

——— & MCKENZIE, E.D. 2002. *Idanoceras*, a new ammonite heteromorph genus from the Late Albian of eastern Australia. *Journal of Paleontology* **76**: 906-909.

HENGSBACH, R. 1981. Zur Evolution der Grundpläne von Cephalopoda, Ammonoidea und Neoammonoidea. Teilt. Zur Evolution des Ammoniten-Zweiges der Cephalopoda. *Zoologische Beiträge* (N.F.) **27**: 223-266.

HENNIG, E. 1932. *Wesen und Wege der Paläontologie.* Gebrüder Borntraeger: Berlin.

HERRICK, C.L. & JOHNSON, D.W. 1900. The geology of the Albuquerque sheet. *Bulletin of the Scientific Laboratories of Denison University* **11**: 175-240.

HIRANO, H. 1979. Importance of transient polymorphism in systematics of Ammonoidea. *Gakujutsu Kenkyu School of Education, Waseda University*, Biology and Geology **28**: 35-42.
———, OKAMOTO, T. & HATTORY, K. 1990. Evolution of some late Cretaceous desmoceratine ammonoids. *Transactions and Proceedings of the Palaeontological Society of Japan*, n.s., **157**: 382-411.
HITZEL, E. 1902. Sur les fossiles de l'étage Albien receullis par M.A. Guebhard dans la région d'Escragnolles (Alpes-Maritimes). *Bulletin de la Société géologique de France* (4) **2**: 874-880.
HOEDEMAEKER, P.J. 1981. The Jurassic-Cretaceous boundary near Miravetes (Caravaca, SE Spain); arguments for its position at the base of the *Occitanica* Zone. *Cuadernos de Geologia* **10**: 235-247.
——— 1982. Ammonite biostratigraphy of the uppermost Tithonian, Berriasian, and lower Valanginian along the Río Argos (Caravaca, SE Spain). *Scripta Geologica* **65**: 1-81.
——— 1995. Ammonite distribution around the Hauterivian-Barremian boundary along the Río Argos (Caravaca, SE Spain). *Géologie Alpine, Mémoire H.S* **20** (for 1994): 219-277 .
———, JANSSEN, N.M.M., CASELLATO, C.E., GARDIN S., REHÁKOVÁ, D. & JAMRICHOVÁ, M. 2016. Jurassic/Cretaceous boundary in the Río Argos succession (Caravaca, SE Spain). *Revue de Paléobiologie* **35** (1): 111-247.
HOEPEN, E.C.N. VAN. 1920. Description of some Cretaceous ammonites from Pondoland. *Annals of the Transvaal Museum* **7**: 142-147.
——— 1921. Cretaceous Cephalopoda from Pondoland. *Annals of the Transvaal Museum* **8**: 1-48.
——— 1931. Die krytfauna van Soeloeland. 2. Voorlopige beskrywing van enige Soeloelandse ammoniete (1). *Lophoceras, Rhytidoceras, Drepanoceras* en *Deiradoceras*. *Palaeontologiese Navorsing van die nasionale Museum, Bloemfontein* **1** (2): 37-54.
——— 1941. Die gekielde ammoniete van die Suid-Afrikaanse Gault. I. Dipoloceratidae, Cechenoceratidae en Drepanoceratidae. *Paleontologiese Navorsing van die nasionale Museum, Bloemfontein* **1** (3): 55-90.
——— 1942. Die gekielde ammoniete van die Suid-Afrikaanse Gault. II. Drepanoceratidae, Pervinquieridae, Arestoceratidae, Cainoceratidae. *Paleontologiese Navorsing van die nasionale Museum, Bloemfontein* **1** (4): 91-157.
——— 1944. Die gekielde ammoniete van die Suid-Afrikaanse Gault. III. Pervinquieridae en Brancoceratidae. *Paleontologiese Navorsing van die Nasionale Museum, Bloemfontein* **1** (5): 159-198.
——— 1946a. Die gekielde ammoniete van die Suid-Afrikaanse Gault. IV. Cechenoceratidae, Dipoloceratidae, Drepanoceratidae, Arestoceratidae. *Paleontologiese Navorsing van die nasionale Museum, Bloemfontein* **1** (6): 199-260.
——— 1946b. Die gekielde ammoniete van die Suid-Afrikaanse Gault. V. Monophyletism or polyphyletism in connection with ammonites of the South African Gault. *Paleontologiese Navorsing van die nasionale Museum, Bloemfontein* **1** (7): 261-271.
——— 1951a. Die gekielde ammoniete van die Suid-Afrikaanse Gault. VI. The so-called old mouth-edges of the ammonites shell. *Paleontologiese Navorsing van die nasionale Museum, Bloemfontein* **1** (8): 273-284.
——— 1951b. Die gekielde ammoniete van die Suid-Afrikaanse Gault. VII. Pervinquieridae, Arestoceratidae, Cainoceratidae. *Paleontologiese Navorsing van die nasionale Museum, Bloemfontein* **1** (9): 285-344, 388-422.
——— 1951c. A remarkable desmoceratid from the South African Albian. *Paleontologiese Navorsing van die nasionale Museum, Bloemfontein* **1** (10): 345-349.
——— 1955a. New and little-known ammonites from the Albian of Zululand. *South African Journal of Science* **51**: 355-361.
——— 1955b. Turonian-Coniacian ammonites from Zululand. *South African Journal of Science* **51**: 361-377.
——— 1955c. A new family of keeled ammonites from the Albian of Zululand. *South African Journal of Science* **51**: 377-382.
——— 1957. The deposits on the Umsinene River. *Compte Rendu, CCTA, Conference géologique, Tananarive* **2**: 349-350.
——— 1965. The Peroniceratinae and allied forms of Zululand, South Africa. *Geological Survey of South Africa, Memoir* **55**: 1-70.
——— 1968a. New and little known Zululand and Pondoland ammonites. *Annals of the geological Survey, South Africa* **4** (1965): 158-181.
——— 1968b. New ammonites from Zululand. *Annals of the geological Survey, South Africa* **4** (1965): 183-191.
HÖFFMAN, R. 2010. New insights on the phylogeny of the Lytoceratoidea (Ammonitina) from the septal lobe and its functional interpretation. *Révue de Paléobiologie, Geneve* **29** (1): 1-156.
——— 2017. Part L, Revised, Volume 3B, Chapter 3: Systematic descriptions of the Lytoceratoidea. *Treatise Online* **70**: 1-34.
———, KEUPP, H. & WIESE, F. 2009. The systematic position of the Lower Cretaceous heteromorphic ammonite *Pictetia* Uhlig, 1883. *Paläontologische Zeitschrift* **83**: 521-531.
HÖLDER, H. 1955. Die Gattung *Taramelliceras* im sudwestdeutschen Unter- und Mittelmalm. *Palaeontographica* (Abt. A) **106**: 37-153.
——— & ZIEGLER, B. 1959. Stratigraphische und faunistische Beziehungen im Weißen Jura (Kimmeridgien) zwischen Süddeutschland und Ardeche. *Neues Jahrbuch für Geologie und Paläontologie, Abhandlungen* **108** (2): 150-214.
HOLLIS, J.D. 1971. Occurrence of the ammonite *Ptychoceras adpressum* (J. Sowerby) in the Upper Albian of Kent. *Palaeontology* **14** (4): 592-594.
HONNORAT-BASTIDE, E.F. 1890. Sur une forme ou peu connues de Céphalopodes du Crétacé inférieur des Basses-Alpes - *Ammonites julianyi* nov. sp. Digne, Réveil des Alpes du 13 avril 1889, 8p. in-12, et A.F.A.S., Limoges, 1890, 1" partie: 189; 2° partie: 367.
HOUDARD, M.J. 1938. Notes sur le Portlandien de l'Yonne et sur quelques especes neocomiennes de l'est du Bassin de Paris. *Memoires de l'Academie des sciences, arts et belles-letteres de Dijon*, Annee **1937**: 106-124.
HOURCQ, V. 1949. Paléontologie de Madagascar. XXVIII. Sur quelques ammonites du Sénonien. *Annales de Paléontologie* **35**: 87-117.
HOUŠA, V. 1965. Podtrída Ammonoidea Zittel, 1884 - Amoniti. *In*, ŠPINAR, Z. (ed.), *Systematická paleontologie bezobratlých*. Academia, Nakladatelství.
——— 1967. *Lewesiceras* Spath (Pachydiscidae, Ammonoidea) from the Turonian of Bohemia. *Sbornik geologickych Ved, paleontologie, Rada P. svazek* **9**: 7-50.
——— & DE LA NUEZ, M.L. 1973. Las faunas de ammonites del Tithoniano y del Infracretaceo de Cuba. *Acadamie de Ciencias de Cuba, Actas Instituto de Geologia* **3**: 18-19.
——— & VAŠÍCEK, N. 2004. Ammonoidea of the Lower Cretaceous deposits (late Berriasian, Valanginian, early Hauterivian) from Štramberk, Czech Republic. *Geolines* **18**: 7-57.
HOWARTH, M.K. 1958. Upper Jurassic and Cretaceous ammonite faunas of Alexander Land and Graham Land. *Falkland Islands Dependencies Survey, Scientific Reports* **21**: 1-16.
——— 1962. The Yorkshire type ammonites and nautiloids of Young and Bird, Phillips, and Martin Simpson. *Palaeontology* **5**: 93-136.
——— 1965. Cretaceous ammonites and nautiloids from Angola. *Bulletin of the British Museum (Natural History)* (Geology) **10**: 335-412.

——— 1992. Tithonian and Berriasian ammonites from the Chia Gara Formation in northern Iraq. *Palaeontology* **35** (3): 597-655.

——— & MORRIS, N.J. 1998. The Jurassic and Lower Cretaceous of Wadi Hajar, southern Yémen. *Bulletin of the Natural History Muséum, London,* Geology Series, **54** (1): 1-32.

HOWLETT, P.J. 1989. Late Jurassic-early Cretaceous cephalopods of eastern Alexander Island, Antarctica. *Special Papers in Palaeontology* **41**: 1-72.

HUMPHREY, W.E. 1949. Geology of the Sierra de Los Muertos area, Mexico (with descriptions of Aptian cephalopods from the La Peña Formation). *Geological Society of America Bulletin* **60**: 89-176.

HUPÉ, L.H. 1854. Mollusca, Echinoderma, Hydrozoa, Foraminifera and Infusoria of Chile. *In*, GAY, G., *Fauna chilena, historia fisica y politica de Chile* (Zoologia) **8**: 7-41.

HUPSCH, J.W.C.A.F. 1768. *Neue in der Naturgeschichte des Niederdeutschlands gemachte Entdeckungen einiger selten und wenig bekannten versteinerten Schaalthiere*. Metternichisschen Buchhandlung: Frankfurt, Leipzig.

HYATT, A. 1867. The fossil cephalopods of the Museum of Comparative Zoology. *Bulletin of the Museum of comparative Zoology, Harvard* **5**: 71-102.

———1875. The Jurassic and Cretaceous ammonites collected in South America by Prof. James Orton, with an appendix upon the Cretaceous ammonites of Prof. Hartt's collection. *Proceedings of the Boston Society of natural History* **17**: 365-372.

——— 1884. Genera of fossil cephalopods [2nd part]. *Proceedings of the Boston Society of natural History* **22** (1883): 73-320.

——— 1889. Genesis of the Arietidae. *Smithsonian Contributions to Knowledge* **673**: xi, 1-238.

——— 1893. Carboniferous cephalopods. Second paper. *Geological Survey of Texas, Annual Report* **4**: 377-474.

——— 1894. Phylogeny of an acquired characteristic. *Proceedings of the American philosophical Society* **32**: 349-647.

——— 1900. Cephalopoda. *In*, ZITTEL, K.A., *Textbook of palaeontology*: 502-592. 1st English ed. (translated by C.R. Eastman). Macmillan: London, N.Y.

——— 1903. Pseudoceratites of the Cretaceous. *Monographs of the United States geological Survey* **44**: 1-352.

IBA, Y. 2009. An early Albian Arctic-type ammonite *Arcthoplites* from Hokkaido, northern Japan, and its paleobiogeographic and paleoclimatological implications. *Journal of Asian Earth Sciences* **34**: 46-50.

ICZN (2000). *The code online*.

IFRIM, C. & STINNESBECK, W. 2007. Early Turonian ammonites from Vallecillo, north-eastern Mexico: taxonomy, biostratigraphy and palaeobiogeographical significance. *Cretaceous Research* **28**: 642-664.

———, ——— & LOPEZ-OLIVA, J.G. 2004. Maastrichtian cephalopods from Cerralvo, north-eastern Mexico. *Palaeontology* **47**: 1575-1627.

———, ——— & SCHAFHAUSER, A. 2005. Maastrichtian shallow-water ammonites of northeastern Mexico. *Revista Mexicana de Ciencias Geológicas* **22**: 48-64.

IGOLNIKOV, A.E. 2006. Revision of *Surites (Caseyiceras) subanalogus* Schulgina, 1972 and stratigraphic implications. *News of Paleontology and Stratigraphy* **9**: 97-103 [in Russian].

——— 2007. A new species of the genus *Boreophylloceras* Alekseev et Repin, 1998 (Ammonitida) from the Berriasian *kochi* Zone of north-central Siberia. *Palaeontological Journal* **4** (2): 128-131.

———2012. *Craspedites (Vitaliites?) sachsi,* a new Boreal Berriasian ammonite species from the north of eastern Siberia (Nordvik Peninsula). *Paleontological Journal* **46** (1): 12-15.

ILOVAISKY, D. 1917. Les ammonites du Jurassique supérieur du pays de Liapine. *Ouvrages de la Section Géologique de la Société Impériale des Amis des Sciences Naturelles de Moscou*, Ouvrage I, Livraisons 1-2: 1-180 [in Russian].

——— 1924. *Pavlovia*, un nouveau genre d'ammonites. *Bulletin of Moscow Society of Naturalists*, Sect. Geol., **2** (4): 329-363.

——— & FLORENSKY, K.R. 1941. Les ammonites du Jurassique supérieur des bassins des rivières Oural et Ilek. *Contribution a la Connaissance de la Géologie de l'USSR*, n.s., **1** (5): 7-195.

ILYIN, V.D. 1957. *Arkhangelkiceras* gen. nov. from the Upper Cretaceous deposits of western Uzbekistan. *Transactions of the Academy of Sciences, USSR* **113**: 425-427 [in Russian].

——— 1958. A new genus of ammonite from the Senonian of southwestern Uzbekistan. *Doklady Akademii nauk SSSR* **121**: 727-729 [in Russian].

——— 1959. Stratigraphy of the Upper Cretaceous deposits of west Uzbekistan and adjacent regions of Turkmenia. *Trudy Vsesoiuznogo Neftianogo Nauchno-Issledovatel'skogo geologicheskogo Neftegazovogo Instituta* **23**: 181-222 [in Russian].

——— 1961. Stratigraphy and fauna of the Albian Stage of western Uzbekistan and adjacent areas of eastern Turkmenia. New data on geology and oil-and-gas content of the Middle Asia. *Transactions VNIGNI* **35**: 36-62.

——— 1975. The ammonite family Placenticeratidae Hyatt in Upper Cretaceous deposits of Central Asia. *In*, BEZNOSOVA, N.B. (ed.), New data on the stratigraphy of Mesozoic deposits of oil bearing regions of southern USSR. *Trudy Vsesoiuznogo Neftianogo Nauchno-Issledovatel'skogo geologicheskogo Neftegazovogo Instituta* **171**: 154-200 [in Russian].

——— 2000. Upper Cretaceous ammonites of the central regions of Middle Asia (Hoplitaceae, Acanthocerataceae, Tissotiaceae, Desmocerataceae). *Bulletin of CF VNIGNI* **4**: 1-157 [in Russian].

IMLAY, R.W. 1938. Ammonites of the Taraises Formation of northern Mexico. *Geological Society of America Bulletin* **49**: 539-602.

——— 1942. Late Jurassic fossils from Cuba and their economic significance. *Geological Society of America Bulletin* **53**(10): 1417-1478.

——— 1956. Stratigraphic and geographic range of the early Cretaceous ammonite *Homolsomites*. *Journal of Paleontology* **30**: 1143-1146.

——— 1957. New genera of Early Cretaceous ammonites from California and Oregon. *Journal of the Washington Academy of Sciences* **47**: 275-277.

——— 1958. *Hertleinites,* new name for the Cretaceous ammonite *Hertleinia,* preoccupied. *Journal of Paleontology* **32**: 1032.

——— 1959. New genera of Early Cretaceous (Albian) ammonites from Alaska. *Journal of Paleontology* **33**: 179-185.

——— 1960a. Ammonites of Early Cretaceous age (Valanginian and Hauterivian) from the Pacific Coast states. *United States Geological Survey Professional Papers* **334-F**: 167-228.

——— 1960b. Early Cretaceous (Albian) ammonites from the Chitina Valley and Talkeetna Mountains, Alaska. *United States geological Survey Professional Papers* **354-D**: 87-114.

——— 1961. Characteristic Lower Cretaceous megafossils from northern Alaska. *United States geological Survey Professional Papers* **335**: 1-74.

——— & JONES, D.L. 1970. Ammonites from the *Buchia* zones in northwestern California and southwestern Oregon. *United States geological Survey Professional Papers* **647-B**: iv, 1-60.

Immel, H. 1987. Die Kreideammoniten der nördlichen Kalkalpen. *Zitteliana* **15**: 3-163.

——— & Seyed-Emami, K. 1985. Die Kreideammoniten des Glaukonitkalkes (O. Alb-O.Cenoman) des Kolah-Qazi-Gebirges südöstlich von Esfahan (Zentraliran). *Zitteliana* **12**: 87-137.

———, Klinger, H.C. & Wiedmann, J. 1982. Die Cephalopoden des unteren Santon der Gosau von Brandenberg/Tirol, Österreich. *Zitteliana* **8**: 3-32.

Indans, J. 1954. Eine Ammonitenfauna aus dem Unter-tithon der argentinischen Kordillere in Süd-Mendoza. *Palaeontographica* **105** (A): 96-132.

Ivanov, A.N. (ed.) 1979a. Late Mesozoic cephalopods of the Upper Volga area. *Inter-college collection of scientific papers* **183**: 1-64 [in Russian].

Ivanov, A.N. 1979b. On the results of the revision of *Laugeites stschurowskii* (Nikitin) species. *In*, Ivanov, A.N. (ed.), Late Mesozoic cephalopods of the Upper Volga area. *Inter-college collection of scientific papers, Issue* **183**: 1-3. Pedagogical University: Yaroslavl [in Russian].

——— & Aristov, V.N. 1969. New ammonite genera from the Lower Cretaceous in the vicinity of Yaroslavl and the origin of simbirskitids. *Biulleten Moskovskogo Obshchestva Ispytatelei Prirody, Otdel Geologicheskii* **44** (6): 84-97 [in Russian].

——— & Idakieva, V. 2013. Lower Aptian ammonite biostratigraphy and potential for further studies of OAE1a in Bulgaria. *Cretaceous Research* **39**: 47-69.

Ivanov, M. 1993. Uppermost Maastrichtian ammonites from the uninterrupted Upper Cretaceous-Paleogene section at Bjala (east Bulgaria). *Geologica Balcanica* **23**: 54.

——— 1995. Amoniti ot gorniia mastrikht v razrezite okolo gr. Biala (Iztochna Bulgariia). *Spisanie na Bulgarskoto Geologichesko Druzhestvo* **56**: 57-73.

Jacob, C. 1905. Étude sur les ammonites et sur l'horizon stratigraphique du gisement de Clansayes. *Bulletin de la Société géologique de France* (4) **5**: 339-432.

——— 1907. Études paléontologiques et stratigraphiques sur la partie moyenne des terrains Crétacés dans les Alpes françaises et les régions, voisines. *Annales de l'Universite de Grenoble* **19** (2): 221-534.

——— 1908a. Études paléontologiques et stratigraphiques sur la partie moyenne des terrains Crétacés dans les Alpes françaises et les régions, voisines. *Travaux du Laboratoire de Géologia de la Faculte des Sciences de l'Universite de Grenoble* **8** (2): 280-590.

——— 1908b. Études paléontologiques et stratigraphiques sur la partie moyenne des terrains crétacés dans les Alpes françaises et les régions, voisines. *Bulletin de la Société de Statistique des Sciences naturelles et des Arts industriels de Departement de l'Isere, Grenoble* (4) **10**: 201-514.

——— 1908c. Études sur quelques ammonites du Crétacé moyen. *Mémoires de la Société géologique de France* (Paléontologie) **15** (38) (for 1907): 1-64.

——— & Tobler, A. 1906. Étude stratigraphique et paléontologique du Gault de la vallée de la Engelberger-Aa. *Mémoires de la Société paléontologique de Suisse* **33**: 3-26.

Jagt, J.W.M. 1989. Ammonites from the Early Campanian Vaals Formation at the CPL quarry (Haccourt, Liège, Belgium) and their stratigraphic implications. *Meded-elingen van de Rijks Geologische Dienst* **43**: 1-33.

——— 2002. Late Cretaceous ammonite faunas of the Maastrichtian type area. *In*, Summesberger, H., Histon, K. & Daurer, A. (eds), Cephalopods - present and past. *Abhandlungen der Geologischen Bundesanstalt Wien* **57**: 509-522.

———, Kennedy, W.J. & Machalski, M. 1999. Giant scaphitid ammonites from the Maastrichtian of Europe. *Bulletin de l'Institut Royal des Sciences Naturelles de Belgique (Sciences de la Terre)* **69**: 133-154.

Jahn, J.J. 1891. Ein Beitrag zur Kenntnis der Fauna der Priesener Schichten der böhmischen Kreideformation. *Jahrbuch der kaiserlich-koeniglichen Geologischen Reichsanstalt, Wien* **41**: 179-186.

Jakowlew, B. 1875. Hemiptera Homoptera of the Russian fauna. *Bulletin de la Societe Imperiale des Naturalistes de Moscou* **49**: 248-285 [in Russian].

Jaubert, J. 1854. Description d'une espèce nouvelle d'*Ancyloceras* de l'étage Néocomien de Castellane (B.-A.). *Annales de la Societe nationale d'Agriculture, d'Histoire naturelle et des Arts utiles de Lyon* (2) **7**: 326-329.

Jaworski, E. 1926a. Bemerkungen zur Paläontologie und Stratigraphie des argentinischen Oberlias und Unter-dogger. *Centralblatt für Mineralogie, Geologie und Paläontologie*, Abteilung B: 273-285.

——— 1926b. Beiträge zur Paläontologie und Stratigraphie des Lias, Doggers, Tithons und der Unterkreide in den Kordilleren im Süden der Provinz Mendoza (Argentinien). Teil I. Lias und Dogger. *Geologische Rundschau* **17**: 373-428.

Jeannet, A. 1908. Sur un genre d'ammonites nouveau de l'Albien du Jura; *Jacobella lugeoni* nov. sp. *Bulletin de la Societe vaudoise des Sciences naturelles, Lausanne* **44**: 205-212.

——— 1951. Stratigraphie und Paläontologie des oolithischen Eisenerzlagers von Herznach und seiner Umgebung. *Beiträge zur Geologie der Schweiz*, Geotechnische Serie **13** (5): xiii, 1-240.

Jeletzky, J.A. 1964. Illustrations of Canadian fossils. Lower Cretaceous marine index fossils of the sedimentary basins of western and arctic Canada. *Geological Survey of Canada Paper* **64-11**: 1-100.

——— 1965. *Thorsteinssonoceras:* a new craspeditid ammonite from the Valanginian of Ellesmere Island, Arctic archipelago. *Geological Survey of Canada Bulletin* **120**: 1-16.

——— 1966. Upper Volgian (latest Jurassic) ammonites and buchias of Arctic Canada. *Geological Survey of Canada Bulletin* **128**: 1-49.

——— 1973. Biochronology of the marine boreal latest Jurassic, Berriasian and Valanginian in Canada. *In*, Casey, R. & Rawson, P.F. (eds), The Boreal Lower Cretaceous. *Geological Journal, Special Issue* **5**: 41-80.

——— 1979. Eurasian craspeditid genera *Temnoptychites* and *Tollia* in the Lower Valanginian of Sverdrup Basin, district of Franklin, with comments on taxonomy and nomenclature of Craspeditidae. *Geological Survey of Canada Bulletin* **299**: 1-89.

——— 1980. New or formerly poorly known, biochronologically and paleobiogeographically important gastroplitinid and cleoniceratinid (Ammonitida) taxa from Middle Albian rocks of mid-western and Arctic Canada. *Geological Survey of Canada Paper* **79-22**: 1-63.

——— 1986. *Pseudoeuryptychites:* a new polyptychitinid ammonite from the Lower Valanginian of the Canadian and Eurasian Arctic. *Current Research B, geological Survey of Canada Paper* **86-1B**: 351-361.

——— & Kemper, E. 1988. Comparative paleontology and stratigraphy of Valanginian Polyptychitinae and Simbirskitinae in Sverdrup Basin (Arctic Canada) and Lower Saxony Basin (northwest Germany). *Geological Survey of Canada Bulletin* **337**: 1-335.

——— & Stelck, C.R. 1981. *Pachygrycia,* a new *Sonneratia*-like ammonite from the Lower Cretaceous (earliest Albian?) of northern Canada. *Geological Survey of Canada Paper* **80-20**: 1-25.

——— & Waage, K.M. 1978. Revision of *Ammonites conradi* Morton, 1834, and the concept of *Discoscaphites* Meek, 1870. *Journal of Paleontology* **52**: 1119-1132.

JIMBO, K. 1894. Beiträge zur Kenntniss der Fauna der Kreide-formation von Hokkaido. *Paläontologische Abhandlungen*, n.s., **2** (heft 3): 147-194 (1-48).
JOHNSON, D.W. 1903. The geology of the Cerrillos Hills, New Mexico. Part 2: Palaeontology. *Columbia University School of Mines Quarterly* **24**: 173-246.
JOLY, B. 1975. Les Phyllocerataceae de l'est et du sud-est du bassin Parisien au Lias. *Bulletin de la Société géologique de France*, Séries 7, **17**: 1084-1091.
——— 1977. Les Phylloceratidae malgaches au Jurassique. Généralités sur quelques Phylloceratidae et quelques Juraphyllitidae. *Documents des Laboratoires de Géologie de la Faculté des Sciences de Lyon* **67**: 1-471.
——— 1993. Les Phyllocerataceae malgaches au Crétacé (Phylloceratina, Ammonoidea). *Documents des Laboratoires de Géologie de la Faculté des Sciences de Lyon* **127**: 1-17.
——— 1998. Les Phyllocerataceae albiens du département de l'Aube. *Bulletin Annuel de l'Association géologique Auboise* **19**: 33-59.
——— 2000a. Les Juraphyllitidae, Phylloceratidae, Neophylloceratidae (Phyllocerataceae, Phylloceratina, Ammonoidea) de France au Jurassique et au Crétacé. *Mémoire de la Société géologique de France* **174**: 1-204.
——— 2000b. Altérations de la croissance chez les Phyllocerataceae Zittel, 1884, conséquence sur la classification. Apports de l'étude des Phyllocerataceae aux modalités de l'évolution. *Comptes Rendus de l'Académie des Sciences* **330**: 709-716.
——— & DELAMETTE, M. 2008. Les Phylloceratoidea (Ammonoidea) aptiens et albiens du bassin vocontien (sud-est de la France). *Carnets de Géologie, Mémoire* **2008/04**: 1-60.
———, PARENT, H. & GARRIDO, A.C. 2023. Aalenian phylloceratid ammonites from Picún Leufú, Neuquén Basin, Argentina. *Revue de Paléobiologie* **42** (1): 143-148.
JONES, D.L., MURPHY, M.A. & PACKARD, E.L. 1965. The Lower Cretaceous (Albian) ammonite genera *Leconteites* and *Brewericeras*. *United States geological Survey Professional Papers* **503-F**: 1-22.
JONES, T.S. 1938. Geology of Sierra de la Peña and paleont-ology of the Indidura Formation, Coahuila, Mexico. *Geological Society of America Bulletin* **49**: 69-150.
JUDD, J.W. 1867. On the strata which form the base of the Lincolnshire Wolds. *Quarterly Journal of the geological Society of London* **23**: 227-251.
JUDOLEY, C.M. & FURRAZOLA-BERMÚDEZ, G. 1968. *Estratigrafia y fauna del jurasico de Cuba*. Instituto Cubano de Recursos Minerales: Havana.
JUIGNET, P., KENNEDY, W.J. & WRIGHT, C.W. 1973. La limite Cénomanien-Turonien dans la région du Mans (Sarthe): Stratigraphie et paléontologie. *Annales de Paléontologie* (Invertebres) **59**: 207-242.
JUKES-BROWNE, A.J. & HILL, W. 1896. A delimitation of the Cenomanian. *Quarterly Journal of the geological Society of London* **52**: 99-177.
KAKABADZÉ, M.V. 1967a. On the systematic position and classification of the colchiditids. *Soobshcheniia Akademii nauk Gruzinskoi SSR* **48**: 437-442 [in Russian].
——— 1967b. On ecology of the genus *Colchidites* Djanelidze. *Bulletin of the geological Society of Georgia* **5** (1): 80-83 [in Russian].
——— 1970. A new genus *Kutatissites* gen. nov. from the Lower Cretaceous deposits of western Georgia. *Soobshcheniia Akademii nauk Gruzinskoi SSR* **58**: 733-736 [in Russian].
——— 1971a. Colchitidids and their stratigraphic significance. *Trudy geologicheskogo Instituta, Akademiia Nauk Gruzinskoi SSR*, n.s., **26**: 1-118 [in Russian].
——— 1971b. On the phylogeny of the Heteroceratidae. *Soobshcheniia Akademii nauk Gruzinskoi SSR* **64** (1): 109-112 [in Russian].
——— 1975. Lower Cretaceous heteroceratids of Georgia and adjacent regions. *Trudy Geologicheskogo Instituta Akademii Nauk Gruzinskoi SSR* **47**: 62-131 [in Russian].
——— 1977. New and little known Ancyloceratidae from the Caucasus. *Trudy geologicheskogo Instituta, Akademiia nauk Gruzinskoi SSR, Tbilisi* **58**: 128-151 [in Russian].
——— 1978. On the systematic and stratigraphical significance of the genera *Pseudocrioceras* Spath, *Audouliceras* Thomel and *Kutatissites* Kakabadze. *Annales du Museum d'Histoire naturelle de Nice* **4** (34): 1-9.
——— 1981. The Ancyloceratidae of the south of the USSR and their stratigraphical significance. *Transactions of the Geological Institute of Georgia, SSR*, n.s., **71**: 1-192 [in Russian].
——— 1988. On the morphological classification of heteromorph ammonites. *In*, WIEDMANN, J. & KULLMANN, J. (eds), *Cephalopods present and past*, 2nd International Cephalopod Symposium: 447-452; Lubrecht & Cramer Ltd.; Stuttgart.
——— 2004. New and little-known genera of the ammonite family Macroscaphitidae Hyatt, 1900 from the Barremian-Aptian deposits of Georgia and the Crimea. *Proceedings of the geological Sciences of Georgia*, n.s., **119**: 360-367 [in Russian].
———, BOGDANOVA, T.N. & MIKHAILOVA, I.A. 1978. Stratigraphy of the middle Aptian in the south of the USSR and some heteromorphous ammonites. *Biulleten Moskovskogo Obshchestva Ispytatelei Prirody, Otdel Geologicheskii* **53** (6): 75-90.
——— & THIEULOY, J.-P. 1991. Ammonites hétéromorphes du Barremien et de l'Aptien de Colombie (Amérique du sud). *Geologie Alpine* **67**: 81-113.
——— & KOTETISHVILI, E. 1995. New data on the Upper Barremian biostratigraphy of the Georgian region (Caucasus). *Memorie descrittive della Carta Geologica d'Italia* **51**: 103-108.
——— & ——— 2003. On the biostratigraphy of the Upper Barremian of Georgia. *Biulleten Moskovskogo Obshchestva Ispytateley Prirody, Otdel Geologicheskii* **78**: 36-45.
———, ——— & MIKHAILOVA, I.A. 1978. On stratigraphy of the Middle Aptian of south of the USSR and some heteromophic ammonites. *Bulletin de la Societe imperi-ale des Naturalistes de Moscow*, Sect.geol., **53** (6): 75-90 [in Russian].
———, KVANTALIANI I.V. & SHARIKADZE M.Z. 1978. On the stratigraphy of the Lower Aptian and adjacent deposits of central Dagestan. *Bulletin of the Academy of Sciences of Georgia, SSR* **92** (1): 121-124 [in Russian]
——— & SHARIKADZE, M.Z. 1991. On mode of life of *Ptychoceras* d'Orbigny (Ammonoidea). *Bulletin of the Academy of Sciences of Georgia, SSR* **141** (2): 345-348.
———, ——— 1993. On the mode of life of heteromorph ammonites (heterocone, ancylocone, ptychocone). *Geobios* **15**: 209-215.
KAPILIMA, S. 1984. Stratigraphische und palaeontologische Untersuchungen im Jura un der Kreide des tansanischen Kustenstreifens im Hinterland von Dar-Es Salaam und Bagamoyo. *Berlin Geowissesnschaftliche Abhandlungen* (Reihe A, Geologie und Paläontologie) **57**: 1-77.
KAPLAN, U. 1988. Die Ammoniten-Subfamilie Collignoniceratinae Wright & Wright 1951 aus dem Turon (Ober-Kreide) von Westfalen und Niedersachsen (NW-Deutschland). *Geologie und Paläontologie in Westfalen* **12**: 5-45.
——— 1989. Die heteromorphe Ammonitengattung *Allocrioceras* Spath aus dem Turon von Nordwest-deutschland. *Geologie und Paläontologie in Westfalen* **15**: 71-105.
——— 1991. Zur Stratigraphie der tiefen Oberkreide im Teutoburger Wald (NW-Deutschland). Teil 2: Turon und Coniac im Steinbruch des Kalkwerks Foerth, Halle/Westfalen. *Berichte des naturwissenschaftlichen Verein Bielefeld und Umgegend* **32**: 125-159.

———, KELLER, S. & WIEDMANN, J. 1984. Ammoniten- und Inoceramen-Gliederung des norddeutschen Cenoman. *Schriftenreihe der Erdwissenschaftlichen Kommission, 1, Österreichische Akademie der Wissenschaften* **7**: 307-347.

——— & KENNEDY, W.J. 1994. Ammoniten des westfälischen Coniac. *Geologie und Paläontologie in Westfalen* **31**: 1-155.

———, ——— & ERNST, G. 1996. Stratigraphy and ammonite fauna of the Campanian in southeastern Munster. *Geologie und Paläontologie in Westfalen* **43**: 1-133.

———, ——— & HISS, M. 2005. Stratigraphie und Ammonitenfaunen des Campan im nordwestlichen und zentralen Münsterland. *Geologie und Paläontologie in Westfalen* **64**: 1-171.

———, ———, LEHMANN, J. & MARCINOWSKI, R. 1998. Stratigraphy and ammonite fauna of the Cenomanian of Westphalia. *Geologie und Paläontologie in Westfalen* **51**: 1-236.

———, ——— & WRIGHT, C.W. 1987. Turonian and Coniacian Scaphitidae from England and north-western Germany. *Geologisches Jahrbuch* **A103**: 5-39.

KARAKASCH, N.I. 1889. Über einige Neokomablagerungen in der Krim. *Sitzungsberichte der kaiserlichen Akademie der Wissenschaften in Wien, Mathematisch-Naturwissenshaftliche Klasse* **98**: 428-438.

——— 1902. Note sur le Crétacé inférieur de Biassala (Crimée). *Annales de l'Universite de Grenoble* **14**: 1-14.

——— 1907. The Lower Cretaceous of the Crimea and its fauna. *Trudy Imperatorskogo S.-Peterburgskogo Obshchestva Estestvoispytatelei, Leningrad* **32** (5): 1-482.

KARSTEN, H. 1858. Über die geognostischen Verhältnisse des westlichen Colombien des heutigen Republiken Neu Granada und Ecuador. *Amtlicher Bericht der Versammlung deutscher Naturforscher und Aerzte* **32** (1856): 80-117.

KAWABE, F. & HAGGART, J.W. 2003. The ammonoid *Desmoceras* in the Upper Albian (Lower Cretaceous) of Japan. *Journal of Paleontology* **77**: 314-322.

KAWADA, M. 1929. On some new species of ammonites from the Naibuchi district, south Saghalien. *Journal of the geological Society of Tokyo* **36**: 1-6.

KAYSER, E. 1924. *Lehrbuch der Geologie. 4. Geologische Formationskunde II.* Ferdinand Enke: Stuttgart.

KAZANSKY, P.A. 1914. Description d'une collection des cephalopodes des terrains Crétacés du Daghestan. *Izvestiia Tomskogo Teckhnologicheskoga Instituta* **32** (4): 1-127.

KELLUM, L.B. & MINTZ, L.W. 1962. Cenomanian ammonites from the Sierra de Tlahualilo, Coahuila, Mexico. *Contributions from the Museum of Paleontology, University of Michigan* **13**: 267-287.

KEMPER, E. 1961. Die Ammonitengattung *Platylenticeras (= Garnieria). Beihefte zum geologischen Jahrbuch, Hannover* **47**: 1-195.

——— 1964. Über eine deutsche *Tollia*-Fauna und den Ursprung der Polyptychiten (Ammonoidea, Mittel-valendis). *Fortschritte in der Geologie von Rheinland und Westfalen* **7**: 15-26.

——— 1966. Beobachtungen an Unterkreideammoniten im deutsch-holländischen Grenzgebiet zwischen Rheine, Bentheim und Winterswijk. *Grondboor en Hamer* **1966** (5): 224-228; (6): 256-258).

——— 1967. Die älteste Ammoniten-Fauna im Aptium Nordwest-Deutschlands. *Paläontologische Zeitschrift* **41**: 119-131.

——— 1968. *Geologischer Führer durch die Grafschaft Bentheim und die angrenzenden Gebiete.* Verlag Heimatverein der Grafschaft Bentheim: Nordhorn.

——— 1971. Zur Abgrenzung und Unterteilung des Valanginium ("Valendis"). *Newsletter on Stratigraphy* **1** (4): 45-58.

——— 1973a. The Valanginian and Hauterivian stages in northwest Germany. *In*, CASEY R. & RAWSON, P.F., (eds), The Boreal Lower Cretaceous. *Geological Journal, Special Issues* **5**: 327-344.

——— 1973b. Die Unterkreide im Untergrund der Gehrdener Berge und in der Deister-Mulde. *Bericht der naturhistorischen Gesellschaft zu Hannover* **117**: 29-54.

——— 1977. Biostratigraphy of the Valanginian in Sverdrup Basin, district of Franklin. *Geological Survey of Canada Paper* **76-32**: 1-6.

——— 1978. Einige neue, biostratigraphisch bedeutsame Arten der Ammoniten-Gattung *Dichotomites* (NW-Deutschland, Obervalangin). *Geologisches Jahrbuch* (A) **45**: 183-253.

——— & JELETZKY, J.A. 1979. New stratigraphically and phylogenetically important olcostephanid (Ammonitida) taxa from the uppermost Lower and Upper Valanginian of Sverdrup Basin, N.W.T. *Geological Survey of Canada Paper* **79-19**: 1-24.

———, RAWSON, P.F. & THIEULOY, J.-P. 1981. Ammonites of Tethyan ancestry in the early Lower Cretaceous of north-west Europe. *Palaeontology* **24**: 251-311.

KENNEDY, W.J. 1971. Cenomanian ammonites from southern England. *Special Papers in Palaeontology* **8**: v, 1-133.

——— 1972. The affinities of *Idiohamites ellipticoides* Spath (Cretaceous Ammonoidea). *Palaeontology* **15**: 400-404.

——— 1977. Ammonite evolution. *Developments in Palaeontology and Stratigraphy* **5**: 251-304.

——— 1985a. A note on *Euhemihoplites paradoxus* Collignon, 1964 (Cretaceous Ammonoidea). *Cretaceous Research* **6**: 307-309.

——— 1985b. *Solgerites* Reeside, 1932 (Cretaceous Ammonoidea) a synonym of *Forresteria* Reeside, 1932, with a revision of *Solgerites brancoi* (Solger, 1904) (Cretaceous Ammonoidea) from Cameroon. *Paläontologische Zeitschrift* **59**: 211-222.

——— 1986a. The ammonite fauna of the Calcaire à *Baculites* (Upper Maastrichtian) of the Cotentin Peninsula (Manche, France). *Palaeontology* **29**: 25-83.

——— 1986b. Observations on *Astiericeras astierianum* (d'Orbigny, 1842) (Cretaceous Ammonoidea). *Geological Magazine* **123**: 507-513.

——— 1986c. Campanian and Maastrichtian ammonites from northern Aquitaine, France. *Special Papers in Palaeontology* **36**: 1-145.

——— 1987. The ammonite fauna of the type Maastrichtian with a revision of *Ammonites colligatus* Binkhorst, 1861. *Bulletin de l'Institut Royal des Sciences naturelles de Belgique* **56**: 151-267.

——— 1988. Late Cenomanian and Turonian ammonite faunas from north-east and central Texas. *Special Papers in Palaeontology* **39**: 1-131.

——— 1989. Thoughts on the evolution and extinction of Cretaceous ammonites. *Proceedings of the Geologists' Association* **100**: 251-279.

——— 1991. Description of the lectotype of *Ammonites deverianus* d'Orbigny, 1841 (Cretaceous, Ammonoidea), *Geobios* **24**: 309-313.

——— 1993a. A note on the lectotype of *Ammonites beaudanti* Brongniart, 1822 (Cretaceous, Albian). *Cretaceous Research* **14**: 235-238.

——— 1993b. Campanian and Maastrichtian ammonites from the Mons Basin and adjacent areas (Belgium). *Bulletin de l'Institut royal des Sciences naturelles de Belgique, Sciences de la Terre* **63**: 99-131.

——— 1994a. Cenomanian ammonites from Cassis, Bouches-du-Rhône, France. *Palaeopelagos, Special Publication* **1**: 209-254.

——— 1994b. Lower Turonian ammonites from Gard (France). *Palaeopelagos, Special Publication* **1**: 255-275.

——— 1995. Maastrichtian ammonites from the United Arab Emirates-Oman border region. *Bulletin of the Natural History Museum* (Geology Series) **51**: 241-250.

——— 2004a. On *Brancoceras* STEINMANN, 1881 (Brancoceratidae) and *Pseudobrancoceras* gen. nov., (type species *Ammonites versicostatus* Michelin, 1838: Lyelliceratinae) from the Albian (Cretaceous) of the western Paris Basin and provence, France. *Acta geologica Polonica* **54**: 251-271.

——— 2004b. Ammonites from the Pawpaw Shale (Upper Albian) in northeast Texas. *Cretaceous Research* **25**: 865-905.

——— 2020. Upper Albian, Cenomanian and Upper Turonian ammonite faunas from the Fahdène Formation of central Tunisia and correlatives in northern Algeria. *Acta geologica Polonica* **70** (2): 147-272.

———, AMÉDRO, F., BADILLET, G., HANCOCK, J.M. & WRIGHT, C.W. 1984. Notes on Late Cenomanian and Turonian ammonites from Touraine, western France. *Cretaceous Research* **5**: 29-45.

———, ——— & COLLETÉ, C. 1986. Late Cenomanian and Turonian ammonites from Ardennes, Aube and Yonne, eastern Paris Basin (France). *Neues Jahrbuch für Geologie und Paläontologie Abhandlungen* **172**: 193-217.

——— & BAYLISS, O.H. 1977. The earliest tissotiid ammonite. *Palaeontology* **20**: 901-906.

———, BILOTTE, M. & HANSOTTE, M. 1998. Albian ammonite faunas from Pech de Foix (Ariège, France). *Bulletin des Centres de Recherches d'Exploration-Production Elf-Aquitaine* **21**: 457-499.

———, ———, LEPICARD, B. & SEGURA, F. 1986. Upper Campanian and Maastrichtian ammonites from the Petites-Pyrénées, southern France. *Eclogae Geologicae Helvetiae* **79**: 1001-1037.

———, ——— & MELCHIOR, P. 1995. Ammonite faunas, biostratigraphy andsequence stratigraphy of the Coniacian-Santonian of the Corbières (NE Pyrénées). *Bulletin des Centres de Recherches d'Exploration-Production Elf-Aquitaine* **19**: 377-499.

———, CHAHIDA, M.R. & DJAFARIAN, M.A. 1979. Cenomanian cephalopods from the glauconitic limestone southeast of Esfahan, Iran. *Acta palaeontologica Polonica* **24**: 3-50.

——— & COBBAN, W.A. 1988a. *Litophragmatoceras incomptum* gen. et sp. nov. (Cretaceous Ammonoidea), a cryptic micromorph from the Upper Cenomanian of Arizona. *Geological Magazine* **125**: 535-539.

———, ——— 1988b. *Nebraskites haresiceratiforme* n. g. n.sp., a new ammonite from the mid-Turonian *Prionocyclus percarinatus* zone in Nebraska, United States. *Neues Jahrbuch für Geologie und Paläontologie, Monatshefte* **1988**: 581-586.

———, ——— 1988c. Mid-Turonian ammonite faunas from northern Mexico. *Geological Magazine* **125**: 593-612.

———, ——— 1990a. Cenomanian ammonite faunas from the Woodbine Formation and lower part of the Eagle Ford Group, Texas. *Palaeontology* **33**: 75-154.

———, ——— 1990b. Cenomanian micromorph ammonites from the Western Interior of the USA. *Palaeontology* **33**: 379-422.

———, ——— 1990c. *Rhamphidoceras saxatilis* n. gen. and sp., a micromorph ammonite from the Lower Turonian of Trans-Pecos Texas. *Journal of Paleontology* **64**: 666-668.

———, ——— 1990d. The madagascan ammonite *Neogauthiericeras* Collignon, 1969 from the Upper Cretaceous (Campanian) of Texas. *Paläontologische Zeitschrift* **64**: 57-61.

———, ——— 1991. Coniacian ammonite faunas from the United States Western Interior. *Special Papers in Palaeontology* **45**: 1-96.

———, ——— 1991. Upper Cretaceous (Upper Santonian) *Boehmoceras* fauna from the Gulf Coast region of the United States. *Geological Magazine* **128**: 167-189.

———, ——— 1993a. Campanian ammonites from the Annona Chalk near Yancy, Arkansas. *Journal of Paleontology* **67**: 83-97.

———, ——— 1993b. Lower Campanian (Upper Cretaceous) ammonites from the Merchantville Formation of New Jersey, Maryland, and Delaware. *Journal of Paleontology* **67**: 828-849.

———, ——— 1993c. The Upper Cretaceous dimorphic pachydiscid ammonite *Menuites* in the Western Interior of the United States. *United States geological Survey Professional Paper* **1533**: 1-14.

———, ——— 1993d. Middle Campanian ammonites and inoceramids from the Wolfe City Sand in northeastern Texas. *Journal of Paleontology* **67**: 71-82.

———, ——— 1993e. Lower Campanian (Upper Cretaceous) ammonites from the Merchantville Formation of New Jersey, Maryland, and Delaware. *Journal of Paleontology* **67**: 828-849.

———, ——— 1993f. Maastrichtian ammonites from the Corsicana Formation in northeast Texas. *Geological Magazine* **130**: 57-67.

———, ——— 1998. 'Erratum: *Chesapeakiceras*, new name for *Chesapeakella* Kennedy and Cobban, 1993 (September 14), not *Chesapeakella* Campbell, 1993 (June 13). *Journal of Paleontology* **72**: 403.

———, ——— & HOOK, S.C. 1988. *Hourcquia* Collignon, 1965 (Cretaceous Ammonoidea) from the Upper Turonian of the southern United States. *Paläontologische Zeitschrift* **62**: 87-93.

———, ——— & LANDMAN, N.H. 1997. Maastrichtian ammonites from the Severn Formation of Maryland. *American Museum Novitates* **3210**: 1-30.

———, ———, ——— 2001. A revision of the Turonian members of the ammonite subfamily Collignoniceratinae from the United States Western Interior and Gulf Coast. *Bulletin of the American Museum of Natural History* **267**: 1-148.

———, ——— & SCOTT, G.R. 2000. Heteromorph ammonites from the middle Campanian *Baculites scotti* Zone in the U.S. Western Interior. *Acta geologica Polonica* **50** (2): 223-241.

——— & COOPER, M.R. 1977. The micromorph Albian ammonite *Falloticeras* Parona and Bonarelli. *Palaeontology* **20**: 793-804.

——— CRAME, A., BENGTSON, P. & THOMSON, M.R.A. 2007. Coniacian ammonites from James Ross Island, Antarctica. *Cretaceous Research* **28**: 509-531.

——— & DELAMETTE, M. 1994. Lyelliceratidae and Flickiidae (Ammonoidea) from the Upper Albian of the Helvetic Shelf (western Alps, France and Switzerland). *Journal of Paleontology* **68**: 1263-1284.

——— & GALE, A.S. 2015. Upper Albian and Cenomanian ammonites from Djebel Mrhila, central Tunisia. *Revue de Paléobiologie, Genève* **34** (2): 235-361.

———, ———, BROWN, P.R., CARON, M., DAVEY, R.J., GRÖCKE, D. & WRAY, D.S. 2000. Integrated stratigraphy across the Aptian-Albian boundary in the Marnes bleues, at Col de Pré-Guittard, Arnayon (Drôme), and at Tartonne (Alpes-de-Haute-Provence), France : a candidate Global Boundary Stratotype Section and Boundary Point for the base of the Albian Stage. *Cretaceous Research* **21**: 591-720.

——— & HANCOCK, J.M. 1970. Ammonites of the genus *Acanthoceras* from the Cenomanian of Rouen, France. *Palaeontology* **13**: 462-490.

———, ——— 1971. *Mantelliceras saxbii,* and the horizon of the *Martimpreyi* Zone in the Cenomanian of England. *Palaeontology* **14**: 437-454.

———,——— & CHRISTENSEN, W.K. 1981. Albian and Cenomanian ammonites from the island of Bornholm (Denmark). *Bulletin of the geological Society of Denmark* **29**: 203-244.

———, ———, COBBAN, W.A. & LANDMAN, N.H. 2004. A revision of the ammonite types described in F. Roemer's "Die Kreidebildungen von Texas und ihre organischen Einschlüsse" (1852). *Acta geologica Polonica* **54**: 433-445

——— & HENDERSON, R.A. 1991a. Revision of *Ammonites gaudama* Forbes, 1846 (Cretaceous Ammonoidea). *Journal of Paleontology* **65**: 891-893.

———, ——— 1991b. A note on *Ammonites sugata* Forbes, 1846 (Cretaceous Ammonoidea). *Neues Jahrbuch für Geologie und Paläontologie, Monatshefte* **1991/8**: 470-476.

———, ——— 1992a. Non-heteromorph ammonites from the Upper Maastrichtian of Pondicherry, south India. *Palaeontology* **35**: 381-442.

———, ——— 1992b. Heteromorph ammonites from the Upper Maastrichtian of Pondicherry, south India. *Palaeontology* **35**: 693-731.

——— & JUIGNET, P. 1977. *Ammonites diartianus* d'Orbigny, 1850, Vascoceratidae du Cénomanien supérieur de Saint-Calais (Sarthe). *Geobios* **10**: 583-595.

———, ——— 1983. A revision of the ammonite faunas of the type Cenomanian. I. Introduction, Ancyloceratina. *Cretaceous Research* **4**: 3-83.

———, ——— 1984. A revision of the ammonite faunas of the type Cenomanian. 2. The families Binneyitidae, Desmoceratidae, Engonoceratidae, Placenticeratidae, Hoplitidae, Schloenbachiidae, Lyelliceratidae and Forbesiceratidae. *Cretaceous Research* **5**: 93-161.

———, ——— 1993. A revision of the ammonite faunas of the type Cenomanian. 4. Acanthoceratinae (*Acompsoceras*, *Acanthoceras*, *Protacanthoceras*, *Cunningtoniceras* and *Thomelites*). *Cretaceous Research* **14**: 145-190.

———, ——— 1994. A revision of the ammonite faunas of the type Cenomanian, 5. Acanthoceratinae *Calycoceras (Calycoceras)*, *C. (Gentoniceras)* and *C. (Newboldiceras)*. *Cretaceous Research* **15**: 17-57.

———, ——— & HANCOCK, J.M. 1981. Upper Cenomanian ammonites from Anjou and the Vendée, western France. *Palaeontology* **24**: 25-84.

———, ——— & WRIGHT, C.W. 1986. A revision of the ammonite faunas of the type Cenomanian. 3. Mantelliceratinae. *Cretaceous Research* **7**: 19-62.

——— & KAPLAN, U. 1995a. *Pseudojacobites farmeryi* (Crick, 1905), ein seltener Ammonit des westfälischen und englischen Ober-Turon. *Berliner geowissenschaftliche Abhandlungen* **E16** (Gundolf-Ernst-Festschrift): 25-43

———, ——— 1995b. *Parapuzosia (Parapuzosia) seppenradensis* (Landois, 1985) and the ammonite fauna of the Lower Campanian Dulmen Beds, Westphalia. *Geologie und Paläontologie in Westfalen* **33**: 1-127.

———, ——— 1997. Ammonites from the Campanian of Stemweder Berg, Damme Upper Cretaceous basin, NW Germany. *Geologie und Paläontologie in Westfalen* **50**: 31-245.

———, ——— 2000. Ammonitenfaunen des hohen Coniac und Santon in Westfalen. *Geologie und Paläontologie in Westfalen* **57**: 1-126.

——— & KLINGER, H.C. 1972. A *Texanites-Spinaptychus* association from the Upper Cretaceous of Zululand. *Palaeontology* **15**: 394-399.

———, ——— 1977a. Cretaceous faunas from Zululand and Natal, South Africa: The ammonite family Tetragonitidae Hyatt, 1900. *Annals of the South African Museum* **73**: 149-197.

———, ——— 1977b. Cretaceous faunas from Zululand and Natal, South Africa: a *Jauberticeras* from the Mzinene Formation (Albian). *Annals of the South African Museum* **74**: 1-12.

———, ——— 1978. Cretaceous faunas from Zululand and Natal, South Africa: a new genus and species of Gastroplitinae from the Mzinene Formation (Albian). *Annals of the South African Museum* **77**: 57-69.

———, ——— 1979a. Cretaceous faunas from Zululand and Natal, South Africa: The ammonite superfamily Haplocerataceae Zittel, 1884. *Annals of the South African Museum* **77**: 85-121.

———, ——— 1979b. Cretaceous faunas from Zululand and Natal, South Africa. The ammonite family Gaudryceratidae. *Bulletin of the British Museum Natural History* (Geology) **31** (2): 121-174.

———, ——— 1985. Cretaceous faunas from Zululand and Natal, South Africa. The ammonite family Kossmaticeratidae Spath, 1922. *Annals of the South African Museum* **95**: 165-231.

———, ——— 1993a. On the affinities of *Cobbanoscaphites* Collignon,1969 (Cretaceous Ammonoidea). *Annals of the South African Museum* **102**: 265-271.

———, ——— 1993b. On the affinities of *Zuluscaphites* van Hoepen 1955 (Cretaceous Ammonoidea) from the Albian of Zululand, South Africa. *Paläontologische Zeitschrift* **67**: 63-67.

———, ——— 2006. Cretaceous faunas from Zululand and Natal, South Africa. The ammonite family Pachydiscidae Spath, 1922. *African Natural History* **2**: 17-166.

———, ——— 2009. The heteromorph ammonite *Ndumuiceras variabile* gen. et sp. nov. from the Albian Mzinene Formation, KwaZulu-Natal, South Africa. *African Natural History* **5**: 43-47.

———, ——— 2010. Cretaceous faunas from Zululand and Natal, South Africa. The ammonite subfamily Acanthoceratinae de Grossouvre, 1894. *African Natural History* **6**: 1-76.

———, ——— 2013. Scaphitid ammonites from the Upper Cretaceous of KwaZulu-Natal and eastern Cape Province, South Africa. *Acta geologica Polonica* **63** (4): 527-543.

——— & KULBROK, F. 2003. An Upper Maastrichtian ammonite faunule from the Eastern Desert, Egypt. *Neues Jahrbuch für Geologie und Paläontologie, Monatshefte* **2003/8**: 449-462.

———, LANDMAN, N.H. & COBBAN, W.A. 1996. The Maastrichtian ammonites *Coahuilites sheltoni* Böse, 1928, and *Sphenodiscus pleurisepta* (Conrad, 1857), from the uppermost Pierre Shale and basal Fox Hills Formation of Colorado and Wyoming. *American Museum Novitates* **3186**: 1-14.

———, ———, ——— & JOHNSON, R.O. 2000a. Additions to the ammonite fauna of the Upper Cretaceous Navesink Formation of New Jersey. *American Museum Novitates* **3306**: 1-30.

———, ———, ———, ——— 2000b. Late Campanian (Cretaceous) heteromorph ammonites from the Western Interior of the United States. *Bulletin of the American Museum of Natural History* **251**: 1-88.

——— & SUMMESBERGER, H. 1979. A revision of *Ammonites mitis* Hauer and *Ammonites glaneggensis* Redtenbacher from the Gosau Beds (Upper Cretaceous) of Austria. *Beiträge zur Paläontologie von Oesterreich* **6**: 71-87.

———, ——— 1986. Lower Maastrichtian ammonites from Neuberg, Steiermark, Austria. *Beiträge zur Paläontologie von Oesterreich* **12**: 181-242.

———, ——— 1987. Lower Maastrichtian ammonites from Neuberg, Steiermark, Austria. *Beiträge zur Paläontologie von Österreich* **12**: 181-242.

———, ——— 1991. A note on the lectotype of *Ammonites galicianus* Favre, 1869. *Annalen des Naturhistorischen Museums in Wien* **92** A: 93-95.

———, ——— 1999. New Late Campanian ammonites from the Gschliefgraben near Gmunden (Ultrahelvetic, Austria). *Beiträge zur Paläontologie von Österreich* **24**: 23-39.

——— & WRIGHT, C.W. 1979. On *Kamerunoceras* Reyment, 1954 (Cretaceous: Ammonoidea). *Journal of Paleontology* **53**: 1165-1178.

———, ——— 1979. Vascoceratid ammonites from the type Turonian. *Palaeontology* **22**: 665-683.

———, ——— 1981. *Euhystrichoceras* and *Algericeras*, the last mortoniceratine ammonites. *Palaeontology* **24**: 417–435.

———, ——— 1983. *Ammonites polyopsis* Dujardin, 1837 and the Cretaceous ammonite family Placenticeratidae Hyatt, 1900. *Palaeontology* **26**: 855-873.

———, ——— 1984. The affinities of the Cretaceous ammonite *Neosaynoceras* Breistroffer, 1947. *Palaeontology* **27**: 159-167.

———, ——— 1985a. *Mrhiliceras* n.g. (Cretaceous Ammonoidea), a new Cenomanian mantelliceratine. *Neues Jahrbuch für Geologie und Paläontologie, Monatshefte* **1985**: 513-526.

———, ——— 1985b. Evolutionary patterns in late Cretaceous ammonites. *Special papers in Palaeontology* **33**: 131-143.

———, ——— 1994a. The affinities of *Nigericeras* Schneegans, 1943 (Cretaceous, Ammonoidea). *Geobios* **27**: 583-589.

———, ——— 1994b. *Baculites (Protobaculites)* Collignon, 1964 (Cretaceous Ammonoidea) is a synonym of *Hemiptychoceras* Spath, 1925. *Journal of Paleontology* **68** (6): 1414-1415.

———, ——— & CHANCELLOR, G.R. 1983. The Cretaceous ammonite *Eopachydiscus* and the origin of the Pachydiscidae. *Palaeontology* **26**: 655-662.

———, ——— & HANCOCK, J.M. 1980a. The European species of the Cretaceous ammonite *Romaniceras* with a revision of the genus. *Palaeontology* **23**: 325-362.

———, ———, ——— 1980b. Collignoniceratid ammonites from the mid-Turonian of England and northern France. *Palaeontology* **23**: 557-603.

———, ———, ——— 1980c. Origin, evolution and systematics of the Cretaceous ammonite *Spathites*. *Palaeontology* **23**: 821-837.

———, ———, ——— 1987. Basal Turonian ammonites from west Texas. *Palaeontology* **30**: 27-74.

———, ——— & KLINGER, H.C. 1979. Cretaceous faunas from Zululand and Natal, South Africa. A new genus and species of tuberculate desmoceratacean ammonite from the Mzinene Formation (Albian). *Annals of the South African Museum* **78**: 29-38.

———, ———, ——— 1983. Cretaceous faunas from Zululand and Natal, South Africa: The ammonite subfamily Barroisiceratinae Basse, 1947. *Annals of the South African Museum* **90**: 241-324.

——— & WALASZCZYK, I. 2004. *Forresteria (Harleites) petrocoriensis* (Coquand, 1859) from the Upper Turonian *Mytiloides scupini* Zone of Slupia Nadbrzezna, Poland. *Acta geologica Polonica* **54**: 55-59.

———, ———, GALE, A.S., DEMBRICZ, K. & PRASZKIER, T. 2013. Lower and Middle Cenomanian ammonites from the Morondava Basin, Madagascar. *Acta geologica Polonica* **63** (4): 625-655.

KEYSERLING, A.G. VON. 1846. *Wissenschaftliche Beobachtungen auf einer Reise in das Petchora-Land, im Jahre 1843.* Carl Kray: St. Petersburg.

KHAKIMOV, F.K. 1972. New ammonites of the genus *Fallotites* from the Lower Turonian of the Tadzhik depression. *Paleontological Journal* **1972** (1): 29-36.

KHIMSHIASHVILI, N.G. 1957. *Upper Jurassic fauna of Georgia (Cephalopoda and Lamellibranchiata)*. Academy of Sciences, Georgian SSR publishing house: Tbilisi [in Russian].

——— 1976. *Tithonian and Berriasian ammonites of the Caucasus.* Metsniereba: Tbilisi [in Russian].

——— 1989. *The Berriasellidae of Abkhazia. The Tithonian fauna of mount Lakorizi-Tau (basin of Bzibi river).* Metsniereba: Tbilisi [in Russian].

KHUDOLEY, K.M. 1960a. New Tithonian phylloceratid species from the Far East. *In*, *New species of ancient plants and animals of the USSR.* Part 2: 159-160. Gosgeoltechizdat: Moscow [in Russian]

——— 1960b. New Tithonian representatives of haploceratids and perisphinctids from South Primorie. *In*, *New species of ancient plants and animals of the USSR.* Part 2: 163-168. Gosgeoltechizdat: Moscow [in Russian].

KIEFFER, J. 1904. Nouvelles cécidomyies xylophiles. *Annales de la Société des Sciences, Bruxelles* **28**: 367-410 (1-44).

KILIAN, C.C.C.W. 1889a. Sur quelques fossiles nouveaux ou peu connus du Crétacé inférieur de la Provence. *Bulletin de la Société géologique de France* (3) **16** (for 1888): 663-691.

——— 1889b. *Description géologique de la Montagne de Lure (Basses-Alpes).* Masson: Paris.

——— 1889c. Études paléontologiques sur les terrains secondaires et tertiaires de l'Andalousie. Le gisement Tithonique de Fuente de los Frailes près de Cabra (province de Cordove). *Mission d'Andalousie Mémoires Presentes pour Divers Savants a l'Academie des Sciences de l'Institut national de France, Paris* **30**: 581-739.

——— 1907-1913. Erste Abteilung: Unterkreide (Palaeocretacicum). Lieferung 1-3. *In*, FRECH, F., *Lethaea Geognostica. II. Das Mesozoicum, Band 3 (Kreide)*: 1-398. Schweizerbart: Stuttgart.

——— 1910a. Sur l'origine du groupe de l'*Ammonites percevali* Uhlig du Barremien. *Bulletin de la Société géologique de France* (4) **10**: 7.

——— 1910b. Un nouvel exemple de phénomènes de convergence chez des ammonitidés. *Comptes Rendus des Seances de l'Academie des Sciences, Paris* **150**: 149-152.

——— 1911. Observations au sujet de la note de H. Douvillé sur l'évolution des pulchelliidés. *Compte Rendu Sommaire des Seances de la Société géologique de France* **11**: 117.

——— & REBOUL, P. 1909. Les céphalopodes néocretacés des îles Seymour et Snow-Hill d'après les matériaux recueillis par l'expédition Antarctique suédoise. *Wissenschaftliche Ergebnisse der schwedischen Süd-polar- expedition, 1901-1903*, III, **6**: 1-75.

———, ——— 1914. Sur la faune du Valanginien moyen du Col de Fréne (Savoie). *Compte Rendu de l'Association Francaise pour l'Avancement des Sciences, Congres Tunis, 1913*, mémoire hors volume: 1-5.

——— 1915. Contribution à l'étude des faunes paléo-crétacés du sud-est de la France. I. La faune de l'Aptien inférieur des environs de Montelimar (Drôme). II. Sur quelques ammonites de l'Hauterivien de la Begude (Basses Alpes). *Mémoires pour servir a l'explication de la carte géologique detaillee de la France* **1915**: 1-288.

KISELEV, D.N. & ROGOV, M.A. 2018. Ammonites and stratigraphy of the terminal part of the Middle Volgian Substage (Upper Jurassic; *Epivirgatites nikitini* Zone and its equivalents) of the Panboreal Realm: 2. *Titanites* and *Glaucolithites*. *Stratigraphy and Geological Correlation* **26** (1): 15-61.

———, ——— & ZAKHAROV, V.A. 2018. The *Volgidiscus singularis* Zone of the terminal horizons of the Volgian Stage of European Russia and its significance for inter-regional correlation and paleogeography. *Stratigraphy and Geological Correlation* **26**: 206-233.

KITCHIN, F.L. 1908. The invertebrate fauna and palaeontological relations of the Uitenhage Series. *Annals of the South African Museum* **7**: 21-250.

Kittl, E. 1908. Beiträge zur Kentniss der Triasbildung der nordöstlichen Dobrudscha. *Denkschriften der kaiserlichen Akademie der Wissenschaften, Wien, Mathematisch-Naturwissenschaftliche Klasse* **81**: 447-532.

Klein, J. 2005. Lower Cretaceous ammonites. I. Perisphinctaceae. 1 (Himalayitidae, Olcostephanidae, Holcodiscidae, Neocomitidae, Oosterellidae. *In*, Riegraf, W. (ed.) *Fossilium Catalogus I: Animalia* pars **139**: 1-484. Backhuys: Leiden.

——— 2013. Lower Cretaceous ammonites. VII. Hoplitoidea & Engonoceratoidea. *Fossilium Catalogus I: Animalia* pars **152**: vi+1-280. Margraf: Weikersheim.

——— 2015. Lower Cretaceous ammonites. VIII. Turrilitoidea 1; Anisoceratidae, Hamitidae, Turrilitidae, including the Upper Cretaceous representatives. *Fossilium Catalogus I: Animalia* pars **154**: 1-265. Backhuys, Margraf: Weikersheim.

——— 2016a. Lower Cretaceous ammonites. IX. Turrilitoidea 2 - Baculitidae, including the Upper Cretaceous representatives. *Fossilium Catalogus I: Animalia* pars **155**: 1-148. Margraf: Weikersheim.

——— 2016b. Lower Cretaceous ammonites. X. Scaphitoidea, including the Upper Cretaceous representatives. *Fossilium Catalogus I Animalia* pars **157**: 1-203. Margraf: Weikersheim.

——— 2018. Lower Cretaceous ammonites. XI. Acanthoceratoidea: Leymeriellidae, Brancoceratidae, Lyelliceratidae, Flickiidae, Forbesiceratidae, including the Upper Cretaceous representatives. *Fossilium Catalogus I Animalia* pars **158**: 1-340. Margraf: Weikersheim .

——— & Bogdanova, T. 2013. Lower Cretaceous ammonites. VI. Douvilleiceratoidea & Deshayesitoidea. *Fossilium Catalogus I: Animalia* pars **151**: vi+1-300. Margraf: Weikersheim.

———, Hoffmann, R., Joly, B., Shigeta, Y. & Vašíček, Z. 2009. Lower Cretaceous ammonites. IV. Boreophylloceratoidea, Phylloceratoidea, Lytoceratoidea, Tetragonitoidea, Haploceratoidea including the Upper Cretaceous representatives. *Fossilium Catalogus I: Animalia* pars **146**: viii+1-416. Margraf: Weikersheim.

——— & Vašíček, Z. 2012. Lower Cretaceous ammonites. V. Desmoceratoidea. *Fossilium Catalogus I: Animalia* pars **148**: vi + 1-311. Margraf: Weikersheim.

——— & Vermeulen, J. 2006. Lower Cretaceous Ammonites II. Perisphinctaceae 2. Polyptychitidae (by Jaap Klein) Endemocerataceae: Pulchelliidae (by Jaap Klein & Jean Vermeulen). *Fossilium Catalogus I: Animalia* pars **151**: vii+1-278. Margraf: Weikersheim.

———, Busnardo, R., Company, M., Delanoy, G., Kaka-badzé, M.V., Reboulet, S., Ropolo, P., Vašíček, Z. & Vermeulen, J. 2007. Lower Cretaceous ammonites III: Bochianitoidea, Protancyloceratoidea, Ancyloceratoidea, Ptychoceratoidea. *In*, Riegraf, W. (ed.), *Fossilium catalogus, I: Animalia* pars **144**: Margraf: Weikersheim.

Kler, M.O. 1908. Neoceratites from eastern Bokhara. *Trudy geologicheskogo Muzeia Petra Velikogo* **2**: 157-173 [in Russian].

Klimova, I.G. 1969. The early Berriasian of western Siberia. *Geologiia i Geofizika* **1969** (4): 128-132.

——— 1972. Ammonites of western Siberia. *In*, Saks, V.N. (ed.), *The Jurassic-Cretaceous boundary and the Berriasian Stage in the Boreal Realm*: 194-204. Nauka: Novosibirsk [in Russian].

——— 1978a. On the systematics of the ammonite *Temnoptychites* (additional new data). *Trudy VNIIG-GIMS: Novye Materialy po Stratigrafii i Paleontologii Sibiri, Vypusk* **260**: 98-114 [in Russian].

——— 1978b. A new genus of ammonite from the Lower Valanginian of north central Siberia. *Geologiia i Geofizika, Novosibirsk* **1978** (12): 50-61 [in Russian].

——— 1983. New ammonites from the Neocomian of northern Siberia. *In*, *New data on the stratigraphy and palaeogeography of the oil and gas basins of Siberia*: 117-124. Sbornik Nauchnykh Trudov SNIIGMS: Novosibirsk [in Russian].

Klinger, H.C. 1976. Cretaceous heteromorph ammonites from Zululand. *Memoir of the geological Survey of South Africa* **69**: 1-142.

——— 1989. The ammonite subfamily Labeceratinae Spath 1925: systematics, phylogeny, dimorphism and distribution (with a description of a new species). *Annals of the South African Museum* **98**: 189-219.

——— & Kennedy, W.J. 1977. Cretaceous faunas from Zululand, South Africa, and southern Mozambique: The Aptian Ancyloceratidae (Ammonoidea). *Annals of the South African Museum* **73**: 215-359.

———, ——— 1978. Turrilitidae (Cretaceous Ammonoidea) from South Africa, with a discussion of the evolution and limits of the family. *The Journal of Molluscan Studies* **44**: 1-48.

———, ——— 1979. Cretaceous faunas from Southern Africa. Lower Cretaceous ammonites, including a new bochianitid genus, from Umgazana, Transkei. *Annals of the South African Museum* **78**: 11-19.

———, ——— 1980a. Cretaceous faunas from Zululand and Natal, South Africa: The ammonite subfamily Texanitinae Collignon, 1948. *Annals of the South African Museum* **80**: 1-357.

———, ——— 1980b. Cretaceous faunas from Zululand and Natal, South Africa: A new sextuberculate texanitid. *Annals of the South African Museum* **82**: 321-331.

———, ——— 1989. Cretaceous faunas from Zululand and Natal, South Africa: The ammonite family Placenticeratidae Hyatt, 1900; with comments on the systematic position of the genus *Hypengonoceras* Spath, 1924. *Annals of the South African Museum* **98**: 241-408.

———, ——— 1993. Cretaceous faunas from Zululand and Natal, South Africa. The ammonite genus *Eubaculites* Spath, 1926. *Annals of the South African Museum* **102** (6): 185-264.

———, ——— 1997. Cretaceous faunas from Zululand and Natal, South Africa. The ammonite family Baculitidae Gill, 1871 (excluding the genus *Eubaculites*). *Annals of the South African Museum* **105** (1): 1-206.

———, ——— 2001. Stratigraphic and geographic distribution, phylogenetic trends and general comments on the ammonite family Baculitidae Gill, 1871 (with an annotated list of species referred to the family). *Annals of the South African Museum* **107** (1): 1-290.

———, ——— 2008. *Mkuzeiella andersoni* gen. et sp. nov. (Cephalopoda, Ammonoidea) from the Albian Mzinene Formation of KwaZulu-Natal, South Africa. *Bulletin de l'Institut Royal des Sciences Naturelles de Belqique, Sciences de la Terre* **78**: 179-191.

———, ——— & Minor, K.P. 2010. *Tarrantites*, a new heteromorph ammonite genus from the Albian of Texas and Pakistan. *African Natural History* **6**: 91-99.

——— & Lunn, G. 2000. Upper Campanian (Cretaceous) ammonites from the Shinarish Formation, Djebel Sinjar, northwest Iraq. *Journal of Paleontology* **74** (3): 464-473.

———, Wiedmann, J. & Kennedy, W.J. 1975. A new carinate phylloceratid ammonite from the early Albian (Cretaceous) of Zululand, South Africa. *Palaeontology* **18**: 657-664.

Klipstein, A. von. 1843. Beiträge zur geologischen Kenntnise der östlichen Alpen. *Mittheilungen aus dem Gebiete der Geologie und Paläontologie, Giessen* **1**: 1-144.

Klug, C., Zaton, M., Parent, H., Hostettler, B. & Tajika, A. 2015. Mature modifications and sexual dimorphism. *In*, Klug *et al.* (eds), Ammonoid paleobiology: from anatomy to ecology. *Topics in Geobiology* **43**: 253-320. DOI 10.1007/978-94-017-9630-9

KNECHTEL, M.N., RICHARDS, E.F. & RATHBUN, M.V. 1947. Mesozoic fossils of the Peruvian Andes. *The Johns Hopkins University Studies in Geology* **15**: 1-150.

KNER, R. 1850. Versteinerungen des Kreidemergels von Lemberg und seiner Umgebung. *Naturwissenschaftliche Abhandlungen (gesammelt und durch subscription herausgegeben von Wilhelm Haidinger)* **3**: 1-42.

KOBAYASHI, T. 1934. The Cambro-Ordovician formations and faunas of South Chosen. Palaeontology, part 1: Middle Ordovician faunas. *Journal of the Faculty of Science, Imperial University of Tokyo* (2, Geology, Mineralogy, Geography, Seismology) **3**: 329-519.

KÜCHLER, T. 2000a. *Nostoceras (Euskadiceras) euskadiense*, a new ammonite subgenus and species from the higher Upper Campanian (Upper Cretaceous) of northern Spain. *Berliner Geowissenschaftliche Abhandlungen*, Reihe E, **34**: 291-307.

——— & ODIN, G.S. 2001. Upper Campanian-Maastrichtian ammonites (Nostoceratidae, Diplomoceratidae) from Tercis les Bains (Landes, France). *In*, ODIN, G.S. (ed.), The Campanian-Maastrichtian stage boundary. Characterisation at Tercis les Bains (France) and correlation with Europe and other continents. *Developments in Palaeontology and Stratigraphy* **19**: 500-528.

KOENEN, A. VON. 1897. Über Fossilien der unteren Kreide am Ufer des Mungo in Kamerun. *Abhandlungen der koeniglichen Gesellschaft der Wissenschaften zu Goettingen*, n.s., **1**: 1-48.

——— 1898. Nachtrag zu Über Fossilien der unteren Kreide am Ufer des Mungo in Kamerun. *Abhandlungen der koeniglichen Gesellschaft der Wissenschaften zu Goettingen*, n.s., **1**: 51-65.

——— 1902. Die Ammonitiden des norddeutschen Neocom (Valanginien, Hauterivien, Barrêmien und Aptien). *Abhandlungen der preussischen Geologischen Landesanstalt und Bergakademie, Berlin*, n.s., **24**: 1-449.

——— 1904. Über die untere Kreide Helgolands und ihre Ammonitiden. *Abhandlungen der koeniglich Gesellschaft der Wissenschaften zu Goettingen*, n.s., **3** (2): 1-62.

——— 1909. Die *Polyptychites*-Arten des Unteren Valanginien. *Abhandlungen der koeniglich preussischen Geologischen Landesanstalt*, n.s., **59**: 1-89.

KOENIG, C.D.E. 1820-1825. *Icones fossilium sectiles*. G.B. Sowerby: London.

KOSSMAT, F. 1895-1898. Untersuchungen über die Süd-indische Kreideformation. *Beiträge zur Paläontologie und Geologie Oesterreich-Ungarns und des Orients* **9** (1895): 97-203 (1-107); **11** (1897): 1-46 (108-153); **11** (1898): 89-152 (154-217).

KOTETISHVILI, E.V. 1970. Stratigraphy and fauna of the colchiditic and adjacent horizons of western Georgia. *Akademiya Nauk Gruzinskoi SSR, Trudy Geologicheskogo Instituta* **25**: 1-115.

——— 1978. On the little-known genus *Torcapella* Busnardo, 1970 from the Lower Cretaceous deposits of Georgia. *Proceedings of the Geological Institute and Academy of Sciences Georgia SSR*, n.s., **59**: 194-197 [in Russian].

——— 1980. Family Pulchelliidae H. Douvillé (from the Lower Cretaceous of the south of USSR). *Transactions of the geological Institute of the Academy of Sciences of Georgia SSR*, n.s., **67**: 1-110.

KOVACS, L. 1939. Bemerkungen zur systematischen Einteilung der Jurassischen Phylloceratan. *Abhandlungen aus dem mineralogisch-geologischen Institut der Stephan Tisza Universität in Debrecen* **13**: 289-320.

——— 1942. Monographie der Liassischen Ammoniten des Nördlichen Bakony. *Geologica Hungarica* (Palaeontologica) **17**: 1-220.

KRANTZ, F. 1926. Die Ammoniten des Mittel- und Ober-tithons. *Geologische Rundschau* Band **17A**: 428-462.

——— 1928. Beiträge zur Paläontologie und Stratigraphie des Lias, Doggers, Tithons und der Unterkreide in den Kordilleren im Süden der Provinz Mendoza (Argentinien). II. Die Ammoniten des Mittel- und Obertithons. *Geologische Rundschau* (Steinmann Festschrift), Sonderband **17A**: 428-462.

KRAUSE, P.G. 1902. Die Fauna der Kreide von Temojoh in West-Borneo. *Sammlung des geologischen Reichs-Museum in Leiden* (1) **7**: 1-28.

KRIMGOL'TS, G.Y., PETROVA, G.T. & PCHELINTSEV, V.F. 1953. Stratigraphy and fauna of the marine Mesozoic slope of northern Siberia. *Trudy Nauchno-Issledovatel'skogo Instituta Geologii Arktiki (NIIGA)* **45**: 1-134 [in Russian].

KÜCHLER, T. & ODIN, G.S. 2001. Upper Campanian-Maastrichtian ammonites (Nostoceratidae, Diplomoceratidae) from Tercis les Bains (Landes, France). *In*, ODIN, G.S. (ed.), The Campanian-Maastrichtian stage boundary: characterisation at Tercis les Bains (France): correlation with Europe and other continents. *Developments in Palaeontology and Stratigraphy* **19**: 500-528.

KUDERNATSCH, J. 1852. Die Ammoniten von Swinitza. *Abhandlungen der koeniglich-kaiserlichen geologischen Reichsanstalt* Band 1, Abt. 2: 1-16.

KULLMANN, J. & WIEDMANN, J. 1970. Significance of sutures in phylogeny of Ammonoidea. *University of Kansas paleontological Contributions Paper* **47**: 1-32.

KUMMEL, B. & DECKER, J.M. 1954. Lower Turonian ammonites from Texas and Mexico. *Journal of Paleontology* **28**: 310-319.

KUTEK, J. 1994. The Sythicus Zone (Middle Volgian) in Poland: its ammonites and biostratigraphic subdivision. *Acta geologica Polonica* **44** (1-2): 1-33.

——— & ZEISS A. 1974. Tithonian-Volgian ammonites from Brzostówka near Tomaszów Mazowiecki, central Poland. *Acta geologica Polonica* **24**: 505-542.

———, ——— 1997. The highest Kimmeridgian and Lower Volgian in central Poland; their ammonites and biostratigraphy. *Acta geologica Polonica* **47** (3-4): 107-198.

KULZH'INSKAYA-VORONETZ, N.S. 1933. Some representatives of the family Lytoceratidae Neumayr emend. Zittel from the Lower Cretaceous deposits of the Crimea. *Transactions of the United Geological and Prospecting Service of USSR* **241**: 1-40 [in Russian].

KVANTALIANI, I.V. 1968. The new genus *Chaschupseceras* from the Clansayesian deposits in western Abkhasiya. *Trudy Gruzhinskogo Politekhnicheskogo Instituta, Tiflis* **2** (122): 62-71 [in Russian].

——— 1971. *The Aptian ammonites of Abkhazia*. Gruzinskii Politekhnicheskii Institut: Tbilisi [in Russian].

——— 1978. On the systematics of the family Parahoplitidae Spath. *Soobshcheniia Akademii nauk Gruzinskoi SSR* **89** (2): 397-400 [in Russian].

——— 1980. On the systematics of the families Deshayesitidae and Leymeriellidae. *Soobshcheniia Akademii nauk Gruzhinskoi SSR* **97**: 121-124 [in Russian].

——— 1999. Berriasian cephalopods of the Crimea and the Caucasus. *Proceedings of the geological Institute of the Georgian Academy of Sciences*, n.s., **112**: 1-189 [in Russian].

——— 2000. Middle Jurassic - Lower Cretaceous ammonites of the Vedi Ophiolotic Zone of the Lesser Caucasus. *Proceedings of the geological Institute of the Georgian Academy of Sciences*, n.s., **116**: 1-71.

———, LOMINADZE, T.A. (1984) On the systematics of the Perisphinctina (Ammonoidea). *Bulletin of the Georgian Academy of Sciences* **116** (3): 553-556 [in Russian].

———, ———, TOPCHISHVILI, I.V. & SHARIKADZE, M.Z. 1999. Systematics and phylogeny of Mesozoic Ammonitida. *Proceedings of the geological Institute of the Georgian Academy of Sciences*, n.s., **113**: 1-160.

——— & LYSENKO, N.I. 1979. A new Berriasian genus *Tauricoceras*. *Soobshcheniia Akademii nauk Gruzinskoi SSR* **93**: 629-632 [in Russian].

——— & SHARIKADZE, M.Z. 1983. The genus *Chaschupseceras* (Ammonoidea), its specific composition and systematic position. *Paleontological Journal* **1983** (2): 39-44.

———, TOPCHISHVILI M., LOMINADZE T. & SHARIKADZE, M. 1999. Upon the systematics of the Mesozoic Ammonitida. *Bulletin of the Georgian Academy of Sciences* **160** (1): 102-105.

LAHUSEN, I. 1874. Fossils from the clays of Simbirsk. *Zapiski imperatorskago S. Peterburgskago mineralogischeskago Obshchestva* (2) **9**: 33-77.

——— 1883. Die fauna der Jurassischen Bildungen des Rjasanschen gouverments. *Trudy geologicheskogo Komiteta* **1** (1): 1-94.

LAMARCK, J.B.P.A. DE M. DE. 1799. Prodrome d'une nouvelle classification des coquilles. *Mémoires du Museum national d'Histoire Naturelle* **1**: 63-90.

——— 1801. *Système des animaux sans vertèbres*. Deterville: Paris.

——— 1822. *Histoire naturelle des animaux sans vertèbres*. **7**. Verdière: Paris.

LANDMAN, N.H. 1988. Heterochrony in ammonites. *In*, *Heterochrony in evolution: multidisciplinary approach*: 159-182. Plenum Press: London.

——— 1989. Iterative progenesis in Upper Cretaceous ammonites. *Paleobiology* **15**: 95-117.

——— & COBBAN, W.A. 2003. Ammonites from the upper part of the Pierre Shale and Fox Hills Formation of Colorado. *American Museum Novitates* **3388**: 1-45.

———, DOMMERGUES, J.L. & MARCHAND, D. 1991. The complex nature of progenetic species - examples from Mesozoic ammonites. *Lethaia* **24** (4): 409-421.

———, GARB, M.P., ROVELLI, R., EBEL, D.S. & EDWARDS, L.E. 2012. Short-term survival of ammonites in New Jersey after the end-Cretaceous bolide impact. *Acta palaeontologica Polonica* **57** (4): 703-715.

———, JOHNSON, R.O. & EDWARDS, L.E. 2004a. Cephalo-pods from the Cretaceous/Tertiary boundary interval on the Atlantic Coastal plain, with a description of the highest ammonite zones in North America. Part 1. Maryland and North Carolina. *American Museum Novitates* **3454**: 1-64.

———, ———, ——— 2004b. Cephalopods from the Cretaceous/Tertiary boundary interval on the Atlantic coastal plain, with a description of the highest ammonite zones in North America. Part 2. Northeastern Monmouth County, New Jersey. *Bulletin of the American Museum of Natural History* **287**: 1-107.

———, ———, GARB, M.P., EDWARDS, L.E. & KYTE, F.T. 2007. Cephalopods from the Cretaceous/Tertiary boundary interval on the Atlantic coastal plain, with a description of the highest ammonite zones in North America. Part III. Manasquan River Basin, Monmouth County, New Jersey. *Bulletin of the American Museum of Natural History* **303**: 1-122.

———, KENNEDY, W.J., COBBAN, W.A. & LARSON, N.L. 2010. Scaphitid ammonites from the Upper Cretaceous Pierre Shale and Bearpaw Shale of the Western Interior of North America. Part 1: *Baculites compressus-B. cuneatus* zones. *Bulletin of the American Museum of Natural History* **342**: 1-242.

——— & WAAGE, K.M. 1993. Scaphitid ammonites of the Upper Cretaceous(Maastrichtian) Fox Hills Formation of South Dakota and Wyoming. *Bulletin of the American Museum of Natural History* **215**: 1-257.

LANDRA, G., CECCA, F. & VAŠÍCEK, Z. 2000. Early Aptian ammonites from the top of the Maiolica and the anoxic "Selli level" (Lombardy, southern Alps). *Bollettino della Societa Paleontologica Italiana* **39**: 29-45.

LAPEYRE, J.-F. 1974. Révision de l'ammonite-index *Pseudothurmannia angulicosta* d'Orbigny. *Annales du Museum d'Histoire naturelle de Nice* **2**: 81-86.

LAPTIKHOVSKY, V.L., ROGOV, M.A., NIKOLAEVA, S.E. & ARK-HIPKIN, A.I. 2013. Environmental impact on ectocochleate cephalopod reproductive strategies and the evolutionary significance of cephalopod egg size. *Bulletin of Geosciences* **88** (1): 83-93.

LARSON, N.L. 2012. The Late Campanian (Upper Cretaceous) cephalopod fauna of the Coon Creek Formation at the type locality. *The Journal of Paleontological Sciences* JPS.H.2012. 01: 1-40.

LASSWITZ, R. 1904. Die Kreide-Ammoniten von Texas (collection F. Roemer). *Geologische und Paläontologische Abhandlungen*, n.s., **6**: 223-259.

LATIL, J.-L. 1989. Les genres *Engonoceras* Neumayr and Uhlig, 1881 et *Hypengonoceras* Spath, 1922 dans l'Albien supérieur (Z. à Dispar) de Salazac, Gard, France. *Revue de Paléobiologie* **8** (1): 51-63.

——— 1990. *Huescarites companyi* nov. gen. nov. sp., a new Albian (early *dispar* Zone) Flickiidae Adkins, 1928 (Ammonoidea, Cephalopoda) from south-west Europe. *Mesogee* **50**: 31-35.

——— 2008. A revision of *Amaltheus ebrayi* de Loriol, 1882, type species of the genus *Parengonoceras* Spath, 1924. *Revue de Paléobiologie* **27** (1): 249-264.

——— 2011. Early Albian ammonites from central Tunisia and adjacent areas of Algeria. *Revue de Paléobiologie* **30** (1): 321-429.

———, ROBERT, E. & BULOT, L.-G. 1990. New insights on the genus *Prolyelliceras* Spath, 1930 and the identity of *Acanthoceras gevreyi* Jacob, 1907 (Cephalopoda, Ammonitina). *Neues Jahrbuch für Geologie und Paläontologie, Abhandlungen* **254**: 337-347.

——— & DOMMERGUES, J.-L. 1997. Variabilité, ontogenèse et phylogenie des Lyelliceratinae (ammonites de l'Albien). *Revue de Paléobiologie* **16** (1): 101-113.

LAUBE, G.C. & BRUDER, G. 1887. Ammoniten der böhmischen Kreide. *Palaeontographica* **33**: 217-239.

LEANZA, A.F. 1945. Ammonites del Jurásico superior y del Cretáceo inferior de la Sierra Azul, en la parte meridional de la provincia de Mendoza. *Anales del Museo de La Plata* (Paleontologia), n.s., **1**: 1-99.

——— 1946. Las opelias de Chacay-Melehue, en el Neuquén, *Strebiles (Pseudopelia) oxynotus* subgen. et sp. nov. *Revista de la Asociación Geológica Argentina* **1** (1): 63-72.

——— 1949. Paleontología mesozoica. I. Sobre *Windhauseniceras humphreyi* n.sp. del Titoniano de Neuquén. *Revista de la Asociación geológica Argentina* **4** (3): 239-242.

——— 1963. *Patagoniceras* gen. nov. (Binneyitidae) y otros ammonites del Cretácico Superior de Chile meridional con notas acerca de su posición estratigráfica. *Boletin de la Academia nacional de Ciencias, Cordoba* **43**: 203-225.

——— 1964. *Parabinneyites,* nuevo nombre generico para *Patagoniceras* Leanza, 1963 *non* Wetzel, 1960. *Revista de la Asociación geológica Argentina* **19**: 84.

——— 1967a. Anotaciones sobre los fósiles jurásico-cretácicos de Patagonia austral (coleccion Feruglio) conservados en la Universidad de Bologna. *Acta geologica Lilloana* **9**: 121-187.

——— 1967c. Algunos ammonites nuevos o poco conocidos del Turoniano de Colombia y Venezuela. *Acta geologica Lilloana* **9**: 189-229.

——— 1967d. Descripcion de la fauna de *Placenticeras* del Cretacico Superior de Patagonia austral con consideraciones acerca de su posicion estratigrafica. *Boletim Academie Nacional de Ciencias* **46**: 5-28.

——— 1967e. *Peroniceras buergli* n.sp. (Amm. Collignoniceratidae) del Coniaciano de Colombia. *Acta geologica Lilloana* **9**: 231-238.

——— 1970. Ammonites nuevos o pocos conocidos del Aptiano, Albiano y Cenomaniano de los Andes australes con notas acerca de su posicion estratigrafica. *Revista de la Asociación geológica Argentina* **25**: 197-261.

——— & CASTELLARO, H.A. 1955. Algunos fosiles Cretacicos de Chile. *Revista de la Asociacion geologica Argentina* **10** (3): 179-213.

——— & LEANZA, H.A. 1973. *Pseudofavrella* gen. nov. (Ammonitina) del Hauteriviano de Neuquen, sus differ-encias con *Favrella* R. Douvillé, 1909, del Aptiano de Patagonia austral y una comparacion entre el geosinclinal andino y el geosinclinal magallanico. *Boletin de la Academia nacional de Ciencias, Cordoba* **50**: 127-145.

LEANZA, H.A. 1972. *Acantholissonia,* nuevo genero de ammonites del Valanginiano de Neuquen, Rep. Argentina y su posicion estratigrafica. *Revista de la Asociacion geologica Argentina* **27** (1): 63-70.

——— 1975. *Himalayites andinus* n.sp. (Ammonitina) del Tithoniano superior de Nequén, Argentina. *Actas del Premer Congreso Argentino de Paleontología y Estratigrafía., Tucuman, 12-16 agosto 1974*, **1**: 581-588.

——— 1980. The Lower and Middle Tithonian ammonite fauna from Cerro Lotena, province of Neuquen, Argentina. *Zitteliana* **5**: 3-49.

——— 1996. The Tithonian ammonite genus *Chigaroceras* Howarth (1992) as a bioevent marker between Iraq and Argentina. *In*, RICCARDI, A.C. (ed.), *Advances in Jurassic research.* GeoResearch Forum **1-2**: 451-458. Transtech Publishing: Uetikon-Zurich.

——— & ZEISS A. 1991. On the ammonite fauna of the lithographic limestones from the Zapala region (Neuquén province, Argentina), with the description of a new genus. *Zentralblatt für Geologie und Paläontologie* **1991** (6): 1841-1850.

——— & WIEDMANN, J. 1980. Ammoniten des Valangin und Hauterive (Unterkreide) von Neuquén und Mendoza, Argentinien. *Eclogae Geologicae Helvetiae* **73** (3): 941-982.

——— & ——— 1992. Nuevos Holcodiscidae (Cephalopoda - Ammonoidea) del Barremiano de la Cuenca Neuquina, Argentina, y su signicado estratigráfico. *Neues Jahrbuch für Geologie und Paläontologie, Monatshefte*, Heft 1: 24-38.

LEHMANN, J. 1998a. Systematic palaeontology of the ammonites of the Cenomanian-Lower Turonian (Upper Cretaceous) of northern Westphalia, northern Germany. *Tübinger Geowissenschaftliche Arbeiten* **A37**: 1-58.

——— 1998b. Taxonomy and nomenclature of large desmoceratids (Ammonoidea) from the Cenomanian and Turonian (Upper Cretaceous) of western Europe. *Neues Jahrbuch für Geologie und Paläontologie, Abhandlungen* **207**: 405-418.

——— 1999. Integrated stratigraphy and palaeoenvironment of the Cenomanian-Lower Turonian (Upper Cretaceous) of northern Westphalia, northern Germany. *Facies* **40**: 25-70.

——— 2011. Ammonite biostratigraphy and correlation of middle/late Albian drilling cores in the Hannover area (northern Germany). *Newsletters on Stratigraphy* **44**: 123-135.

——— & BAYLISS, H.M. 2022. The heteromorph ammonite *Tropaeum (Tropaeum) evolutum* sp. nov. from the Hythe Formation, Lower Greensand Group (basal Upper Aptian, Lower Cretaceous) of Maidstone, Kent, UK. *Cretaceous Research* **132**: 105-110.

——— & HERBIG, H.-G. 2009. Late Cretaceous ammonites from the Bou Angueur syncline (Middle Atlas, Morocco) - stratigraphic and palaeobiogeographic importance. *Palaeontographica* **289**: 45-87.

——— & MURPHY, M.A. 2001. A new placenticeratid ammonite from the Albian (Lower Cretaceous) of California and the inferred history of some placenticeratid and engonoceratid ammonites. *Neues Jahrbuch für Geologie und Paläontologie, Monatshefte* **2001** (6): 348-358.

———, SIMPSON, M.I. & BAYLISS, H.M. 2020. A new ammonite fauna from the *Walpenites* horizon of the upper Aptian (Lower Cretaceous), Isle of Wight, southern England. *Proceedings of the Geologists' Association* **131** (3-4): 321-333.

———, TROGER, K.-A. & OWEN H.G. 2008. Ammonites and associated macrofauna from the early Late Albian of the Zippelsforde 1/64 core, NE-Germany. *Acta geologica Polonica* **58**: 437-453.

LEHMANN, U. 1967. Ammoniten mit Kieferapparat und Radula aus Lias-Geschieben. *Paläontologische Zeitschrift* **41** (1-2): 38-45.

——— & HILLMER, G. 1980. *Wirbellose Tierre der Vorzeit Leitfaden der Systematischen Paläontologie.* Ferdinand Enke Verlag: Stuttgart.

LEMOINE, P. 1906. Études géologiques dans le nord de Madagascar. Contribution à l'histoire géologique de l'Ocean Indien. *Annales Hebert, Paris* **3**: 1-520.

——— 1910-1911. Paléontologie de Madagascar. VIII. Ammonites du Jurassique supérieur du cercle d' Analalava (Madagascar). *Annales de Paléontologie* **5** (4): 1-32 (137-168) (1910); **6**: 33 (45-64) (1911).

LEONHARD, R. 1897. Die Fauna der Kreideformation in Oberschlesien. *Palaeontographica* **44**: 11-70.

LEONOVA, T.B., BARSKOV, I.S. & MITTA, V.V. (eds) 2009. *Contributions to current cephalopod research: morphology, systematics, evolution, ecology and biostratigraphy.* PIN RAS: Moscow [in Russian with English summary].

LESHCHUKH, R.J. 1974. Fauna of ammonites from the Lower Cretaceous volcanic deposits of the basin of Shiben (the Ukrainian Carpathians). *Paleontological Collections* **1** (10): 56-64 [in Russian].

——— 1979. On new ammonite records in the Lower Cretaceous flysh of the south-east of the Ukrainian Carpathicans. *Paleontological Collections* **16**: 59-64 [in Russian].

——— 2014. Some bivalves and cephalopods from the Upper Aptian of the southern Ukraine (Plain Crimea). *Paleontological Review* **46**: 37-51 [in Ukrainian].

——— & HOTSANYUK, H. 2007. New finds of ammonites in the Lower Cretaceous of the Skibian Zone of the Ukrainian Carpathians. *Paleontological Collections* **39**: 31-36 [in Ukrainian].

LÉVEILLÉ, C. 1837. Description de quelques nouvelles coquilles fossiles du département des Basses-Alpes. *Mémoires de la Société géologique de France* (1) **2** (1835): 313-315.

LEWY, Z. 1967. Some Late Campanian nostoceratid ammonites from southern Israel. *Israel Journal of Earth-Sciences* **16**: 165-173.

——— 1969. Late Campanian heteromorph ammonites from southern Israel. *Israel Journal of Earth-Sciences* **18**: 109-135.

LEYMERIE, A. 1841-42. Mémoire sur le terrain Crétacé du Département de l'Aube. *Mémoires de la Société géologique de France* (1) **4** (1841): 291-364; **5** (1842): 1-34.

LINNAEUS, C. 1758. *Systema Naturae per regna tria naturae, secundum classes, ordines, genera, species, cum characteribus, differentiis, synonymis, locis.* [10th revised edition]. Laurentius Salvius: Holmiae.

LISSÓN , C.I. 1907. *Contribución a la geologia de Lima y sus alrededores*. Empresa Gil: Lima.

——— 1908. *Contribución al conocimiento sobre algunos ammonites del Peru.* 4th Congreso Cientifico Latino-Americano, 1st Pan-Americano, Celebrado en Santiago de Chile: 16a-c.

——— 1937. Dos amonitas del Peru. *Boletin de Minas Industrias y Construcciones* **3**: 1-9.

——— & BOIT, B. 1924. *Edad de los fósiles peruenos y distribución de sus depósitos*. 3rd edition. Imprenta Americana: Lima.

LIU, G-F. 1983. Jurassic and Cretaceous ammonites in the Gyangze area, Tibet. *In, Contribution to the geology of the Qinghai-Xizang (Tibet) Plateau*: 131-148. Ministry of Geology and Mineral Resources, Geological Publishing House, Beijing [in Chinese, with English summary].

——— 1988. Jurassic and Cretaceous ammonites from Nyalam Guco area, Xizang (Tibet). *In, Contribution to the geology of the Qinghai-Xizang (Tibet) Plateau*: 131-147. Geological Publishing House: Beijing [in Chinese].

——— & WANG, S.E. 1987. A new advance in study of the Upper Jurassic and Lower Cretaceous of the Himalayas, Xizang (Tibet). *Professional Papers of Stratigraphy and Palaeontology* **17**: 143-166 [in Chinese].

LOMINADZE, T.A. & KVANTALIANI I.V. 1985. On the systematics of family Dorsoplanitidae (Ammonoidea). *Bulletin of the Academy of Sciences, Georgian SSR* **119** (2): 337-340 [in Russian].

———, ——— 1986. The morphogenesis of the shell of the Volgian genus *Dorsoplanites* Semenov, 1898 (Ammonoidea, Cephalopoda). *Paleontological Collection* **23**: 27-32 [in Russian].

———, ——— & SHARIKADZE, M.Z. 1985. Shell morphogenesis of the Volgian ammonite genera *Pavlovia* Ilovaisky and *Laugeites* Spath (Ammonoidea, Cephalopoda). *In*, MENNER, V.V. (ed.), *Fossil Cephalopods*: 121-131. Nauka: Moscow [in Russian].

———, SHARIKADZE, M. & KVANTALIANI, I. 1993. Phylogeny and systematics of perisphinctids as interpreted from suture ontogenies. *Geobios* **15**: 275-286.

LOPEZ-HORGUE, M.A., OWEN, H.G., ARANBURU, A., FERNANDEZ-MENDIOLA, P.A. & GARCIA-MONDEJAR, A. 2009. Early late Albian (Cretaceous) of the central region of the Basque-Cantabrian Basin, northern Spain: biostratigraphy based on ammonites and orbitolinids. *Cretaceous Research* **30**: 385-400.

LORIOL, P. DE. 1882. Études sur la faune des couches du Gault de Cosne (Nièvre). *Mémoires de la Société paléontologique de Suisse* **9**: 1-118.

LORY, P. 1898. Le Crétacé inférieur du Dévolvy et des regions voisines. *Bulletin de la Société géologique de France* (3) **26** (1897): 132-138.

LUGER, P. & GRÖSCHKE, M. 1989. Late Cretaceous ammonites from the Wadi Qena area in the Egyptian Eastern Desert. *Palaeontology* **32**: 355-407.

LUKENEDER, A. 2005. The early Cretaceous *Karsteniceras* Level in the Vienna Woods (northern Calcareous Alps, Lower Austria). *Geologica Carpathica* **56**: 307-315.

——— & LUKENEDER, S. 2014. The Barremian heteromorph ammonite *Dissimilites* from northern Italy: taxonomy and evolutionary implications. *Acta palaeontologica Polonica* **59** (3): 663-680.

LUPPOV, N.P. 1949. New species of the genus *Cicatrites* from the Upper Aptian deposits of northern Caucasus. *Annual of the All-Soviet Paleontological Society* **13**: 71-74.

——— 1961. On ammonites from Lower Cretaceous deposits of south-west spurs of the Gissar range. *Transactions VSEGEI*, n.s., **46**. Problem of oil-and-gas content of the Middle Asia, Issue 2: 175-205.

——— 1963. New Cenomanian and Lower Turonian ammonites of the genus *Placenticeras* from central Asia. *Trudy Vsesoiuznogo Nauchno-Issledovatel'skogo geologicheskogo Instituta (VSEGEI)*, n s., **109**: 135-160.

———, BODYLEVSKII, V.I. & GLAZUNOVA, A.E. 1949. Class Cephalopoda. *In*, LUPPOV, N.P. (ed.), *Atlas of index forms of the fossil faunas of the USSR*. **10**. Gosteolisdat: Moscow, Leningrad [in Russian].

———, BOGDANOVA, T.N., LOBOCHEVA, S.V., AKOPYAN, V.T., DZHALILOV, M.R., KOROTKOV, V.A., MYATLYUK, E.V. & PORETSKAYA, E.S. 1983. Berriasian of Mangyshlak. *Transaction of the Academy of Sciences of the USSR* **17**: 1-235 [in Russian].

——— & DRUSHCHITS, V.V. (eds), 1958. Mollusca-Cephalopoda II. Ammonoidea (ceratites and ammonites) and Endocochlia. Supplement: Coniconchia. *In*, ORLOV, A. (ed.), *Fundamentals of paleontology* **6**. GONTI: Moscow.

LUPU, D. 1965. *Desmoceras (Pseudouhligella) devae*, o noua speciae de amonit Cenomanian de la Cherghes. *Dari de Seama ale sedintelor* **51** (2) (for 1963-64): 19-22.

MACELLARI, C.E. 1986. Late Campanian-Maastrichtian ammonite fauna from Seymour Island (Antarctic Peninsula). *Paleontological Society Memoir* **18**: 1-55.

——— 1988. Late Cretaceous Kossmaticeratidae (Ammonoidea) from the Magallanes Basin, Chile. *Journal of Paleontology* **62**: 889-905

MACHALSKI, M. 2005a. The youngest Maastrichtian ammonite faunas from Poland and their dating by scaphitids. *Cretaceous Research* **26**: 813-836.

——— 2005b. Late Maastrichtian and earliest Danian scaphitid ammonites from central Europe: taxonomy, evolution, and extinction. *Acta palaeontologica Polonica* **50** (4): 653-696.

——— 2012. Stratigraphically important ammonites from the Campanian-Maastrichtian boundary interval of the middle Vistula River section, central Poland. *Acta geologica Polonica* **62** (1): 91-116.

——— & HEINBERG, C. 2005. Evidence for ammonite survival into the Danian (Paleogene) from the Cerithium Limestone at Stevns Klint, Denmark. *Bulletin of the geological Society of Denmark* **52**: 97-111.

MAEDA, H. 1993. Dimorphism of Late Cretaceous false-puzosiine ammonites, *Yokoyamaoceras* Wright & Matsumoto, 1954 and *Neopuzosia* Matsumoto, 1954. *Transactions and Proceedings of the palaeontological Society of Japan*, n.s., **169**: 97-128.

MAGALHÃES, J. 1953. Sobre a posição taxonômica de *Coilopoceras* Hyatt (Ammonoidea - Engonoceratidae). *Revista Cientifica* **4** (1-2): 1-8.

MAHMOUD, I.G. EL-D. 1953. Sur un nouveau genre Albien d'ammonite, *Neposiella* nov. gen. *Compte Rendu Sommaire des Seances de la Société géologique de France* **1953**: 286-287.

——— 1955. Études paléontologiques sur la faune crétacique du massif du Moghara (Sinai, Egypte). *Publications de l'Institut du Desert d'Egypte* **8**: 1-192.

MALINOWSKA, L. 1972. The Middle Oxfordian Perisphinctidae of Zawodzie near Czestochowa. *Acta palaeontologica Polonica* **17** (2): 167-242.

MALLADA, L. 1887. Sinopsis de las especies fósiles que se han encontrado en España. Tomo 3. Terreno Mesozóico. *Boletin de la Comision del Mapa geologico de Espana* **14**: xix, 1-171.

——— 1891. Sinopsis paleontológica. Sistema Cretáceo (Cretáceo Superior). *Boletin de la Comision del Mapa geologico de Espana* **17** (1890): pls 7-12.

MANCINI, E.A. 1978. Origin of the Grayson micromorph fauna (Upper Cretaceous) of north-central Texas. *Journal of Paleontology* **52**: 1294-1314.

MANDOV, G. 1971. Representants du genre *Bochianites* Lory, 1898 (Ammonoidea) dans les depots du Cretace Inferieur en Bulgarie. *Geological Institute, Bulgarian Academy of Sciences*, Series Paleontologie, **20**: 91-106.

——— 1976. L'Etage Hauterivien dans les Balkanides occidentales (Bulgarie de l'ouest) et sa faune d'ammonites. *Annales des Universitie, Sofia* **67**: 11-99.

MANOLOV, J. R. 1962. New ammonites from the Barremian of north Bulgaria. *Palaeontology* **5**: 527-539.

——— 1975. *Barremites balkanicus* sp. n. du Barremien de Bulgarie de nord-ouest. *Spisanie na Bulgarskoto Geologichesko Druzhestvo* **36** (3): 314-317.

——— 1976. Ammonites du Crétacé inférieur du synclinal de Salch (Bulgarie du nord-ouest). *Spisanie na Bulgarskoto Geologichesko Druzhestvo* **37** (2): 139-147.

MANTELL, G.A. 1822. *The fossils of the South Downs, or illustrations of the geology of Sussex*. L. Rolfe: London.

MARCINOWSKI, R. 1974. The transgressive Cretaceous (Upper Albian through Turonian) deposits of the Polish Jura chain. *Acta geologica Polonica* **24**: 117-217.

——— 1979. *Alternacanthoceras* subgen. nov. (Ammonoidea) and some remarks on other Cenomanian representatives of the genus *Acanthoceras* Neumayr, 1875. *Acta geologica Polonica* **29**: 59-65.

——— 1980. Cenomanian ammonites from German Democratic Republic, Poland, and the Soviet Union. *Acta geologica Polonica* **30**: 215-325.

——— & WIEDMANN, J. 1990. The Albian ammonites of Poland. *Palaeontologia Polonica* **50**: 3-94.

MARCOU, J. 1858. *Geology of North America, with two reports on the prairies of Arkansas and Texas, the Rocky Mountains of New Mexico, and the Sierra Nevada of California*. Zürcher & Furrer: Zurich.

MARÉCHAL, M. 1994. Les *Hypacanthoplites* Spath, 1923 des argiles à *Bucaillella cayuexi* de Cauville (76) et leur position stratigraphique. *Bulletin Trimestriel de la Société Géologique de Normandie et des Amis du Muséum du Havre* **81** (2): 83-96.

MARKS, J.G. 1949. Nomenclatural units and tropical American Miocene species of the gastropod family Cancellariidae. *Journal of Paleontology* **23**: 453-464.

MARSHALL, P. 1917. Geology of the central Kaipara. *Transactions and Proceedings of the New Zealand Institute* **49** (for 1916): 433-450.

——— 1926. The Upper Cretaceous ammonites of New Zealand. *Transactions and Proceedings of the New Zealand Institute* **56**: 129-210.

——— 1927. A Kaipara ammonite. *Transactions of the New Zealand Institute* **58**: 357-358.

MARTIN, E. 2003. Intraspecific variation of *Dufrenoyia furcata* Sowerby (Ammonoidea, Ancylocerataceae, Deshayesitidae) from the Gargasian in the Vaucluse (southern France). *Mitteilungen aus dem Geologisch-Paläontologischen Institut der Universität Hamburg*, Heft **87**: 115-124.

MARTIN, W. 1809. *Petrificata derbiensia, or figures and descriptions of petrifactions collected in Derbyshire*. D. Lyon: Wigan.

MARTINEZ, R. 1982. Ammonoideos Cretácicos del Prepirineo de la provincia de Lleida. *Universidad Autónoma de Barcelona, Publicaciones de Geología* **17**: 1-197.

MARYASH, I. & KUREPA, Y. 2011. Acanthoceratid finds from the Cenomanian of Volyn. *Paleontological Collection* **43**: 56-64 [in Ukrainian].

——— & ——— 2014. Heteromorph ammonites from the Middle Cenomanian of the northeastern part of Volyn-Podillya. *Paleontological Review* **46**: 109-118 [in Ukrainian].

MASSE, J.-P. 1976. *Les calcaires urgoniens de Provence; Valanginien - Aptien inférieur. Stratigraphie, paléontologie, paléoenvironnement et leur évolution*. Thèse Sciences, Université d'Aix-Marseille.

MATHERON, P. 1842. *Catalogue méthodique et descriptif des corps organisés fossiles du Département des Bouches-du-Rhône et lieux circonvoisins; précédé d'un mémoire sur les terrains supérieurs au Grès Bigarré du S.E. de la France*. Carnaud Fils: Marseille.

——— 1878-1880. *Recherches paléontologiques dans le Midi de la France, ou étude sur les animaux fossiles découverts dans cette région présentant la description et la figure des espèces nouvelles, douteuses ou peu connues avec l'énumération méthodique des corps organisés fossiles qui le accompagnent dans leurs gisements stratigraphiques*. Marseille: P. Matheron.

MATRION, B., DUBUS, B. & TOUCH, R. 1998. Le genre *Neophlycticeras* (Spath, 1922) dans l'Albien de l'Aube et du Boulonnais. *Bulletin de l'Association géologique Auboise* **19**: 15-28.

MATSUKAWA, M. 1987. Early shell morphology of *Karsteniceras* (ancyloceratid) from the Lower Cretaceous Choshi Group, Japan and its significance to the phylogeny of Cretaceous heteromorph ammonites. *Transactions and Proceedings of the palaeontological Society of Japan*, n.s., **148**: 346-359.

———, SENDON, S.V., MATEER, F.T., SATO, T. & OBATA, I. 2012. Early Cretaceous ammonite fauna of Catanduanes Island, Philippines. *Cretaceous Research* **37**: 261-271.

MATSUMOTO, T. 1938a. Preliminary notes on some of the more important fossils among the Gosyoneura fauna. *Journal of the geological Society of Japan* **45**: 13-24.

——— 1938b. A biostratigraphic study on the Cretaceous deposits of the Naibuti valley, south Karahuto. *Proceedings of the Imperial Academy of Japan* **14**: 190-194.

——— 1938c. *Zelandites*, a genus of Cretaceous ammonite. *Japanese Journal of Geology and Geography* **15**: 137-148.

——— 1942a. A note on the Japanese Cretaceous ammonites belonging to the subfamily Desmoceratinae. *Proceedings of the Imperial Academy of Japan* **18**: 24-29.

——— 1942b. Fundamentals in the Cretaceous stratigraphy of Japan. Part 1: Succession of rocks and fossils at several important Cretaceous localities in Hokkaidô and Karahuto. *Memoirs of the Faculty of Science, Kyushu University* (D, Geology) **1**: 129-280.

——— 1942c. A note on the Japanese ammonites belonging to the Gaudryceratidae. *Proceedings of the Imperial Academy of Japan* **18**: 666-670.

——— 1947. A note on the Japanese Pachydiscinae. *Reports from the Department of Geology, Faculty of Science, Kyushu University* **2** (1): 34-46 (in Japanese).

——— 1951. A note on the Pachydiscinae, a Cretaceous ammonite group. *Transactions and Proceedings of the palaeontological Society of Japan*, n.s., **1**: 19-26.

——— 1953. The ontogeny of *Metaplacenticeras subtilistriatum* (Jimbo). *Japanese Journal of Geology and Geography* **23**: 139-150.

——— (ed.) 1954. *The Cretaceous System in the Japanese islands*. The Japan Society for the Promotion of Science: Tokyo (for 1953).

——— 1955a. Evolution of Peroniceratidae. *Transactions and Proceedings of the palaeontological Society of Japan*, n.s., **18**: 37-44.

——— 1955b. Family Kossmaticeratidae from Hokkaido and Saghalien. *Japanese Journal of Geology and Geography* **26**: 115-164.

——— 1955c. The bituberculate pachydiscids from Hokkaido and Saghalien. *Memoirs of the Faculty of Science, Kyushu University* (D, Geology) **5**: 153-184.

——— 1956. Further notes on the kossmaticeratids from Hokkaido. *Japanese Journal of Geology and Geography* **27**: 173-187.

——— 1959. Cretaceous ammonites from the upper Chitina valley, Alaska. *Memoirs of the Faculty of Science, Kyushu University* (D, Geology) **8**: 49-90.

——— 1960. *Graysonites* (Cretaceous ammonites) from Kyushu. *Memoirs of the Faculty of Science, Kyushu University* (D, Geology) **10**: 41-58.

——— 1965. A monograph of the Collignoniceratidae from

Hokkaido. *Memoirs of the Faculty of Science, Kyushu University* (D, Geology) **16**: 1-80 (part 1), 209-243 (part 2).
——— 1967. Evolution of the Nostoceratidae (Cretaceous heteromorph ammonoids). *Memoirs of the Faculty of Science, Kyushu University* (D, Geology) **18**: 331-347.
——— 1969. A monograph of the Collignoniceratidae from Hokkaido, part 3. *Memoirs of the Faculty of Science, Kyushu University* (D, Geology) **19**: 297-330.
——— 1970a. A monograph of the Collignoniceratidae from Hokkaido, part 4. *Memoirs of the Faculty of Science, Kyushu University* (D, Geology) **20**: 225-304.
——— 1970b. Uncommon keeled ammonites from the Upper Cretaceous of Hokkaido and Saghalien. *Memoirs of the Faculty of Science, Kyushu University* (D, Geology) **20**: 305-317.
——— 1971. A monograph of the Collignoniceratidae from Hokkaido, part 5. *Memoirs of the Faculty of Science, Kyushu University* (D, Geology) **21**: 129-162.
——— 1975. Additional acanthoceratids from Hokkaido. *Memoirs of the Faculty of Science, Kyushu University* (D, Geology) **22**: 99-163.
——— 1977. Some heteromorph ammonites from the Cretaceous of Hokkaido. *Memoirs of the Faculty of Science, Kyushu University* (D, Geology) **23**: 303-366.
——— 1978a. Part 4: Description of some species of *Pseudaspidoceras* and an allied new genus from Hokkaido and Saghalien. *In*, MATSUMOTO, T., KAWASHITA, Y., FUJISHIMA, Y. & MIYAUCHI, T., *Mammites* and allied ammonites from the Cretaceous of Hokkaido and Saghalien. *Memoirs of the Faculty of Science, Kyushu University* (D, Geology) **24**: 16-20.
——— 1984a. Some ammonites from the Campanian (Upper Cretaceous) of northern Hokkaido. *Palaeontological Society of Japan, Special Papers* **27**: 1-93.
——— 1984b. A new tetragonitid ammonite from Hokkaido. *Proceedings of the Japan Academy* (B, Physical and Biological Sciences) **60**: 33-35.
——— 1984c. An aberrant ammonite genus from the Cretaceous of Hokkaido. *Proceedings of the Japan Academy* (B, Physical and Biological Sciences) **60**: 341-344.
——— 1987b. Notes on *Forbesiceras* (Ammonoidea) from Hokkaido [Studies of Cretaceous ammonites from Hokkaido - LX]. *Transactions and Proceedings of the Palaeontological Society of Japan*, n.s., **145**: 16-31.
——— 1988. A monograph of the Puzosiidae (Ammonoidea) from the Cretaceous of Hokkaido. *Palaeontological Society of Japan, Special Papers* **30**: iii, 1-179.
——— 1989. A gigantic ammonite from Hobetsu, Hokkaido. *Bulletin of the Hobetsu Museum* **5**: 1-11.
——— 1991. The mid-Cretaceous ammonites of the family Kossmaticeratidae from Japan. *Palaeontological Society of Japan, Special Papers* **33**: 1-143.
——— 1995. Notes on gaudryceratid ammonites from Hokkaido and Sakhalin. *Palaeontological Society of Japan, Special Paper* **35**: 1-152.
——— & INOMA, A. 1975. Mid-Cretaceous ammonites from the Shumarinai-Soeushinai area, Hokkaido, part 1. *Memoirs of the Faculty of Science, Kyushu University* (D, Geology) **23**: 263-293.
——— & KANIE, Y. 1967. *Ainoceras*, a new heteromorph ammonoid genus from the Upper Cretaceous of Hokkaido. *Memoirs of the Faculty of Science, Kyushu University* (D, Geology) **18**: 349-359.
——— & KAWANO, T. 1975. A find of *Pseudocalycoceras* from Hokkaido. *Transactions and Proceedings of the palaeontological Society of Japan*, n.s., **97**: 7-21.
——— & KAWASHITA, Y. 1984. Two ammonite species of the genus *Sharpeiceras* from the Cretaceous of Hokkaido (Studies of Cretaceous ammonites from Hokkaido and Sakhalin - LXXXI). *Palaeontological Research* **2** (2): 89-91.
———, ———, FUJISHIMA, Y. & MIYAUCHI, T. 1978. *Mammites* and allied ammonites from the Cretaceous of Hokkaido and Saghalien [Studies of the Cretaceous ammonites from Hokkaido and Sakhalin - XXXV]. *Memoirs of the Faculty of Science, Kyushu University* (D, Geology) **24**: 1-24.
———, MOROZUMI, Y., BANDO, Y., HASHIMOTO, H. & MATSU-OKA, A. 1981. Note on *Pravitoceras sigmoidale* Yabe (Cretaceous heteromorph ammonite). *Transactions and Proceedings of the palaeontological Society of Japan*, n.s., **123**: 168-178.
———, MURAMOTO, T. & INOMA, A. 1972. Two small desmoceratid ammonites from Hokkaido. *Transactions and Proceedings of the palaeontological Society of Japan*, n. s., **87**: 377-394.
———, MURAMOTO, T. & TAKAHASHI, T. 1969. Selected acanthoceratids from Hokkaido. *Memoirs of the Faculty of Science, Kyushu University* (D, Geology) **19**: 251-296.
———, ———, ——— 1972. A new gaudryceratine ammonite from the Cenomanian of Hokkaido. *Memoirs of the Faculty of Science, Kyushu University* (D, Geology) **21**: 207-215.
———, NEMOTO, M. & SUZUKI, C. 1990. Gigantic ammonites from the Cretaceous Futaba Group of Fukushima Prefecture. *Transactions and Proceedings of the palaeontological Society of Japan*, n.s., **157**: 366-381.
——— & NISHIDA, T. 2004. A revised taxonomy of the ammonoid genus *Desmoceras* from Japan and southern Sakhalin. *Proceedings of the Academy of Japan*, Series B, **80**: 225-229.
———, ——— & TOSHIMITSU, S. 2003. Early Cenomanian (Cretaceous) ammonoids *Utaturiceras* and *Graysonites* from Hokkaido, north Japan [Studies of Cretaceous ammonites from Hokkaido and Sakhalin - XCV]. *Bulletin of the geological Survey of Japan* **54**: 131-159.
———, ———. ——— 2004. The early Cenomanian (Cretaceous) ammonite fauna from the Soeushinai area of Hokkaido, north Japan [Studies of Cretaceous ammonites from Hokkaido and Sakhalin - XCVII]. *Bulletin of the geological Survey of Japan* **55**: 67-92.
——— & OBATA, I. 1955. Some Upper Cretaceous desmoceratids from Hokkaido and Saghalien. *Memoirs of the Faculty of Science, Kyushu University* (D, Geology) **5**: 119-151.
———, ——— 1981. A new heteromorph ammonite from the Hokkaido in the collection of Yoshitaro Kawashita. *Bulletin of the national Science Museum, Tokyo* (C) **7**: 115-118.
———, ——— 1982. Some interesting acanthocerataceans from Hokkaido. *Bulletin of the national Science Museum, Tokyo* (C) **8**: 67-92.
———, ———, MAEDA, S. & SATA, T. 1964. *Yabeiceras* (Cretaceous ammonites) from Futaba, northeast Japan. *Transactions and Proceedings of the palaeontological Society of Japan*, n.s., **56**: 322-331.
———, SAITO, R. & FUKADA, A. 1957. Some acanthoceratids from Hokkaido. *Memoirs of the Faculty of Science, Kyushu University* (D, Geology) **6**: 1-45.
——— & SARKAR, S.S. 1966. Part 1: *Utaturiceras vicinale* (Stoliczka) from southern India. *In*, MATSUMOTO, T., SASTRY, M.V.A. & SARKAR, S.S., Notes on some Cretaceous ammonites from southern India. *Memoirs of the Faculty of Science, Kyushu University* (D, Geology) **17**: 297-309.
——— & SKWARKO, S.K. 1993. Cretaceous ammonites from south-central Papua New Guinea. *AGSO Journal of Australian Geology & Geophysics* **14/4**: 411-433.
——— & SUEKANE, T. 1987. Some acanthoceratid ammonites from the Yubari Mountains, Hokkaido, part 1. *Science Report of the Yokosuka City Museum* **35**: 1-14.
——— & TAKAHASHI, T. 2000. Further notes on the turrilitid ammonoids from Hokkaido - Part 1 (Studies of the Cretaceous ammonites from Hokkaido and Sakhalin - LXXXIX). *Paleontological Research* **4** (4): 261-273.

——— & Toshimitsu, S. 1984. On the systematic positions of the two ammonite genera *Hourcquia* Collignon, 1965 and *Pseudobarroisiceras* Shimizu, 1932. *Memoirs of the Faculty of Science, Kyushu University* (D, Geology) **25**: 229-246.
——— & Yokoi, K. 1987. Tiny ammonite *Worthoceras* from Hokkaido (Studies of Cretaceous ammonites from Hokkaido. LX). *Transactions and Proceedings of the palaeontological Society of Japan*, n.s., **146**: 42-48.
Matyja, B.A. & Wierzbowski, A. 2016. Ammonites and ammonite stratigraphy of the uppermost Jurassic (Tithonian) of the Owadów-Brzezinki quarry (central Poland. *Volumina Jurassica* **14**: 65-122.
Maury, C.J. 1936. O Cretáceo de Sergipe. *Monographia do Serviço Geológico e Mineralógico do Brasil, Rio de Janeiro* **11**: 1-282.
Mayer-Eymar, C.D.W. 1887. Systematisches Verzeichniss der Kreide und Tertiär Versteinerungen der Umgegend von Thun, nebst Beschreibung der neuen Arten. *Beiträge zur geologischen Karte der Schweiz* (Beilage zur 24 Lieferung) **2**: xxvii, 1-128.
——— 1893. Über Neocomian-Versteinerungen aus dem Somali-Land. *Vierteljahrsschrift der Naturforschenden Gesellschaft in Zürich* **38**: 249-265.
Mayr, E. 1942. *Systematics and the origin of species*. Columbia University Press: N.Y.
——— 1963. *Animal species and evolution*. Belknap Press: Harvard.
Mazenot, G. 1939. Les Palaeohoplitidae Tithoniques et Berriasiens du sud-est de la France. *Mémoires de la Société géologique de France*, n.s., **18** (41): 1-303.
McKenzie, E.D. 1999. A new early to middle Albian (Cretaceous) ammonite fauna from the Great Artesian Basin, Australia. *Proceedings of the Royal Society of Queensland* **108**: 57-88.
———, Rozefelds, A.C. & Deacon, P. 2014. An ancyloceratid ammonite from the Aptian Maryborough Formation, Queensland, Australia, and synonymy of *Australiceras* Whitehouse with *Tropaeum* Sowerby. *Alcheringa: An Australasian Journal of Palaeontology*, DOI: 10.1080/03115518.2014.862062.
McLearn, F.H. 1930. Notes on some Canadian Mesozoic faunas. *Transactions of the Royal Society of Canada* (3, section 4) **24**: 1-11.
——— 1931. The *Gastroplites* and other Lower Cretaceous faunas of the northern Great Plains. *Transactions of the Royal Society of Canada* (3) **25** (section 4): 1-8.
——— 1933. The ammonoid genera *Gastroplites* and *Neogastroplites*. *Transactions of the Royal Society of Canada* (3, section 4) **27**: 13-26.
——— 1945. Revision of the Lower Cretaceous of the Western Interior of Canada. *Geological Survey of Canada Paper* **44-17**: 1-14 (2nd ed.).
——— 1972. Ammonoids of the Lower Cretaceous sandstone member of the Haida Formation, Skidegate Inlet, Queen Charlotte Islands, western British Columbia. *Bulletin of the geological Survey of Canada* **188**: 1-78.
McLellan, R.D. 1927. The geology of the San Juan Islands. *University of Washington Publications in Geology* **2**: 1-185.
McNamara, K.J. 1978. *Myloceras* (Ammonoidea) from the Albian of central Queensland. *Alcheringa* **2**: 231-242.
——— 1980. Heteromorph ammonites from the Albian of South Australia. *Transactions of the Royal Society of South Australia* **104**: 145-159.
——— 1985. A new micromorph ammonite genus from the Albian of South Australia. *Special Publications of the South Australian Department of Mines and Energy* **5**: 263-268.
Medina, F.A. & Riccardi, A.C. 2005. Desmoceratidae, Silesitidae and Kossmaticeratidae (Ammonitida) from the Upper Aptian-Albian of Patagonia (Argentina). *Revue de Paléobiologie* **24** (1): 251-286.
———, ——— 2007. *Tuberosciponoceras*: a new dimorphic ammonoid genus from the Upper Albian of Patagonia, Argentina. *Paläontologische Zeitschrift* **81/2**: 123-130.
——— & Rinaldi, C.A. 1986. Ammonoideos del Albiano Superior en la Formacion Rio Mayer de la estancia La Vega, Provincia de Santa Cruz. *Contribucion Cientifica del Instituto antartico Argentino, Buenos Aires* **324**: 1-34.
Meek, F.B. 1857. Descriptions of new organic remains from the Cretaceous rocks of Vancouver's Island. *Transactions of the Albany Institute* **4**: 37-49.
——— 1863. Descriptions of new Cretaceous fossils from Nebraska territory, collected by the expedition sent out by the government under the command of Lieut. John Mullan, U.S. topographical engineers, for the location and construction of a wagon road from the sources of the Missouri to the Pacific Ocean. *Proceedings of the Academy of natural Sciences of Philadelphia* **1862**: 21-28.
——— 1871a. A preliminary list of fossils, collected by Dr. Hayden in Colorado, New Mexico and California, with brief descriptions of a few of the new species. *Proceedings of the American philosophical Society* **11** (1869-1870): 425-431.
——— 1871b. Preliminary paleontological report, consisting of lists of fossils, with descriptions of some new types, etc. *United States geological and geographical Surveys of the Territories, Annual Report* **4** (1870): 287-318.
——— 1876. A report on the invertebrate Cretaceous and Tertiary fossils of the Upper Missouri country. *In*, Hayden, F.V., *Report of the United States geological and geographical Surveys of the Territories.* **9**: lxiv, 1-629.
——— & Hayden, F.V. 1857. Descriptions of new species of Gasteropoda and Cephalopoda from the Cretaceous formations of Nebraska territory. *Proceedings of the Academy of natural Sciences of Philadelphia* **8** (for 1856): 70-72.
Meister, C. 1989. Les ammonites du Crétacé supérieur d'Ashaka Nigéria. *Bulletin des Centres de Recherche Exploration-Production Elf-Aquitane* **13** (Supplément): 1-84.
——— & Abdallah, H. 1996. Les ammonites du Cénomanien supérieur et du Turonien inférieur de la région de Gafsa-Chotts, Tunisie du centre-sud. *Geobios* **29** (Supplément): 3-49.
———, Buta, A., David, B. & Tavares, T. 2011. Les ammonites de la limite Albien-Cénomanien dans la région de Sumbe (Bassin de la Kwanza, Angola). *Revue de Paléobiologie* **30** (2): 685-781.
——— & Rhalmi, M. 2002. Quelques ammonites du Cénomanien-Turonien de la région d' Errachidia-Boudnid-Erfoud (partie méridionale du Haut Atlas central, Maroc). *Revue de Paléobiologie* **21** (2): 759-779.
———, M'Bina Mounguengui, M. & Lang, J. 2003. Les ammonites du Cénomano-Turonien du bassin côtier nord-gabonais: systématique et intérêt paléogéographique pour la liaison Téthys-Atlantique Sud. *Revue de Paléobiologie* **22** (1): 341-355.
Meledina, S.V., Alifirov, A.S. & Shurygin, B.N. 2010. On the position of the *Praechetaites exoticus* Zone in the Volgian Stage. *Stratigraphy and Geological Correlation* **18** (5): 550-554.
Meléndez, G., Atrops, F., Bello, J., Brochwicz-Lewinski, W., D'Árpa, C., Fozy, I., Pérez-Urresti, I, Ramajo, J. & Sequeiros, L. 2006. On the origin and palaeobiogeographic spread of Oxfordian ammonite genus *Passendorferia* and Tethyan subfamily Passendorferiinae Meléndez. *Volumina Jurassica* **4**: 240-241.
———, ———, ———, ———, ———, ———, ———, —
——— 2009. The Oxfordian ammonite genus *Passendorferia* Brochwicz-Lewinski and the Tethyan subfamily Passendorferiinae Meléndez: origin and palaeobiogeography. *Volumina Jurassica* **7**: 113-134.

MEMMI, L. 1974. *Neohoploceras solignaci*, ammonite nouvelle de l'Hauterivien de Tunisie. *Annales des Mines et de la Géologie, Tunis* **26**: 299-303.

——— 1981. Biostratigraphie du Crétacé inférieur de la Tunisie nordorientale. *Bulletin de la Société géologique de France*, Séries 7, **23**: 175-183.

MENEGHINI, J. 1867-1881. *Monographie des fossiles du calcaire rouge ammonitique (Lias supérieur) de Lombardie et de l'Apennin central*. Bernadoni di C. Rebeschini: Milan.

——— 1875. Nuove specie di *Phylloceras* e di *Lytoceras* del Lias sup. d' Italia. *Atti della Società Toscana di Scienze Naturali, Memorie*, Serie A, **1**: 104-109.

——— 1879. Descrizione dei nuovi Cefalopodi titonici di monte Primo e di Sanvicino. *Atti della Società Toscana di Sciense Naturali residente in Pisa* **4** (1): 3-10.

MERTON, R.K. 1968. The Matthew effect in science. *Science, N.Y.* **159** (3810): 56-63.

MESEZHNIKOV, M.S. 1963. Ammonites of the *Gravesia* genus in the subpolar Urals. *Transactions VNIGRI* **220** (Geological collection no.8): 120-130 [in Russian].

——— 1972. New Arctic genus of the Middle Volgian Dorsoplanitinae. Problems of the paleozoogeography of the Siberian Mesozoic. *Transactions of the Instute of Geology & Geophysics, Siberian Branch of the Academy of Sciences USSR* **111**: 119-131 [in Russian].

———, ALEKSEEV, S.N., KLIMOVA, I.G., SHUL'GINA, N.I. & GIUL'KHADZHAN, L.V. 1983. On the evolution of certain Craspeditidae at the Jurassic-Cretaceous boundary. *In*, ZAKHAROV, V.A. & NAL'NYAEVA, T.I. (eds), Mesozoic of the Soviet Arctic. *Trudy Instituta Geologii i Geofiziki ANSSSR, Sibirskoe Otdelenie Novosibirsk* **555**: 103-125 [in Russian].

MEUNIER, S. 1887. Contribution à la géologie de l'Afrique occidentale. *Bulletin de la Société géologique de France*, 3ème série, **6**: 61-63.

MICHELIN, H. 1838. Note sur une argile dépendant du Gault. *Memoires de la Société géologique de France* (1) **3**: 97-103.

MIKHAILOV, N.P. 1951. Upper Cretaceous ammonites of the south of the European part of the USSR. *Transactions of the Institute of Geological Sciences*, 129, Geological Series, **50**: 1-142 [in Russian].

——— 1962. *Pavlovia* and related ammonite groups. *Bulletin de la Société impériale des naturalistes de Moscou* (Geology) **37** (6): 3-30 [in Russian].

——— 1966. Boreal Jurassic ammonites (Dorsoplanitinae) and zonal subdivision of the Volgian Stage. *Transactions, Geological Institute of the Academy of Sciences of the USSR* **151**: 1-117 [in Russian].

——— & GUSTOMELOV, V. 1964. Boreal Late Jurassic cephalopods (Virgatosphinctinae). *Akademii nauk SSSR* **107**: 1-88 [in Russian].

MIKHAILOVA, I.A. 1957. On the systematics of families Parahoplitidae Spath and Deshayesitidae Stoyanow. *Bulletin of the Moscow State University* **3**: 173-182 [in Russian].

——— 1958. Some data on genera *Acanthohoplites* Sinzow and *Hypacanthoplites* Spath. *Bulletin of the Moscow State University*, Series biology, soil science, geology & geography, **1**: 101-108 [in Russian].

——— 1960. Ontogeny and systematic position of the genus *Colombiceras*. *Bulletin of Moscow Society of the Naturalists*, section Geology, **35** (2): 116-122 [in Russian].

——— 1962. Upper Aptian parahoplitins of the Daghestan and central Ciscaucasia. *In*, *Materials on geology and metallogeny of central and western Caucasus*: 132-140 [in Russian].

——— 1963. On the systematic position and scope of the genus *Diadochoceras*. *Paleontological Journal* **3**: 65-77 [in Russian].

——— 1972. A little known ammonite genus from the Clansayes horizon (Lower Cretaceous). *Paleontological Journal* **1972** (3): 78-86.

——— 1973. Ontogeny of ammonites of the family Leymeriellidae. *Paleontological Journal* **1973** (3): 26-34.

——— 1973. Peculiarities of the ontogeny of *Puzosia mayoriana* Orbigny and possible origin of the Desmocerataceae. *Transactions of the Academy of Sciences, USSR* **212** (1): 217-219 [in Russian].

——— 1974a. The systematic position of the genus *Ptychoceras* Orbigny. *Doklady Akademii nauk SSSR* **214**: 193-196 [in Russian].

——— 1974b. The relationship between early and late Cretaceous Hoplitidae. *Spisanie na Bulgarsko geologichesko Druzhestvo* **35** (2): 117-132 [in Russian].

——— 1974c. Peculiarities of the genus *Sonneratia* Bayle and its systematical position (Hoplitaceae, Ammonoidea). *Bulletin of Moscow Society of the Naturalists*, section Geology, **49** (1): 106-118 [in Russian].

——— 1975. The genus *Sokolovites* Casey and its stratigraphical significance. *Bulletin of Moscow State University* **6**: 38-45 [in Russian].

——— 1976a. New data on the ontogeny of certain parahoplitids. *Paleontological Journal* **1976** (1): 57-66.

——— 1976b. On a correct understanding of the ontogeny and characteristics of the family Deshayesitidae Stoyanow. *Doklady Akademii nauk SSSR* **226**: 444-447 [in Russian].

——— 1976c. Douvilleiceratidae, their systematics and phylogeny. *Spisanie na Bulgarsko geologichesko Druzhestvo* **37** (3): 256-273 [in Russian].

——— 1976d. On true understanding of the ontogeny of Deshayesitidae Stoyanow. *Transactions of the Academy of Sciences USSR* **226** (2): 444-447 [in Russian].

——— 1979. Evolution of the Aptian ammonites. *Paleontological Journal* **1979** (3): 3-11.

——— 1983. *Systematics and phylogeny of the Cretaceous ammonoids*. Nauka: Moscow [in Russian].

——— & BARABOSHKIN, E. 2009. The evolution of the heteromorph and monomorph Early Cretaceous ammonites of the Suborder Ancyloceratina Wiedmann. *Paleontological Journal* **43** (5): 527-536.

——— & NAIDIN, D.P. 2002. On the taxonomic position and distribution of the genus *Borissiakoceras* Arkhangelsky, 1916 (Ammonoidea). *Paleontological Journal* **6**: 46-56.

——— & SAVEL'EV, A.A. 1994. A new genus *Sulcohoplites* (Ammonoidea) from the Albian of the Mangyshlak Peninsula. *Paleontological Journal* **28** (1A): 73-80.

——— & TEREKHOVA, G.P. 1982. Morphogenesis of the *Scaphites* and systematic position of Scaphitaceae. *Bulletin of Moscow Society of the Naturalist*, Ser. geol., **57** (6): 107-112 [in Russian].

MIRZOEV, G.G. 1967a. New species of *Douvilleiceras* from the Lower Albian deposits of the south-west outskirts of the Gissar Range. *Paleontological Journal* **1**: 57-67.

——— 1967b. The interrelationships of the families Hoplitidae and Placenticeratidae. *Paleontological Journal* **1967** (4): 56-70.

——— 1969. New Albian ammonites from the southwestern slopes of the Gissar Mountains. *Paleontological Journal* **1969** (1): 38-50.

——— 1970. New ammonite from the Lower Albian of Turkmenia. *Paleontological Journal* **2**: 113-114.

——— & MELNIKOVA, T.P. 1971. Ontogenetic development of the Albian ammonite *Anahoplites*. *Paleontological Journal* **3**: 116-119.

MITTA, V.V. 1990. Intraspecific variability in the Volgian ammonites. *Paleontological Journal* **24** (3): 10-15.

——— 1993. The systematic composition of the Middle Volgian Virgatitidae (Ammonoidea) of central Russia. *Paleontological Journal* **4**: 41-48.

——— 1994. The systematic composition of the Middle Volgian Dorsoplanitidae (Ammonoidea) from central Russia. *Paleontological Journal* **28** (1): 31-44.

——— 2002. New data on Neocomitidae (Ammonoidea) from the Berriasian of the Moscow region. *Paleontological Journal* **36** (4): 351-355.

——— 2006. Ammonites of the genus *Garniericeras* in the Ryazanian Stage: the end of the myth? *In*, MUSATOV, V.A. (ed.), *Proceedings of the Third All Russia Conference "Cretaceous System of Russia and the adjacent countries: problems of stratigraphy and paleogeography," Saratov, September 26-30*: 98-99. SO EAGO: Saratov.

——— 2007. Ammonite assemblages from basal layers of the Ryazanian Stage (central Russia). *Stratigraphy and Geological Correlation* **15** (2): 80-92 [in Russian].

——— 2009. Ammonites of Tethyan origin from the Ryazanian of the Russian Platform: genus *Subalpinites* Mazenot. *Paleontological Journal* **43** (6): 615-625.

——— 2010. Late Volgian *Kachpurites* Spath (Craspeditidae, Ammonoidea) of the Russian Platform. *Paleontological Journal* **44** (6): 622-631.

——— 2011a. Ammonites of Tethyan origin in the Ryazanian Stage of the Russian Platform: genus *Riasanella* gen. nov. *Paleontological Journal* **45** (1): 13-22.

——— 2011b. Ammonites of Tethyan origin in the Ryazanian Stage of the Russian Platform: genus *Mazenoticeras* and other Neocomitidae. *Paleontological Journal* **45** (2): 143-153.

——— 2018a. Ammonites of Tethyan origin in the Ryazanian Stage of the Russian Platform: genera *Transcaspiites* Luppov and *Karasyazites* gen. nov. *Paleontolological Journal* **52** (3): 245-254.

——— 2018b. The genus *Delphinites* Sayn (Ammonoidea: Neocomitidae) in the Lower Valanginian of the Russian Platform. *Paleontological Journal* **52** (13): 1504-1516.

——— 2019a. Craspeditidae (Ammonoidea) of the Russian Platform at the Jurassic-Cretaceous boundary. I. Genus *Praesurites* Mesezhnikov and Alekseev. *Paleontological Journal* **53** (5): 471-481.

——— 2019b. Craspeditidae (Ammonoidea) of the Russian Platform at the Jurassic-Cretaceous boundary. II. Genus *Hectoroceras* Spath. *Paleontological Journal* **53** (6): 598-610.

——— 2021. Ammonites of the Ryazanian Stage of the Russian Platform: genus *Pronjaites* Sasonova. *Paleontological Journal* **55** (5): 511-519.

———, MIKHAILOVA, I.A. & SUMIN, D.L. 1999. Unusual Volgian scaphitoid ammonites from central Russia. *Paleontological Journal* **6**: 13-17 [in Russian].

——— & SHA, J. 2009. Late Jurassic ammonite evolution and palaeoenvironments of the Russian Platform. *Science in China* (D, Earth Sciences) **52** (12): 2024-2028.

MIZZARO-WIMMER, M. & SALVINI-PLAWEN, L. 2001. *Praktische Malakologie: Beiträge zur vergleichend-anatomischen Bearbeitung der Mollusken: Caudofoveata bis Gastropoda* - "Streptoneura". Springer Verlag: Wien.

MOBERG, J.C. 1885. Cephalopoderna i sveriges Kritsystem. II. Artbeskifning. *Sveriges geologiska Undersoekning* (C, Avhandlingar och Uppsatser) **73**: 1-64.

MOERCH, O.A.L. 1865. Quelques mots sur un arrangement de mollusques pulmonés terrestres (Géophiles, Fer.) basé sur le système naturel. *Journal de Conchyliologie* **13**: 265-283.

MOJSISOVICS, E.V. 1882. Die Cephalopoden der Mediterranen Triasprovinz. *Abhandlungen der geologischen Reichsanstalt* **10**: 1-322.

——— 1893. Das Gebirge um Hallstadt. Abteilung 1: Die Cephalopoden der Hallstatter Kalke. *Abhandlungen der kaiserlich-koeniglichen geologischen Reichsanstalt* **6** (2): 1-835.

MOLINER, L. & OLÓRIZ, F. 2010. New Lower Kimmeridgian ataxioceratin ammonite from the eastern Iberian Chain, Spain: Systematic, biogeographic, and biostratigraphic relevance. *Acta palaeontologica Polonica* **55** (1): 99-110.

MONKS, N. 1999. Cladistic analysis of Albian heteromorph ammonites. *Palaeontology* **42** (5): 907-925.

——— 2000. Functional morphology, ecology and evolution of the Scaphitaceae Gill, 1871 (Cephalopoda). *Journal of molluscan Studies* **66**: 205-216.

MOREMAN, W.L. 1942. Paleontology of the Eagle Ford Group of north and central Texas. *Journal of Paleontology* **16**: 192-220.

MORENO-BEDMAR, J.A., MANZO, R.B., SEMPERE, M.C. & BULOT, L.G. 2013. Aptian (Lower Cretaceous) ammonite biostratigraphy of the Francisco Zarco dam stratigraphic section (Durango State, northeast Mexico). *Journal of South American Earth Sciences* **42**: 150-158.

———, MENDOZA-ROSALES, C.C., MINOR, K.P., DELANOY, G., BARRAGÁN, R. & GONZÁLEZ-LEÓN, O. 2015. Towards an Aptian (Lower Cretaceous) ammonite biostratigraphy of the Mina Texali section, central Atlantic province (Puebla State, central Mexico). *Cretaceous Research* **54**: 203-211.

MORET, L. & MAHMOUD, G.E. 1953. Nouvelles observations stratigraphiques et paleontologiques sur l'Albien du Massif du Moghara (Sinaï-Egypte). *Travaux du Laboratoire de Géologie de la Faculté des Sciences de Grenoble* **31**: 267-272.

MÖRICKE, W. 1894. Versteinerungen des Lias und Unteroolith von Chile. *Neues Jahrbuch für Mineralogie, Geologie und Paläontologie, Beilage-Band* **9**, Heft 1: 1-100.

MORROW, A.L. 1935. Cephalopods from the Upper Cretaceous of Kansas. *Journal of Paleontology* **9**: 463-473.

MORTON, S.G. 1830. Synopsis of the organic remains of the Ferruginous Sand Formation of the United States, with geological remarks. *American Journal of Science* (1) **18**: 243-250.

——— 1834. *Synopsis of the organic remains of the Cretaceous Group of the United States.* Key & Biddle: Philadelphia.

——— 1842. Description of some new species of organic remains of the Cretaceous Group of the United States, with a tabular view of the fossils hitherto discovered in this formation. *Journal of the Academy of natural Sciences of Philadelphia* (1) **8**: 207-227.

——— 1843. Description of several new species of fossil shells from the Cretaceous deposits of the United States. *Proceedings of the Academy of natural Sciences of Philadelphia* **1**: 106-110.

MOSAVJNJA, A., WILMSEN, M., ARYAI, A.A., CHAHIDA, M.R. & LEHMANN, J. 2007. Mortoniceratinae (Amonitina) from the Upper Albian (Cretaceous) of the Atamir Formation, Koppeh Dagh mountains, NE Iran. *Neues Jahrbuch für Geologie und Paläontologie, Abhandlungen* **246**: 83-95.

MOURGUES, A.F., BULOT, L.G. & FRAU, C. 2015. The Valanginian Olcostephaninae Haug, 1910 (Ammonoidea) from the Andean Lower Cretaceous Chañarcillo Basin, northern Chile. *Andean Geology* **42** (2): 213-236.

MÜLLER, G. 1900. Die Versteinerungen des Jura und der Kreide. *In*, BORNHARDT, W., *Zur Oberflächen-gestaltung und Geologie Deutsch-Ostafrika's*: 514-571. Reimer: Berlin.

MÜLLER, G. & WOLLEMANN, A. 1906. Die Molluskenfauna des Untersenon von Braunschweig und Ilsede. II. Die Cephalopoden. *Abhandlungen der koeniglichen preussischen geologischen Landesanstalt*, n.s., **47**: 1-30.

MUNIER-CHALMAS, E.C.P.A. 1882. [Untitled]. *Bulletin de la Société géologique de France* (3) **10**: 425.

——— & Lapparent, A.A. de. 1894. Note sur la nomenclature des terrains sédimentaires. *Bulletin de la Société géologique de France* (3) **21** (for 1893): 438-488.

Muramiya, Y. & Shigeta, Y. 2021. *Sormaites*, a new heteromorph ammonoid genus from the Turonian (Upper Cretaceous) of Hokkaido, Japan. *Paleontological Research* **25** (1): 11-18.

Muravin, E.S. 1979. Ammonites of the genus *Lomonossovella* from the Middle Volgian beds of Glebovo (Yaroslavl area). *In*, Ivanov, A.N. (ed.), Late Mesozoic cephalopods of the Upper Volga area. *Inter-college collection of scientific papers, Issue* **183**: 4-17. Pedagogical University: Yaroslavl [in Russian].

Murphy, M.A. 1967a. The Aptian-Cenomanian members of the ammonite genus *Tetragonites*. *University of California Publications in Geological Sciences* **69**: 1-93.

——— 1967b. The ammonoid subfamily Gabbioceratinae Breistroffer. *Journal of Paleontology* **41**: 595-607.

——— 1967c. Aptian and Albian Tetragonitidae (Ammonoidea) from northern California. *University of California Publications in geological Sciences* **70**: 1-43.

——— 1975. Paleontology and stratigraphy of the Lower Chickabally Mudstone (Barremian-Aptian) in the Ono Quadrangle, northern California. *University of California, Publications in geological Sciences* **113**: 1-52.

——— & Rodda, P.U. 2006. California Early Cretaceous Phylloceratidae (Ammonoidea). *University of California, Riverside, Campus Museum Contribution* **7**: 1-98.

Myczynski, R. 1976. *Organoceras* gen. n. (Ammonoidea) from the Oxfordian of Cuba. *Acta geologica Polonica* **21** (4): 391-394.

——— 1989. Ammonite biostratigraphy of the Tithonian of western Cuba. *Annales Societatis geologorum Poloniae* **59**: 43-125.

——— 1994. Caribbean ammonite assemblages from the Upper Jurassic-Lower Cretaceous sequences. *Studia geologica Polonica* **105**: 91-108.

——— 1999. Some ammonite genera from the Tithonian of western Cuba and their palaeobiogeographic importance. *Studia geologica Polonica* **114**: 93-112.

——— & Iturralde-Vinent, M. 2005. The late Lower Albian invertebrate fauna of the Río Hatillo Formation of Pueblo Viejo, Dominican Republic. *Caribbean Journal of Science* **41** (4): 782-796.

——— & Trif, J. 1986. Los ammonites del Cretacico inferior de las provincias de Pinar del Rio y Matanzas. *Bulletin of the Polish Academy of Sciences,* Earth Science, **34** (1): 113-137.

Nagy, I.Z. 1969. Pulchellidák (Cephalopoda, Ammonoidea) a gerecsei alsókrétából. *Földtani Közlöny* **99**: 206-215 [in Hungerian].

Neave, S.A. 1939. *Nomenclator zoologicus*. **I**. Zoological Society: London.

Neaverson, E. 1925. Ammonites from the Upper Kimmeridge Clay. *Paper of the Geological Department, University of Liverpool* **1**: 1-52.

Neumayr, M. 1871. I. Jurastudien. 3. Die Phylloceraten des Dogger und Malm. *Jahrbuch der kaiserlich-koeniglichen geologischen Reichsanstalt, Wien* **21**: 297-378.

——— 1875a. Über Kreide Ammonitiden. *Sitzungsberichte der kaiserlichen Akademie der Wissenschaften in Wien, Mathematisch-Naturwissenschaftliche Klasse* (Abt. 1) **71**: 639-693.

——— 1875b. Die Ammoniten der Kreide und die Systematik der Ammonitiden. *Zeitschrift der deutschen geologischen Gesellschaft* **27**: 854-942.

——— & Uhlig, V. 1881. Über Ammonitiden aus den Hilsbildungen Norddeutschlands. *Palaeontographica* **27**: 129-203 (1-75).

Nicklès, R. 1890-1894. Contributions à la paléontologie de sud-est de l'Espagne. Terrain crétacé. I. Neocomien. *Mémoires de la Société géologique de France* **1** (2): 1-30 (1890); **4** (3): 31-59 (1894).

——— 1892. Recherches géologiques sur les terrains secondaires et tertiaires de la province d'Alicante et du sud de la province de Valence. *Annales Hebert* **1** (for 1891): 1-220.

Niebuhr, B. 1996. Die Scaphiten (Ammonoidea, Ancyloceratina) des höheren Obercampan der Lehrter Westmulde östlich Hannover (N-Deutschland). *Berliner Geowissenschaftliche Abhandlungen* **E18**: 267-287.

——— 2004. Late Campanian nostoceratid ammonites from the Lehrte West Syncline near Hannover, northern Germany. *Acta geologica Polonica* **54**: 473-487.

Nikitin, S.N. 1881. Die Jura-Ablagerungen zwischen Rybinsk, Mologa und Myschkin. *Mémoires de l'Academie imperiale des Sciences de St.-Petersbourg* (7) **28** (5): 1-98.

——— 1884. Allgemeine geologische Karte von Russland: Fauna der Jurassischen Bildungen des Blattes 56. *Mémoires du Comite géologique St. Petersbourg* **1** (2): 1-136 [in Russian].

——— 1888. Les vestiges de la période Crétacée dans la Russie centrale. *Mémoires du Comite géologique St. Petersbourg* **5** (2): 1-164 [in Russian].

Nikolov, T.G. 1960. The ammonite fauna of the Valanginian of the eastern pre-Balkans. *Trudovevrkhu Geologiiata no Bulgariia* (Paleontologiia) **2**:143-206 [in Bulgarian].

——— 1962. *Dobrogeites* - a new genus of Valanginian ammonites. *Doklady Bolgarskoi Akademii Nauk* **15**: 69-71.

——— 1963. New name for a Valanginian ammonite genus. *Geological Magazine* **100**: 94.

——— 1964. Barremian ammonites from north-eastern Bulgaria. *Comptes Rendus de L'Académie bulgare des Sciences,* Série Paléontologie, **4**: 117-141.

——— 1965. A new Valanginian ammonites (*Dobrodgeiceras benavidesi* sp. nov.) from Peru. *Comptes Rendus de l'Academie bulgare des Sciences* **18** (12): 1153-1155.

——— 1966a. New genera and subgenera of ammonites of family Berriasellidae. *Doklady Bolgarskoi Akademii Nauk* **19**: 639-642.

——— 1966b. *Protoleptoceras* gen. n. - a new genus of Berriasian ammonites. *Doklady Bolgarskoi Akademii Nauk* **19**: 839-841.

——— 1979a. A new ammonite subfamily Argentiniceratinae subfam. n. (Berriasellidae, Tithonian - Berriasian). *Compte Rendu de l'Academie bulgare des Sciences* **32** (2): 197-200.

——— 1979b. A new ammonite subgenus *Jabronella (Erdenella)* subgen. n. (Berriasellidae, Tithonian-Berriasian). *Compte Rendu de l'Academie bulgare des Sciences* **32**: 337-340.

——— & Breskovski, S. 1969. *Abrytusites* - nouveau genre d'ammonites Barrémiennes. *Izvestiya na geologicheskiya Institut, bulgarska Akademiya na Naukite* (Paleontologiya) **18**: 91-96.

——— & Sapunov, I.G. 1977. Sur une nouvelle sous-famille d'ammonites - Pseudosubplanitinae subfam. nov. *Compte Rendu de l'Academie bulgare des Sciences* **30**: 101-103.

Noble, A. H. 1911. On a new species of *Desmoceras* from the Chalk Rock of Buckinghamshire. *Geological Magazine*, n.s., (decade 5) **8**: 398-400.

Noetling, F. 1897. Fauna of the Upper Cretaceous (Maestrichtian) beds of the Mari Hills. *Memoirs of the geological Survey of India, palaeontologia indica* (16) **1** (3): 1-79.

Nolan, H. 1894. Note sur les *Crioceras* du groupe du *Crioceras duvali. Bulletin de la Société géologique de France* (3) **22**: 183-196.

Nowak, J. 1908. Untersuchungen über die Cephalopoden der Oberen Kreide in Polen. I. *Bulletin international de l'Academie des Sciences de Cracovie* (B) **1908**: 327-353.

——— 1911. Untersuchungen über die Cephalopoden der Oberen Kreide in Polen. II. *Bulletin international de l'Academie des Sciences de Cracovie* (B) **1911**: 547-589.

——— 1913. Untersuchungen über die Cephalopoden der Oberen Kreide in Polen. III. *Bulletin international de l'Academie des Sciences de Cracovie* (B) **1913**: 335-415.

——— 1916. Über die bifiden Loben der oberkretazischen Ammoniten und ihre Bedeutung für die Systematik. *Bulletin international de l'Academie des Sciences de Cracovie* (B) **1915**: 1-13.

Obata, I. 1967. Lower Cretaceous ammonites from the Miyako Group. Part 2: Some silesitids from the Miyako Group. *Transactions and Proceedings of the palaeontological Society of Japan*, n.s., **67**: 129-138.

———, Futakami, M., Kawashita, Y. & Takahashi, T. 1978. Apertural features in some Cretaceous ammonites from Hokkaido. *Bulletin of the national Science Museum, Tokyo* (C) **4** (3): 139-155.

——— & Matsukawa, M. 1984. A Barremian occurrence of an olcostephanid, a perisphinctacean ammonite, from the Chosi Group, Japan. *Bulletin of the national Science Museum, Tokyo* (C) **10**: 169-180.

——— & ——— 1988. Some Boreal or Subboreal ammonites in the Japanese Barremian. *In*, Wiedmann, J. & Kullmann, J. (eds), *Cephalopods - present and past:* 469-476. Schweizerbart'sche Verlagsbuchhandlung: Stuttgart.

Ohmert, W. & Zeiss, A. 1980. Ammoniten aus dem Hangenden Bankkalken (Unter-Tithon) der Schwäbischen Alb (Süd-westdeutschland). *Abhandlungen des Geologischen Landesamtes Baden-Württ* Heft **9**: 5-50.

Okamoto, T. 1989. Comparative morphology of *Nipponites* and *Eubostrychoceras* (Cretaceous nostoceratids). *Transactions and Proceedings of the palaeontological Society of Japan*, n.s., **154**: 117-139.

Olivero, E.B. 1988. Early Campanian heteromorph ammonites from James Ross Island, Antarctica. *National Geographic Research* **4**: 259-271.

——— 2012. New Campanian kossmaticeratid ammonites from the James Ross Basin, Antarctica, and their possible relationships with *Jimboiceras? antarcticum* Riccardi. *Revue de Paléobiologie*, Special Volume **11**: 133-149.

——— & Zinsmeister, W.J. 1989. Large heteromorph ammonites from the Upper Cretaceous of Seymour Island, Antarctica. *Journal of Paleontology* **63**: 626-636.

Olóriz, F. 1978a. *Kimmeridgiense-Tithonico inferior en el sector central de las Cordilleras Béticas (Zona Subbética). Paleontologia, bioestratigrafia.* Doctoral thesis, University of Granada.

——— 1978b. *Richterella*, Avram (1974) y *Richteria*, Olóriz (1976) paradigma de supuesta equivalencia. *Tecniterrae, Revista espanola de Geologia y Mineria, Madrid* **23**: 1-6.

——— & Tavera, J.M. 1979. *Simospiticeras* (Ammonoidea) gen. nov. avance sobre nuevas morfologias pertenecientes a la base del Tithonico Superior en las Cordilleras beticas (Zona Subbetica). *Cuadernos de Geologia* **8-9** (for 1977-1978): 181-189.

——— & Tintori, A. 1991. Upper Jurassic (Tithonian) ammonites from the Spiti Shales in western Zanscar (NW Himalaya). *Rivista Italiana di Paleontologia e Stratigrafia* **96** (4): 461-486.

———, Villaseñor, A.B. 2006. *Ceratosphinctes* (Ammonitina, Kimmeridgian) in Mexico: from rare but typical inhabitant of west-Tethyan epioceanic and epicontinental waters to a geographically widespread ammonite genus. *Geobios* **39**: 255-266.

———, ——— & Gonzáles-Arreola, C. 1998. Re-evaluation of *Procraspedites* Spath, 1930 (Ammonitina) from the Upper Kimmeridgian of Mexico. *Bulletin de la Société géologique de France* **169** (2): 243-254.

———, ——— & López Palomino, I. 2008. Middle Oxfordian ammonite biostratigraphy of the Lower Santiago Formation at Tamán, San Luis Potosí, Mexico - with notes on Mexico-Caribbean ochetoceratins. *Revista Mexicana de Ciencias Geológicas* **25** (2): 261-283.

——— & Westermann, G.E.G 1998. The perisphinctid ammonite *Sulaites* n. gen. from the Upper Jurassic of the Indo-Southwest Pacific. *Alcheringa: An Australasian Journal of Palaeontology* **22**: 3, 231-240.

Olsson, A.A. 1944. Contributions to the palaeontology of northern Peru. Part 7: The Cretaceous of the Paita region. *Bulletins of American Paleontology* **28** (111): 159-304 (1-146).

Ooster, W. A. 1860. Catalogue des céphalopodes fossiles des Alpes suisses. V. Céphalopodes tentaculifères, etc. *Neue Denkschriften der Allgemeinen schweizerischen Gesellschaft fuer die Gesamten Naturwissenschaften* **18**: 1-100.

Oppel, A. 1862-1863. Über jurassische Cephalopoden. *Palaeontologische Mitteilungen aus dem Museum des koeniglich-bayerischen Staates, Berlin* **1**, Abt. 1-2: 1-125 (1862); Abt. 3: 127-266 (1863).

——— 1865. Die Tithonische Etage. *Zeitschrift der deutschen geologischen Gesellschaft* **17**: 535-558.

Orbigny, A. d'. 1840-1842. *Paléontologie française. Terrains Crétacés. I. Céphalopodes*: 1-120 (1840); 121-430 (1841); 431-662 (1842). Masson: Paris.

——— 1842b. *Voyage dans l'Amérique méridionale.* **3** (4, Paléontologie). C.F.P. de Martius: Paris, Strasbourg.

——— 1842c. *Paleontologie Francaise, Tome 1. Description zoologique et geologique de tous les animaux mollusques et rayonnes fossiles de France, comprenant leur application a la reconnaissance des couches; terrains oolitiques ou Jurassiques.* Mason: Paris.

——— 1845a. *Paléontologie française. Terrains Jurassiques. I. Cephalopodes*: 313-368. Masson: Paris.

——— 1845b. Mollusques. *In*, Murchison, R.I., de Verneuil, E. de & Keyserling, A. von, *Geologie de la Russie d'Europe. II. Terrains secondaires*: 419-498. R. & J.E. Taylor: London, Paris.

——— 1849a. *Cours élémentaire de paléontologie et de géologie stratigraphiques.* **1**. Masson: Paris.

——— 1849b. *Prodrome de paléontologie stratigraphique universelle des animaux mollusques et rayonnés.* **1**. Masson: Paris.

——— 1850a. *Prodrome de paléontologie stratigraphique universelle des animaux mollusques et rayonnés.* **2**. Masson: Paris.

——— 1850b. Notes sur quelques nouvelles espèces remarquables d'ammonites des étages Néocomien et Aptien de la France. *Journal de Conchyliologie* **1**: 196-201.

——— 1851. Notice sur le genre *Heteroceras* de la classe des céphalopodes. *Journal de Conchyliologie* **2**: 217-222.

——— 1852. Notice sur le genre *Hamulina*. *Journal de Conchyliologie* **3**: 207-228.

Ovando-Figueroa, J.R., Moreno-Bedmar, J.A., Minor, K.P., Franco-Rubio, M., Oviedo, A., Patarroyo, P. & Robert, E. 2018. Ammonite taxonomy and biostratigraphy for the Upper Aptian-Lower Albian (Lower Cretaceous) of Cerro Chino, Chihuahua State, northeast Mexico. *Cretaceous Research* **82**: 109-137.

Owen, D.D. 1852. Description of new and imperfectly known genera and species of organic remains, collected during the geological surveys of Wisconsin, Iowa and Minnesota. *In*, *Report of a geological Survey of Wisconsin, Iowa, and Minnesota*: 571-587. Lippincott: Philadelphia.

Owen, H.G 1971. Middle Albian stratigraphy in the Anglo-Paris Basin. *Bulletin of the British Museum of Natural History* (Geology), Supplement **8**: 1-164.

——— 1988. The ammonite zonal sequence and ammonite taxonomy in the *Douvilleiceras mammillatum* Superzone (Lower Albian) in Europe. *Bulletin of the British Museum (Natural History)* (Geology) **44**: 177-231.

Owen, R. 1832. *Memoir on the Pearly Nautilus (*Nautilus pompilius, *Linn.) with illustrations of its external form and internal structure.* Royal College of Surgeons: London.

Page, K.N. 1993. Mollusca: Cephalopoda (Ammonoidea: Phylloceratina, Lytoceratina, Ammonitina, and Ancyloceratina). *The Fossil Record* **2**: 213-227. Chapman & Hall: London.

Panday, B. 1993. *The Kachchh Jurassic Virgatosphinctinae, high resolution evolutionary succession and chronology.* PhD thesis, Banaras Hindu University, Varanasi.

———, Krishna, J. & Pathak, D.B. 2010. Review of the ammonoid subfamily Virgatosphinctinae with special reference to its evolutionary succession in the Indian subcontinent on the Gondwanian Tethyan margin. *Journal of Scientific Research* **54**: 21-34.

Pandey, D.K., Alberti, M., Fürsich, F.T., Glowniak, E. & Olóriz, F. 2013. Ammonites from the Oxfordian-Kimmeridgian boundary and the Lower-Upper Kimmeridgian of Kachchh, western India. *Volumina Jurassica* **11**: 97-146.

Paquier, V. 1900. Recherches géologiques dans le Diois et les Barronnies orientales. *Bulletin de la Société de Statistique des Sciences naturelles et des Arts industriels de Departement de l'Isere, Grenoble* (4) **5**: 77-476.

Parent, H. 2001. The Middle Tithonian (Upper Jurassic) ammonoid fauna of Cañadón de los Alazanes, southern Neuquén Mendoza Basin (Argentina). [La fauna del Tithoniano Medio de Cañadón de los Alazanes, sur de la cuenca Neuqúen, Mendoza (Argentina)]. *Boletín del Instituto de Fisiografía y Geología* **71** (1-2): 19-38.

——— 2003a. Taxonomic and biostratigraphic re-evaluation of *Perisphinctes internispinosus* Krantz, 1926 (Upper Jurassic, Ammonoidea). *Paläontologische Zeitschrift* **77** (2): 353-360.

——— 2003b. The ataxioceratid ammonite fauna of the Tithonian (Upper Jurassic) of Casa Pincheira, Mendoza (Argentina). *In*, Parent, H., Meléndez, G & Olóriz, F. (eds), Jurassic of South America. *Journal of South American Earth Sciences, Special Issue* **16**: 143-165.

——— & Capello, O.D. 1999. Amonites del Tithoniano inferior de Casa Pincheira, Mendosa (Argentina). *Revue Paléobiologie* **18** (1): 347-353.

——— & Garrido, A.C. 2021. The Tithonian ammonite fauna of the transect Cerro Lotena-Cerro Granito, Vaca Muerta Formation, Argentina. I. Family Himalayitidae. *Boletín del Instituto de Fisiografía y Geología* **91**: 21-60.

———, Scherzinger, A. & Schweigert, G, 2011. The Tithonian-Berriasian ammonite fauna and stratigraphy of Arroyo Cieneguita, Neuquén-Mendoza Basin, Argentina. *Boletín del Instituto de Fisiografía y Geología* **79-81**: 21-94.

———, ———, ——— & Fözy, I. 2015. The Tithonian-Lower Valanginian stratigraphy and ammonite fauna of the Vaca Muerta Formation in Pampa Tril, Neuquén Basin, Argentina. *Boletín del Instituto de Fisiografía y Geología* **86**: 1-96.

———, ———, ——— & Garrido, A.C. 2017a. Additional Tithonian and Berriasian ammonites from the Vaca Muerta Formation in Pampa Tril, Neuquén Basin, Argentina. *Volumina Jurassica* **15**: 139-154.

———, ———, ———, ——— 2017b. Zapaliinae, a new subfamily of Tithonian-Berriasian ataxioceratid ammonites. *Paläontologische Zeischrift* **91**: 507-517.

———, Schweigert, G. & & Melendez, G. 2006. Oxfordian perisphinctid ammonites from Chacay Melehué, Argentina. *Paläontologische Zeitschrift* **80** (4): 307-324.

———, ——— & Scherzinger, A. 2020. A review of the classification of Jurassic aspidoceratid ammonites - the superfamily Aspidoceratoidea. *Volumina Jurassica* **18** (1): 47-52.

———, Myczynski, R., Scherzinger, A. & Schweigert, G 2010. *Cieneguiticeras*, a new genus of Tithonian oppeliids (Ammonoidea, Late Jurassic). *Geobios* **43** (4): 453-463.

Parkinson, J. 1811. The fossil starfish, echini, shells, insects, amphibia, mammalia & c., *In, The organic remains of a former world.* **3**. Sherwood, Neely & Jones: London.

Parnes, A. 1964. Coniacian ammonites from the Negev (southern Israel). *Geological Survey of Israel Bulletin* **39**: 1-42.

Parona, C.F. & Bonarelli, G 1895. Sur la faune du Callovien Inférieur (Chanasien) de Savoie. *Memoires de l'Academie royale des Sciences, Belles-lettres et Arts de Savoie, Chambery* (4) **6**: 1-183.

———, ——— 1897. Fossili Albiani d'Escragnolles del Nizzardo e della Liguria occidentale. *Palaeontographia italica* **2** (for 1896): 53-112.

Passendorfer, E. 1930a. Stratigraphic and paleontologic study of the Kreda Series from a bore in the Tatras. *Prace olskiego Instytutu Geologicznego* **2**: 351-677 (1-327) [in Polish, with French summary].

——— 1930b. Étude stratigraphique et paléontologique du Crétacé de la Série Hauttatrique dans les Tatras. *Travaux de la Société géologique de Pologne* **2**: 511-676.

Patarroyo Gama, P.C.R. 1999. *Entwicklung der Ammoniten der familie Pulchelliidae aus dem Barreme von Zentral-Kolombie.* Inaugural-Dissertation zur Erlangung des Doktorgrades der Naturwissenschaften der Justus-Liebig-Universität Gießen (Fachbereich Geowissenschaften und Geographie).

——— 2000. Distribucion de amonitas del Barremiano de la Formacion Paja en el sector de Villa de Leyva (Boyaca, Colombia). Bioestratigrafia. *Geologia Colombiana* **25**: 149-162.

——— 2011. Sucesion de amonitas del Cretácico Superior (Cenomaniano-Coniaciano) de la parte más alta de la Formación Hondita y de la Formación Loma Gorda en la quebrada Bambucá, Aipe - Huila (Colombia, S.A.). *Boletín de Geología* **33** (1): 69-92.

——— 2020. Barremian deposits of Colombia: a special emphasis on marine successions. *In*, Gómez, J. & Pinilla-Pachon, A.O. (eds), The geology of Colombia. Volume 2, Mesozoic. *Servicio Geológico Colombiano, Publicaciones Geológicas Especiales* **36**: 403-439. doi.org/10.32685/pub.esp.36.2019.12.

Pathak, D.B. 1989. *The Kimmeridgian-Tithonian at Ler-Katrol, Kachchh, India, ammonoid systematics, biochronology and biogeography.* PhD thesis, Banaras Hindu University, Varanasi.

Patrulius, D. & Avram, E. 1976. Les céphalopodes des couches de Carthaga (Tithonique Supérieur-Barrémien Inférieur). *Mémoires de l'Institut de Geologie et de Geophysique, Bucarest* **24**: 153-201.

Paulcke, W. 1907. Die Cephalopoden der Oberen Kreide Südpatagoniens. *Berichte der Naturforschenden Gesellschaft zu Freiburg im Breisgau* **15** (for 1906): 167-248.

Pavlow, A. P. 1890. Études sur les couches Jurassiques et Crétacées de la Russie. I. Jurassique Supérieur et Crétacé Inférieur de la Russie et de l'Angleterre. *Byulleten Moskovskogo Obshchestva Ispytatelei Prirody*, n.s., **3**: 61-127.

——— 1892. Ammonites de Speeton et leurs rapports avec les ammonites des autres pays. *In*, Pavlow, A.P. & Lamplugh, G.W., Argiles de Speeton et leurs équivalents. *Bulletin de la Societe imperiale des Naturalistes de Moscou*, n.s., **5** (4) (for 1891): 181-276, 455-570.

——— 1901. Le Crétacé Inférieur de la Russie et sa faune. I. Aperçu historique des recherches, suivi d'indications sur la distribution des mers et des terres aux différentes époques. II. Céphalopodes du Néocomien Supérieur du type de Simbirsk. *Nouveaux Memoires de la Societe imperiale des Naturalistes de Moscou* **16** (3): 1-87.

——— 1914. Jurassic and Lower Cretaceous cephalopods of northern Siberia. *Zapiski Imperatorskoi Akademii nauk po Fiziko-Matematicheskomu otdeleniyu S. Petersburg* (8) **21** (4): 1-68 [in Russian].

PERGAMENT, M.A. 1965. Inocerams and stratigraphy of the Cretaceous of the Pacific Realm. *Transactions of the geological Institute, Academy of Science USSR* **118**: 1-102 [in Russian].

PERON, A. 1889. *Description des mollusques fossiles des terrains crétacées de la region des Hauts-Plateaux de la Tunisie recueillis en 1885 et 1886 par M. Philippe Thomas: Exploration scientifique de la Tunisie.* Impremier nationale: Paris.

——— 1896-1897. Les ammonites du Crétacé Supérieur de l'Algerie. *Memoires de la Société géologique de France* (Paléontologie) **6-7** (17): 1-24 (1896), 25-88 (1897).

PERVINQUIÈRE, L. 1903a. *Ammonites salmuriensis* Courtiller, 1867. *Palaeontologia Universalis*, Centuria 1, fiche 6, p. 6, 6a, 6b. Congres Geologique International: Laval.

——— 1903b. *Ammonites revelieranus* Courtiller, 1860. *Paléontologia Universalis*, Centuria 1, fiche 7, p. 7, 7a, 7b. Congres géologique International: Laval.

——— 1907. Études de paléontologie tunisienne. I. Céphalopodes des terrains secondaires. *Carte géologique de la Tunisie.* Rudeval: Paris.

——— 1910. Sur quelques ammonites du Crétacé algérien. *Mémoires de la Société géologique de France*, Paléontologie **17** (42): 1-86.

PHILLIPS, J. 1829. *Illustrations of the geology of Yorkshire; or, a description of the strata and organic remains of the Yorkshire coast.* Thomas Wilson & Sons: Highousegate, York.

PICTET, A. 2012. Evolution of the Cheloniceratinae: an alternative to Deshayesitidae for the Lower Aptian biozonation? Preliminary observations. *In*, BERT, D. & BERSAC, S. (eds), First meeting of the research group for paleobiology and biostratigraphy of the ammonites. *Boletin del Instituto de Fisiografia y Geologia, Rosario* **82**: 14-16.

PICTET, F.-J. 1847. Description des mollusques fossiles qui se trouvent dans les Grès Verts des environs de Genève. *Mémoires de la Société de Physique et d'Histoire naturelle de Genève* **11**: 257-412.

——— 1854. Céphalopodes. *In*, *Traité de paléontologie* **2**: 583-716. 2nd ed., J.-B. Bailliere: Paris.

——— 1867. Études paléontologiques sur la faune à *Terebratula diphyoides* de Berrias (Ardèche). *Melanges paléontologiques* **1** (2): 44-130.

——— 1868. Étude provisoire des fossiles de la Portede-France, d'Aizy et de Lémenc. *Melanges paléontologiques* **4**: 207-312.

——— & CAMPICHE, G. 1858-1860. Description des fossiles du terrain Crétacé des environs de Sainte-Croix, part 1. *Materiaux pour la Paléontologie Suisse* (2) **1858**: 1-96; **1859**: 97-208; **1860**: 209-380.

——— 1861-1864. Description des fossiles du terrain Crétacé des environs de Sainte-Croix, part 2. *Materiaux pour la Paléontologie Suisse* (3) **1861**: 1-144 ; **1862**: 145-348; **1863**: 349-524; **1864**: 525-752.

POKHIALAINEN, V.P. 1984. Late Senonian heteromorph ammonites of Anadyr-Koryakia region. Stratigraphy and paleontology of Paleozoic and Mesozoic deposits of north-east USSR. *Collection of scientific Papers, Moscow*: 198-205 [in Russian].

POLKOVOY, K.S. 2021. A new genus *Kislovodskoceras* (Desmoceratidae, Ammonoidea) from the Middle Aptian of the vicinity of Kislovodsk (northern Caucasus). *Palaeontological Journal* **55** (5): 520-528.

——— 2022 Representatives of the subfamily Cheloniceratinae Spath, 1923 (Ammonoidea) from the Aptian of Mangyshlak, western and southern Turkmenistan. Paper 1. General Issues. The genus *Cheloniceras* Hyatt, 1900. *Izvestiya of Saratov University, Earth Sciences* **22** (3): 191-207 [in Russian].

POMEL, A. 1889. Les céphalopodes Néocomiens de Lamoricière. *Materiaux pour la Carte géologique de l'Algerie* (1) **2**: 1-99.

POPPER, K. 1959. *The logic of scientific discovery.* Abingdon-on-Thames: Routledge.

POULTON, T.P., ZEISS, A. & JELETZKY, J.A. 1988. New molluscan faunas from the Late Jurassic (Kimeridgian and Early Tithonian) of western Canada. *Bulletin of the geological Survey of Canada* **373**: 103-116.

POWELL, J.D. 1963. Cenomanian-Turonian (Cretaceous) ammonites from Trans-Pecos Texas and northeastern Chihuahua, Mexico. *Journal of Paleontology* **37**: 309-322.

PRIGIROVSKY, M.M. 1906. Nouvelles donnees sur les ammonites du groupe *Craspedites okensis* du gouvernement de Yaroslavl. *Verhandlungen der Kaiserlichen Russischen Mineralogischen Gesselschaft*, Ser. 2, Band **44**, Lief.1: 483-505 [in Russian, with French summary].

PUGIN, L. 1964. Ammonites préalpines: étude critique des Lytoceratina du Dogger. *Mémoires suisses de Paléontologie* **80**: 1-67.

PUZOS, M. 1832. Sur le *Scaphites yvani*. *Bulletin de la Société géologique de France* (1) **2**: 355-356.

QUENSTEDT, F.A. 1846-1849. *Petrefaktenkunde Deutschlands. Die Cephalopoden.* Fuess: Tübingen.

——— 1856-1858. *Der Jura* [1856: 1-576, 1857: 577-823, 1858: titles, etc.]. H. Lampschen: Tübingen.

RABRENOVIC, D. 1989. *Lepeniceras* n. gen. from the Barremian of Grebena near D. Milanovca, eastern Serbia. *Geoloski anali Balkanskoga poluostrva* **53**: 367-372 (in Serbian with English summary).

——— 1991. Stratigraphy of the Lower Cretaceous with Cephalopoda from Greben on the Danube to Novog Korito E Serbia). Unpublished PhD thesis, University of Belgrade (in Serbian).

RAAB, M. 1962. Jurassic-Early Cretaceous ammonites from the southern coastal plain, Israel. *Bulletin of the geological Survey of Israel* **34**: 24-30

RAISOSSADAT, S.N. 2004. The family Deshayesitidae in the Kopet Dagh Basin, north-east Iran. *Cretaceous Research* **25**: 115-136.

——— 2006. The family Parahoplitidae in the Sanganeh Formation of the Kopet Dagh Basin, north-eastern of Iran. *Cretaceous Research* **27**: 907-922.

——— 2010. Upper Barremian-Lower Albian succession biozonation based on ammonites in the Koppeh Dagh Basin, north east of Iran. *Journal of Science, University of Tehran* **35** (4): 87-96.

——— 2018. Late Barremian Heteroceratidae Hyatt 1900 (Ammonoidea), from the Sarcheshmeh Formation (Koppeh Dagh Basin, northeast Iran): biostratigraphy and paleobiogeography. *Geopersia* **8** (2): 171-197.

RAKOTONIMANANA, R.M. & RANDIANALY, H.N. 2016. Aspect phylogénétique du genre *Cleoniceras* de l'Albien du sud du Bassin de Mahajanga. *Madamines* **1**: 1-35.

RASPAIL, F.V. 1831. Ammonites des Basses-Alpes et des Cevennes. *Annales des Sciences d'Observation* **3**: 115.

RAWSON, P.F. 1971. Lower Cretaceous ammonites from north-east England: the Hauterivian genus *Simbirskites*. *Bulletin of the British Museum (Natural History)*, Geology **20**: 25-86.

——— 1974. Hauterivian (Lower Cretaceous) ammonites from Helgoland. *Geologisches Jahrbuch*, Reiche A, Band **25**: 55-83.

——— 1975a. The interpretation of the Lower Cretaceous heteromorph ammonite genera *Paracrioceras* and *Hoplocrioceras* Spath, 1924. *Palaeontology* **18**: 275-283.

——— 1975b. Lower Cretaceous ammonites from north-east England: the Hauterivian heteromorph *Aegocrioceras*. *Bulletin of the British Museum (Natural History)*, Geology **26**: 129-159.

———, CURRY, D., DILLEY. F.C., HANCOCK, J.M., KENNEDY, W.J., NEALE, J.W., WOOD, C.J. & WORSSAM, B.C. 1978. A correlation of Cretaceous rocks in the British Isles. *Geological Society of London, Special Report* **9**: 1-70.

——— & KEMPER, K. 1978. *Varlheidites*, n. gen. (Ammonoidea, Neocomitinae) aus dem Obervalangin NW-Deutschlands. *Geologisches Jahrbuch* **A45**: 163-181.

REAGAN, A.B. 1924. Cretacic Mollusca of Pacific Slope. *Pan-American Geologist* **41**: 179-190.

REBOULET, S. 1996. L'évolution des ammonites du Valanginien-Hauterivien inférieur du bassin vocontien et de la bordure plateform Provençale (sud-est de la France): relations avec la stratigraphie séquentielle et implications biostratigraphiques. *Documents des Laboratoires de Géologie Lyon* **137**:1-371.

———, HOEDEMAEKER, P.J., AGUIRRE-URRETA, M.B., ALSEN, P., ATROPS, F., BARABOSHKIN, E.Y., COMPANY, M., DELANOY, G., DUTOUR, Y., KLEIN, J., LATIL, J-L., LUKENEDER, A., MITTA, V., MOURGUES, F.A., PLOCH, I., RAISOSSADAT, N., ROPOLO, P., SANDOVAL, J., TAVERA, J.M., VAŠÍCEK, Z. & VERMEULEN, J. 2006. Report on the 2nd International meeting of the IUGS Lower Cretaceous ammonite working group, the "Kilian Group" (Neuchâtel, Switzerland, 8 September 2005). *Cretaceous Research* **27**: 712-715.

———, RAWSON P. F., MORENO-BEDMAR, J.A., AGUIRRE-URRETA, M.B., BARRAGÁN R., BOGOMOLOV, Y., COMPANY, M., GONZÁLEZ-ARREOLA, C., IDAKIEVA STOYANOVA, V., LUKENEDER, A., MATRION, B., MITTA, V., RANDRIANALY, H., VAŠÍCEK, Z., BARABOSHKIN, E.J., BERT, D., BERSAC, S., BOGDANOVA, T.N., BULOT, L.G., LATIL, J-L., MIKHAILOVA, I.A., ROPOLO, P. & SZIVES, O. 2011. Report on the 4th International Meeting of the IUGS Lower Cretaceous ammonite working group, the "Kilian Group" (Dijon, France, 30th August 2010). *Cretaceous Research* **32**: 786-793.

———, SZIVES, O., AGUIRRE-URRETA, B., BARRAGÁN, R., COMPANY, M., IDAKIEVA, V., IVANOV, M., KAKABADZÉ, M.V., MORENO-BEDMAR, J.A., SANDOVAL, J., BARABOSHKIN, E.J., ÇAGLAR, M.K., FŐZY, I., GONZÁLEZ-ARREOLA, C., KENJO, S., LUKENEDER, A., RAISOSSADAT, S.N., RAWSON, P.F. & TAVERA, J.M. 2014. Report on the 5th International Meeting of the IUGS Lower Cretaceous ammonite working group, the "Kilian Group" (Ankara, Turkey, August 2013). *Cretaceous Research* **50**: 126-137.

REDTENBACHER, A. 1873. Die Cephalopodenfauna der Gosauschichten in den nordöstlichen Alpen. *Abhandlungen der geologischen Reichsanstalt, Wien* **5** (5): 91-140.

REESIDE, J. B. 1923. A new fauna from the Colorado Group of southern Montana. *United States geological Survey Professional Paper* **132-B**: 25-32.

——— 1926. A comparison of the genera *Metaplacenticeras* Spath and *Placenticeras* Meek. *United States geological Survey Professional Paper* **147-A**: 1-2.

——— 1927a. Cephalopods from the lower part of the Cody Shale of Oregon basin, Wyoming. *United States geological Survey Professional Papers* **150-A**: iv, 1-19.

——— 1927b. The scaphites, an Upper Cretaceous ammonite group. *United States geological Survey Professional Papers* **150-B**: 21-40.

———1927c. The cephalopods of the Eagle Sandstone and related formations in the Western Interior of the United States. *United States geological Survey Professional Paper* **151**: iv, 1-87.

——— 1932. The Upper Cretaceous ammonite genus *Barroisiceras* in the United States. *United States geological Survey Professional Paper* **170**: 9-20.

——— 1947. Upper Cretaceous ammonites from Haiti. *United States geological Survey Professional Papers* **214-A**: ii, 1-6.

——— & COBBAN, W.A. 1960. Studies of the Mowry Shale (Cretaceous) and contemporary formations in the United States and Canada. *United States geological Survey Professional Papers* **355**: iv, 1-126.

——— & WEYMOUTH, A.A. 1931. Mollusks from the Aspen Shale (Cretaceous) of southwestern Wyoming. *Proceedings of the United States national Museum* **78** (17): 1-24.

RENGARTEN, V. 1926. La faune des dépôts Crétacés de la région d'Assa-Kambileevka, Caucase du nord. *Mémoires du Comité géologique Russe*, n.s., **147**: 1-132.

RENZ, O. 1968. Über die Untergattungen *Venezoliceras* Spath und *Laraiceras* n. subgen. der Gattung *Oxytropidoceras* Stieler (Ammonoidea) aus den venezolanischen Anden. *Eclogae Geologicae Helvetiae* **61**: 615-655.

——— 1970. Über die Gattungen *Parengonoceras* Spath, *Knemiceras* Böhm und *Neophlycticeras* Spath (Ammonoidea) aus den Anden Venezuelas. *Eclogae Geologicae Helvetiae* **63** (3): 1021-1057.

——— 1972. Die Gattungen *Puzosia* Bayle, *Bhimaites* Matsumoto und *Desmoceras* Zittel (Ammonoidea) im Oberen Albien Venezuelas. *Eclogae Geologicae Helvetiae* **65**: 701-724.

——— 1978. Genus *Mathoceras* (Ammonoidea) from the Upper Aptian in the Serranía del Interior, eastern Venezuela. *Eclogae Geologicae Helvetiae* **71**: 677-685.

——— 1982. *The Cretaceous ammonites of Venezuela*. Maraven: Caracas.

——— & ALVAREZ, F.A.G. 1979. Two new ammonite genera from the Lower Turonian of Venezuela. *Eclogae Geologicae Helvetiae* **72**: 973-979.

REPIN, YU. S. 2005. Representatives of the order Phylloceratida (Ammonoidea) from the Middle Jurassic of north-east Asia. *Paleontological Journal* **2**: 38-41.

——— 2012. The endemic branch of the Phylloceratida (Ammonoidea) of Arctic Mesozoic. *In*, LEONOVA, T.B., BARSKOV, I.S. & MITTA, V.V. (eds), Contributions to current cephalopod research: morphology, systematics, evolution, ecology and biostratigraphy. *Proceeding of conference (Moscow, 9–11 April, 2012)*. Moscow: Russian Academy of Sciences, Borissiak Paleontological Institute [in Russian with English summary].

———, MELEDINA, S.V. & ALEXEEV, S.N. 1998. Phylloceratids (Ammonoidea) from the Lower Jurassic of northeastern Asia. *Paleontological Journal* **32** (5): 461-473.

RETOWSKI, O. 1893. Die tithonischen Ablagerungen von Theodosia. Ein Beitrag zur Paläontologie der Krim. *Bulletin de la Societe imperiale des Naturalistes de Moscou*, n.s., **7**: 206-301 (1-95).

REUSS, A.E. VON. 1845. *Die Versteinerungen der böhmischen Kreideformation*, Abt. 1. Schweizerbart: Stuttgart.

REYMENT, R.A. 1954a. New Turonian (Cretaceous) ammonite genera from Nigeria. *Colonial Geology and Mineral Resources* **4**: 149-164.

———1954b. Some new Upper Cretaceous ammonites from Nigeria. *Colonial Geology and Mineral Resources* **4**: 248-270.

——— 1955. The Cretaceous Ammonoidea of southern Nigeria and the southern Cameroons. *Geological Survey of Nigeria, Bulletin* **25**: 1-112.

——— 1956. Über den Bau von *Speetoniceras versicolor* (Trautschold) aus dem Neokom Rußland. *Neues Jahrbuch fur Geologie und Paläontologie Monatshefte* **2**: 101-105.

——— 1957. Über einige wirbellose fossilien aus Nigerien und Kamerun, Westafrika. *Palaeontographica* (Abt. A) **109**: 41-70.

——— 1958a. Über einige Ammoniten aus dem Coniac Kolumbiens und Venezuelas, Südamerika. *Stockholm Contributions* **2**: 1-26.

——— 1958b. Übersichtliche Ergänzung von F. Solgers "Die Fossilien der Mungokreide in Kamerun und ihre geologische Bedeutung" (1904). *Stockholm Contributions in Geology* **2**: 51-72.

——— 1971. Vermuteter Dimorphismus bei der Ammonitengattung *Benueites. Bulletin of the geological Institutions of the University of Uppsala*, n.s., **3** (1972): 1-18.

——— 1972. Some Lower Turonian ammonites from Trinidad and Colombia. *Geologiska Foereningens i Stockholm Foerhandlingar* **94**: 357-368.

——— 1975. Analysis of a generic level transition in Cretaceous ammonites. *Evolution* **28** (1974): 665-676.

——— & KENNEDY, W.J. 2000. Morphometric links in an evolutionary sequence of the Cretaceous ammonite genus *Metoicoceras. Cretaceous Research* 21: doi.10.1006/cres.2000.0234.

REY, J., MORENO-BEDMAR, J.-A., BILOTTE, M. & MARTINEZ, R. 2013. Nouvelles données paléontologiques et stratigraphiques au passage Aptien - Albien dans les Pyrénées ariégeoises (France). *Carnets de Géologie [Notebooks on Geology]* **2013/01** (CG20): 67-81.

REYNÈS, P. 1876. Description de quelques espèces d'ammonites qui se trouvent dans le muséum d'histoire naturelle de la ville de Marseille, accompagnée de quelques observations sur les principes de classification de ces mêmes êtres et sur le classement de quelques céphalopodes déroulés. *Bulletin de la Société scientifique Industrielle, Marseille* **4**: 3-108.

RICCARDI, A.C. 1976. Paleontologia y edad de la formación Springhill. *Congreso geológico Chileno, 1, Santiago, Chile, 2-7 Agosto 1976*: C41-C56.

——— 1977. Berriasian invertebrate fauna from the Springhill Formation of southern Patagonia. *Neues Jahrbuch fur Geologie und Paläontologie Abhandlungen*, Bd.**155**, Hft. 2: 216-252.

——— 1983. Scaphitids from the Upper Campanian-Lower Maastrichtian Bearpaw Formation of the Western Interior of Canada. *Geological Survey of Canada Bulletin* **354**: 1-102.

——— 1988. The Cretaceous System of southern South America. *The geological Society of America, Memoir* **168**: i-v, 1-161.

——— 2008. The Jurassic of Argentina and its ammonites. *Revista de la Asociación Geológica Argentina* **63** (4): 625-643 [in Spanish].

——— 2015. Remarks on the Tithonian-Berriasian ammonite biostratigraphy of west central Argentina. *Volumina Jurassica* **13** (2): 23-52.

———, URRETA, M.B. AGUIRRE & MEDINA, F.A. 1987. Aconeceratidae (Ammonitina) from the Hauterivian-Albian of southern Patagonia. *Palaeontographica* (Abt. A) **196**: 105-185.

——— & MEDINA, F.A. 2002. The Beudanticeratinae and Cleoniceratinae (Ammonitida) from the Lower Albian of Patagonia. *Revue de Paléobiologie* **21** (1): 291-351.

———, WESTERMANN, G.E.G & LEVY, R. 1971. The Lower Cretaceous Ammonitina *Olcostephanus*, *Leopoldia*, and *Favrella* from west-central Argentina. *Palaeontographica*, Abt. A, Bd.136, Lief. 1-6: 83-121.

RIEDEL, L. 1931. Zur Stratigraphie und Faciesbildung im Oberemscher und Untersenon am Südrande des Beckens von Münster. *Jahrbuch der Preussischen geologischen Landesanstalt* **51**: 605-713.

——— 1938. Amonitas del cretácico inferior de la Cordillera Oriental. *Estudios geologicos y paleontologicos sobre la Cordillera Oriental de Colombia* **2**: 7-78.

ROBERT, E. 2002. La transgression albienne dans le Bassin Andin (Pérou): Biostratigraphie, Paléontologie (ammonites) et stratigraphie séquentielle. *Strata* **38**: 1-380.

——— 2004. Origin, phylogeny, faunal composition, and stratigraphical significance of the Albian Engonoceratidae (Pulchelliaceae, Ammonitina) of Peru. *Journal of South American Earth Sciences* **17**: 11-23.

——— & BULOT, L.G., JAILLARD, E., PEYBERNÈES, B. 2003. Revision bioestratigrafica del Albiano inferior a superior basal en la Cuenca Andina peruana (Peru central y norte). *Boletin de la Sociedad geologica del Perú* **95**: 59-74.

———, ——— 2005. Albian ammonite faunas from Peru: the genus *Neodeshayesites* Casey, 1964. *Journal of Paleontology* **79** (3): 611-618.

———, LATIL, J.-L. & BULOT, L.G 2009. Albian ammonite faunas from South America: the genus *Tegoceras* Hyatt, 1903. *Revue de Paléobiologie* **28** (1): 43-51.

ROCH, E. 1927. Etude stratigraphique et paléontologique de l'Aptien inférieur de La Bédoule (prés Cassis) (Bouches du Rhône). *Mémoires de la Société géologique de France*, n.s., **4** (1): 1-37.

RODDA, P. & MURPHY, M. 1991. The Philadelphia syntypes of *Ammonites hoffmanni* Gabb Cretaceous Mollusca: Ammonoidea. *The Veliger* **34** (4): 360-365.

ROEMER, C.F. 1849. *Mit besonderer Rücksicht auf deutsche Auswanderung und die physischen Verhältnisse des Landes nach eigener Beobachtung geschildert; mit einem naturwissenschaftlichen Anhänge, und einer topographisch-geognostischen Karte von Texas.* Adolph Marcus: Bonn.

ROEMER, F.A. 1836. *Die Versteinerungen des nord-deutschen Oolithen-Gebirges.* Hahn'schen Hofbuchhandlung: Hannover.

——— 1840-41. *Die Versteinerungen des Nord-deutsches Kreidegebirges.* **1** (1840): 1-48; **2** (1841): iii-iv, 49-145. Hahn'schen Hofbuchhandlung: Hannover.

——— 1852. *Die Kreidebildungen von Texas und ihre organischen Einschlüsse.* Adolf Marcus: Bonn.

ROGOV, M.A. 2001. Phylogenetic relations within Jurassic Ochetoceratinae (Ammonoidea, Oppeliidae. *Bulletin de la Société impériale des naturalistes de Moscou*, ser. geol., **76** (5): 38-42 [in Russian].

——— 2003. Upper Jurassic Ochetoceratinae (Oppeliidae, Ammonoidea) of central Russia. *Bulletin de la Société impériale des naturalistes de Moscou*, ser. geol., **78** (6): 38-52 [in Russian].

——— 2010. New data on ammonites and stratigraphy of the Volgian Stage in Spitzbergen. *Stratigraphy and Geological Correlation* **18**: 505-531.

——— 2013. Dimorphism and polymorphism of the Boreal and Subboreal Volgian ammonites (Dorsoplanitidae, Craspeditidae, Virgatitidae). *In*, ZAKHAROV, V.A., ROGOV, M.A. & SHURYGIN, B.N. (eds), *Jurassic System of Russia: Problems of stratigraphy and paleogeography*. Fifth All-Russian meeting. September 23-27: 187-190. Yekaterinburg: "ID "Izdat-Nauka-Servis" LLC [in Russian].

——— 2014. *Khetoceras* (Craspeditidae, Ammonoidea) - a new genus from the Volgian Stage of northern Middle Siberia, and parallel evolution of Late Volgian Boreal ammonites. *Paleontological Journal* **48** (5): 457-464.

——— 2017. Ammonites and infrazonal stratigraphy of the Kimmeridgian and Volgian Stages of southern part of the Moscow Syneclise. *Trudy Instituta Geologii Rossiiskaya Akademiya Nauk* **615**: 7-160.

——— 2020. Infrazonal ammonite biostratigraphy, paleobiogeography and evolution of Volgian craspeditid ammonites. *Paleontological Journal* **54** (10): 1189-1219.

——— & MIRONENKO, A.A. 2016. Patterns of the evolution of aptychi of Middle Jurassic to Early Cretaceous Boreal ammonites. *Swiss Journal of Palaeontology* **135** (1): 139-151.

ROLLIER, L. 1909. Phylogénie des principaux genres d'ammonoides de l'Oolithique (Dogger) et de l'Oxfordien. *Archives des Sciences Physiques et Naturelles, Geneve* (4) **28**: 611-623.

——— 1922. Phylogenie des ammonoides. *Eclogae Geologicae Helvetiae* **17**: 358-360.

Roman, F. 1933. Sur quelques formes de céphalopodes de l'Hauterivien, de l'Yonne et des régions voisines. *Travaux du Laboratoire de Géologie de la Faculte des Sciences, Lyon* **22**: 6-22.

——— 1936. Le Tithonique du massif de Djurdjura (Province d'Alger). *Materiaux de la Carte géologique de Algere*, 1 Sér., Paléontologie **7**: 1-43.

——— 1938. *Les ammonites Jurassiques et Crétacées. Essai de genera.* Masson: Paris.

Rominger, C. 1847. Beiträge zur Kenntniss der Böhmischen Kreide. *Neues Jahrbuch für Mineralogie, Geognosie, Geologie und Petrefaktenkunde* **1847**: 641-664.

Ropolo, P. & Gonnet, R. 1998. A propos de quelques Ancyloceratidae nouveaux ou peu connus du Barrémien sommital de La Bédoule (B. du Rh. - France): position stratigraphique et systématique. *Géologie Méditerranéenne* **25**: 117-143.

———, ——— 2003. Les *Dufrenoyia* (Ammonoidea) de la Zone à Furcata (Aptien inférieur) du sud-est de la France. *Annales du Muséum d'Histoire Naturelle de Nice* **18**: 21-55.

———, Conte, G, Gonnet, R., Masse, J.-P. & Moullade, M. 2000. Les faunes d'ammonites du Barrémien supérieur/Aptien inférieur (Bédoulien) dans la région stratotypique de Cassis-La Bédoule (SE France): État des connaissances et propositions pour une zonation par ammonites du Bédoulien-type. *In*, Moullade, M., Masse, J.-P., Tronchetti, G, Kuhnt, W., Ropolo, P., Bergen, J.A., Masure, E. & Renard, M. (eds), Le stratotype historique de l'Aptien inférieur (Bédoulien) dans la région de Cassis-La Bédoule (S.E. France). *Géologie Méditerranéenne* **25** (3-4) (for 1998): 167-175.

———, ———, Moullade, M., Tronchetti, G & Gonnet, R. 2008. The Douvilleiceratidae (Ammonoidea) of the Lower Aptian historical stratotype area at Cassis-La Bédoule (SE France). *Carnets de Géologie/Notebooks on Geology, Memoir* **2008/03**: 1-60.

———, Moullade, M., Gonnet, R., Conte, G & Tronchetti, G 2006. The Deshayesitidae Stoyanov, 1949 (Ammonoidea) of the Aptian historical stratotype region at Cassis-La Bédoule (SE France). *Carnets de Géologie/Notebooks on Geology, Memoir* **2006/01**: 1-42.

———, ——— & Conte, G 1998. Le genre *Pseudocrioceras* dans les couches de passage du Barrémien supérieur / Bédoulien inférieur de Cassis-La Bédoule (SE France). *Géologie Méditerranéenne* **25** (3/4): 85-123.

——— & Salomon, M. 1992. Evolution du déroulement - Passage du stade criocératique au stade subaspinocératique ou protocriocératique - chez certaines populations d'ammonites hétéromorphes de l'Hauterivien moyen (zones à Nodosoplicatum et à Sayni). *Géologie Méditerranéenne* **19** (3): 189-227.

———, Conte, G., Gonnet, R., Masse J.-P. & Moullade, M. 1998. Les faunes d'ammonites du Barrémien supérieur/Aptien inférieur dans la région stratotypique de Cassis-La Bédoule (SE France) état des connaissances et proposition pour une zonation par ammonites du Bédoulien-type. *Géologie Méditerranéenne* **25**: 167-175.

———, ——— & ——— 1999. The '*Pseudocrioceras* interval' and adjacent beds at La Bédoule (SE France, implications to highest Barremian-lowest Aptian biostratigraphy. *Scripta Geologica, Special Issue* **3**: 159-213.

———, Gonnet, R. & Conte, G 1998. Le genre *Pseudocrioceras* dans les couches de passage du Barrémien supérieur/Bédoulien inférieur de Cassis-La Bédoule (SE France). *Géologie Méditerranéenne* **25**: 85-123.

———, Moullade, M., Conte, G & Tronchetti, G 2008. About the stratigraphic position of the Lower Aptian *Roloboceras hambrovi* (Ammonoidea) level. *Carnets de Géologie/Notebooks on Geology, Letter* **2008/03** (CG2008_L03): 1-7.

Rossi, F. 1984. Ammoniti del Kimmeridgiano superiore-Berriasiano inferiore del Passo del Furlo (Appennino Umbro-Marchigiano). *Memorie della Societa Italiana di Scienze Naturali e del Museo civico di storia naturale di Milano* **23** (2): 75-136.

Rouchadzé, J. 1933. Les ammonites aptiennes de la Géorgie occidentale. *Bulletin de l'Institute géologique de Georgie* **1** (3): 165-73.

——— 1938a. Céphalopodes nouveaux ou peu connus de l'Aptien de la Géorgie. *Bulletin de l'Institut géologique de Géorgie* **3** (2): 129-190.

——— 1938b. Les ammonites aptiennes du Caucase du Nord. *Bulletin du Musée d'État de Géorgie* **9A**: 113-209.

Roschen, C.H. 1929. Contributions to the paleontology of Colombia, Peru and Ecuador. The ammonite family Pulchelliidae in Colombia. *Johns Hopkins University Studies in Geology* **10**: 9-60.

Roy, A. & Bardhan, S. 2019. A note on *Placenticeras warthi* Kossmat, 1895 (Ammonoidea), the type species of the genus *Hypengonoceras* Spath, 1922. *Journal of Paleontology* DOI: 10.1017/jpa.2019.49

———, ——— Das, B., Mondal, S. & Mallick, S. 2012. Systematic revision and palaeobiogeography of *Perisphinctes* Waagen (Ammonoidea) from the Oxfordian of Kutch, India: stratigraphic and evolutionary implications. *Palaeoworld* **21**: 167-192.

Royo y Gomez, J. 1945. Fosiles del Barremiense Colombiano. Compilacion de los estudios geologicos oficiales en Colombia. *Servicio geologico Nacional, Bogota* **6**: 459-494.

Rulleau, L. 1998. Évolution et systématique des Phylloceratidae et des Lytoceratidae du Toarcien et du Dogger inférieur de la région lyonnaise. *Documents des Laboratoires de Géologie, Lyon* **149**: 3-167.

Runnegar, B. & Pojeta, J. 1974. Molluscan phylogeny: the paleontological viewpoint. *Science, N.Y.* **186** (4161): 311-317.

Ruppert, E.E., Fox, R.S., & Barnes, R.D. 2004. *Invertebrate zoology: A functional evolutionary approach.* 7th ed. Thomson-Brooks/Cole: Belmont.

Ruzhensev, V.E. 1962. Superorder Ammonoidea. General section. *In*, Ruzhentsev, V.E. (ed.), *Fundamentals of paleontology. 5 (Mollusca-Cephalopoda I.)*: 243-328. Izdatel'stvo Akademi Nauk USSR: Moscow.

Sakharov, A.S. 1982. New ammonitid genus *Gechiceras* from the Late Berriasian of northern Caucasus. *Palaeontological Journal* **3**: 132-136.

——— & Pogodina, I. A. 2003. On the systematics of the ammonite family Berriasellidae Spath, 1922. *Palaeontological Journal* **3**: 36-38 [in Russian].

Sakharova, I.A. 1985. Discussable questions of systematics of the superfamily Douvilleicerataceae Parona et Bonarelli, 1897. *In*, Menner, V.V. (ed.), *Fossil Cephalopoda, Mollusca*: 168-176. Nauka: Moscow [in Russian].

Saks, V.N. & Shul'gina, N.I. 1969. New Neocomian zones and the Berriasian-Valanginian boundary in Siberia. *Geologiia i Geofizika* **1969** (12): 42-52.

———, ——— 1974. Basic problems of the Upper Volgian, Berriasian and Valanginian stratigraphy of the Boreal zone. *Acta geologica Polonica* **24**: 543-560.

Salazar, C. & Stinnesbeck, W. 2015. Tithonian-Berriasian ammonites from the Banos del Flaco Formation, central Chile. *Journal of Systematic Palaeontology*: 1-34. doi.org/10.1080/14772019.2015.1027310

Salfeld, H. 1919. Über die Ausgestaltung der Lobenlinie bei Jura- und Kreide-Ammonoideen. *Nachrichten von der koeniglichen Gesellschaft der Wissenschaften zu Goettingen, Mathematisch-Physikalische Klasse* **1919**: 1-19.

——— 1921. Kiel- und Furchenbildung auf der Schalen-

aussenseite der Ammonoideen in ihrer Bedeutung für die Systematik und Festlegung von Biozonen. *Zentralblatt für Mineralogie, Geologie und Paläontologie* **1921**: 343-347.
——— 1924. Die Bedeutung der Konservativstämme für die Stammesentwicklung der Ammonoideen. *Zentralblatt für Mineralogie, Geologie und Paläontologie* **1924**: 1-16.
SALTER, J.W. & BLANFORD, H.F. 1865. *Palaeontology of Niti in the northern Himalaya; being descriptions and figures of the Palaeozoic and secondary fossils collected by Col. R. Strachey.* Nabu Press: Calcutta.
SAPOZHNIKOV, V.B. 1971. On some Middle Albian hoplitids of Turkmenia. *Scientific Transactions of the Leningrad Institute of Mining* **2**: 55-72 [in Russian].
SAPUNOV, I.G. 1977a. Ammonite stratigraphy of the Upper Jurassic in Bulgaria. IV. Tithonian: substages, zones and subzones. *Geologica Balcanica* **7** (2): 43-64.
——— 1977b. Richterellinae subfam. nov. (Perisphinctidae, Ammonoidea, Tithonian). *Comptes Rendues de l'Ácademie Bulgare de Science* **30**: 103-107.
——— 1979. Jurassique superieur. Ammonoidea. *Les fossiles du Bulgarie.* III (3). Académie bulgare des Sciences: Sofia.
SARASIN, C. 1893. Étude sur les *Oppelia* du groupe du *nisus* et les *Sonneratia* du groupe du *bicurvatus* et du *raresulcatus*. *Bulletin de la Société géologique de France* (3) **21**: 149-164.
——— 1897. Quelques considérations sur les genres *Hoplites, Sonneratia, Desmoceras* et *Puzosia*. *Bulletin de la Société géologique de France* (3) **25**: 760-799.
——— & SCHÖNDELMAYER, C. 1901-02. Étude monograph-ique des ammonites du Crétacique inférieur de Chatel-Saint-Denis. *Memoires de la Société paléontologique Suisse* **28** (1901): 1-91; **29** (1902): 95-195.
ŠARIKADZE, M.Z., LOMINADZE, T.A. & KVANTALIANI, I.V. 1990. Systematische Bedeutung von Muskelabdrücken spätjurassischer und frühkretazischer Ammonoideae. *Zeitschrift für Geologische Wissenschaften*, Band **18**: 1031-1039.
SARKAR, S.S. 1954a. Sur un genre nouveau d'ammonite déroulée. *Compte Rendu Sommaire des Seances de la Société géologique de France* **1954**: 97-98.
——— 1954b. Some new genera of uncoiled ammonites from Lower Cretaceous. *Science and Culture* **19**: 618-621 (1-4).
——— 1955. Révision des ammonites déroulées du Crétacé Inférieur du sud est de la France. *Memoires de la Société géologique de France* (5) **34** (72): 1-176.
——— 1977. Sur *Binelliceras* genre nouveau d'Heteromorpha et sur la validité du genre *Emericiceras* Sarkar, 1954. *Paläontologische Zeitschrift* **51**: 258-261.
SARTI, C. 1988. Due nuove specie di Idoceratinae (Ammonoidea) del Kimmeridgiano delle Prealpi Venete (nord Italia). *Bollettino della Società paleontologica Italiana* **27** (3): 291-299.
——— 1990. Taxonomic revision of the Kimmeridgian (Upper Jurassic) genus *Mesosimoceras* (Ammonoidea) and institution of the new genus *Presimoceras* (Ammonitina, Idoceratinae). *Paläontologisches Zeitschrift* **64** (1/2): 39-55.
——— 1999. *Protancyloceras* (Ammonoidea) in the Lower Tithonian sequences of the Trento Plateau (Venetian Alps, northern Italy). *Profil* **16**: 49-56.
——— 2017. New ammonite genera from the Lower Tithonian (Upper Jurassic) of the southern Alps (northern Italy). *Studi Trentini di Scienze Naturali* **96**: 33-61.
SASONOV, N.T. 1951. On some little-known ammonites of the Lower Cretaceous. *Biulleten Moskovskogo Obshchestva Ispytatelei Prirody, Otdel Geologicheskii* **56**: 57-63 [in Russian].
SASONOVA, I.G. 1958. Lower Cretaceous deposits of the central regions of the Russian Platform. *In*, FLEROVOI, O.V. (ed.), *Mesozoic and Tertiary deposits of the central regions of the Russian Platform*: 31-136. Gostoptekhizdat: Leningrad [in Russian].
——— 1961. Standardized stratigraphical scheme for the Lower Cretaceous deposits of the Russian Platform. *Trudy Vsesoiuznogo Nauchno-Issledovatel'skogo Geologicheskogo Neftegazovogo Insituta* **29** (3): 5-28 [in Russian].
——— 1965. Lower Valanginian ammonites. *In*, SASONOVA, I.G. & SHOOTSKO, E.K. (eds), Mesozoic and Cenozoic faunas of the European parts of the USSR and central Asia. *Trudy Vsesoiuznogo Nauchno-Issledovatel'skogo Geologicheskogo Neftegazovogo Instituta* **44**: 100-106 [in Russian].
——— 1971. Berriasian and Lower Valanginian ammonites of the Russian Platform. *In*, SASONOV, N.T. (ed.), The Berriasian of the Russian Platform (stratigraphy, ammonite fauna and aucellids). *Trudy Vsesoiuznogo Nauchno-Issledovatel'skogo geologicheskogo Neftegazovogo Instituta* **110**: 1-110, 148-157 [in Russian].
——— 1972. Ammonites from the Russian lowland. *In*, SAKS, V.N. (ed.), *The Jurassic–Cretaceous boundary and the Berriasian stage in the Boreal Realm*: 175-194. Novosibirsk: Nauka [in Russian].
——— 1977. Ammonites of the boundary beds of Jurassic and Cretaceous Systems of the Russian Platform. *Trudy Vsesoyuznogo Nauchno-Issledovatel'skogo Geologicheskogo Neftegazovogo Instituta* **185**: 1-128 [in Russian].
——— & SASONOV, N.T. 1983. The Berriasian of the European realm. *Zitteliana* **10**: 439-446.
———, ——— 1991. Stratigraphic classification of the Lower Cretaceous of the east-European Platform (Russian Subboreal Province)]. *Biulleten Moskovskogo Obshchestva Ispytalelei Prirody, Otdel Geologicheskii* **66** (4): 49-63 [in Russian].
SATO, T. 1961. La limite jurassico-crétacée dans la stratigraphie japonaise. *Japanese Journal of Geology and Geography* **32**: 533-541.
SAVEL'EV, A.A. 1960. New Albian hoplitids from Mangyshlak *In, New species of ancient plants and animals of the USSR.* Part 2: 177-188. Gosgeoltechizdat: Moscow [in Russian].
——— 1973. Stratigraphy and ammonites of the Lower Albian of Mangyshlak. *Trudy Vsesoiuznogo Nauchno-Issledovatel 'skogo geologicheskogo Neftegazovogo Instituta* **323**: 1-339 [in Russian].
——— 1974. A new zonal scheme for the Lower Albian of Mangyshlak. *Trudy Vsesoiuznogo Nauchno-Issledovatel-'skogo geologicheskogo Neftegazovogo Instituta* **350**: 116-122 [in Russian].
——— 1992. *Lower Albian ammonites of Mangyshlak, their phylogeny and their significance for the zonal stratigraphy of the Albian of southern SSSR (superzone of* Cleoniceras mangyshlakense*).* Nedra: St. Petersbourg [in Russian].
SAYN, G. 1889. Note sur quelques ammonites nouvelles ou peu connues de néocomien inférieur. *Bulletin de la Société géologique de France* (3) **17**: 679-688.
——— 1891. Description des ammonitides de Barrémien du Djebel Ouach près Constantine. *Annales de la Société nationale d'Agriculture, d'Histoire naturelle et des Arts Utiles de Lyon* (6) **3** (1890): 135-208.
——— 1901. Les ammonites pyriteuses des marnes valanginiennes de sud-est de la France, part 1. *Memoires de la Société géologique de France*, Paléontologie, **9** (2) (23): 1-27.
——— 1904. Review of "*Pseudoceratites* of the Cretaceous" by Alpheus Hyatt. *Revue critique Paleozoologie* **8**: 121-125.
——— 1906. Céphalopodes [Review of "Einige Bemerkungen über Ammoniten-gattung *Hoplites* Neumayr" by V. Uhlig, 1905]. *Revue critique Paleozoologie* **10**: 120-123.
——— 1907a. Les ammonites pyriteuses des marnes valanginiennes de sud-est de la France [part 2]. *Memoires de la Société géologique de France*, Paleontologie **15** (23): 29-68.
——— 1907b. Review of "Fauna der unteren Kreide im westschweizerischen Jura. I & II," by Baumberger. *Revue critique Paleozoologie* **11**: 193-195.

Schairer, G., Fürsich, F.T., Wilmsen, M., Seyed-Emami, K. & Majidifard, M. 2003. Stratigraphy and ammonite fauna of Upper Jurassic basinal sediments at the eastern margin of the Tabas Block (east-central Iran). *Geobios* **36**: 195-222.

Scherzinger, A., Főzy, I. & Parent, H. 2010. The Early Tithonian (Late Jurassic) ammonite genus *Virgatosimoceras* Spath (Ammonoidea: Simoceratidae) - revision and value for correlation. *Neues Jahrbuch für Geologie und Paläontologie, Abhandlungen* **256** (2): 195-212.

——— & Mitta, V. 2006. New data on ammonites and stratigraphy of the Upper Kimmeridgian and Lower Volgian (Upper Jurassic) of the middle Volga region (Russia). *Neues Jahrbuch für Geologie und Paläontologie, Abhandlungen* **241**: 225-251.

———, Parent, H. & Schweigert, G. 2015. A new species of the ammonite genus *Neochetoceras* Spath (Oppeliidae) from the Hybonotum Zone (Lower Tithonian) of southern Germany, with comments on the phylogeny of the genus. *Boletín del Instituto de Fisiografía y Geología* **85**: 1-12.

———, ———, ——— 2018. A new species of the ammonite genus *Physodoceras* Hyatt (Aspidoceratidae) from the Hybonotum Zone (Lower Tithonian) of southern Germany, with comments on the phylogeny of the genus. *Boletín del Instituto de Fisiografía y Geología* **88**: 11-24.

———, Schweigert, G. & Parent, H. 2006. New considerations on dimorphism and aptychus in *Gravesia* Salfeld (Ammonoidea; Late Jurassic). *Neues Jahrbuch für Geologie und Paläontologie, Abhandlungen* **241**: 269-286.

Schindewolf, O.H. 1923. Über die Ausgestaltung der Lobenlinie bei den Neoammonoidea Wdkd. *Zentralblatt für Geologie, Mineralogie und Paläontologie* **1923**: 337-350, 359-370.

——— 1925. Entwurf einer Systematik der Perisphincten. *Neues Jahrbuch für Mineralogie, Geologie und Paläontologie* (Abt. B), Beilage-Band **52**: 309-343.

——— 1933. Vergleichende Morphologie und Phylogenie der Anfangskanunern tetrabranchiater Cephalopoden. Eine Studie über Herkunft, Stammesentwicklung und System der niederen Ammoneen. *Abhandlungen Preußischen Geologischen Landesanstalt*, n.f., Heft **148**: 3-115.

——— 1966. Studien zur Stammesgeschichte der Ammoniten. Lieferung 5. *Abhandlungen der Mathematisch-Naturwissenschaftlichen Klasse, Akademie der Wissenschaften und der Literatur in Mainz* **1966** (3): 325-454 (511-640).

——— 1968. Studien zur Stammesgeschichte der Ammoniten. Lieferung 7. *Abhandlungen der Mathematisch-Naturwissenschaftlichen Klasse, Akademie der Wissenschaften und der Literatur in Mainz* **1968** (3): 41-209 (733-901).

Schlagintweit, O. 1912. Die Fauna des Vracon und Cenoman in Perú. *Neues Jahrbuch für Mineralogie, Geologie und Paläontologie*, Beilage-Band **33**: 43-135.

Schloenbach, U. 1865-1866. Beiträge zur Paläontologie der Jura- und Kreide-Formation im nordwestlichen Deutschland. *Palaeontographica*, Band 13, Lief 4, **6**: 147-192, 267-332.

Schlotheim, E.F. von. 1813. Beiträge zur Naturgeschichte der Versteinerungen in geognostisches Hinsicht. *In*, Leonhard, C.C. von, *Taschenbuch für die gesammte Mineralogie mit Hinsicht auf die neuesten Entdeckungen herausgegeben von Dr. Carl Caesar Leonhard*, Jahrgang 7, Abt. **1**: 3-134.

Schlüter, C. 1867. *Beitrag zur Kenntniss der jungsten Ammoneen Norddeutschlands.* A. Henry: Bonn.

——— 1871-76. Cephalopoden der oberen deutschen Kreide. *Palaeontographica* **21** (1871): 1-24; **22** (1872): 25-120; **24** (1876): 121-264.

Schmid, F. & Ernst, G. 1975. Ammoniten aus dem Campan der Lehrter Westmulde und ihre stratigraphische Bedeutung. Teil 1: *Scaphites, Bostrychoceras* und *Hoplitoplacenticeras. Bericht der naturhistorischen Gesellschaft zu Hannover* **119**: 315-359.

Schneegans, D. 1943. Invertébrés du Crétacé Supérieur du Damergou (territoire du Niger). *Bulletin de la Direction des Mines, Dakar* **7**: 87-166.

Schneid, T. 1915. Die Ammonitenfauna der obertithonischen Kalke von Neuburg a.D. *Geologie und Paläontologie Abhandlungen*, N.F., Band 13, Heft 3: 3-114 (305-416).

——— 1944. Über Ataxioceratiden des nördlichen Frankenjura. *Palaeontographica*, Abt. A, Band **96**: 1-43.

Scholz, G. 1973. Sur l'âge de la faune d'ammonites au Château près de St-Martin-en-Vercors (Drôme) et quelques considerations sur l'évolution des turrilitidés et des hoplitidés vracono-cénomaniens. *Geologie Alpine* **49**: 119-129.

——— 1978. Vracon-Ammoniten (Oberalb, *dispar*-Zone) aus dem Flammenmergel von Salzgitter. *Aspekte der Kreide Europas*, I.U.G.S. (A) **6**: 589-606.

——— 1979. Die Ammoniten des Vracon (Oberalb, *dispar*-Zone) des Bakony-Gebirges (Westungarn) und eine Revision der wichtigsten Vracon-Arten der West-Mediterrannen Faunenprovinz. *Palaeontographica* (Abt. A) **165**: 1-80 (July), 81-136 (August).

Schweigert, G. & Scherzinger, A. 1995. Erstnachweis heteromorpher Ammoniten im Schwäbischen Oberjura. *Jahresberichte und Mitteilungen des Oberrheinische Geologischen Vereins*, N.F., Band **77**: 307-319.

———, ——— & Parent, H. 2002. The *Volanoceras* lineage (Ammonoidea, Simoceratidae) - a tool for long-distance correlations in the Lower Tithonian. *Stuttgarter Beitrage zur Naturkunde*, Serie B (Geologie und Paläontologie) **326**: 1-43.

——— & Zeiss, A. 1998. *Berckhemeria* n.g. (Passendorferinae), eine neue Ammonitengattung aus dem Unter-Tithonium (Hybonotum-Zone) von Süddeutschland. *Neues Jahrbuch für Mineralogie, Geologie und Paläontologie, Monatshefte*, Heft 9: 559-576.

———, ——— 1999. *Lithacoceras ulmense* (Oppel) (Ammonitina) - eine wichtige Leitart des Ober-Kimmeridgiums. *Neues Jahrbuch für Mineralogie, Geologie und Paläontologie, Abhandlungen*, Band **211**, Heft 1/2: 49-73.

———, ——— & Westermann, G.E.G. 2012. The *Gravesia* homeomorphs from the latest Kimmeridgian of Mombasa, Kenya. *Revue de Paléobiologie, Genève,* Special Volume **11**: 13-25.

Scott, G. 1924. Some gerontic ammonites of the Duck Creek Formation. *Texas Christian University Quarterly* **1**: 5-31.

——— 1940. Cephalopods from the Cretaceous Trinity Group of the south-central United States. *University of Texas Publication* **3945**: 969-1106.

——— 1943. Palaeontology of Harrar Province, Ethiopia. 4, Jurassic Cephalopoda and a Cretaceous nautilus. *Bulletin of the American Muséum of natural History* **82** (3): 60-93.

Seeley, H. G. 1865. On ammonites from the Cambridge Greensand. *Annals and Magazine of Natural History* (3) **16**: 225-247.

Seltser, V.B. & Ivanov, A.V. 2010. *Atlas of late Cretaceous ammonites of the Saratov Volga region. Monograph.* Book House «University»: Moscow [in Russian].

Semenov, V.P. 1896. Fauna of the Jurassic formations of Mangyshlak and Tuar-Kyr. *Trudy Imperatorskogo Sankt Peterburgskogo Obshchestva Estestvoispytatelei S.-Peterburg* **24**: 1-112 [in Russian].

——— 1899. La faune des dépôts crétacés de Mangyshlak et de quelques autres localités de la province Transcaspienne. *Trudy Imperatorskogo Sankt Peterburgskogo Obshchestva Estestvoispytatelei S.-Peterburg* **28**: 1-178 [in Russian].

Servais, T., Cecca, F., Harper, D.A.T., Isozaki, Y. & Mac Niocaill, C. 2013. Palaeozoic palaeogeographical and palaeobio-geographical nomenclature. *In*, Harper, D.A.T. & Servais, T. (eds), Early Palaeozoic biogeography and palaeogeography. *Memoir of the Geological Society of London* **38**: doi.org/10.1144/M38.

SEUNES, J. 1887. Sur quelques ammonites du Gault. *Bulletin de la Société géologique de France* (3) **15**: 557-571.
——— 1891. Contributions à l'étude des céphalopodes du Crétacé Supérieur de France. Ammonites du Calcaire à *Baculites* du Cotentin. *Mémoires de la Société géologique de France*, Paléontologie, **2** (1890): 1-22.
SEY, I.I. & KALACHEVA, E.D. 1996. Upper Jurassic-Lower Cretaceous biostratigraphy and fauna of south Primorye (Russian Far East). *Geology of the Pacific Ocean* **12**: 293-312.
——— & ——— 1997. Ammonite assemblages from Middle Tithonian deposits of the southern Primorye, Russian Far East (Haploceratidae, Oppeliidae, Ataxioceratidae). *Regional geology and metallogeny* **6**: 90-102 [in Russian].
——— & ——— 2000. Middle-late Jurassic and early Cretaceous marine fauna evolution in eastern Russia. *Revue de Paléobiologie, Special Volume* **8**: 181-186.
SHA, J. 2019. Opening time of the Hispanic Corridor and migration patterns of pan-tropical cosmopolitan Jurassic and pectinid bivalves. *Palaeogeography, Palaeoclimatology, Palaeoecology* **515**: 34-46.
SHARIKADZE, M.Z. 1979. *Procolombiceras* - a new genus of the family Parahoplitidae from the Lower Aptian of Georgia. *Soobshcheniya Akademii Nauk Gruzinskoi SSR* **94**: 381-384 [in Russian].
——— 1984. On the six-lobed primasuture of the Lower Cretaceous tetragonitids (Ammonoidea). *Doklady Academy of Science, USSR* **275** (5): 1182-1184 [in Russian].
——— 2015. *Aptian ammonites of the Caucasus*. Universal: Tbilisi [in Russian].
——— & GUGUNISHVILI, G.G. 1984. Morphogenesis of *Protacanthoplites* Tovbina (Ammonoidea) genus. Geology and prospecting of mineral deposits of Georgia. *Transactions of the Georgian Polytechnic Institute* **11** (281): 32-39.
——— & KVANTALIANI, I.V. 1975. On unknown Lower Albian lytoceratids of Dzirul massif. *Transactions VNIGRI*, Georgian branch **188**: 12-31 [in Russian].
SHARPE, D. 1853-1857. Description of the fossil remains of Mollusca found in the Chalk of England. Part 1 (1853): 1-26, Part 2 (1855): 27-36, Part 3 (1857): 37-70. Cambridge University Press: Cambridge.
SHATTUCK, G.B. 1903. The Mollusca of the Buda Limestone. *United States geological Survey Bulletin* **205**: 1-36.
SHEVYREV, A.A. 1960. Ontogenetic development of some Upper Jurassic ammonites. *Bulletin de la Societe imperiale des Naturalistes de Moscow* (Geology) **35** (1): 69-78 [in Russian].
——— 2006. Cephalopod macrosystem: a historical review, the present state of knowledge, and unsolved problems: 3. Classification of Bactritoidea and Ammonoidea. *Palaeontological Journal* **2006** (2): 34-46 [in Russian].
SHIGETA, Y. 1989. Systematics of the ammonite genus *Tetragonites* from the Upper Cretaceous of Hokkaido. *Transactions and Proceedings of the Palaeontological Society of Japan*, n.s., **156**: 319-342.
——— 2013. A new species of *Tanabeceras* (Ammonoidea, Gaudryceratidae) from the lowest Cenomanian of Hokkaido, Japan. *Paleontological Research* **17** (4): 312-316.
——— 2014. *Morewites*, a new Campanian (late Cretaceous) heteromorph ammonoid genus from Hokkaido, Japan. *Paleontological Research* **18** (1): 1-5.
———, FUTAKAMI, M. & HOFFMANN, R. 2012. Two new ammonoid genera of the subfamily Gabbioceratinae from the Upper Albian (Lower Cretaceous) of Hokkaido, Japan. *Paleontological Research* **16** (3): 208-218.
———, HOFFMANN, R. & FUTAKAMI, M. 2010. Systematic position and origin of the Cretaceous ammonoid genus *Takahashia*. *Paleontological Research* **14** (3): 196-201.
——— & IZUKURA, M. 2013. The earliest Cenomanian ammonoid *Tanabeceras yezoense* (Shigeta) from the Hobetsu area, Hokkaido. *The Bulletin of the Hobetsu Museum* **28**: 1-6.
———, ——— 2018. Additional specimens of *Pravitoceras sigmoidale* Yabe from the Upper Campanian (Upper Cretaceous) of Hokkaido, Japan. *The Bulletin of the Hobetsu Museum* **33**: 1-10.
———, ——— & NISHIMURA, T. 2019. Campanian (late Cretaceous) ammonoids and inoceramids from the Ribira River area, Hokkaido, northern Japan. *National Museum of Nature and Science Monographs* **50**: ii, 1-139.
———, ———, ———, & TSUTSUMI, Y. 2016. Middle and late Campanian (late Cretaceous) ammonoids from the Urakawa area, Hokkaido, northern Japan. *Paleontological Research* **20**: 322-366.
——— & NISHIMURA, T. 2014. A new species of *Anagaudryceras* (Ammonoidea, Gaudryceratidae) from the lowest Maastrichtian of Hokkaido, Japan. *Paleontological Research* **18** (3): 176-185.
———, TANABE, K. & IZUKURA, M. 2010. *Gaudryceras izumiense* Matsumoto and Morozumi, a Maastrichtian ammonoid from Hokkaido and Alaska and its biostratigraphic implications. *Paleontological Research* **14** (3): 202-211.
SHIGETA, Y. & WEITSCHAT, W. 2004. Origin of the Ammonitina (Ammonoidea) inferred from the internal shell features. *Mitteilungen aus dem Geologisch-Paläontologischen Institut (GPI) der Universität Hamburg* **88**: 179-193.
SHIMIZU, S. 1926. Three interesting Cretaceous ammonites recently acquired from Hokkaido and Saghalin. *Proceedings of the Imperial Academy of Japan, Tokyo* **2**: 547-550.
——— 1931. The marine Lower Cretaceous deposits of Japan, with special reference to the ammonite bearing zones. *Science Reports of the Tohoku Imperial University* (2, Geology) **15**: 1-40.
——— 1932. On a new type of Senonian ammonite, *Pseudobarroisiceras nagaoi* Shimizu, gen. et. sp. nov. from Teshio Province, Hokkaido. *Japanese Journal of Geology and Geography* **10**: 1-4.
——— 1933. Note on two interesting Senonian ammonites from Hokkaido and South Saghalin. *Journal of the Shanghai Science Institute* (2: Geology, Palaeontology, Mineralogy, and Petrology) **1**: 11-15.
——— 1934. Ammonites. *In*, SHIMIZU, S. & OBATA, T., *Ammonites: Iwanami's lecture series of geology and palaeontology*. Tokyo [in Japanese].
——— 1935a. The Upper Cretaceous cephalopods of Japan, Part I. *Journal of the Shanghai Science Institute* (2: Geology, Palaeontology, Mineralogy, and Petrology) **1**: 159-226.
——— 1935b. The Upper Cretaceous ammonites so-called *Hamites* in Japan. *Proceedings of the Imperial Academy, Tokyo* **11**: 271-273.
SHOME, S. & BARDHAN, S. 2007. Genus *Himalayites* (Ammonoidea) from the Upper Tithonian of the Spiti Himalaya - a systematic revision of Uhlig's (1910) material. *Journal of the Palaeontological Society of India* **52** (2): 223-230.
——— 2009a. The genus *Umiaites* Spath, 1931 (Ammonoidea) from the Tithonian (late Jurassic) of Kutch, western India. *Palaeontologia Electronica* **12** (1): 1-10.
———, ——— 2009b. A new late Tithonian ammonite assemblage from Kutch, western India. *Journal of the Palaeontological Society of India* **54** (1): 1-18.
SHUL'GINA, N.I. 1960. Ammonites from the Franz-Joseph Land and Taimyr, and their significance for zonal subdivision of the Kimmeridgian in the Arctic. *Transactions of the NIIGA* **111**: 136-144 [in Russian].
——— 1962. New late Jurassic ammonites of northern Siberia. *Transactions of the NIIGA* **127** (3): 197-202 [in Russian].
——— 1967. Tithonian ammonites of northern Siberia. *Prob-*

lems of the paleontological base of the detailed stratigraphy of Siberia and Far East: 131-149. Nauka Publishing: Leningrad [in Russian].

——— 1968. Ammonites of the *Chetaites* genus from the Jurassic/Cretaceous boundary beds of Siberia. Mesozoic sea faunas of the Soviet Far East and their stratigraphic significance. *Transactions of the Institute of Geology and Geophysics, Siberian Branch Academy of Sciences USSR* **48**: 101-107 [in Russian].

——— 1969. Volgian ammonites. *In*, SAKS, V.N. (ed.), *Key section of the Upper Jurassic deposits of the Kheta River basin:* 125-162. Nauka: Leningrad [in Russian].

——— 1972. Review of the ammonites of the Boreal Realm. *In*, SAKS, V.N. (ed.), *The Jurassic-Cretaceous boundary and the Berriasian stage in the Boreal Realm*: 117-136. Akademiia Nauk SSSR, Instituta Geologii i Geofiziki Sibirskogo Otdeliniia: Novosibirsk [in Russian].

——— 1983. Detailed description of sections in the Valanginian of the R. Boyarka and the Paksa Peninsula. *In*, SHUL'GINA, N.I. & BURDYKINA, M.D., *Paleontological substantiation of the subdivisions of the Palaeozoic and Mesozoic of the Arctic regions of the USSR*: 77-91. Sevmorgeologiya: Leningrad [in Russian].

——— 1985. Boreal basins at the boundary of Jurassic and Cretaceous. *Trudy Vsesoiuznogo nauchnoissle-dovatel'skogo Instituta geologii i mineralnykh resursov Mirovogo Okeana* **193**: 1-164 [in Russian].

——— 1991. New data about ammonites of the Middle Volgian and Berriasian-Valanginian age of Franz Josef Land. *In*, KIRICHKOVA, A.I. & CHIRVA, S.A. (eds), *Stratigraphy and paleogeography of the sedimentary formations of oil-and-gas bearing basin of the USSR*: 127-136. VNIGRI: Leningrad [in Russian].

SHUMARD, B.F. 1854. Appendix E. Palaeontology. Description of the species of Carboniferous and Cretaceous fossils collected. *In*, MARCY, R.B. (ed.), *Exploration of the Red River of Lousiana, in the year 1852*. 32nd U.S. Congress, 2nd session, Senate Executive Document 54 **8**: 197-211.

——— 1860. Descriptions of new Cretaceous fossils from Texas. *Transactions of the Academy of Science of St. Louis* **1**: 590-610.

SIMIONESCU, J. 1899. Note sur quelques ammonites du Néocomien français. *Annales de l'Universite de Grenoble* **11** (3): 1-16.

——— 1900. Synopsis des ammonites néocomiennes [Infravalanginien (Berriasien)-Aptien (incl.)]. *Travaux du Laboratoire de Géologie de Grenoble* (1899-1900) **5** (1): 109-143.

SIMONOVICH, S.E. & BATSEVICH, L.F. 1873. Geological description of the Okriba region. *Materialy dlia Geologii Kavkaza* (1873): 1-36 [in Russian].

———, ——— & SOROKIN, A. 1875. Geological description of parts of the Kutais, Lechkhum, Senaks and Zughdid districts of the Kutais government. *Materialy dlya Geologii Kavkaza* (1874): 1-191 [in Russian].

SIMPSON, G.G. 1944. *Tempo and mode in evolution*. Columbia Uiversity Press: N.Y.

SINZOW, I.T. 1870. A geological sketch of the Saratow region. *Zapiski Imperatorskogo S.-Peterburgskogo Mineralogicheskogo Obshchestva* (2) **5**: 103-161.

——— 1872. Über die jurassische und cretaceische Fossilien aus dem Saratow Gouvernement. *Materialen zur Geologie Russlands* **4**: 1-127.

——— 1906. Die Beschreibung einiger *Douvilléiceras*-Arten aus dem Oberen Neocom Russlands. *Zapiski Imperatorskago S.-Peterburgskago Mineralogicheskago Obshchestva* (2) **44**: 157-198.

——— 1907. Untersuchungen einiger Ammonitiden aus dem Unteren Gault Mangyshlaks und des Kaukasus. *Verhandlungen der russisch-kaiserlichen Mineralogischen Gesellschaft zu St. Petersburg* **45**: 455-519.

——— 1908. Untersuchung einiger Ammonitiden aus dem Unteren Gault Mangyschlaks und des Kaukasus. *Zapiski Imperatorskago S.-Peterburgskago Mineralogicheskago Obshchestva* (2) **45**: 455-519.

——— 1913. Beiträge zur Kenntnis der unteren Kreideablagerungen des Nord-Kaukasus. *Travaux du Musée Géologique Pierre le Grand près Académie Impériale des Sciences* **7**: 93-117.

——— 1915. Über einige Ammonitiden aus dem Gault des Mangyshlaks. *Zapiski Imperatorskago S.-Peterburgskago Mineralogicheskago Obshchestva* (2) **50**: 1-24.

SKWARKO, S.K. & THIEULOY, J.-P. 1989. Early Barremian (early Cretaceous) Mollusca from western Irian Jaya, Indonesia. *Publication of the geological Research and Development Centre, Indonesia* (Paleontology) **6**: 26-43.

SMITH, J.P. 1900. The development and phylogeny of *Placenticeras*. *Proceedings of the California Academy of Sciences* (3, Geology) **1**: 181-240.

SOKOLOV, D.N. 1913. Fossils from boulders on Novaya Zembla. *Trudy Geologicheskogo Muzeia imeni Petra Velikogo Imperatorskoi Akademii Nauk* **7**: 59-92 [in Russian].

SOKOLOV, M.I. 1961. *Karamaites* - a new genus of ammonites from the Vraconian deposits of eastern Mangyshlak. *Biulleten Moskovskogo Obshchestva Ispytatelei Prirody, Otdel Geologicheskii* **36** (5): 152 [in Russian].

——— 1967. *Karamaiceras* gen. nov. from the Vraconian deposits of eastern Mangyshlak. *Paleontological Journal* **4**: 138-139.

SOLGER, F. 1904. Die Fossilien der Mungokreide in Kamerun und ihre geologische Bedeutung, mit besonderer Berücksichtigung der Ammoniten. *Beiträge zur Geologie von Kamerun* **2**: 83-242.

SOMMERMEIER, L. 1910. Die Fauna des Aptien und Albien im nördlichen Perú. Teil 1. Cephalopoden. *Neues Jahrbuch für Mineralogie, Geologie und Paläontologie*, Beilage-Band **30**: 313-382.

SORNAY, J. & TESSIER, F. 1949. Ammonites nouvelles du Maestrichtien du Sénégal. *Bulletin de la Société géologique de France* (5) **19**: 245-249.

SORRENTINO, S. 1942. Osservazioni su forme e gruppi di ammoniti della famiglia Phylloceratidae Zittel. *Rivista italiana di Paleontologia, Milan* **48** (3): 1-19.

SOWERBY, J. & SOWERBY, J. de C. 1812-1846. *The mineral conchology of Great Britain*. **1-7** (Parts 1-117). Meredith: London.

———, ——— 1836. Appendix A. Descriptive notes respecting the shells figured in pls. 11-23. *In*, FITTON, W.H., Observations on some of the strata between the Chalk and the Oxford Oolite in the south-east of England. *Transactions of the geological Society of London* (2) **4**: 335-349.

———, ——— 1837. A communication by James de Carle Sowerby on his new genus of fossil shells, *Tropaeum*. *Proceedings of the geological Society of London* **2**: 535.

———, ——— 1840. Letter to the Secretary on the genus *Crioceratites* and on *Scaphites gigas*. *Transactions of the geological Society of London* (2) **4** (2): 409-411.

SPATH, L.F. 1921a. On Cretaceous Cephalopoda from Zululand. *Annals of the South African Museum* **12**: 217-321.

——— 1921b. On Upper Cretaceous Ammonoidea from Pondoland. *Annals of the Durban Museum* **3**: 39-57.

——— 1922a. On Cretaceous Ammonoidea from Angola, collected by Professor J.W. Gregory, D.Sc., F.R.S. *Transactions of the Royal Society of Edinburgh* **53**: 91-160.

——— 1922b. On the Senonian ammonite fauna of Pondoland. *Transactions of the Royal Society of South Africa* **10**: 113-148.

——— 1923a. Excursion to Folkestone, Saturday, September 30th, 1922, with notes on the zones of the Gault. *Proceedings of the Geologists' Association* **34**: 70-76.

——— 1923b. On the ammonite horizons of the Gault and contiguous deposits. Summary of progress of the Geological Survey for 1922. *Memoirs of the geological Survey of Great Britain* **1923**: 139-149.

——— 1923c. VIII. Appendix. On ammonites from New Zealand. *In*, TRECHMANN, C.T., The Jurassic rocks of New Zealand. *Quarterly Journal of the geological Society of London* **79**: 286-312.

——— 1923d. A monograph of the Ammonoidea of the Gault. Part 1 (1921): 1-72. Palaeontographical Society: London.

——— 1924a. On the ammonites of the Speeton Clay and the subdivisions of the Neocomian. *Geological Magazine* **61**: 73-89.

——— 1924b. On a new ammonite (*Engonoceras iris*, sp. n.) from the Gault of Folkestone. *Annals and Magazine of Natural History* (9) **14**: 504-508.

——— 1924c. On the Blake collection of ammonites from Kachh, India. *Memoirs of the geological Survey of India, palaeontologia indica*, n.s., **9** (1): 1-29.

——— 1924d. The ammonites of the Blue Lias. *Proceedings of the Geologists' Association* **35**: 188-211.

——— 1925a. A monograph of the Ammonoidea of the Gault. 2 (1922): 73-110. Palaeontographical Society: London.

——— 1925b. VII. Ammonites and aptychi. *In*, GREGORY, G.W. (ed.), The collection of fossils and rocks from Somaliland made by Messrs. Wyllie and Smellie. *Monographs of the geological Department of the Hunterian Museum, Glasgow University* **1**: 111-164.

——— 1925c. On Upper Albian Ammonoidea from Portuguese East Africa, with an appendix on Upper Cretaceous ammonites from Maputoland. *Annals of the Transvaal Museum* **11**: 179-200.

——— 1925d. A monograph of the Ammonoidea of the Gault. Part 3 (for 1923): 111-146. Palaeontographic Society: London.

——— 1925e. Sur quelques ammonites de Gault nommées par P. Reynès. *Annales du Musee d'histoire naturelle de Marseille* **20**: 98-105.

——— 1925f. On Senonian Ammonoidea from Jamaica. *Geological Magazine* **62**: 28-32.

——— 1926a. On new ammonites from the English Chalk. *Geological Magazine* **63**: 77-83.

——— 1926b. A monograph of the Ammonoidea of the Gault. Part 4 (for 1924): 147-186. Palaeontographical Society: London.

——— 1927-31. Revision of the Jurassic Cephalopod fauna of the Kachh (Cutch). Pt. I. *Memoirs of the geological Survey of India, palaeontologia indica*, n.s., **9** (2): 1-71 (1927), 73-161 (1928), 163-278 (1928), 79-550 (1931), 551-658 (1931).

——— 1927b. A monograph of the Ammonoidea of the Gault. Part 5 (for 1925): 187-206. Palaeontographical Society: London.

——— 1928a. A monograph of the Ammonoidea of the Gault. Part 6 (for 1926): 207-266. Palaeontographical Society: London.

——— 1928b. Revision of the Jurassic cephalopod fauna of Kachh (Cutch). Part 2. *Memoirs of the geological Survey of India, palaeontologia indica*, n.s., **9** (2): 73-161.

——— 1928c. Revision of the Jurassic cephalopod fauna of Kachh (Cutch). Part 3. *Memoirs of the geological Survey of India, palaeontologia indica*, n.s., **9** (2): 162-278.

——— 1929. Corrections of cephalopod nomenclature. *Naturalist* **871**: 269-271.

——— 1930a. On some Ammonoidea from the Lower Greensand. *Annals and Magazine of natural History* (10) **5**: 417-464.

——— 1930b. The fossil fauna of the Samana Range and some neighbouring areas. V. The Lower Cretaceous Ammonoidea, with notes on Albian Cephalopoda from Hazara. *Memoirs of the geological Survey of India, palaeontologia indica*, n.s., **15**: 50-66.

——— 1930c. A monograph of the Ammonoidea of the Gault. Part 7 (for 1928): 267-312. Palaeontographical Society: London.

——— 1930d. On the Cephalopoda of the Uitenhage Beds. *Annals of the South African Museum* **28**: 131-157.

——— 1931a. A monograph of the Ammonoidea of the Gault. Part 8 (for 1929): 313-378. Palaeontographical Society: London.

——— 1931b. Revision of the Jurassic cephalopod fauna of Kachh (Cutch). Part 4. *Memoirs of the geological Survey of India, palaeontologia indica*, n.s., **9** (2): 279-550.

——— 1931c. Revision of the Jurassic cephalopod fauna of Kachh (Cutch). Part 5. *Memoirs of the geological Survey of India, palaeontologia indica*, n.s., **9** (2): 551-658.

——— 1932. A monograph of the Ammonoidea of the Gault. Part 9 (for 1930): 379-410. Palaeontographical Society: London.

——— 1933a. Revision of the Jurassic cephalopod fauna of Kachh (Cutch). Part 6. *Memoirs of the geological Survey of India, palaeontologia indica*, n. s., **9** (2): vii, 659-945.

——— 1933b. A monograph of the Ammonoidea of the Gault. Part 10 (for 1931): 411-442. Palaeontographical Society: London.

——— 1934. A monograph of the Ammonoidea of the Gault. Part 11 (for 1932): 443-496. Palaeontographical Society: London.

——— 1935. Jurassic and Cretaceous Cephalopoda. *In*, WYLLIE, B.K.N. & SMELLIE, W.R. (eds), On the collection of fossils and rocks from Somaliland. *The Mesozoic palaeontology of British Somaliland*: 205-228. Government Somaliland Protectorate.

——— 1936. The Upper Jurassic invertebrate faunas of Cape Leslie, Milne Land. II. Upper Kimmeridgian and Portlandian. *Meddelelser om Grønland* **99** (3): 1-180.

——— 1937a. The nomenclature of some Lower Chalk ammonites. *Geological Magazine* **74**: 277-281.

——— 1937b. A monograph of the Ammonoidea of the Gault. Part 12 (for 1936): 497-540. Palaeontographical Society: London.

——— 1939a. The Cephalopoda of the Neocomian Belemnite Beds of the Salt Range. *Memoirs of the geological Survey of India, palaeontologia indica*, n.s., **25** (1): 1-154.

——— 1939b. Problems of ammonite nomenclature. 6. The genus *Pachydiscus* Zittel. *Geological Magazine* **76**: 293-296.

——— 1939c. A monograph of the Ammonoidea of the Gault. Part 13 (for 1938): 541-608. Palaeontographical Society: London.

——— 1941. A monograph of the Ammonoidea of the Gault. Part 14: 609-668. Palaeontographical Society: London.

——— 1942. A monograph of the Ammonoidea of the Gault. Part 15: 669-720. Palaeontographical Society: London.

——— 1943. A monograph of the Ammonoidea of the Gault. Part 16: x, 721-787. Palaeontographical Society: London.

——— 1946. Preliminary notes on the Cretaceous ammonite faunas of East Greenland. *Meddelelser om Grønland* **132** (4): 1-12.

——— 1947. Additional observations on the invertebrates (chiefly ammonites) of the Jurassic and Cretaceous of East Greenland. I. The *Hectoroceras* fauna of S.W. Jameson Land. *Meddelelser om Grønland* **133** (4): 1-70.

——— 1950. A new Tithonian ammonoid fauna from Kurdistan, northern Iraq. *Bulletin of the British Museum (Natural History)*, Geology **1**: 95-137.

——— 1952. Additional observations on the invertebrates (chiefly ammonites) of the Jurassic and Cretaceous of East Greenland. II. Some Infra-Valanginian ammonites from Lindemans Fjord, Wollaston Forland; with a note on the base of the Cretaceous. *Meddelelser om Grønland* **133** (4): 1-40.

——— 1953. The Upper Cretaceous cephalopod fauna of Graham Land. *Falkland Islands Dependencies Survey, Science Reports* **3**: 1-60.

STANKIEVICH, E.S. & POJARKOVA, Z.N. 1969. Vascoceratids from the Turonian of southern Kirgisia and the Tadzhiksian depression. *In*, BARKHATOVA, N.N. (ed.), *Continental formations of eastern regions of Soviet Central Asia and Kazakhstan. Lithology and biostratigraphy*: 86-113. Nauka: Leningrad [in Russian].

STANTON, T.W. 1894. The Colorado Formation and its invertebrate fauna. *Bulletin of the United States geological Survey* **106** (for 1893): 1-288.

——— 1896. Contributions to the Cretaceous paleontology of the Pacific Coast: The fauna of the Knoxville Beds. *Bulletin of the United States geological Survey* **133** (1895): 1-132.

——— 1901. The marine Cretaceous invertebrates. *Reports of the Princeton University Expedition to Patagonia, 1896-1899* **4** (Palaeontology 1, 1): 1-43.

STCHIROWSKY, W.A. 1894. Ueber Ammoniten der Genera *Oxynoticeras* und *Hoplites* aus dem nordsimbirsk'schen Neocom. *Bulletin de la Societe imperiale des Naturalistes de Moscou*, n.s., **7** (for 1893): 369-380.

STEFANINI, G. 1933. Molluschi del Giuralias della Somalia. Introdizione, Cephalopodi. *Palaeontographia Italica* **32** (Supplement 1): 97-149.

STEIN, S.F.N. VON. 1878. *Der Organismus der Infusions-thiere nach eigenen Forschungen in systematischer Reihenfolge bearbeitet, Abt. 3*. Engelmann: Leipzig.

STEINMANN, G. 1881. Ueber Tithon und Kreide in den peruanischen Anden. *Neues Jahrbuch für Mineralogie, Geologie und Paläontologie* **1881** (Band 2): 130-153.

——— 1895. Die Cephalopoden der Quiriquina-Schichten. *In*, STEINMANN, G., DEECKE, W. & MÖRICKE, W., Das Alter und die Fauna der Quiriquina-Schichten in Chile. Beiträge zur Geologie und Paläontologie von Südamerika, III. *Neues Jahrbuch für Mineralogie, Geologie und Paläontologie*, Beilage-Band **10**: 64-94.

——— 1925. Beitrage zur Stammesgeschichte der Cephalopoden. *Zeitschrift für induktive Abstammungs- und Vererbungslehre* **36**: 350-416.

——— 1927. Die Methoden der Ammoniten-Phylogenie. *Paläontologische Zeitschrift* **9**: 187-192.

STEPHENSON, L.W. 1941. The larger invertebrate fossils of the Navarro Group of Texas (exclusive of corals & crustaceans and exclusive of the fauna of the Escondido Formation). *University of Texas Publication* **4101**: 1-641.

——— 1955. Basal Eagle Ford fauna (Cenomanian) in Johnson and Tarrant counties, Texas. *United States geological Survey Professional Papers* **274-C**: iv, 53-68.

STEUER, A. 1897. Argentinische Jura-Ablagerungen: Ein Beitrag zur Kenntniss der Geologie und Paläontologie der argentinischen Anden. *Paläontologische Abhandlungen*, n.s., **7** (3): 127-222 (1-96).

——— 1921. Estratos Jurásicos Argentinos. Contribución al conocimiento de la geología y paleontología de los Andes Argentinos entre el río Grande y el río Atuel. *Actas Academia Nacional de Ciencias de Córdoba* **7** (2): 25-128.

STEVENS, G.R. 1997. The late Jurassic ammonite fauna of New Zealand. *Institute of geological and Nuclear Sciences Monograph* **8:** 1-216.

STIELER, C. 1920. Über sogenannte Mortoniceraten des Gault. *Zentralblatt für Mineralogie, Geologie und Paläontologie* **1920**: 345-352, 392-400.

——— 1922. Über Gault- und Cenoman-Ammoniten aus dem Cenoman des Cap Blanc Nez. Mit besonderer Berücksichtung der Gattung *Schloenbachia* Neum. *Neues Jahrbuch für Mineralogie, Geologie und Paläontologie* **47**: 19-44.

STOLICZKA, F. 1863-1866. Ammonitidae, with revision of the Nautilidae. *In*, BLANFORD, H.F. & STOLICZKA, F., The fossil Cephalopoda of the Cretaceous rocks of southern India. *Memoirs of the geological Survey of India, palaeontologia indica* (series 1, 3) 41-56 (1863), 57-106 (1864), 107-154 (1865), 155-216 (1866).

STOLLEY, E. 1907. Ueber ein norddeutsches Aequivalent der Clansayes fauna Südfrankreichs und der Schweiz. *Zentralblatt für Mineralogie, Geologie und Paläontologie* **1907**: 266-270.

——— 1937. Die Gliederung des norddeutschen marinen Unterneocoms. *Zentralblatt für Mineralogie, Geologie, und Paläontologie* (Abt. B) **1937** (11): 434-456; (12): 497-506.

STOYANOW, A. 1949. Lower Cretaceous stratigraphy in southeastern Arizona. *Geological Society of America Memoir* **38**: 1-170.

STOYKOVA, K. 1990. Ammonites of subfamily Venezuellinae Kvantaliani, 1980 (Lower Cretaceous, Aptian) in Bulgaria. *Geologica balcanica* **20**: 19-28.

STRAND, E. 1929. Zoological and palaeontological nomenclatorial notes. *Latvijas Augstskolas Raksti, Riga* **28**: 1-29.

STÜHMER, H.H., SPAETH, C. & SCHMID, F. 1982. *Fossilien Helgolands. Teil 1. Trias und Unter-Kreide. Helgoland.* Niederelbe-Verlag: Ottendorf.

———, ———, ——— 1986. *Fossilien Helgolands. Teil 2. Oberkreide.* Niederelbe-Verlag: Ottendorf.

SUESS, E. 1865. Über Ammoniten. *Sitzungsberichte der kaiserlichen Akademie der Wissenschaften in Wien, Mathematisch-Naturwissenschaftliche Klasse* **52** (Abt. 1): 71-89.

SUMMESBERGER, H. 1979. Eine obersantone Ammonitenfauna aus dem Becken von Gosau (Oberösterreich). *Annalen des naturhistorischen Museums in Wien* **82**: 109-176.

——— & KENNEDY, W.J. 1996. Turonian ammonites from the Gosau Group (Upper Cretaceous; northern Calcareous Alps; Austria) with a revision of *Barroisiceras haberfellneri* (Hauer, 1866). *Beiträge zur Paläontologie* **21**: 105-177.

———, ——— 2004. More ammonites (Puzosiinae, Pachydiscidae, Placenticeratidae, Nostoceratidae, Diplomoceratidae) from the Campanian (late Cretaceous) of the Gschliefgraben (Ultrahelvetic Nappe; Austria). *Annalen des Naturhistorischen Museums in Wien* **106A**: 167-211.

SWINNERTON, H.H. 1935. The rocks below the Red Chalk of Lincolnshire and their cephalopod faunas. *Quarterly Journal of the geological Society of London* **91**: 1-46.

SZAJNOCHA, L. 1885. Zur Kenntniss der mittelcretacischen Cephalopoden-fauna der Inseln Elobi an der Westküste Afrika's. *Denkschriften der kaiserlichen Akademie der Wissenschaften, Wien, Mathematisch-Naturwissenschaftliche Klasse* **49**: 231-238.

SZÀSZ, L. 1982a. Sur la présence des ammonites cénomaniennes dans le bassin de Babadag (Dobrogea de Nord). *Dari de Seama ale Sedintelor* **66**: 27-44.

——— 1982b. Les ammonites hétéromorphes campaniennes des Carpates méridionales (Roumania). *Dari de Seama ale Sedintelor* **66**: 45-56.

——— 1986a. Lower Turonian ammonite assemblage in the Maramures mountains (east Carpathians - Romania). *Dari de Seama ale Sedintelor* **70-71**/3 (for 1983-1984): 17-134.

——— 1986b. Coniacian in Romania; boundaries, subdivisions, ammonite and inoceramid assemblages and their importance for global correlations. *Dari de seama ale sedintelor*. 4. Stratigrafie **70-71** (for 1983-1984): 175-201.

SZIVES, O. 2008. Two new species of *Constrictoceras* nov. gen. (Ammonoidea) from the early Cretaceous (Aptian) of Hungary. *Geobios* **41**: 297-305.

———, Csontos, L., Bujtor, L. & Főzy, I. 2007. Aptian-Campanian ammonites of Hungary. *Geologica Hungarica,* Series Palaeontologica, **57**: 1-187.
——— & Company, M. 2011. Revision of some problematic early Cretaceous ammonite species described by H. Coquand (1880). *Cretaceous Research* **32**: 187-193.
——— & Monks, N. 2002. Heteromorph ammonites from the Tata Limestone Formation (Aptian-Lower Albian), Hungary. *Palaeontology* **45** (6): 1137-1149.
Tajika, A., Tschanz, K. & Klug, C. 2018. New Albian ammonite faunas from Semelenberg (Alpstein, Switzerland) and their paleoecology. *Swiss Journal of Palaeontology* **137**: 65-76.
Tanabe, K. 1977. Functional evolution of *Otoscaphites puerculus* (Jimbo) and *Scaphites planus* (Yabe), Upper Cretaceous ammonites. *Memoirs of the Faculty of Science, Kyushu University* (D, Geology) **23**: 367-407.
——— 2022. Late Cretaceous dimorphic scaphitid ammonoid genus *Yezoites* from the circum-North Pacific regions. *Paleontological Research* **26** (3): 233-269.
———, Obata, I. & Futakami, M. 1981. Early shell morphology in some Upper Cretaceous heteromorph ammonites. *Transactions and Proceedings of the palaeontological Society of Japan*, n.s., **124**: 215-234.
Tarbinskii, S.P. 1932. A contribution to our knowledge of the orthopterous insects of USSR. *Bulletin of the Leningrad Institute for Controlling Farm and Forest Pests* **2**: 181-205 [in Russian with English descriptions].
Taubenhaus, H. 1920. Die Ammoneen der Kreide-formation Palästinas und Syriens. *Deutschen Palästina-Vereins Zeitschrift*, Band **43**: 1-58.
Tavani, G. 1949. Fauna malacologia Cretacea della Somalia e dell'Ogaden, part 2. *Palaeontographia italica* (15) **45**: 1-76.
Teichert, C. 1969. Names and authorship of some cephalopod orders. *Journal of Paleontology* **43** (2): 561-562.
——— 1988. Main features of cephalopod evolution. *In*, Trumann, E.R. (ed.), *The Mollusca: Vol. 12. Paleontology and neontology of cephalopods*: 11-79. Academic Press: N.Y.
Terekhova, G.P. 1980. About some poorly known Cretaceous ammonites from the Anadyr-Koriakia region. *Severo-Vostochnyj Kompleksnyj Nauchno-Issledovatelskij Institut Dalnevostochnogo Nauchnogo Centra Akademii Nauk SSSR*: 94-103 [in Russian].
Thiermann, A. 1964. Die Ammonitengattung *Endemoceras* n. g. aus dem Unter-Hauterive von Nordwest-Europa. *Geologisches Jahrbuch* **81** (for 1963): 345-412.
Thieuloy, J.-P. 1964. *Protanisoceras (Rencurelites)* n. subgen. Nouveau Céphalopode albien de l'horizon des Près de Rencurel (Vercors). *Geologie Alpine, Travaux du Laboratoire de Géologie de la Faculte des Sciences de Grenoble* **40**: 113-120.
——— 1965. Un céphalopode remarquable de l'Hauterivien basal de la Drôme: *Himantoceras* nov. gen. *Bulletin de la Société géologique de France* (7) **6** (1964): 205-213.
——— 1966a. Morphologie de l'ouverture buccale d'un céphalopode microconche: *Saynoceras verrucosum* (d'Orb.). *Bulletin de la Société géologique de France* (7) **7** (for 1965): 839-842.
——— 1966b. Leptocères berriasiens du massif de la Grande-Chartreuse. *Geologie Alpine, Travaux du Laboratoire de Géologie de la Faculte des Sciences de Grenoble* **42**: 281-295.
——— 1969. Sur la présence du genre *Capeloites* Lisson (Ammonoidea) dans le Néocomien des Basses-Alpes et la signification des espèces migratrices trans-atlantiques. *Compte Rendu Sommaire des Seances de la Société géologique de France* **1969**: 256-258.
——— 1971. Réflexions sur le genre *Lyticoceras* Hyatt, 1900 (Ammonoidea). *Comptes Rendus des Seances de l'Academie des Sciences* **272**: 2297-3000.
———1977a. Les ammonites boréales des formations néocomiennes du sud-est français (province subméditerranéenne). *Geobios* **10** (3): 395-461.
——— 1977b. La zone à *callidiscus* du Valanginien Supérieur vocontien (sud-est de la France). Lithostratigraphie, ammonitofaune, limite Valanginien-Hauterivien, corrélations. *Geologie Alpine* **53**: 83-143.
——— 1979. *Matheronites limentinus* n. sp. (Ammonoidea) espèce-type d'un horizon-repère Barremien supérieur du Vercors méridional (Massif Subalpin Français). *Geobios* (mémoir special) **3**: 305-317.
——— & Bulot, L. 1992. Ammonites du Crétacé inférieur du sud-est de la France: 1. Nouvelle èspeces a valeur stratigraphique pour le Valanginien et l'Hauterivien. *Géologie Alpine* **68**: 85-103.
———, Fuhr, M., Bulot, L. 1990. Biostratigraphie du Crétacé inférieur de l'Arc de Castellane (S.E. de la France). 1: Faunes d'ammonites du Valanginien supérieur et âge de l'horizon dit de "La Grande Lumachelle". *Géologie Méditerranéenne* **17** (1): 55-99.
——— & Gazay, M. 1967. Le genre *Dobrodgeiceras* Nikolov en Haute Provence. *Travaux des Laboratoires de Géologie de la Faculte des Sciences de Lyon,* n.s., **14**: 69-78.
Thomel, G. 1964. Contribution à la connaissance des céphalopodes Crétacés du sud-est de la France. Note sur les ammonites déroulées du Crétacé Inférieur vocontien. *Mémoires de la Société géologique de France*, n.s., **43** (101): 1-80.
——— 1969. Réflexions sur les genres *Eucalycoceras* et *Protacanthoceras* (Ammonoidea). *Comptes Rendus hebdomadaires des Seances de l'Academie des Sciences* (D, Sciences Naturelles) **268**: 649-652.
——— 1972. Les Acanthoceratidae cénomaniens des chaines subalpines méridionales. *Mémoires de la Société géologique de France* (5) **51** (116): 1-204.
——— 1980. *Ammonites*. Editions Serre: Nice.
——— 1987. La famille des Tetragonitidae (Ammonoidea) dans le Cénomanien du sud-est de la France. *Annales de Paléontologie* **73** (4): 241-272.
——— 1988. Les ammonites néocrétacées (Coniacien-Santonien-Campanien) des chaînes subalpines méridionales (Alpes-Maritimes et Alpes de Haute-Provence). *Mémoires de la Société géologique de France*, n.s., **153**: 1-79.
———, Coulet, P., Davoux, J., Delanoy, G., Führ, F., Führ, M., Gazel, L., Giverso, J., Jullian, H., Jullian, V., Levan, G., Mascarelli, E., Maurel, L., Poupon, A. & Salomon, M. 1990. Le genre *Acrioceras* Hyatt, 1900 dans le Barrémien du sud-est de la France. *Annales du Muséum d'Histoire naturelle de Nice* **7**: 20-151.
———, Delanoy, G. & Autran, G. 1987. Valeur taxonomique, position stratigraphique et relations phylétiques des genres d'Ammonoidea: *Acrioceras* Hyatt, 1900 et *Aspinoceras* Anderson, 1938, au cours des temps hauteriviens, ainsi que de leurs dérivés. *Comptes Rendus de l'Academie des Sciences* (2) **305**: 215-219.
Thomson, M.R.A. 1974. Ammonite faunas of the Lower Cretaceous of south-eastern Alexander Island. *British Antarctic Survey Scientific Reports* **80**: 1-44.
Tietze, E. 1872. Geologische und palaeontologische Mittheilungen aus dem südlichen Theil des Banater Gebirgsstoches. *Jahrbuch der Kaiserlich-Koeniglichen Geologischen Reichsanstalt* **22**: 35-142.
Tokunaga, S. & Shimizu, S. 1926. The Cretaceous formation of Futaba in Iwaki and its fossils. *Journal of the Faculty of Science, Imperial University of Tokyo*, Section 2: Geology, Mineralogy, Geography, Seismology **1**: 181-212.
Tomlin, J.R. le B. 1930. Some preoccupied generic names. II. *Proceedings of the malacological Society of London* **19**: 22-24.
Torcapel, A. 1884. Quelques fossiles nouveaux de l'Urgonien du

Languedoc. *Bulletin de la Société d'Etude des Sciences naturelles de Nimes* **11** (9, 11): 109-110 (1-2), 133-141 (3-11).

TORNQUIST, A. 1896. Die degenerierten Perisphinctiden des Kimmeridge von Le Havre. *Abhandlungen der schweizerischen paläontologischen Gesellschaft* **23**: 1-44.

TORSVIK, T.H., ROUSSE, S., LABAILS, C. & SMETHURST, M.A. 2009. A new scheme for the opening of the South Atlantic Ocean and the dissection of an Aptian salt basin. *Geophysical Journal International* **177**: 1315-1333.

TOSHIMITSU, S. 1988. Biostratigraphy of the Upper Cretaceous Santonian Stage in northwestern Hokkaido. *Memoirs of the Faculty of Science, Kyushu University* (D, Geology) **26** (2): 125-192.

TOUCAS, A. 1888. Note sur le Jurassique supérieur et le Crétacé inférieur de la vallée du Rhône. *Bulletin de la Société géologique de France* **3**: 903-927.

TOVBINA, S.Z. 1962. *In*, KRIMGOL'TS, G.Y. (ed.), *Field atlas and handbook of the fossils of the Jurassic and Neocomian deposits of western Turkmenistan*: 84-85. Institut zemnoi kory Leningradskogo Gosudarstvennogo Universiteta [in Russian].

——— 1963. Barremian ammonites of Turkmenia. *Trudy Vsesoiuznogo Nauchno-Issledovatel'skogo Geologicheskogo Instituta* **109**: 98-119 [in Russian].

——— 1965. On the ontogeny of the ammonite genus *Colchidites*. *Palaeontological Journal* **3**: 40-48.

——— 1970. A new genus of the family Parahoplitidae. *Paleontological Journal* **1970** (3): 56-65.

——— 1979. The systematics and phylogeny of the family Parahoplitidae: 109-122. *In*, MESEZHNIKOV, M.S. & KIRICHOVKA, A.I., The stratigraphy of Lower Cretaceous deposits of petroleum- and gas-bearing regions of USSR; a collection of works. *Trudy Vsesoiuznogo Neftianogo Nauchno-Issledovatel'skogo Geologorazvedochnogo Neftagazovogo Instituta*: 109-122.

——— 1985. New representatives of family Parahoplitidae of the Turkmenia. *Annual of the All-Union Paleontological Society* **25**: 60-79 [in Russian].

——— 1989. Evolution of ammonites and sedimentation history events in Aptian and Albian of Transcaspian region. *Bulletin de la Societe imperiale des Naturalistes de Moscow*, Sect. geol., **64** (4): 29-41.

TRASK, J.B. 1855. A new ammonite from Arbuckle's diggings, Shasta County. *Proceedings of the California Academy of Sciences* **1** (1): 40.

TRAUTSCHOLD, H. 1861. Recherches geologiques aux environs de Moscou. Fossiles de Kharachovo et supplément. *Bulletin de la Societe imperiale des Naturalistes de Moscou* **34** (3): 267-277.

——— 1866. Zur Fauna des russischen Jura. *Bulletin de la Societe imperiale des Naturalistes de Moscou* **39**: 1-24.

——— 1877. Der Französische Kimmeridge und Portland verglichen mit den gleichaltrigen Moscauer Schichten. *Bulletin de la Societe imperiale des Naturalistes de Moscou* **50** (for 1876): 381-392.

TROUESSART, E.L. 1889. Revue synoptique de la famille des Halacaridae. *Bulletin scientifique de la France et de la Belgique, Paris* **20**: 225-251.

TUOMEY, M. 1856. Description of some new fossils, from the Cretaceous rocks of the southern states. *Proceedings of the Academy of natural Sciences of Philadelphia* **7** (for 1854-1855): 167-172.

TZANKOV, V. 1935. Notes sur le genre *Holcodiscus*. *Annales de Universitie de Sofia* **31** (3): 57-100.

——— 1964. Ammonites from the Maastrichtian near Kladoroub village, Belogradchik region (north-western Bulgaria). *Travaux sur la géologie de Bulgarie*, Série paléontologie, **6**: 143-187.

——— & BRESKOVSKI, S. 1982. Volume et contenu de la famille Holcodiscidae Spath, 1924. *Doklady Bolgarskoi Akademiia Nauk, Sofiia* **35** (4): 491-493.

UHLIG, V. 1882a. Die Wernsdorfer Schichten und ihre Äquivalente. *Sitzungsberichte der kaiserlichen Akademie der Wissenschaften in Wien, Mathematisch-Naturwissenshaftliche Klasse* **86** (1): 86-117.

——— 1882b. Zur Kenntnis der Cephalopoden der Rossfeldschichten. *Jahrbuch der kaiserlich-koeniglichen Geologischen Reichsanstalt, Wien* **32**: 373-396.

——— 1883. Die Cephalopodenfauna der Wernsdorfer Schichten. *Denkschriften der kaiserlichen Akademie der Wissenschaften, Wien, Mathematisch-Naturwissenschaftliche Klasse* **46**: 127-290 (1-166).

——— 1902. Über die Cephalopodenfauna der Teschener und Grodischter Schichten. *Denkschriften der kaiserlichen Akademie der Wissenschaften, Wien, Mathematisch-Naturwissenschaftliche Klasse* **72**: 1-88.

——— 1903-1910. The fauna of the Spiti Shales. *Memoirs of the geological Survey of India, palaeontologia indica* (15, Himalayan fossils) **4** (fascicle 1-3): 1-132 (1903), 133-306 (1910a), 307-395 (1910b).

——— 1905. Einige Bemerkungen über die Ammoniten-gattung *Hoplites* Neumayr. *Sitzungsberichte der kaiserlichen Akademie der Wissenschaften in Wien, Mathematisch-Naturwissenshaftliche Klasse* **114**: 591-636.

——— 1907. Review of "La faune jurassique de Mazapil avec un appendice sur les fossiles du Crétacique Inférieur," by C. Burckhardt (Boletin del Instituto geologico de Mexico. No. 23. 1906. 43 pl.). *Neues Jahrbuch für Mineralogie, Geologie und Paläontologie* **1907** (Band 2): 469-471.

——— 1911. Die marinen Reiche des Jura und die Unterkreide. *Mitteilungen der geologischen Gesellschaft in Wien* **4**: 329-448.

UPTON, B.G.J. 2021. The opening of the North Atlantic. *In*, PANAGIOTAKOPULU, E. & SADLER, J.P. (eds), *Biogeography in the sub-Arctic: the past and future of North Atlantic biota*: 1-27. New Jersey: John Wiley & Sons.

USHER, J.L. 1952. Ammonite faunas of the Upper Cretaceous rocks of Vancouver Island, British Columbia. *Geological Survey of Canada Bulletin* **21**: v, 1-182.

VAŠÍCEK, Z. 1972. Ammonoidea of the Tesin-Hradiste Formation (Lower Cretaceous) in the Moravskoslezke Beskydy Mts. *Rozpravy Ustredniho Ustavu Geologickeho* **38**: 1-104.

——— 1973. Über einige von V. Uhlig (1883) beschriebene Unterkreide-Ammoniten. *Sbornik Geologickych vedeckých Paleontologie* **15**: 49-74.

——— 1979. Zwei neue faunistische Fundorte der Unterkreide in der silesischen Einheit ud Zusammenfassung der Revision von V. Uhlig (1883) Beschreibenen. *Sborník vedeckých prací Vysoké školy bánské v Ostrave Rocník XX, císlo 2, rada hornicko-geologická, clánek* **480**: 119-134.

——— 2005. The oldest (Late Valanginian) Crioceratitinae (heteromorphic ammonoids), central western Carpathians (Slovakia). *Geologica Carpathica* **3**: 245-254.

——— 2006. A remarkable assemblage of early Barremian ammonites in the central western Carpathians (Butkov Quarry, Slovakia). *Acta geologica Polonica* **56** (4): 421-440.

——— 2010. Early Cretaceous ammonites from the Butkov Quarry (Manín Unit, central western Carpathians, Slovakia). *Acta geologica Polonica* **60** (3): 393-415.

——— 2020. *Tescheniceras* gen. nov. (Ammonoidea) and the definition of the Valanginian/Hauterivian boundary in Butkov Quarry (central western Carpathians, Slovakia). *Acta geologica Polonica* **70** (4): 569-584.

——— & FAUPL, P. 1998. Late Valanginian cephalopods in relation to the palaeogeographic position of the Rossfeld and Schrambach Formation of the Reichraming Nappe (northern Calcareous Alps, Upper Austria). *Zentralblatt für Geologie und Paläontologie*, part 1: 1421-1432.

——— & HOEDEMAEKER, PH.J. 2003a. Small Berriasian, lower

Valanginian and Barremian heteromorphic ammonites from the Río Argos succession (Caravaca, southeast Spain). *Scripta Geologica* **125**: 11-33.

———, ——— 2003b. A new *Karsteniceras* from the Barremian (Lower Cretaceous) of Colombia. *Scripta Geologica* **125**: 141-143.

——— & Klajmon, P. 1998. Contribution to the knowledge of some small early Barremian ammonites from Silesian Unit (Outer Carpathians, Czech republic). *Vestník Ceského geologického ústavu* **73**: 331-342.

——— & Klein, J. 2021. On imperfectly known Hauterivian representatives of the families Holcodiscidae Spath, 1923 and Barremitidae Breskovski, 1977 in Butkov Quarry (central western Carpathians, Slovakia). *Acta geologica Polonica* **71** (4): 433-451.

——— & Malek, O. 2017. Pseudothurmanniid ammonites from quarries near Lietavská Lúèka and their stratigraphical significance (late Hauterivian, central western Carpathians, Slovakia). *Geological Quarterly* **61** (3): 613-631.

———, Michalík, J. & Reháková, D. 1994. Lower Cretaceous stratigraphy, paleogeography and life in western Carpathians. *Beringeria* Hft. **10**: 1-170.

———, ———, Radulovic, V.J., Radulovic, B. & Mojsic, I. 2013. Ammonoids (Desmoceratoidea and Silesitoidea) from the Late Barremian of Boljetin, eastern Serbia. *Cretaceous Research* **41**: 39-54.

———, Rabrenovic, D., Skupien, P., Radulovic, V.J., Radulovic, B.V. & Mojsic, I. 2014. Ammonites (Phylloceratina, Lytoceratina and Ancyloceratina) and organic-walled dinoflagellate cysts from the late Barremian in Boljetin, eastern Serbia. *Cretaceous Research* **47**: 140-159.

———, Reháková, D. & Faupl, P. 1999. Zur Biostratigraphie der Schrambachschichten der Oisbergmulde bei Hollenstein a. d. Ybbs (Lunzer Decke, Kalkalpen, Niederösterreich). *Abhandlunger der Geologischen Bundesanstalt, B.-A.* **56** (2): 625-650.

———, Reháková, D. & Skupien, P. 2017. Some perisphinctoid ammonites of the Štramberk Limestone and their dating with associated microfossils (Tithonian to Lower Berriasian, outer western Carpathians, Czech Republic). *Geologica Carpathica* **68** (6): 583-605.

——— & Wiedmann, J. 1994. The Leptoceratoidinae: small heteromorph ammonites from the Barremian. *Palaeontology* **37**: 203-239.

Valduga, A. 1954. Ammoniti ed aptici neogiurassici dell' Ogaden e della Somalia sud-occidentale. *Palaeontographia Italica* **48** (n.s., 18): 1-40.

Vasilievskij, M. 1909. Note sur les couches *Douvilléiceras* dans les environs de la ville de Saratov. *Trudy Geologicheskogo Muzeya Petra Velikago* **2** (1908): 29-52.

Vennari, V.V. 2016. Tithonian ammonoids (Cephalopoda, Ammonoidea) from the Vaca Muerta Formation, Neuquén Basin, west-central Argentina. *Palaeontographica*, Abt. A, **306** (1-6): 85-165.

Vereschagin, V.N., Kinasov, V.P., Paraketsov, K.V., Terekhova, G.P. 1965. *Field atlas of the Cretaceous fauna of north-east of USSR*. Magadan [in Russian].

Verma, H.M. & Westermann, G.E.G. 1973. The Tithonian (Jurassic) ammonite fauna and stratigraphy of Sierra Catorce, San Luis Potosi, Mexico. *Bulletins of American Palaeontology* **63**: 107-314.

———, ——— 1984. The ammonoid fauna of the Kimmeridgian-Tithonian boundary beds of Mombasa, Kenya. *Life Science Contributions, Royal Ontario Museum* **135**: iii-iv, 1-124.

Vermeulen, J. 1972. Contribution à l'étude du genre *Paquiericeras* Sayn, 1901. *Annales du Museum d'Histoire naturelle de Nice* **1**: 41-44.

——— 2003. Etude stratigraphique et paléontologique de la famille des Pulchelliidae (Ammonoidea, Ammonitina, Endemocerataceae). *Géologie Alpine*, Mémoire H.S., **42**: 1-333.

——— 2005. Boundaries, ammonite fauna and main subdivisions of the stratotype of the Barremian. *Géologie Alpine*, Série colloques et excursions, **7**: 147-173.

——— 2006a. Sur quatre especes particulieres d'ammonites du Barremien du sud-est de la France. *Annales du Muséum d'Histoire naturelle de Nice* **20** (2005): 1-24.

——— 2006b . Nouvelle classification à fondement phylogénétique des ammonites hétéromorphes du Crétacé inférieur. *Annales du Muséum d'Histoire naturelle de Nice* **21**: 137-178.

——— 2009a. Ammonites nouvelles (Ancyloceratoidea) du Barrémien inférieur du sud-est de la France. *Riviéra Scientifique* **92**: 49-64.

——— 2009b. Nouvelles données sur la superfamille des Ptychoceratoidea Gill, 1871 *nom. transl.* Klein *et al.*, 2007 (Ammonitida, Turrilitina). *Riviéra Scientifique* **93**: 71-86.

——— 2010. Nouvelle tripartition de la famille des Anahamulinidae Breistroffer, 1952 (Turrilitina, Ptychoceratoidea) et description de quelques taxa du Barrémien inférieur du sud-est de la France. *Annales du Muséum d'Histoire Naturelle de Nice* **25**: 61-91.

——— 2011. Discussion of: Problems in the identity of *"Crioceras" barremense* Kilian, 1895 (Ancyloceratida, Late Barremian), and their proposed resolution, by D. Bert et alii (CG2010_A01) [Alternative title: The nomenclatural status and the acceptation of the genus *Barrancyloceras* Vermeulen, 2000, and of its type species]. *Carnets de Géologie/Notebooks on Geology*, Article 2010/01_Reply (CG2010-A01R).

———, Borro, A., Lazarin, P., Lépinay, P., Leroy, L. & Mascarelli, E. 2020a. Nouveaux taxa d'ammonites du Crétacé inférieur du sud-est de la France. *Riviera Scientifique* **104**: 129-146.

———, ———, ———, ———, ———, ——— 2020b. On fundamental changes to the classification of some ammonite taxa of the Lower Cretaceous (*in press*).

——— & Bulot, L. 2007. Données nouvelles sur la classification phylogénétique des ammonites du Crétacé inférieur. *Annales du Muséum d'Histoire naturelle de Nice* **22**: 101-112.

———, Cecca, F. & Kruta, I. 2007. Nouveaux taxa de la famille des Hamulinidae Gill, 1871 (Ammonitida, Turrilitina, Ptychoceratoidea). *Annales du Muséum d'Histoire naturelle de Nice* **22**: 1-25.

——— & Klein, J. 2006. Endemocerataceae: Pulchelliidae. *In*, Riegraf, W. (ed.), *Fossilium Catalogus I: Animalia* **141**: 187-278. Blackhuys Publishers: Leiden.

———, Lazarin, P., Lépinay, P., Leroy, L. & Mascarelli, E. 2013a. Taxa d'ammonites nouveaux ou peu connus de l'Hauterivien et du Barrémien du sud-est de la France. *Strata*, série 2, mémoires, **48**: 1-58.

———, ——, ——, ——, —— 2013b. Le genre *Defayella* gen. nov. et ses implications taxinomiques (Turrilitina, Ptychoceratoidea, Anahamulinidae). *Riviéra Scientifique* **97**: 83-96.

———, ——, ——, ——, —— 2014. Ammonites du Barrémien du sud-est de la France. *Strata*, série 2, mémoires, **50**: 1-95.

———, ——, ——, ——, —— 2016. Silesitidae du Barremien du sud-est de la France (Ammonitina, Silesitoidea). *Strata*, série 2, **53**: 1-93.

———, ——, ——, ——, ——, Meister, C. & Menkveld-Gfeller, U. 2012. Ammonites nouvelles ou peu connues de l'Hauterivien supérieur. *Annales du Muséum d'Histoire naturelle de Nice* **27**: 271-318.

———, ——, ——, ——, ——, —— & Pacaud, J.-M. 2015. Nouvelles données sur le genre *Ptychoceras* Orbigny, 1842 et sur son espèce-type (Mollusca, Cephalopoda, Ammonoidea).

Annales du Muséum d'Histoire naturelle de Nice **30**: 1-12.

———, ———, LEROY, L. & MASCARELLI, E. 2012. Nouvelles données sur l'évolution et la classification de la sous-famille des Paraspiticeratinae Vermeulen, 2009 (Ancyloceratina, Ancyloceratoidea, Emericiceratidae). *Annales du Muséum d'Histoire naturelle de Nice* **27**: 319-334.

——— & VAŠÍČEK, Z. 2011. Espèces d'ammonites nouvelles ou peu connues de la famille des Anahamulinidae Breistroffer, 1952 (Turrilitina, Ptychoceratoidea). *Annales du Muséum d'Histoire naturelle de Nice* **26**: 47-94.

VETTERS, H. 1905. Die Fauna der Juraklippen zwischen Donau und Thaya. I. Die Tithonklippen von Niederfellabrunn. *Beiträge zur Paläontologie und Geologie Österreich-Ungarns und des Orients* Band **17**, Heft III-IV: 223-259.

VIGH, G. 1984. Die biostratigraphische auswertung einiger Ammoniten-Faunen aus dem Tithon des Bakonygebirges sowie aus dem Tithon-Berrias des Gerecsegebirges. *Annals of the Hungerian geological Institute* **67**: 1-210.

VILLASEÑOR, A.B., GONZÁLEZ-LEÓN, C.M., LAWTON, T.F. & ABERHAN, M. 2005. Upper Jurassic ammonites and bivalves from the Cucurpe Formation, Sonora (Mexico). *Revista Mexicana de Ciencias Geológicas* **22** (1): 65-87.

——— & OLÓRIZ, F. 2006. Ontogeny in a macroconhiate *Idoceras* from Mexico : a key for interpreting suprageneric classification of genus *Idoceras* based on phenotype expression. 7th International Congress on the Jurassic System, Krakow, Abstracts volume. *Volumina Jurassica* **4**: 245-247.

———, ——— 2009. Caribbean Lower Tithonian ammonites from central-east Mexico. *Geobios* **42**: 117-132.

———, ——— & GONZÁLEZ-ARREOLA, C. 2011. Lower Tithonian microconchiate simoceratins from eastern Mexico: taxonomy, biostratigraphy, and palaeobiogeography. *Acta palaeontologica Polonica* **56** (1): 133-158.

VILLOUTREYS, O. DE. 1950. Au sujet de genre *Ptychoceras* d'Orbigny (ammonite). *Bulletin du Museum national d'Histoire naturelle, Paris* (2) **22**: 155-156.

YIN, J. & ÉNAY, R. 2004. Tithonian ammonoid biostratigraphy in eastern Himalayan Tibet. *Geobios* **37**: 667-686.

VOGEL, K.-P. 1959. Zwergwuchs bei Polyptychiten (Ammonoidea). *Geologisches Jahrbuch* **76**: 469-540.

VORONETS, N.S. 1958a. New ammonites from the Pachsa Peninsula on the southern shore of the Laptiev Sea. *Sbornik Statei po Paleontologii i Biostratigrafii, Leningrad* **7**: 22-27 [in Russian].

——— 1958b. The new genus *Virgatoptychites* from the northern regions of Siberia. *Sbornik Statei po Paleontologii i Biostratigrafii, Leningrad* **9**: 68-74 [in Russian].

——— 1962. Stratigraphy and cephalopod molluscs of the Jurassic and Lower Cretaceous deposits of the Leno-Anabar region. *Trudy Nauchno-Issledovetel'skogo Instituta Geologii Arktiki, Moscow* **110**: 1-237 [in Russian].

VREDENBURG, E.W. 1907. The ammonites of the Bagh Beds. *Records of the geological Survey of India* **36** (2): 109-125.

VYALOV, O.S. 1940. *Ilowaiskya* nom. nov. - new genus of Jurassic ammonites. *Doklady AN SSSR* **19** (1): 79-80 [in Russian].

WAAGEN, W. 1873-1875. The Jurassic fauna of Kutch: Cephalopoda. *Palaeontologia indica* (9) **1** (Part 1):1-22 (1873), Parts 2-4: 23-247 (1875).

WALASZCZYKA, I., KENNEDY, W.J., DEMBICZ, K., GALE, A.S., PRASZKIER, T., RASOAMIARAMANANA, A.H. & RANDRIANALY, H. 2014. Ammonite and inoceramid biostratigraphy and biogeography of the Cenomanian through basal Middle Campanian (Upper Cretaceous) of the Morondava Basin, western Madagascar. *Journal of African Earth Sciences* **89**: 79-132.

WARD, P.D. 1976. Upper Cretaceous ammonites (Santonian-Campanian) from Orcas Island, Washington. *Journal of Paleontology* **50**: 454-461.

——— & MALLORY, V.S. 1977. Taxonomy and evolution of the lytoceratid genus *Pseudoxybeloceras* and relationship to the genus *Solenoceras*. *Journal of Paleontology* **51**: 606-618.

——— & WESTERMANN, G.E.G. 1976. Sutural inversion in a heteromorph ammonite and its implication for septal formation. *Lethaia* **9**: 357-361.

———, ——— 1977. First occurrence, systematics, and functional morphology of *Nipponites* (Cretaceous Lytoceratina) from the Americas. *Journal of Paleontology* **51** (2): 367-372.

WARREN, P.S. 1930a. Three new ammonites from the Cretaceous of Alberta. *Transactions of the Royal Society of Canada* (3, section 4) **24**: 21-26.

——— 1930b. New species of fossils from Smoky River and Dunvegan formations, Alberta. Appendix to RUTHERFORD, R.L., Geology and water resources in parts of the Peace River and Grande Prairie districts, Alberta. *Research Council of Alberta Report* **21**: 57-68.

——— 1934. Palaeontology of the Bearpaw Formation. *Transactions of the Royal Society of Canada* (3, section 4) **28**: 81-100.

——— & STELCK, C.R. 1940. Cenomanian and Turonian faunas in the Pouce Coupe district, Alberta and British Columbia. *Transactions of the Royal Society of Canada* (3, section 4) **34**: 143-152.

———, ——— 1958. Lower Cenomanian Ammonoidea and Pelecypoda from Peace River area, western Canada. *Research Council of Alberta, Geological Division Bulletin* **2**: 36-52.

WEAVER, C.E. 1931. Paleontology of the Jurassic and Cretaceous of west central Argentina. *Memoirs of the University of Washington* **1**: 1-469.

WEDEKIND, R. 1916. Über Lobus, Suturallobus und Inzision. *Zentralblatt für Mineralogie, Geologie und Paläontologie* **1916**: 185-195.

——— 1918. Die Genera der Palaeoammonoidea (Goniatiten). *Palaeontographica* **62**: 85-184.

WEERTH, O. 1884. Die fauna des Neocomsandsteins im Teutoberger Walde. *Paläontologische Ablhandlungen* **2**: 1-77.

WEGELE, L. 1929. Stratigraphische und faunistishe Untersuchungen im Oberoxford und Unterkimmeridge Mittelfrankens. II. Paleontologishe Teil. *Palaeontographica* Band **72**: 1-94 (95-188).

WEGNER, R.N. 1909. Uebersicht der bisher bekannten Astieria-Formen der Ammonitengattung *Holcostephanus* nebst Beschreibung zweier neuer Arte. *Neues Jahrbuch für Mineralogie, Geologie und Paläontologie* Band **I**: 77-92.

WEGNER, T. 1905. Die Granulatenkreide des westlichen Münsterlandes. *Zeitschrift der deutschen geologischen Gesellschaft* **57**: 112-232.

WEITER, O.A. 1913. Eine Tithonfauna aus Nord-Peru. *Neues Jahrbuch für Mineralogie, Geologie und Paläontologie* Band **I**: 28-42.

WELLER, S. 1903. The Stokes Collection of Antarctic fossils. *Journal of Geology* **11**: 413-419.

WETZEL, W. 1960. Die Coyhaique-Schichten des patagonischen Neocoms und ihre Ammoniten. *Neues Jahrbuch für Geologie und Paläontologie, Monatshefte* **1960**: 246-254.

WHITE, C.A. 1880. Descriptions of new Cretaceous invertebrate fossils from Kansas and Texas. *Proceedings of the United States national Museum* **2** (for 1879): 292-298.

WHITEAVES, J.F. 1885. Description of a new species of ammonite from the Cretaceous rocks of Fort St. John, on the Peace River. *Proceedings and Transactions of the Royal Society of Canada* (4, Geological and Biological Sciences) **2** (1884): 239-240.

——— 1889. On some Cretaceous fossils from British Columbia, the North West Territory and Manitoba. *Geological Survey of Canada, Contributions to Canadian Palaeontology* **1**: 151-196.

——— 1893a. Notes on the ammonites of the Cretaceous rocks

of the district of Athabasca, with descriptions of four new species. *Proceedings and Transactions of the Royal Society of Canada* (4, Geological and Biological Sciences) **10** (1892): 111-121.

——— 1893b. Descriptions of two new species of ammonites from the Cretaceous rocks of the Queen Charlotte Islands. *Canadian Record of Science* **5**: 441-446.

——— 1900. *Mesozoic fossils*. Volume 1. IV. On some additional or imperfectly understood fossils from the Cretaceous rocks of the Queen Charlotte Islands, with a revised list of the species from these rocks: 263-308. Geological Survey of Canada: Ottawa.

——— 1901. Note on a supposed new species of *Lytoceras* from the Cretaceous rocks at Denman Island, in the strait of Georgia. *Ottawa Naturalist* **15**: 31-32.

WHITEHOUSE, F. 1926. The Cretaceous Ammonoidea of eastern Australia. *Memoirs of the Queensland Museum* **8** (3): 195-242.

——— 1927. Additions to the Cretaceous ammonite fauna of eastern Australia. Part 1 (Simbirskitidae, Aconeceratidae and Parahoplitidae). *Memoirs of the Queensland Museum* **9** (1): 109-120.

——— 1928a. Additions to the Cretaceous ammonite fauna of eastern Australia. Part 2 (Desmoceratidae). *Memoirs of the Queensland Museum* **9** (2): 200-206.

——— 1928b. The correlation of the marine Cretaceous deposits of Australia. *Report of the 18th Meeting of the Australasian Association for the Advancement of Science* **1926**: 275-280.

WHITFIELD, R.P. 1880. Paleontology of the Black Hills of Dakota. *In*, POWELL, J.W. (ed.), *Report on the geology and resources of the Black Hills of Dakota, United States geographical and geological survey of the Rocky Mountain region*: 325-468.

——— 1892. Gasteropoda and Cephalopoda of the Raritan Clays and Greensand Marls of New Jersey. *Monograph of the United States geological Survey* **18**: 1-402.

WHITNEY, F.L. 1911. Fauna of the Buda Limestone. *Bulletin of the University of Texas* **184**: 1-56.

WIEDMANN, J. 1960. Le Crétacé Supérieur de l'Espagne et du Portugal et ses céphalopodes. *Comptes Rendus du Congres des Societes des Savantes-Dijon: Colloque sur le Cretace Superieur Francaise* **1959**: 709-764.

——— 1962a. Ammoniten aus der vascogotischen Kreide (Nordspanien). I. Phylloceratina, Lytoceratina. *Palaeontographica* (Abt. A) **118**: 119-237.

———1962b. Unterkreide-Ammoniten von Mallorca. 1: Lieferung: Lytoceratina, Aptychi. *Abhandlungen der Mathematisch-naturwissenschaftlichen Klasse, Akademie der Wissenschaften und der Literatur in Mainz* **1962**: 1-148.

——— 1962c. Habitus, Skulptur und Sutur bei Kreide-Ammonoideen. *Neues Jahrbuch für Mineralogie, Geologie und Paläontologie, Abhandlungen* **114**: 317-322.

——— 1962d. Die Gabbioceratinae Breistroffer. *Neues Jahrbuch für Geologie und Paläontologie, Abhandlungen* **115**: 1-43.

——— 1963. Unterkreide-Ammoniten von Mallorca. 2. Phylloceratina. *Abhandlungen der mathematisch-naturwissenschaftlichen Klasse der Akademie der Wissenschaften und der Literatur, Mainz* **1963**(4): 151-264.

——— 1965. Origin, limits, and systematic position of *Scaphites*. *Palaeontology* **8**: 397-453.

——— 1966a. Stammesgeschichte und System der posttriadischen Ammonoideen, ein Überblick (1. Teil). *Neues Jahrbuch für Geologie und Paläontologie, Abhandlungen* **125**: 49–79.

——— 1966b. Stammesgeschichte und System der posttriadischen Ammonoideen, ein Überblick (2. Teil). *Neues Jahrbuch für Geologie und Palaeontologie, Abhandlungen* **127**: 13–81.

——— 1969. The heteromorphs and ammonoid extinction. *Biological Reviews, Cambridge* **44**: 563-602.

——— 1970. Über den Ursprung der Neoammonoideen - das Problem einer Typogenese. *Eclogae Geologicae Helvetiae* **63** (3): 923-1020.

——— 1973a. Ancyloceratina (Ammonoidea) at the Jurassic/Cretaceous boundary. *In*, HALLAM, A. (ed.), *Atlas of palaeobiogeography*: 309-316. Elsevier: Amsterdam, London, New York.

——— 1973b. The Albian and Cenomanian Tetragonitidae (Cretaceous Ammonoidea), with special reference to circum-Indic species. *Eclogae Geologicae Helvetiae* **66**: 585-616.

——— 1979. Ammonites. *In*, HERM, D., KAUFFMAN, E.G. & WIEDMANN, J., The age and depositional environment of the "Gosau"-Group (Coniacian-Santonian), Brandenberg/Tirol, Austria. *Mitteilungen der Bayerischen Staatssammlung fuer Paläontologie und historische Geologie* **19**: 41-49.

——— & DIENI, I. 1968. Die Kreide Sardiniens und ihre Cephalopoden. *Palaeontographia italica* **64**: 1-171.

———, KAKABADZE, M.V. & SHARIKADZE, M.Z. 1990. Suture ontogeny in *Ptychoceras* d'Orbigny and its implications for the systematics of Cretaceous heteromorphs. *Neues Jahrbuch für Geologie und Paläontologie, Monatshefte* **1990**: 367-384.

——— & KAUFFMAN, E.G. 1978. Mid-Cretaceous biostratigraphy of northern Spain. *Annales du Museum d'Histoire naturelle de Nice* **4**(1976): III1-III34.

——— & KULLMANN, J. 1979. *Nubidites omarai* n.g. n.sp., ein neuer Ammonit aus dem Nubischen Sandstein Ägyptens. *Palaeontologische Zeitschrift* **53**: 248-258.

——— & MARCINOWSKI, R. 1985. *Scaphamites passendorferi* n.g. n.sp. (Ammonoidea, Cretaceous) - ancestor of Scaphitaceae Meek? *Neues Jahrbuch für Geologie und Paläontologie, Monatshefte* **1985**: 449-463.

WEINKAUF, M.F.G., HOFFMANN, R. & WIEDENROTH, K. 2021. Evolutionary-phylogenetic pathway of the Cretaceous ammonite genus *Aegocrioceras* and its relationship to *Juddiceras* and *Crioceratites* spp. *Papers in Palaeontology* **7** (4): 2113-2139.

WIERZBOWSKI, A. 1976. Oxfordian ammonites of the Pinar del Rio province (western Cuba); their revision and stratigraphical significance. *Acta geologica Polonica* **26** (2): 137-260.

——— 1990. The taxonomy and phylogenetic significance of early Tithonian ammonites of the genus *Protancyloceras* Spath from the Pieniny Klippen Belt (Carpathians, Poland). *Atti Ist Convegno Fossili Evoluzione Ambiente. Pergola, 25-28 ottobre 1984*: 479-489.

WIESE, F. & SCHULZE, F. 2005. The Upper Cenomanian (Cretaceous) ammonite *Neolobites vibrayeanus* (d'Orbigny, 1841) in the Middle East: taxonomic and palaeoecologic remarks. *Cretaceous Research* **26**: 930-946.

WILCKENS, O. 1904. Revision der Fauna der Quiriquina-Schichten. *Neuen Jahrbuch für Mineralogie, Geologie und Paläontologie*, Beilage-band **18**: 181-284.

——— 1947. Paläontologisch und geologische Ergebnisse der Reise von Kohl-Larsen (1928-29) nach Süd-Georgien. *Abhandlungen der senkenbergischen Naturforschenden Gesellschaft* **474**: 1-66.

WILMSEN, M. & NAGM, E. 2013. Upper Cenomanian-Lower Turonian ammonoids from the Saxonian Cretaceous (lower Elbtal Group, Saxony, Germany). *Bulletin of Geosciences* **88** (3): 647-674.

WIMBLEDON, W.A. 1974. *The stratigraphy and ammonite faunas of the Portland Stone, England and northern France*. Unpublished Ph.D. thesis, University of Wales.

——— & COPE, J.C.W. 1978. The ammonite faunas of the English Portland Beds and the zones of the Portlandian Stage. *Journal of the geological Society of London* **135** (2): 183-190.

WITKOWSKI, A. 1967. On the species *Endemoceras (Lyticoceras) noricum* (Roem) of the Lower Cretaceous deposits from the Tomaszow trough. *Kwartalnik Geologiczny* **11** (3): 637-646 (in Polish).

WITTS, J.D., BOWMAN, V.C., WIGNALL, P.B., CRAME, J.A., FRANCIS, J.E. & NEWTON, R.J. 2014. Evolution and extinction of Maastrichtian (late Cretaceous) cephalopods from the López de Bertodano Formation, Seymour Island, Antarctica. *Palaeogeography, Palaeoclimatology, Palaeoecology.*

WOLLEBEN, J.A. 1967. Senonian (Cretaceous) Mollusca from Trans-Pecos Texas and northeastern Chihuahua, Mexico. *Journal of Paleontology* **41**: 1150-1165.

WOLLEMANN, A. 1907. Die Fauna des mittleren Gaults von Algermissen. *Jahrbuch der koeniglich preussischen geologischen Landesanstalt und Bergakademie* **24** (for 1903): 22-42.

WOODS, H. 1896. The Mollusca of the Chalk Rock, part I. *Quarterly Journal of the geological Society of London* **52**: 68-98.

——— 1911. The palaeontology of the Upper Cretaceous deposits of northern Nigeria, Appendix 2. *In*, FALCONER, J.D., *The geology and geography of northern Nigeria*: 273-286. Macmillan: London.

WRIGHT, C.W. 1952. A classification of the Cretaceous ammonites. *Journal of Paleontology* **26**: 213-222.

——— 1953. Notes on Cretaceous ammonites. I. Scaphitidae. *Annals and Magazine of natural History* (12) **6**: 473-476.

——— 1955. Notes on Cretaceous ammonites. II. The phylogeny of the Desmocerataceae and the Hoplitaceae. *Annals and Magazine of natural History* (12) **8**: 561-575.

——— 1956a. Rectification of nomenclature - *Raspailiceras* nom. nov. *Proceedings of the geological Society of London* **1535**: 73-74.

——— 1956b. Notes on Cretaceous ammonites. III. *Utaturiceras* gen. nov. and the Metoicoceratinae. *Annals and Magazine of natural History* (12) **9**: 391-393.

——— 1957a. Some Cretaceous ammonites from New Zealand. *Transaction of the Royal Society of New Zealand* **84**: 805-809.

——— 1957b. [Cretaceous ammonites]. *In*, MOORE, R.C. (ed.), *Treatise on invertebrate paleontology*. Part L. Mollusca 4, Cephalopoda Ammonoidea. Geological Society of America: N.Y.; University of Kansas: Lawrence.

——— 1963. Cretaceous ammonites from Bathurst Island, northern Australia. *Palaeontology* **6**: 597-614.

——— 1975. The Hauterivian ammonite genus *Lyticoceras* Hyatt, 1900 and its synonym *Endemoceras* Thiermann, 1963. *Palaeontology* **18**: 607-611.

——— 1979. The ammonites of the English Chalk Rock (Upper Turonian). *Bulletin of the British Museum (Natural History)*, Geology **31**: 281-332.

——— 1981. Cretaceous Ammonoidea. *In*, HOUSE, M.R. & SENIOR, J.R. (eds), The Ammonoidea, The evolution, classification, mode of life and geological usefulness of a major fossil group. *The Systematics Association, Special Volume* **18**: 157-174.

——— 1996. Cretaceous Ammonoidea (with CALLOMAN (sic), J.H. & HOWARTH, M.K.). *In*, KAESLER, R.L. (ed.), *Treatise on invertebrate paleontology*. Part L. Mollusca 4, Revised. Geological Society of America: N.Y.; University of Kansas: Lawrence.

——— CHANCELLOR, G.R. & KENNEDY, W.J. 1983. The affinities of *Codazziceras* Etayo-Serna, 1979 (Cretaceous Ammonoidea). *Cretaceous Research* **4**: 341-348.

——— & KENNEDY, W.J. 1978. The ammonite *Stoliczkaia* from the Cenomanian of England and northern France. *Palaeontology* **21**: 393-409.

——— & ——— 1979. Origin and evolution of the Cretaceous micromorph ammonite family Flickiidae. *Palaeontology* **22**: 685-704.

———, ——— 1980. Origin, evolution and systematics of the dwarf acanthoceratid *Protacanthoceras* Spath, 1923 (Cretaceous Ammonoidea). *Bulletin of the British Museum (Natural History)*, Geology **34**: 65-108.

———, ——— 1981. The Ammonoidea of the Plenus Marls and the Middle Chalk. *Palaeontographical Society, London* **1981**: 1-148.

———, ——— 1984. The Ammonoidea of the Lower Chalk, part 1. *Palaeontographical Society, London* **1984**: 1-126.

———, ——— 1994. Evolutionary relationships among Stoliczkaiinae (Cretaceous ammonites) with an account of some species from the English *Stoliczkaia dispar* Zone. *Cretaceous Research* **15**: 547-582.

———, ———, HANCOCK, J.M. & GALE, A.S. 2017. The Ammonoidea of the Lower Chalk, Part 7. *Monographs of the Palaeontographical Society* **171**: 648, 461-561,

——— & MATSUMOTO, T. 1954. Some doubtful Cretaceous ammonite genera from Japan and Saghalien. *Memoirs of the Faculty of Science, Kyushu University* (D, Geology) 4: 107-134.

——— & WRIGHT, E.V. 1949. The Cretaceous ammonite genera *Discohoplites* Spath and *Hyphoplites* Spath. *Quarterly Journal of the geological Society of London* **104** (for 1948): 477-498.

———, ——— 1951. A survey of the fossil Cephalopoda of the Chalk of Great Britain. *Palaeontographical Society, London* **1951**: 1-40.

YABE, H. 1902. Note on three Upper Cretaceous ammonites from Japan, outside of Hokkaido. *Journal of the geological Society of Tokyo* **9**: 1-7.

——— 1903. Cretaceous Cephalopoda from the Hokkaido. Part 1: *Lytoceras, Gaudryceras,* and *Tetragonites. Journal of the College of Science, Imperial University of Tokyo* **18** (2): 1-55.

——— 1904. Cretaceous Cephalopoda from Hokkaido, part 2. *Journal of the College of Science, Imperial University of Tokyo* **20**: 1-45.

——— 1909. Zur Stratigraphie und Paläontologie der oberen Kreide von Hokkaido und Sachalin. *Zeitschrift der deutschen geologischen Gesellschaft* **61**: 402-444.

——— 1910. Die Scaphiten aus der Oberkreide von Hokkaido. *Beiträge zur Paläontologie und Geologie Oesterreich-Ungarns und des Orients* **23**: 159-174.

——— 1915. Notes on some Upper Cretaceous fossils from Anaga on the island of Awaji and Toyajo in the province of Kii. *Science Reports of the Tohoku Imperial University, Sendai* (2, Geology) **4**: 13-24.

——— 1926. A study on the genus *Parapachydiscus,* Hyatt. *Proceedings of the Imperial Academy, Tokyo* **2**: 171-173.

——— 1927. Cretaceous stratigraphy of the Japanese islands. *Science Reports of the Tohoku Imperial University, Sendai* (2, Geology) **11**: 27-100.

——— & SHIMIZU, S. 1921. Notes on some Cretaceous ammonites from Japan and California. *Science Reports of the Tohoku Imperial University, Sendai* (2, Geology) **5**: 53-59.

———, ——— 1924. A new species of *Brahmaites* from the Upper Cretaceous of S. Saghalin, with some remarks on the genus *Brahmaites. Japanese Journal of Geology and Geography* **3**: 77-80.

———, ——— 1925. Japanese Cretaceous ammonites belonging to Prionotropidae. I. *Science Reports of the Tohoku Imperial University, Sendai* (2, Geology) **7**: 125-138.

YAKOVLEVA, P.K. 1979. On the *Virgatites* from the upper part of the *Virgatus* Zone from the classical Glebovo outcrop at Volga. *In*, IVANOV, A.N. (ed.), Late Mesozoic cephalopods of the Upper Volga area. *Inter-college collection of scientific papers, Issue* **183**: 18-29. Pedagogical University: Yaroslavl [in Russian].

YOKOYAMA, M. 1890. Versteinerungen aus der japanischen Kreide. *Palaeontographica* **36**: 159-202.

YOUNG, G.M. & BIRD, J. 1828. *A geological survey of the Yorkshire coast: describing the strata and fossils occurring between the Humber and the Tees, from the German Ocean to the plain of York.* 2nd edition. R. Kirby: Whitby.

YOUNG, K. 1957. Upper Albian (Cretaceous) Ammonoidea from Texas. *Journal of Paleontology* **31**: 1-33.

——— 1958a. *Graysonites,* a Cretaceous ammonite in Texas. *Journal of Paleontology* **32**: 171-182.

——— 1958b. Cenomanian (Cretaceous) ammonites from Trans-Pecos Texas. *Journal of Paleontology* **32**: 286-294.

——— 1963. Upper Cretaceous ammonites from the Gulf Coast of the United States. *University of Texas Publication* **6304**: 1-373.

——— 1966. Texas Mojsisovicziinae (Ammonoidea) and the zonation of the Fredericksburg. *Geological Society of America Memoir* **100**: vii, 1-225.

——— 1968. Upper Albian (Cretaceous, *M. romeri* Zone) ammonites in Texas and Mexico. *Journal of Paleontology* **42**: 70-80.

——— 1974. Lower Albian and Aptian (Cretaceous) ammonites of Texas. *Geoscience and Man* **8**: 175-228.

——— 1979. Lower Cenomanian and late Albian (Cretaceous) ammonites, especially Lyelliceratidae, of Texas and Mexico. *Texas Memorial Museum, Bulletin* **26**: vi, 1-100.

——— 1993. Middle Albian ammonites from El Madero, west-central Chihuahua. *The Texas Journal of Science* **44** (4): 165-176.

ZABORSKI, P.M.P. 1982. Campanian and Maastrichtian sphenodiscid ammonites from southern Nigeria. *Bulletin of the British Museum (Natural History)*, Geology, **36**: 303-332.

——— 1987. Lower Turonian (Cretaceous) ammonites from south-east Nigeria. *Bulletin of the British Museum (Natural History),* Geology, **41** (2): 31-66.

——— 1990. The Cenomanian and Turonian (mid-Cretaceous) ammonite biostratigraphy of north-eastern Nigeria. *Bulletin of the British Museum (Natural History).* Geology, **48** (1): 1-18.

——— 1993. Some new and rare Upper Cretaceous ammonites from north-eastern Nigeria. *Journal of African Earth Sciences (and the Middle East)* **17** (3): 359-371.

——— 1995. The Upper Cretaceous ammonite *Pseudaspidoceras* Hyatt, 1903, in north-eastern Nigeria. *Bulletin of the British Museum (Natural History)*, Geology, **51**: 53-72.

——— 1996. The Upper Cretaceous ammonite *Vascoceras* Choffat, 1898 in north-eastern Nigeria. *Bulletin of the natural History Museum, London* **52**: 61-89.

——— 1998. *Bageites bakui* n.g. n.sp., a new ammonite from the Lower Turonian of north-eastern Nigeria. *Neues Jahrbuch für Geologie und Paläontologie Monatshefte* **1998**: 545-558.

——— & MORRIS, N.J. 1999. The late Cretaceous ammonite genus *Libycoceras* in the Iullemmeden Basin (West Africa) and its palaeogeographic significance. *Cretaceous Research* **20**: 63-79.

ZAKHAROV, V.A. & ROGOV, M.A. 2020. High-resolution stratigraphy of buchiid bivalves and ammonites from the Jurassic-Cretaceous boundary beds in the Paskenta area (California). *Cretaceous Research* **110** (10442): 1-27.

ZEISS, A. 1968. Untersuchungen zur Palàontologie der Cephalopoden des Unter-Tithon der südlichen Frankenalb. *Abhandlungen der Bayerischen Akademie fur Wissenschaften, Mathematik-Naturwissenschaft Klasse*, N.S., **132**: 1-191.

——— 2001. Die Ammonitenfauna der Tithonklippen von Ernstbrunn, Niederösterreich. *Denkschriften des Naturhistorischen Museums in Wien* Band **6**: 1-115.

———, BENETTI, A. & PEZZONI, N. 1994. A new ammonite fauna from the Tithonian (Semiforme/Verruciferum Zone) of the Lessinian Alps, Verona Province northern Italy. *Palaeopelagos, Special Publication* **1**, Proceedings of the 3rd Pergola International Symposium: 367-381.

——— & LEANZA, H. 2011. Upper Jurassic (Tithonian) ammonites from the lithographic limestones of the Zapala region, Neuquen Basin, Argentina. *Beringeria* Band **41**: 1-52.

ZEJSZNER, L. 1846. *New and little-known described genera of fossils from the Tatra Mountains. Part 1.* L. Zeuschner: Warsaw [in Polish].

——— 1856. Geognostische Beschreibung des Liaskalkes in der Tatra und in den angrenzenden Gebirgen. *Sitzungsberichte der kaiserlichen Akademie der Wissenschaften in Wien* **19**: 135-182.

ZIEGLER, B. 1958. Monographie der Ammonitengattung *Glochiceras* im epicontinentalen Weissjura Mittel-europas. *Palaeontographica*, Abt. A, Band **110**, Lief. 4-6: 93-164.

——— 1962. Die ammonitengattung *Aulacostephanus* im Oberjura (Taxonomie, Stratigraphie, Biologie). *Palaeontographica*, Abt. A, Band **119**: 1-172.

ZITTEL, K.A. VON. 1868. Palaeontologische studien über Grenzschichten der Jura- und Kreide-Formation im gebiete der Karpaten, Alpen und Apenninen. Abh. I. Die Cephalopoden der Stramberger Schichten. *Paläontologische Mittheilungen aus dem Museum des koeniglich-bayerischen Staates* **2**: 33-118.

——— 1869. Bermerkungen über *Phylloceras tatricum* Pusch sp. und einige andere *Phylloceras*-Arten. *Jahrbuch der kaiserlich-koeniglichen geologischen Reichsanstalt* **19**: 59-68.

——— 1870. Die Fauna der älteren Cephalopoden-führenden Tithonbildungen. *Palaeontographica* (supplement) **1**: 119-310.

——— 1883. Beiträge zur Geologie und Paläontologie der libyschen Wüste und der angrenzenden Gebiete von Aegypten, Theil 1. *Palaeontographica* **30**: 1-127.

——— 1884. Cephalopoda. *In*, ZITTEL, K.A. VON, *Handbuch der Paläontologie*, Band **1**, Abt. 2, Lief 3: 329-522. Oldenbourg: Munich & Leipzig.

——— 1895. *Grundzüge der Paläontologie*. Oldenbourg: Munich & Leipzig.

ZWIERZYCKI, J. 1914. Die Cephalopoden Fauna der Tendaguru-Schichten in Deutsch-Ostafrica. *Berlin Archiv für Biontologie* **3**: 1-96.

List of new genera

Algoiceras - for Algoa Basin, South Africa
Alabushevites - for Dr A.I. Alabushev
Anabaculites - ana (G, towards)
Anatetragonites - ana (G towards)
Anazelandites - ana (G, towards)
Arcturusites - Arcturus, the brightest star
Alekseevites - for Dr S.N. Alekseev
Arkadieviceras - for Prof. V.V. Arkad'ev
Arnouldsagetia - for S. Arnoult-Saget
Askerswellites - Askerswell, English town
Aubersonites - L'Auberson, Swiss village
Baylissites - for O.H. Bayliss
Beaumugnella - Beaumugne, French village
Begastrella - for Begastri, Latin name of Cehegín.
Beloventer - belos (Gr., arrow)

Binckhorstella - for Binckhorst
Brendaites - for Brenda Cooper, my wife
Chernovaites - for Dr E.S. Chernova
Chikkimites - for Chikkim, Indian town
Cinctibrancoceras - *cinctus* (L. girdle)
Cognacella - for Cognac, French town
Companyceras - for Dr M. Company
Delametteiceras - for Dr M. Delamette
Delanoyiceras- for Dr G. Delanoy
Dieniites - for Dr I. Dieni
Drushchitzia - for Dr V.V. Drushchitz
Edwinkemperia - for Dr Edwin Kemper
Eidelweissites - for my mother, Eidelweiss Dorothy Cooper
Elninoceras - for Elnin, town in Madagascar
Emilavramia - for Dr Emil Avram
Epibaculites - epi (Gr., near)
Epikossmaticeras - epi (Gr., near)
Eucyrticeras - eucyrtis (Gr., curved)
Eurokossmaticeras - for Europe
Galloisiceras - for Dr R. Gallois
Gracilicostites - slender ribs (L.)
Greenlandiceras - Greenland
Hanhaites - Hahna, coastal town in Angola
Hattinites - for Dr D.E. Hattin
Hendersoniceras - for Dr R.A. Henderson
Hookiceras - for Dr S.C. Hook
Hughowenites - for Dr H.G. Owen
Iohamitella - *Idiohamites*
Ixioniceras - Ixion (Gr., king of the Lapiths)
Johndixonites - for Dr J.P.G. Dixon
Juigneticeras - for Dr P. Juignet
Karakaschella - for Dr N.I. Karakasch
Katerinella - Katerina (catherine wheel)
Kawashitaiceras - for Dr Y. Kawashita
Keithyoungites - for Prof. K. Young, Texas
Kyushuites - Kyushu, Japanese island
Laeviptychoceras - *laevus* (L., smooth)
Lafresnayella - La Fresnaye, French town
Lambirisites - for Dr A.J. Lambiris
Laqueoceratites - *laqueus* (L.) noose
Lukenederina - for Dr A. Lukeneder
Lytobaculites - *Baculites* with *Lytoceras* suture
Malganisoceras - Madagascan *Anisoceras*
Mangyschlakiceras - Mangyschlak peninsula, Crimea
Manyolaites - Manyola drift, Zululand
Matsumotoceras - for Prof. T. Matsumoto
Matsumotoites - for Prof. T. Matsumoto
Mauryiceras - for Dr C.J. Maury
Mazenotella - for Dr G. Mazenot
Mirapelioides - false *Mirapelia*
Mitraikyella - Mitraiky, Madagascan town
Montanamites - Montana *Hamites*
Neanisoceras - after *Anisoceras*
Neodistoloceras - after *Distoloceras*
Neodouvilleiceras - late *Douvilleiceras*
Neoforbesiceras - after *Forbesiceras*
Neogardeniceras - after *Gardeniceras*
Neoworthoceras - after *Worthoceras*
Neuchateliceras - Neuchâtel, Swiss canton
Nodosoppelia - oppelioid with nodes
Oweniceras - for Dr H.G. Owen
Parafagesia - almost *Fagesia*
Paranaklinoceras - almost *Anaklinoceras*
Paranisoceras - almost *Anisoceras*
Pariatamboites - Pariatambo, Brazilian town
Parvicornutites - small horns
Phindaites - Phinda Game Reserve, Zululand
Plesioscaphites - plesio (G., near)
Powellites - for Dr D. Powell
Prohimalayites - precursor to *Hmalayites*
Prowatinoceras - progenitor to *Watinoceras*
Prozelandites - progenitor to *Zelandites*
Pseudohoepenites - false *Hoepenites*
Pseudomammites - false *Mammites*
Pseudometoicoceras - false *Metoicoceras*
Quimbalaites - Quimbala, coastal town in Angola
Rawsoniceras - for Prof. P.F. Rawson
Regifexiceras - regifex (L., knobbly stick)
Roscheniceras - for Dr C.H. Roschen
Sasonovaites - for Dr I.G. Sasonova
Saxbyites - for Mr S. Saxby, discoverer of the species
Schoendelmayerites - for Dr C. Schoendelmayer
Shulginaella - for Dr N.I. Shul'gina
Skoenbergiceras - Skoenberg, hillock in Zululand
Skoenbergites - Skoenberg, hillock in Zululand
Stephensonites - for Dr L.W. Stephenson
Subtilicostites - gracefully ribbed (L.)
Tanabeites - for Prof. K.Tanabe
Transkiella - Transkei, former republic in Eastern Cape
Tuberohamites - *Hamites* with tubercles
Tuberophylloceras - *Phylloceras* with tubercles
Umzambiceras - Mzamba, town in Eastern Cape
Valbonnetticeras - La Valbonnette, French hamlet
Vermeuleniceras - for Dr J. Vermeulen
Virdeniceras - Virden, town in Montana
Vonkoenenites - for Prof. A. von Koenen
Wakaneneites - Maori chief Waka Nene
Welwitschiella - after Dr Friedrich Welwitsch.
Xenobaculus - *xenos* (Gr., foreigner)

Index

A

F

G

H

M

N

O

P

Q

R

T

U

V

W

X

Y

Z

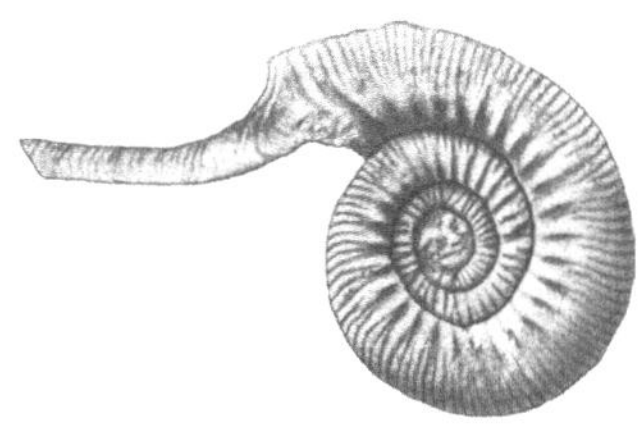